CHEMISTRY

An Introduction to General, Organic, and Biological Chemistry

CHEMISTRY

An Introduction to General, Organic, and Biological Chemistry

JOANNE M. WIDOM
STUART J. EDELSTEIN
Cornell University

W. H. FREEMAN AND COMPANY
San Francisco

To our families

Project Editor: Patricia Brewer
Developmental Editor: Caroline Eastman
Manuscript Editor: Lawrence W. McCombs
Photo Researcher: Elizabeth Whipple
Interior Designer: Marie Carluccio
Cover Designer: Sharon Helen Smith
Production Coordinator: Linda Jupiter
Illustration Coordinator: Cheryl Nufer
Artists: Eric Hieber, John and Jean Foster, Tom Blakeley
Compositor: York Graphic Services
Printer and Binder: Arcata Book Group

Chapter-opening photo credits
Cover, 4, 11, 19, 20: Robert Arnold. 1: Courtesy of the Old Mint Museum. 2, 14: Courtesy of General Motors Research Laboratories. 3: Trumble Photography. 5: Courtesy of the Burndy Library. 6: Courtesy of Cedric L. Chernick and John G. Malm, Argonne National Laboratory. 7, 8: Courtesy of Andrew L. Leed. 9: Courtesy of M. Cukierski, N. Lanza, and J. Maribei, Cornell University. 10: Courtesy of Robert Ishi. 12, 15, 18: Richard Byrnes. 13, 22, 24, 27: From *Tissues and Organs,* by Richard G. Kessel and Randy H. Kardon, W. H. Freeman and Company. Copyright © 1979. 16: Courtesy of BASF Aktiengesellschaft. 17: Ted Streshinsky. 21: Courtesy of Jerome Gross, Massachusetts General Hospital. 23: Courtesy of Robert C. Leif, Papanicolaou Cancer Research Institute. 25: Courtesy of A. K. Kleinschmidt, Elsevier. 26: Courtesy of John Telford, Cornell University. 28: Courtesy of James G. White, University of Minnesota.

Library of Congress Cataloging in Publication Data

Widom, Joanne M 1931–
Chemistry, an introduction to general, organic and biological chemistry.

Includes index.
1. Chemistry. 2. Biological chemistry. I. Edelstein, Stuart J., joint author. II. Title.
QD31.2.W54 540 80-23816
ISBN 0-7167-1224-5

Printed in the United States of America

9 8 7 6 5 4 3 2 1

CONTENTS

4

Liquids and Solids

5

The Divisible Atom

6

The Periodic Table and the Electronic Shell Structure of Atoms

7

The Chemical Bond and the Shapes of Molecules

8

The Structure of Molecules and Crystals

9

Solutions

II

CHEMICAL REACTIONS

10

Chemical Energetics, Chemical Equilibria, and Chemical Kinetics

11

Acids and Bases

12

Oxidation and Reduction

III

ORGANIC CHEMISTRY

13

Introduction to Organic Chemistry

14

Alkanes

15

Unsaturated Hydrocarbons and Aromatic Hydrocarbons

16

Reactions of Alkyl Halides

17

Alcohols, Phenols, and Ethers

18

Aldehydes and Ketones

19

Carboxylic Acids and Esters

20

Amines and Amides

IV

BIOLOGICAL CHEMISTRY

21

Introduction to Peptide and Protein Structure

22

Protein Function

23

Carbohydrates

24

Lipids

25

Nucleic Acids and Protein Synthesis

PREFACE

Chemistry: An Introduction to General, Organic, and Biological Chemistry is written for students who have little or no background in chemistry and whose preparation in mathematics includes only arithmetic and elementary algebra. In writing this book we devoted special attention to clarity and concreteness of explanations, interesting and carefully worked examples, and sound development of principles. These strengths make the book especially appropriate for students in biological and health-related fields of study (nutrition, nursing, ecology, and agriculture) who may take additional courses in chemistry, as well as for students who do not plan to take other courses in science.

The book is divided into four parts: (I) Properties of Matter, (II) Chemical Reactions, (III) Organic Chemistry, and (IV) Biological Chemistry. The first two parts survey the physical principles of chemistry and are illustrated primarily with examples from inorganic chemistry, although examples from organic and biological chemistry are included. The student thus encounters early some topics that are treated in greater detail in the second half of the text. Throughout the book we present chemistry as an experimental science in which theories evolve and change as new information is acquired. Thus we provide some historical development of principles, primarily in the early chapters, because many ideas are best understood by examining the experiments on which they are based. Biomedical examples show students how chemistry is applied to areas of interest to them.

Stoichiometry is introduced in Chapter 2 by using a common-sense approach with many worked examples. This chapter provides the background needed for quantitative work in the laboratory. Gases are presented next because it is our experience that most students find it easier to learn first about macroscopic properties of matter, with which they are already somewhat familiar, before studying atomic structure. Instructors who prefer to start with the structure of the atom can begin with Chapters 5, 6, and 7 (atomic structure, the periodic table, and chemical bonding) and then present stoichiometry and the macroscopic properties of matter.

The third part introduces functional groups of organic compounds and their characteristic reactions. Mechanisms of typical organic reactions provide an understanding of the relationship between molecular structure and chemical reactivity. The most difficult topics—the mechanisms of substitution and elimination reactions—are contained in a single chapter (Chapter 16) that can be skipped in courses for which this detail is not needed. We often use biochemical reactions to illustrate typical organic reactions.

The fourth part introduces proteins, carbohydrates, lipids, and nucleic acids and describes their involvement in fundamental biological processes such as the enzymatic catalysis of reactions, the replication of DNA and RNA, the biosynthesis of proteins, and metabolism. Chapter 22 includes an elementary but sound discussion of simple enzyme kinetics. This topic can be easily omitted, but we feel that the student is well prepared for it by the treatment of reaction kinetics in Chapter 10. Chapter 25 explores recent developments in nucleic acid technology and includes a description of an important new method for sequencing DNA.

The mathematical level of the book is that of arithmetic and of the most elementary algebra. Carefully worked examples of typical problems are included in the text along with simple exercises that provide an immediate test of the student's understanding of the material just presented. Additional problems are provided at the end of each chapter, except Chapters 27 and 28. Many of the worked examples, exercises, and problems are of biological or medical interest. Each chapter also includes an introduction, definitions of key terms, and a summary.

We have primarily used the International System of Units (SI units). In particular, the joule is used as the unit of energy in Parts I, II, and III. In Chapter 4, where units of energy are first introduced, conversions between calories and joules are illustrated with worked examples and exercises. Because biochemists have not adopted the joule to the extent that chemists have, in Part IV the calorie is used as the unit of energy in the text, and energies in both calories and joules are included in tables.

The appendixes include tables of conversion factors and fundamental constants, a review of exponents and logarithms (although logarithms are not required for this book), a discussion of significant figures, and answers to all the exercises and half of the problems.

During the writing we have liberally consulted our colleagues and students. There is almost no member of the Cornell University Chemistry Department who has not contributed to this book in some way—by providing references or answers to particularly difficult questions. We acknowledge with special thanks discussions with Gordon Hammes and Benjamin Widom, and contributions of many interesting problems by Roald Hoffmann (in Parts I and II) and Bruce Ganem (in Parts III and IV). We are also grateful to Scott Mohr (Boston University), Kathleen Richardson (Capital University), Charles Rose (University of Nevada, Reno), and Barbara Schowen (University of Kansas) for their detailed critiques of the manuscript.

We also thank students Andrew Leed and Christopher Akey who read and criticized portions of the manuscript, and Barbara Widom who checked answers to problems. Many of the photographs were taken by Richard Byrnes who cheerfully filled all kinds of unusual requests. Photographs of the bombardier beetle and the paramecium were contributed by Thomas Eisner, the electron micrograph of mitochondria by John Telford, and the scanning electron micrographs of sheep red blood cells by Ronald Minor. Our efforts to achieve clarity and provide interesting biomedical examples were greatly aided by the detailed suggestions of Caroline Eastman during preparation of the final drafts of the text, and we are grateful for her patience, insight, and meticulous care. We also thank Terry Auld for her excellent typing.

The people at W. H. Freeman and Company have, in a real sense, been our collaborators on this project, and it has been a pleasure to work with them. Because they cared about it, the book is much better than it would otherwise have been. In particular, we are grateful to Howard Boyer, Chemistry Editor, for the advice and support he gave us throughout; and to Pat Brewer, Project Editor, for detecting what we would like to imagine were the few remaining errors and inconsistencies, and for cheerfully responding to our interminable requests for changes that would make things a little clearer to the student. We also thank Elizabeth Whipple for her perseverance and imagination in obtaining outstanding chapter-opening photographs, and Ruth Allen for her patience and help with permissions.

Saundra Y. McGuire (University of Tennessee) has prepared a Study Guide to accompany our text. This guide reinforces our attention to mastering the arithmetical aspects of chemistry, a common area of deficiency among beginning students. Each chapter contains learning objectives; a review of the important concepts and terms from the text, with problems and their answers; and self-tests and their answers.

Also accompanying the text are an Instructor's Manual and a set of Transparency Masters. We hope that these materials will ease the instructor's teaching burden.

Joanne M. Widom
Stuart J. Edelstein

August 1980

CHEMISTRY

An Introduction to General, Organic, and Biological Chemistry

I

PROPERTIES OF MATTER

Gold bullion. The bars are poured by hand, using ladles, into standard molds. They are $7 \times 3\frac{5}{8} \times 1\frac{3}{4}$ inches—somewhat smaller than building bricks—and weigh about 400 troy ounces (27.4 lb or 12.4 kg). Each bar bears the mark of the year it was poured, its weight, and its melt (identification) number.

1

The Elements of Chemistry

In this chapter we see how our views of matter have evolved—from 600 B.C. when Greek philosophers tried to identify basic elements by logical reasoning, to the late 1700s when Antoine Lavoisier used careful measurements to develop the law of conservation of mass and to identify 23 of the chemical elements.

1-1
A SEARCH FOR THE ELEMENTS

Consider the many things in the world, each with its own properties. A lump of coal burns with a hot flame, leaving only a small bit of ash. A particular kind of red-brown rock won't burn, but turns to gray iron in the hot coals. Ripe grapes, treated in a certain way, become a fine wine. Over millions of years, humans learned to identify various substances and to make use of their properties: dyeing cloth, fermenting grains and fruits, making pottery and glass, and reducing ores to get metals. Systems of measurement were developed for use in trade. Is there some underlying order to all this diversity of properties?

For a long time, the search for understanding led to various legends about gods who decree the workings of nature. As far as we know, the new approach that eventually led to chemistry began around 600 B.C. in the Greek colonies of Ionia along the shores of Asia Minor. Thinking about changes like that of the red-brown rock to gray metal, Thales began to wonder whether it might be possible to take any one substance and change it into any other. Is all the world really made of only one basic material—some underlying single **element?** The answer, he thought, should be yes; that would give the universe a basic simplicity and order. Thales concluded that the basic element is *water,* the most changeable substance he could imagine.

Over the following centuries, this type of speculation became very popular. Anaximenes argued that the basic element is *air,* which he suggested could be compressed to form liquid and solid substances (we now know he was right about that). Heraclitus suggested that the element is *fire*—always changing yet ever the same, and the key to so many of the transformations from one substance to another.

Perhaps there is more than one element. Around 450 B.C., Empedocles suggested that there are four elements: water, air, fire, and *earth* (his own contribution). Any other substance is made of some combination of differing amounts of these elements, and a change from one substance to another is achieved by adding or removing one or more elements. Aristotle accepted and elaborated this theory (Figure 1-1), and as part of his writings it persisted for 2000 years.

For many centuries after Aristotle, philosophers continued to seek a complete understanding of matter. Alchemists flourished, testing in their laboratories (Figure 1-2) every substance they could lay their hands on, mixing and heating in their crucibles everything from rocks to toads. If a red-brown rock can be changed to gray iron, surely there must be a way to change the gray iron to yellow gold. In their search for the philosopher's stone, a legendary transmuting agent that would enable them to turn base metals into gold, they developed chemical apparatus, experimental procedures, and much practical knowledge of the chemical properties of the substances they tested.

The alchemists discovered and first used the strong acids—sulfuric acid and nitric acid—important contributions though they made little impression at the time. With these acids, the alchemists could dissolve many previously insoluble substances and carry out new reactions. Still clinging to Empedocles' theory of

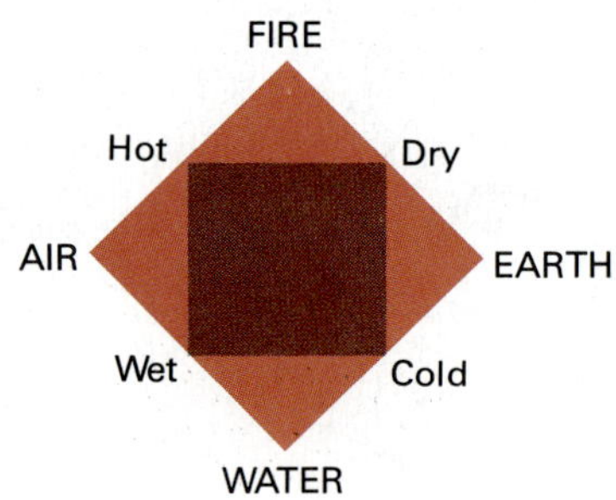

FIGURE 1-1

Elements of ancient times: earth, air, fire, and water. According to Aristotle, each of these elements possesses two of four properties: hot, cold, wet, and dry. Earth, he believed, is cold and dry; fire is dry and hot; air is hot and wet; and water is wet and cold.

FIGURE 1-2

An alchemist. ". . . Surely to alchemy this right is due, that it may be compared to the husbandman whereof Aesop makes the fable; that, when he dies, told his sons that he had left unto them gold buried underground in his vineyard; and they digged over all the ground, and gold they found none; but by reason of their stirring and digging the mould about the roots of their vines, they had a great vintage the year following: so assuredly the search and stir to make gold hath brought to light a great number of good and fruitful inventions and experiments. . . ." [Quotation from Francis Bacon, *The Advancement of Learning,* ed. W. A. Wright, 3d ed. (Oxford: Clarendon Press, 1885), p. 36; as quoted by Mary Elvira Weeks, *Discovery of the Elements* (Easton, Pa.: Journal of Chemical Education, 1939), p. 9. Painting by Cornelis Bega, from the collection of Dr. Alfred Bader.]

the four elements, however, they failed to recognize the new elements they had discovered: antimony, arsenic, bismuth, phosphorus, and zinc.

Modern chemistry grew out of an emphasis on carefully controlled experiments involving precise measurements of weight and volume. One early example of such work is found in the writings of Jan Baptista Van Helmont (1577–1644), who doubted the four-element theory of the Greeks. He asserted that there are only two elements: air and water. To prove that a tree can be formed from water alone, he grew a weighed tree in a weighed container of dirt, adding only water and carefully weighing the tree and the container of dirt at the end of five years (Figure 1-3).

> I took an Earthen Vessel, in which I put 200 pounds of Earth that had been dried in a Furnace, which I moystened with Rain-water, and I implanted therein the Trunk or Stem of a Willow Tree, weighing five pounds; and at length, five years being finished, the Tree sprung from thence, did weigh 169 pounds, and about three ounces: But I moystened the Earthen Vessel with Rain-water, or distilled water (always when there was need) and it was large and implanted into the Earth, and least the dust that flew about should be comingled with the Earth, I covered the lip or mouth of the Vessel, with an Iron-Plate covered with Tin, and easily passable with many holes. I computed not the weight of the leaves that fell off in the four Autumnes. At length I again dried the Earth of the Vessel, and there were found the same 200 pounds, wanting about two ounces. Therefore 164 pounds of Wood, Barks, and Roots, arose out of water only.*

This was a carefully measured and controlled experiment (note the tinplated iron cover, which permitted water to enter but kept out other substances), and Van

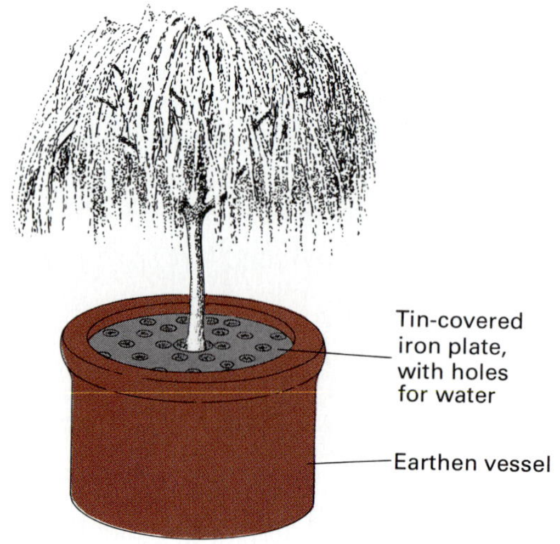

FIGURE 1-3

Van Helmont's willow tree. He grew the weighed tree in a weighed container of dirt, adding only water. After five years the tree had gained 164 lb, but the dirt had lost only about 2 oz.

Helmont was partly right: a tree *is* largely water. Actually carbon dioxide from the air is also consumed in the growth of a tree, along with small amounts of nutrients from the soil.

It was Robert Boyle (1626–1691) who first rejected the Greek method of identifying the elements by reason alone. An element was understood to be a simple substance out of which all matter could be composed. Therefore, according to Boyle, a substance must be tested by experiment; if it can be broken down into simpler substances, or if it can be constructed out of simpler substances, it is not an element. Boyle was the first to emphasize that a scientific theory can be proved or disproved by experiment. Boyle's practical definition of an element as a substance that cannot be broken down into simpler substances stood for over 200 years and will serve us for now. Boyle himself did not attempt to identify which substances are elements.

An important step in identifying elements was the discovery by Antoine Lavoisier (about 1773) of one of

*From J. R. Partington, *A Short History of Chemistry* (London: Macmillan, 1937), pp. 51–52.

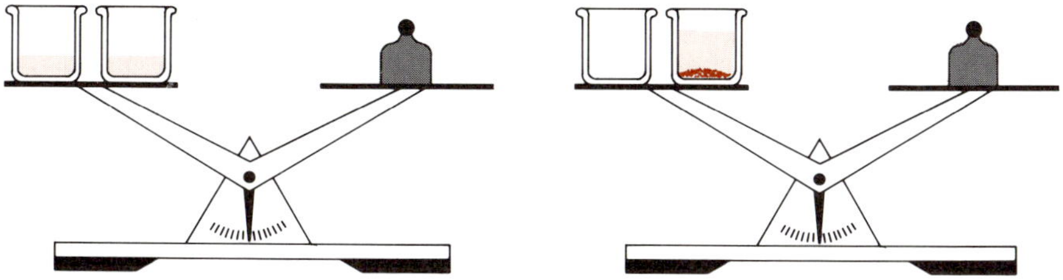

FIGURE 1-4

Conservation of mass. Solutions of silver nitrate and of sodium chloride in water react (when mixed) to produce solid silver chloride and a solution of sodium nitrate in water. The total mass of the beakers plus contents is the same before and after the reaction.

the fundamental laws of chemistry, the **law of conservation of mass:** *In an ordinary chemical reaction, the total mass of the reacting substances equals the total mass of the products formed.* There is no *detectable* change in mass in a chemical reaction (although, under the special conditions of nuclear reactions, measurable amounts of mass may be transformed to energy; see Section 5-11). A natural law describes experimentally observed phenomena; it is a generalization of a large number of observed facts. Thus, the law of conservation of mass is *not* an explanation of the behavior of matter, but a summary of fact, based on experiments carried out by different scientists and obtained under differing conditions. Lavoisier, using special balances more accurate than any ever used before, pioneered in the application of precise measurement to the study of chemical change.

A simple demonstration of the law of conservation of mass is shown in Figure 1-4: two beakers containing clear, colorless solutions—one of silver nitrate in water and the other of sodium chloride in water—are balanced with weights. When the two solutions are mixed, the products are a white solid (silver chloride) and a solution of sodium nitrate in water, but the total mass of the beakers plus the products is the same as it was before the reaction.

Another demonstration of the law of conservation of mass is given by Lavoisier's experiment on the decomposition of mercuric oxide (Figure 1-5). Here the *compound* mercuric oxide, a red powder, is decomposed by heating to yield two *elements:* mercury (a silvery liquid), and oxygen gas. The total mass of the mercury plus oxygen is equal to that of the original mercuric oxide.

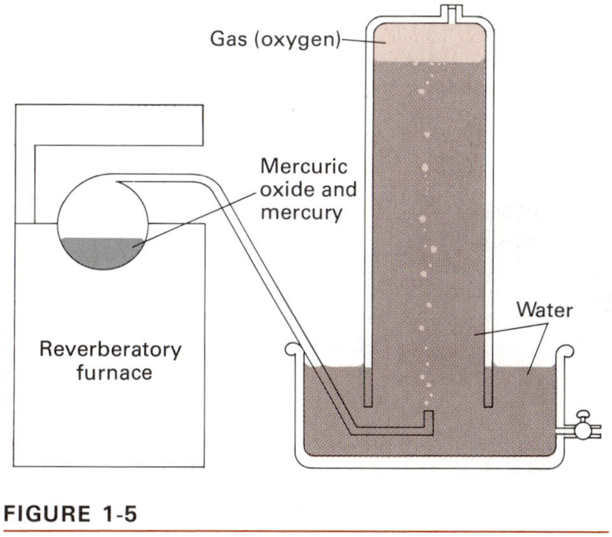

FIGURE 1-5

Lavoisier's apparatus for decomposing mercuric oxide. The mass of the mercuric oxide is equal to the sum of the masses of the mercury and oxygen.

Systematically pursuing experiments like those of Lavoisier, we can determine which substances are elemental and which composite. Elements are those that cannot be formed by combination of other substances and therefore cannot be broken down into simpler substances. **Compounds** are pure substances that are formed by the combination of two or more elements and that have constant compositions. **Mixtures** are aggregates of two or more substances that are not chemically combined; mixtures have no fixed chemical composition. *All* matter is either an element; a compound; or a mixture of elements, of compounds, or of both.

Most things around us are mixtures—a brick, the

Table of simple substances belonging to all the Kingdoms of Nature, which may be considered as the elements of bodies

Names	
Light	forms of energy
Caloric (heat)	
Oxygen	elements
Azote (nitrogen)	
Hydrogen	

Oxidable and acidifiable simple substances not metallic

Names	
Sulfur	elements
Phosphorus	
Charcoal (carbon)	
Muriatic radical	compound containing chlorine
Fluoric radical	compound containing fluorine
Boracic radical	compound containing boron

Table of simple substances, continued oxidable and acidifiable simple metallic bodies

Names	
Antimony	elements
Arsenic	
Bismuth	
Cobalt	
Copper	
Gold	
Iron	
Lead	
Manganese	
Mercury	
Molybdenum	
Nickel	
Platinum	
Silver	
Tin	
Tungsten	
Zinc	

Salifiable Simple Earthy Substances

Names	
Lime	compound containing calcium
Magnesia	compound containing magnesium
Barytes	compound containing barium
Argill	compound containing aluminum
Silex	compound containing silicon

FIGURE 1-6

Lavoisier's table of simple substances. Lavoisier had some doubts about this compilation of elements: ". . . as chemistry advances towards perfection, by dividing and subdividing, it is impossible to say where it is to end; and these things we at present suppose simple may soon be found quite otherwise. All we dare venture to affirm of any substance is that it must be considered as simple in the present state of our knowledge and so far as chemical analysis has been able to show." [Quotation and table from Lavoisier's *Elements of Chemistry* (1789); annotations are in color.]

air, a moon rock; a few are compounds—water, sugar, aspirin; and fewer are elements—a diamond, iron, oxygen. An element cannot be identified at a glance. A list of 33 elements prepared by Lavoisier in 1789 included 23 substances we now know to be elements (Figure 1-6). Eight of the substances were compounds that have since been broken down into simpler substances (as Lavoisier suspected they might be); each of the eight contained a new element. Only two on the list were entirely wrong: light and caloric (heat), which are not substances but forms of energy.

Element hunting became a popular nineteenth-century sport, and new elements were discovered rapidly, with 65 elements identified by 1870. There are now more than 100 known elements; 90 occur naturally, and the rest have been produced in nuclear reactions. Everything in our world, from a grain of salt to an elephant, however differing in appearance and use, is composed of combinations of the naturally occurring elements. Thousands of compounds—including the main components of natural gas, gasoline, petroleum, fuel oils, and asphalt—are composed of only two elements: carbon and hydrogen.

To simplify the writing of chemical formulas and equations, symbols are used to represent the elements. In many cases, the symbol is the first letter of the element's English name: thus, H for hydrogen, N for nitrogen, C for carbon. If more than one element begins with the same letter, two letters may be used in the symbol: Al for aluminum, Ar for argon, As for arsenic. The first letter is capitalized, but the second not. Some symbols are derived from Latin: lead is Pb, from the Latin word *plumbum* (from which the word *plumbing,* which in ancient Rome involved lead pipes, also derives); Au, from *aurum,* for gold; Sn, from *stannum,* for tin. Others are derived from other languages, as in W (from the German *Wolfram*) for tungsten. A list of elements with their symbols and atomic weights is given inside the cover of this book. These symbols may seem difficult to remember, but they are much simpler than those used in earlier times (Figure 1-7).

FIGURE 1-7

Chemical symbols from alchemists' notebooks. [From Oswald Croll, *Basilica Chymica,* 1609.]

1-2 UNITS OF MEASURE

Progress in chemistry accelerated with the introduction of techniques for careful measurement. To describe chemical changes, it is necessary to measure accurately length, volume, mass, temperature, and time. All over the world, the International System of Units (designated SI, for *Système International*) is used in scientific research, and it is used almost exclusively in this book. The SI units are based on multiples of 10, making conversion to different units within the system simply a matter of shifting the decimal point. Remem-

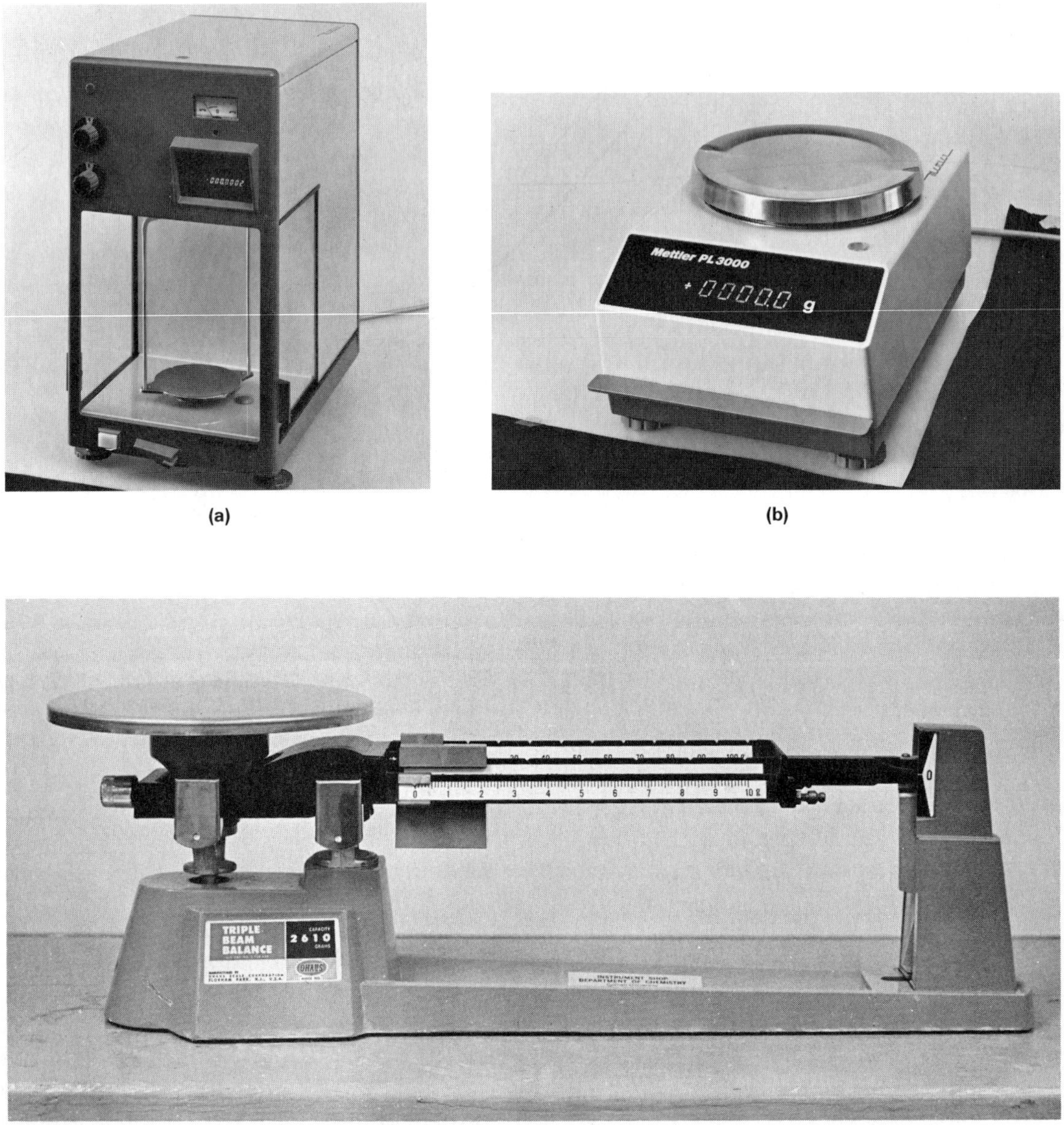

(a) (b) (c)

FIGURE 1-8

Three common types of laboratory balance. **(a)** Electronic analytical balance. **(b)** Electronic top-loading balance. **(c)** Triple-beam platform balance. [(a), (b) courtesy of Mettler Instrument Corporation, Hightstown, New Jersey; (c) by Richard Byrnes.]

bering the size of units is simplified by the use of prefixes, some of which are listed here:

mega- (M)	= million	(10^6)
kilo- (k)	= thousand	(10^3)
deci- (d)	= tenth	(10^{-1})
centi- (c)	= hundredth	(10^{-2})
milli- (m)	= thousandth	(10^{-3})
micro- (μ)	= millionth	(10^{-6})
nano- (n)	= billionth	(10^{-9})
pico- (p)	= millionth millionth	(10^{-12})

The basic unit of length is the *meter* (m), which was originally defined as one ten-millionth of the distance from the equator to the north pole. (The meter is about 39 inches.)

$$1 \text{ kilometer (km)} = 1000 \text{ meters (m)} = 1 \times 10^3 \text{ m}$$
$$1 \text{ centimeter (cm)} = 0.01 \text{ m} = 1 \times 10^{-2} \text{ m}$$
$$1 \text{ millimeter (mm)} = 0.001 \text{ m} = 1 \times 10^{-3} \text{ m}$$

The unit of **mass** is the *kilogram* (kg)—mass refers to the quantity of matter in an object. **Weight** refers to the gravitational attraction of the earth for an object. At a given location on the earth, two objects that have the same mass also have the same weight, and so these terms are often used interchangeably. (The kilogram is about 2.2 pounds.) Figure 1-8 shows three common types of balance used in chemistry laboratories to measure mass.

$$1 \text{ kilogram (kg)} = 1000 \text{ grams (g)} = 1 \times 10^3 \text{ g}$$
$$1 \text{ milligram (mg)} = 0.001 \text{ g} = 1 \times 10^{-3} \text{ g}$$

We will often find it convenient to use the *liter,* which is defined as 1000 cubic centimeters, as a unit of volume. (One liter is roughly 1.1 quarts.) Figure 1-9 shows some devices used in the laboratory to measure volume.

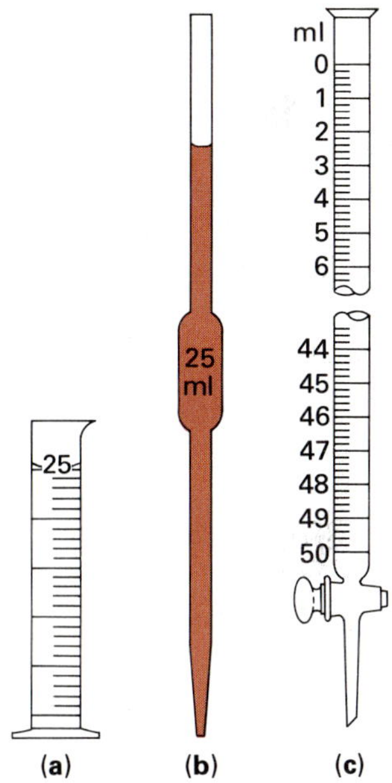

FIGURE 1-9

Devices commonly used in the laboratory to measure volume. **(a)** Graduated cylinder (25 ml). **(b)** Pipet (25 ml). **(c)** Buret (50 ml).

$$1 \text{ liter } (\ell) = 1000 \text{ cubic centimeters (cm}^3) = 1 \times 10^3 \text{ cm}^3$$

$$1 \text{ liter } (\ell) = 1000 \text{ milliliters (ml)} = 1 \times 10^3 \text{ ml}$$

Appendix 1 lists selected conversions between English and SI units of measurement. Appendix 2 contains a review of the use of exponential numbers, which are almost indispensable in chemistry for expressing very large or small numbers. Appendix 3 contains a discussion of significant figures, which will help you decide how many digits to keep in your answer when solving problems. If you keep too few or too many significant figures, you give a misleading impression of the accuracy of the measured quantities you have used.

EXAMPLE 1-1

How many centimeters are in 6.00 inches?

Conversion factor needed: 1 in. = 2.54 cm (1-1)

$$\begin{array}{c}\text{Number of}\\ \text{centimeters}\\ \text{in 6.00 inches}\end{array} = 6.00 \text{ in.} \times \frac{2.54 \text{ cm}}{1 \text{ in.}} = 15.2 \text{ cm}$$

The factor 2.54 cm/1 in. is a conversion factor, which does not change the value but converts the number to new units.

We can see this if we take Equation 1-1 and divide both sides by the same factor, 1 in.:

$$\frac{1 \text{ in.}}{1 \text{ in.}} = \frac{2.54 \text{ cm}}{1 \text{ in.}} = 1$$

Because 2.54 cm *is* 1 in., when we multiply by 2.54 cm/1 in., we are multiplying by 1.

There should be 3 significant figures in the answer, 15.2 cm, because there are 3 in each of the numbers used in the calculation.

In any problem, all units should be written out and manipulated mathematically, like numbers. If Example 1-1 is done correctly, the units of inches will cancel out and give the answer in centimeters, as desired. If, in Example 1-1, you divide by 2.54 cm/in. instead of multiplying, the inches will not cancel out, and you will catch your mistake.

The last step in solving a problem is to check to make sure your answer is reasonable. In this problem, we are converting from larger units (inches) to smaller units (centimeters), so the answer in centimeters should be larger than the original number of inches. Each inch is slightly over $2\frac{1}{2}$ cm, so the answer should be a bit larger than $2\frac{1}{2} \times 6$, or 15 cm.

EXAMPLE 1-2

How many liters is 30 milliliters?

$$\text{Number of liters} = 30 \text{ ml} \times \frac{1 \ \ell}{1000 \text{ ml}}$$

$$= 0.030 \ \ell = 3.0 \times 10^{-2} \ \ell$$

The units cancel out correctly to give the correct units, liters, for the answer. The answer is reasonable because a liter is a large unit of volume and a milliliter a small one, and we expect the number of liters to be much *smaller* than the number of milliliters. There are 1000 ml in 1 ℓ; 30 ml is *much* less than 1 ℓ. If we had multiplied by 1000 instead of dividing, the units would not have worked out correctly, and the answer would have been ridiculous!

EXERCISE 1-1 (a) Mt. Everest is almost 30,000 feet (3×10^4 ft) high. How many centimeters is this?

(b) The volume of blood in an adult human body is approximately 6 ℓ. How many milliliters is this? Give your answer in scientific notation (see Appendix 2).

1-3 DENSITY

The **density** of a substance, a property often measured by chemists, is defined as the mass of a unit volume of that substance. Mercury is a dense, silvery liquid at room temperature. Its density (13.6 grams per cubic centimeters, or 13.6 g/cm^3) is a characteristic property of mercury. One cubic centimeter of mercury weighs 13.6 g; 5.00 cm^3 of mercury weigh 5.00 × 13.6 or 68.0 g; but the density is always 13.6 g/cm^3:

$$\text{Density} = \frac{\text{Mass (g)}}{\text{Volume (cm}^3\text{)}} = \frac{68.0 \text{ g}}{5.00 \text{ cm}^3} = 13.6 \text{ g/cm}^3$$

The density of water at room temperature is 1.0 g/cm^3, only 1/13.6 as great as that of mercury.

EXERCISE 1-2 A cube of pure silver 3 cm on a side weighs 283.5 g. What is the density of silver?

1-4 TEMPERATURE

It is important to have an instrument for measuring temperature because many properties of matter are **functions** of temperature; that is, many properties change as the temperature changes. For example, the state of a substance—whether it is solid, liquid, or gas—depends on its temperature.

We detect temperature differences by observing the direction of heat flow. Heat flows spontaneously from a hotter to a colder substance. Thus, when we hold a cup of hot soup, heat flows from the soup to the cup and our hands; the soup is at a higher temperature than our hands. And when we pick up an ice cube, heat flows from our hands to the ice cube; the ice cube is at a lower temperature. To measure temperature *precisely,* we use a thermometer. If a thermometer is immersed in the substance whose temperature is to be measured, heat flows from the hotter thermometer to the colder substance, or from the hotter substance to the colder thermometer, until the two are at the same temperature. Any substance having a measurable property that is sensitive to changes in temperature can be used to make a thermometer. Gases, liquids, and solids all change volume with temperature—usually expanding when heated and contracting when cooled. This property of volume change with temperature enables all three—gases, liquids, and solids—to be used in thermometers.

The familiar thermometers containing mercury, or alcohol with red dye added to make the alcohol column more easily visible, depend on the expansion of a liquid as the temperature rises. In these thermometers a glass bulb is sealed to a capillary tube (Figure 1-10). The volume of the capillary is much smaller than that of the bulb; therefore, a small change in the volume of the liquid in the bulb, caused by a change in the temperature, produces a large change in the height of the liquid in the capillary.

Two temperature scales in common use are the Celsius or centigrade temperature scale, on which water freezes at 0° and boils at 100°; and the Fahrenheit temperature scale on which water freezes at 32° and boils at 212°. A degree Fahrenheit is only $^5/_9$ as large as a degree Celsius. The following equations give the conversions from degrees Fahrenheit to degrees Celsius, and from degrees Celsius to degrees Fahrenheit:

$$°C = \frac{5}{9}(°F - 32)$$

$$°F = \frac{9}{5} \times °C + 32$$

Figure 1-11 shows several temperatures on the Fahrenheit scale and the corresponding temperatures on the Celsius scale.

EXERCISE 1-3 (a) The element mercury, used in thermometers, is a liquid at room temperature. It freezes at −39°. What temperature is this on the Fahrenheit scale?
(b) A seriously ill person has a temperature of 105°F. What temperature is this on the Celsius scale?

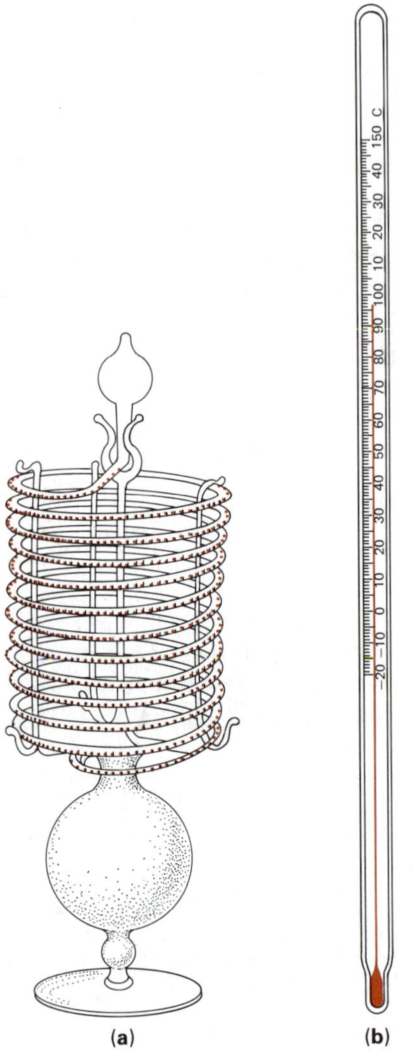

FIGURE 1-10

Thermometers. **(a)** An early thermometer. The temperature scale is marked with colored glass beads on the outside of the glass tube. **(b)** A mercury-in-glass thermometer. The volume of the mercury in the bulb is much greater than that of the mercury in the capillary tube.

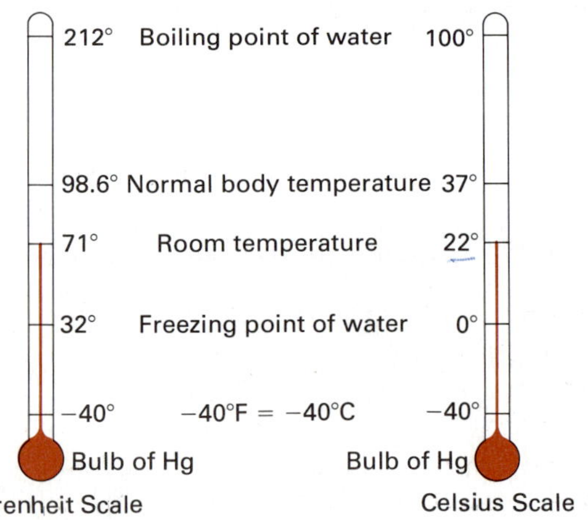

FIGURE 1-11

The Fahrenheit and Celsius temperature scales.

1-5
LAW OF DEFINITE COMPOSITION

Along with the law of conservation of mass, one of the fundamental laws leading to our present understanding of the microscopic nature of matter is the **law of definite composition.** Formulated early in the nineteenth century by the Frenchman Joseph Louis Proust, this law states that *a compound always consists of the same proportions by weight of all its elements.* Thus pure water always consists of 89% oxygen and 11% hydrogen by weight. This is true whether the water is rainwater, distilled seawater, or water prepared in the laboratory by burning hydrogen in oxygen.

$$\underset{\text{water}}{100\text{ g}} \xrightarrow{\text{(yields)}} \underset{\text{oxygen}}{89\text{ g}} + \underset{\text{hydrogen}}{11\text{ g}}$$

$$\underset{\text{oxygen}}{89\text{ g}} + \underset{\text{hydrogen}}{11\text{ g}} \longrightarrow \underset{\text{water}}{100\text{ g}}$$

We can *mix* hydrogen and oxygen in all proportions, but they will *react* only in definite proportion. This derives from the facts of the conservation of mass noted above and the law of definite composition. Thus,

$$\underbrace{\underset{\text{oxygen}}{89\text{ g}} + \underset{\text{hydrogen}}{13\text{ g}}}_{\text{102 g total}} \longrightarrow \underbrace{\underset{\text{water}}{100\text{ g}} + \underset{\text{hydrogen}}{2\text{ g}}}_{\text{102 g total}}$$

There is enough oxygen to react with only 11 g of hydrogen, producing 100 g of water. Two grams of hydrogen remain when the reaction is over. Note that in this example we have used the law of conservation of mass in determining that there was unreacted hydrogen left over at the end of the reaction. If the composition of the mixture is changed,

$$\underbrace{\underset{\text{oxygen}}{95\text{ g}} + \underset{\text{hydrogen}}{11\text{ g}}}_{\text{106 g total}} \longrightarrow \underbrace{\underset{\text{water}}{100\text{ g}} + \underset{\text{oxygen}}{6\text{ g}}}_{\text{106 g total}}$$

There is enough hydrogen to react with only 89 g of oxygen, producing 100 g of water, and leaving the other 6 g of oxygen unreacted.

EXAMPLE 1-3

Calcium carbonate is a major constituent of chalk, marble, eggshells, and seashells. When a 50 g sample of calcium carbonate is decomposed, it yields 20 g of calcium, 6 g of carbon, and 24 g of oxygen.

(a) What is the percentage of calcium, of carbon, and of oxygen in calcium carbonate?

$$\%\text{ calcium} = \frac{20\text{ g}}{50\text{ g}} \times 100 = 40\%\text{ Ca}$$

(The percentage calcium is the fraction of calcium times 100.

$$\%\text{ carbon} = \frac{6\text{ g}}{50\text{ g}} \times 100 = 12\%\text{ C}$$

$$\%\text{ oxygen} = \frac{24\text{ g}}{50\text{ g}} \times 100 = 48\%\text{ O}$$

If correct, the percentages should total 100% (the whole sample), and they do: 40% + 12% + 48% = 100%.

(b) What mass of calcium is contained in a 20 g sample of calcium carbonate? From (a) we know that 40% (or 40/100) of the sample is Ca.

$$\text{Mass Ca} = 20\text{ g} \times 0.40 = 8.0\text{ g Ca}$$

EXERCISE 1-4 A well-balanced daily diet consists of approximately 1 g of protein, 1 to 2 g of fat, and 5 to 7 g of carbohydrate for each kilogram (kg) of body weight. A man weighing 70 kg ate 70 g of protein, 200 g of fat, and 280 g of carbohydrate during a 24 hour period. What percentage of his diet consisted of protein, of fat, and of carbohydrate? Is this a well-balanced diet?

1-6
LAW OF MULTIPLE PROPORTIONS

In some cases, two elements brought together under differing conditions may form two or more different compounds. The relative masses of the two elements in the different compounds confirm the **law of multiple proportions:** *If two elements, A and B, combine in different ways to form two or more different compounds, the ratios of grams of A per gram of B* (g A/g B) *in the series of compounds are simple integer or fractional multiples of each other.* That is, the mass ratio g A/g B in one compound is a simple multiple or simple fraction of the mass ratio g A/g B in the other compounds. For example, hydrogen (H) reacts with oxygen (O) to form water, with a mass ratio g O/g H = 8.0/1. These two elements also form the different substance hydrogen peroxide, with a mass ratio g O/g H= 16.0/1. The mass ratio for hydrogen peroxide is exactly twice that for water.

Table 1-1 gives additional examples of pairs of compounds that illustrate the law of multiple proportions. Note that the mass ratio for cuprous oxide is twice that for cupric oxide (7.94/3.97 = 2/1). Similarly, the mass ratio for mercurous chloride is twice that for mercuric chloride (11.32/5.66 = 2/1). And the mass ratio for ferrous oxide is 3/2 that for ferric oxide (3.49/2.32 = 1.5/1, or 3/2). In each case, the amounts of A (here, Cu, Hg, or Fe) per unit mass of B (O or Cl) are simple integer or fractional multiples of each other.

TABLE 1-1
Law of multiple proportions

Compound	Mass ratio
Cuprous oxide	$\frac{\text{Grams copper (Cu)}}{\text{Grams oxygen (O)}} = \frac{7.94}{1}$
Cupric oxide	$\frac{\text{Grams Cu}}{\text{Grams O}} = \frac{3.97}{1}$
Mercurous chloride	$\frac{\text{Grams mercury (Hg)}}{\text{Grams chlorine (Cl)}} = \frac{11.32}{1}$
Mercuric chloride	$\frac{\text{Grams Hg}}{\text{Grams Cl}} = \frac{5.66}{1}$
Ferrous oxide	$\frac{\text{Grams iron (Fe)}}{\text{Grams O}} = \frac{3.49}{1}$
Ferric oxide	$\frac{\text{Grams Fe}}{\text{Grams O}} = \frac{2.32}{1}$

EXAMPLE 1-4

Hydrogen and nitrogen combine to form several different substances. Two of them are ammonia, used in the production of explosives and fertilizers, which is 82.4% nitrogen and 17.6% hydrogen; and hydrazine, a rocket propellant, which is 87.5% nitrogen and 12.5% hydrogen. Show how these data illustrate the law of multiple proportions.

In 100 g of ammonia, 82.4 g = N and 17.6 g = H:

$$\frac{\text{Mass N}}{\text{Mass H}} = \frac{82.4 \text{ g}}{17.6 \text{ g}} = 4.68$$

In 100 g of hydrazine, 87.5 g = N and 12.5 g = H:

$$\frac{\text{Mass N}}{\text{Mass H}} = \frac{87.5 \text{ g}}{12.5 \text{ g}} = 7.00$$

The ratios (Mass N/Mass H) for the two compounds are related to each other in the proportion: 4.68 : 7.00. After dividing each number by the smaller one, 4.68, the proportion is: 1 : 1.5, or 2 : 3. These ratios are *simple* integer or fractional multiples of each other, confirming the law of multiple proportions.

EXERCISE 1-5 When carbon burns in a limited quantity of air, it forms a poisonous gas that is 42.9% C and 57.1% O. When carbon burns in an unlimited quantity of air, it produces a gas that is 27.3% C and 72.7% O. Show how these data illustrate the law of multiple proportions.

KEY TERMS

The *law of conservation of mass* states that in an ordinary chemical reaction, the total mass of the reacting substances equals the total mass of the products formed.

Elements are substances that cannot be formed by combination of other substances and that cannot be broken down into simpler substances.

Compounds are distinct, reproducible, pure substances composed of two or more elements combined in a fixed mass ratio.

Mixtures may be of elements, of compounds, or both; the proportions of different substances in the mixture can be varied.

The *density* of a substance is the mass of a unit volume of that substance.

Temperature is measured by observing the direction of heat flow, with reference to some standards. On the Celsius scale, the reference points are the freezing and boiling points of water, 0°C and 100°C, respectively.

The *law of definite composition* states that a compound always consists of the same proportions by weight of all its elements.

The *law of multiple proportions* states that if two elements A and B combine in different ways to form two or more compounds, the amounts of A per unit mass of B in the series of compounds are simple integer or fractional multiples of each other.

CHAPTER SUMMARY

The Greek philosophers attempted to identify the *elements* of matter by a process of questioning, reason, and debate. It was not until the seventeenth century that Robert Boyle introduced the idea that only by *experiment* could one determine whether or not a substance was an element. The development of accurate balances was of crucial importance in the identification of elements and in the discovery of the *laws of conservation of mass, definite composition,* and *multiple proportions.* A natural law describes, and is a summary of, a large number of experimentally observed phenomena; it is not an explanation of the behavior of matter.

To describe chemical changes, it is necessary to measure accurately length, volume, mass, temperature, and time. The International System of Units is used in scientific research. In solving problems, it is important to keep the correct number of significant figures. Units should be written out and manipulated mathematically like numbers, and *all answers should be checked to see whether they are reasonable or nonsensical.*

PROBLEMS

1-1. Which of the following are elements, and which compounds?
(a) ozone, O_3 (b) ammonia, NH_3 (c) iodine, I_2 (d) hydrogen cyanide, HCN (e) phosphorus, P_4

1-2. Write the chemical symbol for each of the following:
(a) helium (b) nitrogen (c) bromine (d) gold (e) silver (f) arsenic (g) copper (h) potassium

1-3. Name the element represented by each of the symbols:
(a) Li (b) Be (c) B (d) Fe (e) Ne
(f) Na (g) Mg (h) Al (i) Si

1-4. Name some common substances that are mixtures, some that are compounds, and some that are elements.

1-5. Make the following conversions:
(a) 250 milliliters into liters
(b) 30 grams into kilograms
(c) 5 milligrams into grams
(d) 3500 centimeters into meters
(e) 6 liters into milliliters
(f) 820 kilogams into grams
(g) 3 kilograms into milligrams
(h) 16 meters into millimeters.

1-6. The speed of light is 3.0×10^8 meters per second (m/sec). How many miles per hour is this?

1-7. A 20 ml sample of alcohol weighs 15.8 g. What is the density of alcohol?

1-8. A flask has a volume of 250 ml and weighs 100 g.
(a) How much would the flask weigh when filled with water? (The density of water is 1.00 g/ml.)
(b) How much would the flask weigh when filled with mercury? (The density of mercury is 13.6 g/ml.)
(c) If the flask weighs 277 g when filled with ether, what is the density of ether?

1-9. The density of gold is 19.3 g/cm^3. What is the volume of a gold coin that weighs 2.0 g?

1-10. The density of air is 1.3 g/ℓ at 0°C. What would be the weight of the air at 0°C in a room that is 4 m square and 3 m high?

1-11. Air is liquefied at −193°C. What temperature is this on the Fahrenheit scale?

1-12. Lead melts at 621°F. What temperature is this on the Celsius scale?

1-13. Alcohol colored with red dye is used in thermometers to measure temperatures below the freezing point of mercury. An alcohol thermometer can be used to measure temperatures down to −117°C, the freezing point of the alcohol. What temperature is this on the Fahrenheit scale?

1-14. On a hot day, the temperature reaches 100°F. What temperature is this on the Celsius scale?

1-15. The phosphorus content of cow manure is 5.4 lb of P in 5000 lb manure.
(a) What is the percentage of phosphorus in cow manure?
(b) A corn crop removes 22 lb of phosphorus from 1 acre of soil. The cobs, leaves, and stalks contain 10 lb of phosphorus. Assuming that the cobs, leaves, and stalks are ground up and returned to the soil, what weight of cow manure must be added to restore the soil to its original phosphorus content? (A cow produces about 15 *tons* of manure per year.)

1-16. The stomach secretes about 2.5 ℓ of gastric juice every day. How many milliliters is this? How many quarts?

1-17. The table below gives typical blood counts for a normal person and for a sickle-cell anemia patient. (Assume a total blood volume of 6 ℓ.)

Count	*Normal*	*Sickle-cell patient*
Red-cell	5.4×10^6 cells/ml	2.6×10^6 cells/ml
White-cell	7×10^3 cells/ml	15×10^3 cells/ml
Hemoglobin	0.16 g/ml	0.08 g/ml

(a) For the normal person and for the sickle-cell patient, calculate the number of red cells per liter and the total number of red cells in the body.

(b) For the normal person and for the sickle-cell patient, calculate the number of white cells per liter and the total number of white cells in the body.
(c) For the normal person and for the sickle-cell patient, calculate the number of grams of hemoglobin per liter, and the total number of grams of hemoglobin in the body. What percentage is this of *your* body weight?

1-18. (a) The slightly salty water characteristic of living tissue freezes at about −1.5°C. What temperature is this on the Fahrenheit scale?
(b) Many proteins are denatured (changed in such a way as to destroy their biological functions) at about 45°C. What temperature is this on the Fahrenheit scale?

1-19. If the heart rate is 72 beats per minute, and 70 ml of blood are ejected into the aorta with each beat,
(a) what is the volume of blood pumped in liters per minute?
(b) how many liters are pumped in a 24 hour period?

1-20. If a piece of tinfoil is covered with concentrated nitric acid (HNO_3), and carefully heated, a white oxide of tin is produced. The other products of the reaction are water vapor and gaseous oxides of nitrogen. Outline the measurements you would make and the calculations you would perform to determine the percentage of oxygen in tin oxide. (The atomic weights of tin and oxygen are known.)

1-21. Many salts that have been crystallized from an aqueous solution appear dry, yet yield large quantities of water when heated. Copper sulfate is such a salt, called a hydrate, changing color from deep blue to white when all the "water of hydration" is driven off. Describe an experiment by which you could determine what percentage of the hydrated salt is water. Outline the measurements you would make and the calculations you would perform.

1-22. Sodium metal reacts violently with chlorine gas to form table salt, sodium chloride (NaCl). A 23.0 g sample of sodium reacts with an excess of chlorine to produce 58.5 g of salt?
(a) What percentage of salt is sodium?
(b) What weight of salt can be made from 39.3 g of sodium and 70.0 g of chlorine?
(c) How many grams of sodium are needed to make 2.2 kg (1 pound) of salt?

1-23. Sulfur burns in air to produce an irritating gas that is 50% S and 50% O, by weight. This compound can be made to combine with more oxygen, forming another gas that is 40% S and 60% O, by weight. Show how these data illustrate the law of multiple proportions.

1-24. Two oxides of nitrogen are produced in car exhausts. The one that predominates is 46.7% nitrogen, and the other is 30.4% nitrogen. Show that these data illustrate the law of multiple proportions.

White aluminum oxide (Al_2O_3, see Example 2-4) serves as the catalyst support material in catalytic converters. The oxide's surface is incredibly rough and porous; thus, although the largest of the spherical beads shown here is only a little over 6 mm in diameter, each bead contains about 25 square meters of surface area. That means about 200 beads provide the same surface area as a football field.

2

Stoichiometry: Chemical Arithmetic

The laws of conservation of mass, definite composition, and multiple proportions lead to and support Dalton's atomic hypothesis. Careful analysis of chemical compounds permitted Dalton to establish a table of relative atomic masses. Using an accepted table of atomic masses, along with chemical formulas and balanced chemical equations, we can solve a variety of chemical problems.

2-1
ATOMIC THEORY

From the time of the ancient Greeks, some scientists believed that matter is composed of small, discrete particles. Democritus (about 400 B.C.) argued that atoms of different elements differ in size and shape, giving each element unique properties, and that matter is composed of mixtures of these atoms. Democritus' ideas were unpopular at the time, largely because Aristotle attacked them and promoted the four-element theory instead. Because there were no experiments to confirm or to challenge these ideas, the argument whether matter is continuous or atomic raged even after the English schoolmaster John Dalton formulated his atomic theory in 1803. The laws of conservation of mass, definite composition, and multiple proportions lead to, and support, Dalton's **atomic hypothesis,** which most simply accounts for them. The hypothesis (or theory) is summarized as follows.

1. Matter is composed of discrete, indestructible particles called atoms, which remain intact throughout a chemical change.

2. All atoms of a given element are identical in every respect. (We shall see, in Chapter 5, that most elements are mixtures of what are called isotopes, which vary slightly in mass, but are identical in chemical behavior.)

3. The atoms of one element differ from those of another element in mass and in chemical behavior.

4. Chemical reactions are the combination and separation of atoms.

According to this theory, a compound is formed when two or more atoms combine to form a **molecule.** The molecule carbon monoxide is composed of one atom each of carbon and oxygen. When carbon monoxide (whatever its source and however large the sample) is decomposed into its elements, the weight of oxygen is always 1.33 times that of carbon. This example of the law of definite composition can be explained by assuming that a single atom of oxygen weighs 1.33 times as much as a single atom of carbon.

Then, if one carbon atom combines with one oxygen atom,

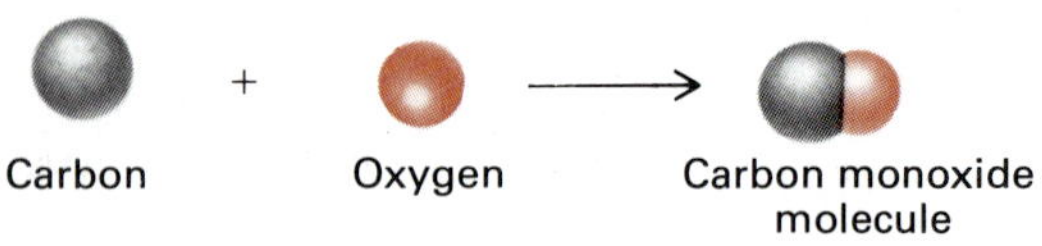

$$\frac{\text{Mass oxygen}}{\text{Mass carbon}} = \frac{1.33}{1}$$

Ten molecules of carbon monoxide would have a mass ratio

$$\frac{\text{Mass oxygen}}{\text{Mass carbon}} = \frac{10 \times 1.33}{10 \times 1} = \frac{1.33}{1}$$

and one hundred molecules of carbon monoxide would have a mass ratio

$$\frac{\text{Mass oxygen}}{\text{Mass carbon}} = \frac{100 \times 1.33}{100 \times 1} = \frac{1.33}{1}$$

Mass is conserved in the reaction because atoms are conserved (Figure 2-1). The atoms combine according to the law of definite composition because they combine atom by atom; only whole atoms combine, and the atoms of each element have their own characteristic atomic mass. The *laws of chemical combination are statements of observed experimental facts.* Dalton's *theory* explains the facts contained in the laws of conservation of mass and definite composition.

If one carbon atom combines with two oxygen atoms, the compound carbon dioxide is formed:

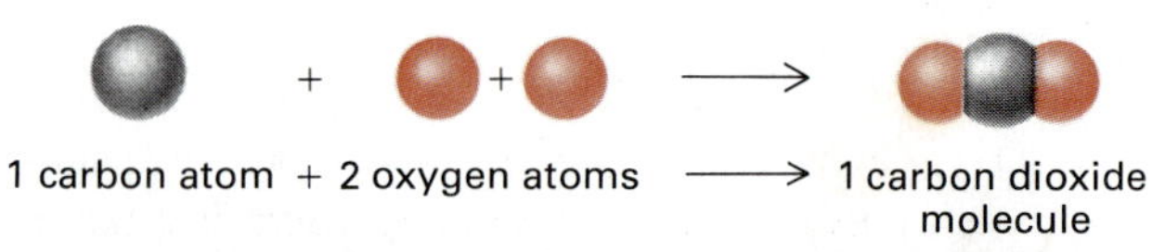

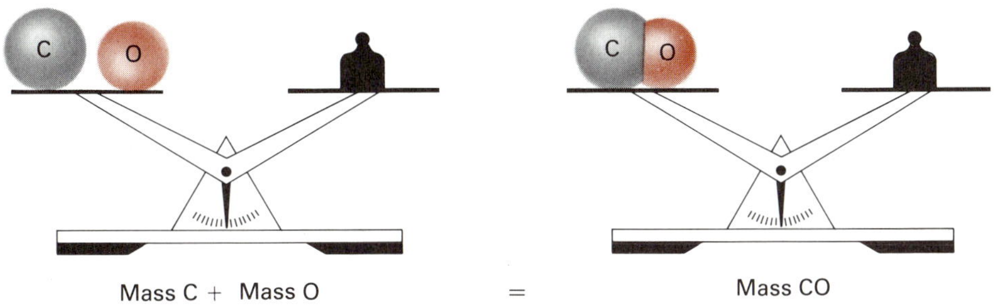

FIGURE 2-1

Mass is conserved because atoms are conserved. (It is not actually possible to weigh individual atoms on a balance; the diagram is symbolic.)

$$\frac{\text{Mass oxygen}}{\text{Mass carbon}} = \frac{2 \times 1.33}{1} = 2.66$$

$$\text{For carbon monoxide, } \frac{\text{Mass O}}{\text{Mass C}} = 1.33$$

$$\text{For carbon dioxide, } \frac{\text{Mass O}}{\text{Mass C}} = 2.66$$

The mass ratio for carbon dioxide is twice that for carbon monoxide because a molecule of carbon dioxide contains twice as many oxygen atoms as does a molecule of carbon monoxide. Because atoms are indivisible, only whole numbers of atoms can react to form a compound; this accounts for the law of multiple proportions.

The law of multiple proportions was actually *predicted* by Dalton as a consequence of his atomic theory and demonstrated in experiments performed by himself and others. The atomic theory beautifully and simply *explained* the facts summarized in the laws of conservation of mass and definite composition, and it *predicted* a new law—the law of multiple proportions. These are important requirements for a theory: simple explanations of known facts and prediction of new observations.

2-2 CHEMICAL FORMULAS

For pure substances (those with a constant composition) there is a handy shorthand way of specifying that composition. The **formula** of a compound tells how many of each kind of atom are present in a molecule of the compound. Each element (each kind of atom) is represented by its symbol. The number of atoms of that kind in the molecule is shown by a subscript to the right of the symbol. If there is only one atom of a kind, the subscript "1" is omitted. The formula for carbon monoxide, CO, indicates that the molecule contains one C atom and one O atom. The formula for carbon dioxide, CO_2, indicates that the molecule contains one C atom and two O atoms. The formula for common sugar (sucrose), $C_{12}H_{22}O_{11}$, means that a molecule of sucrose contains 12 C atoms, 22 H atoms, and 11 O atoms. A slightly more complicated formula is $Pb(C_2H_5)_4$, for the gasoline additive tetraethyl lead; this formula means one atom of Pb and four ethyl groups—each ethyl group (C_2H_5) consisting of two atoms of C and five of H. All the numbers inside the parentheses are multiplied by the subscript 4. One molecule of $Pb(C_2H_5)_4$, therefore, contains one atom of Pb, $4 \times 2 = 8$ atoms of C, and $4 \times 5 = 20$ atoms of H.

EXERCISE 2-1 How many atoms of carbon and how many of hydrogen are in one molecule of octane, a component of gasoline, $CH_3(CH_2)_6CH_3$?

2-3 RELATIVE ATOMIC WEIGHTS

Atoms are much too small to be weighed or measured directly. Later we shall see how much individual atoms and molecules weigh and how large they are. In the meantime, however, we can set up a scale of *relative* weights, called **atomic weights** (or atomic masses). These are simply ratios of the weights (or masses) of atoms of different elements. In Section 2-1, we saw that an atom of oxygen is 1.33 times as heavy as an

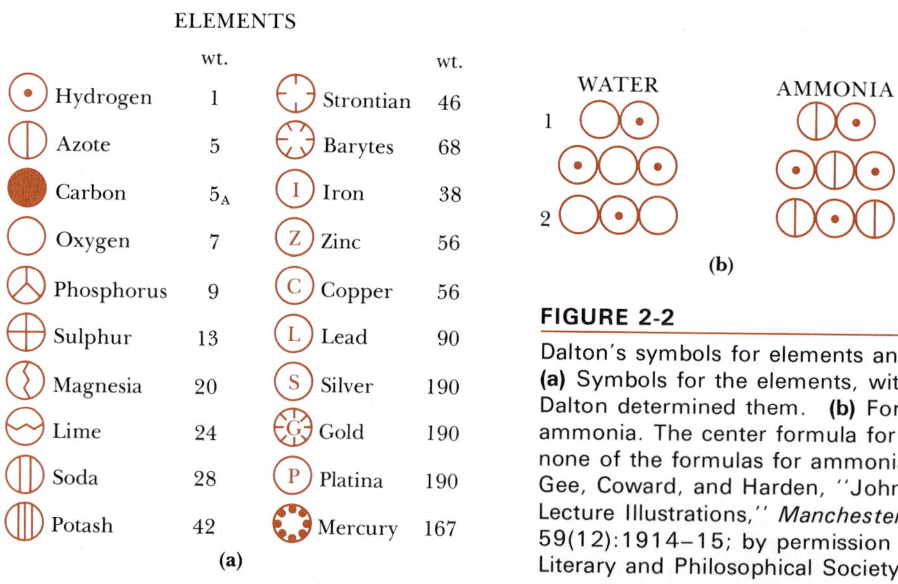

FIGURE 2-2

Dalton's symbols for elements and compounds. **(a)** Symbols for the elements, with atomic weights as Dalton determined them. **(b)** Formulas for water and ammonia. The center formula for water is correct, but none of the formulas for ammonia is correct. [From Gee, Coward, and Harden, "John Dalton's Lectures and Lecture Illustrations," *Manchester Memoirs* 59(12):1914–15; by permission of the Manchester Literary and Philosophical Society.]

atom of carbon. Similar ratios can be computed for other elements, using results experimentally determined by analysis of macroscopic samples—samples large enough to be weighed by the chemists of Dalton's day. Figure 2-2 shows Dalton's symbols for the elements, as well as the atomic weights he computed and his formulas for water and ammonia. As the first table of atomic weights, this table was an important contribution to the progress of chemistry, although some of the atomic weights Dalton reported were incorrect.

Using Dalton's reasoning (with additional checks) and other methods that we shall see in later chapters, we can set up a self-consistent set of relative weights. An arbitrary value is assigned to one element, and all others are fixed relative to that assigned weight. The lightest element is hydrogen (H). Using an arbitrary scale with H assigned weight equal to 1 (on a modern scale it is not exactly 1, but extremely close to it), we find, to a close approximation, the atomic weights shown in Table 2-1.

TABLE 2-1
Atomic weights

Element	Symbol	Approximate atomic weight (at. wt.)
Hydrogen	H	1
Helium	He	4
Carbon	C	12
Nitrogen	N	14
Oxygen	O	16
Calcium	Ca	40

Note: Modern tables of atomic weights actually use as reference standard an isotope (explained in Chapter 5) of carbon that has been assigned the atomic weight of exactly 12 (that is, 12.000 . . .). The table inside the cover of this book is based on this convention. On this scale, the atomic weight of hydrogen is 1.00797.

2-4 MOLECULAR WEIGHTS

The sum of the atomic weights of all the atoms in a molecule is the **molecular weight** (or molar mass). Using the atomic weights listed in Table 2-1, we can calculate some molecular weights:

For carbon monoxide (CO),

1 carbon atom + 1 oxygen atom ⟶ 1 carbon monoxide molecule

$$\text{at. wt. C} + \text{at. wt. O} = \text{mol. wt. CO}$$

$$12 + 16 = 28$$

The molecular weight of carbon monoxide is the atomic weight of carbon plus the atomic weight of oxygen: 12 + 16 = 28.

The molecular weight of carbon dioxide (CO_2) is the atomic weight of carbon plus twice the atomic weight of oxygen: 12 + 2(16) = 44.

Oxygen gas is known to be composed of diatomic molecules, O_2. The molecular weight of O_2 is therefore twice the atomic weight of oxygen: 2(16) ≐ 32.

The molecular weight of water (H_2O) is twice the atomic weight of hydrogen, plus the atomic weight of oxygen: 2(1) + 16 = 18.

The molecular weight of sucrose (table sugar, $C_{12}H_{22}O_{11}$) is 12 times the atomic weight of carbon plus 22 times the atomic weight of hydrogen, plus 11 times the atomic weight of oxygen:

$$12(12) + 22(1) + 11(16) = 342$$

EXERCISE 2-2 Find the weight of one mole of each of the following:

(a) the gasoline additive tetraethyl lead, $Pb(C_2H_5)_4$ (at. wt. Pb = 207);

(b) the simple sugar, glucose, $C_6H_{12}O_6$

(c) aspirin, $C_9H_8O_4$

(d) caffeine, $C_8H_{10}N_4O_2$

So far we have not assigned any units to the atomic and molecular weights. One molecule of carbon monoxide doesn't weigh 28 g; it weighs *much less*. But one molecule of carbon monoxide does weigh 28 times as much as one hydrogen atom; 10 molecules of carbon monoxide weigh 28 times as much as 10 hydrogen atoms; and 10^{23} molecules of carbon monoxide weigh 28 times as much as 10^{23} hydrogen atoms.

It is common to refer to the weights of atoms and molecules as **atomic mass units** (amu). The isotope (Section 5-5) of carbon used as reference standard in modern tables of atomic weight has an atomic weight of exactly 12 amu. An atom of hydrogen has an atomic weight of approximately 1 amu, and a molecule of carbon monoxide has an atomic weight of approximately 28 amu.

2-5 AVOGADRO'S NUMBER AND THE MOLE

The use of relative atomic and molecular weights provides an indirect way of counting atoms and molecules. We are now going to see how the relative weights are related to *actual* weights of matter, and how numbers of atoms and molecules are related to the atomic and molecular weights. We saw that a molecule of carbon monoxide weighs 28 times as much as an atom of hydrogen, and that whatever number of molecules of carbon monoxide we take, they weigh 28 times as much as that same number of atoms of hydrogen. It follows that the *number* of molecules in 28 grams of carbon monoxide is the same as the *number* of atoms in 1 gram of hydrogen. (Convince yourself that this is true.)

If 1 g of hydrogen atoms contains N atoms, then 28 g of carbon monoxide also contain N molecules. And 44 g of carbon dioxide and 18 g of water also contain N molecules. The mass of material that weighs, in grams, a number equal to the atomic or molecular weight, is called a **mole** of that material. The number of atoms or molecules in a mole is called **Avogadro's number,** and is symbolized N. The value of Avogadro's number is enormous:

$$N = 602{,}250{,}000{,}000{,}000{,}000{,}000{,}000$$
$$\cong 6.02 \times 10^{23}$$

The symbol $\cong$ used here means "is approximately equal to." Thus N is approximately 6.02×10^{23}. This is more than 600,000 billion billion. Avogadro's number is so immense that this number of sugar cubes, one-half inch on a side, tightly packed, would cover the surface of the entire earth to a depth of over a mile and a half! This tremendous number was determined by experiment: ingenious techniques were invented to count these particles too small to be seen directly. In a later chapter we shall see how to count the number of molecules in a mole.

A mole of any substance is a *typical* sample, of a size that we can see and handle and work with in the laboratory.

One mole of H atoms weighs 1.0 g
and contains 6.02×10^{23} atoms.

One mole of C atoms weighs 12.0 g
and contains 6.02×10^{23} atoms.

One mole of O atoms weighs 16.0 g
and contains 6.02×10^{23} atoms.

One mole of CO weighs 28.0 g
and contains 6.02×10^{23} molecules.

Each CO molecule contains one C atom and one O atom.

Each mole of CO (28 g) contains one mole of C atoms (12 g) and one mole of O atoms (16 g).

In one *molecule* of CO there are two *atoms;* in one *mole* of CO there are two *moles of atoms,* or 12.04×10^{23} atoms.

In order to convince yourself that these statements are true, see if you can fill in the blanks in the following exercise.

EXERCISE 2-3 (a) One mole of H_2O weighs 18.0 g and contains _____ molecules.
(b) Each *molecule* of H_2O contains _____ atoms of H and _____ atoms of O.
(c) Each *mole* of H_2O contains _____ moles of H atoms and _____ moles of O atoms; also, _____ grams of H and _____ grams of O.
(d) In one *molecule* of H_2O there is a total of _____ atoms (H + O).
(e) In one *mole* of H_2O there is a total of _____ moles of atoms (H + O), or _____ atoms.

Equal numbers of moles of different substances contain equal numbers of particles: 1 mole of *any substance* contains 6.02×10^{23} particles of that substance; $\frac{1}{2}$ mole of any substance contains 3.01×10^{23} particles; 0.1 mole contains 6.02×10^{22} particles; 100 moles contain 6.02×10^{25} particles; and so on.

Counting atoms and molecules in moles amounts to counting them in *bunches* with 6.02×10^{23} of them to a bunch. Counting in moles is like counting eggs by the dozen or the United States budget by the billions of dollars.

To calculate the number of moles in a given sample of any substance, divide the mass (in grams) of the sample by the molecular weight of the substance, the weight of one mole:

$$\text{Moles} = \frac{\text{Wt. of substance (g)}}{\text{Mol. wt. (g/mole)}}$$

$$\text{In 36 g of water, number of moles} = \frac{36\ \text{g}}{18\ \text{g/mole}} = 2.0\ \text{moles}$$

$$\text{In 1.8 g of water, number of moles} = \frac{1.8\ \text{g}}{18\ \text{g/mole}} = 0.10\ \text{mole}$$

To find the weight of a given number of moles of a substance, multiply the number of moles by the molecular weight:

$$\text{Weight (g)} = \text{Moles} \times \text{Mol. wt.} \left(\frac{\text{g}}{\text{mole}}\right)$$

6.0 moles of water weigh

$$6.0\ \text{moles} \times 18\ \text{g/mole} = 108\ \text{g}$$

0.40 moles of water weigh

$$0.40\ \text{moles} \times 18\ \text{g/mole} = 7.2\ \text{g}$$

Knowing the value of Avogadro's number N (the number of molecules in a mole), we are in a position to calculate the size and weight of a typical molecule and to find the number of molecules or atoms in a given sample, as shown in the following two examples.

EXAMPLE 2-1

A mole of water weighs 18 g and occupies a volume of 18 cm^3 (under ordinary conditions—say at room temperature and normal pressure).

(a) How many molecules are in 1 cm^3 of water?

$$\text{Number of molecules} = \frac{1 \text{ mole}}{18 \text{ cm}^3} \times 6.02 \times 10^{23} \frac{\text{molecules}}{\text{mole}}$$

$$= 0.33 \times 10^{23} \frac{\text{molecules}}{\text{cm}^3}$$

$$= 3.3 \times 10^{22} \frac{\text{molecules}}{\text{cm}^3}$$

(b) What volume is occupied by one molecule? If there are 3.3×10^{22} molecules in one cm^3, then

$$\text{Volume occupied by one molecule} = \frac{1.0 \text{ cm}^3}{3.3 \times 10^{22} \text{ molecules}}$$

$$= 0.30 \times 10^{-22} \frac{\text{cm}^3}{\text{molecule}}$$

$$= 3.0 \times 10^{-23} \frac{\text{cm}^3}{\text{molecule}}$$

If we assume that a water molecule is roughly a cube of side s, then

$$\text{Volume of molecule} = s^3 = 3.0 \times 10^{-23} \text{ cm}^3$$

$$= 30 \times 10^{-24} \text{ cm}^3$$

(Here the volume has been rewritten in such a way as to make it easy to take the cube root.)

$$s = (30 \times 10^{-24})^{1/3} \cong 3 \times 10^{-8} \text{ cm}$$

The "side" of a water molecule is about 3×10^{-8} cm; the molecule is so tiny that one hundred million of them, laid down side by side in a line would cover a distance of 3 cm, only a little over an inch. Molecules are not really cubes—they are more nearly spherical (in Chapter 7 we shall see what shapes they actually have)—but this calculation gives a good approximation of the size of a water molecule.

(c) What is the mass of one water molecule?

$$\text{Mass of one molecule} = 18 \frac{\text{g}}{\text{mole}} \times \frac{1 \text{ mole}}{6.02 \times 10^{23} \text{ molecules}}$$

$$= 3.0 \times 10^{-23} \frac{\text{g}}{\text{molecule}}$$

EXAMPLE 2-2

One gallon of octane (C_8H_{18}), a component of gasoline, weighs about 2850 g.

(a) How many moles of octane are there in one gallon?

$$\text{Mol. wt. octane} = 8(\text{at. wt. C}) + 18(\text{at. wt. H})$$

$$= 8(12.0) + 18(1.0)$$

$$= 114 \text{ g/mole}$$

Thus, 1 mole of octane weighs 114 g.

$$\text{Moles per gallon} = \frac{2850 \text{ g/gallon}}{114 \text{ g/mole}} = 25.0 \frac{\text{moles}}{\text{gallon}}$$

(b) How many molecules are there in one gallon of octane?

$$\text{Number of molecules} = 25.0\,\frac{\cancel{\text{moles}}}{\text{gallon}} \times 6.02 \times 10^{23}\,\frac{\text{molecules}}{\cancel{\text{mole}}}$$

$$= 150 \times 10^{23} = 1.50 \times 10^{25}\,\frac{\text{molecules}}{\text{gallon}}$$

(c) How many atoms of carbon are in a gallon of octane?

$$\text{Number of C atoms} = 1.50 \times 10^{25}\,\frac{\cancel{\text{molecules}}}{\text{gallon}} \times 8\,\frac{\text{atoms of C}}{\cancel{\text{molecule}}}$$

$$= 12.0 \times 10^{25} = 1.20 \times 10^{26}\,\frac{\text{atoms of C}}{\text{gallon}}$$

(d) How many atoms of hydrogen are in a gallon of octane?

$$\text{Number of H atoms} = 1.50 \times 10^{25}\,\frac{\cancel{\text{molecules}}}{\text{gallon}} \times 18\,\frac{\text{atoms of H}}{\cancel{\text{molecule}}}$$

$$= 27.0 \times 10^{25} = 2.70 \times 10^{26}\,\frac{\text{atoms of H}}{\text{gallon}}$$

EXERCISE 2-4 A sample of gold weighs 1.97 g.
(a) How many moles of gold atoms is this?
(b) How many atoms of gold is this?
(c) What is the mass of one atom of gold?
(The atomic weight of gold is given in the table inside the cover of this book.)

2-6 CHEMICAL EQUATIONS

A chemical equation is a shorthand statement of what occurs in a chemical reaction. Usually the formulas of the reacting substances (**reagents** or **reactants**) are written on the left and the formulas of the **products** on the right, with an arrow between them indicating that reactants change to products.

For example, carbon reacts with a limited amount of oxygen gas (which has the molecular formula O_2) to form a poisonous gas, carbon monoxide:

$$C + O_2 \longrightarrow CO$$

This equation shows us the reactants (C and O_2) and the product (CO). However, it is not a complete equation, because it is not **balanced.** That is, it does not have the same number of each kind of atom on both sides of the equation. On the left are two oxygen atoms, but on the right is only one oxygen atom. *In a chemical reaction, atoms are conserved.* In going from reactants to products, the atoms combine in new ways, but no atom is created or destroyed or transmuted (changed) into a different kind of atom.

In the equation above, we can conserve O atoms by placing a coefficient 2 before the CO:

$$C + O_2 \longrightarrow 2CO$$

The 2 in front of the CO means 2 *molecules* of CO, not 2C atoms only. Now the O is balanced, but there are 2C atoms on the right and only 1 on the left. Next, then, we should place a 2 in front of the C on the left:

$$2C + O_2 \longrightarrow 2CO$$

Now the equation is correctly balanced, with 2C atoms and 2O atoms on each side of the equation. The equation $2C + O_2 \rightarrow 2CO$ symbolizes the diagram:

2 C atoms + 1 O_2 molecule ⟶ 2 CO molecules

Most chemical equations can be balanced by using such a common sense approach, following these steps.

1. Identify the reactants and products.
2. Write the correct formula for each reactant and each product.
3. Add coefficients, as required, to conserve atoms. When adding coefficients, it is usually best first to choose the reactant that has the greatest number of *different* atoms and to start by balancing an atom that is in only *one* reactant and *one* product.

We'll try this method in the following examples.

EXAMPLE 2-3

Balance the equations for the following chemical reactions.

(a) When acetylene (C_2H_2) is burned in oxygen gas (O_2) as in an acetylene torch, the products are carbon dioxide (CO_2) and water (H_2O).

1. First write the correct formulas for reactants and products:

$$C_2H_2 + O_2 \longrightarrow CO_2 + H_2O$$

Not yet balanced

2. Next balance the C atoms (choosing an atom from the C_2H_2, the reactant with the most different kinds of atoms). We could just as well start with H, but we would run into problems if we start with O, which appears in two different products (making it hard to know where to put coefficients).

$$C_2H_2 + O_2 \longrightarrow 2CO_2 + H_2O$$

Balanced for C; also balanced for H

3. Now add a coefficient to balance the O:

$$C_2H_2 + \tfrac{5}{2}O_2 \longrightarrow 2CO_2 + H_2O$$

Balanced for O; balanced for everything

4. Now all atoms are balanced, but we have $\frac{5}{2}$ oxygen molecules as reactants. Multiplying both sides by two, clears the equation of fractions and puts it in a more conventional form, although the form above is also correct.

$$2C_2H_2 + 5O_2 \longrightarrow 4CO_2 + 2H_2O$$

Balanced; coefficients are whole numbers

5. As a final step, check again to be sure that all atoms are balanced:

$$4C \text{ and } 4H \text{ and } 10O \longrightarrow 4C \text{ and } 4H \text{ and } 10O$$

It checks.

(b) An effective treatment for an acid burn is to wash the exposed area well, first with water and then with a solution of sodium bicarbonate ($NaHCO_3$, baking soda). For example, sulfuric acid (H_2SO_4) reacts with sodium bicarbonate to yield harmless sodium sulfate (Na_2SO_4), carbon dioxide (CO_2), and water (H_2O) as products.

1. The correct formulas for reactants and products are

$$H_2SO_4 + NaHCO_3 \longrightarrow Na_2SO_4 + CO_2 + H_2O$$

Not yet balanced

2. The $NaHCO_3$ has the most different kinds of atoms; we'll start with the Na.

$$H_2SO_4 + 2NaHCO_3 \longrightarrow Na_2SO_4 + CO_2 + H_2O$$

Na balanced; S also balanced

3. Next balance the C (which appears in only one molecule on each side):

$$H_2SO_4 + 2NaHCO_3 \longrightarrow Na_2SO_4 + 2CO_2 + H_2O$$

C balanced

4. Hydrogen appears in only one molecule on the right side; balance it next.

$$H_2SO_4 + 2NaHCO_3 \longrightarrow Na_2SO_4 + 2CO_2 + 2H_2O$$

H balanced; O also balanced

5. Check to make sure that all the atoms are balanced:

4H and 1S and 10O and 2Na and 2C $\longrightarrow$
4H and 1S and 10O and 2Na and 2C

It checks. Note that some judgment is required at each step to choose the element to balance in the next step. In general, you look for an element that can be balanced by introducing some coefficient where one had not yet been added. (Of course, if you *change* a coefficient already added, you will disrupt the balance of some element already considered.)

EXERCISE 2-5 Write a balanced equation for each of the following chemical reactions.

(a) When the zinc ore, zinc sulfide (ZnS) is "roasted" by heating it with oxygen (O_2) in the air, the products are zinc oxide (ZnO) and sulfur dioxide gas (SO_2).

(b) Ammonia gas (NH_3) is produced by reaction between hydrogen gas, (H_2) and nitrogen gas (N_2).

(c) In the process of photosynthesis, the sun's energy is used to convert carbon dioxide (CO_2) and water (H_2O) into glucose ($C_6H_{12}O_6$) and oxygen (O_2).

(d) The poisonous gas phosgene ($COCl_2$) reacts with water (H_2O) in the lungs, to produce hydrogen chloride (HCl, an acid that damages the lungs) and carbon dioxide (CO_2).

(e) Carbon tetrachloride (CCl_4) reacts with oxygen in the air on the surface of a hot metal, to produce the poisonous gas phosgene ($COCl_2$) and chlorine gas (Cl_2). This is one of the reasons that CCl_4 is no longer used in fire extinguishers. Write an equation for the reaction of CCl_4 with O_2 to produce $COCl_2$. (The metal does not change during the reaction, and so does not appear in the equation.)

What kind of information can we get from a balanced equation?

$$\begin{array}{ccccc} 2C & + & O_2 & \longrightarrow & 2CO \\ 2 \text{ atoms} & + & 1 \text{ molecule} & \longrightarrow & 2 \text{ molecules} \end{array}$$

Multiplying both sides by Avogadro's number, N:

$$\begin{array}{rcl} 2N \text{ atoms} + N \text{ molecules} & \longrightarrow & 2N \text{ molecules} \\ 2 \text{ moles} + 1 \text{ mole} & \longrightarrow & 2 \text{ moles} \\ 2 \times 12 \text{ g} + 32 \text{ g} & \longrightarrow & 2 \times 28 \text{ g} \\ 24 \text{ g} + 32 \text{ g} & \longrightarrow & 56 \text{ g} \end{array}$$

From the balanced equation we can determine the relative numbers of molecules or moles, and the relative masses of the reactants and the products of a chemical reaction.

The equation does *not* imply that the reaction is complete. For example, $2C + O_2 \rightarrow 2CO$ does *not* mean that if we mix 2 moles of C with 1 mole of oxygen, all of the C and O_2 will necessarily react. Perhaps they will not react at all—we can leave a lump of coal (carbon) exposed to oxygen in the air, and nothing much will happen until we apply some heat. Even then, the reaction may stop while some of both the carbon and the oxygen remain unreacted. Or some of the carbon might react with oxygen to produce carbon dioxide (CO_2) instead of, or in addition to, CO. The equation *does* mean that for every 2 atoms of C that *do* react with O_2 to form CO, 1 molecule of O_2 is used and 2 molecules of CO are produced. Also, for every 2 molecules of CO produced, we know that 2 atoms of C reacted with 1 molecule of O_2.

The equation also does *not* tell in what proportions the reactants are mixed, but only in what proportions they can react. We can *mix* them in any amounts, but they will *react* only in definite proportion. For example, $2C + O_2 \rightarrow 2CO$ does *not* mean that we must mix 2 moles of C with 1 mole of O_2. We can mix 200 moles of C with 1 of O_2, if we like, but only 2 moles of C can *react* with 1 mole of O_2. The rest will remain as unreacted carbon. In this case, the O_2 is called the **limiting reagent** (or **reactant**), the reagent that determines (limits) the amount of product that can form. The carbon is said to be **in excess** because there is some left over at the end of the reaction—in this case, there will be 198 moles of C left. The amount of product that can, theoretically, be produced from the given amounts of reactants, is called the **theoretical yield**—in this case, the theoretical yield is 2 moles of CO:

	$2C$	$+\ O_2$	$\longrightarrow$	$2CO$
To start:	200 moles	1 mole		none
After reaction:	198 moles	none		2 moles
	(in excess)	(limiting reagent)		(theoretical yield)

In real life, the theoretical yield is approached more or less closely depending on the specific reaction and on reaction conditions, as we shall see in later chapters.

EXERCISE 2-6 Given a reaction mixture of 8 moles of C and 3 moles of O_2, and assuming the only reaction that occurs is the formation of CO,

(a) What is the limiting reagent?

(b) One substance is present in excess. How many moles of this substance are left after reaction?

(c) What is the theoretical yield of CO?

2-7 MORE STOICHIOMETRY

This section consists of examples that apply the ideas we have discussed to a wide variety of chemical problems. Using a table of atomic weights and the information that can be obtained from a balanced equation, we can

1. determine the simplest formula of a compound, given the percent composition of the compound and the requisite atomic weights;
2. calculate the theoretical yield of one or more products of a reaction, given the weight of the limiting (or scarce) reactant;
3. calculate how much of each reactant is needed to produce a specified amount of one of the products of a reaction;
4. identify the limiting reagent, given the starting weights of all the reactants.

The first two examples illustrate how to determine the simplest formula of a chemical compound.

EXAMPLE 2-4

The corrosion of aluminum (Al) is limited by a thin, tough surface coat of aluminum oxide that slowly forms when aluminum is exposed to the air. Upon analysis, this oxide is found to consist of 52.9% by weight Al and 47.1% by weight O. What is the simplest formula of aluminum oxide?

If we choose (arbitrarily but conveniently) to consider a sample of aluminum oxide weighing 100 g, then in that sample

$$\text{Weight of Al} = 52.9\text{ g} \qquad \text{At. wt. Al} = 27.0\text{ g/mole}$$

$$\text{Weight of O} = 47.1\text{ g} \qquad \text{At. wt. O} = 16.0\text{ g/mole}$$

$$\text{Moles of Al atoms} = \frac{52.9\text{ g}}{27.0\text{ g/mole}} = 1.96\text{ moles Al}$$

$$\text{Moles of O atoms} = \frac{47.1\text{ g}}{16.0\text{ g/mole}} = 2.94\text{ moles O}$$

$$\text{The ratio } \frac{\text{moles O}}{\text{moles Al}} = \frac{2.94}{1.96} = \frac{1.5}{1} = \frac{3}{2}$$

Because atoms combine in simple number ratios, we are interested in the simplest whole number ratio, (Moles O)/(Moles Al) = 3/2. In aluminum oxide there are 3 moles of O atoms for every 2 moles of Al atoms. The simplest formula for aluminum oxide, therefore, is Al_2O_3. From the information here, there is no way for us to tell whether aluminum oxide is really Al_2O_3, or Al_4O_6, or Al_6O_9, or some other formula with the same ratio of moles of O to moles of Al. We do know that Al_2O_3 is the *simplest* formula for aluminum oxide consistent with the data. In Chapter 3 we shall see how to determine the real molecular formula of a compound.

Instead of assuming an arbitrary sample of 100 g of aluminum oxide to solve this problem, we could have let X = weight of aluminum oxide (arbitrary). Then,

$$\text{Weight Al} = 0.529\,X\text{ g}$$

$$\text{Weight O} = 0.471\,X\text{ g}$$

$$\text{Moles Al} = \frac{0.529\,X\text{ g}}{27.0\text{ g/mole}} = 0.0196\,X\text{ moles Al}$$

$$\text{Moles O} = \frac{0.471\,X\text{ g}}{16.0\text{ g/mole}} = 0.0294\,X\text{ moles O}$$

$$\text{The ratio } \frac{\text{moles O}}{\text{moles Al}} = \frac{0.0294\,X}{0.0196\,X} = \frac{1.5}{1} = \frac{3}{2}$$

The weight of the sample, X, cancels out, and we get the same ratio of moles O to moles Al that we got before.

EXERCISE 2-7 Ethyl ether (a colorless, volatile, flammable liquid used as an anesthetic and as a solvent) consists of 64.8% by weight carbon, 13.6% by weight hydrogen, and 21.6% by weight oxygen. What is the simplest formula of ethyl ether?

EXAMPLE 2-5

A typical experiment in introductory chemistry is to determine the formula of a metal sulfide. The following data might come from such an experiment: a piece of silver wire weighing 0.540 g is heated with a large quantity of sulfur until all of the silver is converted to silver sulfide. The excess sulfur is then burned off as the noxious gas sulfur dioxide, SO_2 (Figure 2-3). The black sulfide of silver weighs 0.620 g.

What is the simplest formula of silver sulfide?

$$\text{Wt. of Ag} = 0.540\text{ g} \qquad \text{At. wt. of Ag} = 108\text{ g}$$

$$\text{Moles of Ag} = \frac{0.540\text{ g}}{108\text{ g/mole}} = 0.00500\text{ mole Ag}$$

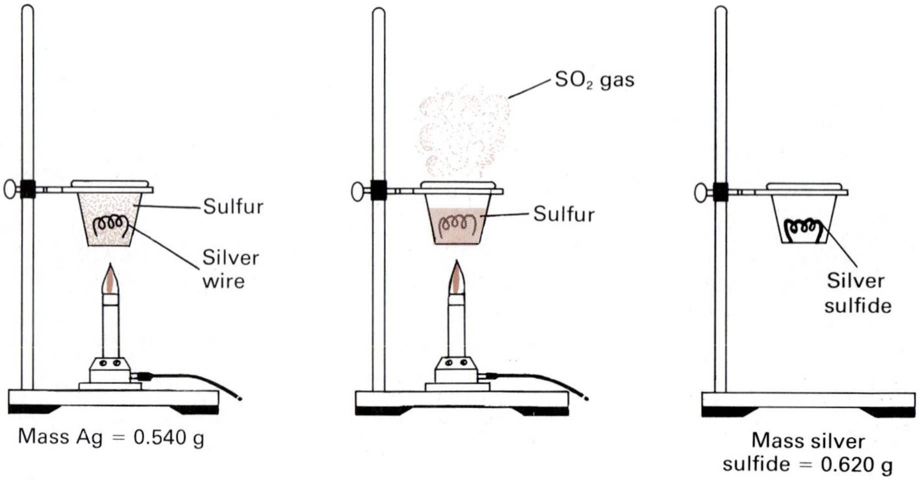

FIGURE 2-3

Experimental determination of the formula of silver sulfide (see text for explanation).

$$\begin{aligned}\text{Wt. of S in silver sulfide} &= \text{Wt. of silver sulfide} - \text{Wt. of silver}\\ &= 0.620\text{ g} - 0.540\text{ g} = 0.080\text{ g S}\end{aligned}$$

$$\text{At. wt. of S} = 32.0\text{ g/mole}$$

$$\text{Moles of S} = \frac{0.080\text{ g}}{32.0\text{ g/mole}} = 0.0025\text{ moles S}$$

$$\text{The ratio }\frac{\text{moles Ag}}{\text{moles S}} = \frac{0.0050}{0.0025} = \frac{2}{1}$$

There are 2 Ag atoms for every 1 S atom in silver sulfide. The simplest formula, therefore, is Ag_2S.

In the following example we are given the weight of one reactant, the limiting reactant, and will calculate the weight of one of the products obtained. We will also calculate the weight needed of a second reactant.

EXAMPLE 2-6

When carbohydrates or sugars react with oxygen, either in the body or in the laboratory, the products are carbon dioxide and water.

(a) Write a balanced equation for the combustion of sucrose ($C_{12}H_{22}O_{11}$).

$$C_{12}H_{22}O_{11} + O_2 \longrightarrow CO_2 + H_2O$$

Correct formulas for reactants and products

$$C_{12}H_{22}O_{11} + O_2 \longrightarrow 12CO_2 + H_2O$$

C balanced

$$C_{12}H_{22}O_{11} + O_2 \longrightarrow 12CO_2 + 11H_2O$$

H balanced

$$C_{12}H_{22}O_{11} + 12O_2 \longrightarrow 12CO_2 + 11H_2O$$

O balanced; all atoms balanced

(b) If 1.00 kg of sucrose is burned, what weight of water is produced?

Mol. wt. sucrose

$$= 12\,(\text{at. wt. C}) + 22\,(\text{at. wt. H}) + 11\,(\text{at. wt. O})$$
$$= 12\,(12.0) + 22\,(1.01) + 11\,(16.0)$$
$$= 342\ \text{g/mole}$$

$$\text{Moles } C_{12}H_{22}O_{11} = \frac{1000\ \cancel{\text{g}}}{342\ \cancel{\text{g}}/\text{mole}} = 2.92\ \text{mole } C_{12}H_{22}O_{11}$$

The balanced equation shows that, for each mole of $C_{12}H_{22}O_{11}$ that reacts, 11 moles of H_2O are produced.

$$\text{Moles } H_2O \text{ produced} = 2.92\ \cancel{\text{moles } C_{12}H_{22}O_{11}} \times \frac{11\ \text{moles } H_2O}{1\ \cancel{\text{mole } C_{12}H_{22}O_{11}}}$$
$$= 32.1\ \text{moles } H_2O$$

$$\text{Wt. } H_2O \text{ produced} = 32.1\ \cancel{\text{moles } H_2O} \times 18.0\ \frac{\text{g } H_2O}{\cancel{\text{mole } H_2O}}$$
$$= 578\ \text{g } H_2O$$

(c) What weight of O_2 is needed to burn all of the sucrose?

$$\text{Mol. wt. } O_2 = 2\ (\text{at. wt. O}) = 2(16.0) = 32.0\ \text{g/mole}$$

For each mole of $C_{12}H_{22}O_{11}$, 12 moles of O_2 are needed

$$\text{Moles } O_2 \text{ needed} = 2.92\ \text{moles}\ \cancel{C_{12}H_{22}O_{11}} \times \frac{12\ \text{moles } O_2}{1\ \text{mole}\ \cancel{C_{12}H_{22}O_{11}}}$$
$$= 35.0\ \text{moles } O_2$$

$$\text{Wt. } O_2 \text{ needed} = 35.0\ \cancel{\text{moles}} \times 32.0\ \text{g}/\cancel{\text{mole}}$$
$$= 1121\ \text{g} = 1.12\ \text{kg}$$

EXERCISE 2-8 Hematite, an iron ore now in scarce supply, has the formula Fe_2O_3. When this ore is heated in a blast furnace with carbon monoxide (produced in the blast furnace from a mixture of coke, C, and air), the products are iron (Fe) and carbon dioxide (CO_2).

(a) Write a balanced equation for the reaction. (This equation is a little trickier than some to balance. The simplest way is to note that each CO is converted to CO_2 by taking one O atom away from Fe_2O_3.)

(b) If 8.0×10^4 kilograms of hematite is reduced to iron in a blast furnace, what weight of iron is produced? (At. wt. Fe = 55.8.)

(c) What weight of CO is needed to convert all of the hematite to iron?

In the next example, we calculate the weight of each reactant necessary to produce a given amount of one of the products of a reaction.

EXAMPLE 2-7

A paraffin candle burns in air, consuming oxygen and producing water and carbon dioxide. Paraffin is really a mixture of compounds, each with its own formula, but of these $C_{25}H_{52}$ is typical.

(a) If 9.0 g of water are produced when a candle burns, what weight of candle was consumed? Assume that the candle is pure $C_{25}H_{52}$. First we should write a balanced equation for the reaction.

$$C_{25}H_{52} + O_2 \longrightarrow CO_2 + H_2O$$

Equation not yet balanced

$$C_{25}H_{52} + O_2 \longrightarrow 25CO_2 + 26H_2O$$

C and H atoms balanced

$$C_{25}H_{52} + 38O_2 \longrightarrow 25CO_2 + 26H_2O$$
O atoms balanced

$$\text{Mol. wt. water} = 2\,(\text{at. wt. H}) + \text{at. wt. O} = (2 \times 1) + 16 = 18 \text{ g/mole}$$

$$\text{Mol. wt. } C_{25}H_{52} = 25\,(\text{at. wt. C}) + 52(\text{at. wt. H}) = (25 \times 12) + (52 \times 1) = 352 \text{ g/mole}$$

$$\text{Moles of } H_2O \text{ produced} = \frac{9.0 \text{ g}}{18 \text{ g/mole}} = 0.50 \text{ mole } H_2O$$

$$C_{25}H_{52} \text{ burned} = 0.50 \text{ moles } H_2O \times \frac{1 \text{ mole } C_{25}H_{52}}{26 \text{ moles } H_2O} = 0.019 \text{ mole } C_{25}H_{52} \text{ burned}$$

$$\text{Wt. of } C_{25}H_{52} \text{ burned} = 0.019 \text{ mole} \times 352 \text{ g/mole} = 6.7 \text{ g of candle burned}$$

(b) What weight of oxygen from the atmosphere reacting with paraffin was needed to produce the 9.0 g of water?

$$\text{Mol. wt. of } O_2 = 2\,(\text{at. wt. O}) = 2\,(16) = 32 \text{ g/mole}$$

$$\text{Moles of } O_2 \text{ needed} = 0.50 \text{ mole } H_2O \times \frac{38 \text{ moles } O_2}{26 \text{ moles } H_2O} = 0.73 \text{ mole } O_2$$

$$\text{Wt. of } O_2 \text{ needed} = 0.73 \text{ mole} \times 32 \text{ g/mole} = 23 \text{ g } O_2 \text{ needed}$$

This last example illustrates a variety of information that can be obtained from a balanced chemical equation, including the determination of the limiting reagent.

EXAMPLE 2-8

Plaster of Paris is a white, powdery substance with the formula $(CaSO_4)_2 \cdot H_2O$. The symbol $\cdot H_2O$ means that the compound is a **hydrate**—it contains combined water that can be driven off by heating (here, *one* molecule of water for every *two* units of $CaSO_4$). When mixed with water, plaster of Paris combines with additional water to form a different hydrate, gypsum ($CaSO_4 \cdot 2H_2O$), a hard substance used to make plaster walls and plaster casts for broken bones. The balanced equation for the reaction is

$$Ca(SO_4)_2 \cdot H_2O + 3H_2O \longrightarrow 2CaSO_4 \cdot 2H_2O$$

(a) How many moles of H_2O are needed to convert 1.5 moles of $(CaSO_4)_2 \cdot H_2O$ to $CaSO_4 \cdot 2H_2O$?

The balanced equation shows that three moles of H_2O are needed for every mole of $(CaSO_4)_2 \cdot H_2O$.

Moles H_2O needed

$$= 1.5 \text{ moles } (CaSO_4)_2 \cdot H_2O \times \frac{3 \text{ moles } H_2O}{1 \text{ mole } (CaSO_4)_2 \cdot H_2O}$$

$$= 4.5 \text{ moles } H_2O$$

(b) If 5.0 moles of H_2O are mixed with 1.5 moles of $(CaSO_4)_2 \cdot H_2O$ to make plaster of Paris, which substance is the limiting reagent, and which is in excess?

We saw in (a) that 4.5 moles of H_2O are needed for the complete reaction of 1.5 moles of $(CaSO_4)_2 \cdot H_2O$. Therefore, $(CaSO_4)_2 \cdot H_2O$ is the limiting reagent, and H_2O is in excess.

(c) Starting with the amounts of reactants given in part b, what is the theoretical yield in grams of $CaSO_4 \cdot 2H_2O$?

For every mole of $(CaSO_4)_2 \cdot H_2O$ (the limiting reagent) that reacts, 2 moles of $CaSO_4 \cdot 2H_2O$ are produced.

Moles $CaSO_4 \cdot 2H_2O$

$$= 1.5 \text{ moles } \cancel{(CaSO_4)_2 \cdot H_2O} \times \frac{2 \text{ moles } \cancel{CaSO_4 \cdot 2H_2O}}{1 \text{ mole } (CaSO_4)_2 \cdot H_2O}$$

$$= 3.0 \text{ moles } CaSO_4 \cdot 2H_2O \text{ (theoretical yield)}$$

$$\begin{aligned}\text{Wt. of 1 mole } CaSO_4 \cdot 2H_2O &= (\text{At. wt. Ca}) + (\text{At. wt. S}) \\ &\quad + (4 + 2)(\text{At. wt. O}) + 4(\text{At. wt. H}) \\ &= 40 + 32 + 6(16) + 4(1) \\ &= 172 \text{ g/mole}\end{aligned}$$

$$\text{Wt. } CaSO_4 \cdot 2H_2O = 3 \text{ moles} \times 172 \text{ g/mole} = 516 \text{ g}$$

EXERCISE 2-9 Pure iron (Fe) may be obtained from the oxide (Fe_3O_4) by passing hydrogen gas (H_2) over it while it is hot. Water (H_2O) and iron are the only products. Fe_3O_4 (magnetite) is different from the iron oxide hematite (Fe_2O_3) mentioned in Exercise 2-8.

(a) Write a balanced equation for the reaction.

(b) How many *molecules* of H_2 are needed to produce 9 atoms of Fe?

(c) How many moles of Fe_3O_4 will react with 1.2 moles of H_2?

(d) If 1.2 moles of H_2 are passed over 0.80 mole of hot Fe_3O_4 to provide pure iron, which substance will be the limiting reagent, and which substance will be in excess?

(e) What is the theoretical yield of pure iron that could be obtained from the amounts of H_2 and Fe_3O_4 given in part (d)?

(f) How many *grams* of Fe is this?

KEY TERMS

A *molecule* is the smallest physical unit of an element or a compound. A molecule of an element consists of one or more like atoms bonded together in a particular arrangement. A molecule of a compound consists of two or more unlike atoms bonded together in a particular arrangement.

The *formula* of a compound tells how many of each kind of atom is in a molecule of the compound. The *simplest formula* gives only the relative numbers of each kind of atom in a molecule.

A modern *atomic weight table* is a table of relative atomic weights that has as reference standard an isotope (Chapter 5) of carbon that has been assigned atomic weight 12 exactly. The *molecular weight* of a compound is the sum of the atomic weights of all the atoms in a molecule of the compound.

The mass of material that weighs, in grams, a number equal to the atomic or molecular weight is called a *mole* of that material. The number of atoms or molecules in a mole is called *Avogadro's number* and is symbolized N. ($N = 6.02 \times 10^{23}$.)

In a *chemical reaction*, atoms combine or are separated, or both. A *chemical equation* is a symbolic representation of what occurs in a chemical reaction.

The *limiting reagent* (or *reactant*) is the reagent that determines the amount of product that can be formed in a chemical reaction. A reagent is *in excess* if some of it is left over at the end of the reaction. The amount of product that can, theoretically, be produced from the given amounts of reagents is called the *theoretical yield*.

CHAPTER SUMMARY

According to *Dalton's atomic theory,* matter is composed of discrete particles called atoms; all atoms of a given element are identical; the atoms of different elements differ in mass and in chemical behavior; and chemical reactions are the combination and separation of atoms. This theory explained the laws of conservation of mass and definite composition, and predicted the law of multiple proportions.

In a chemical reaction, atoms are conserved. A *chemical equation* is a symbolic representation of a chemical reaction, and a balanced equation has the same number of each kind of atom on both sides of the equation.

From a balanced equation, we can determine the relative numbers of molecules or moles, and the relative masses, of the reactants and products of a chemical reaction. The equation does *not* tell in what proportions the reactants are mixed, or whether or not the reaction will occur.

Using a table of *relative atomic weights* and the information contained in a balanced equation, we can solve a wide variety of chemical problems including the determination of the *simplest formula* of a compound, and the calculation of the weights of reactants and products involved in chemical reactions.

PROBLEMS

2-1. (a) Calculate the weight of one mole of each of these substances:

(1) carbon tetrachloride (a solvent, CCl_4)
(2) phosphorus molecules (P_4)
(3) glycine (an amino acid, CH_2NH_2COOH)
(4) codeine hydrochloride ($C_{18}H_{21}NO_3 \cdot HCl \cdot 2H_2O$)
(5) hydrated copper sulfate ($CuSO_4 \cdot 5H_2O$)
(6) nicotine ($C_{10}H_{14}N_2$)
(7) ascorbic acid (vitamin C, $C_6H_8O_6$)

(b) Calculate the number of moles in each of the following, using the molecular weights calculated in part a:

(1) 38.5 g carbon tetrachloride (CCl_4)
(2) 248 g phosphorus (P_4)
(3) 1.5 mg glycine (CH_2NH_2COOH)
(4) 18.6 g codeine hydrochloride ($C_{18}H_{21}NO_3 \cdot HCl \cdot 2H_2O$)
(5) 1.5 kg copper sulfate ($CuSO_4 \cdot 5H_2O$)
(6) 0.405 g nicotine ($C_{10}H_{14}N_2$)
(7) 1.76 g ascorbic acid ($C_6H_8O_6$)

2-2. An important copper ore, chalcocite, has the formula Cu_2S.

(a) How many moles of Cu_2S are contained in a sample of chalcocite weighing 8.0 g?
(b) How many atoms of Cu are contained in the sample?
(c) How many atoms of S are contained in the sample?

2-3. What is the mass of

(a) 3×10^{23} molecules of water (H_2O)?
(b) 2×10^{23} molecules of table sugar ($C_{12}H_{22}O_{11}$)?
(c) 6×10^{21} atoms of mercury?

2-4. Which of the following contains the most atoms: 0.1 mole of aluminum, 8 g of calcium, 20 g of mercury, or 2 g of helium?

2-5. Washing soda, sodium carbonate, is found by chemical analysis to consist of 43.4% Na, 11.3% C, and 45.3% O by weight. What is the simplest formula of sodium carbonate?

2-6. A compound contains 3.0 g of carbon for each 1.0 g of hydrogen. What is the simplest formula of the compound?

2-7. Vitamin B_{12} is found by chemical analysis to have the percentage composition by weight: 2.33% phosphorus; 4.44% cobalt; 16.8% oxygen; 14.7% nitrogen; 6.32% hydrogen; and 56.8% carbon.
(a) Find the simplest formula of vitamin B_{12}.
(b) The recommended daily allowance* of vitamin B_{12} for an adult human is 3.0 μg. How many moles of Co atoms are in 3.0 μg of vitamin B_{12}? (Mol. wt. of vitamin B_{12} = 1330.)
(c) How many Co atoms are in 3.0 μg of vitamin B_{12}?
(d) How many grams of cobalt are in 3.0 μg vitamin B_{12}?

2-8. Hemoglobin has a molecular weight of 64,500. Each hemoglobin molecule contains 4 atoms of iron (Fe).
(a) What is the percentage of Fe in hemoglobin?
(b) An adult has 1.0 kg of hemoglobin in his blood. What is the weight of the iron in 1.0 kg hemoglobin?
(c) How many moles of Fe atoms are in 1.0 kg hemoglobin?
(d) How many Fe atoms are in 1.0 kg hemoglobin?

*_Recommended Dietary Allowances_, 8th rev. ed. (Food and Nutrition Board, National Research Council–National Academy of Sciences, 1974).

2-9. Ascorbic acid (vitamin C) has the molecular formula $C_6H_8O_6$. The recommended daily allowance of ascorbic acid in the diet of a human adult is 45 mg.
(a) How many moles of ascorbic acid is this?
(b) What is the number of molecules of ascorbic acid in 45 mg?

2-10. (a) Methyl mercaptan (CH_3SH), responsible for the characteristic odor of garlic, can be smelled at a concentration of 4×10^{-7} mg per liter of air. How many molecules of methyl mercaptan per liter of air is this?
(b) Strychnine hydrochloride ($C_{21}H_{22}N_2O_2 \cdot HCl \cdot 2H_2O$) has a bitter taste that can be detected at a concentration of less than 2×10^{-6} moles per liter of solution. How many strychnine hydrochloride molecules are there in a liter of such a solution?

2-11. The iron ore called iron pyrite ("fool's gold," FeS_2), reacts with oxygen in the air (an industrial process called roasting) to convert the FeS_2 to the oxide hematite (Fe_2O_3), which can then be used in a blast furnace to produce iron. The other product of the roasting operation is sulfur dioxide gas (SO_2), which is used to make sulfuric acid.
(a) Write a balanced equation for the reaction between FeS_2 and oxygen (O_2) to produce Fe_2O_3 and SO_2 gas.
(b) How many kilograms of FeS_2 are used to produce 256 kg of SO_2 gas?
(c) What mass of Fe_2O_3 is produced?

2-12. The main culprit in a London-type smog is the poisonous gas sulfur dioxide (SO_2), derived largely from the combustion of sulfur-containing coal and oil in power and industrial plants. The percentage of sulfur in coal varies considerably from negligible amounts to over 10% by weight. If a coal containing 2% by weight of sulfur is burned, converting all the sulfur to SO_2 what mass of SO_2 is produced from 1800 kg of coal (2 tons)? (Actually, only about one-fourth of the sulfur burns to form SO_2, the rest is converted to solid sulfates, and is found in the ash.)

2-13. Purple sulfur bacteria are photosynthetic bacteria that use energy from the sun to convert carbon dioxide (CO_2) and hydrogen sulfide (H_2S, a poisonous gas responsible for the characteristic odor of rotten eggs), to water (H_2O), elemental sulfur (S_8), and carbohydrates. A balanced equation for a reaction in which the carbohydrate produced is glucose, a simple sugar of formula $C_6H_{12}O_6$ is

$$12CO_2 + 24H_2S \longrightarrow 2C_6H_{12}O_6 + 12H_2O + 3S_8$$

(a) How many moles of H_2S does the bacterium use to convert 0.020 mole of CO_2 to $C_6H_{12}O_6$?
(b) How many moles, and how many grams, of sulfur (S_8) are produced in the reaction of part a?
(c) How many moles, and how many grams, of $C_6H_{12}O_6$ are produced?
(d) How many moles, and how many grams, of water are produced?

2-14. Swamp gas, produced by the decomposition of organic matter in the absence of oxygen, is methane (CH_4). Swamp gas burns in atmospheric oxygen (O_2) to produce water (H_2O) and carbon dioxide (CO_2).
(a) Write a balanced equation for the combustion of swamp gas.
(b) What mass of methane must be burned to produce 9 g of water?
(c) What mass of oxygen is consumed?
(d) What mass of carbon dioxide is produced?

2-15. When octane (C_8H_{18}), a typical component of gasoline, burns in a *limited* amount of oxygen (O_2), the products are the poisonous gas carbon monoxide (CO) and water (H_2O).
(a) Write a balanced equation for the reaction.
(b) A gallon of gasoline contains 25 moles of octane. What mass of carbon monoxide is produced for each gallon of octane burned in a limited amount of oxygen?
(c) What mass of water is produced?
(The incomplete combustion of gasoline in automobile engines is the major source of CO as a pollutant in the atmosphere. *Complete* combustion of octane in the air, however, produces CO_2 rather than CO, and most of the gasoline in a car *does* burn completely. The federal government's regulations on automobile emissions permitted a maximum of 23 g of CO emitted per vehicle mile in 1971.)

2-16. When octane (C_8H_{18}) burns in an *excess* of oxygen (O_2), the products are carbon dioxide gas (CO_2) and water (H_2O).
(a) Write a balanced equation for the complete combustion of octane to produce water and carbon dioxide.
(b) What mass of oxygen is needed for the complete combustion of one gallon of octane? (A gallon of octane is 25 moles of octane.)
(c) What mass of oxygen is needed for the complete combustion of 1.14 g of octane?

2-17. In an industrial operation, P_4S_3 (tetraphosphorus trisulfide) is produced by a reaction between elemental phosphorus and sulfur:

$$8P_4 + 3S_8 \longrightarrow 8P_4S_3$$

The product is obtained in 80% yield (that is, 80% of the theoretical amount of P_4S_3). What weight of phosphorus would be needed to produce 16 tons of P_4S_3 by this process?

2-18. Ammonia (NH_3) can be decomposed to yield nitrogen gas (N_2) and hydrogen gas (H_2).
(a) Write a balanced equation for the decomposition of ammonia.
(b) How many moles of hydrogen gas are produced for each mole of ammonia decomposed?
(c) N_2 and H_2 can be made to react to form ammonia (NH_3). How many moles of NH_3 could, theoretically, be produced from a mixture of 2.0 moles of N_2 and 4.0 moles of H_2? (Assume here that the reaction goes to completion, though as we shall see in Chapter 10, this is not really so.)

2-19. In the preparation of ammonia gas (NH_3) from N_2 and H_2, 28 g of N_2 are allowed to react with 8.0 g of H_2.
(a) One reactant is present in excess. Identify that reactant and determine how many grams of it are left over.
(b) What is the theoretical yield of NH_3, in grams?
(c) If 17 g of NH_3 are produced, what is the percentage yield?

2-20. Nitroglycerin, $C_3H_5(NO_3)_3$, which is commonly used in the treatment of hypertension, is prepared by reaction between glycerin, $C_3H_5(OH)_3$, and nitric acid, HNO_3. Water is also produced in the reaction.
(a) Write a balanced equation for the reaction.
(b) If 2.0 moles of glycerin and 4.0 moles of nitric acid are mixed and allowed to react, what is the limiting reagent?
(c) What is the theoretical yield of nitroglycerin, in grams, from the amounts of reactants given in part b?
(d) If the actual yield of nitroglycerin is 200 g, what is the percentage yield?

2-21. During strenuous exercise, oxygen may not be supplied to the muscles fast enough to meet the demand for it. Then, pyruvic acid ($CH_3COCOOH$) in the muscles is converted to lactic acid ($CH_3COHCOOH$). When the oxygen supply becomes sufficient, the lactic acid is reconverted to pyruvic acid:

$$2CH_3COHCOOH + O_2 \longrightarrow 2CH_3COCOOH + 2H_2O$$

(a) How many grams of oxygen are required to convert 1 mole of lactic acid to pyruvic acid?
(b) How many grams of water are also produced?

3

Gases

Gases seem insubstantial and elusive. Surprisingly, the gaseous state is the simplest state of matter—the first whose behavior and properties were investigated and understood.

In this chapter we see that the experimentally observed behavior of gases leads to an *ideal gas law*—an equation that relates temperature, volume, pressure, and number of molecules. This equation is used to solve a variety of problems involving gases. The *kinetic theory of gases* describes the microscopic structure of the gaseous state and offers an explanation for the observed behavior of gases.

3-1
PROPERTIES OF GASES

Many properties of gases are familiar to us. We know that they can intermix in all proportions, diffusing through each other. This accounts for the formation of liquid water on a cold window when we boil a kettle of water across the room. It accounts for the pervasiveness of a skunk's odor—and for the homogeneity of air, which is almost the same mixture whether sampled over Siberia or over Cape Horn.

We know that gases expand to fill their containers: an "empty" bottle, a balloon, and a tire are all *filled* with air. And gases are elastic: although they expand to fill their containers, they can also be compressed. A compressed gas is like a compressed spring. It shoots the cork out of a champagne bottle, and the spray from an aerosol can, as it expands and escapes from its container.

When a gas expands rapidly, it usually cools. We can observe this when we use a carbon dioxide fire extinguisher. A cloud appears—a snow of solid water and solid carbon dioxide (Dry Ice) is formed as the expanding gas cools itself and water vapor in the air to freezing (Figure 3-1). When compressed rapidly, on the other hand, a gas usually heats. A bicycle pump gets hot when it is used, and the tire gets hot too.

A gas, like a liquid, flows. The natural gas used for heating and cooking flows through pipes from the gas fields, where it is produced, to our homes.

A gas also has weight and occupies volume. Galileo, in the seventeenth century, was the first to establish that air has weight, by performing a simple experiment in which he balanced a container full of air against a heap of sand. Then he pumped more air into

FIGURE 3-1

A carbon dioxide fire extinguisher. The gas cools and freezes as it expands. Water vapor in the air also freezes. [Photograph by Richard Byrnes.]

the container, sealed it, and found that it now weighed more than the sand. That a gas occupies volume is demonstrated when a narrow-mouthed bottle is immersed in a dishpan of water; the bottle cannot be completely filled with water unless the air is allowed to bubble out. Figure 3-2 illustrates an experiment showing that a gas occupies volume—and, incidentally, that the volume occupied by a gas is much greater than that of the same weight of liquid. Figure 3-3 presents a common laboratory experiment demonstrating that a gas occupies volume.

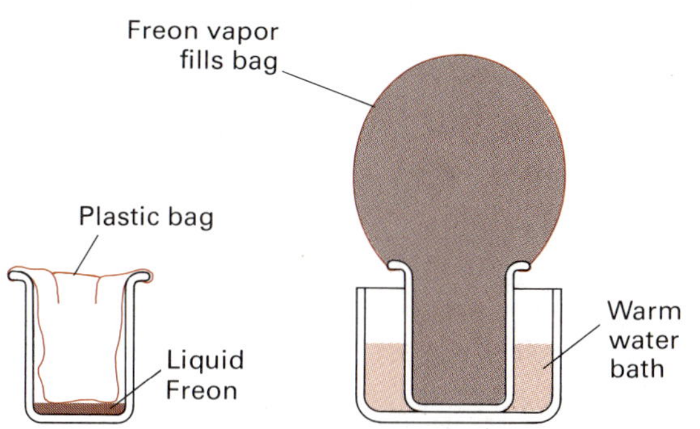

FIGURE 3-2

The volume occupied by Freon as a liquid and as a gas. A few milliliters of liquid Freon are placed in the bottom of a jar. A plastic bag is placed inside the jar and folded over its mouth as shown. As much air as possible is squeezed out of the jar, and the bag is sealed to the jar with tape. When the jar is immersed in a warm water bath, the liquid evaporates and the bag is inflated by the gas.

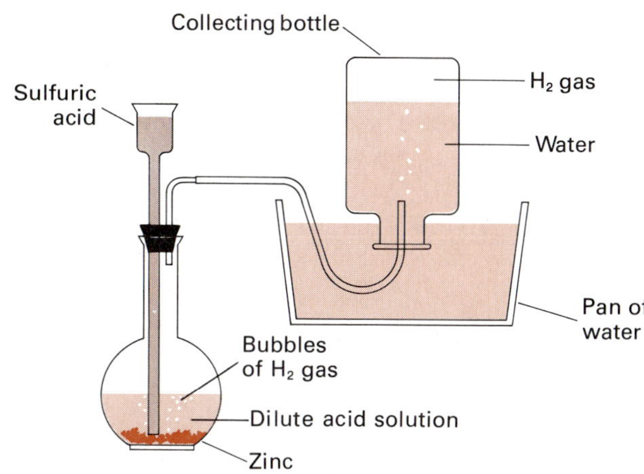

FIGURE 3-3

Collecting a gas by the displacement of water. In this experiment, hydrogen gas is being generated and collected by displacement of water. The gas-collecting bottle was originally filled with water, but the gas has pushed some water out, and the gas now occupies space formerly occupied by water.

How can we most simply explain the properties of gases? To answer this question, we must first see how quantitative studies lead to laws that summarize the behavior of gases. Then we shall see how the kinetic theory of gases explains their observed behavior.

3-2 PRESSURE

What is pressure? **Pressure is defined as the amount of force (or weight) on each unit of area.** The pressure exerted by an object depends both on its weight and on the area over which its weight is spread. Consider a block 2 cm × 3 cm × 6 cm, and weighing 120 g (Figure 3-4).

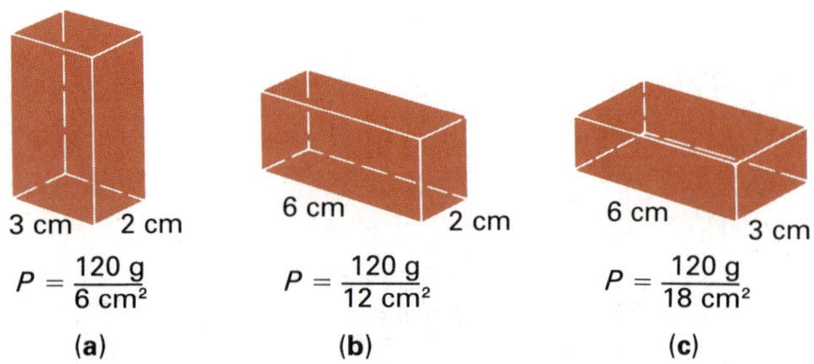

FIGURE 3-4

Pressure exerted by a block. The weight of the block is unchanged, but the pressure can be varied by changing the area on which that weight presses.

If it is lying, as shown in Figure 3-4a, such that its weight is supported by a 6 cm^2 area, then the pressure it exerts is 120 g/6 cm^2 = 20 g/cm^2.

If it is lying as in Figure 3-4b, such that its weight is supported by a 12 cm^2 area, then the pressure it exerts is 120 g/12 cm^2 = 10 g/cm^2.

If it is lying as in Figure 3-4c, such that its weight is supported by an 18 cm^2 area, then the pressure it exerts is 120 g/18 cm^2 = $6\frac{2}{3}$ g/cm^2.

Gases exert pressure, pushing out on the walls of a tire or a balloon. The gases in the earth's atmosphere exert a pressure (**atmospheric pressure**) on the surface of the earth. "We live submerged at the bottom of an ocean of air," wrote Evangelista Torricelli (1608–1647). The pressure of the atmosphere can be regarded as arising from the weight of a column of air extending from the surface of the earth to the outer limits of the atmosphere. The pressure of the atmosphere is equal to that exerted by a weight resting on a horizontal surface with about 14.7 pounds of weight per square inch of surface at sea level (Figure 3-5).

A dramatic demonstration of the existence and magnitude of air pressure was given in 1654 by Otto von Guericke, who invented the vacuum pump, with which he was able to remove the air from any container. He took a hollow bronze sphere, made of two hemispheres about 20 inches in diameter and fitted together along a greased rim and, using his pump, removed the air from the sphere (Figure 3-6). He then hitched a team of horses to each hemisphere, but though the horses were whipped into pulling with all their might, they could not pull apart the two hemispheres. The hemispheres were held together only by the force of the outside air. When air was allowed to reenter the sphere, equalizing the pressure inside and outside the sphere, the two halves easily came apart.

How is the pressure of the earth's atmosphere measured? Torricelli, reasoning that the weight of the atmosphere pushes water from a well as high as 34 ft up an evacuated pipe, invented the **barometer,** a simple instrument that can measure atmospheric pressure. To make a barometer (Figure 3-7), we take a glass tube about 1 m long closed at one end, fill it with mercury, and then invert it in a dish of mercury. The column then falls until its surface is about 750 mm above the surface of the mercury in the dish. Exactly how high it is depends on the air pressure at that time in that

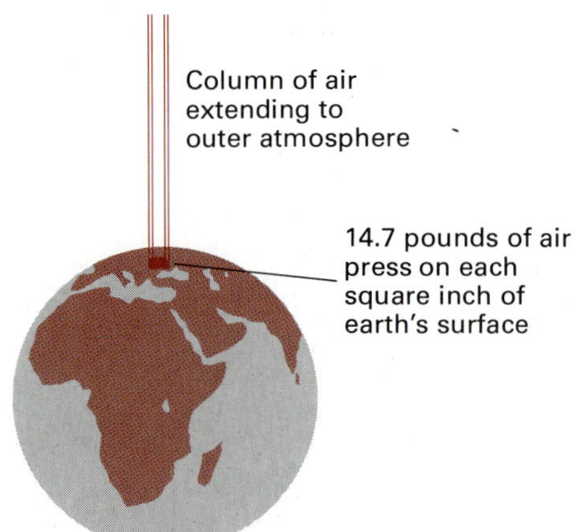

FIGURE 3-5

Pressure of the atmosphere on the earth's surface. The observed air pressure of 14.7 lb/in^2 at sea level can be regarded as the total weight of the column of air above the square inch of surface.

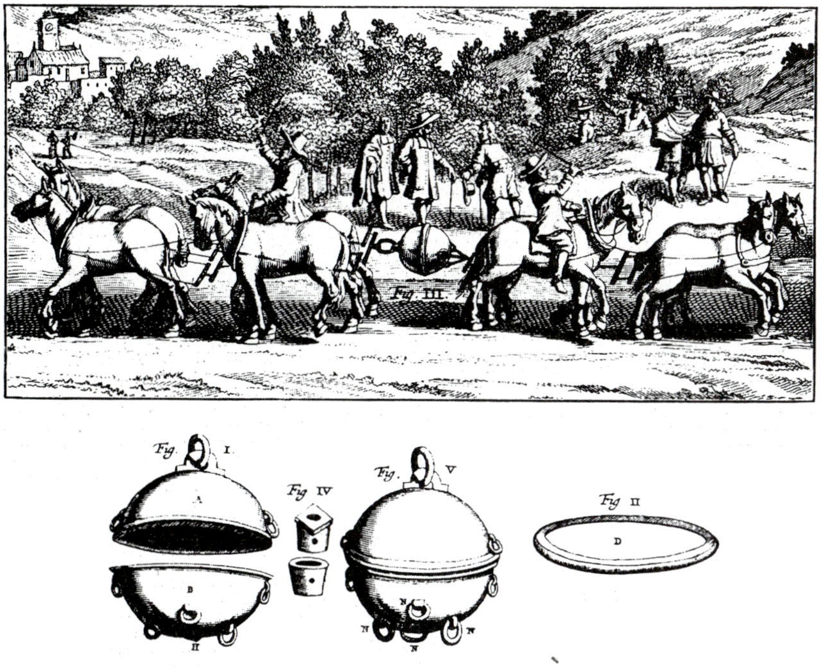

FIGURE 3-6

Von Guericke's sphere. Although the two halves of the sphere were held together only by air pressure, two teams of horses could not pull them apart. [Courtesy of Culver Pictures, Inc.]

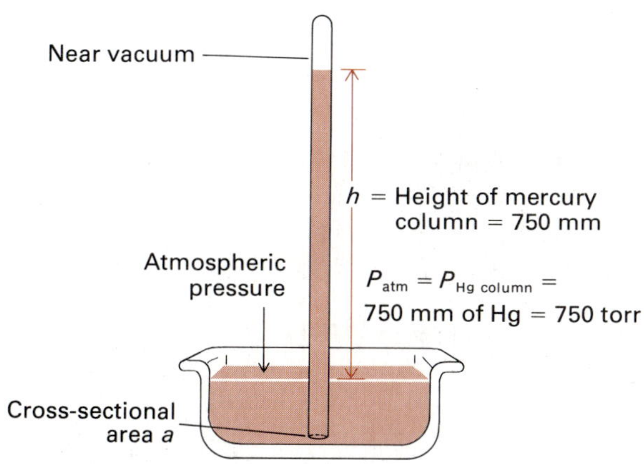

FIGURE 3-7

A mercury barometer.

place. Over the surface of the mercury column in the tube a near vacuum is left. (A little mercury vapor and some gaseous impurities are in that space, but they may be ignored.) Let h be the height of the column above the surface of the mercury in the dish, as shown in the figure. The weight of the mercury column is balanced by the push of air on the surface of the mercury in the dish. The greater the atmospheric pressure, the taller the column of mercury it will support; so the atmospheric pressure may be measured by the height (h) of that column. In Figure 3-7, the atmospheric pressure is 750 mm of mercury (Hg). If the air pressure decreases, the level of mercury in the tube falls, and h decreases; if the air pressure increases, the level of mercury in the tube rises, and h increases.

The height h depends *only* on the atmospheric pressure. In particular, it is independent of the diameter of the tube (Figure 3-8). The weight of the column increases proportionally to its cross-sectional area (a), but the *weight per unit area* (the pressure) is independent of the area itself.

The barometers in Figure 3-8 all show an atmospheric pressure of 760 mm Hg. This would be the pressure on a different day or at a different place from that of Figure 3-7. There the pressure was slightly less (750 mm Hg) as measured by the slightly lower $h = 750$ mm. Atmospheric pressure depends upon the weather and so it varies some from day to day (or even from hour to hour) in any one place. Atmospheric pressure also varies with altitude; there is less air above a mountaintop than above a valley. Figure 3-9 shows the variation in atmospheric pressure with altitude. The air pressure on top of Mount Everest (29,028 ft or 8.85 km above sea level) is only about 250 mm Hg, or about one-third of the pressure at sea level.

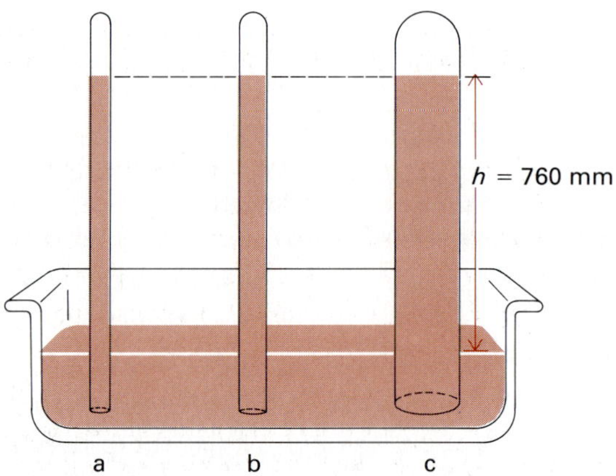

FIGURE 3-8

Barometers with different cross sections. The pressure is the same in all the barometers. The cross-sectional area of tube b is twice that of tube a; therefore the weight of mercury in column b is twice that in column a. Thus,

$$P_b = \frac{\text{Wt.}_b}{\text{Area}_b} = \frac{2 \times \text{Wt.}_a}{2 \times \text{Area}_a} = \frac{\text{Wt.}_a}{\text{Area}_a} = P_a$$

The mercury in column c, whatever its cross-sectional area, must exert the same pressure as that in column a or b.

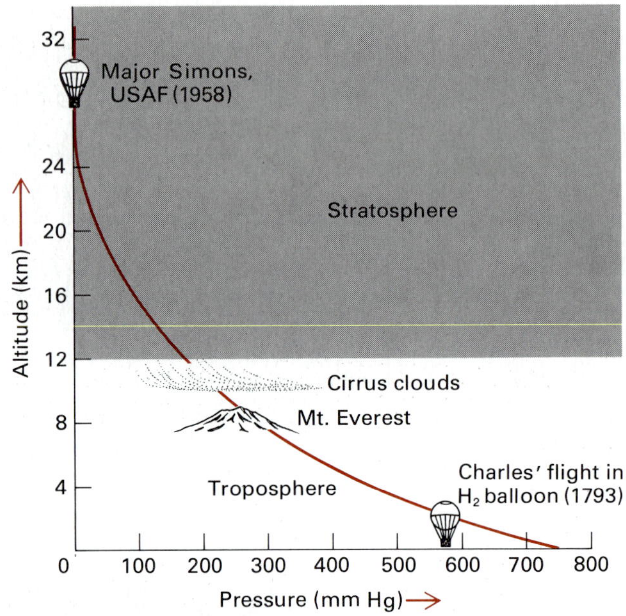

FIGURE 3-9

Air pressure as a function of altitude. [After A. Strahler, *Physical Geography,* p. 119. By permission of John Wiley & Sons, Inc., Copyright © 1952.]

A standard unit of pressure called the *atmosphere* (atm) has been defined: one **standard atmosphere** is the pressure exerted by a column of mercury 760 mm high, measured at 0°C at sea level.

$$1 \text{ atm} = 760 \text{ mm Hg}$$

These conditions, 0°C and 1 atm of pressure, are called **standard temperature and pressure** (abbreviated STP). This is the standard reference point in investigations of gases.

The atmosphere supports a column of mercury about 30 in. (760 mm) high, but pushes water as high as 34 ft. The difference in the heights of the columns supported by the atmosphere is caused by the difference in the densities of the two liquids. At room temperature the density of mercury is 13.6 g/cm^3; this means that 1 cm^3 of mercury weighs 13.6 g. The density of water is 1.00 g/cm^3 at room temperature, 1/13.6 as great as the density of mercury. The *pressure* of the column of liquid that can be supported by one atmosphere is the same for the two liquids. Because water is only 1/13.6 as dense as mercury, the water colunmn will be 13.6 times the height of the mercury column. The atmosphere that supports a column of mercury 760 mm high would, therefore, support a column of water 13.6 times as high:

$$760 \text{ mm} \times 13.6 \cong 10300 \text{ mm}$$
$$\cong 34 \text{ ft}$$

A water barometer requires a tube over 34 ft long. Torricelli realized this and decided to use mercury in his barometer for this reason.

A water barometer, however, was actually constructed by a French mathematician, Blaise Pascal. After hearing of Torricelli's experiment, he obtained (with great difficulty) two 46-ft-long glass tubes and filled one with water and the other with red wine. The atmosphere, he found, supported about 33 ft of water and slightly more of the wine (wine is a little less dense than water).

EXAMPLE 3-1

Air reaching the lungs at sea level contains oxygen gas (O_2) at a **partial pressure** of 149 mm Hg. (The term partial pressure means that the oxygen contributes only part of the total pressure of the air.) How many atmospheres is 149 mm Hg?

$$760 \text{ mm Hg} = 1.00 \text{ atm}$$
$$P_{O_2} = 149 \text{ mm Hg}$$
$$= 149 \cancel{\text{mm Hg}} \times \frac{1.00 \text{ atm}}{760 \cancel{\text{mm Hg}}}$$
$$= 0.196 \text{ atm}$$

EXERCISE 3-1 In addition to oxygen, air reaching the lungs at sea level contains nitrogen, carbon dioxide, and water vapor. (Other gases are present only in trace amounts.)

(a) Given the following partial pressures in mm Hg, find the pressures in atmospheres: $P_{N_2} = 564$ mm Hg; $P_{CO_2} = 0.3$ mm Hg; $P_{H_2O \text{ vapor}} = 47$ mm Hg.

(b) The total pressure of a mixture of gases is the sum of the partial pressures of the gases in the mixture. What is the total pressure (in atmospheres) of the air reaching the lungs at sea level?

For measuring the pressure of gas samples, an instrument called a **manometer** is used. In a closed-end manometer (Figure 3-10a), there is a vacuum above the mercury in the righthand arm, as in a barometer. The downward pressure (measured by the extra height h) exerted by the mercury column at level A in the righthand arm is exactly balanced by the pressure of the gas sample in the lefthand arm. The pressure of the gas sample here is 280 mm.

In an open-end manometer (Figure 3-10b), the righthand arm is open to the atmosphere. The sum of the pressure of the gas plus the downward pressure exerted by the mercury column (measured as the extra height h above level B) is exactly balanced by the atmospheric pressure;

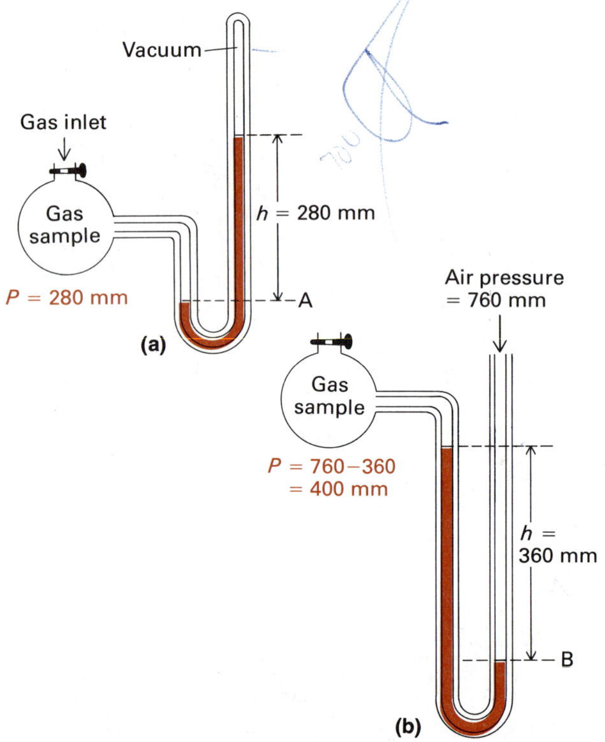

FIGURE 3-10

Manometers. **(a)** A closed-end manometer. **(b)** An open-end manometer.

$$P_{gas} + P_{Hg} = P_{atm}$$

$$P_{gas} = P_{atm} - P_{Hg} = 760 \text{ mm} - 360 \text{ mm} = 400 \text{ mm}$$

The pressure of the gas sample here is 400 mm.

With these instruments for measuring pressure, we can observe some significant laws describing the behavior of macroscopic samples of gases. We can then use these laws to arrive at a theory of the microscopic structure of gases that can simply explain their observed behavior. By "microscopic" we do not mean something merely so small that it can be seen only with an optical microscope, but rather something much smaller than that. (Later in this chapter we'll see just how small atoms and molecules are.) "Macroscopic" is the opposite—it means large.

EXERCISE 3-2 Sketch a closed-end manometer with a gas pressure of 700 mm. Show the extra height h of the mercury column.

EXERCISE 3-3 Sketch open-end manometers with gas pressures of (a) 760 mm (b) 380 mm (c) 1520 mm. Show the extra height h of the mercury column.

3-3 BOYLE'S LAW

His curiosity aroused by von Guericke's air pump and Torricelli's barometer, Robert Boyle (in 1662) performed a series of experiments whose results are summarized in **Boyle's law:** *The pressure of a fixed sample of a gas is inversely proportional to its volume if its temperature is held constant.* Boyle took a J-shaped tube closed at one end, and dropped mercury into the open end until the mercury was at the same level in both arms of the tube. (Figure 3-11a). This is an open-end manometer with a fixed sample of air now trapped in the closed end of the tube. At this point the pressure of the trapped air was equal to the atmospheric pressure, 1 atm:

$$P_a = P_{atm} = 1 \text{ atm}$$

$$V_a = \text{volume of air trapped at 1 atm}$$

Adding mercury until the enclosed volume of air

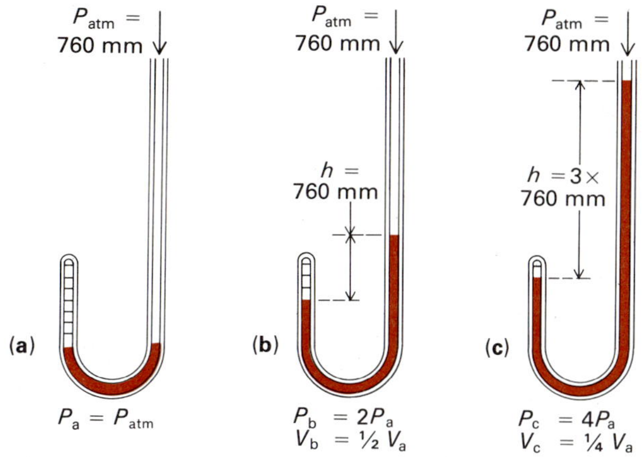

FIGURE 3-11

Boyle's J-tube experiment. This experiment demonstrates that $P \times V$ is constant.

was halved (Figure 3-11b), he found that the extra height h (of the mercury column in the open arm above that of the surface of the mercury in the closed arm) was now 760 mm, equivalent to an additional pressure of 1 atm. This made a total pressure of 2 atm on the trapped air. He had doubled the pressure while compressing the volume to one-half its original value:

$$P_b = P_{atm} + P_{Hg} = 1 \text{ atm} + 1 \text{ atm} = 2 \text{ atm}$$

$$V_b = \frac{1}{2} V_a$$

To halve the volume again, so that the new volume was one-fourth the original volume, required a column of mercury in the open arm that stood 3×760 mm = 2280 mm higher than the surface of the mercury in the closed arm. This made a total pressure of 4 atm on the trapped air (Figure 3-11c). The original pressure was quadrupled while compressing the air to one-fourth its original value:

$$P_c = P_{atm} + P_{Hg} = 1 \text{ atm} + 3 \text{ atm} = 4 \text{ atm}$$

$$V_c = \frac{1}{4} V_a$$

In this classic experiment then, Boyle found that as the pressure was increased, the volume decreased proportionally. Figure 3-12 shows a similar experiment.

Boyle's law is also summarized by the graph shown in Figure 3-13 for a hypothetical case. It shows that as

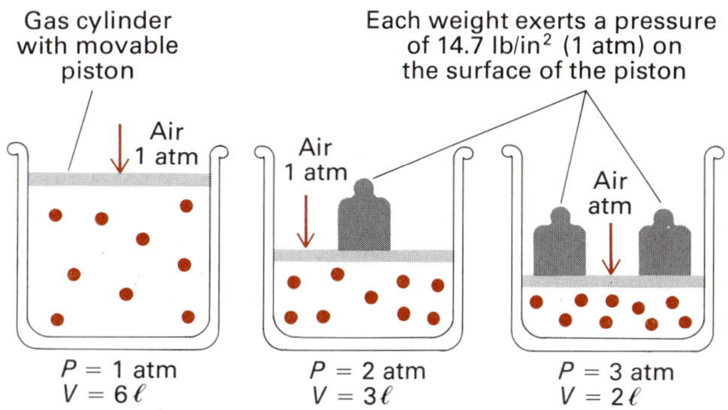

FIGURE 3-12

Boyle's law. The relationship between volume and pressure for a particular gas sample at fixed temperature. $P \times V = 6$ ℓ atm for each cylinder.

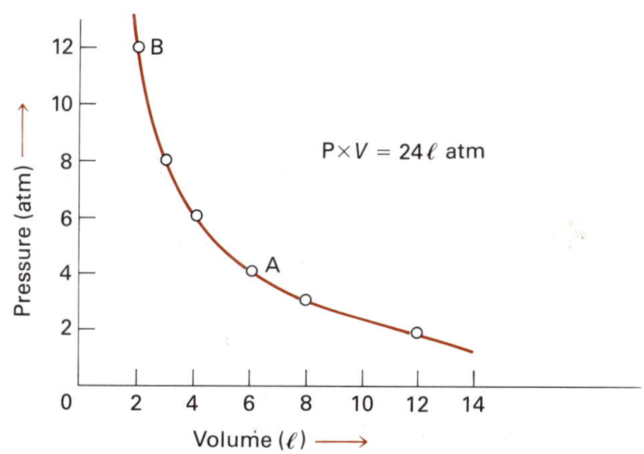

FIGURE 3-13

Boyle's law. The relationship between volume and pressure for a particular gas sample at 0°C.

the pressure is increased, the volume decreases proportionally. The graph shows the relation of **inverse proportionality,** which can also be stated mathematically in the following, equivalent ways:

$P \propto 1/V$: The pressure (P) is proportional to $1/V$; if the volume (V) is doubled, then P is halved.

$V \propto 1/P$: The volume is proportional to $1/P$; if P is doubled, then V is halved.

If a proportionality constant, k_1, is introduced, then Boyle's law becomes

$$P = k_1(1/V) \quad \text{or} \quad PV = \text{constant} = k_1$$

On the graph shown in Figure 3-13, PV = constant = 24 liter-atmospheres.

For example, at point A,

$$P = 4.0 \text{ atm}$$

$$V = 6.0 \text{ ℓ}$$

$$PV = 4.0 \text{ atm} \times 6.0 \text{ ℓ} = 24 \text{ ℓ atm}$$

At point B, the values of P and V are quite different from these. At point B,

$$P = 12 \text{ atm}$$

$$V = 2.0 \text{ ℓ}$$

But the product, PV = 24 ℓ atm, is still the same; and it is the same, 24 ℓ atm, for every point on the curve. In the experiment shown in Figure 3-12 the gas has a constant PV product of 6 ℓ atm.

The behavior summarized by Boyle's law is not always observed with real gases, but when Boyle's law is obeyed, the gas is said to be an **ideal gas.**

Figure 3-14 shows experimental pressure–volume measurements for one mole of carbon monoxide gas at three temperatures. The PV product is plotted against pressure. For a fixed sample of gas at a fixed temperature, the PV product of an ideal gas is constant for all pressures. The solid lines in the figure show the constant values of the PV product for one mole of ideal gas at the three temperatures. The dashed lines show the actual measured values. As can be seen, the behavior summarized by Boyle's law is followed closely by

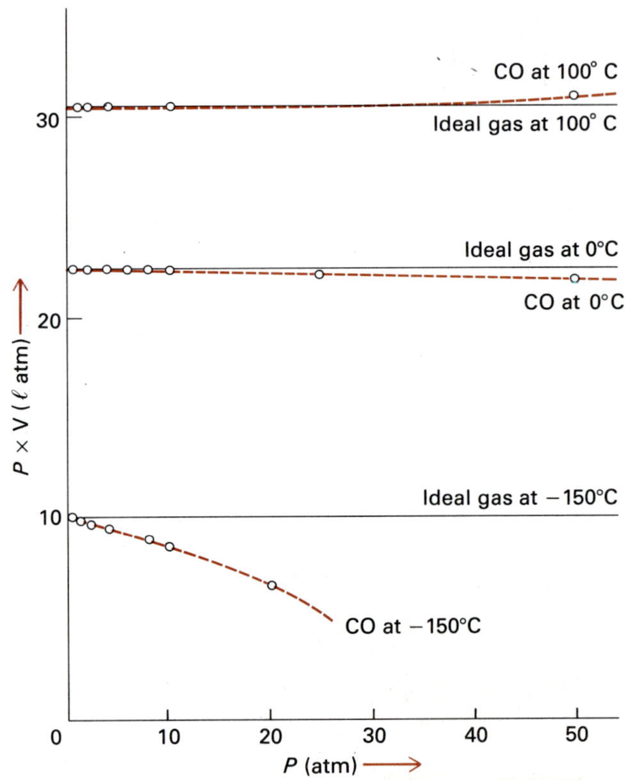

FIGURE 3-14

Pressure–volume measurements for carbon monoxide at 100°C, 0°C, and −150°C. The solid lines represent an ideal gas (one mole); the dashed lines represent actual measurements for CO (one mole).

carbon monoxide at all three temperatures at sufficiently low pressures. At the highest temperature, the *PV* product for carbon monoxide deviates only slightly from the ideal, even at very high pressure, (50 atm). At the lowest temperature, however, the *PV* product for carbon monoxide deviates slightly from the ideal even at 1 atm, and the deviation increases as the pressure increases.

Boyle's law holds for a real gas up to moderately high pressures, though deviations occur increasingly as the pressure is increased. We attempt to account for these deviations later in the chapter (after we have developed a theory explaining the behavior of gases) but, for now, the important point is that Boyle's law holds quite accurately if the pressure is not too high.

EXAMPLE 3-2

Boyle's law calculation: Compressed nitrous oxide gas (N_2O), an anesthetic commonly used by dentists, is sold in 40 ℓ steel cylinders at a pressure of 50 atm. To what volume would the gas in such a cylinder expand if the pressure on it were decreased to 1.0 atm? Assume the temperature remains constant.

The pressure *decreases* from 50 atm to 1.0 atm; thus the final pressure is 1/50 of the initial pressure. The volume *increases* proportionally:

$$V_{\text{final}} = 40\ \ell \times \frac{50\ \text{atm}}{1.0\ \text{atm}} = 2000\ \ell$$

This common-sense method of solving a gas problem is usually the most satisfactory method. Alternatively, we can use the equation that states Boyle's law: $PV = \text{constant}$, and therefore $P_{\text{initial}}V_{\text{initial}} = P_{\text{final}}V_{\text{final}}$. Solving this equation for V_{final},

$$V_{\text{final}} = V_{\text{initial}} \times \frac{P_{\text{initial}}}{P_{\text{final}}} = 40\ \ell \times \frac{50\ \text{atm}}{1.0\ \text{atm}}$$

$$= 2000\ \ell$$

This, of course, is just what the common-sense method amounts to, but there we need only remember the principle of inverse proportionality, whereas the second method requires remembering the equation.

EXERCISE 3-4 If the volume of the gas cylinder of Example 3-2 were increased to 60 ℓ, what would be the pressure of the gas? Assume that the amount of gas in the cylinder and the temperature remain constant.

3-4 CHARLES' LAW

Gases expand when heated and contract when cooled. Balloonists make use of this principle when they heat the air in a balloon, causing it to expand and become less dense than the surrounding air, and thereby allowing the balloon to be lifted and buoyed by the surrounding air (Figure 3-15).

The relationship between temperature and volume was first studied (in 1787) by J. A. C. Charles, who was led to study behavior of gases by his interest in balloons. Charles was on board the second balloon to ascend from the earth, a balloon filled with hydrogen gas. While investigating hot gases, Charles discovered

FIGURE 3-15

A hot-air balloon over the Wisconsin River. The heated air in the balloon is less dense than the surrounding air, so the balloon floats in the air. [© Trumble Photography, Wisconsin Dells.]

the law of gas behavior that relates volume and temperature; a simple experiment illustrates this relationship (Figure 3-16).

A small-bore glass tube sealed at one end and plugged with a drop of mercury has a sample of air in the bottom of the tube. The mercury plug can move up and down as the gas expands and contracts. The pressure on the air sample is the sum of the atmospheric pressure and the slight extra pressure due to the weight of the mercury plug, and the pressure remains

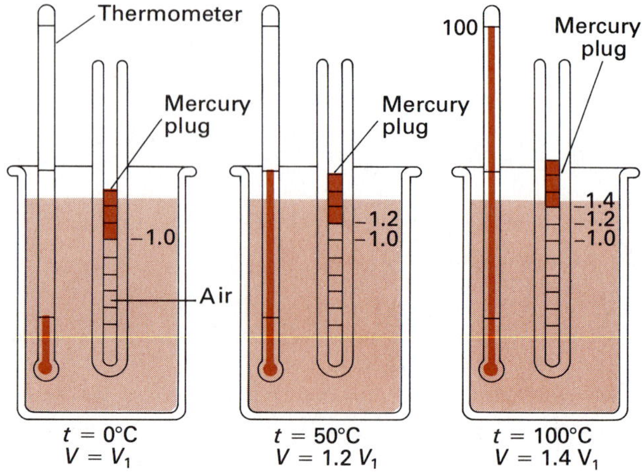

FIGURE 3-16

Charles' law. The relationship between volume and temperature for a particular gas sample at fixed pressure.

constant throughout the experiment. Because the tube has a uniform bore, the volume of the air sample is proportional to the length of the air column. We can cool the sample to below 0°C with an ice and salt bath, and we can warm it to 100°C in a boiling water bath, measuring the volume at these and at intermediate temperatures.

The graph shown in Figure 3-17 summarizes the results of such an experiment. If volume is plotted as a function of temperature in degrees Celsius, a straight line is obtained. If the line is extended to the point of zero volume, the corresponding temperature is −273.15°C at a pressure of one atmosphere. The data for a real gas would deviate from this line long before absolute zero was reached. Nevertheless, the hypothetical point at which the extrapolated volume diminishes to zero (that is, it *vanishes*), is a temperature of special significance: the **absolute zero** temperature, −273°C. An experiment relating the temperature and volume of any sized sample of any gas, at any constant pressure, *always* gives an extrapolated temperature at

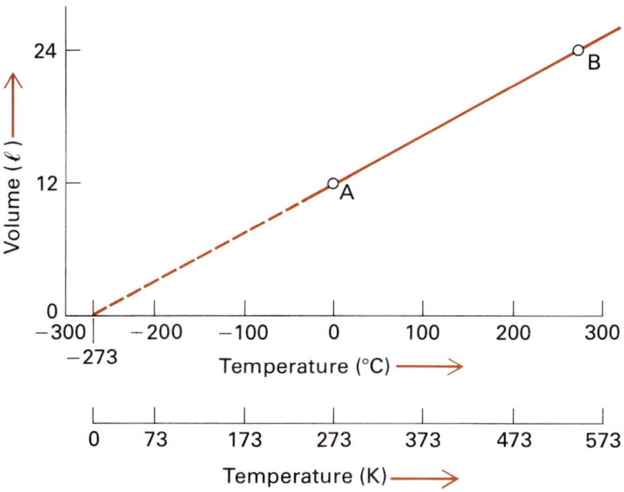

FIGURE 3-17

Charles' law. The relationship between volume and temperature for a particular gas sample at fixed pressure. The dashed portion of the line is an extrapolation from the measurements made at higher temperatures. The data for a real gas would deviate from this straight line well before a temperature of absolute zero was reached.

zero volume of -273°C. Temperatures T on that scale (which has its zero at -273°C)—called the **kelvin** scale after Lord Kelvin who proposed it—are called **absolute** temperatures. The size of the kelvin temperature unit is the same as the Celsius degree; therefore, we can convert from degrees Celsius to kelvins (symbolized K) by adding 273.15 to the value of the temperature in degrees Celsius,

$$T(\text{K}) = t(°\text{C}) + 273.15$$

$$t(°\text{C}) = T(\text{K}) - 273.15$$

Figure 3-18 shows temperatures on the kelvin scale at which some familiar phenomena occur. By convention, the term "degree" and the degree symbol are not used in expressing temperatures in kelvins.

Using the kelvin scale to express temperatures, we can state mathematically the observations summarized in the graph of Figure 3-17:

$V \propto T$: The volume (V) is proportional to the absolute (kelvin) temperature (T); if T is doubled, V is doubled.

Introducing a proportionality constant k_2, we can express this relationship as

$$V = k_2 T$$

We can now state **Charles' law:** *For a fixed sample of gas at constant pressure, the volume is directly proportional to the absolute temperature.* If the absolute temperature is

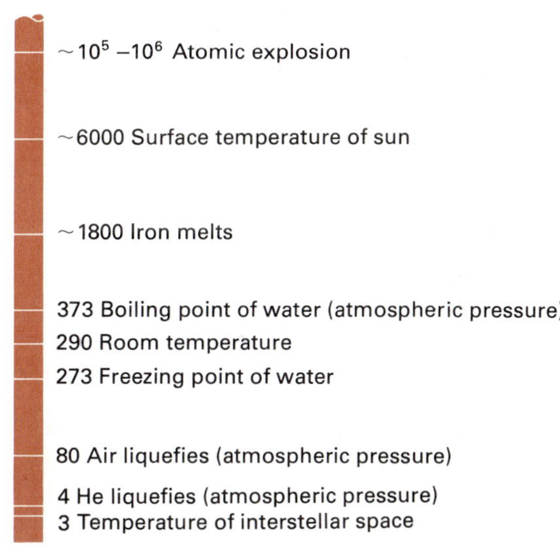

FIGURE 3-18

Temperatures of some phenomena on the kelvin scale. Scientists in low-temperature laboratories can work conveniently with temperatures as low as 0.003 K. Temperatures as low as 10^{-5} or 10^{-6} K have been attained.

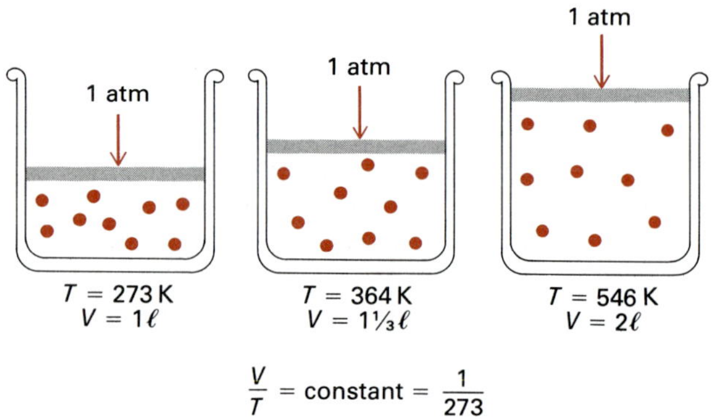

FIGURE 3-19

Charles' law. The relationship between absolute temperature and volume for a particular gas sample at fixed pressure.

doubled, the volume (at constant pressure) is doubled. At point A in Figure 3-17, the temperature is 273 K and the volume is 12 ℓ. At point B, the temperature has been doubled to 546 K and the volume has doubled to 24 ℓ.

Figure 3-19 shows another experiment (essentially identical to that performed by Charles) that demonstrates Charles' law. The absolute temperature in Figure 3-19b is 4/3 that in Figure 3-19a; accordingly, the volume in Figure 3-19b is 4/3 that in Figure 3-19a. The absolute temperature in Figure 3-19c is twice that in Figure 3-19a; and the volume in Figure 3-19c is twice that in Figure 3-19a.

As with Boyle's law, Charles' law is said to hold always for an ideal or perfect gas. For a real gas it holds at low or moderate pressures. At high pressures, deviations from Charles' law are observed. Charles' law is applied in the following example.

EXAMPLE 3-3

Assume you have 3 ℓ of helium gas at a temperature of 27°C. What volume would the helium occupy if cooled to −263°C (6°C above its normal boiling point) at fixed pressure?

$$\text{Initial temperature} = 27°\text{C} = 300\ \text{K}$$

$$\text{Final temperature} = -263°\text{C} = 10\ \text{K}$$

The temperature *decreases* from 300 K to 10 K; the final temperature is 10/300 of the initial temperature. The volume must *decrease* proportionally:

$$V_{\text{final}} = V_{\text{initial}} \times \frac{10\ \text{K}}{300\ \text{K}} = 3\ \ell \times \frac{10\ \cancel{\text{K}}}{300\ \cancel{\text{K}}} = 0.1\ \ell$$

EXERCISE 3-5 In a gas thermometer, a gas occupies a volume of 25 ml at a temperature of 27°C. The gas is then heated (at constant pressure), expanding to a final volume of 50 ml. What is the final temperature of the gas?

3-5 GAY-LUSSAC'S LAW RELATING PRESSURE AND TEMPERATURE

We have already determined how the volume of a gas sample varies with pressure at fixed temperature (Boyle's law), and how the volume varies with temperature at fixed pressure (Charles' law). We now examine the relationship between pressure and temperature at fixed volume. This is **Gay-Lussac's law:** *The pressure*

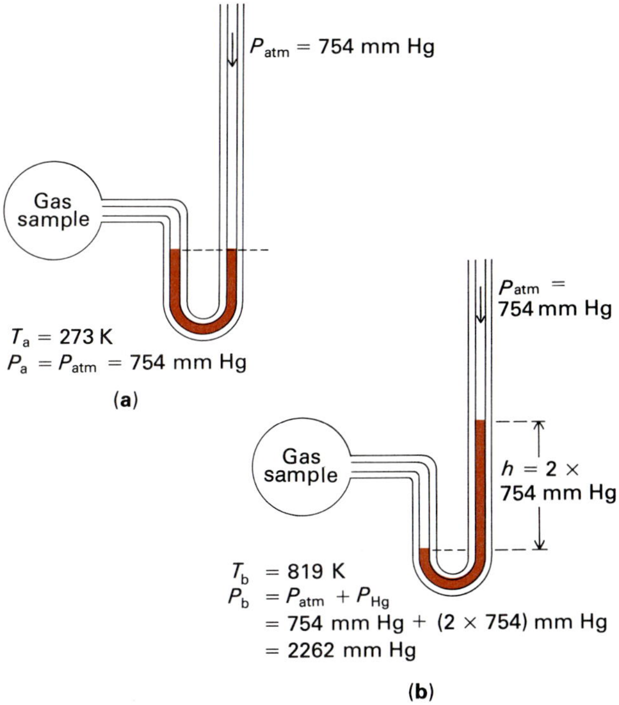

FIGURE 3-20

Gay-Lussac's law. The relationship between pressure and temperature for a particular gas sample at fixed volume.

of a fixed sample of gas at constant volume varies directly with the absolute temperature. (This is why car tires have a higher pressure at the end of a long trip on a hot day than at the beginning.)

The relationship between pressure and temperature of a gas sample held at constant volume can be investigated with the apparatus shown in Figure 3-20. A large glass globe is connected to an open-end manometer. The manometer is a capillary tube of volume negligible compared to the volume of the globe. Therefore, to a good approximation, the volume occupied by the gas is the volume of the globe, and thus it remains essentially constant throughout the experiment. It can expand only by an insignificant volume into the capillary tube. An increase in pressure is shown by a change in the height of the mercury column.

In Figure 3-20a, the mercury is at the same level in the two arms of the manometer; therefore, the pressure of the gas is equal to the atmospheric pressure, which in this example is supposed equal to 754 mm of Hg, a little less than one standard atmosphere.

$$P_a = P_{atm}$$
$$T = 273\ \text{K}$$

In Figure 3-20b, the absolute temperature is 819K, three times that in Figure 3-20a, and the pressure has increased proportionally, to 2262 mm Hg (three times what it was to start).

$$\begin{aligned} P_b &= P_{atm} + P_{Hg} \\ &= 754\ \text{mm Hg} + (2 + 754)\ \text{mm Hg} \\ &= 2262\ \text{mm Hg} \\ T &= 819\ \text{K} \end{aligned}$$

The relationship between pressure and temperature for a fixed sample of gas is shown graphically in Figure 3-21. If the line is extended to the point of zero pressure, the corresponding temperature (once again) is −273.15°C, the absolute zero.

The observation summarized in Gay-Lussac's law can be stated mathematically:

$P \propto T$: Pressure is proportional to absolute temperature; if T is doubled, P is doubled.

Or, introducing a proportionality constant k_3,

$$P = k_3 T$$

As expected, Gay-Lussac's law holds for a real gas up to moderate pressures, but deviations occur at high pressures. Gay-Lussac's law is applied in the following example.

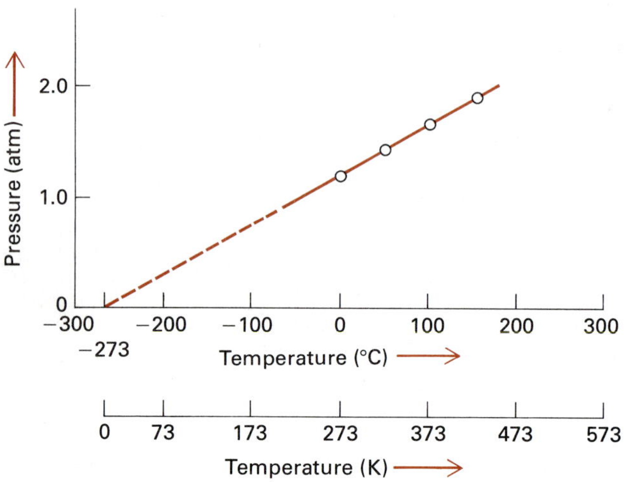

FIGURE 3-21

Gay-Lussac's law. The relationship between pressure and temperature for a particular gas sample at fixed volume. The dashed portion of the line is an extrapolation from the measurements at higher temperatures. The data from a real gas would deviate from this straight line well before absolute zero was reached.

EXAMPLE 3-4

A cylindrical steel tank contains oxygen gas at a pressure of 2000 pounds per square inch (lb/in^2, or psi) at 25 °C, somewhat above room temperature. The tank is guaranteed to stand a pressure of 6000 psi, but it may rupture at higher pressures. What is the highest temperature to which the tank may be safely heated?

The pressure *increases* from 2000 psi to 6000 psi; the final pressure is 6000/2000 times the initial pressure. To cause this pressure increase, the *absolute* temperature must increase proportionally. Therefore,

$$T_{\text{initial}} = 25\,°\text{C} = 298\ \text{K}$$

$$T_{\text{final}} = T_{\text{initial}} \times \frac{6000\ \cancel{\text{psi}}}{2000\ \cancel{\text{psi}}}$$

$$= 298\ \text{K} \times \frac{6000\ \cancel{\text{psi}}}{2000\ \cancel{\text{psi}}} = 894\ \text{K}$$

$$894\ \text{K} = 621\,°\text{C}$$

This temperature at which the highest safe pressure is reached is only about 1150 °F. A wood fire burns at a temperature well *above* that.

EXERCISE 3-6 To what pressure would the gas in Example 3-4 be reduced if the tank were stored at 0 °C?

3-6 COMBINED GAS LAWS

Boyle's law and Charles' law are

$V = k_1/P$ for a fixed sample of gas at constant temperature (Boyle's law)

$V = k_2T$ for a fixed sample of gas at constant pressure (Charles' law)

The dependence of volume on pressure *and* temperature must be as given in these two equations and, therefore, both these laws can be expressed in a single equation:

$$V \propto T/P \qquad \text{or} \qquad V = k_4T/P$$

If T is constant, this equation shows that V is proportional to $1/P$ (Boyle's law). If P is constant, this equation shows that V is proportional to T (Charles' law). We can deduce Gay-Lussac's law from it also. Because $V = k_4T/P$, then if V is to remain constant, whenever T changes, P must change proportionally.

If two of the three variables (P, V, and T) are changed *simultaneously,* we can calculate the effect on the third using the relation $V \propto T/P$. For example, if the pressure is doubled (an effect which by itself would reduce the volume by one half), and the absolute temperature is doubled simultaneously (an effect which by itself would double the volume), then there is no observed change in the volume:

$$V_{\text{final}} = V_{\text{initial}} \times (\text{factor for pressure change}) \times (\text{factor for temperature change})$$

$$= V_{\text{initial}} \times \frac{P}{2P} \times \frac{2T}{T} = V_{\text{initial}}$$

A typical problem in which two variables are changed simultaneously is given in Example 3-5.

EXAMPLE 3-5

A scuba diver carries three tanks of air. Each has a capacity of 7.00 ℓ and is at a pressure of 150 atm at 25°C. What volume of air does this correspond to at STP (0°C and 1 atm of pressure)?

The pressure *decreases* from 150 atm to 1 atm; the final pressure is 1/150 of the initial pressure. This causes a proportional *increase* in the volume; the pressure change causes the volume to increase by the factor, 150/1.

$$\text{Initial temperature} = 25\,°\text{C} = 298\text{ K}$$

$$\text{Final temperature} = 0\,°\text{C} = 273\text{ K}$$

The absolute temperature *decreases* from 298 K to 273 K; the final temperature is 273/298 the initial temperature. This causes a proportional *decrease* in the volume; the temperature change causes the volume to *decrease* by the factor 273/298.

$$V_{\text{final}} = V_{\text{initial}} \times (\text{factor for pressure change}) \times (\text{factor for temperature change})$$

$$= 3 \times 7.00\ \ell \times \frac{150}{1} \times \frac{273}{298}$$

$$= 2886\ \ell \text{ at STP}$$

$$= 2.89 \times 10^3\ \ell \text{ at STP}$$

The pressure effect accounts for most of the volume change; the temperature effect is much less significant.

The rate of air consumption depends on the depth at which the diver is submerged, as well as on his activities. At a depth of 20 m, a diver uses approximately 69 ℓ of air (measured at STP) per minute. The three tanks, therefore, would permit him to stay under

$$\frac{2.89 \times 10^3\ \ell}{69\ \ell/\text{minute}} \cong 42\text{ min}$$

EXERCISE 3-7 What is the effect on the volume of a gas of doubling the pressure, and simultaneously halving the absolute temperature?

3-7
GAY-LUSSAC'S LAW OF COMBINING VOLUMES

Some gases react chemically with each other when mixed. A mixture of hydrogen gas and oxygen gas can exist without reacting at normal temperatures and pressures, but if ignited by a spark, the mixture will react explosively to form water. At a temperature above the boiling point of water, two volumes of hydrogen will react with one volume of oxygen to form two volumes of water vapor, if all volumes are measured at the same temperature and pressure:

hydrogen gas + oxygen gas ⟶ water vapor
2 volumes + 1 volume ⟶ 2 volumes

Similarly, a mixture of hydrogen gas and chlorine gas can exist without reacting until exposed to ultraviolet light. The gases then react to form the gaseous compound, hydrogen chloride. If all volumes are measured at the same temperature and pressure, one volume of hydrogen reacts with one volume of chlorine to produce two volumes of hydrogen chloride:

hydrogen gas + chlorine gas ⟶ hydrogen chloride
1 volume + 1 volume ⟶ 2 volumes

This behavior observed in mixtures of gases involved in chemical reaction is summarized in **Gay Lussac's law of combining volumes:** *If all volumes are measured under the same conditions of temperature and pressure, then the volumes in which gases combine in a chemical reaction are related to each other, and to the volumes of any gaseous products, in simple ratios.*

3-8
AVOGADRO'S HYPOTHESIS

Dalton had hypothesized that atoms react in simple whole-number ratios to form compounds, and Gay-Lussac had observed that gases combine in volumes that are simply related to each other. Avogadro then reasoned that there must consequently be a simple relationship between gas volume and number of particles. This simple relationship is summarized in **Avogadro's hypothesis:** *Equal volumes of any gases at the same temperature and pressure contain equal numbers of molecules* (or, equivalently, *equal numbers of moles*). That is, 1 ft^3 of hydrogen gas contains the same number of molecules (moles) as 1 ft^3 of oxygen gas, and 73 ℓ of chlorine gas contain the same number of molecules (moles) as 73 ℓ of ammonia gas, if all gases are measured at the same temperature and pressure.

An important fact that Avogadro was the first to recognize is that many of the common gaseous elements including hydrogen, oxygen, and chlorine, are diatomic—for example, H_2, O_2, and Cl_2. They occur naturally as the diatomic molecules rather than as the separate atoms H, O, and Cl. The reasoning by which he concluded this can be illustrated as follows. We have seen that when hydrogen gas and chlorine gas react to form hydrogen chloride gas, *one volume* of hydrogen reacts with *one volume* of chlorine to produce *two volumes* of hydrogen chloride. If the hypothesis that equal volumes contain equal numbers of molecules is right, it means that *one molecule* of hydrogen reacts with *one molecule* of chlorine to produce *two molecules* of hydrogen chloride. *Then if hydrogen chloride is assumed to be simply HCl* (the simplest assumption—and, as we now know, the correct one), it follows that the molecules of hydrogen and chlorine must be H_2 and Cl_2:

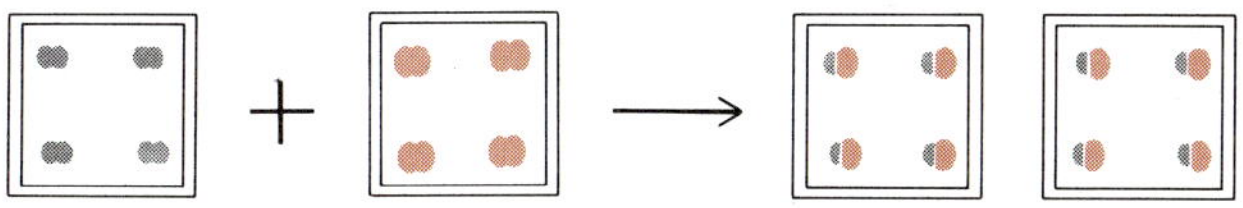

H_2 molecules	+	Cl_2 molecules	⟶	HCl molecules
1 volume	+	1 volume	⟶	2 volumes
n molecules	+	*n* molecules	⟶	2*n* molecules

The chemical equation that describes the reaction correctly is:

$$H_2(g) + Cl_2(g) \longrightarrow 2HCl(g)$$

The symbol (g) means that the substance is in the gaseous state.

Similarly, when two volumes of hydrogen gas combine with one volume of oxygen gas to form two volumes of steam, there is no problem if we assume that hydrogen and oxygen molecules each consist of two atoms (H_2 and O_2) and that each water molecule contains two hydrogen and one oxygen atom (H_2O):

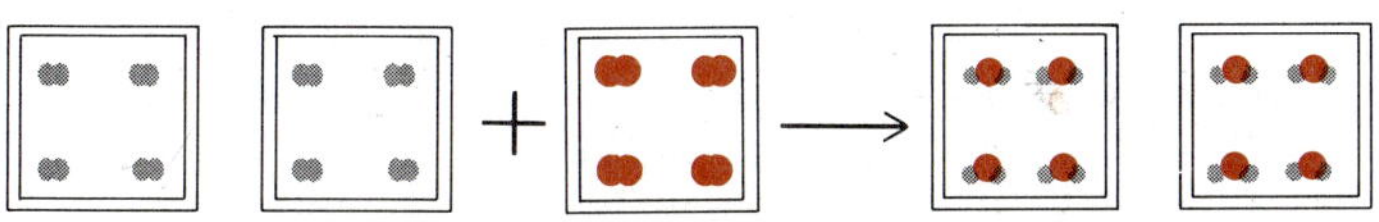

H_2 molecules	+	O_2 molecules	⟶	H_2O molecules
2 volumes	+	1 volume	⟶	2 volumes
2*n* molecules	+	*n* molecules	⟶	2*n* molecules

The chemical equation for the above reaction is

$$2H_2(g) + O_2(g) \longrightarrow 2H_2O(g)$$

EXERCISE 3-8 In the Haber process (of great industrial importance in "fixing" the nitrogen of the air for use in chemical fertilizers and other nitrogen-containing compounds), 1 volume of N_2 gas reacts with 3 volumes of H_2 gas to produce 2 volumes of ammonia gas. What is the chemical formula of ammonia?

EXAMPLE 3-6

A gas space heater, of a size to heat one room, might burn 100 ℓ (at STP) of natural gas per hour. What volume of air, at STP, is required to provide enough oxygen to operate the heater for 8 hours? Assume that natural gas is methane, CH_4 (it *is* mostly methane); that the products of the reaction are carbon dioxide gas (CO_2) and water vapor (H_2O); and that air is 1/5 oxygen by volume. (The rest of the air is almost all nitrogen, which is chemically quite inert and does not take part in the reaction.)

The balanced equation for the reaction is

$$CH_4(g) + 2O_2(g) \longrightarrow CO_2(g) + 2H_2O(g)$$

In 8 hours we will burn 100 ℓ/hour × 8 hours = 800 ℓ of CH_4. From the balanced equation we see that 2 *moles* of O_2 are needed for each *mole* of CH_4. Because the reactants and products are all gases, we know (by Avogadro's hypothesis) that we need 2 *volumes* of O_2 for each *volume* of CH_4. Therefore, we need

$$800\ \ell\ CH_4 \times \frac{2\ \ell\ O_2}{1\ \ell\ CH_4} = 1600\ \ell \text{ of } O_2$$

Because air is only one-fifth oxygen, we need 5 × 1600 ℓ = 8000 ℓ of air to operate the heater for 8 hours.

A gas-burning heater must be properly vented.

Used in a closed room, it would gradually consume enough oxygen that significant amounts of the poisonous gas carbon monoxide would be produced along with the carbon dioxide. Occasionally a person sleeping in a closed room in which a gas heater is operating is asphyxiated when the carbon monoxide concentration reaches a lethal level.

3-9 IDEAL GAS LAW

We can now consider Avogadro's hypothesis along with the relationship $V \propto T/P$ that we developed earlier (the combined gas laws). This relationship holds for any fixed amount of the gas. But Avogadro's hypothesis tells us that (at fixed pressure and temperature) the volume occupied by a gas is directly proportional to the number of molecules, and therefore to the number (n) of moles. We already know that

$$V \propto T \qquad \text{(at fixed } n \text{ and } P\text{)}$$

$$V \propto \frac{1}{P} \qquad \text{(at fixed } n \text{ and } T\text{)}$$

Now we can add a third statement:

$$V \propto n \qquad \text{(at fixed } P \text{ and } T\text{)}$$

We can combine all three statements into one:

$$V \propto nT/P$$

Introducing a new proportionality constant, R, we obtain

$$V = RnT/P$$

This equation is more usually (but equivalently) expressed as

$$PV = nRT$$

and is known as the **ideal gas law.**

In this equation, n is the number of moles of the gas. The gas constant R has an invariable and universal value for any gas that behaves ideally. We shall calculate the value of R shortly.

If we consider the equation for constant P and n, we see that $V \propto T$, which is Charles' law. At constant T and P, we see that $V \propto n$, which is Avogadro's hypothesis.

At standard temperature and pressure (0°C and 1 atm), the volume occupied by 32.00 g (one mole) of oxygen gas has been experimentally determined to be 22.4 ℓ. This is the **molar volume** of oxygen at STP. And, because equal volumes of all gases are supposed (by Avogadro's hypothesis) to contain equal numbers of molecules, this should be the volume occupied by one mole of any gas at STP. Table 3-1 gives observed molar volumes of some gases.

Real gases indeed have molar volumes equal or very nearly equal to 22.4 ℓ/mole at STP. The molar volume for an ideal gas at STP is 22.414 ℓ/mole. This value is determined by measuring the volumes of real gases at low pressure (where their behavior approximates ideality) and then calculating the molar volume at STP using the gas laws.

TABLE 3-1
Molar volumes of gases

Gas	Molar volume (ℓ/mole)
Air	22.401
Argon (Ar)	22.390
Nitrogen (N_2)	22.404
Ammonia (NH_3)	22.080
Carbon dioxide (CO_2)	22.262
Carbon monoxide (CO)	22.400
Acetylene (C_2H_2)	22.157
Ethylene (C_2H_4)	22.246
Propane (C_3H_8)	21.990
Ethane (C_2H_6)	22.179
Methane (CH_4)	22.360

Using this ideal value for the molar volume at STP we can calculate R:

$$R = PV/nT$$

At 0°C and 1 atm, the volume occupied by one mole is 22.4 ℓ.

$$R = \frac{1 \text{ atm} \times 22.4 \ \ell}{1 \text{ mole} \times 273 \text{ K}} = 0.082 \ \ell \text{ atm/mole K}$$

This is the constant and universal value of R for all gases, at all temperatures and pressures for which the gases remain ideal. The units of R are liter atmospheres per mole per kelvin. Because of the way it is calculated, when this value of R is used in the ideal gas law, *P must be expressed in atmospheres, V in liters, n in moles, and T in kelvins.*

If three out of the four variables (P, V, T, and n) are known the fourth can be calculated using the ideal gas law. Several examples follow.

EXAMPLE 3-7

The total lung capacity of a woman is about 4.2 ℓ. What, approximately, is the mass of the air in the lungs? Assume that the pressure is 1.0 atm and the temperature is body temperature, 37°C. Assume, also, that the average molecular weight of air is about 29 g/mole (air is about 80% N_2 and 20% O_2). We know the following:

$$P = 1 \text{ atm}$$

$$T = 37°\text{C} = 310 \text{ K}$$

$$V = 4.2 \ \ell$$

$$R = 0.082 \ \ell \text{ atm/mole K}$$

We can calculate n, the number of moles of air in the lungs, by rearranging the ideal gas law $PV = nRT$:

$$n = \frac{PV}{RT} = \frac{1.0 \text{ atm} \times 4.0 \ \ell}{0.082 \dfrac{\ell \text{ atm}}{\text{mole K}} \times 310 \text{ K}} = 0.17 \text{ moles}$$

One mole of air $\cong$ 29 g. Therefore, the mass of air in the lungs is

$$0.17 \text{ moles} \times 29 \text{ g/mole} = 4.8 \text{ g}$$

EXAMPLE 3-8

Natural gas is mostly (80% to 95%) methane. It is obtained from wells, where it exists as a gas at high pressure and also dissolved in oil. What volume of a well would contain one mole of gas at a pressure of 1500 pounds per square inch (lb/in^2) and a temperature of 77°C? (Approximate 1 atm as 15 lb/in^2, which is close enough.)

We know the following:

$$n = 1 \text{ mole}$$
$$T = 77^\circ\text{C} = 350 \text{ K}$$
$$P = 1500 \text{ lb/in}^2$$
$$\cong \frac{1500 \cancel{\text{lb/in}^2}}{(15 \cancel{\text{lb/in}^2})/\text{atm}} = 100 \text{ atm}$$

Therefore, we can calculate V:

$$V = nRT/P$$
$$= \frac{1 \cancel{\text{mole}} \times (0.082\ \ell\ \cancel{\text{atm}}/\cancel{\text{mole}}\ \cancel{\text{K}}) \times 350 \cancel{\text{K}}}{100 \cancel{\text{atm}}}$$
$$= 0.29\ \ell$$

A mole of methane at STP occupies 22.4 ℓ. The tremendous decrease in the volume occupied by a mole of methane in the well is due almost entirely to the great pressure. At such high pressures, the ideal gas law is not highly accurate, but it is accurate enough so that this estimate of the volume is realistic.

When a hole is drilled into an oil pool that is under high pressure from gas and water, the pressure is relieved and the gas expands. Sometimes, when the pressure is high enough, tremendous quantities of gas and oil erupt through the well opening. Modern drilling techniques make these "blow-outs" or "gushers" rare, but they still occur occasionally. In June 1979, an exploratory well off the coast of Mexico exploded. More than three million barrels of crude oil were spilled before the well was successfully capped nine months later.

EXAMPLE 3-9

An aerosol can that squirts cheese spread uses nitrogen gas (N_2) as a propellant. Assume that the gas occupies a volume of 300 ml and weighs 2.8 g.

(a) Calculate the pressure of the N_2 at 27°C (about 81°F).

We know the following:

$$V = 300 \text{ ml} = 0.300\ \ell$$
$$n = 2.8 \cancel{\text{g}}/(28 \cancel{\text{g}}/\text{mole}) = 0.10 \text{ mole}$$
$$T = 27^\circ\text{C} = 300 \text{ K}$$

In calculating n, we used the fact that the molecular weight of $N_2 = 2 \times 14 = 28$. We are now ready to calculate P:

$$P = nRT/V$$
$$= \frac{0.10 \cancel{\text{mole}} \times 0.082 \cancel{\ell}\ \text{atm}/\cancel{\text{mole}}\ \cancel{\text{K}} \times 300 \cancel{\text{K}}}{0.30 \cancel{\ell}}$$
$$= 8.2 \text{ atm}$$

(b) What pressure would develop if the can were thrown into a campfire and heated to 900 K (about 1100°F)? (Neglect any effect the heat might have on the cheese spread, and assume that the can does not melt at this temperature.)

The values of n and V are unchanged; T increases from 300 K to 900 K, so the absolute temperature increases by the factor 900/300 = 3/1. The pressure increases proportionally, by the factor 3/1:

$$P_{\text{final}} = 8.2 \text{ atm} \times 900 \cancel{\text{K}}/300 \cancel{\text{K}} = 25 \text{ atm}$$

We can also solve the problem by again using the ideal gas law:

$$P = nRT/V$$
$$= \frac{0.10 \cancel{\text{mole}} \times 0.082 \cancel{\ell}\ \text{atm}/\cancel{\text{mole}}\ \cancel{\text{K}} \times 900 \cancel{\text{K}}}{0.30 \cancel{\ell}}$$
$$= 25 \text{ atm}$$

The can might well explode before reaching this pressure. You will note that aerosol spray cans carry explicit warnings against incinerating or heating above 120°F.

An important application of the ideal gas law is in the determination of molecular weights, which permits one to ascertain the correct molecular formula of a compound whose *simplest* formula is known from chemical analysis. This application is illustrated in the two following examples.

EXAMPLE 3-10

Acetylene, a hydrocarbon gas used in welding, is 7.7% hydrogen and 92.3% carbon, by weight. It has a density of 4.33 g/ℓ at a pressure of 4.1 atm and temperature of 300 K.

(a) What is the simplest formula of acetylene?

Of 100 g of acetylene, 92.3 g are carbon and 7.7 g are H.

$$\text{Moles of C} = \frac{92.3\,\cancel{g}}{12.0\,\cancel{g}/\text{mole}} = 7.7 \text{ moles C}$$

$$\text{Moles of H} = \frac{7.7\,\cancel{g}}{1.0\,\cancel{g}/\text{mole}} = 7.7 \text{ moles H}$$

There is one hydrogen atom for each atom of carbon. Therefore, the simplest formula is CH, but the actual formula might be C_2H_2, C_3H_3, C_4H_4, or any formula with one atom of H for each C atom.

(b) Calculate the molecular weight of acetylene.

We know

$$T = 300 \text{ K}$$

$$P = 4.1 \text{ atm}$$

We do not know n or V, but we can calculate n/V, the number of moles in one liter. By rearranging the equation $PV = nRT$, we get

$$\frac{P}{RT} = \frac{n}{V}$$

$$\frac{n}{V} = \frac{P}{RT} = \frac{4.1\,\cancel{\text{atm}}}{(0.082\ \ell\,\cancel{\text{atm}}/\text{mole}\,\cancel{\text{K}}) \times 300\,\cancel{\text{K}}}$$

$$= 0.167 \text{ mole}/\ell$$

But, we also know (from the gas density measurement) that 1 ℓ of acetylene weighs 4.33 g at this temperature and pressure. Therefore, 0.167 mole weighs 4.33 g. We know that

$$\text{Moles} = \text{g/molecular weight}$$

so $\quad \text{Molecular weight} = \text{g/moles}$

$$\text{Molecular weight of acetylene} = 4.33 \text{ g}/0.167 \text{ mole} = 26.0 \text{ g/mole}$$

Equivalently, we might have reasoned as follows:

$$0.167 \text{ mole} = 4.33 \text{ g}$$

Dividing both sides by 0.167 gives

$$1 \text{ mole} = 4.33 \text{ g}/0.167 = 26.0 \text{ g}$$

(c) What is the actual formula of acetylene?

The simplest formula, CH, has a formula weight of 12 + 1 = 13 g/mole. The observed molecular weight is 26.0 g/mole. The observed molecular weight of acetylene is 26.0/13 = 2 times as great as the simplest formula weight for CH. Thus, the acetylene molecule must contain two CH groups. The actual chemical formula of acetylene is $(CH)_2$ or C_2H_2.

EXAMPLE 3-11

A 200 ml sample of phosphorus vapor at a pressure of 0.41 atm and temperature of 227°C weighs 0.248 g.

What is the formula of the phosphorus molecule? (The atomic weight of phosphorus is 31.)

We know the following:

$$T = 227°\text{C} = 500 \text{ K}$$

$$P = 0.41 \text{ atm}$$

$$V = 200 \text{ ml} = 0.200 \; \ell$$

We can calculate the number of moles of phosphorus vapor in the 200 ml sample:

$$n = PV/RT = \frac{0.41 \text{ atm} \times 0.200 \; \ell}{(0.082 \; \ell \text{ atm/mole K}) \times 500 \text{ K}}$$

$$= 1/500 \text{ mole} = 0.0020 \text{ mole}$$

We also know that this 200 ml sample weighs 0.248 g. Therefore,

$$0.0020 \text{ mole} = 0.248 \text{ g}$$

$$\text{Molecular weight} = 0.248 \text{ g}/0.0020 \text{ mole}$$

$$= 124 \text{ g/mole}$$

The atomic weight of phosphorus is 31 g/mole. Therefore, there are 124/31 = 4 moles of phosphorus atoms in a mole of phosphorus vapor, or 4 atoms of phosphorus in a molecule of phosphorus vapor. Thus, the formula of the molecule of phosphorus vapor must be P_4.

3-10 KINETIC THEORY OF GASES

The ideal gas law applies to all gases, whatever their molecular complexity, and summarizes the physical behavior observed by Boyle, Charles, and Gay-Lussac. Because gas behavior does not depend on the *kind* of gas, studies of the macroscopic behavior of gases—as observed in measurements of pressure, volume, mass, and temperature—lead us to a relatively simple microscopic picture of gases: the **kinetic theory of gases.** The diffusion of gases through each other suggests that gases consist of particles in motion. This idea of molecules in motion is the basis of the kinetic theory of gases, and it can be used to explain the observed properties of gases.

The kinetic theory describes molecules as moving rapidly and randomly in straight lines, colliding with each other and with the walls of the container, and changing direction only upon such a collision (Figure 3-22).

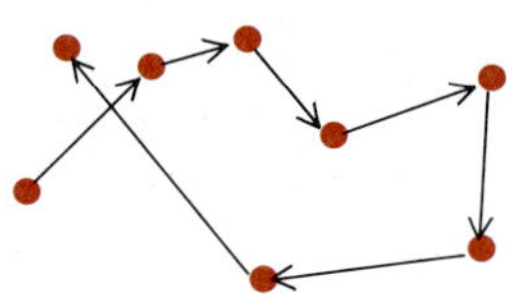

FIGURE 3-22

Random motion of one molecule in a gas. There is a sudden change of direction at each collision with another molecule or with a container wall.

The pressure of a gas is caused by molecules banging into the walls of the container. The more collisions per unit time, the greater the pressure. If we take a fixed sample of gas and reduce the size of its container (its volume) at constant temperature, then there will be more collisions per unit time—and thus greater pressure—because the molecules have a shorter average distance to travel between successive collisions

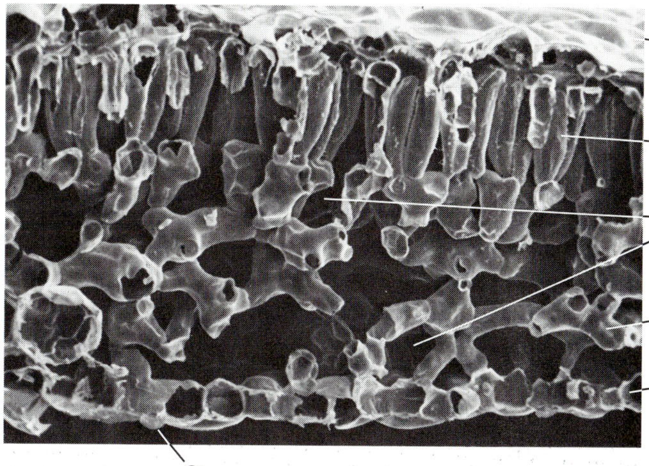

FIGURE 3-23

Cross section of a yellow birch leaf. Carbon dioxide enters through stomata (openings) on the lower surface of the leaf. The carbon dioxide diffuses through the intercellular spaces into the parenchymal cells, where it is consumed in the process of photosynthesis. Oxygen, produced during photosynthesis, diffuses out into the atmosphere. [Courtesy of Ellen T. Paparozzi, Cornell University.]

with the walls. If the volume is increased, the molecules will have to travel farther, on the average, between two successive collisions with the walls of the container, and therefore the pressure will decrease. This picture is consistent with Boyle's law: $P \propto 1/V$.

The observed gas properties of diffusion and compressibility are consequences of the large amount of free space existing in a gas. Figure 3-23 illustrates an example of diffusion that is of great consequence to life on earth—the diffusion of carbon dioxide and oxygen into and out of the photosynthetic cells of a plant. When a gas is compressed, the molecules are squeezed closer together, reducing the amount of free space. A comparison of the density of a gas with the density of a liquid or a solid shows that the volume occupied by a mole of gas is many times greater than the volume occupied by a mole of liquid or solid. In Example 3-12, it proves to be over 800 times as great.

EXAMPLE 3-12

(a) The density of solid nitrogen (N_2) is about 1.03 g/cm^3. (Nitrogen exists in the solid form at 1 atm pressure only at temperatures below $-210°C$.) What is the molar volume of solid N_2?

$$\text{Molar volume of solid N}_2 = \frac{28.0\ \cancel{g}/\text{mole}}{1.03\ \cancel{g}/\text{cm}^3} = 27.2\ \text{cm}^3/\text{mole}$$

(b) At $-210°C$ and 1 atm pressure, the density of liquid N_2 is 0.81 g/cm^3. What is the molar volume of liquid N_2?

$$\text{Molar volume of liquid N}_2 = \frac{28.0\ \cancel{g}/\text{mole}}{0.81\ \cancel{g}/\text{cm}^3} = 34.6\ \text{cm}^3/\text{mole}$$

The liquid is somewhat less dense than the solid—that is, one mole of the liquid occupies a volume somewhat greater than the volume occupied by one mole of the solid.

(c) We know from our earlier discussion that the molar volume of *any* gas at STP must be 22.4 ℓ/mole = 22,400 cm^3/mole. (As an exercise, confirm this value for the molar volume of N_2 gas at STP, given that the measured density of N_2 gas at STP is 0.00125 g/cm^3.) Compare the molar volumes of solid, liquid, and gaseous N_2.

The volume occupied by one mole of liquid N_2 is about 34.6/27.2 = 1.27 times as great as the volume occupied by one mole of solid N_2. The volume occupied by one mole of gaseous N_2 (at STP) is about 22,400/27.2 = 824 times as great as the volume occupied by one mole of solid N_2 (which cannot actually exist at STP). Figure 3-24 illustrates this dramatic change in molar volume between the solid or liquid states and the gaseous state. The molecules themselves are the same size in the solid, liquid, and gas states, but the space between the molecules in the gas state is very much greater than that in the liquid or the solid.

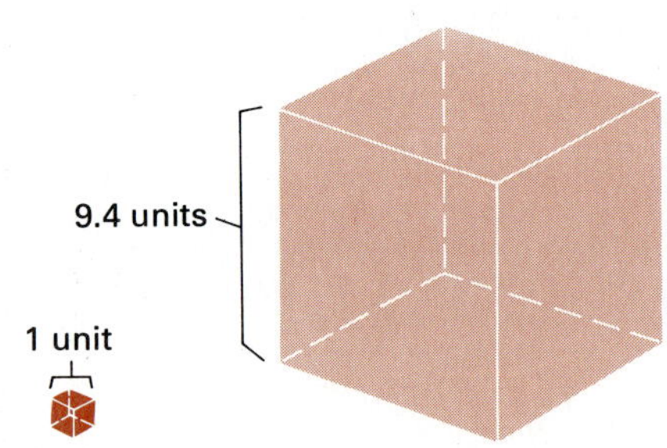

FIGURE 3-24

Comparison of the volumes of solid and gas. The volume occupied by a certain mass of N_2 gas at STP is about 800 times as great as the volume occupied by the same mass of solid N_2. The number of N_2 molecules is the same in each case.

In this example, the volume occupied by the gas is almost 1000 times as great as the volume occupied by an equal mass of the solid. If we picture the molecules as touching in the dense solid, then the distance between neighboring molecules in the gas must be about 10 times the width of a molecule. This is because a tenfold increase in distance produces a thousandfold increase in volume ($10^3 = 1000$).

We have seen (in Figure 3-14) that real gases deviate from ideal behavior at high pressures. What causes these deviations? At low pressures, 99.9% or more of the volume of a gas may be free space, and the actual volume of the gas molecules is negligible. As the pressure is increased, however, the molecules are squeezed closer together, and eventually a pressure is reached where the actual volume of the molecules is no longer negligible, but is comparable to the volume of free space available to them (Figure 3-25).

The deviation from ideal behavior at high pressure shown in Figure 3-14 results from the fact that molecules exert forces on each other. The origin of these forces is investigated in Chapter 7. The forces between molecules are attractive at large distances, tending to pull them together. At sufficiently high pressure and low temperature, these forces cause the gas to condense to a liquid. In a gas, however, the molecules are so far apart that these attractive forces, which decrease rapidly with increasing distance between the molecules, are weak and readily overcome by the motion of the molecules. The molecules therefore move independently of one another. When the pressure is increased, the molecules are squeezed together until, finally, they begin to feel the attractions of their near neighbors. The attractive forces then pull the molecules closer together. Eventually, if the pressure is great enough and the temperature low enough, the gas molecules will coalesce and form a liquid or a solid.

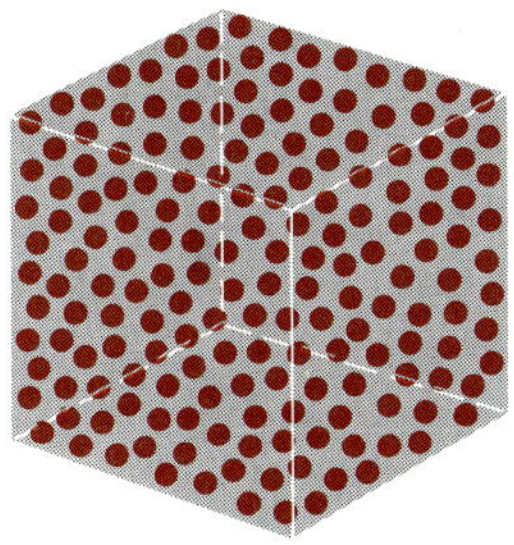

FIGURE 3-25

Kinetic-molecular concept of a gas at high pressure. The free space between molecules is comparable to the volume of the molecules themselves.

The kinetic theory hypothesizes further that the average energy of motion of gas molecules depends *only* on the absolute temperature and is directly proportional to it. This energy of motion, called the **kinetic energy** (K.E.) of a particle, is given by the expression K.E. $= \frac{1}{2}mv^2$, where m is the mass of the particle and v is the velocity. It is the energy of the molecules' motion. When a gas is heated, the average kinetic energy of its molecules increases, and the molecules, on the average, move faster. When a gas cools, the average kinetic energy decreases, and the molecules, on the average, move more slowly.

At any *one* temperature, the average kinetic energy of the molecules of a gas is the same in any sample of any gas at any pressure. Thus the hydrogen molecules (H_2) in a 7 ℓ tank of hydrogen gas at 100 atm pressure and 20°C, have the *same average kinetic energy* as the oxygen molecules (O_2) in a roomful of air at 1/5 atmosphere pressure and 20°C. Both gases are at the same temperature, and therefore the average kinetic energy of the molecules of the two gases is the same. Because the average $\frac{1}{2}mv^2$ is the same for the two gases, the hydrogen molecules, with molecular weight 2, will be traveling at a greater average velocity than the oxygen molecules with molecular weight 32.

The hypothesis that the kinetic energy is proportional to the absolute temperature is consistent with Gay-Lussac's and Charles' laws. If the temperature of a gas is increased, the average velocity increases and causes the molecules to hit the walls of the container *harder* and *more frequently*. At constant volume, both of these effects lead to an increase in pressure. This is consistent with Gay-Lussac's law: $P \propto T$ (for constant n and V). If the pressure, instead, is to remain constant, the volume must increase. Then the frequency of collisions with the container decreases and counteracts the greater force of the collisions. This is consistent with Charles' law: $V \propto T$ (for constant n and P).

We have spoken of the *average* kinetic energy at any temperature because the individual molecules do not all have the same kinetic energy. There is instead a **distribution** of velocities, with many molecules traveling at a velocity close to the average, but some traveling much faster and some much more slowly. We can compare this situation to the distribution of speeds among a group of marathoners running a race. The majority run at a moderate rate, and only a few run very fast or very slowly. Figure 3-26 shows a distribution for a group of 99 runners in a marathon. There is a wide range of average velocities, with more runners in the velocity interval 7.5 to 8.0 miles per hour than in any other interval. The most probable velocities are those in this interval. Most of the runners had an average velocity close to this. Only 2 ran more slowly (on the average) than 6.0 miles per hour, and only 2 ran faster than 9.5 miles per hour.

Figure 3-27 shows *distributions of molecular velocities* for samples of nitrogen at two temperatures. The area of the shaded rectangle is proportional to the number of molecules that have velocities in that little interval, just as a bar in Figure 3-26 represents the number of runners with velocities in a small interval. The most probable velocity at 273 K is about 375 m/sec. At 1273 K the most probable velocity is much greater—about 800 m/sec. A wide range of velocities exists at

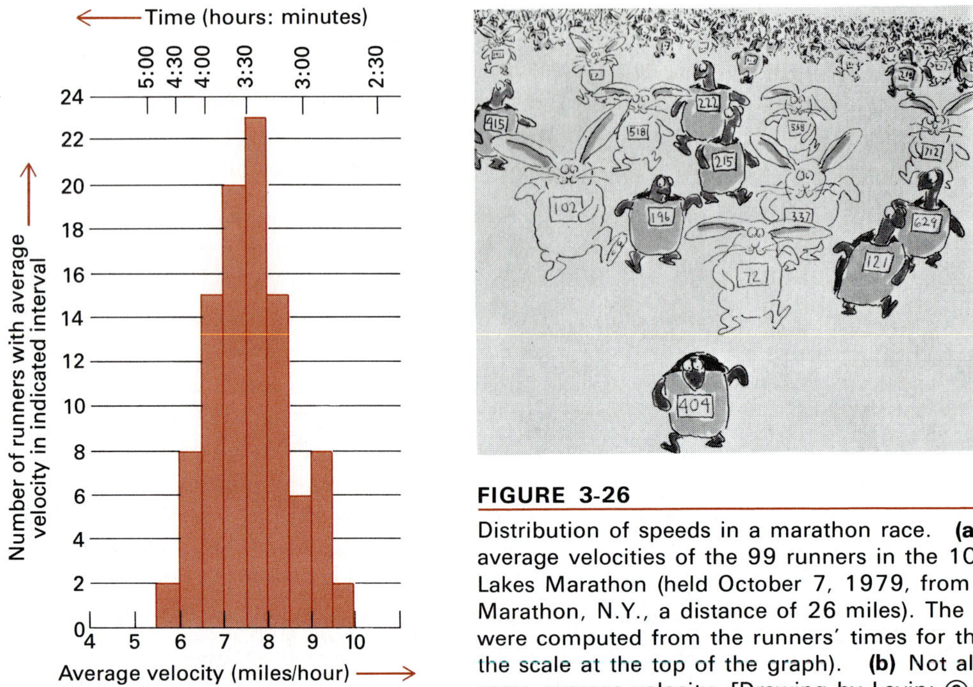

(a)

(b)

FIGURE 3-26

Distribution of speeds in a marathon race. **(a)** Distribution of average velocities of the 99 runners in the 10th Annual Finger Lakes Marathon (held October 7, 1979, from Ithaca, N.Y. to Marathon, N.Y., a distance of 26 miles). The average velocities were computed from the runners' times for the race (as shown on the scale at the top of the graph). **(b)** Not all runners have the same average velocity. [Drawing by Levin; © 1979 The New Yorker Magazine, Inc.]

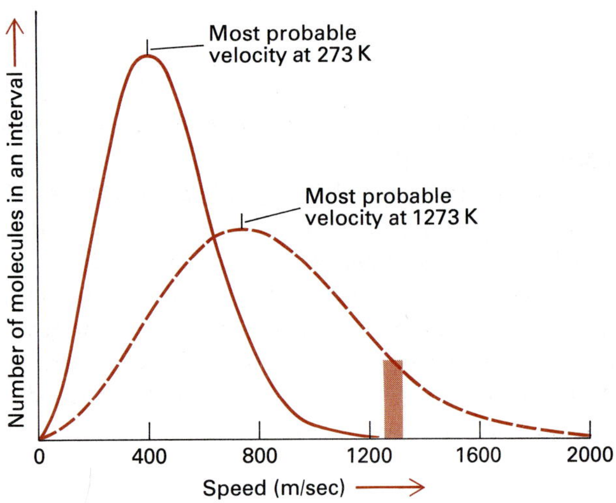

FIGURE 3-27

Distribution of molecular velocities in a gas. Data for nitrogen gas (N_2) at two temperatures. The area under a section of the curve is proportional to the number of molecules having velocities within the corresponding interval.

TABLE 3-2
Most probable velocities for gas molecules

Molecule	Molecular weight	Temperature (K)	Most probable velocity (m/sec)
H_2	2	273	15×10^2 (3400 miles/hour)
H_2	2	1000	29×10^2
H_2O	18	273	5×10^2
CO_2	44	273	4×10^2
Sn	119	273	1.9×10^2
Hg	201	273	1.5×10^2
Hg	201	1000	2.9×10^2
Hg	201	1500	3.5×10^2

both temperatures, but at the higher temperature *more* molecules move at a high velocity. And although there are always *some* slow-moving molecules, there are fewer slow molecules at 1273 K than at 273 K. This principle is crucial to an understanding of the rates of chemical reactions; when we come to study that subject in Chapter 10, we shall refer again to Figure 3-27. Table 3-2 gives the most probable molecular velocities of several gases at different temperatures.

KEY TERMS

Pressure is the amount of force, or weight, on a unit area. The *pressure of the atmosphere* arises from the weight of a column of air extending from the earth's surface to the outer limits of the atmosphere. A *barometer* measures the pressure of the atmosphere.

A *standard atmosphere* is the pressure exerted by a column of mercury 760 mm high at 0°C at sea level: 1 atm = 760 mm Hg. *Standard temperature and pressure* (*STP*) refer to a temperature of 0°C and a pressure of 1 atmosphere. A *manometer* is used to measure the pressure of a sample of gas.

Temperatures on the *kelvin* scale (K) are called *absolute temperatures,* and symbolized T. The *absolute zero temperature* is -273°C or 0 K (K = °C + 273).

An *ideal gas* is a gas that obeys the *ideal gas law,* $PV = nRT$, incorporating Boyle's law, Charles' law, and Avogadro's hypothesis. R is the universal gas constant.

According to *Avogadro's hypothesis,* equal volumes of all gases at the same temperature and pressure contain equal numbers of molecules.

The *kinetic theory of gases* is a molecular theory that explains the observed macroscopic behavior of gases. *Kinetic energy* is energy of motion. The average kinetic energy of a sample of gas is proportional to the absolute temperature.

CHAPTER SUMMARY

We have investigated the following laws summarizing the observed behavior of gases:

1. $P \propto \frac{1}{V}$ for a fixed sample of gas at constant T (Boyle's law);

2. $V \propto T$ for a fixed sample of gas at constant P (Charles' law);

3. $P \propto T$ for a fixed sample of gas at constant V (Gay-Lussac's law);

4. At a given temperature and pressure, gases react in simple proportion by volume, and the volume of any gaseous product is related simply to the volume of any gaseous reactant (Gay-Lussac's law of combining volumes).

These laws, together with Avogadro's hypothesis that equal volumes of gases at the same temperature and pressure contain equal numbers of molecules, lead us to the ideal gas law:

5. $PV = nRT$.

The kinetic theory of gases explains their observed behavior. The principal ideas included in this theory are the following.

1. A gas is composed of molecules that are widely separated and are in rapid, random motion—colliding with each other and with the walls of the container. Pressure results from collisions of gas molecules with the walls of the container.

2. The average kinetic energy of the molecules depends *only* on the absolute temperature and is proportional to it. At any temperature, there is a distribution of velocities, with some molecules moving very slowly and some very rapidly, but with most moving at a velocity close to the average.

3. Except at high pressures or low temperatures, the attractive and repulsive forces between the molecules in a gas are negligible.

PROBLEMS

3-1. Blood pressure is the force of the blood pushing against the walls of blood vessels. It varies at different points in the circulatory system. Normal blood pressure in a young woman is about 110 mm Hg when the ventricles contract (systolic blood pressure) and 70 mm Hg when the ventricles relax (diastolic pressure). What are these pressures in atmospheres?

3-2. What is the pressure (in atmospheres) exerted by
(a) a column of mercury 380 mm high?
(b) a column of mercury 38 cm high?
(c) a column of mercury 70 cm high?
(d) a column of water 17 ft high?
(e) a column of mercury 800 mm high?
(f) a column of mercury 1140 mm high?

3-3. Discuss the following phenomena in terms of the gas laws.
(a) The rising of dough.
(b) The "popping" of a paper bag.
(c) The explosion of an "empty" spray can when incinerated.

(d) The greater difficulty in breathing at higher altitudes.
(e) The expansion of a hydrogen-filled weather balloon upon rising. (A weather balloon appears shriveled when it takes off, and expands as it rises.)
(f) The pressure in an automobile tire increasing on a hot day or on a long trip.

3-4. Figure 3-28a shows air trapped above the surface of mercury in a tube, the other end of which is open to the atmosphere. More mercury is then poured into the open end of the tube, whereupon the condition of the tube becomes that shown in Figure 3-28b. In terms of the original volume V of the trapped air, what is the final volume?

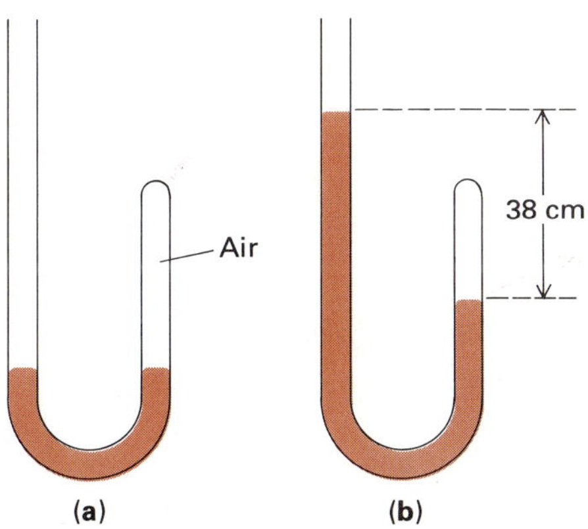

FIGURE 3-28

3-5. At what temperature is the volume of a sample of gas five times what it is at standard temperature? Assume that the pressure remains constant.

3-6. Solve the following problems about gases.
(a) A gas occupies 2.0 ℓ at STP. What is its volume at a pressure of 0.5 atm and a temperature of 546 K?
(b) A gas occupies 0.50 ℓ at a temperature of 27°C and a pressure of 2.0 atm. What is the final pressure, if the temperature is increased to 327°C and the volume is increased to 4.0 ℓ?
(c) A gas occupies 10 ℓ at a pressure of 4.0 atm and a temperature of 819°C. What would its volume be at STP?
(d) Which contains more molecules—2 ℓ of carbon dioxide gas at STP, or 2 ℓ of ammonia gas at STP?
(e) A sealed vial contains 6.72 ml of carbon monoxide gas (CO) at STP. How many moles of CO does it contain?

3-7. What is the effect on the pressure of a gas of doubling the volume of its container and simultaneously doubling its absolute temperature?

3-8. When gasoline burns in a cylinder of an automobile, the pressure increases, thus forcing a piston outward. The motion of the piston is transmitted to the wheels of the car. In the cylinder, gasoline vapor reacts with oxygen (O_2) in the air to produce carbon dioxide gas (CO_2) and water vapor (H_2O).
(a) Taking octane (C_8H_{18}) as a representative component of gasoline, find the effect on the pressure in the cylinder of the change in the number of moles of gas as a result of the chemical reaction.
(b) Heat produced in the reaction causes the temperature to increase from about 300 K to about 1500 K. What is the effect on the pressure in the cylinder of this change in temperature?
(c) By what factor does the pressure change as a result of the change in the number of moles of gas and the change in the temperature, taken together?

3-9. Each of two identical 1.12 ℓ flasks contains *one* of the gases nitrogen (N_2) and carbon tetrachloride (CCl_4) at STP.
(a) How can you determine, without opening the flasks, which flask contains which gas?
(b) Which flask contains the greater number of moles of gas?
(c) Which flask contains the greater number of molecules of gas?
(d) How many molecules are in the nitrogen-containing flask?

3-10. Water can be decomposed to hydrogen gas (H_2) and oxygen gas (O_2). If 36 g of water are decomposed, what is the total volume of gas (hydrogen plus oxygen) that is produced, if the gases are collected at STP?

3-11. (a) The oxygen content of arterial human blood is about 200 ml (at STP) of O_2 per liter of blood. How many moles of O_2 are in 1 ℓ of arterial blood? Most of the O_2 (98%) is combined with hemoglobin.
(b) Oxygen is removed from the blood as it circulates through the tissues. The oxygen content of venous blood for a person at rest is 150 ml (at STP) of O_2 per liter of blood. How many moles of O_2 are in 1 ℓ of venous blood?
(c) How many moles of oxygen per liter of blood are removed by the tissues of a person at rest?
(d) If the blood flows at a rate of 5.4 ℓ per minute, how many moles (and how many grams) of O_2 are used by the tissues of a person at rest during a 24 hour period?

3-12. Because carbon dioxide is produced in the tissues, venous blood contains more CO_2 than does arterial blood. The amount of CO_2 removed from the tissues of a person at rest is 37 ml CO_2 (at STP) per liter of blood. If the blood flows at a rate of 5.4 liters per minute, how many moles (and how many grams) of CO_2 are excreted in a 24 hour period?

3-13. During strenuous exercise, as much as 4 ℓ (at STP) of oxygen per minute might be removed from arterial blood by the tissues, and as much as 8 ℓ (at STP) per minute of CO_2 be excreted into the venous blood. How many grams of O_2 are used (and how many grams of CO_2 excreted) per hour?

3-14. An improperly adjusted stove is leaking carbon monoxide (CO) into a closed room at a rate of 0.60 moles of CO per minute. The volume of the room is 6.0×10^5 ℓ, and the temperature is 0°C. Assume that all the CO remains in the room and that the total pressure remains constant at 1 atm. How long will it take for the CO to reach the level of 0.1% CO by volume? (At this point, 0.1% of all the gas molecules in the room are CO. This concentration of CO causes unconsciousness in 1 hour and death in 4 hours.)

3-15. Methane gas (CH_4) burns in air to produce carbon dioxide (CO_2) and water.
(a) How many liters of air are needed to burn each liter of CH_4? Assume that air is 20% oxygen by volume, and that the volumes of all of the gases are measured at the same pressure and temperature.
(b) If 11.2 ℓ of oxygen are used, how many liters of CH_4 are burned? How many moles of CH_4 are burned?
(c) If 10 ℓ of CH_4 and 10 ℓ of O_2 are mixed, how many liters of water vapor can be produced by the combustion? Which gas (CH_4 or O_2) and how much remains at the end of the reaction?

3-16. Find the densities at STP (in grams per liter) of the following gases:
(a) nitrous oxide (N_2O)

FIGURE 3-29

Weather balloon during launch. This balloon has a volume of 5,660,000 cubic feet; it carries instruments for experiments by researchers at Harvard University. [Photograph by Mack Gore, National Scientific Balloon Facility, Palestine, Texas.]

(b) hydrogen (H_2)
(c) air (80% N_2 and 20% O_2)
Which gas would you choose to fill a weather balloon? Explain.

3-17. A weather balloon is filled with 300 ℓ of hydrogen gas at STP. The balloon is designed to rise to a height of 10 miles (Figure 3-29), at which altitude the temperature is $-55\,°C$ and the atmospheric pressure is 0.1 atm.
(a) What volume will the balloon have at the 10 mile altitude?
(b) How many moles of hydrogen gas are in the balloon?
(c) If the hydrogen is produced by a reaction between solid calcium hydride (CaH_2) and water—producing solid calcium hydroxide, $Ca(OH)_2$, and hydrogen gas, H_2—what mass of calcium hydride is needed to produce the required amount of H_2?

3-18. On being strongly heated, solid calcium carbonate decomposes into solid calcium oxide and gaseous carbon dioxide: $CaCO_3 \rightarrow CaO + CO_2$. When 10.0 g of $CaCO_3$ are decomposed, 5.6 g of CaO remain. The CO_2, collected at a pressure of 1 atm and a temperature of 300 K, occupies a volume of 2.46 ℓ. Calculate the gas constant R from these data.

3-19. Bromine vapor has a density of 1.00 g/ℓ at a pressure of 0.41 atm and a temperature of 800 K. Show from these data that the formula of the bromine molecule is Br_2.

3-20. An unknown organic compound is determined by chemical analysis to have the simplest formula $C_6H_{12}O$. At normal atmospheric pressure and a temperature of 227°C, the vapor of the substance has a density of about 2.5 g/ℓ. Of the various possibilities $C_6H_{12}O$, $C_{12}H_{24}O_2$, $C_{18}H_{36}O_3$, and so on, which is the true formula of the compound?

3-21. A closed container with a volume of 1.0 ℓ contains 8.0 g of oxygen gas (O_2) and an unknown amount of carbon dioxide gas (CO_2). If the overall pressure in the container is 8.2 atm and the temperature is 27°C, what is the weight (in grams) of the CO_2?

3-22. Atmospheric pressure in outer space is about 1×10^{-14} mm Hg, with the gas mainly hydrogen atoms (H). The temperature is about 2 K. Find the density of H (in *atoms* per liter) in outer space.

3-23. Dry Ice (solid carbon dioxide, CO_2) is provided in 100 kg blocks that are cut into quarters for distribution. Each quarter is a cube of about 25 cm on a side. The Cornell University chemistry department uses about 100 kg of Dry Ice every day.

(a) What volume does one 25 kg cube of Dry Ice occupy as a gas at STP?
(b) Find the volume of the 25 kg cube of the solid (Dry Ice) and compare it with the volume of the gas calculated in part a.
(c) What volume of CO_2 gas, measured at STP, does the Cornell University chemistry department use in one year (365 days)?

3-24. Propane is composed *only* of the elements carbon and hydrogen. In the complete combustion of propane gas, 1 volume of propane reacts with 5 volumes of oxygen gas (O_2) to produce 3 volumes of carbon dioxide gas (CO_2) and 4 volumes of water vapor (H_2O). What is the formula of propane? (All volumes are given at the same temperature and pressure.)

3-25. A large lecture hall (about 40′ × 50′ × 60′) has a volume of 120,000 ft^3, or about 3×10^6 ℓ. What, approximately, is the weight of the air in the lecture hall at 27°C and 1 atm pressure? Assume that the molecular weight of air is about 30 g/mole (it is closer to 29, but 30 is a good approximation that will make the calculation simpler.)

4

Liquids and Solids

The kinetic theory of matter also helps explain some observed properties of liquids and solids as well as changes from one state to another. A liquid evaporating in a closed container furnishes an example of dynamic equilibrium—although no observable change is taking place, there is a continuous exchange of individual molecules between the liquid and vapor phases. A situation that seems static on the macroscopic level is in fact dynamic on the microscopic level.

4-1
CONDENSATION OF A GAS

We have seen that a gas under high pressure deviates from ideal behavior; the molecules do not move independently but exert significant forces on each other. At moderate distance, the forces between molecules are attractive, pulling them together, and causing a gas at sufficiently low temperature or high pressure to condense to a liquid or solid. At very short distances the forces become repulsive, causing the molecules to have definite and characteristic *sizes* or *volumes*. The repulsive forces keep the atoms from penetrating each other, and thus resist the compression of liquids and solids. (We examine the origin of these forces in Chapter 7.)

We think of the molecules in a liquid or solid as touching each other so that they cannot be squeezed closer together without interpenetrating. A 1.00 ml sample of *liquid* water is only slightly compressed to 0.96 ml when the pressure is increased from 1 atm to 1000 atm. But the volume of water *vapor* is halved when the pressure is merely doubled. Gases, which are mostly free space, have densities that are directly proportional to the pressure. Because liquids and solids have little free space, their densities do not change very much when the pressure is increased. The density of a vapor reflects the *distance between* the molecules, whereas the density of a liquid or a solid reflects the *sizes of* the molecules. We saw this in Example 3-12, where we found the volume occupied by one mole of N_2 gas to be over 800 times that occupied by one mole of the liquid or the solid.

In a gas the forces of attraction are overcome by the kinetic energy. When a gas at a given pressure is cooled, it eventually reaches a temperature at which the average kinetic energy of the molecules is low enough compared to the attractive forces that the gas condenses. Table 4-1 shows the temperatures at which some gases liquefy at a pressure of 1 atm. The familiar phenomenon of water vapor from the air condensing on the outside of a glass of ice water is an example of a gas condensing to a liquid. This is a **change of state,** or a **phase change:** water in the gaseous state (or phase) has changed to the liquid state (or phase). As the temperature of the liquid is lowered further, the kinetic energy of the molecules decreases still more, and eventually the attractive forces between the molecules become important enough that the fluid liquid freezes to a rigid solid (Figure 4-1). Water freezes at 0°C, changing from liquid water to ice. In a phase change, only the *state* of the substance (whether solid or liquid or gas) changes, not its chemical identity. Ice, water, and steam are three different forms of the same chemical substance; all consist of the same H_2O molecules.

TABLE 4-1
Liquefaction temperature of some gases (at 1 atm pressure)

Gas	Liquefaction temperature (°C)	(K)
Chlorine (Cl_2)	−34	239
Oxygen (O_2)	−183	90
Argon (Ar)	−186	87
Fluorine (F_2)	−188	85
Nitrogen (N_2)	−196	77
Neon (Ne)	−246	27
Hydrogen (H_2)	−253	20
Helium (He)	−268.9	4.2

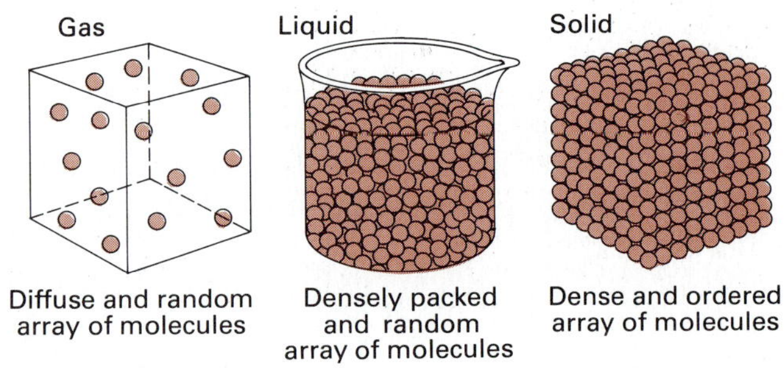

FIGURE 4-1
Molecules in the gaseous, liquid, and solid states.

4-2
PROPERTIES OF LIQUIDS

Only two liquids are prevalent in nature: water, which covers three-fourths of the earth's surface, filling over 324 million cubic miles of rivers, lakes, and seas; and petroleum, which is stored as a liquid in the earth's crust. Some other liquids—such as milk or maple syrup—are mainly water. And many—such as ethylene glycol (used as a permanent antifreeze) or trichloroethane (used as a spot remover)—are synthetic. Only two elements—mercury (which is used in barometers) and fuming, red-brown bromine—are liquids at room temperature and atmospheric pressure. Neither is found free in nature in any quantity.

The molecules of a gas move chaotically and randomly, so that a gas expands to fill and take on the shape of its container. The molecules of a liquid, on the contrary, are confined by attractive forces. The molecules cling together, the liquid then occupying a fixed volume—though, like a gas, a liquid is shapeless, adapting to the shape of its container. The liquid water in a glass remains in the bottom, but if the glass is overturned, the water flows, spreading over the table and the floor and losing its shape. The *volume* of a liquid remains the same, whatever the size and shape of its container—it does not expand to fill its container as a gas does.

Like gases, liquids diffuse—though much more slowly than gases. There is much less empty space in a liquid than in a gas, and therefore the molecules in a liquid cannot move about and diffuse so freely as those in a gas.

Another characteristic property of liquids is **surface tension.** It is caused by the pull of the molecules that are below the surface of a liquid sample on those molecules that are at the surface. This occurs whether the surface is that of a small spherical rain drop or that of a large lake. The surface molecules are pulled together by each other and by the molecules underneath, making a tough surface film that can support a water strider (Figure 4-2).

FIGURE 4-2

A water strider is supported by surface tension. The terminal segment of each of the strider's six legs rests on the surface of the water, making a shallow depression. [Photograph by Richard Byrnes.]

Another property of liquids is **wetness.** Spilled water wets almost everything it touches—clothes, a table, a rug—as the forces of attraction of water molecules for each other give way to the stronger attractive forces between water molecules and the molecules of the substances they wet. (An exception is that water does not wet greasy or oily substances, for reasons that we discuss in Section 9-4.) The atoms of liquid mercury are more strongly attracted to each other than to the molecules of most other substances, so mercury does not ordinarily wet other substances but, when spilled, beads up into droplets (Figure 4-3).

The attraction of a liquid for other substances is also illustrated by the phenomenon called **capillarity,** which draws water up a fine glass tube or into a blotter, a towel, or a sponge, and allows melted wax to travel up a wick where it is vaporized and burns. It is capillary action that moves water upward through the pores of the soil to reach the roots of plants.

We are all familiar with the fact that liquids **evaporate**—the water from an open glass gradually evaporates, and eventually all of the liquid escapes into the

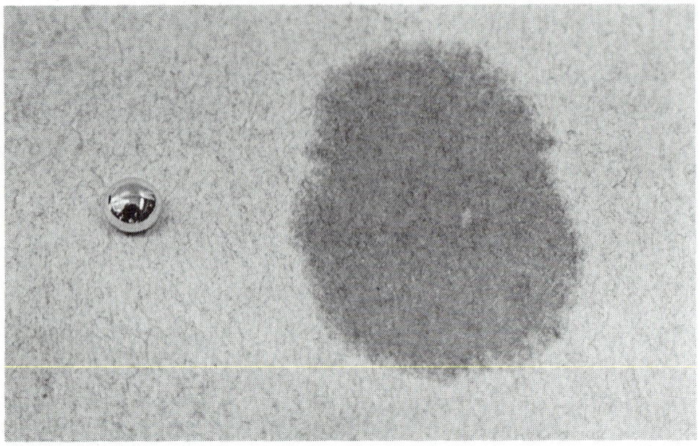

FIGURE 4-3
Drops of water and of mercury on a paper towel. Water wets the towel, but mercury beads up. [Photograph by Richard Byrnes.]

gas phase. Like the molecules of a gas, the molecules of a liquid have a *distribution* of kinetic energies. Some of the molecules at the surface of a liquid have enough kinetic energy to overcome the forces of attraction and escape from the liquid into the gas. Loss of the energetic molecules lowers the average kinetic energy of the remaining molecules, and the temperature drops. If the liquid is in an uninsulated container (such as a glass of water), heat will flow from the surroundings to the liquid, maintaining the liquid at the same temperature as the surroundings, and eventually all of the liquid will evaporate. However, if the liquid is in an insulated container such as a Dewar flask or thermos bottle (Figure 4-4), its temperature drops while the liquid evaporates because the liquid loses its hottest molecules—those with the greatest kinetic energy. At the lower temperature, fewer molecules have enough kinetic energy to escape, so the rate of evaporation decreases—the small heat flow from the surroundings permits only a slow rate of evaporation to occur. Even liquid nitrogen, which boils at $-196°C$, can be kept for many hours at room temperature in an open Dewar flask.

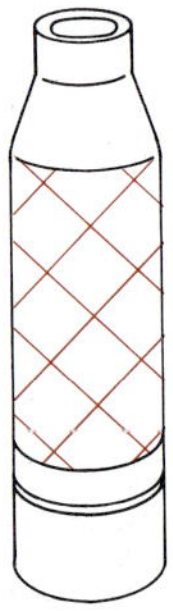

FIGURE 4-4
A Dewar flask. This container is similar to a thermos bottle, with an evacuated space between two reflecting walls. The contents of a Dewar flask remain at a nearly constant temperature over a long period of time.

4-3 PROPERTIES OF SOLIDS

A solid is firm and sticks together compactly, whether held in a container or let loose. When a glass of ice water is overturned, the ice does not flow as the liquid does. The water molecules in the ice cube are held tightly in a rigid, ordered structure. (The orderly arrangement of the atoms in a solid is discussed in Section 8-6.) The idea that the solid state is highly ordered is an old one, and it accounts for the observed fact that many solids occur in beautiful crystalline forms. A solid can be grasped and held firmly in the hand, unlike a gas or a liquid. It has its own fixed shape and does not take on the shape of its container.

We have seen that liquids diffuse more slowly than gases. Solids diffuse much more slowly even than liquids—so slowly, in fact, that ordinarily we do not think of solids as diffusing at all. The rock layers in the Grand Canyon (Figure 4-5) have been lying in contact with each other for millions of years (some for over 600 million years) without significant diffusion from one to another, so the sharp boundaries between the layers are maintained. In some solids, however, at elevated temperatures below the melting point, a measurable extent of diffusion occurs in a few hours. Because there is essentially no free space in a solid, it may be surprising that diffusion occurs at all. Imperfections in solids leave occasional empty spaces that permit diffusion.

FIGURE 4-5

The Grand Canyon. The rock layers have been in contact for millions of years without significant diffusion of solid rock from one layer to another. [Drawing by W. H. Holmes, from Clarence E. Dutton, *Tertiary History of the Grand Cañon District* (Washington, D.C.: U.S. Government Printing Office, 1882).]

A solid can be as hard as a block of oak or a diamond, or as soft as wax or graphite. It can be as brittle as a sheet of ice, or as pliable and elastic as a rubber band. Some solids (such as butter) melt in a warm room; iron doesn't melt until it is heated to 1530°C. Many solids (such as salt, sugar, and copper sulfate) are soluble in water, but others (such as butter, silver, and plastic) are insoluble in water. We shall learn more in Chapters 8 and 9 about the structure of solids, their solubilities, and the properties of their solutions.

4-4 LIQUID–VAPOR EQUILIBRIUM, AND VAPOR PRESSURE OF LIQUIDS

The evaporation of liquids from open containers is a familiar phenomenon, but what happens to a liquid in a closed container? If some alcohol is placed in the bottom of a cylinder and the cylinder covered, the liquid level drops a little and then remains constant. How can we explain this phenomenon? Some of the molecules near the liquid surface have enough kinetic energy to overcome the forces of attraction and escape from the liquid to the vapor phase. At first, evaporation occurs. Then, as more molecules collect in the vapor, some of them collide with the surface and reenter the liquid. Finally a situation is reached where

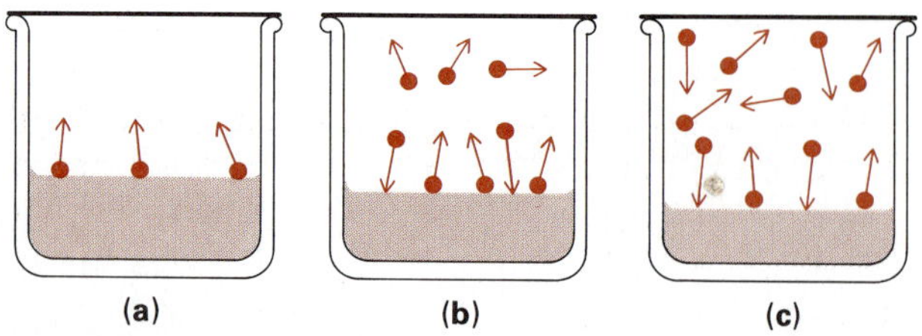

FIGURE 4-6

Evaporation of a liquid in a closed container. **(a)** Initially, molecules leave the surface of the liquid to enter the vapor phase. **(b)** Eventually, some vapor molecules collide with the surface of the liquid and reenter the liquid. **(c)** Finally, molecules leave and reenter the liquid at the same rate.

molecules are leaving the liquid at the same rate at which other molecules from the vapor are reentering it (Figure 4-6). This is a state of **dynamic equilibrium.** The level of liquid in the container remains constant. If we could count the water molecules in the vapor phase at equilibrium, we would find that their number also remains constant. Individual molecules continue to escape from the liquid to the vapor phase, but during the same time an equal number of others condense from the vapor to the liquid phase.

There are many familiar examples of equilibrium situations. The number of books on the shelves of a public library remains almost constant from one day to the next, for example, although books are continually withdrawn and returned; and cars are parked solidly all day along a city street, although individual cars remain only a short time. An important example of a nonequilibrium situation is the world population which is continually increasing; people die, but at present others are born at a more rapid rate.

A liquid in a closed container, then, is at equilibrium with its vapor when the number of molecules in the vapor phase no longer changes with time, as in Figure 4-6c. We cannot count individual molecules, but we can determine their number by measuring the vapor pressure. We know (from the ideal gas law, $PV = nRT$) that for constant volume and temperature, the pressure is proportional to the number of moles of gas: $P \propto n$. Therefore, when the pressure of the vapor over a liquid in a closed container remains constant, the number of molecules in the vapor is constant, and the liquid and vapor are at equilibrium. Figure 4-7 shows how one might measure the **equilibrium vapor pressure** of a liquid. In this illustration, we measure the vapor pressure of three liquids (water, carbon tetrachloride, and diethyl ether), all at a temperature of 20°C.

The barometer in Figure 4-7a shows an atmospheric pressure, at the time and place of the experiment, of 750 mm Hg. In Figure 4-7b, a small volume of water has been inserted into the barometer; the water is lighter than mercury, and it floats to the top of the mercury column. To measure the equilibrium vapor pressure, some excess liquid must be present to insure that the liquid and vapor are at equilibrium, so enough water is added that some remains as a thin

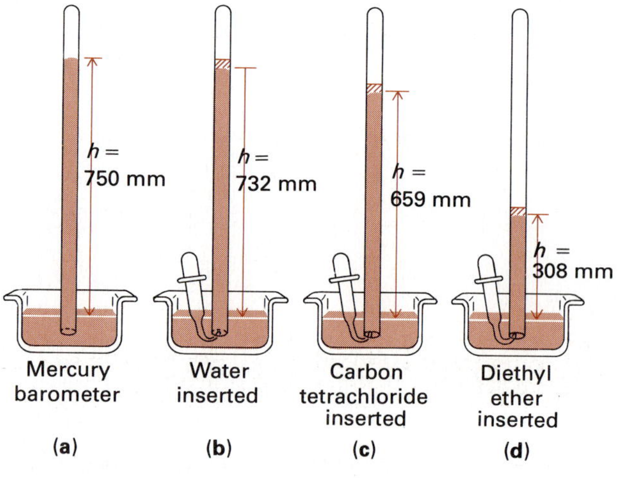

FIGURE 4-7

Measurements of equilibrium vapor pressure for three different liquids at 20°C.

layer of liquid on top of the mercury. The pressure of the water vapor quickly depresses the mercury column, which then remains at a constant height. We can calculate the pressure P_{vap} of the water vapor as follows:

$$P_{atm} = P_{Hg} + P_{vap} + P_{liq}$$

The layer of liquid water is only about 1 mm high, and the pressure due to its weight is only 1/13.6 that of an equal height of mercury. In mm Hg, the pressure of the thin layer of liquid water is

$$P_{liq} = 1 \text{ mm } H_2O \times \frac{1 \text{ mm Hg}}{13.6 \text{ mm } H_2O} = 0.07 \text{ mm Hg}$$

Because it is so small, P_{liq} can be neglected. Therefore,

$$P_{vap} = P_{atm} - P_{Hg} = 750 \text{ mm} - 732 \text{ mm}$$
$$= 18 \text{ mm Hg}$$

In Figure 4-7c, carbon tetrachloride has been inserted into the barometer until again a thin layer of liquid lies on top of the mercury column. This time, P_{vap} is the pressure of the carbon tetrachloride vapor:

$$P_{vap} = P_{atm} - P_{Hg} = 750 \text{ mm} - 659 \text{ mm}$$
$$= 91 \text{ mm Hg}$$

if we ignore the pressure due to the liquid carbon tetrachloride layer.

In Figure 4-7d, diethyl ether has been inserted into the barometer until a thin layer of liquid lies on top of the mercury column. This time, P_{vap} is the pressure of the ether vapor:

$$P_{vap} = P_{atm} - P_{Hg} = 750 \text{ mm} - 308 \text{ mm}$$
$$= 442 \text{ mm Hg}$$

again ignoring the pressure due to the thin liquid layer.

When equilibrium is attained, as in Figure 4-7b,c,d, no further observable changes take place, but there is a continuing exchange of individual molecules between the liquid and vapor phases; evaporation and condensation occur at equal rates. In Sections 7-9 and 7-10 we investigate the intermolecular forces that account for the great differences in the vapor pressures of these three liquids.

EXERCISE 4-1 If chloroform ($CHCl_3$) is inserted into a barometer until a thin layer of liquid chloroform floats on top of the mercury, the pressure of the chloroform vapor depresses the mercury column. The vapor pressure of $CHCl_3$ at room temperature (20°C) is 160 mm Hg. Sketch the barometer, giving the new height h of the mercury column. Assume that the atmospheric pressure is standard (760 mm Hg).

4-5 CHANGE OF EQUILIBRIUM VAPOR PRESSURE WITH TEMPERATURE

Table 4-2 shows the equilibrium vapor pressures of five liquids at different temperatures. We see that as the temperature is raised, the vapor pressure also increases. This is true for all liquids. Figure 4-8 gives vapor pressure curves for water and diethyl ether. The vertical scale represents pressure and the horizontal scale temperature. To each temperature corresponds some equilibrium concentration of molecules in the

TABLE 4-2
Equilibrium vapor pressure of liquids (in mm Hg)

°C	Diethyl ether	Carbon tetrachloride	Water	Octane	Mercury
0	185	32.9	4.6	2.9	1.9×10^{-4}
20	442	91.0	17.5	10.5	12.0×10^{-4}
40	920	216	55.3	30.9	60.8×10^{-4}
60	1730	451	149	77.6	252×10^{-4}
80	3000	838	355	175	888×10^{-4}
100	4865	1457	760	354	2729×10^{-4}

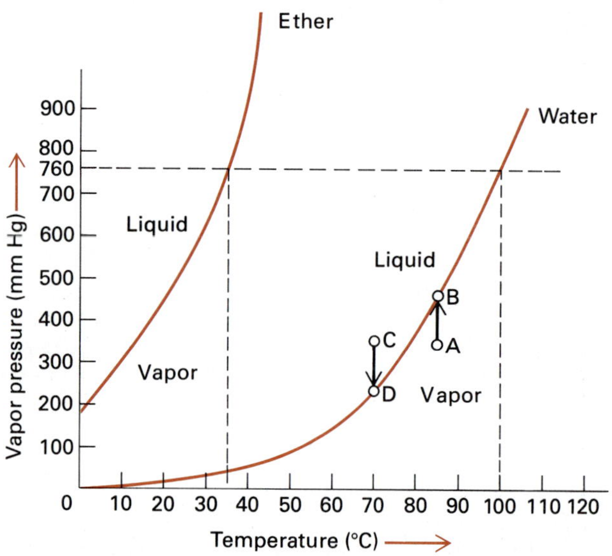

FIGURE 4-8

Vapor pressure curves for liquid water and for diethyl ether.

vapor phase, reflected in the vapor pressure of the liquid at that temperature. *At any temperature and pressure on the curve, the two phases (liquid and vapor) are in equilibrium.* For example, at a temperature of 85 °C, when the amount of substance in the vapor phase is so small that the vapor is at a pressure *less* than the equilibrium vapor pressure at that temperature (as is the case for point A in Figure 4-8, the rate of evaporation is greater than the rate of condensation. More liquid evaporates until the equilibrium vapor pressure is reached at point B. At B the system is at equilibrium. Similarly, at a temperature of 75 °C, when the amount of substance in the vapor phase is such that the vapor is at a pressure *greater* than the equilibrium vapor pressure at that temperature (as is the case for point C), the rate of condensation is greater than the rate of evaporation. Vapor condenses until the equilibrium vapor pressure is reached at point D. At D the system is at equilibrium. When the amount of substance in the vapor is such that its pressure is *equal* to the equilibrium vapor pressure at that temperature, the system is at equilibrium; evaporation and condensation occur at equal rates. The vapor pressure of a substance depends *only* on the temperature, not on the amount of the substance present (so long as *some* liquid is present to insure that liquid and vapor are at equilibrium) nor on the size of the container.

We saw that liquid water at point A in Figure 4-8 was *not* stable and would evaporate. The stable phase at point A is the vapor; liquid water at that temperature and pressure evaporates. At point C, the stable phase is the liquid; vapor at that temperature and pressure is unstable and condenses. The region above the equilibrium curve corresponds to temperatures and pressures where the liquid is the stable phase, and the region below the curve to temperatures and pressures where the vapor is the stable phase.

EXERCISE 4-2 Using Figure 4-8, determine which phase of ether (liquid or vapor) is stable at 0°C and 1 atm of pressure. Which phase is stable at 40°C and 1 atm of pressure?

4-6 BOILING POINT

We see in Figure 4-8 that water has an equilibrium vapor pressure of 18 mm Hg at 20°C, and 355 mm Hg at 80°C. The temperature at which the vapor pressure is 760 mm Hg (1 standard atmosphere) is 100°C. **Boiling** *occurs at a temperature where the vapor pressure of a liquid equals the external pressure.* Below this temperature, if bubbles form in the interior of a liquid, they are suppressed by the pressure of the atmosphere because the pressure of the vapor inside the bubbles is less than atmospheric pressure. Evaporation then occurs quietly from the surface of the liquid. (Any tiny bubbles seen rising from the water *before* it boils are bubbles of dissolved air that is driven off as the water is heated.) But, when the pressure of the vapor inside the bubbles is the same as atmospheric pressure, the bubbles are no longer suppressed by the pressure of the atmosphere. Evaporation then occurs in the *interior* of the liquid; and the liquid boils. Figure 4-9 shows a schematic diagram of the boiling process.

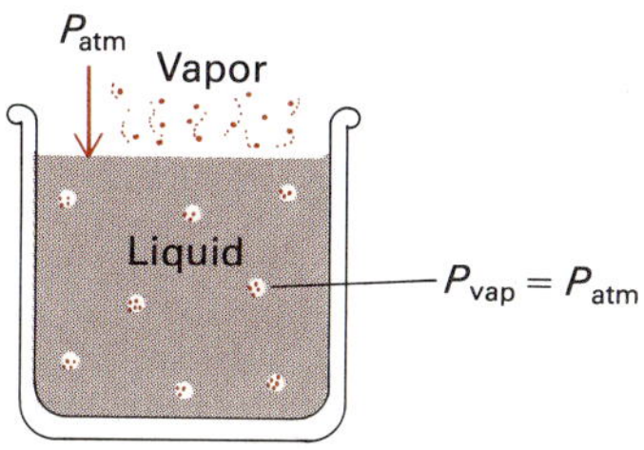

FIGURE 4-9
Schematic diagram of boiling. At the boiling point, the pressure of the vapor inside the bubbles is equal to atmospheric pressure.

The temperature at which the vapor pressure of a liquid is equal to 1 standard atmosphere (760 mm Hg) is called the **normal boiling point** *of the liquid.* Figure 4-8 shows that the normal boiling point of water is 100°C and the normal boiling point of diethyl ether is 35°C. Suppose we are near the top of Mount Everest, where the atmospheric pressure might be only 280 mm Hg. According to Figure 4-8, the boiling temperature of water at that pressure is about 74°C. The effect of altitude on the boiling temperature of water is illustrated in Figure 4-10.

Producers of frozen concentrated orange juice take advantage of the fact that water boils at lower temperatures at reduced pressures. By reducing the pressure, they are able to remove water from the juice by boiling it off at a temperature low enough that the vitamins and natural flavorings are not destroyed. If, on

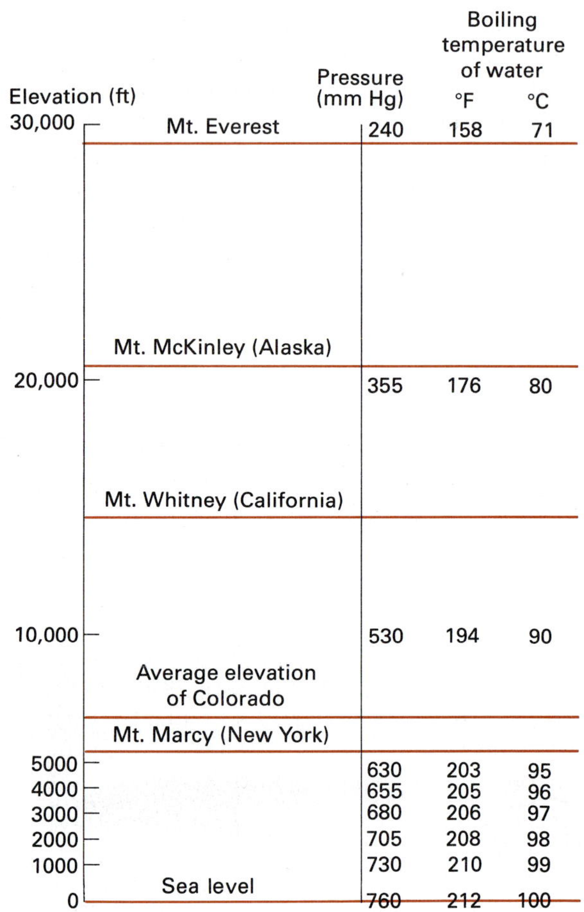

FIGURE 4-10

Effect of altitude on boiling temperature.

EXERCISE 4-3 Figure 4-11 shows the vapor pressure of a certain liquid as a function of temperature.

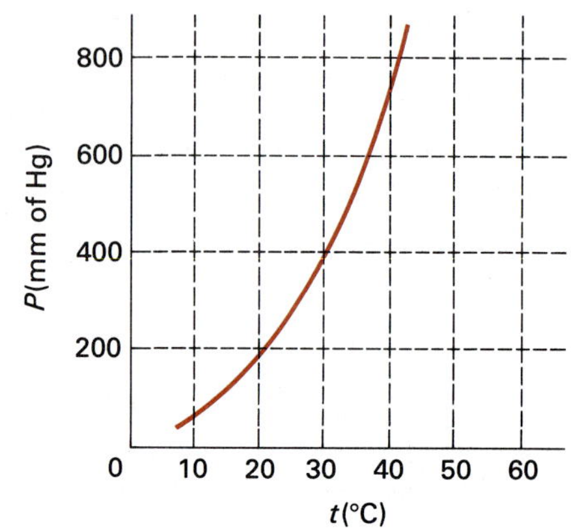

FIGURE 4-11

(a) What (approximately) is the normal boiling point of the liquid?

(b) To what value (approximately) must the external pressure be reduced in order that the liquid boil at 10°C?

(c) Is it the liquid or the vapor form of the substance that is stable at $P = 800$ mm Hg and $t = 20°C$? Explain.

(d) Is it the liquid or the vapor form of the substance that is stable at $P = 100$ mm Hg and $t = 30°C$? Explain.

the other hand, we use a pressure cooker at a pressure of 1039 mm Hg (20 lb/in.^{2}), the boiling temperature is 109°C, and food will cook much more rapidly than at atmospheric pressure. In Section 10-6, we see that the rates of many chemical reactions, including those of cooking, roughly double with an increase of 10°C in temperature. (What change in temperature in degrees Fahrenheit is required to produce this same increase in rate?)

Figure 4-8 also illustrates the meaning of the term **relative humidity,** used in weather studies. At any temperature, there is a limit to the amount of water that can be held as vapor in the air; this limit is represented by the equilibrium vapor pressure of water at that temperature. If the pressure of the water vapor in the air is only half the equilibrium vapor pressure at that temperature, then the relative humidity is 50%. If the pressure of the water vapor in the air is nine-tenths the equilibrium pressure at that temperature, the relative humidity is 90%. If the temperature suddenly rises, increasing what would be the *equilibrium* vapor pressure but without any increase in the actual amount of water vapor present, then the relative humidity drops. If instead the temperature suddenly falls, reducing what would be the equilibrium vapor pressure but with a fixed actual pressure, then the relative humidity increases. And if the actual pressure of the water vapor in the air is greater than this now lowered equilibrium vapor pressure (making the relative humidity greater than 100%), dew, fog, or frost forms until the pressure of the water vapor in the air is equal to or less than the equilibrium vapor pressure.

The **dew point** is the temperature at which the air, with whatever pressure of water vapor it contains at the time, is saturated with that water vapor. At a temperature below the dew point, condensation occurs. Thus on a humid summer day, the air near cold-water pipes often cools below its dew point, causing water vapor from the air to condense on the pipes.

4-7 VAPOR PRESSURE OF SOLIDS, AND SOLID–VAPOR EQUILIBRIUM

Though we are generally more familiar with evaporation of liquids, solids also evaporate. The process is called **sublimation.** One well-known example is that of Dry Ice (solid carbon dioxide) which evaporates without melting—it sublimes. When wet clothes are hung out to dry at a temperature below freezing, the clothes first freeze and then the ice evaporates, or sublimes, leaving the clothes dry. And snow covering the ground sometimes gradually disappears by subliming when the temperature remains below freezing. Producers of freeze-dried coffee and other foods take advantage of the fact that solids evaporate. They use vacuum pumps to remove the water as a vapor from the frozen food, permitting it to retain vitamins and natural flavorings that might be destroyed by heat. Figure 4-12 shows iodine crystals subliming as they are warmed.

FIGURE 4-12

Iodine crystals sublime when heated. **(a)** Schematic diagram of sublimation. **(b)** Iodine sublimes from the crystals at the bottom of the beaker, and then recrystallizes on the cool surface at the top. [Photograph by Richard Byrnes.]

Thus solids also have equilibrium vapor pressures that can be measured in experiments similar in principle to those used to determine the vapor pressure of liquids. Like that of a liquid, the vapor pressure of a solid increases as the temperature is raised. At a higher temperature, more molecules have enough kinetic energy to escape from the solid to the vapor. Figure 4-13 shows vapor pressure curves for water and carbon dioxide below their freezing points (where each is a solid). At each temperature there is a unique (single) equilibrium concentration of molecules in the vapor phase, reflected in the vapor pressure of the solid at that temperature. At absolute zero temperature, the vapor pressure of all substances goes to zero. *Each point*

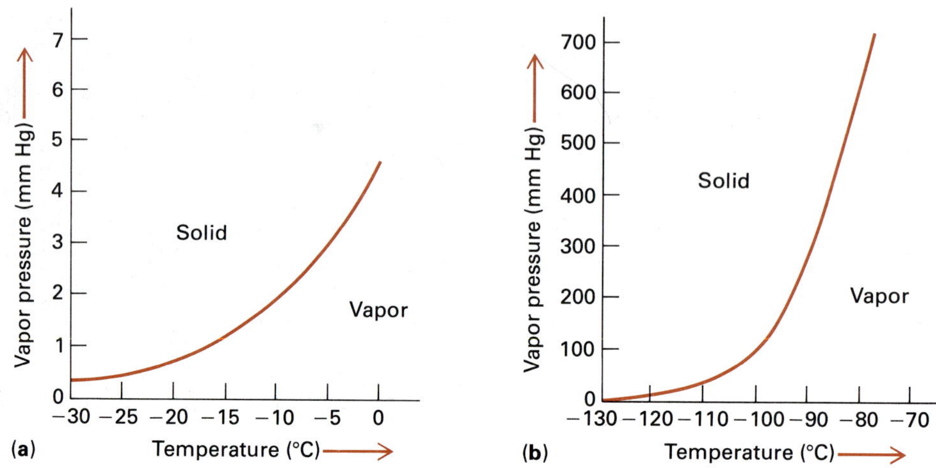

FIGURE 4-13

Vapor pressure curves for ice **(a)** and for solid carbon dioxide (Dry Ice) **(b)**.

on the curve represents the equilibrium vapor pressure for the solid at that temperature. Thus, at any point on the curve, liquid and vapor are at equilibrium, and no measurable change occurs. At temperatures and pressures in the region above the curve, the solid is the stable phase; in the region below the curve, the vapor is the stable phase.

EXERCISE 4-4 What change in vapor pressure would you observe if you took an ice cube at a pressure of 1 atm and let it warm up from −30°C to −5°C?

4-8 MELTING AND FREEZING

We are all familiar with many solid-to-liquid and liquid-to-solid changes: the thawing of the ice on a roadside puddle on a warm spring day, and the freezing of the puddle on a cold day; the melting of butter left out in the sun, and its resolidifying when returned to the refrigerator; and the melting of wax near the flame of a candle, while the melted wax that drips down the sides of the candle, away from the heat, resolidifies. Just as a liquid or a solid can be in equilibrium with a vapor, a solid can also be in equilibrium with a liquid—for example, ice in equilibrium with liquid water at 0°C.

Again, there is one temperature at which any particular solid and its liquid form are in equilibrium (melting point, freezing point) at any given pressure. Just as the liquid–vapor equilibrium temperature depends on pressure, so does the solid–liquid equilibrium temperature depend on pressure, but to a much smaller degree. Thus there is a "melting line" (or solid–liquid equilibrium line) for each substance, showing the unique temperature at which the solid and liquid are in equilibrium at each pressure. The melting lines for benzene and water are shown in Figure 4-14. Each line is almost vertical, showing that the melting point does not change much with pressure. For water the melting point changes by only 0.0098 °C when the pressure increases from 0.006 atm to 1 atm. *The* **normal melting (freezing) point** *is the temperature at which the solid melts (the same also as that at which the liquid freezes) at a pressure of 1 standard atmosphere.*

Most substances are denser as solids than as liquids. Benzene is in that respect typical. Figure 4-15 shows a cube of solid benzene sunk in liquid benzene—when the solid is more dense than the liquid, the solid does not float on the liquid but sinks to the bottom. An outstanding exception to this typical behavior is that of water, which expands (becoming less dense) when it freezes. Figure 4-15 also shows the atypical (but very familiar) phenomenon of an ice cube floating on liquid water. It is this expansion on freezing that causes bursting of pipes when plumbing freezes during a prolonged spell of cold weather.

FIGURE 4-15

Densities of solids and liquids. Ice cubes float in water (left), but benzene cubes sink in liquid benzene (right). [Photograph by Richard Byrnes.]

EXERCISE 4-5 Consult Figure 4-14a to determine whether the melting point of benzene increases or decreases when the pressure is raised. Does the melting point of water increase or decrease when the pressure is increased?

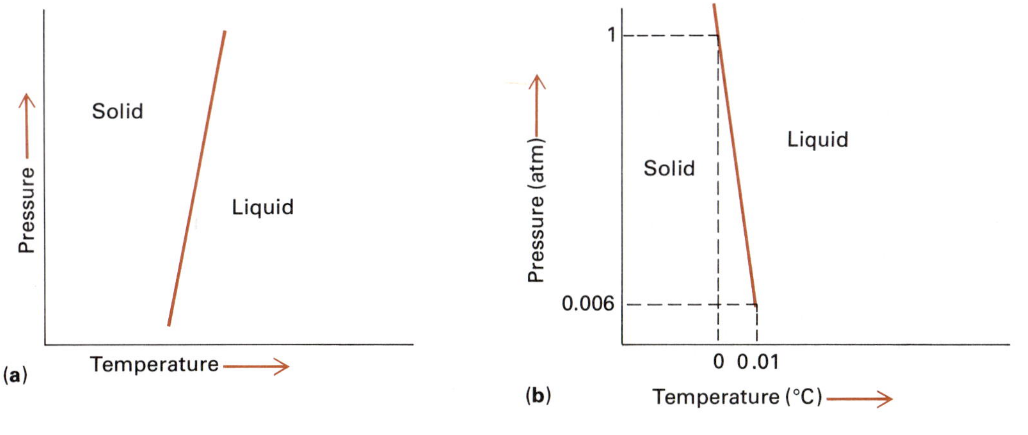

FIGURE 4-14

Solid–liquid equilibrium lines (melting lines). **(a)** Benzene. **(b)** Water.

4-9
PHASE DIAGRAMS

For any substance, the three equilibrium curves (for liquid–vapor, solid–vapor, and solid–liquid equilibrium) meet at the **triple point,** a unique temperature and pressure where all three phases can exist simultaneously at equilibrium. Figure 4-16 shows phase diagrams for water and carbon dioxide. The three curves in the phase diagram for water are shown separately in Figures 4-8, 4-13, and 4-14. When they are combined, we see that the triple point for water is about 0.01 °C and 0.006 atm. When liquid water, ice, and water vapor are placed in a closed, insulated container, equilibrium will be at that temperature of 0.01 °C and pressure of 0.006 atm. In each of the three regions of the phase diagram, one phase of the substance (solid, liquid, or gas) is the *stable* phase. Each region of the diagram is labeled to show which is the stable phase there. For all values of the temperature and pressure that lie inside one region, the substance is stable only in that one form.

Figure 4-16a shows that at a pressure of 1 atm, ice is the stable phase below 0 °C; liquid water is the stable phase between 0 °C and 100 °C; and water vapor is the stable phase above 100 °C. (What happens at a pressure of 1 atm at 0 °C, and what happens at 100 °C?) The triple point for carbon dioxide (Figure 4-16b) is at 5.2 atm and −57 °C. Line A shows that as solid carbon dioxide (Dry Ice) warms at 1 atm, the solid sublimes when the temperature reaches −78 °C, going directly to the vapor without melting.

Figure 4-17 shows a phase diagram for carbon, a substance that has more than one crystalline form in the solid state. There are two "solid" regions in the phase diagram: diamond and graphite. In principle, there also is a vapor region on the phase diagram, but it is not visible on this scale. Notice that at normal pressures and temperatures, graphite is the stable form of carbon. Diamond is stable at normal temperatures only at pressures above about 20,000 atm. Naturally

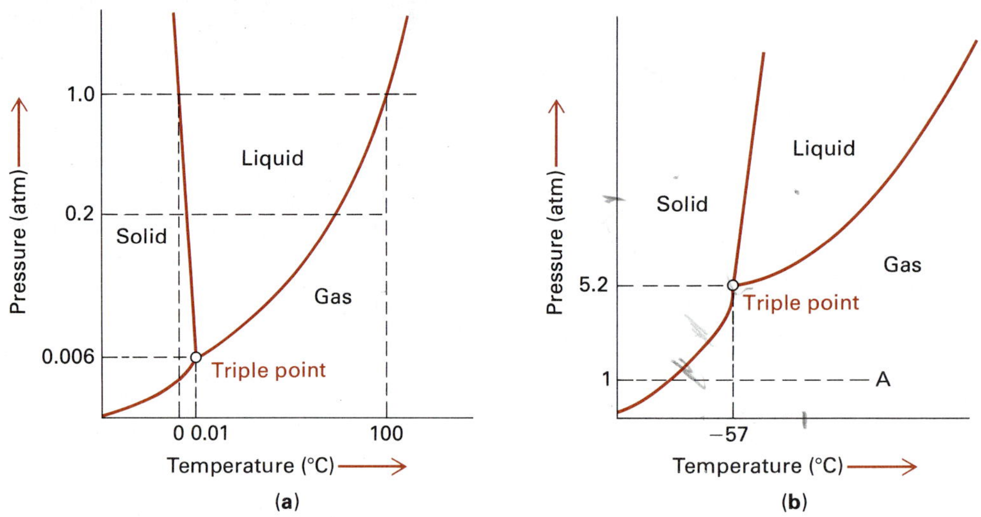

FIGURE 4-16

Phase diagrams. Note that the triple point of H_2O is at a pressure below 1 atm, and the triple point of CO_2 is above 1 atm. As a result, at 1 atm ice melts, but Dry Ice sublimes. **(a)** Water. **(b)** Carbon dioxide.

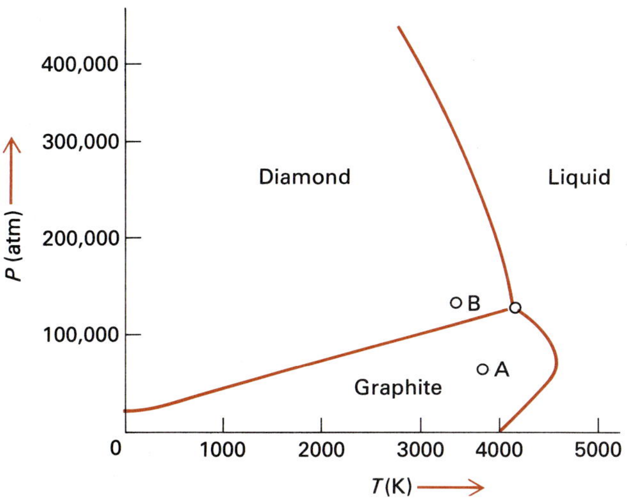

FIGURE 4-17

Phase diagram for carbon. (Recent studies show that the carbon phase diagram may be more complex than shown here. A metallic phase of carbon may exist between the solid and liquid phases.)

occurring diamonds are formed from graphite under conditions of tremendous pressure.

Just as liquid water at 1 atm spontaneously freezes at any temperature below 0°C, diamond may be expected to convert to graphite at normal temperatures and pressures. But in the case of diamond, the *rate* of conversion is too slow to observe. Diamond has been *rapidly* converted to graphite at pressures of 20,000 to 80,000 atm and a temperature of 3800 K (in the region of point A in Figure 4-17). And in the reverse transition, graphite has been rapidly converted to diamond in the region of point B, at pressures a little over 130,000 atm and temperatures of about 3500 K, some 500 to 600 K below the melting point of carbon at that pressure. Figure 4-18 is a photograph showing top and side views of a graphite rod that was subjected to a pressure of 130,000 atm. The center section, which was flash-heated to 3500 K, was completely converted to diamond. These artificially produced diamonds are not of gemstone quality, but they have many industrial applications including use as phonograph needles and abrasives. Large, perfect diamond crystals *can* be grown artificially, but a 1 carat (0.2 g, 6 mm) crystal takes several days to grow. Because the high-pressure apparatus that is required is so expensive, this process does not compete economically with digging naturally formed diamonds from the ground.

EXERCISE 4-6 Using Figure 4-16a, determine the boiling point of water at a pressure of 0.20 atm.

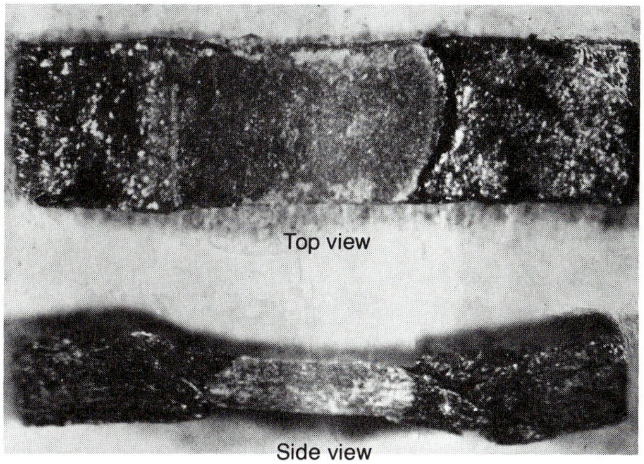

FIGURE 4-18

Diamond formed from the center section of a graphite bar. The graphite bar was subjected to a pressure of about 140,000 atm and then flash-heated by a very strong pulse of electric current. The instant the middle section reached a temperature of about 3300 K, it converted spontaneously to polycrystalline diamond. This section contracted as it formed the more dense diamond (3.5 g/cm^3). The ends of the bar did not get as hot, and they remained in the form of graphite (density 2.3 g/cm^3). [Photograph courtesy of F. P. Bundy, General Electric Research Laboratories, Schenectady, N.Y.]

4-10 HEAT AND TEMPERATURE

Before we can fully understand what happens when a solid or a liquid is heated, it is necessary to have a unit of heat to measure the heat added. **Heat** is defined as a *quantity of energy* and is distinguished from temperature, which is the *degree of hotness.* A kettle full of boiling water contains a much greater quantity of heat than a few drops of boiling water, and can burn one much more severely, though both are at the same temperature. Until recently, the unit used to measure quantity of heat was the calorie (cal). The **calorie** is the quantity of heat required to raise the temperature of 1 g of liquid water from 14.5 °C to 15.5 °C. In SI units, the **joule** (J) is the unit of heat, and the joule will be used in most of this book. One calorie is the same amount of heat as 4.184 joules. Because we shall use the calorie in the biochemistry section of this book, and because you may encounter the calorie in nutrition tables and older books, it is important to be able to convert from calories to joules and the reverse:

$$1 \text{ cal} = 4.184 \text{ J}$$

Thus, 4.184 J is the amount of heat required to raise the temperature of 1 g of water from 14.5 °C to 15.5 °C. The following example illustrates these conversions.

EXAMPLE 4-1

The "Calorie" used in nutrition is really the *kilocalorie* (kcal), which is equivalent to 1000 calories. Energy is produced by the chemical reaction (essentially combustion) of food with oxygen. This energy is used to maintain body temperature, to build new cells, and for mechanical energy with which to do work. An adult requires approximately 2000 to 3000 kcal per day for these purposes. The statement that a 13 ounce malted milkshake contains 514 Calories means that on combustion it yields 514,000 cal, or 514 kcal.

(a) How many joules—and how many kilojoules (kJ)—are equivalent to 514 kcal? (1 kJ = 1000 J).

$$1 \text{ cal} = 4.184 \text{ J}$$

$$514 \times 10^3 \text{ cal} \times 4.18 \frac{\text{J}}{\text{cal}} = 2.15 \times 10^6 \text{ J}$$

$$2.15 \times 10^6 \text{ J} \times \frac{1 \text{ kJ}}{1000 \text{ J}} = 2.15 \times 10^3 \text{ kJ}$$

(b) The energy needed to raise the temperature of 1 ℓ of water from 60 °C to 65 °C is 20.9×10^3 J (20.9 kJ). How many calories (and how many kilocalories) is this?

$$20.9 \times 10^3 \text{ J} \times \frac{1 \text{ cal}}{4.18 \text{ J}} = 5.00 \times 10^3 \text{ cal}$$

$$5.00 \times 10^3 \text{ cal} \times \frac{1 \text{ kcal}}{1000 \text{ cal}} = 5.00 \text{ kcal}$$

EXERCISE 4-7 (a) A medium slice of meatloaf yields 100 kcal on combustion. How many joules is this?
(b) The heat released when 2 ℓ of water are cooled from 100 °C to 0 °C is 8.36×10^5 joules. How many kilocalories is this?

The **specific heat** of a substance is the amount of heat required to raise the temperature of 1 g of that substance by 1 °C (1 deg). The specific heat of water is 4.18 J/g deg (or 1.0 cal/g deg). Table 4-3 lists the specific heats of some substances.

More heat is required to raise the temperature of water one degree than to raise the temperature of the same mass of many other materials, as the table shows. The high heat capacity of water has important consequences for climate. Cities on the coast of a large lake or a sea tend to have more even climates than inland

TABLE 4-3
Specific heats of some substances at 20°C

Substance	Specific heat (cal/g deg)	Specific heat (J/g deg)
Water (liquid)	1.00	4.18
Aluminum	0.21	0.88
Cast iron	0.12	0.50
Normal thermometer glass	0.20	0.84
Wood	0.42	1.76
Dry clay	0.22	0.92
Rock (typical)	0.21	0.88

cities. Large amounts of heat are absorbed by a lake or a sea during the warm summer months. The heat stored in the water is released slowly, warming the cold winter winds as they pass over, and thus warming the coastal regions. By the time summer returns, the water has cooled; it now absorbs heat from the warm winds as they pass over, producing cool ocean breezes over the coast. The same feature makes water a good heat-storage medium in solar heating systems.

EXAMPLE 4-2

How many joules are required to raise the temperature of 1.0 mole of water from 0°C to its boiling point (100°C)? The specific heat of water is 4.18 J/g deg.

$$\text{Molecular weight of water} = 18 \text{ g/mole}$$

$$\text{Change in temperature} = 100° - 0° = 100°\text{C}$$

$$\text{Heat required} = 4.18 \text{ J/g deg} \times 18 \text{ g} \times 100 \text{ deg} = 7524 \text{ J} = 7.5 \text{ kJ}$$

EXAMPLE 4-3

If a cast-iron bar weighing 2 kg is heated to 500°C and then cooled by immersing it in an insulated bucket containing 2 ℓ of water at 20°C, what will be the common final temperature of the water and the iron bar? The density of water is 1.0 g/ml, the specific heat of water is 4.18 cal/g deg, and the specific heat of cast iron is 0.50 J/g deg.

Let the common final temperature of the water and iron bar be t_{final}.

$$\text{Decrease in temperature of iron} = (500 - t_{\text{final}}) \text{ deg}$$

$$\text{Heat lost by iron} = 0.50 \text{ J/g deg} \times 2000 \text{ g} \times (500 - t_{\text{final}}) \text{ deg}$$

$$\text{Gain in temperature of water} = (t_{\text{final}} - 20) \text{ deg}$$

$$\text{Heat gained by water} = 4.18 \text{ J/g deg} \times 2000 \text{ g} \times (t - 20) \text{ deg}$$

But the heat lost by the iron equals the heat gained by the water (because the container is insulated, no heat is lost to the surroundings). So,

$$0.50 \text{ J/g deg} \times 2000 \text{ g} \times (500 - t_{\text{final}}) \text{ deg} = 4.18 \text{ J/g deg} \times 2000 \text{ g} \times (t_{\text{final}} - 20) \text{ deg}$$

$$250 - 0.50\, t_{\text{final}} = 4.18\, t_{\text{final}} - 83.6$$

$$-4.68\, t_{\text{final}} = -334$$

$$t_{\text{final}} = 71°\text{C}$$

The final temperature is 71°C. Thus, because of the great specific heat of water, the water has increased in temperature by only about 52°C while an equal mass of iron has decreased in temperature by about 429°C.

EXERCISE 4-8 A 5.0 kg nickel bar is heated to 300°C and then dropped into an insulated bucket containing 10 ℓ of water at 10°C. The final temperature of the bar and water is 25°C. What is the specific heat of nickel?

4-11
HEAT EFFECTS AND CHANGES OF STATE

Now we can consider how the temperature of a solid changes as it absorbs heat. Imagine a lump of ice at −20°C. The molecules are arranged in an orderly network, a crystalline structure. At any temperature there is some molecular motion, so the molecules in the ice are vibrating about their fixed positions in the solid. As heat is added, the temperature rises, and the molecular motion (the kinetic energy) increases. Figure 4-19 shows a heating curve for a mole of ice heated from a temperature of −20°C. The temperature rises until the normal melting point (0°C) is reached.

At the melting point, the temperature and the average kinetic energy of the molecules remain constant. The energy added to melt the ice at 0°C is used to overcome the forces of attraction between the molecules, breaking down the crystalline structure of the ice to form the liquid (Figure 4-20), and is thereafter stored as potential energy. **Potential energy** is energy stored in an object by virtue of its position or arrangement with respect to other objects (Figure 4-21). In the

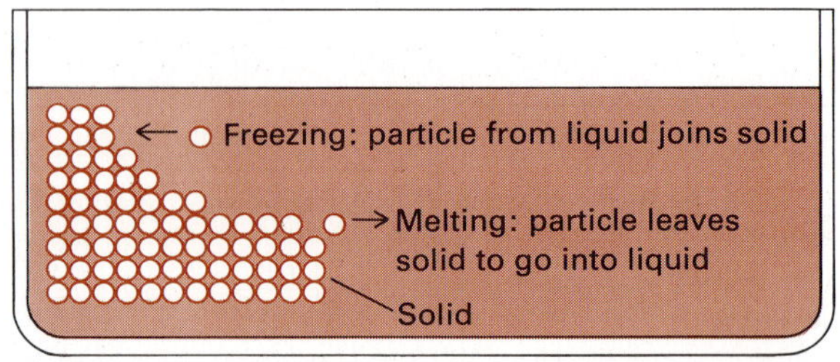

FIGURE 4-20

Melting and freezing.

FIGURE 4-21

Potential energy. The water at the top of the falls, by virtue of its position, is greater in potential energy than the water at the bottom of the falls. [Photograph by Russ Hamilton, courtesy of Cornell University.]

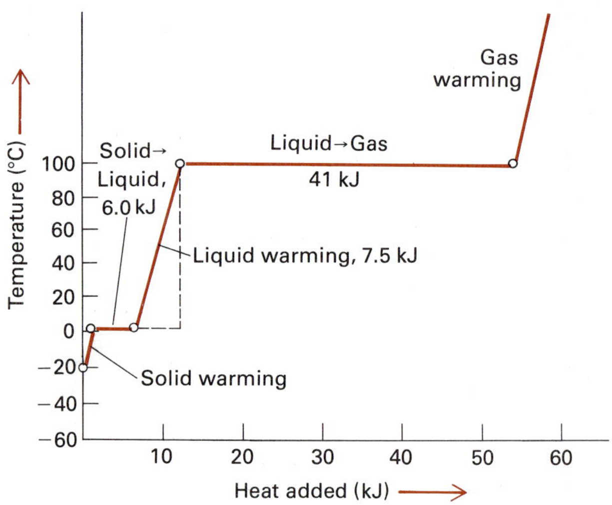

FIGURE 4-19

Heating curve for 1 mole of ice at −20°C to steam above 100°C.

TABLE 4-4
Melting points and heats of fusion of some substances

Substance	Melting point* (°C)	Melting point* (K)	Molar heat of fusion (kJ/mole)	Heat of fusion (kJ/g)
Helium (He)	−273	0	0.014	0.0035
Hydrogen (H_2)	−259	14	0.12	0.059
Chlorine (Cl_2)	−101	172	6.40	0.092
Water (H_2O)	0	273	6.0	0.33
Sodium (Na)	98	371	2.6	0.11
Sodium chloride (NaCl)	808	1081	29	0.49
Copper (Cu)	1083	1356	13	0.20

*For all entries except helium, the normal melting point (at 1 atm) is given. Solid helium, however, cannot exist at pressures less than 25 atm. The melting point of He is 0 K at 25 atm; at any temperature above the absolute zero, solid helium can exist only at pressures greater than 25 atm.

case of a melting solid, the heat energy added to the system increases the potential energy while the kinetic energy remains constant. The molecules in the liquid are at a higher potential energy than those in the solid. That same potential energy is released as heat when the process is reversed and the liquid freezes.

Figure 4-19 shows that 6.0 kJ are needed to melt one mole of ice at 0°C (334 J for 1 g of ice). When the ice has completely melted, the temperature again rises as heat is added and the kinetic energy of the molecules increases. Ice cubes cool a drink, not so much because they are cold, but because they melt (absorbing, in the process, 334 J of heat per gram of ice that melts).

For one mole of ice at 0°C:

$$H_2O\text{ (solid)} + 6.0\text{ kJ} \longrightarrow H_2O\text{ (liquid)} \quad (4\text{-}1)$$

The amount of energy required to melt a solid is called the **heat of fusion;** that required to melt one mole is the **molar heat of fusion.** Table 4-4 gives the melting points and heats of fusion of some substances.

EXAMPLE 4-4

How much heat is required to change 50 moles of ice at 0°C to liquid water at 100°C?

Heat required to melt ice to liquid water at 0°C

$= 50\text{ moles} \times 6.0\text{ kJ/mole}$

$= 3.0 \times 10^2\text{ kJ}$

Heat required to raise temperature of liquid from 0°C to 100°C

$= 50\text{ mole} \times 18\text{ g/mole} \times 4.18\text{ J/g deg} \times 100\text{ deg}$

$= 3.8 \times 10^5\text{ J} = 3.8 \times 10^2\text{ kJ}$

Total energy required $= 3.0 \times 10^2 + 3.8 \times 10^2$

$= 6.8 \times 10^2\text{ kJ}$

Roughly equal amounts of heat are required to melt the ice and to heat the liquid from the freezing to the boiling point. In Exercise 4-10 we find that a much greater amount of heat is needed to *vaporize* the liquid at its boiling point.

EXERCISE 4-9 How much heat is required to melt 500 g of sodium chloride (NaCl) at 808°C, its normal melting point?

When one mole of water freezes at 0°C, the reaction shown in Equation 4-1 is reversed, and the same quantity of heat is *released.* Tubs of water are sometimes placed in orchards or greenhouses on a freezing night. As the water freezes, it releases heat, thus preventing the plants from freezing.

For one mole of water at 0°C:

$$H_2O\text{ (liquid)} \longrightarrow H_2O\text{ (solid)} + 6.0\text{ kJ}$$

Figure 4-19 also shows how the temperature of liquid water changes as it absorbs heat from 0°C to 100°C. Initially, as heat is added, the temperature rises. When the water reaches a temperature of 100°C, the water boils, and the temperature remains constant while all the liquid boils away. This constant temperature, 100°C, is the normal boiling point. When all of the liquid has evaporated, the temperature again rises as the vapor is heated.

The 7.5 kJ calculated in Example 4-2 as the heat required to raise the temperature of one mole of water from 0°C to 100°C is the same as the value shown on the heating curve of Figure 4-19. On the portion of the heating curve between 0°C and 100°C, the average kinetic energy of the molecules is increasing. This is reflected in the increase in temperature. At the boiling point, however, the temperature and the average kinetic energy of the molecules remain constant. The energy required to vaporize the liquid at the boiling temperature (100°C) is used to overcome the attractive forces between the molecules; instead of increasing the kinetic energy of the molecules, this energy is stored as potential energy. The more widely separated molecules in the gas are at a higher potential energy than the molecules in the liquid. That energy is released as heat when the molecules come together again—that is, when the vapor recondenses to a liquid. Figure 4-19 shows that 41 kJ are needed to vaporize one mole of liquid water at its boiling point (about 2.3 kJ per gram).

For one mole of water at 100°C:

$$H_2O\text{ (liquid)} + 41\text{ kJ} \longrightarrow H_2O\text{ (gas)} \qquad (4\text{-}2)$$

The amount of energy required to vaporize a liquid is called the **heat of vaporization;** that required to vaporize one mole is the **molar heat of vaporization.** The energy is used to overcome attractive forces and separate the molecules from each other. Because energy is required to convert a liquid into a vapor, evaporation is a cooling process. The heat used to evaporate perspiration is absorbed from the skin, cooling the skin in the process. Table 4-5 gives the normal boiling points and the molar heats of vaporization for some substances.

EXERCISE 4-10 How much heat is required to vaporize 50 moles of liquid water at its boiling point, 100°C? (Compare with the amount of heat required to melt 50 moles of ice, calculated in Example 4-4.)

When one mole of water condenses from a vapor to a liquid, the reaction shown in Equation 4-2 is reversed, and the same quantity of heat is released.

For one mole of water at 100°C:

$$H_2O\text{ (gas)} \longrightarrow H_2O\text{ (liquid)} + 41\text{ kJ}$$

TABLE 4-5
Normal boiling points and heats of vaporization of some substances

Substance	Boiling point (°C)	Boiling point (K)	Molar heat of vaporization (kJ/mole)	Heat of vaporization (kJ/g)
Helium (He)	−269	4	0.079	0.020
Hydrogen (H_2)	−253	20	0.90	0.45
Chlorine (Cl_2)	−34	239	20	0.29
Water (H_2O)	100	373	41	2.3
Sodium (Na)	883	1156	89	3.9
Sodium chloride (NaCl)	1473	1746	171	2.9
Copper (Cu)	2582	2855	304	48

This reverse reaction accounts for the severity of a steamburn and for the formation of raindrops in a cloud during a storm, releasing tremendous quantities of energy as heat. The amount of heat released in a storm is calculated in the following example.

EXAMPLE 4-5

Calculate how much energy is released when there are 1.8 cm (about $\frac{3}{4}$ in.) of rain over an area 10^{10} cm^2 (about 1 square mile). The density of water is 1 g/cm^2:

$$\text{Volume of rain} = 1.8 \text{ cm} \times 10^{10} \text{ cm}^2 = 1.8 \times 10^{10} \text{ cm}^3$$

$$\text{Mass of rain} = 1.8 \times 10^{10} \text{ cm}^3 \times 1 \text{ g/cm}^3 = 1.8 \times 10^{10} \text{ g}$$

$$\text{Mol. wt. of water} = 18 \text{ g/mole}$$

$$\text{Moles of water} = \frac{1.8 \times 10^{10} \text{ g}}{18 \text{ g/mole}} = 1.0 \times 10^{9} \text{ moles}$$

$$\text{Heat released} = 1 \times 10^{9} \text{ mole} \times 41 \text{ kJ/mole} = 4.1 \times 10^{10} \text{ kJ}$$

(This is about the same energy as that produced in burning 1300 tons of coal.) Our calculation is not entirely accurate because the water vapor is not condensing at its boiling point, but at some lower temperature, at which the heat of vaporization is slightly different. However, the error is small.

EXERCISE 4-11 Ethanol (common alcohol) melts at −114°C and boils at +78°C. The specific heat of solid ethanol is 0.84 J/g deg and that of liquid ethanol is 2.1 J/g deg. The heat of fusion of ethanol is 0.11 kJ/g and its heat of vaporization is 0.85 kJ/g.

(a) How much heat would be required to transform 10 g of ethanol from the solid at −194°C (liquid air temperature) to a vapor at 78°C?

(b) Sketch a heating curve for 1 g of alcohol from −194°C to just above its boiling point. For each portion of the heating curve, indicate what phase is stable or what phase transition is occurring.

4-12
WHAT DETERMINES THE STABLE PHASE?

It is a familiar fact that a system tends to change spontaneously to a state of lower energy. (We shall call any region or object we are studying at the moment a **system.**) Thus a walnut falls from its branch to the ground, water flows downhill, and a wound clock runs down. In all these cases, the potential energy decreases. But in some familiar processes, a system goes spontaneously to a state of *higher* energy. Ice spontaneously melts to liquid water at temperatures above 0°C, although liquid water is higher in energy than ice. And at 1 atm, water spontaneously vaporizes above 100°C, although the vapor is higher in energy than liquid water. Why is it that the liquid and vapor states, both of which are higher in energy than the solid, are *ever* the stable phase?

Imagine a beaker of water colored with a blue dye and a separate beaker of water colored with a red dye, both at the same temperature. When poured together, the solutions quickly mix, yielding a violet-colored solution. The energy of the two solutions is the same, so this spontaneous mixing is not caused by a decrease in energy. The driving force in this case is the tendency for a system to achieve the most random, disordered molecular arrangement possible. *Disorder tends to increase spontaneously, just as energy tends to decrease spontaneously.* The fact that disorder tends to increase spontaneously is amply illustrated by familiar examples. The chairs in a classroom might be arranged in neat rows in the morning but not at the end of the day. A dormitory room is austere in its order when its occupant arrives in the fall but not when he leaves. A stack of plates is compactly arranged before a meal but not at its end. In all these examples, there is a rapid change to a state of greater disorder with use. It requires work (hence energy) to maintain or to restore order.

Molecules move randomly. Random motion is much more likely to result in disorder than in order, because there are many more disordered arrangements possible than ordered ones. Thus the pink and blue molecules, when poured together, arrange themselves in a random, disordered pattern (yielding a violet mixture) rather than arranging themselves neatly in an ordered pattern (to yield a blue region and a pink region side by side).

P	B	P	P	B	P	B	B
P	P	B	B	P	B	P	B
B	P	B	P	P	B	B	
P	B	B	B	B	P	P	B

Violet mixture

rather than

P	P	P	P	B	B	B	B
P	P	P	P	B	B	B	B
P	P	P	P	B	B	B	B
P	P	P	P	B	B	B	B

Pink plus blue

The degree of randomness or disorder of a system is called its **entropy.** (Entropy also has a precise, mathematical definition that we shall not discuss.) We see that the spontaneous tendency for entropy to increase is just as important as that for energy to decrease. When a change involves both an energy and an entropy change (as when ice melts), there may be competition between the natural tendencies toward lower energy and higher entropy. Liquid water is higher in energy than ice. But liquid water is the more disordered, its molecules sliding over each other, whereas those of ice are fixed in a regular, crystalline structure. So the liquid is also higher in entropy. There is competition between the lower energy of the ice and the higher entropy of the liquid.

Above 0°C (at 1 atm) the state of higher entropy, liquid water, is the stable phase. *Below* 0°C the state of lower energy, ice, is the stable phase. And *at* 0°C the two phases, liquid and solid, are equally stable. At 0°C and 1 atm, the tendencies toward minimum energy and maximum entropy are precisely balanced, resulting in an equilibrium between solid and liquid. At 100°C and 1 atm, another such balance leads to equilibrium between liquid water and its vapor.

It is, then, not only low energy and not only high entropy that determines what phase is stable. It is a *combination* of energy and entropy that is important. The competition between energy and entropy is expressed quantitatively by the particular combination

$\Delta H - T\Delta S$. Here T is the absolute temperature, S is the entropy, and H represents a quantity closely related to the energy, called the **heat content** of the substance. The quantity ΔH represents the difference between the final heat content and the initial heat content when the substance changes its state. For example, when a solid melts at constant pressure, ΔH is the same as the heat absorbed. The Greek letter Δ (delta) is a symbol for *difference* or *change.*

We have already seen ΔH in our heating curve for water (Figure 4-19), but there we called it the heat of vaporization and the heat of fusion. One horizontal portion of the heating curve shows the heat required to vaporize one mole of water at 100°C. This amount of heat (41 kJ/mole) is the ΔH for the vaporization of water (the heat of vaporization):

$$H_2O \text{ (liquid)} \longrightarrow H_2O \text{ (gas)}$$

Heat of vaporization

$$= \Delta H$$
$$= \begin{pmatrix}\text{Final heat content} \\ \text{(gas)}\end{pmatrix} - \begin{pmatrix}\text{Initial heat content} \\ \text{(liquid)}\end{pmatrix}$$
$$= 41 \text{ kJ/mole}$$

The heat content of the vapor is 41 kJ/mole greater than the heat content of the liquid. This difference arises from, and is very nearly the same as, the higher potential energy of the molecules in the vapor and their lower potential energy in the liquid.

The other horizontal portion of the curve shows the heat required to melt one mole of ice at 0°C. This amount of heat (6.0 kJ/mole) is the ΔH for the melting of ice (the heat of fusion):

$$H_2O \text{ (solid)} \longrightarrow H_2O \text{ (liquid)} \qquad \Delta H = 6.0 \text{ kJ/mole}$$

The heat content (essentially the potential energy of the molecules) of the liquid is 6.0 kJ/mole greater than that of the solid.

The quantity ΔS is the difference between the final entropy (the entropy of the substance after its change of phase) and the initial entropy (the entropy of the substance before its change of phase):

$$\text{Difference in entropy} = \Delta S$$
$$= \text{Final entropy} - \text{Initial entropy}$$

In Example 4-6 the symbol Δ (for difference) means the difference between the maximum and minimum temperatures. This illustrates the precise, quantitative meaning of the symbol Δ.

EXAMPLE 4-6

An outstanding extreme temperature rise was recorded in Granville, North Dakota, in February 1918, when the temperature (t) rose from −36°C (−33°F) to +10°C (50°F) in a 12 hour period. What was the Δt for this period?

Difference in temperature

$$= \Delta t$$
$$= \text{Final temperature} - \text{Initial temperature}$$
$$= 10°\text{C} - (-36)°\text{C} = 46°\text{C}$$

The temperature difference or change was 46°C. (What was Δt in degrees Fahrenheit?)

The important combination $\Delta H - T\Delta S$, which we shall later see is also important in chemical changes, is symbolized by ΔG. For all changes at constant temperature,

$$\Delta G = \Delta H - T\Delta S$$

This quantity G is called the **Gibbs free energy,** and ΔG is the change in free energy (the difference between the free energy of the substance after its change of phase and the free energy of the substance before its change of phase). The value ΔG expresses the net driving force for a change in phase (or a chemical change); it results from the competition between the two opposite tendencies toward lower heat content and higher entropy. *The direction of a spontaneous change at any fixed temperature and pressure is always toward a condition of lower free energy.* Thus for a change to occur spontaneously, the combination $\Delta H - T\Delta S$ (or ΔG) must be *negative* (final G *less than* initial G).

For a spontaneous change,

$$\Delta G = \Delta H - T\Delta S$$

is negative.

At absolute zero, only ΔH (the change in heat content) counts because $T\Delta S$ is equal to zero. The higher the temperature, the more important $T\Delta S$ becomes. Whatever the temperature, at equilibrium there is an exact balance between these two factors. At equilibrium, ΔH and $T\Delta S$ are equal, and there is no tendency for change.

$$\text{At equilibrium,} \quad \Delta H = T\Delta S$$

$$\text{so} \quad \Delta G = 0$$

That is, the condition that there be no tendency for spontaneous change from one phase to another is that the free energies of the two phases must be equal. Illustrations and applications of this condition for equilibrium are given in Examples 4-7 and 4-8.

EXAMPLE 4-7

In the melting of ice at 0°C, H_2O (solid) → H_2O (liquid), with $\Delta H = 6.0$ kJ/mole. Calculate ΔS for this phase change.

At the melting point, where solid and liquid are at equilibrium, there is no tendency for change—neither for the ice to melt nor for the water to freeze. Thus,

$$\Delta G = 0,$$

and

$$\Delta H = T_{\text{melting}}\Delta S$$

$$T_{\text{melting}} = 0°\text{C} + 273 = 273\ \text{K}$$

$$\Delta S = \frac{\Delta H}{T} = \frac{6.0\ \text{kJ/mole}}{273\ \text{K}} = 0.022\ \text{kJ/mole K}$$

$$= 22\ \text{J/mole K}$$

At temperatures *above* 273 K, the entropy factor ($T\Delta S$) is greater than the change in heat content (ΔH), so the combination $\Delta H - T\Delta S$ is negative; that is, ΔG is then negative, and so ice spontaneously melts at those higher temperatures (as of course we knew).

At temperatures below 273 K, ΔH is greater than $T\Delta S$, so the combination $\Delta H - T\Delta S$ is positive. Then ΔG is positive, so at these lower temperatures ice does not melt spontaneously. In this familiar case we know that the reverse change, liquid water freezing, occurs spontaneously below 273 K (0°C). When ΔG for a particular change is positive, that change does not occur spontaneously; but then the reverse change does occur.

EXAMPLE 4-8

When the red-brown liquid bromine (Br_2) vaporizes at its normal boiling point, Br_2 (liquid) → Br_2 (gas), $\Delta H = 31$ kJ/mole and $\Delta S = 0.093$ kJ/mole K. What is the normal boiling point of bromine?

At the boiling point, liquid and vapor are at equilibrium, so

$$\Delta G = 0$$

and $$\Delta H = T_{\text{boiling}}\,\Delta S$$

$$T_{\text{boiling}} = \frac{\Delta H}{\Delta S} = \frac{31 \text{ kJ/mole}}{0.093 \text{ kJ/mole K}} = 333 \text{ K}$$

$$t_{\text{boiling}} = 333 - 273 = 60^\circ\text{C}$$

We use these ideas about heat content and entropy again when we study equilibrium in chemical reactions in Chapters 9 and 10.

KEY TERMS

Matter can exist in three *states* or *phases:* gas, liquid, and solid.

Surface tension is caused by the pull of molecules below the surface of a liquid on the surface molecules.

Capillarity, the phenomenon of a liquid climbing up a narrow tube, results from the attraction of the liquid for other substances.

Dynamic equilibrium is a condition where no observable changes are taking place—the *rates* of two opposing processes are equal.

The *equilibrium vapor pressure* of a liquid is the pressure of the vapor over a liquid in a closed container when the liquid and vapor are in equilibrium.

Boiling occurs at a temperature where the vapor pressure of the liquid is equal to the external pressure. The *normal boiling point* of a substance is the temperature at which its vapor pressure is equal to one standard atmosphere (760 mm Hg).

Sublimation is the process of a solid evaporating without melting.

The *normal melting (freezing) point* is the temperature at which a solid melts (freezes) at a pressure of one atmosphere.

The *triple point* is a unique temperature and pressure where the three phases of a substance can coexist at equilibrium.

The *joule* (J) is the SI unit used to measure heat: 4.18 J is the amount of heat required to raise the temperature of 1 g of water from 14.5°C to 15.5°C. The *calorie* (cal) is the quantity of heat required to raise the temperature of 1 g of water from 14.5°C to 15.5°C, so 1 cal = 4.18 J.

The *specific heat* of a substance is the amount of heat required to raise the temperature of 1 g of that substance by 1°C.

Potential energy is the capacity to do work. It is the energy stored in an object by virtue of its position or arrangement with respect to other objects.

The *molar heat of vaporization* is the amount of energy required to vaporize one mole of liquid at its boiling point. The *molar heat of fusion* is the amount of energy required to melt one mole of solid at its melting point.

The change in *heat content* (ΔH) for a reaction is the difference between the final and the initial heat content; it essentially represents the difference in potential energy between the products and the reactants.

The change in *free energy* (ΔG) for a reaction is the difference between the free energy of the products and that of the reactants.

Entropy (S) is the degree of randomness or disorder in a system, and ΔS is the difference in disorder between products and reactants.

CHAPTER SUMMARY

The kinetic theory of gases can be extended to explain some of the observed properties of liquids and solids.

A liquid evaporating in a closed container may reach a state of *dynamic equilibrium* where molecules are leaving the liquid at the same rate at which molecules from the vapor are reentering the liquid. The equilibrium vapor pressure of a liquid increases as the temperature is raised. Boiling occurs when the vapor pressure of the liquid is equal to the external pressure. The *normal boiling point* is the temperature at which the vapor pressure is 1 atm. Just as a liquid can be in equilibrium with its vapor, a solid can also be in equilibrium with its vapor; and a solid can be in equilibrium with the liquid.

Every point on an *equilibrium vapor pressure curve* represents a unique temperature and pressure at which two phases of a substance are in equilibrium. For all values of temperature and pressure inside a given region of a *phase diagram,* only one phase is stable. Given a substance at a certain temperature and pressure, we can use a phase diagram to predict what changes will occur as the temperature or pressure is changed.

A *heating curve* shows how the temperature of a substance changes as it absorbs heat. During a phase change (solid to liquid, or liquid to vapor), the temperature remains constant while heat is absorbed—the energy is used to overcome the forces of attraction between the molecules, and it is stored as potential energy.

It is not only the tendency toward lower *energy* and not only the tendency toward greater *entropy* that determines what phase of a substance is stable. Rather, the determining factor is the special combination of energy and entropy that is called *free energy.* The direction of spontaneous change is always toward a condition of lower free energy.

PROBLEMS

4-1. Ethyl chloride (C_2H_5Cl), which has a boiling point of 12.3°C, is used as a local anesthetic. When sprayed on the skin, it cools the skin and reduces the rates of the chemical reactions that transmit pain. Explain how ethyl chloride cools the skin.

4-2. In December, a ten-year-old boy put an unwrapped snowball in the freezer. When he went to get it in August, it was only one-half its original size. What happened to the rest of the snowball?

4-3. A student needed some boiling water for an experiment. He filled a flask half full with water, sealed it with a rubber stopper, and began to heat it. What do you think happened? Explain.

4-4. In a laboratory demonstration, a flask half full of water was brought to boiling, sealed with a rubber stopper, and cooled under running water. The water continued to boil as it cooled. Why?

4-5. How much heat must be removed to freeze the water in an ice-cube tray of 300 ml volume? Where does this heat go? (The heat of fusion of water is 0.33 kJ/g.)

4-6. (a) A person at rest loses about 50 ml/hour of water by evaporation from the skin and the mucous membranes of the mouth and respiratory passages. How much heat does the body lose in a 24 hour period as a result of this evaporation? Assume that the heat of vaporization of water at skin temperature is about the same as it is at 100°C (2.3 kJ/g).
(b) During heavy exercise in a hot, dry climate, as much as 1600 ml/hour of water might evaporate. How much body heat is used to vaporize 1600 ml of water? Compare your answer to that of part a.

4-7. When preparing crepes suzette, why does one warm the brandy before igniting it?

4-8. The vapor pressure of ethanol is given as a function of temperature in the following table:

t(°C)	0	20	40	60	80	100	120
P(mm of Hg)	10	40	150	350	800	1800	3550

(a) What, approximately, is the normal boiling point of the alcohol?
(b) What, approximately, is the vapor pressure of the alcohol near room temperature?
(c) Describe an experiment by which the result in part b could have been determined.

4-9. A quartered apple left uncovered overnight is shriveled and dry in the morning, but one that is wrapped in plastic remains moist. Explain.

4-10. Liquid sodium is heated at its boiling temperature, 1156 K. Although heat energy is being absorbed by the liquid, its temperature does not increase. Explain.

4-11. A drop (0.050 ml, or 0.050 g) of water is placed in the bottom of a 1 ℓ flask, which is then stoppered. Will a state of equilibrium be reached between liquid water and water vapor
(a) at room temperature, 20°C? Explain.
(b) at 60°C? Explain.
(Use data given in Table 4-2.)

4-12. A beaker of diethyl ether and a beaker of mercury were left open in a locked laboratory overnight. In the morning one beaker was empty, but the other appeared unchanged. Which beaker was empty? Explain.

4-13. Dichlorobenzene ($C_6H_4Cl_2$) mothballs have a strong odor, whereas table salt (NaCl) is odorless. Which would you expect to have the higher vapor pressure? Explain.

4-14. Which would cause the more severe burn,
(a) 100 g of liquid water at 100°C or 100 g of steam at 100°C?
(b) 500 g of liquid water at 100°C or 100 g of steam of 100°C?
(Assume that the final temperature of the water is close to room temperature, 20°C. The heat of vaporization of water is 2.3 kJ/g.)

4-15. What is the minimum amount of ice one could use to cool 400 g of water from room temperature (20°C) to 0°C?
(Assume that the temperature of the ice is 0°C. The heat of fusion of ice is 0.33 kJ/g.)

4-16. An iron bar at 1000°C is plunged into an insulated container of water and ice. When the system reaches equilibrium, there is still ice remaining in the container.
(a) What is the final temperature of the iron bar?
(b) What happened to the heat lost by the iron bar?
(c) If 300 g of ice are melted, how many joules are lost by the iron bar?
(d) If the iron bar weighs 200 g, what is the specific heat of the iron?

4-17. Use the diagrams in Figure 4-22 to answer the following questions.

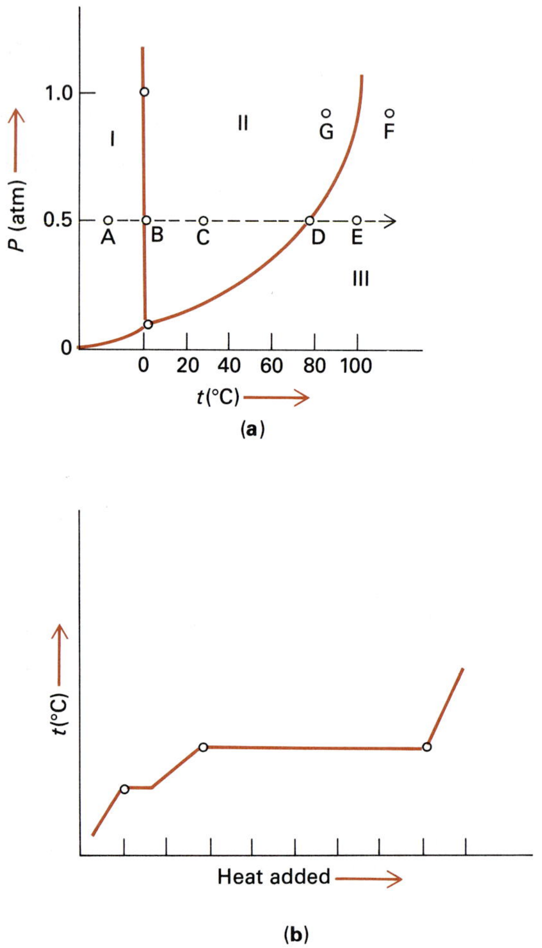

FIGURE 4-22

(a) Phase diagram for H_2O. **(b)** Heating curve for 1 g of H_2O at 0.5 atm.

(a) Identify as solid, liquid, or gas the regions I, II, and III on the phase diagram for H_2O.
(b) Reproduce the heating curve for water. Examine points A, B, C, D and E on the phase diagram. On the *heating curve* you have drawn, identify the points A, B, C, D, and E that correspond to those points on the *phase diagram.*
(c) Identify on your heating curve the points at which melting and boiling occur as the system moves along the line ABCDE in the phase diagram. What is the boiling point of the water?
(d) Explain why one of the horizontal portions of the heating curve is longer than the other.
(e) If your answer in part c for the boiling point is different from 100°C (the normal boiling point of water), explain why.

4-18. Consider solid, liquid, and gaseous water at the triple point (approximately 0°C). Compare the average kinetic energy, the average potential energy, and the entropy per mole for the three states.

4-19. Give several examples (other than those in the text) of dynamic equilibrium.

4-20. Give several examples (other than those in the text) of spontaneous changes in which there is an increase in entropy.

4-21. Iodine (I_2) is a solid that sublimes when warmed.
(a) Does the tendency to achieve maximum entropy favor solid or gas?
(b) Does the tendency to achieve minimum energy favor solid or gas?

4-22. The energies released by the burning of foods in the body are 4.1 kcal per gram of carbohydrate or protein, and 9.3 kcal per gram of fat. What are these energies in kilojoules per gram?

4-23. The metabolic rate is the amount of energy liberated per unit of time by the burning of foods in the body. When the metabolic rate is determined at rest, 12 to 14 hours after the last meal, it is called the "basal" metabolic rate, and is

usually given in kilocalories per 24 hour day. Some typical basal metabolic rates are
(a) mouse, 3.82 kcal/day;
(b) pig, 2444 kcal/day;
(c) horse, 4983 kcal/day;
(d) human, 2054 kcal/day.
Convert these metabolic rates to kilojoules per day.

4-24. The energy content of some foods is given in kilojoules. Convert the kilojoules to kilocalories.
(a) Chocolate fudge (1 in. square), 418 kJ
(b) Yoghurt (1 cup), 1,100 kJ
(c) Zucchini (1 cup), 105 kJ
(d) Beer (12 oz), 585 kJ
(e) Cantaloupe ($\frac{1}{2}$ average melon), 209 kJ
(f) Chicken ($\frac{1}{2}$ medium), 525 kJ

4-25. A dehumidifier in a basement collected 4 ℓ of water in one day. How many calories were released by the condensation of the water?

5

The Divisible Atom

At the end of the nineteenth century, scientists were convinced that the atom is the fundamental, indestructible unit of which all matter is composed. They very soon discovered, however, that the atom is in fact divisible. Here we see that an atom is composed of smaller particles—including negatively charged electrons, positively charged protons, and uncharged neutrons. This model of the divisible atom provides explanations of radioactivity and of the huge amounts of energy released in nuclear reactions.

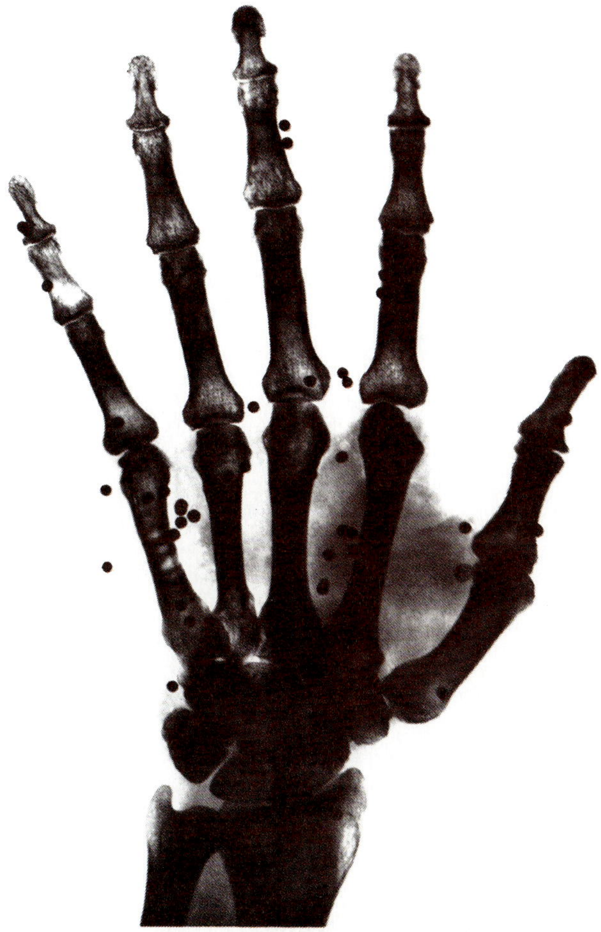

5-1
THE ELECTRON

A simple observation illustrates some properties of electric charge. When you pass a comb through clean hair, the hair is attracted to the comb and clings to it, while individual hairs repel each other and refuse to lie flat. This happens because the comb and the hair have acquired electric charges. There are two kinds of electric charge: **positive charge** and **negative charge.** The unlike charges on the comb and the hair attract each other. The like charges on the individual hairs repel each other.

In a battery, positive and negative charges are separated. An electric charge may be placed on an object by connecting it to the positive or the negative terminal of a battery. (We'll learn more about the construction of batteries and how they work in Chapter 12.)

Electric current is the movement of electrically charged particles. When particles carrying positive or negative charge flow through empty space or through a conducting material such as a metal wire, we have an electric current. The electron is the fundamental particle that carries negative electric charge. It is this particle that conducts electric current through wires.

Figure 5-1 shows a diagram of a gaseous discharge tube (also called a cathode ray tube), an instrument used by J. J. Thomson (1856–1940) to study the charged particles that carry electric current through a rarefied gas. Metal plates called electrodes are sealed in each end of a long glass tube. The electrodes are connected to opposite terminals of a high-voltage source so that one electrode (the **anode**) acquires a positive electric charge, and the other electrode (the **cathode**) acquires a negative charge. A vacuum pump removes gas from the tube. No electricity flows through the tube until the gas pressure is reduced to about 0.01 atm. Then an electric current flows, and the gas glows—red if the gas is neon. The now familiar orange-red neon (and green helium) advertising signs and the fluorescent light bulb all make use of the passage of electric current through a gas at low pressure.

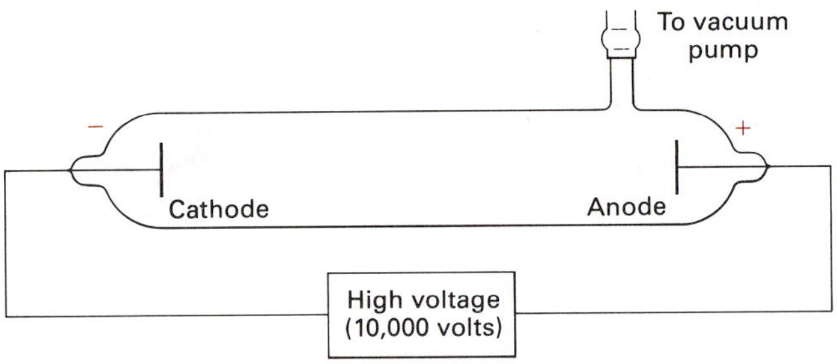

FIGURE 5-1

Cathode ray tube.

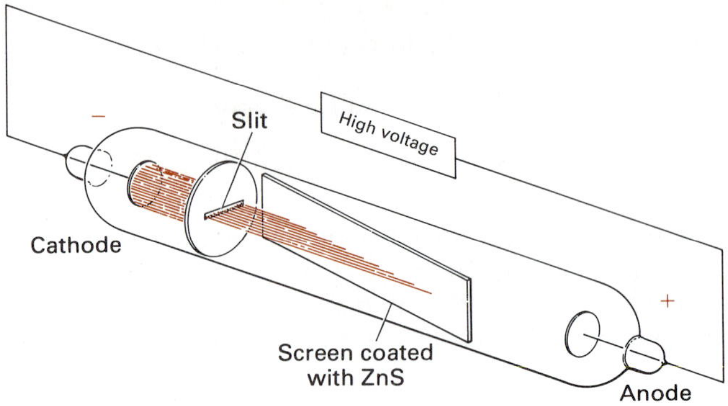

FIGURE 5-2

Cathode ray tube with a fluorescent screen. The rays pass through the slit and move in straight lines toward the anode.

In the experiment shown in Figure 5-2, a metal screen with a narrow slit is placed in front of the cathode. A second screen, coated with a fluorescent material such as zinc sulfide (ZnS), is placed diagonally across the tube. When a voltage is applied, rays can be seen to emanate from the cathode, pass through the slit, and travel in a straight line until they strike the fluorescent screen—which glows when struck by the rays, just as a television screen does. These rays, which originate at the negatively charged cathode and travel in a straight line toward the positively charged anode, were given the name cathode rays. The following experiments provided additional information about these rays.

A gaseous discharge tube is constructed as in Figure 5-3b, with a perforated anode to form a narrow beam of cathode rays, and a screen coated with zinc sulfide to detect the beam. A second set of electrodes is sealed in the tube, and a voltage is applied. The electric charge on these electrodes exerts an attractive or repulsive force on any other electric charge in the surrounding space. This force is called an **electric field.** The beam of cathode rays is deflected toward the positive plate of the deflecting electrodes, again indicating that cathode rays are negatively charged.

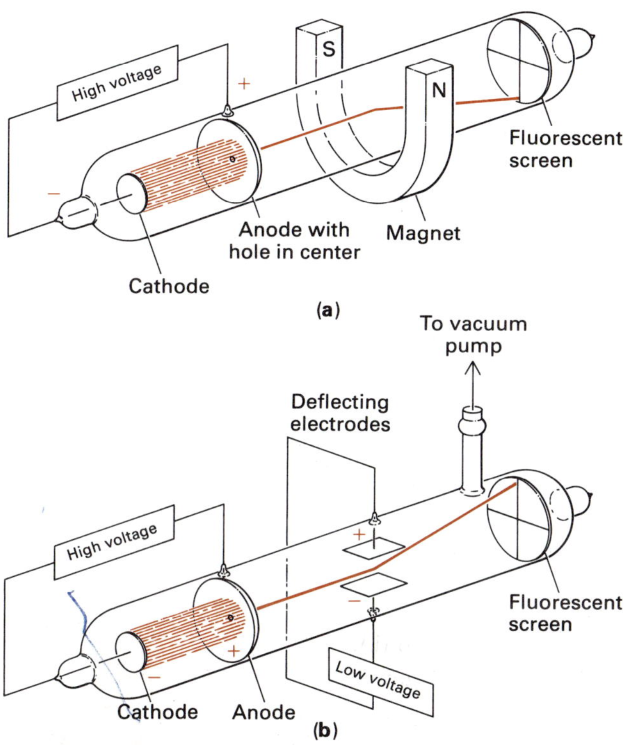

FIGURE 5-3

Deflection of cathode rays. **(a)** Cathode rays are deflected by a magnetic field. **(b)** Cathode rays are deflected by an electric field.

If, instead of applying an electric field, we bring a magnet near (Figure 5-3a), the beam is again deflected. A magnet exerts a force on another magnet or on a moving charge in the surrounding space. This force is called a **magnetic field.** The direction of deflection of the beam by the magnet is the same as it would be for any negatively charged particles.

The *amount* of deflection by an electric or a magnetic field depends on the following three factors:

1. The strength of the field—the stronger the magnet, or the greater the voltage on the deflecting electrodes, the greater the deflection.

2. The magnitude of the charge on the deflected particle—the larger the charge on the particle, the greater the attractive or repulsive forces on it, and the greater the deflection.

3. The mass of the particle—the heavier the particle, the smaller the deflection. (Just as it is more difficult to deflect a Greyhound bus than a Volkswagen, it is more difficult to deflect a heavy than a light particle from its straight-line path.)

J. J. Thomson combined the principles of the experiments shown in Figure 5-3 in the apparatus illustrated in Figure 5-4. The magnetic field bends the beam down, and the charged electrodes bend the beam up. By balancing the strengths of the magnetic and electric fields, it is possible to make the beam of cathode rays strike the fluorescent screen at its center. Using the known strengths of the magnetic and electric fields, Thomson was able to calculate the ratio of the particle's charge to its mass (q/m) for the cathode rays, with the result

$$q/m = -1.76 \times 10^8 \text{ coulombs/g}$$

The minus sign indicates that the cathode ray particle

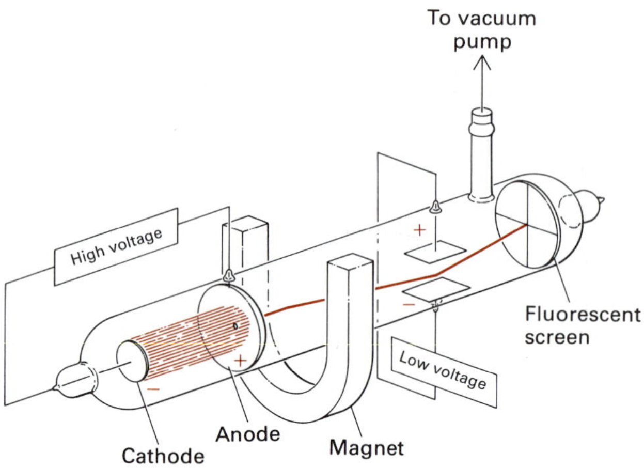

FIGURE 5-4

Determination of charge/mass ratio with a cathode ray tube.

has a negative charge. The coulomb (C) is the unit of electrical charge.

The charged particles that form cathode rays are called **electrons.** The same value of q/m was obtained in every experiment with cathode rays, independent of the cathode material used and of the gas in the tube. This observation suggested that the electron is one of the fundamental constitutents of all atoms.

The *actual* charge (q) of the particle—and therefore, indirectly, its mass (m)—was measured in an ingenious experiment designed by the American physicist Robert A. Millikan, who won a Nobel Prize in 1923 for this work. Millikan's experiment is diagrammed in Figure 5-5. Electrodes are connected to opposite terminals of a battery so that one plate is positively charged and the other is negatively charged. Tiny droplets of oil are sprayed in the upper section of the apparatus. As a droplet falls through the hole, its mass M can be determined by the rate at which it falls due to the force of gravity. If high-energy light shines on molecules of air in the container, electrons are occasionally removed from the molecules, and some oil drops acquire one or more of these electrons.

When a droplet of oil thus negatively charged comes between the electrodes, it is attracted to the positively charged plate, and it can be kept suspended,

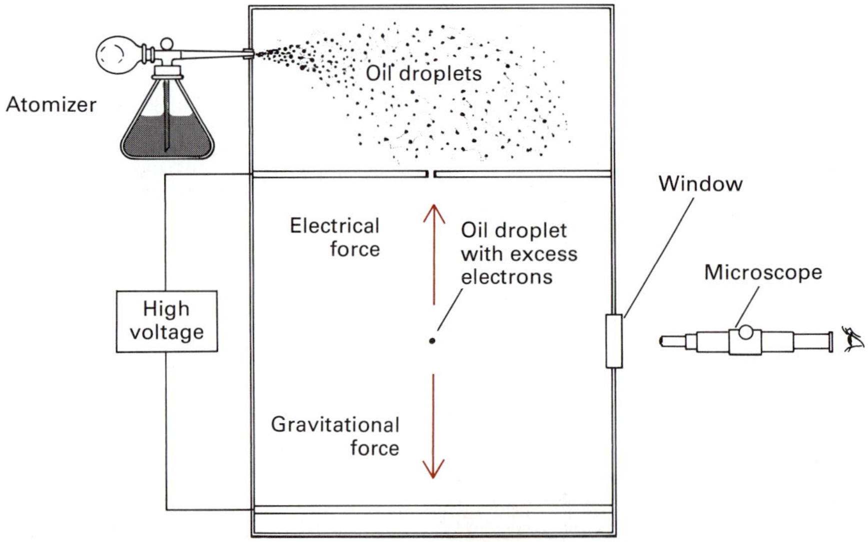

FIGURE 5-5

Millikan's oil-drop experiment.

without rising or falling, by adjusting the charge on the plates. The gravitational and electrical forces on the droplet are then exactly balanced. The mass of the oil droplet with electrons attached is practically the same as the mass of the oil droplet alone, because the mass of an electron is negligible compared to the mass of an oil droplet. Using the droplet mass M and the known value of the strength of the electric field, Millikan calculated the charge on the oil droplet. It turned out that over thousands of such experiments, the charge on the oil droplet was *always* a small integral multiple of a quantity that, Millikan reasoned, must be the charge on a single electron:

$$q = -1.60 \times 10^{-19}\,\mathrm{C}$$

The charge on the oil drop was always a small whole-number multiple of -1.6×10^{-19} C. No droplet carrying $\frac{1}{2}$, $\frac{3}{4}$, or any other fraction of this unit of charge was ever found.

The mass of the electron could then be determined by combining Millikan's value for q with Thomson's value for the ratio q/m:

From Thomson's experiment, $q/m = -1.76 \times 10^{8}\,\mathrm{C/g}$

From Millikan's experiment, $q = -1.60 \times 10^{-19}\,\mathrm{C}$

$$m = \frac{q}{q/m} = \frac{-1.6 \times 10^{-19}\,\mathrm{C}}{-1.76 \times 10^{8}\,\mathrm{C/g}} = 9.11 \times 10^{-28}\,\mathrm{g}$$

5-2 THE PROTON

We saw that negatively charged electrons in a gas discharge tube move from the cathode toward the anode. If a gas discharge tube is set up with a perforated cathode, as shown in Figure 5-6, other rays, called canal rays, can be detected moving from the anode to the cathode, with some rays passing through holes in the cathode. The deflection of these rays by an electric or a magnetic field is opposite to that for electrons, showing that these rays consist of *positively* charged particles. Their charge/mass ratio (q/m) is also much smaller than that for electrons (and opposite in sign). Further, the value of q/m for these positively charged particles, unlike that for the negatively charged particles (electrons) in the cathode rays, depends on what gas is present in the discharge tube.

To summarize, in a gas discharge tube, electrons and positively charged particles are formed. The movement of these electrically charged particles constitutes an electric current. Electrons have negative charge, have the same value of q/m whatever the gas

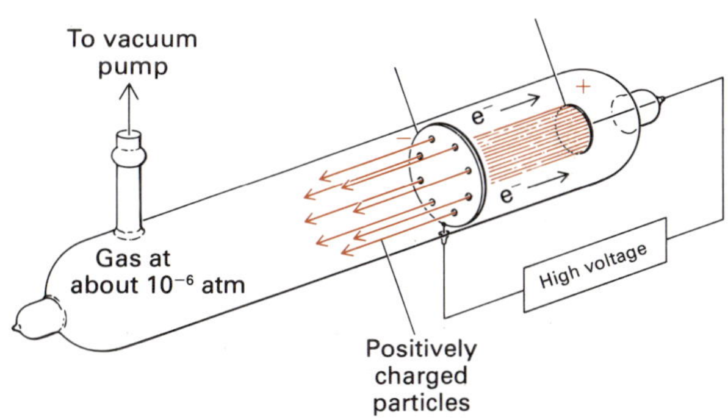

FIGURE 5-6
Canal ray experiment.

in the tube, and move toward the anode. The positive particles have a smaller numerical value of q/m than that of electrons, with the value of q/m dependent upon the nature of the gas in the tube, and they move toward the cathode.

How can we explain the presence of these positively charged particles in a discharge tube? The originally neutral atoms of the gas contained in the tube are not indivisible; rather they are a combination of negatively and positively charged particles. In a gas discharge tube, high-velocity electrons from the cathode strike the neutral atoms (or molecules) of the gas. A neutral atom may lose an electron when struck by a cathode electron, thereby acquiring a positive charge. The following equations illustrate the formation of positive particles and electrons in a gas discharge tube:

$$He \longrightarrow He^+ + e^-$$

$$Na \longrightarrow Na^+ + e^-$$

The superscript + represents a positive charge, and − represents a negative charge; e^- represents an electron. For all atoms except hydrogen atoms (H), it is possible to remove more than one electron (a hydrogen atom, as we shall later see, *has* only one electron):

$$He \longrightarrow He^+ + e^-$$

$$He^+ \longrightarrow He^{++} + e^-$$

The helium atom has now lost two electrons and has two units of positive charge. These positively charged particles are called positive **ions,** after the Greek word for "wanderer."

In terms of electron charge units, if a particle carries one unit of negative electric charge (the charge on one electron), it is said to have a 1− charge; if it carries two units of negative charge it is said to have a 2− charge. If a particle carries one unit of positive charge (the charge on the hydrogen ion H^+), it is said to have 1+ charge. The He^{++} ion then has a 2+ charge, and is usually written He^{2+}. A particle carrying no electric charge is said to have a charge of 0 (zero).

When a diatomic gas such as hydrogen gas (H_2) is present in the gas discharge tube, the process is more complicated, but the end result is to produce H_2^+ and H^+ ions in a series of steps such as

$$H_2 \longrightarrow H_2^+ + e^-$$

$$H_2 \longrightarrow 2H$$

$$H \longrightarrow H^+ + e^-$$

The charge/mass ratio for the negative particle is always the same because it is always an electron that is produced. Electrons removed from the neutral atom join the stream of cathode-ray electrons moving toward the anode.

The value of the charge/mass ratio of the positive ion depends on the gas in the discharge tube because the identity, and therefore the mass, of the ion depends upon what the gas is. Thus, from He we get He^+ ions (or He^{2+} ions) of atomic mass 4, and from H_2 gas we get H_2^+ ions of mass 2 and H^+ ions of mass 1. The positive ions are so much more massive than the electrons that, to a first approximation, the positive ions have the same masses as the corresponding neutral atoms or molecules. So we can say that He atoms, He^+ ions, and He^{2+} ions each have atomic mass 4, and that H atoms and H^+ ions each have atomic mass 1.

The charge/mass ratio (q/m) for the lightest ion, the hydrogen ion (H^+), is the largest value found for any positive ion. The charge on this ion is opposite in sign, but equal in magnitude to that on the electron (-1.60×10^{-19} C). Therefore, the charge on the H^+ ion is $+1.60 \times 10^{-19}$ C/ion. The mass of a single H^+ ion, calculated from its q/m ratio, is 1.67×10^{-24} g, about 1840 times the mass of an electron. This particle, the H^+ ion, is called the **proton.** If two electrons are removed from an atom, as in the case of He^{2+}, the remaining positive ion has a charge of $2(1.60 \times 10^{-19}) = 3.20 \times 10^{-19}$ C/ion. The He^{2+} ion is called the **alpha particle** and is one product of natural radioactivity.

From the known mass of the proton, it is possible to determine Avogadro's number. Because the mass of an electron is so much smaller than the mass of the proton (only about 1/1840 as great), the mass of the neutral H atom, which is composed of one proton and one electron, is essentially identical to the mass of the proton. We can then say that the mass of a single H atom is 1.67×10^{-24} g, the mass of a proton. We can use this value, and the known mass of a mole of H atoms (the atomic weight), to determine Avogadro's number, N:

$$N \cong \frac{\text{Mass of one mole of H atoms}}{\text{Mass of one H atom}}$$

$$= \frac{1.0\ (\text{g/mole})}{1.67 \times 10^{-24}\ (\text{g/atom})}$$

$$= 6.0 \times 10^{23}\ \text{atoms/mole}$$

This method of determining Avogadro's number is based on the measured values of the charge of the electron, and the charge/mass ratio of the proton.

5-3 X RAYS AND RADIOACTIVITY

The glass wall behind the anode of a gas discharge tube glows with a greenish light when struck by cathode rays. In 1895 W. K. Roentgen, while experimenting in a darkened room with a discharge tube completely enclosed in a black box, discovered that a fluorescent screen far across his laboratory was glowing, though no visible light could reach it. It glowed whenever his discharge tube was turned on. How could he account for this remarkable phenomenon? Cathode rays (electrons) could not be responsible because they cannot penetrate a box nor more than a few inches of air. Roentgen reasoned that when cathode rays struck the glass wall of the tube and made it glow, new rays that he called **X rays** were produced. He determined that X rays can penetrate wood, thin sheets of aluminum, and human flesh, but are partially stopped by bones and completely stopped by thin lead plates. The discovery that X rays darken photographic plates led to their immediate application in medicine (Figure 5-7). We now know that X rays are a form of light with very high energy.

Shortly after Roentgen's discovery, Henry Becquerel (a French scientist) discovered that a mineral containing the element uranium causes photographic plates wrapped in black paper to darken. Roentgen's X rays were produced by using high voltages in a gaseous discharge tube, but the radiation from the uranium flows spontaneously without the application of any outside energy. Studies of this phenomenon, called **radioactivity,** show that nothing—not heating, nor cooling, nor chemical changes such as dissolving the radioactive material in acid or digesting it in the stomach—affects the intensity of the radiation. The radiation flows out at a rate that depends only on the type and amount of radioactive material present.

The radiation emanating from radioactive materials may consist of three different components. These can be distinguished and separated by passing the

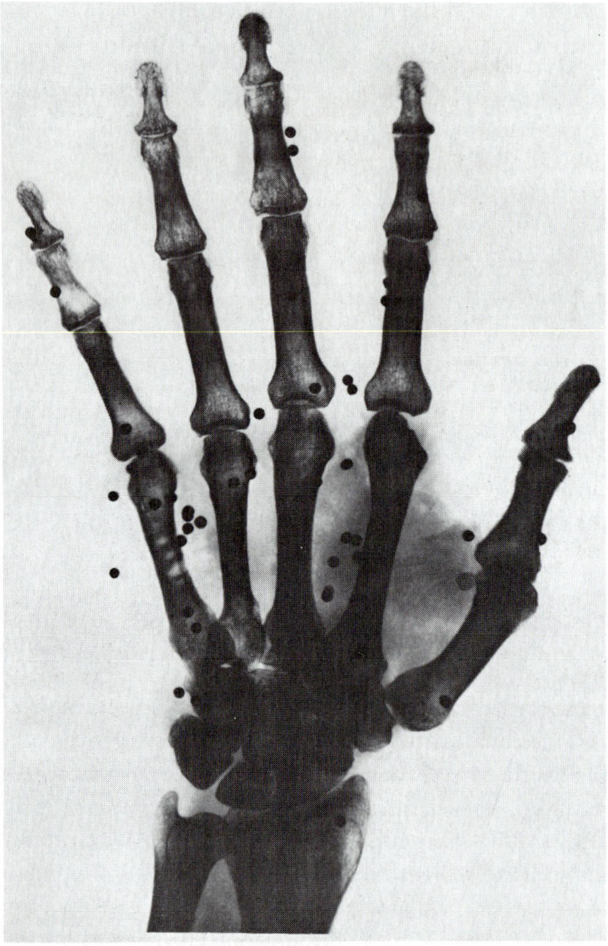

FIGURE 5-7
An early X-ray photograph of a hand, showing gunshot embedded in the flesh and in the bone. [Photograph, courtesy of Burndy Library, Norwalk, Conn.]

radiation through a strong magnetic field as shown in Figure 5-8. A tiny sample of radium, a naturally occurring radioactive metal, is placed inside a hole in a lead block. The radiation emerges in a beam through a small opening in the lead block. The beam passes through a strong magnetic field where it splits into three separate streams of radiation, which are called alpha, beta, and gamma rays.

Gamma (γ) **rays** are undeflected by the magnetic field. They are a high-energy form of light, even more penetrating than X rays. Gamma rays have a velocity equal to the speed of light and will penetrate several inches of lead.

Beta (β) **rays** are high-speed electrons. They are identical to cathode rays, except for their greater velocity. Beta particles have velocities up to nine-tenths of the speed of light and will penetrate a few millimeters of human tissue.

Alpha (α) **particles** are massive, positively charged particles. An alpha particle, as we have seen, is the He^{2+} ion (and we'll see that it is the nucleus of the helium atom). Alpha particles have a velocity of about one-tenth of the speed of light and have little penetrating power. They are stopped by paper or human skin.

Thus, radioactive atoms *spontaneously* disintegrate, producing positively and negatively charged particles. This is convincing evidence that the atom *is* divisible.

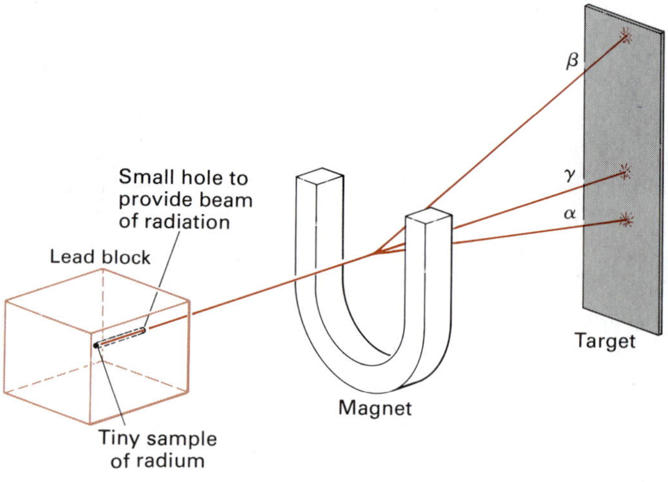

FIGURE 5-8
Effect of a magnetic field on radiation from radium.

5-4
THE RUTHERFORD ATOM

The next step forward in establishing a picture of the atom resulted from experiments directed by Ernest Rutherford in the early twentieth century. In one experiment, diagrammed in Figure 5-9, he investigated the deflection of alpha particles by a thin metal foil. A sample of a radioactive substance such as polonium, known to emit alpha particles, was placed in a lead block, and a beam of alpha particles was permitted to escape through a small hole. A thin metal foil (say, gold) was placed in the path of the alpha-particle beam, and a fluorescent detecting screen set up around the foil. Whenever an alpha particle hit the screen, a flash of light was observed.

If the positive charges of the atoms of the metal foil were distributed uniformly over its volume, then we would expect all of the speeding alpha particles to pass through the thin foil essentially undeflected. Figure 5-9 shows that most of the alpha particles did pass through the foil undeflected or with small deflections (A, B, C, and D), but a few were deflected through large angles (E and F). Rutherford was astonished that these relatively massive, speeding particles should be so deflected: "It was almost as incredible as if you had fired a 15 inch shell at a piece of tissue paper and it came back and hit you."

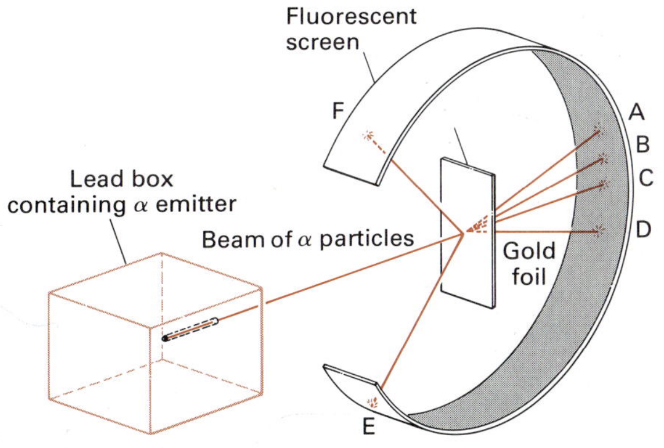

FIGURE 5-9

The Rutherford experiment.

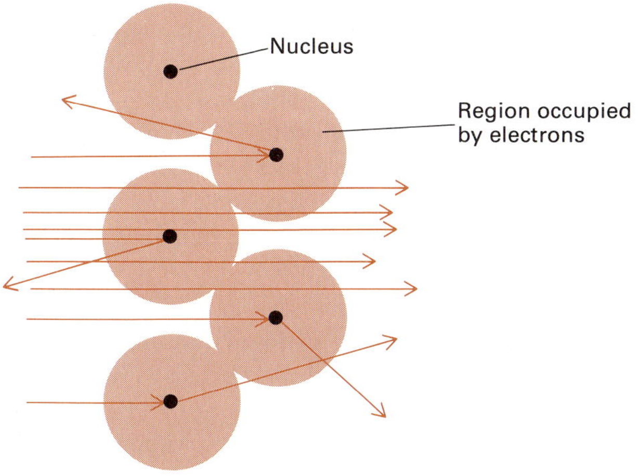

FIGURE 5-10

The scattering of α particles by gold foil (based on Rutherford's model of the atom).

How could Rutherford account for this experimental fact? A speeding alpha particle of molecular weight 4 could not be deflected by an electron weighing only about 1/7000 as much, just as a billiard ball is not deflected by a particle of dust. The simplest explanation, it seemed to Rutherford, was that the gold atom has a large central positive charge contained in a very small volume. This central core of the atom he named the **nucleus.**

Figure 5-10 illustrates how Rutherford's view of the atom, the beginning of the modern view, accounts for the results of his alpha-scattering experiments. The atom has a very small nucleus in which is located all the positive charge of the atom (in the form of protons) and most of the mass of the atom. The positive charge on the nucleus is always an integral multiple of the charge on the proton. The electrons, of negligible mass, are equal in number to the number of protons in the nucleus, and they surround the nucleus. Because they are in rapid motion, they effectively occupy most of the volume of an atom. It is the repulsive forces between the electrons of neighboring atoms that prevent interpenetration of the atoms and account for the volume of the atom. As shown in Figure 5-10, most of the alpha particles pass undeflected through the large volume occupied by the electrons. An occasional alpha particle (which is positively charged) passes close to the nucleus of a gold atom and is deflected slightly. Very rarely an alpha particle strikes the tiny, massive, positively charged nucleus of a gold atom directly and is deflected sharply.

By determining the fraction of alpha particles deflected through particular angles, Rutherford was able to calculate the diameter of the nucleus. The nuclear diameter is 1/10,000 to 1/100,000 of the diameter of the atom. (If a nucleus were pictured as being an *inch* in diameter, the entire atom would have a diameter of about a *mile.*) A typical atomic diameter is 2×10^{-8} cm, and a nuclear diameter between 10^{-13} and 10^{-12} cm. Atomic and molecular sizes are usually expressed in nanometers (1 nanometer = 1 nm = 10^{-9} m). A typical atomic diameter is then 0.2 nm, and the corresponding nuclear diameter is 2×10^{-6} nm.

The protons in a nucleus account for its positive charge, but how many protons are there in a given nucleus? The answer to this question resulted from experiments performed in 1913 by Henry Moseley, who found that each element, when bombarded by cathode rays, emits X rays of characteristic energy. From the differences in X-ray energies, each element was shown to have associated with it an **atomic number** (Z) *equal to the number of positive charges on the nucleus—that is, to the number of protons in the nucleus.* In Chapter 6 we see that it is useful to arrange the elements in order of increasing atomic number. Such an arrangement leads to the periodic table. The atomic number of hydrogen (H) is 1, and its nucleus carries a 1+ charge; the atomic number of helium (He) is 2, and its nucleus carries a 2+ charge; and the atomic number of gold (Au) is 79, and its nucleus carries a 79+ charge.

The atomic number of an element is often written at the lower left corner of the symbol for the element ($_Z$Element): thus, $_1$H, $_2$He, and $_{79}$Au. *Because in an electrically neutral atom the number of electrons is equal to the number of protons, the atomic number also gives the number of electrons present in a neutral atom.* The neutral H atom has 1 electron, the neutral He atom has 2 electrons, and the neutral Au atom has 79 electrons.

If an electron is removed from a neutral atom, the remaining particle has a positive charge. The neutral hydrogen atom has only one electron. If it is removed, the remaining positively charged particle is the nucleus of the hydrogen atom, the proton H^+. The neutral helium atom has two electrons. If both are removed, the remaining positively charged particle is the nucleus of the helium atom, the alpha particle He^{2+}. If one electron is removed from a neutral gold atom, the remaining positively charged particle still has 78 electrons. This particle is a gold ion with a single positive charge, Au^+.

According to Rutherford's model of the atom, the tiny nucleus contains most of the mass of the atom and all of its positive charge. But the mass of the protons does not entirely account for the mass of the nucleus. This fact led Rutherford to postulate the existence in the nucleus of a particle with mass equal to that of the proton but with no electric charge. This particle, the **neutron,** was discovered about 20 years later (in 1932) by James Chadwick. The neutron is contained in the nucleus, is almost identical in mass to the proton, and carries no charge. Each atom then (except for H) has a nucleus composed of protons and neutrons, which together constitute almost the entire mass of the atom. Table 5-1 compares the charges and masses of the proton, neutron, and electron.

TABLE 5-1
Comparison of proton, neutron, and electron

Particle	Charge	Mass (g)
Proton	1+	1.67×10^{-24}
Neutron	none	1.67×10^{-24}
Electron	1−	9.11×10^{-28}

5-5
ISOTOPES

Not all atoms of an element have the same mass. Atoms that have the same atomic number (and thus the same number of protons), but that have different nuclear masses, are called **isotopes.** Isotopes of an element have the same numbers of protons and electrons, but different numbers of neutrons, leading to different nuclear masses.

The masses of the isotopes and their relative abundances are determined using an instrument called a **mass spectrometer.** Figure 5-11a shows one type of mass spectrometer—similar to the instrument used by Thomson to measure q/m for positive particles. A gas—say, pure neon—is subjected to bombardment by high-speed electrons. An electron is knocked off the neutral Ne atom, forming the positive ion Ne^+. These Ne^+ ions are accelerated and passed through a slit to produce a beam of high-speed positive ions, and the beam is then passed through a strong magnetic field.

In the case of neon, the beam is separated into three streams corresponding to neon ions (Ne^+) of three different masses. The heaviest particles are hardest to deflect from their original path and so are least affected by the magnetic field (B in Figure 5-11a). The lightest particles are the easiest to deflect and so are deflected the most from their original path (A in the figure). From the position at which an ion emerges, its charge/mass ratio can be calculated. Because the charge is known, the mass of the ion (essentially the mass of the atom) can be accurately calculated. Relative abundances of the isotopes are determined by measuring the intensity of the beam of ions for each isotope (Figure 5-11b). Neon isotopes of atomic masses 20.0, 21.0, and 22.0 are found, with relative abundances of 90.92%, 0.257%, and 8.82%, respectively.

Almost every element exists in nature as a mixture of isotopes. Table 5-2 lists the isotopes of several elements with their relative natural abundances. The scale of atomic weights is based on the most abundant isotope of carbon, which has been assigned the mass of exactly 12, or 12.000. . . . Atomic weights are weighted

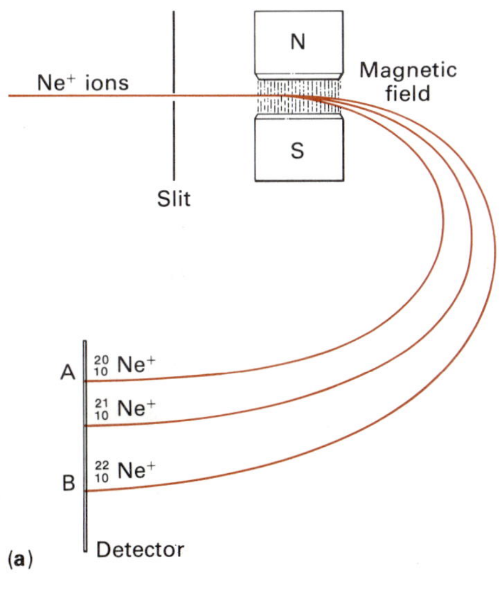

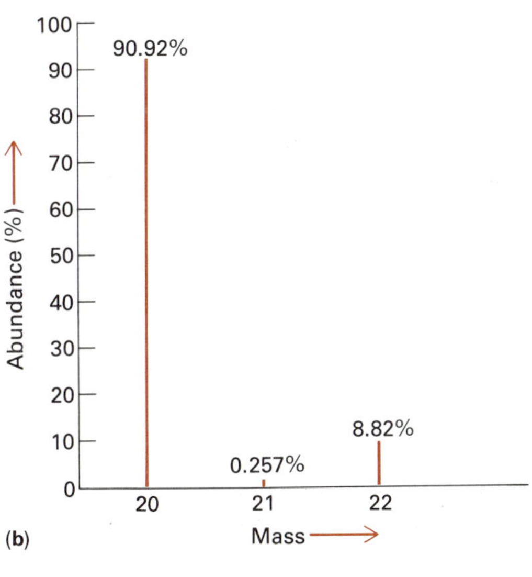

FIGURE 5-11

The mass spectrometer. **(a)** Schematic diagram of mass spectrometer. **(b)** A computer-plotted bar graph showing relative abundances of the three isotopes of neon.

TABLE 5-2
Isotopes of some elements

Isotope		Natural abundance (%)	Atomic weight	Atomic number	Mass number	Protons	Neutrons	Electrons in neutral atom
Hydrogen	$^{1}_{1}H$	99.985	1.0079	1	1	1	0	1
	$^{2}_{1}H$	0.015		1	2	1	1	1
Helium	$^{3}_{2}He$	1.3×10^{-4}	4.0026	2	3	2	1	2
	$^{4}_{2}He$	100		2	4	2	2	2
Carbon	$^{12}_{6}C$	98.892	12.011	6	12	6	6	6
	$^{13}_{6}C$	1.108		6	13	6	7	6
Oxygen	$^{16}_{8}O$	99.759	15.9984	8	16	8	8	8
	$^{17}_{8}O$	0.037		8	17	8	9	8
	$^{18}_{8}O$	0.204		.8	18	8	10	8
Fluorine	$^{19}_{9}F$	100	18.9984	9	19	9	10	9
Neon	$^{20}_{10}Ne$	90.92	20.179	10	20	10	10	10
	$^{21}_{10}Ne$	0.257		10	21	10	11	10
	$^{22}_{10}Ne$	8.82		10	22	10	12	10
Chlorine	$^{35}_{17}Cl$	75.53	35.453	17	35	17	18	17
	$^{37}_{17}Cl$	24.47		17	37	17	20	17

averages that allow for the relative abundances of the naturally occurring isotopes of each element that would actually be found in a typical sample on the earth. As an example, we'll calculate the atomic weight of neon. The atomic weight of neon is the sum of the masses of the three isotopes, each multiplied by its relative abundance expressed as a decimal.

Atomic weight	Natural abundance
20.0	90.92% = 0.9092
21.0	0.257% = 0.00257
22.0	8.82% = 0.0882

$$\text{At. wt.} = 20.0\,(0.9092) + 21.0\,(0.00257) + 22.0\,(0.0882) = 20.2$$

EXERCISE 5-1 The element thallium, Tl, has atomic weight 204.4. It is a mixture of the naturally occurring isotopes with masses 203 and 205, respectively. Which isotope is more abundant? Can you calculate the relative abundance?

All three isotopes of neon have the same number (10) of protons in the nucleus. The isotope of mass 20 has 10 protons and 10 neutrons; the isotope of mass 21 has 10 protons and 11 neutrons; and the isotope of mass 22 has 10 protons and 12 neutrons. The **mass number** (A) *of an atom is the sum of the number of protons and the number of neutrons in the nucleus.* The two kinds of nuclear particles, proton and neutron, are given the general name **nucleon.** Thus *the mass number is equal to the number of nucleons.* The mass number is placed above and to the left of the symbol for the atom: $^{20}_{10}Ne$, $^{21}_{10}Ne$, and $^{22}_{10}Ne$ are the three isotopes of neon. (Remember that the lower number represents the atomic number.) An isotope may also be designated by its written name and mass: neon-20, neon-21, and neon-22.

To determine the number of neutrons in an atom, we subtract the atomic number (the number of protons) from the mass number (number of protons plus neutrons):

$$\text{Number of neutrons} = \text{Mass number } (A) - \text{Atomic number } (Z)$$

Thus, in $^{21}_{10}Ne$ there are $21 - 10 = 11$ neutrons;
in $^{40}_{20}Ca$ (ordinary calcium) there are $40 - 20 = 20$ neutrons;
in $^{2}_{1}H$ (the isotope of hydrogen called deuterium) there is $2 - 1 = 1$ neutron;
and in $^{3}_{1}H$ (tritium) there are $3 - 1 = 2$ neutrons.

The number of electrons in a neutral atom equals the atomic number (Z), the number of protons in the nucleus. In the neutral atom $^{22}_{10}Ne$ there are 10 electrons, in $^{40}_{20}Ca$ there are 20 electrons, and in each of $^{1}_{1}H$, $^{2}_{1}H$, and $^{3}_{1}H$ there is only 1 electron.

EXERCISE 5-2 Give the numbers of protons, neutrons, and electrons in the neutral atoms (a) $^{235}_{92}U$ and (b) $^{238}_{92}U$.

Chemists are concerned primarily with electrons, because it is primarily the configuration of the electrons that determines the chemical behavior of an atom. In a chemical reaction, the nucleus remains intact. Isotopes of an element, because they have the same number of electrons, differ very little chemically. Thus, a basic assumption of Dalton's atomic theory, that all atoms of an element are identical, although not literally true—we now know that isotopes exist—is accurate enough as far as chemical reactions are concerned.

5-6 STABLE AND RADIOACTIVE NUCLEI

We saw earlier in the chapter that certain atoms are radioactive, spontaneously disintegrating to produce alpha or beta particles, usually accompanied by gamma radiation. Later, a fourth type of radiation was discovered: the β^+ *particle,* or **positron.** The positron is a high-velocity particle equal in mass but opposite in charge to the electron, a positive counterpart to the electron. These particles come from the nucleus of the atom. Radioactivity is then a *nuclear* disintegration. The radioactive nucleus is unstable and thus spontaneously breaks apart.

Some combinations of neutrons and protons are

stable; other combinations are unstable and lead to radioactive atoms. A stable nucleus of low mass has approximately equal numbers of neutrons and protons and thus has a mass number about twice its atomic number. Examples are $^{4}_{2}He$ (helium) with 2 protons and 2 neutrons, $^{16}_{8}O$ (oxygen) with 8 protons and 8 neutrons, and $^{19}_{9}F$ (fluorine) with 9 protons and 10 neutrons. (The hydrogen isotope $^{1}_{1}H$ with 1 proton and no neutrons also is stable.)

As the atomic number increases, the stable nuclei tend to have more neutrons than protons, so the mass number is more than twice the atomic number. Examples are $^{197}_{79}Au$ (gold) with 79 protons and 118 neutrons, and $^{206}_{82}Pb$ (lead) with 82 protons and 124 neutrons. Somehow neutrons help bind the nucleus together against the repulsive forces between the positively charged protons. The forces that hold the nucleus together involve other particles in addition to the proton and the neutron, and are not completely understood. We shall not discuss nuclear forces here.

In Figure 5-12, each stable nucleus is represented by a dot. The region in which the stable nuclei occur is called the **belt of stability.** The belt of stability ends with bismuth, atomic number 83; all elements of higher atomic number undergo natural radioactive breakdown. Apparently, if there are more than 83 protons in a nucleus, that nucleus is unstable no matter how many neutrons are present. Nuclei that lie outside the belt of stability (whether naturally occurring or artificial, and whether above, below, or beyond it) are radioactive, and disintegrate spontaneously so as to achieve a stable nucleus. (They may pass through other unstable nuclear configurations on the way.)

A nucleus outside the belt of stability decomposes into two or more smaller pieces in a **nuclear reaction,** at the same time emitting large quantities of energy. If a nucleus lies *beyond* the belt of stability (with too many nucleons), stability can be achieved by the emission of an alpha particle, $^{4}_{2}He$. An example of this process is the decay of polonium-212:

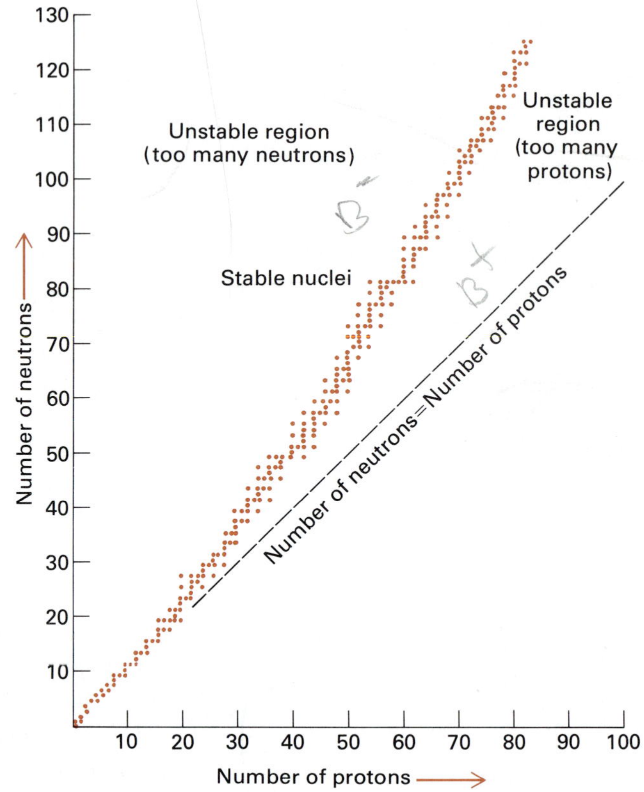

FIGURE 5-12

Stable nuclei.

$$^{212}_{84}Po \longrightarrow \, ^{208}_{82}Pb + \, ^{4}_{2}He^{2+} \text{ (}\alpha\text{ particle)} + 2e^-$$

The unstable nucleus of polonium ejects an alpha (α) particle, thereby achieving a stable nucleus of lead-208. The total nuclear charge, represented by the atomic numbers, has been conserved: $84 = 82 + 2$. The total number of nucleons (protons plus neutrons), represented by the mass numbers, has also been conserved: $212 = 208 + 4$. The chemical identity of the atoms and the number of atoms has *not* been conserved: a single Po atom has been transformed into the two atoms Pb and He. *In a nuclear reaction, charge and mass number (total number of nucleons) are conserved.* If we consider the numbers of electrons involved in the reaction above, we see that the polonium originally has 84, and the neutral lead atom needs only 82, so there are two electrons available to combine with the alpha particle (a helium nucleus) to form helium atoms. Gaseous helium is a product of radioactive disintegration by alpha emission.

EXERCISE 5-3 Naturally occurring radium ($^{226}_{88}Ra$) disintegrates to radon (Rn) with the emission of an α particle. Write a balanced equation for the disintegration of $^{226}_{88}Ra$.

A nucleus that lies *above* the belt of stability (with too many neutrons for the number of protons) can achieve stability by the process of β^- decay (the ejection of a β^- particle—that is, a high-energy electron). For example,

$$^{14}_{6}C \longrightarrow \, ^{14}_{7}N + \, ^{0}_{-1}e \; (\beta^- \text{ particle})$$

This electron is from the *nucleus,* not from those surrounding the nucleus. The radioactive isotope carbon-14 lies above the belt of stability; it achieves a stable nucleus (that of nitrogen-14) by ejecting a β^- particle from the nucleus and thereby converting a neutron to a proton. Charge is conserved: $6 = 7 - 1$. Mass number also is conserved: $14 = 14 + 0$.

EXERCISE 5-4 A radioactive isotope of nitrogen, $^{16}_{7}N$, disintegrates to oxygen with the emission of a β^- particle. Write a balanced equation for the disintegration of $^{16}_{7}N$.

If a nucleus lies *below* the belt of stability, with too many protons for the number of neutrons, it can achieve stability by gaining neutrons or by losing protons. In β^+ (positron) emission, a proton is lost and a neutron gained simultaneously by conversion of a proton to a neutron. For example,

$$^{11}_{6}C \longrightarrow \, ^{11}_{5}B + \, ^{0}_{+1}e \; (\beta^+ \text{ particle})$$

The radioactive isotope carbon-11 (with 6 protons and 5 neutrons) lies below the belt of stability. It achieves a stable nucleus by ejecting a positron, producing a stable isotope boron-11 (with 5 protons and 6 neutrons). The $^{11}_{6}C$ has lost a proton and, simultaneously, gained a neutron. Again, mass number and charge are conserved in the reaction.

5-7 RADIOACTIVE ISOTOPES AND HALF-LIFE

Radioactive nuclei—whether naturally occurring or artificially produced—have definite, measurable lifetimes. Some exist for only a small fraction of a second before undergoing reaction; others last for thousands, millions, or even billions of years. Most of the naturally occurring radioactive elements (such as uranium, radium, polonium, and thorium) are almost stable—that is, they have long lifetimes. This accounts for their continued existence as natural radioactive substances. The shorter-lived radioisotopes formed with the earth have now all but disappeared.

The rate at which alpha, beta, and gamma radiation is emitted from a particular radioactive isotope depends only on the amount of radioactive material (that is, on the number of such radioactive nuclei in the sample). The rate is independent of any physical or chemical changes the material undergoes. Carbon-14 ($^{14}_{6}C$), for example, decays at the same rate whether it is incorporated in human muscle, in ethyl alcohol, or in carbon dioxide gas. In a definite interval of time, characteristic of the particular nucleus, a constant fraction of the nuclei decompose. That length of time during which one-half of the nuclei disintegrate is known as the **half-life** ($t_{1/2}$) for that nucleus. For example, the half-life of the isotope thorium-90 is 24 days. After 24 days, one-half of the nuclei in *any* sample of thorium-90 will have disintegrated, no matter how large a sample we began with. There is no way to know when any *one* nucleus will disintegrate, but one-half of them will have disintegrated after 24 days.

If there are originally 200 atoms of thorium-90, then after 24 days (or one half-life), one-half will be lost and one-half ($\frac{1}{2} \times 200$ atoms $=$ 100 atoms) will remain. And after 24 days more (a total of two half-lives), one-half of those 100 atoms of thorium-90 will be lost, and only 50 atoms will remain. How many atoms of thorium-90 will remain after another 24 days (3 half-lives)? The change in the size of a sample with time is shown in Table 5-3 for a sample that initially contains 1250 atoms; this change is shown graphically in Figure 5-13.

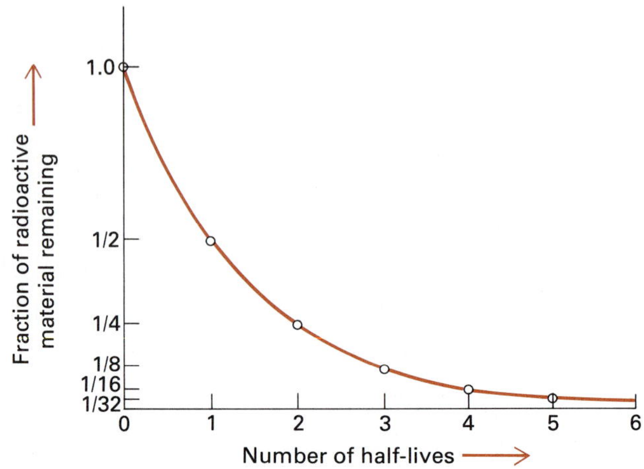

FIGURE 5-13
Change in the size of a radioactive sample with time.

TABLE 5-3
The change in size of a radioactive sample with time

Number of half-lives	Sample size (atoms)
0	1250
1	625
2	313
3	156
4	78
5	39
6	20
7	10
8	5
9	2
10	1

EXERCISE 5-5 Iodine-131 (^{131}I), widely used in the treatment of cancer of the thyroid, has a half-life of 8 days. If at some time you had a 100 mg sample of ^{131}I, how much ^{131}I would you have after 40 days? What are the advantages of this isotope in the treatment of cancer over one with a short half-life (say, a few seconds) or one with a long half-life (say, 100 years)?

The tremendous variation in the half-lives of radioactive nuclei is seen in the stepwise natural decay of uranium-238 to produce a stable isotope of lead, lead-206. In the first step of the series (Figure 5-14),

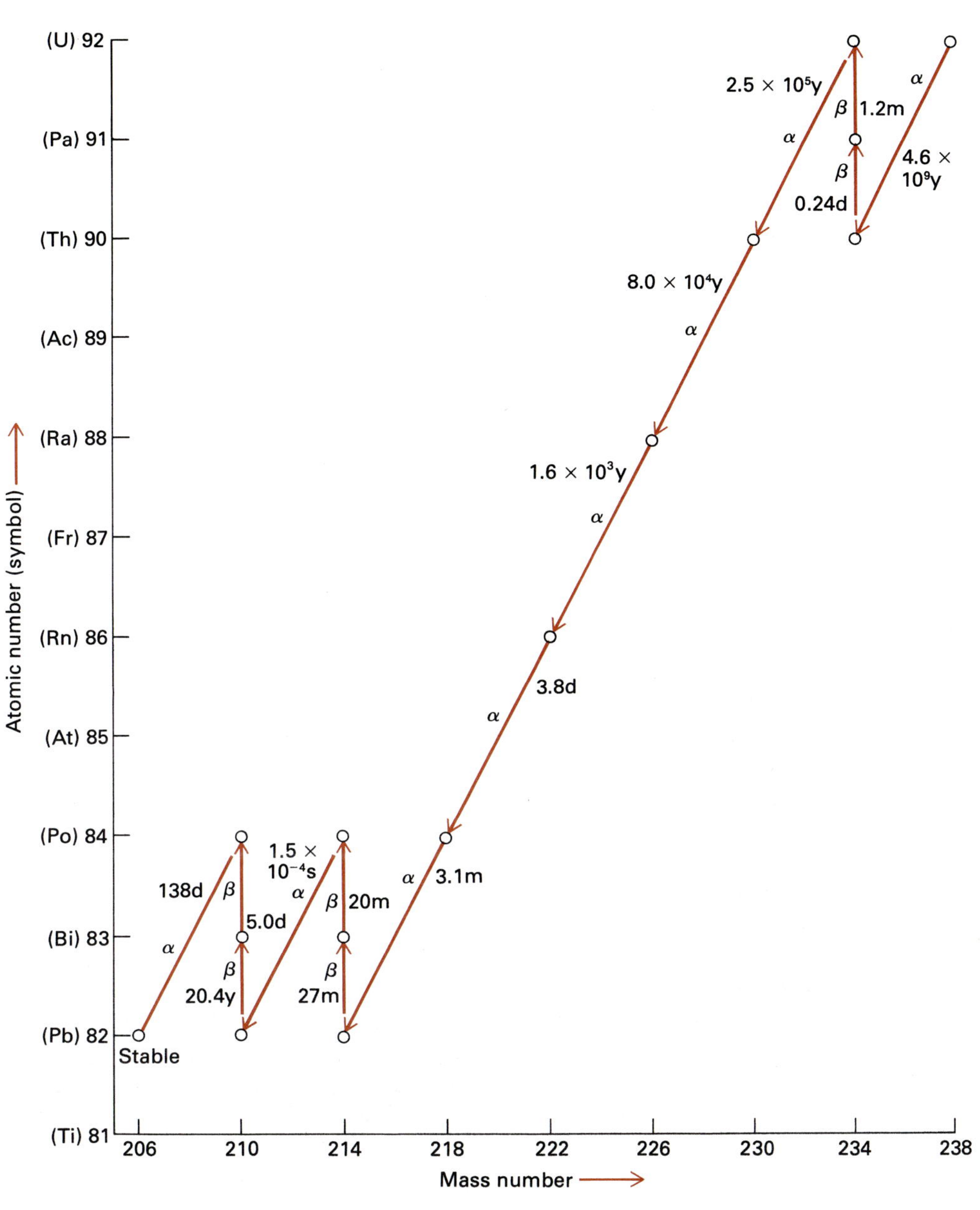

FIGURE 5-14

The radioactive decay of uranium-238. The half-life for each nucleus is given; s = seconds, m = minutes, d = days, y = years.

^{238}U ejects an alpha particle to produce ^{234}Th. The half-life of ^{238}U is 4.6×10^9 years.

$$^{238}_{92}\text{U} \xrightarrow{t_{1/2} = 4.6 \times 10^9 \text{ yr}} {}^{234}_{90}\text{Th} + {}^{4}_{2}\text{He}$$

Later in the series, there is a half-life of only 1.5×10^{-4} seconds. That is the half-life of polonium-214, which decays by alpha emission to yield lead-210:

$$^{214}_{84}\text{Po} \xrightarrow{t_{1/2} = 1.5 \times 10^{-4} \text{ sec}} {}^{210}_{82}\text{Pb} + {}^{4}_{2}\text{He}$$

The series ends with the production of lead, because ^{206}Pb is not radioactive. In this decay series, some of the isotopes eject alpha particles and others beta particles; each type of emission in most cases is accompanied by gamma radiation.

5-8 DATING BY RADIOACTIVITY, AND ARTIFICIAL ISOTOPES

How can we measure the age of the earth, a moon rock, or an old scroll? An ingenious and reliable method of establishing the age of old objects depends on the presence of naturally occurring radioactivity. It is based on the fact that the rate of decay depends only on the particular radioactive nucleus and is independent of its history. For example, we can use the known half-life of uranium to determine the age of a rock. We know that uranium-238 has a half-life of 4.6×10^9 years, and that the final product of the natural decay series is lead-206. The half-life of the ^{238}U is by far the longest half-life in the series. So, as an approximation we can say that of any sample of ^{238}U, in about 5 billion years one-half will have been transformed to ^{206}Pb.

A pure crystal of a uranium ore (say, pitchblende, U_2O_5) always contains some lead. If we assume that *all* the lead comes from the radioactive decay of ^{238}U (and this assumption can be tested by means that we shall not consider here), we can determine the age of the sample by measuring the relative amounts of uranium and lead. When, for example, there are equal numbers of lead and uranium atoms, then one-half of the original uranium has decayed, and the sample is about 5 billion years old. This method of dating shows that the oldest known terrestrial rock samples are about 3.4 billion years old. The oldest meteorites, which may have been formed at the same time as the earth, are about 4.5 billion years old. These measurements lead to the conclusion that the earth is not much less than 5 billion years old.

Similar estimates of the earth's age are obtained by measurements based on the radioactivity of an isotope of potassium, ^{40}K, which has a half-life of 1.26×10^9 years and decays to argon. Other measurements are based on the radioactivity of an isotope of rubidium ^{87}Rb, which has a half-life of 4.8×10^{10} years and decays to produce strontium. Potassium–argon dating of moon rocks brought back by Apollo astronauts shows the oldest to be 4.2 billion years old.

A radioactive isotope of carbon, ^{14}C, is used to date objects that are thousands (rather than billions) of years old. Carbon-14 has a half-life of 5570 years, decaying by β^- emission to yield nitrogen:

$$^{14}_{6}\text{C} \longrightarrow {}^{14}_{7}\text{N} + {}^{0}_{-1}\text{e} \ (\beta^- \text{ particle})$$

Carbon dioxide gas formed in the atmosphere thus contains a certain fraction of ^{14}C. The concentration of ^{14}C has remained approximately constant over the ages at one part ^{14}C to 10^{10} parts stable carbon (^{12}C), because new atoms of ^{14}C are produced at about the same rate that others decay.

Plants, in the process of photosynthesis, absorb CO_2 from the atmosphere, and therefore have the same ratio of ^{14}C to stable ^{12}C atoms as that present in the atmosphere. Animals obtain their carbon from plants so they contain the same fraction of ^{14}C. This ratio remains constant as long as the organism lives, but from the moment it dies no new carbon is absorbed. The carbon-14 already in the organism decays at its characteristic rate. So an old block of wood, a piece of paper, or a piece of leather will be less radioactive than a fresh sample.

The amount of radioactivity in the sample can be measured using a device such as a Geiger counter. Such a counter detects radioactive emissions by the positive and negative ions they produce in a gas. It indicates them with clicks, lights, or a print-out of the number per time interval. The carbon in a fresh sample of wood gives a count of 15.3 disintegrations per gram per minute. If a sample of carbon from a piece of old cloth is found to have an activity of 7.7 counts per gram per minute, the activity (and thus also the ^{14}C concentration) has decayed to one-half its original value. We thus know that the plant from which the cloth was made died about 5600 years ago. (How old is a sample that has an activity of about 1.9 counts per gram per minute?) Because it measures age in thousands of years, ^{14}C plays the same role in the dating of human history that ^{238}U plays in the dating of the earth's history. In Figure 5-15, the dates obtained for old objects by ^{14}C analysis agree remarkably well with the dates estimated from historical evidence.

Some radioactive isotopes, as we have seen, occur naturally. Others are artificially produced by bombarding stable nuclei with high-energy particles. The earliest induced nuclear reactions were brought about by bombarding atoms with alpha particles from naturally occurring radioactive nuclei. The first such reaction was performed by Rutherford in 1919. By bombarding nitrogen with alpha particles, he produced a heavy isotope of oxygen:

$$^{14}_{7}N + ^{4}_{2}He \longrightarrow ^{17}_{8}O + ^{1}_{1}H$$

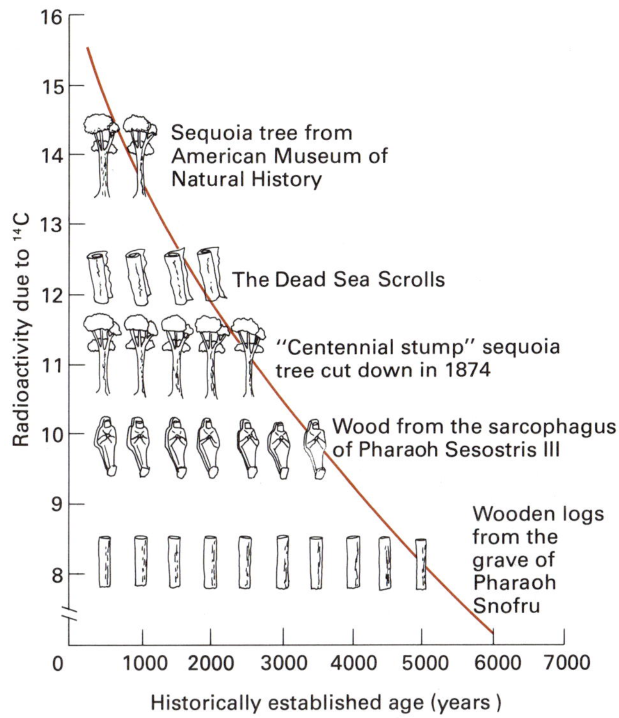

FIGURE 5-15

Dating by the radioactivity of carbon-14. Each symbol represents 500 years of age as determined by historical evidence. The curve is a plot of the decay with time of radioactivity due to ^{14}C. (The same kind of plot as shown in Figure 5-13.) [From G. Gamow, *The Atom and Its Nucleus,* (Englewood Cliffs, N.J.: Prentice-Hall, 1961), with permission of W. F. Libby.]

In this reaction, the nitrogen atom first captures an alpha particle and then ejects a proton.

Thousands of different radioactive isotopes have now been produced by bombarding stable nuclei with charged particles (protons or electrons) that have been accelerated to high velocities. Radioactive isotopes of almost every element have been prepared, and we shall consider a few of their many uses shortly.

All of the transuranium elements (those with atomic number greater than 92) are artificially produced; none occurs naturally. The first of the transuranium elements, neptunium, is produced by bombarding uranium with neutrons:

$$^{238}_{92}\mathrm{U} + ^{1}_{0}\mathrm{n} \longrightarrow ^{239}_{92}\mathrm{U}$$

Uranium-239 is unstable and decays by emitting a β^- particle to yield neptunium, of atomic number 93:

$$^{239}_{92}\mathrm{U} \longrightarrow ^{239}_{93}\mathrm{Np} + ^{0}_{-1}\mathrm{e}.$$

Neptunium also decays by emitting a β^- particle to produce the second transuranium element, plutonium, with atomic number 94:

$$^{239}_{93}\mathrm{Np} \longrightarrow ^{239}_{94}\mathrm{Pu} + ^{0}_{-1}\mathrm{e}.$$

This isotope of plutonium is relatively stable ($t_{1/2} = 24{,}390$ years) and is used as a fuel in nuclear reactors.

EXERCISE 5-6 Element 98, californium ($^{246}_{98}\mathrm{Cf}$) is made by bombarding $^{238}_{92}\mathrm{U}$ with carbon nuclei, $^{12}_{6}\mathrm{C}$. Neutrons are also produced. Write a balanced equation for the reaction.

5-9 USES OF RADIOACTIVE ISOTOPES

Radioactive isotopes are useful in a variety of ways, far too many to enumerate here. Chemists use them primarily as **tracers,** to establish the mechanism, or pathway, of a reaction; we shall see examples of this use in later chapters. The characteristic that permits radioactive isotopes to be used as tracers is their emission of easily detected radiation. No matter what chemical compound the isotope is incorporated in, or what reaction it undergoes, the radiation continues. However, a radioactive isotope reacts chemically like other atoms of the same element. Thus, if we want, for example, to study how quickly sodium ingested as table salt (NaCl) is circulated through the body, we can prepare salt that contains a **trace** (a small amount) of radioactive sodium ($^{24}\mathrm{Na}$) along with the normal isotope ($^{23}\mathrm{Na}$). Such a compound containing a small amount of radioactive material is called a **tagged** compound. The $^{24}\mathrm{Na}$ will circulate through the body along with the ordinary $^{23}\mathrm{Na}$, and its location and progress can be easily detected from outside the body with a radiation counter. Tagged calcium salts have been used to study the assimilation of calcium in the bones and teeth. Sugar tagged with radioactive carbon is useful for study of sugar metabolism by diabetics.

Tracers have been used in agriculture to determine the most effective method of applying fertilizer, the best time to apply particular minerals, and the correct amounts to add, thus improving the efficiency of fertilizer application. A specific example is the use of zinc to increase the productivity of fruit trees. By using a tagged zinc-containing compound, it was shown that zinc absorption through the leaves is much more efficient than through the roots, and that zinc is quickly absorbed when the branches of trees are sprayed during below-freezing weather. This method of supplying zinc to fruit trees is now widely used.

Industrial applications of radioactive tracers include their use in locating leaks or blockages in underground pipes, to avoid unnecessary digging. Such tracers are also used to determine wear of machine tools and parts. For example, if piston rings in a car engine are made radioactive, wear on the rings can be determined after a very short running time by measuring the amount of radioactive material in the lubricating oil. The effect of different oils and of different metal combinations in the piston rings can then be evaluated.

The biological effect of radioactivity, the damage to or destruction of tissue, also leads to many uses of radioactive isotopes in medicine and agriculture. Alpha and beta particles and gamma rays destroy tissue (whether diseased or not) by ionizing some molecules:

$$\alpha + \text{molecule} \longrightarrow \alpha + \text{molecule}^+ + e^-$$

The alpha particle knocks an electron out of the molecule, leaving the molecule positively charged and causing it to decompose. This property of radiation can be used to destroy cancerous tissue, but it also produces the undesirable side effects of radiation exposure.

The standard unit used to measure radioactivity is the **curie.** A sample of one curie undergoes 3.7×10^{10} disintegrations per minute, the approximate rate of decay of 1 g of radium. (A cubic centimeter of air naturally has a radioactivity of about 10^{-16} curie, mostly from radon.)

The biological effect of a given dose of radiation depends on the energy and kind of radiation involved, the organism it affects, and the particular type of tissue exposed. Thus, other units are needed to measure the amount of radiation required to produce a given effect in matter. The **rad** (*r*adiation *a*bsorbed *d*ose) measures *absorbed* radiation. One rad is the amount of radiation that delivers 1×10^{-5} J of energy to 1 g of matter. A single dose of several hundred rads delivered to a human body causes death in a few weeks. One rad of alpha particles causes approximately ten times as much damage as one rad of beta particles or gamma radiation. A dose in **rems** (*r*adiation *e*quivalent *m*an) is equal to the absorbed dose in rads times a factor that takes into account the relative biological effectiveness of the radiation.

Any dose of radiation destroys *some* tissue, though a dose of less than about 25 rem delivered over a short time produces no discomfort. A single dose of 100 to 200 rems causes nausea and a marked decrease in the number of white blood cells. A dose of 500 rems would within one month cause the death of one-half the people exposed. Exposure to even small doses of radiation may increase the chance of cancer.

There is no way of avoiding radiations entirely. Naturally occurring radioactive materials are all around us—in soil, food, and water. This unavoidable radiation is called **background radiation.** The average individual exposure of a United States citizen due to background radiation is from 70 to 200 millirems per year. The *additional* average exposure inside a stone house (due to natural radioactive materials) is about 100 millirems per year. Federal radiation standards at present allow an average individual maximum dose of 170 millirems per year from all sources, exclusive of exposure from medical treatments and background radiation. (These "allowable" doses are still the subject of much controversy.) An average chest X-ray dose is 200 millirems, and a gastro-intestinal tract examination dose is 22,000 millirems.

Nuclear power plant operations are required to keep their releases of radioactivity "as low as practicable," which means that they must limit releases to a small fraction of the allowed 170 millirems. In 1970, the actual exposure to radiation of persons living at a plant boundary was only about 5 millirems per year, on the average, for all the nuclear power plants then in operation. A malfunction at a nuclear power plant in Three Mile Island, Pennsylvania, during the spring of 1979 resulted in the release of some radioactivity to the environment. An estimated maximum dosage of 70 to 80 millirems would have been received by a person remaining at the fence during the entire incident—although of course no one did this! Plant workers, of course, received higher dosages; these figures have not yet been made public.

Radiation damage is much greater to the rapidly multiplying cells of fetuses, children, and cancerous growths than it is to the normal cells of an adult. Therefore, radiation can be used for the destruction of cancerous tissue in doses that do not destroy many normal cells. The long-lived radioactive isotope of cobalt ^{60}Co (with a half-life of 5.3 years) produces gamma radiation that can be aimed directly at cancerous tissue, delivering a large dose of radiation to it for short periods of time.

Many short-lived radioactive isotopes suitable for internal use have been prepared. Iodine-131 (with a half-life of 8 days) has been probably the most widely and successfully used of these. It can be administered in a glass of water, and its relatively short half-life means that after a short time the radiation essentially stops. The radioactive iodine is selectively absorbed in the thyroid gland, particularly in cancerous tissue in the thyroid and in cancer offshoots (called metastases) of the thyroid. In small doses, it thus acts as a tracer to locate cancers for further treatment with larger therapeutic doses of ^{131}I or for surgical removal.

When plants or animals are exposed to light doses of radiation, one result is an increased rate of production of mutants—individuals that differ from their parents because of a change in gene. Most, but not all, mutations are harmful. Many mutants survive only briefly. Some desirable mutants have been produced by irradiating plants, mainly with X rays. Among these are an early-maturing, disease-resistant bush-type pea-bean; disease-resistant oats; a tough-hulled, disease-resistant peanut; and a wide variety of new ornamental flowers.

Radiation is also the basis of a promising method of insect control. Insects are reared, made sterile by exposure to radiation, and then released. If a sterile male mates with a normal female, she lays sterile eggs. This method has had dramatic success in the control of the screwworm fly in the United States. The female screwworm fly lays eggs on the edges of an open cut on livestock. The larvae burrow, tearing away the flesh and often killing untreated animals. Damage in the southeastern United States from this insect at one time amounted to 15–25 million dollars annually. After successful tests on the island of Curaçao, where the screwworm was eliminated in one year, a fly-breeding plant was set up in Florida. Flies were raised to the pupal stage (consuming more than 40 tons of ground horse and whale meat). They were exposed to gamma radiation from a ^{60}Co source to cause sexual sterilization, and were then allowed to mature. During the attack, 50 million sterile flies per square mile were released per week from airplanes over Florida, Georgia, and Alabama. Within two years, the screwworm was eliminated from the southeast. Billions of sterile adult flies have now been released in the southwestern United States and Mexico in a project to eliminate the screwworm from that area. Other potential targets of this method of insect control are the fruit fly, the boll weevil, the gypsy and codling moths, the tsetse fly, and the anopheles mosquito.

This small discussion gives only an inkling of the uses to which radioactive isotopes can be put. In all cases, however, the advantages to be gained must be weighed against the damage that may result from *any* exposure to radiation.

5-10 NUCLEAR ENERGY AND BINDING ENERGY

When a radioactive nucleus disintegrates, a great deal of energy is released. This energy takes the form of gamma radiation and kinetic energy of the products. Kinetic energy equals heat, and in a nuclear power plant this heat is used to generate electricity. Where does this energy come from? It comes from the conversion of tiny amounts of mass into energy according to the Einstein formula: $E = mc^2$. According to this equation, the energy (E) contained in mass (m) is equal to the product of m and the square of the velocity of light (c). If the mass is expressed in kilograms, and the velocity of light in meters per second ($c = 3 \times 10^8$ m/sec), the energy is given in joules. If one gram of matter were converted to energy, the energy produced would be

$$\begin{aligned} E = mc^2 &= 1 \times 10^{-3}\,\text{kg} \times (3 \times 10^8\,\text{m/sec})^2 \\ &= 9 \times 10^{13}\,\text{kg m/sec}^2 \\ &= 9 \times 10^{13}\,\text{J} = 9 \times 10^{10}\,\text{kJ} \end{aligned}$$

A joule is a very small unit of energy, but 9×10^{10} kJ is enough energy, if converted to food, to feed 10 million people for one day. This is the same energy as that contained in about 25,000 tons of TNT. One pound of any substance, if it could be completely converted into energy, would yield as much as is produced by burning 15 million tons of coal. This amount of energy could keep a home furnace operating for tens of thousands of years, or could fuel a car for 3 billion miles. In Chapter 10 we shall see that the energy absorbed or released in a chemical reaction corresponds to an infinitesimal change in mass, a change much too small to detect. (To break all the bonds in a mole of H_2 molecules, for example, requires only about 400 kJ.)

Mass is converted to energy in the spontaneous decay of the naturally occurring radioactive isotope of carbon, ^{14}C

$$^{14}_{6}\text{C} \longrightarrow {}^{14}_{7}\text{N}^+ + {}^{0}_{-1}\text{e}^- \text{ (beta particle)}$$

(The ^{14}C on the left side of the equation is a neutral carbon atom. The ^{14}N on the right side is first a singly charged N^+ ion. This ion together with the β^- particle is equivalent to a neutral nitrogen atom.)

$$\begin{aligned} \text{Mass of } ^{14}\text{C atom} &= 14.003242\text{ g/mole} \\ \text{Mass of } ^{14}\text{N atom} &= 14.003074\text{ g/mole} \\ \hline \text{Mass lost in reaction} &= 0.000168\text{ g/mole} \\ &= 1.68 \times 10^{-7}\text{ kg/mole} \\ \text{Energy produced} = mc^2 &= 1.68 \times 10^{-7}\text{ kg/mole} \\ &\quad \times (3 \times 10^8\text{ m/sec})^2 \\ &= 1.5 \times 10^{10}\text{ J/mole} \\ &= 1.5 \times 10^{7}\text{ kJ/mole} \end{aligned}$$

When one mole of ^{14}C decays according to the above reaction, 1.5×10^7 kJ/mole energy is released; 0.000168 g of matter is converted to energy. The energy appears as kinetic energy of the products (the nitrogen atom and the β^- particle). In this particular reaction, no gamma radiation is emitted.

Energy is converted to mass in Rutherford's original nuclear reaction:

$$^{14}_{7}\text{N} + {}^{4}_{2}\text{He} \longrightarrow {}^{17}_{8}\text{O} + {}^{1}_{1}\text{H}$$

$$\begin{aligned} \text{Mass of } ^{14}_{7}\text{N atoms} &= 14.003074\text{ g/mole} \\ \text{Mass of } ^{4}_{2}\text{He atom} &= 4.002603\text{ g/mole} \\ \hline \text{Total mass of reactants} &= 18.005677\text{ g/mole} \\ \\ \text{Mass of } ^{17}_{8}\text{O atom} &= 16.999133\text{ g/mole} \\ \text{Mass of } ^{1}_{1}\text{H atom} &= 1.007825\text{ g/mole} \\ \hline \text{Total mass of products} &= 18.006958\text{ g/mole} \end{aligned}$$

$$\begin{aligned} \text{Mass } \textit{gained} \text{ in reaction} &= 18.006958 - 18.005677 \\ &= 0.001281\text{ g/mole} \\ &= 1.28 \times 10^{-6}\text{ kg/mole} \end{aligned}$$

$$\begin{aligned} \text{Energy } \textit{consumed} \text{ in reaction} &= mc^2 \\ &= 1.28 \times 10^{-6}\text{ kg/mole} \\ &\quad \times (3 \times 10^8\text{ cm/sec})^2 \\ &= 1.15 \times 10^{11}\text{ J/mole} \\ &= 1.15 \times 10^{8}\text{ kJ/mole} \end{aligned}$$

When 1 mole of nitrogen is converted to oxygen according to the above reaction, 1.15×10^8 kJ of energy is consumed and is converted to 0.001281 g of matter. This energy comes from the kinetic energy of the alpha particle, which is greater than that of the ejected proton.

The masses of the different isotopes of the elements (in atomic mass units, or amu) are close to whole numbers. (We would expect integers because the atom is composed of neutrons and protons, both of atomic mass 1 amu, and of electrons, which are very light. One atomic mass unit = 1 amu = 1/12 the mass of a carbon-12 atom.) For example, the mass of an oxygen-16 atom is 15.99492 amu. But the sum of the masses of the 8 neutrons, 8 protons, and 8 electrons that constitute the ^{16}O atom is greater: 16.13192 amu. The ^{16}O atom is too light by 0.13700 amu (16.13192 − 15.99492). This mass is lost in the form of energy when an oxygen-16 nucleus is formed from neutrons and protons. The amount of energy produced when one mole of ^{16}O is formed in this way is

$$E = mc^2 = 0.137 \times 10^{-3}\,\frac{\text{kg}}{\text{mole}} \times \left(3.0 \times 10^8\,\frac{\text{m}}{\text{sec}^2}\right)^2$$
$$= 1.23 \times 10^{13}\ \text{J}$$

This same amount of energy would be required to decompose a mole of ^{16}O nuclei into protons and neutrons. The energy required to decompose a nucleus into protons and neutrons is called the **binding energy.**

Figure 5-16 graphs the binding energy per nucleon against mass number. The most stable nuclei, those with the greatest binding energy, are those of medium mass. Light and heavy nuclei, which have lower binding energies, are less stable. In the following section, we see that large amounts of energy are released when two light nuclei fuse to yield a nucleus of medium mass; and also when a heavy nucleus splits to yield two nuclei of medium mass.

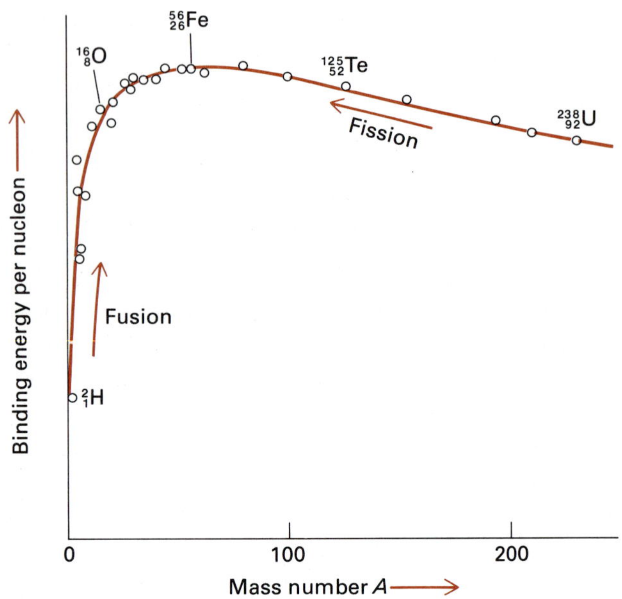

FIGURE 5-16
Binding energy per nucleon as a function of mass number.

EXERCISE 5-7 The deficiency in mass that results from the formation of ^{4}He atoms is 0.03040 g/mole. Calculate the energy released when one mole of ^{4}He atoms is formed from neutrons, protons, and electrons. (Assume that *all* the energy results from the formation of the nucleus. The contribution due to the addition of electrons is negligible.)

5-11
NUCLEAR FISSION AND NUCLEAR FUSION

> Atom-Powered World Absurd, Scientists Told—Lord Rutherford Scoffs at Theory of Harnessing Energy in Laboratories.
>
> —Headline of front-page story, *New York Herald Tribune,* September 12, 1933

The conversion of mass to energy represents a source of energy for use on the earth. As we saw in Section 5-10, very light nuclei and very heavy nuclei are less stable than those of medium mass. Thus, energy is released both when a heavy atom such as uranium is split into two atoms of medium weight and also when very light atoms such as hydrogen are combined to form a heavier atom. In a **fission reaction,** a large nucleus is split, and mass is converted to energy. In a **fusion reaction,** two small nuclei are combined, and again mass is converted to energy.

A typical fission reaction results when certain uranium atoms are bombarded with neutrons:

$$^{235}_{92}U + ^{1}_{0}n \longrightarrow ^{236}_{92}U \longrightarrow ^{90}_{38}Sr + ^{143}_{54}Xe + 3^{1}_{0}n + \gamma$$

The fissionable ^{235}U nucleus first absorbs a neutron, and the atom then splits into the two medium-weight atoms ^{90}Sr and ^{143}Xe, emitting three neutrons and gamma radiation in the process. About 1.9×10^{10} kJ of energy is released in this reaction for each mole of uranium that reacts. This corresponds to about 3.6×10^{10} kJ for a pound of uranium. In this reaction we have invested only one neutron but have produced three. Figure 5-17 illustrates a different way in which ^{235}U can split, in this case to produce tellurium-137, zirconium-97, and two neutrons. These are only two of many ways the ^{235}U nucleus splits in nuclear reactions. Hundreds of isotopes of more than 30 different elements have been found in the products of fission reactions, and most of them are radioactive. On the average, each fission of ^{235}U produces 2.4 neutrons.

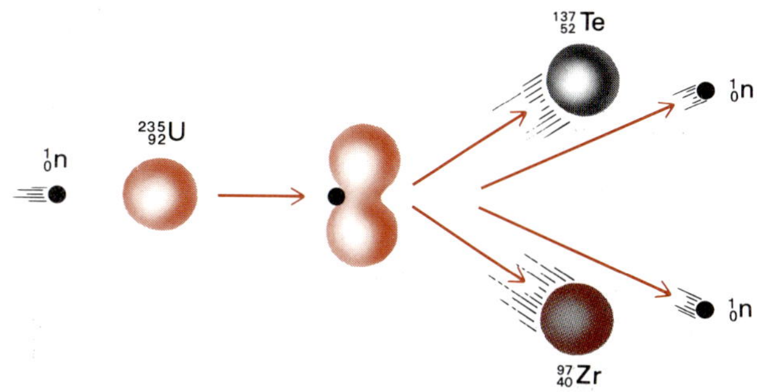

FIGURE 5-17

Uranium-235 is split into tellurium-137, zirconium-97, and two neutrons.

If some of the neutrons produced in fission reactions are absorbed by other ^{235}U nuclei, a **chain reaction** (self-propagating reaction) is initiated as shown schematically in Figure 5-18. In a nuclear reactor, energy is released slowly, under carefully controlled conditions; in an atomic bomb, there is an uncontrolled release of energy. The energy released by the fission of one pound of uranium is equivalent to that obtained by burning over 2 million pounds of coal. In the nuclear reactor diagrammed in Figure 5-19, heat is absorbed from the reactor by large quantities of cooling water that reaches a temperature of about 300°C under pressure. This water then converts a second system of water to steam that powers electricity-generating turbines.

Radioactive waste is produced when the spent fuel from a nuclear (fission) reactor is recycled. The spent fuel is dissolved in acid, and the usable uranium and plutonium separated out. What remains is a rusty-brown liquid that boils for years on heat generated by the remaining radioactive waste substances.

Some of the isotopes in the waste decay quickly to harmless levels. Zirconium-95, for example, has a half-life of only 65 days. After about a year, its concentration is reduced to less than 1/32 of its original

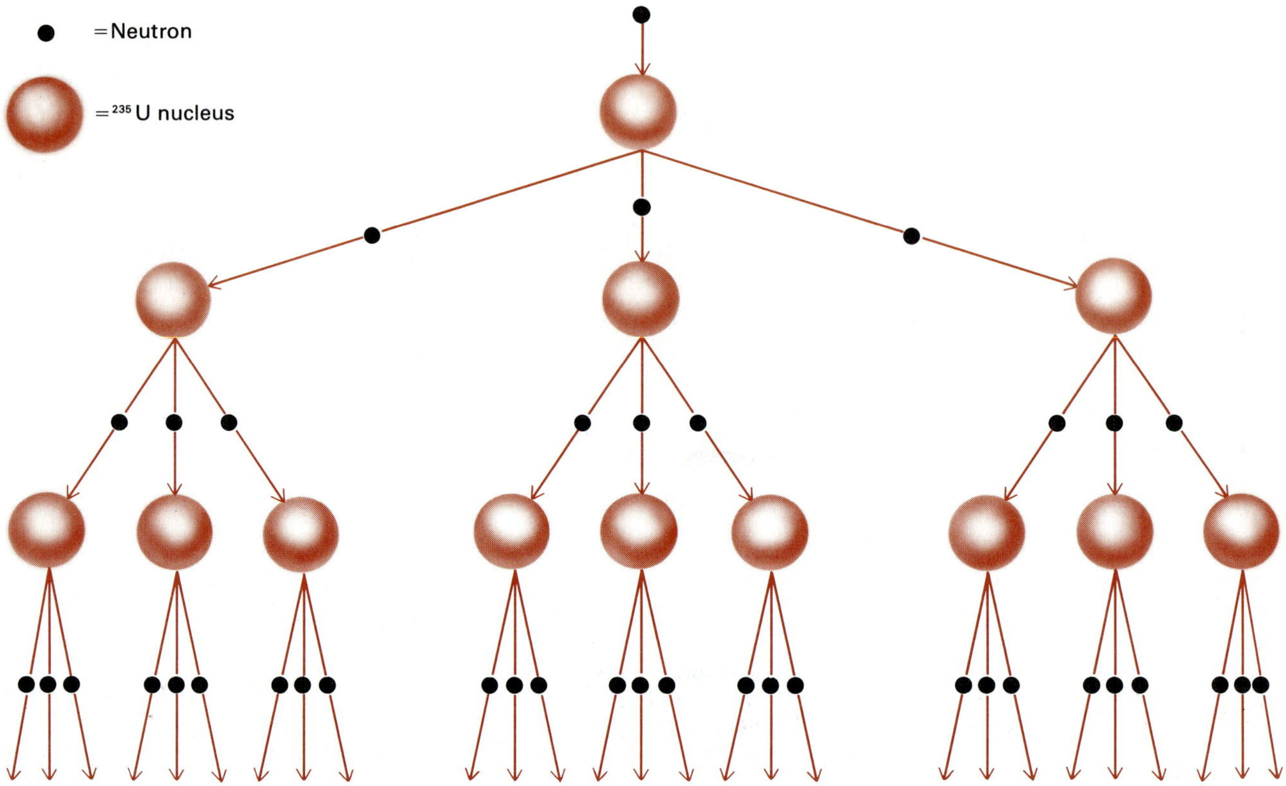

FIGURE 5-18

Schematic diagram of a chain reaction. Each red circle represents a ^{235}U nucleus; each black circle represents a neutron.

level. But other isotopes are very long-lived. Strontium-90 and cesium-137 have half-lives of 30 years, and there is so much of them present in the waste material that they must be isolated from the environment for 600 to 1000 years. Cesium-137 emits highly penetrating gamma radiation, but if ingested it is excreted from the body within a few weeks. Strontium-90 also emits highly penetrating gamma radiation, but because it is chemically similar to calcium, it is incorporated into the bones and continues to bombard the surrounding tissue with radiation for years. Strontium-90 is therefore, the most deadly of the radioactive waste products.

The waste products must be contained, sealed away, for 250,000 years. (Remember that Neanderthal man appeared only 75,000 years ago.) In the United States, some wastes are being temporarily stored buried in the ground in tanks that have useful lives of only about 30 years; others have been solidified into radioactive salt cakes, granular substances, or a glass or ceramic-like

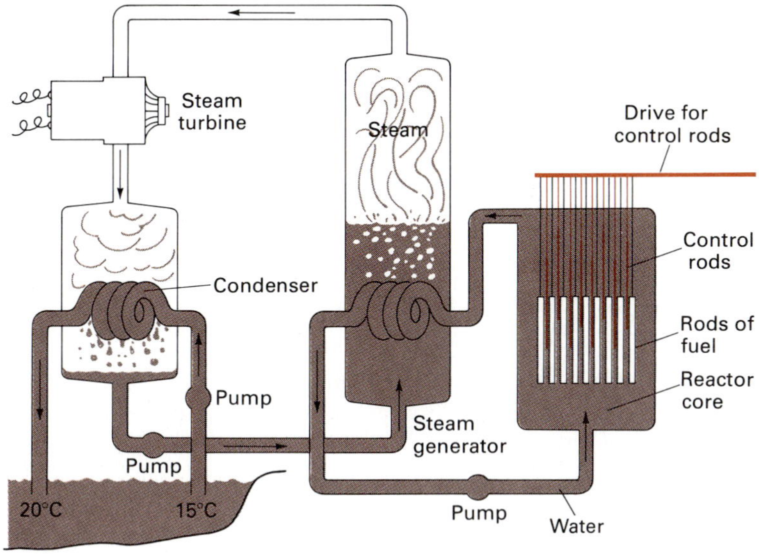

FIGURE 5-19

Schematic diagram of a nuclear fission reactor. The control rods (of cadmium or boron), which absorb neutrons, can be raised or lowered to control the rate of the reaction. The water that circulates through the core is in a closed system to prevent escape of radioactive products. The steam is condensed by cooling water obtained from a river or lake. This cooling water is returned to the river or lake at a higher temperature and thus is a source of thermal pollution.

material, and sealed inside lead casks. Permanent solutions to the problem of waste disposal that have been proposed include shooting them off into outer space, depositing them on ice in Greenland or Antarctica, depositing them on the sea floor in areas where they would soon be buried in sand, transmuting them by neutron bombardment to shorter-lived radioactive isotopes or to harmless stable materials, and burying them in salt caverns or artificial caverns in bedrock.

Fusion reactors are a promising source of energy and are almost free of radioactive waste products, but it may be decades before they are put into operation. Nuclear fusion involves combining two or more light elements to form a heavier one. Per pound of reacting material, even more energy is produced in a fusion reaction than in a fission reaction. The tremendous quantity of energy our sun radiates into space results from the fusion of four protons into a helium nucleus and two positrons. The overall reaction can be expressed as

$$4\,{}^{1}_{1}\text{H} \longrightarrow {}^{4}_{2}\text{He} + 2\,{}^{0}_{1}\text{e}\,(\beta^{+}\text{ particle}) + \gamma$$

The energy released (by conversion of mass to energy)

is 2.4×10^9 kJ per mole of helium produced, or about 2.8×10^{11} kJ per pound of hydrogen atoms fused. This is almost ten times the amount of energy released in the fission of one pound of uranium. The sun burns 657 million tons of hydrogen each second, converting it into 653 million tons of helium. The missing 4 million tons of mass are radiated into space as γ rays and β^+ particles. There are hydrogen reserves in the sun to last 5 billion years.

Because of the tremendous amount of energy released and the great abundance of hydrogen on the earth, fusion reactions are a promising potential source of energy. The seas contain enough hydrogen atoms to provide fuel for millions of years if an economical method can be found to fuse their nuclei. The major obstacle in obtaining a successful fusion reaction is the requirement to maintain the sample for one second at a temperature of over 100 million kelvins. This temperature is required to provide the kinetic energy necessary to overcome the repulsive forces between the positively charged nuclei.

Other potential fuels for a fusion reaction are the isotopes of hydrogen, deuterium ^2_1H (which occurs only as a trace in nature) and tritium ^3_1H (which must be artificially produced in a nuclear reaction). The following are some fusion reactions involving deuterium and tritium:

$$^2_1H + ^2_1H \longrightarrow ^3_2He + ^1_0n + \gamma$$

(Energy released = 3.2×10^8 kJ per mole of ^3_2He formed)

$$^2_1H + ^2_1H \longrightarrow ^3_1H + ^1_1H + \gamma$$

(Energy released = 3.8×10^8 kJ per mole of ^3_1H formed)

$$^2_1H + ^3_1H \longrightarrow ^4_2He + ^1_0n + \gamma$$

(Energy released = 17×10^8 kJ per mole of ^4_2He formed)

The third reaction, which releases the greatest amount of energy, is that used in the hydrogen bomb. In this case, the required temperature is produced by an atomic (fission) reaction, which yields temperatures of hundreds of millions of kelvins just after detonation.

When the controlled release of energy by the fusion process is accomplished, this process will have several important advantages over the fission process. Among these advantages are the abundance of the fuel (hydrogen) and the relative absence of radioactive waste products. Both the United States and the Soviet Union have begun construction of test plants for nuclear fusion, and both hope by the mid-1980s to have controlled reactions in which as much energy is produced as is consumed. Controlled fusion reactions have been achieved on a small scale, using droplets of frozen deuterium as fuel and a laser or high-energy electron beam as fuel source.

KEY TERMS

Electric current is the movement of charged particles.

An *electron* is a negatively charged particle of mass 9.11×10^{-28} g found outside the nucleus of an atom. A *positive ion* is formed when a neutral atom loses one or more electrons. A *proton* (H^+) is the nucleus of the hydrogen atom. It is a positively charged particle of mass 1.67×10^{-24} g found in the nucleus of all atoms.

Radioactive nuclei disintegrate spontaneously, emitting particles and radiation. An *alpha* (α) *particle* (He^{2+}) is the nucleus of the helium atom. A *beta* (β)

particle is a high-energy electron emitted by a radioactive nucleus. A *gamma* (γ) *ray* is a high-energy form of light emitted by a radioactive nucleus.

The *neutron* is a neutral particle found in the nucleus of an atom, with a mass almost the same as that of the proton.

Each element has an *atomic number* (Z) equal to the number of protons in the nucleus. The *mass number* (A) of an atom is the sum of the number of protons and the number of neutrons in the nucleus.

Isotopes are atoms with the same atomic number but different masses.

The length of time during which one-half of the nuclei of a radioactive sample disintegrate is known as the *half-life* ($t_{1/2}$) for that nucleus.

A *curie* is the standard unit for measurement of radioactivity; a sample with 1 curie of radioactivity produces 3.7×10^{10} disintegrations per minute. A *rad* is the amount of radiation that delivers 1×10^{-5} J of energy to 1 gram of matter. A dose in *rems* is a measure of the biological effectiveness of a given dose of radiation. *Background radiation* is naturally occurring radiation that is all around us.

Nuclear energy results from the conversion of mass into energy according to the Einstein formula, $E = mc^2$. *Binding energy* is energy that is released when an atom is formed from protons, neutrons, and electrons.

In a *fission reaction*, a large nucleus is split, and mass is converted to energy. In a *fusion reaction*, two small nuclei are combined, and mass is converted to energy.

CHAPTER SUMMARY

An atom is composed of smaller particles including *protons* and *neutrons* (which are in the *nucleus*) and *electrons* (which surround the nucleus). The protons account for the positive charge on the nucleus, and the protons and neutrons together account for almost the entire mass of the atom. The electrons occupy most of the volume of the atom, but are negligible in mass—an electron has a mass only about 1/1840 of that of a proton.

The experiments on which this model of the atom are based include Thomson's cathode-ray-tube experiments, which led to the discovery of the electron and the determination of its charge/mass ratio; Millikan's oil-drop experiment, which led to the determination of the charge on the electron; and Rutherford's alpha-scattering experiments on metal foils, which demonstrated that the positive charge and most of the mass of an atom are contained in a tiny nucleus.

The *atomic number* (Z) gives the number of protons in the nucleus, and also the number of electrons in a neutral atom. The *mass number* (A) is the sum of the number of protons and the number of neutrons. *Isotopes* are atoms with the same atomic number but different masses. A mass spectrometer is used to separate isotopes and to determine their masses. Most elements in nature consist of a mixture of isotopes.

The nuclei of *radioactive* elements disintegrate spontaneously by emitting α, β^-, or β^+ particles (and usually γ radiation) to achieve a *stable nucleus*. Radioactive nuclei, whether naturally occurring or artificially produced, have definite and measurable *half-lives*. The rate at which a radioactive nucleus decays is independent of any chemical or physical changes the material undergoes. This leads to many constructive uses of radioactive isotopes, as well as to problems. Radiation damages tissue by ionizing molecules and causing them to decompose.

Mass is converted to energy in the spontaneous decay of radioactive nuclei according to the Einstein equation, $E = mc^2$. The amounts of energy produced in nuclear reactions are tremendous. The energy released by the fission of one pound of uranium, for example, is equivalent to that obtained by burning over 2 million pounds of coal.

PROBLEMS

5-1. (a) Given the mass of the electron (9.11×10^{-28} g), calculate the mass of a mole of electrons.
(b) Given the mass of the proton (1.67×10^{-24} g), calculate the mass of a mole of protons.
(c) Compare the results of parts a and b with the atomic mass of H.

5-2. (a) Given the charge on a single electron (-1.60×10^{-19} coulombs), calculate the charge on a mole of electrons.
(b) What is the charge on a mole of protons?

5-3. Radioactive isotopes of radium ($^{226}_{88}Ra$), cobalt ($^{60}_{27}Co$), and iodine ($^{131}_{53}I$) are used in the treatment of cancer. How many protons, neutrons, and electrons are in an atom of each isotope?

5-4. A medical technician working with radioactive material wears protective clothing containing a thin layer of lead. Why?

5-5. How can α, β, and γ rays be distinguished from each other?

5-6. What evidence leads to the model that pictures the atom as mostly empty space?

5-7. The radius of a gold atom is about 1.5×10^{-8} cm. How many gold atoms does an α particle pass through when it goes through gold foil 0.0002 cm thick (about 1/10,000 of an inch)? Assume that the atoms are tightly packed in even rows.

5-8. A typical nuclear diameter is 1×10^{-13} cm, and a typical atomic diameter is 1×10^{-8} cm.
(a) Find the ratio of atomic diameter to nuclear diameter.
(b) Find the ratio of atomic volume to nuclear volume.

5-9. Give the numbers of protons, electrons, and neutrons in each of the following neutral atoms: (a) $^{9}_{4}Be$ (b) $^{24}_{12}Mg$ (c) $^{127}_{53}I$ $^{239}_{93}Np$.

5-10. Give the numbers of protons, electrons, and neutrons in each of the following ions: (a) $^{23}_{11}Na^{+}$ (b) $^{27}_{13}Al^{3+}$ (c) $^{35}_{17}Cl^{-}$ (d) $^{32}_{16}S^{2-}$.

5-11. The atom of an element has 34 neutrons in its nucleus and a mass number of 63. What is the element?

5-12. What is the mass number of the isotope of lead that has 126 neutrons in the nucleus?

5-13. Copy the following table and fill in the blank spaces.

Element	Atomic number	Number of: Protons	Electrons	Neutrons	Mass number
Nitrogen (N)	7				14
Phosphorus (P)		15			31
Chromium (Cr)			24		52
Arsenic (As)			33	42	
Strontium (Sr)		38		50	
Silver (Ag)	47			60	
Iodine (I)			53		127
Lead (Pb)				126	208

5-14. Two atoms each have mass number 40. Are they necessarily atoms of the same element? Explain.

5-15. Naturally occurring carbon contains 98.89% carbon-12 and 1.11% carbon-13. What is the atomic weight of carbon? The atomic mass of carbon-13 is 13.0 amu.

5-16. Argon has naturally occurring isotopes of mass numbers 36, 38, and 40. The atomic weight of argon is 39.948. Which of the three isotopes of argon is the most abundant? Explain.

5-17. The isotope of polonium with mass number 198 ($^{198}_{84}Po$) disintegrates to Pb with the emission of an α particle, with a half-life of 1.7 minutes.
(a) Write a balanced equation for the disintegration of $^{198}_{84}Po$.
(b) How might one identify an α particle?

5-18. Tritium, a radioactive isotope of hydrogen ($^{3}_{1}H$), disintegrates to helium with the emission of a β^- particle, with a half-life of 12.3 years.
(a) Write a balanced equation for the disintegration of tritium.
(b) How might one identify a β^- particle?

5-19. The nucleus of nitrogen-18 ($^{18}_{7}N$) lies *above* the belt of stability (Figure 5-15). Write an equation for a nuclear reaction by which $^{18}_{7}N$ can achieve stability.

5-20. The nucleus of radon-212 ($^{212}_{86}Rn$) lies *beyond* the belt of stability (Figure 5-15). Write an equation for a nuclear reaction by which $^{212}_{86}Rn$ can achieve stability.

5-21. The nucleus of nitrogen-13 ($^{13}_{7}N$) lies *below* the belt of stability (Figure 5-15). Write an equation for a nuclear reaction by which $^{13}_{7}N$ can achieve stability.

5-22. Uranium-235 ($^{235}_{92}U$) undergoes a decay series in which α, β^-, α, α, β^-, α, α, α, β^-, β^-, and α particles are given off successively. What are (a) the atomic number and (b) the mass number of the final stable product of this radioactive decay series?

5-23. A $^{235}_{92}U$ nucleus, upon being struck by a neutron, sometimes splits into two medium-sized nuclei and three neutrons. If one of the nuclei is $^{141}_{56}Ba$, what is the other?

5-24. Element 100, fermium, was first identified as the isotope $^{255}_{100}Fm$. If the fermium was produced from $^{238}_{92}U$ by the capture of 17 neutrons, how many β^- particles were emitted?

5-25. Astatine was first produced as the isotope $^{211}_{85}At$ by bombarding bismuth ($^{209}_{83}Bi$) with α particles. Write a balanced nuclear equation for this reaction.

5-26. A piece of charcoal from a cave in France has a β^- count of 3.8 particles per minute per gram, produced by the disintegration of ^{14}C, which has a half-life of 5570 years. A fresh sample of wood gives a count of 15.3 particles per minute. How old, approximately, is the charcoal?

5-27. Strontium (^{90}Sr) is one of the products of the fission of ^{235}U. This isotope of strontium has a half-life of 28.1 years, decaying by the emission of two β^- particles to the stable isotope of zirconium, ^{90}Zr. Suppose you want to store the ^{90}Sr in some secure way until 98% to 99% of it has disintegrated (less than 2% of the ^{90}Sr remains). For how many years, approximately, must the storage container last?

5-28. Radioactive isotope A decays with a half-life of 4 seconds to B. B is also radioactive, decaying with a half-life of 15 days to C. C decays with a half-life of 1 second to D, which is stable.

$$\text{A} \xrightarrow{4\text{ sec}} \text{B} \xrightarrow{15\text{ days}} \text{C} \xrightarrow{1\text{ sec}} \text{D}$$

Suppose we start out with 1 mole of A, and none of B, C, or D. After 30 days, how many moles of A, B, C, and D would we have?

5-29. One method used for dating the rock samples brought back from the moon by Apollo 11 used the relative abundance of $^{40}_{19}$K and $^{40}_{18}$Ar. The assumption is that initially there was no $^{40}_{18}$Ar (a stable isotope) in the rocks, and that it was produced entirely by the decay of $^{40}_{19}$K (a positron emitter) present at the time of the moon's formation. It is also assumed that no further ^{40}K or ^{40}Ar was deposited in the rock after the initial process of formation.

A sample of moon rock containing these two isotopes has 16% ^{40}K and 84% ^{40}Ar. The half-life of ^{40}K is approximately 1.0×10^9 years. *Estimate,* to within 10%, the age of this moon rock. (Because we are quoting a real measurement here, the answer is not as simple as most textbook examples.)

5-30. *The alchemist's dream revived.* Mercury (Hg) is element number 80; gold (Au) is number 79. Here's a scheme for making gold. Take one of the isotopes of mercury $^{196}_{80}$Hg and bombard it with neutrons released by a nuclear reactor. The aim is to achieve absorption of a neutron to give $^{197}_{80}$Hg, which would then decay instantaneously to the stable gold isotope $^{197}_{79}$Au.

(a) What particle would be emitted in the decay of $^{197}_{80}\text{Hg} \rightarrow {}^{197}_{79}\text{Au}$?

(b) The scheme works. The only problem is that the best nuclear reactors produce something like 10^{14} neutrons per second. Making the far-fetched assumption that each of these neutrons is absorbed by ^{196}Hg, *how long* would it take to make 1 g of gold this way? Give the answer in seconds.

Crystals of xenon tetrafluoride, the first noble gas compound to be synthesized ($\times 10$).

6

The Periodic Table and the Electronic Shell Structure of Atoms

By 1830, some 55 different elements had been discovered. Each had been assigned a characteristic atomic weight, though not all were accurate. The elements vary widely in physical and chemical properties; to simplify their study, chemists tried to organize them in groups. When arranged in order of atomic number, the elements show a periodic recurrence of properties. This observation led to the grouping of elements in a periodic table. Models of the electronic structure of the atom provide an explanation of this periodicity.

6-1
THE PERIODIC TABLE

Many scientists contributed to the important achievement of arranging the elements in a **periodic table** that organizes elements with similar properties into groups. The major contributions, however, were made independently by the Russian chemist Dmitri Mendeléev and the German Lothar Meyer. Working in the third quarter of the nineteenth century, each man found that when the elements are arranged in order of increasing atomic weight, they show a definite and regular recurrence of chemical properties. Mendeléev's periodic table is shown in Figure 6-1. In this table, elements having similar properties lie in vertical columns.

Mendeléev believed so strongly in the periodic recurrence of properties of the elements that he reversed the atomic-weight ordering of some element pairs to make them better fit the similarity of properties. And when he came to a spot in his table for which no element then known had the proper chemical properties, he decided to leave the space blank, predicting that it would eventually be filled by the discovery of a new element. He predicted that elements with atomic weights of about 44, 68, and 72 would be discovered, and he even predicted their properties. Within the next fifteen years, scandium (atomic weight 45.0), gallium (atomic weight 69.7), and germanium (atomic weight 72.6) were discovered. Table 6-1 compares Mendeléev's predictions of the properties of the hypo-

Reihen	Gruppe I. — R^2O	Gruppe II. — RO	Gruppe III. — R^2O^3	Gruppe IV. RH^4 RO^2	Gruppe V. RH^3 R^2O^5	Gruppe VI. RH^2 RO^3	Gruppe VII. RH R^2O^7	Gruppe VIII. — RO^4
1	H=1							
2	Li=7	Be=9,4	B=11	C=12	N=14	O=16	F=19	
3	Na=23	Mg=24	Al=27,3	Si=28	P=31	S=32	Cl=35,5	
4	K=39	Ca=40	—=44	Ti=48	V=51	Cr=52	Mn=55	Fe=56, Co=59, Ni=59, Cu=63.
5	(Cu=63)	Zn=65	—=68	—=72	As=75	Se=78	Br=80	
6	Rb=85	Sr=87	?Yt=88	Zr=90	Nb=94	Mo=96	—=100	Ru=104, Rh=104, Pd=106, Ag=108.
7	(Ag=108)	Cd=112	In=113	Sn=118	Sb=122	Te=125	J=127	
8	Cs=113	Ba=137	?Di=138	?Ce=140	—	—	—	— — — —
9	(—)	—	—	—	—	—	—	
10	—	—	?Er=178	?La=180	Ta=182	W=184	—	Os=195, Ir=197, Pt=198, Au=199.
11	(Au=199)	Hg=200	Tl=204	Pb=207	Bi=208	—	—	
12	—	—	—	Th=231	—	U=240	—	— — — —

FIGURE 6-1

Mendeléev's periodic table. To permit elements with similar properties to be grouped together, Mendeléev placed tellurium before iodine despite tellurium's higher atomic weight. Iodine would then be grouped with chlorine and bromine, two elements with which it has many properties in common. We now know that it is atomic number, not atomic weight, that properly determines the order of elements in the periodic table. [From *Annalen der Chemie,* supplemental Vol. 8 (1872), p. 151.]

TABLE 6-1
Predicted and actual properties of germanium

	Mendeléev's predictions for eka-silicon, Es (1871)	Winkler's data for germanium, Ge (1886)
Atomic weight	About 72	72.6
Source	Will be obtained by reaction of Na with EsO_2 or K_2EsF_6	Obtained by reaction of Na with K_2GeF_6
Physical properties	Dark gray metal; high melting point; density about 5.5 g/cm^3	Gray metal; melting point = 958°C; density = 5.36 g/cm^3
Oxide	Will react with oxygen to form EsO_2 with a high melting point and density 4.7 g/cm^3	Reacts with oxygen to form GeO_2; melting point = 1100°C; density = 4.7 g/cm^3
Sulfide	EsS_2 will be insoluble in water, but soluble in ammonium sulfide	GeS_2 is only slightly soluble in water but dissolves readily in ammonium sulfide

thetical element "eka-silicon" with the actual properties of this element, which was discovered by a German chemist and named germanium.

The X-ray experiments of Henry Moseley (Section 5-5) showed that each element has associated with it an atomic number Z equal to the number of protons in the nucleus. *When the elements are arranged in order of increasing atomic number, there is a periodic recurrence of properties, leading to identifiable families (or groups) in a periodic table.* A study of the properties of the 105 presently known elements is greatly simplified by considering them in such groups.

For example, Table 6-2 lists some characteristics of elements with atomic numbers 1 through 20; groups of elements that show similar chemical behavior are evident. The elements helium, neon, and argon stand out—all are monatomic gases that form no hydrides. The elements lithium, sodium, and potassium—which immediately follow these monatomic gases—are all soft, reactive, metallic solids that form hydrides of general formula MH (where M represents the metallic element). The elements fluorine and chlorine, which immediately precede two of the monatomic gases, are diatomic gases that form hydrides of general formula HX (where X represents F or Cl). It is not clear where to put hydrogen—it could be placed with fluorine and chlorine, or with lithium, sodium, and potassium. In Section 6-13 we shall see that it belongs with neither group.

In Table 6-3, the first 20 elements are arranged in order of increasing atomic number, so that elements having similar properties fall in vertical columns. The atomic number is given in parentheses beside the symbol for each element, and the formula of the chloride of each element is shown below its symbol to illustrate a similarity in group properties. (Hydrogen is set off separately.) Table 6-3 is a periodic table for the first 20 elements.

TABLE 6-2
Some characteristics of the first 20 elements

Element (symbol)	Atomic number	State at 25°C and 1 atm	Formula of hydride
Hydrogen (H)	1	Diatomic gas (H_2)	H_2
Helium (He)	2	Monatomic gas (He)	——
Lithium (Li)	3	Soft metallic solid; reactive	LiH
Beryllium (Be)	4	Metallic solid	BeH_2
Boron (B)	5	Semimetallic solid	B_2H_6
Carbon (C)	6	Nonmetallic solid	CH_4
Nitrogen (N)	7	Diatomic gas (N_2)	NH_3
Oxygen (O)	8	Diatomic gas (O_2)	H_2O
Fluorine (F)	9	Diatomic gas (F_2)	HF
Neon (Ne)	10	Monatomic gas (Ne)	——
Sodium (Na)	11	Soft metallic solid; reactive	NaH
Magnesium (Mg)	12	Metallic solid	MgH_2
Aluminum (Al)	13	Metallic solid	AlH_3^*
Silicon (Si)	14	Semimetallic solid	SiH_4
Phosphorus (P)	15	Nonmetallic solid	PH_3
Sulfur (S)	16	Nonmetallic solid	H_2S
Chlorine (Cl)	17	Diatomic gas (Cl_2)	HCl
Argon (Ar)	18	Monatomic gas (Ar)	——
Potassium (K)	19	Soft metallic solid; reactive	KH
Calcium (Ca)	20	Metallic solid	CaH_2

*The hydride of aluminum is an unstable compound having three H atoms for each Al atom. It is analogous to the hydride of boron, B_2H_6, which has three H atoms for each B atom.

TABLE 6-3
Periodic table for first 20 elements

H (1) HCl							He (2) ——
Li (3) LiCl	Be (4) $BeCl_2$	B (5) BCl_3	C (6) CCl_4	N (7) NCl_3	O (8) OCl_2	F (9) ClF	Ne (10) ——
Na (11) NaCl	Mg (12) $MgCl_2$	Al (13) $AlCl_3$	Si (14) $SiCl_4$	P (15) PCl_3	S (16) SCl_2	Cl (17) Cl_2	Ar (18) ——
K (19) KCl	Ca (20) $CaCl_2$	. . . .					

1A	2A	3B	4B	5B	6B	7B	8B			1B	2B	3A	4A	5A	6A	7A	8A
			1 **H** 1.008														2 **He** 4.003
3 **Li** 6.941	4 **Be** 9.01	←				Transition metals					→	5 **B** 10.81	6 **C** 12.011	7 **N** 14.01	8 **O** 16.00	9 **F** 19.00	10 **Ne** 20.18
11 **Na** 22.99	12 **Mg** 24.31											13 **Al** 26.98	14 **Si** 28.09	15 **P** 30.97	16 **S** 32.06	17 **Cl** 35.45	18 **Ar** 39.95
19 **K** 39.10	20 **Ca** 40.08	21 **Sc** 44.96	22 **Ti** 47.90	23 **V** 50.94	24 **Cr** 52.00	25 **Mn** 54.94	26 **Fe** 55.85	27 **Co** 58.93	28 **Ni** 58.70	29 **Cu** 63.55	30 **Zn** 65.38	31 **Ga** 69.72	32 **Ge** 72.59	33 **As** 74.92	34 **Se** 78.96	35 **Br** 79.90	36 **Kr** 83.80
37 **Rb** 85.47	38 **Sr** 87.62	39 **Y** 88.91	40 **Zr** 91.22	41 **Nb** 92.91	42 **Mo** 95.94	43 **Tc** (98)	44 **Ru** 101.07	45 **Rh** 102.91	46 **Pd** 106.4	47 **Ag** 107.87	48 **Cd** 112.40	49 **In** 114.82	50 **Sn** 118.69	51 **Sb** 121.75	52 **Te** 127.60	53 **I** 126.90	54 **Xe** 131.30
55 **Cs** 132.91	56 **Ba** 137.33	57 **La** 138.91	72 **Hf** 178.49	73 **Ta** 180.95	74 **W** 183.85	75 **Re** 186.2	76 **Os** 190.2	77 **Ir** 192.22	78 **Pt** 195.09	79 **Au** 196.97	80 **Hg** 200.59	81 **Tl** 204.37	82 **Pb** 207.2	83 **Bi** 208.98	84 **Po** (209)	85 **At** (210)	86 **Rn** (222)
87 **Fr** (223)	88 **Ra** (226)	89 **Ac** (227)	104 **(Rf)** (261)	105 **(Ha)** (262)	106												

Lanthanides	58 **Ce** 140.12	59 **Pr** 140.91	60 **Nd** 144.24	61 **Pm** (145)	62 **Sm** 150.4	63 **Eu** 151.96	64 **Gd** 157.25	65 **Tb** 158.93	66 **Dy** 162.50	67 **Ho** 164.93	68 **Er** 167.26	69 **Tm** 168.93	70 **Yb** 173.04	71 **Lu** 174.97
Actinides	90 **Th** 232.04	91 **Pa** (231)	92 **U** 238.03	93 **Np** (237)	94 **Pu** (244)	95 **Am** (243)	96 **Cm** (247)	97 **Bk** (247)	98 **Cf** (251)	99 **Es** (252)	100 **Fm** (257)	101 **Md** (258)	102 **No** (259)	103 **Lr** (260)

(available radioactive isotope of longest half-life)

Metals

Metalloids and nonmetals

FIGURE 6-2

Periodic table of the elements.

A complete periodic table, including all the known elements, is shown in Figure 6-2 and repeated inside the cover of this book. Elements that lie in a vertical column constitute a **group** (or **family**) and have similar properties. By studying groups of elements we can organize and greatly simplify the study of chemistry. From top to bottom of a group, most properties change in a regular way. Table 6-4 shows properties of the elements in group 1A of the periodic table, the **alkali metal** group. From top to bottom of the group, there is a gradual increase in atomic weight and atomic radius, and a decrease in melting point, boiling point, and ionization potential.

The **atomic radius** measures the size of the atom. In Section 6-10 we shall see that the radius of an atom cannot be defined precisely and depends in any case on the surrounding atoms. The values given in Table 6-4 for the atomic radii are given for all the atoms under the same conditions, so that they may be compared. We see that, from top to bottom of the

TABLE 6-4
Properties of the alkali metals

Element	Atomic weight	Melting point (°C)	Boiling point (°C)	Atomic radius (nm)	Ionization potential (kJ/mole)
Li	6.9	179	1336	0.123	520
Na	23.0	97.8	883	0.157	496
K	39.1	63.5	758	0.203	419
Rb	85.5	39.0	700	0.216	403
Cs	132.9	28.6	670	0.235	376
Fr*	(223)	(27)			

*Francium is a short-lived radioactive element whose properties are largely unknown.

group, the size of the atom gradually increases (Figure 6-3a).

The **ionization potential** is the minimum energy required to remove one electron from a neutral atom in the gaseous state—that is, the energy required for the reaction

$$\text{M (gas)} \longrightarrow \text{M}^{+} \text{ (gas)} + e^{-}$$

Ionization potential is expressed here in kilojoules/mole. The energy required to remove one electron from each atom of a mole of sodium atoms in the gas state is 520 kJ. (That corresponds to only 9×10^{-19} J to remove one electron from a single atom of sodium.) From top to bottom of the group, the ionization potential gradually decreases. That is, it becomes progressively easier to remove an electron from the neutral atom. The variation of ionization potential with atomic number is graphed in Figure 6-3b.

EXERCISE 6-1 (a) Write an equation for the ionization of potassium gas (K) to the K^+ ion.

(b) Which of the alkali metals is the most difficult (requires the most energy) to ionize?

Because of the gradual and regular change of properties of the elements from top to bottom of a group in the periodic table, Mendeléev was able to predict the properties of the undiscovered elements for which he left blank spaces in his table. He left a blank space in the carbon group for an element with atomic weight, density, and melting point between those of silicon (Si) and tin (Sn). The properties of both these elements were well known. Mendeléev knew also that the new element should form compounds with the same general formulas and properties as those of the known members of the carbon group: C, Si, Sn, and lead (Pb). Thus he predicted the properties listed earlier (Table 6-1) for germanium.

From left to right across a horizontal row, or **period**, in the periodic table, the properties of the elements often change in an irregular way, as illustrated in Table 6-5 for the elements in the second period. Of the properties shown in Table 6-5, the atomic radius and the ionization potential vary in a generally regular way across the period, but the melting point shows irregular variations.

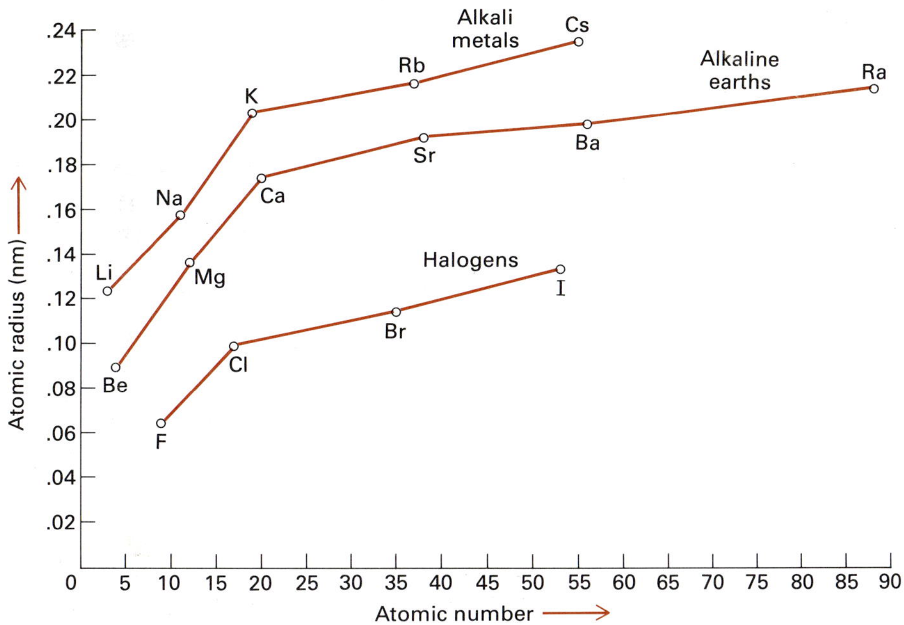

(a)

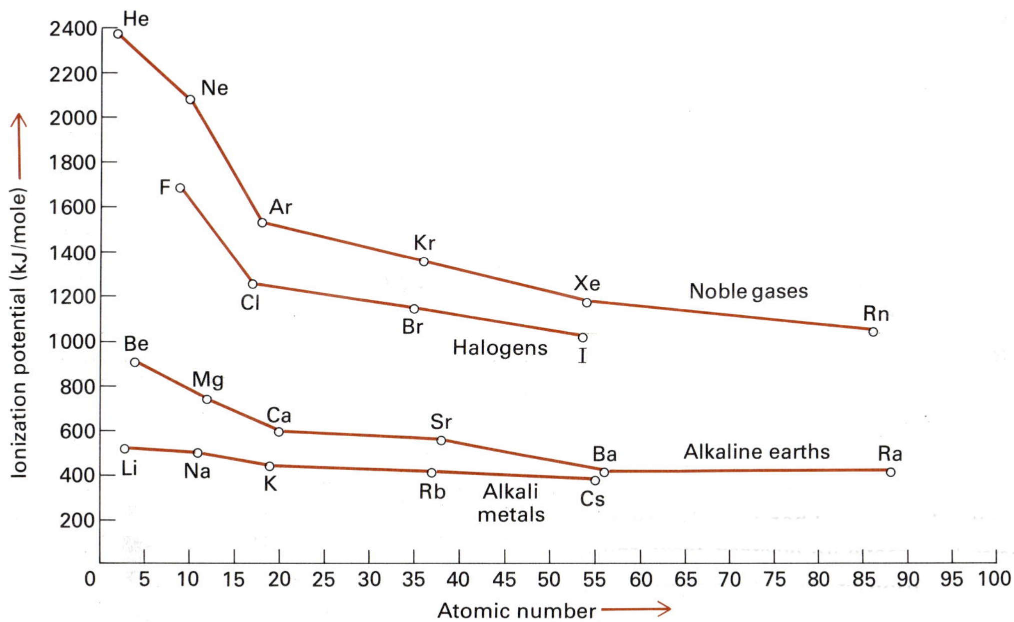

(b)

FIGURE 6-3

Variation in properties with atomic numbers. **(a)** Variation in atomic radius with atomic number for the alkali metals, alkaline earths, and halogens. **(b)** Variation in ionization potential with atomic number for the alkali metals, alkaline earths, halogens, and noble gases.

TABLE 6-5
Some properties of elements in the second period

Property	Li	Be	B	C	N	O	F	Ne
	Element							
Atomic number	3	4	5	6	7	8	9	10
Melting point at 1 atm (in °C)	179	1285	2074	3850 (sublimes)	−210	−219	−220	−249
Atomic radius* (nm)	0.123	0.125	0.090	0.077	0.075	0.073	0.064	——
Ionization potential (kJ/mole)	520	900	8010	1090	1400	1310	1680	2080

*These radii have been determined for the atoms in compounds. Neon does not form compounds, so no radius is given for neon.

EXERCISE 6-2 Plot graphs of (a) ionization potential, (b) atomic radius, and (c) melting point of the elements in the second period against atomic number, and observe the regularities and irregularities.

We next briefly consider some properties of four prominent groups in the periodic table: the noble gases, the alkali metals, the alkaline earth metals, and the halogens.

6-2 THE NOBLE GASES

The noble gas group (group 8A in the periodic table), whose existence was unknown at the time of Mendeléev, provides the clearest example of the periodic recurrence of properties of the elements. The **noble gas group** consists of the elements helium (He), neon (Ne), argon (Ar), krypton (Kr), xenon (Xe), and radon (Rn). All are monatomic, colorless gases with the outstanding property of an almost total lack of chemical reactivity. For this reason, they often are called the inert gases. These elements are most stable as single, neutral atoms. Until 1962 no compounds of them had ever been prepared, and they were considered to be totally unreactive. Compounds between Xe and F, Xe and O, Xe and N, and Kr and F have now been prepared, but even these compounds readily decompose on heating to yield the uncombined noble gas atoms.

The first clue to the existence of the noble gases was detected in 1785 by Henry Cavendish, who noticed that a small component of the air ("not more than 1/120 part of the whole") would not react under any conditions. This gas was identified as an element, argon, by Lord Rayleigh and William Ramsay in 1894, and was assigned to a new group in the periodic table. By 1898, neon, krypton, and xenon had been isolated from the atmosphere by the liquefaction of air; as the liquid air evaporates, each gas boils off at a different temperature and can thus be obtained in pure form. Helium was first detected by the characteristic bright yellow color of its vapor in light emitted from the sun during an eclipse in 1868. The yellow light was attributed to a new element that was named helium (after *helios,* the Greek word for sun). Helium was first found on earth in 1895 in a uranium-bearing mineral. (As we saw in Chapter 5, helium is produced on earth in radioactive decay by alpha emission, and in the sun by the fusion of hydrogen.) Radon, the densest gaseous element, is a decay product of radium and is itself radioactive.

The noble gases are rare. Together they comprise only about 1% by volume of the earth's atmosphere. In addition to being present in the atmosphere, helium is found also in some rocks and in some natural gas wells. Table 6-6 summarizes the properties of the noble gases.

There is a gradual increase from top to bottom of the group in melting point, boiling point, and atomic radius, and a gradual decrease in ionization potential. If we compare the high ionization potentials of the elements in this group with those shown in Table 6-4 for the elements in the alkali metal group, we see that it is much more difficult to remove an electron from a noble gas atom than from an atom of any alkali metal. Among the noble gases, the lower ionization potentials of xenon and krypton account for the fact that compounds of these elements have been prepared; no compounds have yet been prepared of He, Ne, and Ar, which have higher ionization potentials. The ionization potentials of the noble gases are graphed in Figure 6-3b together with those of the alkali metals.

Helium is used in gas thermometers, and in liquid form is used as a coolant to produce very low temperatures. Because of its low density, it is also used in balloons; it is safer than the even-less-dense hydrogen gas (which forms an explosive mixture with oxygen), yet more buoyant than hot air, which is denser. Neon's chief use is in advertising signs. Argon, the most abundant of the noble gases, is often used when an inert atmosphere is required—for example, it is used in welding to provide an atmosphere in which metals will not burn.

TABLE 6-6
Some properties of the noble gases

Element	Atomic weight	Melting point (°C)	Boiling point at 1 atm (°C)	Ionization potential (kJ/mole)
He	4.00	−273	−269	2370
Ne	20.2	−249	−246	2080
Ar	39.9	−189	−186	1520
Kr	83.8	−157	−153	1350
Xe	131.3	−112	−108	1170
Rn	(222)	−71	−62	1040

Note: The melting (freezing) points are given at 1 atm pressure for all the elements except helium. Helium freezes at −273°C (absolute zero) at 25 atm; below this pressure it cannot exist as a solid.

6-3
THE ALKALI METALS

The **alkali metal group** (group 1A in the periodic table) consists of the elements lithium (Li), sodium (Na), potassium (K), rubidium (Rb), cesium (Cs), and the rare, short-lived, radioactive element francium (Fr), which is a decay product of actinium. Because the most stable isotope of Fr decays quite rapidly, its properties are not well known, and we do not discuss Fr further here.

The alkali metals are all silvery metals, soft enough to cut with a knife. Lithium, sodium, and potassium are less dense than water; lithium has a density of only 0.53 g/cm^3. The alkali metals have low melting points; the melting point of Cs is 29°C, and Na melts at 98°C (below the boiling point of water). They are highly reactive, and are good conductors of electricity and heat. Because they are so reactive, they are rarely seen in pure form. They are never found free in nature, but always in combination with other elements. The alkali metals react rapidly with oxygen in the air, and violently with water, and so must be stored sealed in an atmosphere of an inert gas such as argon or immersed in an unreactive substance such as kerosene.

The reaction of an alkali metal with water produces hydrogen gas and a bitter-tasting, slippery feeling, corrosive solution. For example, reaction of sodium and water yields hydrogen gas and a solution of sodium hydroxide:

$$Na(s) + H_2O(l) \longrightarrow Na^+ + OH^- + \frac{1}{2}H_2(g)$$

The symbols (s), (*l*), and (g) indicate the solid, liquid, and gas states, respectively. The sodium hydroxide (NaOH) is present in the solution not as the compound, but as separate Na^+ and OH^- ions. (The OH^- ion is a stable ion composed of an oxygen atom and a hydrogen atom, with one extra electron.) In Chapter 11 we see that the alkali metal hydroxides are **bases;** they neutralize acids. The familiar substance lye is impure sodium hydroxide.

EXERCISE 6-3 Write an equation for the reaction that occurs when potassium metal is put into water.

Sometimes the reaction between an alkali metal and water is so violent that heat given off in the reaction causes the hydrogen gas to ignite and burn with the oxygen in the air to form water.

The alkali metals also react vigorously with chlorine to produce crystalline chlorides with the general formula MCl, where M represents any alkali metal. For example, potassium and chlorine react to form potassium chloride:

$$2K(s) + Cl_2(g) \longrightarrow 2KCl(s)$$

Sodium chloride (NaCl) is ordinary table salt.

Large quantities of alkali metal chlorides are found in the oceans, inland seas, and in salt deposits that originated from the drying of former seas. Fifty million billion tons of NaCl are still dissolved in the earth's oceans. Table 6-7 gives the relative abundances of elements in the universe, in the earth's crust, in seawater, and in humans. Sodium ranks fourth in abundance in seawater (in percentage of total atoms) and tenth in the human body. Potassium ranks eighth in seawater and eighth in the human body.

The properties of the alkali metals are summarized in Table 6-4. As with the noble gases, there is a gradual decrease from top to bottom of the group in ionization potential, and a gradual increase in atomic radius. The low ionization potentials of the alkali metals show that it is relatively easy to remove an electron from an alkali metal atom. In their compounds the alkali metals usually occur as ions with a 1+ charge: Li^+, Na^+, K^+, Rb^+, and Cs^+.

Because the alkali metals are so soft and so reactive, they do not find many uses in the pure form. Liquid

TABLE 6-7
Abundances of elements in the universe, earth's crust, seawater, and the human body

Composition of universe		Composition of earth's crust		Composition of seawater		Composition of human body	
H	91	O	47	H	66	H	63
He	9.1	Si	28	O	33	O	25.5
O	0.057	Al	7.9	Cl	0.33	C	9.5
N	0.042	Fe	4.5	Na	0.28	N	1.4
C	0.021	Ca	3.5	Mg	0.033	Ca	0.31
Si	0.003	Na	2.5	S	0.017	P	0.22
Ne	0.003	K	2.5	Ca	0.006	Cl	0.03
Mg	0.002	Mg	2.2	K	0.006	K	0.06
Fe	0.002	Ti	0.46	C	0.0014	S	0.05
S	0.001	H	0.22	Br	0.0005	Na	0.03
		C	0.19			Mg	0.01
All others	< 0.01	All others	< 0.1	All others	< 0.1	All others	< 0.01

Note: Abundances are expressed as percentages of the total number of atoms in each case. Note that phosphorus (P) is the only element in the fourth column that does not appear in any other column; thus the sixth most common element in the body is rare in inanimate nature. Carbon, the third most plentiful element in the body, also is very scarce elsewhere.
Source: After Earl Frieden, "The Chemical Elements of Life."

sodium, because of its high heat conductivity, is used as a coolant in nuclear reactors, conducting the heat to the exterior of the reactor where it converts water into useful steam. Cesium—which has so low an ionization potential that an electron can be removed by the energy in visible light—is used in photocells, devices for converting the light energy to electricity.

Lithium ion (Li^+) has not been found in the body, but is used as lithium carbonate (Li_2CO_3) to treat the extremes of excitement and depression suffered by manic depressives. The mechanism by which it works, however, is not yet well understood. Sodium ion (Na^+) is the principal positive ion in the fluids surrounding the cells in our bodies, where it is needed for water retention and muscle action. Na^+ is, therefore, sometimes administered intravenously in "physiological saline solution" (which has a Na^+ concentration the same as that of Na^+ in the blood) to patients who cannot eat. Too much Na^+, however, appears to be undesirable. A number of strains of rats develop hypertension (high blood pressure) when fed a high-sodium diet. Because Na^+ is suspected of being partly responsible for hypertension in humans, people suffering from hypertension are usually put on a low-sodium diet. The body has mechanisms for retaining Na^+, and it is almost impossible to prepare a sodium-free diet, so salt depletion is rarely a problem. Potassium ion (K^+) is the principal positive ion in the cells of the body. The transport of Na^+ and K^+ ions across nerve cell membranes is of crucial importance in the conduction of impulses along nerve cell fibers.

6-4 THE ALKALINE EARTH METALS

The **alkaline earth group** (group 2A in the periodic table) consists of the elements beryllium (Be), magnesium (Mg), calcium (Ca), strontium (Sr), barium (Ba), and radium (Ra). The alkaline earth elements are reactive, silvery metals. Each is somewhat harder than the neighboring alkali metal and has higher melting and boiling points. Each element in this group is less reactive than the alkali metal that immediately precedes it in the periodic table. The alkaline earth elements are always found chemically combined with other elements and, in the pure form, all except Be and Mg must be stored protected from the atmosphere and moisture.

The alkaline earths form hydroxides of general formula $M(OH)_2$. Calcium, strontium, and barium react vigorously with cold water, as do the alkali metals:

$$Ca(s) + 2H_2O(l) \longrightarrow Ca^{2+} + 2OH^- + H_2(g)$$

As with the alkali metals, the calcium hydroxide is present in solution as ions (Ca^{2+} and OH^-).

The alkaline earths react with chlorine to produce crystalline chlorides of general formula MCl_2:

$$Sr(s) + Cl_2(g) \longrightarrow SrCl_2(s)$$

EXERCISE 6-4 Write an equation for the reaction that occurs when barium metal reacts with chlorine gas (Cl_2) to form barium chloride, a white solid.

Beryllium occurs in trace amounts in the earth's crust in the mineral beryl ($Be_3Al_2Si_6O_{18}$). (The gems emerald and aquamarine are beryl with trace impurities that give the color.) Magnesium is abundant in seawater, salt beds, and in the mineral dolomite ($CaCO_3 \cdot MgCO_3$). Calcium, the most abundant element of the group, is found principally as calcium carbonate ($CaCO_3$) in the form of limestone, marble, chalk, seashells, and the mineral calcite. Table 6-7 shows that calcium ranks fifth and magnesium eleventh in percentage of total atoms in the body. Strontium occurs in nature mainly as the carbonate ($SrCO_3$). Barium is found chiefly as the sulfate ($BaSO_4$). Radium is a very rare, radioactive metal, a decay product of uranium.

Some properties of the alkaline earths are summarized in Table 6-8. As with the noble gases and the alkali metals, there is an increase from top to bottom of the group in atomic radius and a decrease in ioniza-

TABLE 6-8
Some properties of the alkaline earth metals

Element	Atomic weight	Melting point (°C)	Boiling point (°C)	Atomic radius (nm)	Ionization potential (kJ/mole)
Be	9.0	1285	2970	0.125	900
Mg	24.3	650	1117	0.145	738
Ca	40.1	851	1487	0.174	590
Sr	87.6	774	1366	0.191	550
Ba	137.3	850	1537	0.198	503
Ra	226.0	~700	~1525		509

tion potential (see Figure 6-3). The ionization potentials of the alkaline earths are lower than those of the noble gases, but somewhat higher than those of the nearest alkali metals. In their compounds, the alkaline earths are usually present as ions with a 2^+ charge: Be^{2+}, Mg^{2+}, Ca^{2+}, Sr^{2+}, and Ba^{2+}.

EXERCISE 6-5 From which metal atom is it easier to remove an electron? (See Tables 6-4 and 6-9.) (a) Na or Mg? (b) Li or Ba?

Beryllium is used in small amounts as an alloying metal to impart hardness. Because of its low density, magnesium is combined with aluminum to produce a hard, light alloy for use in aircraft. Magnesium powder or ribbon burned in air generates intense heat and light, so Mg is used in flashbulbs and fireworks.

Calcium is present in the skeletons of animals—the human adult contains about two pounds, mostly in the bones and teeth. Calcium is also essential to the clotting of blood. Because of the prevalence of calcium-containing minerals, Ca^{2+} ions are often present in water (causing "hard" water), where they form an insoluble "scum" with soap.

Radioactive strontium (^{90}Sr) is a deadly product of the fission of $^{235}_{92}U$ in nuclear reactors, where it is produced in large quantity. Strontium is chemically so similar to calcium that it enters the food chain in grass, is concentrated in milk, and replaces some of the calcium in bone tissue. There it may destroy bone marrow and cause cancer if it is present in large enough amounts.

The compound barium sulfate ($BaSO_4$) is a dense white substance, insoluble in water, and opaque to X rays. These properties enable it to be used to outline the stomach and intestines in X-ray photographs. The $BaSO_4$ is administered by mouth and is so insoluble that it passes through the body unchanged. Its passage is followed by X rays, and any blockages can be detected. Before the dangers of its radiation were recognized, radium was used to make luminous watch dials. The same destructive capacity is now used in the treatment of cancer.

6-5 THE HALOGENS

The **halogen group** (group 7A in the periodic table) consists of the elements fluorine (F), chlorine (Cl), bromine (Br), iodine (I), and astatine (At), Astatine is radioactive, so its properties are not well known, and it is not included in this discussion. All of the halogens are nonmetallic, and in the pure (elemental) form occur at room temperature as the diatomic molecules: fluorine (F_2) as a pale yellow gas; chlorine (Cl_2) as a yellow-green gas; bromine (Br_2) as a fuming, red-brown liquid; and iodine (I_2) as a blue-black solid.

The halogens have pungent, irritating odors, and all are poisonous to breathe, although iodine is less dangerous than the others. Like the alkali metals and the alkaline earths, the halogens are highly reactive, and thus are not found free in nature, but always chemically combined with other substances. Fluorine is the most reactive element of the group and iodine the least reactive. Fluorine reacts with almost every element (except He, Ne, and Ar): wood and rubber burn in fluorine; it attacks platinum, and even glass in the presence of moisture. Thus fluorine is difficult to handle and to store. It can be stored in Teflon, a synthetic plastic made of carbon and fluorine.

We have seen that chlorine combines with the alkali metals and alkaline earths to form crystalline compounds. This behavior is typical of all the halogens:

$$2Na(s) + F_2(g) \longrightarrow 2NaF(s) \text{ (sodium fluoride)}$$

$$Mg(s) + Br_2(g) \longrightarrow MgBr_2(s) \text{ (magnesium bromide)}$$

The halogens react vigorously with hydrogen to produce a gas of general formula HX, where X is a halogen atom:

$$H_2(g) + Cl_2(g) \longrightarrow HCl(g) \text{ (hydrogen chloride)}$$

The reactions between H_2 and F_2 and between H_2 and Cl_2 are explosive. Even turpentine ($C_{10}H_{16}$) reacts violently with Cl_2 gas as the chlorine removes hydrogen from the turpentine according to the following reaction:

$$8Cl_2(g) + C_{10}H_{16}(l) \longrightarrow 16HCl(g) + 10C(s)$$

producing hydrogen chloride and a black soot of carbon. The HX compounds are all gases at room temperature, and they dissolve in water to produce acidic solutions, of which the best known is hydrochloric acid.

Fluorine occurs primarily in the minerals fluorspar (CaF_2) and cryolite (Na_3AlF_6). Large quantities of chlorides and bromides are dissolved in the oceans and are also found in salt deposits. Iodine is found in seaweed. Chlorine is the most abundant of the halogens, ranking seventh in percentage of total atoms in the body, where it is the principal intracellular and extracellular anion.

Some properties of the halogens are summarized in Table 6-9. The halogens have low melting and boiling points, only slightly higher than those of the noble gases. Even at room temperature, liquid bromine fumes to form reddish Br_2 vapor, and solid iodine sublimes to violet I_2 vapor. As with the noble gases, the alkali metals, and the alkaline earths, there is a decrease from top to bottom of the group in ionization potential; and there is an increase in atomic radius (see Figure 6-3). The ionization potentials of the halogens are very high, only slightly lower than those of the noble gases; it is difficult to remove an electron from a halogen atom. Halogens thus rarely occur as the positive ions F^+, Cl^+, Br^+, or I^+. Instead, they pick up electrons easily, so in their compounds the halogens

TABLE 6-9
Some properties of the halogens

Element	Molecular formula	Molecular weight	Melting point (°C)	Boiling point (°C)	Atomic radius (nm)	Ionization potential (kJ/mole)
F	F_2	38.0	−220	−188	0.071	1680
Cl	Cl_2	70.9	−101	−34	0.099	1250
Br	Br_2	159.8	−7	59	0.114	1140
I	I_2	253.8	114	184	0.133	1010
At		(210)	302	334		

usually occur as ions with a 1− charge: F^-, Cl^-, Br^-, and I^-. These ions are **halide** ions: fluor*ide,* chlor*ide,* brom*ide,* and iod*ide.*

EXERCISE 6-6 Write equations for the reactions that occur

(a) between barium metal and fluorine gas
(b) between cesium metal and chlorine gas
(c) between hydrogen gas and solid iodine

EXERCISE 6-7 From which atom is it easier to remove an electron? (a) Cl or Ar? (b) Cl or Xe? (c) Cl or Na?

Fluorine is used to make fluorocarbons (compounds of carbon and fluorine), which are extremely inert, corrosion-resistant materials that resist attack by strong acids and other corrosive substances. Teflon is an example of a fluorocarbon. Fluoride ions (F^-) present in water or toothpaste in trace amounts form a hard, protective coating of fluorapatite on the teeth and bones. In fluorapatite, F^- replaces the OH^- of calcium hydroxyapatite, $Ca_{10}(PO_4)_6(OH)_2$, the principal bone mineral. Too high a concentration of F^-, however, causes stains on the teeth and makes teeth and bones more brittle.

Chlorine is used as a bleach and as a disinfectant to purify drinking water. Hydrochloric acid is consumed in large quantities in the preparation of plastics and other compounds, and it is a normal secretion of the human stomach. Bromine is also a disinfectant, often used to purify water in swimming pools. Bromides are added to leaded gasolines to keep lead from accumulating in the engine. The lead combines with the bromide to form lead bromide ($PbBr_2$), which boils at a low temperature and escapes in the exhaust. Iodine is commonly added to table salt as sodium iodide (NaI) to supplement the diet. Trace amounts are needed for the production of the hormone thyroxine in the thyroid gland. In the absence of iodine, no thyroxine is produced, and a goiter (an enlargement of the thyroid gland) may result. Radioactive iodine is used in the diagnosis and treatment of thyroid diseases.

In our consideration of these four groups of elements—the alkali metals, the alkaline earths, the halogens, and the noble gases—we have illustrated the periodic recurrence of chemical properties, but we have yet to account for it. We shall see that much of this behavior can be explained by the electronic shell structure of atoms. Before considering the electronic structure of atoms, however, we shall investigate some properties of light, because it is the interaction between light and matter that gives the best evidence for the electronic shell structure of atoms.

6-6 LIGHT

Radiation is a form of energy that can be transmitted through empty space. Among the forms of radiation are gamma rays, X rays, ultraviolet light, visible light, infrared radiation, and radio waves. The familiar energy carried by sunlight warms the land and the seas in summer and melts glaciers. Sunlight provides the energy with which chlorophyll breaks down H_2O molecules into protons, electrons, and O_2 molecules in the

FIGURE 6-4

Waves in water. Raindrops falling in a pond send out ripples in all directions. [Photograph by Richard Byrnes.]

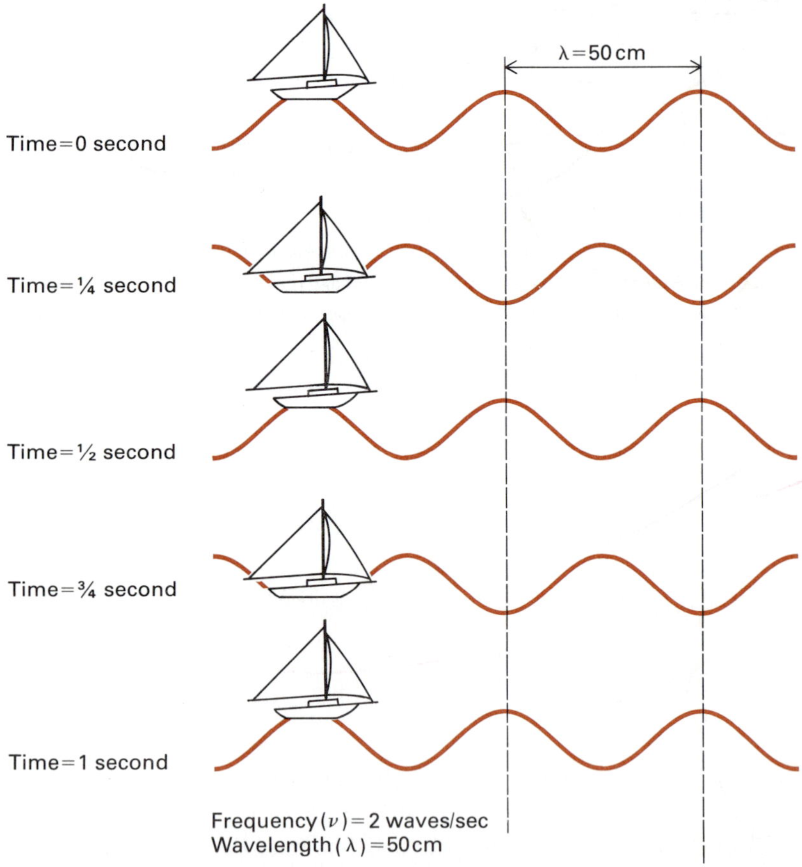

FIGURE 6-5

Wavelength and frequency of waves on a pond.

first step of photosynthesis. The term **electromagnetic radiation** is used to describe all forms of radiant energy because all have both an electrical and magnetic nature. Electromagnetic radiation consists of electric and magnetic fields traveling together through space.

Light is often characterized as a kind of wave; in many ways it is like waves on the ocean or ripples on a pond (Figure 6-4). Imagine a toy boat bobbing up and down on the ripples on a pond into which a stone has been tossed. The number of times per second that the boat rises to a maximum (or the wave crests) at a fixed point is called the **frequency** of the wave (Figure 6-5), and is usually represented by the Greek letter ν (*nu*). The frequency of the waves shown in Figure 6-5 is two waves every second; $\nu = 2$ waves/sec, or 2 hertz (Hz), the SI unit for frequency: 1 Hz $= 1\ \text{sec}^{-1}$.

The **wavelength,** λ (*lambda*), is the distance from the top of one wave to the top of the next wave, at any one instant. In Figure 6-5 the wavelength is 50 centimeters; $\lambda = 50$ cm/wave. Figure 6-6 shows waves of different wavelengths and frequencies.

Waves in water can have different speeds. The speed or **velocity** (v) of a wave (the rate at which a wave front moves, say, from the center of the pond toward the shore) is determined by multiplying its wavelength by its frequency:

$$\text{velocity (cm/sec)} = \text{wavelength (cm/wave)} \times \text{frequency (waves/sec)}$$

or

$$v = \lambda\nu \text{ cm/sec}$$

In Figure 6-6, all the waves are traveling at the same velocity of 120 cm/sec.

All forms of electromagnetic radiation travel in a vacuum at the same velocity of 3×10^{10} cm/sec. The **velocity of light** (c) in a vacuum is then 3×10^{10} cm/sec. If we know the frequency of light, we can then determine its wavelength from the relation $c = \lambda\nu$, just as with water waves. Figure 6-7 shows the **electromagnetic spectrum,** the frequency and wavelength for some forms of electromagnetic radiation. Visible light (radiation to which the human eye is sensitive) has wavelengths between 380 nm (blue) and 700 nm (red)—only a very small part of the entire electromagnetic spectrum. (Nanometers: 1 nm $= 10^{-9}$ m)

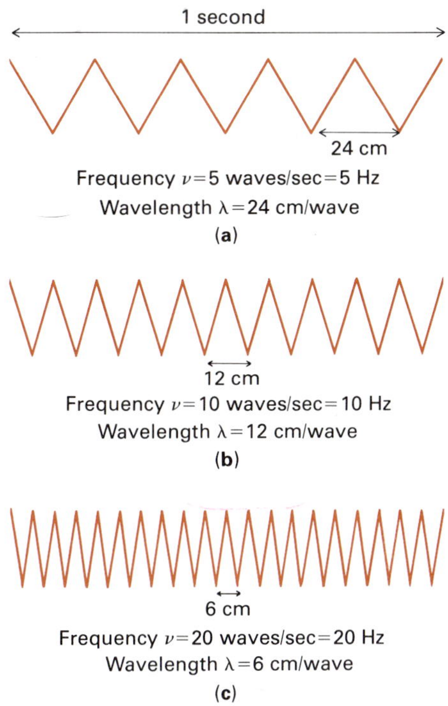

FIGURE 6-6

Waves of different wavelengths and frequencies. In all three cases, velocity $= \lambda \times \nu = 120$ cm/sec.

EXERCISE 6-8 The electromagnetic waves used in standard AM radio broadcasting have frequencies of about 1 megahertz (1,000,000 waves/sec). What is the wavelength of the radiation?

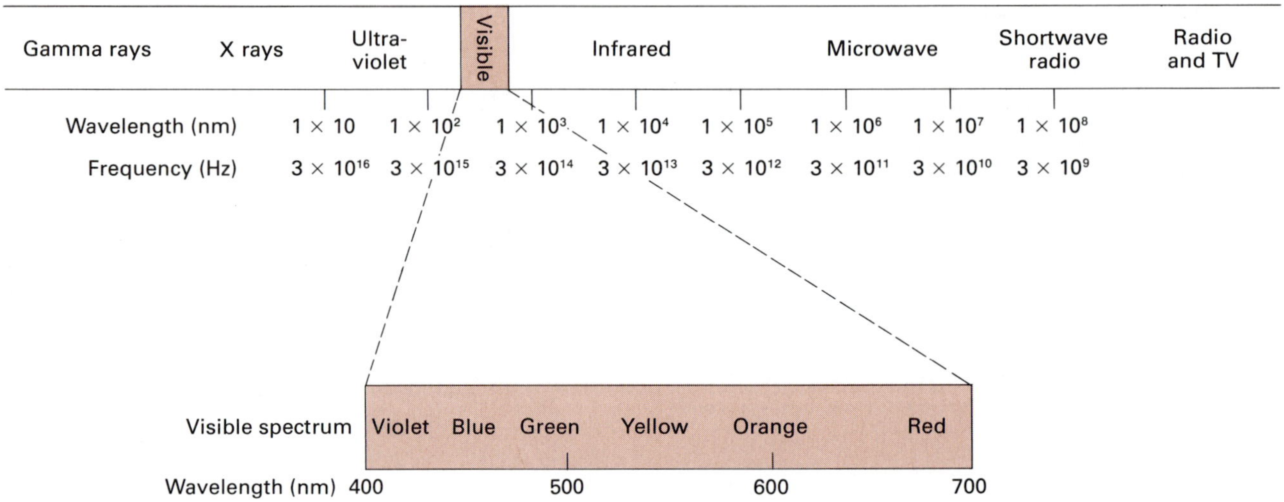

FIGURE 6-7

The electromagnetic spectrum.

6-7 SPECTRA

When sunlight shines through a diamond, a prism, or a raindrop, it is separated into a rainbow of colors. White light is composed of all these colors. All are present in the original light, but they are separated by the prism, the diamond, or the raindrop. The simple spectroscope shown in Figure 6-8 uses a prism to separate light into its components. White light from a hot tungsten wire passes through a slit, producing a narrow beam of light. When the light enters the prism, the light waves are **refracted,** or bent. The components of the white light with different frequencies are bent through different angles. The blue light is bent most and the red light least. The light from the original narrow beam is thus spread out into the **continuous spectrum** or rainbow of colors.

Up to now, we have characterized light as a wave, but in some ways it is also like a particle. The energy carried by light comes in "packets" called **quanta** (singular, **quantum**) or **photons,** just as electric charge comes in units of the charge on the electron. The energy and the frequency of these photons, or quanta, of light are related by the equation $E = h\nu$. The symbol h, called Planck's constant (after the German scientist who first introduced it), here plays the role of a conversion factor that converts frequency units to energy units. The value of h is 6.6256×10^{-34} J sec. Using the equation $E = h\nu$, we can calculate the energy of quanta of light. For a quantum of red light of frequency 4.3×10^{14} Hz,

$$E = 6.6 \times 10^{-34}\ \text{J sec} \times 4.3 \times 10^{14}\ \text{sec}^{-1}$$
$$= 2.8 \times 10^{-19}\ \text{J}$$

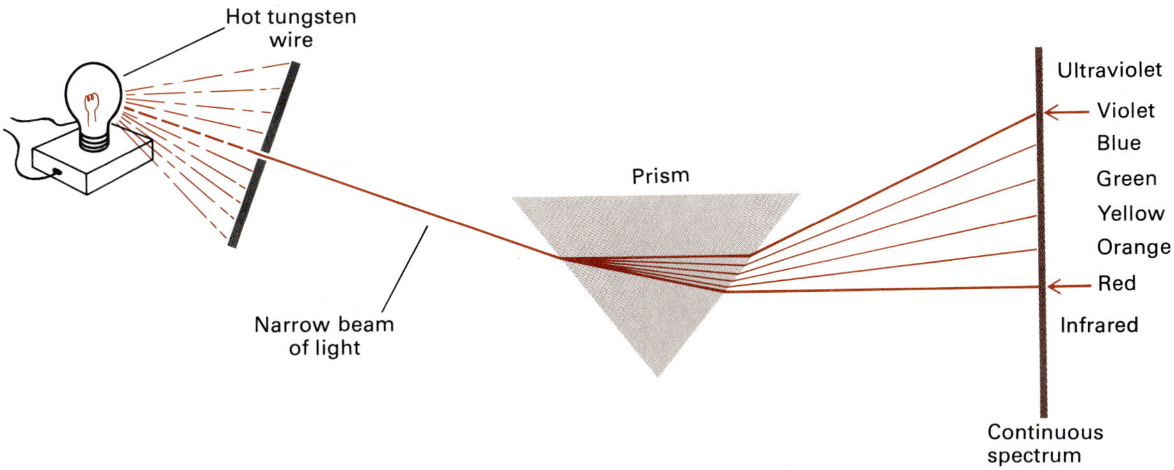

FIGURE 6-8

Diagram of prism spectroscope.

The different colors in the spectrum of white light correspond to different, characteristic wavelengths or frequencies of light, and thus to different energies.

If the tungsten wire in the spectroscope of Figure 6-8 is replaced with a discharge tube containing hydrogen gas, the spectrum produced at the right is not continuous as with the tungsten wire, but instead a series of sharp lines corresponding to definite frequencies (and thus also to definite wavelengths and to definite photon energies). Some of these lines are in the visible portion of the electromagnetic spectrum, and others are in the ultraviolet or infrared regions. Each line is the image, in a different color, of the original beam. This kind of spectrum is a **line spectrum.** Figure 6-9a shows a photograph of a line spectrum for hydrogen. Each line represents a definite frequency and energy of light emitted by hydrogen atoms. The photograph is a negative (it is light where the photographic plate was *not* exposed). The light spaces thus correspond to frequencies of light that are never emitted by hydrogen atoms.

Line spectra of several other elements, and of a mixture of elements, are shown in Figure 6-9b. The set of spectral lines is unique for each element and can therefore be used for identification. We noted earlier that helium was first detected on the sun; a spectrum of the sun's light contained the lines of this then-unknown element. These bright spectral lines can be seen when a substance is subjected to a high voltage in a gas discharge tube, or heated to a high temperature. Some substances also emit characteristic colors of light when heated to a high temperature in a flame: lithium compounds burn with a red flame; sodium, yellow; potassium, violet; calcium, orange-red, strontium, deep red; and barium, yellow-green. Some of these elements can be readily identified by their flame colors, but others (such as Ca and Sr) have colors so nearly alike that it is difficult to distinguish them by their flame colors, viewed by eye. Their line spectra, however, are easily distinguished, as can be seen in Figure 6-9b.

Hospital laboratories use spectral lines to determine

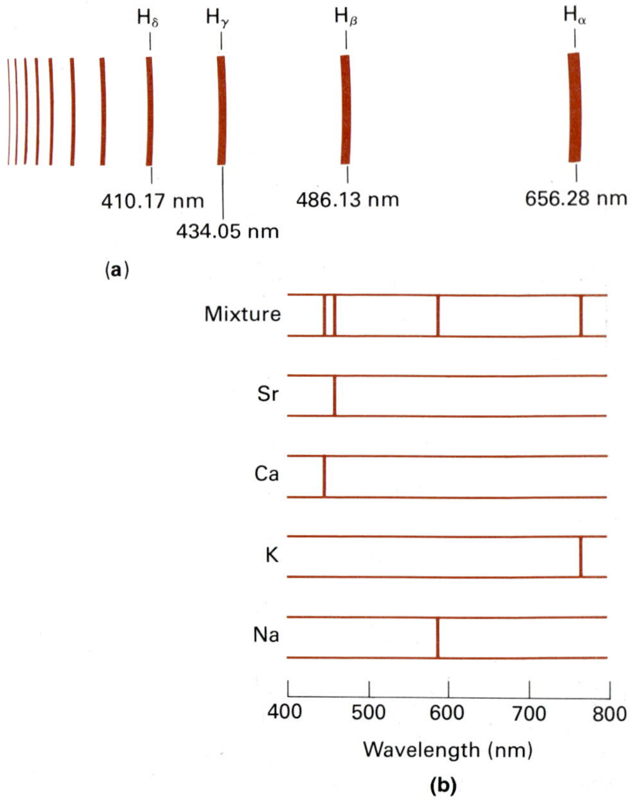

FIGURE 6-9

Line spectra. **(a)** The emission spectrum of the hydrogen atom in the visible and near ultraviolet regions. The dark lines represent wavelengths of light emitted by the atom. The intensity is indicated roughly by the thickness of the lines. [Courtesy of Gerhard Herzberg.] **(b)** Schematic representation of the most intense lines in the visible spectra of Na, K, Ca, Sr, and a mixture of all four. (Note that Ca and Sr do *not* emit light in the orange or red regions of the spectrum. Those flame colors, observed when Ca or Sr salts are heated in a flame, result from electron transitions in the molecules CaOH and SrOH, and not from the atoms.)

the concentrations of sodium, potassium, and calcium in blood serum. In a recently developed technique, serum samples are injected into a flame without pretreatment, and the spectral lines viewed on a television screen. The concentration of each element is determined by the intensity of its spectral line.

6-8 THE BOHR MODEL OF THE HYDROGEN ATOM

How can we explain characteristic spectra of the kind seen in Figure 6-9? The Rutherford model of the atom pictures the electrons whirling rapidly around a minute nucleus like planets around the sun, occupying by their rapid motion most of the volume of the atom, just as a swarm of bees effectively "occupies" a picnic area. But moving charged particles are known to radiate and lose energy. The Rutherford model of the atom failed to account for the existence of line spectra, and also failed to explain why the negative electrons do not lose energy through radiation and fall toward the positive nucleus.

These phenomena were accounted for by the Danish physicist Niels Bohr, who (in 1913) made the bold *assumption* that in a stable atom the electron can circle the nucleus forever without losing energy and falling into the nucleus. But an electron cannot circle the nucleus in just any orbit, rather only in certain specific orbits. We shall soon see that even Bohr's picture of the atom is not completely accurate, because an electron does not move in a well-defined path. Nevertheless, we shall briefly consider Bohr's model of the hydrogen atom because many of the ideas included in this model are still helpful.

The hydrogen atom consists of a nucleus and a single electron which, according to the Bohr model, can orbit the nucleus only in paths confined to certain **shells.** Each shell has associated with it a definite energy, so the electron can have *only* these energy values. The energy of the electrons in these shells increases with the shell number in the order 1, 2, 3, 4, 5, ... The shells are often called **energy levels;** shell number 1 is then energy level 1, shell number 2 is energy level 2, and so on. The evidence for these discrete energy levels is provided by line spectra.

In the stable hydrogen atom at room temperature, the electron is in the lowest energy level and the atom is said to be in its **ground state.** The electron can absorb a discrete amount of energy and jump to a higher energy level. The atom is then said to be in an **excited state.** The energy absorbed is exactly equal to the difference between energy levels. This energy may be supplied in the form of a photon of light, or as heat, or as electricity.

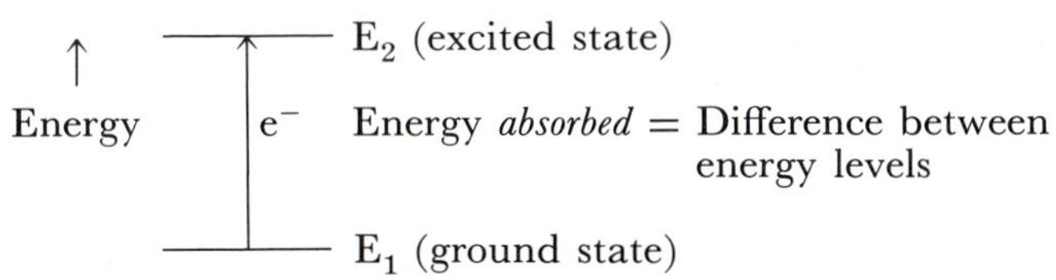

When an electron falls from a higher to a lower energy level, the atom radiates energy, as seen in a spectral line or flame color. The amount of energy emitted is exactly equal to the difference between energy levels.

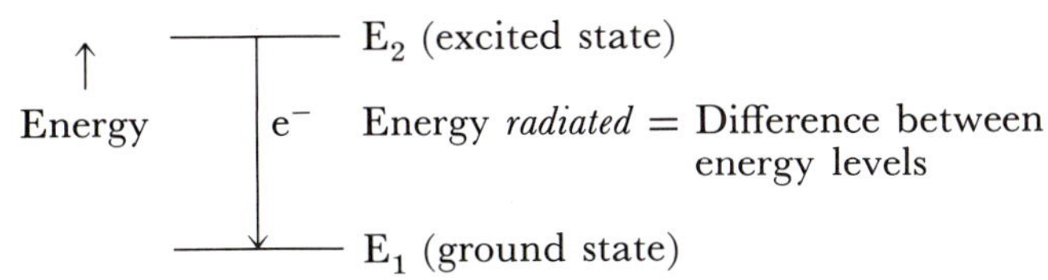

Energy is absorbed or emitted only in definite packets, or quanta, corresponding exactly to differences between the energy levels that are available to the electron. Thus, when an electron makes a transition from energy level 2 to energy level 1, the energy of the photon of light emitted is equal to the energy difference between energy level 2 and energy level 1.

The energy ($h\nu$) of any photon of light emitted by an atom is exactly the difference in energy between two such energy levels. By determining the frequency of a line in the hydrogen spectrum, we can find the difference in energy levels associated with that line. From the frequencies of all the lines in the hydrogen spectrum, the specific energy levels permitted to the hydrogen electron can then be determined.

Figure 6-10 shows an energy level diagram for the hydrogen atom. Arrows indicate some electron transitions from higher to lower energy levels. On each arrow, the energy change for that transition is given. The difference in energy between level 2 and level 1 is 1.63×10^{-18} J. When one electron falls from level 2 to level 1, a photon of light with energy 1.63×10^{-18} J is emitted (or 986 kJ per mole of photons). As seen in Table 6-10, this light is the lowest energy light in the ultraviolet region of the hydrogen spectrum, at a frequency of 24.7×10^{14} Hz or a wavelength of 122 nm. All the lines in the ultraviolet region correspond to electron transitions between higher energy levels and level 1, the ground state. All the lines in the visible region of the spectrum correspond to electron transitions from higher energy levels to level 2. Another series of spectral lines, in the infrared region, corresponds to electron transitions between higher energy levels and level 3.

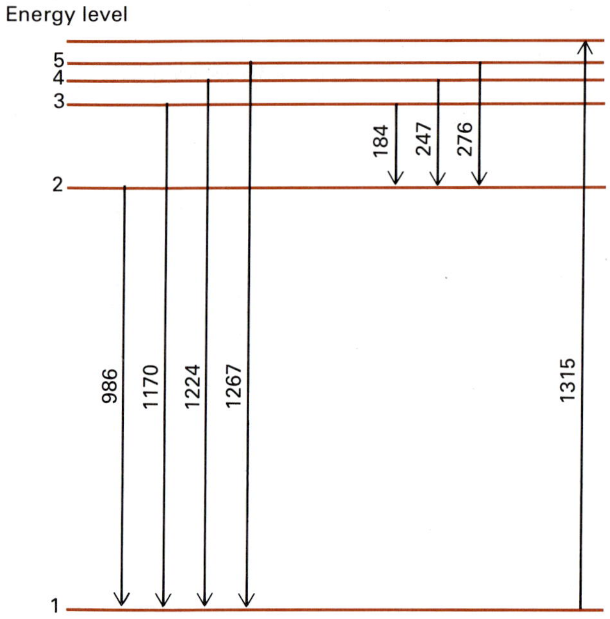

FIGURE 6-10
Energy levels of the hydrogen atom (kJ/mole).

The number of the energy level is designated n; thus $n = 1$ for the ground state. As n increases, the energy *difference* between successive levels becomes progressively smaller. Above $n = 6$ or 7, the energies are so close together they are almost indistinguishable. (It takes almost as much energy to move an electron from $n = 1$ to $n = 7$ as it does to remove the electron from the atom completely.) The highest energy shown in Figure 6-10 corresponds to an ionized state of the hydrogen atom, where the electron has just enough energy to escape from the attractive force of the nucleus. For the hydrogen atom, the minimum amount of energy that must be added to the atom in the ground state to ionize it is 2.184 J/atom or 1315 kJ/mole:

For one mole: $$H + 1315\ kJ \longrightarrow H^+ + e^-$$

This amount of energy is the ionization energy (or ionization potential) of the hydrogen atom, that same quantity we considered earlier in the chapter when describing the properties of groups in the periodic table.

TABLE 6-10
Some lines in the hydrogen spectrum

Region of spectrum	Frequency (ν) (Hz)	Wavelength (λ) (nm)	Energy $= h\nu$ (J/atom)	Energy $= h\nu$ (kJ/mole)
Visible	7.55×10^{14}	397.1	4.98×10^{-19}	301
	7.31×10^{14}	410.1	4.82×10^{-19}	290
	6.91×10^{14}	434.0	4.56×10^{-19}	276
	6.17×10^{14}	486.1	4.06×10^{-19}	247
	4.57×10^{14}	656.3	3.02×10^{-19}	184
Ultraviolet	32.2×10^{14}	93.1	21.3×10^{-19}	1283
	32.0×10^{14}	93.8	21.1×10^{-19}	1275
	31.6×10^{14}	95.0	21.0×10^{-19}	1267
	30.8×10^{14}	97.3	20.3×10^{-19}	1224
	29.2×10^{14}	102.6	19.4×10^{-19}	1170
	24.7×10^{14}	121.6	16.3×10^{-19}	986

6-9
ELECTRONIC ENERGY LEVELS

Bohr's simple model of the atom accounted for line spectra, and it even allowed accurate calculation of the energy levels of the hydrogen atom. However, this theory could not account for chemical bonding, or for details in the spectra of many-electron atoms. The years following Bohr's revolutionary proposals in 1913 were a period of great excitement and activity in physics and chemistry. By the late 1920s, Bohr's theory had been modified by Werner Heisenberg, Erwin Schrödinger, and others, who developed a mathematical theory of atomic structure called the **quantum theory.** This theory, which we shall not discuss in detail, is based on the idea that electrons have characteristics of waves as well as of particles, just as photons have properties both of waves and of particles.

For the hydrogen atom, this theory agrees exactly with the older theory of Bohr, and gives energy levels by the simple formula

$$\begin{aligned}\Delta E &= E_{\text{final}} - E_{\text{initial}}\\ &= h\nu\\ &= 1315\left(\frac{1}{n^2_{\text{initial}}} - \frac{1}{n^2_{\text{final}}}\right)\frac{\text{kJ}}{\text{mole}}\end{aligned}$$

Note that, as n_{final} grows very large, corresponding to the ionized state seen in Figure 6-10, $1/n^2$ approaches zero. If now n_{initial} is 1 (so that the electron transition to the ionized state is from $n = 1$), $\Delta E =$ 1315 kJ/mole. This is the same ionization energy we saw in Section 6-8 and in Figure 6-10.

EXERCISE 6-9 Calculate ΔE for the electron transition from $n = 1$ to $n = 2$, and compare it with that shown in Figure 6-10.

There are, however, important differences between Bohr's model and the quantum theory model. According to the quantum theory, electrons do not move in fixed orbits, like planets around the sun, as Bohr had assumed. Neither do they just move at random. Instead, as we shall see in the following section, there are regions of space around the nucleus in which the electrons are most likely to be found. We cannot precisely locate the electron in space; we can only describe the probability, or chance, of finding the electron at various positions around the nucleus.

The region of space in which an electron is most likely to be found is called an **orbital;** the shape and size of the orbital depends on the energy of the electron. We shall consider the shapes and sizes of orbitals shortly. The orbitals of an atom are quantized. They come in definite intervals of fixed energy—in definite **energy levels**—as we have already seen for the hydrogen atom. The names, or symbols, of the orbitals are:

1*s*					
2*s*	2*p*				
3*s*	3*p*	3*d*			
4*s*	4*p*	4*d*	4*f*		
5*s*	5*p*	5*d*	5*f*	5*g*	
6*s*	6*p*	6*d*	6*f*	6*g*	6*h*
7*s*	7*p*	. . .			

The orbitals are in shells and subshells. The number in the symbol for the orbital specifies the **shell;** it is the **principal quantum number** (n). The letters *s, p, d, f,* . . . specify the **subshells** within each shell. *For each principal quantum number n, there are n subshells.* Thus, for principal quantum number $n = 1$, there is one subshell, the 1*s* subshell. For $n = 2$, there are two subshells, the 2*s* and 2*p* subshells. For $n = 3$, there are three subshells, the 3*s,* 3*p,* and 3*d* subshells; etc. A subshell itself may consist of more than one orbital. We shall soon learn more about these orbitals, but for now we give only their count:

The *s* subshell consists of 1 orbital;

the *p* subshell consists of 3 orbitals;

the *d* subshell consists of 5 orbitals;

the *f* subshell consists of 7 orbitals.

Notice that for each value of the principal quantum number n, there are n^2 orbitals:

For $n = 1$, there is one subshell (1*s*), consisting of a single *s* orbital ($n^2 = 1$).

For $n = 2$, there are two subshells (2*s* and 2*p*), with a total of $n^2 = 4$ orbitals: one *s* orbital and three *p* orbitals.

For $n = 3$, there are three subshells (3*s*, 3*p*, and 3*d*), with a total of $n^2 = 9$ orbitals: one *s* orbital, three *p* orbitals, and five *d* orbitals.

For $n = 4$, there are four subshells (4*s*, 4*p*, 4*d*, and 4*f*), with a total of $n^2 = 16$ orbitals: one *s* orbital, three *p* orbitals, five *d* orbitals, and seven *f* orbitals.

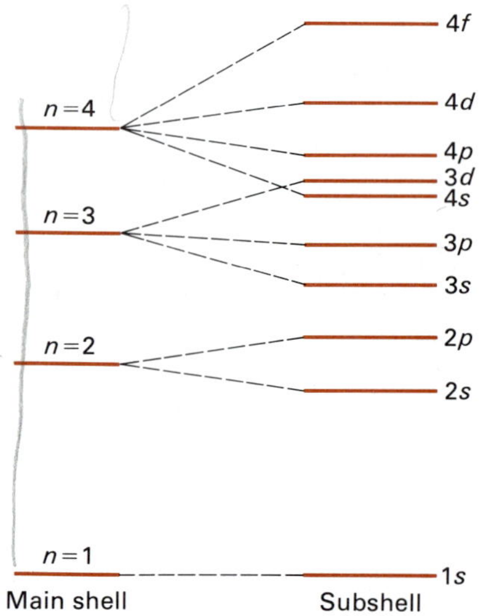

FIGURE 6-11

Electrons in different subshells differ in energy.

In the hydrogen atom, which has only one electron, all the orbitals of any given principal quantum number are of the same energy. Thus Figure 6-10 shows energies only for the principal quantum numbers, or main shells. In a many-electron atom, however, the electrons in the different subshells of any main shell differ somewhat in energy. The *s* subshell is of slightly lower energy than the *p*, the *p* subshell of lower energy than the *d*, and the *d* subshell of lower energy than the *f* (Figure 6-11). No scale is given for the energy levels in the figure because the energy is characteristic of the element. As the figure shows, the 4*s* orbital is of lower energy than the 3*d* orbital; and both the 5*p* and the 6*s* orbitals are of lower energy than the 4*d* orbital.

The diagram shown in Figure 6-12 is a device for remembering the order of energy of the orbitals in a typical many-electron atom. The energy of the subshells increases in the direction of the arrows. Thus, the energies of the orbitals increase in the order 1*s*, 2*s*, 2*p*, 3*s*, 3*p*, 4*s*, 3*d*, 4*p*, 5*s*, . . . , in agreement with Figure 6-11.

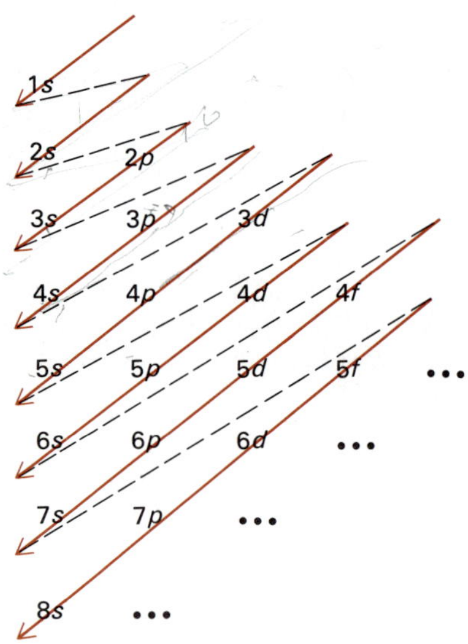

FIGURE 6-12

Order of orbital energy levels.

6-10
SHAPES OF ORBITALS

Electrons move in such a way that we cannot know exactly where they are at any instant, but we know where they are on the average. The probability, or chance, of finding an electron at a specific point can be determined. The shape of this electron probability distribution in space depends on the orbital the electron is in. With each orbital there is associated some definite spatial distribution for the electrons that occupy that orbital.

All the *s* orbitals are spherically symmetric about the nucleus: at a *given distance* from the nucleus, the chance of finding the electron is the same *in any direction.* That is, the probability of finding the electron at any point depends only on the distance of that point from the nucleus, not on the direction. Figure 6-13 is an electron "cloud" picture of a 1*s* orbital. It shows the shape of the orbital—a spherical region around the nucleus in which there is a strong chance of finding the electron. The electron can be imagined as spread out like that cloud, smeared throughout a spherical region around the nucleus. The entire sphere represents the region around the nucleus where the electron would be found most (say, 95%) of the time, but there is no sharply defined limit to the size of the atom. There is some slight chance that the electron will be found outside this region, far from the nucleus, though that chance becomes rapidly smaller as the distance from the nucleus increases.

The 2*s* orbital, like the 1*s* orbital, is spherically symmetric about the nucleus. The 2*s* orbital is of higher energy than a 1*s* orbital, and a 2*s* electron is, on the average, farther from the nucleus than a 1*s* electron.

Figure 6-14 shows the spatial distributions of three *p* orbitals. Unlike *s* orbitals, the *p* orbitals have angular, or directional, properties. (We shall see in the next chapter that these directional properties are important in determining the shapes of molecules.) Each *p* orbital consists of two dumbbell-shaped regions (one on either side of the nucleus) concentrated about one direction, or line, through the nucleus. An electron in a *p* orbital is concentrated within the two lobes of the orbital. As shown in the figure, the three orbitals are at

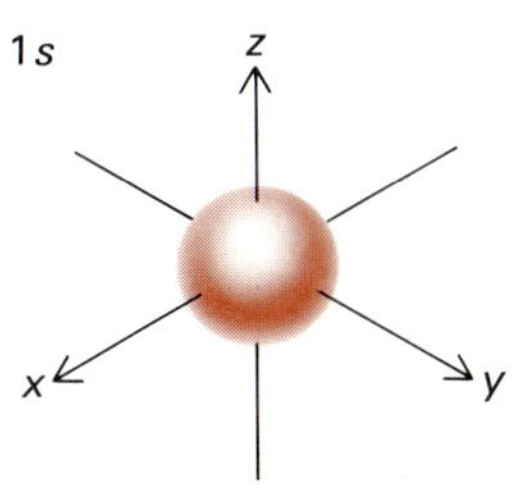

FIGURE 6-13

The shape of a 1*s* orbital. There is a high probability of finding the electron within the sphere.

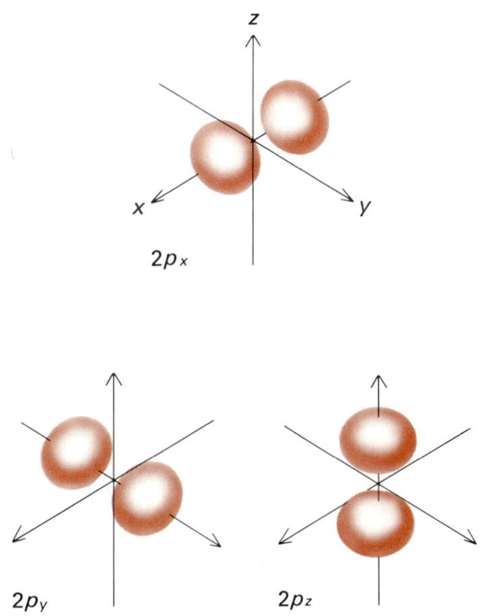

FIGURE 6-14

Shapes of *p* orbitals. An electron in a *p* orbital is concentrated within the two lobes of the orbital.

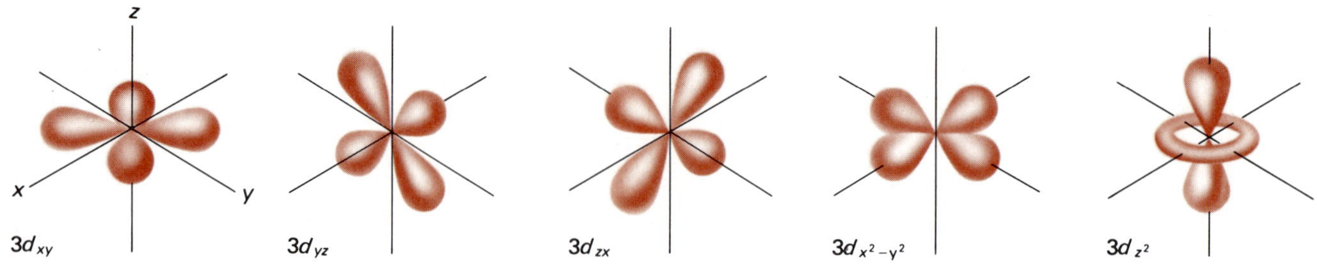

FIGURE 6-15

Shapes of *d* orbitals.

90° angles to each other; these directions are called x, y, and z, and the corresponding p orbitals are called p_x, p_y, and p_z. For every principal quantum number n greater than 1, there are three p orbitals of equal energy. Electrons in p orbitals (like those in s orbitals) are, on the average, farther from the nucleus for a larger principal quantum number of the orbital. Thus a $3p$ electron is, on the average, farther from the nucleus than a $2p$ electron.

Figure 6-15 shows the spatial distribution of the five d orbitals. For every principal quantum number greater than $n = 2$, there are five d orbitals of equal energy. The d orbitals, like the p orbitals, have directional properties.

6-11 ELECTRONIC STRUCTURE OF ATOMS

We are now almost ready to describe the electronic structure of an atom—that is, the distribution of the electrons among the atomic orbitals. We know how many orbitals there are, but we do not yet know how many electrons there can be in each orbital. All electrons are negatively charged, and hence repel each other. But electrons are not all identical; there are two kinds, which differ in a property called "spin." We can think of the electron as rotating about its own internal axis, just as the earth rotates about its axis every 24 hours. The electron can spin in two possible directions, say in a clockwise direction or in a counterclockwise direction. These are often called "spin up" and "spin down," and designated ↑ and ↓. *In any one orbital there can be at most two electrons, one of each spin.*

We have seen that for each principal quantum number n there are n subshells, comprising n^2 orbitals altogether. Each orbital can have in it a *maximum* of 2 electrons. Thus, there can be at most $2n^2$ electrons with the same principal quantum number n. Each s orbital can accommodate 2 electrons:

$$\underline{\downarrow\uparrow}$$
s orbital

Each group of three p orbitals can accommodate 6 electrons:

↓↑ ↓↑ ↓↑
p orbitals

Each group of five *d* orbitals can accommodate 10 electrons:

↓↑ ↓↑ ↓↑ ↓↑ ↓↑
d orbitals

Each group of seven *f* orbitals can accommodate 14 electrons:

↓↑ ↓↑ ↓↑ ↓↑ ↓↑ ↓↑ ↓↑
f orbitals

So, for $n = 1$, a maximum of $2n^2 = 2$ electrons is permitted:

↓↑
1*s*
2 electrons

For $n = 2$, a maximum of $2n^2 = 8$ electrons is permitted:

↓↑	↓↑ ↓↑ ↓↑
2*s*	2*p*
2 electrons	6 electrons

8 electrons

For $n = 3$, a maximum of $2n^2 = 18$ electrons is permitted:

↓↑	↓↑ ↓↑ ↓↑	↓↑ ↓↑ ↓↑ ↓↑ ↓↑
3*s*	3*p*	3*d*
2 electrons	6 electrons	10 electrons

18 electrons

We can now describe the lowest-energy (that is, the most stable, or ground-state) electronic configuration of atoms of all the elements. To see this, we can imagine assigning the electrons in an atom to orbitals, always placing an electron in the lowest energy level available to it. (Remember, the energy levels of the orbitals increase in the order shown by the arrows in Figure 6-12: 1*s*, 2*s*, 2*p*, 3*s*, 3*p*, 4*s*, 3*d*, . . .). From this order, we can deduce the ordinary, stable electron configurations of the first 20 elements (Table 6-11). In the filling of atomic orbitals, when there are two or more electrons in a *p, d,* or *f* subshell, those electrons remain unpaired as long as possible, because of the repulsive forces between electrons. In nitrogen, for example, the three *p* electrons have the same spin and are distributed one to each of the three different *p* orbitals (see Table 6-11). In oxygen there are four *p* electrons, so two must be paired in the same orbital, but the other two are unpaired.

The electronic configurations are often represented as shown in the righthand column of Table 6-11, where the number and letter name the orbital, and the superscript gives the number of electrons in that orbital. Thus, hydrogen ($1s^1$) means that there is one electron in the 1*s* orbital; and boron ($1s^2 2s^2 2p^1$) means that there are two electrons in the 1*s* orbital, two electrons in the 2*s* orbital, and one electron in the 2*p* orbital. The ground-state electronic configurations of the atoms of all the elements are given in Table 6-12.

EXERCISE 6-10 How many electrons can be accommodated in the fourth shell ($n = 4$)? Show how many electrons are in each subshell.

EXERCISE 6-11 How many electrons are in the calcium atom, which has the electronic configuration $1s^2\ 2s^2\ 2p^6\ 3s^2\ 3p^6\ 4s^2$?

TABLE 6-11
Ground state electronic configurations of the first 20 elements

Element	Atomic number	$1s$	$2s$	$2p$	$3s$	$3p$	$3d$	$4s$	Shorthand form of writing electronic configuration
H	1	↓							$1s^1$
He	2	↓↑							$1s^2$
Li	3	↓↑	↓						$1s^2 2s^1$
Be	4	↓↑	↓↑						$1s^2 2s^2$
B	5	↓↑	↓↑	↓					$1s^2 2s^2 2p^1$
C	6	↓↑	↓↑	↓ ↓ —					$1s^2 2s^2 2p^2$
N	7	↓↑	↓↑	↓ ↓ ↓					$1s^2 2s^2 2p^3$
O	8	↓↑	↓↑	↓↑ ↓ ↓					$1s^2 2s^2 2p^4$
F	9	↓↑	↓↑	↓↑ ↓↑ ↓					$1s^2 2s^2 2p^5$
Ne	10	↓↑	↓↑	↓↑ ↓↑ ↓↑					$1s^2 2s^2 2p^6$
Na	11	↓↑	↓↑	↓↑ ↓↑ ↓↑	↓				$1s^2 2s^2 2p^6 3s^1$
Mg	12	↓↑	↓↑	↓↑ ↓↑ ↓↑	↓↑				$1s^2 2s^2 2p^6 3s^2$
Al	13	↓↑	↓↑	↓↑ ↓↑ ↓↑	↓↑	↓ — —			$1s^2 2s^2 2p^6 3s^2 3p^1$
Si	14	↓↑	↓↑	↓↑ ↓↑ ↓↑	↓↑	↓ ↓ —			$1s^2 2s^2 2p^6 3s^2 3p^2$
P	15	↓↑	↓↑	↓↑ ↓↑ ↓↑	↓↑	↓ ↓ ↓			$1s^2 2s^2 2p^6 3s^2 3p^3$
S	16	↓↑	↓↑	↓↑ ↓↑ ↓↑	↓↑	↓↑ ↓ ↓			$1s^2 2s^2 2p^6 3s^2 3p^4$
Cl	17	↓↑	↓↑	↓↑ ↓↑ ↓↑	↓↑	↓↑ ↓↑ ↓			$1s^2 2s^2 2p^6 3s^2 3p^5$
Ar	18	↓↑	↓↑	↓↑ ↓↑ ↓↑	↓↑	↓↑ ↓↑ ↓↑			$1s^2 2s^2 2p^6 3s^2 3p^6$
K	19	↓↑	↓↑	↓↑ ↓↑ ↓↑	↓↑	↓↑ ↓↑ ↓↑		↓	$1s^2 2s^2 2p^6 3s^2 3p^6 4s^1$
Ca	20	↓↑	↓↑	↓↑ ↓↑ ↓↑	↓↑	↓↑ ↓↑ ↓↑		↓↑	$1s^2 2s^2 2p^6 3s^2 3p^6 4s^2$

TABLE 6-12
Electronic configurations of the elements in the ground state

Element	Atomic number	$1s$	$2s$	$2p$	$3s$	$3p$	$3d$	$4s$	$4p$	$4d$	$4f$	$5s$	$5p$	$5d$	$5f$	$5g$	$6s$	$6p$	$6d$	...	$7s$
H	1	1																			
He	2	2																			
Li	3	2	1																		
Be	4	2	2																		
B	5	2	2	1																	
C	6	2	2	2																	

TABLE 6-12 (continued)

Element	Atomic number	1*s*	2*s*	2*p*	3*s*	3*p*	3*d*	4*s*	4*p*	4*d*	4*f*	5*s*	5*p*	5*d*	5*f*	5*g*	6*s*	6*p*	6*d*	. . .	7*s*
N	7	2	2	3																	
O	8	2	2	4																	
F	9	2	2	5																	
Ne	10	2	2	6																	
Na	11	2	2	6	1																
Mg	12	2	2	6	2																
Al	13	2	2	6	2	1															
Si	14	2	2	6	2	2															
P	15	2	2	6	2	3															
S	16	2	2	6	2	4															
Cl	17	2	2	6	2	5															
Ar	18	2	2	6	2	6															
K	19	2	2	6	2	6		1													
Ca	20	2	2	6	2	6		2													
Sc	21	2	2	6	2	6	1	2													
Ti	22	2	2	6	2	6	2	2													
V	23	2	2	6	2	6	3	2													
Cr	24	2	2	6	2	6	5	1													
Mn	25	2	2	6	2	6	5	2													
Fe	26	2	2	6	2	6	6	2													
Co	27	2	2	6	2	6	7	2													
Ni	28	2	2	6	2	6	8	2													
Cu	29	2	2	6	2	6	10	1													
Zn	30	2	2	6	2	6	10	2													
Ga	31	2	2	6	2	6	10	2	1												
Ge	32	2	2	6	2	6	10	2	2												
As	33	2	2	6	2	6	10	2	3												
Se	34	2	2	6	2	6	10	2	4												
Br	35	2	2	6	2	6	10	2	5												
Kr	36	2	2	6	2	6	10	2	6												
Rb	37	2	2	6	2	6	10	2	6			1									
Sr	38	2	2	6	2	6	10	2	6			2									
Y	39	2	2	6	2	6	10	2	6	1		2									
Zr	40	2	2	6	2	6	10	2	6	2		2									
Nb	41	2	2	6	2	6	10	2	6	4		1									
Mo	42	2	2	6	2	6	10	2	6	5		1									
Tc	43	2	2	6	2	6	10	2	6	5		2									
Ru	44	2	2	6	2	6	10	2	6	7		1									
Rh	45	2	2	6	2	6	10	2	6	8		1									
Pd	46	2	2	6	2	6	10	2	6	10											
Ag	47	2	2	6	2	6	10	2	6	10		1									

(continued)

TABLE 6-12 (continued)

Element	Atomic number	1*s*	2*s*	2*p*	3*s*	3*p*	3*d*	4*s*	4*p*	4*d*	4*f*	5*s*	5*p*	5*d*	5*f*	5*g*	6*s*	6*p*	6*d*	. . .	7*s*
Cd	48	2	2	6	2	6	10	2	6	10		2									
In	49	2	2	6	2	6	10	2	6	10		2	1								
Sn	50	2	2	6	2	6	10	2	6	10		2	2								
Sb	51	2	2	6	2	6	10	2	6	10		2	3								
Te	52	2	2	6	2	6	10	2	6	10		2	4								
I	53	2	2	6	2	6	10	2	6	10		2	5								
Xe	54	2	2	6	2	6	10	2	6	10		2	6								
Cs	55	2	2	6	2	6	10	2	6	10		2	6				1				
Ba	56	2	2	6	2	6	10	2	6	10		2	6				2				
La	57	2	2	6	2	6	10	2	6	10	1	2	6				2				
Ce	58	2	2	6	2	6	10	2	6	10	2	2	6				2				
Pr	59	2	2	6	2	6	10	2	6	10	3	2	6				2				
Nd	60	2	2	6	2	6	10	2	6	10	4	2	6				2				
Pm	61	2	2	6	2	6	10	2	6	10	5	2	6				2				
Sm	62	2	2	6	2	6	10	2	6	10	6	2	6				2				
Eu	63	2	2	6	2	6	10	2	6	10	7	2	6				2				
Gd	64	2	2	6	2	6	10	2	6	10	7	2	6	1			2				
Tb	65	2	2	6	2	6	10	2	6	10	9	2	6				2				
Dy	66	2	2	6	2	6	10	2	6	10	10	2	6				2				
Ho	67	2	2	6	2	6	10	2	6	10	11	2	6				2				
Er	68	2	2	6	2	6	10	2	6	10	12	2	6				2				
Tm	69	2	2	6	2	6	10	2	6	10	13	2	6				2				
Yb	70	2	2	6	2	6	10	2	6	10	14	2	6				2				
Lu	71	2	2	6	2	6	10	2	6	10	14	2	6	1			2				
Hf	72	2	2	6	2	6	10	2	6	10	14	2	6	2			2				
Ta	73	2	2	6	2	6	10	2	6	10	14	2	6	3			2				
W	74	2	2	6	2	6	10	2	6	10	14	2	6	4			2				
Re	75	2	2	6	2	6	10	2	6	10	14	2	6	5			2				
Os	76	2	2	6	2	6	10	2	6	10	14	2	6	6			2				
Ir	77	2	2	6	2	6	10	2	6	10	14	2	6	7			2				
Pt	78	2	2	6	2	6	10	2	6	10	14	2	6	9			1				
Au	79	2	2	6	2	6	10	2	6	10	14	2	6	10			1				
Hg	80	2	2	6	2	6	10	2	6	10	14	2	6	10			2				
Tl	81	2	2	6	2	6	10	2	6	10	14	2	6	10			2	1			
Pb	82	2	2	6	2	6	10	2	6	10	14	2	6	10			2	2			
Bi	83	2	2	6	2	6	10	2	6	10	14	2	6	10			2	3			
Po	84	2	2	6	2	6	10	2	6	10	14	2	6	10			2	4			
At	85	2	2	6	2	6	10	2	6	10	14	2	6	10			2	5			
Rn	86	2	2	6	2	6	10	2	6	10	14	2	6	10			2	6			
Fr	87	2	2	6	2	6	10	2	6	10	14	2	6	10			2	6			1
Ra	88	2	2	6	2	6	10	2	6	10	14	2	6	10			2	6			2
Ac	89	2	2	6	2	6	10	2	6	10	14	2	6	10			2	6	1		2

TABLE 6-12 (continued)

Element	Atomic number	1s	2s	2p	3s	3p	3d	4s	4p	4d	4f	5s	5p	5d	5f	5g	6s	6p	6d	. . .	7s
Th	90	2	2	6	2	6	10	2	6	10	14	2	6	10			2	6	2		2
Pa	91	2	2	6	2	6	10	2	6	10	14	2	6	10	2		2	6	1		2
U	92	2	2	6	2	6	10	2	6	10	14	2	6	10	3		2	6	1		2
Np	93	2	2	6	2	6	10	2	6	10	14	2	6	10	4		2	6	1		2
Pu	94	2	2	6	2	6	10	2	6	10	14	2	6	10	6		2	6			2
Am	95	2	2	6	2	6	10	2	6	10	14	2	6	10	7		2	6			2
Cm	96	2	2	6	2	6	10	2	6	10	14	2	6	10	7		2	6	1		2
Bk	97	2	2	6	2	6	10	2	6	10	14	2	6	10	8		2	6	1		2
Cf	98	2	2	6	2	6	10	2	6	10	14	2	6	10	10		2	6			2
Es	99	2	2	6	2	6	10	2	6	10	14	2	6	10	11		2	6			2
Fm	100	2	2	6	2	6	10	2	6	10	14	2	6	10	12		2	6			2
Md	101	2	2	6	2	6	10	2	6	10	14	2	6	10	13		2	6			2
No	102	2	2	6	2	6	10	2	6	10	14	2	6	10	14		2	6			2
Lr	103	2	2	6	2	6	10	2	6	10	14	2	6	10	14		2	6	1		2
Ku	104	2	2	6	2	6	10	2	6	10	14	2	6	10	14		2	6	2		2
Ha	105	2	2	6	2	6	10	2	6	10	14	2	6	10	14		2	6	3		2

Note: There are occasional irregularities in the filling of *d* and *f* orbitals. Usually these are such as to produce a half-filled or filled orbital earlier than would otherwise be accomplished. Filled and half-filled orbitals are especially stable.

6-12 ELECTRONIC CONFIGURATION AND THE PERIODIC TABLE

The first horizontal row (period) of the periodic table consists of two elements: hydrogen (H) with the electronic configuration $1s^1$, and helium (He), with the electronic configuration $1s^2$. In this period, the $1s$ orbital, which can hold a maximum of two electrons, is half-filled at H and filled at He.

The second row consists of eight elements. It begins with the alkali metal lithium (Li), with the electronic configuration $1s^22s^1$, and ends with the noble gas neon (Ne), with the electronic configuration $1s^22s^22p^6$. The $2s$ and $2p$ orbitals, which together can accommodate eight electrons, are increasingly occupied and finally filled across this period of eight elements (see Table 6-12).

The third row also consists of eight elements. It begins with the alkali metal sodium (Na), with the electronic configuration $1s^22s^22p^63s^1$, and ends with the noble gas argon (Ar), with the electronic configuration $1s^22s^22p^63s^23p^6$. The $3s$ and $3p$ orbitals, which together can accommodate eight electrons, are gradually filled across this row.

In the fourth row there are 18 elements. The additional electrons of these elements, beyond those of Ar, are accommodated first in the $4s$ orbital, which is lower in energy than the $3d$ orbital (Figure 6-11):

Potassium (K): $1s^22s^22p^63s^23p^64s^1$
or $[Ar]4s^1$ an alkali metal

Calcium (Ca): $1s^22s^22p^63s^23p^64s^2$
or $[Ar]4s^2$ an alkaline earth

Here the symbol $[\text{Ar}]4s^1$ means that K has the electronic configuration of argon, plus an extra electron in the 4*s* orbital. Then, over the next 10 elements, 10 electrons are added in the 3*d* orbitals. These 10 elements constitute the **first transition series:** scandium (Sc), titanium (Ti), vanadium (V), chromium (Cr), manganese (Mn), iron (Fe), cobalt (Co), nickel (Ni), copper (Cu), and zinc (Zn). They are all metals, and are good conductors of heat and electricity. After the 3*d* orbitals are filled, electrons go into the 4*p* orbitals (Figure 6-11), completing the row:

Gallium (Ga): in the group of B and Al	$[\text{Ar}]3d^{10}4s^24p^1$
Germanium (Ge): in the group of C and Si	$[\text{Ar}]3d^{10}4s^24p^2$
Arsenic (As): in the group of N and P	$[\text{Ar}]3d^{10}4s^24p^3$
Selenium (Se): in the group of O and S	$[\text{Ar}]3d^{10}4s^24p^4$
Bromine (Br): a halogen	$[\text{Ar}]3d^{10}4s^24p^5$
Krypton (Kr): a noble gas	$[\text{Ar}]3d^{10}4s^24p^6$

This row, like the second and third, begins with an alkali metal (K) and ends with a noble gas (Kr). The 4*s*, 3*d*, and 4*p* orbitals, which together can accommodate 18 electrons, are filled across this row of 18 elements.

In the fifth period there are again 18 elements. The additional electrons, beyond those of Kr, are accommodated first in the 5*s* orbital:

Rubidium (Rb):	$[\text{Kr}]5s^1$	an alkali metal
Strontium (Sr):	$[\text{Kr}]5s^2$	an alkaline earth

Then, over the next ten elements, the 4*d* orbitals are filled. These ten elements constitute the **second transition series:** yttrium (Y), zirconium (Zr), niobium (Nb), molybdenum (Mo), technetium (Tc), ruthenium (Ru), rhodium (Rh), palladium (Pd), silver (Ag), and cadium (Cd). Technetium was the first "artificial" element, produced in 1937 by bombarding molybdenum (Mo) with neutrons. After the 4*d* orbitals are filled, electrons go into the 5*p* orbitals, completing the row:

Indium (In): in the group of B, Al, and Ga	$[\text{Kr}]4d^{10}5s^25p^1$
Tin (Sn): in the group of C, Si, and Ge	$[\text{Kr}]4d^{10}5^25p^2$
Antimony (Sb): in the group of N, P, and As	$[\text{Kr}]4d^{10}5s^25p^3$
Tellurium (Te): in the group of O, S, and Se	$[\text{Kr}]4d^{10}5s^25p^4$
Iodine (I): a halogen	$[\text{Kr}]4d^{10}5s^25p^5$
Xenon (Xe): a noble gas	$[\text{Kr}]4d^{10}5s^25p^6$

This row also begins with an alkali metal (Rb) and ends with a noble gas (Xe). The 5*s*, 4*d*, and 5*p* orbitals, which together can accommodate 18 electrons, are filled across this row of 18 elements.

Period 6 has a total of 32 elements. The 6*s* orbital is filled first:

Cesium (Cs):	$[\text{Xe}]6s^1$	an alkali metal
Barium (Ba):	$[\text{Xe}]6s^2$	an alkaline earth

Next filled are, primarily, the 4*f* orbitals, which can accommodate 14 electrons (though as seen in Table 6-12, lanthanum, La, and gadolinium, Gd, each have one electron in a 5*d* orbital). These elements, beginning with La and ending with lutetium (Lu), are called the **lanthanides,** or **rare earths.** They are all soft, easily worked metals, so similar in their properties that separating them is extremely difficult. Naturally occurring mixtures of the rare earths constitute an alloy called "Misch metal," which is used in the production of iron and steel to absorb impurities. Next to be filled are the 5*d* orbitals; this is the **third transition series,** which ends with mercury (Hg). Finally, the 6*p* orbitals are filled, completing the row:

Thallium (Tl): $[Xe]4f^{14}5d^{10}6s^{2}6p^{1}$
in the group of B, Al, Ga, and In

Lead (Pb): $[Xe]4f^{14}5d^{10}6s^{2}6p^{2}$
in the group of C, Si, Ge, and Sn

Bismuth (Bi): $[Xe]4f^{14}5d^{10}6s^{2}6p^{3}$
in the group of N, P, As, and Sb

Polonium (Po): $[Xe]4f^{14}5d^{10}6s^{2}6p^{4}$
in the group of O, S, Se, and Te

Astatine (At): $[Xe]4f^{14}5d^{10}6s^{2}6p^{5}$
a halogen

Radon (Rn): $[Xe]4f^{14}5d^{10}6s^{2}6p^{6}$
a noble gas

This row also begins with an alkali metal (Cs) and ends with a noble gas (Rn). The 6*s*, 4*f*, 5*d*, and 6*p* orbitals, which together can accommodate 32 electrons, are filled across this row of 32 elements.

The seventh row begins with the filling of the 7*s* orbital:

Francium (Fr):	$[Rn]6s^{1}$	an alkali metal
Radium (Ra):	$[Rn]6s^{2}$	an alkaline earth

Next filled are, primarily, the 5*f* orbitals, which can accommodate 14 electrons. These elements, beginning with actinium (Ac) and ending with lawrencium (Lr) are called the **actinides.** They are all radioactive. All the elements having atomic number greater than 92 (uranium) are artificial. (They may once have existed in nature and vanished because of their short half-lives.) The next orbitals to be filled are the 6*d* orbitals, beginning a new transition series, which thus far includes only two elements: elements 104, kurchatorium (Ku), and 105, hahnium (Ha).

We see that the number of elements in each row of the periodic table is related to the electronic configurations of the atoms. Figure 6-16 shows which orbitals are filled in each row of the periodic table. The number of electrons that can be accommodated in the orbitals filled in any row is equal to the number of elements in that row.

In the electronic configurations of the atoms, we can see a basis for the periodic recurrence of chemical

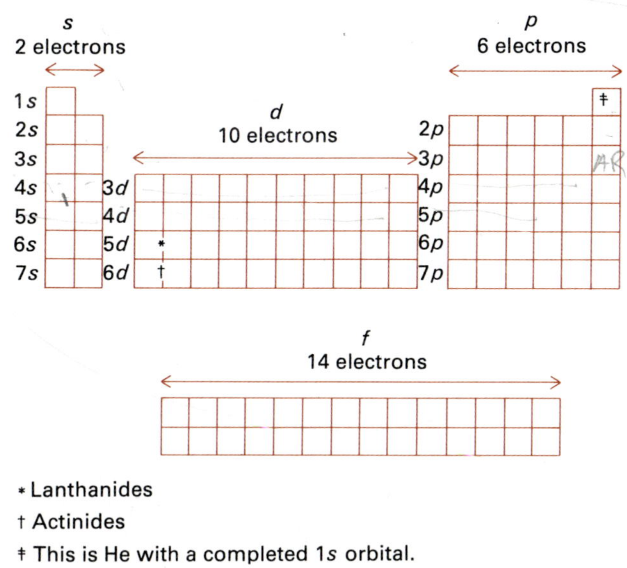

FIGURE 6-16

Orbitals filled in each period of periodic table.

properties of the elements: the chemical properties of an element depend primarily on the electronic configuration of its outer or **valence** shell. Elements with the same outer-shell electronic configuration have similar chemical properties, and fall into the same group in the periodic table.

The noble gases:

He: $1s^2$
Ne: $1s^2\,2s^2\,2p^6$
Ar: $1s^2\,2s^2\,2p^6\,3s^2\,3p^6$
Kr: . . . . $4s^2\,4p^6$
Xe: . . . . $5s^2\,5p^6$
Rn: . . . . $6s^2\,6p^6$

Helium has a completed first shell. Each of the other noble gases has completed the inner two subshells (the *s* and *p* subshells) of its outer main shell, and thus has 8 electrons in its outer shell.

The alkali metals:

Li: $[He]2s^1$
Na: $[Ne]3s^1$
K: $[Ar]4s^1$
Rb: $[Kr]5s^1$
Cs: $[Xe]6s^1$
Fr: $[Rn]7s^1$

Each of the alkali metals has only a single *s* electron in its outer shell; a lone electron outside a core that has the electronic configuration of the preceding noble gas atom.

The alkaline earths:

Be: $[He]2s^2$
Mg: $[Ne]3s^2$
Ca: $[Ar]4s^2$
Sr: $[Kr]5s^2$
Ba: $[Xe]6s^2$
Ra: $[Rn]7s^2$

Each of the alkaline earths has two electrons in its outer shell; two electrons outside a core that has the electronic configuration of the preceding noble gas atom.

The halogens:

F: $1s^2\,2s^2\,2p^5$
Cl: $1s^2\,2s^2\,2p^6\,3s^2\,3p^5$
Br: . . . . $4s^2\,4p^5$
I: . . . . $5s^2\,5p^5$
At: . . . . $6s^2\,6p^5$

Each of the halogens has seven electrons in its outer shell, lacking only one electron of the electronic configuration of a noble gas atom.

To sum up, we have seen that elements with the same electronic configurations in their outer shells form groups in the periodic table. It is these outer electrons that principally determine the chemical behavior of an element.

6-13 IONIZATION POTENTIAL AND THE PERIODIC TABLE

The noble gases are almost totally unreactive. They occur as monatomic gases and, as recently as 1962 (the year of the discovery of xenon tetrafluoride, XeF_4) were thought to form no compounds. (See Figure 6-17.) This inertness can be explained in terms of their electronic configurations, which are of special stability

FIGURE 6-17

Crystals of xenon tetrafluoride (XeF_4). The container is 22 mm in diameter. [Photograph courtesy of Cedric L. Chernick and John G. Malm, Argonne National Laboratory.]

(as seen, for example, in the ionization potential—the energy required to remove an electron from the neutral atom).

Of the elements in the second period, the noble gas neon has the highest ionization potential (Table 6-5). This means that it is more difficult to remove an electron from a neon atom than from any other element in this period. The same is true of the noble gas atom in each period, as shown graphically in Figure 6-18. The special characteristic of the electronic configuration of the noble gases is that the *s* and *p* orbitals of the outer shell are complete, with eight electrons (He, which has only two electrons, has a completed 1*s* shell). A pair of electrons in the $n = 1$ shell or an octet of electrons in higher shells is a particularly stable electronic configuration.

The alkali metals, as we have seen, are highly reactive. The ionization potentials of this group are, as a whole, lower than those of any other group (Figure 6-18). It is relatively easy to remove an electron from an alkali metal atom. Each alkali metal atom has a single electron in the *s* orbital of its outer shell. When one electron is removed, the ion formed (M^+) has the electronic configuration of the immediately preceding noble gas atom:

$$\underset{1s^22s^1}{\text{Li}} \longrightarrow \underset{1s^2}{\text{Li}^+ + e^-}$$

(the He electronic configuration)

$$\underset{1s^22s^22p^63s^1}{\text{Na}} \longrightarrow \underset{1s^22s^22p^6}{\text{Na}^+ + e^-}$$

(the Ne electronic configuration)

In their compounds the alkali metals usually occur as ions with a 1+ charge. This last fact reflects the low ionization potentials of the alkali metals; their low ionization potentials are another manifestation of the stability of the noble gas configuration.

Each of the alkaline earths has two electrons in the *s*

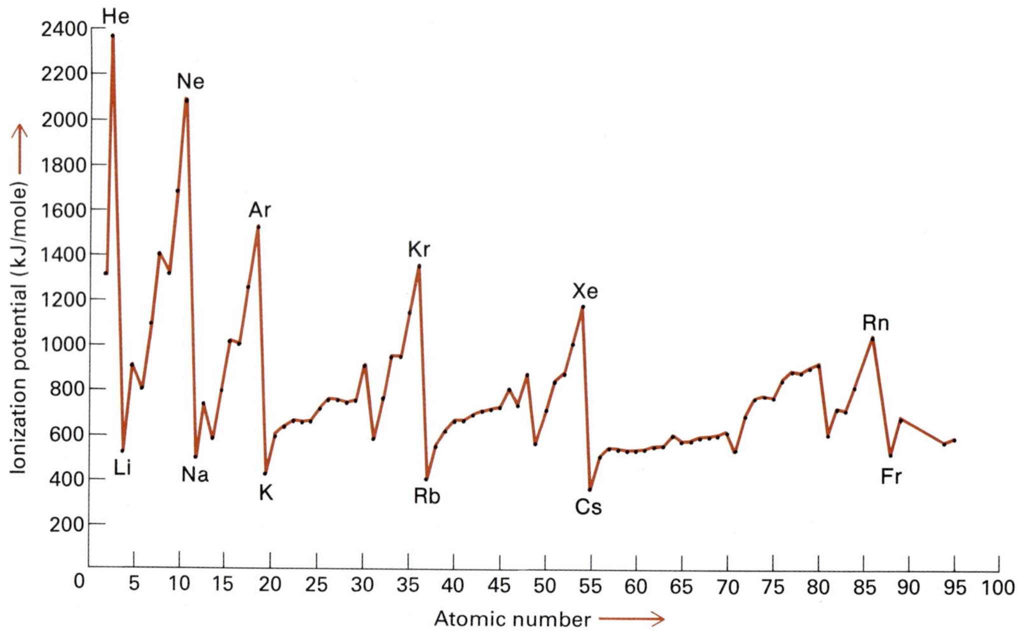

FIGURE 6-18

Ionization potential as a function of atomic number.

orbital of its outer shell. In their compounds, the alkaline earths usually lose two electrons to form ions with a 2+ charge (M^{2+}), thereby acquiring the electronic configuration of the closest preceding noble gas atom:

$$\underset{1s^22s^2}{\text{Be}} \longrightarrow \underset{1s^2}{\text{Be}^{2} + 2e^-}$$

(the He electronic configuration)

$$\underset{1s^22s^22p^63s^2}{\text{Mg}} \longrightarrow \underset{1s^22s^22p^6}{\text{Mg}^{2+} + 2e^-}$$

(the Ne electronic configuration)

Be^{2+}, Li^+, and He are **isoelectronic**—that is, they have the same number of electrons. Mg^{2+}, Na^+, and Ne also are isoelectronic.

The halogens lack one electron of a noble gas configuration, having seven electrons in their outer shells. The ionization potentials of the halogens are high, and they rarely occur as positive ions—instead, they pick up electrons easily, and usually occur as ions with a 1− charge. The neutral halogen atom gains an electron (from another atom such as Na, as we shall see in the following chapter), thereby acquiring the electronic configuration of the immediately following noble gas atom:

$$\underset{1s^22s^22p^5}{\text{F}} + e^- \longrightarrow \underset{1s^22s^22p^6}{\text{F}^-}$$

(the Ne electronic configuration)

$$\underset{1s^22s^22p^63s^23p^5}{\text{Cl}} + e^- \longrightarrow \underset{1s^22s^22p^63s^23p^6}{\text{Cl}^-}$$

(the Ar electronic configuration)

F^- is isoelectronic with Ne, Na^+, and Mg^{2+}.

EXERCISE 6-12 Give the formula for an alkali metal ion, an alkaline earth ion, and a halide ion that is isoelectronic with argon.

Hydrogen, with the electronic configuration $1s^1$, resembles the alkali metals in having a single electron in its outer shell, and resembles the halogens in lacking one electron of a noble gas configuration. The fact that the ionization potential of hydrogen is high (1315 kJ/mole) shows that H is not an alkali metal. The chemical behavior of hydrogen is such that in some ways, however, it is like an alkali metal, forming compounds of general formula HX with the halogens; and in other ways it is like a halogen, forming compounds of general formula MH with the alkali metals.

The ionization potential decreases from top to bottom of any group in the periodic table, as shown in Figure 6-3 for the alkali metals, alkaline earths, halogens, and noble gases. Going down a group, the outer electrons, which are the most easily removed, are of increasing principal quantum number, and thus of increasing average distance from the nucleus. The farther an electron is from the nucleus the weaker the attractive force of the nucleus on the electron, and the more easily the electron can be removed. Furthermore, in a many-electron atom, the inner electrons (those with lower principal quantum numbers than the outermost electrons) partially "shield" the outer electrons from the nucleus. The negative charges of the inner electrons effectively counteract part of the positive charge of the nucleus. Thus the outer electrons do not feel the full attractive force of the positively charged nucleus.

We saw in Figure 6-18, as we did earlier in Table 6-5, that there is a general trend of increasing ionization potential from left to right across a row in the periodic table. In each row, the alkali metal has the lowest ionization potential and the noble gas the highest. This increase in ionization potential across a row results from an increase in the nuclear charge without a corresponding increase in the number of inner electrons that shield the outer electrons from the nucleus. In any row, the alkali metal has the smallest nuclear charge and the lowest ionization potential; and the noble gas has the greatest nuclear charge and the highest ionization potential.

6-14 ATOMIC RADIUS AND THE PERIODIC TABLE

We have seen that the size of an atom is not sharply defined. The quantum mechanical model of the atom gives only a fuzzy picture of it. It is possible, however, to determine the apparent sizes of the atoms or ions in a crystal by thinking of them as touching. In this way we can compare the sizes of atoms under similar conditions.

Figure 6-3a shows an increase in atomic radius from top to bottom of any group. This increase in size reflects the fact that the principal quantum number of the outer electrons increases by one from one period to the next. And the greater the principal quantum number of an orbital, the farther (on the average) electrons in that orbital are from the nucleus.

In Table 6-5, we saw that there is a decrease in atomic radius from left to right across a row in the periodic table. This decrease in atomic radius, like the increase in ionization potential, can be attributed to the increase in positive charge on the nucleus across the row while the number of inner shielding electrons remains the same. The outermost electrons, those that determine the size of the atom, thus feel an increasing nuclear charge and are pulled in closer to the nucleus. Table 6-5 shows that Li is the largest atom in the second period, with an atomic radius of 0.123 nm, and that F is the smallest atom, with a radius of only 0.064 nm.

In the following chapter we see how the electronic configurations of atoms determine both the kinds of bonds formed between them and the shapes of the resulting molecules.

EXERCISE 6-13 Which would you expect to be larger? (a) Na or P? (b) Cl or Br? (c) Cl^- or K^+?

KEY TERMS

Elements having similar chemical properties are grouped together in a *periodic table*. Elements that lie in a vertical column in the periodic table constitute a *group* or *family* of elements with similar properties. A horizontal row in the periodic table is called a *period*.

The *atomic radius* is a measure of the size of an atom. The *ionization potential* is the minimum energy required to remove an electron from an atom in the gaseous state.

The *noble gases* constitute group 8A in the periodic table. They are monatomic gases with an almost total lack of chemical reactivity. The *alkali metals* constitute group 1A in the periodic table. They are soft, highly reactive metals. The *alkaline earths* constitute group 2A in the periodic table. They are reactive metals, somewhat harder than the alkali metals. The *halogens* constitute group 7A in the periodic table. They are highly reactive nonmetals that occur as diatomic molecules at room temperature.

Electromagnetic radiation, which includes visible light, is a form of energy characterized as a wave. All types of electromagnetic radiation move with a velocity of 3.0×10^{10} cm/sec.

A *continuous spectrum* contains light of all possible wavelengths (or frequencies, or energies). A *line spectrum* consists of radiation of only certain, definite wavelengths (or frequencies, or energies).

The energy carried by light comes in "packets" called *quanta* or *photons*.

Electrons in atoms, ions, or molecules can be in only certain, discrete, *energy levels*. An atom is in its *ground state* when all its electrons are in the lowest possible energy levels. When one electron (or more) of an atom in the ground state absorbs energy and jumps to a higher energy level, the atom is in an *excited state*.

An *orbital* is a region of space around the nucleus in which an electron is likely to be found. The ground-state *electronic configuration* of an atom shows what orbitals all the electrons are in when an atom is in its ground state.

Two or more atoms or ions that have the same number of electrons are *isoelectronic*.

CHAPTER SUMMARY

When the elements are arranged in order of increasing atomic number, we find a periodic recurrence of properties leading to identifiable families or groups in a *periodic table.* We considered the properties of four important groups in the periodic table: the alkali metals, the alkaline earths, the halogens, and the noble gases.

Every element has a characteristic *line spectrum,* in which each line corresponds to a definite energy of light. We can account for these discrete energies emitted or absorbed by atoms by assuming that electrons are allowed only certain specific *energy levels.* The energy associated with each spectral line is equal to the difference between two energy levels in the atom. By measuring the frequency of an atom's spectral lines, it is possible to determine the energies allowed its electrons, using the relation $E = h\nu$.

The *quantum theory* of atomic structure is based on the idea that electrons have characteristics of waves as well as of particles. According to this theory, we cannot locate an electron precisely, but can only describe the probability of finding the electron at various positions around the nucleus. The regions of space in which electrons are most likely to be found are called *orbitals.* Each orbital has a definite energy and a characteristic probability distribution in space, which gives it a size and a shape. A maximum of two electrons, one of each spin, can be in an orbital.

Ground-state *electronic configurations* can be written for atoms by assigning the electrons to orbitals, always placing an electron in the orbital of lowest energy. When there are two or more electrons in a p, d, or f *subshell,* the electrons remain unpaired as far as possible.

Elements with the same outer-shell electronic configuration have similar properties and are in the same group in the periodic table. It is these valence electrons that determine the chemical behavior of an atom. The alkali metals have a single s electron in the outer main shell; the alkaline earths two s electrons; the halogens two s and five p electrons; and the noble gases two s and six p electrons in the outer main shell (except for He which has a two $1s$ electrons). The electronic configuration of the noble gases—filled s and p orbitals in the outer shell—is of special stability.

Ionization potentials increase across a row in the periodic table, with alkali metals having the lowest, and noble gases the highest, ionization potential. Ionization potentials decrease down a group in the periodic table.

Atomic radii decrease across a row in the periodic table, and increase down a group.

PROBLEMS

6-1. Which two of the elements in each set would you expect to be most similar in chemical and physical properties?

(a) Ar, S, Al, Se, and Ba
(b) Mg, K, Si, I, and Sn
(c) Cl, P, As, Pb, and O
(d) I, Na, Ca, Rb, and Bi

6-2. State whether each of the following elements is an alkali metal, an alkaline earth metal, a halogen, or a noble gas: (a) Cs (b) Be (c) Br (d) Ne (e) F (f) Li (g) Ca

6-3. Predict the charges on the ions of (a) Li (b) Ra (c) S (d) F (e) Br (f) Rb (g) Ba (h) O.

6-4. (a) Use data given in Table 6-4 to predict the melting and boiling points of the radioactive element 87, francium. (Draw graphs to help answer the questions.)
(b) Write an equation for one chemical reaction involving francium.

6-5. (a) Use data given in Table 6-9 to predict the atomic radius and ionization potential of the radioactive element 85, astatine. (Draw graphs to help answer the questions.)
(b) Write an equation for one chemical reaction involving astatine.

6-6. From the experimental results given, identify the group or family and the specific element described. (Consult data in Tables 6-4, 6-6, 6-8, and 6-9.)
(a) Element X (1) has an ionization potential of less than 418 kJ/mole; (2) reacts with chlorine to form an ionic compound; (3) reacts with water to liberate hydrogen gas; and (4) melts about 10°C above room temperature.
(b) Element Y (1) has an ionization potential of more than 1.7×10^3 kJ/mole; (2) has its boiling point below room temperature; (3) has a density lower than that of air; and (4) does not form a hydride.
(c) Element Z (1) has an ionization potential of less than 1.3×10^3 kJ/mole; (2) is a solid at room temperature; (3) does not react with water; (4) forms a gaseous hydride; and (5) reacts with sodium to form an ionic compound.

6-7. Of the atoms in the third period of the periodic table (Na, Mg, Al, Si, P, S, Cl, and Ar), which would you expect to have the highest ionization potential, and which the lowest?

6-8. Of the atoms in group 4A of the periodic table (C, Si, Ge, Sn, and Pb), (a) which would you expect to have the lowest ionization potential, and (b) which would you expect to have the largest radius?

6-9. In the ground state, lithium has the electronic configuration $1s^2\,2s^1$. How can the electron in the 2*s* orbital be raised to the 3*s* orbital?

6-10. Identify the group in the periodic table and the specific element that has each of the following electronic configurations.
(a) $1s^2\,2s^2\,2p^6\,3s^2\,3p^4$
(b) $1s^2\,2s^2\,2p^6\,3s^2\,3p^6\,4s^1$
(c) $1s^2\,2s^2\,2p^6\,3s^2\,3p^2$
(d) $1s^2\,2s^2\,2p^6\,3s^2\,3p^5$
(e) $1s^2\,2s^2\,2p^6\,3s^2\,3p^6\,4s^2$
(f) $1s^2\,2s^2\,2p^6\,3s^2\,3p^6\,3d^{10}\,4s^2\,4p^3$
(g) $1s^2\,2s^2\,2p^6\,3s^2\,3p^6\,3d^{10}\,4s^2\,4p^6\,4d^{10}\,5s^2\,5p^4$

6-11. Give the electronic configuration and the number of valence electrons for each of the following elements.
(a) ${}_{13}Al$ (b) ${}_{16}S$ (c) ${}_{20}Ca$ (d) ${}_{32}Ge$
(e) ${}_{35}Br$ (f) ${}_{37}Rb$ (g) ${}_{56}Be$ (h) ${}_{83}Bi$

6-12. When calcium chloride is heated in a flame, and the flame is observed through a spectroscope, distinct lines are seen. Why do just lines appear, instead of a continuous spectrum?

6-13. (a) Why do sodium chloride and potassium chloride produce spectra with different lines?
(b) Compounds containing the alkali and alkaline earth metals are often used to impart color to fireworks or a wood fire. Explain.

6-14. How does an electron in a 2*p* orbital differ from an electron in a 1*s* orbital (a) in energy, and (b) in characteristic distribution in space?

6-15. Nitrogen has the electronic configuration $1s^2\,2s^2\,2p^3$. What is the electronic configuration of phosphorus, which lies just below nitrogen in the periodic table? (Try to answer without consulting tables.)

6-16. (a) Two atoms have the electronic configurations $1s^2\ 2s^2\ 2p^6$ and $1s^2\ 2s^2\ 2p^6\ 3s^1$. One has the ionization potential 497 kJ/mole and the other 2100 kJ/mole. Which ionization potential goes with which electronic configuration? Explain. (Try to answer without consulting tables.)
(b) One of the atoms is chemically similar to potassium, and the other is chemically similar to argon. Which is which?

6-17. Which *one* of the following elements, specified by their atomic numbers, would have very different chemical and physical properties from the other elements in the set: elements 8, 16, 34, 50, and 84? (Try to answer by writing out electronic configurations; not by using tables.)

6-18. Which *two* of the following elements, specified by their atomic numbers, would be similar in chemical and physical properties: elements 3, 12, 36, 56, and 83? (Try to answer by writing out electronic configurations; not by using tables.)

6-19. (a) Give the electronic configuration of the element having atomic number 33.
(b) Give the atomic numbers of three or four other elements expected to have the same general chemical and physical properties as element 33.
(Try to answer both parts without using tables.)

6-20. (a) Give symbols for, and charges on, one negatively charged ion and two positively charged ions that are isoelectronic with Xe.
(b) Give symbols for, and charges on, two negatively charged ions and two positively charged ions that are isoelectronic with Ar.

Halite crystals (NaCl). The shape of a crystal reflects the microscopic structure of the substance.

7

The Chemical Bond and the Shapes of Molecules

Most of the matter around and within us consists not of single atoms, but of groups of atoms bonded in molecules. In this chapter we first look at chemical bonds—the forces that hold atoms together in a molecule. But molecules are not just unstructured clumps of atoms. They have consistent organization and three-dimensional shapes. We study here these shapes and how they affect the properties of the substance. Finally, we investigate the relatively weak forces that hold molecules together in a liquid or solid.

7-1
THE CHEMICAL BOND: IONIC BONDS

Chemical reactions are the making and breaking of the **bonds** that hold atoms together in a molecule. There are two extremes of chemical bonds, called *ionic* and *covalent* bonds, that confer different properties on compounds. We first look at the **ionic bond**—a bond that results from the attraction between ions of opposite charge. In forming an ionic bond, one atom *loses* one or more electrons while another atom *gains* one or more electrons; each atom thereby attains a noble gas electronic configuration:

$$2Na + Cl_2 \longrightarrow 2Na^+ + 2Cl^-$$

As we saw in Chapter 6, both Na^+ and Cl^- have inert gas electronic structures.

The kind of change represented in the equation shown above is often written in terms of **electron-dot** (or **Lewis**) **structures:**

$$2Na\cdot + :\ddot{\underset{\cdot\cdot}{Cl}}:\ddot{\underset{\cdot\cdot}{Cl}}: \longrightarrow 2Na^+ + 2:\ddot{\underset{\cdot\cdot}{Cl}}:^-$$

In this type of notation, the nucleus and inner electrons are represented by the symbol for the element, and the electrons *in the outermost main shell* are represented by dots. Electrons in the outer main shell are often called **valence electrons.** Thus Na, which has the electronic configuration $1s^22s^22p^63s^1$, has one electron in the third main shell (one valence electron) represented by one dot, and Cl, which has the electronic configuration $1s^22s^22p^63s^23p^5$, has seven valence electrons represented by seven dots in the electron-dot structure.

When silvery metallic sodium and gaseous chlorine are brought together, there is a violent reaction, forming a white crystalline compound (table salt, NaCl). As shown in the equation, each sodium atom loses one electron to form Na^+, and each chlorine atom gains one electron to form Cl^-. An ionic bond between Na^+ and Cl^- results from the strong electrostatic forces between the positively charged Na^+ and the negatively charged Cl^- ions. A crystal of salt (NaCl) is composed of Na^+ and Cl^- ions arranged in a lattice, as shown in Figure 7-1. Each ion is surrounded by six ions of the opposite charge. Because the crystal is electrically neutral, the numbers of Na^+ and Cl^- ions must be the same. Solids such as NaCl are called **ionic solids.** (In Chapter 8, we shall see how the geometrical structures of solids—the arrangements of the atoms in space—are determined.)

Ionic compounds are formed, typically, by the combination of elements on the left side of the periodic table—those that lose electrons most readily (say, the alkali metals and the alkaline earths)—with the strongly electron-attracting nonmetals on the right side of the periodic table—which are one or two electrons away from a noble gas structure (say, the halogens and oxygen and sulfur). Here are further exam-

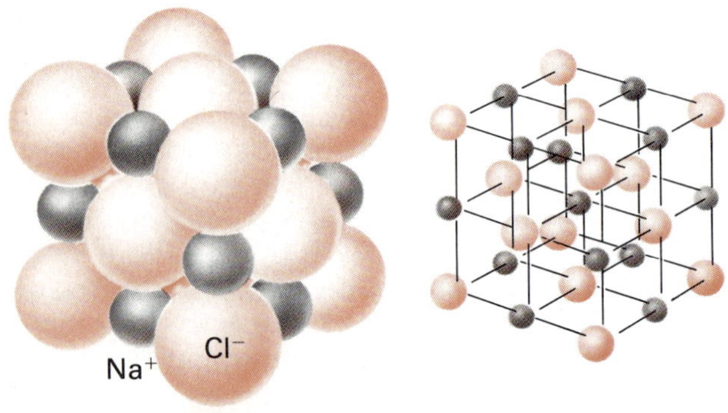

(a) (b)

FIGURE 7-1

The crystal structure of sodium chloride. **(a)** Relative sizes of the atoms. **(b)** The crystal lattice.

ples of the formation of ionic compounds (for simplicity, we represent fluorine and oxygen as single atoms rather than as the diatomic molecules):

$$\ddot{\text{Mg}} + 2\cdot\ddot{\underset{\cdot\cdot}{\text{F}}}: \longrightarrow \text{Mg}^{2+} + 2\,:\ddot{\underset{\cdot\cdot}{\text{F}}}:^{-}$$

(magnesium fluoride)

$$\ddot{\text{Ba}} + \cdot\ddot{\underset{\cdot\cdot}{\text{O}}}\cdot \longrightarrow \text{Ba}^{2+} + :\ddot{\underset{\cdot\cdot}{\text{O}}}:^{2-}$$

(barium oxide)

All the ions Na^+, Cl^-, Mg^{2+}, F^-, Ba^{2+}, and O^{2-} have achieved noble gas electronic structures. In the case of magnesium fluoride (MgF_2), for each magnesium atom that loses two electrons, two fluorine atoms gain one electron each. Thus, the product is electrically neutral—there is no net gain or loss of electrons in the reaction. In a crystal of MgF_2, there are twice as many F^- ions as there are Mg^{2+} ions.

EXERCISE 7-1 Write electron-dot structures that represent the formation (a) of the ionic solid, calcium bromide; (b) of the ionic solid, potassium oxide.

7-2 THE CHEMICAL BOND: COVALENT BONDS BETWEEN LIKE ATOMS

Ionic bonding is one extreme, resulting from the complete transfer of electrons from one atom to another. The **covalent bond** between two atoms of the same element—as in the diatomic gas H_2—is at the other extreme: entirely equal sharing of electrons. Although ionic bonding is common between elements at opposite sides of the periodic table, most bonding is covalent, with a sharing of electrons between the two atoms participating in the bond.

The electron-dot representation of the reaction between two H atoms to form a molecule of H_2 is

$$\text{H}\cdot + \cdot\text{H} \longrightarrow \text{H:H}$$

Here, the two dots in H:H represent the electron pair shared by the two atoms. Alternatively, a **line structure,** or **structural formula,** for the H_2 molecule may be given as H—H, where the line representing the bond stands for the shared electron pair. In H_2, each H atom has a share in two electrons and thus has completed its 1*s* orbital—each H atom has acquired the electronic structure of helium.

Another example of completely equal sharing of electrons between two atoms is chlorine gas (Cl_2):

$$:\ddot{\underset{\cdot\cdot}{\text{Cl}}}\cdot + \cdot\ddot{\underset{\cdot\cdot}{\text{Cl}}}: \longrightarrow :\ddot{\underset{\cdot\cdot}{\text{Cl}}}:\ddot{\underset{\cdot\cdot}{\text{Cl}}}:, \quad \text{or Cl—Cl}, \quad \text{or } \text{Cl}_2$$

In the Cl_2 molecule, each atom has acquired an octet of electrons by the sharing of a pair, each atom having contributed one electron to the shared pair. The shared electron pair constitutes the covalent bond that binds the two atoms together and is, in this case and in hydrogen (H_2), shared equally by the two atoms.

EXERCISE 7-2 Using electron-dot structures, show how each iodine (I) atom in the molecule I_2 has acquired a noble gas electronic configuration.

7-3 THE CHEMICAL BOND: POLAR COVALENT BONDS

We have seen examples of two *extreme* types of chemical bonds: the ionic bond, in which one atom loses one or more electrons while another atom gains them; and the covalent bond between like atoms, in which a pair of electrons is shared equally by the two atoms. The most common type of bond, however, is a covalent bond between unlike atoms, in which electrons are shared between the two atoms but are shared unequally—the **polar covalent bond.**

In hydrogen chloride gas (HCl), for example, a pair of electrons is shared by the H and Cl atoms:

$$\mathrm{H}\cdot + \cdot\ddot{\underset{\cdot\cdot}{\mathrm{Cl}}}: \longrightarrow \mathrm{H}:\ddot{\underset{\cdot\cdot}{\mathrm{Cl}}}:, \quad \text{or H—Cl,} \quad \text{or HCl}$$

By sharing a pair of electrons, both the H and the Cl acquire noble gas electronic structures. In this case, however, the electrons are not shared equally by the two atoms. The Cl has a much greater tendency to attract electrons than does the H, so the Cl has a greater share of the shared electron pair. That is, the electrons spend more time close to the Cl atom than close to the H atom.

This tendency to attract shared electrons is reflected in an atom's electronegativity. **Electronegativity** is a measure of the ability of an atom to attract shared electrons. The Cl atom is more electronegative than the H atom. The relative electronegativities of the elements, on an arbitrary scale, are shown in Figure 7-2. On this scale, fluorine, which has the greatest attraction for electrons, has an electronegativity of 4.0. Cesium, which has the least attraction for electrons (it has a great tendency to *lose* an electron to a more electronegative atom), has an electronegativity of only 0.7.

In HCl we can think of the Cl atom, which has the greater share of the shared electrons, as having a partial negative charge (where "partial" means less than the full charge of an electron) and the H atom as having a partial positive charge. The HCl molecule is said to have some ionic character, with a partial positive charge at one end of the molecule and a partial negative charge at the other. Such a molecule, which has a positive and a negative end, is called a **polar molecule.** It is said to have a **dipole,** a separation of charge, and the H–Cl bond is said to be **polar.** Polar bonds are often represented as follows:

$$\overset{+\!\!\longrightarrow}{\text{H—Cl}} \quad \text{or} \quad \overset{\delta^+}{\text{H}}\text{—}\overset{\delta^-}{\text{Cl}}$$

where the + tail on the arrow or δ^+ shows that the H has a partial positive charge, and the head of the arrow or δ^- shows that the Cl has a partial negative charge.

The greater the difference in the electronegativities of two bonded atoms, the more ionic (or polar) the

1A	2A	3B	4B	5B	6B	7B	8B	8B	8B	1B	2B	3A	4A	5A	6A	7A	8A
				H 2.1													He
Li 1.0	Be 1.5											B 2.0	C 2.5	N 3.0	O 3.5	F 4.0	Ne
Na 0.9	Mg 1.2											Al 1.5	Si 1.8	P 2.1	S 2.5	Cl 3.0	Ar
K 0.8	Ca 1.0	Sc 1.3	Ti 1.5	V 1.6	Cr 1.6	Mn 1.5	Fe 1.8	Co 1.8	Ni 1.8	Cu 1.9	Zn 1.6	Ga 1.6	Ge 1.8	As 2.0	Se 2.4	Br 2.8	Kr
Rb 0.8	Sr 1.0	Y 1.2	Zr 1.4	Nb 1.6	Mo 1.8	Tc 1.9	Ru 2.2	Rh 2.2	Pd 2.2	Ag 1.9	Cd 1.7	In 1.7	Sn 1.8	Sb 1.9	Te 2.1	I 2.5	Xe
Cs 0.7	Ba 0.9	La-Lu 1.1-1.2	Hf 1.3	Ta 1.5	W 1.7	Re 1.9	Os 2.2	Ir 2.2	Pt 2.2	Au 2.4	Hg 1.9	Tl 1.8	Pb 1.8	Bi 1.9	Po 2.0	At 2.2	Rn
Fr 0.7	Ra 0.9	Ac-Lr 1.1-															

FIGURE 7-2

The Pauling scale of electronegativities.

bond between them. When there is a large difference in electronegativities (as there is, say, between Cs and F, or between Na and Cl), the bond is essentially completely ionic. When the difference in electronegativities is smaller (as in HCl), the bond is polar covalent. At the extreme of two identical atoms, where there is no difference in electronegativity (as in H_2 or Cl_2), the bond is nonpolar covalent. Any degree of polarity is possible. Ionic and nonpolar bonds represent the two extremes, with most chemical bonds lying somewhere between these extremes in polarity.

EXERCISE 7-3 In the following pairs, which molecule is more polar? For each show the direction of the dipole, using the δ^+, δ^- or $+\!\!\longrightarrow$ notation.
(a) HCl or HBr (b) ClBr or BrI

We shall now construct electron-dot and line structures for several molecules that have polar covalent bonds. These are all shown in two dimensions in the plane of the page. (In Section 7-7 we consider three-dimensional shapes and polarities of molecules.) In drawing electron-dot structures, it is important to use the correct number of valence electrons for each atom, to make sure that electrons are conserved in the molecule, and to be sure that each atom in the molecule has an octet of electrons (a pair, for H).

We first consider the structure of methane (CH_4), representing the formation of the molecule simply as the combination of one C and four H atoms:

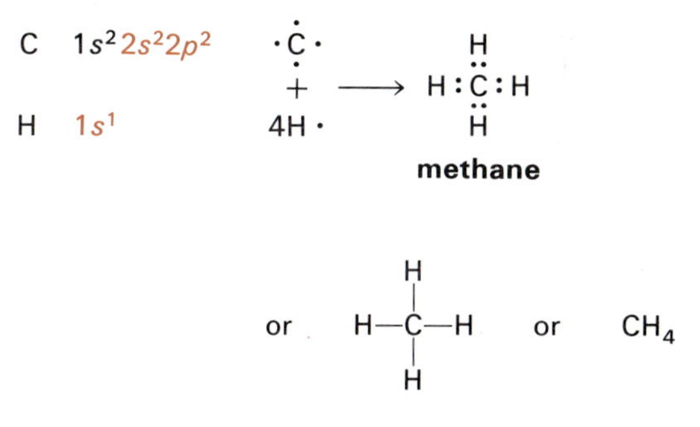

(Remember that of carbon's six electrons, four are valence electrons.) In CH_4, each H atom has acquired its complement of two electrons, and C its complement of eight. Note that in molecules, *carbon typically forms four covalent bonds and hydrogen forms one.*

Now consider carbon tetrachloride (CCl_4):

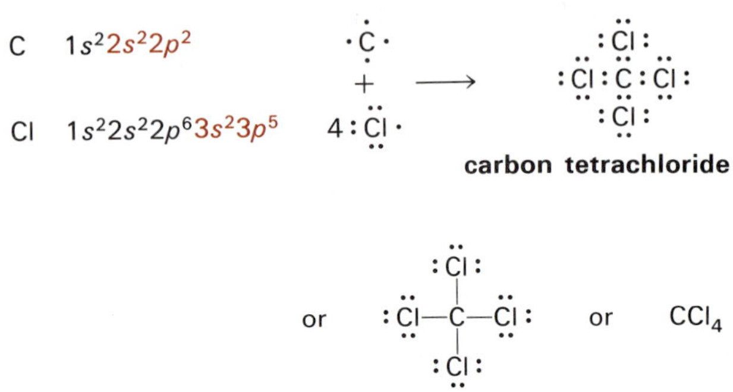

carbon tetrachloride

In line structures, *unshared* pairs of electrons are sometimes shown and sometimes not shown. Here they are shown on the Cl atoms. Again, the C atom has formed four covalent bonds, while each Cl atom has formed one covalent bond. All the atoms have acquired noble gas structures by sharing electrons.

Next we look at water (H_2O):

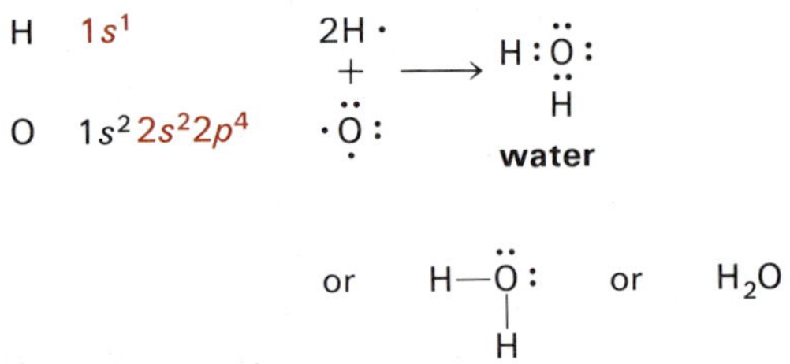

water

The O atom has formed two covalent bonds, and each H atom one, each atom thereby acquiring a noble gas structure. *Typically an O atom forms two covalent bonds.*

Now we consider the structure of ammonia (NH_3):

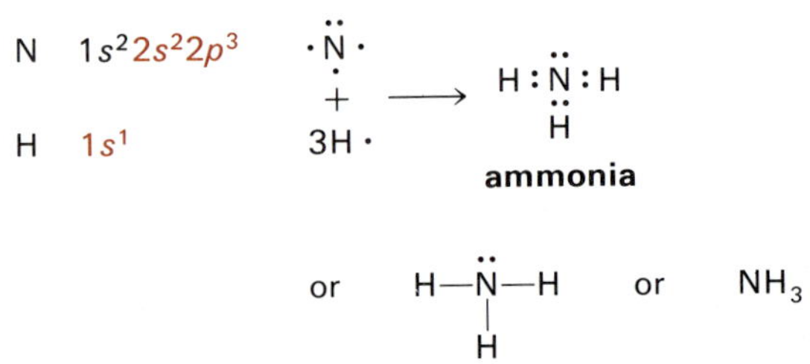

ammonia

The N atom has formed three covalent bonds, and each H atom one. Each atom has acquired a noble gas structure. *An N atom typically forms three covalent bonds.*

The structure of methyl alcohol, or methanol (CH_3OH), is more complex:

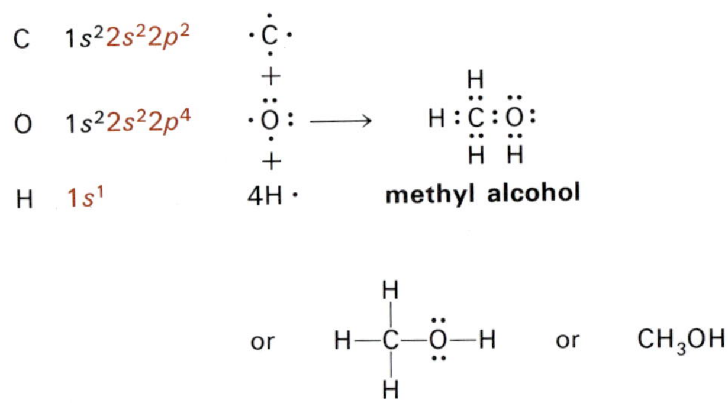

methyl alcohol

The C atom has formed four covalent bonds, the O atom two, and each H atom one. Each atom has acquired a noble gas structure by sharing electrons.

Ethane (C_2H_6) has a very regular structure:

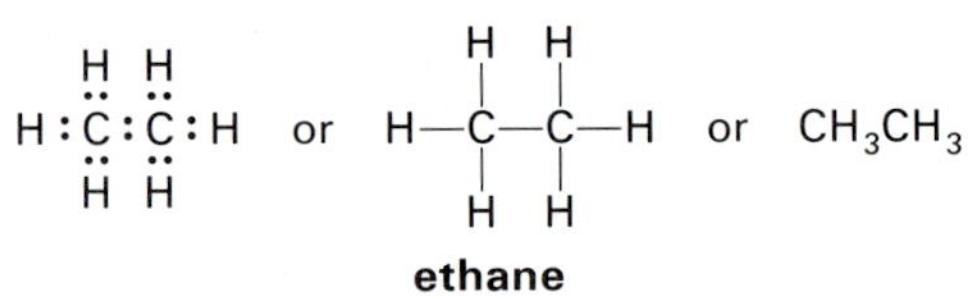

ethane

Tetraethyl lead, $Pb(C_2H_5)_4$, provides an example of a fairly complex structure:

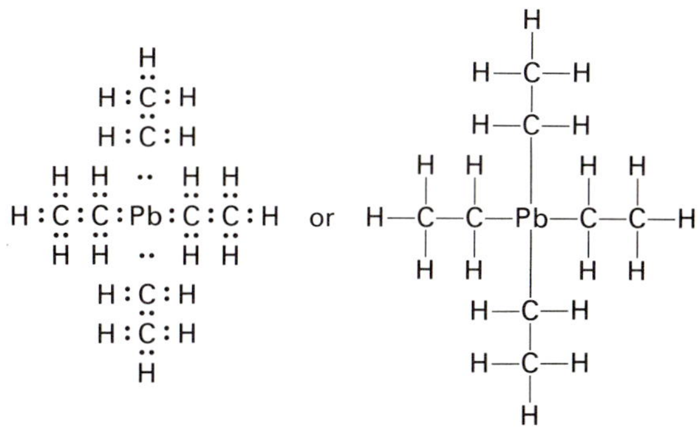

or $Pb(C_2H_5)_4$
tetraethyl lead

Sometimes both electrons of a shared pair come originally from the same atom. This occurs, for example, in the formation of the ammonium ion, NH_4^+, from NH_3 and H^+.

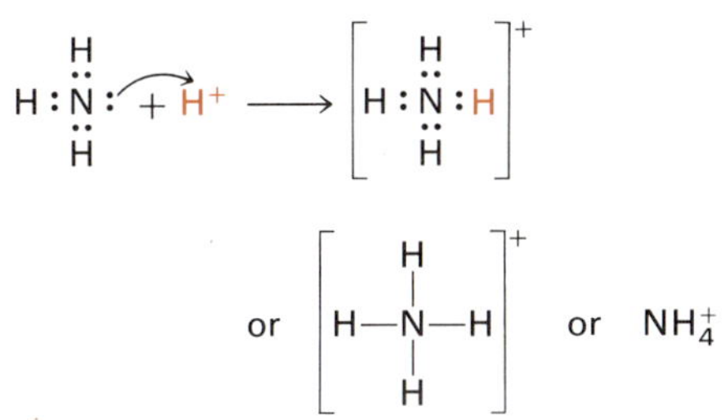

(7-1)

The unshared pair of electrons on the N atom in ammonia is shared with a proton. In NH_4^+, all the N—H bonds are identical, and the entire ion has a 1+ charge. The curved arrow is a convention we shall use often. The tail of the arrow indicates the origin of the electron pair, and the head points to the atom to be bonded.

The NH_4^+ looks like an exception to the rule of thumb stated earlier, that N typically forms three covalent bonds. Then we were considering covalent bonds where the N atom contributes *one* electron and accepts one electron of each shared pair. In NH_4^+, the N atom does form three covalent bonds in this way. In the fourth covalent bond, N contributes *both* electrons of the shared pair—but this bond, once formed, is identical to the others.

In the formation of the hydronium ion (H_3O^+), a pair of unshared electrons on the O atom in water is shared with a proton:

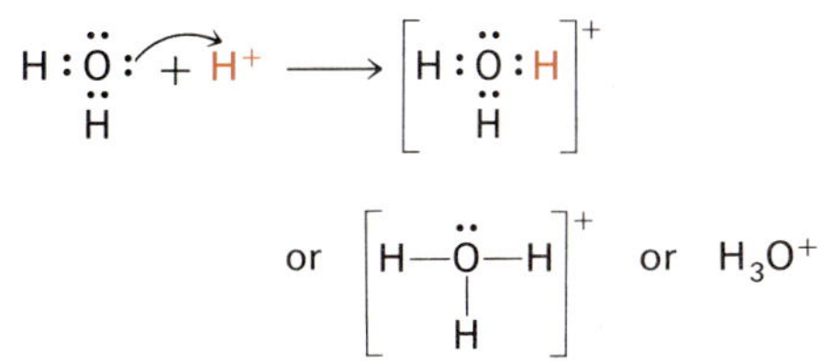

(7-2)

In H_3O^+, all the O-H bonds are identical, and the entire ion has a 1+ charge.

The hydroxide ion OH^- consists of one oxygen and one hydrogen atom, with an extra electron. If, for example, a proton H^+ is removed from a water molecule, an OH^- ion is formed in which both the O and the H have noble gas structures:

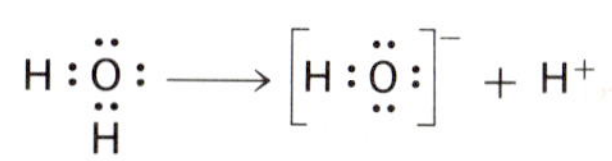

(The electron that initially belonged to the proton has been left on the hydroxide ion.)

The ions just considered (NH_4^+, H_3O^+, and OH^- are **complex ions**—ions composed of two or more atoms. The group as a whole has lost or gained one or more electrons, leaving it either positively or negatively charged. A complex ion usually acts as a unit in a chemical reaction, just as a simple ion does. The formula for a complex ion is like that of any molecule, with a superscript to indicate the charge. Thus, the sulfate ion, SO_4^{2-}, is composed of a sulfur atom bonded (by covalent bonds) to four oxygen atoms; two extra electrons have been gained in the formation, which accounts for the 2− charge on the ion. Table

7-1 gives the formulas of some common complex ions.

The bonding *within* each of the complex ions is covalent (as we saw for NH_4^+, H_3O^+, and OH^-), but, like simple ions, complex ions form ionic compounds. Washing soda (Na_2CO_3) is an ionic compound composed of sodium ions Na^+ and carbonate ions CO_3^{2-} arranged in a crystal lattice. Because the compound is neutral, there are *two* Na^+ ions for every CO_3^{2-} ion.

TABLE 7-1
Complex ions

Name	Formula
Ammonium	NH_4^+
Hydronium	H_3O^+
Acetate	CH_3COO^-
Bicarbonate	HCO_3^-
Bisulfate	HSO_4^-
Carbonate	CO_3^{2-}
Cyanide	CN^-
Hydroxide	OH^-
Nitrate	NO_3^-
Nitrite	NO_2^-
Phosphate	PO_4^{3-}
Sulfate	SO_4^{2-}
Sulfite	SO_3^{2-}

EXERCISE 7-4 Draw line structures for the following compounds: (a) dichlorodifluoromethane (Freon 12, F_2CCl_2); (b) phosphine (PH_3), a foul smelling toxic gas; and (c) hydrogen sulfide (H_2S), the poisonous gas responsible for the odor of rotten eggs. Show lone pairs of electrons.

7-4 MULTIPLE BONDS

Sometimes more than one pair of electrons is shared between two atoms, resulting in a double or triple bond. We now consider some examples of molecules with a **double bond,** in which two electron pairs are shared.

Consider ethylene (C_2H_4). If we try to draw an electron-dot structure with only single bonds, we get the structure

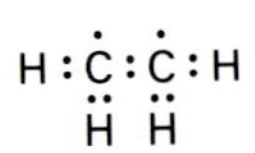

in which there are two unpaired electrons, and in which each C atom has only seven electrons. The only arrangement of the electrons that gives noble gas configurations about each atom is

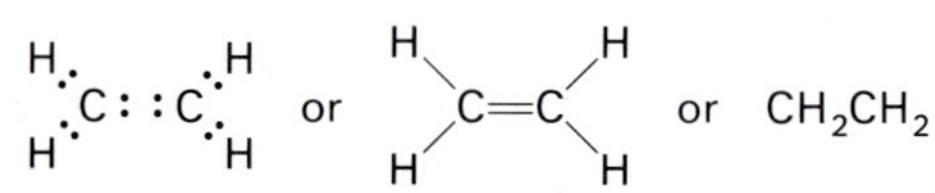

Each C atom now has its typical arrangement of four covalent bonds (two single bonds and one double bond), and each H atom has one.

Similarly, for carbon dioxide (CO_2), we obtain

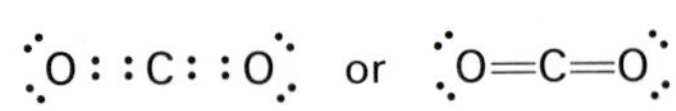

The structure of acetic acid (CH_3COOH) is

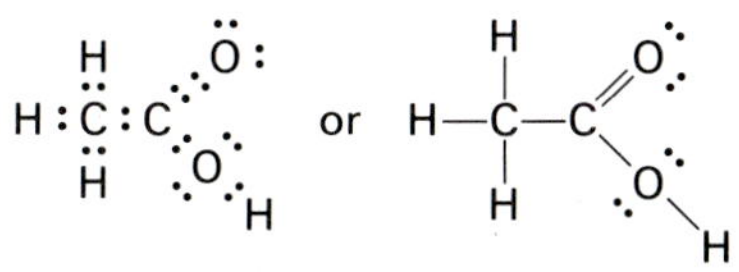

An important exception to the octet rule is oxygen gas (O_2), which, according to the octet rule, we would expect to have the electronic structure :O::O:. In this structure, each O has an octet and the atoms are double bonded. This structure cannot be correct for the O_2 molecule, however, because magnetic experiments show that O_2 has unpaired electrons. (We saw in Chapter 6 that electrons have "spin." A spinning electron is magnetic, like a small bar magnet with a north and a south pole. When two electrons of opposite spin are paired, there is no net magnetism, but an unpaired electron is magnetic.) Thus O_2 is more accurately :Ö:Ö:, with two unpaired electrons, and therefore is an exception to the octet rule (each O atom has a share in only *seven* electrons).

Nitrogen gas (N_2) provides an example of a **triple bond:**

:N⋮⋮N: or :N≡N:

Each N atom in N_2 has its typical arrangement of three covalent bonds.

Acetylene (C_2H_2) also has a triple bond:

H:C⋮⋮C:H or H—C≡C—H

As usual, C has formed four covalent bonds.

Consider hydrogen cyanide (HCN):

H:C⋮⋮N: or H—C≡N:

Again, N has formed three covalent bonds, and C has formed four.

EXERCISE 7-5 Draw the line structure for formaldehyde (H_2CO). Show lone pairs of electrons.

7-5 RESONANCE HYBRID STRUCTURES

In some cases, the electronic structure of a molecule cannot be represented by a *single* electron-dot structure that conforms both to the octet rule and to the experimentally determined properties of the molecule. For example, in the compound sulfur dioxide (SO_2), a possible electron-dot structure that satisfies the octet rule is

S $1s^22s^22p^63s^23p^4$:S̈·

:O::S̈:Ö: or O=S̈—Ö:

According to this structure, SO_2 has one sulfur–oxygen double bond and one single bond. However, experiments show that the two bonds in sulfur dioxide are equal in bond length (the distance between the two nuclei), and in bond energy (the energy required to break the bond). That is, the two sulfur–oxygen bonds are identical. The real structure of SO_2 is a **resonance hybrid** of the two structures, neither of which by itself is correct:

SO_2 is a resonance hybrid of

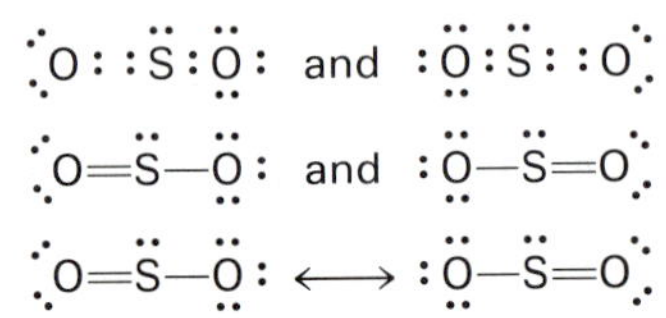

The one actual structure of SO_2 is intermediate between that of the two resonance hybrids. The two sulfur–oxygen bonds are equivalent, and both are intermediate between double and single bonds. The problem is that a single electron-dot representation is inadequate to describe this molecule. The double-headed arrow is often used to indicate resonance hybrid structures.

There are many other molecules that, like SO_2, cannot be described adequately by electron-dot structures, and that are commonly described as resonance hybrids of two or more structures. Benzene is a notable example (discussed in Chapter 15). Still other molecules, like O_2, do not satisfy the octet rule. But the great majority of molecules have electronic structures that can be described satisfactorily by electron-dot structures.

EXERCISE 7-6 Sulfur trioxide (SO_3) is another example of a molecule that cannot be adequately described by a single electron-dot structure. Write electron-dot structures for the resonance hybrids of SO_3.

7-6 BOND ENERGY AND BOND LENGTH

Although the **bond energy,** the energy required to break a chemical bond such as the C—H bond, depends to some extent on the specific molecule in which the bond occurs, its percentage variation is not very large. For example, for most molecules having a C—H bond, the bond energy lies between 375 and 430 kilojoules per mole, with the average being 414 kJ per mole. (This is equivalent to about 100 kcal per mole.) We can think of the bond energy as being close to constant. Average bond energies for some common chemical bonds are given in Table 7-2.

Like the bond energy, the length of a specific chemical bond (say, the C—C bond) is largely independent of the molecule in which the bond occurs. Thus the C—C bond length in ethane (C_2H_6) and in propane (C_3H_8) is 0.154 nm, and in ethanol (C_2H_5OH) it is 0.155 nm. Average bond lengths for some common chemical bonds are given in Table 7-2 together with the bond energies. Note that the C=C double bond is both shorter and stronger than the C—C single bond, and the C≡C triple bond is both shorter and stronger than the double bond.

The lengths of the bonds in the compounds acetic acid (the acid in vinegar) and acrolein (an irritating

TABLE 7-2
Average bond energies and bond lengths for some common chemical bonds

Bond	Average bond energy* (kJ/mole)	Average bond length (nm)
O—H	464	0.096
C—H	414	0.110
C—Cl	339	0.177
C—N	305	0.147
C—C	347	0.154
C=C	610	0.134
C≡C	836	0.120
C—O	355	0.143
C=O	744	0.123
N—H (in ammonia)	389	0.107

*This is the energy required to *break* the bond. To convert to kilocalories, divide the number of kilojoules by 4.18.

substance found in automobile exhaust) are very close to those given in Table 7-2. Compare the bond lengths (in nm) shown in the structures with those in the table:

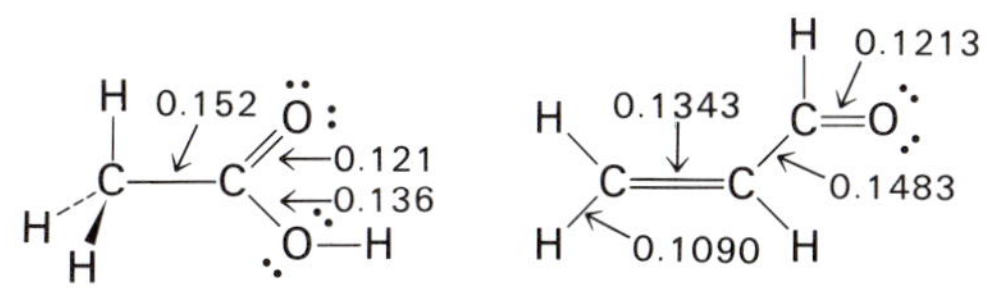

EXERCISE 7-7 Give the expected bond lengths of all the bonds in the following structures. In each case, indicate which carbon–carbon bond is the strongest and give its approximate bond energy.

(a) 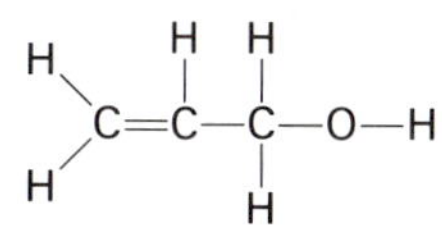

(b) 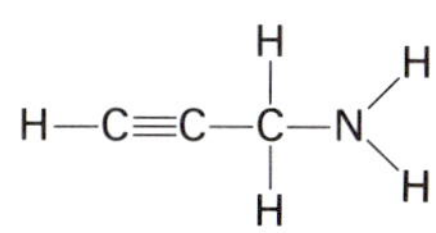

7-7 THE THREE-DIMENSIONAL SHAPES OF MOLECULES

Thus far we have discussed the arrangement of electrons in molecules as though the molecules were flat, two-dimensional objects, like thin pancakes. This is not the case, and the actual three-dimensional shape of a molecule is of great importance in determining its chemical behavior. We shall see in later chapters, for example, that the three-dimensional structure of molecules is a dominant feature in the fields of organic chemistry and biochemistry. We now consider the most common shapes of simple molecules composed of up to five atoms. In considering the shapes of molecules, we are interested primarily in the positions of the *nuclei* in space—less in the location of electrons within the molecule.

We have seen that a covalent bond between unlike atoms is polar. A diatomic molecule composed of unlike atoms thus is a polar molecule. A molecule composed of three or more atoms may be polar, or the individual bond dipoles may cancel each other and yield a nonpolar molecule. In the following discussion, we shall see that, in a molecule composed of three or more atoms, whether individual polar bonds lead to a polar or to a nonpolar molecule depends on the shape of the molecule.

A molecule composed of two atoms is necessarily linear (both nuclei lie on a line), and a molecule composed of three atoms is necessarily planar (all the nuclei lie in a plane). Although a molecule composed of more than three atoms could also be linear or planar, there are many other shapes it could have as well. We now consider some of the common three-dimensional structures of molecules.

A molecule composed of two atoms is linear.

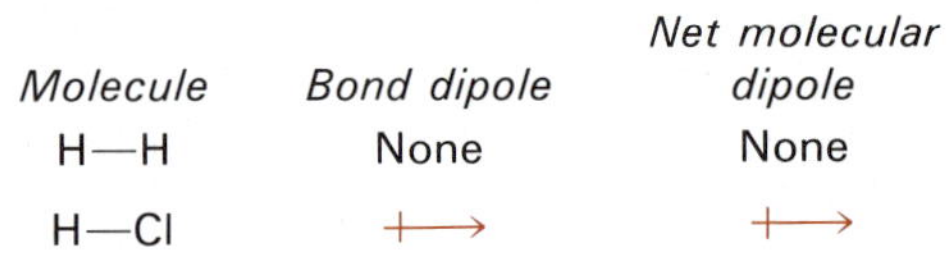

A molecule composed of three atoms is planar and may be linear or nonlinear. For example, CO_2 is a linear molecule:

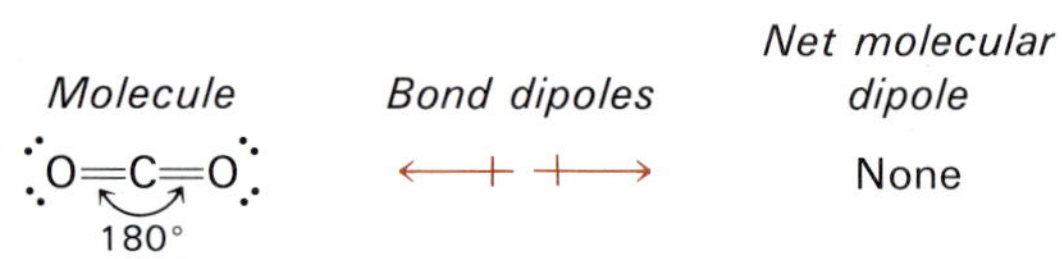

In this linear triatomic molecule, the two bond dipoles are symmetrical and cancel each other so that the molecule as a whole is nonpolar.

H_2O is a nonlinear molecule:

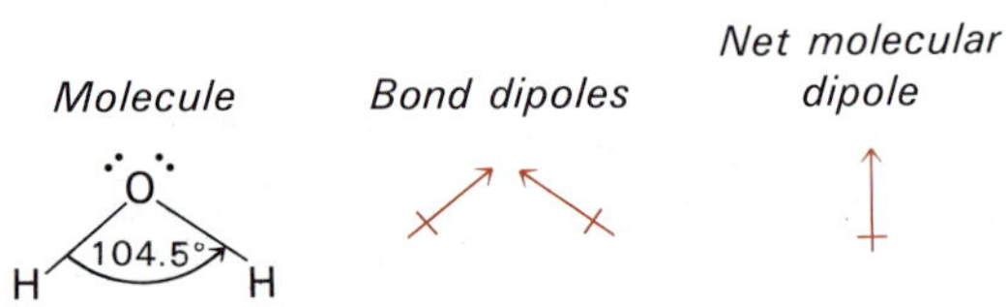

There are covalent bonds between hydrogen and oxygen in H_2O, but no chemical bond between the two H atoms. The molecule is angular, with an H—O—H bond angle of 104.5°, so the two bond dipoles do not cancel each other but yield a net molecular dipole as indicated above.

A molecule composed of four atoms is, typically, either triangular within a plane or pyramidal. Boron trifluoride (BF_3) is a *planar triangular* molecule:

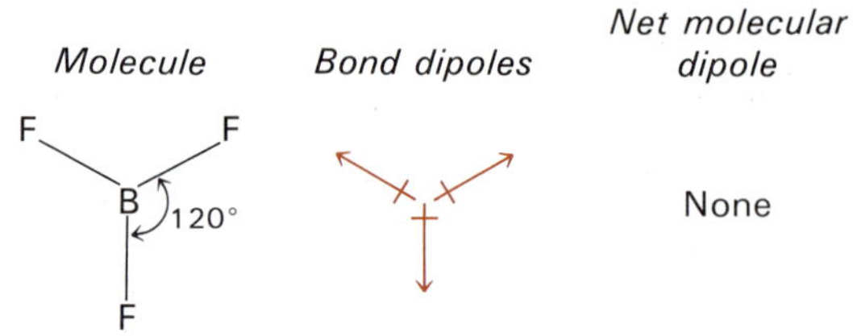

The F atoms are at the corners of an equilateral triangle, with all three F—B—F bond angles equal to 120°. The three bond dipoles cancel each other and yield a nonpolar molecule.

Ammonia (NH_3) is a pyramidal molecule:

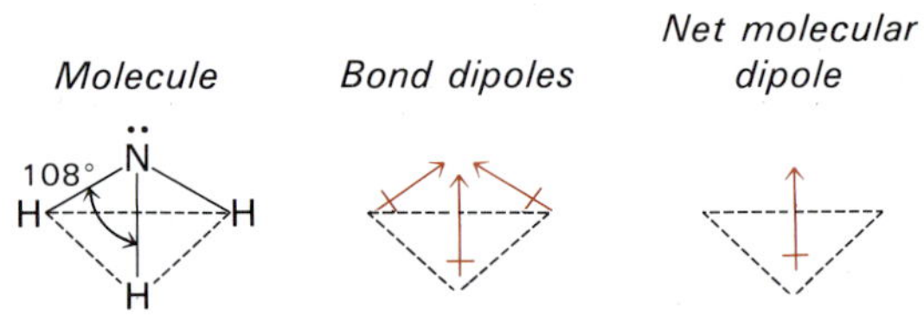

This molecule is pyramidal, with the N atom above a triangular base formed by the three H atoms. This triangular base is shown by the dashed lines. The H—N—H bond angles are about 108°. Because of the pyramidal shape of the molecule, there is a net molecular dipole with its positive end in the plane of the hydrogen atoms and its negative end at the nitrogen atom.

Elemental phosphorus, in the form called white phosphorus, consists of P_4 molecules arranged in a tetrahedron, a pyramid with a triangular base.

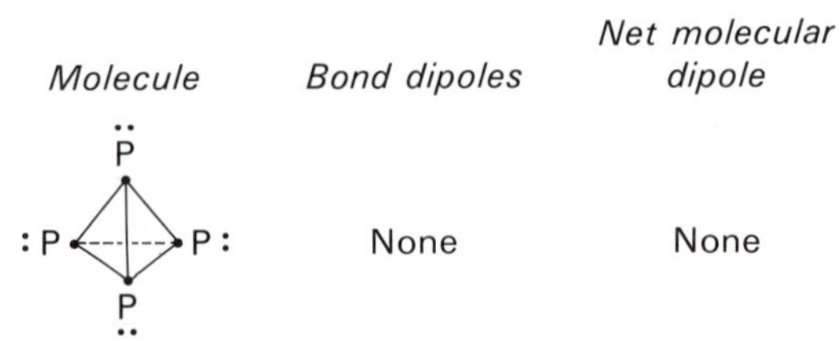

Each P atom is covalently bonded to three other P atoms. Each face of the tetrahedron is an equilateral triangle. All the bonds are between like atoms and so are nonpolar.

A molecule composed of five atoms is commonly in the shape of a tetrahedron of four atoms around a central atom. For example, consider methane (CH_4):

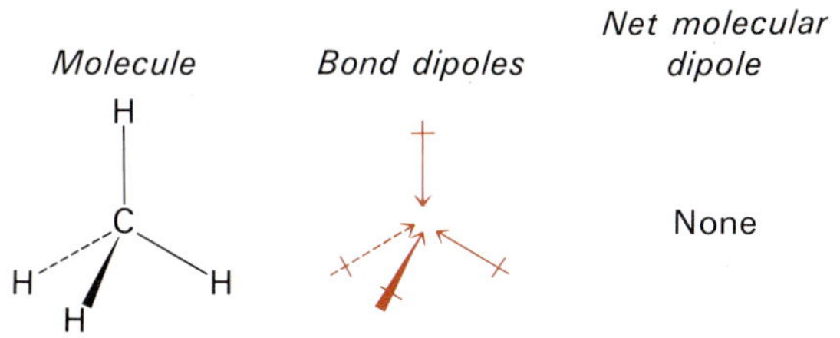

The C atom is at the center of a tetrahedron, with H atoms at the four corners. The individual CH bond dipoles (which are quite small) cancel each other out in the symmetrical, tetrahedral molecule. The tetrahedral bond angle is 109.5°; all the H—C—H bond angles are the same, 109.5°. The drawing of the tetrahedron above introduces several conventions we shall use often. A bond extending out from the plane of the paper is indicated by a wedge, a bond directed behind the plane of the paper is indicated by a dashed line, and a bond concentrated in the plane of the paper is indicated by a solid line.

If a Cl atom is substituted for one of the H atoms in CH_4, to form chloromethane (CH_3Cl), the individual bond dipoles no longer cancel each other. CH_3Cl is a polar molecule:

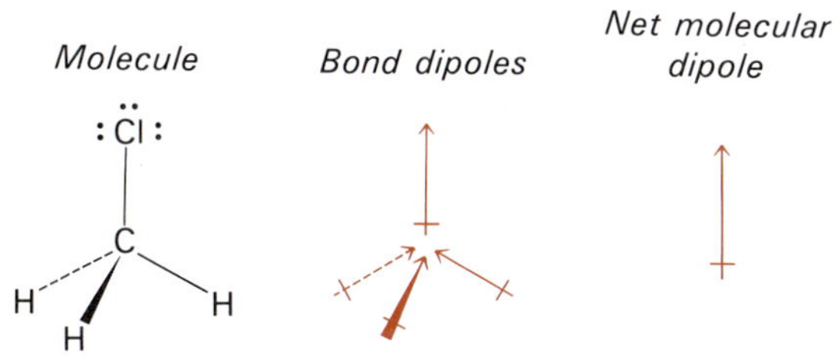

Both methane and chloromethane are tetrahedral molecules. Many other molecules have approximately this same shape. Most atoms, in forming chemical bonds, acquire an octet of electrons (four electron pairs) in their outer shells. *Because of their mutual repulsion, the electron pairs arrange themselves in a way that places them as far from each other as possible. The optimal arrangement for four pairs of electrons (that separates them to the greatest extent) is a tetrahedral arrangement, with the bonds at angles of 109.5° to each other.*

Several of the molecules we have already considered can be regarded as tetrahedra:

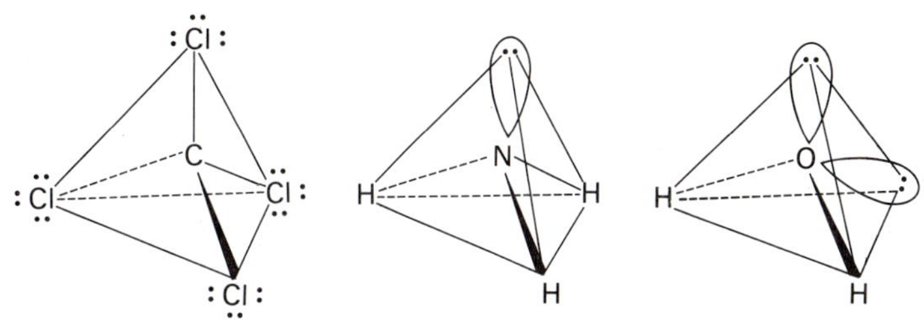

carbon tetrachloride **ammonia** **water**

In CCl_4, the carbon is at the center of a tetrahedron, with Cl atoms at the four corners. The NH_3 molecule, which we pictured earlier as pyramidal, can be regarded as a tetrahedron with the N atom at the center, H atoms at three corners, and an unshared pair of electrons at the fourth corner. The H_2O molecule, which we pictured earlier as angular, can also be visualized as a tetrahedron with the O atom at the center, H atoms at two corners, and unshared pairs of electrons at the other two corners.

Some of the molecules that we have considered have a net molecular dipole because they have individual bond dipoles that do *not* cancel each other: HCl, H_2O, NH_3, and CH_3Cl. These molecules are *polar.* Other molecules have no net molecular dipoles, and so are nonpolar molecules. The phosphorus molecule (P_4) is nonpolar because the bonds, which are between like atoms, are nonpolar. And CO_2, BF_3, CH_4, and CCl_4

are nonpolar molecules because they are symmetric molecules in which the individual bond dipoles cancel.

EXERCISE 7-8 (a) $BeCl_2$ is a nonpolar molecule. What is its shape?
(b) OF_2 is a polar molecule. What is its shape?

7-8 MOLECULAR SHAPES AND HYBRID ORBITALS

We'll soon see how we can explain the shapes of molecules, but first we'll look again at a covalent bond. As we saw earlier, covalent bonding occurs when atoms *share* electrons. We can think of this as happening when an atomic orbital of one atom overlaps with an atomic orbital of another atom. The formation of H_2 from two H atoms, for example, results from the *overlap* of the $1s$ orbitals of the separate H atoms. Figure 7-3 plots potential energy against distance between the two nuclei. At A in the figure, the two atoms are weakly attracting. The electron of each H atom is attracted to the nucleus of the other H atom. The $1s$ orbitals of the two H atoms overlap. As the atoms are drawn closer together, the amount of overlap increases. The observed equilibrium bond distance of H_2 corresponds to a minimum in the potential energy. This distance is a compromise between the attractive forces that draw the atoms together and the repulsive forces between the two nuclei that push the atoms apart.

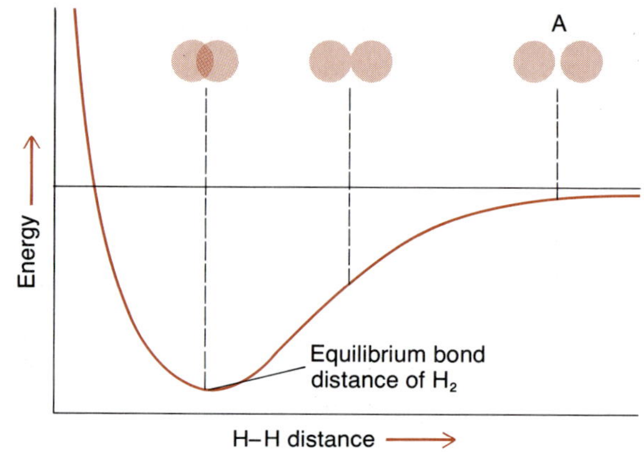

FIGURE 7-3
Formation of an H_2 molecule from two H atoms. The equilibrium H—H bond distance r corresponds to a minimum in the potential energy.

In H_2, the shared electron pair occupies a region that consists in part of the atomic orbital of each atom. Both electrons can occupy both orbitals, but the electron pair is most likely to be found in the region of overlap (the darkened area in the diagram). An electron in this region is simultaneously attracted to both nuclei.

Now we'll look at the bonding in methane (CH_4) and see how the tetrahedral shape of molecules can be explained in terms of the atomic orbitals of the carbon atom. Experiments show that the bond angle of all the H—C—H bonds in methane is the same tetrahedral angle (109.5°), and all four H—C—H bonds are equivalent in length and in strength. The isolated C atom has the electronic configuration

$$\text{C} \quad 1s^2 2s^2 2p^2 \qquad \underset{1s}{\underline{\downarrow\uparrow}} \quad \underset{2s}{\underline{\downarrow\uparrow}} \quad \underline{\uparrow}\ \underset{2p}{\underline{\uparrow}}\ \underline{\quad}$$

In forming compounds, carbon gains a share in four more electrons to complete an octet in its second main shell. If these four additional electrons were shared in the unfilled *p* orbitals of the carbon atom, we would expect to find bond angles of 90° in carbon compounds, because *p* orbitals are oriented at 90° to each other (Figure 6-14). Experimentally, the bond angles are found to be 109.5° instead. This observation is explained by a theory that says the four orbitals of the second shell of the carbon atom (the single 2*s* orbital and the three 2*p* orbitals) are **hybridized,** or mixed, to form four new **hybrid orbitals** called sp^3 hybrid orbitals. The term sp^3 means that one *s* and three *p* orbitals have been mixed to form the four new orbitals. Note the difference in meaning of the symbol sp^3 and the electronic configuration $1s^2 2s^2 2p^6$. . . . The carbon 2*s* and its three 2*p* orbitals do not retain their original atomic character. Instead, in CH_4 they form four new sp^3 hybrid orbitals that are equal in energy and are directed toward the corners of a tetrahedron (Figure 7-4):

↓↑	↑	↑	—	⟶	↑ ↑ ↑ ↑
$2s$	$2p_x$	$2p_y$	$2p_z$		sp^3

one *s* + three *p* orbitals — four sp^3 hybrid orbitals
isolated C atom — **C atom in CH_4**

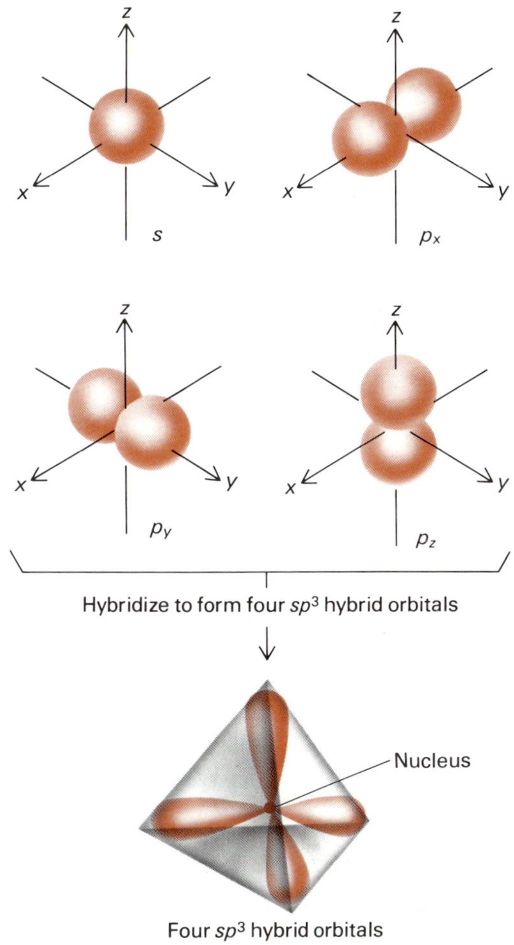

FIGURE 7-4
Formation of four sp^3 hybrid orbitals from one *s* and three *p* orbitals.

In CH_4 the four valence electrons of the C atom are distributed one to each of the four sp^3 hybrid orbitals. Each C—H bond consists of a pair of electrons, one electron of the pair from an sp^3 orbital of the C atom and the other from the 1*s* orbital of the H atom (Figure 7-5). The electrons are concentrated in the region of overlap of the two orbitals. This kind of bond, concentrated at and symmetric about the axis joining the two nuclei, is called a **σ bond** (sigma bond).

In ammonia, the H—N—H bond angle is about 108°, close to the tetrahedral angle. The 2*s* and the three 2*p* orbitals of the N atom are hybridized to form four new sp^3 hybrid orbitals, and the five valence electrons are distributed in these orbitals as follows:

↓↑	↑	↑	↑		↓↑ ↑ ↑ ↑
$2s$	$2p_x$	$2p_y$	$2p_z$		sp^3

isolated N atom — **N atom in NH_3**

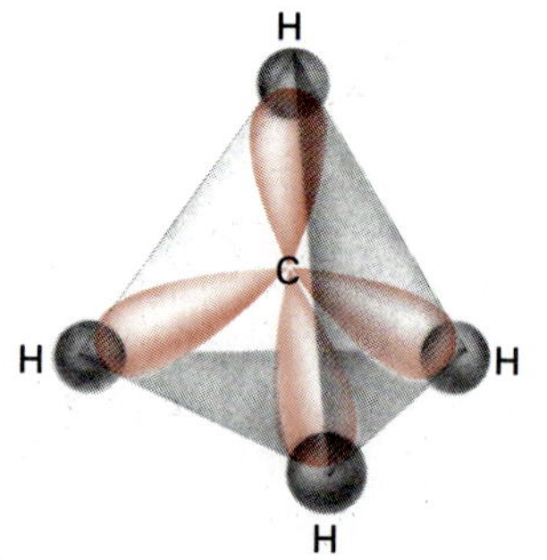

FIGURE 7-5

Formation of methane by the overlap of the four sp^3 hybrid orbitals of carbon with the $1s$ orbitals of four hydrogen atoms.

In the NH_3 molecule, one sp^3 orbital is occupied by a pair of unshared electrons. Each of the three N—H bonds consists of a pair of electrons, one electron from an sp^3 orbital of the N atom and the other from the $1s$ orbital of an H atom (Figure 7-6):

$$\underline{\downarrow\uparrow}\ \underline{\downarrow\uparrow}\ \underline{\downarrow\uparrow}\ \underline{\downarrow\uparrow}$$
$$sp^3$$

NH_3: N atom + three H atoms

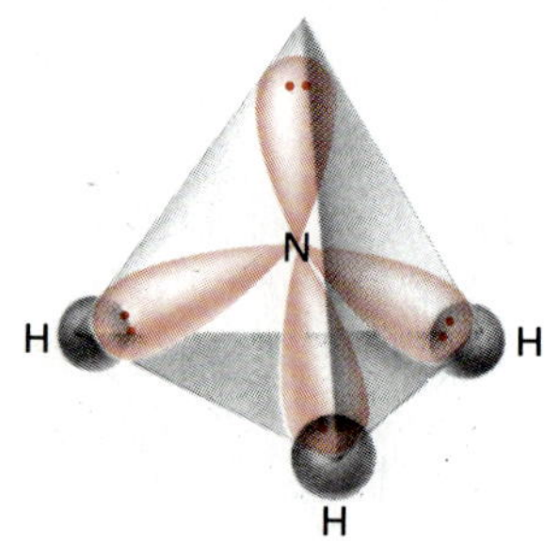

FIGURE 7-6

sp^3 hybrid orbitals of nitrogen with the $1s$ orbitals of three hydrogen atoms.

When ammonium chloride (NH_4Cl) is formed—say, from NH_3 and HCl—a proton (H^+) is transferred from the HCl molecule to the NH_3 molecule. There the proton acquires its complement of two electrons by occupying the fourth corner of the tetrahedron about the N atom, thus sharing nitrogen's lone pair of electrons:

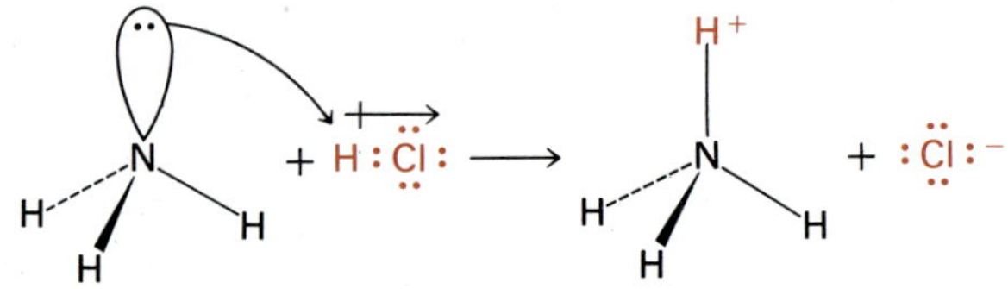

The ammonium ion has a positive charge because the hydrogen is transferred from the HCl molecule to the NH_3 molecule as a proton (H^+), leaving its electron on the Cl atom to yield a Cl^- ion. All four N—H bonds are equivalent, and they are directed toward the corners of a tetrahedron, with H—N—H bond angles of 109.5°.

In water, the H—O—H bond angle is about 105°, close to the tetrahedral angle. The $2s$ and the three $2p$ orbitals of the O atom are hybridized to form four new sp^3 hybrid orbitals, and the six valence electrons are distributed in these orbitals as follows:

$\underline{\downarrow\uparrow}$ $2s$	$\underline{\downarrow\uparrow}$ $2p_x$	$\underline{\uparrow}$ $2p_y$	$\underline{\uparrow}$ $2p_z$		$\underline{\downarrow\uparrow}\ \underline{\downarrow\uparrow}\ \underline{\uparrow}\ \underline{\uparrow}$ sp^3
isolated O atom					**O atom in H_2O**

In the H_2O molecule, each of two sp^3 orbitals is occupied by a pair of unshared electrons. Each O—H bond consists of a pair of electrons, one electron from an sp^3 orbital of the O atom, and the other from the $1s$ orbital of an H atom (Figure 7-7):

$$\underline{\downarrow\uparrow}\ \underline{\downarrow\uparrow}\ \underline{\downarrow\uparrow}\ \underline{\downarrow\uparrow}$$
$$sp^3$$

H_2O: O atom + 2 H atoms

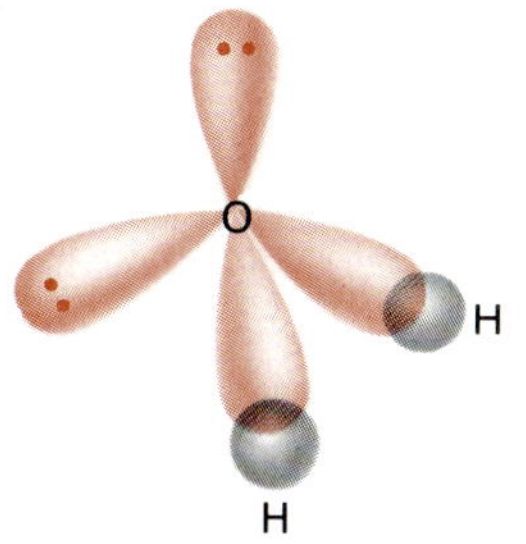

FIGURE 7-7

Formation of water by the overlap of two of the sp^3 hybrid orbitals of oxygen with the 1s orbitals of two hydrogen atoms.

When HCl gas is dissolved in water, hydronium ions (H_3O^+) and Cl^- ions are formed by the transfer of a proton from an HCl molecule to a water molecule where it shares a lone pair of electrons:

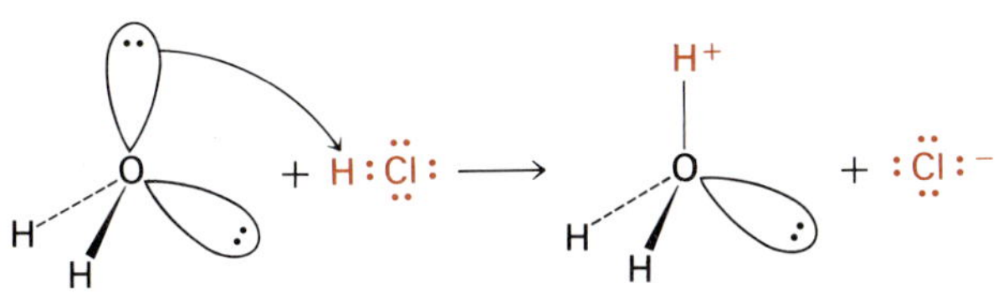

The H atoms occupy three of the four corners of a tetrahedron about the O atom, and a lone pair of electrons occupies the fourth.

The shapes of larger molecules that contain chains of carbon atoms—such as ethane (C_2H_6), propane (C_3H_8), and butane (C_4H_{10})—can also be explained by assuming that the carbon electrons are in sp^3 hybrid orbitals:

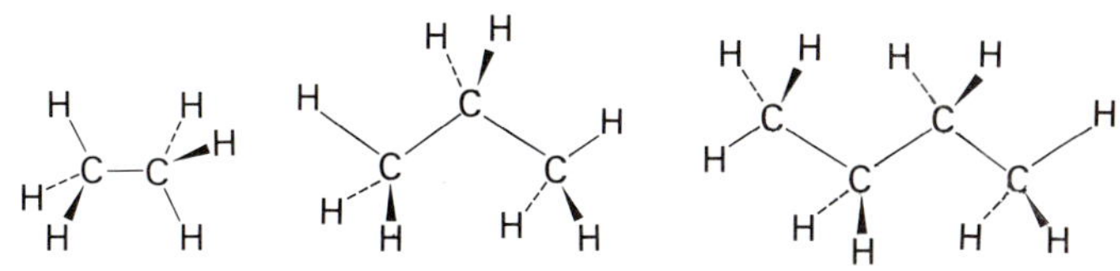

ethane **propane** **butane**

In these compounds containing carbon atoms linked together in chains, each carbon atom uses four sp^3 hybrid orbitals directed toward the corners of a tetrahedron. Each C—H bond is formed by pairing a 1s electron of the H atom with an sp^3 electron of the C atom. Each C—C bond is formed by pairing two sp^3 electrons, one from each C atom.

7-9 MULTIPLE BONDS: sp^2 AND sp HYBRID ORBITALS

Ethylene (C_2H_4) is a planar molecule with a carbon-carbon *double* bond, and all bond angles close to 120°:

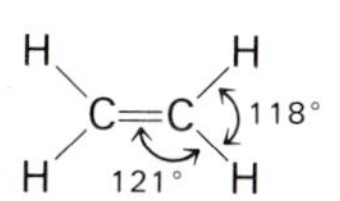

The nuclei of all six atoms lie in one plane, shown here as the plane of the paper. The shape of this molecule is explained by assuming a different kind of hybridization of the atomic orbitals of carbon—a mixing of the 2s orbital with only *two* of the 2p orbitals to yield three new hybrid orbitals, symbolized by sp^2. The third 2p orbital is unhybridized. The four valence electrons of the C atom are distributed as follows:

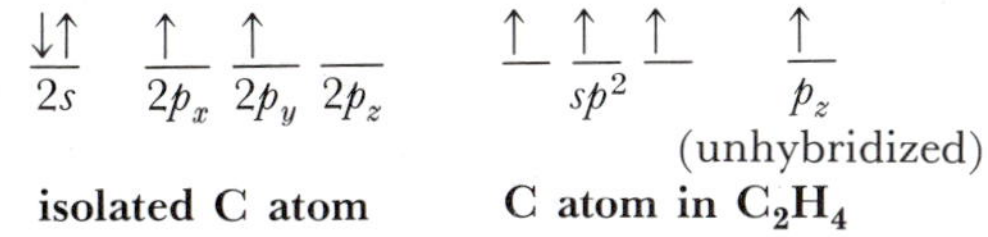

The three sp^2 orbitals lie in a plane and are, theoretically, at 120° to each other. An sp^2 orbital on one C atom overlaps with an sp^2 orbital on the other to form a new orbital that is occupied by a pair of electrons. This shared electron pair constitutes one of the two bonds between the two C atoms (Figure 7-8a). This bond is a σ bond—like the bonds in CH_4, NH_3, and H_2O. Each of the other two sp^2 orbitals of each C atom overlaps with a 1*s* orbital of an H atom, as can be seen in the figure. The C—H bonds are also σ bonds.

The second of the two bonds joining the two C atoms (the first was the σ bond described above) results from a sidewise overlap of the unhybridized *p* orbitals (one on each C atom), which gives two sausage-shaped lobes, one above and one below the plane of the molecule (Figure 7-8b). A shared pair of electrons (one electron from the unhybridized *p* orbital of each C atom) is contained in this two-lobed orbital. The bond thus formed, which is not concentrated on the axis joining the two carbon nuclei, is called a **π bond** (pi bond). When we study organic chemistry, in later chapters, we shall see that a π bond is weaker than a σ bond, and is more easily attacked and broken in a chemical reaction. As shown in Table 7-2, the C=C double bond (the *combination* of one σ and one π bond) is shorter and stronger than a C—C single bond.

Acetylene (C_2H_2) is a linear molecular, with a carbon–carbon triple bond:

H—C≡C—H

180°

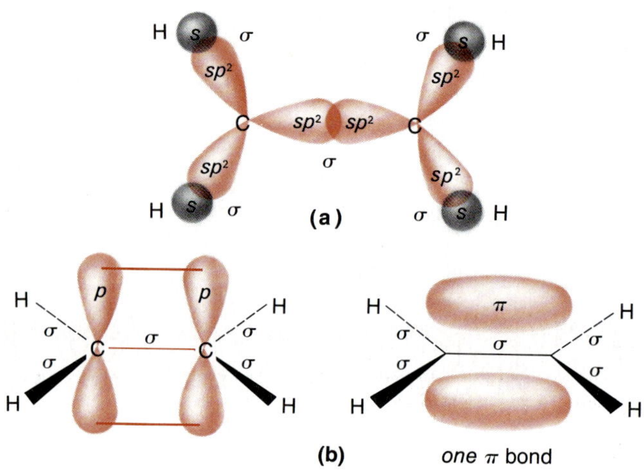

FIGURE 7-8

Formation of ethylene. **(a)** The σ bond framework in ethylene is formed from the sp^2 hybrid orbitals on the carbon atoms and the 1*s* orbitals of the hydrogen atoms. **(b)** The π bond in ethylene is formed by the overlap of the unhybridized *p* orbitals (one from each carbon atom).

This shape is explained by assuming that the 2*s* orbital and *one* of the 2*p* orbitals of the C atom are hybridized to yield two new *sp* hybrid orbitals. The remaining two *p* orbitals of each C atom are unhybridized. The four valence electrons of the carbon atom are distributed as follows:

↓↑	↑	↑			↑ ↑	↑	↑
2*s*	$2p_x$	$2p_y$	$2p_z$		*sp*	p_y	p_z
						(unhybridized)	
isolated C atom					**C atom in C_2H_2**		

The two *sp* orbitals point in opposite directions along a line—that is, they are directed at 180° to each other. An *sp* orbital on one carbon atom overlaps with an *sp* orbital on the other (Figure 7-9a) to form a new orbital that is occupied by a pair of electrons. This shared electron pair constitutes a σ bond between the two carbon atoms, one of the three bonds that join

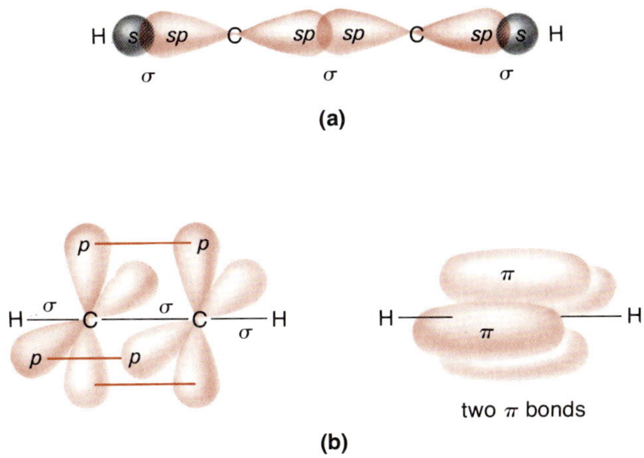

FIGURE 7-9

Formation of acetylene. **(a)** The σ bond framework in acetylene is formed from the *sp* hybrid orbitals of the carbon atoms and the 1*s* orbitals of the hydrogen atoms. **(b)** The π bonds in acetylene are formed by the overlap of pairs of parallel, unhybridized *p* orbitals of the two carbon atoms.

them. The remaining *sp* orbital of each C atom overlaps with a 1*s* orbital of an H atom, as can be seen in the figure; the C—H bonds are also σ bonds. The second and third bonds connecting the two C atoms are π bonds formed by the overlap of the p_y and p_z orbitals of the two carbon atoms (Figure 7-9b). One π bond consists of lobes above and below the molecule, and the other of lobes in front and in back of the molecule. Each pair of lobes contains one pair of electrons. As shown in Table 7-2, the C≡C triple bond (the *combination* of *one* σ and *two* π bonds) is shorter and stronger than either the C—C single or the C═C double bond.

EXERCISE 7-9 Describe the hybrid orbitals used by each C atom, and estimate the bond angles in the following compounds. Indicate which bonds are σ bonds, and which are π bonds.

(a) $CH_3-C\equiv C-H$ (b) $CH_3-\underset{}{\overset{H}{\overset{|}{C}}}=\overset{H}{\overset{|}{C}}-CH_3$

7-10 INTERMOLECULAR FORCES: VAN DER WAALS FORCES

So far we have discussed the strong forces that hold atoms together in a molecule—**intramolecular forces.** But we have not yet considered the weaker forces between molecules, those that hold molecules together in the liquid or the solid state—**intermolecular forces.** If a molecule is polar, with a permanent dipole, small attractions between the negative end of one molecule and the positive end of another help to hold the molecules together in the liquid or solid. Although the attractions between permanent dipoles are of great importance in many reactions of organic and biological chemistry, they are often weak, and in any case cannot by themselves account for the liquid and solid states of nonpolar molecules. What forces account for this?

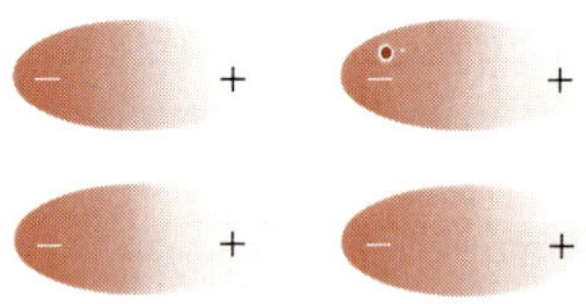

FIGURE 7-10
Induced dipoles in iodine molecules (I_2).

And how can we explain the existence of liquid and solid states for a nonpolar molecule like I_2, or for the noble gas Ne?

The forces that hold I_2 molecules or Ne atoms together in the liquid or the solid phase result from the formation of temporary or instantaneous dipoles that produce attractive forces between the particles. Consider an I_2 molecule in the gas phase. On the average, the electronic charge is distributed uniformly between the two nuclei—that is, the electrons are as close to one nucleus as to the other. At any one instant, however, the charge distribution might be such that the electron cloud is closer to one nucleus than to the other, giving the molecule a *temporary* negative charge on that side and a positive charge on the other—a temporary dipole. The effect of this dipole is to **polarize** a neighboring molecule, inducing a dipole in it, too; and so on (Figure 7-10). The resulting attractive forces between the neighboring instantaneous dipoles, called **van der Waals forces** after the Dutch scientist who investigated them, pull the molecules together to form a liquid—or, if the temperature is low enough (so that the kinetic energy of the molecules is small enough), a solid.

Van der Waals forces are typically of the order of a few kilojoules per mole, three to four times as strong as the forces between permanent dipoles, but very much weaker than covalent bonds (only of the order of 1/100 as great). The strength of van der Waals forces depends on the number of electrons in the molecule and on its size and shape. The more electrons there are in a molecule, the more *polarizable* it is, and the more easily a dipole can be induced in it by a neighboring electrical charge. Every electron contributes to the polarizability, but the least tightly bound (those farthest from the nucleus) contribute the most.

The boiling temperature of a substance reflects the strength of the forces holding the molecules together in the liquid—the stronger the van der Waals forces, the higher the boiling point—because, in order to escape from the liquid, the molecules must have enough kinetic energy to overcome the forces of attraction. (Remember that kinetic energy is proportional to absolute temperature; Section 3-10). Table 7-3 gives the normal (1 atm) boiling temperatures of several molecules of different sizes, shapes, and numbers of electrons. As the number of electrons and the size of the molecule increases, from methane to pentane, the boiling temperature increases. Neopentane, which has the same atoms and number of electrons as *n*-pentane, but which is more compactly arranged (see Table 7-3) with less surface exposed to neighboring molecules, has a lower boiling temperature than *n*-pentane. Temporary dipoles are less easily induced in neopentane, so the van der Waals forces between the neopentane molecules are weaker than those between molecules of *n*-pentane. Note that all the substances in Table 7-3 are nonpolar, so it is *only* the van der Waals forces that hold the molecules together in the liquid or the solid.

The attractive forces between nonpolar gas molecules that help account for deviations from ideality (and eventually lead to liquefaction; Chapter 3) are van der Waals forces. The heat needed to melt a nonpolar solid (heat of fusion) or to vaporize a liquid (heat of vaporization) is used to overcome the van der Waals forces of attraction. Table 7-3 also gives the heats of vaporization of several substances. As we would expect, they are in the same order as the boiling temperatures: the greater the van der Waals forces, the higher the heat of vaporization.

TABLE 7-3
Boiling temperatures (at 1 atm) of some hydrocarbons

Compound	Formula	Line structure	Condensed structural formula	Boiling point (°C)	Heat of vaporization (kJ/mole)
Methane	CH_4	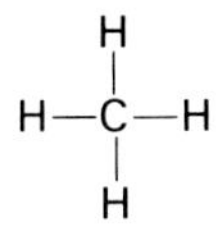	CH_4	−161.5	8.15
Ethane	C_2H_6	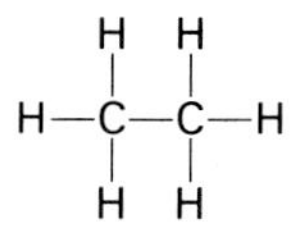	CH_3CH_3	−88.6	14.7
Propane	C_3H_8	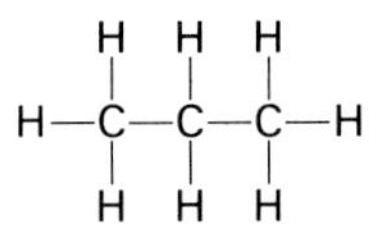	$CH_3CH_2CH_3$	−42.1	18.8
n-Butane (normal butane)	C_4H_{10}	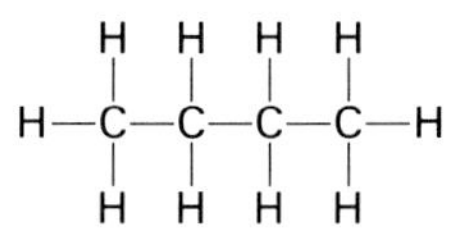	$CH_3(CH_2)_2CH_3$	−0.5	22.4
n-Pentane (normal pentane)	C_5H_{12}	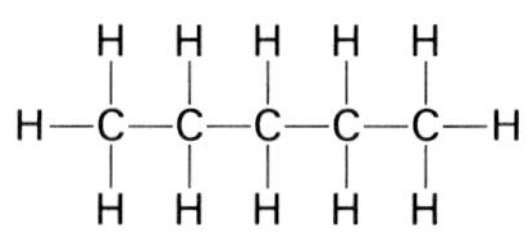	$CH_3(CH_2)_3CH_3$	36.1	26.5
Neopentane (dimethylpropane)	C_5H_{12}	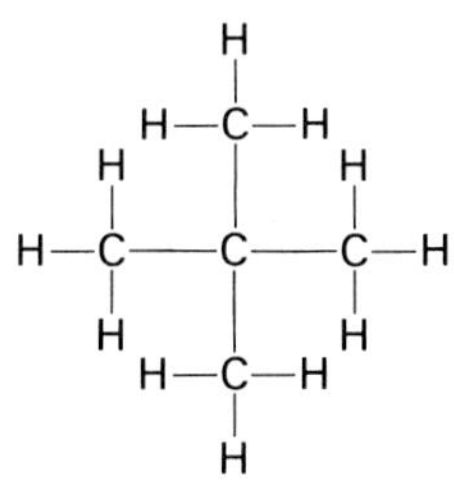	$C(CH_3)_4$	9.5	22.7

7-11
INTERMOLECULAR FORCES: HYDROGEN BONDING

The freezing and boiling temperatures of substances containing hydrogen bonded to nitrogen, oxygen, or fluorine are often found to be unexpectedly high. This can be seen in Figure 7-11, where the boiling temperature is plotted against row in the periodic table for the hydrides of elements in the C, N, O, and F groups (groups 4, 5, 6, and 7). The van der Waals forces should be progressively larger going down a group in the periodic table because of the increasing number of electrons and the increasing distance of the outer electrons from the nucleus. Thus, the boiling points of the hydrides should increase with the row in the periodic table, if the only attractive forces present between molecules are van der Waals forces. The unusually high boiling points of NH_3, H_2O, and HF give evidence that another attractive force (stronger than van der Waals forces) is present, holding the molecules together in the liquid.

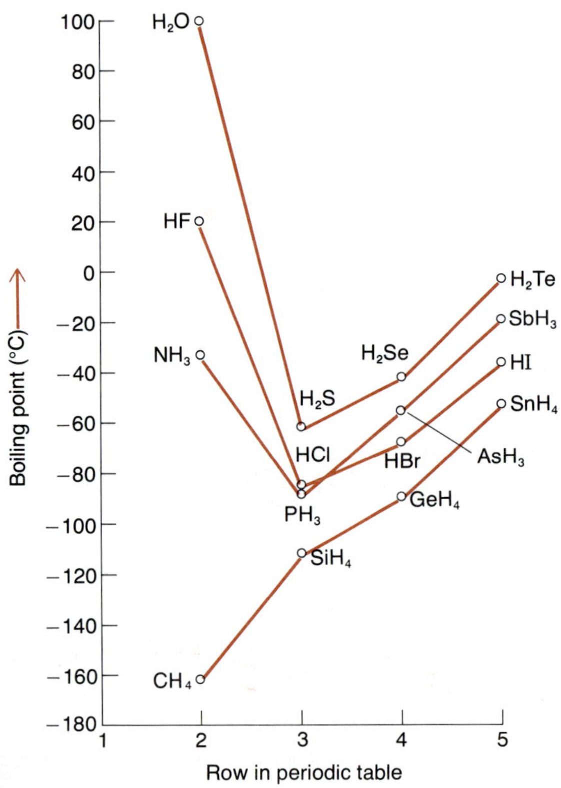

FIGURE 7-11

Normal boiling points of the hydrides of four families.

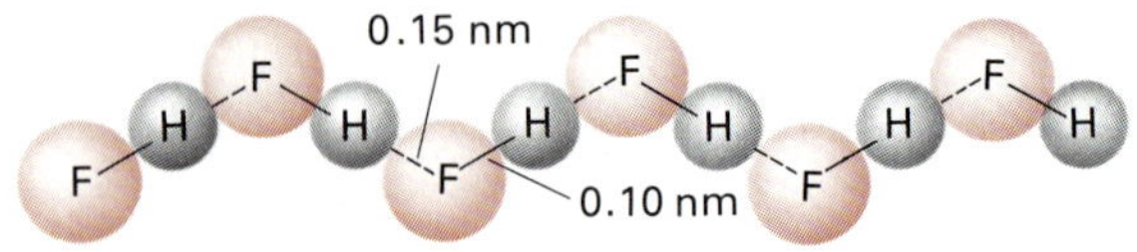

FIGURE 7-12

Association of molecules by hydrogen bonding. The structure of crystalline HF.

A hydrogen atom bonded to one of the small, highly electronegative atoms N, O, or F is simultaneously attracted to an unshared pair of electrons on an N, O, or F atom of a neighboring molecule, and can form a "bridge," called a **hydrogen bond,** between the two molecules (Figure 7-12). The strength of hydrogen bonds is of the order of 40 kJ per mole, about 10 times that of van der Waals forces, though only about 1/10 that of covalent bonds.* Hydrogen bonds thus account for the unexpectedly high boiling points of NH_3, H_2O, and HF.

In liquid water, clusters of water molecules are held together by hydrogen bonds. The hydronium ion (H_3O^+) discussed in Section 7-3 is just one of many protonated species in the liquid; $H_5O_2^+$, $H_7O_3^+$, $H_9O_4^+$, and larger clusters exist. Hydrogen bonds account for the many special properties of water described in Chapter 4: the high surface tension, the high molar heats of fusion and vaporization, and the exceptionally large specific heat. Because of the large specific heat of water, organisms that live in the oceans exist in an environment of relatively constant temperature.

*The corresponding bond strengths in kilocalories are: covalent bond, ~100 kcal/mole; hydrogen bond, ~10 kcal/mole; van der Waals forces, ~1 kcal/mole.

Animals and plants that live on land, where the temperature fluctuates widely, have a high water content (50% to 95% of their body weight) that helps them maintain a relatively constant internal temperature. This is important because biological reactions take place in a narrow temperature range.

Hydrogen bonds also account for the unusual ability of water to climb in narrow tubes—the phenomenon called capillarity. In capillary action, hydrogen bonds formed between water molecules and solid surfaces such as glass or clay (both of which contain oxygen atoms) pull water up through a glass capillary tube or through porous soil to the roots of plants.

In ice, hydrogen bonding leads to the formation of a giant, three-dimensional network in which each H atom forms a bridge between two O atoms, and each O is surrounded tetrahedrally by four H atoms. Each water molecule is thus bonded to four other water molecules. The resulting open crystalline structure (Figure 7-13a) is responsible for the low density of ice. When ice is heated, some hydrogen bonds break, and the ice melts. The crystalline structure partially breaks down to form the denser liquid water which has fewer hydrogen bonds and a less open structure (Figure 7-13b). Thus, hydrogen bonding gives water the unusual property of being less dense as a solid than as a liquid.

Hydrogen bonds also are formed in alcohols such as ethanol (CH_3CH_2OH) and (just as we saw earlier in NH_3, H_2O, and HF) cause unexpectedly high boiling points. Table 7-4 gives the boiling temperatures of ethanol and several other alcohols and, for comparison, chloroethane. Chloroethane is about the same size as ethanol, but has more electrons, so should have stronger van der Waals forces between the molecules. Thus chloroethane might be expected to have a somewhat higher boiling point than ethanol. However, the fact that chloroethane does not form hydrogen bonds clearly outweighs the effect of the van der Waals forces. The presence of two OH groups, as in ethylene glycol, allows the formation of a greater number of hydrogen bonds and markedly increases the boiling temperature—hence the use of ethylene glycol as the principal ingredient of permanent antifreeze (it does not boil off at normal radiator temperatures). Glycerol, with three OH groups, can form still more hydrogen bonds, which results in an even higher boiling point.

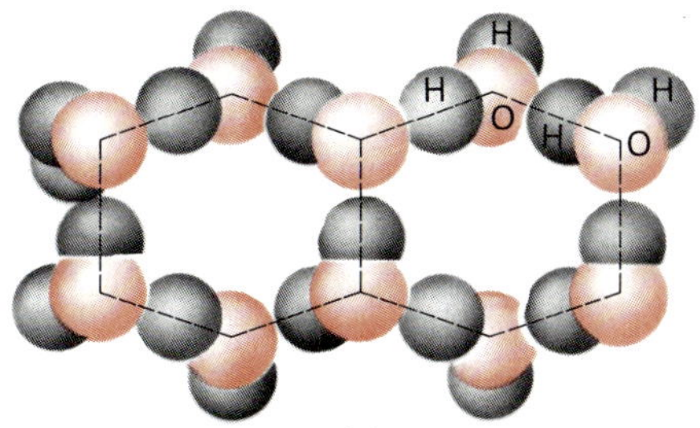

(a)

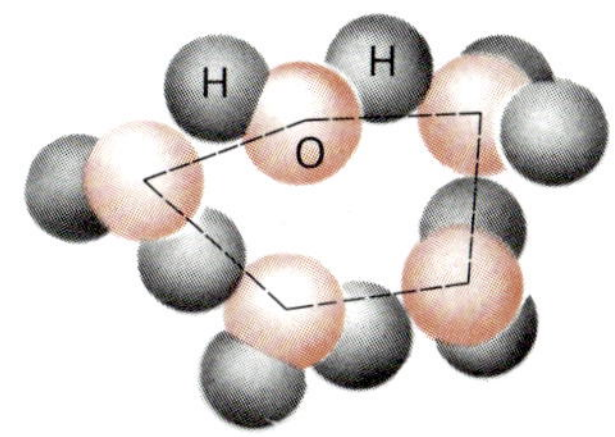

(b)

FIGURE 7-13

Structure of H_2O. **(a)** The crystal structure of ice. Note the hexagonal pattern and the "open" structure. [After L. Pauling and P. Pauling, *Chemistry,* W. H. Freeman and Company. Copyright © 1975.] **(b)** The structure of liquid water. Note that the liquid has a more random and less open structure than does the ice.

Many molecules of biological importance contain NH or OH groups that can participate in hydrogen bonds with water or other biologically important molecules. In proteins, *intramolecular* hydrogen bonds be-

TABLE 7-4
Boiling temperatures of some alcohols and chloroethane

Compound	Condensed structural formula	Line structure	Boiling point (°C)
Chloroethane	CH_3CH_2Cl	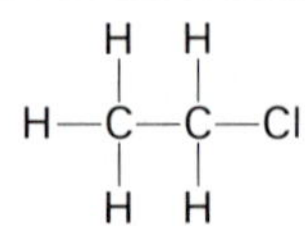	12.3
Ethanol	CH_3CH_2OH	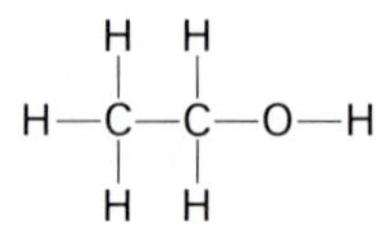	78.4
Ethylene glycol	$HOCH_2CH_2OH$	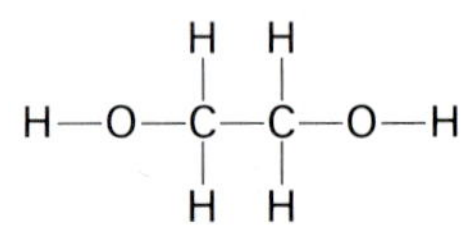	197.9
n-Propanol	$CH_3CH_2CH_2OH$	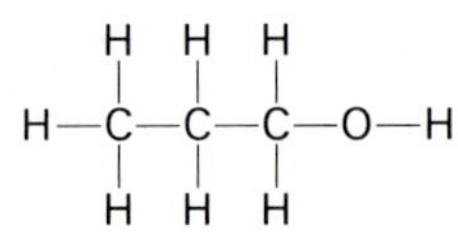	97.2
Glycerol	$HOCH_2CHOHCH_2OH$	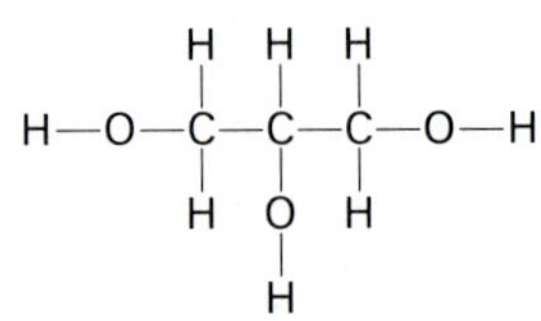	290
n-Butanol	$CH_3(CH_2)_2CH_2OH$	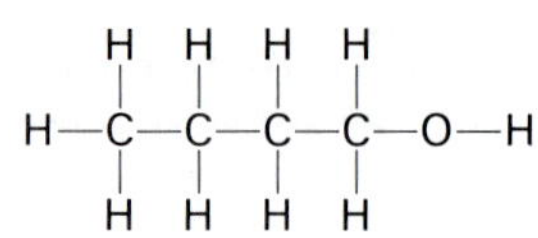	118.0

tween the H atom of an NH group and an O atom of a C=O group of the same molecule may stabilize the long chain in a definite helical shape (see Figure 21-5, for example). And hydrogen bonds tie together the two strands of the double helix of DNA—the molecule that stores genetic information (Figure 7-14). We shall learn more about DNA and its remarkable properties in Chapter 25. When a protein is denatured (as, for example, when egg white is heated and turns from a clear, viscous liquid to an opaque, white solid), the change in its properties is due partly to the destruction of hydrogen bonds and to the consequent breakdown of the previously fixed, three-dimensional shape of the molecule. In later chapters, we shall see many other examples of the role of hydrogen bonds in determining properties of biologically important compounds.

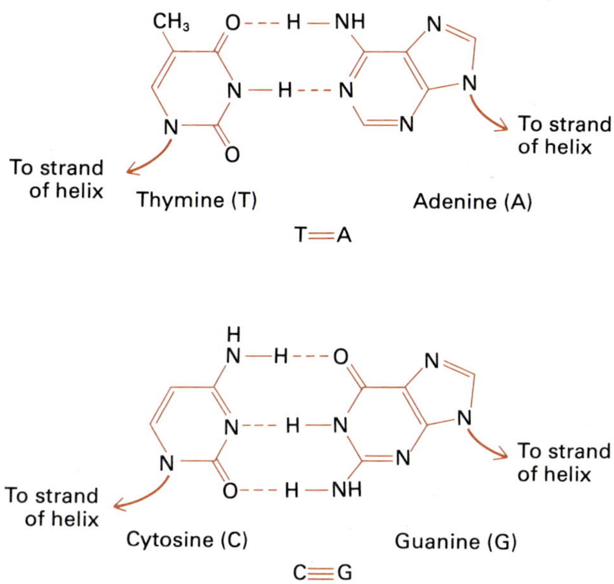

(a)

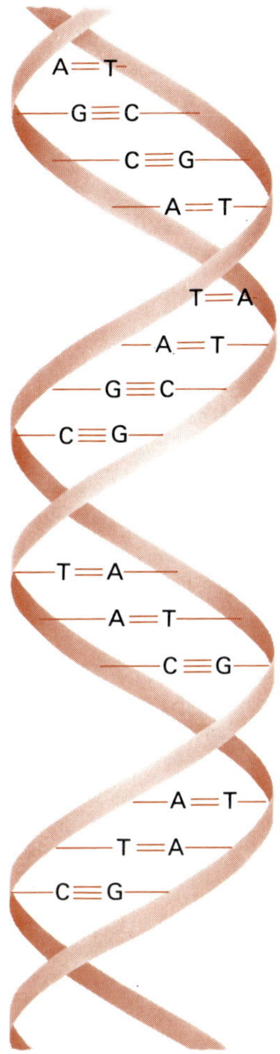

(b)

FIGURE 7-14

Hydrogen bonds in DNA. **(a)** Hydrogen bonds tie together pairs of structures (bases). **(b)** The paired bases tie together the two strands of the DNA double helix.

KEY TERMS

In a *Lewis* (or *electron-dot*) *structure,* the nucleus and inner electrons are represented by the symbol for the element, and the electrons in the outermost main shell are represented by dots. *Valence electrons* are electrons in the outermost mainshell.

In forming an *ionic compound,* one atom loses one or more electrons, while another atom gains one or more electrons. An *ionic bond* between a positive ion and a negative ion results from the strong electrostatic forces between the oppositely charged particles.

Most bonding is *covalent,* with a sharing of electrons between the two atoms participating in the bond. In a *nonpolar covalent bond,* there is equal sharing of electrons between the two atoms. In a *polar covalent bond,* there is unequal sharing of electrons between the two atoms. In a *line structure,* a line represents a chemical bond and stands for a shared electron pair.

Electronegativity is a measure of the ability of an atom to attract electrons.

A diatomic molecule composed of unlike atoms is a *polar molecule.* In a molecule composed of three or more atoms, whether individual polar bonds lead to a polar or nonpolar molecule depends on the shape of the molecule.

When two pairs of electrons are shared by two atoms, the bond is a *double bond.* When three pairs of electrons are shared by two atoms, the bond is a *triple bond.*

When a molecule cannot be adequately represented by a single electron-dot structure, the real structure of the molecule is said to be a *resonance* hybrid of two or more structures.

Bond energy (or bond dissociation energy) is the energy required to break a bond.

Hybrid orbitals are orbitals formed by mixing atomic orbitals. For example, four sp^3 hybrid orbitals are formed by mixing one s and three p orbitals.

A σ *bond* (sigma bond) is concentrated at, and cylindrically symmetric about, the axis joining the two nuclei. A π *bond* (pi bond) results from the sidewise overlap of two p orbitals, and consists of two lobes that lie above and below the axis joining the two nuclei.

Intramolecular forces are the strong forces that hold atoms together in molecules. *Intermolecular forces* are weaker forces that hold molecules together in the liquid or solid state. The *van der Waals forces* are weak attractive forces between neighboring instantaneous dipoles. A hydrogen atom bonded to N, O, or F can form a "bridge" to a lone pair of electrons on an N, O, or F atom of another molecule; this bridge between two molecules is called a *hydrogen bond.*

Polarizability is the ability of a molecule to have a dipole induced in it.

CHAPTER SUMMARY

A *chemical bond* forms between two atoms because electrons of each atom are attracted simultaneously to both nuclei. The electronic shell structure of atoms can be used to help us understand the nature of the chemical bond. The special stability of the noble gas electronic configuration is seen in the formation of chemical bonds: *atoms give up or accept electrons, or share them, to reach an inert gas configuration.*

The two extreme examples of chemical bonds are the *ionic bond* and the *nonpolar covalent bond* between like atoms. Most chemical bonds are *polar covalent bonds* in which the shared electron pair is shared unequally by the two atoms. Electron-dot and line structures symbolize the bonding in molecules. Carbon typically forms four covalent bonds; nitrogen, three; oxygen, two; and hydrogen and the halogens, one. Sometimes two or three electron-pairs are shared by two atoms, resulting in a double or a triple bond.

Most atoms, in forming bonds, acquire an octet of electrons in their outer shells. The optimal arrangement for these four electron pairs is *tetrahedral,* and many molecules have a tetrahedral shape. This shape is explained in the methane molecule (and others) by a theory that says the $2s$ and the three $2p$ orbitals of carbon are hybridized to yield four new sp^3 hybrid orbitals that are directed toward the corners of a tetrahedron. In ethylene (C_2H_4), which is a planar molecule, the carbon is in the sp^2 hybridization state, and the double bond consists of one σ and one π bond. In acetylene (C_2H_2), which is a linear molecule, the carbons are in the *sp* hybridization state, and the triple bond consists of one σ and two π bonds.

The *intermolecular* forces that hold molecules together in a liquid or a solid are dipole–dipole forces, van der Waals forces, and hydrogen bonds. The *van der Waals forces* result from instantaneous, induced dipoles, and are only about 1/100 as strong as a covalent bond. Boiling points and heats of vaporization give an indication of the relative strengths of van der Waals forces in different molecules—as the size of the molecule increases, the van der Waals forces also increase.

A hydrogen atom bonded to one of the small, highly electronegative atoms N, O, or F is simultaneously attracted to an unshared pair of electrons on an N, O, or F atom of a neighboring molecule. This hydrogen bridge between the two molecules is called a *hydrogen bond.* Hydrogen bonds are about 1/10 as strong as covalent bonds, and account for the high boiling point and heat of vaporization, and for many of the other properties of water.

PROBLEMS

7-1. Draw the electron-dot structures (a) of strontium and bromine in the compound strontium bromide, and (b) of lithium and sulfur in the compound lithium sulfide.

7-2. Give the formulas of the compounds formed between the following pairs of elements or ions. In each case, assume that the bonding is ionic.

(a) Al and O	(b) K and N
(c) Ba and N	(d) Rb and S
(e) Ca and At	(f) Ca^{2+} and No_3^-
(g) Ba^{2+} and PO_4^{3-}	(h) Na^+ and SO_4^{2-}
(i) NH_4^+ and CO_3^{2-}	(j) Cr^{3+} and OH^-
(k) K^+ and HCO_3^-	(l) Zn^{2+} and CH_3COO^-

7-3. Draw the electron-dot structures for NH_4^+ and I^- in the ionic compound ammonium iodide (NH_4I).

7-4. Draw the electron-dot structures for the ions H^+, H^-, and CN^-.

7-5. Draw the electron-dot or the line structures for the following compounds: phosphorus trichloride (PCl_3); hydroxylamine (NH_2OH); hypochlorous acid (HOCl), the active ingredient in laundry bleaches; the poison gas phosgene (Cl_2CO); acetonitrile (CH_3CN), a useful solvent.

7-6. The element with atomic number 12 (call it X) reacts with the element of atomic number 9 (call it Y) to give a stable compound XY_n. What is the number *n*? Assume that the bonding in the compound is ionic. (Try to answer without consulting tables.)

7-7. The triatomic molecule ozone (O_3), a form of oxygen that is produced in the stratosphere and that protects the earth from the sun's ultraviolet light, is angular in shape, with two of the oxygen atoms not directly bound to each other (but with both bound to the third, or central, oxygen atom). Draw an electron-dot and a line structure for the ozone molecule. Is it possible adequately to represent this molecule by only one such structure?

7-8. (a) Ethylene (C_2H_4), is the first in a family of hydrocarbons called alkenes, each of which has a double bond. Draw the electron-dot or the bond structure of the second member of the family, propene (C_3H_6).
(b) Acetylene (C_2H_2) is the first in a family of hydrocarbons called alkynes, each of which has a triple bond. Draw the electron-dot or the line structure of the second member of the family, propyne (C_3H_4).

7-9. Cyclopentane (C_5H_{10}) is a molecule with five carbon atoms joined in a ring, with each carbon bonded also to two hydrogen atoms. Draw the line structure for cyclopentane.

7-10. Recall that the electron configurations of the carbon and nitrogen atoms in their states of lowest energy are $1s^2\,2s^2\,2p^2$ and $1s^2\,2s^2\,2p^3$, respectively. Just below carbon in the periodic table is silicon (Si) and just below nitrogen is phosphorus (P).
(a) Write the electron configurations of the silicon and phosphorus atoms in their states of lowest energy. (Be careful to specify which is which. Try to answer without consulting tables.)
(b) What are the atomic numbers of silicon and phosphorus?
(c) The compound silane is like methane, but with silicon in place of carbon. Write the chemical formula, the electron-dot structure, and the corresponding line structure of the silane molecule.
(d) Describe in words (with the help also of a figure, if you wish) the three-dimensional geometrical structure of the silane molecule. What hybrid orbitals is the Si atom using?
(e) The compound phosphine is like ammonia, but with phosphorus in place of nitrogen. Write the chemical formula, the electron-dot structure, and the corresponding line structure of the phosphine molecule.
(f) Describe in words (with the help also of a figure, if you wish) the three-dimensional geometrical structure of the phosphine molecule. What hybrid orbitals is the P atom using?

7-11. Prostaglandins are biological molecules that modulate the action of hormones. The structure of the prostaglandin $PGF_{2\alpha}$ is

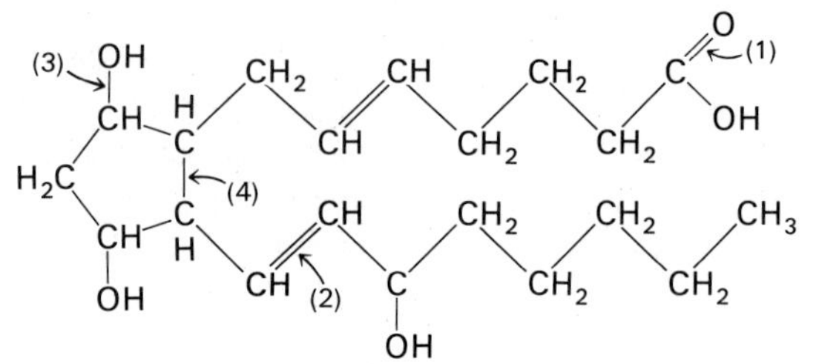

$PGF_{2\alpha}$

(a) Which, if any, carbon–carbon bond(s) should be the shorter one(s) in the molecule?
(b) Which, if any, carbon–carbon bond(s) should be the stronger one(s) in the molecule?
(c) Tell what orbitals are being used by each carbon atom participating in the bonds labeled (1), (2), (3), and (4) in the structure.

7-12. Tell what is wrong with each of the following structures:

(a) 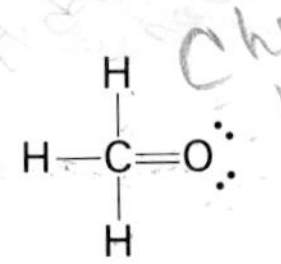

(b) 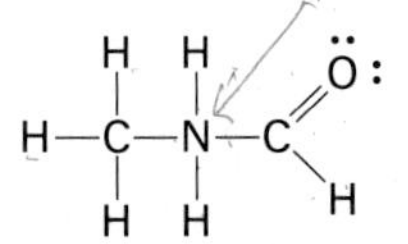

(c) 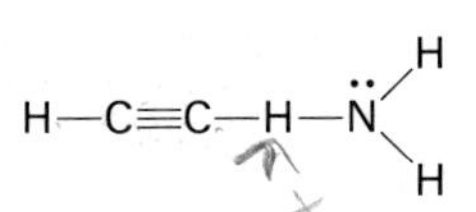

(d) 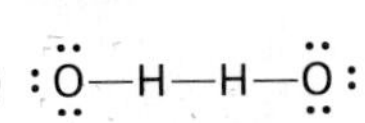

7-13. Which of the following are polar molecules: N_2, HF, Cl–F, and S_8? For each polar molecule, show the direction of the dipole. Use the δ^+, δ^- or the $+\!\!\longrightarrow$ notation.

7-14. Which of the following are polar molecules: CF_4, CHF_3, CH_2F_2, CH_3F? For each polar molecule, show the direction of the bond dipoles, using the δ^+,δ^- or the $+\!\!\longrightarrow$ notation.

7-15. H_2S is a polar molecule. Write a line structure for H_2S, showing its shape and the directions of the bond dipoles.

7-16. What are the hybridization states of the two carbons in the following compound:

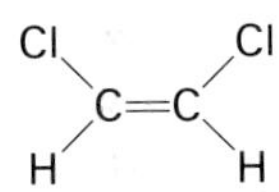

Draw a picture showing the bonds and orbitals. Label the σ and π bonds.

7-17. What are the hybridization states of the two carbons that are joined by the triple bond in the compound $CH_3—C{\equiv}C—CH_3$? Draw a picture showing the bonds and orbitals. Label the σ and the π bonds.

7-18. Which of these two molecules is a polar molecule? Explain your answer.

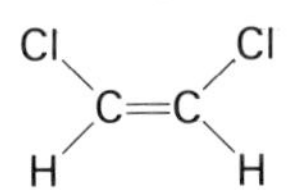

or

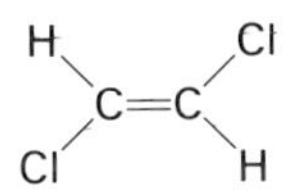

7-19. Which has the higher boiling point,

$$CH_3—CH_2—CH_2—OH$$

or

$$HO—CH_2—CH_2—CH_2—OH$$

Explain your answer.

7-20. Of the three molecules, C_2H_6, C_2H_5Br, and C_2H_5OH, which would you expect to have the highest, and which the lowest, boiling point?

7-21. In each of the following pairs, which compound would you expect to have the lower boiling point? Explain your answers.

(a) Silane (SiH_4) or silicon tetrabromide ($SiBr_4$).

(b) Normal hexane,

$$CH_3—CH_2—CH_2—CH_2—CH_2—CH_3$$

or 2,2-dimethyl butane,

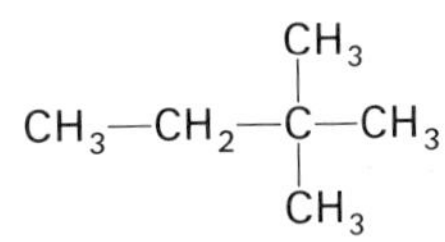

(c) 1-Fluoropropane, $CH_3—CH_2—CH_2—F$, or 1-propanol, $CH_3—CH_2—CH_2—OH$.

7-22. Account for the fact that HF is a liquid at room temperature (below 20°C), whereas HCl is a gas.

7-23. Acetic acid forms dimers (two molecules held together by hydrogen bonds). Draw a diagram showing how hydrogen bonds account for the existence of acetic acid dimers. Acetic acid has the following structure:

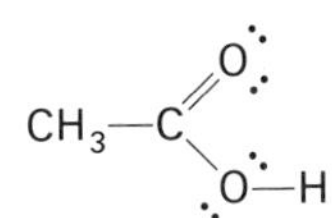

7-24. Explain how hydrogen bonding accounts for the large specific heat of water.

Quartz crystals (SiO_2). Crystals of a given substance have the same characteristic shapes.

8

How does a chemist identify a substance? What elements are present in it and in what amounts? How are the atoms connected in the molecule? What is the three-dimensional arrangement of these atoms in space? In this chapter, we see how chemists answer such questions.

The Structure of Molecules and Crystals

8-1
METHODS OF PURIFICATION

Purity is a relative matter—nothing is *absolutely* pure. For example, in 18.0 g of water (one mole), there are 6.02×10^{23} molecules of H_2O. If one molecule is not H_2O, but instead is sugar ($C_{12}H_{22}O_{11}$), we would not be able to detect this impurity by tasting because our detector (the tongue) is not sensitive enough. Many laboratory instruments can't detect impurities of less than about 0.1% by weight—say, 1 g of impurity in 1000 g of substance. One exception is if the impurity is radioactive. As we saw in Chapter 5, tiny amounts of radioactive material can be detected with a Geiger counter. A biological detector may also be extremely sensitive; the male silk moth can detect with its antennae, and respond to, a stimulus of only 200 molecules of the female sex attractant, bombykol. (This stimulus is received when an airstream contains about 1000 bombykol molecules per cubic centimeter of air, or a concentration of one molecule of bombykol among 10^{16} molecules of other components of the air.) So purity is defined arbitrarily, depending on the degree of purity required. Isopropyl alcohol used for sterilizing hypodermic needles is only 90% pure, refined sugar is 99.90% pure sucrose, copper used in electrical wires is more than 99.95% pure, and silicon used to make transistors is over 99.9999999% pure.

Some methods of separating a mixture into its components are **mechanical:** panning for gold, picking objects of iron and steel out of a canal with a magnet, or separating sand from sugar by dissolving the sugar in water and then filtering off the sand.

Another physical method of purification is **distillation,** which separates mixtures on the basis of differences in boiling points. Figure 8-1 shows a simple distillation apparatus. The solution in the distilling flask is salt water. What we would like to do is to purify the water—as in desalinization to prepare drinkable water from brackish water or seawater. When the solution is boiled, water vaporizes and the vapor (steam) flows

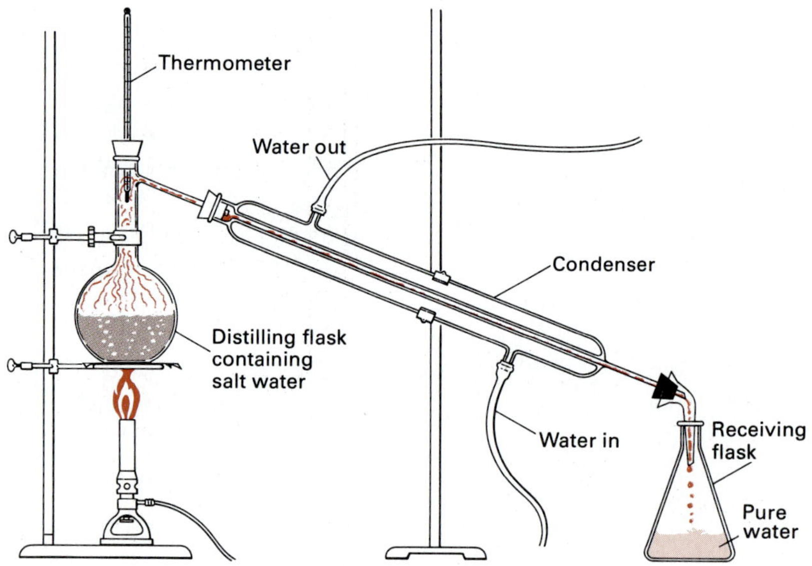

FIGURE 8-1

A simple distillation apparatus. Cool water circulating through the jacket of the condenser causes the vapor to condense.

into the condenser. There it is cooled by water circulating through the jacket of the condenser, and returns to the liquid state (condenses). Pure water thus collects in the receiving flask, while the salt—which is essentially nonvolatile at the temperature of the boiling solution (the boiling point of the salt is 1743°C)—remains in the distilling flask. To separate mixtures of liquids whose boiling points are close together, the distillation may have to be repeated many times to achieve the desired degree of purity. Distillation is used to separate the alcohol from a fermentation mixture, to desalinate sea water, and to refine crude petroleum (see Chapter 14).

Extraction separates substances by taking advantage of differences in solubility of a substance in two different solvents. Figure 8-2a shows a separatory funnel containing a brownish solution of iodine (I_2) and nickel chloride ($NiCl_2$) in water. We would like to separate the $NiCl_2$ from the I_2. Another solvent is added (say, carbon tetrachloride, CCl_4) that is immiscible with (insoluble in) water, and in which the I_2, but not the $NiCl_2$, is soluble. The flask is shaken and, when the two liquid layers are allowed to settle, a purple solution of I_2 in the denser CCl_4 settles out at the bottom, with a green solution of $NiCl_2$ in water floating on top (Figure 8-2b). The CCl_4 layer can then be drained off the bottom, thus completing the separation of the I_2 from the $NiCl_2$ (Figure 8-2c). By evaporating off the solvents, pure I_2 and $NiCl_2$ are obtained. Repeated extractions are often necessary for complete separation of a mixture.

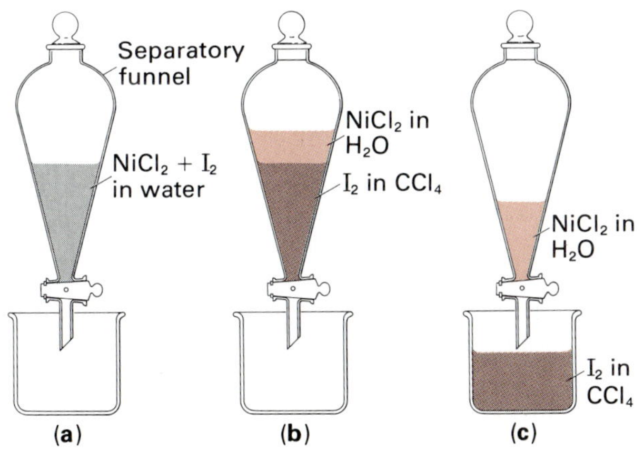

FIGURE 8-2

Extraction. The I_2 is extracted from the water solution by the carbon tetrachloride. The $NiCl_2$ remains in the water layer.

Extraction is used in the preparation of instant decaffeinated coffee. A solution of strong coffee is extracted with methylene chloride (CH_2Cl_2), in which the caffein, but not the coffee, is very soluble. By repeated extractions, as much as 97% of the caffein is removed from the coffee. Instant decaffeinated coffee is obtained by freeze-drying the aqueous (water) layer; and the caffein is obtained by evaporating the methylene chloride. The caffein, recovered here as a byproduct, is sold for use in cola drinks and "pep" pills.

Fractional crystallization can be used to purify a solid, crystalline substance. The impure solid is dissolved in as small a volume as possible of a hot solvent—say, unrefined dark sugar in hot water. The solution is then permitted to cool slowly. As it cools, pure, white sugar crystals are formed, leaving the impurities behind in solution. Or, if there are two substances in the same solution, the two may crystallize out at different times. In the mineral sample shown in Figure 8-3a, the bottom layer of crystals is dolomite, $CaMg(CO_3)_2$, and the top layer is chalcopyrite, $CuFeS_2$. At one time, both were together in solution and, as the solution slowly cooled over millions of years, first the dolomite crystallized out and then the chalcopyrite. In Figure 8-3b, three different minerals crystallized out separately.

Zone refining is a modification of the technique of fractional crystallization. In zone refining, only a part of the material to be purified is melted at any one time. As shown in Figure 8-4, a short length of molten material (in this case, silicon) passes slowly through a relatively long, solid silicon rod. The impurities re-

(a)

(b)

FIGURE 8-3

Fractional crystallization. **(a)** As a molten mixture cooled, dolomite (*white*) crystallized out first, and then chalcopyrite. **(b)** In this rock, the small white crystals are quartz; the large, flat white ones are albite; and the dark ones are microcline. [Photographs by Richard Byrnes.]

main in the molten silicon and are carried upward, and pure silicon crystallizes below as the heater is moved along the rod from bottom to top. To obtain Si of greater than 99.9999999% purity for use in transistors, the heater is passed 67 times over a silicon rod of 1.3 cm diameter, at a rate of 3.6 inches per hour. Zone refining is applied to other metals when a high degree of purity is required, but it is a time-consuming and expensive process.

Chromatography is a method of purification and identification that has special importance in organic and biological chemistry, where it is often necessary to separate compounds that differ only slightly in composition and structure, and that are very similar in solubility, boiling point, and melting point. According to the classical technique, a solution of the mixture is introduced onto a column containing a powdered

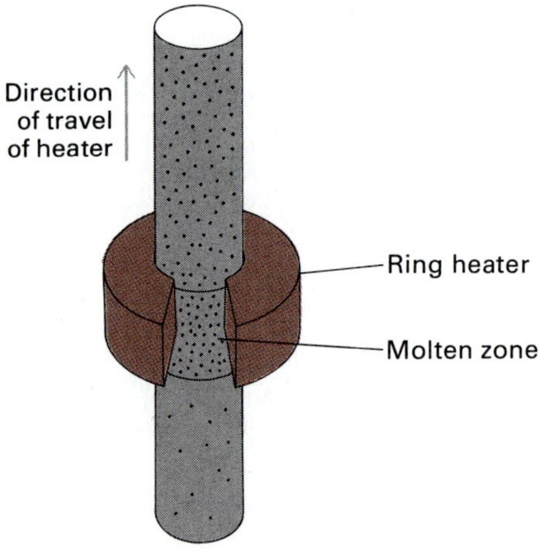

FIGURE 8-4

Zone refining. As the heater is raised, a short length of molten silicon passes through a silicon rod, carrying the impurities (black dots) along and leaving purer silicon behind.

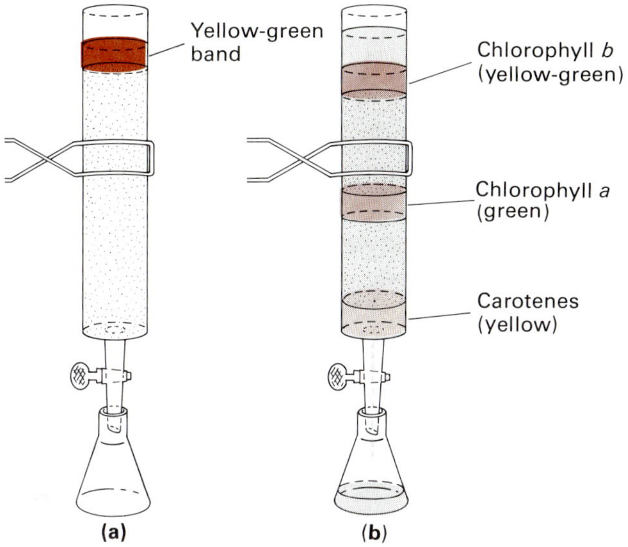

FIGURE 8-5

Column chromatography. The column is packed with powdered sugar. **(a)** An extract of dried leaves in hexane (C_6H_{12}) is introduced at the top of the column. **(b)** The column is washed with hexane containing a small amount of ethyl alcohol. This resolves the mixture into three components that can be eluted (washed from the column) separately and collected in separate test tubes.

material composed of polar molecules—for example, silica gel (SiO_2). The packing material has a weak attraction for organic molecules—one that is strong enough to hold, or **adsorb,** the organic molecules, but weak enough to be broken again, allowing adsorbed molecules to be washed through the column with solvent. The rate of movement of a compound through the column depends on the strength of its adsorption by the packing material, its solubility in the solvent, and the temperature. The mixture is thus divided into various components, which can be collected separately in different containers.

Figure 8-5 shows the separation of pigments from plant leaves. The mixture of pigments is separated on a column packed with powdered sugar, using a mixture of hexane (C_6H_{12}) and ethyl alcohol (C_2H_6O) as solvent. An experiment similar to this one led to the development of chromatography by Mikhail Tswett, a Russian botanist who separated the pigments in chlorophyll in about 1900.

Many additional chromatographic techniques have since been developed. The mixture to be analyzed can be in the gas or liquid phase, and the adsorbing material can be a solid or a liquid supported on an inert solid such as paper. The technique called thin-layer chromatography, which uses silica gel on plastic strips, is illustrated in Figure 8-6. Chromatography has also

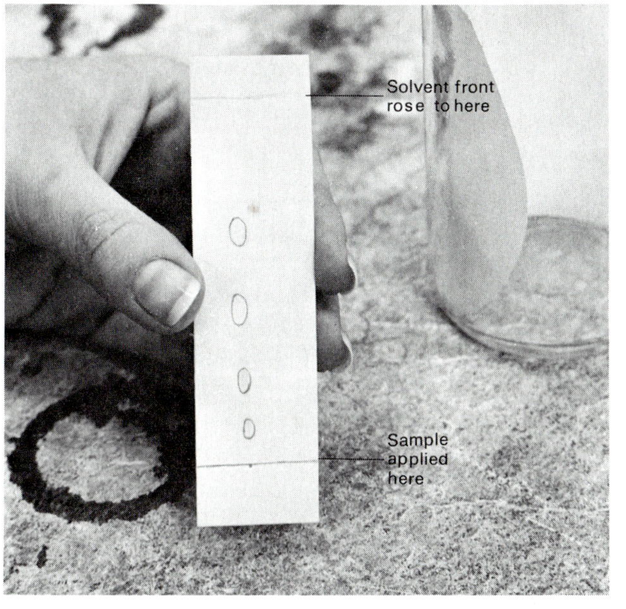

FIGURE 8-6

Thin-layer chromatography. The plate consists of a thin layer of SiO_2 on plastic. The bottle contains a shallow layer of the solvent (ethyl alcohol). A microdrop of an extract of spinach was applied at the center of the line near the bottom of the chromatography plate. As the solvent front moved up the strip, the mixture separated into four components. Because the spots fade rapidly, they have been outlined in pencil for a permanent record. [Photograph by Richard Byrnes.]

been extended to separate *colorless* components of a mixture. Various methods are used to locate the colorless compounds of the mixture. In thin-layer chromatography, these include converting the colorless substance to a colored one—say, by reaction with iodine, which forms a brown complex with many organic molecules. Fluorescent components of a mixture can be seen under ultraviolet light. Alternatively, if the components of the mixture are *not* fluorescent, a fluorescent substance can be added to the adsorbing material. Then the components show up as dark spots under ultraviolet light.

Other methods of detection, used with column chromatography where the components are collected separately in test tubes, include incorporating a radioactive isotope in the compound of interest—say, carbon-14, for example, in an organic molecule. Then that compound can be detected with a Geiger counter. Or, if the substance desired is an enzyme, it can be detected by testing for reaction with its substrate (the molecule on which the enzyme acts). Chromatography plays an extremely important role in the purification of enzymes and in the determination of protein structure, as we shall see in Chapter 21.

8-2 DETERMINATION OF MOLECULAR FORMULA AND STRUCTURAL FORMULA

Suppose that we have obtained a pure sample of the compound ethyl alcohol (ethanol), the common alcohol produced by fermentation. How can we determine what elements ethanol is composed of, and in what amounts? One method, **combustion analysis,** uses the apparatus diagrammed in Figure 8-7. When ethanol is burned in oxygen in the combustion apparatus, the *only* products of the combustion are carbon dioxide (CO_2) and water (H_2O). From this we can conclude that carbon and hydrogen are definitely present in ethanol, and that oxygen may be present. (We have used oxygen from the air in the combustion process,

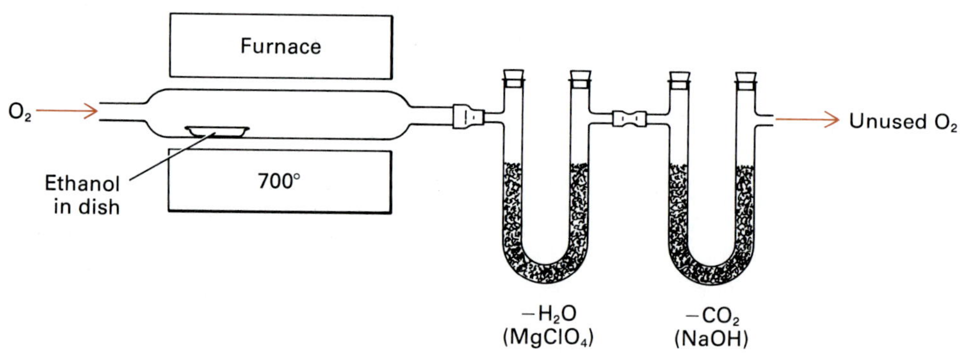

FIGURE 8-7

Apparatus for combustion analysis. Magnesium perchlorate absorbs the water, and sodium hydroxide absorbs the carbon dioxide.

but some oxygen may also have been present in the ethanol to begin with.)

The mass of CO_2 and H_2O produced in the combustion of a known mass of ethanol can be determined by collecting the products (as shown in Figure 8-7) and weighing them. It is then possible to calculate how much C, H, and O were present in the sample and to find the simplest formula, using methods described in Chapter 2. In Example 8-1, we calculate the simplest formula of ethanol.

EXAMPLE 8-1

Suppose 92 g of ethanol are burned in a combustion apparatus, producing 176 g of CO_2 and 108 g of H_2O. What is the simplest formula of ethanol? (Compare with Example 3-11.)

1. First calculate the mass of carbon in the CO_2 and the mass of hydrogen in the H_2O:

$$\text{Mol. wt. of } CO_2 = 12 + 2 \times 16$$
$$= 44 \text{ g/mole} \quad \left(\frac{12 \text{ g}}{44 \text{ g}} \text{ of } CO_2 \text{ is C}\right)$$

$$\text{Mol. wt. of } H_2O = 2 \times 1.0 + 16$$
$$= 18 \text{ g/mole} \quad \left(\frac{2.0 \text{ g}}{18.0 \text{ g}} \text{ of } H_2O \text{ is H}\right)$$

$$\text{Mass of C in 176 g of } CO_2 = \frac{12}{44} \times 176 \text{ g}$$
$$= 48 \text{ g C}$$

$$\text{Mass of H in 108 g of } H_2O = \frac{2.0}{18} \times 108 \text{ g}$$
$$= 12 \text{ g C}$$

2. All the carbon in the CO_2 came from the ethanol. All the hydrogen in the H_2O came from the ethanol. Thus, in 92 g of ethanol, 48 g are carbon and 12 g are hydrogen. The remainder of the 92 g is oxygen.

Mass of oxygen in ethanol

$$= \text{Mass of ethanol} - (\text{Mass of C} + \text{Mass of H})$$

Mass of oxygen in 92 g of ethanol

$$= 92 - (48 + 12) = 32 \text{ g O}$$

3. To find the simplest formula, calculate the number of moles of C, H, and O in the sample:

$$\text{Moles C in sample} = \frac{48 \text{ g}}{12 \text{ g/mole}} = 4.0 \text{ moles C}$$

$$\text{Moles H in sample} = \frac{12 \text{ g}}{1.0 \text{ g/mole}} = 12 \text{ moles H}$$

$$\text{Moles O in sample} = \frac{32 \text{ g}}{16 \text{ g/mole}} = 2.0 \text{ moles O}$$

The ratio of C:H:O atoms in ethanol is 4:12:2, or 2:6:1. The simplest formula is C_2H_6O.

The molecular formula of ethanol might be the same as the simplest formula (C_2H_6O) or it might be some multiple (such as $C_4H_{12}O_2$ or $C_6H_{18}O_3$). In Chapter 3, we saw that gas density can be used to determine molecular weight and thus establish the molecular formula. In Example 8-2, the density of ethanol vapor is used to establish its molecular formula.

EXAMPLE 8-2

The density of ethanol vapor is 1.15 g/ℓ at a pressure of 0.82 atm and a temperature of 400 K. What is the molecular formula of ethanol if its simplest formula is C_2H_6O?

We know

$$T = 400 \text{ K}$$

$$P = 0.82 \text{ atm}$$

We do not know n or V, but we can calculate n/V, the number of moles in one liter, by rearranging the ideal gas equation $PV = nRT$:

$$\frac{n}{V} = \frac{P}{RT} = \frac{0.82 \text{ atm}}{0.082 \ \ell \text{ atm/mole K} \times 400 \text{ K}}$$
$$= 0.025 \text{ mole}/\ell$$

But, we also know from the gas density, that 1 ℓ weighs 1.15 g at this temperature and pressure. Therefore, 0.025 mole = 1.15 g, and

$$\text{Mol. wt.} = \frac{1.15 \text{ g}}{0.025 \text{ mole}} = 46 \text{ g/mole}$$

The molar mass of the simplest formula C_2H_6O

$$= 2 \times 12 + 6 \times 1.0 + 16$$
$$= 46 \text{ g/mole}$$

Therefore, the molecular formula of ethanol is the same as the simplest formula, C_2H_6O. (Now that we know the formula, we can find the molecular weight of ethanol even more precisely: mol. wt. $C_2H_5OH = 2 \times 12.0000 + 6 \times 1.0080 + 15.9994 = 46.0474$.)

EXERCISE 8-1 When 30 g of acetic acid are burned in a combustion apparatus, the products are 44 g of CO_2 and 18 g of H_2O. What is the simplest formula of acetic acid?

EXERCISE 8-2 The density of acetic acid vapor is 1.00 g/ℓ at a pressure of 0.41 atm and a temperature of 300 K. What is the molecular formula of acetic acid? Its simplest formula is CH_2O.

We now know the formula of the ethanol molecule, but how are the atoms connected in the molecule? Remembering that typically each C atom forms four covalent bonds, each O atom two, and each H atom one, we can still write two different possible structures for ethanol:

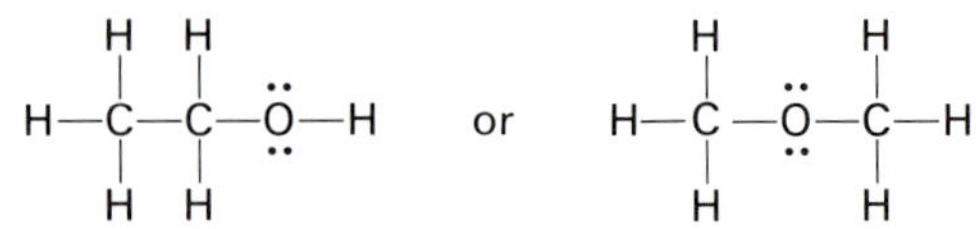

CH_3CH_2OH structure A or **CH_3OCH_3 structure B**

Both structures satisfy the octet rule, but which one is ethanol? Early chemists used a detailed knowledge of chemical behavior to differentiate between structural formulas such as these, but now a large variety of chemical and physical methods are available. We shall first consider some examples of chemical behavior that can be used to establish whether ethanol is structure A or structure B.

Sodium metal reacts with ethanol to liberate hydrogen gas and yield a basic solution. We have already seen (Chapter 6) that sodium reacts with water, breaking an O—H bond, to produce H_2 gas and a basic solution:

$$\text{Na(s)} + \text{H—}\ddot{\underset{..}{\text{O}}}\text{—H(l)} \longrightarrow \text{Na}^+ + \ddot{\underset{..}{\text{O}}}\text{H}^- + \tfrac{1}{2}\text{H}_2 \text{ (gas)}$$

Formula A contains an O—H bond that could be attacked by sodium to liberate H_2 gas:

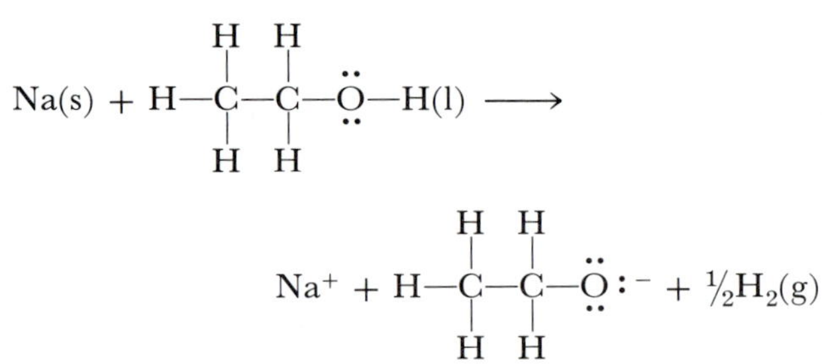

Formula B, however, has all the H atoms bonded to C,

and it is known that sodium does not attack C—H bonds (in fact, sodium is often stored in hydrocarbon liquids, which contain only C—C and C—H bonds). This experiment, then, indicates that ethanol has structure A.

In another experiment, ethanol reacts with aluminum oxide (Al_2O_3) a known **dehydrating agent,** to produce ethylene (C_2H_4) and water. A dehydrating agent removes water (H_2O) from another substance. The Al_2O_3 removes H_2O from ethanol to form hydrated aluminum oxides, $Al_2O_3 \cdot H_2O$ and $Al_2O_3 \cdot 2H_2O$ and $Al_2O_3 \cdot 3H_2O$. Ethylene has the structure

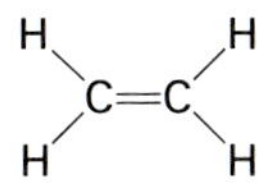

with a double bond between the two C atoms. It is much easier to see how structure A, which already has a single C—C bond, could be converted to ethylene and water (by removing H from one C and OH from the other), than to see how structure B, which has no C—C bond and no OH group, could be so converted. The reaction between structure A and Al_2O_3 involves breaking one C—H and one C—O bond:

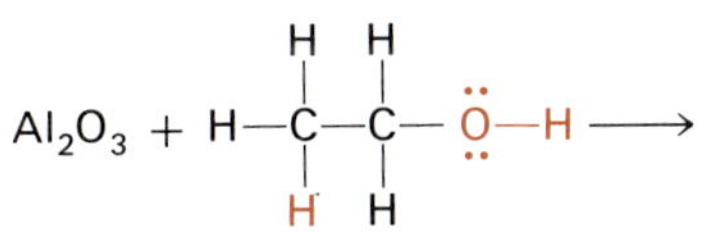

structure A

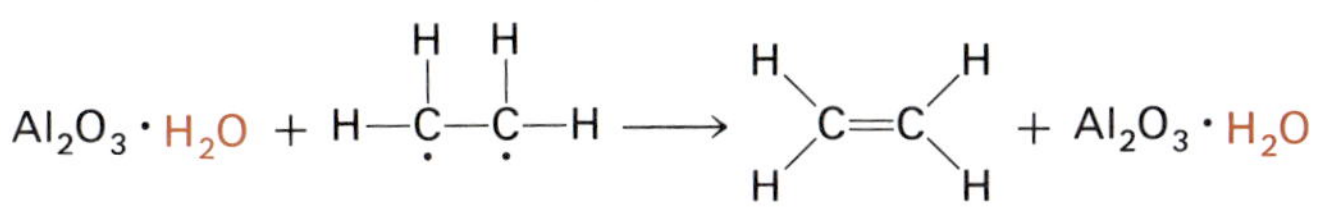

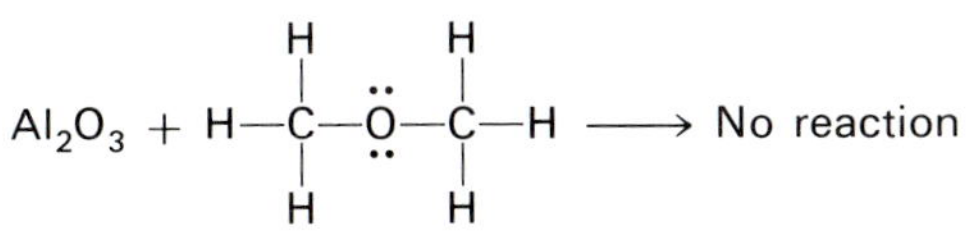

structure B

To remove H_2O from structure B would require breaking two C—O bonds and two C—H bonds, a much more difficult reaction than that shown for structure A. This experiment gives additional confirmation that ethanol has structure A.

A final confirmation that structure A is ethanol results from a **synthesis** of structure A, and a comparison of the properties of this substance with those of ethanol. A known reaction of organic chemistry (that we shall see again in Chapter 15) is that HB (where B is an atom or group of atoms other than H) adds to ethylene (CH_2CH_2) to yield CH_3CH_2B in the presence of sulfuric acid.

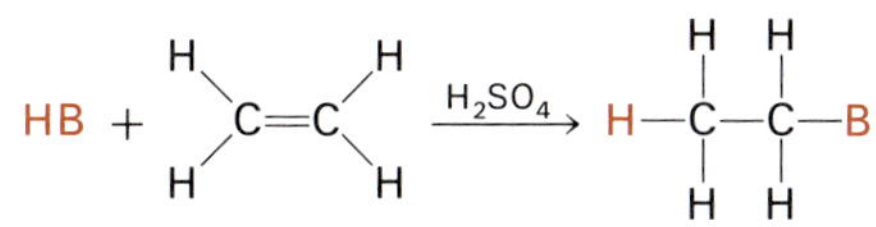

(The H_2SO_4 is written over the arrow because it must be present for the reaction to occur, but we are not now interested in it.) Structure A can therefore be prepared by combining ethylene and water in the presence of sulfuric acid:

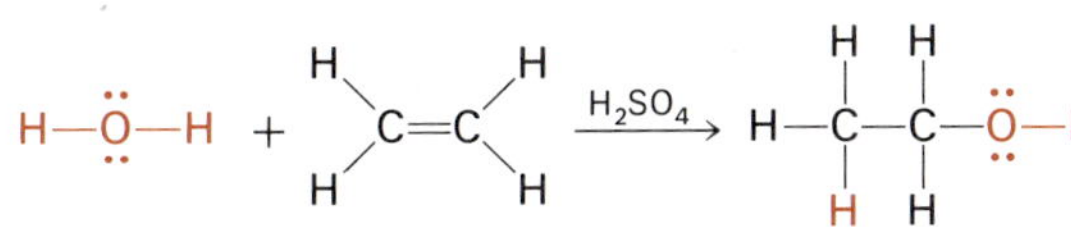

water **ethylene** **structure A**

If this product and ethanol have the same physical and chemical properties, we know that ethanol is *definitely* structure A.

A comparable set of chemical reactions can be used to show that another substance, dimethyl ether, which is a gas at room temperature, has structure B. Two such compounds, that have the same molecular formula, but different molecular structures and different chemical and physical properties, are called **isomers.** Isomers, which are important in organic chemistry, are discussed in detail in Chapter 13.

8-3

PHYSICAL METHODS OF STRUCTURE DETERMINATION: SPECTROSCOPY

Several methods of determining molecular structure are based on the interaction of matter with electromagnetic energy. Light (radiation) can be *absorbed*, *transmitted* undisturbed, or *scattered* by a sample of matter. We first consider **spectroscopy,** which depends on the absorption or transmission of energy.

A variety of spectroscopic techniques—differing not in principle, but only in the wavelength of the radiation used—are used to study the structures of molecules. In Chapter 6 we saw that *atoms* have characteristic spectra—each atom absorbs and may later radiate energy only in definite "packets," or quanta. All other wavelengths are transmitted unchanged. The spectra we saw in Chapter 6 were all **emission spectra,** in which each line corresponds to a wavelength of light *radiated* (or emitted) in the transition of electrons from higher to lower energy levels. **Absorption spectra,** in which each line corresponds to a wavelength of light *absorbed* in the transition of electrons from lower to higher energy levels, can also be obtained.

Many *molecules* absorb radiation in the infrared region of the spectrum, light of wavelengths from 10^3 to 10^5 nm. Both dimethyl ether and ethanol are transparent to visible light but absorb infrared light. The absorption of infrared light is associated with vibrations of atoms in the molecule about their equilibrium positions (Figure 8-8). The energies of these vibrations, like electronic energies, are quantized. Thus substances only absorb infrared light of definite wavelengths. All other wavelengths are transmitted. As a result the absorption spectrum is characteristic of the kinds of bonds in the molecules. For example, when an OH group vibrates by alternate expansion and contraction of the O—H bond, the wavelength of infrared light absorbed is almost the same no matter what molecule the OH group is in. From measurements made on simple molecules of known structure, it is known that a molecule that has an OH group absorbs light of wavelength 3.2×10^3 nm. A CH group absorbs light of 3.4×10^3 nm. Infrared spectra of ethyl alcohol and dimethyl ether are shown in Figure 8-9, along with a schematic diagram of a spectrophotometer. Light of wavelength 3.4×10^3 nm is strongly absorbed by *both* ethyl alcohol and ether—both molecules have C—H bonds. There is said to be an **absorption peak** at this wavelength. The characteristic absorption due to O—H vibration is seen in the strong absorption peak at 3.2×10^3 nm in the spectrum of ethyl alcohol. Dimethyl ether, on the contrary, which has no O—H group, does not absorb strongly at this wavelength. Because infrared spectroscopy is quickly and easily done (and because it can be used on gases, liquids, or solids), it is routinely used to help identify substances and to check on purity.

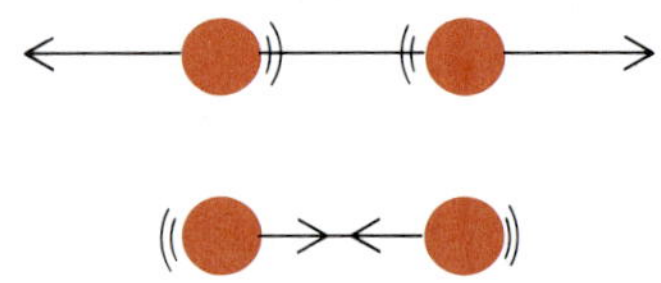

FIGURE 8-8

Atoms in a molecule vibrate about their equilibrium positions.

EXERCISE 8-3 A compound of molecular formula $C_5H_{12}O$, known to be either an alcohol or an ether, absorbs light of frequency 9.3×10^{13} Hz. Using this information, can you tell whether the compound is an alcohol or an ether? (Refer to Chapter 6, if necessary, for the relationship between frequency and wavelength.)

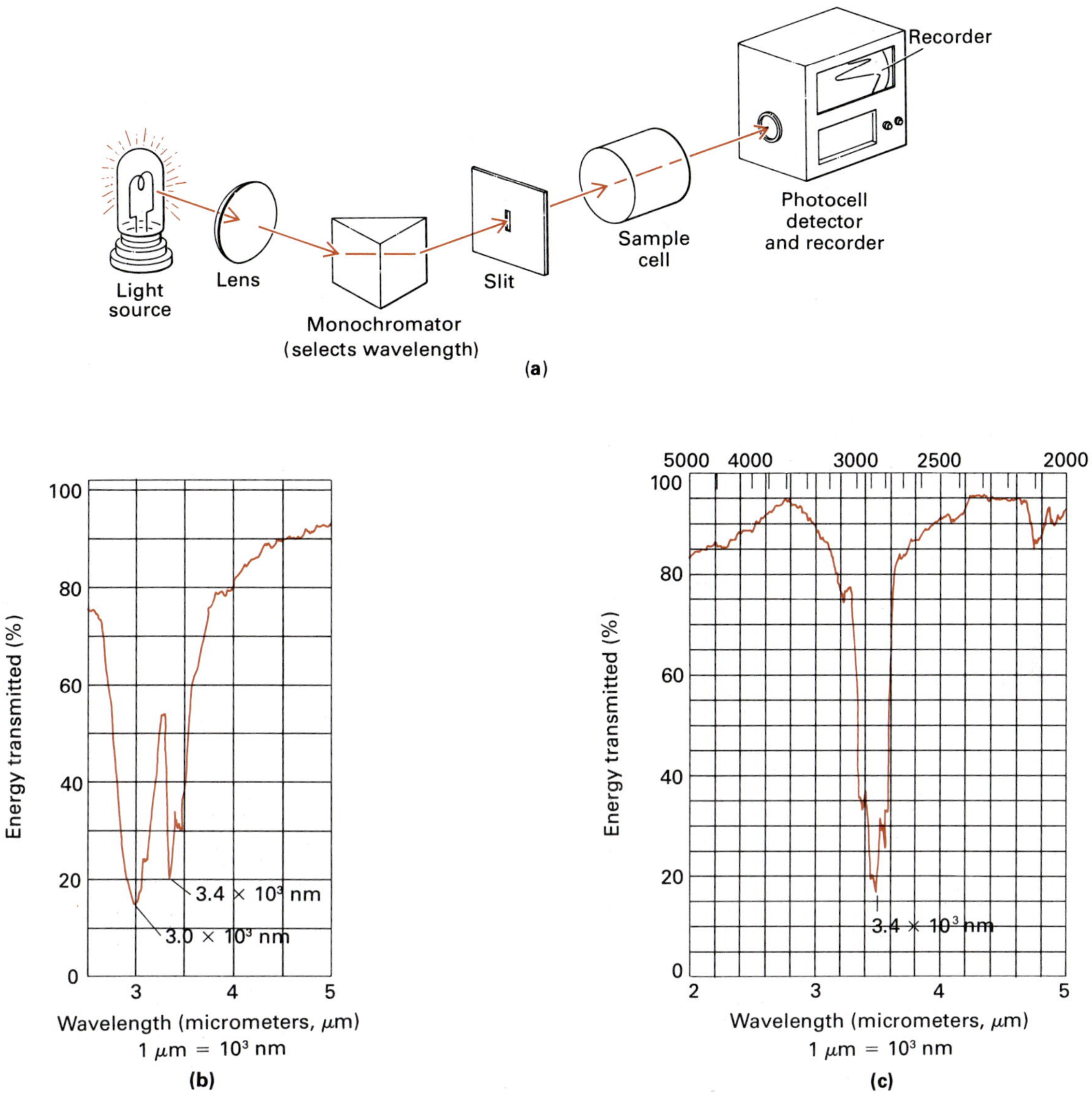

FIGURE 8-9

Spectrophotometry. **(a)** Schematic diagram of the essential parts of a spectrophotometer. The light source produces a definite number of photons of known wavelengths. The monochromator can vary the wavelength of light that passes through the sample. The photocell detector measures the percentage of light of each wavelength that passes through the sample (the transmittance). **(b)** Part of the infrared spectrum of ethyl alcohol. **(c)** Part of the infrared spectrum of dimethyl ether.

8-4
NUCLEAR MAGNETIC RESONANCE

Another spectroscopic method of determining molecular structure, which also depends on the absorption of electromagnetic radiation—this time in the form of radio waves (radiation of very low energy)—is **nuclear magnetic resonance** (abbreviated **NMR**). The nuclei of some atoms (for example, hydrogen, ${}^{1}_{1}H$) have "spin" as electrons do. They act like tiny magnets with north and south poles. In the absence of a magnetic field, these magnets are oriented at random, in all directions. In a strong magnetic field, each proton lines up with the field.

If radio waves are passed through a sample that contains protons, most frequencies are transmitted. However, certain definite frequencies are absorbed, causing some protons to turn in the magnetic field. The frequency of the radiation absorbed depends on the strength of the magnetic field. In a molecule, the strength of the magnetic field "felt" by each proton depends on the electronic charge distribution near that proton. The strength of the magnetic field at a proton in an OH group is different, therefore, from that felt by a proton of a CH_3 group. As a result, different frequencies of radiation are absorbed by the protons of different groups. Figure 8-10 shows the NMR spectra of ethanol and dimethyl ether. The energy absorbed is plotted against the frequency. In ethanol there are three absorption peaks, corresponding to the three different kinds of protons. The areas under the peaks are in the proportion 1:2:3, corresponding to the numbers of protons in each group (one OH proton, two CH_2 protons, and three CH_3 protons). In dimethyl ether there is only one kind of proton (that of a CH_3 group), and there is only one absorption peak in the NMR spectrum. This method of determining molecular structure is applied primarily to substances containing hydrogen atoms, but it can also be used in molecules containing ^{19}F, ^{13}C, ^{15}N, ^{17}O, and ^{31}P.

EXERCISE 8-4 The isomers ethyl mercaptan, CH_3CH_2SH, and dimethyl sulfide, $(CH_3)_2S$, can be distinguished by their NMR spectra. How many peaks would you expect to find in the NMR spectrum of each? If there is more than one peak in a spectrum, give the relative areas under the peaks.

(The nose is actually the most sensitive instrument for distinguishing between these two compounds. Both have disagreeable odors, but that of ethyl mercaptan is so intense that humans can detect 1 part ethyl mercaptan in 50 billion parts of air. For this reason, very small amounts of ethyl mercaptan are added to natural gas, which is almost odorless, to warn of any leaks.)

FIGURE 8-10

NMR spectra. **(a)** Ethyl alcohol. **(b)** Dimethyl ether.

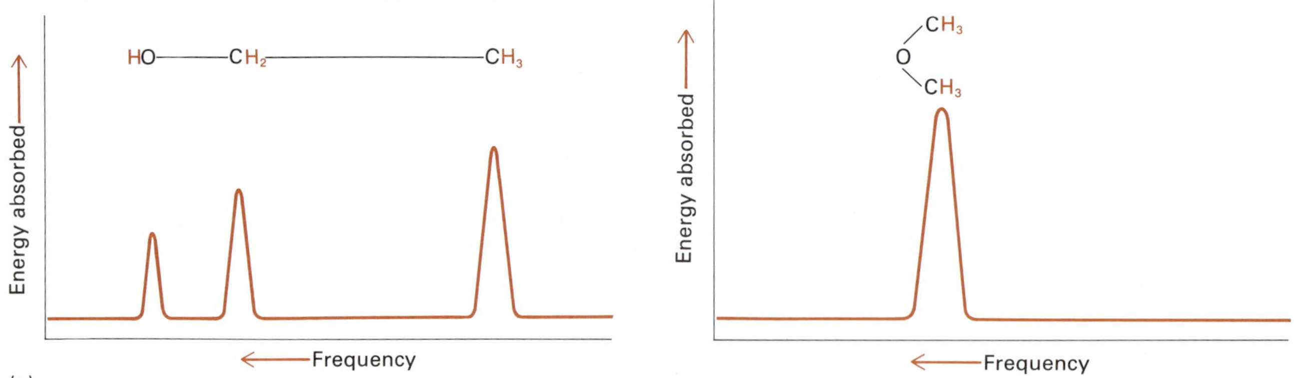

(a) (b)

8-5 DIFFRACTION

Thus far we have considered methods of determining structure that depend on the *absorption* of electromagnetic radiation. Other information can be obtained through methods that depend on **diffraction,** or ordered **scattering,** of radiation. The diffraction of X rays is used to study crystal structures, diffraction of electrons to determine the structures of molecules in gases, and diffraction of neutrons to study the structures of liquids and solids. (Neutron diffraction can readily locate hydrogen atoms, which X-ray diffraction cannot do.) The principles involved in diffraction are illustrated in the following discussion of **X-ray diffraction.**

The wavelengths of visible light (about 400 to 700 nm) are much too long to study the submicroscopic structure of a crystal made up of atoms approximately 0.1 nm in diameter. Instead, X rays (with wavelengths about 0.1 nm) are used. When X rays strike a crystal, some of the radiation is scattered by atoms (or molecules, or ions) in the crystal, and the pattern of diffraction can be recorded on a photographic plate (Figure 8-11a). The spots of high radiation intensity result from the interactions of the waves that bounce off atoms in the crystal lattice. The angles at which these rays emerge, as well as their intensities (and hence the pattern of the diffracted radiation that is recorded) are characteristic of the crystalline substance, and depend on the spacing of the atoms in the crystal. X-ray diffraction patterns showing a regular arrangement of spots, which reflect the regular spacing of atoms in a crystal lattice, are shown in Figure 8-11b for powdered copper and tungsten. Each substance has a unique diffraction pattern. Working backward from the diffraction pattern, the submicroscopic, three-dimensional structure of the crystal can be reconstructed. The arrangement of the atoms and the distances between them can be calculated. Figure 8-12 shows the crystalline structure of vitamin B_{12}, first described in 1959 by Dorothy Hodgkin and a team of colleagues, who had invested eight years in the elucidation of this structure by X-ray analysis. Although use of modern, high-speed digital computers has now greatly shortened the time required for structure determination by X-ray diffraction, it is still a long and complicated process.

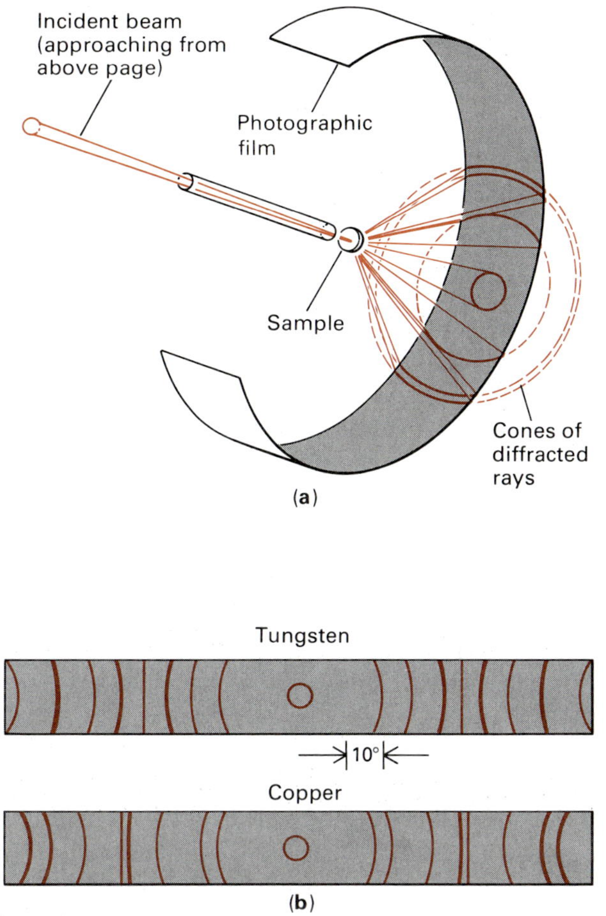

FIGURE 8-11

X-ray diffraction. **(a)** An X-ray diffraction experiment. **(b)** X-ray diffraction patterns of powdered crystals of copper and tungsten.

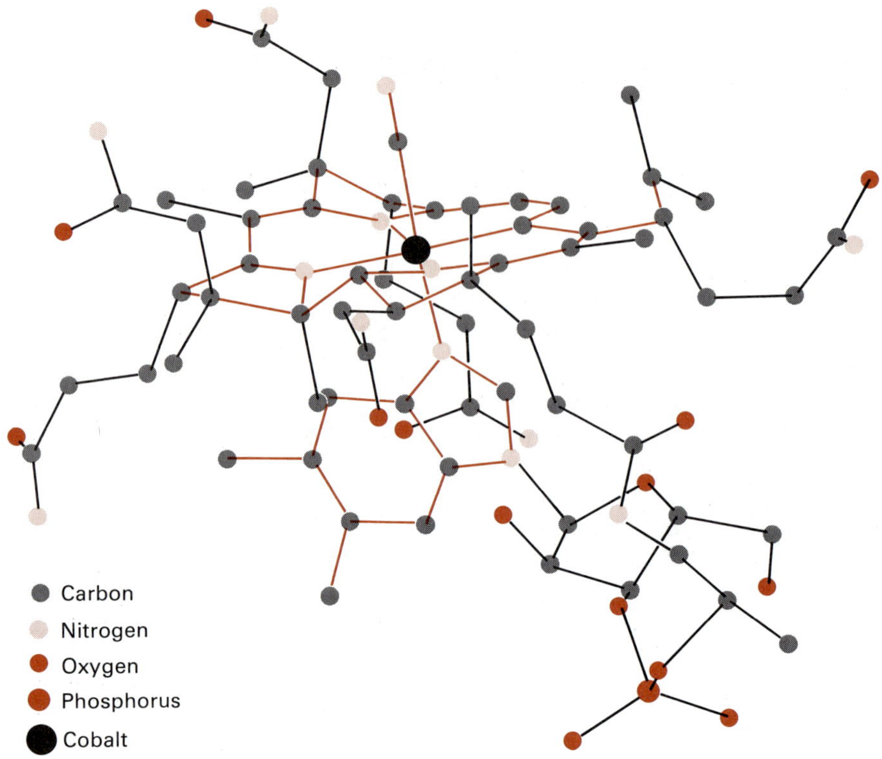

(a)

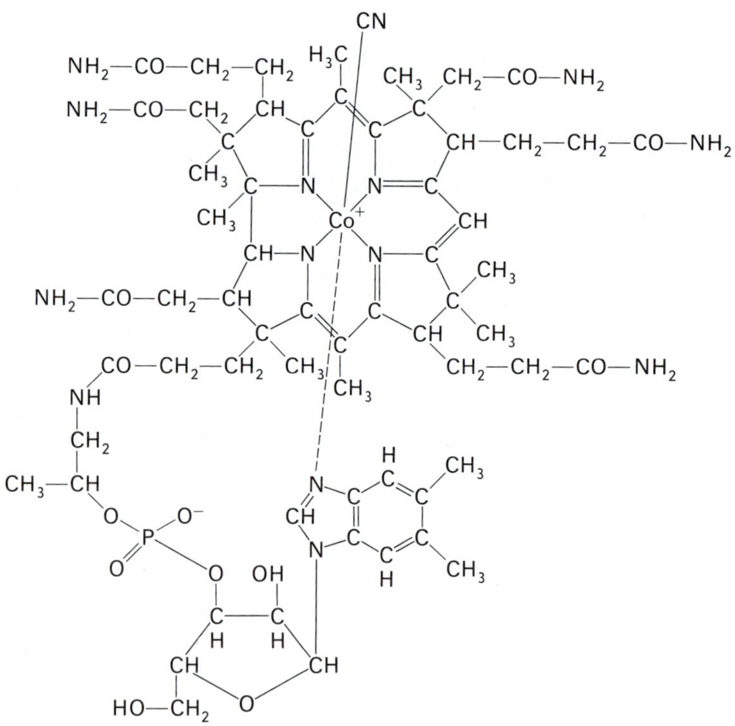

(b)

FIGURE 8-12

(a) The crystalline structure of vitamin B_{12} ($C_{63}H_{84}N_{14}O_{14}PCo$). Three-dimensional structure, with H atoms omitted. **(b)** Vitamin B_{12} shown flattened, in two dimensions, with the H atoms included. [(a) from Sir Lawrence Bragg, "X-ray Crystallography." Copyright © 1968 by Scientific American, Inc. All rights reserved.]

8-6
CRYSTAL STRUCTURE: METALLIC SOLIDS

Long before the days of X-ray diffraction, scientists interested in crystals knew that crystals of a given substance (say, quartz) always have the same characteristic shape—that is, the same angles between corresponding faces. This can be seen in the crystals shown in Figure 8-13. (Some substances, when crystallized under different conditions, have more than one crystalline shape, but each shape is recognizable and reproducible.) In 1781 Abbé Hauy, professor of humanities at the University of Paris, dropped a calcite crystal and noticed that it broke along planes in such a way that the pieces maintained the characteristic crystalline form. Gem cutters had long made use of this phenomenon when splitting gem crystals along "preferred cleavage planes." Hauy reasoned that the regu-

(a)

(b)

(c)

FIGURE 8-13

Crystals of a given substance have the same characteristic shapes. The shape reflects the microscopic structure of the substance. **(a)** Halite (NaCl). **(b)** Quartz (SiO_2). **(c)** Aragonite ($CaCO_3$). [Photographs by Andrew Leed.]

lar crystalline shape and the existence of cleavage planes reflect a regularity in the stacking of the particles of which the crystal is composed. A model of a crystal structure proposed by Hauy is shown in Figure 8-14, along with an electron micrograph of a crystal of tobacco necrosis virus. It was the application of X-ray diffraction to the study of crystal structure (from 1912), however, that first showed conclusively that *the characteristic shapes of crystals can be attributed to their microscopic structure.* The atoms, ions, or molecules of which the crystal is composed are arranged in a latticelike structure, each particle occupying a fixed position about which it can vibrate only a bit.

In pure metallic crystals, where all the atoms in the crystal lattice are identical, and so, in particular, are all the same size, the ions are usually packed together as closely as possible, like the grapefruit in Figure 8-15. X-ray analysis gives the size and shape of the **unit cell**—the smallest unit, which, repeated over and over in three dimensions, yields the crystal lattice. Figure 8-16 shows packing characteristic of metallic crystals: the hexagonal closest-packed structure, exhibited by magnesium; the cubic closest-packed (or face-centered cubic) structure, exhibited by silver and aluminum; and the body-centered cubic structure (which is slightly less efficiently packed—leaves more empty space—than the other two structures) exhibited by sodium and iron.

Metals have many properties in common; most are lustrous and malleable, and they are good conductors of heat and electricity. These similarities in properties can be explained by similarities in their electronic

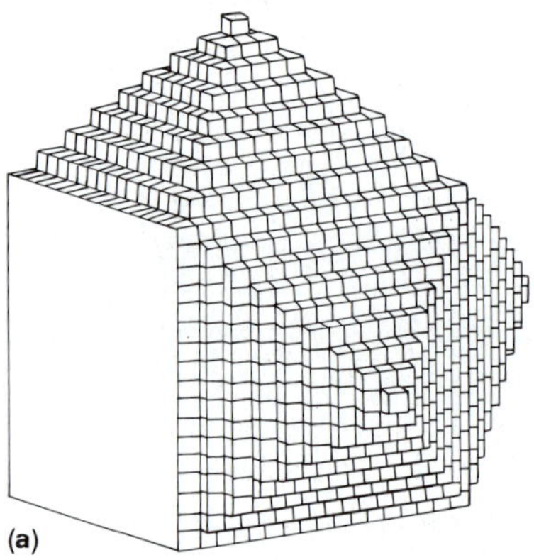
(a)

(b)

FIGURE 8-14

(a) Model of a crystal structure proposed by Hauy (1781). [From *Traite elementaire de Physique,* Vol. 1 (Paris: De L'Imprimerie de Delance et Lesur, 1803).] **(b)** The packing of protein molecules in a crystal of tobacco necrosis virus. [Electron micrograph courtesy of R. W. G. Wyckoff.]

FIGURE 8-15

Closest packing. In a metallic crystal, the metal ions usually are packed as closely as possible, like these grapefruit. [Photograph by Richard Byrnes.]

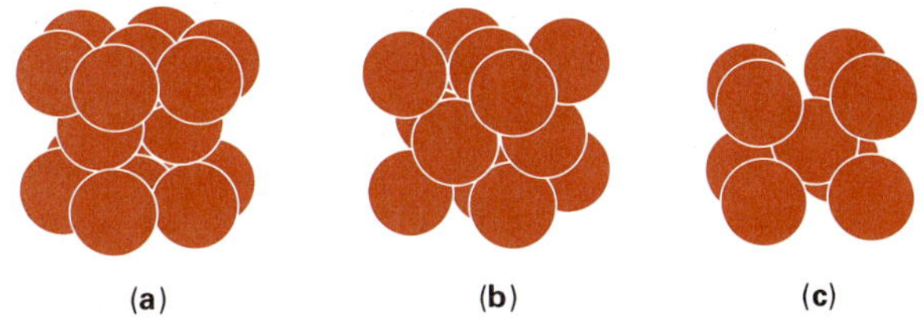

FIGURE 8-16

Unit cell structures. **(a)** Hexagonal closest-packed. **(b)** Face-centered cubic. **(c)** Body-centered cubic.

structures, which result in relatively low ionization potentials. A metal can be pictured as a three-dimensional lattice of positive ions embedded in a "sea" or "cloud" of mobile electrons. In the case of silver, for example, the crystal is composed of Ag^+ ions—the single valence electron of each atom has been contributed to the mobile electron "cloud" that belongs to the entire crystal. Every electron in the cloud can move from one Ag^+ to another throughout the crystal. These electrons are said to be **delocalized** because they are not each confined to a single atom. The crystal is held together by the attractive forces between the positively charged ions and the cloud of negatively charged electrons.

Many metals are soft and malleable because the bonding is nondirectional (the ions are bound together by the electron cloud but are not directly linked to each other), allowing layers of ions to be pushed over each other. (We shall see, shortly, that this is in contrast to ionic and network solids, where the ions or atoms are tightly held in their lattice positions.) In some cases—for example, in the transition metals vanadium, chromium, and tungsten—there is some covalent bonding between the positive ions in addition to the metallic bonding, making the metal hard and brittle.

Metals are good conductors of electricity because of the electron cloud which is free to travel throughout the crystal. Irregularities in the crystal structure, caused by impurities, interfere with the mobility of the electrons and cause a decrease in conductivity. The conductivity of pure copper, for example, is reduced by a factor of 10 in an alloy with about 4% manganese. This accounts for the high degree of purity (over 99.95%) required for copper used in electrical wires.

Heat is carried through solids by the vibrations of atoms, ions, or molecules about their equilibrium lattice positions. In metals, energy is additionally transported by the mobile, delocalized electrons; so metals are especially good conductors of heat.

8-7
CRYSTAL STRUCTURE: IONIC SOLIDS AND MOLECULAR SOLIDS

In crystals of **ionic solids** such as NaCl, the ions pack together closely, with positive ions surrounded by negative ions, and negative ions by positive ones. In NaCl, each Na^+ ion is surrounded by six Cl^- ions, and each Cl^- ion by six Na^+ ions (Figure 8-17). In the crystal, both the Na^+ ions separately, and the Cl^- ions separately, form face-centered cubic patterns. Most ionic solids have high melting and boiling temperatures because of the strong electrostatic forces holding the ions together. NaCl, for example, melts at 800°C and boils at 1400°C. The electrons are localized around individual ions and, unlike in metals, are not free to move through the crystal. Thus, ionic solids are poor conductors of heat and electricity. When melted, however, the ions themselves can conduct an electric current, though much less efficiently than metals do.

Molecular solids are those in which the lattice points are occupied by molecules. Within each molecule the bonding is covalent, but the solid is held together primarily by relatively weak dipole–dipole forces, van der Waals forces, or hydrogen bonds. Because the forces between the molecules are weak, most molecular substances are soft (the molecules can be pushed over one another) and have low melting and boiling temperatures. Because the valence electrons are associated with specific covalent bonds, rather than being delocalized as those in metals are, most molecular solids are poor conductors of heat and electricity.

Examples of substances that form molecular solids are N_2, O_2, CO_2, and I_2. All of these substances are held together weakly in the solid or liquid, and go to the gaseous state at relatively low temperatures. N_2, O_2, and CO_2 are gases at room temperature, and I_2 sublimes when warmed. The boiling point of I_2 is only 180°C, much lower than that of NaCl (1400°C).

Sulfur and phosphorus are also molecular solids: phosphorus (as we saw in Example 3-11 and Section 7-7) is composed of P_4 molecules, with the P atoms at the corners of a tetrahedron; and sulfur is composed of S_8 molecules (Figure 8-18). In Part III of this book, we shall see that most solid organic substances are molecular solids—aspirin is an example.

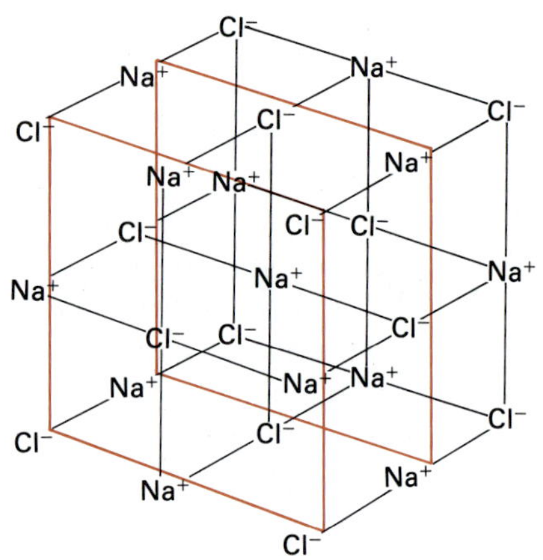

FIGURE 8-17

Crystal structure of NaCl.

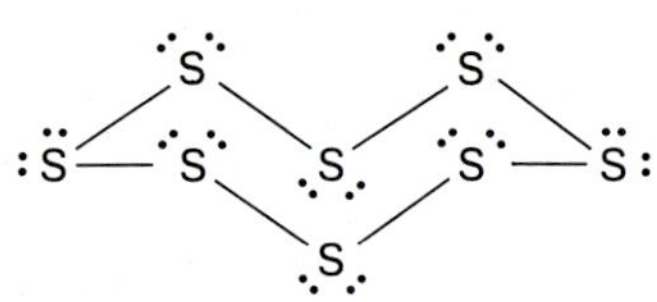

FIGURE 8-18

The sulfur molecule (S_8). In solid sulfur, S_8 molecules occupy the crystal lattice points.

8-8
CRYSTAL STRUCTURE: COVALENT OR NETWORK SOLIDS

In a **covalent** or **network solid** the lattice points are occupied by atoms that are covalently bonded to the other atoms of the crystal in a three-dimensional network. Each such crystal may be considered to be a single giant molecule. Silicon and carbon are examples of substances that form network crystals. In diamond, each C atom is bonded tetrahedrally (using sp^3 hybrid orbitals) to four other C atoms in a structure extended indefinitely in three dimensions (Figure 8-19). The many strong covalent bonds account for the extreme hardness of diamond, and for its high melting point (4800°C). Many covalent bonds must be broken to break down the crystal lattice and thus melt the solid. Because all the valence electrons are localized in the covalent bonds, diamond is a poor conductor of heat or electricity. Silicon, which is directly below carbon in the periodic table, has the same structure as diamond. In graphite (Figure 8-20), there are strong covalent bonds between C atoms in a sheet, but only weak van der Waals forces hold the sheets together. (Each C atom is using sp^2 hybrid orbitals, giving the planar hexagonal arrangement shown in the figure.) The layers can be easily separated, and they readily slide over each other if entrapped gas molecules are present to act like ball bearings—a property that accounts for the use of graphite as a lubricant.

Other network compounds contain silicon and oxygen bonded together in long chains, extended planes, or three-dimensional networks. Sand and quartz are **silica** (SiO_2), whose structure is similar to that of diamond and pure silicon. Each silicon atom is at the center of a tetrahedron, with oxygen atoms at the four corners, and each oxygen bridges two silicon atoms (Figure 8-21). In this structure, both Si and O have stable octets of electrons. (Convince yourself that this is so.) The giant network of strong covalent bonds ac-

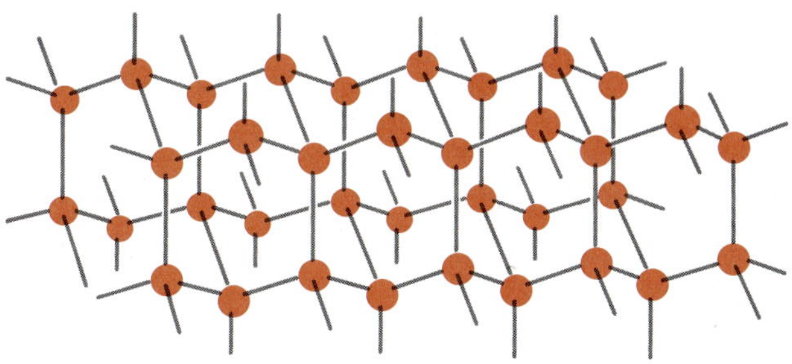

FIGURE 8-19

Structure of diamond.

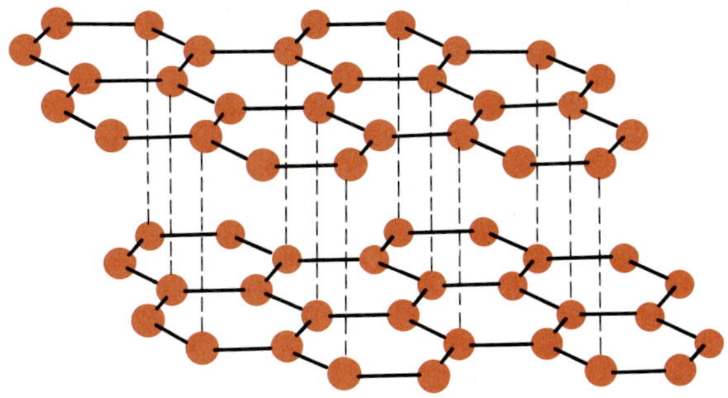

FIGURE 8-20

Structure of graphite.

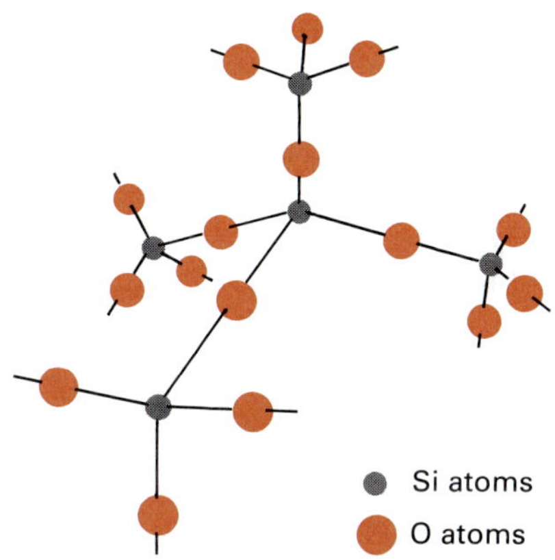

FIGURE 8-21

Structure of silica (SiO_2). In this three-dimensional network, each Si is bonded to four O atoms, and each O is bonded to two Si atoms.

counts for the hardness and high melting point (about 1700°C) of silica.

When silica melts, some of the Si–O bonds break, and the giant crystalline structure partially breaks down. As the liquid cools, the large chains cannot reorganize themselves into the original network; they become entangled, and a **glass** results. A glass is a noncrystalline solid—a solid with the random structure of a liquid. The irregular structure of glass is reflected in the irregular patterns in which it shatters—there are no preferred cleavage planes as there are with crystalline solids. Glass is an example of an **amorphous** solid—one with the rigidity of a solid, but the random structure of a liquid. Glass flows very slowly, and so is also considered a supercooled liquid (in the liquid state at a temperature at which it should be solid).

Other compounds of silicon and oxygen are the **silicates.** Silicates contain the same tetrahedra as silica, but with negative charges on one or more oxygen atom of each tetrahedron. In micas and clays, three of the four corners of each tetrahedron are "bridge" oxygens like those in silica, but the fourth oxygen is *not* bonded to a second silicon (Figure 8-22). This oxygen has acquired its eighth electron from a metal atom (aluminum, magnesium, or calcium, for example), and so has a negative charge. As shown in the figure, these silicates consist of extended planes of tetrahedra.

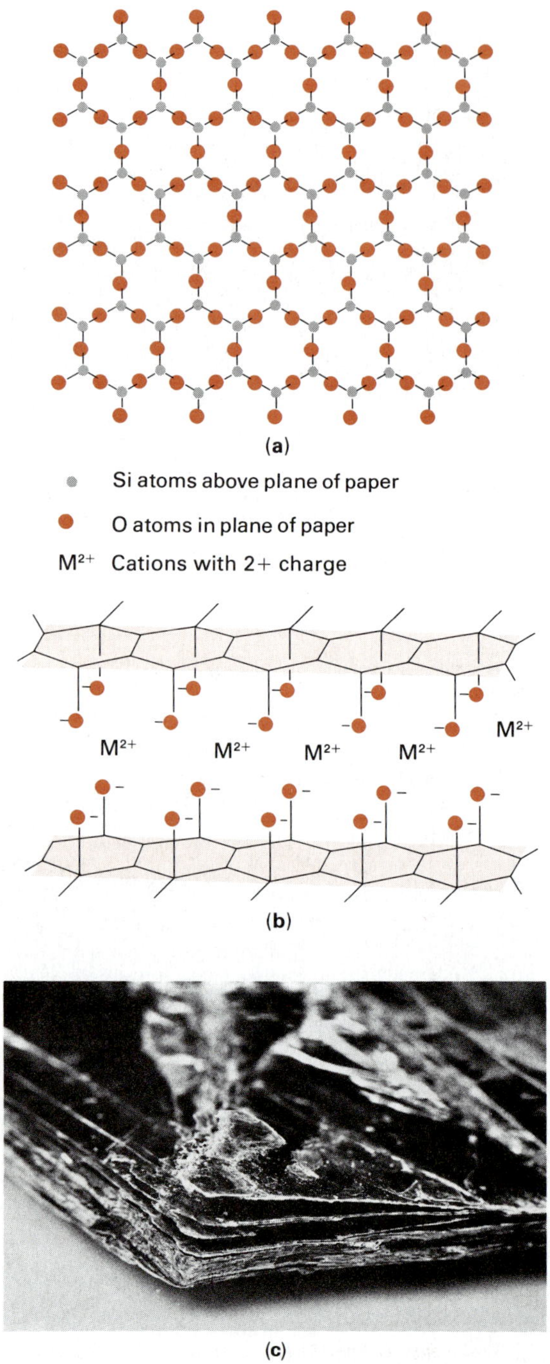

FIGURE 8-22

Structure of micas. Micas are composed of layers of silicate sheets, with positively charged ions sandwiched between layers. **(a)** View of a sheet in the plane of the paper. The red circles represent O atoms in the plane of the paper; the gray circles represent Si atoms slightly above the plane of the paper. Each Si atom is bonded to a fourth O atom (not shown) projecting up from the paper; these O atoms are negatively charged. **(b)** Some of the negatively charged oxygens are shown. Cations are attracted to the negative charges, and the cations in turn attract another such sheet turned upside down. **(c)** Photograph of muscovite, a mica mineral. [Photograph by Richard Byrnes.]

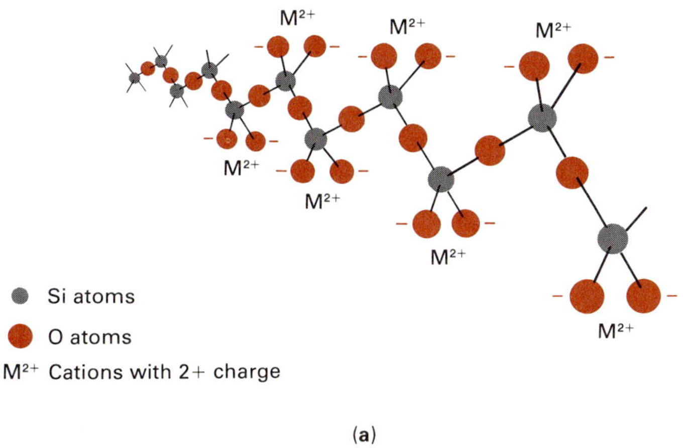

FIGURE 8-23

Structure of asbestos-like minerals. **(a)** This structure is composed of silicate chains. The O atoms bonded to only one Si atom are negatively charged. Cations are attracted to the negative charge. **(b)** Photograph of amosite, an asbestos-like mineral. [Photograph by Andrew Leed.]

The sheets are arranged in layers with positively charged metal ions sandwiched between them, making the compound electrically neutral. As with graphite, thin sheets of mica can be split off; and layers of clay slide over each other when wet.

If only two of the oxygen atoms of each tetrahedron are "bridge" oxygens, and the other two oxygens have negative charges, the resulting structure is a long chain of tetrahedra (Figure 8-23), with metal ions to make the compound electrically neutral. The asbestos-like minerals have this structure, and threadlike strands (each consisting of many parallel chains) can be peeled off. The known carcinogenicity of asbestos may be due to these extremely fine fibers, which natural defense mechanisms cannot handle.

EXERCISE 8-5 Given the following substances in the *solid* state, classify each as metallic, ionic, molecular, or network, and predict some properties of each.

(a) chlorine (Cl_2) (b) silicon (Si)
(c) zinc (Zn) (d) potassium bromide (KBr)

KEY TERMS

Distillation is a method of purification that separates mixtures on the basis of differences in the boiling points of the substances to be separated. *Extraction* is a method of purification that is based on differences in solubility of a substance in two different solvents that are insoluble in each other. *Fractional crystallization* is a method of purification based on differences in solubilities of the desired substance and the impurities.

Isomers are compounds that have the same molecular formula, but that have different structures and different chemical and physical properties.

Chromatography is the separation of mixtures into their constituents by differences in their adsorption by a solid and their solubility in a solvent.

A *dehydrating agent* is a substance that removes H_2O from another substance.

In an *emission spectrum,* each line corresponds to a wavelength of light *radiated* in the transition of electrons from higher to lower energy levels. In an *absorption spectrum,* each line corresponds to a wavelength of light *absorbed* in the transition of electrons from lower to higher energy levels.

Transmitted light is light that passes through a substance unchanged. *Diffraction of light* is the bending or breaking up of light waves by obstacles (atoms, for example) in their paths.

Spectroscopy is the science that deals with the use of instruments that measure absorbance and transmittance of electromagnetic radiation, and with the interpretation of spectra. *Nuclear magnetic resonance* (NMR) is a spectroscopic method of determining molecular structure that depends on the absorbance of radio waves.

Delocalized electrons are not confined to a single bond, but are shared among three or more atoms.

A *metallic solid* can be pictured as a three-dimensional lattice of positive ions in a "sea" of mobile electrons. An *ionic solid* consists of positive and negative ions arranged in a crystal lattice. A *molecular solid* consists of molecules arranged in a crystal lattice. A *network solid* consists of covalently bonded atoms arranged in a crystal lattice—the entire crystal is one molecule.

CHAPTER SUMMARY

The first step in the determination of chemical composition is to obtain the substance in pure form. Methods of purification include mechanical separation, distillation, extraction, fractional crystallization, zone refining, and chromatography.

Once a pure sample has been obtained, the simplest formula and the molecular formula are determined. In the case of molecules that can exist as *isomers*—compounds with the same molecular formula, but different chemical and physical properties—it is necessary to distinguish which, of all possible structures, corresponds to a given substance. Both chemical and physical tests can be used to establish the correct structure. A final confirmation of structure results from a synthesis of the molecule.

The characteristic shapes of *crystals* can be attributed to their microscopic structures. The atoms, ions, or molecules of which the crystal is composed are arranged in a latticelike structure, with each particle occupying a fixed position about which it can vibrate only a bit. A *metallic crystal* can be pictured as a three-dimensional lattice of positive ions embedded in a "cloud" of electrons. This model accounts for many of the properties of metals—for example, their malleability and the fact that they are good conductors of heat and electricity.

In *ionic crystals,* each ion is surrounded by ions of the opposite charge. Because the electrons are localized around individual ions, ionic solids are poor conductors. Because of the strong electrostatic forces between ions of opposite charge, ionic solids have high melting and boiling points.

Molecular solids are those in which the lattice points are occupied by molecules. The solid is held together only by weak van der Waals forces. Thus, most molecular solids are soft and have low melting and boiling points. They are poor conductors of heat or electricity.

A *covalent or network crystal* may be pictured as a single molecule, with the atoms covalently bonded in a giant, three-dimensional network. These solids are extremely hard, are generally poor conductors, and have high melting and boiling points.

PROBLEMS

8-1. In the disease called phenylketonuria (one of the first genetic defects to be discovered in humans), an excess of the compound phenylpyruvic acid is produced. This causes severe, irreversible mental retardation. Phenylpyruvic acid is composed of carbon, hydrogen, and oxygen only. When 1.64 g of phenylpyruvic acid is burned in a combustion apparatus, the products are 3.96 g CO_2 and 0.72 g H_2O. What is the simplest formula of phenylpyruvic acid?

8-2. Nicotine has the simplest formula C_5H_7N. How many grams of CO_2, and how many grams of H_2O, would be produced in the combustion of 1.62 g of nicotine?

8-3. When 45 g of glyceraldehyde (an intermediate produced in the metabolism of glucose) are burned in a combustion apparatus, the *only* products are 66 g of CO_2 and 27 g of H_2O. (Metabolism is the sum of all chemical reactions within a living unit.)
(a) Find the simplest formula of glyceraldehyde.
(b) The density of glyceraldehyde vapor is 2.0 g/liter at a pressure of 0.82 atm and a temperature of 450 K. What is the molecular formula of glyceraldehyde?

8-4. When 8.1 g of quinine (a substance found in the bark of the cinchona tree and once widely used as an antimalarial drug) are burned in a combustion apparatus, the *only* products are 22.0 g of CO_2, 5.4 g of H_2O, and 2.3 g of NO_2.
(a) Find the simplest formula of quinine.
(b) The molecular weight of quinine is found, by some method, to be 324 g. What is the molecular formula of quinine?

8-5. You are given unlabeled samples of two different pure substances, each of which has the molecular formula C_3H_8O. The names and structural formulas of the two substances are known:

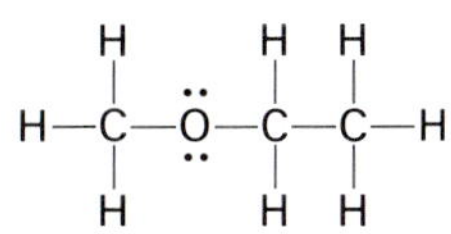

methyl ethyl ether

and

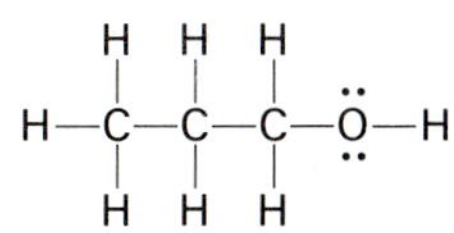

1-propanol

Suggest some methods by which you might determine which sample has which stucture.

8-6. How many peaks would you expect to find in the NMR spectrum of each of the following substances, and what would be the relative areas under the peaks?
(a) formic acid (HCOOH)
(b) methanol (CH_3OH)
(c) trichloromethane (chloroform, $CHCl_3$)
(d) propane ($CH_3CH_2CH_3$)

8-7. The NMR spectrum of hypophosphoric acid (H_3PO_2) has two peaks; that of phosphoric acid (H_3PO_4) has only one. Using this information, draw structural formulas (or electron-dot structures) for these two compounds.

8-8. Two bottles in a chemical laboratory, each containing a white powder, have lost their labels. One bottle contains aspirin, a molecular solid with the formula $C_9H_8O_4$; the other contains the ionic solid lithium chloride, LiCl. Suggest some methods by which you might distinguish between these two solids.

8-9. Given a sample of an unknown solid, describe some tests you might perform in order to classify the solid as metallic, ionic, molecular, or network.

8-10. Consider the following substances in the *solid* state: PH_3, Hg, BaS, and Ge.
(a) Which one of these substances conducts an electric current?
(b) Which one of these substances is a network solid?
(c) Which one of these substances is held together by van der Waals forces, and boils below room temperature?
(d) Which of these substances is a nonconductor of electricity in the solid state, but a conductor in the liquid state?

8-11. The distance between the centers of neighboring Na^+ and Cl^- ions in solid NaCl is determined by X ray diffraction to be 0.282 nm. The density of the solid is easily measured and is found to be 2.16 g/cm^3. The atomic weights of Na and Cl are known to be 23.0 and 35.5. From these data, calculate Avogadro's number, and note that the value obtained is the same as that found by other methods.

Hints: (a) Suppose that a sodium chloride crystal that contains Avogadro's number (N) of Na^+ ions, and an equal number of Cl^- ions, is of volume V. If the measured distance between neighboring Na^+ and Cl^- ions is s, the volume per ion is s^3; then $V = 2Ns^3$. (Why? This may take a little thought.) (b) From the known atomic weights and the measured density, find the volume of solid sodium chloride that contains one mole of NaCl. This is the molar volume V. (c) Combine the results of parts a and b to find N.

8-12. Silicon carbide, or carborundum (SiC), is almost as hard as diamond. Account for this by describing the kind of bonding and the structure you would expect SiC to have.

8-13. Boron nitride, BN, has a structure similar to that of graphite.
(a) Explain why this is a reasonable structure to expect for BN.
(b) Suggest another structure that might be expected for BN, and tell how you might be able to convert BN from the graphite structure to this other structure. (It might help to consult Section 4-9.)

8-14. Why do no pure elements crystallize as ionic solids?

8-15. Sketch the NMR spectra of the following cyclic compounds:

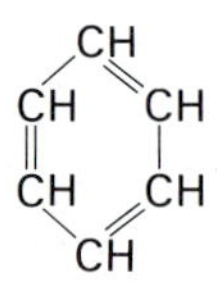
benzene

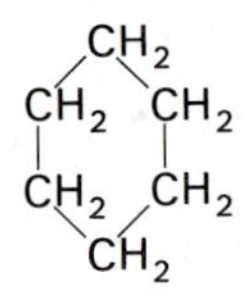
cyclohexane

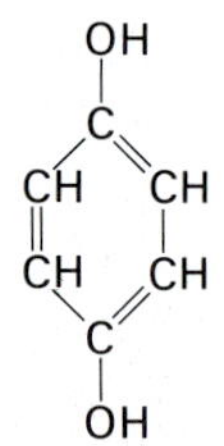
hydroquinone

8-16. Unlike metallic and molecular crystals, ionic and network crystals are very difficult to deform. Explain why this should be so. (Base your explanation on differences in bonding.)

Sheep red blood cells in physiological saline solution. Scanning electron micrograph, $\times 10{,}000$.

9

Solutions

Life on earth began in water, and from 50% to 95% of the weight of any living organism is water. It is primarily the solvent properties of water—its ability to dissolve many substances—that make it so important to life. Cells take up nutrients and discharge waste products into a sea of slightly salty extracellular fluids, and all the reactions of the cell take place in water solution. In this chapter we study properties of solutions and see how they depend on the nature of the solvent, the substance that does the dissolving, and the concentration of solute particles, the substance that is dissolved.

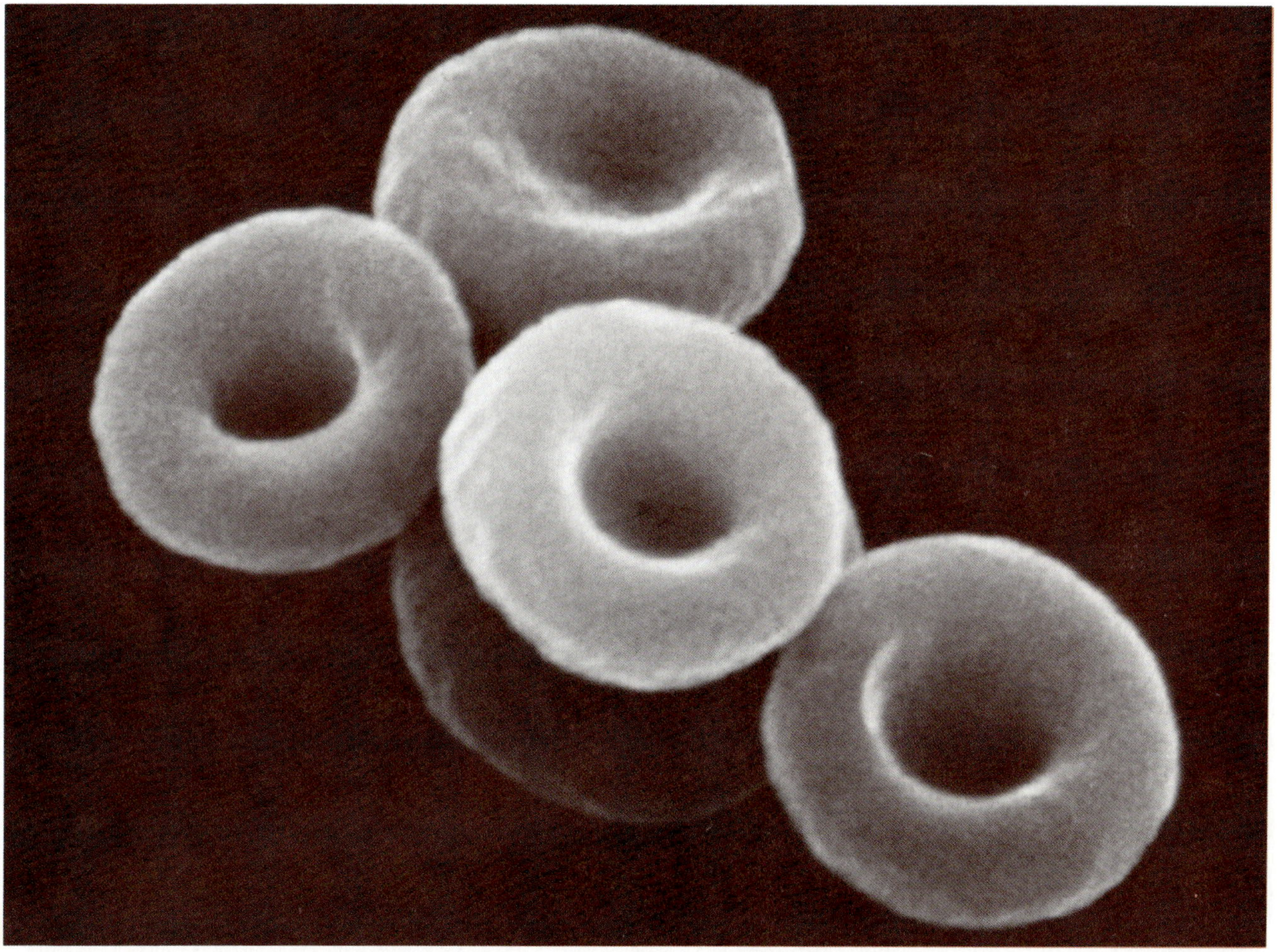

9-1 LIQUID, SOLID, AND GASEOUS SOLUTIONS

A **solution** is a homogeneous mixture of two or more substances. By *homogeneous,* we mean that a sample taken from one part of the solution has the same chemical composition and the same properties as a sample taken from any other part of the solution. The composition of a solution, unlike that of a compound, can be continuously varied. Thus sugar syrup (a solution of sugar in water) can be made up in a wide range of strengths, but any *one* such solution has fixed taste, density, and composition, whatever sample of it we take.

Although gaseous and solid solutions are not uncommon, here we are interested mainly in solutions in which a liquid is the **solvent**—*the substance that is present in large excess and does the dissolving*, and another liquid or a solid is the **solute**—*the substance that is dissolved.* The major theme of our discussion is how the nature of the solute particles affects the properties of the solvent. The most important solvent in everyday life is, of course, water, which dissolves and transports the nutrients required for plant and animal growth and the waste products of animals.

A liquid solution might be a gas dissolved in a liquid—for example, oxygen gas in water, which is of great importance in marine biology. It might also be a liquid dissolved in a liquid—an example is rubbing alcohol, a solution of isopropyl alcohol in water. Or a liquid solution might be a solid dissolved in a liquid—seawater is a solution of sodium chloride and other salts in water.

Because of the relatively large distances between molecules, gases readily mix in all proportions to yield gaseous solutions. Table 9-1 shows the percentages by volume of the gases present in a typical sample of dry air, the most familiar gaseous solution.

TABLE 9-1
Components of atmospheric air (exclusive of water vapor)

	Content	
Constituent	(percentage by volume)	(parts per million by volume)*
N_2	78.08	
O_2	20.95	
CO_2	0.033	
Ar	0.934	9340
Ne	0.00182	18.2
He		5.24
Kr		1.14
Xe		0.087
H_2		0.5
CH_4		2
N_2O		0.5

*This value tells the number of volumes of the substance in one million volumes of air. (Percentage by volume is the number of volumes of the substance in 100 volumes of air.)

A solid solution of a metal dissolved in a metal is an **alloy.** The 16 karat gold used in jewelry and other applications is a solution of copper in gold. Some of the atoms in the crystal lattice of gold have been replaced by copper atoms. Steel is a solution of carbon in iron, in which iron atoms occupy their regular crystalline lattice positions, and carbon atoms occupy some of the spaces between iron atoms.

9-2 CONCENTRATION OF SOLUTIONS

The properties of a solution vary with the relative amounts of solute and solvent present—the taste and color of a cup of coffee depend upon the strength or concentration of the solution. We look now at several common methods of expressing concentration.

Molarity is the number of moles of solute per unit

volume of solution, generally the number of moles per liter of solution. This method of expressing concentration is the one most often used by chemists. A 1 molar (1 M) solution contains 1 mole of solute in 1 ℓ of *solution*; a 2 M solution contains 2 moles of solute in 1 ℓ of solution; and a 0.5 M solution contains 0.5 moles of solute in 1 ℓ of solution.

$$\text{Molarity} = \frac{\text{Number of moles of solute}}{\text{Volume of solution}}$$

To prepare 1 ℓ of a 1.000 M solution of copper chloride ($CuCl_2$), we would weigh out 1 mole, or 134.5 ($63.5 + 2 \times 35.5$) g of $CuCl_2$, and then dissolve it in enough water to make a final volume of 1.000 ℓ of solution (Figure 9-1).

$$\text{Molarity } CuCl_2 = \frac{1.000 \text{ mole}}{1.000\ \ell} = 1.000 \text{ M } CuCl_2$$

To prepare 1 ℓ of a 0.100 M solution of $CuCl_2$, we would take only 0.100 moles (13.45 g) of the $CuCl_2$, and dissolve it in enough water to make a final volume of 1.000 ℓ of solution. For this solution,

$$\text{Molarity } CuCl_2 = \frac{0.100 \text{ mole}}{1.000\ \ell} = 0.100 \text{ M } CuCl_2$$

EXAMPLE 9-1

What is the molarity of a solution prepared by dissolving 26.9 g of $CuCl_2$ in a total volume of 100 ml of solution?

$$\text{Moles of } CuCl_2 = \frac{26.9 \text{ g}}{134.5 \text{ g/mole}} = 0.200 \text{ mole}$$

$$\text{Volume of solution} = 100 \text{ ml} = 0.100\ \ell$$

$$\text{Molarity } CuCl_2 = \frac{0.200 \text{ mole}}{0.100\ \ell} = 2.00 \text{ M } CuCl_2$$

EXERCISE 9-1 What is the molarity of an aqueous solution of glucose ($C_6H_{12}O_6$) that contains 90 g of glucose in each 2 ℓ of solution?

If the molarity of a solution is known, the number of moles of solute present in any volume of solution can be calculated by rearranging the equation

$$\text{Molarity} = \frac{\text{Moles of solute}}{\text{Volume of solution}}$$

to give

$$\text{Moles of solute} = \text{Molarity} \times \text{Volume of solution}$$

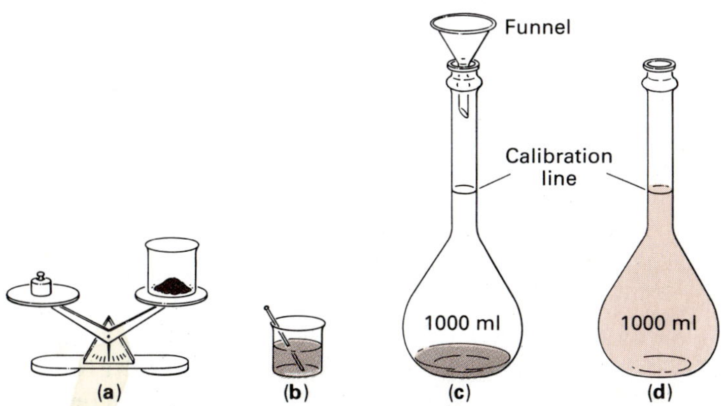

FIGURE 9-1

Preparation of 1.000 M $CuCl_2$. **(a)** Weigh out 134.5 g (one mole) $CuCl_2$. **(b)** Dissolve in water. **(c)** Transfer the solution to a 1.000 ℓ volumetric flask. **(d)** Add water until the *solution* reaches the calibration line on the neck of the flask.

Thus, in a 50 ml (0.050 ℓ) sample of 2.00 M $CuCl_2$,

$$\text{Moles } CuCl_2 = 2.00\,\frac{\text{moles}}{\ell} \times 0.050\ \ell$$
$$= 0.100 \text{ mole } CuCl_2$$

EXAMPLE 9-2

What mass of glucose ($C_6H_{12}O_6$) is needed to prepare 100 ml of a 0.200 M solution?

$$\text{Moles } C_6H_{12}O_6 \text{ required} = 0.200\,\frac{\text{moles}}{\ell} \times 0.100\ \ell$$
$$= 0.200 \text{ mole}$$

$$\text{Mol. wt. } C_6H_{12}O_6 = 6 \times 12.0 + 12 \times 1.0 + 6 \times 16.0$$
$$= 180 \text{ g/mole}$$

$$\text{Mass } C_6H_{12}O_6 \text{ required} = 0.200 \text{ moles} \times 180 \text{ g/mole}$$
$$= 36.0 \text{ g}$$

EXERCISE 9-2 (a) How many moles of glucose are needed to prepare 2.00 ℓ of a 0.250 M solution?
(b) How many grams of glucose is this?

One reason working with solutions is so convenient is that when a sample containing a certain number of *moles* of solute is needed, the sample can be obtained relatively easily and accurately by measuring out a known volume of a solution of known concentration. The required volume can be calculated by again rearranging the equation

$$\text{Molarity} = \frac{\text{Moles of solute}}{\text{Volume of solution}}$$

to give

$$\text{Volume of solution} = \frac{\text{Moles of solute}}{\text{Molarity}}$$

Thus, if 0.300 moles of NaCl are needed, and a solution of 1.50 M NaCl is available, the *volume* of solution required is

$$\text{Volume of solution} = \frac{0.300 \text{ moles}}{1.50 \text{ moles}/\ell}$$
$$= 0.200\ \ell = 200 \text{ ml}$$

Graduated cylinders and pipets are available to measure liquid volumes from fractions of a milliliter to several liters. Some of these are shown in Figure 1-9.

Similarly, it is often convenient to prepare a dilute solution from a more concentrated one. For example, suppose you have a 5.00 M solution of NaCl, and you need 250 ml of a 1.00 M solution. How could you prepare such a solution? First, we'll calculate the number of *moles* of NaCl required, and then the *volume* of the 5.00 M solution that contains that number of moles.

$$\text{Moles of NaCl} = \text{Molarity} \times \text{Volume of solution}$$
$$= 1.00 \text{ mole}/\ell \times 0.250\ \ell$$
$$= 0.250 \text{ mole}$$

$$\text{Volume of 5.00 M NaCl} = \frac{0.250 \text{ mole}}{5.00 \text{ moles}/\ell}$$
$$= 0.0500\ \ell$$

You could prepare the required solution by withdrawing 50.0 ml of the 5.00 M solution by pipet, transferring it to a 250 ml volumetric flask, and adding water to the calibration line.

EXERCISE 9-3 (a) What volume of 0.500 M NaCl is needed to provide 2.00×10^{-3} moles of NaCl?
(b) What volume of 0.50 M glucose solution could be prepared from 45 g of glucose ($C_6H_{12}O_6$)?

Molality is the number of moles of solute per unit mass of solute, generally per 1000 g (1 kg) of solvent. A solution that is 1.00 molal in glucose can be prepared by adding 1000 g of water to 1.00 moles (180 g) of glucose. A 0.50 molal glucose solution is prepared by adding 1000 g of water to 0.50 moles (90 g) of glucose.

Molality is sometimes used instead of molarity in making very precise measurements because molality does not change with temperature. If a solution contains 1 mole of solute per kilogram of water, then each gram of solution contains a fixed amount of solute, whatever the temperature. Molarity, on the contrary, changes with temperature because the volume of the solution changes with temperature. For small changes in temperature, the volume change is not significant, and we shall usually use molarities.

Percentage by weight of solute is the percentage contributed by the pure solute to the total mass of the solution. Thus, if a solution is 2% by weight glucose, there are 2 g of glucose in every 100 g of solution.

Percentage by volume of solute is the percentage contributed by the pure solute to the total volume of the solution. If a solution is 60% by volume alcohol, it contains 60 ml of pure alcohol in every 100 ml of solution.

EXERCISE 9-4 (a) How many grams of glucose must be added to 0.500 ℓ (500 g) of water to yield a 0.20 molal solution? (The molecular weight of glucose is 180 g/mole.)

(b) What is the percentage by weight of glucose in a solution that contains 5 g of glucose in 200 g of solution?

9-3 ELECTROLYTES AND NONELECTROLYTES

Many ionic solids **dissociate** (split into their component ions) when they dissolve in water. Thus, a solution of table salt in water contains Na^+ ions and Cl^- ions rather than NaCl units. A few covalently bonded substances, such as hydrogen chloride (which at room temperature is a gas composed of HCl molecules) also dissociate into ions when dissolved in water. Most covalently bonded substances, however, do not dissociate into ions when they dissolve in water.

The simple experiment in electrical conductivity diagrammed in Figure 9-2 demonstrates the presence or absence of ions in solution. In the figure, a pair of electrodes (metal or graphite rods) is connected to a wall socket and a light bulb. As long as the electrodes

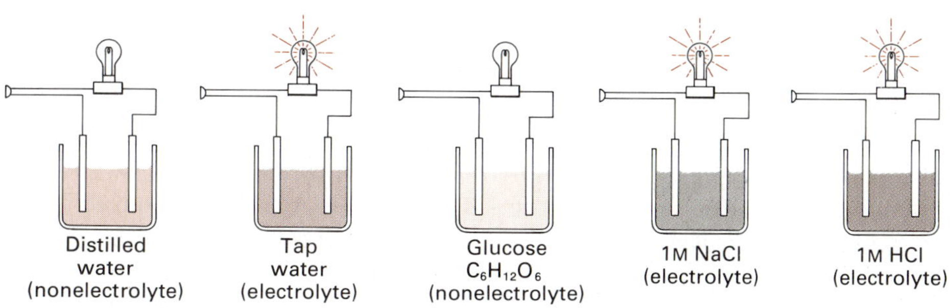

FIGURE 9-2

Electrolytes and nonelectrolytes. Electrolytes conduct an electric current.

are kept separate, no electrons can flow, and the bulb remains dark. When the two electrodes are connected by an electrical conductor, a current is set up in the circuit, and the bulb lights. Thus, if the electrodes are dipped into a solution that contains ions (which carry an electrical charge, and so can conduct an electric current through the solution), the bulb lights. If the electrodes are dipped into a solution that contains only molecules, which do not carry electrical charge, the bulb fails to light. As seen in the figure, when the beaker contains only the molecular substances distilled water and glucose, the bulb remains dark. However, when the beaker is filled with ordinary tap water (which contains dissolved minerals), or with hydrochloric acid or sodium chloride, then the bulb lights, showing that *ions are present in the solution to conduct the current*. An ionic *solid*, such as a sodium chloride crystal, does not conduct an electric current because the ions are in fixed positions and are not free to move, but if such an ionic solid is melted, the liquid melt can conduct a current because the ions in the liquid are mobile.

Substances that dissolve in water to yield conducting solutions are called ***electrolytes.*** *Substances that dissolve in water to yield nonconducting solutions are called* ***nonelectrolytes.*** Nonelectrolytes retain their identity as molecules in solution. If the light bulb is replaced by a meter that measures electrical conductivity, a quantitative determination of the concentration of ions in solution can be made. When the conductivity of 1 M acetic acid (CH_3COOH) is compared with that of 1 M hydrochloric acid, it is found that the conductivity of the acetic acid is much lower than that of the hydrochloric acid. We conclude that the acetic acid provides fewer ions to conduct the current—acetic acid is an example of a **weak electrolyte,** *a substance that is only slightly dissociated into ions in solution.* Most of the acetic acid molecules remain as undissociated molecules in solution. Strong electrolytes such as hydrochloric acid are essentially 100% dissociated into ions in solution.

Electrolytes in the blood and other body fluids have important roles in various body functions. The body electrolytes include Na^+, K^+, Mg^{2+}, Ca^{2+}, Cl^-, HCO_3^-, HPO_4^{2-}, and $H_2PO_4^-$. These ions maintain the electrical neutrality of the cells and extracellular fluids. They are important in the maintenance of protein structure, membrane permeability, and muscle contraction. In Section 9-9, we shall see that the concentrations of electrolytes in the cells and in the body fluids determine osmotic pressure, thus helping to maintain the proper liquid volume of the blood and extracellular fluids.

9-4 SOLUBILITY

The solutions considered thus far have been **unsaturated** solutions, whose concentrations could be increased by adding more solute. For any given solute and solvent, a wide range of concentrations of unsaturated solutions can be prepared. However, for most solutes (especially solids) with most solvents, as more solute is added, a point eventually is reached where the addition of still more solute does not increase the concentration of the solution. Instead, the additional material remains undissolved. At the point where no more material can be dissolved, the solution is said to be **saturated.** The **solubility** of the solute in that solvent is its concentration in the saturated solution. To ensure having a saturated solution, an excess of solid solute is left in contact with the solution. The total amount of excess solid solute then remains constant

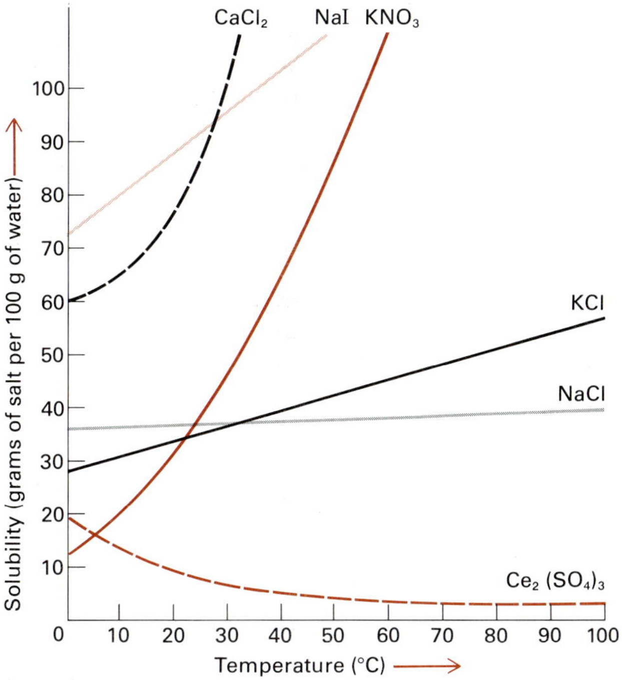

FIGURE 9-3

Solubilities of some ionic solids in water as a function of temperature.

with time, though this equilibrium between the solid solute and the solute particles in solution is dynamic, not static. Solid is continually dissolving, but dissolved solute particles are crystallizing out onto the surfaces of the solid at the same rate. In Chapter 4 we saw similar examples of dynamic equilibrium between a liquid and its vapor, a solid and its melt, and a solid and its vapor.

Solubility is affected by temperature. Commonly the solubility of a solid in a liquid increases with a rise in temperature. The graphs in Figure 9-3 show the variation of solubility in water with temperature for several substances. As shown in the figure, the solubility of cerium sulfate, $Ce_2(SO_4)_3$, is atypical. The solubility of $Ce_2(SO_4)_3$ decreases as the temperature is raised. The solubility of NaCl is almost independent of temperature. In the case of a gas dissolved in water, the solubility usually decreases as the temperature is raised. An important consequence of this is the effect that thermal pollution of lakes (say, by nuclear power plants) can have on life in the lake. If lake water is used for cooling and the hot water is immediately returned to the lake without prior cooling, the lake temperature will increase in the vicinity of the plant. This will result in a decrease in the concentration of oxygen gas dissolved in the lake and thus seriously affect the aquatic animal life.

A rough rule of thumb that describes the effect on solubility of the relative natures of the solvent and solute is "like dissolves like." Ionic solids dissolve only in strongly polar solvents—those that have partial ionic character. As we saw in Chapter 7, water is a polar molecule, with a partial negative charge on the oxygen atom and a partial positive charge on each hydrogen:

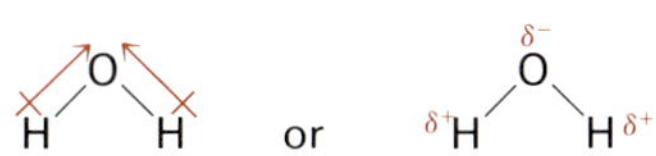

When an ionic solid such as NaCl dissolves in water, negatively charged chloride ions (Cl^-) on the surface of the crystal attract and become surrounded by the positively charged ends of water molecules. Such a chloride ion might then be torn from the crystal lattice and carried through the solution as a **hydrated** Cl^- ion. At the same time, the positively charged sodium ions on the surface of the lattice attract the negative ends of water molecules and go into solution, as shown schematically in Figure 9-4.

Table 9-2 summarizes the solubilities of some common ionic solids in water. Here, "soluble" means that we can make a solution of greater than 0.1 M concentration at room temperature. We see that all common compounds of the alkali metals and ammonium ion are soluble, as are all common nitrates and acetates. Differences in the solubilities of ionic solids are used as

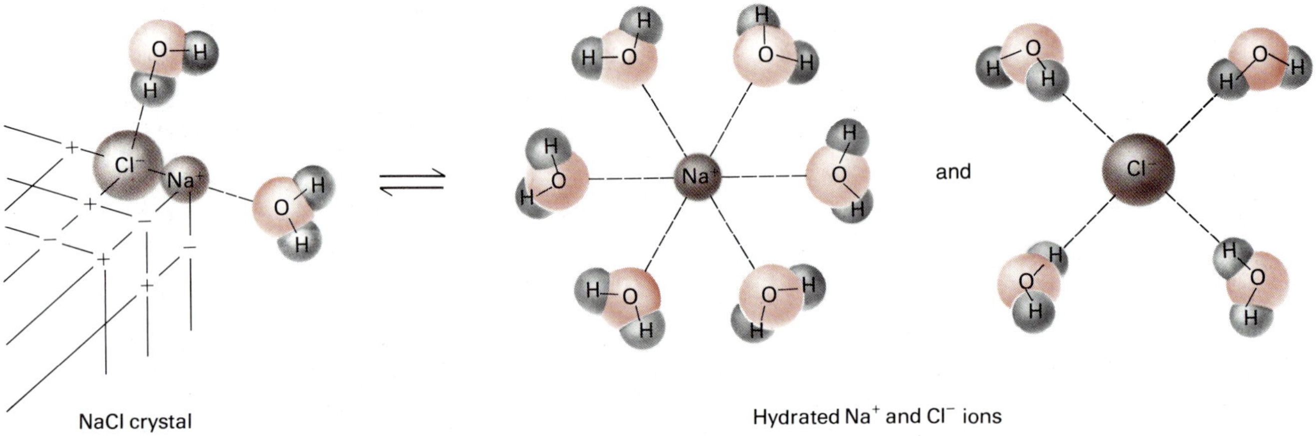

FIGURE 9-4

A sodium chloride crystal dissolves in water. There are six water molecules about each Na^+ ion. We do not know how many water molecules surround a Cl^- ion.

TABLE 9-2
Solubilities of some ionic compounds in water

Rule	Exceptions
1. All compounds of the alkali metal ions (Li^+, Na^+, K^+, Rb^+, Cs^+) and the ammonium ion (NH_4^+) are soluble.*	No important ones
2. All nitrates (NO_3^-) and acetates (CH_3COO^-) are soluble.	No important ones
3. The halides Cl^-, Br^-, and I^- are soluble.	Those of Ag^+, Hg_2^{2+}, and Pb^{2+}
4. Sulfates (SO_4^{2-}) are soluble.	Those of Ca^{2+}, Sr^{2+}, Ba^{2+}, and Pb^{2+}
5. Phosphates (PO_4^{3-}), carbonates (CO_3^{2-}), and sulfites (SO_3^{2-}) are *not* soluble.	Those of the alkali metal ions and NH_4^+
6. Hydroxides (OH^-) are *not* soluble.	Those of the alkali metal ions and NH_4^+ (Those of Ca^{2+}, Sr^{2+} and Ba^{2+} are moderately soluble.)
7. Sulfides (S^{2-}) are *not* soluble.	Those of the alkali metal ions, alkaline earth ions (Be^{2+}, Mg^{2+}, Ca^{2+}, Sr^{2+}, Ba^{2+}), and NH_4^+

*"Soluble" means that more than 0.1 mole of solute will dissolve per liter of solution.

the basis of qualitative analysis, the separation and identification of positive and negative ions in the laboratory.

As an example, suppose we would like to separate the silver ions from the sodium ions in a solution containing silver nitrate ($AgNO_3$) and sodium nitrate ($NaNO_3$). We see in Table 9-2 that all compounds of sodium are soluble, but that silver chloride (AgCl) is insoluble. This means that, when a solution contains both Ag^+ and Cl^-, these ions immediately combine to form solid AgCl, which is insoluble. Thus, Ag^+ can be separated from Na^+ by adding a chloride, such as hydrochloric acid; this causes the Ag^+ to precipitate out of solution as solid AgCl, while the Na^+ remains in solution.

Or suppose we know that a certain solution consists of either lead nitrate, $Pb(NO_3)_2$, or calcium nitrate, $Ca(NO_3)_2$. We can establish which it is by adding sulfide ions (say, as Na_2S). If the solution is of $Pb(NO_3)_2$, a precipitate of insoluble lead sulfide (PbS) will form. If the solution is of $Ca(NO_3)_2$, no precipitate will form because calcium sulfide is soluble.

We saw in Chapter 7 that when strongly polar hydrogen chloride dissolves in water, the positive end of an HCl molecule is transferred to the negative end of a water molecule, yielding a hydronium ion (H_3O^+) and a Cl^- ion:

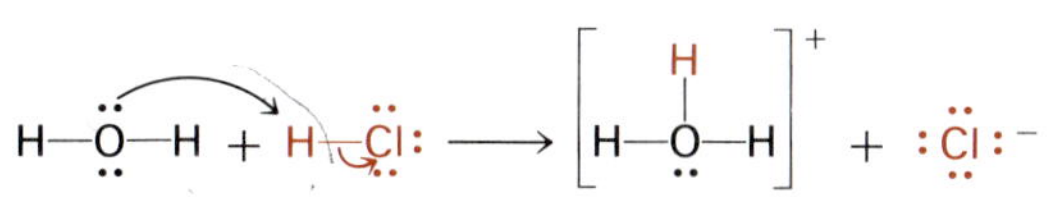

$$H_2O \quad + \quad HCl \longrightarrow H_3O^+ \quad + \quad Cl^-$$

Both the hydronium and the chloride ions are then further hydrated. They are themselves surrounded by additional water molecules to which they are more or less tightly bound.

Water is also a good solvent for ethyl alcohol (ethanol, C_2H_5OH). Here, both the alcohol and the water molecules have the hydroxyl group, OH. (Remember the "like dissolves like" rule.) The nonpolar hydrocarbon end of the alcohol molecule is quite foreign to the water, but in this case it is not too large a part of the ethanol molecule to prevent its solubility in water. Water and ethyl alcohol are miscible in all proportions. As shown in Table 9-3, as the length of the nonpolar hydrocarbon chain increases, the solubility of the alcohol in water decreases.

TABLE 9-3
Solubilities of some alcohols in water

Name	Formula	Solubility (grams per 100 g H_2O)
Methanol	CH_3OH	Infinitely soluble
Ethanol	CH_3CH_2OH	Infinitely soluble
n-Propanol	$CH_3(CH_2)_2OH$	Infinitely soluble
n-Butanol	$CH_3(CH_2)_3OH$	7.9
n-Pentanol	$CH_3(CH_2)_4OH$	2.3
n-Hexanol	$CH_3(CH_2)_5OH$	0.6
n-Heptanol	$CH_3(CH_2)_6OH$	0.2
n-Octanol	$CH_3(CH_2)_7OH$	0.05

Likewise, ethyl alcohol and diethyl ether ($C_2H_5OC_2H_5$) are completely miscible because large parts of both molecules are hydrocarbons. Ether and water, on the contrary, are hardly miscible at all because too large a part of the ether molecule consists of nonpolar hydrocarbon chains.

Figure 9-5 shows three cylinders, each containing three layers of solvent: ether (the lightest of the three) floating on top, water in the middle, and carbon tetrachloride (the densest of the three) on the bottom. The nonpolar carbon tetrachloride layer and the ether layer are both immiscible with water. Into the cylinder of Figure 9-5b, a bit of powdered copper sulfate (an ionic solid) has been dropped. Into the cylinder of Figure 9-5c, some crystals of iodine (a molecular solid) have been dropped. As can be seen, the ionic solid dissolves only in the polar solvent, whereas the molecular solid dissolves only in the nonpolar solvents.

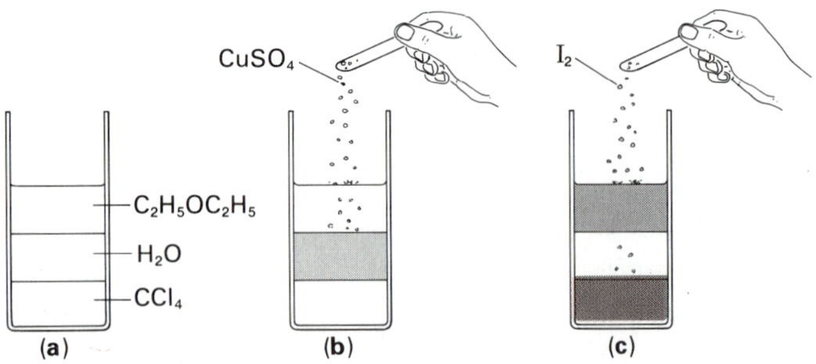

FIGURE 9-5

"Like dissolves like." **(a)** Three solvents floating in distinct layers. **(b)** The polar solvent water dissolves the ionic solid $CuSO_4$. **(c)** The nonpolar solvents carbon tetrachloride and diethyl ether dissolve the nonpolar solid I_2.

9-5 DISTILLATION AND VAPOR PRESSURE OF A SOLVENT

The vapor in equilibrium with a solution is richer in the more volatile component of the solution. An extreme example is the case considered in the discussion of distillation in Chapter 8—that of the nonvolatile solute NaCl in water. The vapor in equilibrium with an aqueous solution of NaCl is entirely water. We saw that the NaCl and water can be effectively separated by distillation, where the water is evaporated from the solution, conducted away as steam, and then the steam condensed to give pure water, leaving pure NaCl behind in the distilling flask (Figure 8-1). In general—say, in distilling a mixture of ethyl alcohol (C_2H_5OH), which boils at 78.5°C, and diethyl ether ($C_2H_5OC_2H_5$), which boils at 34.6°C—the vapor contains some of both components, but the first fraction of the solution to distill over is richer in the more volatile ether. If the vapor is then condensed, a distillate richer in ether is obtained. The liquid remaining in the first flask is richer in alcohol, the less volatile component. We can now start over with each solution separately (the distillate and the residue). By repeating this process many times, we can separate the mixture into its two components: pure alcohol and pure ether. This can be accomplished in a single operation with the distilling flask shown in Figure 9-6.

The experiment shown in Figure 9-7 demonstrates the effect of a solute on the vapor pressure of a solvent. At the start, beaker A contains 100 ml of pure water, and beaker B contains 100 ml of a sugar solution. The beakers are covered with a plastic dome, so that no water can enter or leave the system. Water molecules can, however, escape from and reenter each of the beakers. The level of the water in beaker A *decreases* with time, while that of the solution in beaker B *increases*. Eventually all of the water from A is transferred to B. How can we explain this?

This phenomenon is a result of the fact that when a nonvolatile solute is dissolved, *the vapor pressure of the solvent in solution is less than that of the pure solvent.* (Even if the solute is volatile, the **partial pressure** of the solvent in the vapor—that part of the total pressure that is due just to the solvent molecules in the vapor—is less than that of the pure solvent.) As shown in Figure 9-8, the concentration of solvent molecules on the surface of a solution (those most likely to escape to the vapor) is reduced by the presence of solute particles. In Figure 9-8a, pure solvent is at equilibrium with its vapor. The rate at which liquid molecules enter the vapor equals the rate at which vapor molecules return to the liquid. When solute is added (Figure 9-8b), the *rate* at which

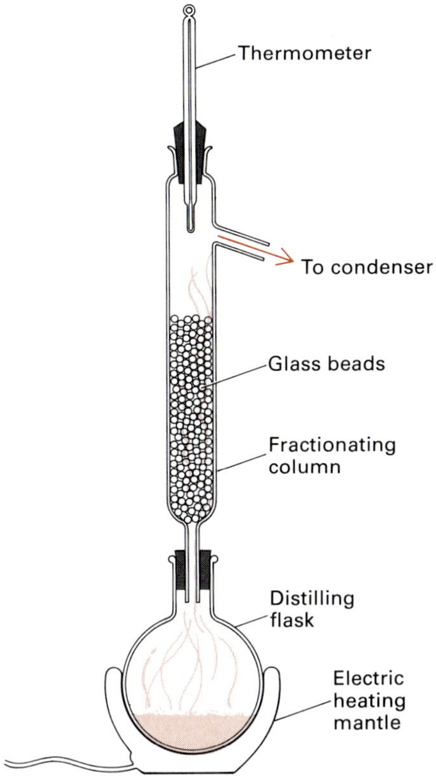

FIGURE 9-6

Separation of dimethyl ether from ethyl alcohol by fractional distillation. The column is packed with glass beads, on which the vapor cools and condenses as it passes through the column. The condensed liquid is richer in the alcohol, the less volatile component. As the liquid drips back down toward the distilling flask, it comes in contact with fresh vapor, which is hotter. An exchange of heat between liquid and vapor causes the ether in the condensed liquid to vaporize, and the alcohol in the vapor to condense. This process is repeated again and again as the vapor moves up the column, and it results in a separation of the ether from the alcohol.

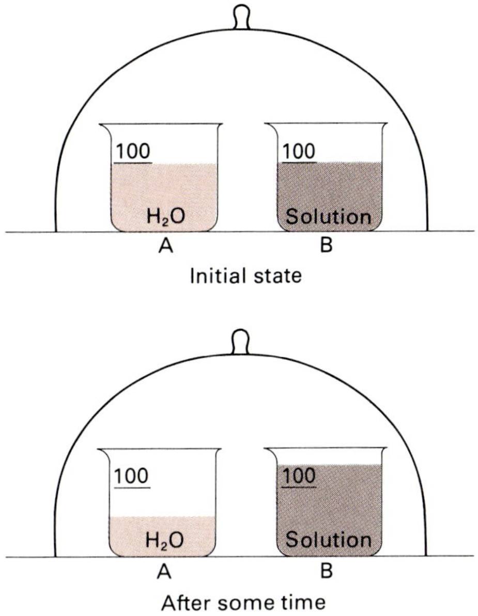

FIGURE 9-7

Effect of a solute on the vapor pressure of the solvent. Beaker A contains pure water, and beaker B contains a sugar solution.

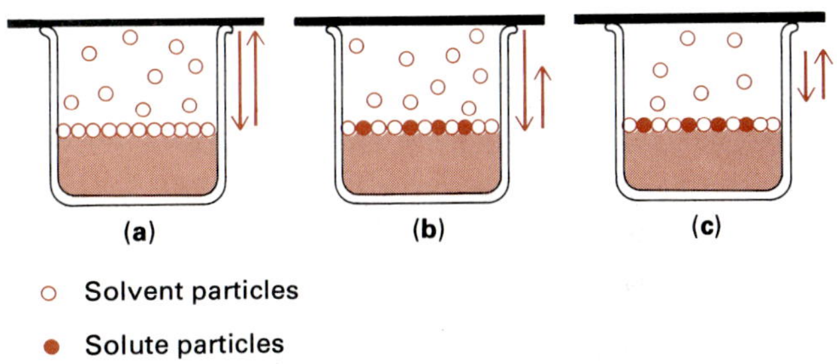

FIGURE 9-8

Adding a solute lowers the vapor pressure of the solvent. The arrows represent evaporation of the liquid and condensation of the vapor, and the lengths of the arrows represent the relative rates. **(a)** There is equilibrium between liquid and vapor. **(b)** The equilibrium is momentarily disrupted by the addition of solute. **(c)** The equilibrium is reestablished, with a lower vapor pressure of the solvent.

liquid solvent molecules enter the vapor decreases while, momentarily, the rate at which solvent molecules in the vapor return to the liquid remains the same as it was for pure solvent. Finally, a new equilibrium is established (Figure 9-8c) with fewer solvent molecules in the vapor—that is, with the vapor pressure of the solvent lowered.

In the experiment of Figure 9-7, water molecules escape from the pure water in beaker A faster than they escape from the solution in B, but they return to the beakers at equal rates. Finally, after all the water has been transferred to B, a state of equilibrium is reached, with water molecules entering and leaving the solution at equal rates.

The extent to which the vapor pressure of the solvent is lowered is determined entirely by the concentration of solute particles. The extent of the vapor pressure lowering does *not* depend on the kind of particles. Thus a 1 M solution of glucose and a 1 M solution of sucrose have the same vapor pressure because each contains the same concentration of molecules. A 1 M solution of NaCl, however, gives a 2 M solution of ions (1 M in Na^+ and 1 M in Cl^-), and causes the same reduction of the vapor pressure of water as a 2 M solution of glucose. (The 2 M glucose solution contains 2 moles of glucose molecules per liter of solution, and the 1 M NaCl solution contains 2 moles of ions per liter of solution.)

To summarize, *the depression of the vapor pressure of the solvent in a dilute solution below that of the pure solvent is proportional to the total concentration of solute particles but is independent of the kind of solute particles.* This vapor-pressure lowering leads to several closely related phenomena that we next consider: boiling-point elevation, freezing-point depression, and osmosis.

EXERCISE 9-5 (a) What concentration of KCl would cause the same lowering of the vapor pressure of water as a 0.050 M solution of sucrose?

(b) What concentration of KCl would cause the same lowering of the vapor pressure of water as a 0.200 M solution of NaCl?

9-6 BOILING-POINT ELEVATION

Figure 9-9 shows phase diagrams of pure water and of an aqueous solution of a nonvolatile solute. At every temperature, the vapor pressure of the solution is lower than that of pure solvent. From the diagram, we can see why the boiling point of the solution must be higher than that of the pure solvent. In the figure, 100°C is the boiling point of the pure water—the point at which the vapor pressure is 1 atm. Because the vapor pressure of the solution is lower, its temperature must be raised higher (to t_2) before its vapor pressure reaches 1 atm.

Just as the extent of lowering of the vapor pressure of the solvent in a dilute solution depends on the concentration of solute particles, so also does the extent of the elevation of the boiling point. *The boiling point of a solution is elevated above that of pure solvent by an amount that is proportional to the concentration of solute particles.* In studying boiling-point elevation, concentrations are conventionally expressed in *molalities*. A 1.0 molal solution of glucose (or of sucrose) molecules causes the boiling point of water to increase by 0.52°C. The solution boils at 100.52°C. However, a 1.0 molal solution

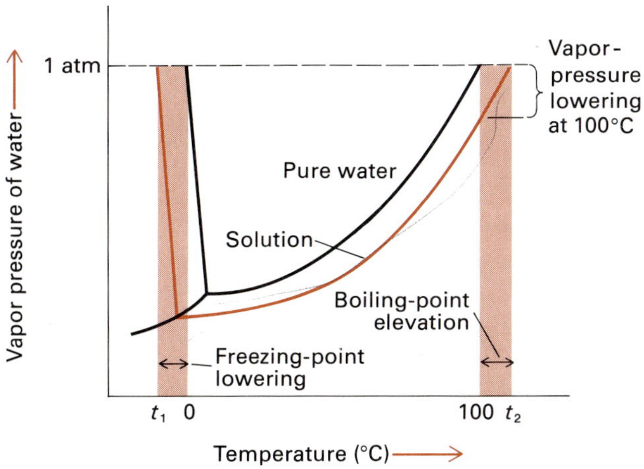

FIGURE 9-9

A comparison of the phase diagrams of water and of an aqueous solution of a nonvolatile solute. (The temperature scale is distorted.)

of NaCl, which is 2.0 molal in ions, causes the boiling point to increase by 2 × 0.52°C, or 1.04°C. This solution boils at 101.04°C.

The elevation of the boiling point also depends on the solvent. We saw that the boiling temperature of a water solution is raised 0.52°C by a 1 molal solution. The boiling temperature of a solution in ethyl alcohol is raised 1.24°C by a 1 molal solution, and the boiling temperature of a solution in carbon tetrachloride is raised 5.05°C by a 1 molal solution.

For any solvent with a nonvolatile solute, we can write:

$$\Delta T_b = K_b \times m$$

where ΔT_b is the elevation of the boiling point, K_b is called the **molal boiling-point elevation constant;** and m is the molality of all solute particles in the solution.

The molal boiling-point elevation constant K_b has the numerical value of 0.52 °C/molal for water; 1.24 °C/molal for ethanol; and 5.05 °C/molal for carbon tetrachloride.

EXAMPLE 9-3

(a) Physiological saline solution is 0.9% NaCl or 0.15 molal NaCl, in water. What is the boiling point of physiological saline solution?

A 0.15 molal solution of NaCl is 2 × 0.15 = 0.30 molal in ions (0.15 molal in Na^+ and 0.15 molal in Cl^-).

$$\Delta T_b = 0.52 \text{ °C/molal} \times 0.30 \text{ molal} = 0.16\text{°C}$$

The boiling point of the solution is 100.16°C.

(b) What is the boiling point of an aqueous solution that is 0.15 molal in glucose?

$$\Delta T_b = 0.52 \text{ °C/molal} \times 0.15 \text{ molal} = 0.08\text{°C}$$

The boiling point of the solution is 100.08°C. (The boiling-point elevation of the solution in (a) is twice as great as that in (b) because the NaCl is dissociated into ions in solution, but the glucose exists as molecules in solution. It is the concentration of *particles* that is important in determining the boiling-point elevation.)

EXERCISE 9-6 Iodine molecules (I_2) are a beautiful violet color when dissolved in carbon tetrachloride (CCl_4). (This color is the basis of a laboratory test for I_2.) What is the boiling point of a 0.20 molal solution of I_2 in CCl_4? The boiling point of CCl_4 is 76.54°C, and K_b for CCl_4 is 5.05 °C/molal.

The magnitude of the boiling-point elevation is small, but it is very important. In the following example, it allows the determination of the molecular weight and the molecular formula of fructose.

EXAMPLE 9-4

The simplest formula of fructose is found by combustion analysis to be CH_2O. A solution of 18.0 g fructose in 500 g ethanol has a boiling point of 78.60°C. What is the molecular formula of fructose? The normal boiling point of ethanol is 78.35 °C, and the boiling-point elevation constant for ethanol is 1.24 °C/molal.

$$\Delta T_b = 78.60° - 78.35° = 0.25°C$$

From ΔT_b and K_b we can calculate the molality of the solution:

$$\text{Molality} = \frac{\Delta T_b}{K_b} = \frac{.25°C}{1.24\ °C/\text{molal}} = 0.20 \text{ molal}$$

From the molality, we know that the solution contains 0.20 moles fructose for every 1000 g (1 kg) alcohol. Therefore, in 500 g alcohol, there is one-half this number of moles, or 0.10 moles of fructose. We also know that this solution has 18.0 g fructose in 500 g alcohol. Therefore,

$$0.10 \text{ moles} = 18.0 \text{ g}$$

$$\text{Mol. wt. fructose} = \frac{18.0 \text{ g}}{0.10 \text{ moles}} = 180 \text{ g/mole}$$

The formula weight of the simplest formula (CH_2O) is $12.0 + 2 \times 1.0 + 16.0 = 30.0$ g. Six of these formula weights are required to give the correct molecular weight of 180 g/mole.

Molecular formula of fructose = $(CH_2O)_6$, or $C_6H_{12}O_6$

9-7 FREEZING-POINT DEPRESSION

Another consequence of the lowering of the vapor pressure of the solvent in a solution is a *depression* of the freezing point of the solvent, when it is pure solvent that crystallizes out as the solution freezes. This phenomenon is seen in the lowering of the freezing point of water by the addition of ethylene glycol ("antifreeze") to car radiators, and in the use of salt to melt ice on the roads. We shall see later that if an icy road is salted enough to yield a solution 1 molal in calcium chloride, then the freezing point of the solution (at which pure ice crystallizes out) is lowered to 22°F.

We concluded, from Figure 9-9, that the boiling point of a solution of a nonvolatile solute is raised above that of pure solvent as a result of a depression of the vapor pressure of the solvent. We can use this same figure to see that the freezing point of the solution is depressed below that of the pure solvent. In the figure, t_1 is the freezing point of the solution, and 0°C is the freezing point of pure water. Like the elevation of the boiling point, *the depression of the freezing point of the solvent in a dilute solution is proportional to the total concentration of all solute particles, with a proportionality constant that depends only on the solvent:*

$$\Delta T_f = K_f \times m$$

Here ΔT_f is the depression of the freezing point, K_f is the **molal freezing-point depression constant,** and m is the molality of all solute particles in the solution. The value of K_f for water is 1.86 °C/molal; for benzene, 4.90 °C/molal; and for naphthalene, 6.9 °C/molal.

A 1.0 molal solution of $CaCl_2$ in water (3.0 molal in

ions, because 1.0 molal in Ca^{2+} and 2.0 molal in Cl^-) lowers the freezing point of water by 3.0 molal × 1.6 °C/molal = 5.58°C. When an icy road is salted to yield a 1.0 molal solution of $CaCl_2$, the freezing point is lowered to −5.6°C, or 22°F.

EXAMPLE 9-5

How many grams of ethylene glycol ($C_2H_6O_2$) must be added to each kilogram (1000 g) of water to protect a car radiator down to −29.0°C (−20°F)? (K_f for water is 1.86 °C/molal.)

$$\Delta T_f = K_f \times m$$

$$m = \frac{\Delta T_f}{K_f} = \frac{29.0\,^\circ\mathrm{C}}{1.86\,^\circ\mathrm{C/molal}} = 15.6 \text{ molal}$$

For each 1 kg of water, 15.6 moles of ethylene glycol are needed.

$$\text{Mol. wt. } C_2H_6O_2 = 2 \times 12.0 + 6 \times 1.0 + 2 \times 16.0 = 62.0 \text{ g/mole}$$

$$\text{Mass } C_2H_6O_2 \text{ needed} = 15.6 \text{ mole} \times 62.0 \text{ g/mole} = 967 \text{ g}$$

(This is about 880 ml of ethylene glycol for each 1000 ml of water. This is much too concentrated for the relationship $\Delta T_f = K_f \times m$ to be very accurate, but the order of magnitude is right.)

EXERCISE 9-7 If the freezing point of a solution of LiBr is −3.72°C, what is the molality of LiBr in the solution?

9-8 OSMOSIS

The earliest living organisms probably had salt and mineral concentrations very similar to those of the seawater in which the organisms lived. Most organisms living today, on the contrary, have chemical compositions quite different from those of their environments. In humans, the blood plays a major role in maintaining a constant internal chemical environment, delivering nutrients to the cells, and removing waste products. The blood can function in this way because the kidneys constantly remove the cellular waste products from the blood and excrete them, while returning most of the salts and minerals and water to the body.

Water moves freely through the cell membranes, into and out of the three major water compartments in the body: (1) the blood plasma (which contains 7% of the body fluid); (2) the interstitial fluid, the fluid between the cells (28%); and (3) the intracellular fluid, the fluid within the cells (65%). Although water moves freely through cell membranes, the membranes prevent the passage of most molecules. Such a membrane is a **semipermeable membrane,** and the movement of water through the membrane is osmosis.

Osmosis *is the movement of water across a semipermeable membrane from a region of higher water concentration (lower solute concentration) to one of lower water concentration (higher solute concentration).*

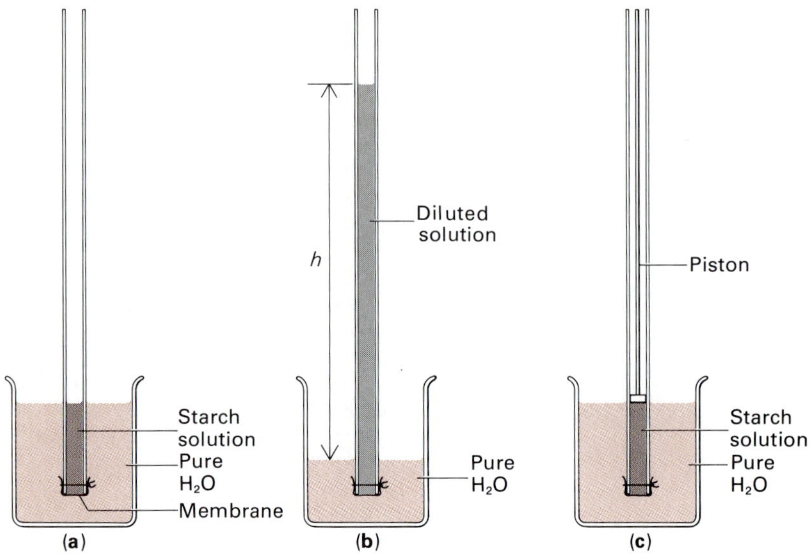

FIGURE 9-10

Osmotic pressure. **(a)** Initial state. **(b)** The excess pressure h exerted by the column of solution at equilibrium is the osmotic pressure. **(c)** The osmotic pressure is measured by applying just enough force with a piston to prevent the rise of solution in the tube.

In the experiment illustrated in Figure 9-10, a semipermeable cellophane membrane covers one end of the glass tube. The tube contains a concentrated starch solution. The beaker contains pure water. The cellophane is permeable to water but not to the much larger starch molecules. The water and the starch solution are *not* in equilibrium because of the unequal concentrations on the two sides.

Complete equilibrium would be complete uniformity. Because the starch molecules cannot move through the cellophane, this uniformity can be approached only by making the solution inside the sack more dilute. Thus water molecules move into the sack from the outside region of higher water concentration, seeking to reach an equilibrium in which the concentrations on the two sides of the cellophane are the same. The starch solution rises in the glass tube. Eventually the weight of the solution in the tube exerts a downward pressure that is just enough to balance the tendency of the water molecules to enter the sack, thereby preventing the further *net* entrance of water. (Water molecules still move in and out, but at the same rate.) This *excess* pressure exerted by the column of solution (over the atmospheric pressure exerted both on the water in the beaker and on the solution in the tube) is the **osmotic pressure** of the solution.

Osmotic pressure is measured by applying enough force to a piston to prevent the rise of the solution into the tube. Just as with vapor pressure, freezing-point lowering, and boiling-point elevation, *osmotic pressure in a dilute solution is proportional to the total concentration of all solute species, but is independent of their kind.*

The relationship between osmotic pressure and number of solute particles is identical in form to the ideal gas law:

$$\pi V = nRT$$

Here, π is the osmotic pressure, V is the volume of the solution, n is the number of moles of *all* solute particles in volume V, R is the gas constant, and T is the temperature in kelvins.

This relationship can be used to determine molecular weights, as shown in the following example.

EXAMPLE 9-6

At room temperature, the osmotic pressure of a solution containing 20 g of the protein lactoglobulin (a milk protein) per liter is 13.6 cm of water. Find the approximate molecular weight of lactoglobulin. (Use the rough values of 0.08 ℓ atm/mole deg for R, and 300 K for room temperature.)

The osmotic pressure of the solution is 13.6 cm H_2O. This corresponds to a pressure of 1.00 cm of Hg, because Hg is 13.6 times as dense as water (see Section 3-2):

$$13.6 \text{ cm } H_2O \times \frac{1.00 \text{ cm Hg}}{13.6 \text{ cm } H_2O} = 1.00 \text{ cm Hg}$$

$$\text{Osmotic pressure} = 1.00 \text{ cm Hg} = \frac{1.00 \text{ cm}}{76 \text{ cm/atm}} = \frac{1}{76} \text{ atm}$$

We can now calculate the number of moles of lactoglobulin per liter:

$$n = \frac{\pi V}{RT} = \frac{1/76 \text{ atm} \times 1 \ \ell}{0.08 \ \ell \text{ atm/mole K} \times 300 \text{ K}} = \frac{1}{1800} \text{ mole}$$

We know that 1/1800 mole weighs 20 g. Thus one mole weighs 1800 × 20 g = 36,000 g. The molecular weight of lactoglobulin is approximately 36,000 g/mole. Although this value seems very large in comparison to inorganic molecules, lactoglobulin is a relatively small protein.

Measurement of osmotic pressure is especially useful for determining molecular weights of proteins because it is very sensitive, and because it is a gentle procedure that most proteins can tolerate without undergoing chemical change. The sensitivity of osmotic pressure measurements is illustrated in the following calculation of the osmotic pressure of physiological saline solution.

EXAMPLE 9-7

Physiological saline solution is 0.15 M NaCl. What is the osmotic pressure of physiological saline solution at 27°C?

We'll use the equation $\pi = nRT/V$. Here, n/V is the concentration in moles/liter of *all* solute particles. Because each NaCl dissociates to form two ions,

$$\frac{n}{V} = 2 \times 0.15 \frac{\text{mole NaCl}}{\ell} = 0.30 \frac{\text{mole ions}}{\ell}$$

$$\pi = 0.30 \frac{\text{mole}}{\ell} \times 0.082 \frac{\ell \text{ atm}}{\text{mole K}} \times 300 \text{ K}$$

$$= 7.4 \text{ atm (equivalent to } 5.6 \times 10^3 \text{ mm of Hg)}$$

(The boiling-point elevation of this same solution, calculated in Example 9-3, is only 0.08°C.)

We see that a 0.15 M NaCl solution has an osmotic pressure of over 7 atm. Because pressures of as little as 10^{-3} atm can be measured precisely, the osmotic pressure of a protein solution of concentration as low as 1×10^{-4} M can be readily determined (see Problems

9-25, 9-26, and 9-28). Because many proteins have low solubilities, and because often only small amounts of protein are available, this method has been much used in determining molecular weights of proteins.

In working this problem, we used molarity, as usual, to express concentration, but we had to remember that here it is the number of moles of *ions* per liter of solution that is important. Physiologists sometimes use, instead, a concentration called osmolarity. One **osmole** is the molecular weight of the substance in grams divided by the number of particles each molecule provides in solution, and **osmolarity** is the number of osmoles per liter of solution:

$$1 \text{ mole of NaCl} = 58.5 \text{ g}$$

$$1 \text{ osmole of NaCl} = \frac{58.5}{2} \text{ g} = 29.3 \text{ g}$$

A 1 M solution of NaCl contains 58.5 g/ℓ of NaCl and gives 2 moles/ℓ of ions. A 1 osmolar solution of NaCl contains 29.3 g/ℓ of NaCl and gives 1 mole/ℓ of ions. A 2 osmolar solution of NaCl contains 58.5 g/ℓ of NaCl and gives 2 mole/ℓ of ions. (This is the same as the 1 M solution.)

There are 2 osmoles in 1 mole of NaCl, and the osmolarity of an NaCl solution is 2 times the molarity. The osmolarity of the 0.15 M NaCl solution is 0.30 osmolar.

The concentration in *osmolarity* can be used *directly* for the concentration n/V in the equation

$$\pi = \frac{n}{V} RT = 0.030\, RT = 7.4 \text{ atm}$$

Special terms are used to compare the osmotic pressures of two solutions: solutions that have equal concentrations of solute particles, and thus equal osmotic pressures, are **isotonic.** There is no *net* movement of water molecules across a membrane separating the two solutions. A **hypotonic** solution has a lower concentration of solute particles than the solution with which it is being compared; and a **hypertonic** solution has a higher concentration of solute particles than the solution with which it is being compared. The prefix *iso* means "the same"; *hypo* means "less" (solute particles); and *hyper* means "more" (solute particles).

Figure 9-11 shows red blood cells of a sheep in an isotonic solution, in a hypotonic solution, and in a hypertonic solution. In the hypotonic solution, where the salt concentration outside the cell is lower than that inside (the water concentration is higher), there is a net movement of water into the cells by osmosis. The cells swell, and some even burst. In the hypertonic solution, the situation is reversed; water moves out of the cell, causing the cell to shrivel. In the isotonic solution, whose osmotic pressure is identical to that of the cells, the cells retain their normal size.

Figure 9-12 illustrates how a paramecium manages to maintain an internal environment that is hypertonic with respect to the pond water in which it lives. Water moves into the cell by osmosis. Too much water would interfere with the biochemical reactions in the cell and might rupture the cell membrane. As shown in the photograph, water is expelled through a special organelle (called a contractile vacuole) that collects excess water and empties it through a small central pore.

EXERCISE 9-8 What is the osmotic pressure at body temperature (37°C) of a 5% by weight glucose solution? (The 5% glucose solution is 5 g of glucose in 100 g of solution. To make the calculation easier, assume that there are 5 g of glucose in 100 ml of solution, which is a close approximation.) The molecular weight of glucose is 180 g/mole.

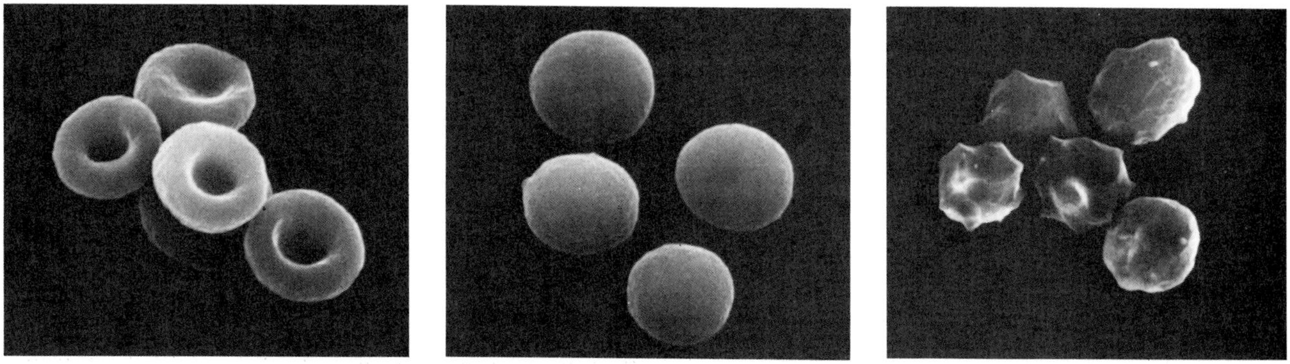

(a) (b) (c)

FIGURE 9-11

Red blood cells of a sheep. **(a)** In an isotonic solution. **(b)** In a hypotonic solution. **(c)** In a hypertonic solution. [Scanning electron micrographs by M. Cukierski, N. Lanza, and J. Maribei, New York State College of Veterinary Medicine, Cornell University.]

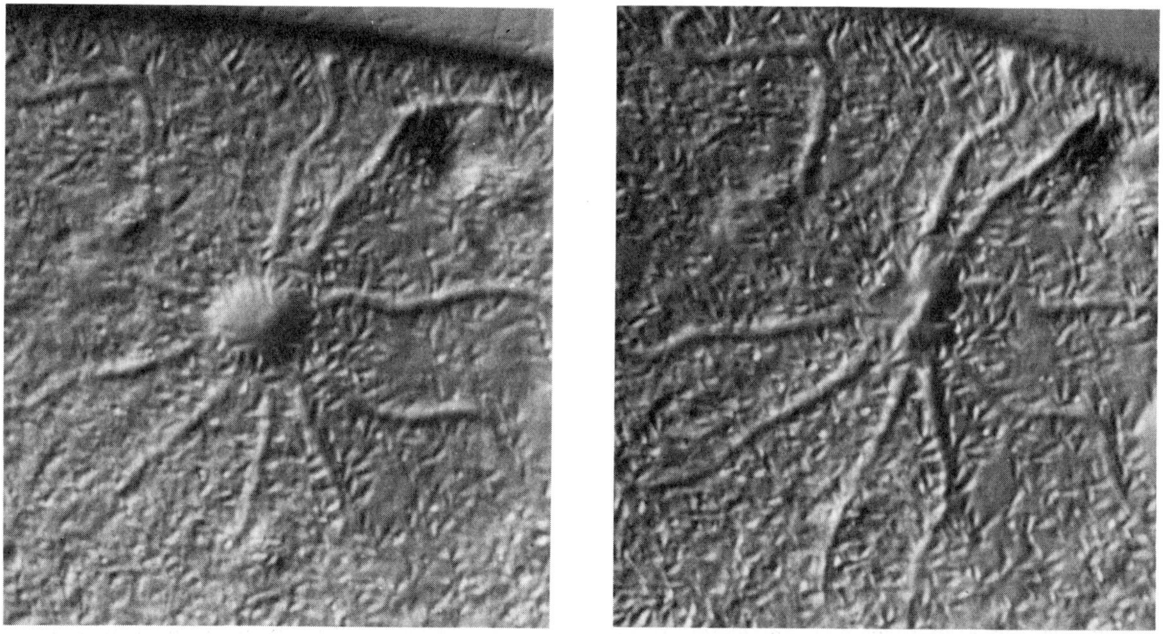

(a) (b)

FIGURE 9-12

The problem of osmotic pressure in living cells. The fluid inside the paramecium is hypertonic with respect to the pond water, so water flows into the paramecium by osmosis. Excess water in the paramecium is collected in tubules and is squeezed out by contraction through a small hole in the contractile vacuole. **(a)** A full contractile vacuole. **(b)** An empty contractile vacuole. [Photographs by Thomas Eisner, Cornell University.]

KEY TERMS

In a solution, if one substance is present in large excess, it is called the *solvent* (the substance that does the dissolving). The *solute* is the substance that is dissolved.

An *alloy* is a solid solution of a metal dissolved in a metal.

Molarity is the number of moles of solution in 1 ℓ of solution. *Molality* is the number of moles of solute in 1000 g of solvent.

Electrolytes are substances that dissolve in water to yield conducting solutions. Electrolytes dissociate to form ions in solution. *Nonelectrolytes* are substances that dissolve in water to yield nonconducting solutions. Nonelectrolytes retain their identity as molecules in solution. A *weak electrolyte* is only slightly dissociated into ions in solution.

A solution is *unsaturated* if its concentration can be increased by adding more solute. At the point where solution and excess solid are in equilibrium, and no additional solute can be dissolved, a solution is said to be *saturated*.

The *partial pressure* of the solvent in the vapor is that part of the total vapor pressure that is due just to the solvent molecules in the vapor.

A *semipermeable* membrane permits the passage of water molecules, but blocks the passage of most other molecules. *Osmosis* is the movement of water across a semipermeable membrane, from a region of higher water concentration to a region of lower water concentration.

If two solutions are *isotonic,* they have the same *osmotic pressure,* the pressure generated by the flow of water in osmosis. A *hypertonic* solution has a greater osmotic pressure than the solution with which it is being compared. A *hypotonic* solution has a lower osmotic pressure than the solution with which it is being compared.

CHAPTER SUMMARY

The properties of a solution vary with the concentration. Some common methods of expressing concentration are *molarity* (the method most often used by chemists), *molality,* percentage by weight, percentage by volume, and parts per million.

A rough rule of thumb that describes the effect on solubility of the relative natures of solvent and solute is "like dissolves like." Water, which is a polar solvent, is a good solvent for ionic solids and for polar molecular substances, especially those with which it can form hydrogen bonds.

Substances that dissolve in water to yield conducting solutions are *electrolytes,* and substances that dissolve to yield nonconducting solutions are *nonelectrolytes. Strong electrolytes* are essentially completely dissociated into ions in solution. *Weak electrolytes* are only slightly dissociated.

When excess solid is in contact with a saturated solution, there is a state of dynamic *equilibrium* between the solid and the dissolved solute particles.

In a solution containing a nonvolatile solute, the vapor pressure of the solvent is depressed. For a dilute solution, the depression of the vapor pressure below that of the pure solvent is proportional to the total concentration of solute particles, but is independent of their kind. This depression of the vapor pressure leads to several closely related phenomena: *boiling-point elevation, freezing-point depression,* and *osmosis*. These phenomena are important because they permit the determination of the molecular weight, and thus often also determination of the molecular formula of the solute.

PROBLEMS

9-1. What is the molarity of each of the following solutions?
(a) 2.0 moles of glucose in a volume of 6.0 ℓ
(b) 0.100 mole of NaCl in a volume of 250 ml
(c) 1.71 g of sucrose, $C_{12}H_{22}O_{11}$ (mol. wt. = 342), in 50.0 ml of solution
(d) 11.7 g of NaCl in 0.500 ℓ of solution

9-2. (a) How would you prepare 2.0 ℓ of physiological saline solution, 0.15 M NaCl? (What measurements of weight and volume would you make?)
(b) How would you prepare 2.0 ℓ of 1.5 M NaCl?
(c) How many liters of physiological saline solution could you make from 100 ml of 1.5 M NaCl? Describe how you would prepare the physiological saline solution.
(d) How many milliliters of 3.0 M NaCl solution would you need to make 1.0 ℓ of physiological saline solution?

9-3. How many moles, and how many grams, of solute are in each of the following solutions?
(a) 300 ml of 0.200 M glucose ($C_6H_{12}O_6$)
(b) 2.00 ml of 1.50 M potassium chloride (KCl)
(c) 2.00 ℓ of 0.350 M sulfuric acid (H_2SO_4)
(d) 5.0 ml of 18 M H_2SO_4
(e) 10 ml of 12 M hydrochloric acid (HCl)
(f) 25.0 ml of 6.00 M sodium hydroxide (NaOH)

9-4. (a) What mass of glucose ($C_6H_{12}O_6$, mol. wt. = 180 g/mole) is needed to prepare 5.00 ℓ of 0.28 M glucose solution? (This solution is 5% by weight glucose.)
(b) What mass of glucose must be taken to prepare 1.00 ℓ of 2.8 M glucose solution?

9-5. (a) What volume of 18 M H_2SO_4 is needed to prepare 1.0 ℓ of 6.0 M H_2SO_4?
(b) What volume of 12 M HCl is needed to prepare 0.50 ℓ of 1.0 M HCl?
(c) What volume of 6.0 M NaOH is needed to prepare 100 ml of 1.0 M NaOH?
(d) What volume of 2.8 M glucose is needed to prepare 2.0 ℓ of 0.28 M glucose solution?

9-6. (a) How many grams of glucose (mol. wt. = 180 g/mole) must be added to 1.00 kg of water to make a 2.00 molal solution?
(b) How many grams of sucrose (mol. wt. = 342 g/mole) must be added to 0.500 g of water to make a 1.0 molal solution?
(c) What is the percentage by weight of NaCl in a solution that contains 4.4 g NaCl in 500 g of solution?
(d) There are about 10 g of calcium in 1 ℓ of milk. What is the molarity of Ca^{2+} in milk?
(e) Urea, $(NH_2)_2CO$, collects in the blood in cases of kidney malfunction. In one case of uremic poisoning, the concentration of urea in the blood was 0.030 M. How many grams of urea were contained in a total blood volume of 5.5 ℓ?

9-7. (a) The solubility of sodium carbonate ("washing soda," Na_2CO_3) in water is 7.1 g per 100 ml of solution at 0°C. What is the molarity of a saturated solution of Na_2CO_3 at 0°C?
(b) What volume of 0.050 M Na_2CO_3 could you prepare using 454 g (1 lb) of washing soda?
(c) If you wanted to prepare 5.0 ℓ of 0.30 M Na_2CO_3, what mass of Na_2CO_3 would you need?

9-8. (a) What is the molality of a solution that contains 6.84 g of sucrose ($C_{12}H_{22}O_{11}$) in 400 g of water?
(b) What mass of sucrose is contained in 2000 g of water if the solution is 0.2 molal in sucrose?

9-9. A bottle of sherry is 20% ethanol (ethyl alcohol) by volume.
(a) What volume of pure ethanol (C_2H_5OH) is contained in 1.0 ℓ (just over 1 quart) of sherry?
(b) What is the molarity of ethanol in sherry? The density of pure ethanol is 0.785 g/ml.

9-10. Which of the following compounds would you expect to be electrolytes and which nonelectrolytes?
(a) sucrose, $C_{12}H_{22}O_{11}$
(b) barium hydroxide, $Ba(OH)_2$
(c) methanol, CH_3OH
(d) glycerol, $CH_2OHCHOHCH_2OH$
(e) sodium sulfide, Na_2S

9-11. A saturated solution of potassium nitrate (KNO_3) is prepared in 100 g of water at 50°C. The solution is allowed to cool. What mass of potassium nitrate will have precipitated out when the solution reaches room temperature (20°C)? (Consult Figure 9-3 to obtain the required data.)

9-12. Using information given in Table 9-2, suggest one method by which you might
(a) separate Na^+ from Ag^+
(b) separate Ba^{2+} from Cu^{2+}
(c) separate Pb^{2+} from Ag^+
(d) separate Mg^{2+} from K^+.

9-13. Use information given in Table 9-2 to answer these questions.
(a) When HCl is added to a solution of unknown composition, a precipitate forms. What cations might be present in the precipitate?
(b) When H_2SO_4 is added to another solution of unknown composition, no precipitate forms. What cations are definitely not present in the solution?
(c) When Na_2CO_3 is added to still another solution of unknown composition, no precipitate forms. What cations may be present in the original solution?
(d) When HCl is added to yet another solution of unknown composition, a precipitate forms. When H_2SO_4 is added to another portion of this same solution, a precipitate forms. What positive ion is definitely present in the original solution? What negative ions may be present in the original solution?

9-14. (a) Would you expect liquid ammonia to be a good solvent for ionic solids? Explain your answer.
(b) Would you expect water or *n*-hexane (C_6H_{14}) to be the better solvent for stearic acid, $CH_3(CH_2)_{16}COOH$, a fatty acid? Explain your answer.

9-15. (a) Blood cells are sometimes stored in physiological saline solution, but never in distilled water. Explain.
(b) If normal red blood cells are placed in 0.44% saline solution, they swell, and if they are placed in 0.25% saline solution they burst. Explain.
(c) What would happen to normal red blood cells if they were placed in 2% saline solution?

9-16. The root cells of plants contain a solution that is hypertonic in relation to soil water, so water moves into the roots by osmosis. Explain why salt spread on the roads to melt ice in winter sometimes kills nearby trees.

9-17. Some mammals, including humans, have developed the ability to excrete hypertonic urine in order to conserve water. In humans, just before the urine leaves the kidney, the collecting ducts of the kidney (which contain the urine) pass through a fluid that contains many times the salt concentration normally found in the blood and tissues. Explain how this helps conserve water in the body.

9-18. Some raw carrots were prepared for a picnic. Half of the carrots were stored in water, and because many people like their carrots salted, the other half were stored in salt water. When it was time to eat, the carrots stored in salt water were limp and rubbery but those stored in water were crisp. Explain.

9-19. You are given two unlabeled bottles, each containing a colorless liquid. One bottle contains 1-butanol, $CH_3(CH_2)_2CH_2OH$, and the other has 1,4-butanediol, $HOCH_2(CH_2)_2CH_2OH$. Suggest some methods by which you might determine which bottle contains which substance.

9-20. Draw a phase diagram (plot pressure against temperature) for pure water. On the *same* diagram, sketch in phase diagrams for
(a) a 1.0 molal solution of glucose
(b) a 1.0 molal solution of sodium chloride.

9-21. A certain aqueous solution freezes at $-4.65\,°C$. What is the boiling point of the solution? The freezing-point depression constant for water is 1.86°C/molal, and the boiling-point elevation constant is 0.52°C/molal.

9-22. The freezing point of a solution of 10 g of a molecular solid dissolved in 500 g of water is $-0.372\,°C$. What is the molecular weight of the solid? The freezing-point depression constant for water is 1.86°C/molal.

9-23. Pure camphor melts at 179°C. A mixture consisting of 5 g of camphor and 0.312 g of the organic compound *n*-decylcaprate ($C_{20}H_{40}O_2$) melts at 171°C. A mixture consisting of 5 g of camphor and 0.193 g of cholesterol melts at 175°C. What is the molecular weight of cholesterol?

9-24. At room temperature, the osmotic pressure of a solution containing 7.5 g of insulin per liter is 24.5 mm Hg at room temperature. Find the approximate molecular weight of insulin. Use the value 0.082 ℓ atm/mole K for R, and 300 K for the temperature.

9-25. Amylose, a soluble starch, has a molecular weight of about 30,000 g/mole. What, roughly, would be the osmotic pressure (at room temperature) of a solution containing 3.0 g of amylose dissolved in 1 ℓ of water? Use the value 0.082 ℓ atm/mole K for R, and 300 K for room temperature.

9-26. (a) What is the osmotic pressure at 0°C of a solution that is simultaneously 0.10 M in NaCl and 0.3 M in glucose?
(b) What is the osmotic pressure at 0°C of a solution that is 0.20 M in $CaCl_2$?
(c) Of the solutions in parts a and b, which has the higher boiling point, and which has the lower freezing point?
(d) What is the osmotic pressure at 0°C of a solution that is 1 M in solute particles?

9-27. A solution containing 2.00 g of human hemoglobin in 100 ml of water has an osmotic pressure at 0°C of 5.28 mm of Hg. What is the molecular weight of hemoglobin?

II

CHEMICAL REACTIONS

10

Chemical Energetics, Chemical Equilibria, and Chemical Kinetics

Energy changes accompany chemical changes. Some reactions give off energy; others absorb energy. Some chemical reactions are fast—like these fireworks; others are slow—like a car rusting. Every chemical reaction, however, reaches a state of dynamic equilibrium, with some of each product and some of each reactant present.

10-1
HEATS OF REACTION

All the energy we use to heat homes and power airplanes, as well as that to run a marathon and build new body cells, is derived from chemical reactions—from the breaking and making of chemical bonds. Energy is required to break chemical bonds, and energy is released when new bonds form.

In discussing changes of state in Chapter 4, we defined a quantity called the heat content H of a substance; H is related to, and very nearly the same as, the energy. The quantity ΔH for a change of state is defined as the difference between the final heat content and the initial heat content when a substance changes its state—say, from solid to liquid. We shall apply these same ideas to chemical reactions, using ΔH to represent the difference between the heat content of the products and that of the reactants:

$$\Delta H = \text{Heat content of products} - \text{Heat content of reactants}$$

The heat of reaction (ΔH) is the heat absorbed or given off in a reaction.

A significant chemical reaction in which energy is absorbed is the process of photosynthesis: chlorophyll molecules use the sun's energy to convert carbon dioxide and water into sugar and oxygen:

$$6H_2O + 6CO_2 + 2.82 \times 10^3\,\text{kJ} \longrightarrow \underset{\text{glucose}}{C_6H_{12}O_6} + 6O_2$$

The heat content of the products ($C_6H_{12}O_6$ and O_2) is greater than that of the reactants, so ΔH is positive. Such a reaction (one that absorbs heat and stores it as energy in the products) is called an **endothermic reaction.** In an endothermic reaction, heat is absorbed. The heat content of the products is greater than that of the reactants, and ΔH is positive (Figure 10-1). For this reaction, $\Delta H = 2.82 \times 10^3$ kJ/mole glucose.

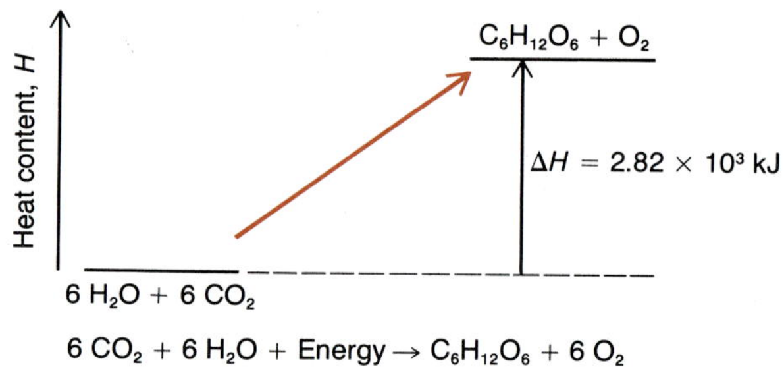

FIGURE 10-1

An endothermic reaction.

In the combustion of coal, energy is given off as heat and light. The chemical reaction is

$$C + O_2 \longrightarrow CO_2 + 393\,\text{kJ}$$

For each mole of coal burned, 393 kJ of energy are *released* (almost 17 million kJ per ton). The heat content of the product (CO_2) is less than that of the reactants (C and O_2), so ΔH is negative. A reaction that releases energy is an exothermic reaction. In an **exothermic reaction,** the heat content of the products is less than that of the reactants, and ΔH is negative (Figure 10-2). ΔH for the combustion of coal is -393 kJ/mole C. The reverse reaction, the formation of C and O_2 from CO_2, is endothermic; ΔH is $+393$ kJ/mole C.

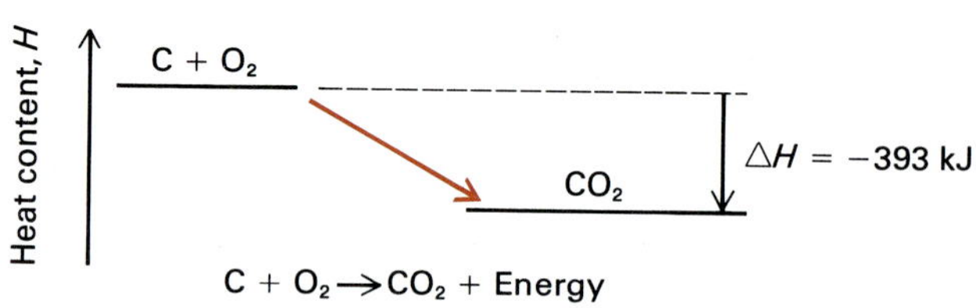

FIGURE 10-2

An exothermic reaction.

10-2
MEASURING HEATS OF REACTION

The energy of a chemical reaction is measured using a device like that shown in Figure 10-3, a simple **calorimeter.** A reaction between known amounts of reactants takes place in the reaction chamber. The reaction vessel is immersed in a large insulated container of water. Any heat absorbed or released in the reaction changes the temperature of the water. The amount of heat (the number of joules or calories) involved in the reaction can be calculated if the mass of the water and its temperature change are known, as shown in the following example.

EXAMPLE 10-1

One mole of $H_2(g)$ and one-half mole of $O_2(g)$ are introduced into the reaction vessel of a calorimeter. The insulated reservoir of the calorimeter contains 10.00 ℓ of water at an initial temperature of 21.50°C. The hydrogen and oxygen are then ignited with a spark to produce water, $H_2O(l)$. The final temperature of the water in the reservoir is 28.33°C. Calculate ΔH for the reaction

$$H_2(g) + \frac{1}{2}O_2(g) \longrightarrow H_2O(l)$$

The specific heat of water is 4.18 J/g deg, and its density is 1.00 g/ml.

Change in temperature of water = 28.33°C − 21.50°C = 6.83°C.

One liter of water has a volume of 1000 ml and, thus, a mass of 1000 g. The mass of 10.00 ℓ of water is 10,000 g.

$$\begin{aligned}\text{Calories } \textit{absorbed} \text{ by water} &= 10{,}000 \text{ g} \times 4.18 \frac{\text{J}}{\text{g deg}} \times 6.83 \text{ deg} \\ &= 2.85 \times 10^5 \text{ J} \\ &= 285 \text{ kJ}\end{aligned}$$

Because the temperature of the reservoir water has been raised, we know that 285 kJ are *released* per mole of $H_2O(l)$ formed. The reaction is exothermic.

$$\Delta H = -285 \text{ kJ/mole } H_2O(l)$$

The ΔH determined in Example 10-1, the heat content of $H_2O(l)$ relative to that of the elements $H_2(g)$ and $O_2(g)$, is called the **heat of formation** of liquid water from its elements. We'll use the symbol ΔH_f to represent the heat of formation of a substance from its elements:

$$\Delta H_{f,H_2O(l)} = -285 \text{ kJ/mole } H_2O(l)$$

Table 10-1 gives the heats of formation (ΔH_f) of several substances from their component elements. The data are given for all substances at 25°C and 1 atm. By convention, *the heat of formation of an element in its normal state at 25°C and 1 atm is defined as zero.*

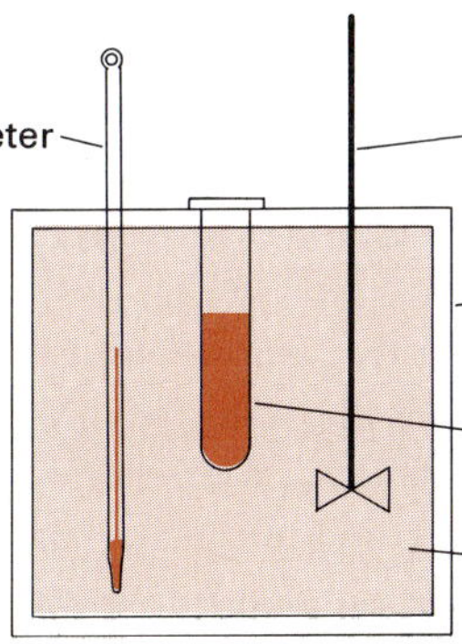

FIGURE 10-3
A simple calorimeter.

TABLE 10-1
Heats of formation of some substances

Name	Formula	ΔH_f (kJ/mole)	Name	Formula	ΔH_f (kJ/mole)
Acetic acid	$CH_3COOH(l)$	−484.5	Ethane	$C_2H_6(g)$	−84.68
Acetylene	$C_2H_2(g)$	+226.7	Ethylene	$C_2H_4(g)$	+52.26
Ammonia	$NH_3(g)$	−46.11	Glucose	$C_6H_{12}O_6(s)$	−1260
Calcium carbonate	$CaCO_3(s)$	−1207	Glutaric acid	$CH_2(CH_2CO_2H)_2$	−2154
Calcium oxide	$CaO(s)$	−635.1	Methane	$CH_4(g)$	−74.81
Carbon (diamond)	C(diamond)	+1.90	Methanol	$CH_3OH(l)$	−238.7
Carbon dioxide	$CO_2(g)$	−393.5	Sucrose	$C_{12}H_{22}O_{11}(s)$	−2221
Carbon monoxide	$CO(g)$	−110.5	Urea	$CO(NH_2)_2(s)$	−332.9
Chloral hydrate	$CCl_3CH(OH)_2(s)$	−576.1	Water	$H_2O(g)$	−241.8
Dimethyl ether	$(CH_3)_2O(g)$	−184.1	Water	$H_2O(l)$	−285.8

For example, the heats of formation of $H_2(g)$, $Cl_2(g)$, $Br_2(l)$, and Ca(s) are all zero. We are free to do this because the measured quantities are heats of *reaction,* ΔH, which, we shall see, are *differences* of heats of formation. Adopting this conventional reference point does not affect these differences.

10-3 CALCULATING HEATS OF REACTION

Given the heats of formation, we can calculate the heats of many chemical reactions, using the following generalization: *the heat of a chemical reaction is the same whether the reaction takes place in one or in several steps.* Then, if a reaction can be expressed as the sum of several reactions, the heat of the overall reaction is the sum of the heats of reaction of the individual steps. Figure 10-4 shows that the heat of combustion of carbon to carbon dioxide is the same, whether or not carbon monoxide is formed in an intermediate step.

We can apply this principle, using data given in Table 10-1, to find the heat of combustion of methane (CH_4) to carbon dioxide and water:

$$CH_4(g) + 2O_2(g) \longrightarrow CO_2 + 2H_2O(g)$$

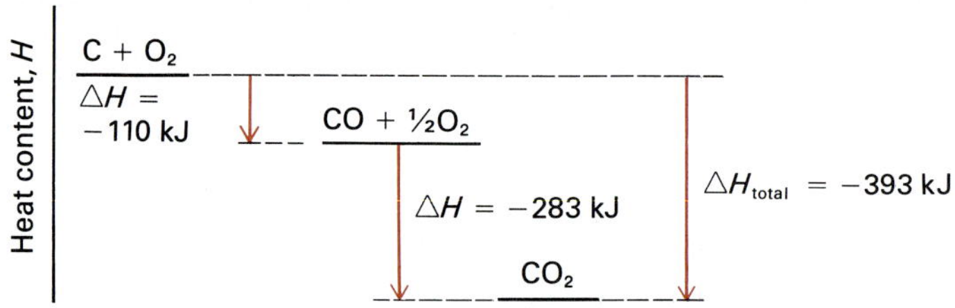

FIGURE 10-4
The heat of a reaction is the same whether the reaction takes place in one or several steps. For the reaction $C + O_2 \longrightarrow CO_2$, the value of $\Delta H = -393$ kJ/mole, whether the reaction takes place in one step or in two.

To begin, we find (in Table 10-1) the heat of formation of CH_4 from the elements:

$$\mathrm{C(s) + 2H_2(g) \longrightarrow CH_4(g)} \qquad \Delta H_{\mathrm{f,CH_4}} = -74.8\ \mathrm{kJ/mole\ CH_4} \qquad (10\text{-}1)$$

In the combustion reaction, however, the CH_4 appears as a reactant, on the left side of the equation. We can reverse the direction of Equation 10-1 to get the CH_4 on the left, *changing the sign of* ΔH. If 74.8 kJ/mole are released when CH_4 is formed from its elements, then that same 74.8 kJ/mole must be supplied to break CH_4 into its elements:

$$\mathrm{C(s) + 2H_2(g) \longrightarrow CH_4(g) + 74.8\ kJ}$$
$$\Delta H_{\mathrm{f,CH_4}} = -74.8\ \mathrm{kJ/mole\ CH_4}$$
$$\mathrm{CH_4(g) + 74.8\ kJ \longrightarrow C(s) + 2H_2(g)}$$
$$\Delta H = 74.8\ \mathrm{kJ} = -\Delta H_{\mathrm{f,CH_4}}$$

We can now obtain the desired overall reaction by adding the following three reactions; and the ΔH for the overall reaction by adding the heats of the individual reactions:

$$\mathrm{CH_4(g) \longrightarrow \cancel{C(s)} + \cancel{2H_2(g)}} \qquad \Delta H = +75\ \mathrm{kJ}$$
$$\mathrm{\cancel{C(s)} + O_2(g) \longrightarrow CO_2(g)} \qquad \Delta H_{\mathrm{f,CO_2}} = -394\ \mathrm{kJ}$$
$$\mathrm{\cancel{2H_2(g)} + O_2(g) \longrightarrow 2H_2O(g)} \qquad \Delta H = -484\ \mathrm{kJ} \qquad (10\text{-}2)$$

Net reaction:

$$\mathrm{CH_4(g) + 2O_2(g) \longrightarrow CO_2(g) + 2H_2O(g)} \qquad \Delta H_{\mathrm{net}} = -803\ \mathrm{kJ}$$

Equation 10-2 is written for the formation of *two* moles of water from the elements because we need two moles of water in the balanced equation. The $\Delta H_{\mathrm{f,H_2O}}$ also is doubled to give the heat produced in the formation of *two* moles of water:

$$2\ \mathrm{moles\ H_2O} \times (-242)\ \mathrm{kJ/mole\ H_2O} = -484\ \mathrm{kJ}$$

The net reaction for the combustion of methane is highly exothermic, as would be expected because methane is a good fuel. The heat of combustion calculated here is the same as the experimentally determined heat of combustion of methane (−803 kJ per mole of CH_4).

It is not necessary to go through all of the above steps to calculate the ΔH for a reaction. For any reaction,

$$\Delta H = \Delta H_{\mathrm{f,products}} - \Delta H_{\mathrm{f,reactants}}$$

where $\Delta H_{\mathrm{f,products}}$ and $\Delta H_{\mathrm{f,reactants}}$ are heats of formation, of *all* the products and *all* the reactants. In the case of the above reaction,

$$\begin{aligned}\Delta H &= \Delta H_{\mathrm{f,products}} - \Delta H_{\mathrm{f,reactants}} \\ &= \Delta H_{\mathrm{f,CO_2}} + 2\Delta H_{\mathrm{f,H_2O}} - \Delta H_{\mathrm{f,CH_4}} - 2\Delta H_{\mathrm{f,O_2}} \\ &= -394 + 2(-242) - (-75) - 2(0) \\ &= -803\ \mathrm{kJ\ per\ mole\ CH_4}\end{aligned}$$

Both methods give the same value for ΔH. (If you compare the two methods carefully, you can convince yourself that they are really the same.)

EXAMPLE 10-2

Using data in Table 10-1, determine ΔH for the combustion of sucrose ($C_{12}H_{22}O_{11}$) to carbon dioxide (CO_2) and liquid water, $H_2O(l)$. (This is the same reaction that occurs in our bodies in the process of catabolism, the breaking down of large food molecules into CO_2 and H_2O. But in our bodies, in contrast to the calorimeter, the process occurs at 37°C, in a series of small steps that liberate energy in usable amounts.)

First the equation must be balanced:

$$C_{12}H_{22}O_{11}(s) + 12O_2(g) \longrightarrow 12CO_2(g) + 11H_2O(l)$$

$$\begin{aligned} \Delta H &= 12\,\Delta H_{f,CO_2} + 11\,\Delta H_{f,H_2O} \\ &\quad - \Delta H_{f,C_{12}H_{22}O_{11}} - 12\,\Delta H_{f,O_2} \\ &= 12(-393.5\text{ kJ}) + 11(-285.8\text{ kJ}) \\ &\quad - (-2221\text{ kJ}) - 12(0) \\ &= -5.645 \times 10^3\text{ kJ/mole sucrose} \end{aligned}$$

The reaction is highly exothermic, as we might have expected because sucrose is a good source of energy.

EXERCISE 10-1 (a) Using data from Table 10-1, determine ΔH for the combustion of glucose ($C_6H_{12}O_6$) to carbon dioxide and liquid water. (Be sure to balance the equation.) Compare your answer with the heat of combustion of sucrose calculated in Example 10-2.

(b) Now calculate ΔH of combustion per *gram* of glucose and ΔH per *gram* of sucrose, and compare them. The molecular weight of sucrose is 342 g/mole, and that of glucose is 180 g/mole. (Both glucose and sucrose are carbohydrates. Typically, the energy obtained from 1 g of carbohydrate is about 17 kJ. This same amount of energy is produced when 1 g of carbohydrate is catabolized, or broken down, in the human body.)

10-4 CHEMICAL EQUILIBRIUM

We have already seen that a state of dynamic equilibrium can exist between different phases of a pure substance, between a solution and its vapor, and between solid solute and solute particles in solution. We have not yet considered equilibrium in connection with chemical reactions. Instead, we have thought of most chemical reactions as going to completion.

We shall now see that a chemical reaction never goes to completion but always goes to a condition of equilibrium, although it may sometimes *appear* to go to completion. Consider the reaction

$$2H_2(g) + O_2(g) \rightleftharpoons 2H_2O(g)$$

where the double arrows represent forward and backward reactions. At room temperature, the forward reaction ($2H_2 + O_2 \rightarrow 2H_2O$) appears to go to completion. If hydrogen and oxygen are mixed in correct proportions and the reaction initiated (say, by a spark), *all* the hydrogen and oxygen *seem* to be converted (explosively) to water. At very high temperatures, however, there is clearly a condition of dynamic equilibrium. Hydrogen, oxygen and water are all present simultaneously. Water decomposes to hydrogen and oxygen, and hydrogen and oxygen are at the same time combining to form water, at the same rate. Such an equilibrium also exists at lower temperatures, but then the *position* of equilibrium lies so far to the right—products are present in much greater amounts than reactants—that the amount of uncombined hydrogen and oxygen present is imperceptible.

We can see this fact more easily in the equilibrium between the brown gas nitrogen dioxide (NO_2) and the colorless gas dinitrogen tetroxide (N_2O_4), as shown

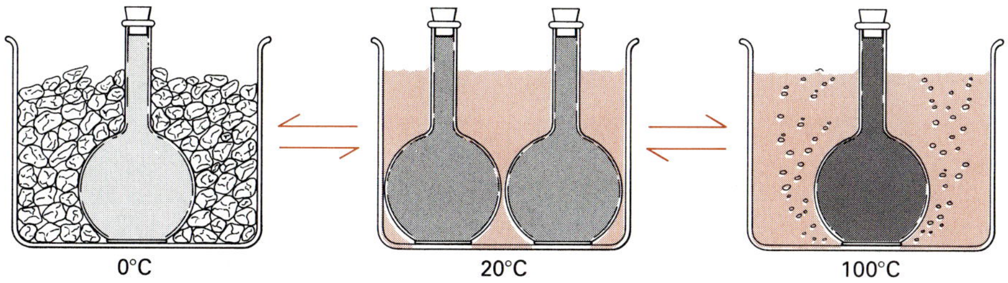

FIGURE 10-5

Effect of temperature on equilibrium for the reaction

$$\underset{\text{(colorless)}}{N_2O_4\ (g)} \rightleftharpoons \underset{\text{(brown)}}{2NO_2\ (g)}$$

in the experiment illustrated in Figure 10-5. Suppose an equilibrium mixture of the two gases is prepared at room temperature, and two identical flasks are filled with the same amount of the mixture. The flasks are then sealed. The color of the mixture is pale brown. One flask is then immersed in a water bath at 100°C, and the other in a water bath at 0°C.

The intensity of the color of the gas changes with the changes in temperature. Equilibrium has been reached when the intensity of the color of the gases in the flasks no longer changes with time but remains constant. At equilibrium, the gas at 0°C is almost colorless, showing that there is less NO_2 in the flask than there was at room temperature. The gas at 100°C is a darker brown than it was at room temperature, showing that more NO_2 is present. The reactions that are occurring here are

$$N_2O_4(g) \rightleftharpoons 2NO_2(g)$$

At the lower temperature the reverse reaction is favored, and the equilibrium lies far to the left (toward the N_2O_4), but *some* NO_2 is still present, as shown by the pale color. At the higher temperature, the forward reaction is favored, and the equilibrium lies far to the right (toward the NO_2), but *some* N_2O_4 is still present. We see here that temperature has an effect on the position of equilibrium. We shall later see why that is so.

The position of equilibrium also shifts with changes in concentration, as illustrated in the experiment shown in Figure 10-6. The beakers contain aqueous solutions of the ionic solids ferric nitrate, $Fe(NO_3)_3$, and sodium thiocyanate, NaSCN. These solids dissolve in water to yield Fe^{3+}, NO_3^-, Na^+, and SCN^- ions. The Fe^{3+} and the SCN^- react, when mixed, to yield a deep red substance in solution:

$$\underset{\text{(pale yellow)}}{Fe^{3+}} + \underset{\text{(colorless)}}{3SCN^-} \rightleftharpoons \underset{\text{(deep red)}}{Fe(SCN)_3}$$

The Na^+ and NO_3^- ions do not react, but remain as ions in solution. The red color is actually a mixture of species, ranging from the complex ion $FeSCN^{2+}$ to $Fe(SCN)_6^{3-}$, but $Fe(SCN)_3$ is an adequate representative. The red color of $Fe(SCN)_3$ is used as a qualitative test to detect the Fe^{3+} ion.

To start, a red solution containing some $Fe(NO_3)_3$ and NaSCN is distributed among four beakers. Beaker A contains only this solution, and it is set aside as a standard for purposes of comparison.

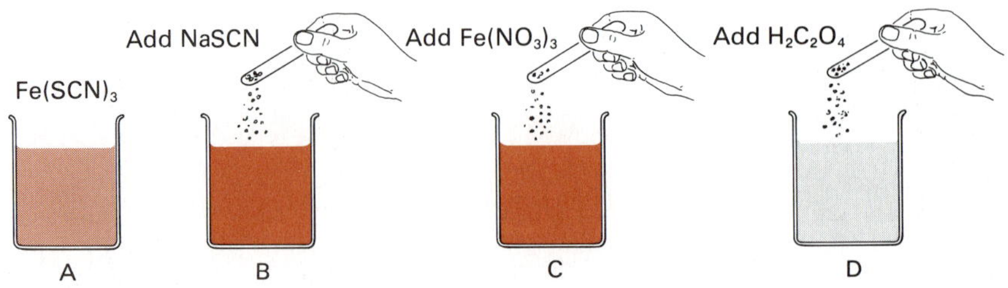

FIGURE 10-6

Effect of concentration on equilibrium for the reaction

$$\underset{\text{(pale yellow)}}{Fe^{3+}} + \underset{\text{(colorless)}}{3SCN^-} \rightleftharpoons \underset{\text{(red)}}{Fe(SCN)_3}$$

To beaker B we add NaSCN, but no more $Fe(NO_3)_3$. The solution in beaker B turns a deeper red than the standard, showing that more $Fe(SCN)_3$ has been formed (and that some Fe^{3+} remained free in the standard to react with this added SCN^-). The position of equilibrium shifts to the right, toward more $Fe(SCN)_3$.

To beaker C we add $Fe(NO_3)_3$, but no more NaSCN. Again, the solution turns a deeper red than the standard, showing that more $Fe(SCN)_3$ has been formed (and that some SCN^- remained free in the standard to react with this added Fe^{3+}). The position of equilibrium shifts to the right, toward more $Fe(SCN)_3$.

To beaker D we add crystals of oxalic acid, $H_2C_2O_4$. The solution turns almost colorless, showing that very little $Fe(SCN)_3$ remains. The oxalate ion, $C_2O_4^{2-}$, reacts with Fe^{3+} in solution to form a colorless complex ion, $Fe(C_2O_4)_3^{3-}$, that is extremely stable. This has the effect of removing Fe^{3+} from the solution (most of the Fe^{3+} is no longer free in solution to combine with SCN^-). The position of equilibrium shifts to the left, toward more of the reactants and *less* $Fe(SCN)_3$.

This experiment shows that *increasing* the concentration of a reactant (Fe^{3+} or SCN^-) shifts the position of equilibrium to the right, toward formation of more product. *Decreasing* the concentration of a reactant (Fe^{3+}) shifts the position of equilibrium to the left, toward the reactants. We see that in the standard solution (and the same is true in all the solutions), some of *each* of the reactants is present along with the product. There is always an equilibrium between reactants and products, but the position of equilibrium shifts with changes in concentration.

EXERCISE 10-2 In the experiment shown in Figure 10-5, there is still some N_2O_4 present at 100°C. Describe an experiment that would demonstrate the presence of N_2O_4 at 100°C.

EXERCISE 10-3 In the experiment shown in Figure 10-6, how could you demonstrate that there is *some* Fe^{3+} and *some* SCN^- present in each of the solutions?

10-5
CHEMICAL EQUILIBRIUM AND LE CHÂTELIER'S PRINCIPLE

If a system at equilibrium is subjected to some sort of stress—a change in temperature, pressure, or concentration, for example—the position of equilibrium changes. The direction of the change can be predicted on the basis of **Le Châtelier's principle:** *when a solution at equilibrium is subjected to a stress, the equilibrium shifts in such a way as to relieve the stress.* We shall consider the effects of changes of concentration, pressure, and temperature on reactions at equilibrium.

We saw in Figure 10-6 that when one of the reactants (Fe^{3+} or SCN^-) is added to the system at equilibrium, the position of equilibrium shifts to the right:

$$Fe^{3+} + 3SCN^- \rightleftharpoons Fe(SCN)_3$$

The shift in equilibrium is in the direction of *consuming* the added substance and restoring the system to equilibrium, as predicted on the basis of Le Châtelier's principle. On the other hand, when Fe^{3+} is removed by the addition of oxalate ion, the position of equilibrium shifts to the left, producing more Fe^{3+}—thus partially restoring the depleted Fe^{3+} concentration, again in accord with Le Châtelier's principle.

If one or more of the reactants or products in a chemical reaction is gaseous, a change in the pressure of the gas may also affect the position of equilibrium. Increasing the pressure by decreasing the volume of the reaction container shifts the equilibrium in the direction of *fewer* gas molecules, thereby relieving the stress of the increased pressure. Thus, in the reaction $N_2O_4 \rightleftharpoons 2NO_2$, an increase in pressure shifts the position of equilibrium to the left, in the direction of fewer gas molecules (and of lower pressure). A decrease in pressure shifts the position of equilibrium to the right, in the direction of more gas molecules and of higher pressure. (Increasing the total gas pressure by adding an inert gas to the reaction container, however, has *no effect* on the equilibrium between N_2O_4 and NO_2.)

The reaction $N_2O_4 \rightleftharpoons 2NO_2$ is endothermic, with $\Delta H = 59.0$ kJ/mole N_2O_4. That is

$$N_2O_4 + 59.0 \text{ kJ} \rightleftharpoons 2NO_2$$

We saw that, when the temperature is raised (when heat is added to the system), more NO_2 is produced (Figure 10-5). The position of equilibrium shifts in the direction of *consuming* the added heat, thereby relieving the stress. *If a reaction is endothermic (heat is absorbed), the position of equilibrium shifts toward products (to the right) with increasing temperature. If a reaction is exothermic (heat is released), the position of equilibrium shifts toward reactants (to the left) with increasing temperature.*

In the following example, the Haber process (the tremendously important reaction in which atmospheric nitrogen is converted to ammonia) is used to illustrate the application of Le Châtelier's principle to chemical equilibria.

EXAMPLE 10-3

(a) The Haber process is summarized by the following equation:

$$3H_2(g) + N_2(g) \overset{\text{Fe}}{\rightleftharpoons} 2NH_3(g) + 92 \text{ kJ}$$

$$\Delta H = -46 \frac{\text{kJ}}{\text{mole } NH_3 \text{ produced}}$$

What would be the effect on the equilibrium concentration of NH_3 if the temperature were increased?

Because the reaction is exothermic (negative ΔH), increasing the temperature shifts the position of equilibrium to the left, decreasing the equilibrium concentration of NH_3. Low temperature favors the production of NH_3. The effect of increasing temperature on

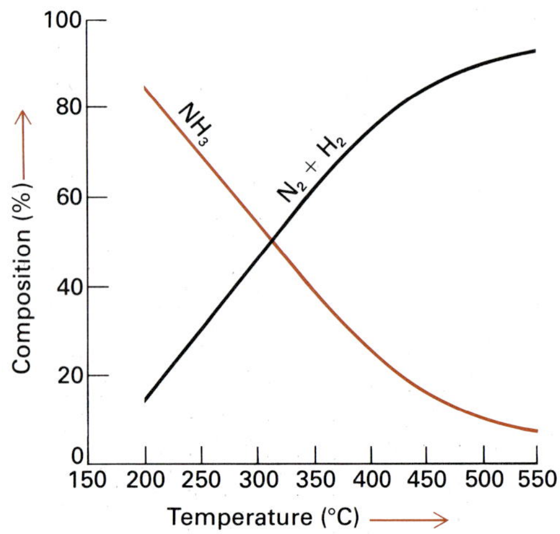

FIGURE 10-7

Effect of temperature on the position of equilibrium for the reaction $3H_2 + N_2 \rightleftharpoons 2NH_3$. The higher the temperature, the lower the percentage of NH_3 at equilibrium. [From V. Haensel and R. Burwell, Jr., "Catalysis." Copyright © 1971 by Scientific American, Inc. All rights reserved.]

an equilibrium mixture of H_2, N_2, and NH_3 is shown graphically in Figure 10-7. Increasing the temperature increases the *rate* of the reaction, which is desirable in an industrial process, but decreasing the temperature increases the equilibrium concentration of ammonia. Therefore, in practice, a compromise is struck at a temperature of about 500°C. This gives both a reasonable rate and a reasonable yield of NH_3.

(b) What would be the effect on the equilibrium concentration of NH_3 of increasing the pressure by decreasing the volume of the container?

Increasing the pressure will shift the position of equilibrium in the direction of *fewer* gas molecules—that is, toward more NH_3. Thus, high pressure favors the production of NH_3. A pressure of about 350 atm is used in the industrial process.

(c) What would be the effect on the total amount of NH_3 produced if NH_3 is removed from the reaction mixture by liquefying it at a temperature of -34°C, just below its boiling point? (H_2 has a boiling point of -253°C, and N_2 of -196°C.)

At -34°C, when the NH_3 has been liquefied and can be drained from the reaction vessel, H_2 and N_2 are still gases, so the forward reaction is favored, and more N_2 and H_2 will be converted to NH_3.

EXERCISE 10-4 Because of the ability of hemoglobin in the red blood cells to bind oxygen, whole blood absorbs about 250 ml of O_2 per liter. Blood plasma (without the red cells) can dissolve only 5 ml of O_2. The manner in which hemoglobin binds O_2 involves a complex series of equilibria (discussed in Section 22-7), but some information can be obtained by summarizing the equilibria by the equation

$$HHb^+ + O_2(g) \rightleftharpoons HbO_2 + H^+$$

where Hb represents one of four subunits of the hemoglobin molecule, HbO_2 is the oxygenated form of hemoglobin, and HHb^+ is a protonated form (it has an extra proton) of deoxygenated hemoglobin. The structure of the hemoglobin molecule is given in Figure 21-7.

(a) How would decreasing the partial pressure of O_2 in the lungs (say, by climbing Mount Everest) affect the concentration of HbO_2 formed in the lungs?

(b) Protons (H^+) are formed in the tissues as a product of metabolism. How does the increase in concentration of H^+ in the tissues affect the concentrations of oxygenated and deoxygenated hemoglobin?

(c) Oxygen is used for chemical reactions in the tissues. Is the concentration of HbO_2 greater in the tissues or in the lungs? Is the concentration of HHb^+ greater in the tissues or in the lungs?

10-6
LAW OF MASS ACTION

We have observed, qualitatively, the changeability of chemical equilibria, and we have seen that the position of equilibrium is shifted by changes in temperature or concentration. There is a *quantitative* expression that relates the amounts of reactants and products present in a system at equilibrium.

The N_2O_4–NO_2 equilibrium has been studied quantitatively in a series of experiments performed at the same temperature. In each trial a different *initial* concentration of N_2O_4 and NO_2 was used. The initial and final concentrations of each substance are shown in Table 10-2. We use the symbol [A] to indicate the concentration of a substance A.

The ratio $[NO_2]^2/[N_2O_4]$, in the last column of the table, has been calculated for the *final*, or equilibrium, concentrations of NO_2 and N_2O_4. Note that it is constant for all the trials. In fact, it is found to remain constant for every initial concentration of NO_2 and N_2O_4 at 8°C. Thus, we can write

$$\frac{[NO_2]^2}{[N_2O_4]} = \text{Constant} = 1.6 \times 10^{-3}\,\frac{\text{mole}}{\text{liter}} \quad \text{at } 8°\text{C}$$

The balanced equation for the reaction is

$$N_2O_4 \rightleftharpoons 2NO_2$$

The constant ratio is the square of the equilibrium concentration of the product, divided by the equilibrium concentration of the reactant.

Experiments show that for any reaction at a given temperature, there is a definite relationship between the equilibrium concentrations of the reactants and products. Consider the general reaction

$$aA + bB + \cdots \longrightarrow xX + yY + \cdots$$

in which a molecules of A react with b molecules of B, etc., to yield x molecules of X and y molecules of Y, etc. The letters A, B, . . . , X, Y, . . . represent formulas of reactants and products, and a, b, . . . , x, y, . . . represent the coefficients of the corresponding molecules in the balanced equation. The ratio that is constant at equilibrium is

$$\frac{[X]^x[Y]^y \cdots}{[A]^a[B]^b \cdots}$$

where [A], [B], . . . , [X], [Y], . . . are the *equilibrium* concentrations of reactants and products. This ratio is called the **mass action expression.**

At equilibrium, at a given temperature T, the law of mass action states that

$$\frac{[X]^x[Y]^y \cdots}{[A]^a[B]^b \cdots} = \text{Constant} = K_T$$

TABLE 10-2
Equilibrium concentrations of N_2O_4 and NO_2 at 8°C

	Initial concentrations		Final concentrations		
Experiment	$[N_2O_4]$(M)	$[NO_2]$(M)	$[N_2O_4]$(M)	$[NO_2]$(M)	Final $\frac{[NO_2]^2}{[N_2O_4]}$ (M)
1	0.100	0	0.094	0.012	1.6×10^{-3}
2	0.030	0	0.027	0.0066	1.6×10^{-3}
3	0	0.08	0.036	0.0076	1.6×10^{-3}
4	0.040	0.020	0.046	0.0086	1.6×10^{-3}

where K_T is the equilibrium constant at temperature T. In the $N_2O_4 \rightleftharpoons 2NO_2$ reaction, $x = 2$ and $a = 1$, and the equilibrium constant expression is

$$K_T = \frac{[X]^x}{[A]^a} = \frac{[NO_2]^2}{[N_2O_4]}$$

When the concentrations are given in moles/liter, the equilibrium constant for the N_2O_4–NO_2 reaction has the units of moles per liter. An equilibrium constant usually has units associated with it. But if $x + y + \cdots = a + b + \cdots$, then it is dimensionless. All the units cancel out.

The following example illustrates applications of the law of mass action.

EXAMPLE 10-4

(a) Hydrogen and iodine react slowly to produce hydrogen iodide according to the equation,

$$H_2(g) + I_2(g) \rightleftharpoons 2HI(g)$$

When a mixture of H_2, I_2, and HI is permitted to come to equilibrium at 475°C, the equilibrium concentrations are $[H_2] = 4.53 \times 10^{-3}$ M, $[I_2] = 0.74 \times 10^{-3}$ M, and $[HI] = 12.8 \times 10^{-3}$ M. What is the value of the equilibrium constant K for the reaction at 475°C?

$$K_{475°C} = \frac{[HI]^2}{[H_2][I_2]}$$

$$= \frac{(12.8 \times 10^{-3})^2}{(4.53 \times 10^{-3})(0.74 \times 10^{-3})} = 49$$

This is *the* value of K for this reaction at 475°C, and it applies however the equilibrium may be reached. In particular, then, this is the value of K that applies in the following problem.

(b) If, to begin with, there were 0.0090 moles of HI in a one-liter vessel, but no H_2 or I_2, how many moles of H_2 would be present by the time equilibrium was reached at 475°C?

From the balanced equation $H_2 + I_2 \rightleftharpoons 2HI$, we see that for each molecule of H_2 produced in the reverse reaction, *one* molecule of I_2 is also produced, and *two* molecules of HI have decomposed. Therefore, at equilibrium,

$$[H_2] = [I_2]$$

$$[HI] = 0.0090 - 2[H_2]$$

$$K_{475°C} = \frac{(0.0090 - 2[H_2])^2}{[H_2][I_2]}$$

$$= \frac{(0.0090 - 2[H_2])^2}{[H_2]^2} = 49$$

Taking the square root of both sides of the equation, we get

$$\frac{0.0090 - 2[H_2]}{[H_2]} = 7.0$$

$$0.0090 = 9.0[H_2]$$

$$[H_2] = 0.0010 \text{ M}$$

At equilibrium, there are 0.0010 moles of $H_2(g)$ in the one-liter vessel.

EXERCISE 10-5 In still another experiment involving the reaction discussed in Example 10-4, a flask initially filled with pure HI is allowed to come to equilibrium at 475°C. The equilibrium concentration of H_2 is 0.20 M. What are the equilibrium concentrations of I_2 and HI?

We now consider the effect on the mass action expression of applying a stress to a system at equilibrium. For the reaction

$$Fe^{3+} + 3SCN^- \rightleftharpoons Fe(SCN)_3$$

the law of mass action is

$$\frac{[Fe(SCN)_3]}{[Fe^{3+}][SCN^-]^3} = K_T$$

If this system is initially at equilibrium and then the equilibrium is disrupted by adding more of one of the reactants (say, Fe^{3+}), the mass action expression $[Fe(SCN)_3]/[Fe^{3+}][SCN^-]^3$ momentarily becomes smaller than K_T (because of the larger $[Fe^{3+}]$ in the denominator). The system returns to equilibrium as some Fe^{3+} reacts with SCN^- to form more $Fe(SCN)_3$. This increases the concentration of $Fe(SCN)_3$ and decreases that of SCN^- (the reaction shifts to the right). Finally, once again, a condition of equilibrium is reached, and the mass action expression is once more equal to K_T.

If the equilibrium is disrupted by adding oxalate ion ($C_2O_4^{2-}$) to *remove* Fe^{3+}, the mass action expression is momentarily larger than K_T (because of the lower concentration of Fe^{3+}). The system returns to equilibrium as $Fe(SCN)_3$ dissociates to provide more Fe^{3+}, thereby increasing the concentration of SCN^- and decreasing the concentration of $Fe(SCN)_3$. That is, the reaction shifts to the left. Finally, once again, a condition of equilibrium is reached, and the mass action expression equals K.

If a system at equilibrium is disturbed in such a way that, momentarily, the mass action expression exceeds K, reaction occurs spontaneously to the left to return to equilibrium.

If a system at equilibrium is disturbed in such a way that, momentarily, the mass action expression is less than K, reaction occurs spontaneously to the right to return to equilibrium.

Table 10-3 lists equilibrium constants for some reactions, each at a specified temperature.

TABLE 10-3
Equilibrium constants for some reactions

Reaction	Equilibrium constant, K_T
$2H_2O(g) \rightleftharpoons 2H_2(g) + O_2(g)$	$K_{1727°C} = 5.31 \times 10^{-10}$ mole/ℓ
$3H_2(g) + N_2(g) \rightleftharpoons 2NH_3(g)$	$K_{727°C} = 2.37 \times 10^{-3}$ (ℓ/mole)^2
$H_2(g) + I_2(g) \rightleftharpoons 2HI(g)$	$K_{475°C} = 49$
$2CO(g) + O_2(g) \rightleftharpoons 2CO_2(g)$	$K_{727°C} = 2.24 \times 10^{22}$ ℓ/mole
$N_2O_4(g) \rightleftharpoons 2NO_2(g)$	$K_{8°C} = 16 \times 10^{-4}$ mole/ℓ
Ca-citrate(aq) $\rightleftharpoons Ca^{2+}$ + citrate^{2-}	$K_{25°C} = 2.1 \times 10^{-5}$ mole/ℓ
Pb-EDTA^{2-}(aq) $\rightleftharpoons Pb^{2+}$ + EDTA^{4-}	$K_{25°C} = 5.0 \times 10^{-19}$ mole/ℓ
Sucrose(aq) + $H_2O \rightleftharpoons$ glucose(aq) + fructose(aq)	$K_{25°C} = 1.36 \times 10^{5}$ mole/ℓ
β-D-glucose(aq) $\rightleftharpoons$ α-D-glucose(aq)	$K_{25°C} = 0.57$

Note: Here "(aq)" means that the substance is in water solution. The citrate ion is the anion of citric acid, the acid in lemons. The structure of Pb-EDTA is shown in Figure 11-6. The structures of sucrose, glucose, and fructose are given in Chapter 23.

EXAMPLE 10-5

Consider again the following reaction at 475°C:

$$H_2(g) + I_2(g) \rightleftharpoons 2HI \qquad K_{475^\circ C} = 49$$

If momentarily the concentrations are $[H_2] = 6.50 \times 10^{-3}$ M, $[I_2] = 2.80 \times 10^{-3}$ M, and $[HI] = 27.0 \times 10^{-3}$ M, will a reaction occur spontaneously? If so, what will be the direction of the reaction?

The mass action expression at these concentrations has the value

$$\frac{[HI]^2}{[H_2][I_2]} = \frac{(27.0 \times 10^{-3})^2}{(6.50 \times 10^{-3})(2.80 \times 10^{-3})} = 40$$

The system is not at equilibrium because the mass action expression is not equal to K_T. Therefore a reaction will occur.

The mass action expression has the value 40, which is *smaller* than K_T ($K_{475^\circ C} = 49$). Therefore, reaction occurs spontaneously to the right, increasing the concentration of HI, and decreasing those of H_2 and I_2, until the mass action expression is equal to 49. That is, a spontaneous reaction occurs in the forward direction.

EXERCISE 10-6 In the same reaction, if momentarily the concentrations are $[H_2] = 1.0 \times 10^{-3}$ M, $[I_2] = 1.0 \times 10^{-3}$ M, and $[HI] = 7.5 \times 10^{-3}$ M, will a reaction occur spontaneously? If so, what will be the direction of the reaction?

10-7 EQUILIBRIA INVOLVING PURE SOLIDS, PURE LIQUIDS, OR THE SOLVENT IN A DILUTE SOLUTION

We have written mass action expressions for reactions involving gases or solutes in dilute solutions, but we have not yet considered a reaction in which one or more of the reactants is a pure solid, a pure liquid, or the *solvent* in a dilute solution. An example of such a reaction is that between hydrogen gas and solid sulfur to produce hydrogen sulfide gas:

$$H_2(g) + S(s) \rightleftharpoons H_2S(g)$$

The mass action expression for this reaction is

$$\frac{[H_2S]}{[H_2]} = K$$

Why is the solid sulfur not included in the expression? The concentration (essentially the density) of a pure solid, pure liquid, or the solvent in a dilute solution depends only on the temperature. Thus, at any one temperature, the *effective* concentration of the pure solid, pure liquid, or the solvent is just a constant. By universally accepted convention, this constant is combined with the equilibrium constant. So *there is no explicit reference to pure solids, pure liquids, or solvents in the mass action expression.*

When limestone ($CaCO_3$) is heated, it decomposes to lime (CaO) and carbon dioxide gas:

$$CaCO_3(s) \rightleftharpoons CaO(s) + CO_2(g)$$

The mass action expression for the reaction is simply

$$[CO_2(g)] = K$$

Again, there is no explicit reference to the pure solids ($CaCO_3$ and CaO) in the mass action expression; their effective concentrations are included in K. The equilibrium constant for the decomposition of limestone is equal to the concentration (or the pressure) of the carbon dioxide at equilibrium. This means that, at any fixed temperature, there is a constant concentration of $CO_2(g)$ in any equilibrium mixture of $CaCO_3(s)$, CaO(s), and $CO_2(g)$.

When chlorine gas dissolves in water, the products are hypochlorous acid (HOCl) and H^+ and Cl^-, all in solution:

$$Cl_2(g) + H_2O(l) \rightleftharpoons H^+ + Cl^- + HOCl$$

The mass action expression for this reaction (in which the solvent, water, is a reactant) is

$$\frac{[H^+][Cl^-][HOCl]}{[Cl_2(g)]} = K$$

Some water reacts, but because water is the solvent, it is present in high concentration; the small amount that reacts is negligible in comparison. Thus, there is no explicit reference to the solvent (water) in the mass action expression; its constant effective concentration is included in K.

EXERCISE 10-7 (a) Write the mass action expression for the reaction between sucrose and water to yield fructose and glucose (see Table 10-3).

(b) Write the mass action expression for the reaction

$$4Fe(s) + 3O_2(g) \rightleftharpoons 2Fe_2O_3(s)$$

10-8 SOLUBILITY PRODUCTS

We saw in Section 9-4 that a saturated solution is a solution in which excess solute is in a state of dynamic equilibrium with solute particles in solution. For example, a saturated solution of the ionic solid barium sulfate ($BaSO_4$) has excess solid $BaSO_4$ in equilibrium with Ba^{2+} and SO_4^{2-} ions in solution:

$$BaSO_4(s) \rightleftharpoons Ba^{2+} + SO_4^{2-}$$

The mass action expression for the dissolving of $BaSO_4$ has the form

$$K_{sp} = [Ba^{2+}][SO_4^{2-}] = 1.5 \times 10^{-9}$$

where K_{sp} is the equilibrium constant, which is here called the **solubility product** constant. The product $[Ba^{2+}][SO_4^{2-}]$ is called the **ion product.** There is no explicit reference to the solid ($BaSO_4$) in the mass action expression; its constant effective concentration is included in K_{sp}.

Table 10-4 gives solubility products of some slightly soluble ionic solids at 25°C.

TABLE 10-4
Solubility products

Substance	Formula	K_{sp}
Barium sulfate	$BaSO_4$	1.5×10^{-9} (mole/ℓ)^2
Calcium carbonate	$CaCO_3$	2.8×10^{-9} (mole/ℓ)^2
Calcium oxalate	CaC_2O_4	4×10^{-9} (mole/ℓ)^2
Calcium phosphate	$Ca_3(PO_4)_2$	2.0×10^{-29} (mole/ℓ)^5
Lead chloride	$PbCl_2$	1.6×10^{-5} (mole/ℓ)^3
Magnesium carbonate	$MgCO_3$	3.5×10^{-8} (mole/ℓ)^2
Magnesium phosphate	$Mg_3(PO_4)_2$	10^{-23} to 10^{-27} (mole/ℓ)^5
Mercuric sulfide	HgS	1.6×10^{-54} (mole/ℓ)^2
Silver bromide	AgBr	5.0×10^{-13} (mole/ℓ)^2
Silver chloride	AgCl	1.7×10^{-10} (mole/ℓ)^2
Silver iodide	AgI	8.5×10^{-17} (mole/ℓ)^2

EXAMPLE 10-6

Write the equilibrium expression for the dissolving of the ionic solid $PbCl_2$ to yield Pb^{2+} and Cl^- ions in solution.

The equation for the reaction is

$$PbCl_2(s) \rightleftharpoons Pb^{2+} + 2Cl^-$$

The equilibrium expression is, therefore,

$$K_{sp} = [Pb^{2+}][Cl^-]^2$$

EXERCISE 10-8 Write the solubility product expression for the solution of aluminum hydroxide, $Al(OH)_3$, in water to yield Al^{3+} and OH^- ions.

Sodium chloride is very soluble in water. At room temperature, a saturated solution of NaCl is 5.3 M (that is, 5.3 M in Na^+ and 5.3 M in Cl^-). The K_{sp} for NaCl is therefore greater than one:

$$K_{sp} = [Na^+][Cl^-] = 5.3\text{ M} \times 5.3\text{ M} = 28\left(\frac{\text{mole}}{\ell}\right)^2$$

If we compare substances that have solubility product constants of the same dimensions, we can say that a substance with a low K_{sp} is less soluble than one with a higher K_{sp}. For example, referring to Table 10-4, we can say that AgI is less soluble than AgBr, which in turn is less soluble than AgCl. Of all the substances listed in Table 10-4 that have (mole/ℓ)^2 for the dimensions of K_{sp}, $CaSO_4$ is the most soluble, and HgS is the least soluble. In the following example, we calculate the solubility of HgS in water.

EXAMPLE 10-7

Mercuric sulfide (HgS) dissolves in water to yield Hg^{2+} and S^{2-} ions. How many grams of HgS can be dissolved in 1 ℓ of solution at 25°C? The K_{sp} for HgS is 1.6×10^{-54} (mole/ℓ)^2.

The equation for HgS dissolving in water is

$$HgS(s) \rightleftharpoons Hg^{2+} + S^{2-}$$

From the balanced equation, we see that for each HgS that dissolves there is *one* Hg^{2+} and *one* S^{2-} ion in solution.

Thus, if s = the number of moles of HgS that dissolve to form 1 ℓ of saturated solution, then at equilibrium,

$$[Hg^{2+}] = [S^{2-}] = s$$

$$K_{sp} = [Hg^{2+}][S^{2-}] = s \times s = 1.6 \times 10^{-54} \left(\frac{\text{mole}}{\ell}\right)^2$$

$$s^2 = 1.6 \times 10^{-54} \left(\frac{\text{mole}}{\ell}\right)^2$$

$$s = 1.3 \times 10^{-27} \text{ mole}/\ell$$

The molecular weight of HgS is 201 + 32 = 233 g/mole.

$$\text{Mass of HgS in saturated solution} = 1.3 \times 10^{-27} \frac{\text{mole}}{\ell} \times 233 \frac{\text{g}}{\text{mole}} = 3.0 \times 10^{-25} \frac{\text{g}}{\ell}$$

The solubility of HgS is so low that there are only 1.3×10^{-27} moles/ℓ $\times 6.0 \times 10^{23}$ ions/mole $\cong 8 \times 10^{-3}$ ions/ℓ—or one of each kind of ion per 10^3 ℓ.

EXAMPLE 10-8

Given that K_{sp} of calcium sulfate ($CaSO_4$) is 2.4×10^{-5} (mole/ℓ)^2, calculate its solubility (a) in pure water, and (b) in 0.10 M Na_2SO_4 solution.

(a) Let s = solubility of $CaSO_4$ in pure water (moles of $CaSO_4$ that dissolve to form 1 ℓ of solution).

$$CaSO_4(s) \rightleftharpoons Ca^{2+} + SO_4^{2-}$$

Then, in the saturated solution,

$$[Ca^{2+}] = [SO_4^{2-}] = s$$

$$K_{sp} = [Ca^{2+}][SO_4^{2-}] = s \times s = 2.4 \times 10^{-5} \left(\frac{\text{mole}}{\ell}\right)^2$$

$$s^2 = 2.4 \times 10^{-5} \left(\frac{\text{mole}}{\ell}\right)^2$$

$$s = 4.9 \times 10^{-3} \text{ mole}/\ell$$

The solubility of $CaSO_4$ in water is 4.9×10^{-3} mole/ℓ.

(b) Let s' = solubility of $CaSO_4$ in 0.10 M Na_2SO_4 solution. Then, in the saturated solution,

$$[Ca^{2+}] = s'$$

In this case, however, $[Ca^{2+}]$ is not the same as $[SO_4^{2-}]$ because of the extra Na_2SO_4 in solution. Instead,

$$[SO_4^{2-}] = 0.10 + s'$$

(0.10 M SO_4^{2-} from the Na_2SO_4 solution, and s' from the dissolved $CaSO_4$). Therefore,

$$K_{sp} = [Ca^{2+}][SO_4^{2-}] = s' \times (0.10 + s') = 2.4 \times 10^{-5}$$

Because K_{sp} is small, not much $CaSO_4$ dissolves (we saw in part a that only 0.0049 mole dissolves per liter, even in pure water), so we can consider s' negligible compared to the 0.10 M.

That is, we can assume, $0.10 + s' \cong 0.10$, and

$$K_{sp} = s' \times 0.10 = 2.4 \times 10^{-5}$$

$$s' = 2.4 \times 10^{-4} \text{ M}$$

Note that s' is in fact small enough that $0.10 + s'$ is very closely approximated by 0.10, because $0.10 + s' = 0.10 + 0.00024 \cong 0.10$. Thus the solubility of $CaSO_4$ in 0.10 M Na_2SO_4 is 2.4×10^{-4} M, only

1/200 as great as its solubility in pure water. This is an example of the **common ion effect,** by which the solubility of a slightly soluble salt is greatly decreased by adding a different solute that has an ion in common with the slightly soluble salt. Here the solubility of $CaSO_4$ was decreased by the addition of SO_4^{2-} ions. It would also be decreased by the addition of Ca^{2+} ions instead—say, as $CaCl_2$ (Ca^{2+} and $2Cl^-$).

EXERCISE 10-9 Would you expect calcium phosphate, $Ca_3(PO_4)_2$, to be more soluble in pure water or in a dilute solution of sodium phosphate, Na_3PO_4? Use a chemical equation and Le Châtelier's principle to explain your answer.

Just as we used the law of mass action to predict the direction of a reaction, we can use the solubility product to predict whether a **precipitate** (an insoluble solid) will separate out when two solutions are mixed. If, when two solutions are mixed, the ion product is less than K_{sp} (that is, less than the ion product in a saturated solution of the substance), then the solution is unsaturated, and precipitation will not occur (a precipitate will not form). If the ion product in the mixture is, on the other hand, momentarily greater than K_{sp}, then solid will precipitate until the ion product of the ions remaining in solution has been reduced to K_{sp}.

EXAMPLE 10-9

Will a precipitate of the slightly soluble salt $CaCO_3$ form when equal volumes of the following two solutions are mixed? Solutions: 4.0×10^{-5} M Na_2CO_3 and 5.0×10^{-4} M $CaCl_2$. The K_{sp} for $CaCO_3$ is 4.7×10^{-9} (mole/ℓ)^2.

After mixing, each ion is in a volume twice as great as the original volume, so the concentration of each is one-half what it was in the original solution:

$$\text{After mixing,}\quad [Ca^{2+}] = \frac{5.0 \times 10^{-4}}{2}\ \text{M} = 2.5 \times 10^{-4}\ \text{M}$$

$$[CO_3^{2-}] = \frac{4.0 \times 10^{-5}}{2}\ \text{M} = 2.0 \times 10^{-5}\ \text{M}$$

The ion product is

$$[Ca^{2+}][CO_3^{2-}] = (2.5 \times 10^{-4})(2.0 \times 10^{-5}) = 5.0 \times 10^{-9}$$

The ion product is greater than K_{sp}, so $CaCO_3$ will precipitate until the ion product is equal to K_{sp}. The Na^+ and Cl^- ions will not form a precipitate, because NaCl is a highly soluble salt with a large K_{sp}. Sodium carbonate (Na_2CO_3) often is used to soften "hard" water for washing. The CO_3^{2-} ions cause the Ca^{2+} ions to precipitate from the water as $CaCO_3$ (and similarly with some other metal ions), thereby preventing these ions from forming an insoluble "scum" with the soap.

The formation of soap "scums" is the result of an ion product exceeding the K_{sp}. Sodium laurate, $CH_3(CH_2)_{10}COONa$, for example, is a soap whose calcium salt is relatively insoluble. If this soap is used in water that contains a high enough concentration of Ca^{2+}, the slightly soluble salt calcium laurate, $(CH_3(CH_2)_{10}COO)_2Ca$, precipitates out of solution as a scum.

A biological process related to solubility products is that of bone formation. Details of the process of bone formation are not well understood, but a contributing factor may be the precipitation of $Ca_3(PO_4)_2$ near bone-forming cells. Bone is living tissue composed of collagen fibers (Section 21-8) impregnated with minerals—primarily hydroxyapatite, $Ca_{10}(PO_4)_6(OH)_2$. Bone matter is constantly being absorbed and reformed; part of bone is readily exchangeable, and a larger part only slowly exchangeable. The Ca^{2+} in the blood plasma is in equilibrium with Ca^{2+} in the bone. If we assume, for simplicity, that it is $Ca_3(PO_4)_2$ that precipitates as bone, the equilibrium expression for the formation and resorption of bone is

$$Ca_3(PO_4)_2(s) \rightleftharpoons 3Ca^{2+} + 2PO_4^{3-}$$
$$K_{sp} = [Ca^{2+}]^3[PO_4^{3-}]^2$$

This K_{sp} is the constant for the reaction in blood at body temperature, and $[Ca^{2+}]$ and $[PO_4^{3-}]$ are the concentrations in the blood. One factor in the formation of bone may be that the bone-forming cells release extra phosphate, so that the K_{sp} for $Ca_3(PO_4)_2$ is exceeded in their vicinity. The reverse process of bone absorption may result in part from the secretion of an acid that dissolves $Ca_3(PO_4)_2$ by lowering the concentration of PO_4^{3-} (by the reaction $H^+ + PO_4^{3-} \rightarrow HPO_4^{2-}$). Alternatively, some substance may be secreted that binds to the Ca^{2+} and lowers its concentration in the blood.

EXERCISE 10-10 The K_{sp} for $Ca_3(PO_4)_2$ in water at 25°C is 2.0×10^{-29}. Would you expect a precipitate of $Ca_3(PO_4)_2$ to form in a solution that contains Ca^{2+} at a concentration of 1.2×10^{-3} M and PO_4^{3-} at a concentration of 6.7×10^{-9} M? (These are approximately the concentrations of free Ca^{2+} and free PO_4^{3-} in blood.)

10-9 FREE ENERGY AND CHEMICAL EQUILIBRIUM

What determines, absolutely, the position of equilibrium in any given set of conditions? This question is similar to that (considered in Chapter 4) of what determines the stable phase of a substance. For phase equilibria, we saw that a combination of energy and entropy determines which phase is stable—a competition between low energy and high entropy that is expressed quantitatively as $\Delta H - T\,\Delta S$ and is symbolized ΔG (see Section 4-12).

In the case of a chemical reaction at constant temperature, it is this same combination

$$\Delta H - T\,\Delta S = \Delta G$$

that determines the position of equilibrium. Here ΔH represents the heat of the reaction; T is the absolute temperature; and ΔS represents the difference between the final entropy (or disorder) of the system and the initial entropy—the entropy of all products minus entropy of all reactants. The difference in free energy (ΔG) represents the net driving force of the reaction, the combined effect of the two opposing tendencies toward lower energy and higher entropy.

As with phase changes, a chemical reaction occurs spontaneously in the direction of lower free energy. When ΔG is negative, reaction occurs spontaneously in the forward direction. When ΔG is zero, the system is at equilibrium, and no net reaction occurs. When ΔG is positive, reaction does not occur spontaneously in the forward direction, but the reverse reaction occurs spontaneously.

In the reaction

$$59.0\ \text{kJ} + N_2O_4(g) \rightleftharpoons 2NO_2(g)$$

$$\Delta H = 59.0 \frac{\text{kJ}}{\text{mole } N_2O_4}$$

the N_2O_4 is lower in energy (the reaction is endothermic), but the $2NO_2$ are higher in entropy—the larger *number* of the NO_2 molecules (though smaller in size) means that they are more disordered than the smaller number of the N_2O_4 molecules. At high temperatures, the tendency toward greater entropy, as expressed by the factor $T\,\Delta S$, wins (because of the large value of T), and the NO_2 is favored. At low temperatures, the tendency toward lower energy is more important, and the N_2O_4 is favored. This is in accord with the experimental results shown in Figure 10-5.

10-10 CHEMICAL KINETICS: RATES OF REACTIONS

We have looked at some chemical reactions and have considered factors that affect the position of equilibrium, but we have not yet asked how a chemical reaction comes about or what affects its rate.

Chemical reactions can be slow (as silver tarnishing), medium (as blood clotting or a plaster cast setting), or rapid (as an explosion). Quantitative statements about reaction rates are made by choosing either *one* reactant or *one* product from a reaction,

$$A + B + \cdots \longrightarrow M + N + \cdots$$

and defining the **reaction rate** as the rate at which the concentration of that substance changes. For reactant A, this is

$$\text{Rate} = \frac{\text{Concentration of A that disappeared}}{\text{Time interval}}$$

and for product M, the reaction rate is

$$\text{Rate} = \frac{\text{Concentration of M that appeared}}{\text{Time interval}}$$

where the time interval (Δt) is the length of time required for the concentration change.

When hydrogen iodide (HI) is heated to 400°C, it decomposes to yield hydrogen and iodine:

$$2HI(g) \longrightarrow H_2(g) + I_2(g)$$

Hydrogen iodide and hydrogen are colorless gases, but iodine gas is purple. As the reaction proceeds, the intensity of the purple color increases (Figure 10-8). The amount of I_2 present at any time can be determined by measuring the intensity of the purple color. If the amount of I_2 is determined at two times, t_1 and t_2, the rate of the reaction is

$$\text{Rate} = \frac{[I_2] \text{ at } t_2 - [I_2] \text{ at } t_1}{t_2 - t_1} = \frac{\Delta[I_2]}{\Delta t}$$

It is reasonable to assume that, for a reaction to occur, the molecules of the different reactants must come close to each other. The more *collisions* between reactant molecules, the more reaction that can occur. The collision frequency, in turn, depends on the concentrations of the reactant molecules. Thus, one factor on which the rate depends is the concentrations of the

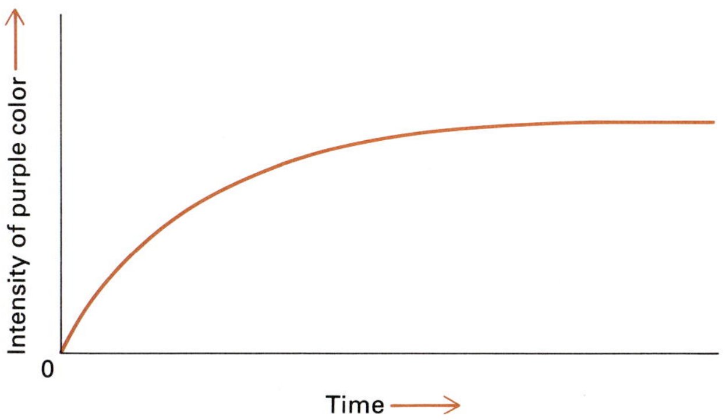

FIGURE 10-8

Rate of decomposition of hydrogen iodide:

$$\underset{\text{(colorless)}}{2HI\ (g)} \rightleftharpoons \underset{\text{(colorless)}}{H_2\ (g)} + \underset{\text{(purple)}}{I_2\ (g)}$$

reactants. Consider the decomposition of N_2O_5 to yield NO_2 and O_2 according to the equation

$$N_2O_5 \longrightarrow 2NO_2 + \tfrac{1}{2}O_2$$

The **rate law,** a mathematical expression determined by experiment, is

$$\text{Rate of disappearance of } N_2O_5 = k[N_2O_5]$$

The **rate constant** k at 25°C is $3.38 \times 10^{-5}\ \text{sec}^{-1}$. When the concentration of N_2O_5 is given in moles/ℓ, the rate of disappearance of N_2O_5 is in moles/ℓ sec. At another temperature, k would have a different value. We shall see why this is so in Section 10-11.

The general form of a **reaction rate law** (or expression) for the reaction

$$A + B + \cdots \longrightarrow \text{Products}$$

is

$$\text{Rate} = k[A]^m[B]^n \cdots$$

where [A] and [B] are the *concentrations* of the reactants (usually expressed in moles/ℓ); m and n are exponents that may be simple integers, zero, or nonintegers; and k is a rate constant that (as we shall see later) depends on the temperature.

We have said that any of the powers of the concentrations in the expression for the reaction rate might even be zero. Anything raised to the zero power is just one. If m were zero, that would mean that the rate is independent of the concentration of reactant A:

$$\begin{aligned}\text{Rate} &= k[A]^m[B]^n \cdots \\ &= k \times [A]^0 \times [B]^n \times \cdots \\ &= k[B]^n \cdots\end{aligned}$$

How could that come about? To answer this question, consider (as an analogy) a group of three students putting together a magazine. The magazine will have 50 pages, and the pages have been printed and stacked in piles, one pile for each page. Student X assembles the pages in order; Y straightens the pages; and Z staples them. Student X can assemble 20 magazines per hour, but Y can straighten 120 per hour, and Z can staple 200 per hour. The first step is the "bottleneck" in the process—almost as soon as a magazine is assembled, it is straightened and stapled. We cannot speed up the rate of production by assigning more people to straighten or staple. The rate depends only on the number of students assembling the magazine, which is the slowest step in the process. Thus, if a second student is set to assembling, the rate doubles—but if another is set to stapling, the rate is unchanged.

Many chemical reactions occur in a series of steps in which one step is a similar "bottleneck." The slowest step is the **rate-determining, rate-limiting,** or **rate-controlling** step. The rate of reaction depends on the concentrations of any reactants participating in the rate-determining step, and the rate is independent of the concentrations of reactants involved only in other steps. Reaction rates are not, generally, predictable from theory; they must be determined experimentally.

As an example, we'll consider the reaction in which hydrogen and iodine are produced by heating hydrogen iodide:

$$2HI(g) \rightleftharpoons H_2(g) + I_2(g)$$

Experimentally it is found that when the concentration of HI is doubled, the rate of the reaction is quadrupled (multiplied by four). The rate law has the form Rate $= k[HI]^m$. We would like to determine the exponent m. One way is to try some numbers and see which best fits the observed data. The exponents in a rate law are never very large, so we'll first consider $m = 0$.

$$\text{If } m = 0, \qquad \text{Rate} = k[HI]^0 = k$$

According to this expression, doubling [HI] leaves the rate unchanged, because any number raised to the zero power is one. Therefore, we know that $m \neq 0$ (m is not equal to 0).

$$\text{If } m = 1, \qquad \text{Rate} = k[HI]^1 = k[HI]$$

This expression predicts that if [HI] is doubled, then the rate also is doubled (not quadrupled as experiment indicates):

$$[HI]_{new} = 2[HI]_{old}$$

$$\text{Old rate} = k[HI]_{old}$$

$$\text{New rate} = k[HI]_{new} = 2k[HI]_{old} = 2 \times \text{Old rate}$$

Therefore, we know that $m \neq 1$.

$$\text{If } m = 2, \qquad \text{Rate} = k[HI]^2$$

Now if [HI] is doubled, the rate is quadrupled because

$$[HI]_{new} = 2[HI]_{old} \quad \text{and} \quad [HI]^2_{new} = 2^2[HI]^2_{old}$$

$$\text{Old rate} = k[HI]^2_{old}$$

$$\text{New rate} = k[HI]^2_{new} = 2^2k[HI]^2_{old} = 4 \times \text{Old rate}$$

Therefore, we know that $m = 2$ and that the rate law is

$$\text{Rate} = k[HI]^2$$

because the value $m = 2$ fits the experimental data.

Example 10-10 illustrates how a rate law for a more complicated reaction is determined from experimental data.

EXAMPLE 10-10

In a rate study of the reaction A + B + C → Products, seven experiments were run, using various initial concentrations of A, B, and C:

Experiment	[A](M)	[B](M)	[B](M)	Rate*
1	1.0	1.0	1.0	x
2	2.0	1.0	1.0	$4x$
3	4.0	1.0	1.0	$16x$
4	1.0	2.0	1.0	$2x$
5	1.0	4.0	1.0	$4x$
6	1.0	1.0	2.0	x
7	1.0	1.0	4.0	x

*Here x represents the magnitude of the rate observed in experiment 1. As a reaction progresses, the reactants are gradually consumed, so the rate decreases as the concentrations of the reactants decrease. The rates we refer to thus are *initial* rates, measured while the concentrations are essentially at the initial values.

(a) What is the experimental rate law for this reaction?

The rate law has the form Rate $= k[A]^m[B]^n[C]^p$.

To find the exponent m, consider experiments 1, 2, and 3, in which the concentrations of B and C are held constant while that of A is varied. The data show that when A is doubled, the rate is multiplied by 4; when the concentration of A is quadrupled, the rate is multiplied by 16. These three experiments show that the reaction rate is proportional to the square of the concentration of A; that is, $m = 2$ in the rate equation.

To find the exponent n, consider experiments 1, 4, and 5, in which the concentrations of A and C are held constant while that of B is varied. We see that when the concentration of B is doubled, the rate is multiplied by 2; when the concentration of B is quadrupled, the rate is multiplied by 4. These three experiments show that the reaction rate is directly proportional to the concentration of B; that is $n = 1$ in the rate equation.

To find the exponent p, consider experiments 1, 6, and 7, in which the concentrations of A and B are held constant while that of C is varied. When the concentration of C is doubled, and also when it is quadrupled, the rate remains unchanged. The reaction rate is independent of the concentration of C; that is, $p = 0$ in the rate equation.

Therefore, the rate law is

$$\text{Rate} = k[\text{A}]^2[\text{B}]^1[\text{C}]^0 = k[\text{A}]^2[\text{B}]$$

We conclude that C does not appear in the rate-determining step of this reaction.

(b) In this reaction, what would be the effect on the reaction rate of tripling the concentration of A while, simultaneously, doubling the concentrations of B and C?

$$\text{Original rate} = k[\text{A}]^2[\text{B}]$$

$$\text{New rate} = k3^2[\text{A}]^2 2[\text{B}] = 18k[\text{A}]^2[\text{B}]$$

The reaction rate is multiplied by 18 if the concentration of A is tripled while those of B and C are doubled. Note that the doubling of the concentration of C had no effect on the reaction rate.

EXERCISE 10-11 We found that the rate law for $2\text{HI(g)} \rightleftharpoons \text{H}_2\text{(g)} + \text{I}_2\text{(g)}$ is Rate $= k[\text{HI}]^2$. What would be the effect on the reaction rate of tripling the concentration of HI?

EXERCISE 10-12 Assume that A, B, and C of Example 10-10 are gases.

(a) If the initial amounts of these gases are 1 mole of A, 2 moles of B, and 1 mole of C in a 1 ℓ flask, what would be the initial concentrations of A, B, and C?

(b) If these same amounts of A, B, and C are confined in a 500 ml flask, what would be the concentrations of A, B, and C?

(c) In which flask would the reaction be more rapid, and how many times more rapid would it be?

Generally we know only the starting materials and products of a reaction, and not the **reaction mechanism,** the entire series of steps by which the reaction occurs. It is usually difficult to determine the mechanism of a reaction. The steps in any proposed mechanism must be consistent with the experimentally determined rate law, and they must add to give the correct net equation. Consider, for example, the reaction of hydrogen bromide with oxygen:

$$4\text{HBr} + \text{O}_2 \longrightarrow 2\text{H}_2\text{O} + 2\text{Br}_2$$

Experimentally it is found that the rate doubles when the concentration of HBr is doubled while that of O_2 is held constant. The rate also doubles when the concentration of O_2 is doubled while that of HBr is constant. The rate is directly proportional to both reactants. The rate law is, therefore,

$$\text{Rate} = k[\text{HBr}][\text{O}_2]$$

The rate law implies that in the rate-controlling step of the reaction, *one* HBr molecule reacts with *one* O_2 molecule. Then, doubling the number of HBr molecules doubles the frequency of collisions between O_2 and HBr molecules, and the rate doubles. And if the number of O_2 molecules is doubled, there are twice as many collisions between HBr and O_2 molecules, and again the rate doubles. The actual mechanism of the reaction is now believed to be

$$HBr + O_2 \xrightarrow{\text{slow}} HOOBr$$
$$HOOBr + HBr \xrightarrow{\text{fast}} 2HOBr$$
$$(HOBr + HBr \xrightarrow{\text{fast}} H_2O + Br_2)$$
$$\text{Net reaction:} \quad 4\,HBr + O_2 \longrightarrow 2H_2O + 2Br_2$$

The separate steps of the mechanism add to give the correct net equation for the reaction. The molecules HOOBr and HOBr, which are produced in the course of the reaction and later consumed, are called **reaction intermediates.** The rate-controlling step involves only one molecule each of HBr and O_2, and thus it is consistent with the rate law. When the exponents in a rate law are greater than one, or when the rate law involves more than two subtances, it still is possible to deduce mechanisms consistent with it, but we shall not consider these more complicated mechanisms here.

EXERCISE 10-13 A possible mechanism for the decomposition of hydrogen iodide to produce hydrogen and iodine consists of the following steps:

$$HI(g) + HI(g) \xrightarrow{\text{slow}} H_2I_2(g)$$
$$H_2I_2(g) \xrightarrow{\text{fast}} H_2(g) + 2I(g)$$
$$2I(g) \xrightarrow{\text{fast}} I_2(g)$$

(a) What is the overall, or net, reaction?
(b) Which species are reactive intermediates?
(c) Write a rate law that is consistent with this mechanism.

10-11 ENERGY OF ACTIVATION, AND THE EFFECT OF TEMPERATURE ON REACTION RATE

A mixture of methane and oxygen in a Bunsen burner is stable at room temperature but burns rapidly when ignited with a lighted match, to form carbon dioxide and water:

$$CH_4(g) + 2O_2(g) \longrightarrow CO_2(g) + 2H_2O(g)$$

Why is this mixture stable until the match is introduced, and how does the lighted match initiate the reaction? At room temperature there are many collisions between CH_4 and O_2 molecules, but no observable reaction occurs—the collisions are not effective. For reactions to occur, the molecules must collide with enough energy to overcome the repulsive forces that exist between the electron clouds of two molecules as they approach each other closely. That is, the colliding molecules must have enough energy to lead to a rearrangement of chemical bonds.

This critical energy, called the **activation energy,** is shown schematically in Figure 10-9 for an exothermic reaction, where it is seen to be an energy barrier between reactants and products. Just as a ball rolled up a hill may not have enough energy to get over the top but instead rolls back down the hill, so two colliding molecules may not have enough energy to get over the activation-energy barrier. Even if the colliding molecules do have the critical energy, the collision may not be effective—the molecules may collide and fly apart without reacting. In the case of the combustion of methane, the activation energy is supplied by a lighted match. We shall now see how.

In Chapter 3 we saw that the average kinetic energy (which is proportional to the square of the speed) of the molecules is proportional to the absolute temperature. We saw further that at any temperature, there is a distribution of molecular velocities, with many molecules traveling at a velocity close to the average but some traveling faster and some more slowly. Figure 10-10 shows a typical distribution of molecular velocities at two temperatures. At the higher temperature, more molecules move at a high velocity. If v_c is the critical velocity the molecules need in order to react (the velocity that corresponds, in energy, to the activation energy), we see that *at the higher temperature there are many more molecules that can react.* Thus, at the higher temperature, more colliding pairs have enough energy to react, and the rate of the reaction increases. In the combustion of methane, heat from the match raises the temperature of the mixture locally and thus supplies the activation energy. Once the reaction has been initiated in this way, heat produced in the exothermic combustion reaction provides energy to keep the reaction going.

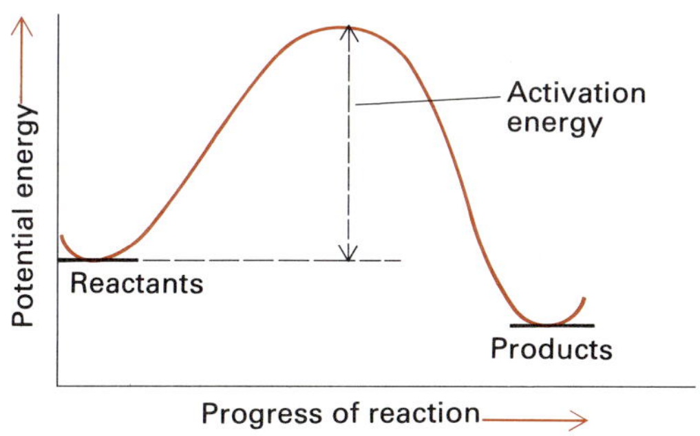

FIGURE 10-9

Activation energy. Only molecules that have enough energy to get over the activation energy barrier can react.

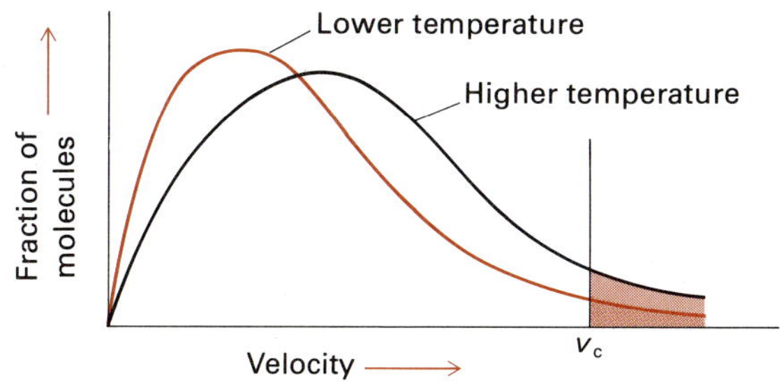

FIGURE 10-10

Distribution of molecular velocities in a sample of gas at two different temperatures.

The rate of a chemical reaction increases rapidly with temperature. It would be quite typical, say, for a temperature increase of 10°C to double or triple the rate of reaction. Near room temperature (about 300 K), an increase in temperature of 10°C is only a 3% increase. Nonetheless, it has a tremendous effect on the rate of reaction, because two or three times as many molecules now have enough energy to react.

We see the effect of temperature on reaction rate when using a pressure cooker, where increased pressure leads to an increase in the boiling temperature and thus to more rapid cooking; and in the slowing of biochemical processes when the body temperature of a hibernating animal decreases. Sterilization takes place when proteins in bacteria are altered, killing the bacteria. The rate of sterilization in an autoclave depends on the temperature of the steam. Figure 10-11 shows the *time* required for sterilization as a function of temperature in an autoclave. The time required to kill the bacteria decreases from 11 minutes to 3 minutes when the temperature is raised from 125°C to 135°C (the *rate* of sterilization is increased by a factor of almost 4).

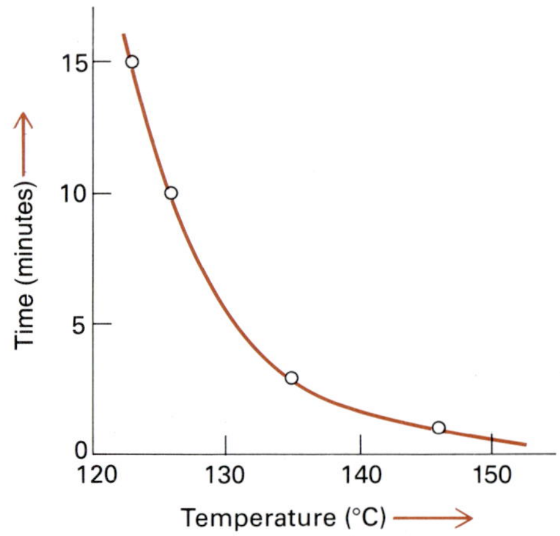

FIGURE 10-11

Time required for sterilization as a function of autoclave temperature.

10-12 CATALYSIS

Many reactions that normally proceed slowly can be made to go much more rapidly by the introduction of a **catalyst,** a substance that *participates in the reaction but is not consumed by it.* Consider the reaction

$$A + B \xrightarrow{\text{slow}} \text{Products}$$

If a catalyst for the reaction can be found, a possible mechanism might be

$$A + \text{Catalyst} \xrightarrow{\text{fast}} A \cdot \text{catalyst}$$

$$A \cdot \text{Catalyst} + B \xrightarrow{\text{fast}} \text{Catalyst} + \text{Products}$$

where A · Catalyst is a reaction intermediate that is a complex molecule formed by reaction of A with the catalyst. The catalyst takes part in the reaction but is regenerated at the end, and so it is not consumed.

Enzymes, large protein molecules that perform biological functions, are remarkably efficient catalysts. The enzyme α-amylase in saliva has the specific function of speeding the breakdown of large starch molecules to sugar. This reaction uncatalyzed (that is, in the absence of α-amylase) requires weeks at body temperature. It occurs within seconds in the presence of α-amylase. Breaking down large proteins into small fragments in the laboratory requires digestion for long times in hot, concentrated hydrochloric acid. In the body this reaction is catalyzed by pepsin and other enzymes in the gastric juices and occurs rapidly at body temperature. Enzymes are tremendously efficient, speeding up reactions typically by factors of 10^8 to 10^{11}. In one minute, a single enzyme molecule may catalyze the reaction of as many as 10 million (10^7) reactant molecules, though 10 to 1000 is a more typical range. In Chapter 22 we shall learn that the *shape* of the enzyme molecule is of crucial importance in its ability to act as a catalyst.

In some laboratory reactions, a solid catalyst is used. A mixture of hydrogen and oxygen gases is stable at room temperature (unless ignited), but when the mixture is exposed to platinum powder, a rapid reaction forming water molecules occurs on the surface of the metal particles:

$$2H_2(g) + O_2(g) \xrightarrow{Pt} 2H_2O(g)$$

Platinum is also used as a catalyst in the "contact process" for the production of sulfuric acid:

$$2SO_2(g) + O_2(g) \xrightarrow{Pt} 2SO_3(g)$$

$$SO_3(g) + H_2O \longrightarrow H_2SO_4$$

and in the hydrogenation of compounds with carbon–carbon double or triple bonds:

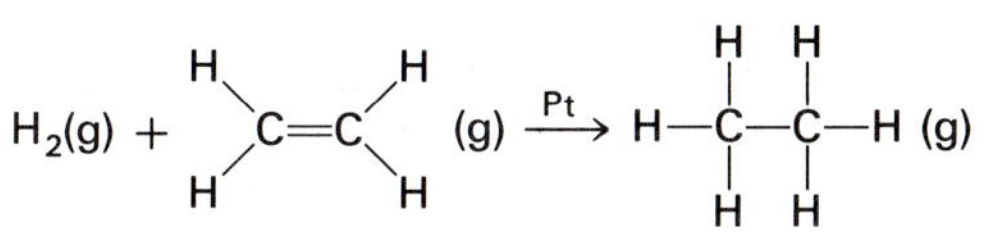

ethylene **ethane**

How does a catalyst increase the rate of a chemical reaction? A catalyst provides an alternative mechanism for the reaction—one that has a lower activation energy than the uncatalyzed reaction. Figure 10-12 shows the lowering of the activation energy of a hypothetical reaction by a catalyst.

Table 10-5 gives activation energies and relative rates of reactions for the decomposition of hydrogen peroxide to water and oxygen,

$$2H_2O_2 \longrightarrow 2H_2O + O_2(g)$$

with no catalyst, with powdered platinum as catalyst, and with the enzyme catalase as catalyst. Platinum speeds the rate of reaction by a factor of 26,000, and the enzyme catalase increases the rate by a factor of 10^{10}!

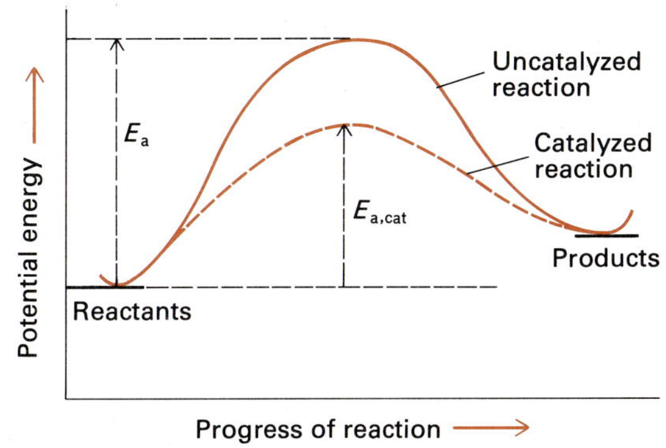

FIGURE 10-12

The activation energy (E_a) is lowered by a catalyst.

A catalyst increases the rate of a chemical reaction by providing an alternative mechanism with a lower activation energy. What effect does it have on the position of equilibrium? If you look at Figure 10-12 again, you will see that both the activation energy of the forward reaction and that of the reverse reaction are lowered by the same amount. The result is that *when a catalyst is added to a system that is already at equilibrium, there is no observable effect.* The *rates* of both forward and reverse reactions are increased equally, so equilibrium is reached faster, but there is no effect on the *position* of equilibrium.

TABLE 10-5

Activation energies and relative rates for uncatalyzed and catalyzed reactions

Reaction	Activation energy (kJ/mole)	Relative rate
$2H_2O_2 \rightarrow 2H_2O + O_2(g)$	75	1
$2H_2O_2 \xrightarrow{Pt} 2H_2O + O_2(g)$	50	26,000
$2H_2O_2 \xrightarrow{catalase} 2H_2O + O_2(g)$	0.6	10^{10}

EXERCISE 10-14 Two gases, A and B, are introduced into a container at room temperature. The rate equation for the reaction between A and B is

$$\text{Rate} = k[\text{A}]^3[\text{B}]$$

What effect will each of the following procedures have on the reaction rate?

(a) The volume of the container is doubled, at constant temperature.
(b) The concentration of A is halved.
(c) The concentration of A is doubled while that of B is halved.
(d) The temperature is decreased at constant volume.
(e) A catalyst is added to the system.

KEY TERMS

An *endothermic reaction* is a reaction in which heat is absorbed from the surroundings; ΔH is positive. An *exothermic reaction* is a reaction in which heat is released; ΔH is negative.

A *calorimeter* is used to measure heats of reaction.

The *heat of formation* of a substance is the heat absorbed or released when one mole of the substance (at 25°C and 1 atm) is formed from its elements (also at 25°C and 1 atm).

The *rate* of a chemical reaction is defined in terms of the increase in concentration of one product, or the decrease in concentration of one reactant, with time.

Some chemical reactions occur in a series of steps called a *reaction mechanism*. The *rate-determining step* determines the overall rate of a reaction. *Reaction intermediates* are molecules that are produced in the course of a reaction and later are consumed.

Activation energy is the critical energy that colliding molecules must have for reaction to occur.

A *catalyst* speeds the rate of a reaction. The catalyst participates in the reaction but is not consumed.

When a chemical reaction reaches *dynamic equilibrium,* the rate of the forward reaction is equal to the rate of the reverse reaction.

The *law of mass action* is a quantitative expression that relates the amounts of reactants and products in a system at equilibrium.

CHAPTER SUMMARY

Given tables of *heats of formation,* it is possible to calculate the heats of many reactions: $\Delta H = \Delta H_{\text{f,products}} - \Delta H_{\text{f,reactants}}$.

A chemical reaction does not go to completion. Instead, it reaches a state of *dynamic equilibrium* in which some of each of the products and some of each of the reactants are present. Le Châtelier's principle can be used to predict the effects of changes of pressure, concentration, and temperature on the position of equilibrium.

The *law of mass action* is a quantitative expression that relates the concentrations of reactants and products in a system at equilibrium in terms of an equilibrium constant K that, for a given reaction, depends only on the temperature. The value of K is not affected by changes in the concentrations of the reacting substances or by the presence of a catalyst. The law of mass action can be used to predict the direction of reaction and to calculate the equilibrium concentrations of all the reacting substances, given their initial concentrations.

It is a universally accepted convention to omit the concentrations of pure liquids or solids, or that of the solvent in a dilute solution, from the mass action expression. Their effective concentrations are constant and are incorporated into the equilibrium constant.

For chemical equilibria, as for phase equilibria, it is the competition (between low energy and high entropy) expressed by ΔG that determines the position of equilibrium.

In general, the rates of chemical reactions are not predictable from theory but must be determined experimentally. The *rate of reaction* depends on the concentrations of any reactants participating in the *rate-determining step* and on the temperature. For a reaction to occur, molecules must collide with a certain critical energy called the *activation energy*. The higher the temperature, the greater the number of molecules with enough energy to react. Thus, the rate of a chemical reaction increases rapidly with the temperature. A *catalyst* speeds a reaction by providing an alternate mechanism that has an activation energy lower than that of the uncatalyzed reaction.

PROBLEMS

10-1. Which would give more heat upon combustion, 1 mole of ethane (C_2H_6) or 1 mole of acetylene (C_2H_2)? Assume that the products are carbon dioxide gas and water vapor in both cases. Use data given in Table 10-1.

10-2. Explain why the temperature inside a compost pile is higher than that of the surroundings.

10-3. (a) How much energy is liberated when 2 moles of hydrogen gas combine with enough oxygen gas to convert the hydrogen completely to liquid water?
(b) How much energy is required to break down 36 g of liquid water into hydrogen gas and oxygen gas?

10-4. The temperature of a certain calorimeter increases by 1.00°C when 4.18×10^3 kJ are absorbed. The combustion of 1.71 g of sucrose ($C_{12}H_{22}O_{11}$) in the calorimeter results in a temperature rise of 6.75°C. What is the heat of combustion per mole of sucrose?

10-5. By consulting Table 10-1, tell whether the reaction in which each of the following substances is formed from the elements is endothermic or exothermic: glucose, urea, diamond, acetylene.

10-6. Using data given in Table 10-1, calculate ΔH for
(a) the decomposition of limestone ($CaCO_3$) to calcium oxide (CaO) and carbon dioxide (CO_2)
(b) the oxidation of carbon monoxide to carbon dioxide

(c) the oxidation of diamond to carbon dioxide.

10-7. ΔH for the combustion of ethanol (C_2H_5OH) to $CO_2(g)$ and $H_2O(l)$ is 1.369×10^3 kJ per mole of ethanol burned. What is the heat of formation of ethanol?

10-8. Chloral hydrate, $CCl_3CH(OH)_2$, has a narcotic effect that leads to its illegal use as "knockout drops." Determine the heat of combustion *per mole* of chloral hydrate to carbon dioxide, liquid water, and chlorine gas. (Be sure to balance the equation first—you'll need 2 molecules of chloral hydrate.)

10-9. (a) A cold pack used in ambulances and hospitals is a vial of ammonium nitrate (NH_4NO_3) in a water pack. When the pack is hit, the vial breaks, and the NH_4NO_3 dissolves (endothermically) and cools the pack. The reaction is

$$NH_4NO_3(s) + H_2O \longrightarrow NH_4^+(aq) + NO_3^-(aq)$$
$$\Delta H = 26 \text{ kJ/mole}$$

If 1 mole of NH_4NO_3 dissolves in 1 ℓ of water, by how many degrees is the temperature of the water lowered?

(b) A hot pack is a vial of calcium chloride ($CaCl_2$) in a water pack. When the vial in the pack breaks, the $CaCl_2$ dissolves (exothermically) and heats the pack. The reaction is

$$CaCl_2(s) + H_2O \longrightarrow Ca^{2+}(aq) + 2Cl^-(aq)$$
$$\Delta H = -73 \text{ kJ/mole}$$

If 1 mole of $CaCl_2$ dissolves in 1 ℓ of water, by how many degrees is the temperature of the water raised?

10-10. Palmitic acid is $CH_3(CH_2)_{14}CO_2H$. For this fatty acid, $\Delta H_f = -902.1$ kJ/mole.
(a) What is the heat of combustion of palmitic acid (in kJ/mole)?
(b) What is the heat of combustion of palmitic acid (in kJ/g)?
This energy is typical for the catabolism of fats in the body.
(c) Compare your answer to part b with the heats of combustion per gram of sucrose and glucose calculated in Exercise 10-1.

10-11. When glucose is dissolved in water, an equilibrium is established between the two forms α-D-glucose and β-D-glucose (see Section 23-3):

$$\beta\text{-D-glucose(aq)} \rightleftharpoons \alpha\text{-D-glucose(aq)}$$
$$K_{25°C} = 0.57$$

What is the concentration of each form at equilibrium in a 1 M solution of glucose?

10-12. (a) Ca^{2+} is essential to the clotting of blood. To prevent clotting in samples of whole blood, potassium oxalate ($K_2C_2O_4$) is sometimes added. It reacts with Ca^{2+} to form insoluble calcium oxalate and prevents coagulation. If enough potassium oxalate is added to give $[C_2O_4^{2-}] = 0.01$ M at equilibrium, what is the equilibrium Ca^{2+} concentration? The K_{sp} for $CaC_2O_4 = 4 \times 10^{-9}$ (mole/ℓ)^2.
(b) Another way to prevent clotting of blood is to add a solution of sodium citrate to the sample—say, to the syringe into which the blood is drawn. The citrate ion binds to the Ca^{2+} to form a soluble complex, Ca-citrate, which prevents the Ca^{2+} from entering into the clotting process. The equilibrium expression is

$$\text{Ca-citrate(aq)} \rightleftharpoons \text{Ca}^{2+} + \text{citrate}^{2-}$$
$$K = 2.1 \times 10^{-5}\ \text{mole}/\ell$$

If the equilibrium concentration of citrate ion is 0.05 M, and that of Ca-citrate is 0.025 M, what is the concentration of Ca^{2+}?
(c) Do you think that oxalate and citrate would be effective as water softeners?

10-13. Cases of lead poisoning are sometimes treated with EDTA. EDTA forms a soluble complex with the Pb^{2+} which is then excreted in the urine. If after treatment, the concentration of Pb-EDTA^{2-} is 1×10^{-4} M, and that of EDTA^{4-} is 0.01 M, what is the concentration of free Pb^{2+} that remains in the body fluids? The equilibrium expression is

$$\text{Pb-EDTA}^{2-}\text{(aq)} \rightleftharpoons \text{Pb}^{2+} + \text{EDTA}^{4-}$$
$$K = 5.0 \times 10^{-19}\ \text{mole}/\ell$$

10-14. Suppose that the following reaction is at equilibrium at 2000 K:

$$H_2(g) + CO_2(g) \rightleftharpoons H_2O(g) + CO(g)$$
$$\Delta H = +41.8\ \frac{\text{kJ}}{\text{mole CO}}$$

(a) What would be the effect on the amount of CO present at equilibrium if the total pressure is decreased by decreasing the volume of the container?
(b) Would more or less CO be present at equilibrium if the reaction were run at 1000 K instead of at 2000 K?
(c) What would be the effect on the amount of CO present at equilibrium if a catalyst is added?
(d) What would be the effect on the amount of CO present at equilibrium if more $H_2O(g)$ is added to the reaction mixture?

10-15. At 475°C, the equilibrium constant for the gas-phase reaction $H_2 + I_2 \rightleftharpoons 2HI$ is 49.
(a) If initially 0.018 mole HI is put into a 1 ℓ vessel, how many moles of I_2 are produced as the gas comes to equilibrium at 475°C?
(b) Heat is evolved when the reaction $H_2 + I_2 \rightarrow 2HI$ occurs. At 400°C, is the equilibrium constant for this reaction greater than, equal to, or less than 49? Explain.
(c) When at 475°C the concentrations of H_2, I_2, and HI are momentarily all equal, which reaction will then occur spontaneously: $H_2 + I_2 \rightarrow 2HI$, or $2HI \rightarrow H_2 + I_2$? Explain.
(d) For the reaction that occurs in part c, is ΔH greater than, equal to, or less than $T\Delta S$? Explain.
(e) When H_2, I_2, and HI are initially at equilibrium, and the volume of the container is then increased at fixed temperature, does the amount of HI increase, decrease, or remain unchanged? Explain.
(f) Suppose that $H_2 + I_2 \rightarrow 2HI$ and $2HI \rightarrow H_2 + I_2$, exactly as written, are the rate-determining steps in the forward and reverse reactions of $H_2 + I_2 \rightleftharpoons 2HI$. (1) What effect does doubling the concentration of H_2 have on the rate of the reaction $H_2 + I_2 \rightarrow 2HI$? (2) What is the effect of doubling the concentration of I_2? (3) What effect does *tripling* the concentration of HI have on the rate of the reaction $2HI \rightarrow H_2 + I_2$? (4) Is the rate of $H_2 + I_2 \rightarrow 2HI$ increased or decreased by increasing the temperature? Explain. (5) Answer part 4 for the reaction $2HI \rightarrow H_2 + I_2$.

10-16. Phosgene ($COCl_2$) is produced by reaction of carbon monoxide and chlorine: $CO(g) + Cl_2(g) \rightleftharpoons COCl_2(g)$. The forward reaction is exothermic. What effect does each of the following operations have on the amount of phosgene present at equilibrium?
(a) Adding more Cl_2 at constant volume.
(b) Increasing the volume of the container.
(c) Removing the CO by adding a substance with which it forms a very stable complex.
(d) Adding a catalyst.
(e) Lowering the temperature.

10-17. At 100°C the equilibrium constant for the gas-phase reaction $CO + Cl_2 \rightleftharpoons COCl_2$ is 4.6×10^9 ℓ/mole.
(a) Heat is evolved when the reaction $CO + Cl_2 \rightarrow COCl_2$ occurs. At 150°C, is the equilibrium constant for this reaction greater than, equal to, or less than 4.6×10^9 ℓ/mole? Explain.
(b) When at 100°C the concentrations of CO, Cl_2, and $COCl_2$ are momentarily all equal, with the common value 1.0×10^{-3} mole/ℓ, which reaction will then occur spontaneously: $CO + Cl_2 \rightarrow COCl_2$, or $COCl_2 \rightarrow CO + Cl_2$? Explain.
(c) For the reaction that occurs in part b, is ΔH greater than, equal to, or less than $T\Delta S$? Explain.
(d) When CO, Cl_2, $COCl_2$ are initially at equilibrium, and the volume of the container is then decreased at fixed temperature, does the amount of $COCl_2$ increase, decrease, or remain unchanged? Explain.

10-18. Consider the oxidation of iron to ferric oxide:

$$4Fe(s) + 3O_2(g) \rightleftharpoons 2Fe_2O_3(s)$$

What would be the effect on the equilibrium concentration of oxygen if more iron is added to the reaction vessel? Explain.

10-19. Calcium hydroxide, $Ca(OH)_2$, dissolves to a small extent in water to yield Ca^{2+} and OH^- ions.
(a) A saturated solution of $Ca(OH)_2$ at 25°C contains 0.051 g of $Ca(OH)_2$ in 100 ml of solution. What is the K_{sp} of $Ca(OH)_2$ at 25°C?
(b) How many grams of $Ca(OH)_2$ would dissolve in 100 ml of 1.0 M NaOH? Assume that the volume of the solution remains constant, 100 ml.

10-20. Write solubility constant expressions for (a) $Fe(OH)_3$, (b) $Mg_3(PO_4)_2$, and (c) PbI_2.

10-21. (a) Seawater contains about 19 g of Cl^- per liter. What is the maximum concentration of $AgNO_3$ solution you could prepare using seawater? The K_{sp} for AgCl is 1.7×10^{-10} (mole/ℓ)^2.
(b) Suppose you would like to prepare a 0.5 M solution of $AgNO_3$ using distilled water. What is the maximum concentration of Cl^- that can be tolerated in the distilled water?

10-22. If you add 10 ml of a solution of 0.01 M $BaCl_2$ to 90 ml of a solution of 1 M NaOH, will $Ba(OH)_2$ precipitate? K_{sp} for $Ba(OH)_2$ is 5.0×10^{-3} M^3.

10-23. For each of the following reactions, consider (a) whether the tendency toward minimum energy favors the forward or the backward reaction, and (b) whether the tendency toward maximum entropy favors the forward or the backward reaction.
(1) The decomposition of limestone:

$$CaCO_3(s) \rightleftharpoons CaO(s) + CO_2(g)$$
$$\Delta H = 176 \text{ kJ}$$

(2) The formation of salt from the elements:

$$2Na(s) + Cl_2(g) \rightleftharpoons 2NaCl(s)$$
$$\Delta H = -820 \text{ kJ}$$

(3) The decomposition of ammonium nitrite:

$$NH_4NO_2(s) \rightleftharpoons N_2(g) + 2H_2O(g)$$
$$\Delta H = -224 \text{ kJ}$$

(4) The production of glucose in photosynthesis:

$$6H_2O(l) + 6CO_2(g) \rightleftharpoons C_6H_{12}O_6(s) + 6O_2(g)$$
$$\Delta H = 2.82 \times 10^3 \text{ kJ}$$

(c) Which of the above reactions would occur spontaneously at every temperature?
(d) Which of the above reactions would *not* occur spontaneously at any temperature?

10-24. Graphite is the stable form of carbon at normal temperatures and pressures. The reaction $C_{graphite} \rightleftharpoons C_{diamond}$ has $\Delta H = 1.9$ kJ/mole and $\Delta S = -3.3$ kJ/mole.
(a) Calculate ΔG for the conversion of graphite to diamond near room temperature (300 K).
(b) If ΔH and ΔS are independent of temperature, is there some temperature at which graphite should spontaneously convert to diamond, and if so, what is the temperature? (As we saw in Chapter 4, at high *pressures* diamond becomes stable because its density is greater than that of graphite, but here we are asking for the effect of *temperature,* at a fixed pressure of 1 atm.)
(c) If graphite is the stable form of carbon at normal temperatures and pressures, how do you account for the fact that we never observe diamonds changing into graphite?

10-25. Ethane is decomposed by ultraviolet light to ethylene and hydrogen:

$$H_3C—CH_3 \xrightarrow{h\nu} H_2C{=}CH_2 + H—H$$

Three possible mechanisms are suggested for the reaction.

Mechanism A: "One fell swoop." One hydrogen from each carbon pops off simultaneously, forming H_2 and leaving behind ethylene:

$$H_3C—CH_3 \longrightarrow H_2C{=}CH_2 + H—H$$

Mechanism B: "Radical chain."

$$H_3C—CH_3 \longrightarrow H_3C—CH_2\cdot + H\cdot$$
$$H\cdot + H_3C—CH_3 \longrightarrow H_2 + H_3C—CH_2\cdot$$
$$H_3C—CH_2\cdot \longrightarrow H_2C{=}CH_2 + H\cdot$$
$$H\cdot + H\cdot \longrightarrow H_2$$

The dot on $H\cdot$ and on $CH_3CH_2\cdot$ means that each has an unpaired electron. Such species, which are highly reactive, are called "free radicals."

Mechanism C: "Elimination at one carbon, followed by rearrangement." The initial step is the formation of the hydrogen molecule from two hydrogens bonded to the same ethane carbon.

$$H_3C—CH_3 \longrightarrow H_3C—\ddot{C}H + H_2$$
$$H_3C—\ddot{C}H \longrightarrow H_2C{=}CH_2$$

The following two experiments were done (around 1960). Here, D stands for deuterium, the ${}^2_1H$ isotope.

(a) A mixture of equal amounts of $H_3C—CH_3$ and $D_3C—CD_3$ was decomposed by ultraviolet light. The hydrogen recovered was an equal mixture of H_2 and D_2, but it contained no HD. Which mechanism(s) of the above three does this experiment eliminate? Why?
(b) A clever synthetic chemist made $H_3C—CD_3$. On decomposition, it also gave only H_2 or D_2, and it yielded no HD at all. Which mechanism(s) of the above three does this experiment eliminate? Why?
(c) Do the results in parts a and b unambiguously demonstrate the "real mechanism" of this process? Discuss.

10-26. Iron reacts with oxygen gas to form ferric oxide (Fe_2O_3). For which of the following iron objects would the reaction between the iron and oxygen be most rapid: an iron nail, iron wire, or iron filings? Explain your answer.

10-27. Given the equation for the combustion of sucrose,

$$C_{12}H_{22}O_{11}(s) + 12O_2(g) \longrightarrow 12CO_2 + 11H_2O$$

Is it possible to determine from it the mechanism of the reaction? Explain your answer.

10-28. Why do fruits ripen more slowly in the refrigerator than at room temperature?

10-29. The kinetics of the reaction $A_2 + B + C \rightarrow AB + AC$ are being studied. It is found that doubling the concentration of A_2 doubles the reaction rate, tripling the concentration of B triples the rate, and doubling the concentration of C has no effect.
(a) Write the rate law for this reaction.
(b) How can it be that the concentration of C has no effect on the rate of the reaction, although its presence is necessary for the reaction to occur?
(c) Suggest a mechanism for the reaction that is consistent with the observed rate law.

10-30. Hydrogen gas reacts with nitric oxide to produce water and nitrogen:

$$2H_2(g) + 2NO(g) \longrightarrow 2H_2O(g) + N_2(g)$$

The data shown in the table below are obtained in an investigation of this reaction.

Experiment	Initial concentrations		Initial rate
	[NO](M)	[H](M)	
1	1×10^{-3}	1×10^{-3}	X
2	2×10^{-3}	1×10^{-3}	4X
3	3×10^{-3}	1×10^{-3}	9X
4	3×10^{-3}	2×10^{-3}	18X
5	3×10^{-3}	3×10^{-3}	27X

(a) Write the rate law for the reaction.
(b) What would be the rate when $[H_2] = 4 \times 10^{-3}$ M and $[NO] = 6 \times 10^{-3}$ M?

10-31. The proposed mechanism for a certain reaction is given by the following sequence:

Step 1: $Cl_2 + AlCl_3 \xrightarrow{\text{fast}} AlCl_4^- + Cl^+$

Step 2: $Cl^+ + C_6H_6 \xrightarrow{\text{slow}} C_6H_6Cl^+$

Step 3: $C_6H_6Cl^+ \xrightarrow{\text{fast}} C_6H_5Cl + H^+$

Step 4: $AlCl_4^- + H^+ \xrightarrow{\text{fast}} AlCl_3 + HCl$

(a) What is the overall reaction being studied?
(b) Which species is a catalyst?

(c) Which species is (are) reactive intermediates?

10-32. In the "contact process" for the production of the important industrial chemical sulfuric acid, the first step is the oxidation of sulfur dioxide to sulfur trioxide in the presence of a platinum catalyst:

$$2SO_2(g) + O_2(g) \xrightleftharpoons{Pt} 2SO_3(g)$$
$$\Delta H = -96 \text{ kJ/mole } SO_3$$

When the concentrations are given in moles/ℓ, the equilibrium constant for the reaction at 727°C is 261 ℓ/mole.

(a) For reactions involving gaseous species alone, the mass action expression often is written in terms of the partial pressures of the substances instead of their concentrations. (The partial pressure of each constituent is the pressure it would exert if it were alone in the container.) For the present reaction this alternative form of the mass action expression is $P^2_{SO_3}/P^2_{SO_2}P_{O_2} = K_p$. What is the value of the equilibrium constant K_p for this reaction at 727°C? Assume that the gases are ideal.

(b) If, momentarily, the gases are all at equal partial pressures of 1 atm, will a reaction occur spontaneously? If so, in which direction?

(c) What experimental conditions would lead to a high yield (that is, a high equilibrium concentration) of SO_3?

Bubbles of carbon dioxide gas form when baking soda ($NaHCO_3$) is sprinkled on an acid spill.

11

Acids and Bases

We have used the terms "acid" and "base" in earlier chapters without defining them in detail. We now study in some detail these important reagents, found in every chemical laboratory as well as in homes. The sour taste of many foods is due to acids—acetic acid in vinegar, citric acid in lemon juice, and lactic acid in sour milk. Solutions of bases, on the contrary, are bitter tasting and slippery feeling—sodium hydroxide as "lye" and ammonia as household ammonia are examples of bases. In this chapter we consider the properties of acids and bases.

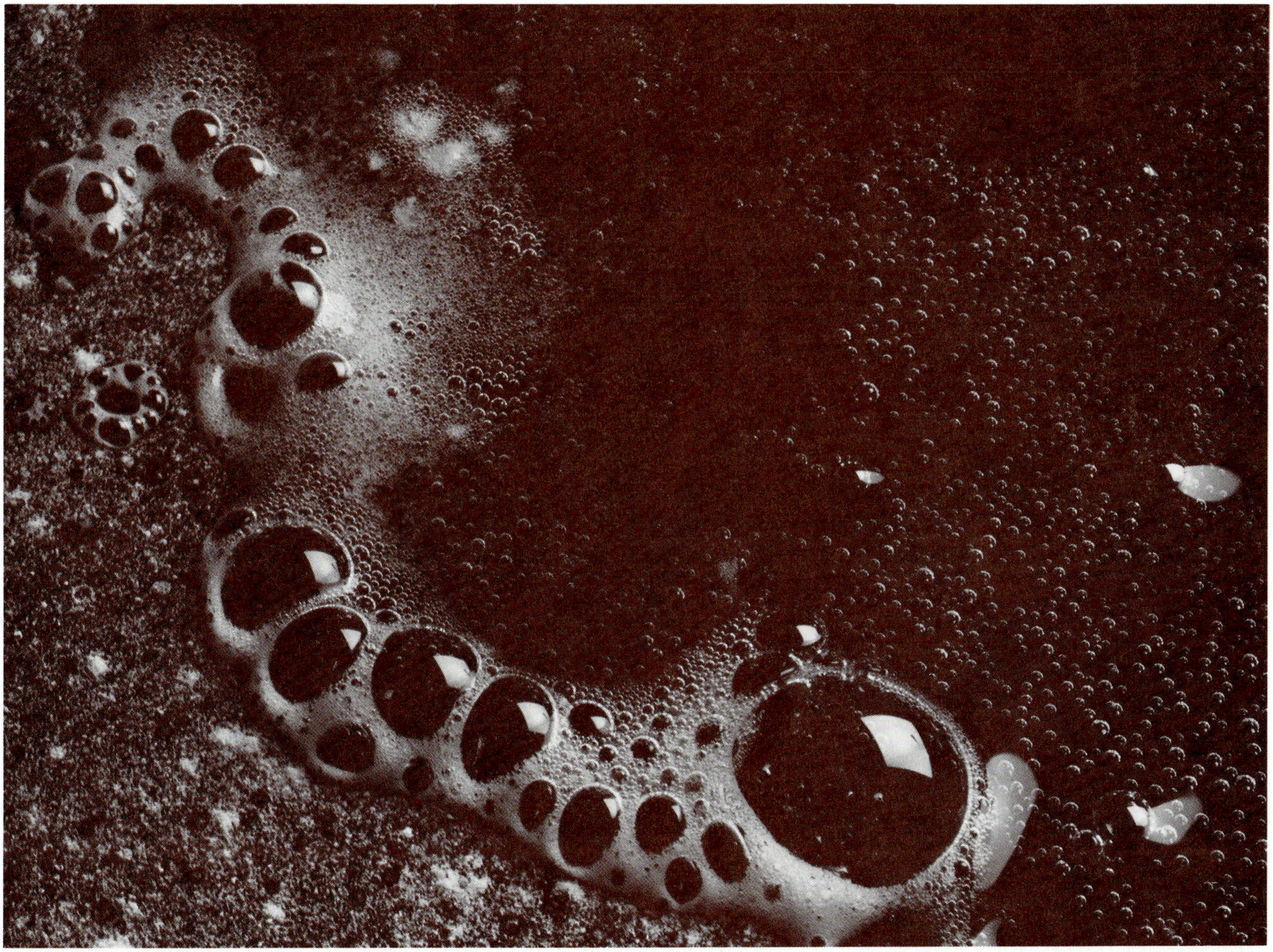

11-1
SIMPLE TESTS FOR ACIDS AND BASES

In the laboratory, acids and bases are often identified by **indicators,** dyes that have different colors in the presence of acids than they do with bases. The dye litmus is one of the best known indicators. It is red in acidic solution and blue in basic solution. Litmus is usually used in the form of litmus paper, paper that has been impregnated with the dye. We shall see other examples of indicators and how they work, later in this chapter.

Acids can also be identified by the reaction of their aqueous solutions with such metals as magnesium to form hydrogen gas:

$$2HCl(aq) + Mg \longrightarrow H_2(g) + MgCl_2(aq)$$

$$2CH_2COOH(aq) + Mg \longrightarrow H_2(g) + (CH_2COO)_2(aq)$$

Here "(aq)" means that the acid is in water solution. The reaction between magnesium and acetic acid (CH_3COOH) is slower than that between magnesium and hydrochloric acid (Figure 11-1). We shall see why shortly.

The experiment in electrical conductivity discussed in Chapter 9 demonstrates the presence of ions in some solutions. A similar experiment (Figure 11-2) measures the conductivity of solutions of acids and bases. Instead of a light bulb, a meter is used. All of the solutions conduct electricity, but some are better conductors than others. Table 11-1 summarizes the results of the experiment with several acids and bases. All of the solutions conduct electricity. They are all **electrolytes,** though some are better conductors than others, as measured by the meter. Acetic acid and ammonia are weaker electrolytes than the other acids and bases listed.

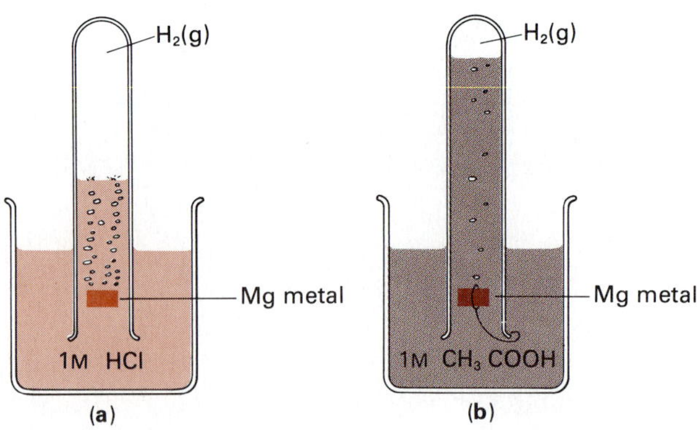

FIGURE 11-1

Reaction of acid and metal. Magnesium metal reacts much more rapidly **(a)** with hydrochloric acid than **(b)** with acetic acid.

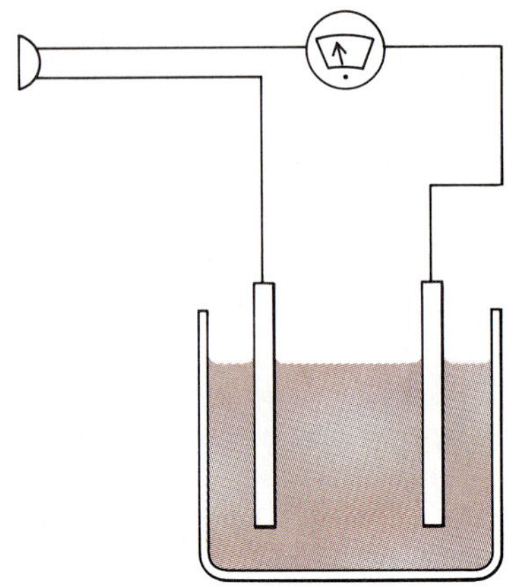

FIGURE 11-2

The conductivity of solutions of acids and bases can be measured.

TABLE 11-1
Strong and weak acids and bases

Strong acids and bases	Weak acids and bases
Hydrochloric acid (HCl)	Acetic acid (CH_3COOH)
Nitric acid (HNO_3)	Boric acid ($HB(OH)_4$)
Sulfuric acid (H_2SO_4)	Hydrocyanic acid (HCN)
Potassium hydroxide (KOH)	Ammonia (NH_3)
Sodium hydroxide ($NaOH$)	

11-2
ARRHENIUS DEFINITION OF ACIDS AND BASES

Why is it that some acids and bases are better conductors than others? And why does acetic acid, which is a weak electrolyte, react more slowly with magnesium than does the strong electrolyte, hydrochloric acid? To answer these questions, the Swedish chemist, Svante Arrhenius, proposed the following definitions of acids and bases.

Acids are substances that, in water solution, provide hydrogen ions (H^+). (Actually, the isolated proton cannot exist independently in aqueous solution, but is always **hydrated**—associated with at least one water molecule.)

Bases are substances that, in water solution, provide hydroxide ions (OH^-).

The chemical activity and electrical conductivity of acids and bases are directly related to the extent to which these H^+ and OH^- ions are present in the solution.

When hydrogen chloride gas dissolves in water, it is almost completely ionized. The equation for the reaction is usually written

$$HCl \longrightarrow H^+ + Cl^- \qquad (K \text{ is very large})$$

with the understanding that *all* the ions are really hydrated, and that the reaction goes essentially to completion in the forward direction.

When sulfuric acid (H_2SO_4) dissolves in water, it is almost 100% ionized to H^+ and HSO_4^- (bisulfate ion):

$$H_2SO_4 \longrightarrow H^+ + HSO_4^- \qquad (K \text{ is very large})$$

The bisulfate ion itself (HSO_4^-) is a fairly strong acid, ionizing (to a moderate extent) to yield H^+ and SO_4^{2-}. Acids such as HNO_3, HCl, and H_2SO_4, which are essentially completely ionized to H^+ and a negative ion (NO_3^-, Cl^-, HSO_4^-) in aqueous solution, are called **strong acids.**

When acetic acid dissolves in water, most of it remains as the undissociated molecule. Only a small fraction dissociates, and the concentration of H^+ is small:

$$CH_3COOH \rightleftharpoons H^+ + CH_3COO^-$$

$$K_a = \frac{[H^+][CH_3COO^-]}{[CH_3COOH]} = 1.8 \times 10^{-5} \frac{\text{moles}}{\text{liter}}$$

Here K_a is the equilibrium constant for the dissociation of acetic acid. In any but an extremely dilute solution of the acid, the position of equilibrium lies far to the left, toward the undissociated acid. Acetic acid is called a **weak acid**—it is low in conductivity and in its reactivity as an acid, as we have seen in its reaction with magnesium. Note that the words *strong* and *weak* refer to the ability of the acid to furnish H^+, and not to the concentration of the acid in solution. Thus, a 1 M solution of HCl provides 1 mole of H^+ per liter, but a 1 M solution of CH_3COOH provides only 0.004 moles of H^+ per liter. (Later in the chapter, we shall see how to calculate the concentration of H^+ in a solution of acetic acid.)

Sodium hydroxide, potassium hydroxide, and barium hydroxide are **strong bases,** almost 100% ionized in solution:

$$NaOH \longrightarrow Na^+ + OH^-$$

$$KOH \longrightarrow K^+ + OH^-$$

$$Ba(OH)_2 \longrightarrow Ba^{2+} + 2OH^-$$

Barium hydroxide is only moderately soluble in water (0.1 M at room temperature), but to the extent that it dissolves, it is 100% ionized.

Ammonia, on the contrary, is a weak electrolyte. The equilibrium in an aqueous solution of NH_3 is

$$NH_3 + H_2O \rightleftharpoons NH_4^+ + OH^-$$

$$K = \frac{[NH_4^+][OH^-]}{[NH_3]} = 1.8 \times 10^{-5} \frac{\text{moles}}{\text{liter}} \qquad (11\text{-}1)$$

Most of the ammonia remains as NH_3 molecules dissolved in water, but some reacts (according to Equation 11-1) to yield a weakly basic solution. Ammonia is a **weak base.**

A solution of NH_3 in water is often called "ammonium hydroxide, NH_4OH," which is said to dissociate, to a small extent, to NH_4^+ and OH^-. But there is probably no real chemical species NH_4OH present to any significant extent. The solution should just be pictured as NH_3 molecules dissolved in water, providing OH^- by the reaction shown in Equation 11-1.

11-3 NEUTRALIZATION

Acids and bases neutralize each other, yielding water and a salt:

Acid	+ *Base*	$\longrightarrow$	*Water*	+ *Salt*
HCl	+ $NaOH$	$\longrightarrow$	H_2O	+ $NaCl$ (sodium chloride)
HNO_3	+ KOH	$\longrightarrow$	H_2O	+ KNO_3 (potassium nitrate)
H_2SO_4	+ KOH	$\longrightarrow$	H_2O	+ $KHSO_4$ (potassium hydrogen sulfate, or potassium bisulfate)
H_2SO_4	+ $2KOH$	$\longrightarrow$	$2H_2O$	+ K_2SO_4 (potassium sulfate)
CH_3COOH	+ $NaOH$	$\longrightarrow$	H_2O	+ $NaCH_3COO$ (sodium acetate)

If these solutions are boiled long enough to drive off the water, what remains are the salts formed in the neutralization reactions. Salts taste salty and are electrolytes in aqueous solution. If formed from *strong* acids and bases, the salts usually form solutions that are neutral, or close to neutral. (Later in this chapter we shall see examples of salts whose solutions are not neutral.)

The above equations should, more accurately, be written so as to show that most of the substances are present as hydrated ions. For example,

$$H^+(aq) + Cl^-(aq) + Na^+(aq) + OH^-(aq) \longrightarrow H_2O + Na^+(aq) + Cl^-(aq) \quad (11\text{-}2)$$

The Na^+ and Cl^- are called **spectator ions.** They do not take part in the reaction, and so we do not need to include them in the equation for the reaction. Equation 11-2 is more simply written as

$$H^+ + OH^- \longrightarrow H_2O \quad (11\text{-}3)$$

Equation 11-3 is a **net ionic equation.** For the reaction between any *strong* acid and *strong* base, the net ionic equation is always the same: $H^+ + OH^- \rightarrow H_2O$. The reaction between a weak acid (say, CH_3COOH) and a strong base (say, $NaOH$) is written

$$CH_3COOH + OH^- \longrightarrow H_2O + CH_3COO^-$$

The acetic acid is present mainly as the undissociated molecules in solution, so its molecular formula is used in the equation (the concentration of H^+ in solution is small); the $NaOH$, on the other hand, is essentially

100% dissociated into Na^+ and OH^- ions. The reaction then is between CH_3COOH molecules and OH^- ions.

Evidence that all strong acid–strong base neutralization reactions are similar in character comes from their heats of reaction. The ΔH for neutralization is the same for different combinations of acid and base. As seen in Table 11-2, neutralization reactions are exothermic. The ΔH is -56.1 kJ per mole of water formed for every combination of a strong acid and a strong base. With the weak acetic acid or the weak base ammonia, the heat of reaction is slightly lower—some heat is absorbed in removing H^+ from the CH_3COOH molecule, and in removing OH^- from the H_2O by the NH_3.

TABLE 11-2
Heats of neutralization

Acid	Base	Heat of neutralization (kJ per mole of H_2O formed)
HCl	NaOH	−56.1
HCl	KOH	−56.1
HCl	$Ba(OH)_2$	−56.1
HCl	NH_3	−52.7
HNO_3	NaOH	−56.1
$HClO_4$	KOH	−56.1
CH_3COOH	NH_3	−51.5

EXERCISE 11-1 (a) Write a balanced net ionic equation for the reaction of KOH with HNO_3.
(b) Write a balanced net ionic equation for the reaction of CH_3COOH with KOH.

11-4 WATER EQUILIBRIUM AND pH

The equation for an acid–base neutralization should be written so as to show that an equilibrium exists between the reactants (H^+ and OH^-) and the product (H_2O). However, we saw in Chapter 9 that pure water is nonconducting, a nonelectrolyte. The neutralization reaction thus goes virtually to completion, and its reverse (the dissociation of H_2O into H^+ and OH^-) occurs only to a slight extent:

$$H_2O \rightleftharpoons H^+ + OH^- \qquad (K \text{ is very small})$$

Because very little water dissociates, its concentration is (to a close approximation) constant, both in the pure liquid and in dilute aqueous solution. We can, therefore, include this virtually constant concentration of the water in the equilibrium constant, as we did in Chapter 10:

$$K_w = [H^+][OH^-] = 1.0 \times 10^{-14} \left(\frac{\text{mole}}{\ell}\right)^2 \text{ at } 25\,°C$$

where K_w represents the dissociation constant for water. *This relation between H^+ and OH^- always holds in water,* whether it be pure water or a dilute aqueous solution—and, in the latter case, whether the solutes be ionized or not. Thus:

In pure water at 25°C, $[H^+][OH^-] = 1.0 \times 10^{-14}$
In 0.1 M NaCl solution, $[H^+][OH^-] = 1.0 \times 10^{-14}$
In 6.0 M HCl solution, $[H^+][OH^-] = 1.0 \times 10^{-14}$

In 1 M sucrose solution, $[H^+][OH^-] = 1.0 \times 10^{-14}$
And so on.

For each water molecule that dissociates, *one* H^+ and *one* OH^- are formed. In pure water, or in neutral solution,

$$[H^+] = [OH^-]$$

Therefore, *in pure water or in neutral solution* at 25°C,

$$K_w = [H^+][OH^-] = [H^+][H^+] = 1.0 \times 10^{-14} \left(\frac{\text{mole}}{\ell}\right)^2$$

$$[H^+]^2 = 1.0 \times 10^{-14}$$

$$[H^+] = 1.0 \times 10^{-7}\ \text{M}$$

$$[OH^-] = 1.0 \times 10^{-7}\ \text{M}$$

The concentration of H^+ and OH^- ions in pure water is always 1×10^{-7} M. The dissociation constant for water applies also in acid solution, but there the concentrations of H^+ and OH^- are no longer equal. The concentration of H^+ is now *much* greater than that of OH^-, with most of the H^+ coming from the acid and only a negligible amount from the dissociation of water. But the water dissociation is the *only* source of OH^-. We can use the water equilibrium relationship, $[H^+][OH^-] = K_w$, to determine the concentration of OH^- in acid solution. For example, we can calculate the concentration of OH^- in 0.10 M HCl as follows.

The HCl dissociates to yield 0.10 M H^+. The total concentration of H^+ is 0.10 M, plus a contribution from the dissociation of water. The concentration of H^+ from the dissociation of *pure* water is only 1.0×10^{-7} M, and it will be even smaller here because (in accord with Le Châtelier's Principle) the added H^+ from the HCl will repress the dissociation of the water. So, the contribution of H^+ from the dissociation of water is negligible and, to a close approximation,

$$[H^+] = 0.10\ \text{M}$$

We know that $[H^+][OH^-] = 1.0 \times 10^{-14}$, so

$$[OH^-] = \frac{K_w}{[H^+]} = \frac{1.0 \times 10^{-14}}{0.10} = 1.0 \times 10^{-13}\ \text{M}$$

We see now that only 1.0×10^{-13} mole of water per liter dissociates in 0.10 M HCl solution, producing 1.0×10^{-13} M OH^-, and also adding 1.0×10^{-13} M H^+ to the very much larger H^+ concentration contributed by the strong acid. Our assumption (that the H^+ from the water is negligible compared to the 0.10 M H^+ from the HCl) has proved to be a good one!

In basic solution, the concentration of OH^- is greater than that of H^+. Essentially *all* of the OH^- comes from the base, and very little comes from the dissociation of water. But the water dissociation is the *only* source of H^+. In 0.010 M NaOH, for example, because NaOH is a *strong* base, (completely dissociated into Na^+ and OH^-),

$$[OH^-] = 0.010\ \text{M (plus a negligible amount from the dissociation of water)}$$

$$[H^+] = \frac{K_w}{[OH^-]} = \frac{1.0 \times 10^{-14}}{0.010} = 1.0 \times 10^{-12}\ \text{M}$$

A convenient way of expressing concentrations of H^+, even very small concentrations, is through the use of pH, which is defined as follows:

$$[H^+] = 10^{-\text{pH}}\ \frac{\text{moles}}{\ell}$$

or

$$\text{pH} = -\log [H^+]$$

We can use the first definition to find the pH of the following solutions.

Neutral solution: $[H^+] = 1.0 \times 10^{-7}$ M, and pH = 7.

Acidic solution, 0.10 M HCl: $[H^+] = 1.0 \times 10^{-1}$ M, and pH = 1. Solutions of pH less than 7 are acidic.

Basic solution, 0.010 M NaOH: $[OH^-] = 0.010$ M, so $[H^+] = 1.0 \times 10^{-12}$ M, and pH = 12. Solutions of pH greater than 7 are basic.

If $[H^+] = 6 \times 10^{-4}$ M, the pH lies *between* 3 and 4 (because 6×10^{-4} is more than 10^{-4} but less than 10^{-3}). To find the pH more precisely, it is necessary to use logarithms and the second definition of pH: $pH = -\log[H^+] = -\log(6 \times 10^{-4}) = -(0.8 - 4) = 3.2$. Logarithms and pH are discussed (with examples) in Appendix 2.

If $[H^+] = 1.2 \times 10^{-8}$ M, the pH lies between 7 and 8, but close to 8. The precise pH, obtained with logarithms, is 7.92.

Table 11-3 gives concentrations of H^+ with the corresponding pH values, and the concentrations of OH^- at those same pH values; Figure 11-3 gives the pH values of some familiar substances. The following three exercises will help you see whether you understand these important ideas.

EXERCISE 11-2 (a) If the pH of a solution is 5, is the solution acidic or basic?

(b) If the pH of a solution is 11, is the solution acidic or basic?

(c) What is the OH^- concentration in a solution of pH 9?

EXERCISE 11-3 Find the concentrations of H^+ and OH^- and the pH of each of the following solutions: (a) 0.001 M HNO_3 (b) 1.0 M KOH (c) 10 M HCl

EXERCISE 11-4 (a) What is the approximate pH of a 1.3×10^{-5} M solution of HCl?

(b) What is the approximate $[H^+]$ of a solution whose pH is 10.8?

TABLE 11-3
H^+ Concentration, pH, and OH^- concentration

Concentration of H^+ (moles/ℓ)	pH	Concentration of OH^- (moles/ℓ)	
1×10^{0} (1 M)	0	1×10^{-14}	
1×10^{-1} (0.1 M)	1	1×10^{-13}	
1×10^{-2} (0.01 M)	2	1×10^{-12}	
1×10^{-3}	3	1×10^{-11}	Acidic
1×10^{-4}	4	1×10^{-10}	
1×10^{-5}	5	1×10^{-9}	
1×10^{-6}	6	1×10^{-8}	
1×10^{-7}	7	1×10^{-7}	Neutral
1×10^{-8}	8	1×10^{-6}	
1×10^{-9}	9	1×10^{-5}	
1×10^{-10}	10	1×10^{-4}	
1×10^{-11}	11	1×10^{-3}	Basic
1×10^{-12}	12	1×10^{-2} (0.01 M)	
1×10^{-13}	13	1×10^{-1} (0.1 M)	
1×10^{-14}	14	1×10^{-0} (1 M)	

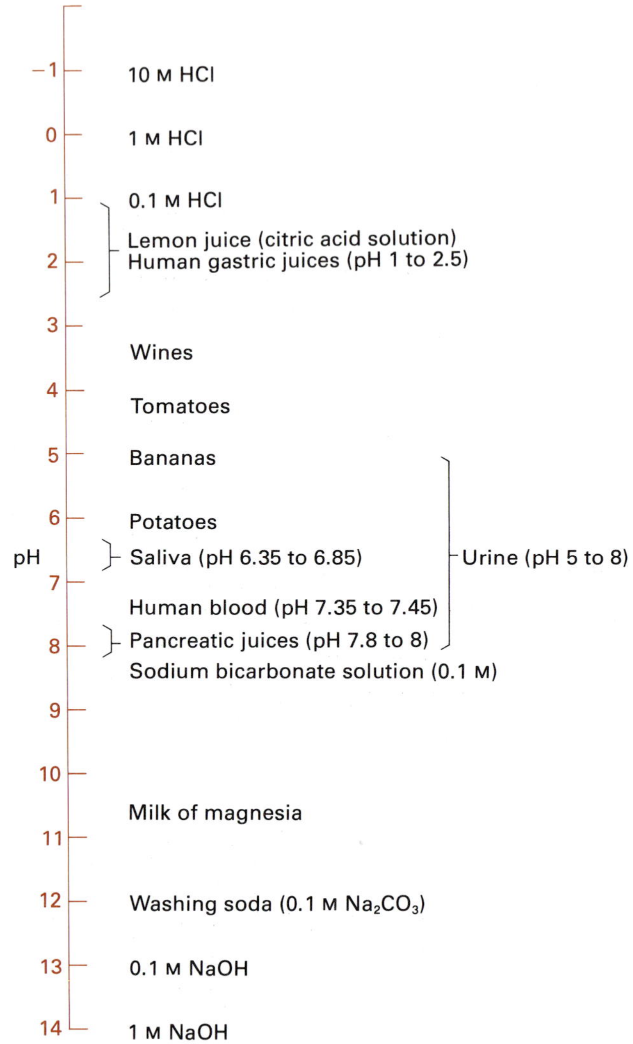

FIGURE 11-3

The pH values of some familiar substances.

11-5 TITRATION

Titration is an experimental method of determining concentration that is commonly used to determine the concentrations of acidic or basic solutions.

In the following discussion, which illustrates the principles of titration for an acid–base neutralization reaction, we see how the concentration of H^+ changes in the titration of a strong acid (HCl) with a strong base (NaOH). To simplify the example, we use *solid* NaOH here, although in a real titration we would use NaOH in solution.

TABLE 11-4
Titration of 100 ml of 1.00 M HCl with solid NaOH

Moles NaOH added	Excess moles H^+	Excess moles OH^-	Concentration (moles/ℓ) H^+	Concentration (moles/ℓ) OH^-	pH
0	0.100		1.0 (1×10^0)	1×10^{-14}	0
0.050	0.050		0.50	2×10^{-14}	0.3
0.090	0.010		0.10 (1×10^{-1})	1×10^{-13}	1
0.099	0.001		0.01 (1×10^{-2})	1×10^{-12}	2
0.100	0	0	1×10^{-7}	1×10^{-7}	7
0.101		0.001	1×10^{-12}	0.01	12
0.110		0.010	1×10^{-13}	0.10	13
0.200		0.100	1×10^{-14}	1.0	14

We start out with 100 ml of 1.00 M HCl, and add solid NaOH, bit by bit. (We assume that the addition of the NaOH does not change the volume of the solution significantly, so the total volume remains 100 ml throughout the titration.) Table 11-4 shows the results of the experiment.

$$\text{Initial number of moles of H}^+ = 1.00\,\frac{\text{moles}}{\ell} \times 0.100\ \ell$$

$$= 0.100 \text{ mole}$$

As NaOH is added, it neutralizes the H^+, by the reaction

$$H^+ + OH^- \longrightarrow H_2O$$

And so the solution becomes progressively less strongly acid. In Table 11-4, we see that after the first three additions of NaOH, there is still excess H^+ left in the solution. When 0.100 mole of NaOH has been added, just enough to neutralize *all* the H^+, the solution is neutral (the solution now is 0.100 M in NaCl). The H^+ and OH^- concentrations are now the same and are 1.0×10^{-7} M. Then, as more NaOH is added, the solution becomes more and more basic.

In Example 11-1, we see how to calculate the pH of the 1.00 M HCl solution after the addition of solid NaOH.

EXAMPLE 11-1

As in Table 11-4, 0.090 mole of solid NaOH is added to 100 ml of 1.00 M HCl. What is the pH of the resulting solution? (Assume no change in volume.)

$$\text{Moles HCl} = 1.00\,\frac{\text{mole}}{\ell} \times 0.100\ \ell$$

$$= 0.100 \text{ mole HCl}$$

$$\text{Moles NaOH added} = 0.090 \text{ mole NaOH}$$

A neutralization reaction occurs in which *all* of the NaOH reacts with the HCl (that is, 0.090 mole NaOH reacts with 0.090 mole HCl):

$$HCl + NaOH \longrightarrow H_2O + NaCl$$

$$\text{Moles HCl after reaction} = 0.100 - 0.090$$

$$= 0.010 \text{ mole HCl}$$

$$\text{Moles H}^+ \text{ in solution} = 0.010 \text{ mole H}^+$$

This number of moles of H^+ is contained in 100 ml of solution.

$$\text{Concentration of } H^+ = \frac{0.010 \text{ mole}}{0.100\ \ell} = 0.10\ \text{M}$$
$$= 1.0 \times 10^{-1}\ \text{M}$$
$$\text{pH} = 1.0$$

EXERCISE 11-5 As in Table 11-4, 0.110 mole of solid NaOH is added to 100 ml of 1.00 M HCl. What is the pH of the resulting solution? (Assume no change in volume.)

The data from Table 11-4 are plotted in Figure 11-4, with $[H^+]$ plotted against the number of moles of added NaOH in Figure 11-4a, and pH against the number of moles of added NaOH in Figure 11-4b. On the scale of Figure 11-4a, H^+ concentrations of less than about 0.01 M are indistinguishable from zero, but on the pH scale of Figure 11-4b, very low concentrations are just as easily distinguished as high concentrations. It is for that very reason that the pH scale is so frequently used as a measure of H^+ concentration.

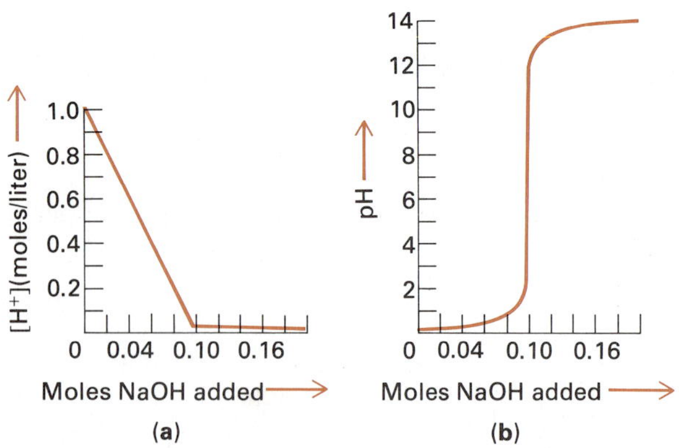

FIGURE 11-4
Titration of 100 ml of 1 M HCl with *solid* NaOH.

The abrupt rise in pH at the **equivalence point** (the point where the number of moles of NaOH added is exactly equal to the number of moles of HCl in the original solution) is characteristic of a titration involving a strong acid and strong base. The equivalence point is at a pH of 7. If either the acid or the base is weak, the rise in pH at the equivalence point is not so sharp, and the pH at the equivalence point may not be 7. (The value depends on the particular acid or base used. We shall see the reason for this later in the chapter.)

By locating the point in a titration where the pH changes rapidly, we can determine the equivalence point. For example, if a few drops of the indicator phenolphthalein (which is colorless in acidic solution but pink in basic solution) are added at the beginning of the titration, it will indicate the point where the pH changes rapidly by changing from colorless to pink. The solution is colorless through the point where 0.099 mole of NaOH is added, but pink by the point where 0.101 mole of NaOH has been added. (In this region, the addition of only 0.002 mole of NaOH changes the pH from 2 to 12.) Thus phenolphthalein indicates the equivalence point by a change of color. Alternatively, a pH meter—a device that measures $[H^+]$ electronically—can be used to plot a pH curve. The equivalence point is the midpoint of the section of the curve where the pH increases most rapidly.

In actual practice, the base with which the acid is titrated would also be in solution, and would be added a drop at a time. Figure 11-5 shows an experimental set-up for the titration of a sample of vinegar, of *unknown* concentration of acetic acid, with a solution of sodium hydroxide of *known* concentration, a *standardized* solution of NaOH.

In the following examples, titration is used to determine the concentration of an acidic (or a basic) solution of unknown concentration. In each case, the reaction that occurs is an acid–base neutralization where one of the solutions is of known, and the other of unknown, concentration.

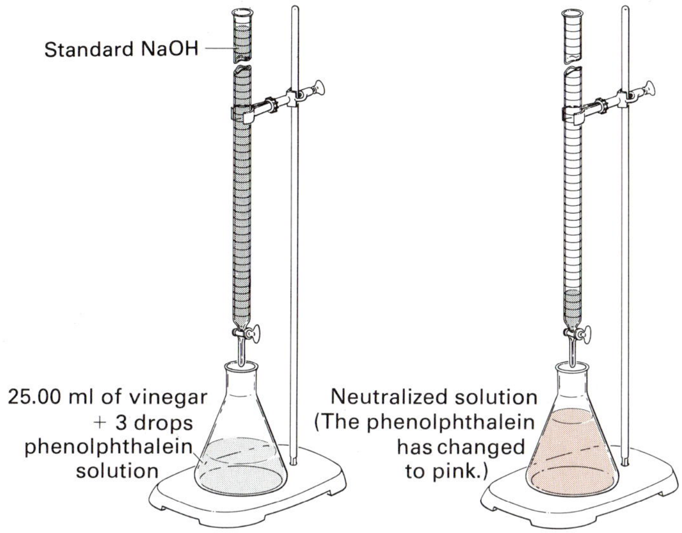

FIGURE 11-5

Titration of vinegar with standardized sodium hydroxide, using an indicator to determine the equivalence point.

EXAMPLE 11-2

A 0.100 M solution of NaOH is used to titrate 100 ml of an HCl solution of unknown concentration, using phenolphthalein as an indicator. If 75 ml of the NaOH are needed to reach the equivalence point (where the solution turns from colorless to pink), what is the concentration of the HCl solution?

$$\text{Moles of NaOH added} = 0.100\,\frac{\text{mole}}{\ell} \times 0.075\ \ell$$

$$= 0.00750\ \text{mole}$$

At the equivalence point, the number of moles of NaOH added is exactly equal to the number of moles of HCl in the 100 ml sample:

$$NaOH + HCl \longrightarrow H_2O + NaCl$$

(*One* mole of NaOH neutralizes *one* mole of HCl.)

Therefore,

$$\text{Moles of HCl} = 0.00750\ \text{mole NaOH} \times \frac{1\ \text{mole HCl}}{1\ \text{mole NaOH}}$$

$$= 0.00750\ \text{mole HCl}$$

This number of moles of HCl is contained in 100 ml of the HCl solution.

$$\text{Concentration of HCl solution} = \frac{0.00750\ \text{mole}}{0.100\ \ell}$$

$$= 0.0750\ \text{M HCl}$$

This answer is reasonable—because the number of moles of NaOH in only 75 ml of solution is equal to the number of moles of HCl in 100 ml of solution, the NaOH must be more concentrated than the HCl.

EXAMPLE 11-3

A solution of barium hydroxide, $Ba(OH)_2$, of unknown concentration is used to titrate 100 ml of the HCl solution whose concentration was determined (in Example 11-2) to be 0.0750 M. Phenolphthalein is used as an indicator. If 50.0 ml of the $Ba(OH)_2$ is needed to reach the equivalence point, what is the concentration of the $Ba(OH)_2$ solution?

$$\text{Number of moles of HCl} = 0.0750\,\frac{\text{mole}}{\ell} \times 0.100\ \ell$$

$$= 0.00750\ \text{mole}$$

At the equivalence point, the number of moles of $Ba(OH)_2$ added is *one-half* the number of moles of HCl in the 100 ml sample:

$$Ba(OH)_2 + 2\,HCl \longrightarrow 2H_2O + BaCl_2$$

(*One* mole of $Ba(OH)_2$ neutralizes *two* moles of HCl. Note that the number of moles of OH^- added is equal to the number of moles of H^+ provided by the acid.)

Therefore,

$$\text{Moles Ba(OH)}_2 = 0.00750 \text{ mole HCl} \times \frac{1 \text{ mole Ba(OH)}_2}{2 \text{ mole HCl}}$$

$$= 0.00375 \text{ mole Ba(OH)}_2$$

This number of moles of $Ba(OH)_2$ is contained in 50.0 ml of solution.

$$\text{Concentration of solution} = \frac{0.00375 \text{ mole}}{0.0500 \text{ } \ell}$$

$$= 0.0750 \text{ M Ba(OH)}_2$$

This answer is reasonable—we have used only *half* as much of the $Ba(OH)_2$ solution as we did of the HCl solution, but each $Ba(OH)_2$ furnishes *two* OH^- ions.

EXERCISE 11-6 Find the H^+ concentrations of the solutions prepared by mixing

(a) 50 ml of 0.200 M NaOH and 40 ml of 0.200 M HCl
(b) 50 ml of 0.200 M NaOH and 50 ml of 0.200 M HCl
(c) 50 ml of 0.200 M NaOH and 60 ml of 0.200 M HCl.

(First determine how many moles of H^+ or OH^- remain after reaction. Then, taking account of the volume change, calculate their concentrations. Assume that the final volume is equal to the sum of the two separate volumes.)

In the examples just given, the concentrations of the acids and bases are expressed in molarities. You may encounter, elsewhere, acid and base concentrations in terms of normality. **Normality** (N) is the number of equivalents of solute per liter of solution, where an **equivalent** of an acid is the amount that furnishes one mole of H^+ in a neutralization, and an **equivalent** of a base is the amount that furnishes one mole of OH^-. One equivalent of an acid always reacts with one equivalent of a base.

One mole of HCl furnishes one mole of H^+. There is, then, one equivalent in one mole of HCl, and the normality is equal to the molarity:

$$1 \frac{\text{mole}}{\ell} \text{HCl} = 1 \frac{\text{equiv}}{\ell} \text{H}^+$$

$$1 \text{ M HCl} = 1 \text{ N HCl}$$

But one mole of H_2SO_4 can furnish two moles of H^+ in a neutralization. Then there are two equivalents in one mole of H_2SO_4, and the normality is twice the molarity:

$$1 \frac{\text{mole}}{\ell} \text{H}_2\text{SO}_4 = 2 \frac{\text{equiv}}{\ell} \text{H}^+$$

$$1 \text{ M H}_2\text{SO}_4 = 2 \text{ N H}_2\text{SO}_4$$

11-6 BRØNSTED–LOWRY DEFINITION OF ACIDS AND BASES

The Arrhenius model has the advantage of simplicity, but it does not give any information about the nature of the reaction between an acid or a base and water. Furthermore, there are many substances (such as sodium carbonate, Na_2CO_3) that do not provide OH^- ions but that are bases in aqueous solution. In 1923, Danish chemist J. N. Brønsted and (independently) the English chemist T. M. Lowry proposed to define acids and bases more broadly.

An **acid** is a substance that *donates* a proton.

A **base** is a substance that *accepts* a proton.

This definition of acids and bases is the one that we shall find most useful.

Any acid or base according to the Arrhenius definition also is one according to the Brønsted–Lowry definition. Although the Brønsted–Lowry definition of an acid is similar to that of Arrhenius, the definition of a base is more general, as can be seen in the following reactions. In each the first reactant acts as an acid, and the second as a base:

Acid	*+ Base*		*Acid*	*+ Base*	
$HCl(g)$	$+ H_2O$	$\rightleftharpoons$	H_3O^+	$+ Cl^-$	(11-4)
CH_3COOH	$+ H_2O$	$\rightleftharpoons$	H_3O^+	$+ CH_3COO^-$	(11-5)
H_2O	$+ NH_3(g)$	$\rightleftharpoons$	NH_4^+	$+ OH^-$	(11-6)
CH_3COOH	$+ NH_3(g)$	$\rightleftharpoons$	NH_4^+	$+ CH_3COO^-$	(11-7)

In the above reactions, each acid (HCl, CH_3COOH, and H_2O) *donates* a proton, and each base (H_2O and NH_3) *accepts* a proton. The H_2O in Equation 11-5 and the NH_3 in Equation 11-7 are bases in the Brønsted–Lowry system, but would not be bases in the Arrhenius system because they do not furnish hydroxide ions. Note that H_2O acts as a base in Equations 11-4 and 11-5, but as an acid in Equation 11-6; H_2O can either accept or donate a proton.

The *products* in the above reactions also are acids and bases, as can be seen in the reaction between CH_3COOH and NH_3:

$$\underset{\text{acid 1}}{CH_3COOH} + \underset{\text{base 2}}{NH_3(g)} \rightleftharpoons \underset{\text{acid 2}}{NH_4^+} + \underset{\text{base 1}}{CH_3COO^-}$$

The CH_3COOH (acid 1) can donate a proton to NH_3 (base 2), but the reaction is reversible, and NH_4^+ (acid 2) can donate a proton to CH_3COO^- (base 1). Each pair, CH_3COOH—CH_3COO^- (acid 1–base 1), and NH_4^+—NH_3 (acid 2–base 2), is called a **conjugate acid–base pair;** its two members differ only by a proton. Each acid–base reaction is the transfer of a proton from an acid to a base, producing thereby the conjugate base and acid.

We have already seen that acids have different strengths, varying in their tendency to lose a proton. A strong acid necessarily has a weak conjugate base—for example, the strong acid HCl is entirely dissociated into H^+ and Cl^- in aqueous solution. Its conjugate base, Cl^-, has almost no tendency to accept protons, and so is a weak base. A relatively weak acid (such as CH_3COOH) has a relatively strong conjugate base (CH_3COO^-). The CH_3COO^- has a much greater tendency to accept a proton than does Cl^-, so the CH_3COO^- is a stronger base. The stronger an acid, the weaker its conjugate base; and the weaker an acid, the stronger its conjugate base.

In Table 11-5, acids and their conjugate bases are listed in order of *decreasing* acid strength (with the strongest acid at the top of the table, and the weakest at the bottom). The conjugate bases are, consequently, in order of *increasing* strength (with the weakest base at the top of the table, and the strongest at the bottom). *Any acid has a great tendency to donate a proton to a base below it in the table, but only a small tendency to donate a proton to a base above it.* For example, if HF is added to a solution containing CO_3^{2-} (say, from Na_2CO_3), which is a base below HF in the table, then there is a strong tendency for the HF to donate its proton to the CO_3^{2-}. The following proton transfer will occur:

$$\cancel{Na^+} + HF + CO_3^{2-} \rightleftharpoons HCO_3^- + F^- + \cancel{Na^+}$$

(The Na^+ is a spectator ion.) In this reaction, products are favored over reactants. The reaction goes, to a great extent, in the forward direction—though, of course, all four species (HF, CO_3^{2-}, HCO_3^-, and F^-) are present at equilibrium.

TABLE 11-5
Brønsted–Lowry conjugate acid–base pairs

	Acid	Base	
Strongest			Weakest
	$HClO_4$	ClO_4^-	
	H_2SO_4	HSO_4^-	
	HCl	Cl^-	
	HNO_3	NO_3^-	
	H_3O^+	H_2O	
	HSO_4^-	SO_4^{2-}	
	H_3PO_4	$H_2PO_4^-$	
	HF	F^-	
	CH_3COOH	CH_3COO^-	
	$CO_2 + H_2O$	HCO_3^-	
	H_2S	HS^-	
	$H_2PO_4^-$	HPO_4^{2-}	
	NH_4^+	NH_3	
	HCO_3^-	CO_3^{2-}	
	HPO_4^{2-}	PO_4^{3-}	
	H_2O	OH^-	
	HS^-	S^{2-}	
	OH^-	O^{2-}	
Weakest			Strongest

If, however, HF is added to a solution containing Cl^- (say, from NaCl), there is *little* tendency for the HF to donate its proton to Cl^-, which is a base above HF in the table. The reaction

$$\cancel{Na^+} + HF + Cl^- \rightleftharpoons HCl + F^- + \cancel{Na^+}$$

occurs only to a very small extent. In this case, reactants are favored over products.

We see in Table 11-5 that H_2O is a very weak acid, having a great tendency to donate a proton only to the strongest bases, S^{2-} and O^{2-}. Only $HClO_4$, H_2SO_4, HCl, and HNO_3 are strong acids in aqueous solution, with a great tendency to donate a proton to water (which is a weak base). All other acids are weak, with little tendency to transfer a proton to water. Note that HSO_4^- has only a small tendency to donate a proton to water, so 1 M H_2SO_4 yields only a little over 1 mole of H^+ per liter in aqueous solution. (If 2 moles of a strong base, such as NaOH, are added to 1 ℓ of 1 M H_2SO_4, however, *both* of the H^+ ions in the H_2SO_4 will be neutralized.)

With the Brønsted–Lowry theory, it is easy to see why some salts dissolve in water to give acidic or basic solutions. Sodium carbonate (Na_2CO_3, "washing soda") yields a basic solution:

$$\cancel{Na^+} + \underset{\text{base 2}}{CO_3^{2-}} + \underset{\text{acid 1}}{H_2O} \rightleftharpoons \underset{\text{base 1}}{OH^-} + \underset{\text{acid 2}}{HCO_3^-} + \cancel{Na^+}$$

The base CO_3^{2-} lies above the acid H_2O in the table, and so reactants are favored over products. Still, the forward reaction occurs to *some* extent, and solutions of Na_2CO_3 are basic. The Na^+ does not combine with OH^- because NaOH is a strong base. *Salts of a weak acid and a strong base* (such as Na_2CO_3) *dissolve in water to give basic solutions.*

Ammonium chloride (NH_4Cl), on the contrary, dissolves in water to yield an acidic solution:

$$\cancel{Cl^-} + \underset{\text{acid 2}}{NH_4^+} + \underset{\text{base 1}}{H_2O} \rightleftharpoons \underset{\text{acid 1}}{H_3O^+} + \underset{\text{base 2}}{NH_3} + \cancel{Cl^-}$$

The base H_2O lies above the acid NH_4^+ in the table, so reactants are favored over products. But the forward reaction occurs to a small extent, and it accounts for the fact that solutions of NH_4Cl are acidic. The Cl^- does not react with the H_3O^+ because HCl is a strong acid. *Salts of a strong acid and a weak base* (such as NH_4Cl) *dissolve in water to give acidic solutions.*

Sodium chloride (NaCl) gives a neutral solution in water because NaOH is a strong base and HCl is a strong acid. The following reactions *do not* occur:

$$Na^+ + H_2O \xrightarrow{\text{does not occur}} NaOH + H^+$$

$$Cl^- + H_2O \xrightarrow{\text{does not occur}} HCl + OH^-$$

As we have already seen in our discussion of acid–base neutralization, the *reverse* of these reactions goes essentially to completion. *Salts of a strong acid and a strong base* (such as NaCl) *give neutral solutions when dissolved in water.*

EXERCISE 11-7 (a) Write the equation for the reaction of H_2S with CO_3^{2-}. Indicate the conjugate acid–base pairs. Are reactants or products favored?

(b) Write the equation for the reaction of H_2S with CH_3COO^-. Indicate the conjugate acid–base pairs. Are reactants or products favored?

(c) The salt trisodium phosphate (Na_3PO_4) gives a basic solution in water, so it is often used in cleaning solutions. Write an equation for the principal reaction that occurs when Na_3PO_4 is dissolved in water.

11-7 LEWIS DEFINITION OF ACIDS AND BASES

At about the same time that Brønsted and Lowry introduced their definition of acids and bases, the American chemist G. N. Lewis proposed an even more general definition.

An **acid** is a substance that can accept an electron pair.

A **base** is a substance that can donate an electron pair.

In an acid–base neutralization reaction, a **coordinate covalent bond** is formed—a bond in which the shared electron pair is contributed by only one atom. It is easy to demonstrate that any Arrhenius or Brønsted–Lowry acid is also a Lewis acid. And any Arrhenius or Brønsted–Lowry base is also a Lewis base—for example, ammonia, water, the cyanide ion, and the hydroxide ion:

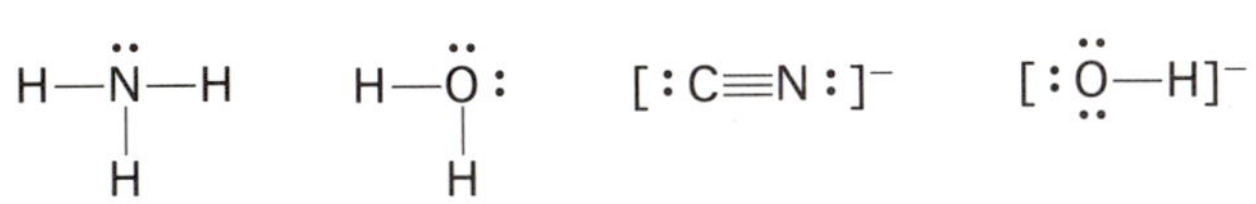

Each of these can accept a proton and is therefore a Brønsted–Lowry base, and each also can donate an electron pair and is therefore a Lewis base.

The Lewis definition, however, greatly increases the number of substances that can be classified as acids. The substance that accepts the electron pair may be a proton:

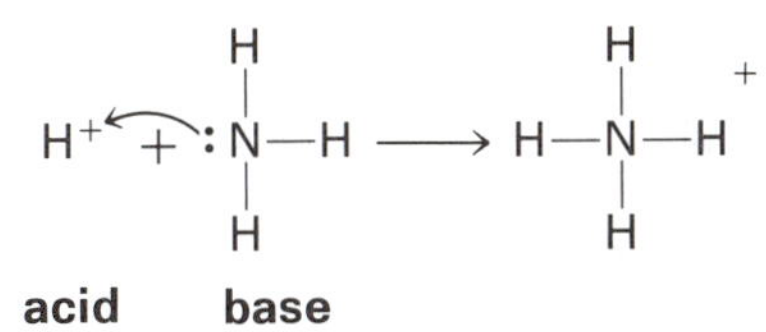

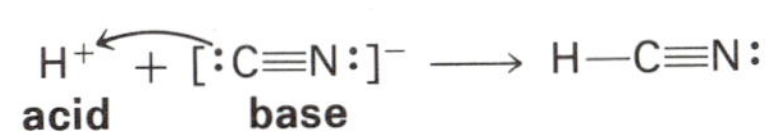

But, in the Lewis definition, the electron-pair acceptor need not be a proton. For example, it might be a molecule with an incomplete octet of electrons. Boron trifluoride (BF_3) is an example of a Lewis acid that is not a Brønsted–Lowry acid:

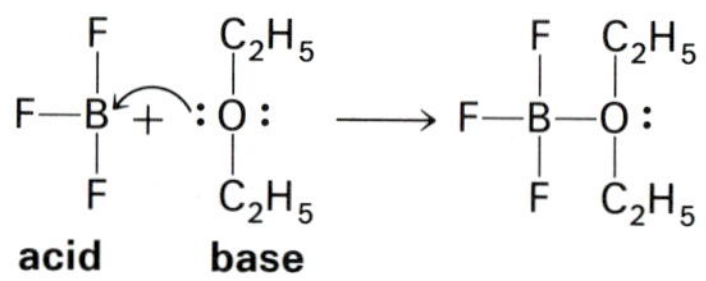

The molecule formed in the reaction between BF_3 and diethyl ether is called an etherate—it is used as an acid catalyst in organic reactions. The etherate is a convenient form in which to use the BF_3, and when a stronger Lewis base (a better electron-pair donor) than the ether is present, the BF_3 releases the ether molecules and combines with the stronger base instead.

Metal cations and atoms can also act as Lewis acids. In the reaction between Cu^{2+} and NH_3 to form the copper–ammonia complex ion, $Cu(NH_3)_4^{2+}$, Cu^{2+} is a Lewis acid. The complex ion, which is a beautiful deep blue color, is used to identify the Cu^{2+} ion.

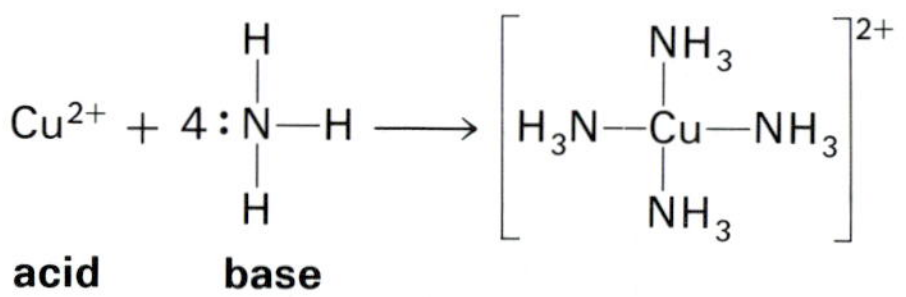

In water solution, all metal ions undergo acid–base reactions with water as do the Zn^{2+} ions shown here, though the number of water molecules in the hydrated complex ion (the water of hydration) is not always six:

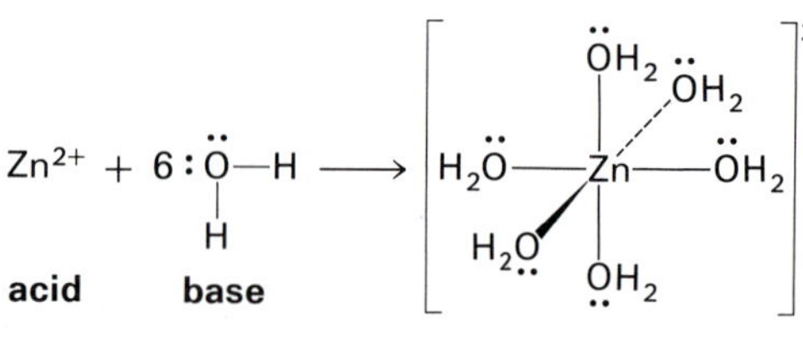

(Here the zinc has acquired twelve electrons (instead of eight) in its valence orbitals. In forming six bonds, the zinc is using *d* orbitals in addition to *s* and *p* orbitals.)

Another Lewis base—used as an antidote for poisoning by heavy metals such as lead and mercury, and also as a food preservative—is the ethylenediaminetetraacetate ion, EDTA:

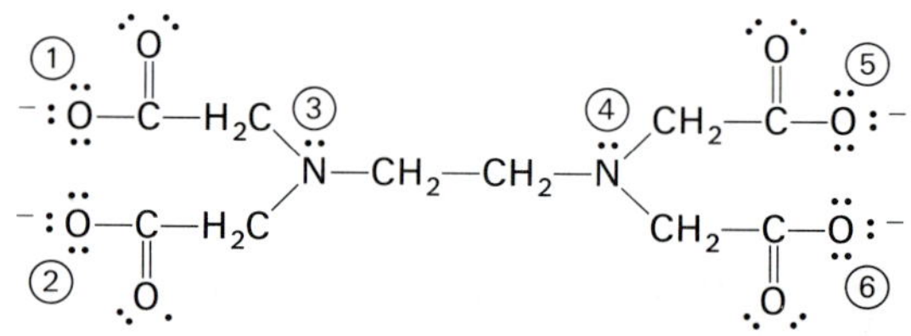

(The commonly used chemicals are Na_4 EDTA and Na_2Ca EDTA.) The ion has six positions (numbered on the structural formula) where it can act as a Lewis base. It grabs a metal ion "like a claw" (Figure 11-6), and hence is called a chelating agent (from the Greek word for "claw"). When administered as an antidote

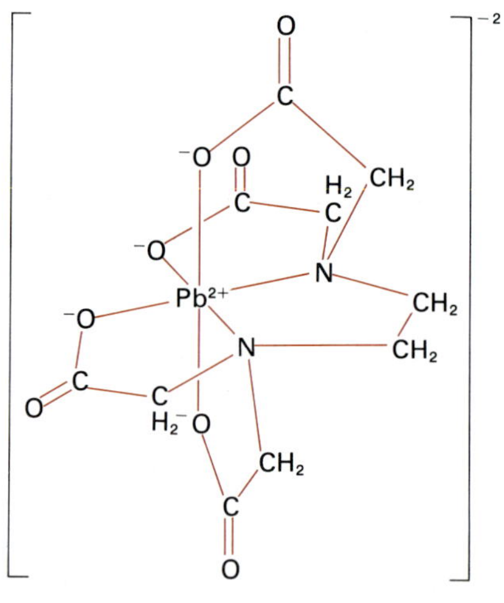

FIGURE 11-6
Structure of a lead–EDTA complex.

for lead or mercury poisoning, EDTA forms soluble complexes with the Pb^{2+} or Hg^{2+} ions, which are then excreted in the urine. EDTA works in the same way when it acts as a food preservative—it complexes with, and so renders inactive, metal ions that catalyze the reactions involved in the spoiling process.

EXERCISE 11-8 Which of the following are Lewis bases? CH_4, NH_3, NH_4^+, CH_3OCH_3, CH_3CH_2OH, $CH_3CH_2NH_2$, CH_3CH_3, PH_3, H_2S.

EXERCISE 11-9 (a) The silver ion (Ag^+) forms a complex with two ammonia molecules: $Ag(NH_3)_2^+$. Write an equation that shows that the reaction between Ag^+ and NH_3 is a Lewis acid–base reaction.

(b) The nickel ion (Ni^{2+}) forms a complex with four Cl^- ions: $NiCl_4^{2-}$. Write an equation that shows that the reaction between Ni^{2+} and Cl^- is a Lewis acid–base reaction.

11-8 DISSOCIATION OF WEAK ACIDS

We have seen that strong acids are almost completely dissociated in aqueous solution and that weak acids have only a small tendency to dissociate. A weak acid, HA, dissociates in water to yield some H^+ and A^- ions, but undissociated HA is also present at equilibrium:

$$HA \rightleftharpoons H^+ + A^- \qquad K_a = \frac{[H^+][A^-]}{[HA]}$$

Table 11-6 gives equilibrium constants for the dissociation of some weak acids. Also given are values for pK_a where

$$K_a = 10^{-pK_a} \text{ moles}/\ell$$

or

$$pK_a = -\log K_a$$

The definition of pK_a is analogous to that of pH, and we can calculate pK_a from K_a just as pH is calculated from $[H^+]$. For example, if $K_a = 4 \times 10^{-9}$, pK_a is *between* 8 and 9 (because 4×10^{-9} is more than 10^{-9}, but less than 10^{-8}). The precise pK_a, obtained using logarithms, is 8.4.

TABLE 11-6
Dissociation constants and pK_a values of weak acids

Acid	K_a (moles/ℓ)	pK_a
HSO_4^-	1.3×10^{-2}	1.89
H_3PO_4	7.5×10^{-3}	2.12
HF	6.6×10^{-4}	3.18
CH_3COOH	1.8×10^{-5}	4.74
$CO_2 + H_2O$	4.3×10^{-7}	6.37
H_2S	1.3×10^{-7}	6.88
$H_2PO_4^-$	6.2×10^{-8}	7.21
HCN	4.0×10^{-10}	9.40
HCO_3^-	4.8×10^{-11}	10.32
HPO_4^{2-}	2.2×10^{-13}	12.66
HS^-	7.1×10^{-15}	14.15

Note: The acids are arranged in order of decreasing acid strength.

The acids in Table 11-6 are arranged in order of decreasing acid strength. If you compare this table with Table 11-5, you will see that what we have done here is merely to express the Brønsted–Lowry ranking of acids in quantitative terms.

Because all the dissociation constants given in Table 11-6 have the same dimensions (moles per liter), we can compare the strengths of the acids on the basis of their dissociation constants: the greater the K_a, the greater the tendency of the acid to furnish H^+, and the stronger the acid. We also know (from Section 10-9) that the requirement for a reaction to occur spontaneously is a negative ΔG. In fact, ΔG and the equilibrium constant K are related:

$$\Delta G = -2.303\, RT \log K$$

According to the equation, the greater the value of K, the more negative ΔG is, and the greater is the tendency for reaction to occur.

In the following examples, we use the dissociation constant to determine the concentration of H^+ in a solution of given concentration of a weak acid or a weak base.

EXAMPLE 11-4

What is the concentration of H^+ in 0.10 M CH_3COOH? The K_a for acetic acid is 1.8×10^{-5} mole/ℓ.

$$CH_3COOH \rightleftharpoons H^+ + CH_3COO^-$$

$$K_a = \frac{[H^+][CH_3COO^-]}{[CH_3COOH]} = 1.8 \times 10^{-5}\,\frac{\text{mole}}{\ell}$$

According to the equation for the dissociation, for each H^+ formed, one CH_3COO^- is also produced, so

$$[CH_3COO^-] = [H^+]$$

For each H^+ formed, one CH_3COOH has dissociated, so

$$[CH_3COOH] = 0.10\text{ M} - [H^+]$$

But, because K_a is small, we know that *most* of the CH_3COOH remains undissociated—that is, $[H^+]$ is much less than 0.10 M. We can therefore take

$$[CH_3COOH] \cong 0.10\text{ M}$$

to a good approximation.

Substituting these concentrations in the equilibrium expression,

$$K_a = \frac{[H^+][CH_3COO^-]}{[CH_3COOH]} = \frac{[H^+][H^+]}{0.10} = 1.8 \times 10^{-5}$$

$$[H^+]^2 = 1.8 \times 10^{-6}$$

$$[H^+] = 1.3 \times 10^{-3}\text{ M}$$

(Note that $[H^+]$ *is* small compared to 0.10 M, so our assumption, that the concentration of CH_3COOH is close to 0.10 M, is good.) The pH of this solution is between 2 and 3, but close to 3. In fact, the pH is 2.9.

EXAMPLE 11-5

What is the concentration of H^+ in a 1.0 M solution of NH_3? The principal reaction occurring in the solution is

$$NH_3 + H_2O \rightleftharpoons NH_4^+ + OH^-$$

$$K = \frac{[NH_4^+][OH^-]}{[NH_3]} = 1.8 \times 10^{-5}\,\frac{\text{mole}}{\ell}$$

For each OH^- formed, one NH_4^+ is also produced, so

$$[NH_4^+] = [OH^-]$$

For each OH^- formed, one NH_3 has reacted, so

$$[NH_3] = 1.0\text{ M} - [OH^-]$$

Because K is small, the concentration of OH^- is low, and we will assume it is negligible with respect to 1.0 M, so

$$[NH_3] \cong 1.0\ \text{M}$$

and

$$K = \frac{[NH_4^+][OH^-]}{[NH_3]} = \frac{[OH^-][OH^-]}{1.0} = 1.8 \times 10^{-5}$$

$$[OH^-]^2 = 1.8 \times 10^{-5} = 18 \times 10^{-6}$$

$$[OH^-] = 4.2 \times 10^{-3}\ \text{M}$$

(Note that our assumption that $[OH^-]$ is negligible compared to 1.0 M is good.)

The OH^- is provided almost entirely by the NH_3. Only a small amount comes from the dissociation of water. At the same time, there is also H^+ in the solution that comes entirely from the dissociation of water; its concentration is related to that of the OH^- by the dissociation constant for water:

$$[H^+] = \frac{K_w}{[OH^-]} = \frac{1.0 \times 10^{-14}}{4.2 \times 10^{-3}} = 2.4 \times 10^{-12}\ \text{M}$$

The pH of this solution of a weak base is therefore between 11 and 12, but close to 12. The exact pH is 11.6.

EXERCISE 11-10 Of the weak acids HCN, HF, and H_2S, which is weakest and which is strongest?

EXERCISE 11-11 What is the concentration of H^+ in 1.0 M CH_3COOH? The K_a for acetic acid is 1.8×10^{-5} M.

EXERCISE 11-12 (a) Salicylic acid, which is used in the preparation of aspirin, has $pK_a = 3.00$. What is the K_a of salicylic acid?

(b) Ascorbic acid (vitamin C) has $K_a = 5 \times 10^{-5}$ M. What is the pK_a of ascorbic acid?

11-9 ACID IN RAIN

In the last 20 years, rain falling on the eastern United States, Europe, and Japan has increased in acidity to 100–1000 times the previously normal levels. Under normal conditions, rainwater is only *slightly* acid due to the CO_2 in the atmosphere:

$$CO_2(g) + H_2O \rightleftharpoons H^+ + HCO_3^-$$

This normal acidity is about that of a potato, with a pH of 5.6.

In recent years, however, the average acidity of rainwater in industrialized regions has increased to that of a tomato (pH 4) and occasionally the acidity has been that of pure lemon juice (pH 2). The main source of the acid (60%) is sulfur dioxide gas (SO_2), an atmospheric pollutant resulting primarily from the combustion of sulfur-containing fuels. It yields acidic solutions by the reaction

$$SO_2 + H_2O \rightleftharpoons H^+ + HSO_3^-$$

and is thus, although technically a "weak" acid, actually moderately strong (a 0.01 M solution of SO_2 has a pH of 2.2).

Nitrogen oxides, which account for about 35% of the acid in rain, originate primarily in automobile exhausts. A complicated series of reactions is involved. One of these reactions produces nitric acid:

$$NO_2(g) + H_2O \longrightarrow HNO_3$$

The United States and Canada discharge about 50 million tons of sulfur and nitrogen oxides into the air each year.

Acid rain leaches essential plant nutrients from the soil, stunting the growth of farm crops and forests. It is also estimated that 50,000 lakes in the Adirondacks and Canada have become so acidic that the fish population, particularly trout and salmon, has seriously declined or been destroyed. Furthermore, acids in the lakes and streams extract toxic metals such as mercury and lead from the bottom sediment. These metals can also destroy fish and contaminate drinking water. Acid rain can also dissolve limestone and marble buildings and sculptures by the reaction

$$CaCO_3(s) + H^+(aq) \longrightarrow HCO_3^-(aq) + Ca^{2+}(aq)$$

11-10 INDICATORS

Various dyes can be used as indicators. We have already encountered litmus, a natural dye that is red in acidic solution and blue in basic. Other familiar indicators are cranberry juice (which is red in acid and green in base) and tea (which is red-brown in neutral solution and yellow-brown in acid—when lemon juice is added, for example). The juice of a red cabbage goes through several color changes: it is red at pH 1, purple at pH 7, blue at pH 9, green at pH 12, and yellow at pH 14.

How do indicators work? These dyes are weak acids, of general formula HIn, that dissociate slightly in aqueous solution to yield H^+ and In^- ions:

$$HIn \rightleftharpoons H^+ + In^- \qquad K_a = \frac{[H^+][In^-]}{[HIn]} \tag{11-8}$$

where HIn and In^- are of different colors. The formula of the indicator HIn may be very complex, as, for example, in phenolphthalein, which has the structure shown at the bottom of the page.

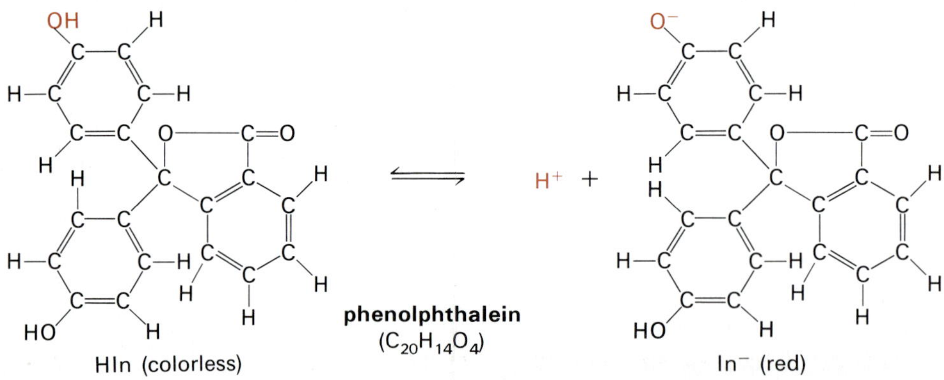

phenolphthalein
($C_{20}H_{14}O_4$)

TABLE 11-7
Indicator colors

Indicator	Approximate pH at which color changes	Color at lower pH	Color at higher pH
Methyl green	1	Yellow	Blue
Thymol blue	2	Red	Yellow
Erythrosine	3	Orange	Red
Congo red	4	Blue	Red
Methyl red	5	Red	Yellow
Alizarin	6	Yellow	Red
Bromothymol blue	7	Yellow	Blue
α-Naphtholphthalein	8	Orange-yellow	Green-blue
Phenolphthalein	9	Colorless	Red
Thymolphthalein	10	Colorless	Blue
Alizarin yellow	11	Yellow	Red

So HIn represents the indicator, and In^- represents the ion formed when HIn loses a proton. With phenolphthalein, HIn is colorless and In^- is red. If H^+ is added to a solution containing an indicator, the position of equilibrium in Equation 11-8 shifts to the left, in accord with Le Châtelier's principle, increasing the concentration of HIn. If OH^- is added to a solution containing an indicator, it neutralizes the HIn, increasing the concentration of In^-.

With the indicator bromthymol blue, HIn is yellow. Acidic solutions, with a high concentration of HIn and a low concentration of In^-, are yellow. Here In^- is blue, so basic solutions, where the concentration of In^- is large and that of HIn small, are blue. When the concentrations of the yellow and the blue species, HIn and In^-, are about equal, the solution is green.

We can see the effect of the H^+ concentration on the concentrations of HIn and In^- more clearly if we rewrite Equation 11-8 as follows:

$$\frac{[In^-]}{[HIn]} = \frac{K_a}{[H^+]}$$

In acidic solution, where the concentration of H^+ is high, the ratio $[In^-]/[HIn]$ is small, and the species HIn is favored. In basic solution, where the concentration of H^+ is low, the ratio $[In^-]/[HIn]$ is large, and the species In^- is favored. The indicator changes color when the concentration of H^+ is approximately equal to K_a. Then the ratio $[In^-]/[HIn] = 1$, with HIn and In^- present in equal concentration. Table 11-7 lists some indicators, with the pH at which a color change occurs, and gives the color at higher and at lower pH. We saw (Figure 11-4b) that in a strong acid–strong base titration, the pH changes rapidly near the equivalence point, over a very narrow range of added base. In such a titration, any indicator that changes color in the pH range from around 4 to 10 is satisfactory.

EXERCISE 11-13 Suppose you want to titrate a basic solution, containing a certain antacid tablet dissolved in water, to a pH of 2 or 3 (the pH of stomach acid) in order to test the effectiveness of the tablet. You titrate with a solution of HCl of known concentration. Name two indicators from Table 11-7 that would be suitable for this titration, and tell what color change you would see at the endpoint of the titration.

Universal indicator solutions that indicate pH variations over a range of pH can be prepared from *mixtures* of indicators. One such mixture is composed of methyl red, bromthymol blue, α-naphtholphthalein, phenolphthalein, and thymolphthalein (see Table 11-7). This mixture is red at pH 4, orange at pH 5, yellow at pH 6, yellow-green at pH 7, green at pH 8, blue-green at pH 9, blue-violet at pH 10, and red-violet at pH 11. This mixture can be used to determine the approximate pH of solutions whose pH lies between 4 and 11. For precise determination of pH, a pH meter would be used instead. A pH paper is paper that has been impregnated with mixtures of indicators. The papers are supplied with color charts that show the color at each pH. The pH of a solution is estimated by matching the color of the paper when wet by the solution with a color on this chart.

11-11 BUFFERS

Many chemical reactions and most biological processes occur only when the pH is close to some fixed value. Most of the chemistry of living things occurs at a pH of from 6 to 8, a notable exception being gastric juices (which are strongly acid, with a pH of about 2 in humans). Human blood maintains a pH between 7.35 and 7.45, although it transports chemicals and waste products, many of which by themselves are acids or bases.

This almost constant pH is maintained through **buffer systems** that prevent appreciable changes of pH in solutions to which small amounts of acids or bases are added. Buffer systems are composed of conjugate acid–base pairs—for example, CH_3COOH and CH_3COO^-, a weak acid and a salt of that weak acid. It is the salt, such as $NaCH_3COO$, that provides the conjugate base. The equilibrium in this system is

$$CH_3COOH \rightleftharpoons CH_3COO^- + H^+ \quad (11\text{-}9)$$

with mass action expression

$$K_a = \frac{[H^+][CH_3COO^-]}{[CH_3COOH]} = 1.8 \times 10^{-5} \frac{\text{mole}}{\ell} \quad (11\text{-}10)$$

If we rewrite Equation 11-10 with $[H^+]$ on the left, we get

$$[H^+] = K_a \frac{[CH_3COOH]}{[CH_3COO^-]}$$

$$= 1.8 \times 10^{-5} \frac{[CH_3COOH]}{[CH_3COO^-]} \quad (11\text{-}11)$$

The concentration of H^+ depends on K_a and on the ratio $[CH_3COOH]/[CH_3COO^-]$. If the solution is moderately concentrated (say, 1 M) in CH_3COOH and the same concentration in CH_3COO^-, the ratio $[CH_3COOH]/[CH_3COO^-]$ is unity, and $[H^+]$ is approximately equal to K_a:

$$[H^+] = K_a \frac{[CH_3COOH]}{[CH_3COO^-]} = K_a \frac{(1)}{(1)}$$

$$= 1.8 \times 10^{-5}\ \text{M}$$

and

$$\text{pH} = 4.7$$

(To be precise, $[CH_3COOH] = 1 - [H^+]$, and

$[CH_3COO^-] = 1 + [H^+]$, but $[H^+]$ is small compared to 1, and so, to a close approximation, $[CH_3COOH] \cong 1$ and $[CH_3COO^-] \cong 1$.)

If small amounts of H^+ are added to this buffer solution, the reaction of Equation 11-9 is driven to the left to restore equilibrium, consuming most of the added H^+. If, instead, a small amount of OH^- is added, it reacts with CH_3COOH and is mostly consumed. The ratio $[CH_3COOH]/[CH_3COO^-]$ changes only slightly when H^+ or OH^- is added because both CH_3COOH and CH_3COO^- are present in *large* concentration, and we are adding only *small* amounts of H^+ or OH^-. Thus the H^+ concentration of the buffer remains fixed (close to 1.8×10^{-5} M) in spite of the addition of *small* amounts of strong acids or bases. (If such a *large* amount of a strong acid or base is added as to use up either the CH_3COO^- or the CH_3COOH, the buffer will be ineffective against further addition of acid or base.)

A buffer system can also be prepared from a conjugate acid–base pair composed of a weak base and a salt of that weak base—for example, NH_3 and NH_4^+, which might come from ammonium chloride (NH_4Cl). In this system, the NH_3 is the weak base, and the salt provides the conjugate acid, NH_4^+. The equilibrium in this system is

$$NH_3 + H_2O \rightleftharpoons NH_4^+ + OH^-$$

with mass action expression

$$K = \frac{[NH_4^+][OH^-]}{[NH_3]} = 1.8 \times 10^{-5} \qquad (11\text{-}12)$$

Rewriting Equation 11-12 with $[OH^-]$ on the left, we get

$$[OH^-] = K\frac{[NH_3]}{[NH_4^+]}$$

If the concentrations of NH_3 and NH_4^+ are large and equal, then the concentration of OH^- is buffered at 1.8×10^{-5} M, and the H^+ concentration at $(1 \times 10^{-14})/(1.8 \times 10^{-5}) \cong 5 \times 10^{-10}$ M.

In the examples given here, we have assumed that the buffers are composed of *equal* concentrations of CH_3COOH and CH_3COO^- or of NH_3 and NH_4^+, so that the ratios $[CH_3COOH]/[CH_3COO^-]$ and $[NH_3]/[NH_4^+]$ are equal to 1. The resulting H^+ concentration in the CH_3COOH–CH_3COO^- buffer is then equal to K_a for acetic acid; and the OH^- concentration in the NH_3–NH_4^+ buffer is equal to K for ammonia. These are only examples—the pH at which the solution is buffered can be *tailored* by varying the relative amounts of the two constituents of the buffer.

A convenient form of Equation 11-11 is obtained by taking the negative logarithm of both sides, and then rewriting the equation in terms of pH and pK:

$$[H^+] = K_a\frac{[CH_3COOH]}{[CH_3COO^-]} \qquad (11\text{-}11)$$

$$\log [H^+] = \log K_a + \log\frac{[CH_3COOH]}{[CH_3COO^-]}$$

$$-\log [H^+] = -\log K_a - \log\frac{[CH_3COOH]}{[CH_3COO^-]}$$

or

$$pH = pK_a - \log\frac{[CH_3COOH]}{[CH_3COO^-]}$$

The equation can be rewritten (because $\log (a/b) = -\log (b/a)$):

$$pH = pK_a + \log\frac{[CH_3COO^-]}{[CH_3COOH]}$$

In general terms, for the acid HA,

$$pH = pK_a + \log\frac{[A^-]}{[HA]} \qquad (11\text{-}13)$$

Equation 11-13 is widely used in the quantitative

treatment of acid–base equilibria in biological systems. With this equation, which is known as the Henderson-Hasselbalch equation, it is easy to calculate the pH of any buffer of known concentrations of HA and A^-. The equation can also be used to determine K_a for an acid if the pH is measured with known concentrations of HA and A^-.

EXERCISE 11-14 (a) Use Table 11-6 to choose a buffer that would maintain a pH of about 2. Tell what acid and what salt you would use. Write an equation that can be used to explain how this buffer maintains a constant pH when small amounts of acid or base are added.

(b) If the buffer is 1 M in the acid component and 1 M in the salt, what is the H^+ concentration? What is the pH?

The pH of the blood is kept between 7.35 and 7.45, despite the production of over 12 moles of CO_2 and other acids every day as a result of the metabolism of food. Several buffers help to maintain this constant pH, but the principal buffer system is the carbonic acid–bicarbonate system:

$$H_2O + CO_2(aq) \xrightleftharpoons{\text{carbonic anhydrase}} H_2CO_3(aq) \rightleftharpoons H^+(aq) + HCO_3^-(aq)$$

The CO_2 in solution is also in equilibrium with carbon dioxide gas in the atmosphere. The enzyme carbonic anhydrase catalyzes the formation of H_2CO_3 in solution (and the reverse reaction), but the concentration of H_2CO_3 in solution is nevertheless very small—only 0.37% of the dissolved CO_2 is present as H_2CO_3. The equilibrium is usually summarized more simply:

$$H_2CO_3 \rightleftharpoons H^+ + HCO_3^- \qquad (11\text{-}14)$$

with the H_2CO_3 representing the *total* carbon dioxide dissolved in the blood (that is, H_2CO_3 *and* CO_2).

When H^+ is added, the equilibrium shifts to the left, and most of the added H^+ is removed from solution. When OH^- is added, H^+ and OH^- combine, and more H_2CO_3 dissociates to maintain a constant pH.

The equilibrium constant expression for the carbonic acid–bicarbonate buffer system in blood at body temperature (37°C) is

$$K_a = \frac{[H^+][HCO_3^-]}{[H_2CO_3]} = 7.9 \times 10^{-7}$$

or

$$\begin{aligned} pH &= pK_a + \log\frac{[HCO_3^-]}{[H_2CO_3]} \\ &= 6.1 + \log\frac{[HCO_3^-]}{[H_2CO_3]} \end{aligned} \qquad (11\text{-}15)$$

The normal concentration of HCO_3^- is 25×10^{-3} M, and that of H_2CO_3 is 1.25×10^{-3} M; the ratio $[HCO_3^-]/[H_2CO_3]$ is 20. This gives $[H^+] = 4.0 \times 10^{-8}$ M or pH = 7.40, the pH of normal blood.

Given the large amounts of CO_2 produced in metabolism, buffers alone would not be able to maintain a constant pH in blood. The buffering capacity would soon be exhausted if it were not for the rapid excretion of CO_2 through the lungs in respiration and of H^+ through the kidneys and urine (the pH of urine is normally between 5 and 6).

Acidosis, in which the pH of the blood may drop, can result from taking large amounts of acid drugs such as aspirin, or from a reduced ability to excrete CO_2. This happens in emphysema when the lungs lose elasticity, and in hyaline membrane disease in newborns, when the CO_2 does not diffuse out of the blood into the lungs rapidly enough.

An accumulation of lactic acid in the blood may occur temporarily during strenuous exercise, but this has no lasting effect.

Alkalosis, in which the pH of the blood may rise, can result from excessive vomiting (with loss of HCl from the stomach), or from the ingestion of too much HCO_3^- from fruits, or from taking $NaHCO_3$ ("bicarbonate of soda") for indigestion.

KEY TERMS

Strong acids and *strong bases* are almost completely dissociated in aqueous solution. *Weak acids* and *weak bases* are only slightly dissociated in aqueous solution.

Acids and bases *neutralize* each other to yield water and a salt.

Spectator ions are ions that do not take part in the reaction; they are not included in the *net ionic equation* for the reaction.

Titration is an experimental method of determining the concentration of a solution, commonly used to determine the concentrations of acidic or basic solutions.

$$\text{pH} = -\log[\text{H}^+]; \quad \text{or} \quad [\text{H}^+] = 10^{-\text{pH}} \frac{\text{moles}}{\ell}$$

$$\text{pK}_\text{a} = -\log K_\text{a}; \quad \text{or} \quad K_\text{a} = 10^{-\text{pK}_\text{a}} \frac{\text{moles}}{\ell}$$

The *equivalence point* in an acid–base titration is the point where the number of moles of OH^- provided by the base is exactly equal to the number of moles of H^+ provided by the acid.

The two members of a *conjugate acid–base pair* differ only by a proton.

An *indicator* is a dye that changes color at some characteristic pH.

Buffers prevent appreciable changes of pH in solutions to which small amounts of acids or bases are added.

A *coordinate covalent bond* is a bond in which the shared electron pair is contributed by only one atom.

CHAPTER SUMMARY

Acids and bases can be defined in various ways. According to the *Arrhenius definition,* acids are substances that in water provide hydrogen ions (H^+), and bases are substances that in water provide hydroxide ions (OH^-). Acids and bases neutralize each other to yield a salt and water.

According to the *Brønsted–Lowry definition,* which is more general than the Arrhenius definition, an acid is a substance that donates a proton, and a base is a substance that accepts a proton. Each acid has a *conjugate base* and each base a *conjugate acid,* that differ only by a proton. The stronger an acid is, the weaker its conjugate base. If acids and bases are listed in order of decreasing acid strength (and thus increasing base strength) as in Table 11-5, any acid has a great tendency to donate a proton to a base that is lower in the table, but only a small tendency to donate a proton to a base that is higher in the table.

According to the *Lewis definition,* which is even more general than the Brønsted–Lowry definition, an acid is a substance that accepts an electron pair, and a base is a substance that donates an electron pair.

In pure water, or in dilute aqueous solution, the relationship

$$[\text{H}^+][\text{OH}^-] = K_\text{w} = 1.0 \times 10^{-14}\,(\text{mole}/\ell)^2 \text{ at } 25\,^\circ\text{C}$$

always holds, whatever the solutes. In neutral solution, $[\text{H}^+] = [\text{OH}^-] = 1.0 \times 10^{-7}\,\text{M}$, and $\text{pH} = 7$. In acidic solution, the pH is less than 7; in basic solution, the pH is greater than 7.

Acid–base *titration,* which depends on the controlled neutralization of an acid by a base, is used to determine the concentration of a solution. The concentration of either the acid or the base must be known, and the volumes of both solutions must be measured.

There are strong and weak acids and bases. Because the dissociation constants of the acids given in Table 11-6 have the same dimensions, we can compare the strengths of the acids on the basis of their dissociation constants—the larger the K_a, the stronger the acid. The dissociation constants of weak acids and weak bases can be used to calculate the concentration of H^+ in a solution of known concentration of the acid or base. The pH of buffer solutions can also be calculated, and buffers can be tailored to maintain any pH.

PROBLEMS

11-1. This problem reviews some ideas about grams, moles, and molarities (Chapter 9).

(a) Calculate the number of moles and grams of solute in each of the following solutions:
(1) 50 ml of 0.50 M H_2SO_4
(2) 2.0 ml of 18 M H_2SO_4
(3) 100 ml of 6.0 M HCl
(4) 1.0 ℓ of 0.050 M KOH
(5) 6.0 ℓ of 0.10 M $NaHCO_3$

(b) What volume of each of the following solutions is required to give 1.0 mole of solute?
(1) 0.50 M H_2SO_4 (2) 6.0 M HCl
(3) 0.025 M KOH (4) 18 M H_2SO_4
(5) 2.0 M HNO_3

(c) What volume of each of the solutions in part b is required to give 0.25 mole of solute?

11-2. This problem gives practice in the use of *normality* and *equivalents*.

(a) What is the normality of each of the following solutions?
(1) 2 M H_2SO_4 (2) 2 M HNO_3
(3) 18 M H_2SO_4 (4) 0.5 M HCl
(5) 0.010 M $Ca(OH)_2$, a strong base

(b) What is the molarity of each of the following solutions?
(1) 3 N NaOH (2) 1 N H_2SO_4
(3) 0.050 N H_2SO_4 (4) 0.010 N $Ba(OH)_2$
(5) 0.1 N KOH

(c) How many equivalents, and how many moles, of solute are in each of the following?
(1) 50 ml of 3 N HCl
(2) 100 ml of 0.50 N H_2SO_4
(3) 1.0 ml of 5.0 N H_2SO_4
(4) 0.10 ml of 2 N NaOH
(5) 500 ml of 0.010 N $Ba(OH)_2$

11-3. Suggest several tests you might perform to determine whether a solution is acidic.

11-4. Sucrose ($C_{12}H_{22}O_{11}$) contains hydrogen, yet is not an acid. State one or more properties of acids that sucrose does not possess.

11-5. When vinegar, which contains acetic acid (CH_3COOH), is poured onto baking soda ($NaHCO_3$), bubbles of carbon dioxide gas are given off. Write a balanced net ionic equation for the reaction.

11-6. Write a balanced net ionic equation for the reaction of each of the following pairs of acids and bases in water solution:
(a) HNO_3 and NH_3
(b) CH_3COOH and NH_3
(c) HCl and $Ba(OH)_2$
(d) CH_3COOH and $Ba(OH)_2$
(Barium hydroxide is a strong base.)

11-7. If the OH^- concentration in a certain aqueous solution is 1.0×10^{-5} M, what is the concentration of H^+? Is the solution acidic or basic?

11-8. Rank the following equimolar solutions in order of decreasing H^+ concentration: NaCN, H_2SO_4, Na_2S, $NaHSO_4$. (Use data in Tables 11-5 and 11-6.)

11-9. (a) How many moles of NaOH are needed to neutralize 100 ml of 0.10 M HCl?
(b) How many moles of NaOH are needed to neutralize 100 ml of 0.10 M CH_3COOH?
(c) How many moles of NaOH are needed to neutralize 100 ml of 0.10 M H_2SO_4?

11-10. (a) How many ml of 0.200 M NaOH are needed to neutralize 50 ml of 0.300 M HCl?
(b) When the two solutions of part a are mixed, what species are present in significant concentration in the resulting solution?
(c) What is the pH of the resulting solution?

11-11. Find the approximate pH of
(a) 0.020 M HCl (b) 2 M HNO_3
(c) 0.020 M NaOH (d) 5×10^{-6} M NaOH.

11-12. Given the pH, find the approximate $[H^+]$ for each of the following solutions, and tell which are acidic and which are basic.
(a) pH = 4.10 (b) pH = 8.20
(c) pH = 3.95 (d) pH = −1.

11-13. Determine the approximate K_a for each of the following substances:
(a) Boric acid, $HB(OH)_4$, used in eye drops ($pK_a = 9.24$)
(b) Acetylsalicylic acid (aspirin) ($pK_a = 4.56$)
(c) Dichloroacetic acid ($Cl_2CHCOOH$), a moderately strong acid ($pK_a = 1.30$)
(d) Glycerol, a product of the partial breakdown of fats ($pK_a = 14.15$)
(e) Lactic acid, produced in the muscles during exercise ($pK_a = 3.86$)

11-14. Determine the approximate pK_a of each of the following substances:
(a) Butyric acid, the acid in rancid butter ($K_a = 1.5 \times 10^{-5}$ M)
(b) Formic acid, the acid transmitted in ant bites ($K_a = 1.8 \times 10^{-4}$ M)
(c) Glucose, a simple sugar ($K_a = 3.7 \times 10^{-13}$ M)
(d) Trichloroacetic acid, a strong organic acid ($K_a = 0.23$ M)
(e) Vanillin, a flavoring ($K_a = 4.0 \times 10^{-8}$ M)

11-15. Normal gastric juice at the height of digestion is typically of pH 2. About 2.5 ℓ are secreted into the stomach daily. How many moles of HCl is this per day?

11-16. The CO_2 excreted in respiration in 24 hours is equivalent to more than 12.5 moles of HCl. If this many moles of HCl were contained in 6 ℓ of water, what would be the pH of the solution? (6 ℓ is the approximate volume of blood in an adult.)

11-17. Normal saline solution (0.9% NaCl) has a pH of 7.00. What is the pH of 6 ℓ of saline solution to which 1 ml of 6 M HCl has been added? (This amount of HCl added to 6 ℓ of blood changes the pH from 7.40 to about 7.38.)

11-18. A change in the pH of the blood from 7.4 to 7.0 can make the difference between life and death. What is the *change* in H^+ concentration that corresponds to this change in pH?

11-19. A 0.50 M solution of KOH is used to titrate 50 ml of vinegar, which contains acetic acid of unknown concentration. If 80 ml of the KOH solution are needed to reach the equivalence point, what is the concentration of the CH_3COOH in the vinegar?

11-20. What is the pH of each of the following solutions?
(a) 0.01 M HNO_3 (b) 0.001 M KOH
(c) 0.01 M KNO_3
(d) 0.01 M HX, where HX is a weak acid with $K_a = 1.0 \times 10^{-8}$ mole/ℓ.

11-21. A solution of KOH of unknown concentration is used to titrate 100 ml of 0.100 M H_2SO_4. If 150 ml of the KOH is needed to neutralize the acid, what is the concentration of the KOH solution?

11-22. What is the concentration of H^+ in 0.010 M HCN? The K_a for HCN $= 4.0 \times 10^{-10}$ mole/ℓ.

11-23. (a) If K_a for the dissociation of HF is 6.7×10^{-4} mole/ℓ, what is the equilibrium constant for the reaction $H^+ + F^- \rightleftharpoons HF$?
(b) What is the general relation between the equilibrium constant for a reaction written in one direction and the equilibrium constant for the same reaction written in the opposite direction?

11-24. Write an equation for the reaction of each of the following pairs of substances in water solution:
(a) HSO_4^- and F^- (b) H_2S and SO_4^{2-}
(c) HCO_3^- and OH^- (d) NH_4^+ and H_2O
For each reaction, indicate the conjugate acid–base pairs, and state whether reactants or products are favored. (Use Table 11-5.)

11-25. What would be the color of litmus indicator in the water solution of each of the following? KNO_3, KF, NH_3, $NaCH_3COO$, K_2O, NH_4NO_3.

11-26. A 0.10 M solution of boric acid, $HB(OH)_4$, has a H^+ concentration of 7.8×10^{-6} M. What is K_a for boric acid? Assume that only *one* H^+ of the $HB(OH)_4$ dissociates.

11-27. Suppose you want to prepare a buffer solution with pH between 3 and 4 (H^+ between 1×10^{-4} and 1×10^{-3} mole/ℓ). Suggest chemicals you could use to prepare the buffer solution, and give their approximate concentrations. Write equations that show how the solution is buffered against added OH^- and against added H^+. (Use data given in Table 11-6.)

11-28. Containers of sodium bicarbonate ($NaHCO_3$) are often kept in chemical laboratories so that the $NaHCO_3$ may be readily available to neutralize any acids or bases that are spilled. Write a net ionic equation that shows how the $NaHCO_3$ neutralizes NaOH, and write one that shows how it neutralizes HCl.

11-29. Which of the following are Brønsted–Lowry acids, and which are bases, in water solution? H_2O, HCl, NH_3, OH^-.

11-30. We have seen that the neutralization of a strong acid with a strong base can be represented by the net ionic equation $H^+ + OH^- \rightarrow H_2O$. The heat of reaction is 56.1 kJ per mole of water formed, independent of the acid and base used. Here we shall see the same principle illustrated for the reaction $Ag^+ + Cl^- \rightarrow AgCl(s)$.

The ionic compounds $AgNO_3$ and NaCl are separately very soluble in water, where the first is completely dissociated into the ions Ag^+ and NO_3^-, and the second into Na^+ and Cl^-. When solutions of the two compounds are mixed, the very insoluble AgCl precipitates out, and *heat is evolved in the process.* The reaction that forms the insoluble AgCl is $Ag^+ + Cl^- \rightarrow AgCl$. The Na^+ and NO_3^- remain in solution unchanged. ($NaNO_3$ is very soluble in water.)

When 50 ml of a solution of $AgNO_3$ of concentration 0.2 mole/ℓ, and an equal volume of

a solution of NaCl of the same concentration, both at the same temperature, are mixed in a polystyrene cup, the temperature goes up by 1.55 °C. Assume that *all* of the heat is absorbed by the solution (none is absorbed by the cup or the thermometer), and that the specific heat of the reaction mixture is essentially the same as that of pure water, 4.18 J/g deg.
(a) Find the heat evolved in the reaction $Ag^+ + Cl^- \rightarrow AgCl$, and express it in *kilojoules per mole*.
(b) $CaCl_2$ and $Ca(NO_3)_2$ are also very soluble in water, where they are completely dissociated into Ca^{2+} and $2Cl^-$, and into Ca^{2+} and $2NO_3^-$, respectively. What is the temperature rise that occurs when 50 ml of a solution of $AgNO_3$ of concentration 0.2 mole/ℓ is mixed in a polystyrene cup with 50 ml of a solution of $CaCl_2$ of concentration 0.2 mole/ℓ? Explain.

11-31. Al^{3+} forms a hydrate with six H_2O molecules: $Al(H_2O)_6^{3+}$. Write an equation showing that the reaction between Al^{3+} and H_2O is a Lewis acid–base reaction.

11-32. Ethylenediamine, abbreviated (*en*), has the formula $H_2N–CH_2–CH_2–NH_2$. It can donate two pairs of electrons when acting as a Lewis base, and so it is called bidentate (for two-toothed). Draw a diagram of the complex ion that Ni^{2+} forms with *two* (*en*) molecules.

11-33. The principal intracellular buffer is an $H_2PO_4^-$ and HPO_4^{2-} buffer.
(a) Using data from Table 11-6, calculate the ratio $[HPO_4^{2-}]/[H_2PO_4^-]$ required to maintain the fluid in the cell at a pH of 7.40, or $[H^+] = 4.0 \times 10^{-8}$ M.
(b) If the *total* concentration of phosphate ($H_2PO_4^- + HPO_4^{2-}$) in the cell is 38 mM (38×10^{-3} M), what is $[H_2PO_4^-]$ and what is $[HPO_4^{2-}]$?

11-34. With Equation 11-13, it is possible to calculate the change in pH caused by the addition of acid or base to the blood. In this problem you will determine the change in pH caused by the addition of 10 ml of 6 M HCl. Assume that the HCl is distributed through 6 ℓ of blood. The H_2CO_3 is in equilibrium with CO_2 in solution, which in turn is in equilibrium with CO_2 in the atmosphere. Because of the enzyme carbonic anhydrase and the rapid excretion of CO_2 through the lungs, the concentration of H_2CO_3 is almost constant at 1.25×10^{-3} M. The initial concentration of HCO_3^- is 25×10^{-3} M. The added acid reacts completely with HCO_3^-. The concentration of HCO_3^- is thus reduced, and it takes several hours for it to be replenished via the kidneys.
(a) What is the concentration of HCO_3^- after the addition of 10 ml of 6 M HCl to 6 ℓ of blood?
(b) What is the pH of the blood after the addition of the HCl?
(c) Compare your answer to part b with the pH of a solution of 10 ml of 6 M HCl added to 6 ℓ of water.

Crystals of metallic silver are deposited on a copper wire immersed in a solution of silver nitrate.

12

Oxidation and Reduction

Oxidation, which originally meant a combination with oxygen, is now more generally defined as a loss of electrons. Reduction, which originally meant a removal of oxygen, is more generally defined as a gain of electrons. Oxidation and reduction always occur together—when one substance loses electrons, another substance must gain them.

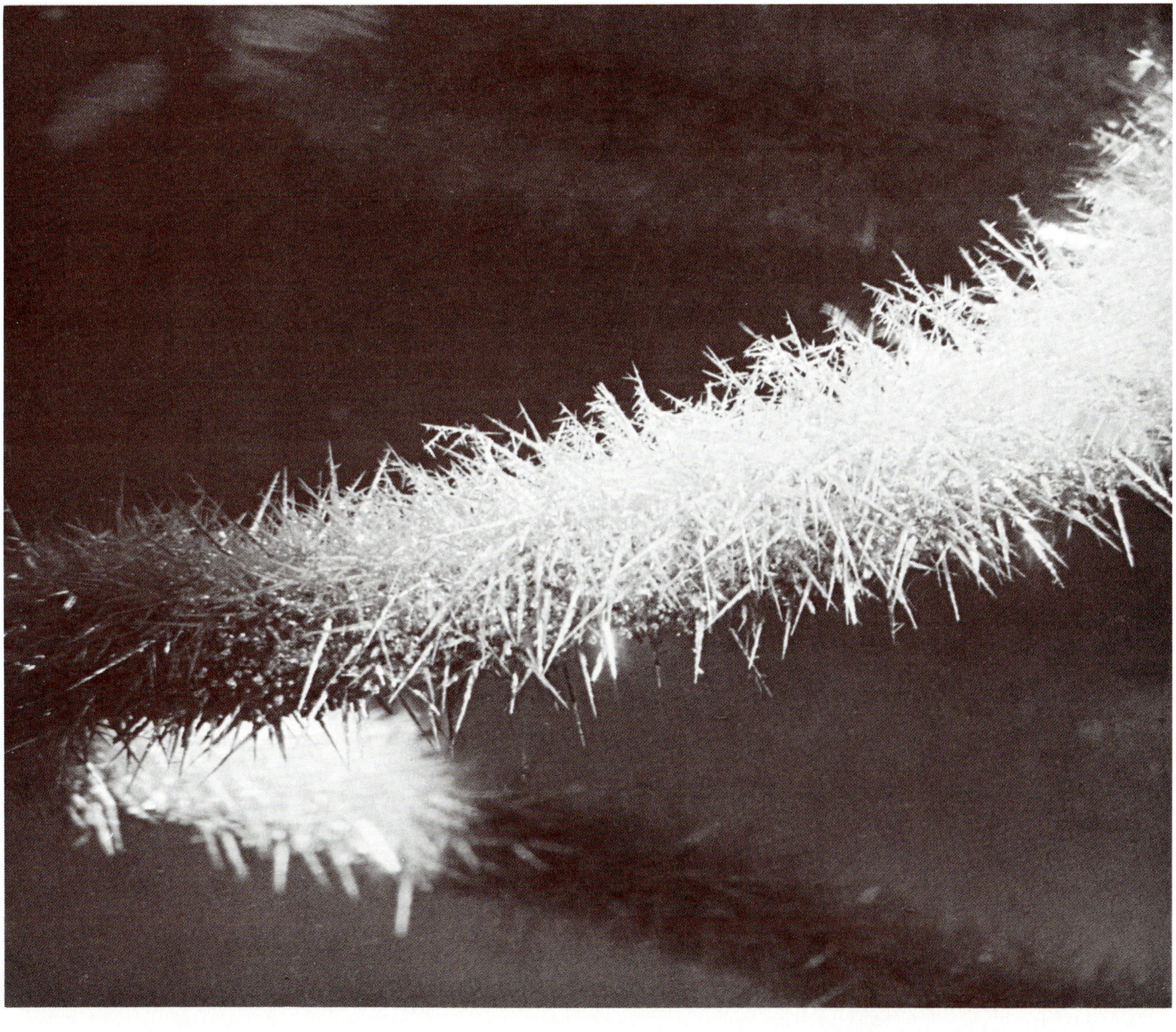

12-1
DEFINING OXIDATION AND REDUCTION

Originally, the term *oxidation* just meant a combination with oxygen, as in the burning of coal,

$$C + O_2 \longrightarrow CO_2$$

the corrosion of a metal,

$$4Fe + 3O_2 \longrightarrow 2Fe_2O_3 \qquad (12\text{-}1)$$

or the conversion of ethyl alcohol in wine to the acetic acid in vinegar.

$$C_2H_5OH + O_2 \longrightarrow CH_3COOH + H_2O$$

The term *reduction* just meant a removal of oxygen, as in the reduction of iron ore with carbon to produce metallic iron:

$$2Fe_2O_3 + 3C \longrightarrow 3CO_2 + 4Fe$$

After years of studying similar reactions, chemists realized that it would be convenient to have a more general definition of oxidation and reduction because some of these reactions do not involve oxygen. The reaction between iron and chlorine, for example,

$$2Fe(g) + 3Cl_2(g) \longrightarrow 2FeCl_3(s) \qquad (12\text{-}2)$$

is in some respects similar to that between Fe and O_2. In what way are these reactions similar? We can think of a metal oxide such as Fe_2O_3 as ionic, composed of Fe^{3+} and O^{2-} ions, and of the oxidation of iron by oxygen (Equation 12-1) as a transfer of electrons from iron atoms to oxygen atoms. $FeCl_3$ is also ionic, composed of Fe^{3+} and Cl^- ions. And the reaction between Fe and Cl_2 is a transfer of electrons from iron atoms to chlorine atoms. The chlorine in Equation 12-2 plays the same role as oxygen plays in Equation 12-1. Viewing these reactions in this more general way, we see that *it is the transfer of electrons that is important.* In the processes of oxidation and reduction, oxygen is only incidental and is not necessary. This observation leads to a more general definition of oxidation and reduction as processes in which electrons are transferred.

Oxidation *is the loss of electrons by the substance being oxidized.* The electrons are gained by the **oxidizing agent.** In Equation 12-1, the Fe is being oxidized by (loses electrons to) the oxidizing agent O_2. And in Equation 12-2, Fe is also oxidized, this time by the Cl_2.

Reduction *is the gain of electrons by the substance being reduced.* The electrons are lost by the **reducing agent.** In Equation 12-1, the O_2 is being reduced by (gains electrons from) the reducing agent Fe. In Equation 12-2, the Cl_2 is reduced by the Fe. Oxidation and reduction always occur together—the reducing agent is oxidized (loses electrons), and at the same time the oxidizing agent is reduced (gains electrons).

EXERCISE 12-1 For each of the following reactions, name the substance oxidized, the substance reduced, the oxidizing agent, and the reducing agent:

(a) $C_2H_5OH + O_2 \rightarrow CH_3COOH + H_2O$
(b) $2Al(s) + 3Fe^{2+} \rightarrow 2Al^{3+} + 3Fe(s)$
(c) $Fe(s) + 2Ag^+ \rightarrow 2Ag(s) + Fe^{2+}$
(d) $Mg(s) + 2H^+ \rightarrow H_2(g) + Mg^{2+}$

12-2
OXIDATION NUMBERS

Metallic iron (Fe) can readily be oxidized to the ferrous ion, (Fe^{2+})—say, by dipping a strip of iron into a hydrochloric acid solution:

$$Fe + 2H^+ \longrightarrow H_2(g) + Fe^{2+}$$

The Fe^{2+} ion can itself be further oxidized to the ferric ion (Fe^{3+}) with a strong oxidizing agent such as chlorine gas:

$$2Fe^{2+} + Cl_2(g) \longrightarrow 2Fe^{3+} + 2Cl^-$$

To help identify oxidation–reduction reactions and as an aid to balancing their equations, **oxidation numbers** are assigned to atoms that enter into oxidation–reduction reactions. (Oxidation–reduction reactions are also commonly known as redox reactions.) Metallic iron, which is electrically neutral, is said to be in the oxidation state zero. The Fe^{2+}, which has lost two electrons and has a +2 charge, is said to be in the oxidation state +2. And Fe^{3+}, which has lost three electrons and has a +3 charge, is said to be in the oxidation state +3. If we think of the compounds ferrous chloride ($FeCl_2$) and ferric chloride ($FeCl_3$) as being ionic, it is relatively easy to assign oxidation numbers to the iron and the chlorine. In $FeCl_2$, iron is in the +2 oxidation state, and chlorine is in the −1 oxidation state. In $FeCl_3$, iron is in the +3 oxidation state, and chlorine is in the −1 oxidation state. In a compound such as potassium permanganate ($KMnO_4$) however, it is more difficult to assign oxidation numbers, and so the following conventions are widely accepted.

1. The oxidation number of a neutral atom is zero.

2. The oxidation number of a monatomic ion is equal to the charge on the ion. The alkali metals form *only* +1-charged ions in their compounds and in solution, and so always have oxidation numbers of +1, except when they are uncombined metals. The alkaline earths form *only* +2-charged ions in their compounds and in solution, and so have oxidation numbers of +2. Except when combined with oxygen, the halogens form *only* −1-charged ions in their compounds and in solutions, and so have oxidation numbers of −1. (In combination with oxygen, the oxidation number of a halogen is found by applying rules 3 and 5.)

3. Oxygen, in compounds, has oxidation number −2 except in peroxides. In peroxides (molecules containing two oxygen atoms bonded together) such as H_2O_2 or Na_2O_2, oxygen has oxidation number −1.

4. Hydrogen, in compounds, has oxidation number +1 (except in metallic hydrides such as NaH, where hydrogen has oxidation number −1).

5. All other oxidation numbers are assigned so that the sum of all the oxidation numbers gives the net charge on the molecule or ion.

The following example illustrates the application of these rules.

EXAMPLE 12-1

Assign oxidation numbers to all the atoms in the following substances: potassium chloride (KCl); permanganate ion (MnO_4^-); methyl alcohol (CH_3OH); formaldehyde (CH_2O); formic acid (HCO_2H); chlorine (Cl_2); and perchloric acid ($HClO_4$).

Molecule or ion	*Oxidation numbers*	*Comments*
KCl	K: +1	By rule 2
	Cl: −1	By rule 2
MnO_4^-	O: −2	By rule 3
	Mn: +7	By rule 5

CH_3OH	O:	-2	By rule 3
	H:	$+1$	By rule 4
	C:	-2	By rule 5
CH_2O	O:	-2	By rule 3
	H:	$+1$	By rule 4
	C:	0	By rule 5
HCO_2H	O:	-2	By rule 3
	H:	$+1$	By rule 4
	C:	$+2$	By rule 5
Cl_2	Cl:	0	By rule 1
$HClO_4$	O:	-2	By rule 3
	H:	$+1$	By rule 4
	Cl:	$+7$	By rules 2 and 5

The information given in Example 12-1 is often summarized by writing oxidation numbers *above* the symbols in a formula:

$$\overset{+1}{K}\overset{-1}{Cl} \quad \overset{+7}{Mn}\overset{-2}{O_4^-} \quad \overset{-2}{C}\overset{+1}{H_3}\overset{-2}{O}\overset{+1}{H} \quad \overset{0}{C}\overset{+1}{H_2}\overset{-2}{O}$$

$$\overset{+1}{H}\overset{+2}{C}\overset{-2}{O_2}\overset{+1}{H} \quad \overset{0}{Cl_2} \quad \text{and} \quad \overset{+1}{H}\overset{+7}{Cl}\overset{-2}{O_4}$$

The oxidation number $+7$ does *not* mean that the Mn in MnO_4^- is an ion with a $7+$ charge. Because the conventions for assigning oxidation numbers are arbitrary, oxidation numbers of polyatomic species often do not represent the real charge on an atom. Fractional oxidation numbers (or an oxidation number of zero for a combined atom) are possible.

A Roman numeral placed in parentheses after the symbol often is used to indicate the oxidation number of a metal. Thus Fe(III) is Fe^{3+}, and Fe(II) is Fe^{2+}; similarly, Fe(III) oxide is Fe_2O_3.

EXERCISE 12-2 (a) Determine the oxidation number of manganese in the compounds Mn_3O_4, MnO_2, $Mn(OH)_2$, MnF_3, $KMnO_4$, and K_2MnO_4.

(b) Assign oxidation numbers to all the atoms in the following equations, and tell which substance is being oxidized and which reduced:

(1) $5CH_3OH(aq) + 6NO_3^-(aq) + 6H^+(aq) \longrightarrow 5CO_2(aq) + 3N_2(g) + 13H_2O(l)$

(2) $H_2S(aq) + 2O_2(g) \longrightarrow SO_4^{2-}(aq) + 2H^+(aq)$

(3) $SO_4^{2-}(aq) + 2H^+(aq) + 3H_2S(g) \longrightarrow 4S(s) + 4H_2O(l)$

12-3 GALVANIC CELLS

Figure 12-1 shows several oxidation–reduction reactions that occur spontaneously, with a *direct* transfer of electrons from the reducing agent to the oxidizing agent. In Figure 12-1a, iron metal dissolves in copper sulfate to yield Fe^{2+} in solution, and metallic copper. In Figure 12-1b, copper metal dissolves in silver nitrate to yield Cu^{2+} in solution and metallic silver. In Figure 12-1c, zinc metal dissolves in acid to yield Zn^{2+} in solution, and bubbles of hydrogen gas.

These spontaneous reactions can be made to occur *indirectly*, and so to be a source of electric current. In a **galvanic cell**, a spontaneously occurring chemical reaction is used as a source of electrical energy, with the electron transfer occurring through a wire, rather than

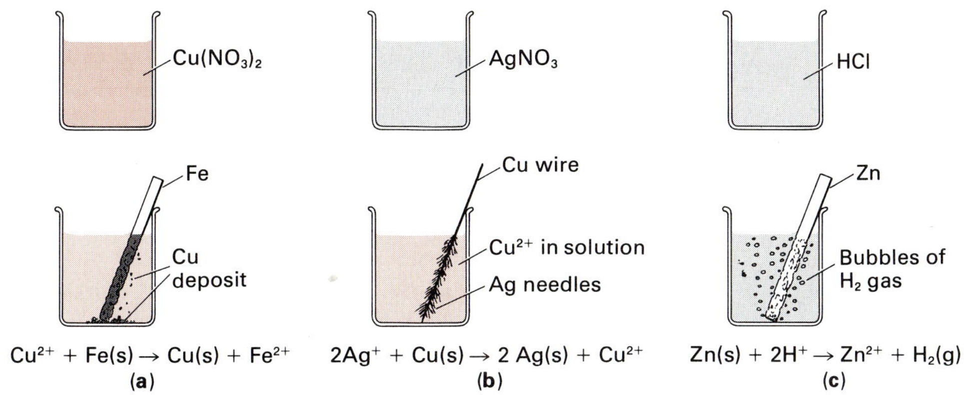

FIGURE 12-1

Oxidation–reduction reactions.

directly as in the earlier examples. In Figure 12-2a, the lefthand beaker contains a strip of metallic copper dipping into a solution of copper nitrate, and the righthand beaker contains a silver wire dipping into a solution of silver nitrate. The metal strips, called **electrodes,** are connected together by wires that pass through an ammeter, an instrument that measures the flow of electrons, or electrical current. The ammeter shows that no electrons are flowing.

But when a **salt bridge** containing an electrolyte such as ammonium chloride connects the two beakers (Figure 12-2b), there is a current. The salt bridge permits ions to flow from one beaker to the other, conducting the current through the solution. As time goes on, the copper strip dissolves, and the silver wire gains in mass. If the silver and copper electrodes are weighed before and after the reaction, it is found that two moles of silver are produced for each mole of copper that dissolves, as would be expected from the balanced equation for the overall reaction:

$$Cu(s) + 2Ag^+ \longrightarrow 2Ag(s) + Cu^{2+}$$

In the galvanic cell, this reaction has been split into two **half-reactions,** with the electron transfer occurring through the wire:

$$Cu(s) \longrightarrow Cu^{2+} + 2e^- \quad \text{(Oxidation half-reaction)}$$
$$Ag^+ + e^- \longrightarrow Ag(s) \quad \text{(Reduction half-reaction)}$$

There is no *net* gain or loss of electrons in the reaction. Whatever number of electrons is given up by the Cu *must* be gained by the Ag^+. *In any oxidation–reduction reaction, the number of electrons lost by the substance that is oxidized* (here the Cu), *must equal the number gained by the substance that is reduced* (here the Ag^+). The equation for the overall reaction is obtained by adding the equations for the two half-reactions. But first the equation for the Ag^+–Ag half-reaction must be multiplied by 2, so that the number of electrons lost equals the number gained, and no electrons appear in the overall equation:

$$Cu(s) \longrightarrow Cu^{2+} + 2e^- \quad \text{(Anode)}$$
$$\underline{2Ag^+ + 2e^- \longrightarrow 2Ag(s)} \quad \text{(Cathode)}$$

Net reaction:

$$Cu(s) + 2Ag^+ \longrightarrow 2Ag(s) + Cu^{2+}$$

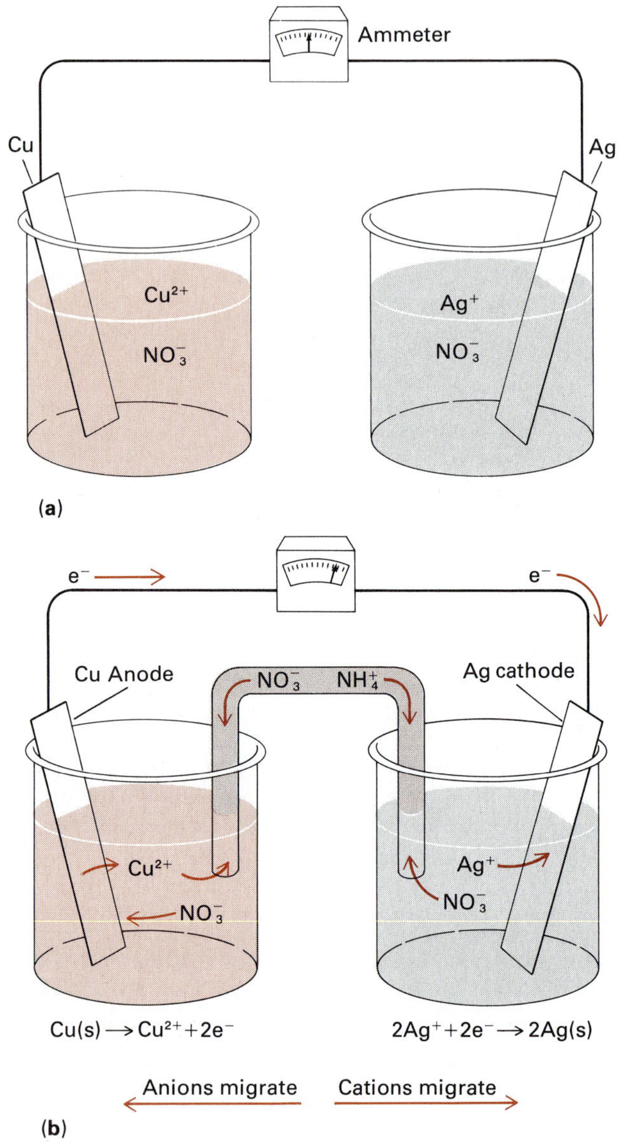

FIGURE 12-2

Galvanic cells. **(a)** An incomplete galvanic cell—the ammeter shows that no current is flowing. **(b)** A galvanic cell—the ammeter shows that current is flowing.

The copper electrode is called the **anode**—oxidation occurs at the anode. Because oxidation is the loss of electrons, electrons are released at the anode. The silver electrode is called the **cathode**—reduction occurs at the cathode, and electrons are "consumed" in the process. Electrons move from the anode, where they are given up by the copper atoms, through the external circuit (including the ammeter) to the cathode, where they are taken up by the Ag^+. The solution in each beaker is kept electrically neutral by the migration of ions through the salt bridge. Negative ions, called **anions** (NO_3^- and Cl^- in Figure 12-2), move toward the anode, where positive ions (Cu^{2+}) are formed by the oxidation of Cu. Positive ions, or **cations** (Ag^+, NH_4^+, and Cu^{2+} in Figure 12-2), move toward the cathode to replace the positively charged ions (Ag^+) that have been reduced at the cathode. Negative ions are called anions because they migrate toward the anode, and positive ions are called cations because they migrate toward the cathode.

EXERCISE 12-3 A galvanic cell consists of (1) a strip of zinc metal dipping into a solution of zinc nitrate, $Zn(NO_3)_2$, solution and (2) a strip of silver metal dipping into a solution of $AgNO_3$. The salt bridge contains NH_4Cl. The half reactions are

$$Zn(s) \longrightarrow Zn^{2+} + 2e^-$$

$$e^- + Ag^+ \longrightarrow Ag(s)$$

(a) Which electrode gains in weight as the reaction proceeds?

(b) Which electrode is the anode; which the cathode?

(c) Toward which electrode do the NO_3^- and Cl^- ions migrate?

12-4
REDUCTION POTENTIALS

Just as acids have different strengths, with some acids having a great tendency to donate protons and others having only a small tendency, so metals and other reducing agents vary in their tendency to lose electrons. We saw (Figure 12-1) that iron spontaneously transfers electrons to Cu^{2+}, and that Cu spontaneously transfers electrons to Ag^{+}. The reverse reactions do not occur spontaneously. For example, if a copper strip is dipped into a solution of ferrous nitrate, $Fe(NO_3)_2$, or if a silver wire is dipped into a solution of copper nitrate, no reaction is observed. Copper does not spontaneously displace iron from solution (does not transfer electrons to Fe^{2+}), and silver does not spontaneously displace copper from solution (does not transfer electrons to Cu^{2+}). Consider the three metals iron, copper, and silver; iron has the greatest tendency to lose electrons, and silver has the least tendency. We say that iron is the strongest reducing agent of the three, and that silver is the weakest.

These tendencies to gain and lose electrons can be measured—numbers can be assigned to them. The apparatus that is used consists of a galvanic cell with a voltmeter (instead of an ammeter) in the external circuit (Figure 12-3). The voltage, or **potential,** of a cell (symbolized $\mathcal{E}$) indicates the ability of the cell to transfer electrons to Cu^{2+}. The voltage of a galvanic cell depends on the particular oxidation–reduction reaction occurring in the cell, on the cell temperature, and on the concentrations of the species taking part in the net cell reaction. A reference temperature of 25°C and ionic concentrations of 1 M (and gas pressures of 1 atm) have been agreed upon as the standards for measuring cell potentials. The symbol $\mathcal{E}^0$ represents the cell potential for standard reference conditions.

The cell shown in Figure 12-3 has a standard cell potential $\mathcal{E}^0$ of 1.10 volts (1.10 V). If a galvanic cell is constructed using as one half-cell a zinc electrode and Zn^{2+} ions at a concentration of 1 M, and as the other half-cell a copper electrode and Cu^{2+} ions at a concentration of 1 M, then the potential of that cell is 1.10 V.

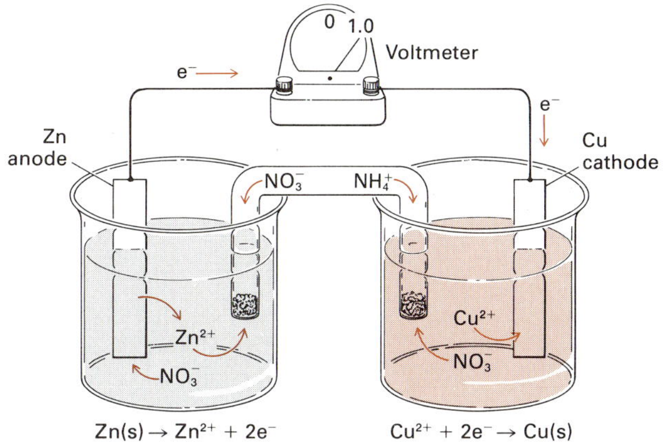

FIGURE 12-3

A galvanic cell with the net reaction

$$Zn(s) + Cu^{2+} \longrightarrow Cu(s) + Zn^{2+}$$

The potential is 1.10 V.

The voltage of such a cell is $\mathcal{E}^0 = 1.10$ V, no matter how large the volume of the cell or the area of the electrodes.

$$\begin{array}{llll}
\text{Anode:} & & & \\
& Zn(s) \longrightarrow Zn^{2+} + 2e^- & \mathcal{E}^0_{ox} & = ? \\
\text{Cathode:} & & & \\
& 2e^- + Cu^{2+} \longrightarrow Cu(s) & \mathcal{E}^0_{red} & = ? \\
\hline
\text{Net reaction:} & & & (12\text{-}3) \\
& Zn(s) + Cu^{2+} \longrightarrow Cu(s) + Zn^{2+} & \mathcal{E}^0 & = 1.10\text{ V}
\end{array}$$

We cannot isolate a single half-reaction—if one species loses electrons, some other species must gain them. So, we cannot measure the voltage of a single half-cell (the $\mathcal{E}^0$ of *one* half-reaction). We can measure the potential only of an overall reaction, the *combination* of two half-reactions. But it would be convenient, in order to compare relative strengths of oxidizing and reducing agents, to be able to attribute part of the measured voltage to each half-reaction. A way to do this has been worked out: we can assign to *one* particular half-reaction the voltage zero, and then measure all others relative to it.

The **standard half-reaction,** to which has been assigned the voltage zero, is the reduction of H^+ to hydrogen gas (H_2):

$$2H^+ + 2e^- \longrightarrow H_2(g) \qquad \mathcal{E}^0_{red} = 0 \quad (12\text{-}4)$$

If a galvanic cell is constructed with the hydrogen gas–hydrogen ion half-reaction as one half-cell (Figure 12-4), the voltage of the whole cell will be that of the *other* half-cell. (To measure the *standard* cell potential, $\mathcal{E}^0$, ion concentrations must be 1 M, the pressure of H_2 must be 1 atm, and the temperature must be 25°C.)

The cell diagrammed in Figure 12-4 is composed of zinc–zinc ion and hydrogen gas–hydrogen ion half-cells. The cell potential is 0.76 V, and the cell reactions are

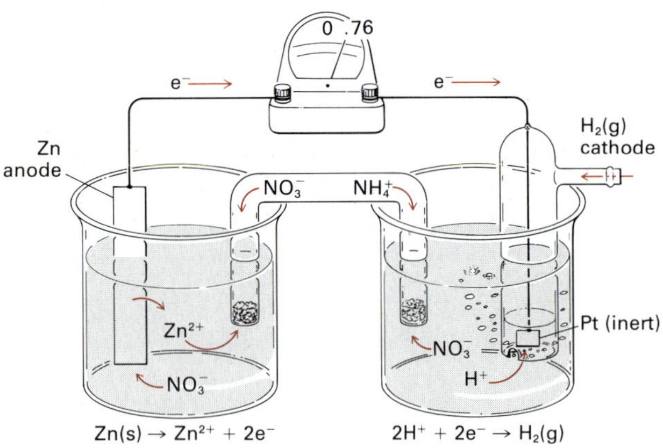

FIGURE 12-4

A galvanic cell with a hydrogen reference half-cell. In the hydrogen gas–hydrogen ion half-cell, $H_2(g)$ at a pressure of 1 atm is bubbled over an electrode that does not react (an inert electrode). The $[H^+] = 1$ M.

Anode:

$$Zn(s) \longrightarrow Zn^{2+} + 2e^- \qquad \mathcal{E}^0_{ox} = ? \quad (12\text{-}5)$$

Cathode:

$$2e^- + 2H^+ \longrightarrow H_2(g) \qquad \mathcal{E}^0_{red} = 0 \text{ (Standard)}$$

Net reaction: (12-6)

$$Zn(s) + 2H^+ \longrightarrow H_2(g) + Zn^{2+} \qquad \mathcal{E}^0 = 0.76\text{ V (Measured)}$$

The $\mathcal{E}^0 = 0.76$ V for the whole cell is an absolute, *measured* quantity. The $\mathcal{E}^0_{red} = 0$ for the hydrogen half-cell is our arbitrary convention. The $\mathcal{E}^0_{ox}$ for the zinc half-cell must be 0.76 V:

$$\mathcal{E}^0 = \mathcal{E}^0_{ox} + \mathcal{E}^0_{red}$$

$$0.76\text{ V} = \mathcal{E}^0_{ox} + 0 = \mathcal{E}^0_{ox}$$

When the cell operates, zinc dissolves and hydrogen gas is produced. Electrons are transferred, through the external circuit, from Zn to H^+. The H^+ has a greater tendency to gain electrons than does Zn^{2+}, and 0.76 V is the quantitative measure of this difference.

A cell composed of a copper and cupric ion (Cu^{2+}) half-cell and a hydrogen gas and hydrogen ion half-cell has a measured potential of 0.34 V. In this case, the needle on the voltmeter moves in the direction opposite to that shown on the voltmeter in Figure 12-4, because electrons in this cell are transferred from H_2 to Cu^{2+}. When the cell operates, H_2 is converted to H^+, and more Cu metal plates out on the Cu cathode. The cell reactions are

Anode:

$$H_2(g) \longrightarrow 2H^+ + 2e^- \qquad \mathcal{E}^0_{ox} = 0 \text{ (Standard)}$$

Cathode: (12-7)

$$2e^- + Cu^{2+} \longrightarrow Cu(s) \qquad \mathcal{E}^0_{red} = 0.34\text{ V (For consistency)}$$

Net reaction: (12-8)

$$H_2(g) + Cu^{2+} \longrightarrow Cu(s) + 2H^+ \quad \mathcal{E}^0 = 0.34\ V \ \text{(Measured)}$$

As before, the potential of the hydrogen gas–hydrogen ion half-cell is zero, so the overall cell potential can be ascribed entirely to the copper–cupric ion half-cell. In this case, however, the H^+ has a *smaller* tendency to gain electrons than does Cu^{2+}, with 0.34 V the measure of this difference. The hydrogen gas–hydrogen ion half-reaction used here is the reverse of Equation 12-4.

So far we have considered only the *magnitudes* of the standard cell potentials, and not their signs. By universally accepted convention, a positive sign is given to the $\mathcal{E}^0$ of any reaction that proceeds spontaneously from left to right as written. Thus, the reactions of Equations 12-6 and 12-8 are given positive $\mathcal{E}^0$ values *because* they proceed spontaneously as written, from left to right.

If the spontaneous reaction is instead from right to left, the $\mathcal{E}^0$ is given a negative sign. Thus, for the *reverse* of the reactions of Equations 12-6 and 12-8, the $\mathcal{E}^0$ values are negative:

$$H_2 + Zn^{2+} \longrightarrow Zn(s) + 2H^+ \quad \mathcal{E}^0 = -0.76\ V$$

$$Cu(s) + 2H^+ \longrightarrow H_2(g) + Cu^{2+} \quad \mathcal{E}^0 = -0.34\ V$$

Reversing an equation changes the sign of the $\mathcal{E}^0$ but does not change its magnitude. Similarly, when a half-reaction is reversed, the sign of the $\mathcal{E}^0$ is changed, as seen here for the half-reactions of Equations 12-4, 12-5, and 12-7:

$$H_2(g) \longrightarrow 2H^+ + 2e^- \quad \mathcal{E}^0 = -0 = 0$$

$$Zn^{2+} + 2e^- \longrightarrow Zn \quad \mathcal{E}^0 = -0.76\ V \quad (12\text{-}9)$$

$$Cu(s) \longrightarrow Cu^{2+} + 2e^- \quad \mathcal{E}^0 = -0.34\ V \quad (12\text{-}10)$$

The voltage associated with an oxidation half-reaction such as Equation 12-10 is called an **oxidation potential,** and the voltage associated with a reduction half-reaction such as Equation 12-9 is called a **reduction potential.** Standard reduction potentials are given in Table 12-1 for some common half-reactions. The reduction potential gives a measure of the tendency of an oxidizing agent to gain electrons. Using these reduction potentials, we can determine the voltage of a

TABLE 12-1
Standard reduction potentials

Half-reaction	$\mathcal{E}^0$ (V)
$F_2 + 2e^- \rightarrow 2F^-$	+2.87
$8H^+ + MnO_4^- + 5e^- \rightarrow Mn^{2+} + 4H_2O$	+1.52
$Au^{3+} + 3e^- \rightarrow Au(s)$	+1.50
$Cl_2(g) + 2e^- \rightarrow 2Cl^-$	+1.36
$14H^+ + Cr_2O_7^{2-} + 6e^- \rightarrow 2Cr^{3+} + 7H_2O$	+1.33
$4H^+ + O_2(g) + 4e^- \rightarrow 2H_2O$	+1.23
$Br_2(l) + 2e^- \rightarrow 2Br^-$	+1.06
$4H^+ + NO_3^- + 3e^- \rightarrow NO(g) + 2H_2O$	+0.96
$Ag^+ + e^- \rightarrow Ag(s)$	+0.80
$Hg^{2+} + 2e^- \rightarrow Hg(l)$	+0.78
$Fe^{3+} + e^- \rightarrow Fe^{2+}$	+0.77
$2H^+ + O_2(g) + 2e^- \rightarrow H_2O_2$	+0.68
$I_2(s) + 2e^- \rightarrow 2I^-$	+0.53
$Cu^{2+} + 2e^- \rightarrow Cu(s)$	+0.34
$2H^+ + S(s) + 2e^- \rightarrow H_2S(g)$	+0.14
$2H^+ + 2e^- \rightarrow H_2(g)$	0 (Standard)
$2H^+ + CO_2(g) + 2e^- \rightarrow CO(g) + H_2O$	−0.12
$Sn^{2+} + 2e^- \rightarrow Sn(s)$	−0.14
$Co^{2+} + 2e^- \rightarrow Co(s)$	−0.28
$Fe^{2+} + 2e^- \rightarrow Fe(s)$	−0.44
$Zn^{2+} + 2e^- \rightarrow Zn(s)$	−0.76
$Al^{3+} + 3e^- \rightarrow Al(s)$	−1.66
$Mg^{2+} + 2e^- \rightarrow Mg(s)$	−2.37
$Na^+ + e^- \rightarrow Na(s)$	−2.71
$Ca^{2+} + 2e^- \rightarrow Ca(s)$	−2.87
$K^+ + e^- \rightarrow K(s)$	−2.93
$Li^+ + e^- \rightarrow Li(s)$	−3.05

galvanic cell. As an example, we can use Table 12-1 to calculate the voltage of the cell shown in Figure 12-3, which has a measured $\mathcal{E}^0$ of 1.10 V:

Anode:

$$Zn(s) \longrightarrow Zn^{2+} + 2e^- \qquad \mathcal{E}^0 = 0.76\ V \text{ (From table)}$$

Cathode:

$$2e^- + Cu^{2+} \longrightarrow Cu(s) \qquad \mathcal{E}^0 = 0.34\ V \text{ (From table)}$$

Net reaction:

$$Zn(s) + Cu^{2+} \longrightarrow Zn^{2+} + Cu(s) \qquad \mathcal{E}^0 = 1.10\ V \text{ (Calculated)}$$

The calculated cell voltage, 1.10 V, agrees with the measured voltage for the cell at 25°C and ionic concentrations of 1 M. (Note that because the direction of the Cu—Cu^{2+} half-reaction was reversed, the sign of the $\mathcal{E}^0$ was changed.)

Any oxidation–reduction reaction for which the cell potential ($\mathcal{E}^0$) is positive occurs spontaneously as written. A reaction for which the cell potential is negative does not occur spontaneously as written, but instead it goes spontaneously in the opposite direction.

An example of a reaction that does not go spontaneously as written is the oxidation of metallic silver by Zn^{2+} ions to yield metallic zinc and Ag^+ ions:

Anode: (12-11)

$$2Ag(s) \longrightarrow 2Ag^+ + 2e^- \qquad \mathcal{E}^0 = -0.80\ V \text{ (From table)}$$

Cathode:

$$2e^- + Zn^{2+} \longrightarrow Zn(s) \qquad \mathcal{E}^0 = -0.76\ V \text{ (From table)}$$

Net reaction:

$$2Ag(s) + Zn^{2+} \longrightarrow Zn(s) + 2Ag^+ \qquad \mathcal{E}^0 = -1.56\ V \text{ (Calculated)}$$

This reaction, for which the $\mathcal{E}^0$ is negative, does not occur spontaneously. (If a silver wire is dipped into a solution of zinc nitrate, no reaction is observed.) The reverse reaction, however, has a positive $\mathcal{E}^0$ and does occur spontaneously. (If a zinc strip is dipped into a solution of silver nitrate, zinc dissolves, and metallic silver plates out.)

Note that Equation 12-11 for the oxidation of Ag to Ag^+ has been multiplied by two, but the $\mathcal{E}^0$ *was not doubled.* The voltage of a cell does not depend on the amounts of reacting substances present, but only on their concentrations (amounts per unit volume). The $\mathcal{E}^0$ is a measure of the relative tendencies to lose or gain electrons. Doubling the total *amounts* of Ag and Ag^+ (as distinct from concentrations) has no effect on the tendency of Ag to lose electrons—that is, no effect on the $\mathcal{E}^0$.

EXERCISE 12-4 A chromium bar dissolves in $Fe(NO_3)_2$ solution, and a zinc bar dissolves in $CrCl_3$ solution. List chromium, iron, and zinc in order of decreasing strength as reducing agents.

EXERCISE 12-5 (a) Would you expect a magnesium bar to dissolve in a solution of $CuCl_2$? (Consult Table 12-1.)

(b) Would you expect a Cu bar to dissolve in a solution of $MgCl_2$?

EXERCISE 12-6 (a) Diagram a galvanic cell composed of magnesium–magnesium ion (Mg^{2+}) and cobalt–cobalt ion (Co^{2+}) half-cells. Label the anode and the cathode. Indicate the direction of electron flow in the external circuit, and the direction of anion and cation flow in the cell. Use data given in Table 12-1.

(b) What is the $\mathcal{E}^0$ of the cell?

(c) Write the chemical equation for the net reaction that occurs when the cell operates.

(d) Cobalt pellets are dropped into a $Mg(NO_3)_2$ solution. Does a reaction occur? If so, write an equation for the reaction.

(e) A strip of magnesium ribbon is dipped into a $Co(NO_3)_2$ solution. Does a reaction occur? If so, write an equation for the reaction.

We saw in Chapter 4 that the criterion for a reaction to occur spontaneously is a negative free energy, ΔG. And we saw in Chapter 11 that the equilibrium constant K is related to ΔG. The standard cell potential $\mathcal{E}^0$ is also related to ΔG^0 (where the superscript zero indicates standard conditions). The equation that expresses this relationship is $\Delta G^0 = -\text{Constant} \times \mathcal{E}^0$. Thus, the more positive $\mathcal{E}^0$, the more negative ΔG^0, and the greater the tendency for the reaction to occur.

Galvanic cells are familiar to us in the form of flashlight and transistor batteries, and in the lead storage battery used in automobiles. The lead storage battery is composed of cells in which a lead anode and a lead dioxide cathode are immersed in a solution of sulfuric acid (Figure 12-5). When the cell operates, the reactions are

Anode: $\mathcal{E}^0 = 0.36\ \text{V}$

$$\text{Pb(s)} + \text{HSO}_4^- \longrightarrow \text{PbSO}_4\text{(s)} + \text{H}^+ + 2\text{e}^-$$

Cathode: $\mathcal{E}^0 = 1.68\ \text{V}$

$$\text{PbO}_2\text{(s)} + \text{HSO}_4^- + 3\text{H}^+ + 2\text{e}^- \longrightarrow \text{PbSO}_4\text{(s)} + 2\text{H}_2\text{O}$$

Net reaction: $\mathcal{E}^0 = 2.04\ \text{V}$

$$\text{Pb(s)} + \text{PbO}_2\text{(s)} + 2\text{HSO}_4^- + 2\text{H}^+ \underset{\text{charging}}{\overset{\text{discharging}}{\rightleftharpoons}} 2\text{PbSO}_4\text{(s)} + 2\text{H}_2\text{O}$$

The insoluble $PbSO_4$ formed at each electrode remains in place, stuck to the lead grids. The battery is charged by the automobile's generator, which sets up a potential greater than 2.04 V opposing that of the spontaneously occurring reaction—all the cell reactions are reversed, restoring the cell to its original condition. When the battery is instead providing electricity, sulfuric acid is consumed and water is produced. The concentration of the sulfuric acid decreases. Because the density of a sulfuric acid solution depends upon the concentration of H_2SO_4 (high density means a high concentration of H_2SO_4, and low density means a low concentration of H_2SO_4), measuring the density of an acid solution in a battery indicates how fully charged it is. In fact, mechanics measure the density to determine whether a car battery is fully charged.

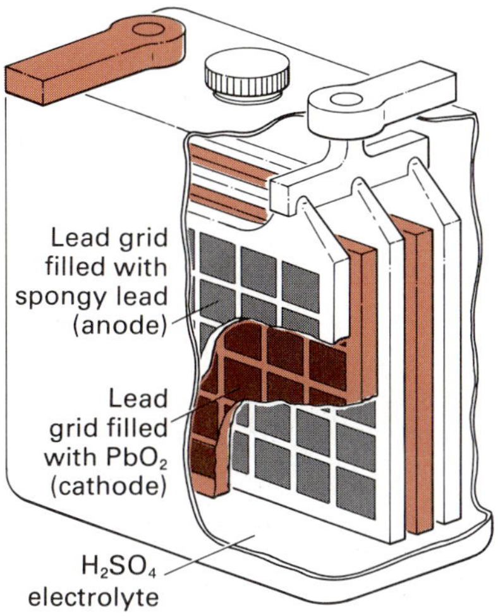

FIGURE 12-5

A lead storage battery.

So far we have considered the potentials of inorganic reactions only. Many reactions of organic and biological chemistry also are oxidation–reduction reactions, and thus have potentials associated with them. We have already seen that the process of photosynthesis—the capture of the sun's energy and its conversion to chemical energy—is an oxidation–reduction reaction:

$$6\overset{4+}{\text{C}}\text{O}_2 + 6\text{H}_2\overset{-2}{\text{O}} \xrightarrow{\text{light}} \overset{0}{\text{C}}_6\text{H}_{12}\text{O}_6 + 6\overset{0}{\text{O}}_2$$

In photosynthesis, the carbon in carbon dioxide accepts electrons, and is reduced from the 4+ oxidation state to the zero oxidation state in sugar. The oxygen in water is oxidized to molecular oxygen. When the chemical energy stored in the sugar is used by the cells of a living organism, the carbon atoms in the sugar are reoxidized to carbon dioxide, and molecular oxygen is reduced to water. The reduction of carbon dioxide to sugar requires energy; the oxidation of sugar to carbon dioxide releases energy.

The processes by which cells break down food and convert the chemical energy stored by plants to forms that can participate in cell reactions are complicated. But the principle involved is the oxidation of energy-rich compounds to energy-poorer compounds plus energy. Part of the energy is stored temporarily in molecules such as NADH (nicotinamide adenine dinucleotide), the reduced form of NAD^+ (see Section 26-2). Table 12-2 lists the standard reduction potentials for some biological half-reactions. The potentials are given the symbol $\mathcal{E}^{0\prime}$ instead of $\mathcal{E}^0$ because the half-reactions are given for the physiological pH of 7. (Remember that the $\mathcal{E}^0$ values of Table 12-1 are for all ions, including H^+, at a concentration of 1 M.) The potentials for the reduction of H^+ and of O_2 in Table 12-2 are therefore different from those given in Table 12-1. Using Table 12-2, we can calculate the $\mathcal{E}^{0\prime}$ for a biochemical reaction just as we did before for other chemical reactions. As an example, we'll calculate the $\mathcal{E}^{0\prime}$ for the reduction of O_2 to H_2O by NADH, the final stage in the conversion of sugar to carbon dioxide and water. The reaction occurs in a series of steps, called the electron-transport chain (see Section 26-3). The unbalanced net reaction can be written

$$\text{NADH} + O_2 \longrightarrow \text{NAD}^+ + H_2O$$

The $\mathcal{E}^{0\prime}$ is obtained by combining the half-reactions:

$$O_2 + 4H^+ + 4e^- \longrightarrow 2H_2O \qquad \mathcal{E}^{0\prime} = +0.82\ \text{V}$$

$$2\text{NADH} + 2H^+ \longrightarrow 2\text{NAD}^+ + 4H^+ + 4e^- \qquad \mathcal{E}^{0\prime} = +0.32\ \text{V}$$

Net reaction: $\mathcal{E}^{0\prime} = 1.14$ V

$$O_2 + 2\text{NADH} + 2H^+ \longrightarrow 2\text{NAD}^+ + 2H_2O$$

As expected (because this is a spontaneous, energy-releasing reaction), the standard potential is positive. Note that when a half-reaction is multiplied by two, the $\mathcal{E}^{0\prime}$ is *not* doubled. (Chapter 26 discusses some of the oxidation–reduction reactions involved in photosynthesis and metabolism.)

TABLE 12-2
Standard reduction potentials of biochemical substances

Half-reaction	$\mathcal{E}^{0\prime}$ (V)
$O_2 + 4H^+ + 4e^- \rightarrow 2H_2O$	+0.82
Cytochrome *c* (+3) + $e^- \rightarrow$ Cytochrome *c* (+2)	+0.22
Oxaloacetate + $2H^+ + 2e^- \rightarrow$ Malate	−0.17
Pyruvate + $2H^+ + 2e^- \rightarrow$ Lactate	−0.19
Acetaldehyde + $2H^+ + 2e^- \rightarrow$ Ethyl alcohol	−0.20
$NAD^+ + 2H^+ + 2e^- \rightarrow NADH + H^+$	−0.32
$NADP^+ + 2H^+ + 2e^- \rightarrow NADPH + H^+$	−0.32
α-Ketoglutarate + $CO_2 + 2H^+ + 2e^- \rightarrow$ Isocitrate	−0.38
$2H^+ + 2e^- \rightarrow H_2(g)$	−0.42
Acetic acid + $2H^+ + 2e^- \rightarrow$ Acetaldehyde + H_2O	−0.58

Note: These standard reduction potentials $\mathcal{E}^{0\prime}$ are measured at pH = 7 and a temperature of 25 to 30°C.

12-5
CORROSION

Corrosion is an important (and usually undesirable) consequence of oxidation–reduction reactions. It is most familiar in the form of the rusting of iron—say, in automobile tail pipes.

The rusting of iron is a complicated process, not yet completely understood. It requires the presence of iron, oxygen, and water. Furthermore, it is known that strains in the metal cause more rapid rusting, that acids catalyze the reaction, and that the presence of stronger reducing agents than iron (such as zinc or magnesium) retards rusting. A mechanism has been proposed consisting of the following steps:

$$(1) \quad Fe(s) \longrightarrow Fe^{2+} + 2e^-$$

$$(2) \quad H^+ + e^- \longrightarrow H$$

$$(3) \quad 4H + O_2(g) \longrightarrow 2H_2O$$

$$(4) \quad 4Fe^{2+} + O_2(g) + 4H_2O \longrightarrow 2Fe_2O_3 + 8H^+$$

In the first step, Fe acts as an anode, becoming oxidized to Fe^{2+}. Electrons move through the iron bar to a moist surface, where H^+ ions in water (or in acidic solution) accept them, forming neutral H atoms. The H atoms react with O_2 in the air to produce H_2O. In the last step, Fe^{2+} ions are oxidized to Fe^{3+} in the presence of water to form rust, a hydrated form of ferric oxide ($Fe_2O_3 \cdot xH_2O$). Pitting—a phenomenon in which the metal corrodes at one spot, but the rust is found at another—occurs when the Fe^{2+} does not come into contact with O_2 where the Fe^{2+} is formed, but instead diffuses along a film of moisture to a distant point where it then comes in contact with O_2. Pitting is seen, for example, when a steel pipe is partially immersed in water. The corrosion may occur below the water line where there is a strain in the metal, but the rust appears at the water line, where O_2 is available (Figure 12-6).

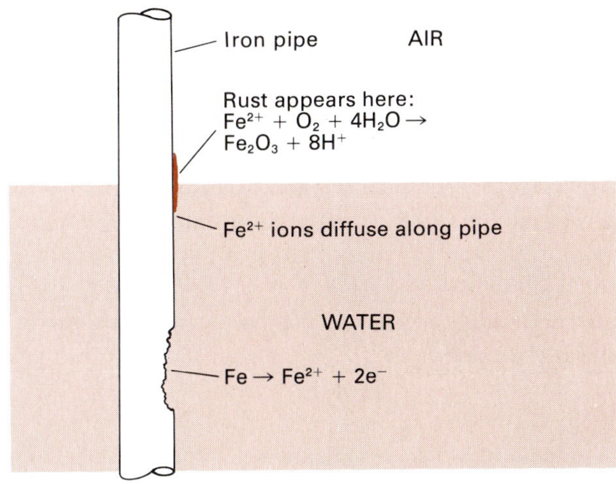

FIGURE 12-6

Pitting. The iron pipe has corroded below the water line, but rust appears at the water line.

Several ways to minimize rusting are suggested by the proposed mechanism. If water is excluded, step 4 is prevented. The exclusion of O_2 prevents steps 3 and 4. Thus, coating iron completely with grease, tar, paint, or tin (to exclude water and oxygen) inhibits rusting. If iron is coated with tin, a tough and inert coat of tin oxide (which is resistant to corrosion) forms on the surface. But if the tin coat is punctured, the corrosion of the steel is accelerated. This may be observed in "tin" cans, which are tin-coated iron. If the tin coat is scratched through, the iron rusts more rapidly than an uncoated iron can would do. The iron is a stronger reducing agent than tin, and so it is preferentially oxidized. The mechanism also accounts for the fact that strong reducing agents such as magnesium retard rusting. If magnesium is placed in contact with iron, the magnesium acts as an anode, forcing the iron to act as a cathode, and, thus, preventing step 1 in the rusting mechanism:

$$\text{Anode:} \quad Mg(s) \longrightarrow Mg^{2+} + 2e^-$$

$$\text{Cathode:} \quad 2e^- + Fe^{2+} \longrightarrow Fe(s)$$

The magnesium is preferentially oxidized, instead of the iron. The magnesium is called a **sacrificial metal,** and this method of protecting iron is called **cathodic protection.** Ship hulls and underground pipes are protected in this way (Figure 12-7). When iron is coated with a thin layer of zinc (galvanized iron), the iron is protected in two ways. A strong, protective coat of zinc oxide forms on the zinc, sealing off the iron and the zinc from water and oxygen. Furthermore, if the zinc is punctured, it acts as a sacrificial metal and is preferentially oxidized.

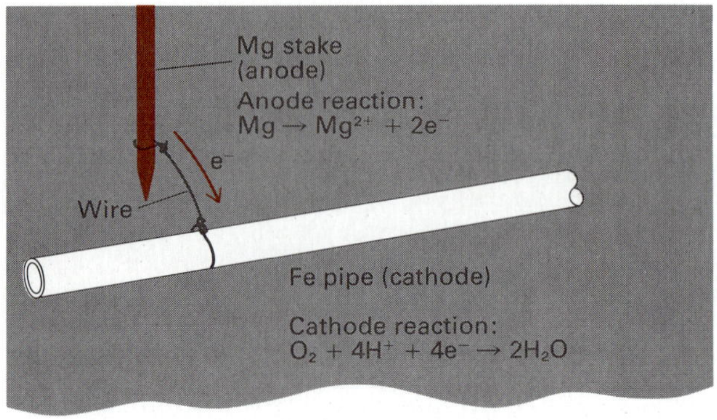

FIGURE 12-7

Cathodic protection. Electrons are carried from the magnesium to the iron by ions in the moist soil. As a result, the magnesium stake is corroded instead of the iron pipe.

12-6 ELECTROLYSIS

Galvanic cells, as we have seen, are based on chemical reactions that proceed spontaneously. We can, however, force a reaction to go in the direction *opposite* to that in which it goes spontaneously. This is done by placing in the circuit a generator or battery with a potential greater than, and opposed to, the potential of the galvanic cell. This is the process of **electrolysis.** The recharging of a lead storage battery is an example we have already mentioned.

We know that hydrogen gas and oxygen gas react, spontaneously, to produce water. But, we can force the reaction to occur in the opposite direction—that is, we can decompose water to hydrogen and oxygen, by electrolysis (Figure 12-8). The electrodes are an inert substance (here carbon), and the solution is water with a small amount of an electrolyte such as sodium sulfate (Na_2SO_4). The Na_2SO_4 is required to furnish ions to conduct the current through the solution, with the cations (Na^+) moving toward the cathode, and the anions (SO_4^{2-}) moving toward the anode. The cell reactions are

Anode:

$$H_2O \longrightarrow \tfrac{1}{2}O_2(g) + 2H^+ + 2e^-$$

Cathode:

$$2e^- + 2H_2O \longrightarrow H_2(g) + 2OH^-$$

Net reaction:

$$3H_2O \xrightarrow{\text{electrolysis}} H_2(g) + \tfrac{1}{2}O_2(g) + 2H^+ + 2OH^-$$

Both H^+ and OH^- appears as products. If a few drops of litmus are added to each half-cell, the solution near the anode will turn red, and that near the cathode will turn blue. If the H^+ (at the anode) and the OH^- (at

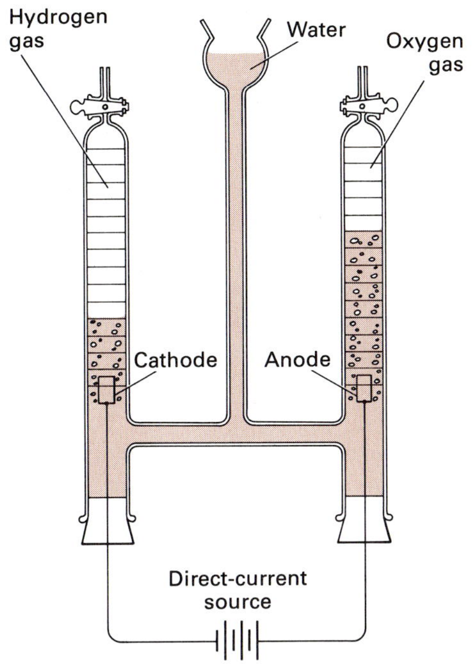

FIGURE 12-8

Electrolysis of water.

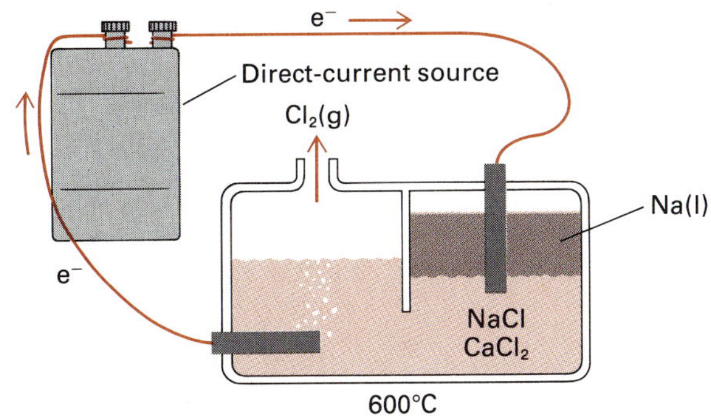

FIGURE 12-9

Preparation of sodium by the electrolysis of sodium chloride.

the cathode) are allowed to mix, they will neutralize each other, forming H_2O. The overall equation then becomes

$$H_2O \xrightarrow{\text{electrolysis}} H_2(g) + \tfrac{1}{2}O_2(g)$$

Electrolysis is used commercially in the preparation of active metals such as aluminum, magnesium, and sodium. Sodium is prepared by the electrolysis of a molten mixture of sodium chloride and calcium chloride at 600°C (Figure 12-9). The calcium chloride is added to lower the melting point of the sodium chloride (the melting point of pure NaCl is 801°C). The cell reactions are

$$\text{Anode:} \qquad 2Cl^- \longrightarrow Cl_2(g) + 2e^-$$

$$\text{Cathode:} \quad 2Na^+ + 2e^- \longrightarrow 2Na(l)$$

Net reaction:

$$2Na^+ + 2Cl^- \xrightarrow{\text{electrolysis}} 2Na(l) + Cl_2(g)$$

The pure liquid sodium collects on top of the molten mixture.

EXERCISE 12-7 What would be the minimum voltage required for the electrolysis of $FeBr_2$ to yield Fe(s) and $Br_2(l)$? Write equations for the two half-reactions that would occur.

12-7 BALANCING EQUATIONS FOR OXIDATION–REDUCTION REACTIONS

Many equations for oxidation–reduction reactions, like those for other chemical reactions, can be balanced by inspection—that is, by trial and error as described in Chapter 2. We use this method here to balance the equation for a reaction in which a strip of aluminum dissolves in a solution of copper nitrate, $Cu(NO_3)_2$, to yield metallic copper and aluminum ions, Al^{3+}:

$$Al(s) + Cu^{2+} \longrightarrow Cu(s) + Al^{3+}$$

(Correct formulas of reactants and products)

$$Al(s) + 3Cu^{2+} \longrightarrow Cu(s) + 2Al^{3+}$$

(Charge balanced; atoms not balanced)

$$2Al(s) + 3Cu^{2+} \longrightarrow 3Cu(s) + 2Al^{3+}$$

(Charge and atoms balanced)

An oxidation–reduction reaction that is difficult to balance by inspection can often be balanced by using a table of half-reactions such as that given in Table 12-1. For example, when iodide ion (I^-) is added to an acidic solution of the salt potassium permanganate ($KMnO_4$), the color of the solution turns from purple to pale yellow, and the products of the reaction are iodine (I_2) and manganous ions (Mn^{2+}). The reactants and products are shown in the unbalanced equation:

$$I^- + MnO_4^- \longrightarrow Mn^{2+} + I_2(s) \qquad (12\text{-}12)$$

This equation is difficult to balance by inspection, but it can be balanced easily by adding the appropriate half-reactions from Table 12-1, which are already balanced as to atoms and charge.

The half-reactions in which we are interested are those involving I^- and I_2, and MnO_4^- and Mn^{2+}:

$$2I^- \longrightarrow I_2(s) + 2e^- \qquad (12\text{-}13)$$

$$Mn^{2+} + 4H_2O \longrightarrow MnO_4^- + 8H^+ + 5e^- \qquad (12\text{-}14)$$

These two half-reactions involve the reactants and products that we want, but Equation 12-14 must be reversed to give the MnO_4^- on the left and the Mn^{2+} on the right, as in the unbalanced Equation 12-12:

$$MnO_4^- + 8H^+ + 5e^- \longrightarrow Mn^{2+} + 4H_2O \qquad (12\text{-}15)$$

Note that the balanced half-reaction for the manganese also involves H_2O and H^+. For reactions that take place in aqueous solution, it sometimes is necessary to add H_2O—and H^+ if the solution is acidic, or OH^- if the solution is basic—in order to balance the equation.

So that no electrons appear in the equation for the overall reaction when the two appropriate half-reactions are added together (in this case Equations 12-13 and 12-15), the numbers of electrons must be made the same in the two half-reactions. Therefore, we will multiply Equation 12-13 by five, and Equation 12-15 by two, so that there are ten electrons in each half reaction. The resulting half-reactions and their sum are

$$10I^- \longrightarrow 5I_2(s) + 10e^-$$

$$2MnO_4^- + 16H^+ + 10e^- \longrightarrow 2Mn^{2+} + 8H_2O$$

Net reaction:

$$10I^- + 2MnO_4^- + 16H^+ \longrightarrow 5I_2(s) + 2Mn^{2+} + 8H_2O \qquad (12\text{-}16)$$

Equation 12-16 is a balanced equation for the overall reaction, with both atoms and charge balanced.

EXAMPLE 12-2

Balance the following equation, using half-reactions from Table 12-1:

$$Ag(s) + Cr_2O_7^{2-} \longrightarrow Ag^+ + Cr^{3+}$$

First, choose the two half-reactions involving the reactants and products:

$$Ag(s) \longrightarrow Ag^+ + e^-$$
$$2Cr^{3+} + 7H_2O \longrightarrow 14H^+ + Cr_2O_7^{2-} + 6e^-$$

The chromium reaction must be reversed:

$$Cr_2O_7^{2-} + 14H^+ + 6e^- \longrightarrow 2Cr^{3+} + 7H_2O$$

So that no electrons appear in the overall equation, the silver half-reaction must be multiplied by six. The two half-reactions are then added together:

$$6Ag(s) \longrightarrow 6Ag^+ + 6e^-$$
$$Cr_2O_7^{2-} + 14H^+ + 6e^- \longrightarrow 2Cr^{3+} + 7H_2O$$

Net reaction:

$$6Ag(s) + Cr_2O_7^{2-} + 14H^+ \longrightarrow 6Ag^+ + 2Cr^{3+} + 7H_2O$$

Finally, check to be certain the equation is balanced as to both atoms and charge.

EXERCISE 12-8 Balance the following equation by combining the appropriate equations from Table 12-1:

$$Cr_2O_7^{2-} + Br^- \longrightarrow Cr^{3+} + Br_2(l)$$

Reaction occurs in acid solution, so you may add H^+ and H_2O, if needed.

You know how to balance equations for oxidation–reduction reactions using tables of half-reactions, but what happens when no table is available? Then you must write your own half-reactions. The following example shows how to do it.

EXAMPLE 12-3

Microorganisms in the sea use dissolved oxygen to convert ammonium ion (NH_4^+) to nitrite ion (NO_2^-). Water is also produced. Write a balanced equation for the reaction, adding H^+ and H_2O if needed.

The first step is to write an unbalanced *net* equation, giving reactants and products:

$$NH_4^+ + O_2(g) \longrightarrow NO_2^- + H_2O$$

The second step is to write separate unbalanced equations for the oxidation and reduction processes. These are then balanced, by inspection, with respect to atoms and charge. For the oxidation half-reaction,

$$\overset{-3}{NH_4^+} \longrightarrow \overset{+3}{NO_2^-}$$
(N balanced)

$$2H_2O + NH_4^+ \longrightarrow NO_2^-$$
(Balance O by adding H_2O)

$$2H_2O + NH_4^+ \longrightarrow NO_2^- + 8H^+$$
(Balance H by adding H^+)

$$2H_2O + NH_4^+ \longrightarrow NO_2^- + 8H^+ + 6e^-$$
(Balance charge by adding electrons)

Check to make sure the equation is balanced with respect to atoms and charge. For the reduction half-reaction,

$$\overset{0}{O_2} \longrightarrow 2H_2\overset{-2}{O}$$
(Balance O by adding H_2O)

$$4H^+ + O_2 \longrightarrow 2H_2O$$
(Balance H by adding H^+)

$$4e^- + 4H^+ + O_2 \longrightarrow 2H_2O$$
(Balance charge by adding electrons)

Check to make sure the equation is balanced with respect to atoms and charge.

Note that the equation for the reduction half-reaction is the same as that given in Table 12-1 for the reduction of oxygen to water. Using this method, you can construct for yourself all the half-reactions given in the table!

Now that we have the balanced half-reactions, it is easy to finish balancing the equation as we did in Example 12-2. Finish balancing this one, for practice.

To summarize, use the following procedure to balance equations for oxidation–reduction reactions when no tables are available.

1. Write an unbalanced equation for the net reaction.

2. Write unbalanced equations for the two half-reactions.

3. Balance each half-reaction separately by inspection, with respect to both atoms and charge.
   a. First balance the atom that is oxidized or reduced.
   b. Then balance the O atoms by adding H_2O, as required.
   c. Next balance the H atoms by adding H^+, as required.
   d. Finally, balance the charge by adding electrons, as required.
   e. Check to make sure that the equations are properly balanced.

4. Combine the equations for the half-reactions so that the electrons cancel out, thus obtaining the net equation.

5. Check the final equation to make sure that atoms and charge are balanced.

These rules are for balancing equations in acid or neutral solutions. When the solution is basic, the concentration of H^+ is small, so OH^- and H_2O are added when required rather than H^+ and H_2O. An example of such a reaction follows.

EXAMPLE 12-4

In basic solution, permanganate ion (MnO_4^-) reacts with sulfide ion (S^{2-}) to yield solid manganese dioxide (MnO_2) and free sulfur (S). Write a balanced equation for the reaction. (The reaction is in basic solution, so OH^- and H_2O can be added if needed to balance the equation.)

First, write an unbalanced equation giving reactants and products:

$$MnO_4^- + S^{2-} \longrightarrow MnO_2(s) + S(s)$$

Next, write balanced equations for the half-reactions. For oxidation,

$$\overset{-2}{S^{2-}} \longrightarrow \overset{0}{S}(s) + 2e^-$$

(Atoms balanced; add electrons to balance charge)

For reduction,

$$\overset{+7}{MnO_4^-} \longrightarrow \overset{+4}{MnO_4}(s) \qquad \text{(Mn balanced)}$$

To balance O atoms, first determine how many are needed (here two O atoms are required on the right). Then add *two* OH^- ions for every O required. (Adding $2OH^-$ to one side of the equation, and $1H_2O$ to the other, has the net effect of adding 1 oxygen atom to the side to which OH^- was added.)

We need 2 oxygen atoms on the right, so we add 4 OH^- on the right, and balance them by adding 2 H_2O on the left:

$$\overset{+7}{} \qquad \overset{+4}{}$$
$$2H_2O + MnO_4^- \longrightarrow MnO_2(s) + 4OH^-$$
(O balanced; H balanced)

$$3e^- + 2H_2O + MnO_4^- \longrightarrow MnO_2(s) + 4OH^-$$
(Add electrons to balance charge)

Now that you have the two half-reactions, finish balancing the equation, for practice.

EXERCISE 12-9 Fixed nitrogen in the sea is returned to the atmosphere as $N_2(g)$ by nitrate-reducing bacteria. In a typical reaction, a substance such as methyl alcohol (CH_3OH) is oxidized by nitrate ion (NO_3^-) to produce carbon dioxide gas and nitrogen gas. Write a balanced equation for the reaction, adding H^+ and H_2O if needed.

KEY TERMS

Oxidation is the loss of electrons. The *oxidizing agent* gains electrons. *Reduction* is the gain of electrons. The *reducing agent* loses electrons.

In a *galvanic cell,* a spontaneously occurring chemical reaction is used as a source of electrons. Oxidation occurs at the *anode* and reduction occurs at the *cathode.* Negative ions are called *anions*—they migrate toward the anode. Positive ions are called *cations*—they migrate toward the cathode.

The voltage or *potential* of a galvanic cell is a measure of the ability of the cell to do electrical work.

In the process of *electrolysis,* electrical energy is used to force a reaction to go in the direction opposite to that in which it goes spontaneously.

CHAPTER SUMMARY

We have seen that *oxidation* and *reduction* can be regarded as processes in which electrons are transferred. Oxidation is a loss of electrons, and reduction is a gain of electrons. Oxidation and reduction always occur together. Electrons are gained by the *oxidizing agent,* and electrons are lost by the *reducing agent.*

Oxidation numbers can be assigned to atoms in compounds using widely accepted conventions.

In a *galvanic cell,* chemical energy is converted to electrical energy. A spontaneous oxidation–reduction reaction is made to take place indirectly, with the electron transfer occurring through a wire. In an *electrolytic cell,* on the contrary, electrical energy is converted to chemical energy. A battery or generator is used to force a chemical reaction to go in the direction opposite to that in which it would go spontaneously.

An oxidation–reduction reaction can be regarded as the sum of two *half-reactions:* one an oxidation half-reaction and the other a reduction half-reaction. In the net reaction, the number of electrons gained by the oxidizing agent and the number of electrons lost by the reducing agent must be the same. Equations for oxidation–reduction reactions can be balanced by combining the appropriate half-reactions in such a way that the electrons cancel out.

Each half-reaction has associated with it a *standard reduction potential* $\mathcal{E}^0$ based on a reference temperature of 25°C and ionic concentrations of 1 M (or gas pressures of 1 atm). The hydrogen half-reaction, $2H^+ + 2e^- \rightarrow H_2(g)$, is arbitrarily assigned the potential zero, and the potentials of all other half-reactions are measured relative to it. Using half-cell potentials, we can calculate the voltage of a galvanic cell. An oxidation–reduction reaction for which the cell potential is positive occurs spontaneously as written. A reaction for which the cell potential is negative does not occur spontaneously as written, but goes spontaneously in the opposite direction.

PROBLEMS

12-1. Sodium metal reacts with bromine to produce sodium bromide. In this reaction, what substance is oxidized and what substance is reduced? What is the oxidizing agent, and what is the reducing agent?

12-2. Chlorine gas reacts with Br^- to produce $Br_2(l)$ and Cl^-. In this reaction, what is the oxidizing agent and what is the reducing agent?

12-3. Purple sulfur bacteria use energy from the sun to convert carbon dioxide (CO_2) and hydrogen sulfide (H_2S) to carbohydrates $(CH_2O)_x$ and sulfur. An unbalanced equation for the reaction is

$$CO_2(g) + H_2S(g) \xrightarrow{\text{light}} (CH_2O)_x + 2S(s)$$

In this reaction, what is being oxidized, and what is being reduced? What is the oxidizing agent, and what is the reducing agent?

12-4. Assign oxidation numbers to all the atoms in the following equations, and tell which substance is being oxidized, and which reduced.
(a) $C_6H_{12}O_6(aq) + 3SO_4^{2-}(aq) + 6H^+(aq) \rightarrow 3H_2S(g) + 6H_2O(l) + 6CO_2(g)$
(b) $Cl_2(aq) + H_2O(l) \rightarrow HOCl(aq) + H^+(aq) + Cl^-(aq)$
(c) $Cd(s) + Ni_2O_3(s) + 3H_2O(l) \rightarrow Cd(OH)_2(s) + 2Ni(OH)_2(s)$

12-5. (a) Give the oxidation number of each nitrogen atom in the compound ammonium nitrate (NH_4NO_3). This compound, which contains nitrogen in two different oxidation states, is unstable. When jarred, it decomposes to yield nitrous oxide (laughing gas, N_2O) and water. (NH_4NO_3, used in blasting explosives, caused a disastrous explosion of a ship laden with it in Texas City in 1947.)
(b) Give the oxidation number of nitrogen in the compounds HNO_3, $NaNO_2$, N_2H_4 (hydrazine), NH_2OH (hydroxylamine), N_2O_4, NO, N_2O, and Na_3N.

12-6. The following compounds each contain carbon in a different oxidation state: methane (CH_4); methyl alcohol (CH_3OH); formaldehyde (CH_2O); formic acid (HCO_2H); and carbon dioxide (CO_2). Determine the oxidation number of carbon in each of these compounds.

12-7. Which of the following are oxidation–reduction reactions?

(a) $2Na(s) + 2H_2O(l) \rightarrow 2Na^+(aq) + 2OH^-(aq) + H_2(g)$

(b) $Ba(NO_3)_2(aq) + H_2SO_4(aq) \rightarrow BaSO_4(s) + 2HNO_3(aq)$

(c) $2Ag^+(aq) + H_2S(aq) \rightarrow Ag_2S(s) + 2H^+(aq)$

(d) $2KClO_3(s) \rightarrow 2KCl(s) + 3O_2(g)$

(e) $C_2H_4(g) + Br_2(l) \rightarrow C_2H_4Br_2(l)$

12-8. The "thermite reaction" is a spectacular exothermic reaction in which metallic aluminum reacts with ferric oxide (Fe_2O_3) to produce metallic iron and aluminum oxide (Al_2O_3).
(a) In this reaction, what substance is oxidized, and what substance is reduced?
(b) How many grams of aluminum metal are needed to react with 40 g of ferric oxide?

12-9. Strontium dissolves in a solution of $MnCl_2$. Nickel dissolves in a solution of $Pb(NO_3)_2$, but not in one of $MnCl_2$. From this information, arrange the four elements strontium, nickel, manganese, and lead in order of decreasing strength as reducing agents.

12-10. The two half-reactions occurring in a galvanic cell are

$$Fe(s) \longrightarrow Fe^{2+} + 2e^-$$

$$Ag^+ + e^- \longrightarrow Ag(s)$$

(a) Which reaction occurs at the anode, and which at the cathode?
(b) When 112 g of iron have dissolved, what mass of silver has been deposited?
(c) If you wanted to plate a copper ring with silver, would you make the ring the anode or the cathode in a galvanic cell?

12-11. Which of the following reactions will proceed spontaneously to give products?
(a) $Mg^{2+} + Fe \rightarrow Mg + Fe^{2+}$
(b) $Cu + Zn^{2+} \rightarrow Cu^{2+} + Zn$
(c) $Hg + 2Ag^+ \rightarrow 2Ag + Hg^{2+}$

12-12. Explain why silver, gold, and copper can be found in nature as the uncombined metals, whereas potassium, magnesium, and zinc are always found combined (in compounds).

12-13. Which of the following metals would you expect to dissolve in acidic solutions: Sn, Ag, Al, Zn, and Hg? Explain.

12-14. (a) Would you expect a gold container to be suitable for storing fluorine gas? Explain.
(b) Would a silver container be suitable for storing a solution of $ZnCl_2$? Explain.

12-15. A student in a chemistry course was analyzing an unknown solid for metals. In the process of the analysis, he dissolved the unknown in hydrochloric acid in a Pyrex beaker, stirring with a steel spatula to help dissolve the solid. He reported silver, tin, lead, and iron in the unknown, but he was told that the unknown did not contain iron. What was the error in the student's procedure?

12-16. One method of removing the tarnish of silver sulfide (Ag_2S) from silver is to put the silver in an aluminum pan that has first been polished with steel wool to remove the protective oxide coating, and then to fill the pan with water, adding a little salt. After a short time the silver is bright and shiny. Explain how this is possible.

12-17. Name one or two metals (other than magnesium and zinc) that might be used as sacrificial metals to protect iron from corroding.

12-18. During the operation of a galvanic cell, electrons are given up at the anode and are consumed at the cathode. Does half of the cell then acquire a positive, and the other half a negative, charge? Explain.

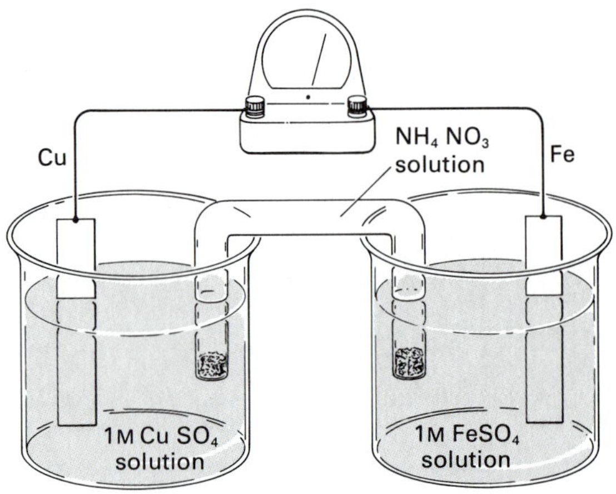

FIGURE 12-10
Galvanic cell for Problem 12-19.

12-19. (a) In the cell shown in Figure 12-10, do the electrons flow in the external circuit from the copper electrode to the iron, or from the iron to the copper? (Use data given in Table 12-1.)
(b) What voltage would be read on the voltmeter?
(c) Write the chemical equation for the net reaction that occurs as the result of the operation of this galvanic cell.
(d) An iron strip is dipped into a $CuSO_4$ solution, and a copper strip is dipped into an $FeSO_4$ solution. In which case does a reaction occur, and what is the reaction?

12-20. Metallic sodium is usually produced by electrolysis of molten sodium chloride. Name two or three substances that could be used to reduce the sodium in NaCl chemically (instead of electrolytically) to metallic sodium.

12-21. Silver dissolves in nitric acid (HNO_3) to yield nitric oxide gas (NO), Ag^+, and water. What volume of 5.0 M HNO_3 is required to dissolve 108 g of Ag?

12-22. An aqueous solution of Br_2 can be used to oxidize Fe^{2+} to Fe^{3+}. If 1.0 mole of Br_2 reacts, what mass of Fe^{2+} is oxidized?

12-23. Lactic acid is a product of glycolysis, which is the anaerobic (without O_2) breakdown of sugars in the body. In the presence of oxygen, lactic acid is oxidized to carbon dioxide and water. Large amounts of energy for life processes are also produced. The unbalanced equation for the net reaction is

$$\underset{\text{lactic acid}}{C_3H_6O_3} + O_2 \longrightarrow CO_2 + H_2O$$

(a) Write balanced equations for the two half-reactions. (Use H^+ and H_2O if needed.)
(b) Write a balanced net equation for the reaction.

12-24. One step in the Krebs tricarboxylic acid cycle (or citric acid cycle, the major energy-producing series of reactions in living organisms) is the oxidation of isocitrate ion to α-ketoglutarate ion and CO_2. The oxidizing agent is NAD^+. Using information given in Table 12-2, write a balanced equation for the reaction, and calculate the standard potential, $\mathcal{E}^{0\prime}$.

12-25. In photosynthesis, electrons flow in a series of reactions from water to the oxidizing agent $NADP^+$ (see Chapter 26). The unbalanced equation for the net reaction is

$$H_2O + NADP^+ \longrightarrow NADPH + O_2$$

(a) Using data from Table 12-2, write a balanced net equation, and calculate the standard potential $\mathcal{E}^{0\prime}$.
(b) Look at your answer to part a carefully, and explain what it means.

12-26. Some microorganisms (such as brewers' yeast) ferment glucose to ethyl alcohol and carbon dioxide. The final step of the process is the reduction of acetaldehyde (CH_3CHO) to ethyl alcohol (CH_3CH_2OH). The reducing agent is NADH. Using data from Table 12-2, write a balanced net equation for the reaction, and calculate the standard potential $\mathcal{E}^{0\prime}$.

12-27. Balance the equation for each of the following reactions by any method you choose (assume that all are in acid solution):
(a) $Cr^{3+} + Fe^{3+} \rightarrow Cr_2O_7^{2-} + Fe^{2+}$
(b) $Cu(s) + NO_3^- \rightarrow Cu^{2+} + NO(g)$
(c) $MnO_4^- + C_2O_4^{2-} \rightarrow Mn^{2+} + CO_2(g)$
(d) $Al(s) + H^+ \rightarrow Al^{3+} + H_2(g)$
(e) $Cr_2O_7^{2-} + H_2S \rightarrow Cr^{3+} + S(s)$
(f) $C(s) + NO_3^- \rightarrow NO_2(g) + CO_2(g)$
(g) $Ag_2S(s) + NO_3^- \rightarrow Ag^+ + NO(g) + S(s)$

12-28. (a) Permanganate ion (MnO_4^-) oxidizes methyl alcohol (CH_3OH) to formic acid (HCO_2H). The manganese in MnO_4^- is reduced to Mn^{2+}. Write a balanced equation for the reaction (in acid solution).
(b) Oxidation of ethyl alcohol (CH_3CH_2OH) to acetic acid (CH_3CO_2H) by oxygen in the air is responsible for the souring of wine. Water is also produced in the reaction. Write a balanced equation for the reaction (in acid solution).
(c) Acetic acid (CH_3CO_2H) can be reduced to ethyl alcohol (CH_3CH_2OH) by reaction with hydrogen gas in the presence of a platinum catalyst. Water is also a product. Write a balanced equation for the reaction.

12-29. The reduction half-reactions of Table 12-1 are arranged in order of decreasing strength of oxidizing agent. The *second* of the following two half-reactions occurs lower in the table than the first:

$$(1)\ n\,e^- + \text{Oxidizing agent}_1 \longrightarrow \text{Reducing agent}_1$$

$$(2)\ n\,e^- + \text{Oxidizing agent}_2 \longrightarrow \text{Reducing agent}_2$$

Which reaction occurs spontaneously,

$$\text{Oxidizing agent}_1 + \text{Reducing agent}_2 \longrightarrow \text{Reducing agent}_1 + \text{Oxidizing agent}_2$$

or the reverse of this reaction? (Does this remind you of the Brønsted–Lowry acid–base table, Table 11-5? Explain.)

III

ORGANIC CHEMISTRY

Long slender rod cells from the retina of the eye. When light falls on these cells, the compound retinal, an oxidized form of vitamin A, changes its shape and initiates a sequence of events that results in vision (see page 363) ($\times$2000).

13

Introduction to Organic Chemistry

Until the nineteenth century, the special properties of living things were thought to be the consequence of a vital force. Substances were divided into two classes: organic compounds (those obtained from living things) and inorganic compounds (from nonliving sources). This system of classification broke down when chemists began to synthesize organic substances in the laboratory—today organic chemistry is defined as the chemistry of carbon compounds.

This chapter includes a review of the bonding patterns of carbon and the shapes of carbon-containing molecules, and an introduction to several kinds of isomers—compounds that have the same molecular formula, but are nevertheless different compounds.

13-1
ORGANIC CHEMISTRY: THE COMPOUNDS OF CARBON

The ability of plants and animals to grow, to reproduce, and to adapt to changing conditions clearly distinguishes them from inanimate objects. Therefore it is not surprising that the special properties of living things were long thought to be the consequence of a special life energy, or **vital force.** Moreover, because many of the substances isolated from plants and animals have properties not found in samples from nonliving sources, the vital force was thought to be reflected in the compounds derived only from living things, the so-called **organic compounds.** Substances available from nonliving sources were called **inorganic** by contrast. This division of matter into two classes persisted until a series of developments, beginning in 1828, drastically altered the early view. In that year, the German chemist Friedrich Wöhler accidentally synthesized the organic chemical urea (naturally found in urine) by heating the inorganic substance ammonium cyanate, a salt of ammonium ions (NH_4^+) and cyanate ions (OCN^-):

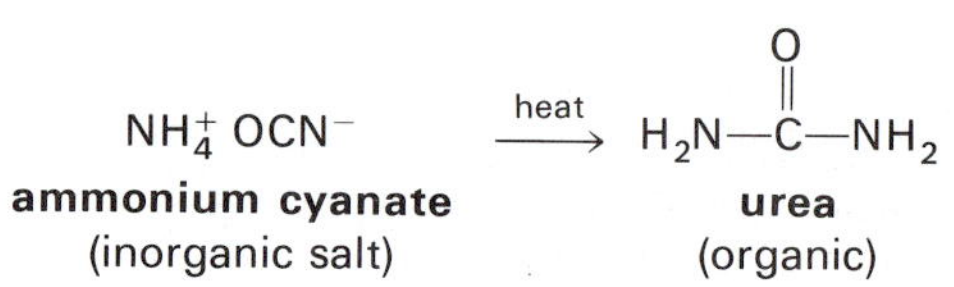

The idea that organic substances can be synthesized in the laboratory was not immediately accepted, but the possibility stimulated further experiments. In 1844, Adolph Kolbe (a student of Wöhler) prepared acetic acid, known to be a product of fermentation, starting with mineral calcium carbide.

In subsequent years, chemists synthesized in the laboratory many other organic compounds indistinguishable from material isolated from plants and animals. By the 1960s, the laboratory synthesis of organic molecules included such important examples as chlorophyll (an essential agent in plant photosynthesis), insulin (a hormone whose absence can give rise to the disease diabetes), ribonuclease (an enzyme involved in digestion), vitamin B_{12}, and even fragments of DNA (the material of genetic inheritance). The structures of chlorophyll and vitamin B_{12} are shown in Figure 13-1. A molecule of ribonuclease has 1876 atoms: 587 of carbon, 909 of hydrogen, 197 of oxygen, 171 of nitrogen, and 12 of sulfur—$C_{587}H_{909}O_{197}N_{171}S_{12}$! All of these important molecules are discussed in more detail in following chapters.

There is no basic chemical difference between organic and inorganic compounds, and organic chemistry is now defined as the chemistry of the compounds of carbon. There are a few exceptions to this rule, including carbon dioxide (CO_2), calcium carbide (CaC_2), and salts such as ammonium cyanate (NH_4OCN), sodium cyanide ($NaCN$), or potassium carbonate (K_2CO_3)—these compounds are considered to be inorganic.

The compounds of carbon play a predominant role in life processes. Furthermore, artificial organic compounds (materials that don't occur in nature but whose existence is due entirely to laboratory chemistry) have had a major impact on modern civilization. Examples of the organic products created by chemists include nylon, the plastics, pesticides such as DDT, and many drugs for the treatment of disease. Many organic products are obtained from petroleum, the "fossil fuel" derived from living matter.

The fact that carbon is the central element in so many important compounds suggests that it must have some special properties. Carbon is, in fact, unique among the elements by virtue of the following:

1. Carbon can form long chains of strong covalent carbon–carbon bonds, with an almost infinite variety of arrangements. Carbon can bond to two, three, or four other carbon atoms.

2. Carbon has the ability to form single, double, or triple bonds.

3. Carbon has the ability to form strong covalent bonds with many other atoms such as H, O, S, N, P, Cl, Br, and I.

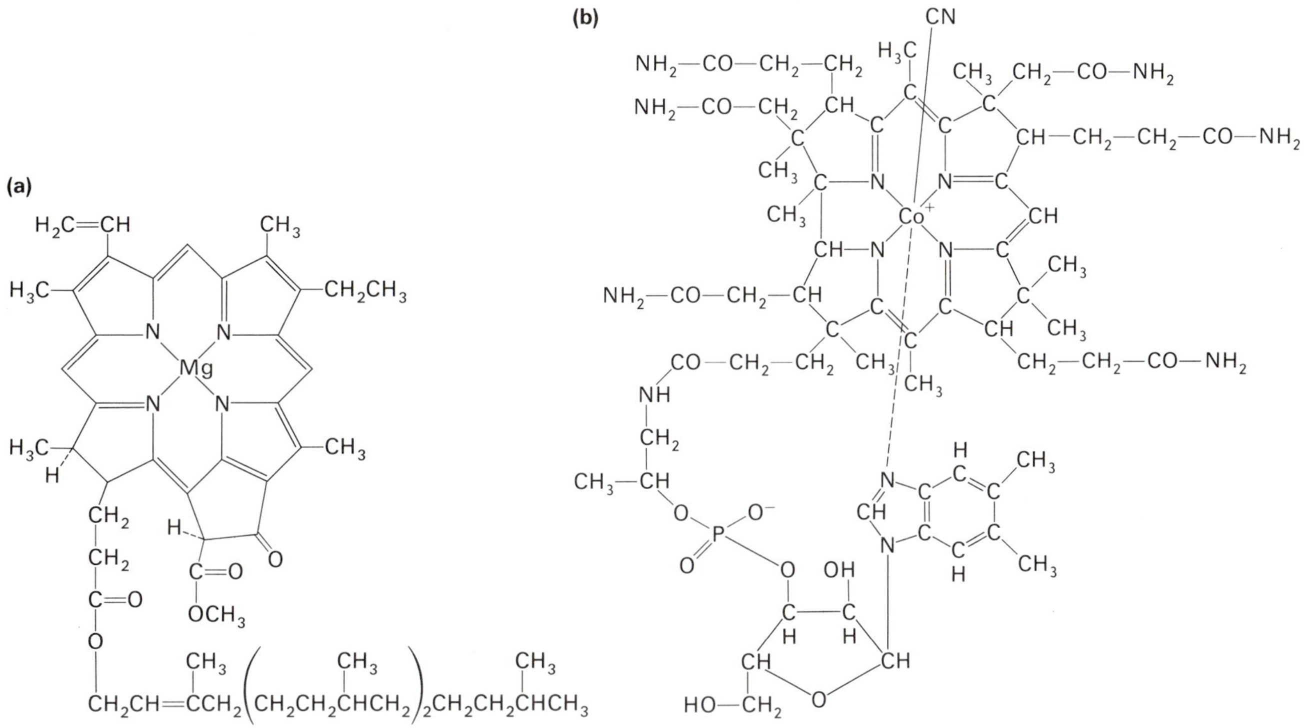

FIGURE 13-1

Structures of chlorophyll *a* and vitamin B_{12}. In these structures, every line represents a bond, and a carbon atom is at every corner where no atom is shown. All bonds not shown—to make a total of four bonds for each carbon—are to hydrogen. Chlorophyll *a* was synthesized in 1956 by R. B. Woodward of Harvard (and others). Woodward (who died in 1979) received the Nobel Prize in 1966 for his synthetic work. The synthesis of vitamin B_{12} by Woodward and Albert Eschenmoser of Switzerland (and others) was completed in 1973.

13-2 BONDING IN CARBON COMPOUNDS

Carbon has atomic number 6 and has the ground-state electronic configuration, $1s^22s^22p^2$. A carbon atom achieves a noble gas configuration by participating in four covalent bonds.

Typical bonding patterns of carbon and the actual shapes of carbon-containing molecules are described in Chapter 7. We saw there that when carbon forms four single bonds, it is sp^3 hybridized; when it forms a double bond (composed of one σ bond and one π bond), it is sp^2 hybridized; and when it forms a triple bond (composed of one σ bond and two π bonds), it is sp hybridized. Table 13-1 summarizes the bonding

TABLE 13-1
Bonding features in single-, double-, and triple-bonded carbon compounds

Example	Number of atoms bonded to carbon	Hybridization state of C atom	Orbitals hybridized	Unhybridized orbitals on carbon	Kinds of bonds to each carbon	Geometry
$CH_3—CH_3$	4	sp^3	$2s + 2p_x + 2p_y + 2p_z$	None	Four σ	Tetrahedral
$CH_2═CH_2$	3	sp^2	$2s + 2p_x + 2p_y$	$2p_z$	Three σ + one π	Triangular planar
$CH≡CH$	2	sp	$2s + 2p_x$	$2p_y + 2p_z$	Two σ + two π	Linear

features of ethane, ethylene, and acetylene; their structures are:

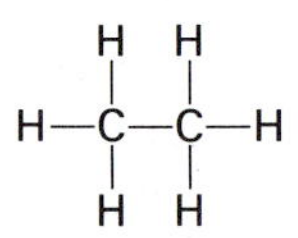
ethane

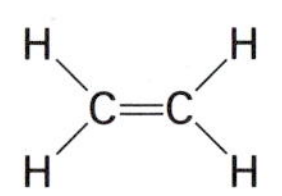
ethylene

H—C≡C—H
acetylene

TABLE 13-2
Length and strength of carbon–carbon bonds

Compound	Average carbon–carbon bond length (nm)	Carbon–carbon bond energy (kJ/mole)
$CH_3—CH_3$	0.154 (σ bond)	347
$CH_2═CH_2$	0.133 ($\sigma + \pi$ bonds)	611
$CH≡CH$	0.120 (σ + two π bonds)	837

The arrangement of the four atoms bonded to an sp^3 hybridized carbon atom is tetrahedral, with bond angles of 109.5°. The arrangement of the three atoms bonded to an sp^2 hybridized carbon atom is planar triangular, with bond angles of 120°. When two atoms are bonded to an sp hybridized carbon atom, the arrangement is linear, with bond angles of 180°.

As more electrons are drawn into the region between two carbon nuclei, the two nuclei are drawn closer together, and the bond between them becomes stronger. Thus, as seen in Table 13-2, carbon–carbon triple bonds are shorter and stronger than double bonds, which in turn are shorter and stronger than single bonds. We can also deduce from the table that a σ bond is stronger than a π bond—the strength of a carbon–carbon σ bond is 347 kJ/mole, whereas that of a π bond is only about 264 kJ/mole (611 kJ − 347 kJ). (The corresponding bond strengths in kilocalories are: σ bond ~ 83 kcal/mole, and π bond ~ 63 kcal/mole.)

13-3 STRUCTURAL ISOMERS AND CONFORMERS

We saw in earlier chapters that most pairs of elements form only one compound, or at most a small number of different compounds. For example, sodium and oxygen, when combined alone, form only the two compounds sodium oxide (Na_2O) and sodium peroxide (Na_2O_2). Carbon and hydrogen, on the contrary, form different compounds in countless numbers. The number of possible compounds of molecular formula

$C_{40}H_{82}$, for example, is 62,491,178,805,831 (and a chemist could synthesize any one of them)! Collectively these compounds of carbon and hydrogen are known as **hydrocarbons.** The existence of such a large number of isomers is the result of the special ability of carbon atoms to bond together in long chains, branched chains, and rings—and with single, double, or triple bonds between adjacent carbon atoms. We shall consider some of these more complicated molecules shortly, but first we look at the configurations (the arrangements of the atoms in space) of the simple molecules methane (CH_4), ethane (C_2H_6), and propane (C_3H_8).

These hydrocarbon molecules are not flat; the arrangement of atoms about each carbon is tetrahedral. Figure 13-2 shows the three-dimensional structures of methane, ethane, and propane with two kinds of models. The **ball-and-stick models** show the framework of the molecule—the nuclei and the bonds. The **space-filling models** show the full bulk of the atoms (including the electron clouds) of each molecule, but they do not show the skeleton clearly. On paper, the three-dimensional shapes of molecules are commonly represented by perspective diagrams such as those shown in Figure 13-3. (We have seen similar diagrams in Section 7-7.) In interpreting these **wedge-dash-line diagrams,** certain conventions are essential. A bond extending out from the plane of the paper toward the reader is indicated by a wedge; a bond directed behind the plane of the paper is indicated by a dashed line; and a bond lying in the plane of the paper is indicated by a solid line.

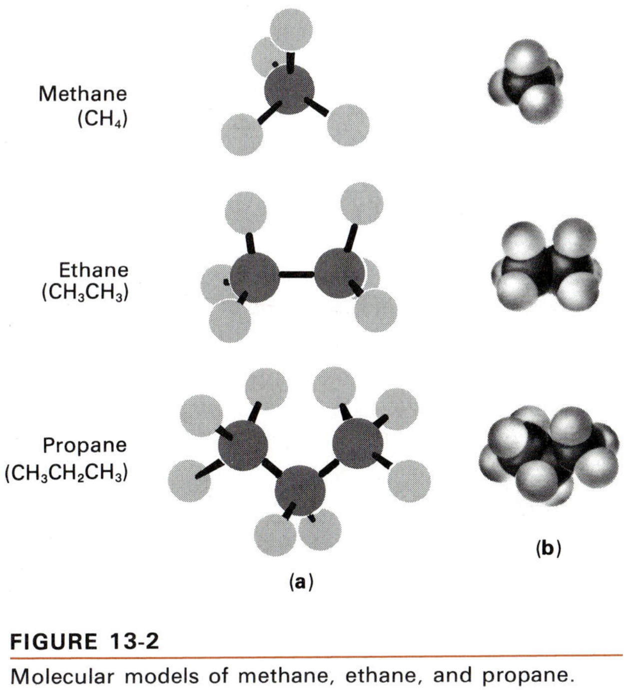

FIGURE 13-2
Molecular models of methane, ethane, and propane. **(a)** Ball-and-stick models. **(b)** Space-filling models.

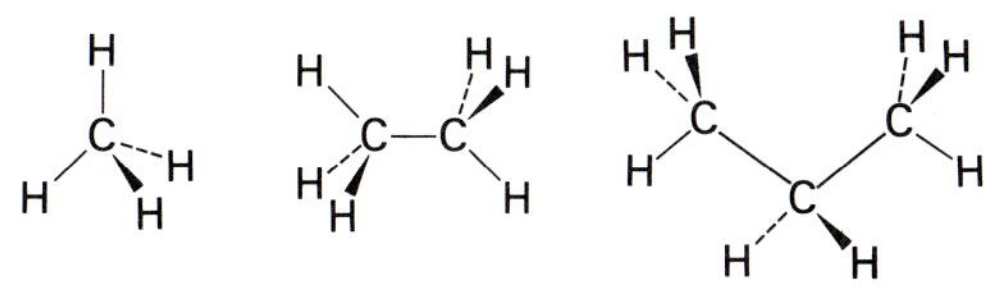

FIGURE 13-3
Wedge–dash–line representations of methane, ethane, and propane. A solid line represents a bond in the plane of the paper; a dashed line represents a bond directed behind the plane of the paper; and a wedge represents a bond extending out from the plane of the paper.

Figure 13-4 shows the formation of σ bonds in ethane from the overlap of the atomic orbitals of the separate atoms. As we saw in Chapter 7, a σ bond is cylindrically symmetric about the internuclear axis. You can see from the figure that twisting or rotating a hydrogen atom about a C—H bond or a $-CH_3$ group about a C—C bond does not reduce the overlap of the atomic orbitals, and thus does not break the bond.

Only a small amount of energy is required to *rotate* groups about a carbon–carbon single bond—13 kJ/mole for ethane, compared to the 347 kJ/mole required to *break* the carbon–carbon single bond.

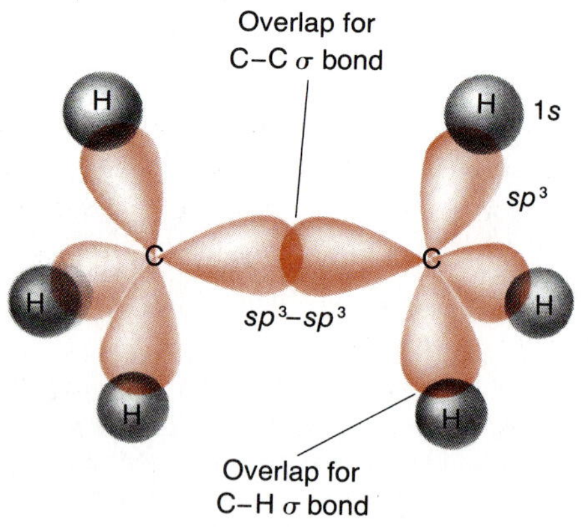

FIGURE 13-4

Formation of a σ bond. In ethane, σ bonds are formed by the overlap of the 1*s* orbitals of the hydrogen atoms with sp^3 orbitals of the carbon atoms, and by the overlap of sp^3 orbitals of the two carbon atoms.

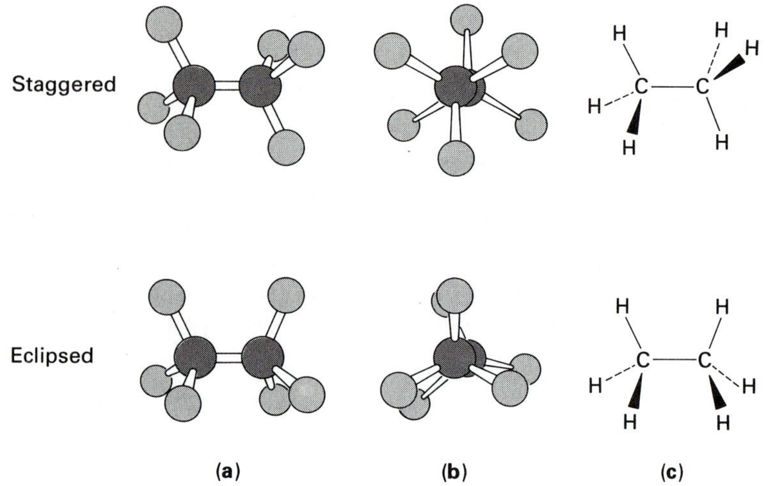

FIGURE 13-5

Staggered and eclipsed conformations of ethane **(a)** Side view. **(b)** Head-on view. **(c)** Wedge–dash–line representation.

Thus, an infinite number of arrangements of the atoms in a given molecule are possible at normal temperatures. These arrangements are called **conformations.** The individual molecules corresponding to these arrangements are called **conformers.** Figure 13-5 shows models of two conformers of the ethane molecule.

The **staggered conformation,** in which the hydrogen atoms (and thus their electron clouds, which repel each other) are as far apart as possible, is the most stable (or preferred) conformation. In any sample of ethane molecules, more are in the staggered than in any other conformation.

The **eclipsed conformation,** in which the hydrogen atoms are closest to each other (and thus the repulsions between the electron clouds are greatest), is the least stable. Because only a small amount of energy is required to rotate from the staggered to the eclipsed conformation, there is essentially free rotation about carbon–carbon single bonds, and it is impossible to isolate or distinguish between two conformers at ordinary temperatures. Perspective diagrams of the two extreme conformations of ethane are also given in Figure 13-5.

The formulas of simple hydrocarbons such as methane (CH_4), ethane (C_2H_6), and propane (C_3H_8) correspond to only *one* structure. The situation becomes more complicated for compounds with four or more carbon atoms. For example, the next larger member of the series, C_4H_{10}, can occur in *two* structural forms:

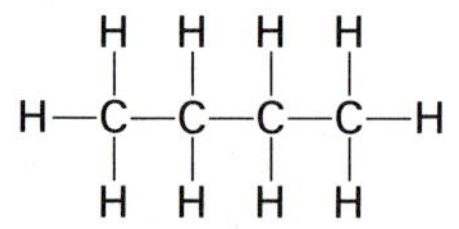

C_4H_{10}
***n*-butane**

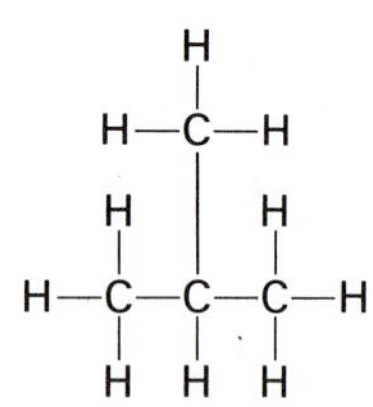

C_4H_{10}
isobutane

Although in isobutane and in the following structures some C—C bonds are shown longer than others for clarity, all C—C bonds are, in fact, very close to the same length. The unbranched form is called normal butane (abbreviated *n*-butane), and the branched form is called isobutane. The prefix "normal" or "*n*-" designates a continuous, unbranched carbon chain. Although both butanes have the same molecular formula, they are distinctly different compounds with different properties. Table 13-3 lists some of the properties of *n*-butane and isobutane.

TABLE 13-3
Some properties of n-butane and isobutane

Compound	Melting point (°C)	Boiling point (°C)	Density at −20°C (g/ml)
n-Butane	−138	−0.5	0.662
Isobutane	−160	−12	0.604

In trying to visualize which structures represent different conformations of the same compound and which represent isomers, it is important to remember that the molecules are three-dimensional, and that groups can be rotated about single bonds. Thus, the following three structures all represent the *same* molecules (*n*-butane) with four carbon atoms connected in an unbranched chain:

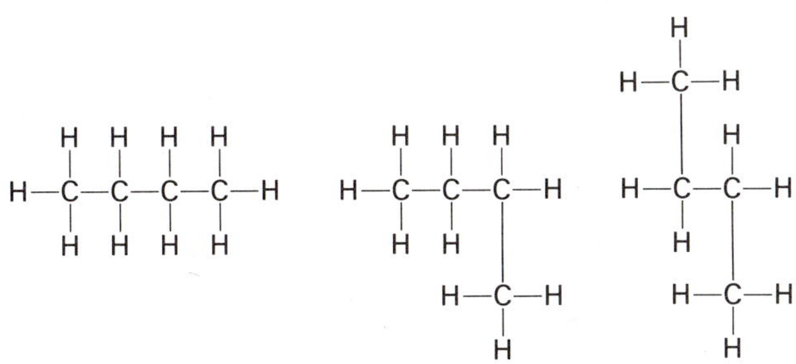

These structures represent different conformations of *n*-butane. They do not represent isomers because any one of them can be converted into the others by rotating $—CH_3$ and $—CH_2$ groups about carbon–carbon single bonds. (To see that these structures represent the same molecule, it might help to imagine the carbon chain as being made of links. If you pull on the ends of the bent chains, they straighten out and are identical to the straight chain shown on the left.)

Isobutane, on the contrary, has only three carbon atoms connected in an unbranched chain; the fourth carbon branches from the center carbon. The following structures all represent isobutane:

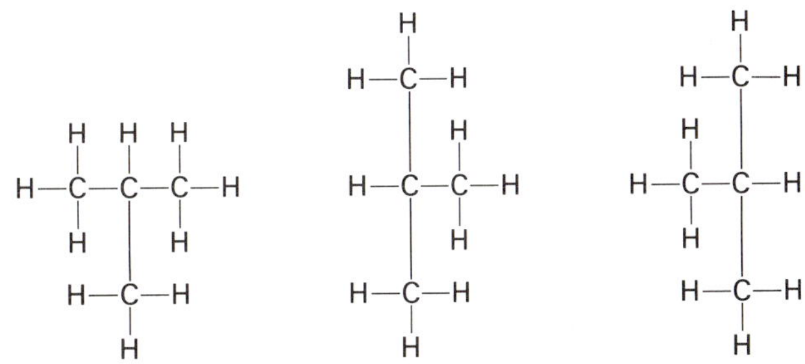

By rotating $—CH_3$ and $—CH$ groups about the carbon–carbon bonds, any one of these structures can be converted into any other.

But *n*-butane cannot be converted into isobutane (nor isobutane into *n*-butane) by rotation about single bonds. To convert from one compound to the other requires breaking and reforming chemical bonds. Compounds such as *n*-butane and isobutane, which have the same molecular formula (C_4H_{10} in this case), but different carbon skeletons and different chemical and physical properties, are called **structural isomers.** Table 13-4 gives the formulas of some hydrocarbons, along with the numbers of structural isomers that exist for each.

In Chapter 8 we saw another example of structural isomerism. The two compounds ethyl alcohol and dimethyl ether have the same molecular formula (C_2H_6O) but different structures and different chemical and physical properties. Their structures are

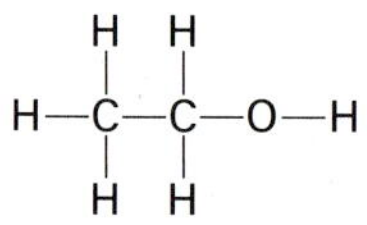

ethyl alcohol
CH_3CH_2OH

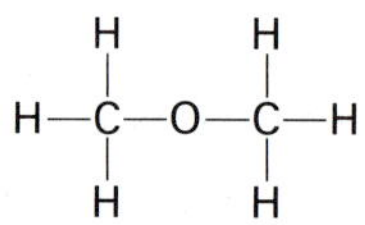

dimethyl ether
CH_3OCH_3

To convert from one of these compounds to the other requires breaking and reforming chemical bonds.

To summarize, (1) a molecule may have an infinite number of conformations that differ from one another only by rotation about single bonds. The following perspective diagrams and line structures represent different conformations of the same molecule, and not isomers:

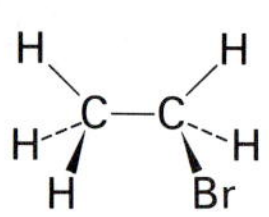 is the same molecule as 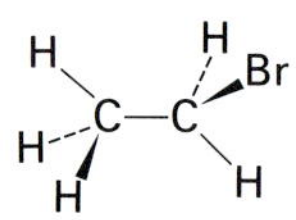

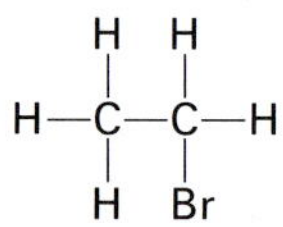 is the same molecule as 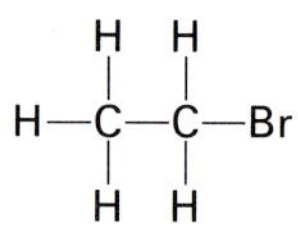

bromoethane
Condensed formula: CH_3CH_2Br

TABLE 13-4
Numbers of isomers of some hydrocarbons

Formula	Number of isomers
CH_4	1
C_2H_6	1
C_3H_8	1
C_4H_{10}	2
C_5H_{12}	3
C_6H_{14}	5
C_7H_{16}	9
C_8H_{18}	18
C_9H_{20}	35
$C_{10}H_{22}$	75

(2) Two compounds are structural isomers if they have the same molecular formula, but different skeletal structures and different chemical and physical properties. To go from one isomer to another, a chemical reaction must occur—bonds must be broken and reformed. The following compounds are structural isomers:

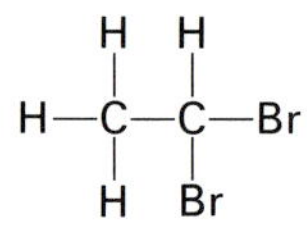

Condensed formula:
CH_3CHBr_2
1,1-dibromoethane
(boiling point: 108°C)

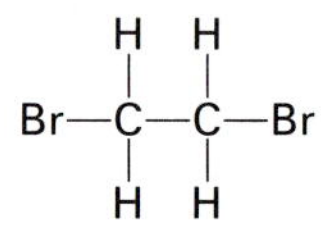

Condensed formula:
$BrCH_2CH_2Br$
1,2-dibromoethane
(boiling point: 132°C)

In the *condensed* structural formulas, which we often use, all atoms or groups following a given carbon in the formula are bonded to that carbon. Thus,

CH_3 means 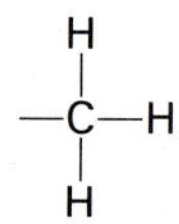

CH_2 means 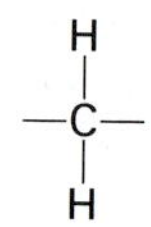

CH means 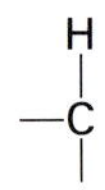

$(CH_2)_2$ means 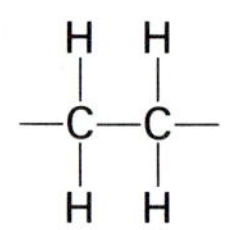

CH_2Cl means 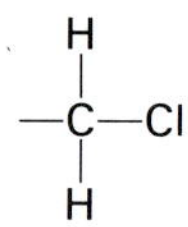

$CHCl_2$ means 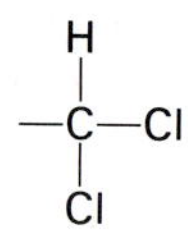

The structural formula of the compound isooctane is

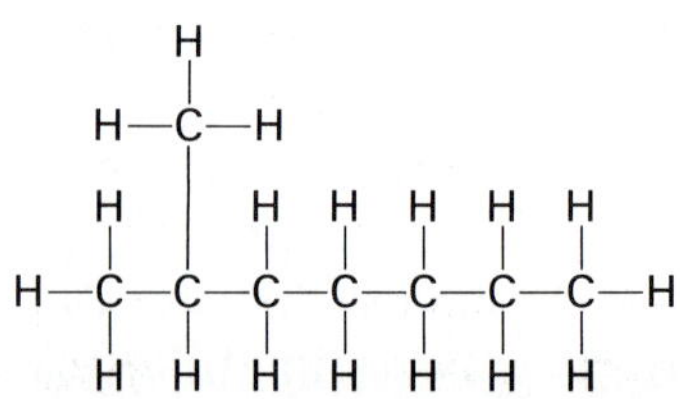

Its condensed structural formula is

$$\begin{array}{c}CH_3\\|\\CH_3CHCH_2CH_2CH_2CH_2CH_3\end{array} \quad \text{or} \quad \begin{array}{c}CH_3\\|\\CH_3CH(CH_2)_4CH_3\end{array}$$

If a chain is branched, in this book we usually show the branch with a vertical bond as we have here.

EXERCISE 13-1 (a) Draw the most stable and the least stable conformations of 1,1-difluoroethane (CH_3CHF_2). Use the wedge–dash–line representation.

(b) Write the structural formula and condensed structural formula of an isomer of 1,1-difluoroethane.

EXERCISE 13-2 Write structural formulas and condensed structural formulas for the three compounds having molecular formula C_5H_{10}.

13-4 STEREOISOMERISM: GEOMETRIC ISOMERISM

In addition to structural isomerism, organic compounds display **stereoisomerism,** in which molecules with the *same* connections of atoms (that is, compounds that are *not* structural isomers) can be distinguished by the different three-dimensional arrangements of their atoms in space. We shall consider two types of stereoisomerism—first geometric isomerism, and then (in Section 13-5) optical isomerism.

We have noted that at normal temperatures there is free rotation about single bonds so that we can't distinguish conformers. But if rotation is blocked, then it sometimes becomes possible to distinguish separate compounds. **Geometric isomerism** *results from a restriction of rotation about double bonds, or about single bonds in cyclic compounds.* As shown in Figure 13-6, when one $—CH_2$ group of the carbon–carbon double bond in ethylene is rotated through 90°, the σ bond remains intact, but the π bond is broken. There is no longer an overlap of the unhybridized p orbitals that constituted the π bond, and no further sharing of the π bond electrons between the two carbon atoms. A considerable amount of energy (about 250 kJ/mole) is required to break a π bond, in contrast to the energy of rotation about a carbon–carbon single bond (only about

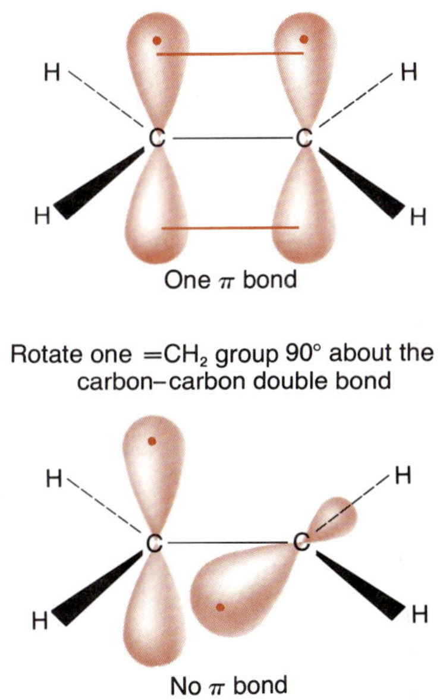

FIGURE 13-6

Rotation of one $=CH_2$ group of ethylene about the carbon–carbon double bond. The overlap of the p orbitals that form the π bond is lost in the rotation—the π bond is broken.

13 kJ/mole). Thus, under ordinary circumstances, when two carbon atoms are connected by a double bond, one =C< group cannot rotate in relation to the other. Therefore, the **substituents** (atoms or groups of atoms) attached to the doubly-bonded carbon atoms are held in a rigid configuration. As a result, 1,2-dichloroethylene can exist in two different spatial arrangements:

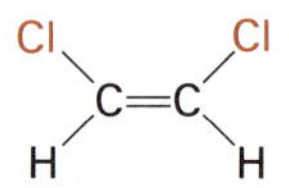

***cis*-1,2-dichloroethylene**
(boiling point: 60°C)

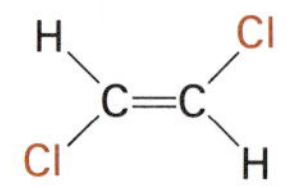

***trans*-1,2-dichloroethylene**
(boiling point: 47°C)

These two distinct compounds are geometric isomers—different compounds with different properties. The *cis* isomer has the two H atoms on the same side of the double bond, and the *trans* isomer has the two H atoms on opposite sides of the double bond. Converting one isomer into the other requires rotating one =CHCl group through 180° about the carbon–carbon bond, breaking and reforming the π bond in the process.

For geometric isomers to exist for compounds containing a double bond, *each of the doubly-bonded carbon atoms must have two different groups attached to it.* Thus, in the two isomers of 1,2-dichloroethylene, *one* of the carbons has two different groups—an H and a Cl. And the *other* carbon also has two different groups—an H and a Cl.

A third compound, 1,1-dichloroethylene, has the same molecular formula ($C_2H_2Cl_2$) as *cis*- and *trans*-1,2-dichloroethylene, but a different structural formula:

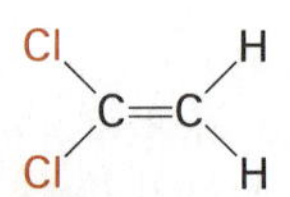

1,1-dichloroethylene
(boiling point: 37°C)

This compound is a structural isomer of the two considered earlier. It does not have a geometric isomer because one of the carbons has two chlorine atoms (and the other has two hydrogen atoms). If even *one* of the doubly-bonded carbon atoms has two identical groups attached, the compound does not have a geometric isomer. Thus, for example, chloroethylene (CH_2CHCl) does not have a geometric isomer:

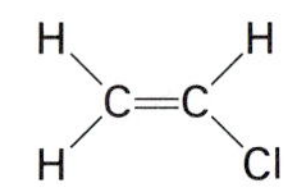

chloroethylene

If the $=CH_2$ group of chloroethylene were rotated 180°, about the double bond, the structure would be *identical* to that shown. (Convince yourself that this is so.)

The compound with molecular formula C_4H_8 consists of six isomers:

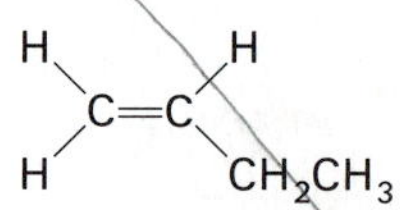

1-butene

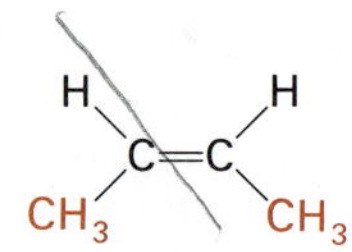

***cis*-2-butene**

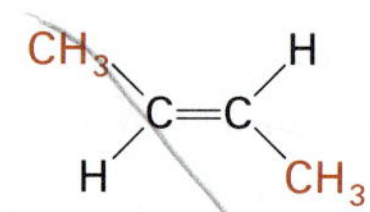

***trans*-2-butene**

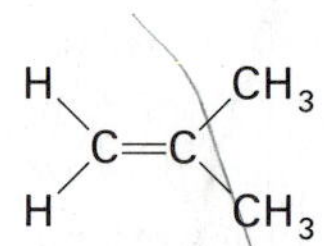

methylpropene

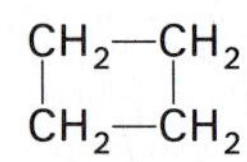

cyclobutane

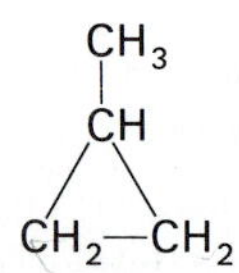

methylcyclopropane

Two of these compounds (*cis*-2-butene and *trans*-2-butene) are geometric isomers.

EXERCISE 13-3 Explain why 1-butene and methylpropene do not have geometric isomers.

An example of geometric isomerism of importance in biology occurs in the compound retinal, an oxidized form of vitamin A present in the retina of the eye. Retinal combines with proteins called opsins to form visual pigments. The initial chemical event responsible for vision is the isomerism of *cis*-retinal to *trans*-retinal; light is the source of the energy required to break the π bond.

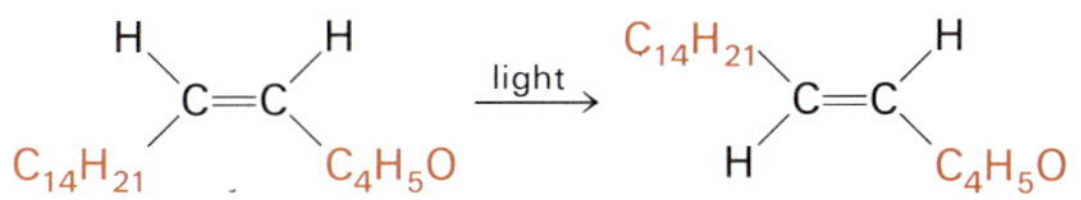

***cis*-retinal**   ***trans*-retinal**

All further changes that occur in vision are consequences of this photochemical *cis–trans* conversion. It has been shown that the human eye can detect a flash of light when as few as five molecules of *cis*-retinal each absorb just one quantum of light. Figure 13-7 shows how the resulting change in shape of the retinal molecule, in going from the *cis* to the *trans* form, causes the retinal to be released by the opsin, initiating the sequence of changes that results in vision.

EXERCISE 13-4 (a) Draw the structures of the geometric isomers of $CH_3CH{=}CH{-}CH_2{-}CH_3$.

(b) Show why $CH_2{=}CH{-}CH_2{-}CH_2{-}CH_3$ has only one possible structure.

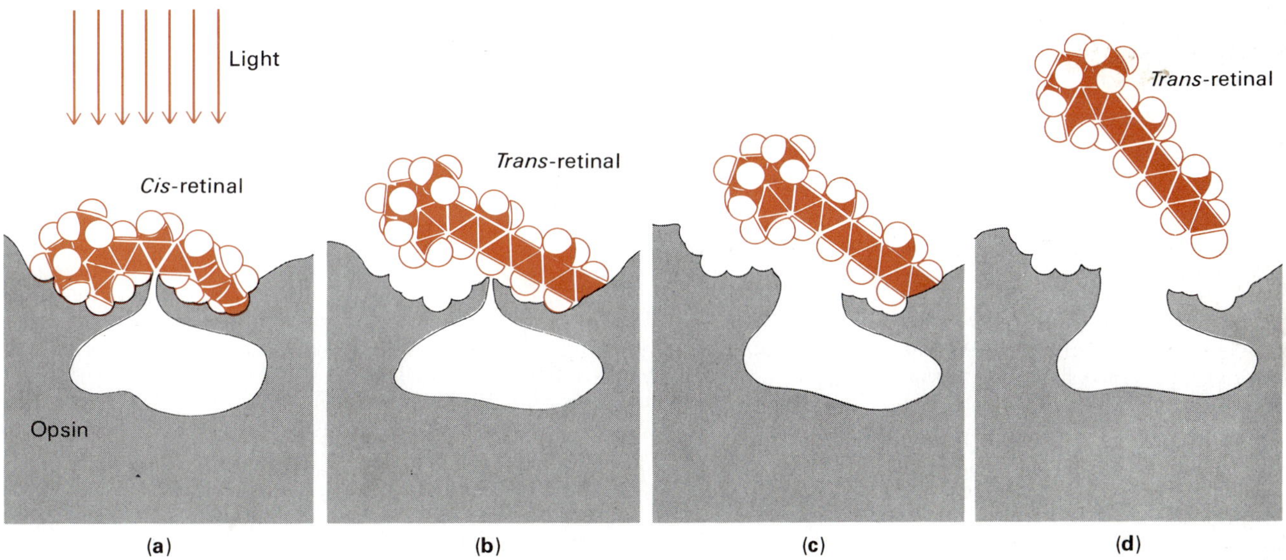

FIGURE 13-7

Molecular events in vision. **(a)** A molecule of *cis*-retinal is shown attached to its binding site in the large protein molecule, opsin. **(b)** After absorbing light, the bent *cis* isomer is converted to the straight *trans* isomer. **(c), (d)** A change in the shape of the opsin may facilitate the release of the *trans*-retinal. [From Ruth Hubbard and Allen Kropf, "Molecular Isomers in Vision." Copyright © 1967 by Scientific American, Inc. All rights reserved.]

Cyclic compounds also may exhibit geometric isomerism because of the rigidity of the ring structure. The cyclopropane ring, for example, lies in a plane with attached groups held above and below the plane. Thus, 1,2-dichlorocyclopropane exists as *cis* and *trans* isomers as shown in Figure 13-8. In the *cis* isomer, the chlorine atoms are on the same side of the plane of the ring; in the *trans* isomer, the chlorine atoms are on opposite sides of the plane of the ring. These are not just two conformations of the same molecule, but are geometric isomers; it is impossible to rotate a —CHCl— group to go from one structure to the other without breaking a carbon–carbon bond.

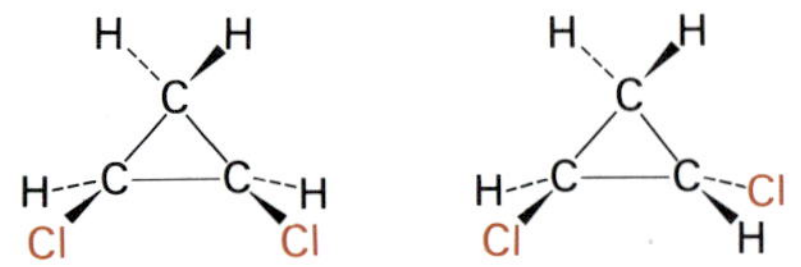

FIGURE 13-8

Cis and *trans* isomers of 1,2-dichlorocyclopropane.

13-5 STEREOISOMERISM: OPTICAL ISOMERISM

One of the more interesting chapters in the development of modern chemistry concerns the interaction of matter with light. Important contributions to our understanding of this subject were made by Louis Pasteur, the French scientist who is best known for his brillliant contributions to biology in the nineteenth century. One of Pasteur's greatest achievements resulted from a repetition of experiments that had been performed by others on the interaction with light of tartaric acids and their salts. Pasteur noticed that crystals of sodium ammonium tartrate, $NaNH_4(C_4H_4O_6)$, occur in two forms, one a mirror image of the other (Figure 13-9). By carefully sorting the crystals one by one, he was able to place the "lefthanded" crystals in one pile and the "righthanded" ones in another. Testing the effect of solutions of the two forms of crystal on light, he discovered that the righthanded and lefthanded crystals rotate light in opposite directions. This discovery laid the foundation for the field of stereochemistry—the study of the spatial arrangement of atoms in a molecule—and led to the idea that a molecule is three-dimensional.

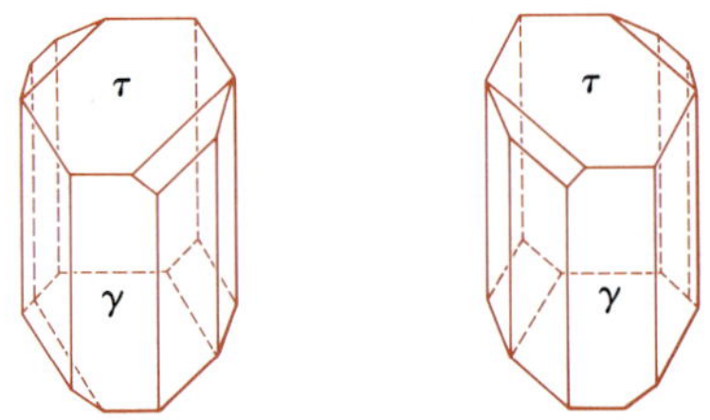

FIGURE 13-9

Crystals of sodium ammonium tartrate. The drawings are by Pasteur. [From George B. Kauffman and Robin D. Myers, "The Resolution of Racemic Acid," *J. Chem. Ed.* 52:778 (1975).]

In order to understand what property of a molecule causes the rotation of light, we must first see how this property is measured. Measurement of the rotation of light by a compound, its **optical activity**, requires **plane-polarized light.** This can be produced by passing light through certain types of crystal such as calcite ($CaCO_3$) or through synthetic materials such as the Polaroid filter used in the lenses of Polaroid sunglasses. The electric field of light vibrates in all possible planes perpendicular to the light path. After passage through a polarizer, only one plane of waves remains; the light is said to be plane-polarized (Figure 13-10). When a second polarizer is placed in the light path, the light is transmitted through the second one also only if the two polarizers are aligned in the same direction (Figure 13-11a). When, instead, the second filter is at right angles to the first, the plane-polarized light cannot get through, and the light is completely extinguished (Figure 13-11b). (You can demonstrate these effects with two lenses from Polaroid sunglasses.

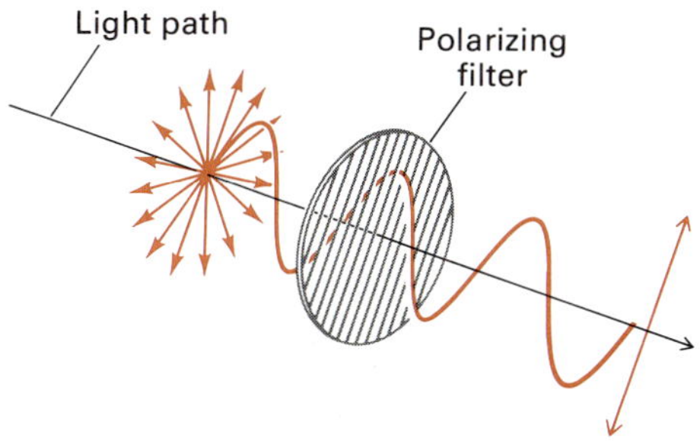

FIGURE 13-10

Polarization. The electric field of polarized light vibrates in all possible planes perpendicular to the light path. After passing through a polarizer, it vibrates in only one plane.

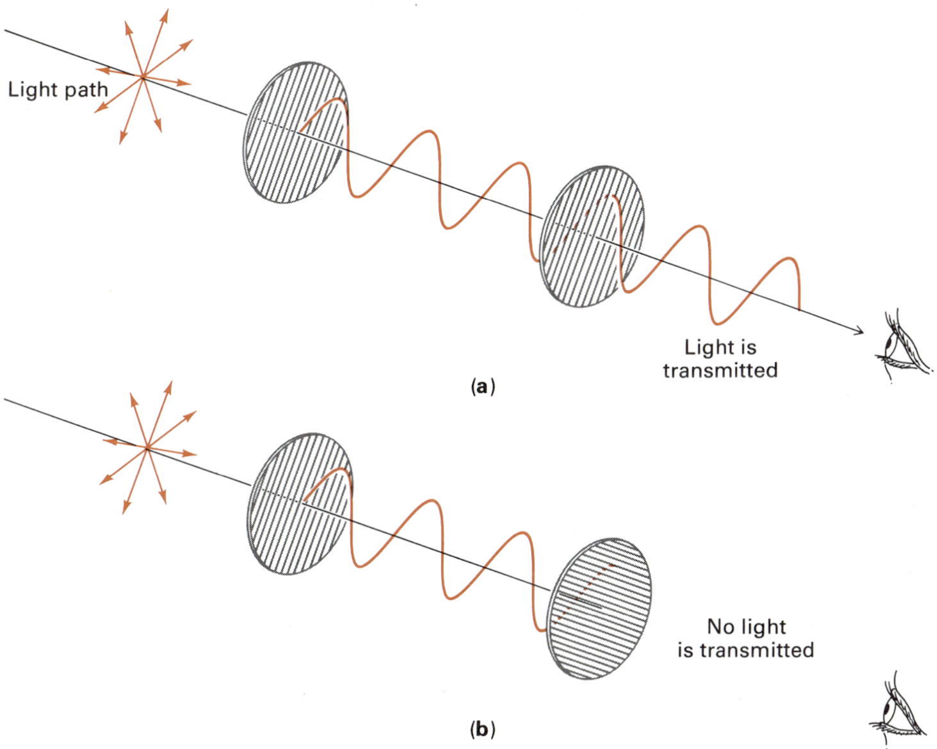

FIGURE 13-11

Detecting polarization. **(a)** When a second polarizer is aligned in the same direction as the first, the plane of polarized light is transmitted through the second filter also. **(b)** When the second polarizer is aligned at 90° to the first, no light is transmitted.

A maximum amount of light gets through when the lenses are aligned in the same direction; no light gets through when they are at right angles to each other.)

Figure 13-12 is a schematic diagram of a **polarimeter,** an instrument that measures the rotation of polarized light by a compound in solution. In Figure 13-12a, the sample cell is empty, and the second polarizer (the analyzer in the figure) is at right angles to the first, so the light is extinguished. In Figure 13-12b, the sample cell contains a compound that rotates the plane of polarized light before it reaches the second polarizer. The second polarizer now needs to be set at an angle other than 90° with respect to the first to extinguish the light. If the second polarizer must be turned from its right-angle position in a clockwise direction, the chemical substance interposed is said to have a plus (+) rotation and is called **dextrorotatory.** If the second polarizer must be moved counterclockwise, a minus (−) rotation is indicated, and the substance is called **levorotatory.** Substances that thus rotate the plane of polarized light are said to be **optically active.**

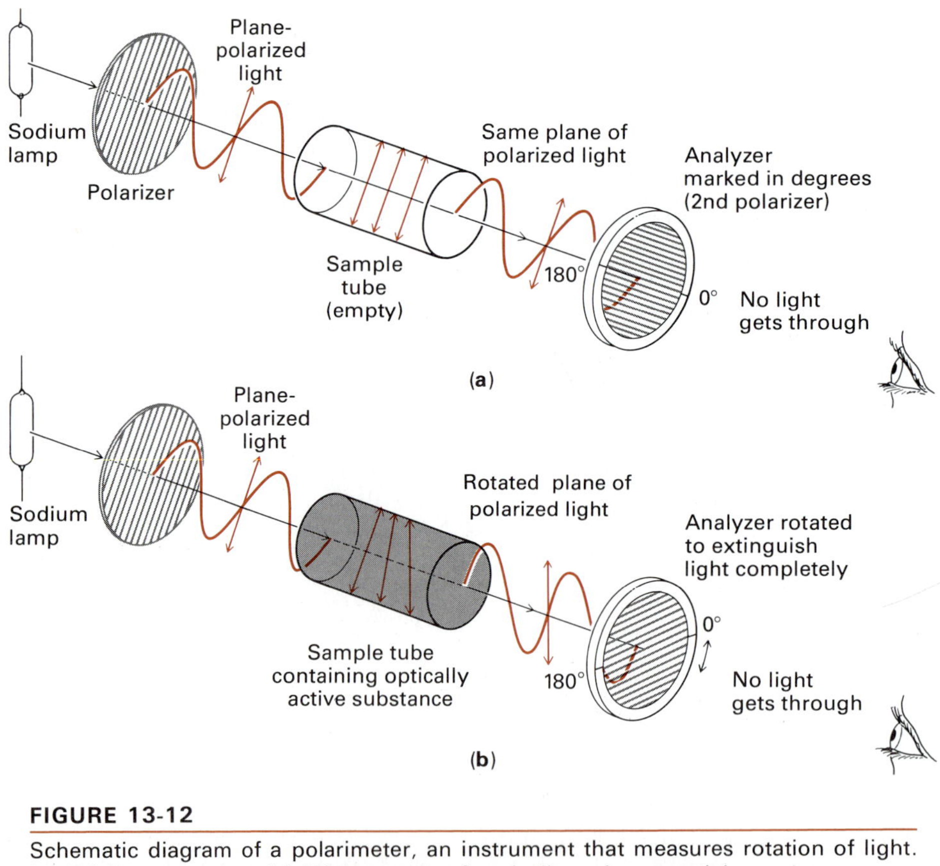

FIGURE 13-12

Schematic diagram of a polarimeter, an instrument that measures rotation of light. **(a)** Without sample. **(b)** With sample of optically active material.

(a)

(b)

FIGURE 13-13

Handedness. **(a)** The left and right hands are not superimposable. **(b)** Is this person superimposable on her mirror image? [(a), Albrecht Dürer, *Disciple's Hands,* with permission of Graphische Sammlung Albertina, Vienna. (b), cartoon by Remo, with permission of Willy Breinholst's Forlag.]

What is it that causes this optical activity? To answer this question we first consider the relationship of an object to its mirror image. A left hand and its mirror image (the right hand) are *not* identical. This is evident when one places the right hand on top of the left, or when one tries to fit a right hand into a left-hand glove! The right and left hands are mirror images that are not superimposable (Figure 13-13). *We shall see that there are also molecules that are not superimposable on their mirror images—they have a "handedness"—and it is these molecules that are optically active.*

Consider the molecule CH_3X and its mirror image:

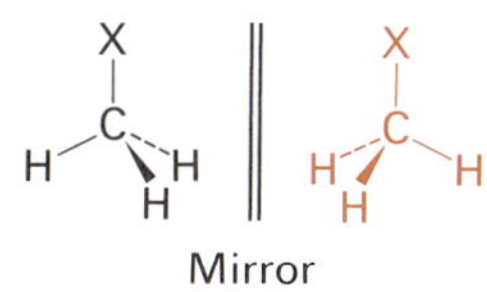

The two are identical—by rotating one to a new position, we can superimpose it upon the other. For example, if we rotate the righthand structure 60° around the C—X bond, we obtain the lefthand structure. (It is *very* helpful to construct molecules using a molecular model set in order to see whether or not a molecule is identical to its mirror image. If you do not have a molecular model set, marshmallows or gumdrops and toothpicks will do.)

Next, consider the molecule CH_2XY and its mirror image:

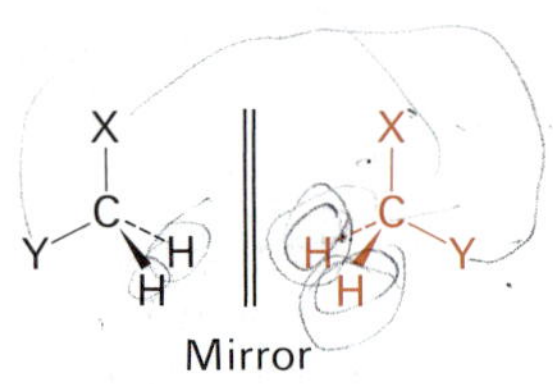

Again, these two are superimposable—this time by rotating the righthand structure 180° about the C—X bond.

Finally, consider the molecule CHXYZ and its mirror image:

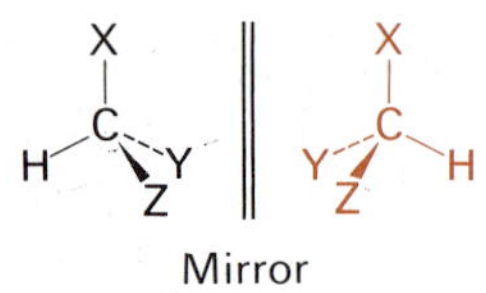

These molecules are *nonsuperimposable*. To convert one form to the other, two of the groups on the central carbon must be interchanged while the other two are fixed in position. That cannot be done merely by rotating one of the molecules into a new position; it can be done only by breaking and reforming chemical bonds. It takes a lot of energy to break chemical bonds, so such isomers are *different* compounds.

A carbon atom that is bonded to four different atoms or groups of atoms is called a **chiral** (or asymmetric) carbon atom. We have seen that a molecule that contains a chiral carbon atom (such as CHXYZ above) can exist in two forms that are nonsuperimposable mirror images of each other. *It is this property—that a molecule and its mirror image be nonsuperimposable—that makes a molecule optically active.* The molecule and its nonsuperimposable mirror image are called **optical isomers,** or **enantiomers.**

An example of a molecule that contains a chiral carbon atom and exists as a pair of enantiomers is 2-butanol, $CH_3CH—CH_2—CH_3$:
|
OH

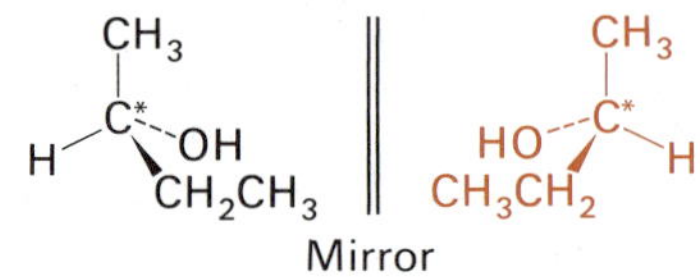

Note that, although there are two carbon atoms attached to the chiral carbon (the starred carbon), the two attached carbon atoms are not the same—one is in a methyl ($—CH_3$) group and the other in an ethyl ($—CH_2CH_3$) group. There are, then, four *different* groups attached to the starred carbon atom. Some properties of the optical isomers of 2-butanol are given in Table 13-5.

One enantiomer rotates the plane of polarized light in *one* direction; its mirror image rotates the plane of polarized light in the *other* direction. A mixture of equal amounts of a pair of enantiomers is not optically active (has no *net* effect on polarized light) and is called a **racemic mixture** of enantiomers. One molecule of the racemic mixture rotates polarized light in one direction, but the next molecule may rotate it back. Thus, if there are equal amounts of the two enantiomers, there is no *net* rotation of the plane of polarization. For example, the two forms of 2-butanol rotate the plane of polarized light equally, but in opposite directions. A mixture of equal amounts of the two enantiomers has no *net* effect on polarized light.

The magnitude of the rotation of plane-polarized light by a given solution (measured in a polarimeter)

TABLE 13-5
Some properties of optically active compounds

Substance	Melting point (°C)	Density (g/cm³)	Rotation (in degrees; sodium light, wavelength 590 nm; cell length = 10 cm; temperature = 20°C)
2-Butanol (C_4H_9OH)	−115	0.808	+13.9 (pure liquid)
	−115	0.807	−13.9
Tartaric acid ($C_4H_6O_6$)	170	1.76	+12.7 (17.4 g solute/100 ml solution,
	170	1.76	−12.7 in water; see structures in
meso-Tartaric acid	150	1.67	0 text)
Ascorbic acid ($C_6H_8O_6$,	192	1.65	+48 (1 g solute/100 g solution, in
vitamin C is the + isomer)	192	1.65	−48 methyl alcohol)
Aspartic acid ($C_4H_7NO_4$,	270–271	1.661	+5.0 (1 g solute/100 g solution, in
an amino acid)	270	1.661	−5.0 water)
Limonene ($C_{10}H_{16}$,	−97	0.841	+126 (pure liquid)
lemon oil)	−97	0.842	−122

Note: Where there are small differences in values of a property of + and − isomers, the difference is due to impurities in the sample or to inaccuracies in measurement.

depends on the particular compound and on the number of optically active molecules in the light path. For a given optically active compound, the greater the concentration of the solution, or the longer the sample cell, the greater the angle of rotation. The wavelength of the light, the temperature, and the solvent also affect the amount of rotation. Table 13-5 lists some properties, including the rotation, of several optically active substances. As you can see, the two members of a pair of enantiomers are identical except in their effect on polarized light.

Pasteur's separation of a racemic mixture of sodium ammonium tartrate crystals into its two component enantiomers was the first experiment to show a clear connection between chemical structure and optical activity. Each of the two enantiomers of the tartrate ion contains two chiral carbon atoms:

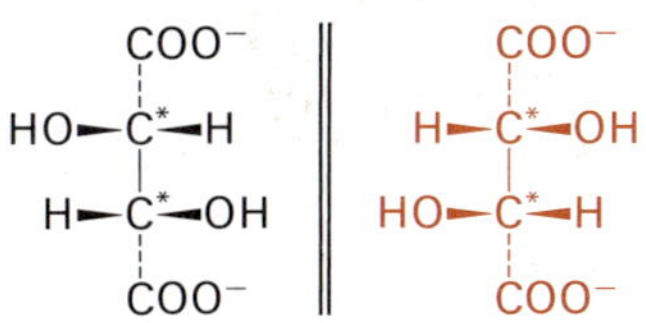

Mirror

tartrate ions

The two crystalline forms of sodium ammonium tartrate in Figure 13-9 result from these differences in molecular structure. The "righthanded" crystals are composed of one enantiomer, and the "lefthanded" crystals are composed of the other. The two crystalline forms, when dissolved separately in water, rotate the plane of polarized light equally but in opposite directions. The optical activity is caused by the asymmetric arrangement of atoms in the molecules. A solution containing equal amounts of the "lefthanded" and "righthanded" crystals—a racemic mixture—has no net effect on polarized light.

Because the tartrate ion contains two identical chiral carbon atoms, it can also exist in a third form:

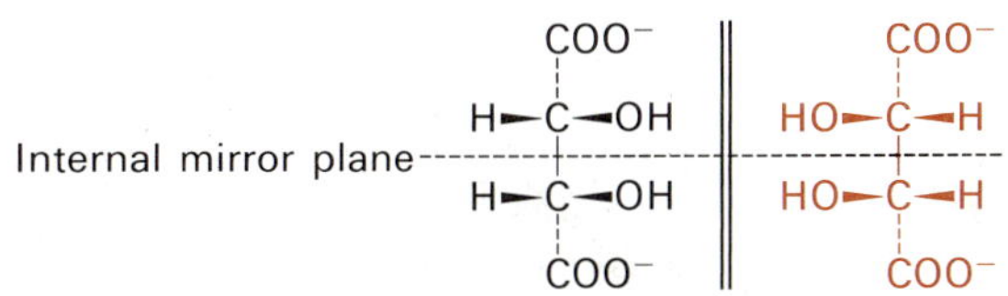

Mirror

This form, known as a *meso* compound, is optically inactive because it has an internal plane of symmetry. One half of the molecule is the mirror image of the other half, and the whole molecule is superimposable on its mirror image. As you can see in Table 13-5, the properties of *meso*-tartaric acid are different from those of the optical isomers.

Pasteur's discovery of optical activity led to the theory of the tetrahedral carbon atom, proposed simultaneously and independently in 1874 by Jacobus van't Hoff in Holland and Joseph le Bel in France. The suggestion that molecules could have such three-dimensional shapes, when first made, was widely ridiculed by eminent chemists. Van't Hoff's paper provoked Hermann Kolbe, one of the most eminent organic chemists of his day, to publish a withering denunciation.

"Not long ago," Kolbe wrote in 1877, "I expressed the view that the lack of general education and of thorough training in chemistry of quite a few professors of chemistry was one of the causes of the deterioration of chemical research in Germany. . . . Will anyone to whom my worries may seem exaggerated please read, if he can, a recent memoir by a Herr van't Hoff on 'The Arrangements of Atoms in Space,' a document crammed to the hilt with the outpourings of a childish fantasy. This Dr. J. H. van't Hoff, employed by the Veterinary College at Utrecht, has, so it seems, no taste for accurate chemical research. He finds it more convenient to mount his Pegasus (evidently taken from the stables of the Veterinary College) and to announce how, on his daring flight to Mount Parnassus, he saw the atoms arranged in space."

In spite of Kolbe's denunciation, experimental evidence in support of their "childish fantasy" soon vindicated van't Hoff and le Bel. By 1900, when van't Hoff received the first Nobel Prize in chemistry, the idea that molecules are three-dimensional was universally accepted.

EXERCISE 13-5 Of the following compounds, which are optically active?

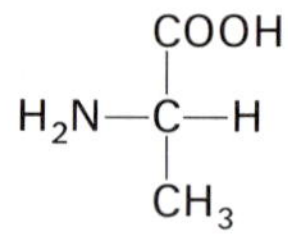

alanine

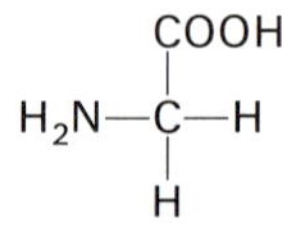

glycine

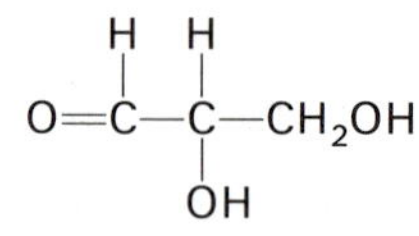

glyceraldehyde

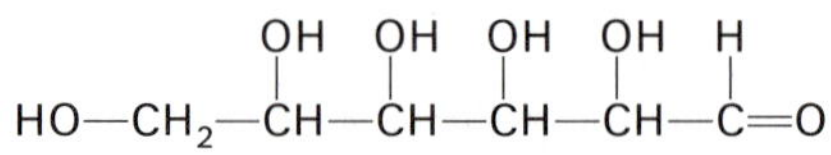

glucose

We have seen that enantiomers have opposite effects on polarized light. Otherwise, their physical properties (melting point, boiling point, density, solubility in water, and so on) are identical. They also behave identically in all chemical reactions with **achiral** substances (those that do not have a handedness). An achiral molecule can react equally with either of a pair of enantiomers, just as a walking stick fits either hand. So, of what consequence or interest are enantiomers? As we shall discuss in greater detail in later chapters, a great many of the compounds in living organisms are enantiomers. Most of the biochemical reactions that take place in the cells and fluids of living things are stereospecific. That is, *generally, only one member of a pair of stereoisomers is able to participate in a given reaction.* Proteins, for example, are composed of amino acids that have the following structure, where R represents some group:

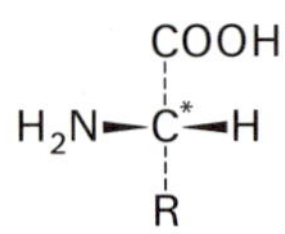

or more accurately

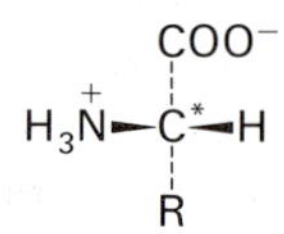

(At pH 7, the acid group —COOH is ionized to COO^-, and the basic amino group $—NH_2$ is protonated to $—NH_3^+$. See Section 21-1.)

Except for glycine (NH_2CH_2COOH), which does

not have a chiral carbon atom, all of the amino acids can exist as either of two enantiomers. But in fact, one form predominates in nature—the one with the structure

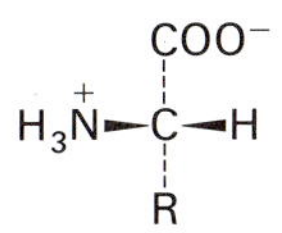

an L-amino acid

which has been designated an L-amino acid. Its mirror image, called a D-amino acid, occurs only rarely in nature. The prefixes D- and L- refer *only* to the stereochemical family of the enantiomer, and *not* to the direction in which it rotates light. The L- family has the structure given above, and the D- family has the mirror image structure:

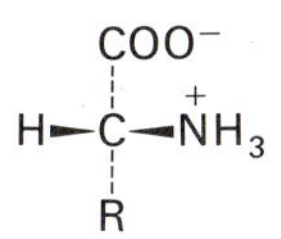

It is not possible to determine whether a molecule is dextrorotatory or levorotatory merely by looking at its structure, but only by doing an actual measurement of its effect on the plane of polarization. When amino acids are prepared in the laboratory, a racemic mixture of the two forms usually is obtained.

How does the body distinguish a molecule from its mirror image? The receptor of an optically active molecule in the body (say, an enzyme; Chapter 22) in virtually every case has a handedness itself. The enzyme fits the molecule like a glove fits a hand. Thus, in many biological reactions involving the amino acids, only the L-enantiomer can participate, and its mirror image (the D-enantiomer) cannot substitute. If an organism is fed a diet containing a racemic mixture of amino acid enantiomers, it will utilize only the one that occurs naturally, while excreting the other.

How did those molecules of nature that have a definite handedness come to have the *same* handedness in all terrestrial organisms—viruses, bacteria, plants, and animals? No one now can answer that question. It is known that gases present in the primitive atmosphere could have been the precursors of organic compounds. In one experiment, CH_4, NH_3, H_2O, and H_2 in a closed flask at 80°C were subjected for a week or more to electric sparking that simulated lightning. The reaction products in this and in other similar experiments included all the amino acids commonly found in proteins—but, so far, only racemic mixtures of enantiomers have been produced in this way.

It seems likely that out of a rich broth of these organic molecules the first living organisms arose. Having all amino acids be of the same handedness is known to be biologically advantageous. This is because a protein, which consists of long chains of amino acids bonded together, can form certain structures such as a stable α helix (see Section 21-8) only if *all* the amino acids are of the *same* form. Thus, biologically active proteins could be constructed from either L-amino acids or from D-amino acids, but not from a mixture of D- and L-isomers. It was, quite likely, chance alone that gave the advantage to some molecule of the L-form. If the stereochemistry of life on earth was determined by chance, will we meet up with life on other planets whose chemistry is the mirror image of ours? One of the experiments of the Viking mission to Mars was designed to test this. *Both* enantiomers of lactic acid and of alanine,

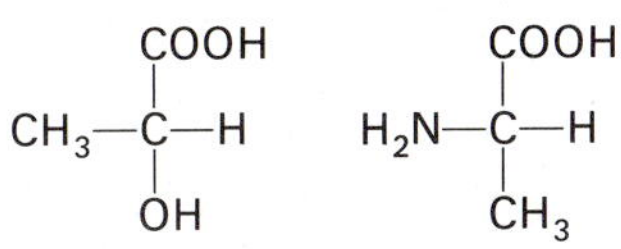

lactic acid **alanine**

were taken along to determine whether living organisms on Mars—if there are any—have a handedness and would metabolize only one of the enantiomers. But up to now, no living organisms have been found.

KEY TERMS

Hydrocarbons are compounds composed of only hydrogen and carbon.

Conformations are different spatial arrangements of the atoms in a molecule resulting from rotation about single bonds. The *configuration* of a molecule is the arrangement in space of the atoms of the molecule.

Wedge–dash–line diagrams are perspective diagrams of molecules; the diagrams represent in two dimensions the three-dimensional shapes of molecules.

Structural isomers are compounds that have the same molecular formula but different skeletal structures and different chemical and physical properties. *Stereoisomers* are isomers that have the same skeletal structures (and so are not structural isomers), but different three-dimensional arrangements of the atoms in space. There are two types of stereoisomerism—geometric and optical. *Geometric isomers* are stereoisomers that result from a restriction of rotation about double bonds or single bonds in cyclic compounds. Geometric isomers are designated by the prefixes *cis*- and *trans*-.

A *chiral* or asymmetric carbon atom is a carbon that is bonded to four different atoms or groups of atoms. A molecule that contains a chiral carbon atom can exist in two forms that are nonsuperimposable mirror images of each other. These two forms are called *optical isomers,* or *enantiomers.*

Plane-polarized light is light whose electrical component lies in a single plane.

A *racemic mixture* is a mixture of equal amounts of a pair of enantiomers. The racemic mixture has no *net* effect on polarized light.

CHAPTER SUMMARY

Organic chemistry deals with the compounds of carbon. Carbon atoms can participate in four covalent bonds and can combine with themselves and with hydrogen to form long-chain compounds known as *hydrocarbons*. For hydrocarbons with more than three carbon atoms, both unbranched and branched chains can occur. Such compounds, which have the same molecular formula but different skeletal structures, are called *structural isomers*. An individual isomer can exist in various *conformations* that result from the free rotation of atoms about single bonds.

In addition to structural isomerism, organic compounds display *stereoisomerism,* in which different compounds with the same chemical structures are distinguished by the different three-dimensional arrangements of their atoms in space. There are two kinds of stereoisomerism: *geometric isomerism* can occur in molecules with a double bond or a ring structure; *optical isomerism* can occur in molecules that contain a carbon atom bonded to four *different* atoms or groups of atoms.

PROBLEMS

13-1. Tell whether the members of each pair are isomers or the same molecule:

(a) $CH_3—CH_2—CH_3$ and 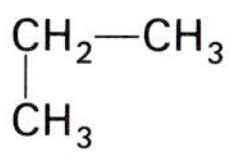

(b) $CH_3CH_2CH_2OH$ and 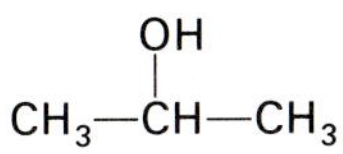

(c) $CH_3CH_2CH_2OH$ and 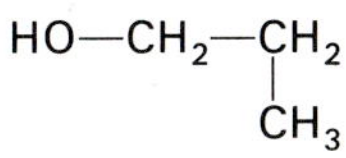

(d) $CH_3—S—CH_3$ and CH_3CH_2SH

(e) 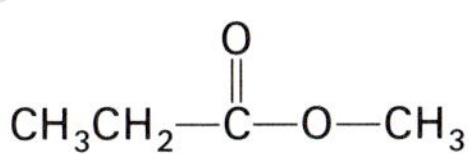

and 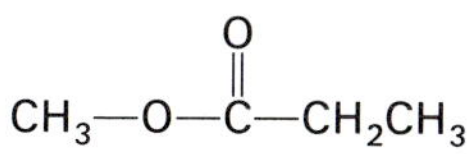

(f) $CH_3—CH_2—CH_2—CH_2—CH_2—CH_3$

and 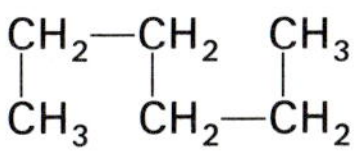

(g) 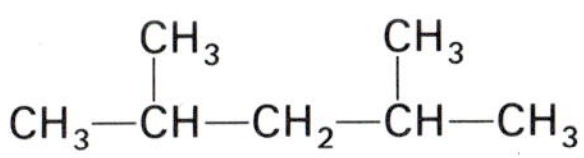

and 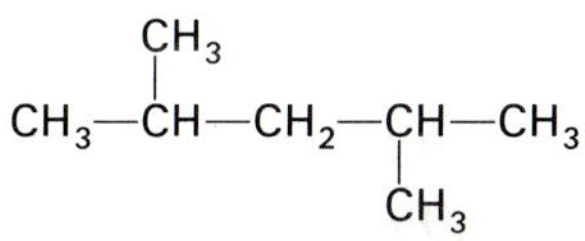

(h) 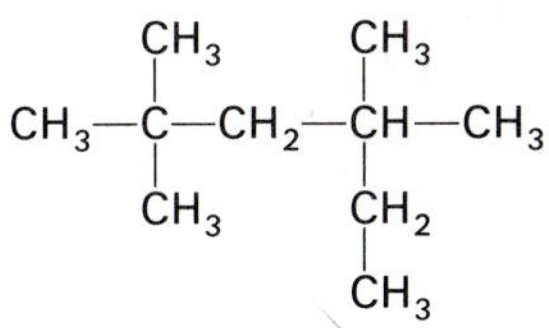

and 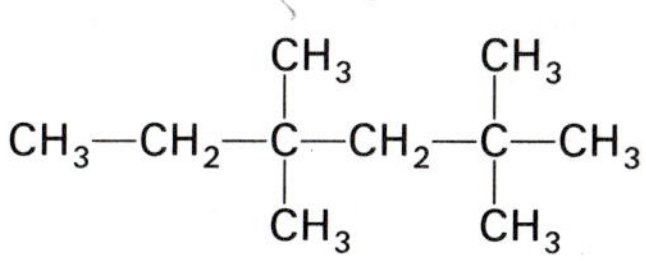

13-2. What are the hybridization states of the two carbon atoms in bromoethylene ($H_2C{=}CHBr$)? Draw a picture of the molecule, showing all σ and π bonds. (Refer to Chapter 7, if necessary.)

13-3. Describe the hybridization state of each carbon atom in 2-butyne ($CH_3—C{\equiv}C—CH_3$). Draw one picture showing all the σ bonds, and another picture showing the π bonds.

13-4. Draw wedge–dash–line structures for
(a) bromoethylene ($H_2C{=}CHBr$)
(b) propyne ($CH{\equiv}C—CH_3$)
(c) propane (C_3H_8) in the staggered conformation
(d) any conformation of *cis*-2-butene ($CH_3—CH{=}CH—CH_3$)
(e) the eclipsed conformation of chloroethane (C_2H_5Cl).

13-5. Write structural formulas for all the isomers of hexane (C_6H_{14}).

13-6. Write structural formulas for all the isomers of chlorobutane (C_4H_9Cl).

13-7. Write structural formulas for all the isomers of dibromopropane ($C_3H_6Br_2$).

13-8. Write the structural formula of a molecule that exhibits geometric isomerism, and explain what feature of the molecule is responsible for the isomerism.

13-9. Which one of the following can exist in *cis* and *trans* forms?

(a) 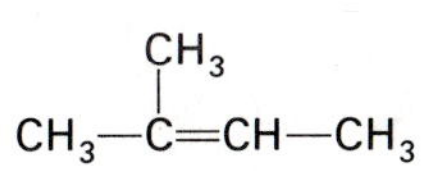
(b) $CH_3—CH{=}CH_2$
(c) $CH_2{=}CH—Cl$
(d) $CH_3—CH{=}CH—Cl$

13-10. Draw the structures of all isomers of C_3H_5Cl. (*Hint:* some have a carbon–carbon double bond, and one is a cyclic compound.)

13-11. Draw the structures of all isomers of bromochlorofluoroethylene ($C_2HBrClF$). (The molecule has a carbon–carbon double bond.)

13-12. Draw the structure of a molecule that exhibits optical isomerism, and explain what feature of the molecule is responsible for the isomerism.

13-13. Which of the following molecules could exhibit optical activity?

(a) HO—CH_2—CH_2—OH

(b) 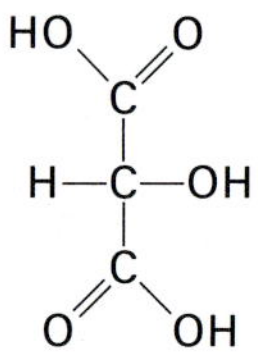 (c) 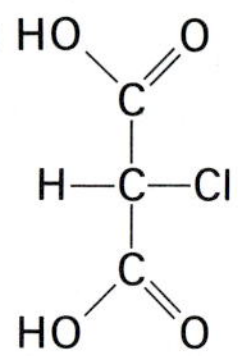 (d) 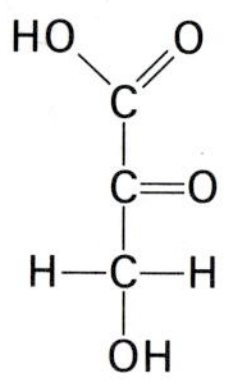

13-14. Define each of the following, and illustrate each with an example:
(a) structural isomers
(b) chiral carbon atom
(c) enantiomers
(d) racemic mixture

13-15. Consider the following structures:

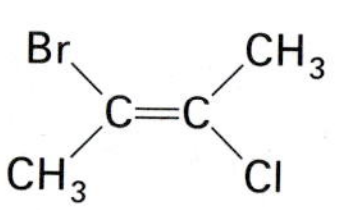
A

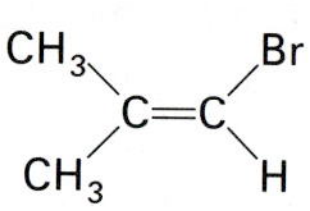
B

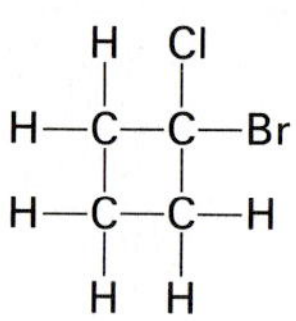
C

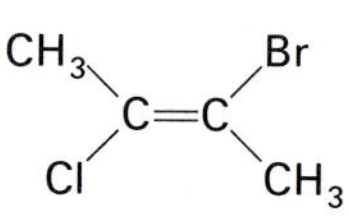
D

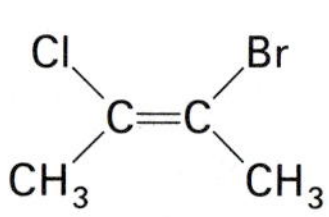
E

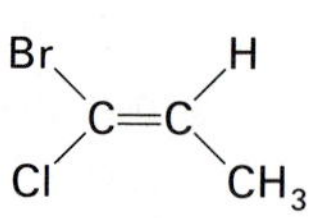
F

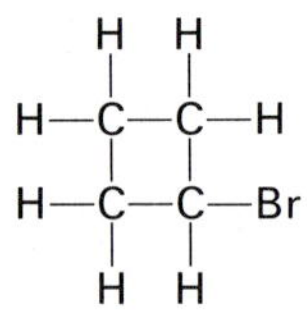
G

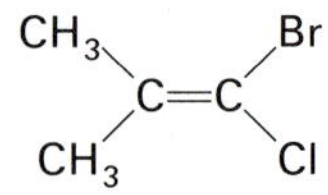
H

(a) Which of these are identical to molecule A?
(b) Which are structural isomers of molecule A?
(c) Which are geometric isomers of molecule A?

13-16. Suppose methane were not tetrahedral, but square planar (with all bonds in a plane) and with hydrogen atoms at the corners of the square:

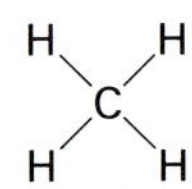

(a) Draw all possible CH_3X isomers if methane were square planar.
(b) Draw all possible CH_2XY isomers if methane were square planar.
(c) Draw all possible CHXYZ isomers if methane were square planar.
(d) How do your conclusions for the hypothetical square planar structure differ from what actually happens for the true (tetrahedral) geometry of methane when it is substituted by X, Y, and Z in successive reactions? That is, when X substitutes for one hydrogen, then Y substitutes for a second hydrogen; and finally, Z substitutes for a third hydrogen. (It was reasoning such as this that led van't Hoff and le Bel to hypothesize the tetrahedral shape of carbon compounds.)

13-17. The three carbon atoms in the simple organic

compound allene ($H_2C{=}C{=}CH_2$) lie in a line. The molecule contains carbon atoms in two different hybridization states.
(a) What is the hybridization state of the "end" carbons?
(b) What is the hybridization state of the central carbon?
(c) Sketch the structure of allene, showing only the σ-bonds in the molecule.
(d) What kind of orbitals (be specific) does the central carbon use for its σ bonds?
(e) What value (in degrees) do you predict for the H—C—H bond angle in allene?
(f) Now draw another picture of allene showing all the π bonds.
(g) Referring to your picture from part f, what do you conclude about the angle between the π bonds in allene?
(h) Draw a wedge–dash–line structure for allene.
(i) Is 1,1-dichloroallene ($Cl_2C{=}C{=}CH_2$) optically active? Support your answer with proof.

13-18. Draw all possible isomers of C_4H_7Cl. (There are a total of thirteen.)

13-19. Draw the structure of the simplest molecule (fewest atoms, only carbon and hydrogen, only σ bonds) that has one asymmetric carbon atom. Use the line–wedge–dash representation for the asymmetric carbon.

13-20. Consider bonding possibilities for boron, which has electronic structure $1s^22s^22p^1$. Answer the following questions about the molecule BH_2CH_3, which is drawn here with all the σ bonds:

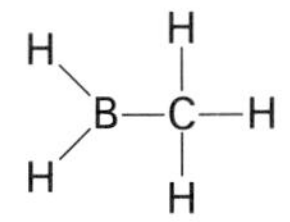

(a) The angle H—B—H is 120°. What is the hybridization state of boron?
(b) What is the H—C—H bond angle?
(c) What is the B—C—H bond angle?
(d) How many π bonds are there?
(e) How many lone pairs of electrons (Chapter 7) are there?
(f) Is rotation about the B—C bond likely to be free or restricted?

13-21. Consider the molecule NH_2—BH_2, which is drawn here with only σ bonds shown. This molecule shows *restricted* rotation about the B—N bond.

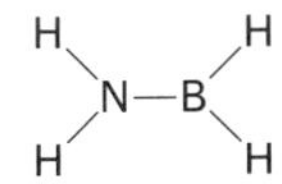

(a) Draw the structure showing all valence orbitals not involved in σ bonds, and label them.
(b) Draw in and label any π bonds and/or unshared pairs of electrons.
(c) Explain how your picture is consistent with restricted rotation about the B—N bond.
(d) What is the hybridization state of B?
(e) What is the hybridization state of N?
(f) Is the B—N bond distance likely to be shorter or longer than the B—C bond distance in BH_2CH_3 (see Problem 13-20)?
(g) Four isomers of BCl_2H_2N can be synthesized. Draw all four isomers.

13-22. You will learn about the following compounds in Part IV of this book.

(a) Which of them is optically active?

(1) 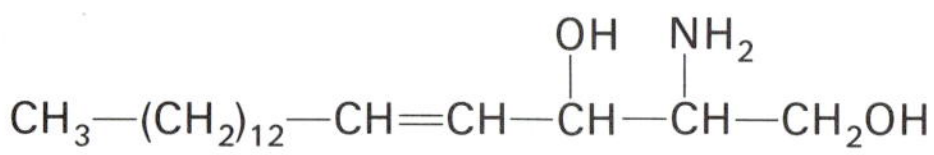

sphingosine

(2) 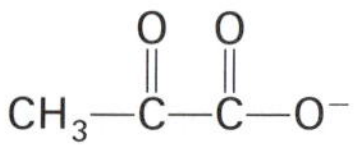

pyruvate

(3) 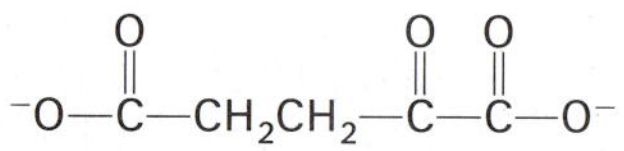

α-ketoglutarate

(4) 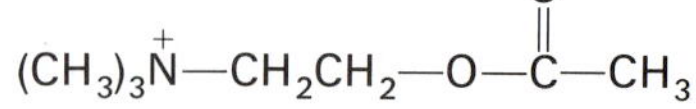

acetylcholine

(5) 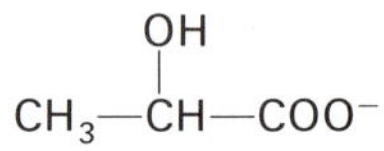

lactate

(6) 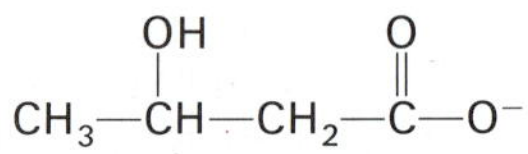

β-hydroxybutyrate

(b) Which of the compounds has a geometric isomer?

A thin film of lubricating oil separates into streamers as it exits from the region of contact between two surfaces in relative motion. This phenomenon was observed on an apparatus used to study conditions typically found in engine bearings.

14

Alkanes

Alkanes, hydrocarbons containing carbon–carbon *single* bonds only, are not commonly found in living organisms, but they are the major components of petroleum and natural gas, both of which were formed from living matter. Although they are valuable fuels, giving off heat in the process of combustion, alkanes are otherwise not very reactive. In this chapter we discuss the structures, properties, and chemical reactions of alkanes.

14-1
STRUCTURE AND PROPERTIES OF ALKANES

The group of hydrocarbons consisting of carbon and hydrogen joined only by *single* covalent bonds is called the **alkane** family. These compounds are also called **saturated hydrocarbons,** because they contain the maximum possible number of hydrogen atoms for a given number of carbon atoms. Table 14-1 gives the names and formulas of the first ten **normal** (unbranched) alkanes. Each successive member of the family differs from the one before by the addition of a $—CH_2—$ group. Such a series of compounds, each of whose members differs from the previous one by the same increment, is called a **homologous series.** The general formula C_nH_{2n+2} represents the open-chain alkane family. (We consider the cyclic alkanes separately in Section 14-6.)

The simplest alkane is methane (CH_4). Methane is the principal constituent of "swamp gas" formed by the decay of vegetation in swamps, and is sometimes itself called swamp gas. Natural gas is over 80% methane, and also contains the larger alkanes ethane, propane, and butane. As we have seen, the carbon atom in methane is in the sp^3 hybridization state, and the molecule is tetrahedral, with bond angles of 109.5°. Because each carbon atom in an alkane is bonded to four other atoms by single bonds, there is always a tetrahedral arrangement of atoms about every carbon.

Carbon and hydrogen are close in electronegativity (C = 2.5 and H = 2.1; see Figure 7-2), so the carbon–hydrogen bond is essentially nonpolar. Thus, hydrocarbons are only slightly soluble in water and other polar solvents, but they are soluble in nonpolar solvents such as other hydrocarbons.

Table 14-1 also gives physical properties of some alkanes. Note that the boiling point of the alkanes rises with increasing size and number of electrons in the alkane molecule. This effect, which is due to an increase in the van der Waals forces—the attractive forces between instantaneous dipoles—is also shown in Figure 14-1, which plots the boiling points against the molecular weights of normal alkanes.

Because hydrocarbons are nonpolar and their intermolecular forces are only of the weak van der Waals

TABLE 14-1
Physical properties of normal alkanes

Name	Formula	Melting point (°C)	Boiling point (°C)	Density of liquid (g/cm^3 at 20°C)	State (at STP*)
Methane	CH_4	−183	−162	0.424	Gas
Ethane	C_2H_6	−183	−89	0.546	Gas
Propane	C_3H_8	−187	−42	0.582	Gas
n-Butane	C_4H_{10}	−138	0	0.579	Gas
n-Pentane	C_5H_{12}	−130	36	0.626	Liquid
n-Hexane	C_6H_{14}	−95	69	0.659	Liquid
n-Heptane	C_7H_{16}	−91	98	0.684	Liquid
n-Octane	C_8H_{18}	−57	126	0.703	Liquid
n-Nonane	C_9H_{20}	−54	151	0.718	Liquid
n-Decane	$C_{10}H_{22}$	−30	174	0.730	Liquid
n-Eicosane	$C_{20}H_{42}$	37	343	0.789	Solid

*Remember that STP (standard temperature and pressure) is 0°C and 1 atm.

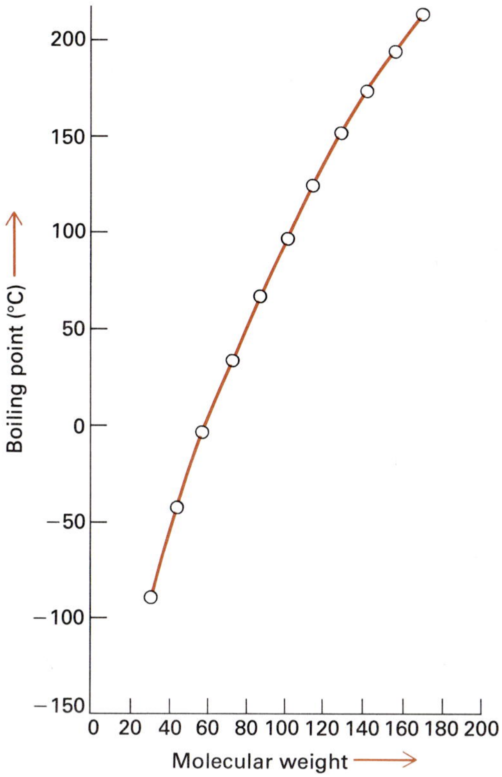

FIGURE 14-1
Boiling points of normal alkanes.

type, their melting points and boiling points are low compared to those of polar compounds of approximately the same size. For example, propane (with molecular weight 44) is a gas at room temperature and atmospheric pressure, whereas acetaldehyde CH_3CHO (a polar molecule with the same molecular weight) is a liquid at this temperature. We shall see many similar examples in later chapters.

The hydrocarbons are all less dense than water and so float on it—a phenomenon we observe, unfortunately rather frequently, in oil spills.

14-2 NOMENCLATURE OF ALKANES

We have already seen that methane, ethane, and propane can have only one structure each, but that butane can have two:

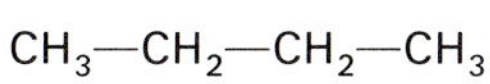

***n*-butane**

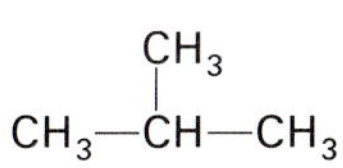

isobutane

And pentane can have three:

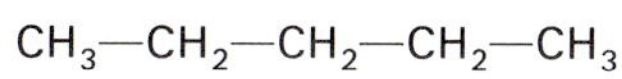

***n*-pentane**

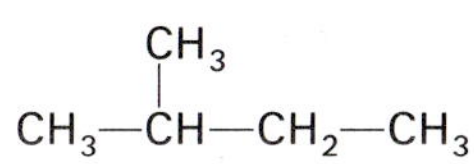

isopentane

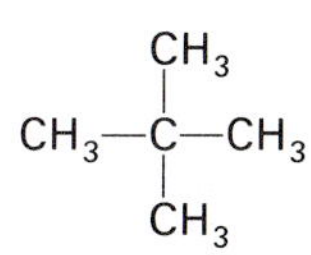

neopentane

The prefix *n-* stands for "normal," which always means an unbranched chain.

Given the tremendous numbers of isomers that exist for the larger hydrocarbons (there are, for example, over four million for $C_{30}H_{62}$), it is clearly impossible to give each a special name. A simple system of nomenclature is imperative! Such a system has been developed by the International Union of Pure and Applied Chemistry (IUPAC), and we now apply it to the alkanes. (In spite of this system, however, chemists often find it convenient to use the common or historic names for many compounds, and we shall use them, too.)

1. The names of the first ten alkanes are given in Table 14-1. For unbranched saturated hydrocarbons, the prefix *n-* is used with a name that designates the number of carbon atoms in the chain and ends in -ane. For example, the continuous-chain alkane with six carbon atoms is *n*-hexane.

2. For branched alkanes, the parent name is that of the longest unbranched chain in the molecule. The parent name is given *last* in the IUPAC name of the molecule. Each branch, or side chain, is designated by a *name* and a *number.*

a. The number identifies the carbon atom of the parent chain to which the branch is attached. The carbon at the end of the parent chain closer to the branch is assigned the number 1:

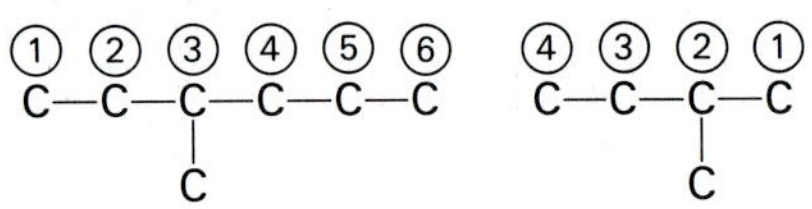

b. Each side chain is an **alkyl group,** an alkane with a hydrogen atom removed. The name of the alkyl group is obtained by dropping the *-ane* ending from the name of the corresponding alkane, and substituting *-yl:*

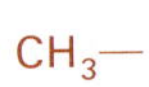
methyl

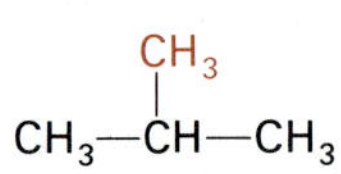
2-methylpropane

ethyl

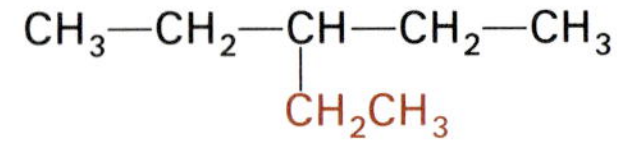
3-ethylpentane

c. The presence of more than one alkyl group with the same structure is indicated by a prefix: *di-* for two; *tri-* for three; *tetra-* for four; and so on. A separate number is used to designate the position of *each* of the groups. As an example, we'll name the following structure, which is an isomer of *n*-nonane (C_9H_{20}):

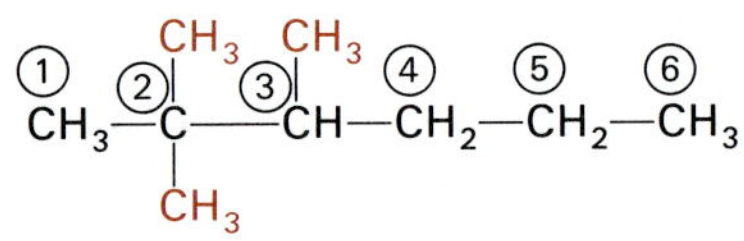
2,2,3-trimethylhexane

The way to name this structure is the following. First, find the longest unbranched chain and name it. Here, the chain is six carbons long, so the compound is a hexane. Next, number the chain in the correct order; assign the number 1 to the end of the parent chain that is closer to a branch. Finally, name and number the substituents. The name trimethylhexane indicates that there are *three* methyl groups attached to an unbranched chain of six carbon atoms. The numerical prefixes show that *two* methyl groups are attached to carbon 2, and *one* methyl group to carbon 3. The complete name, 2,2,3-trimethylhexane, specifies this one particular structure—every other isomer of nonane would have a different name.

d. If more than one kind of alkyl group is attached to the parent chain, the names are given in alphabetical order:

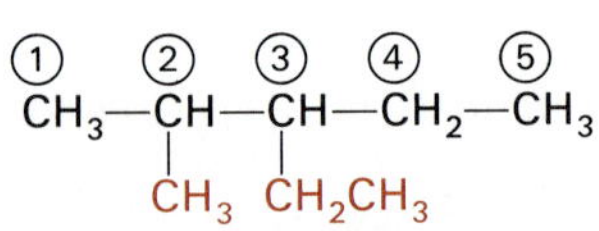
3-ethyl-2-methylpentane

TABLE 14-2
Alkyl groups

Name	Structure
Methyl	$CH_3—$
Ethyl	$CH_3—CH_2—$
n-Propyl	$CH_3—CH_2—CH_2—$
Isopropyl	$CH_3—CH(CH_3)—$
n-Butyl	$CH_3—CH_2—CH_2—CH_2—$
Isobutyl	$CH_3—CH(CH_3)—CH_2—$
Secondary butyl (or *sec*-butyl)	$CH_3—CH_2—CH(CH_3)—$
Tertiary butyl (or *tert*-butyl)	$CH_3—C(CH_3)_2—$

Table 14-2 gives the names of all the alkyl groups of up to four carbon atoms. Larger alkyl groups are named according to the IUPAC system just described, but we shall not encounter any of them in this book.

Here are the structures and names of some other alkanes:

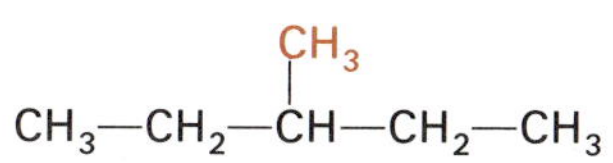

3-methylpentane

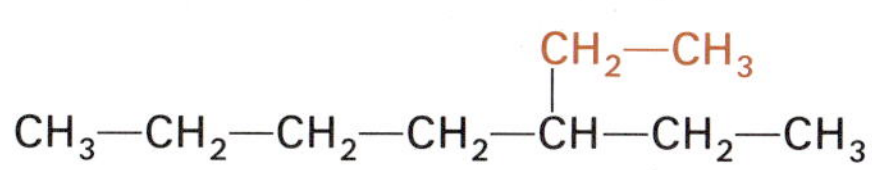

3-ethylheptane

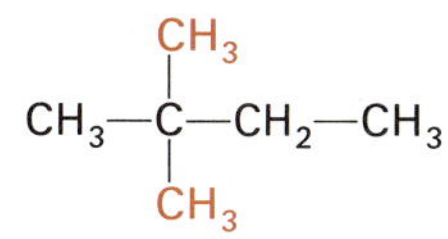

2,2-dimethylbutane

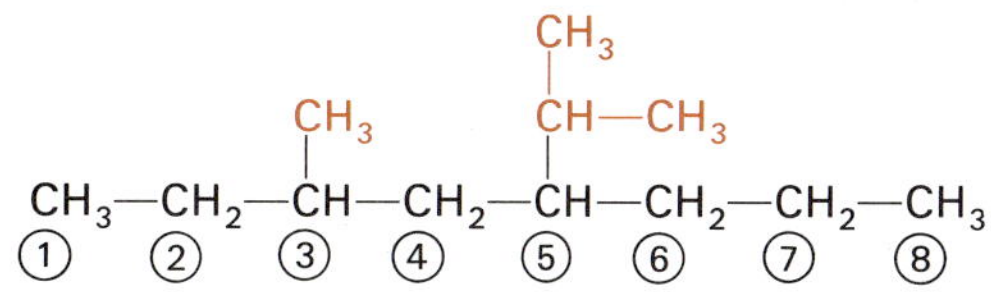

3-methyl-5-isopropyloctane

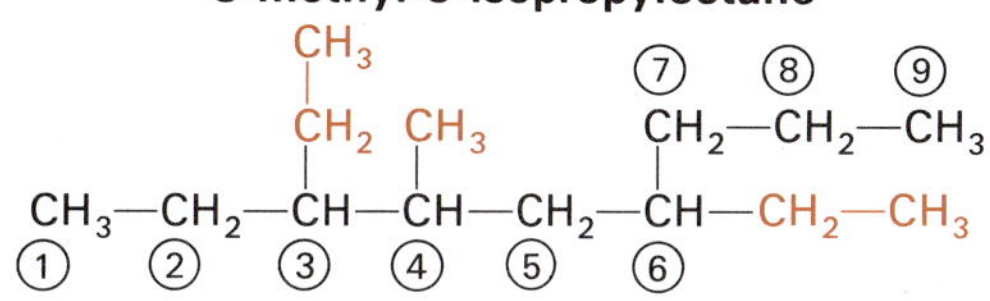

3,6-diethyl-4-methylnonane

EXERCISE 14-1 (a) Name the following compounds:

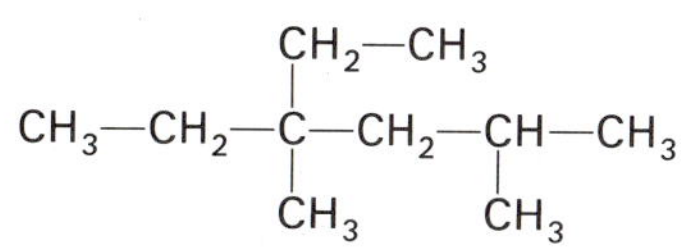

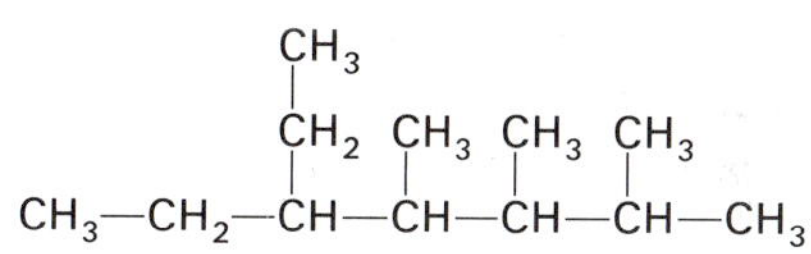

(b) Draw the structural formula of 2,2-dimethylheptane.

The number of carbon atoms bonded to a central carbon atom affects the chemical activity of the central carbon. For this reason we sometimes use a second system of classification that defines atoms as primary (1°), secondary (2°), tertiary (3°), or quaternary (4°), depending on the number of *other* carbon atoms bonded to the given carbon. A *primary* carbon atom is bonded to *one* other carbon; a *secondary* carbon atom is bonded to *two* other carbons, and so on. Thus in propane, the terminal carbons are primary carbons, and the central carbon is a secondary carbon:

$$\underset{1^\circ}{CH_3}—\underset{2^\circ}{CH_2}—\underset{1^\circ}{CH_3}$$

In isobutane, the central carbon is a tertiary carbon:

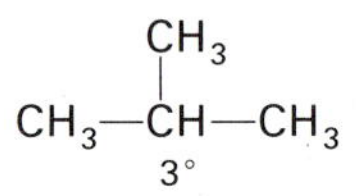

And in neopentane, the central carbon is a quaternary carbon:

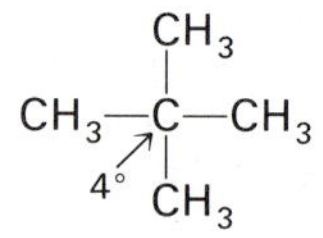

EXERCISE 14-2 Write a structural formula for a molecule that contains one or more primary, secondary, tertiary, and quaternary carbon atoms, and label each carbon 1°, 2°, 3°, or 4°.

14-3 PETROLEUM AND GASOLINE

Alkanes are not commonly found in living organisms, but they are major components of petroleum and of natural gas. The massive structures of pipes, tanks, and fractionating columns known as refineries (Figure 14-2) are responsible for separating petroleum into particular groups of alkanes and other products. This **fractionation** is based on the fact that the boiling points of hydrocarbons vary markedly depending on the size of the molecule, as we saw in Figure 14-1. This difference in properties permits separation by distillation, as discussed in Chapter 8. Some of the major fractions into which petroleum is separated in a refinery are shown in Table 14-3, with their states at room temperature and their distillation temperature ranges.

Gasoline, for example, is a mixture of hydrocarbons consisting primarily of molecules with five to ten carbon atoms. In the internal combustion engine, these compounds are burned with oxygen in the air, giving off heat, and carbon dioxide and water. (Because of incomplete combustion, carbon monoxide is also a product, along with various other pollutants.) The complete combustion of *n*-octane to carbon dioxide and water is represented by the equation

$$\underset{(n\text{-octane})}{2C_8H_{18}} + 25O_2 \longrightarrow 16CO_2 + 18H_2O$$

$$\Delta H = -5452 \text{ kJ/mole } C_8H_{18}$$

When the combustion process is irregular, or too rapid, a small explosion occurs that is heard as a "knock" in the engine. An octane-rating scale has been developed to rate the evenness of combustion (or the knocking properties) of gasolines. This arbitrary scale is based on *n*-heptane, which has been assigned octane number zero, and 2,2,4-trimethylpentane (isooctane), which has been assigned octane number one hundred.

FIGURE 14-2

An oil refinery. Note the size of the people marked with a circle in the center foreground.

TABLE 14-3
Petroleum fractions

Product	Range of hydrocarbon sizes	State at 20°C	Distillation range (°C)
Petroleum gas	C_1 to C_5	Gas	Less than 20°
Petroleum ether	C_5 to C_7	Liquid	30–90
Gasoline	C_5 to C_{10}	Liquid	40–200
Kerosene and aviation fuel	C_{10} to C_{18}	Liquid	175–275
Diesel fuel	C_{12} to C_{20}	Liquid	250–400
Lubricating oil	C_{20} to C_{30}	Liquid	Above 300
Petroleum jelly	C_{22} to C_{40}	Semisolid	Nonvolatile
Paraffin wax	C_{25} to C_{50}	Semisolid	Nonvolatile

To determine the octane number of gasoline, one burns a sample in a standard engine, and compares its knocking properties to those of standard mixtures of *n*-heptane and isooctane. For years tetraethyl lead, $Pb(C_2H_5)_4$, has been added to gas to reduce the knocking without raising the octane number of the gasoline. On burning, the lead combines with bromine from another additive to form lead bromide ($PbBr_2$) in the exhaust. This is the pollutant formed by leaded gasolines. For comparable smoothness of combustion, unleaded gas must be "higher octane" than leaded gas.

14-4 CHEMICAL REACTIONS OF ALKANES

We know that hydrocarbons are valuable fuels, undergoing combustion to yield carbon dioxide and water, and giving off heat in the process. The alkanes are, however, not very reactive. Combustion is the *only* reaction that alkanes undergo with ease, and even this reactivity decreases with increasing size of the alkane—it is much more difficult to burn tar than it is to burn propane. Alkanes do not react with concentrated strong acids such as HCl, strong bases such as NaOH, or with strong oxidizing agents such as $KMnO_4$.

In a process called **halogenation,** however, one or more of the hydrogen atoms in an alkane can be substituted by one or more halogen atoms. For example, when a mixture of methane and chlorine is heated to about 120°C (or is irradiated with ultraviolet light), the following highly exothermic reaction occurs:

$$\underset{}{CH_4} + Cl_2 \xrightarrow[\text{UV light}]{\text{heat or}} \underset{\substack{\text{methyl}\\\text{chloride}}}{CH_3Cl} + HCl$$

$$\Delta H = -103 \text{ kJ/mole}$$

In actual practice, a mixture of products is obtained because the methyl chloride can be further chlorinated to yield dichloromethane (methylene chloride, CH_2Cl_2), trichloromethane (chloroform, $CHCl_3$), and tetrachloromethane (carbon tetrachloride, CCl_4). The series of reactions is

$$\underset{\text{methane}}{CH_4} \xrightarrow{Cl_2 \quad HCl} \underset{\text{methyl chloride}}{CH_3Cl} \xrightarrow{Cl_2 \quad HCl}$$

$$\underset{\substack{\text{methylene}\\\text{chloride}}}{CH_2Cl_2} \xrightarrow{Cl_2 \quad HCl} \underset{\text{chloroform}}{CHCl_3} \xrightarrow{Cl_2 \quad HCl} \underset{\substack{\text{carbon}\\\text{tetrachloride}}}{CCl_4}$$

The symbol $\xrightarrow{Cl_2 \quad HCl}$ means that a Cl_2 is added and an HCl is produced at each step.

Methyl chloride is a gas at room temperature; methylene chloride, chloroform, and carbon tetrachloride are liquid. The three liquids are used as solvents for oils, grease, and other organic substances. Chloroform and methylene chloride are also used as anesthetics. Ethyl chloride (CH_3CH_2Cl), produced by the chlorination of ethane, is used as a local anesthetic. It boils at 12.3°C. Thus, when sprayed on the skin (say, where a ball player has been hit by a pitched ball), it vaporizes; this cools the sprayed area and deadens the pain.

A high concentration of CCl_4 in the air can cause death through respiratory failure. Prolonged exposure to small concentrations can cause damage to the liver and kidneys. Carbon tetrachloride is also suspected to be carcinogenic. Methylene chloride has largely replaced carbon tetrachloride as a cleaning fluid, but all halogenated hydrocarbons are probably toxic. They should be used with care, in well-ventilated areas. These compounds can be absorbed through the skin as well as by inhalation.

During the past few years, concern has developed that halogenated hydrocarbons present as pollutants in the atmosphere might cause a decrease in the amount of ozone (O_3) in the stratosphere (the upper atmosphere). The ozone absorbs ultraviolet light from the sun, thus protecting life on earth from radiation damage. An increase in the amount of ultraviolet light reaching the earth would cause an increase in the incidence of skin cancer in humans and in the number of mutations in DNA, the genetic information carrier.

The major problem is caused by the fully halogenated hydrocarbons dichlorodifluoromethane (CCl_2F_2, Freon-12) and trichlorofluoromethane (CCl_3F, Freon-11), used as propellants in aerosol spray cans and as refrigerants in air conditioners, freezers, and refrigerators. The Freons are very unreactive, and probably exist as long as 50 years in the atmosphere. Thus they can diffuse into the stratosphere, where they are finally broken down by ultraviolet light from the sun to produce free Cl atoms. The Cl atoms are highly reactive because each has an unpaired electron, and they start a series of reactions that result in a decrease in the amount of ozone. The problem is complex, and its magnitude is not yet known with certainty, but it is serious enough that the use of Freons as propellants in spray cans is now being phased out. Substitutes are being sought for use as refrigerants.

The alkyl halides are important starting materials in the synthesis of organic compounds. We shall see other ways to prepare them, and many examples of their reactions, in following chapters.

EXERCISE 14-3 (a) Write an equation for the bromination of ethane (C_2H_6) to produce bromoethane (C_2H_5Br).

(b) If more bromine is permitted to react with bromoethane, two different products with the formula $C_2H_4Br_2$ are obtained. Write their structural formulas.

14-5 NOMENCLATURE OF ALKYL HALIDES

Simple alkyl halides are commonly named as salts of alkyl radicals:

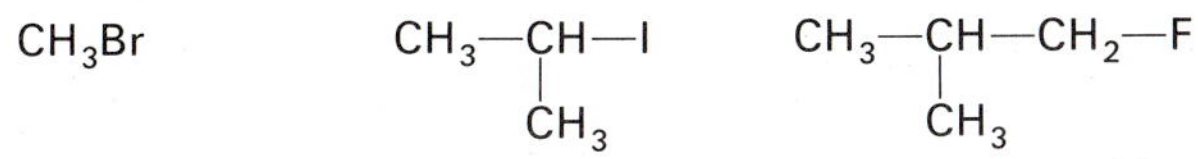

methyl bromide **isopropyl iodide** **isobutyl fluoride**

Although you will often hear these names, they are not the IUPAC designations. In the IUPAC system, the compounds are named as halogen derivatives of a parent hydrocarbon.

The alkane is named first, and a prefix derived from the name of the halogen (fluoro-, chloro-, bromo-, iodo-) is attached. A number is assigned to show the position of the halogen:

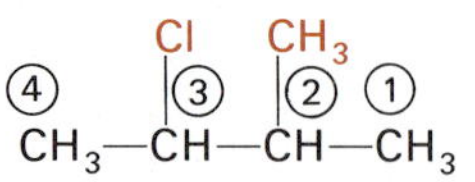

3-chloro-2-methylbutane

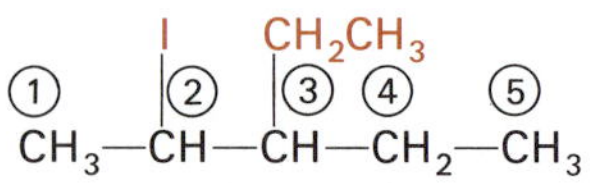

2-iodo-3-ethylpentane

If there are two or more atoms of the same halogen, the prefixes di-, tri-, tetra-, and so on are used:

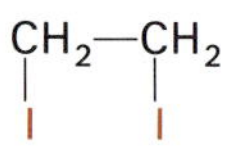

1,2-diiodoethane

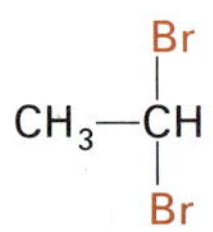

1,1-dibromoethane

$CH_3—CF_3$
1,1,1-trifluoroethane

CF_4
tetrafluoromethane

If there are different halogens, they are named in alphabetical order:

$Cl—CH_2—CH_2—CH_2—Br$
1-bromo-3-chloropropane

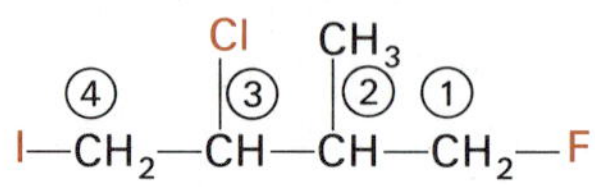

3-chloro-1-fluoro-4-iodo-2-methylbutane

EXERCISE 14-4 (a) Draw structural formulas for the four isomers of dibromopropane ($C_3H_6Br_2$) and name them.

(b) Draw structural formulas for all isomers of bromofluoroethane (C_2H_4BrF) and name them.

14-6 CYCLOALKANES

Cycloalkanes, whose properties are generally similar to those of the alkanes, have the general formula C_nH_{2n}. A cyclic alkane has two hydrogen atoms fewer than the corresponding alkane, and has one additional carbon–carbon bond instead:

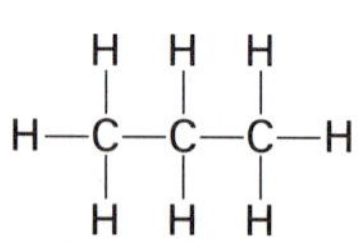

propane (C_3H_8)

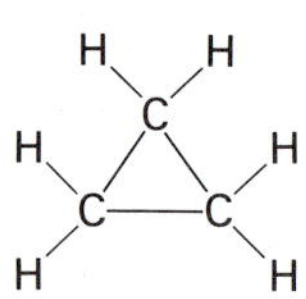

cyclopropane (C_3H_6)

The cyclic alkanes are commonly represented by a polygon that shows only carbon–carbon bonds. Each corner represents a carbon atom, and all bonds not shown (those required to make a total of four bonds for each carbon) are to hydrogen atoms. Substituents other than H are shown explicitly, with bonds.

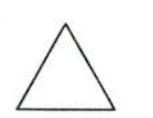

cyclopropane (C_3H_6)

cyclobutane (C_4H_8)

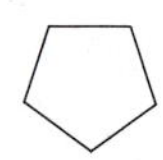

cyclopentane (C_5H_{10})

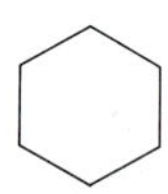

cyclohexane (C_6H_{12})

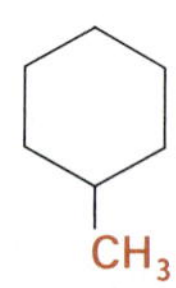

methylcyclohexane (C_7H_{14})

We use this convention frequently in drawing the

structures of complicated organic compounds that contain carbon rings.

The cycloalkanes are named in the IUPAC system according to the number of carbons in the ring, with the prefix *cyclo-*. The names of several cycloalkanes have already been given with their structures. For substituted cycloalkanes, the following rules are used.

1. It is not necessary to designate the position of attachment if there is only one substituent on the ring, because all positions are equivalent:

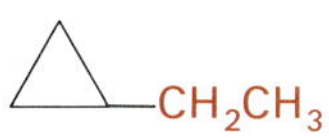

ethylcyclopropane

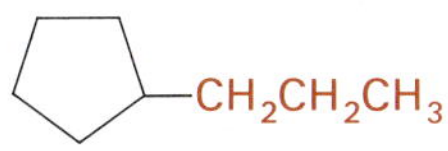

propylcyclopentane

2. When there is more than one substituent on the ring, the lowest possible combination of numbers is assigned:

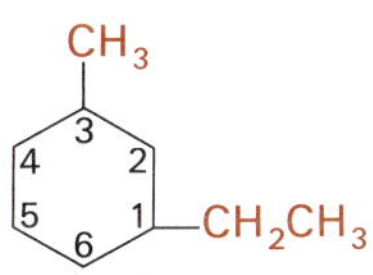

1-ethyl-3-methylcyclohexane

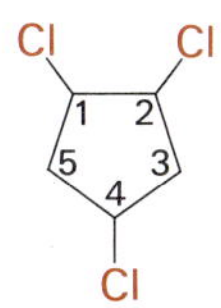

1,2,4-trichlorocyclopentane

Cyclopropane (C_3H_6) and cyclobutane (C_4H_8) are much more reactive than the corresponding open-chain alkanes. This is attributed to their strained geometry:

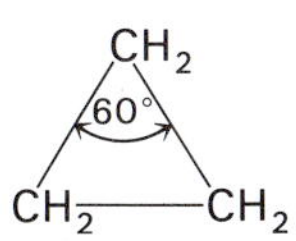

cyclopropane

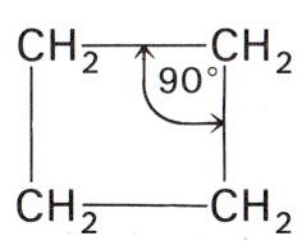

cyclobutane

Cyclopropane, for example, has carbon–carbon bond angles of 60°, far from the normal tetrahedral angle of 109.5° expected for carbon in the sp^3 hybridization state. It is therefore a highly strained and unstable molecule. Cyclobutane, with carbon–carbon bond angles of 90°, is also a strained molecule.

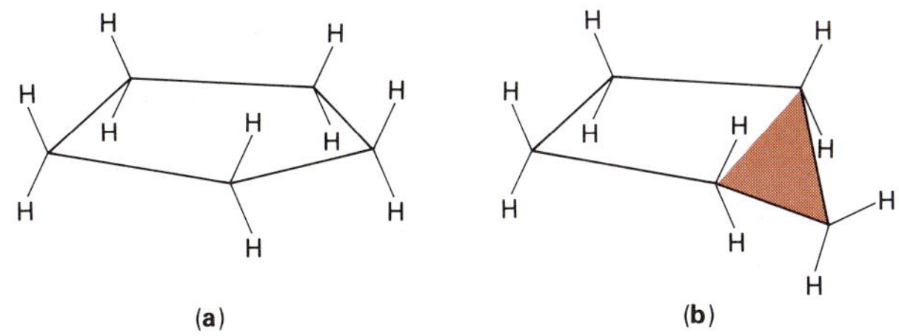

FIGURE 14-3

Structure of cyclopentane. **(a)** Planar conformation of cyclopentane. All of the hydrogens are eclipsed. **(b)** The most stable conformation of cyclopentane is a puckered ring. The out-of-plane CH_2 group is staggered with respect to its nearest neighbors.

Cyclopentane is a stable molecule. In the planar conformation, the bond angles of the ring are 108°, close to the normal angle, but the hydrogen atoms close to each other are in an eclipsed arrangement (Figure 14-3a). The most stable conformation is a slightly puckered ring, in which the hydrogen atoms can be somewhat farther apart in a staggered arrangement.

Cyclohexane is the most stable cyclic alkane. As we shall see in later chapters, its derivatives (compounds that can be derived from it) are very common in nature. Two examples follow, and we shall see many more in later chapters.

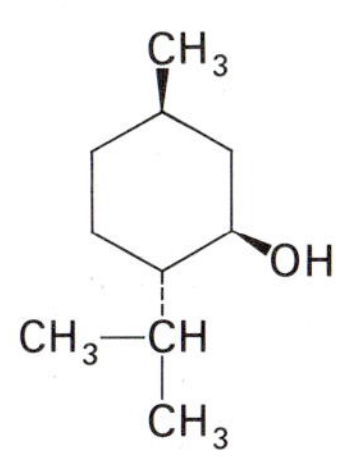

menthol
(mint oil)

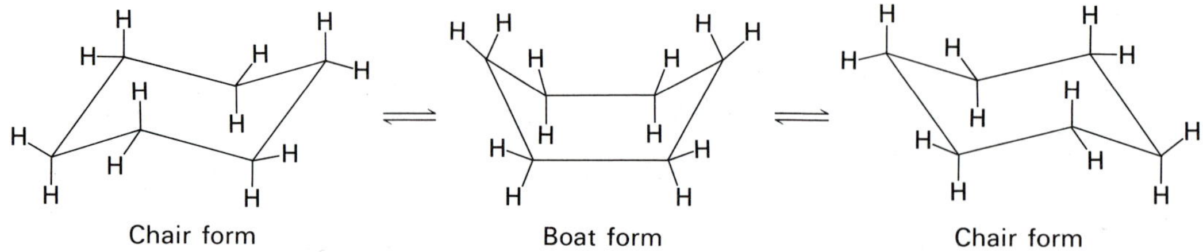

FIGURE 14-4

Chair and boat forms of cyclohexane. The chair forms are the most stable conformations.

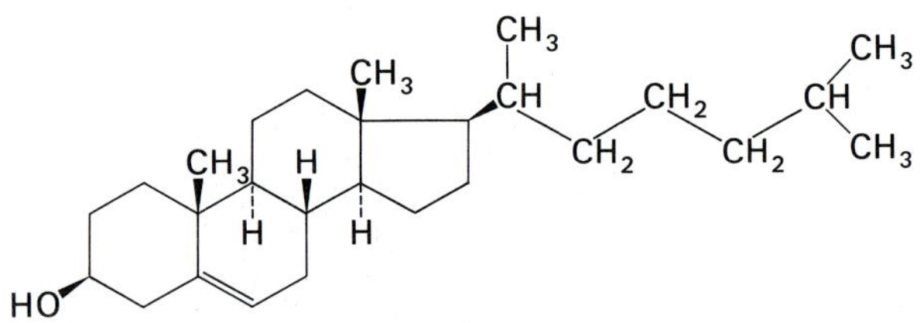

cholesterol
(major steroid in human body and precursor of steroid hormones)

The stability of cyclohexane results from the fact that a conformation exists in which all the bond angles are tetrahedral (in the planar conformation the bond angles would be 120°C) and all the hydrogen atoms are in a staggered arrangement. This stable conformation, called the **chair form,** is shown in Figure 14-4. There are actually two chair forms, as shown in the figure, and the molecule converts rapidly from one to the other at room temperature. An intermediate, somewhat less stable, conformation called the **boat form,** is also shown. In this conformation, the bond angles again are tetrahedral, but some of the hydrogen atoms are in an eclipsed arrangement.

If we consider the carbon ring in the chair form of cyclohexane to be *roughly* in a plane, six of the carbon–hydrogen bonds are perpendicular to that plane, with three extending upward and three downward (see Figure 14-5). These six bonds are called **axial** bonds, and the three on one side are all quite close together. The other six carbon–hydrogen bonds are approximately in the "plane" of the ring, and are called **equatorial** bonds. When one chair form is converted to the other, the bonds that were originally axial become equatorial, and vice versa. This is shown in Figure 14-6 for methylcyclohexane.

When a methyl or some other group is substituted for one of the hydrogen atoms of cyclohexane, the preferred position for the substituent is an equatorial position because this places the larger, bulkier group as far as possible from the close-together axial hydrogen atoms (Figure 14-6). At equilibrium, 95% of the methylcyclohexane has the methyl group in the equatorial position, and only 5% has the methyl group in the axial position. With even larger substituents (say, a butyl group), the ring is effectively locked into the form that has the substituent in the equatorial position and can no longer assume other conformations.

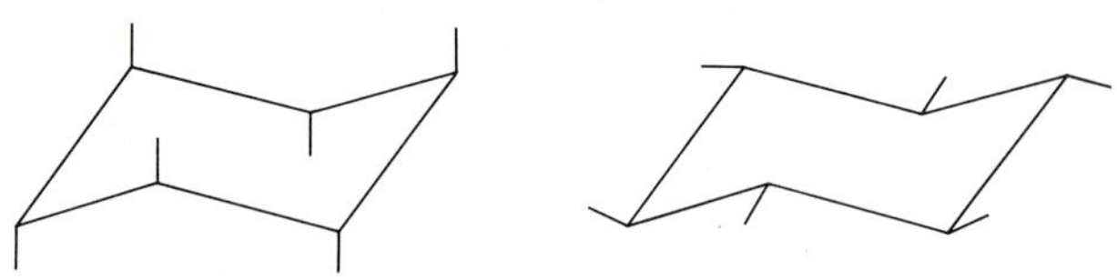

(a) (b)

FIGURE 14-5

Chair form of cyclohexane. **(a)** Only axial bonds are shown. **(b)** Only equatorial bonds are shown.

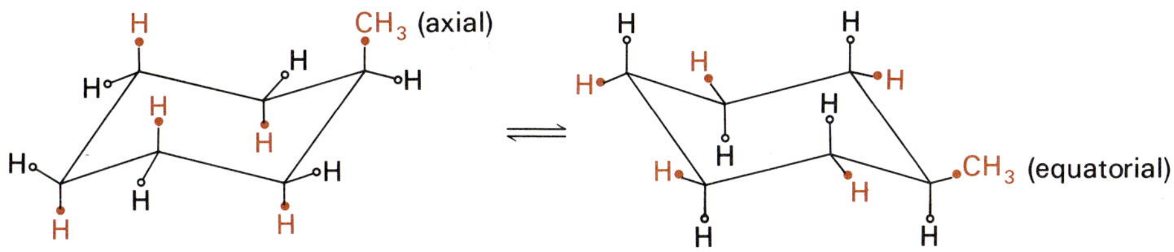

structure A **structure B**

FIGURE 14-6

The two chair forms of methylcyclohexane. The axial bonds in structure A (those perpendicular to the plane of the ring) become equatorial when the chair form A converts to the chair form B. The B form predominates (95%) because in it the bulky methyl group is farther from adjacent hydrogen atoms.

Due to restriction of rotation about the carbon–carbon bonds in cyclic compounds, they can exist as geometric isomers (Chapter 13). There are *cis* and *trans* isomers, for example, of 1,3-dibromocyclopentane:

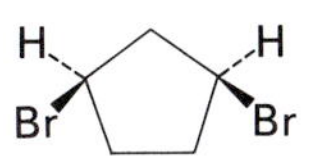

***cis*-1,3-dibromocyclopentane**

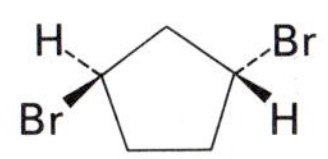

***trans*-1,3-dibromocyclopentane**

Even though the molecule is slightly puckered in its most stable conformation, we can think of the two bromine atoms as being on the same side (*in the cis* isomer) or on opposite sides (in the *trans* isomer) of the approximate plane of the ring.

EXERCISE 14-5 Draw the structural formulas of all *cyclic* isomers of dichlorocyclobutane, and name them.

Carbon compounds containing rings of 15 or more carbon atoms are found in nature. Examples are muscone (with a 15-membered carbon ring) and civetone (with a 17-membered ring), both of which are used in the manufacture of expensive perfumes:

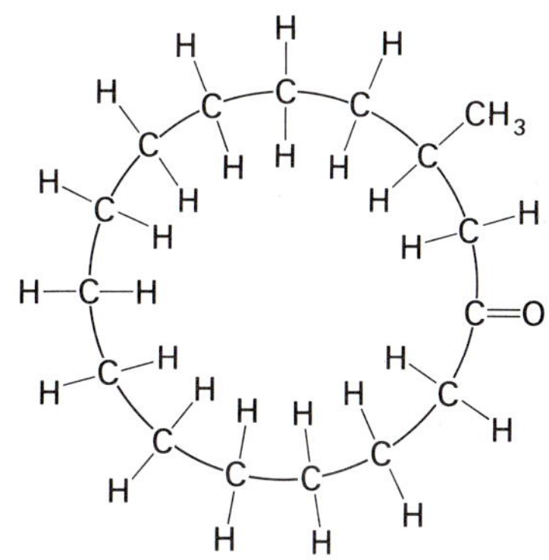

muscone
(sex attractant of the Tibetan musk deer)

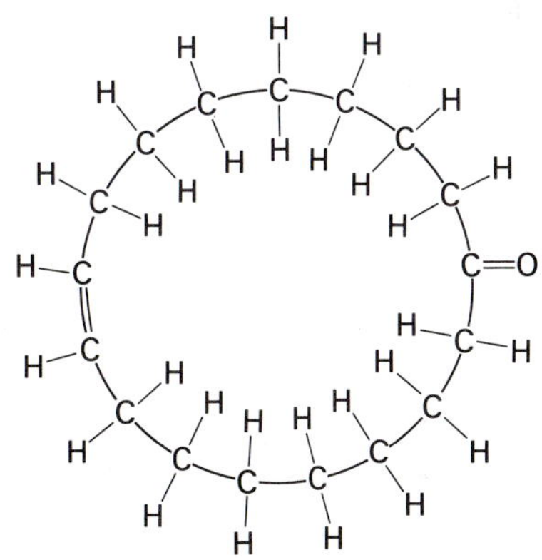

civetone
(sex attractant of the male civet cat)

14-7
FUNCTIONAL GROUPS

The alkanes are stable, relatively unreactive compounds, but the substitution of another atom, or group of atoms, for one of the hydrogens on the chain usually produces a molecule that is more reactive. For example, the compound ethyl chloride (CH_3CH_2Cl) reacts with water or with hydroxide ion to form ethyl alcohol (CH_3CH_2OH) under conditions where ethane itself is unreactive:

$$CH_3{-}CH_2{-}Cl + OH^- \longrightarrow CH_3{-}CH_2{-}OH + Cl^-$$
$$CH_3{-}CH_3 + OH^- \longrightarrow \text{No reaction}$$

If we use the symbol R to represent an alkyl group (a saturated hydrocarbon with a single H atom removed), we can write an equation for the general reaction in which an alkyl chloride is converted to an alcohol:

$$R{-}Cl + OH^- \longrightarrow R{-}OH + Cl^- \quad (14\text{-}1)$$

In Equation 14-1, R could, for example, represent the ethyl group ($CH_3CH_2{-}$). The family of compounds R—Cl, known as the alkyl chlorides, exhibits the behavior summarized in Equation 14-1, in which the chlorine atom is the reactive part of the molecule. The chlorine atom is called a **functional group** (though in this case the "group" is only a single atom).

Before considering the properties of functional groups, we shall look at the three ways in which a carbon–hydrogen bond can be broken.

1. *Carbanion formation. Both* electrons of the shared pair remain with the carbon atom:

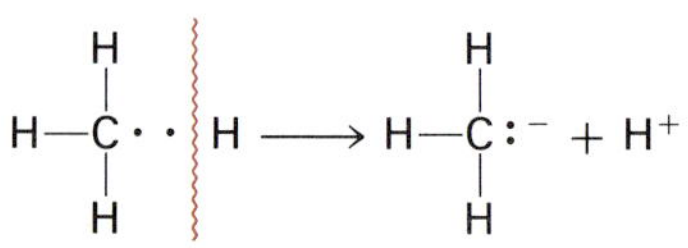

The anion formed when the bond is broken in this way is called a **carbanion,** *an ion having a negatively charged carbon.* This is the most likely way for the C—H bond to break, because carbon is slightly more electronegative than hydrogen (C = 2.5 and H = 2.1).

2. *Free-radical formation.* One electron of the shared pair remains with each atom:

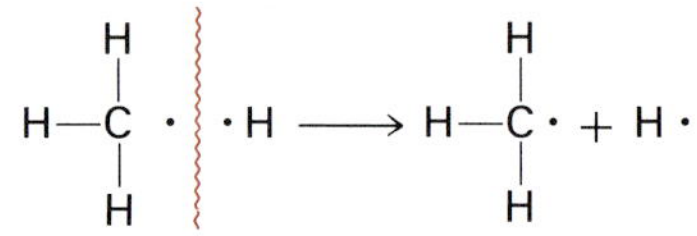

The products, in this case, are free radicals. We won't encounter these frequently; the Cl atom formed in the ozone layer from the Freons (Section 14-4) is one example.

3. *Carbonium-ion formation.* Both electrons of the shared pair remain with the hydrogen:

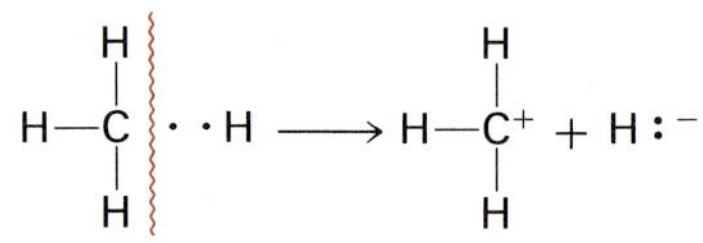

The cation formed when the bond is broken in this way is called a **carbonium ion,** *an ion having a positively charged carbon.* This is the least likely way for the C—H bond to break, because carbon is more electronegative than hydrogen.

Now, instead of the C—H bond, which is not very reactive, we consider the C—X bond, where X is any functional group such as Cl, Br, I, or some combination of atoms with characteristic properties, such as the —OH group of alcohols. These functional groups are generally more electronegative than carbon. Using the symbols δ^+ and δ^- (Chapter 7) to represent a polar bond, we can symbolize the bond $^{\delta+}C—X^{\delta-}$ to show that the functional group X bears a partial negative charge and the carbon atom bears a partial positive charge.

We want to focus on the C—X bond, so it is convenient to represent unreactive groups attached to the carbon by the symbol R, which we have used earlier to represent an alkyl group. If three *different* groups are attached to the carbon, the symbols R, R′, and R″ are used:

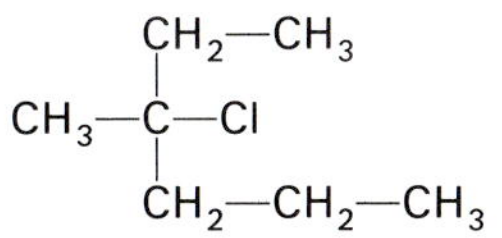

or

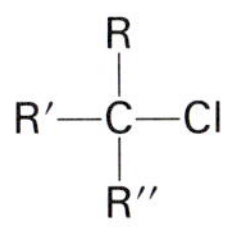

Here, R stands for —CH_2CH_3 (an ethyl group), R′ stands for —CH_3 (a methyl group), and R″ stands for —$CH_2CH_2CH_3$ (a propyl group).

Next we consider the three ways the C—X bond can be broken.

1. *Both* electrons of the shared pair remain with X:

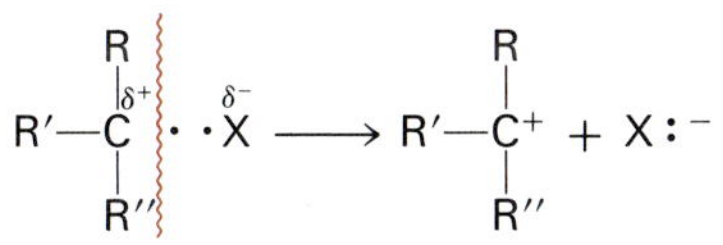

When, as is usual, X is more electronegative than carbon, this is the most likely way for the C—X bond to be broken. This reaction forms a carbonium ion and an anion of X. We shall see many examples.

2. One electron of the shared pair remains with each atom:

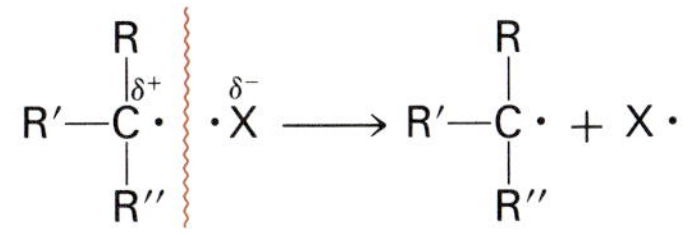

The products are free radicals. We shall see very few examples of this type of reaction.

3. Both electrons of the shared pair remain with the carbon:

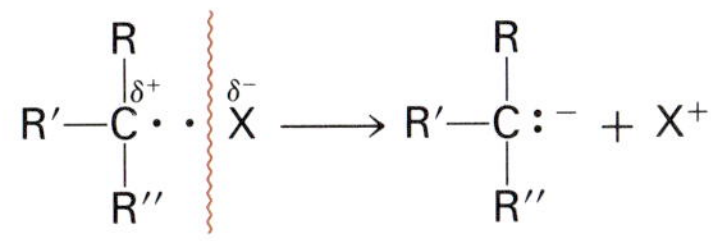

Where X is more electronegative than carbon, this is the least likely way for the bond to be broken. But, if X is *less* electronegative than carbon (say, X is a metal atom, such as lithium), this is a likely way for the bond to be broken. This reaction forms a carbanion and a cation of X. We shall see examples of this in later chapters.

Summary. An alkane is unreactive, but the replacement of one or more hydrogen atoms by a functional group makes the molecule more reactive. Whatever reaction occurs usually occurs at or near the functional group. Table 14-4 lists some of the common functional groups. In following chapters, we discuss the characteristic reactions of the most important functional groups. The basic classes of organic reactions that we shall consider are summarized in the next section.

TABLE 14-4
Functional groups

Family	General formula	Example	Common name
Alcohol	R—OH	CH_3CH_2—OH	Ethyl alcohol
Thiol	R—SH	CH_3CH_2—SH	Ethanethiol
Amine	R—N(R′)(R″)	CH_3CH_2—NH_2	Ethylamine
Aldehyde	R—C(=O)—H	H—C(=O)—H	Formaldehyde
Ketone	R—C(=O)—R′	CH_3—C(=O)—CH_3	Acetone
Carboxylic acid	R—C(=O)—OH	H—C(=O)—OH	Formic acid
Ester	R—C(=O)—O—R′	CH_3—C(=O)—O—CH_2CH_3	Ethyl acetate
Amide	R—C(=O)—N(R′)(R″)	CH_3C(=O)—NH_2	Acetamide
Ether	R—O—R′	CH_3CH_2—O—CH_2CH_3	Diethyl ether
Alkyl halide	R—X	CH_3—Cl	Methyl chloride
Alkene*	R_3R_4C=CR_2R_1	H_2C=CH_2	Ethylene
Alkyne	R—C≡C—R′	H—C≡C—H	Acetylene
Aromatic ring*	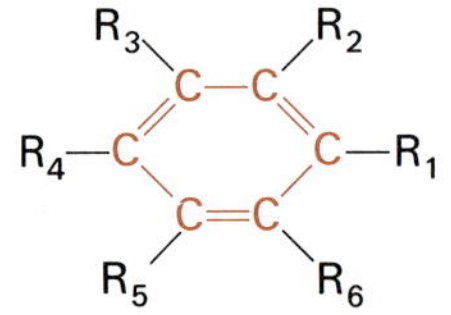	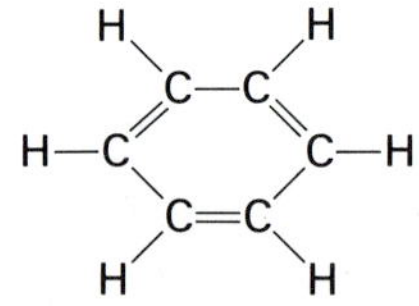	Benzene

*R_1, R_2, R_3, and so on are used instead of R′, R″, and so on.

14-8 REACTIONS OF ORGANIC CHEMISTRY

The number of reactions that have been reported for organic compounds is enormous. Many involve very complicated structures and complex mechanisms. However, certain generalities can be made about organic reactions, and most reactions of organic compounds fall into one of five basic classes.

1. **Addition reactions,** in which two or more reactants combine to form a new molecule—as in the addition of bromine to ethylene to yield 1,2-dibromoethane:

$$H_2C{=}CH_2 + Br_2 \longrightarrow H_2\underset{\displaystyle Br}{\underset{|}{C}}{-}\underset{\displaystyle Br}{\underset{|}{C}}H_2$$

In an addition reaction, there is an *increase* in the number of atoms or groups attached to carbon.

2. **Elimination reactions,** in which two or more covalent bonds are broken, and a new multiple (double or triple) bond is formed—as in the elimination of hydrogen bromide from bromoethane to yield ethylene:

$$\underset{\displaystyle H}{\underset{|}{C}}H_2{-}\underset{\displaystyle Br}{\underset{|}{C}}H_2 \longrightarrow CH_2{=}CH_2 + HBr$$

In an elimination reaction, there is a *decrease* in the number of atoms or groups attached to carbon.

3. **Substitution reactions,** in which one atom or group is *replaced* by another atom or group (one bond is broken, and a new one formed)—as in the substitution of bromine for chlorine in chloromethane:

$$CH_3{-}Cl + Br^- \longrightarrow CH_3{-}Br + Cl^-$$

4. **Oxidation reactions,** in which an organic compound is oxidized by oxygen or by some other oxidizing agent (Chapter 12). An example is the oxidation of ethyl alcohol to acetic acid that is carried out by bacteria in wine:

$$2CH_3CH_2OH + O_2 \longrightarrow 2CH_3CH_2{-}C\begin{matrix}{=}O\\ {-}OH\end{matrix}$$

5. **Reduction reactions,** in which an organic compound is reduced by hydrogen or by some other reducing agent. An example is the reduction of acetic acid to ethyl alcohol:

$$CH_3{-}C\begin{matrix}{=}O\\ {-}OH\end{matrix} + H_2 \xrightarrow{Pt} CH_3CH_2OH + H_2O$$

In the following chapters, we discuss examples of each of these kinds of reaction. We also investigate the mechanisms by which they occur.

KEY TERMS

Alkanes are hydrocarbons that contain the carbon–carbon single bond only, and that have the general formula C_nH_{2n+2}. The alkanes are said to be *saturated* because they contain the maximum number possible of hydrogen atoms.

An *alkyl group* is an alkane with a hydrogen atom removed. The symbol R— is commonly used to represent an alkyl group.

Carbon atoms are classified as *primary* (1°), *secondary* (2°), *tertiary* (3°), or *quaternary* (4°) depending on the number of other carbon atoms bonded to the given carbon.

The *IUPAC system* is a universally accepted system of naming organic compounds developed by the International Union of Pure and Applied Chemistry.

Free radicals are atoms or groups of atoms with an unpaired electron.

In the process of *halogenation* of an alkane, a halogen atom is substituted for a hydrogen atom in the alkane.

Cycloalkanes are cyclic alkanes having the general formula C_nH_{2n}.

The *chair form* and the *boat form* are two conformations of cyclohexane. The chair form is the most stable conformation. In the chair form of cyclohexane there are six *axial* bonds, which are perpendicular to the (approximate) plane of the molecule, and six *equatorial* bonds, which are in the approximate plane of the molecule.

A *carbanion* is an ion with a negatively-charged carbon atom. A *carbonium ion* is an ion with a positively-charged carbon atom.

A *functional group* is an atom or group of atoms that replaces a hydrogen atom in an alkane and makes the molecule more reactive.

In *addition reactions* two or more reactants combine to form a new molecule. In *elimination reactions* two or more covalent bonds are broken and a new multiple bond is formed. In *substitution reactions,* one atom or group is replaced by a different atom or group.

CHAPTER SUMMARY

Alkanes are *hydrocarbons* that contain carbon–carbon single bonds only, and that have the general formula C_nH_{2n+2}. For the higher alkanes, the number of isomers is so great that it becomes imperative to have a systematic way to name them. The *IUPAC system* is a simple, internationally accepted method of naming organic compounds.

Alkanes are generally unreactive, their only important reactions being *combustion* and *halogenation,* a free-radical initiated chain reaction in which a halogen atom is substituted for a hydrogen.

Cycloalkanes are similar to alkanes both in their properties and their reactions. The most important cycloalkanes are cyclopentane and cyclohexane, which are the most stable. The *chair form* is the most stable conformation of cyclohexane. The cycloalkanes are commonly represented by polygons that show only the carbon–carbon bonds.

The alkanes are relatively unreactive, but substitution of a *functional group* X for one of the hydrogen atoms makes the molecule more reactive. When a C—H or C—X bond is broken, a *carbonium ion,* a *carbanion,* or a *free radical* is produced. The basic classes of organic reactions include *addition, elimination, substitution, oxidation,* and *reduction.*

PROBLEMS

14-1. Isooctane is 2,2,4-trimethylpentane. Draw structural formulas for (and name) a substituted heptane, hexane, pentane, and butane that are isomers of isooctane.

14-2. Draw structural formulas for
(a) 3-ethylpentane
(b) 2,4-dichlorohexane
(c) 3,3,5-triiodoheptane
(d) 1-chloro-2-fluoroethane
(e) 1,1,1-trichloroethane
(f) 1,1,2-trichloroethane
(g) 4-isobutyldecane
(h) 4-ethyl-2-methyloctane
(i) 2,2,4-trimethylpentane.

14-3. Name each of the following compounds:

(a) 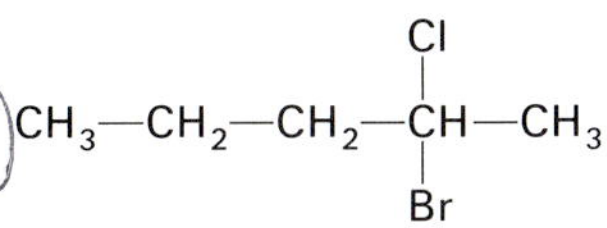

(b) 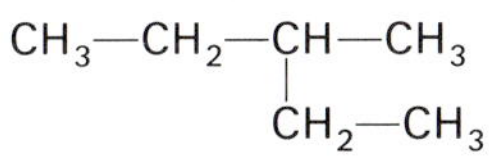

(c) 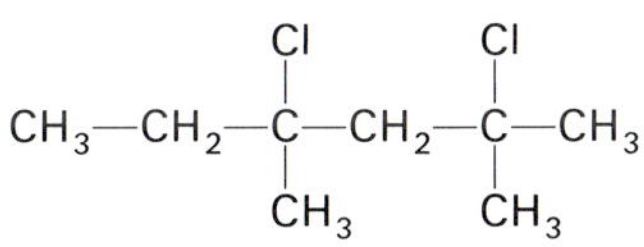

(d) 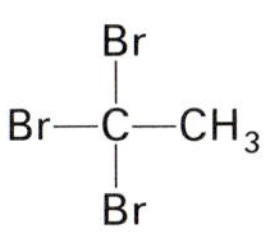

(e) 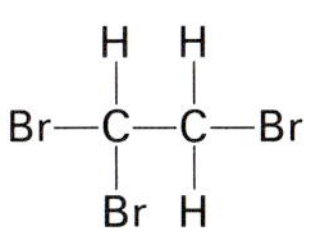

(f) 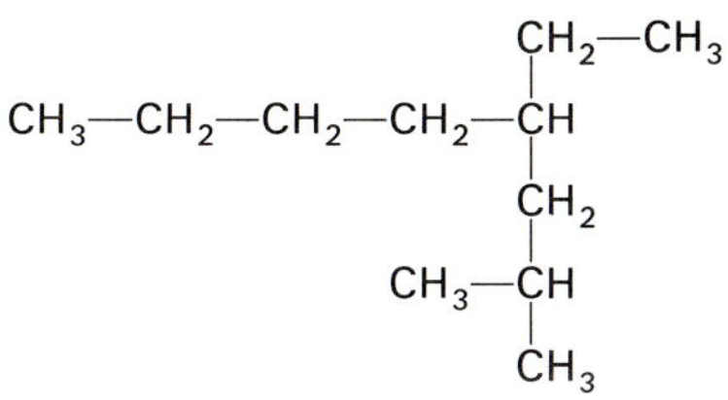

(g) 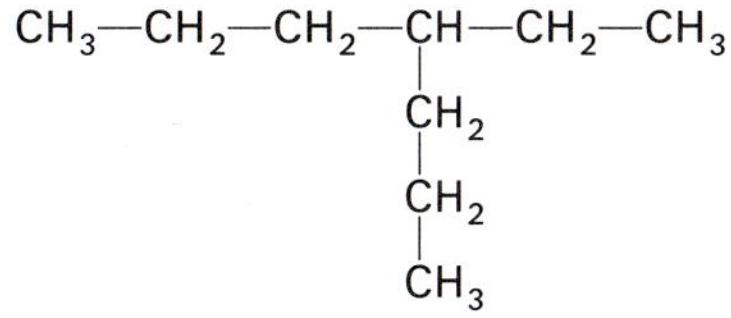

(h) $Cl-CH_2-CH_2-CH_2-CH_2-Cl$

14-4. Draw structural formulas for (and name) all the isomers of hexane (C_6H_{14}).

14-5. Draw structural formulas for (and name) the two cycloalkanes with molecular formula C_4H_8.

14-6. Draw structural formulas for (and name) the five cycloalkanes with molecular formula C_5H_{10}.

14-7. Draw the structural formulas for ethylcyclohexane in the two chair conformations. Which conformation is the more stable?

14-8. Draw the structural formula for bromocyclohexane in its most stable conformation. Label axial and equatorial bonds.

14-9. Draw structural formulas for *cis*- and *trans*-1,3-dibromocyclobutane.

14-10. The following are three representations of 1,2-dibromocyclohexane:

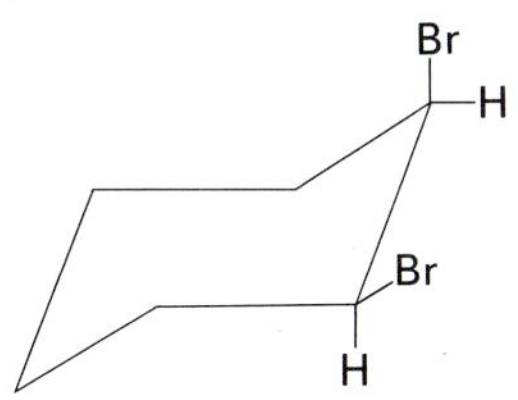

A

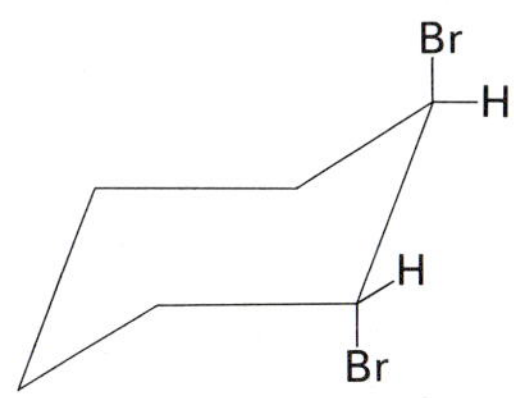

B

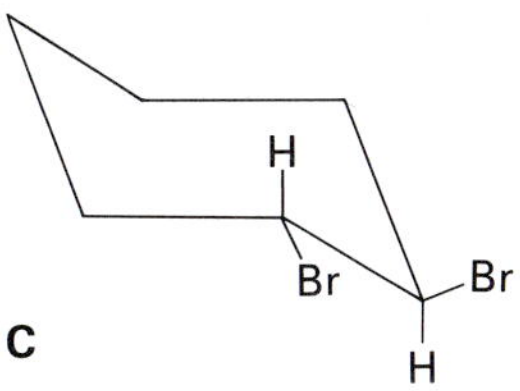

C

(a) Tell which pair or pairs are isomers.
(b) Tell which pair or pairs are different conformations of the same molecule.

14-11. Classify each of the following reactions as to general type (i.e., addition, elimination, or substitution):

(a) 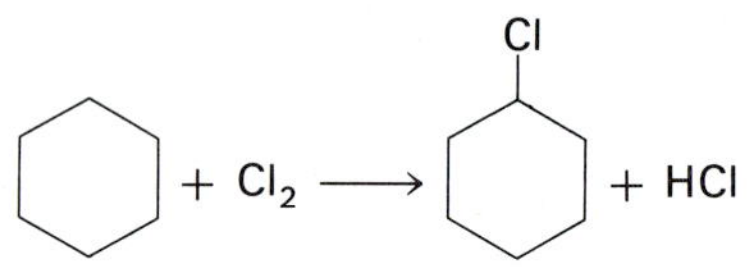

(b) $CH_3—C{\equiv}CH + HBr \longrightarrow CH_3—C(Br){=}CH_2$

(c) $CH_3—C(Br){=}CH_2 \longrightarrow CH_3—C{\equiv}CH + HBr$

14-12. Label and name the functional groups in each of the following molecules.

(a) $CH_3—CH_2—C{\equiv}CH$

(b) 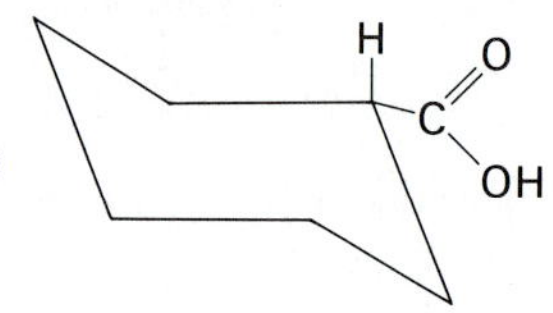

(c) $CH_3—CH(OH)—CH_3$

(d) $CH_2{=}CH—CH_2—CH_2—OH$

(e) 

(f) $CH_3—C(=O)—O—CH_2—CH_3$

(g) $CH_3CH_2OCH_2CH_3$

(h) 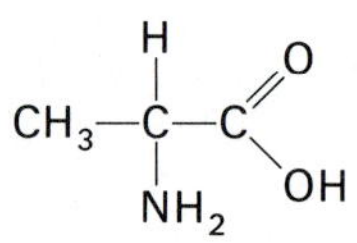

(i) CH_3SH

14-13. The following are structures of some molecules you will learn more about in Part IV of this book. Label and name all the functional groups in each.

(a) 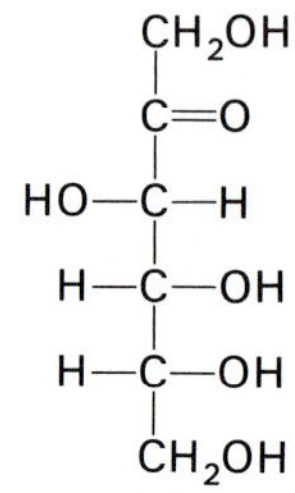

D-fructose
(a simple sugar)

(b) 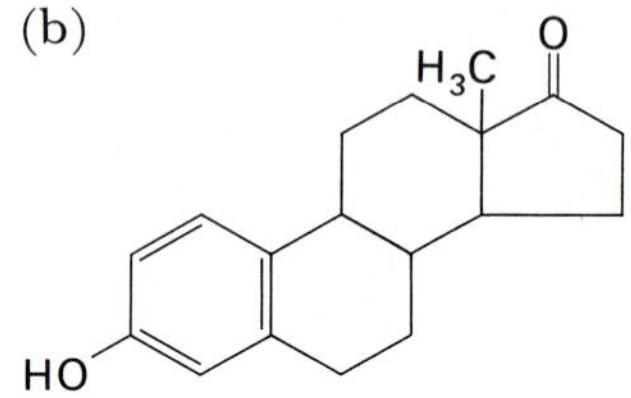

estrone
(a hormone)

(c) 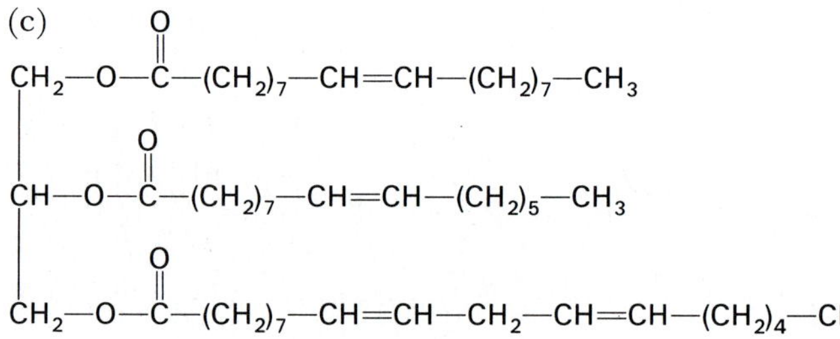

a typical fat

(d) 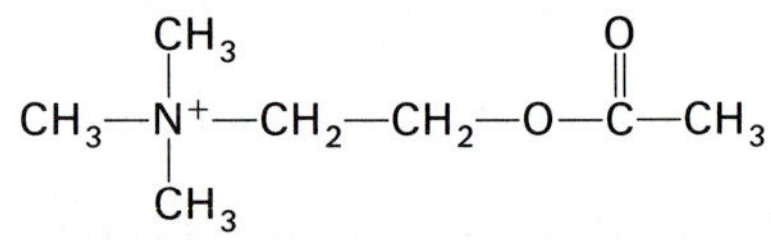

acetylcholine (important in the transmission of signals by the nerves)

(e) 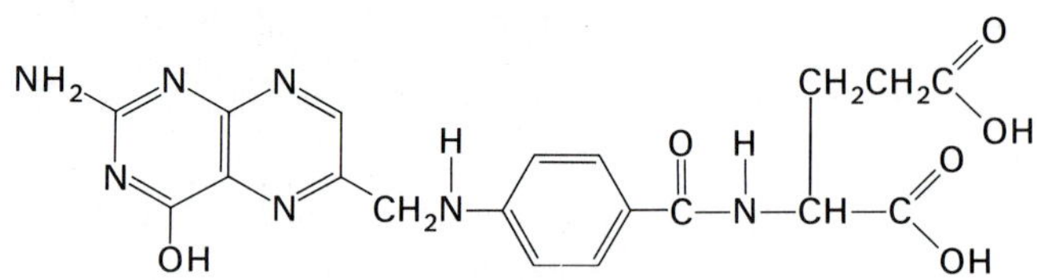

folic acid (a vitamin)

When bromine is added to an unsaturated oil, its red-brown color fades rapidly as it reacts.

15

Unsaturated Hydrocarbons and Aromatic Hydrocarbons

Hydrocarbons that contain one or more carbon–carbon double bonds are called alkenes; those with one or more triple bonds are alkynes. Aromatic hydrocarbons contain the benzene ring, a six-membered carbon ring in which the bonding is of a unique kind. Here we discuss the structures and properties of these compounds, together with two important reactions: addition to the carbon–carbon double or triple bond, and aromatic substitution.

15-1
STRUCTURE AND PROPERTIES OF ALKENES

Alkenes are hydrocarbons that have one or more carbon–carbon double bonds. Each double bond in an alkene reduces the number of hydrogen atoms in the molecule by two. An open-chain alkene with only one double bond has the general formula C_nH_{2n} (the same general formula as that of the cycloalkanes). For example, the simplest alkene is ethylene, C_2H_4. Ethylene was formerly called olefiant (oil-forming) gas because it reacts with chlorine to produce an oily liquid, and the alkenes are often called **olefins.** The alkenes are said to be **unsaturated,** because additional hydrogen atoms can be added to their structures.

As shown in Figure 15-1, the carbon–carbon double bond in ethylene is composed of one σ and one π bond. (See also Section 7-9.) Because the carbon atoms are in the sp^2 hybridization state, ethylene is a planar molecule, with bond angles close to 120°. We saw in Chapter 13 that the carbon–carbon double bond is shorter and stronger than the carbon–carbon single bond, but the π bond is weaker than the σ bond. It is the reactivity of the π bond that leads to the special characteristics of the alkene family. The functional group of the alkenes is the $>C{=}C<$ group. The carbon–carbon double bond often occurs as a functional group in compounds that contain other functional groups as well.

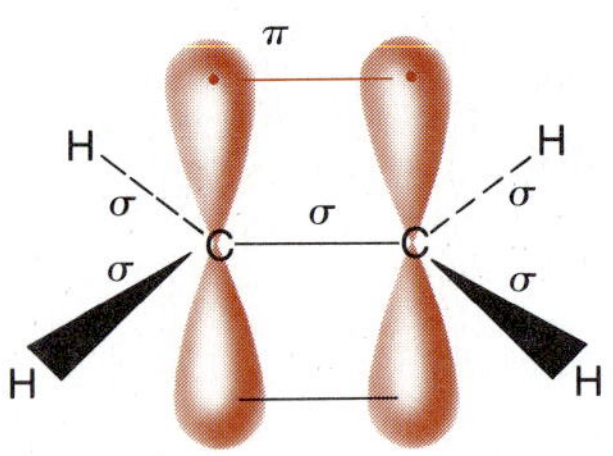

Bond length (nm)	
C–H	0.11
C=C	0.13
Bond angle	
H–C–H	117°
H–C=C	122°

FIGURE 15-1
The structure of ethylene. The σ bonds are formed from the sp^2 hybrid orbitals on the carbon atoms. The π bond is formed by the overlap of the unhybridized p orbitals.

The alkenes are similar to the alkanes in their physical properties—they are nonpolar compounds, are insoluble in water but soluble in nonpolar solvents, and are less dense than water. Ethylene (C_2H_4), propylene (C_3H_6), and the butenes (C_4H_8) are gases at room temperature and atmospheric pressure. Table 15-1 gives the physical properties of some alkenes.

Compounds with this functional group are widely distributed in nature, and their reactions are of great importance in organic and biological chemistry because they are used as starting materials, or occur as intermediates, in the synthesis of a wide variety of compounds. Structures of some naturally occurring molecules that contain one or more carbon–carbon double bonds are shown here:

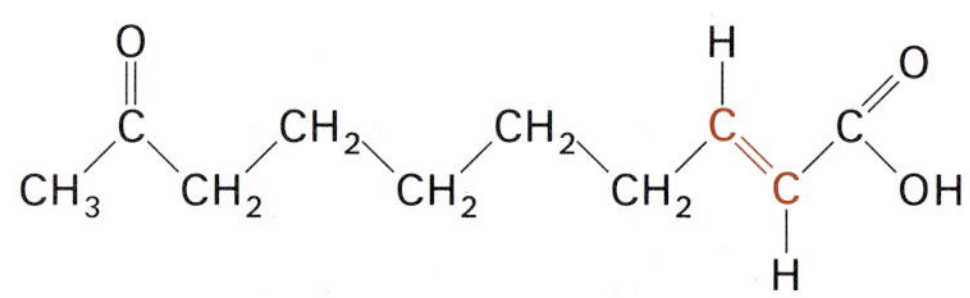

honeybee queen substance
(sex attractant)

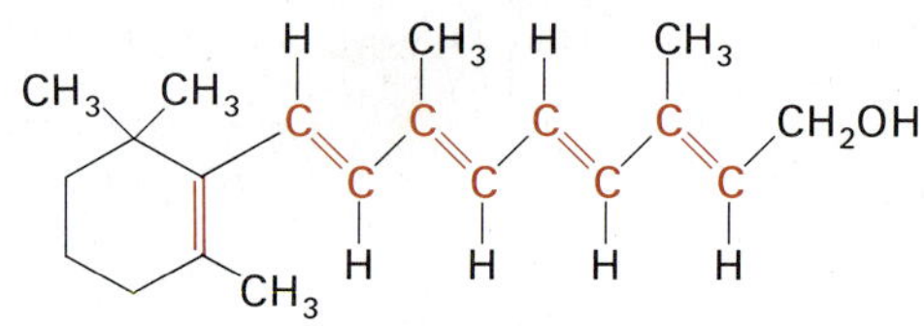

vitamin A

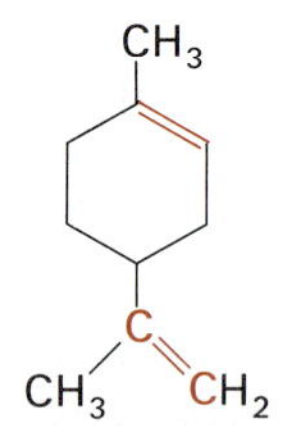

limonene
(lemon oil)

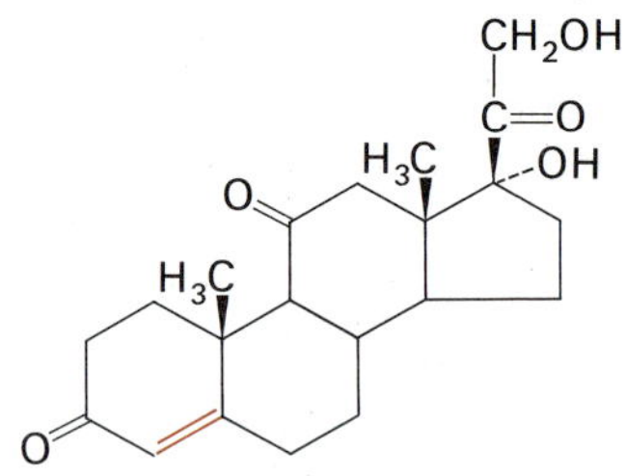

cortisone
(a steroid hormone)

TABLE 15-1
Physical properties of alkenes

Name	Structure	Boiling point (°C)	Density of liquid (g/cm^3 at 20°C)	State (at STP)
Ethylene	$CH_2{=}CH_2$	−104	——	Gas
Propylene	$CH_3{-}CH{=}CH_2$	−47	0.519	Gas
1-Butene	$CH_2{=}CH{-}CH_2{-}CH_3$	−6.3	0.595	Gas
cis-2-Butene	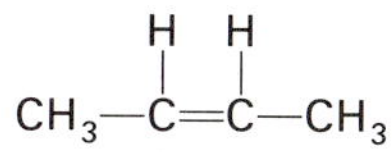	3.7	0.621	Gas
trans-2-Butene	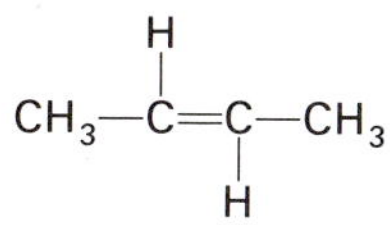	0.9	0.604	Gas
1-Pentene	$CH_2{=}CH{-}CH_2{-}CH_2{-}CH_3$	30.0	0.641	Liquid
cis-2-Pentene	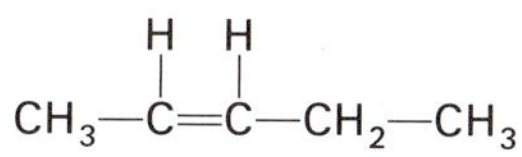	36.9	0.656	Liquid
trans-2-Pentene	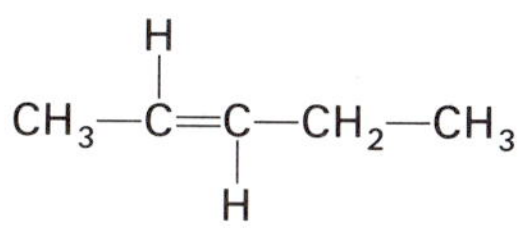	36.4	0.648	Liquid

15-2 NOMENCLATURE OF ALKENES

In the IUPAC system, the following rules are used in naming alkenes.

1. The names of the alkenes are obtained by changing the *-ane* suffix of the corresponding alkane to *-ene*:

$$\underset{\textbf{ethene}}{H_2C{=}CH_2} \qquad \underset{\textbf{propene}}{H_2C{=}CH{-}CH_3}$$

These compounds are more often called ethylene and propylene. The smaller members of each family of compounds were named long before the IUPAC convention was adopted, and the older names are used much more than the systematic ones.

2. When more than one position is possible for the double bond, the numbering begins at the end of the carbon chain closer to the double bond. A number is prefixed to the name to designate the *first* carbon of the double bond. The third member of the alkene series, butene, has *positional* isomers that differ in the location of the double bond, as do all higher alkenes:

$H_2C{=}CH{-}CH_2{-}CH_3$
1-butene

$CH_3{-}CH{=}CH{-}CH_3$
2-butene

$H_2C{=}CH{-}CH_2{-}CH_2{-}CH_3$
1-pentene

$CH_3{-}CH{=}CH{-}CH_2{-}CH_3$
2-pentene

$CH_3{-}CH_2{-}CH_2{-}CH{=}CH{-}CH_3$
2-hexene

(In Section 13-4 we saw that 2-butene and higher alkenes can also exist as *cis* and *trans* isomers.)

3. Branched alkenes are numbered and named according to the same rules as alkanes, but with the numbering always based on the longest unbranched chain *containing the double bond:*

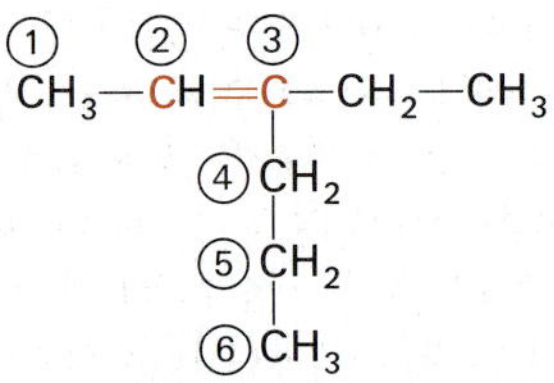

3-ethyl-2-hexene

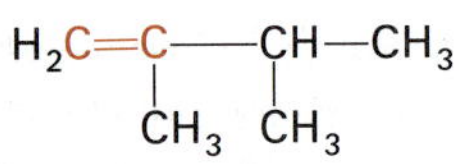

2,3-dimethyl-1-butene

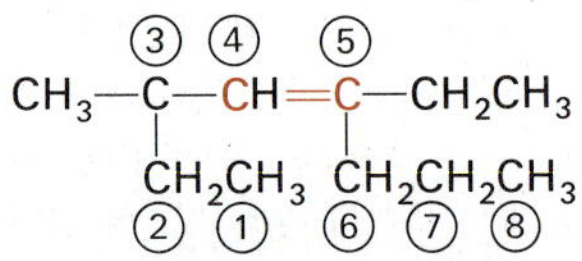

3-methyl-5-ethyl-4-octene

4. If an alkene molecule contains two double bonds, its name ends in *-diene*; for compounds with three double bonds, the ending is *-triene*; and so on:

$CH_2{=}CH{-}CH{=}CH_2$
1,3-butadiene

$CH_2{=}CH{-}CH{=}CH{-}CH{=}CH{-}CH_3$
1,3,5-heptatriene

The compound on the top is butadiene because it has an unbranched chain of four carbons containing two double bonds. The prefixed numbers 1 and 3 designate the first carbons of the double bonds. The compound on the bottom is heptatriene because its longest chain contains seven carbons and three double bonds. The prefixed numbers 1, 3, and 5 designate the first carbons of the double bonds.

EXERCISE 15-1 Write structural formulas for (and name) all six isomers of pentene (C_5H_{10}) that are alkenes. (Neglect the isomers that are cycloalkanes.)

15-3 PREPARATION OF ALKENES

Alkenes are prepared from petroleum in the process called "cracking," in which saturated hydrocarbons are heated in the absence of air, with or without catalysts, to yield smaller hydrocarbons:

$$CH_3CH_2CH_2CH_2CH_3 \xrightarrow{\text{heat}} CH_2{=}CH_2 + CH_3CH_2CH_3$$

This is the primary source of the tremendous quantities of ethylene and propylene used in the polymer industry.

Alkenes are also produced by elimination reactions such as the elimination of a molecule of water from an alcohol (**dehydration**):

$$\underset{\text{H}}{\text{CH}_2}\text{—}\underset{\text{OH}}{\text{CH}_2} \xrightleftharpoons{H_2SO_4} CH_2\text{=}CH_2 + H_2O$$

Concentrated sulfuric acid plays the role of catalyst. It also absorbs large amounts of water, and thereby (according to Le Châtelier's principle) drives the reaction to the right, favoring the production of the alkene.

Another method of preparing alkenes is by the elimination of HBr or HCl from an alkyl halide by reaction with a strong base:

$$CH_3\underset{\text{Br}}{\text{CH}}\underset{\text{H}}{\text{CH}_2} + NaOH \xrightleftharpoons{\text{alcohol}} CH_3CH\text{=}CH_2 + NaBr + H\text{—}OH$$

Here the alcohol is a solvent in which both the NaOH and the alkyl halide are soluble. We shall see the mechanism of this elimination reaction in Chapter 16.

15-4 REACTIONS OF ALKENES: HYDROGENATION

We have seen that alkanes are generally unreactive; the alkenes, on the contrary, with their $\rangle C\text{=}C\langle$ functional group, react readily with strong acids, halogens, some oxidizing and reducing agents, and even with water in the presence of an acid catalyst. *The characteristic reaction of alkenes is addition to the double bond,* and we shall later discuss the mechanism of these addition reactions.

In the process called *hydrogenation,* a hydrogen molecule adds to an unsaturated molecule to form a saturated compound. The reaction takes place in the presence of a catalyst such as finely divided platinum, or palladium, or Raney nickel (a finely subdivided form of nickel). The H_2 molecules bind to the surface of the metal catalyst. This weakens the H—H bond and catalyzes the addition of H_2 to the alkene:

$$\underset{\text{ethylene}}{CH_2\text{=}CH_2} + H_2 \xrightarrow{Pt} \underset{\text{ethane}}{\overset{\text{H}}{\text{CH}_2}\text{—}\overset{\text{H}}{\text{CH}_2}}$$

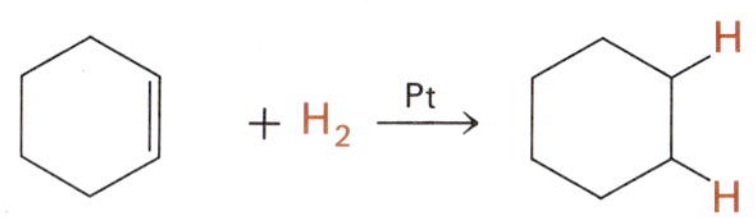

cyclohexene **cyclohexane**

Hydrogenation is commonly used to convert liquid oils (such as peanut, corn, or safflower oil) to margarines (solid fats consisting primarily of saturated compounds):

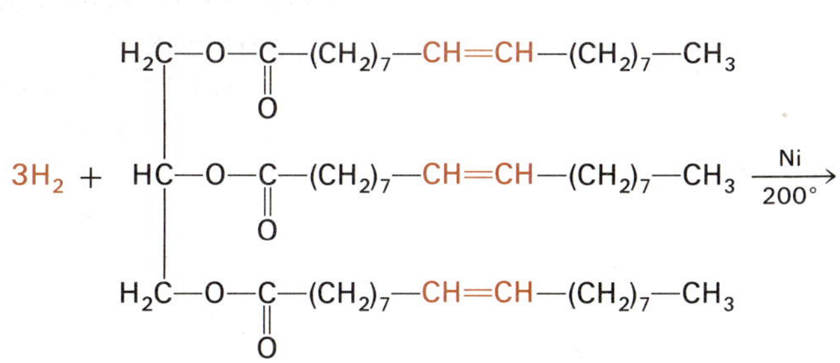

triolein (a liquid oil)

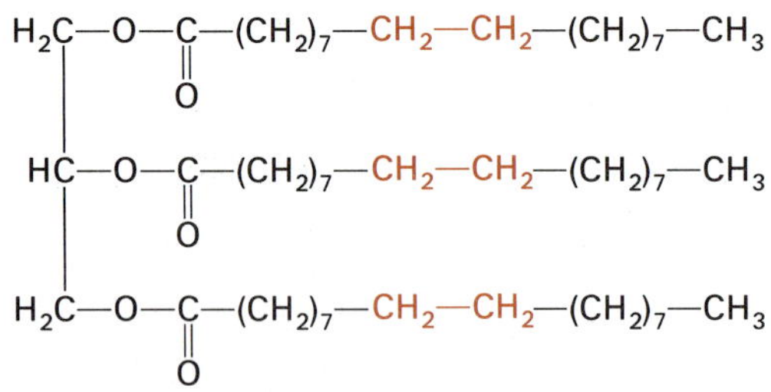

tristearin (a solid fat)

Fats (of which margarine is one) and oils have the same carbon skeletons. They differ only in the number of double bonds. Fats are the more saturated and therefore have higher melting points than the corresponding less-saturated oils. The reason for this difference in melting point is that the long hydrocarbon chains of saturated fats can assume a "straight-chain" conformation. This regular shape packs well to form a stable solid:

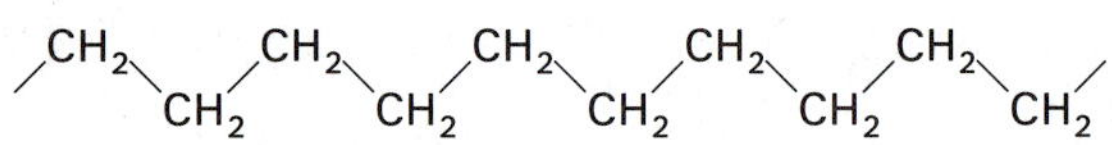

The unsaturated oils essentially always have a *cis* configuration about the double bond. This causes a bend in the hydrocarbon chain that interferes with a regular packing in the solid and results in a less stable solid with a lower melting point:

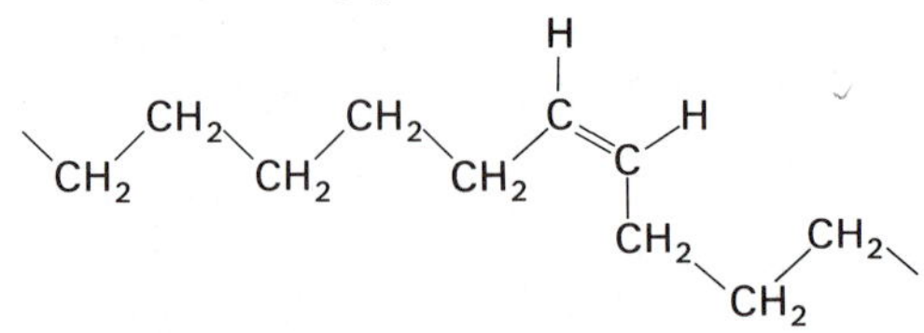

(A "*trans*" unsaturated hydrocarbon would not have such a sharp bend. Note the difference in the shapes of *cis*- and *trans*-retinal in Figure 13-7.)

A study has been made of fats in the copepod—a cold-blooded sea creature (like a tiny shrimp) that constitutes 70% of the animal protein in the sea. The fats of copepods that live in cold water have more double bonds than the fats of those that live in warm water. This has a real advantage for the copepod: cell membranes, which are composed of a kind of fat (see Section 24-5), must have a certain degree of fluidity in order to carry out their functions. The proper degree of fluidity is achieved with more-saturated (higher-melting) fats in copepods living in warm temperatures and less-saturated (lower-melting) fats in cold-water copepods.

15-5 ADDITION OF H—X: MARKOVNIKOV'S RULE

The hydrogen halides (HCl, HBr, and HI) react readily with alkenes by adding to the double bond to produce alkyl halides. The addition of HBr to ethylene is an example:

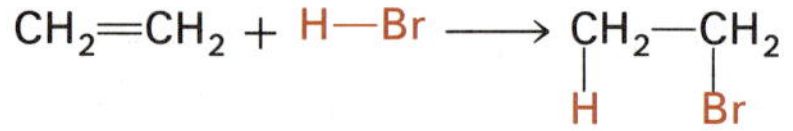

In this case, there is only one possible product, bromoethane, because the two carbon atoms are equivalent. (Note that in this reaction H—X is H—Br.)

From the addition of HBr to 1-butene, however, we might expect two products—1-bromobutane if the bromine adds to carbon-1, and 2-bromobutane if the bromine adds to carbon-2:

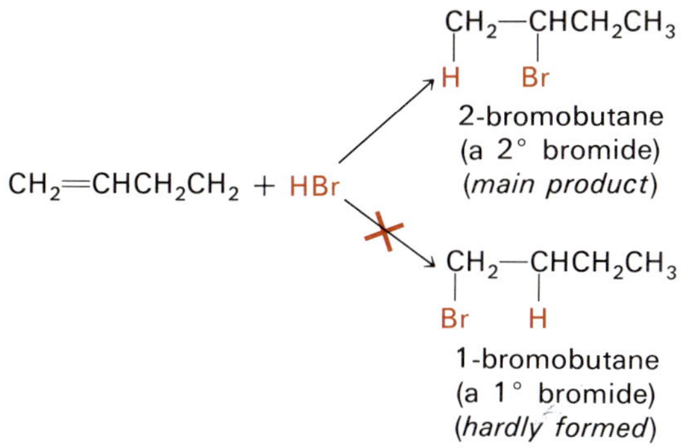

(15-1)

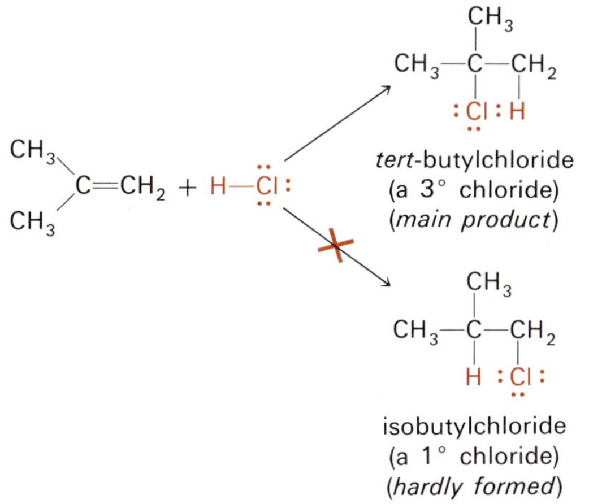

(15-2)

However, when this reaction is carried out, only *one* product (2-bromobutane) is formed in significant quantity. The other product (1-bromobutane) is formed in much smaller amount. Note that 2-bromobutane) is called a secondary (2°) bromide because the carbon attached to the bromine is bonded to *two* other carbons; and 1-bromobutane is a primary (1°) bromide because the carbon attached to the bromine is bonded to only *one* other carbon.

Similarly, when HCl is added to methylpropene, only one of the two possible products is formed in significant amount, *tert*-butylchloride. See Equation 15-2 at the top of the next column.

The results of these and similar experiments are summarized in **Markovnikov's rule,** first stated on empirical grounds in 1871. This rule predicts the *direction* of addition to carbon–carbon double and triple bonds. *In the addition of H—X to an alkene or an alkyne (a compound with a carbon–carbon triple bond), the hydrogen will bond to the carbon atom that already bears the greater number of hydrogen atoms.* Both reactions 15-1 and 15-2 illustrate this rule; in each case, the product formed in greater amount is the one in which the hydrogen adds to the carbon already bearing the greater number of hydrogen atoms.

EXERCISE 15-2 Write the structural formula of the product of the addition of HI to (a) l-butene; (b) propene.

15-6 REACTION MECHANISMS

Before we try to understand the mechanisms of the reactions summarized in Markovnikov's rule, we need to review and extend our ideas about reaction mechanisms beyond those introduced in Chapter 10. Figure 15-2a is the energy diagram for an elementary (one-step) process:

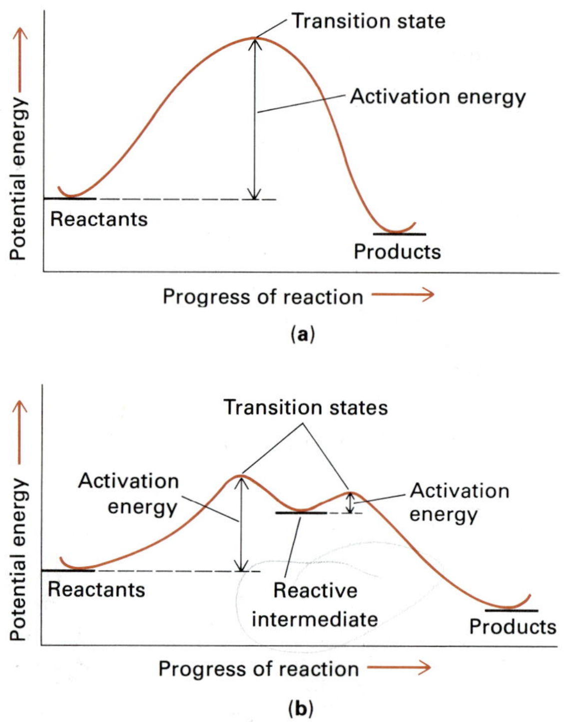

FIGURE 15-2

Energy diagrams. **(a)** A one-step reaction. **(b)** A two-step reaction.

$$\text{Reactants} \rightleftharpoons \text{Products}$$

The activation energy, shown in the figure, is the energy that must be added to the reactants to get them to the top of the energy barrier (see Section 10-11). Figure 15-2b is a diagram for a two-step process:

$$\text{Reactants} \rightleftharpoons \text{Intermediates} \rightleftharpoons \text{Products}$$

The **reactive intermediate,** shown in the diagram as a "well" between two energy "barriers," is a real chemical substance that can be observed (and sometimes isolated) but generally is unstable. A carbonium ion is an example of a reactive intermediate; we shall see other specific examples shortly.

A **transition state** is the chemical species at the point of highest energy for any *elementary* process, and the **activation energy** is the energy required to go from reactants to the transition state. A transition state (in contrast to a reactive intermediate) does *not* correspond to a real chemical substance that can be observed or isolated, but instead represents a fleeting state in which some bonds are partly formed, and others partly broken. Because it corresponds to an energy peak, it will spontaneously change to the lower-energy reactants or products. In later chapters, we shall see examples of likely transition states—inferred from the structures of the reactants and the products, and from studies of the rates of reactions. Figure 15-2b shows that there are two transition states for a two-step process, one for each step of the reaction. Each step has its own activation energy.

We saw in Section 10-11 that each step in a multistep process has its own rate. Just as in an assembly line, one slow step holds up the entire process. Because only molecules that have enough energy to get over the activation energy barrier can react, the overall rate of any reaction is determined by the transition state of highest energy—it is the energy *maximum* for the entire process that is important, just as it is the height of the *highest* mountain that is important when one flies over a mountain range.

15-7 MECHANISM OF ADDITION REACTIONS

Markovnikov's rule *summarizes* the experimental facts, but it does not *explain* them. How can we explain these facts? We must begin by looking at a mechanism for the addition of H—X to a carbon–carbon double bond. The generally accepted mechanism is one in which the first step, which is found to be the rate-determining step, is a reaction of the acid–base type between a proton (H^+) and the π bond of the alkene. The π bond acts as a Lewis base, donating an electron pair to the proton. One carbon of the double bond is then left with only six electrons around it. It is a carbonium ion (Section 13-2). This first step in the addition of hydrogen bromide to 1-butene could produce either (or both) of two reactive intermediates, both of which are carbonium ions:

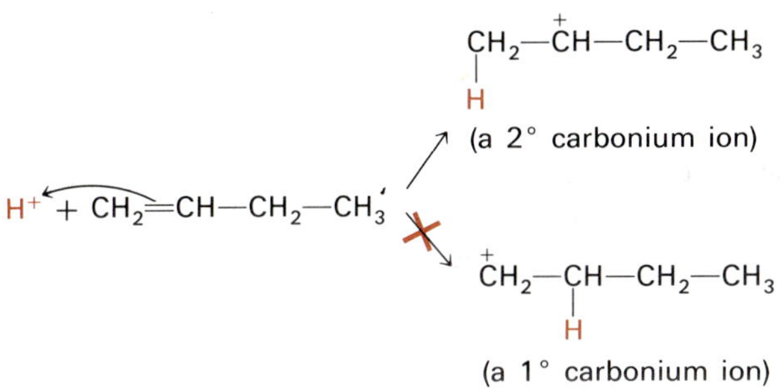

(15-3)

(In carbonium ions, the carbon atom with the positive charge is the one that is bonded to only three other atoms, and so has only six electrons instead of eight, in its valence orbitals. In a secondary carbonium ion, the carbon with the positive charge is bonded to *two* other carbon atoms; and in a primary carbonium ion, the carbon with the positive charge is bonded to *one* other carbon.) We saw in Equation 15-1 that the principal product of this reaction is 2-bromobutane, so we conclude (though we have not yet seen why this should be so) that in the reaction of Equation 15-3, the secondary carbonium ions are produced in much greater amounts than the primary carbonium ion.

The negative bromide ion then adds, in a fast step, to the positive carbon of the secondary carbonium ion. The two-step reaction mechanism is summarized in two equations:

Step 1: Addition of H^+

$$H^+ + CH_2{=}CH{-}CH_2{-}CH_3 \xrightarrow{\text{slow}} CH_2(H){-}\overset{+}{C}H{-}CH_2{-}CH_3 \qquad (15\text{-}4)$$

Step 2: Addition of Br^-

$$:\!\ddot{\underset{..}{Br}}\!:^- + CH_2(H){-}\overset{+}{C}H{-}CH_2{-}CH_3 \xrightarrow{\text{fast}} CH_2(H){-}CH(:\!\ddot{Br}\!:){-}CH_2{-}CH_3 \qquad (15\text{-}5)$$

The net (or overall) reaction is the *sum* of these two steps, and is just the addition of H—Br in accord with Markovnikov's rule:

$$CH_2{=}CH{-}CH_2{-}CH_3 + H^+ + :\!\ddot{\underset{..}{Br}}\!:^- \longrightarrow CH_2(H){-}CH(:\!\ddot{Br}\!:){-}CH_2{-}CH_3$$

The carbonium ion is a reactive intermediate, produced in the first step but consumed in the second, so it does not appear in the net equation. The other possible product (1-bromobutane) is obtained in much smaller quantity because the primary carbonium ion (the reactive intermediate that would yield it) is formed in much smaller quantity. We conclude that secondary carbonium ions are more stable (of *lower* energy) than primary carbonium ions, and so the secondary carbonium ion is formed in preference to the primary. The existence and the relative stabilities of carbonium ions were inferred from such experiments long before there was direct experimental evidence of their existence.

In Equations 15-4 and 15-5, curved arrows are used to show the movement of electron pairs: the tail of the arrow indicates the origin of the electron pair, and the head of the arrow shows where the electrons are going. The use of these curved arrows, called **arrow formalism**, helps us to follow the mechanism of a reaction by showing the movement of electrons. The arrow represents the movement of a *pair* of electrons. In Equation 15-4, it shows an electron pair from the π bond forming a new bond with the proton. In Equation 15-5, the arrow represents the movement of a lone pair of electrons; a lone pair from the bromide ion is donated to the positive carbon of the carbonium ion. We shall use the arrow formalism to follow the mechanisms of reactions throughout our study of organic and biological chemistry. You will find it helpful to use arrows in working out reaction mechanisms for yourself.

As another example, we next consider the mechanism of the reaction of Equation 15-2, the addition of hydrogen chloride to methylpropene. The two reactive intermediates that could be formed are

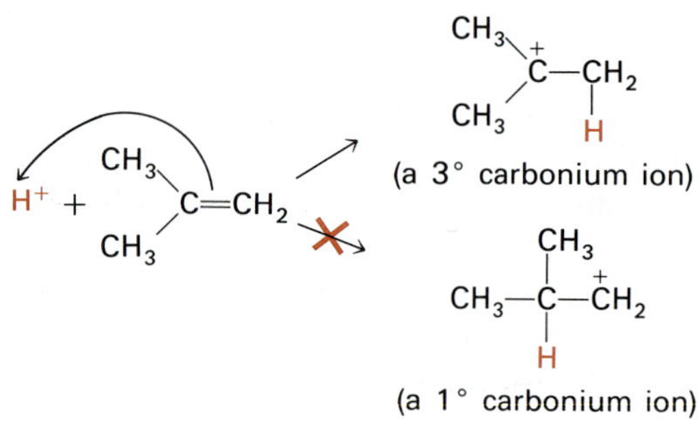

But we saw in Equation 15-2 that the only product is *tert*-butylchloride. This product is formed by reaction between the tertiary carbonium ion and the chloride ion. The mechanism for the reaction that occurs is summarized as follows:

Step 1: Addition of H^+

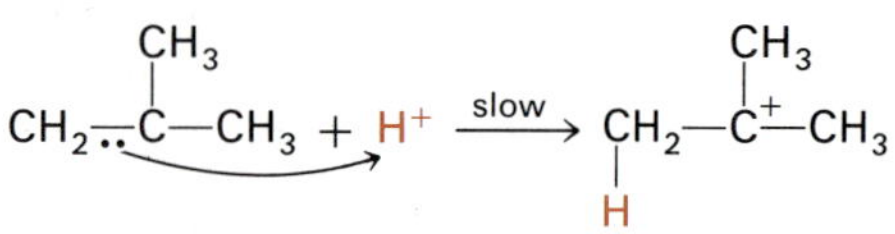

Step 2: Addition of Cl^-

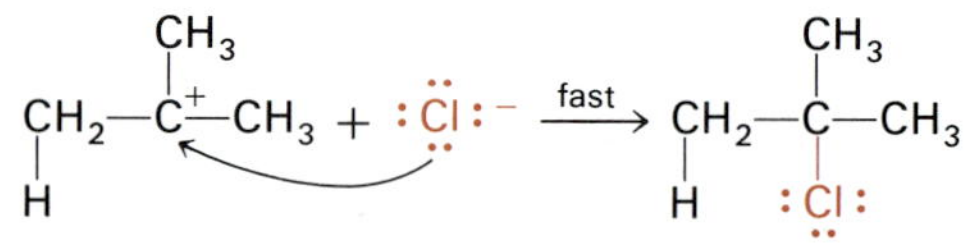

The *net* reaction (the sum of these two steps) is just the addition of HCl to an alkene, in accord with Markovnikov's rule:

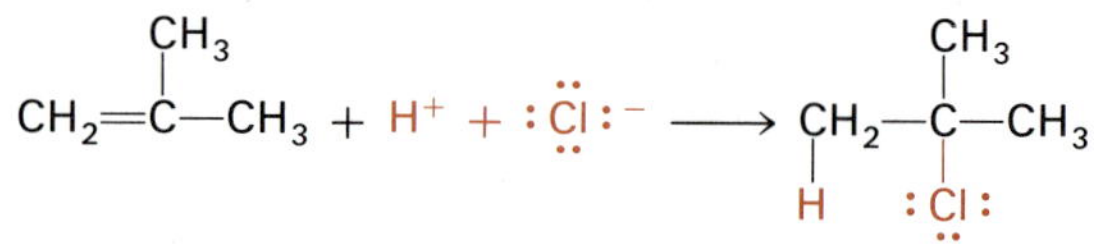

The other possible product (isobutyl chloride) is not formed because the primary carbonium ion that would yield it is not formed. We conclude that the tertiary carbonium ion is more stable (of lower energy) than the primary carbonium ion, and so the tertiary carbonium ion is formed in preference to the primary.

These and other experiments involving carbonium ions show that the stability of carbonium ions, in order of decreasing stability, is

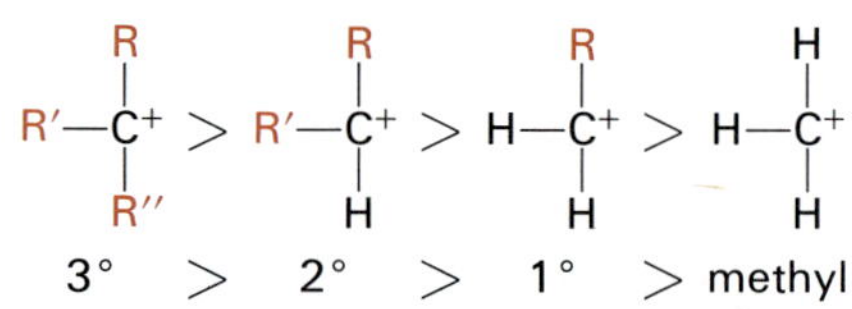

The tertiary carbonium ion is the most stable, and the methyl carbonium ion is the least stable.

We can explain this order by considering the effect of replacing a hydrogen atom of a secondary carbonium ion with a methyl group:

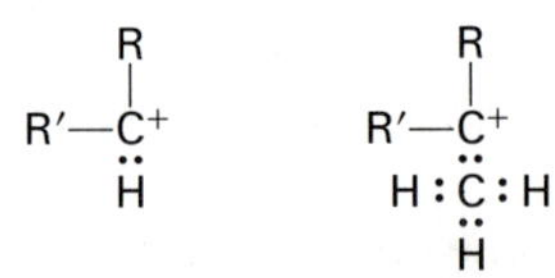

The positively charged carbon attracts toward itself the electrons of the neighboring groups, so there is a migration of electrons through the adjacent σ bonds of the methyl group toward the central positive charge. This partly neutralizes the charge on *that* carbon but, at the same time, leaves a *partial* positive charge spread over the atoms of the methyl group that have lost part of their share in the electrons. The positive charge is thus delocalized over several atoms. The methyl group, with eight electrons around it, is more polarizable (Section 7-10) than the hydrogen atom, which has only two. A charge concentrated in a small volume is less stable than the same charge spread through a larger volume. Thus, the migration of electrons from alkyl groups toward the positively charged carbon *stabilizes* (by diffusing, or making less concentrated) the positive charge on the carbonium ion. This accounts for the fact that the tertiary carbonium ion is the more stable of the two ions. This reasoning can be extended to explain the relative order of stabilities of carbonium ions.

EXERCISE 15-3 Write equations that show the mechanism of the addition of hydrogen bromide to propene. Use the arrow formalism.

15-8 HYDRATION OF ALKENES

The **hydration** of (addition of water to) alkenes in acid solution also follows Markovnikov's rule. The product of the reaction is an alcohol:

$$CH_2{=}\underset{}{\overset{CH_3}{\overset{|}{C}}}{-}CH_3 + H{-}\ddot{\underset{\cdot\cdot}{O}}{-}H \underset{}{\overset{H^+}{\rightleftharpoons}} \underset{H}{\underset{|}{CH_2}}{-}\underset{:\ddot{O}-H}{\underset{|}{\overset{CH_3}{\overset{|}{C}}}}{-}CH_3 \qquad (15\text{-}6)$$

In this addition of $H{-}\ddot{\underset{\cdot\cdot}{O}}{-}H$ (the way we shall write H_2O when we wish to understand how it reacts), the hydrogen adds to the double-bonded carbon that already bears more hydrogen atoms, and the hydroxide adds to the other double-bonded carbon. This reaction is the reverse of one we saw earlier in the chapter, in which alkenes are prepared by the *dehydration* of alcohols. The position of equilibrium—how much of the alcohol, and how much of the alkene, are present at equilibrium—depends to a great extent on the amount of water present. According to Le Châtelier's principle, a large amount of water favors the production of the alcohol.

Although, of course, *all* reactions reach a state of equilibrium—with reactants and products both present, and with forward and reverse reactions occurring at equal rates—we often are interested primarily in the forward reaction (Reactants $\rightarrow$ Products) and not in the reverse reaction (Products $\rightarrow$ Reactants). Thus, we often omit the arrow that signifies the reverse reaction, showing only Reactants $\rightarrow$ Products. Where we are concerned with the conditions under which a reaction can be made to go in one direction or the other, we include arrows for both forward and reverse reactions. But, whether or not we use both arrows, *all reactions are reversible.*

The mechanism for the acid-catalyzed addition of water to alkenes consists of the following three steps:

Step 1: Addition of H^+

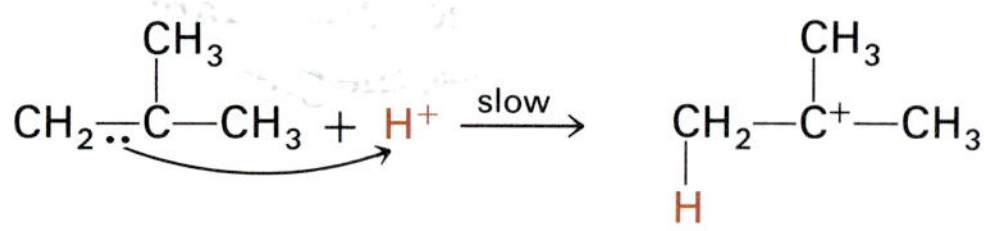

(a 3° carbonium ion)

Step 2: Addition of H_2O

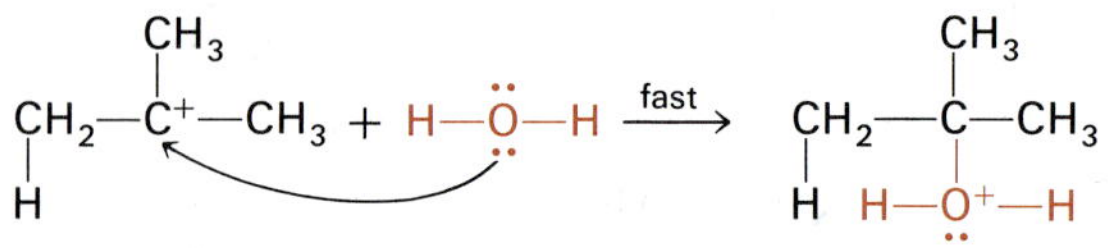

(an oxonium ion)

Step 3: Elimination of H^+

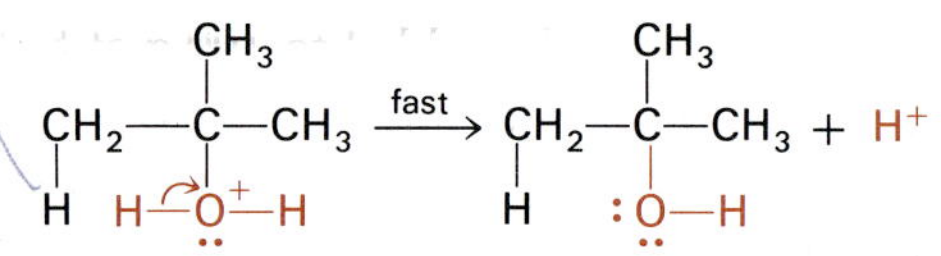

In the first step, a proton from the acid catalyst is added to the double bond to produce a tertiary carbonium ion.

In the second step, a lone pair of electrons from the oxygen of a water molecule bonds to the positively charged carbon of the carbonium ion. The oxygen provides *both* of the electrons for the carbon–oxygen bond, so it has, in effect, lost a half-share of those electrons. In the process, it acquires a positive charge. The product is called an **oxonium ion**—an ion in which the oxygen bears a positive charge.

In the final step of the reaction, a proton leaves the oxonium ion, yielding an alcohol and regenerating the catalyst. The *net* reaction (the sum of the three steps) is just the addition of one molecule of water to one molecule of the alkene (Equation 15-6). The two reactive intermediates (the carbonium ion and the oxonium ion) and the catalyst (the proton) do not appear in the net equation.

The mechanism of this reaction is almost identical to that we saw earlier for the addition of H—X to an alkene. Because H—X is a strong acid that itself provides the proton for the slow first step, there is no need to provide an acid catalyst for the addition of H—X as there is for the addition of H—$\ddot{O}$—H, which is a weak acid.

EXERCISE 15-4 Draw the structural formula of the product of the hydration of 1-butene.

15-9 HALOGENATION OF ALKENES

We have seen that bromine and chlorine can be *substituted* for hydrogen atoms in alkanes, but only through the action of heat or light. With alkenes, however, bromine or chlorine can readily *add* to the double bond:

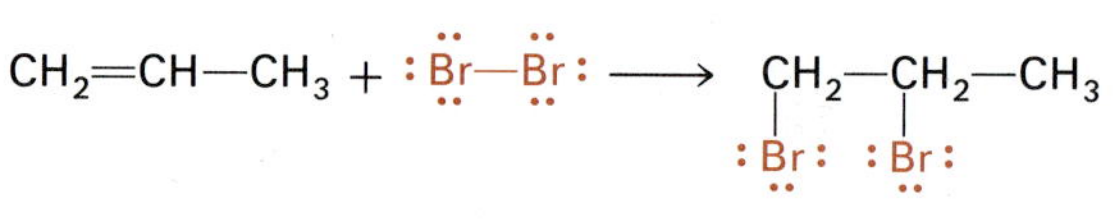

(15-7)

Iodine generally is not reactive enough to add to alkenes, and fluorine reacts explosively.

The generally accepted mechanism consists of two steps, with the formation of a carbonium ion as a reactive intermediate. The first step depends on the breakup of bromine molecules to form ions:

Step 1: Formation of Br^+

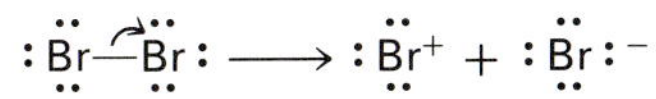

Step 2: Addition of Br^+

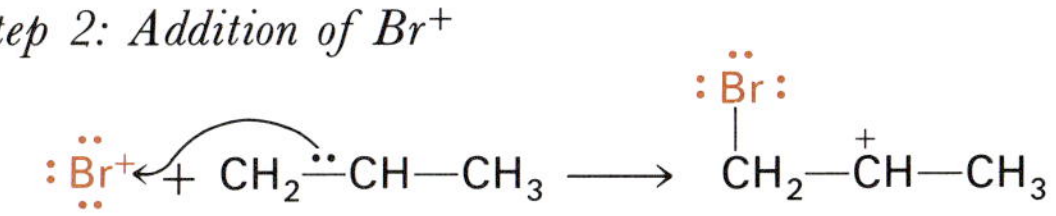

Step 3: Addition of Br^-

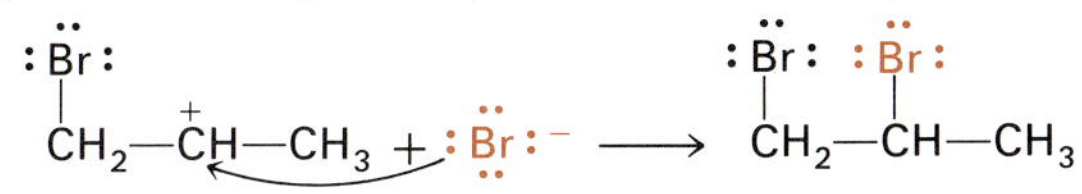

(It is more accurate to say that the bromine molecule becomes polarized to yield $^{\delta+}Br—Br^{\delta-}$ as it approaches the π bond, and that the positive end then adds to the π bond to yield the carbonium ion and the bromide ion, Br^-. The view given in the proposed mechanism, however, is simpler, and it is adequate for our purposes.)

In the second step, Br^+ adds to the double bond, forming a secondary carbonium ion. (The other possible carbonium ion, the primary one, is not formed in significant amount.) Finally, the bromide ion donates a pair of electrons to form a new bond with the positively charged carbon of the carbonium ion. The *net* reaction is just the addition of Br_2 to the double bond (Equation 15-7).

EXERCISE 15-5 Draw the structural formula of the carbonium ion formed as a reactive intermediate in the chlorination of 2-methyl-2-butene.

The addition of bromine can be used as a test for unsaturated oils. If a few drops of red-brown liquid bromine are added to safflower oil (a highly unsaturated oil), the color disappears instantly. However, if a few drops of bromine are added to highly saturated fats such as those in butter, the color fades only slowly. The degree of unsaturation of an oil can be determined by measuring the amount of bromine that a fixed amount of oil will decolorize rapidly—the more bromine the oil decolorizes, the greater its degree of unsaturation.

In fat and oil chemistry, the degree of unsaturation is measured by iodine number: the number of grams of iodine that could be added to 100 g of the fat or oil. The analysis is not actually done with iodine because I_2 is not reactive enough. Iodine chloride (ICl), iodine bromide (IBr), or bromine (Br_2) are used instead, and the calculations are made as if iodine had been used. A completely saturated fat has iodine number zero (no I_2 could be added). Butter, which is composed of highly saturated fats, has an iodine number of about 30—that is, 30 g of iodine could be added to 100 g of butter. Soybean oil, which is highly unsaturated, has an iodine number of about 130. Table 15-2 lists the iodine numbers and solidification points of some fats and oils. The solidification point is the temperature at which the oil or melted fat solidifies as it cools. As we would expect, the highly unsaturated oils (those with high iodine numbers) solidify at lower temperatures than the more saturated fats. We shall learn more about fats and oils in Chapters 19 and 24.

TABLE 15-2
Iodine numbers and solidification points of fats and oils

Name	Iodine number	Solidification point (°C)
Butter fat	26 to 38	20 to 23
Chicken fat	66 to 72	22 to 27
Coconut oil	66 to 100	14 to 22
Cod liver oil	137 to 166	−3
Corn oil	111 to 128	−10 to −20
Human fat	57 to 73	15
Linseed oil	175 to 202	−19 to −27
Olive oil	79 to 88	+2 to −6
Safflower oil	122 to 141	−13 to −18
Sunflower oil	129 to 136	−17
Whale oil	90 to 146	−2 to 0

15-10
ADDITION POLYMERS

We have seen that halogens, hydrogen, hydrogen halides, and water can be added to double bonds. Giant molecules called **polymers** (from the Greek words for *many parts*) are formed when alkene molecules add to the double bonds of each other. Polystyrene and polyethylene are examples, and we shall soon see how they are made. Other polymers (such as Dacron and Nylon) are synthesized in a different process (discussed in Sections 19-10 and 20-7). And still other polymers such as wool, silk, and rubber occur naturally. Other natural polymers—starch and cellulose, proteins, and nucleic acids—are at the heart of biochemistry and are discussed in Part IV of this book.

The individual small molecules—the building blocks—of which polymers are constructed are called **monomers.** A typical molecule of the polymer polyethylene may be composed of a thousand ethylene monomers:

$$\underset{\text{ethylene}}{n\mathrm{H_2C{=}CH_2}} \longrightarrow \underset{\text{polyethylene}}{\mathrm{-(CH_2{-}CH_2{-})}_n}$$

where n may be 1000.

In **addition polymerization,** alkene monomers combine to form long chains, and *every* atom of the monomer is incorporated into the polymer. There are several ways to start the polymerization. One of the simplest in concept is the use of a very reactive free radical—an atom or group of atoms with an unpaired electron. The presence of an unpaired electron makes the free radical very reactive. Examples of free radicals are the chlorine atom (Cl) and the methyl radical (H_3C, methane with a hydrogen atom removed). Free-radical-initiated polymerization is used to produce billions of pounds of polyethylene annually in the United States.

We use the symbol X· to represent the free-radical initiator of the reaction. The dot represents the unpaired electron. The identity of X· is not important. The polymerization is carried out at a temperature of over 100°C and at a pressure of over 1000 atm of ethylene gas. The first step is an attack by X· on the π electrons of ethylene:

$$\mathrm{X\cdot + CH_2\overset{..}{=}CH_2 \longrightarrow X\cdot\cdot CH_2{-}\dot{C}H_2}$$

The product is a new free radical that can attack *another* ethylene molecule, generating still another free radical:

$$\mathrm{X\cdot\cdot CH_2{-}\dot{C}H_2 + CH_2\overset{..}{=}CH_2 \longrightarrow X\cdot\cdot CH_2{-}CH_2\cdot\cdot CH_2{-}\dot{C}H_2}$$

This step is repeated many times:

$$\mathrm{X\cdot\cdot CH_2{-}CH_2\cdot\cdot CH_2{-}\dot{C}H_2 + CH_2\overset{..}{=}CH_2 + CH_2\overset{..}{=}CH_2 + CH_2\overset{..}{=}CH_2 \longrightarrow}$$
$$\mathrm{X\cdot\cdot CH_2{-}CH_2\cdot\cdot CH_2{-}CH_2\cdot\cdot CH_2{-}CH_2\cdot\cdot CH_2{-}CH_2\cdot\cdot CH_2{-}CH_2\cdot\cdot CH_2{-}\dot{C}H_2}$$

At each step, a free radical is produced as well as consumed, so the reaction can continue in this way indefinitely; such a reaction is called a chain reaction. Finally a chain-terminating reaction, such as the combination of two chains, occurs:

$$\mathrm{X{-}(CH_2{-}CH_2{-})_m CH_2{-}\dot{C}H_2 + \dot{C}H_2{-}CH_2{-}(CH_2{-}CH_2{-})_n X \longrightarrow}$$
$$\mathrm{X{-}(CH_2{-}CH_2{-})_m CH_2{-}CH_2{\cdot}{\cdot}CH_2{-}CH_2{-}(CH_2{-}CH_2{-})_n X}$$

Both m and n may be 100 to 1000 or larger, so that the number of terminal free radical groups (X) is negligible, and there is no reference to them in the name of the product. As we noted earlier, the number of ethylene monomers in polyethylene is of the order of 1000. In any synthetic polymer sample, the molecules are of different lengths and thus are not all of the same molecular weight. Only an average molecular weight can be determined for a synthetic polymer. The same is true of the natural polymers starch, glycogen, and rubber. Other natural polymers—for example, proteins (Chapter 22) and nucleic acids (Chapter 25)—can have a precise molecular weight. Table 15-3 lists some synthetic addition polymers.

The polyethylene produced by the free-radical-initiated polymerization does not have the simple linear structure implied so far. Branches consisting of ethyl and butyl groups are known to occur along the chain.

TABLE 15-3
Addition polymers

Monomer name and structure	Polymer name	Uses
Ethylene $CH_2{=}CH_2$	Polyethylene	Films, tubing, molded objects
Propylene $CH_2{=}CHCH_3$	Polypropylene	Fibers for carpeting and clothes, heart valve leaflets, knitted surgical mesh, hip-joint repair
Styrene $CH_2{=}CH{-}$ 	Polystyrene	Styrofoam insulation, molded objects
Vinyl chloride $CH_2{=}CHCl$	Polyvinyl chloride (PVC)	Records, floor coverings
Vinylidene chloride $CH_2{=}CCl_2$	Polyvinylidene chloride	Saran wrap
Acrylonitrile $CH_2{=}CH{-}C{\equiv}N$	Polyacrylonitrile	Fibers—Orlon and Acrilan
Tetrafluoroethylene $CF_2{=}CF_2$	Polytetrafluoroethylene (Teflon)	Chemically resistant films, electrical insulation, housing for heart pacemakers, artificial heart valves
Vinyl acetate $CH_2{=}CH{-}O{-}C({=}O){-}CH_3$	Polyvinyl acetate	Chewing gum, adhesives, fabric coatings
Methyl methacrylate $CH_2{=}C(CH_3){-}C({=}O){-}O{-}CH_3$	Polymethyl methacrylate	Transparent sheets and rods—Plexiglas, Lucite

A true linear polyethylene without branches is produced in a different process, using a special catalyst of trialkyl aluminum (R_3Al) and titanium tetrachloride ($TiCl_4$). Because the unbranched chains of linear polyethylene can lie side by side in a regular arrangement, linear polyethylene is more dense and has a higher melting point than branched polyethylene. Familiar household articles made of polyethylene include refrigerator containers and pails. In the laboratory, squeeze bottles, disposable syringes, and tubing commonly are polyethylene; so are such hospital articles as catheters and bags to hold intravenous solutions. Linear polyethylene is used for artificial sockets in the repair of ball-and-socket joints such as the knee.

Polypropylene is the addition polymer of the monomer propylene:

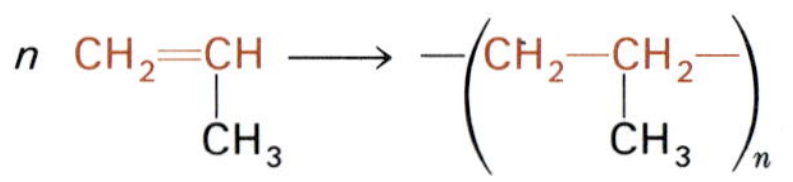

Note that the backbone of the polymer consists of the carbons that are joined by the double bond in the monomer. The CH_3 groups are branches from the main polymer chain. Polypropylene is used to make fibers for carpeting, clothing, and rope. It makes an abrasion-resistant carpet backing. Its uses in medicine include hip-joint repair.

EXERCISE 15-6 Write the structure of polymethylmethacrylate (Lucite or Plexiglas). Include at least three monomers in the structure. The monomer, methyl methacrylate, has the structure

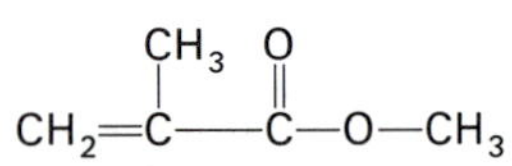

Remember that the backbone of the polymer consists of the double-bonded carbons of the monomer.

EXERCISE 15-7 In a free-radical-initiated addition polymerization, the average length of the chains can be controlled to some extent by varying the concentration of the initiator. A high concentration of initiator leads to short chains, and a low concentration of initiator leads to long chains. Explain why this is true.

Natural rubber (obtained by drying the juice of rubber trees and plants) is a polymer of isoprene. The monomer contains two double bonds:

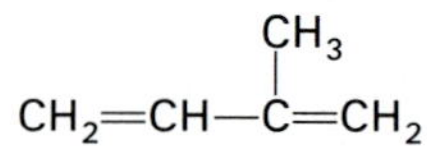

isoprene

One double bond per monomer unit remains after polymerization. A sample of rubber might contain molecules composed of from 1000 to 5000 units of isoprene, with a molecular weight of about 200,000. "Natural" rubber is the all-*cis* isomer of polyisoprene:

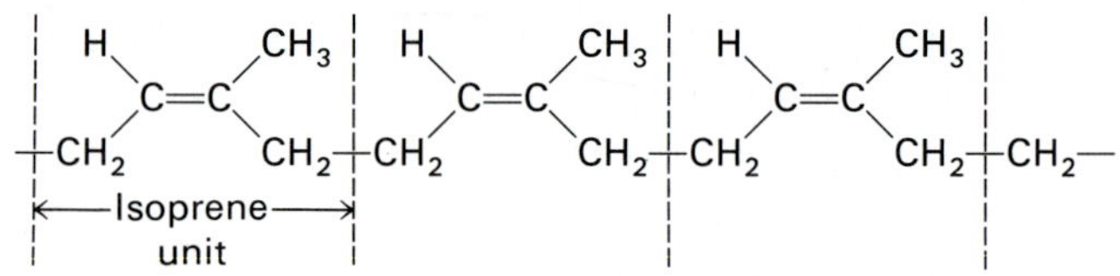

"natural" rubber (*cis*-polyisoprene)
(average molecular weight ≅ 200,000)

The all-*cis* configuration about the double bonds has a critical effect on the properties of rubber. The *trans* isomer, gutta percha (which is obtained from the sap of various Malaysian trees), is much harder and more brittle than rubber:

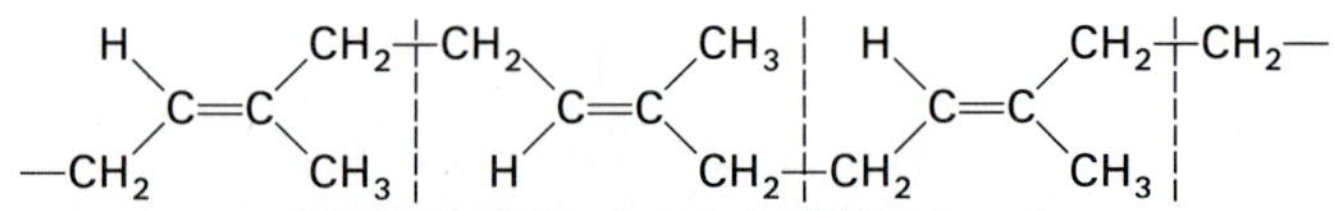

gutta percha (*trans*-polyisoprene)

It was not until 1955 that a synthetic rubber chemically identical to natural rubber was first synthesized from isoprene using special catalysts that led to the all-*cis* polymer. Previous attempts to synthesize "natural" rubber had always led to the all-*trans* form or to a random form of polyisoprene, both of which were poor substitutes for rubber.

EXERCISE 15-8 Neoprene is a tough substitute for natural rubber that is insensitive to heat and acids and is also oil-resistant. It was discovered during World War II. It is a polymer of chloroprene:

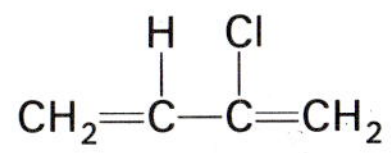

chloroprene

Draw the structure of neoprene, including at least three monomer units.

15-11 OXIDATION OF ALKENES

The π electrons of alkenes are readily attacked by oxidizing agents. For example, potassium permanganate in dilute aqueous solution oxidizes the alkene to a **dialcohol** (an alcohol with two OH groups; also called a **diol** or a **glycol**):

$$3CH_3{-}CH{=}CH{-}CH_3 + \underset{\text{(purple)}}{2KMnO_4} + 3H_2O \longrightarrow$$

$$3CH_3{-}\underset{\displaystyle OH}{\underset{|}{CH}}{-}\underset{\displaystyle OH}{\underset{|}{CH}}{-}CH_3 + \underset{\text{(brown)}}{2MnO_2(s)} + 2KOH$$

In the course of the reaction, the purple color of the potassium permanganate solution disappears, and a brown precipitate of manganese dioxide forms. This reaction, like the addition of bromine, can be used in the laboratory to distinguish between alkenes and alkanes, because alkanes are not oxidized by potassium permanganate.

EXERCISE 15-9 When ethylene is oxidized by $KMnO_4$, the product is ethylene glycol (which is used as a permanent antifreeze). Write the structure of ethylene glycol.

15-12 ALKYNES

The **alkynes** are hydrocarbons that contain a carbon–carbon triple bond. The general formula of the open-chain alkynes containing only one triple bond is C_nH_{2n-2}, and the simplest alkyne is acetylene (C_2H_2).

The carbon–carbon triple bond in acetylene is composed of one σ and *two* π bonds (Figure 15-3, see also Section 7-9). The carbon atoms are in the *sp* hybridization state, so acetylene is a linear molecule. The

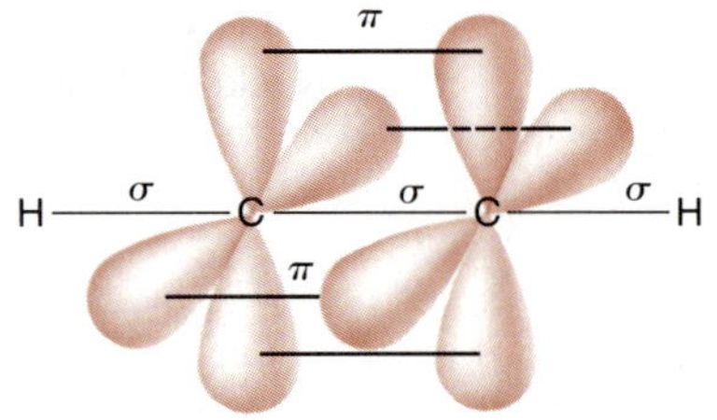

Bond length (nm)		Bond angle	
C≡C	0.12	H−C≡C	180°
H−C	0.11		

FIGURE 15-3

The structure of acetylene. The σ bonds are formed from the *sp* hybrid orbitals on the carbon atoms. The π bonds are formed by the overlap of two sets of unhybridized *p* orbitals.

carbon–carbon triple bond is both shorter and stronger than either the carbon–carbon single or double bond. Because the alkyne functional group (—C≡C—) has π bonds, the chemistry of the alkyne family is similar to that of the alkenes. Unlike the alkenes, the alkynes are not common in nature.

In the IUPAC system, the alkynes are named according to rules analogous to those for the alkenes.

1. The names of the alkynes are obtained by changing the *-ane* suffix of the corresponding alkane to *-yne:*

$$H—C≡C—H$$
ethyne
(acetylene)

$$H—C≡C—CH_3$$
propyne
(methyl acetylene)

Common names are given in parentheses under the IUPAC names.

2. When more than one position of the triple bond is possible, its position is designated by a numerical prefix. As in naming alkenes, the numbering begins at the end of the chain closer to the triple bond, and the number designates the *first* carbon of the triple bond:

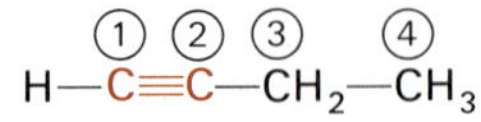
1-butyne

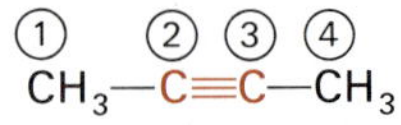
2-butyne

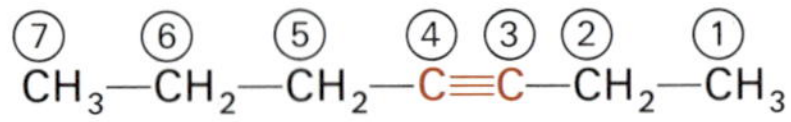
3-heptyne

3. Branched alkynes are numbered and named according to the same rules as alkanes, but with the numbering always based on the longest unbranched chain *containing the triple bond:*

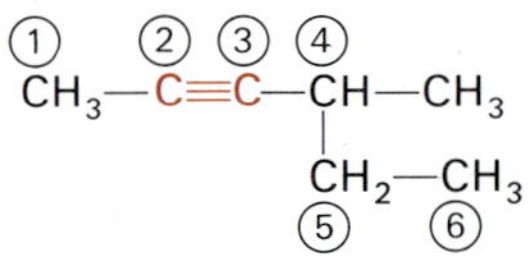
4-methyl-2-hexyne

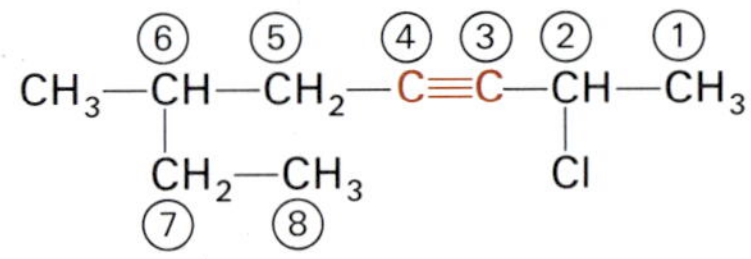
2-chloro-6-methyl-3-octyne

EXERCISE 15-10 Name the following substances:

(a) 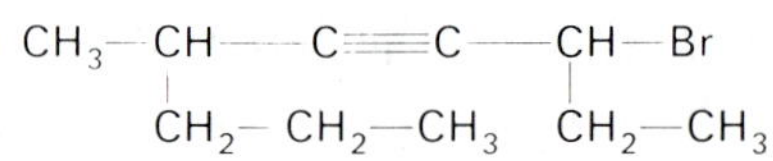

(b) 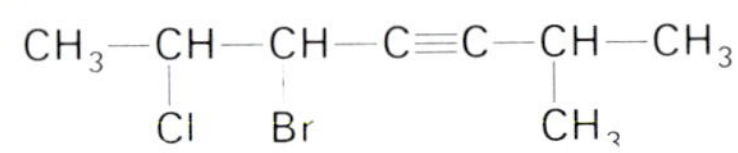

In their physical properties, the alkynes are similar to the alkenes and the alkanes—they are nonpolar molecules, insoluble in water, and less dense than water. Through the butynes, the alkynes are gases at room temperature. Acetylene (widely used in oxyacetylene torches, and important as a starting material in a variety of organic syntheses) is easily prepared by reaction between calcium carbide and water:

$$CaC_2 + 2H_2O \longrightarrow H{-}C{\equiv}C{-}H + Ca(OH)_2$$

The most important reaction of the alkyne family is addition to the π bonds, and the mechanisms of these reactions are identical to those we have already seen for the alkenes. As an example, consider the addition of HCl to 1-butyne:

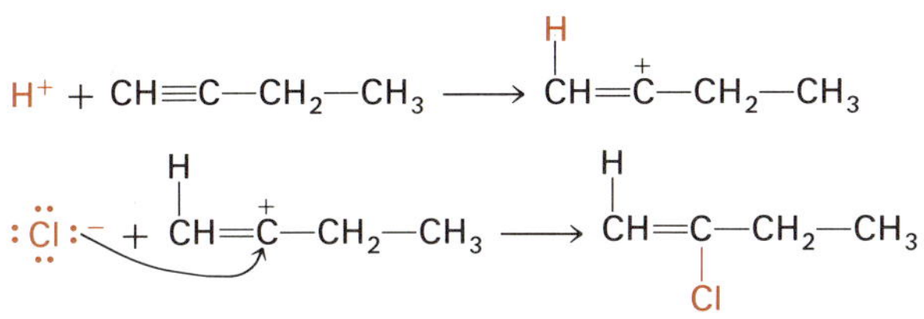

2-chloro-1-butene

When only *one* molecule of HCl is added to a molecule of 1-butyne, the main product (in accord with Markovnikov's rule) is 2-chloro-1-butene. But a second molecule of HCl can be added to the double bond (also in accord with Markovnikov's rule) to produce 2,2-dichlorobutane:

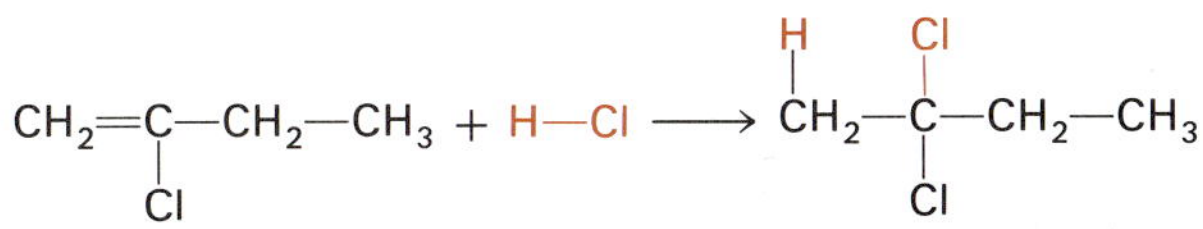

2,2-dichlorobutane

EXERCISE 15-11 (a) Write equations that show the mechanism of the addition of *one* molecule of HBr to *one* molecule of propyne. Use curved arrows to show the movement of electron pairs.

(b) Draw the structure of the product of the addition of *two* molecules of HBr to *one* molecule of propyne.

15-13 AROMATIC HYDROCARBONS

Benzene, the simplest aromatic hydrocarbon, was discovered in 1825 by Michael Faraday, who isolated it from a liquid that condensed out of the gas used in gas lights. (This gas, called illuminating gas, was obtained from whale oil.) Benzene and its derivatives were called **aromatic** because many of them have characteristic aromas. The molecular formula of benzene, C_6H_6, suggests that benzene would have a high degree of unsaturation. (A six-carbon alkane would have the molecular formula C_6H_{14}, and a cycloalkane would be C_6H_{12}.) But benzene does not react like an alkene. For example, it does not undergo the ready addition reactions of alkenes. It was not until 1865, forty years after its discovery, that August Kekulé suggested a possible structure—a ring structure with alternating double and single bonds:

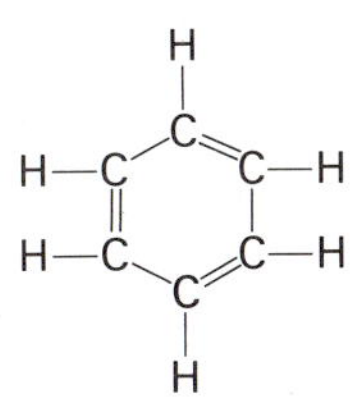

This structure, called a Kekulé structure, is usually written in abbreviated form as

Kekulé's theory is not entirely satisfactory, however, because *benzene does not react like a typical alkene*—and the Kekulé structure does not account for this fact. For example, benzene *can* be hydrogenated to yield cyclohexane, but the reaction requires much higher temperatures and pressures than does the hydrogenation of alkenes:

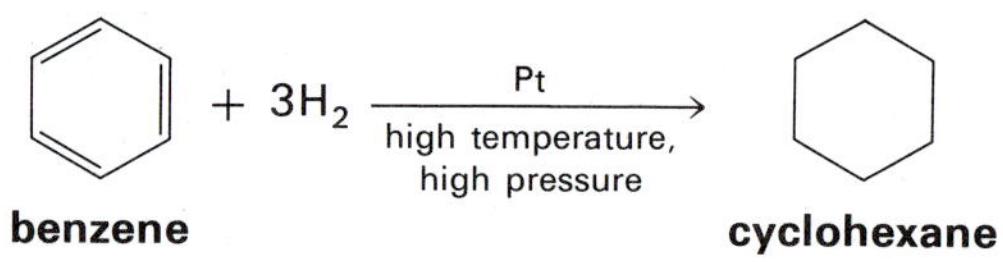

Furthermore, when bromine reacts with benzene, it reacts only slowly, and a bromine *substitutes* for a hydrogen, just as it does with alk*anes*:

This is in contrast to the rapid *addition* of bromine to the double bond of an alkene:

$$CH_2{=}CH_2 + Br{-}Br \longrightarrow \underset{\substack{| \\ Br}}{CH_2}{-}\underset{\substack{| \\ Br}}{CH_2}$$

In fact, in compounds containing both an alkene and a benzene ring, bromine adds readily to the double bond of the alkene, while leaving the benzene ring untouched:

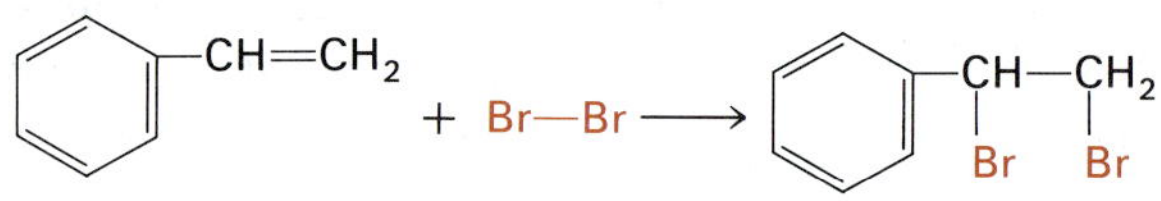

We see from these reactions that *the benzene ring is surprisingly stable.*

15-14 RESONANCE EFFECT AND THE STRUCTURE OF BENZENE

How can we account for the stability of the benzene ring? The special stability of the benzene ring is explained by the **resonance** effect. This theory says that no *single* structure can adequately represent benzene, but that the real molecule is a combination or **hybrid** of two structures:

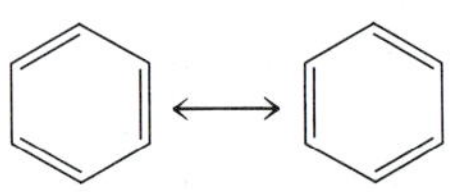

Much as a mule is a hybrid of a horse and a donkey that has some characteristics of each and cannot be accurately represented by either one alone, so benzene is a hybrid of the two Kekulé structures and cannot be represented by either structure alone. The real benzene molecule is said to be a **resonance hybrid** of the two structures, intermediate between them (see Chapter 7 also). The double headed arrow ↔ symbolizes resonance hybrids, and implies that the real molecule is best represented as a combination of these struc-

TABLE 15-4
Length of carbon–carbon bonds

Bond	Length (nm)
C—C	0.154
C⩦C (benzene)	0.140
C=C	0.133

tures. (This symbol is different from the two arrows ⇌ used to represent a reversible chemical reaction.)

In effect, each carbon has one-and-a-half bonds to each of its neighboring carbon atoms; in fact, the length of the carbon–carbon bond in benzene is between that of a single and a double bond (Table 15-4).

Benzene can also be described in terms of hybrid orbitals. Each carbon in benzene is bonded to three other atoms at 120° angles. The molecule is planar. The carbon atoms in benzene are thus in the sp^2 hybridization state. Each carbon is bonded to two other carbon atoms and to one hydrogen atom by σ bonds (Figure 15-4a). The single unshared electron of each carbon is in an unhybridized p orbital—say, the p_z orbital. We saw that the pair of unhybridized parallel p orbitals in ethylene overlap to form the π bond. Similarly, each p_z orbital in benzene can overlap with the p_z orbitals of the neighboring carbon atoms to form the π-bond network illustrated in Figure 15-4b: a set of doughnut-shaped π orbitals lying above and below the plane of the carbon ring. There is a **delocalization** (a spreading or sharing) of the six π electrons (3 electron pairs) around the ring. All six π electrons can move around the carbon ring through the network of π orbitals. It is this delocalization of the π electrons that accounts for the special stability of benzene.

We often use the following structure to represent benzene and to emphasize the delocalization of its π electrons:

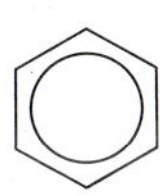

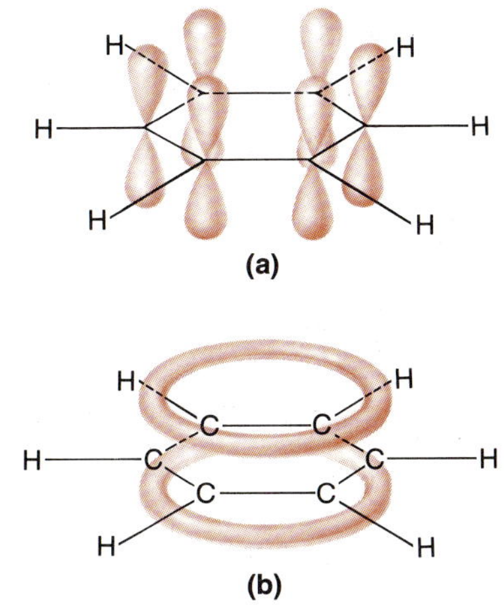

FIGURE 15-4
The structure of benzene. **(a)** The σ bonds and the unhybridized p orbitals. **(b)** The two lobes of the π-bond network.

But we shall also sometimes use one of the resonance structures to represent benzene, understanding that neither one alone accurately represents the bonding in the real molecule.

In later chapters we shall see further examples of the resonance effect, which requires a framework of parallel p orbitals, and involves a flow of electrons through these orbitals. The resonance effect gives a special, extra stability to the molecule—as we have seen for benzene, which has a much greater resistance to reaction than does an ordinary unsaturated molecule.

EXERCISE 15-12 Use curved arrows to show how one resonance structure of benzene can be converted into the other.

15-15
AROMATIC COMPOUNDS

The benzene ring is found in many compounds—both naturally occurring and synthetic. All of these compounds are called **aromatic** (although not all have the characteristic aroma responsible for the name). Compounds with *no* benzene ring are called **aliphatic** (from the Greek for fat) because such molecules characteristically occur in fats. The alkanes, alkenes, and alkynes are all aliphatic.

Examples of aromatic compounds include halogen-substituted benzenes such as chlorobenzene, bromobenzene, or iodobenzene:

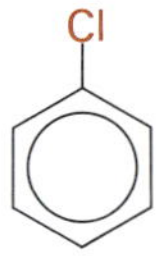

chlorobenzene

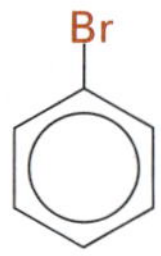

bromobenzene

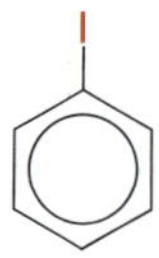

iodobenzene

When two groups are present on benzene, three different arrangements known as *ortho, meta,* and *para* (abbreviated *o*-, *m*-, or *p*-) are possible. Thus dichlorobenzene can occur as three distinct compounds:

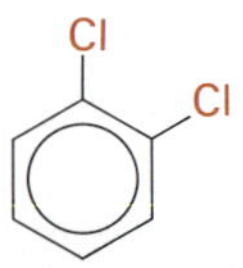

***o*-dichlorobenzene**
Melting point:
−17.2°C

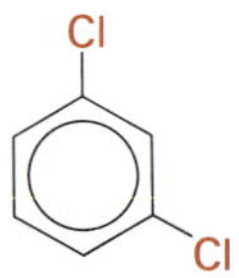

***m*-dichlorobenzene**
−24.8°C

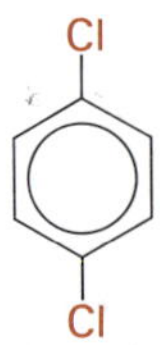

***p*-dichlorobenzene**
53.1°C

In the *ortho* compound, the substituents are on neighboring carbons; in the *meta* isomer, the substituents are once removed from each other; and in the *para* isomer, the substituents are on opposite sides of the ring.

When there are three or more substituents on the benzene ring, a numbering system is used:

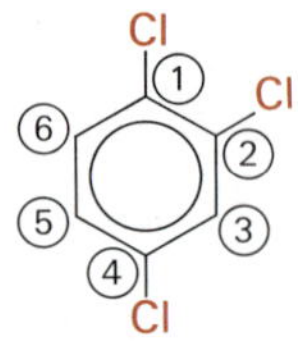

1,2,4-trichlorobenzene

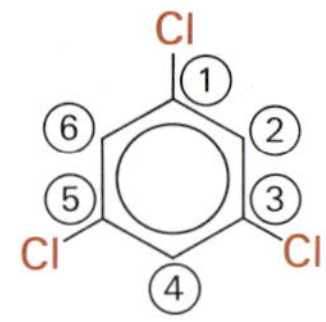

1,3,5-trichlorobenzene

Methylbenzene has the common name toluene:

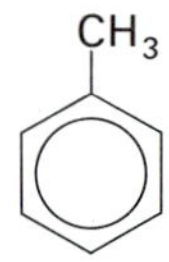

When there are additional substituents on the ring, the compound is named as a toluene:

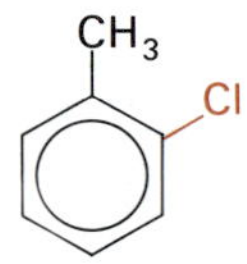

***o*-chlorotoluene**

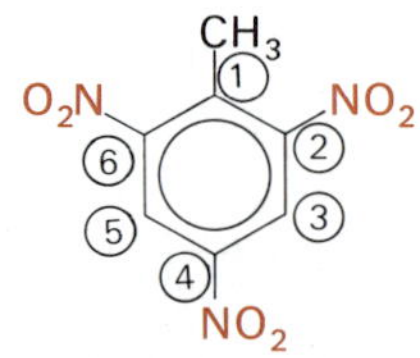

2,4,6-trinitrotoluene

EXERCISE 15-13 Draw the structures of *ortho-, meta-,* and *para-* iodotoluene.

Both benzene and toluene are obtained by the distillation of soft coal in the absence of air. Benzene is extremely toxic—a high concentration of the vapor in the air can cause convulsions or death due to respiratory failure. Benzene is also carcinogenic. Poisoning can occur by absorption through the skin as well as by inhalation. Toluene is less toxic than benzene because small amounts of toluene (in contrast to benzene) can

TABLE 15-5
Structures and uses of some aromatic compounds

Name	Structure	Use
Aniline	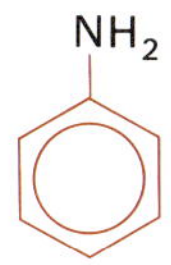	Preparation of dyes and sulfa drugs
Phenol (carbolic acid)	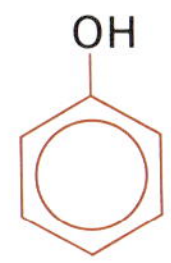	Disinfectant
Novocaine	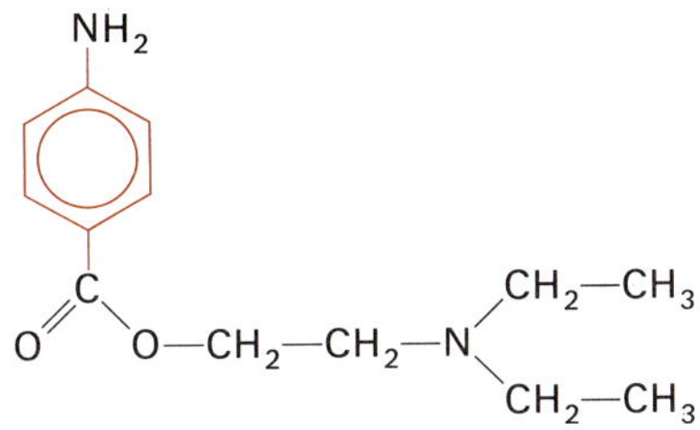	Local anesthetic
Dichlorodiphenyltrichloroethane (DDT)	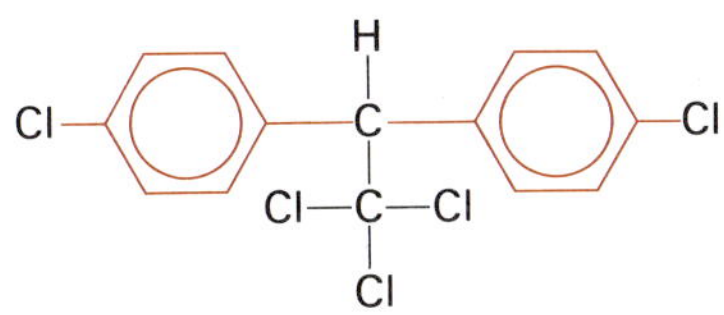	Insecticide
Styrene	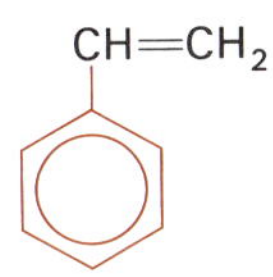	Preparation of polystyrene
Aspirin	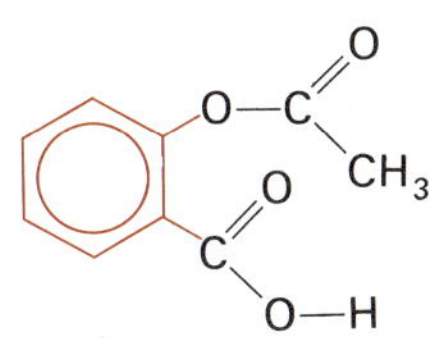	Many medical applications

be metabolized by the body and excreted. Toluene should therefore be substituted for benzene as a laboratory solvent or reagent whenever possible.

Table 15-5 gives the structures and uses of some other benzene derivatives. One of the compounds on the list, the insecticide DDT, has been shown to be toxic not only to insects but also to other animals, particularly fish and birds. Thus, in 1972 the Environmental Protection Agency banned DDT for general use in the United States. DDT may still be used, however, in case of an outbreak of malaria or typhoid, and under certain conditions, it may be used on green peppers, onions, and stored sweet potatoes.

Some more complicated aromatic compounds are composed of **fused** benzene rings (rings that have a side in common):

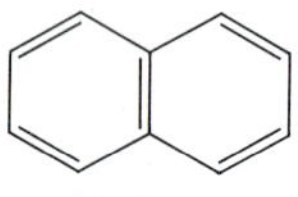

naphthalene

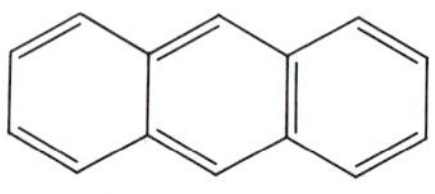

anthracene

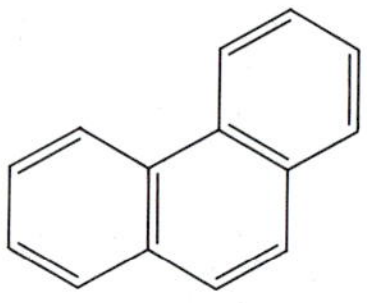

phenanthrene

These compounds, (as well as benzene and toluene) are obtained by the distillation of soft coal in the absence of oxygen. Naphthalene occurs in the fraction distilling at 195 °C to 230 °C; anthracene and phenanthrene occur in the highest-boiling fraction, which is called coal tar.

Some (but not all) polycyclic aromatic hydrocarbons (those composed of three or more fused benzene rings) are highly carcinogenic—small amounts painted on the skin of mice cause skin cancers within a few months. Two of the most potent carcinogens are dibenz[*a*,*h*]anthracene and benzo[*a*]pyrene (the letters *a* and *h* in these names designate the specific arrangement of the benzene rings in the compound by a system we shall not discuss here):

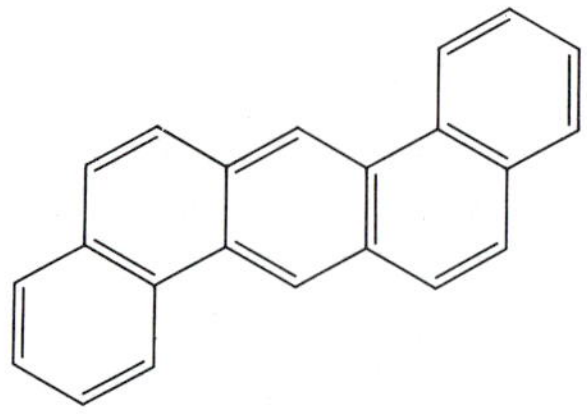

dibenz[*a*,*h*]anthracene

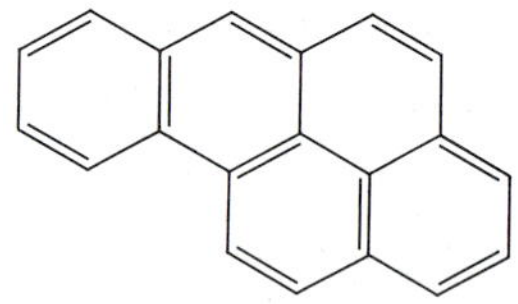

benzo[*a*]pyrene

Both of these compounds can be obtained from coal tar. Polycyclic aromatic hydrocarbons are also present in minute quantity in tobacco smoke, automobile exhaust, and in the effluent from coal-burning power plants. They are considered to be at least partly responsible for the increased incidence of lung cancer among cigarette smokers and among people who live in large cities.

In addition, many important compounds have aromatic-type five- or six-membered ring structures, but with one or more atoms other than carbon in the ring (**heterocyclic** compounds). Examples that are discussed in later chapters are

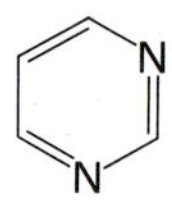

pyrimidine

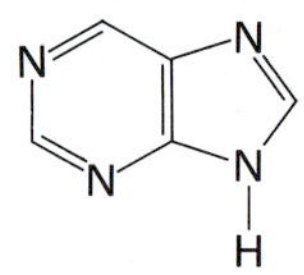

purine

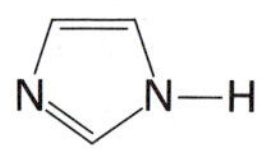

imidazole

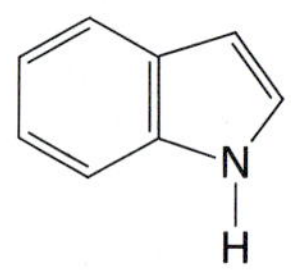

indole

15-16
REACTIONS OF BENZENE: ELECTROPHILIC AROMATIC SUBSTITUTION

We have already seen that benzene can be halogenated—a process in which a halogen atom is substituted for a hydrogen atom. To understand the mechanism of this reaction (and many others), we will find two new terms useful.

A **nucleophile** (from *nucleus* and *philos,* Greek for

"loving") *is any ion or molecule that is attracted to positive charge.* A nucleophile may be negatively charged, or it may be a neutral molecule with a lone pair of electrons. Examples are

$$:\ddot{\underset{\cdot\cdot}{\text{C}}}:^- \qquad :\ddot{\underset{\cdot\cdot}{\text{Br}}}:^- \qquad :\ddot{\underset{\cdot\cdot}{\text{I}}}:^- \qquad \text{H}-\ddot{\underset{\cdot\cdot}{\text{O}}}:^- \qquad \ddot{\text{N}}\text{H}_3$$

$$\text{H}-\ddot{\underset{\cdot\cdot}{\text{O}}}-\text{H} \qquad :\text{N}\equiv\text{C}:^- \qquad \text{R}-\ddot{\underset{\cdot\cdot}{\text{O}}}-\text{H}$$

and carbanions. Electrons in a π bond also are nucleophilic.

An **electrophile** (electron-lover) *is any ion or molecule that seeks out negative charge.* Examples are

$$:\ddot{\underset{\cdot\cdot}{\text{Br}}}^+ \qquad \text{H}^+ \qquad \text{Mg}^{2+}$$

and carbonium ions.

An electrophile *accepts* a pair of electrons from another atom or ion to form a new covalent bond, and a nucleophile *donates* a pair of its electrons to form a new bond. We have already seen examples of reactions between electrophiles and nucleophiles—in the addition of H—X to an alkene, for example:

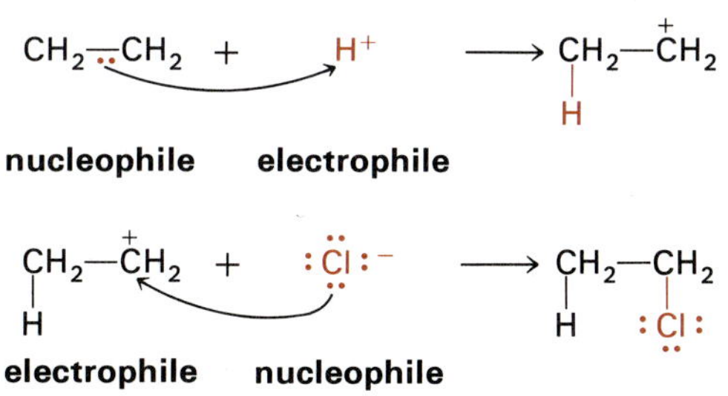

In this reaction, the proton and the carbonium ion are electrophiles, and the alkene and the chloride ion are nucleophiles.

The mechanism by which the benzene ring is halogenated is an example of a general reaction called **electrophilic aromatic substitution.** The reaction involves the *addition* of an electrophile to a nucleophilic π bond of the benzene ring, followed by the *elimination* of a proton:

$$:\ddot{\underset{\cdot\cdot}{\text{Br}}}-\ddot{\underset{\cdot\cdot}{\text{Br}}}: \longrightarrow :\ddot{\underset{\cdot\cdot}{\text{Br}}}:^- + \ddot{\underset{\cdot\cdot}{\text{Br}}}:^+ \qquad (15\text{-}8)$$

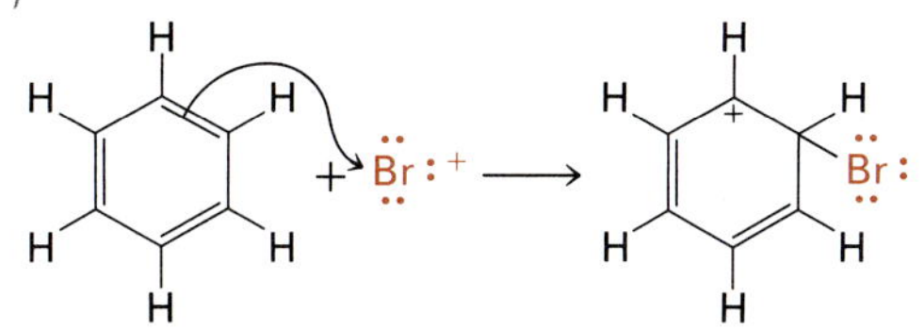

(15-9)

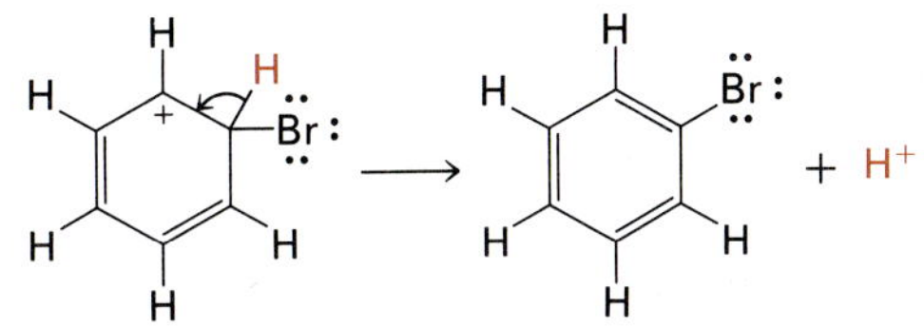

(15-10)

First, Br^+ is generated, as in the halogenation of an alkene. The Br^+ is an electrophile, and it attacks a π bond of the benzene ring, producing a carbonium ion. In the addition of bromine to an *alkene,* the final step would be an attack on the carbonium ion by the nucleophilic bromide ion, Br^-:

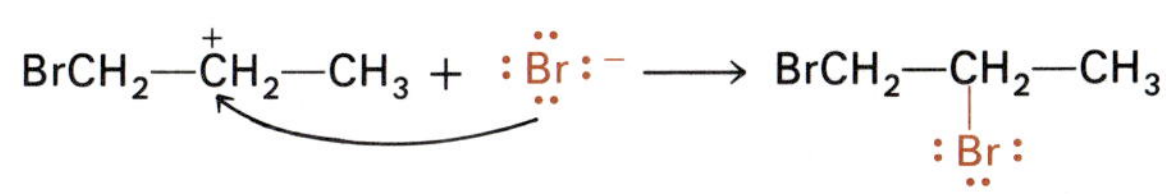

In this case, on the contary, a proton is lost to regenerate the benzene ring and yield a *substituted* benzene compound. *Because of the exceptional stability of the benzene ring,* the final step is the loss of a proton, rather than the addition of a bromide ion. The *net* reaction is the **substitution** of Br for H:

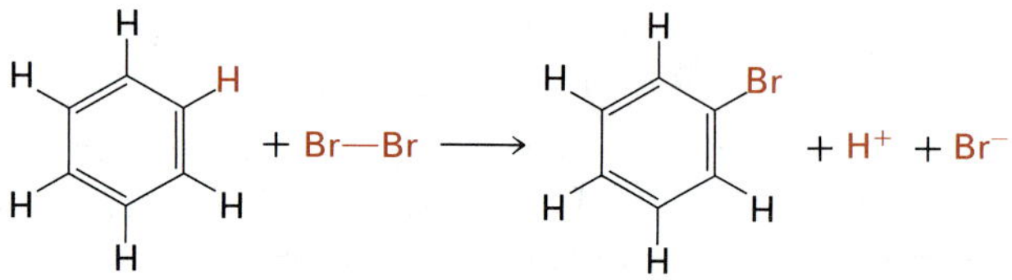

Several other reactions that are important industrially and in the laboratory are also substitution reactions:

Nitration is the addition of NO_2^+ (from a mixture of nitric and sulfuric acids, HNO_3 and H_2SO_4), followed by elimination of a proton:

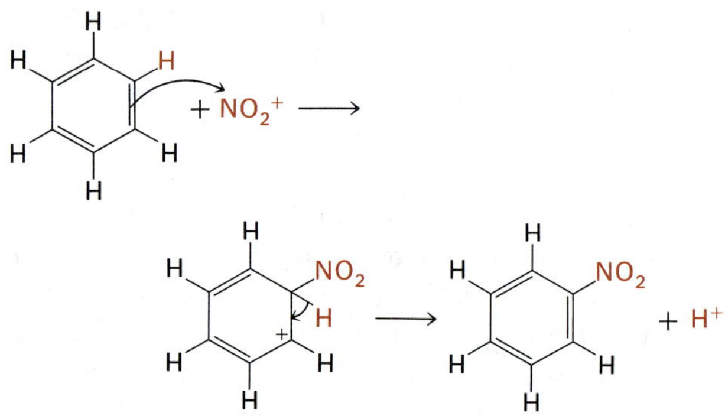

The net reaction is the substitution of NO_2 for H.

Alkylation is addition of a carbonium ion (from an alkyl halide and $AlCl_3$ or $FeCl_3$), followed by elimination of a proton:

Formation of carbonium ion

$$R{-}Cl + AlCl_3 \longrightarrow \underset{\text{carbonium ion}}{R^+} + AlCl_4^-$$

Alkylation

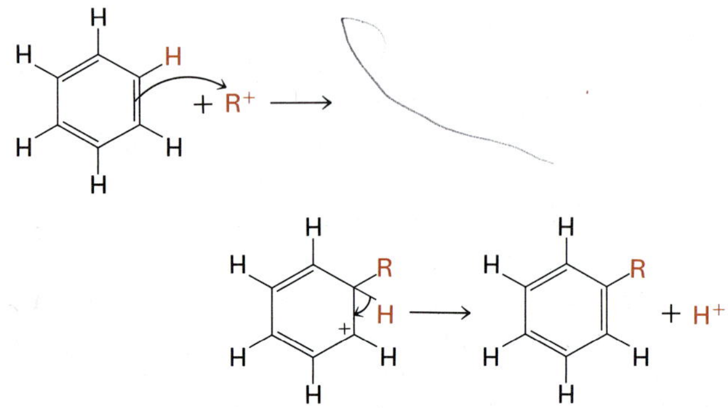

The net reaction is the substitution of R for H. If R is a methyl group, the product is toluene.

Sulfonation is addition of SO_3H^+ (from SO_3 and H_2SO_4), followed by elimination of a proton:

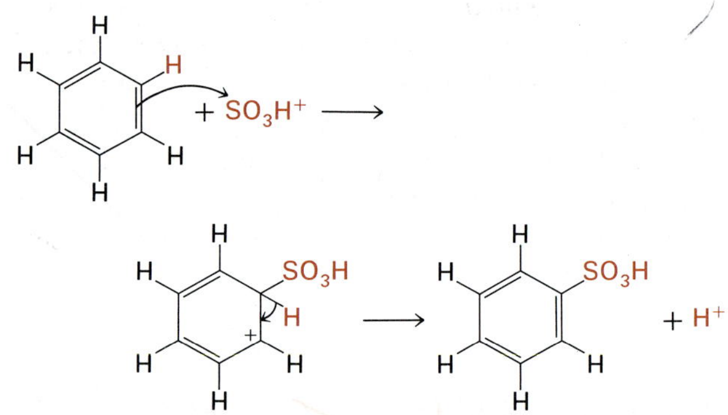

The net reaction is the substitution of SO_3H for H.

KEY TERMS

Alkenes are open-chain hydrocarbons that contain a carbon–carbon double bond and have the general formula C_nH_{2n}. *Alkynes* are open-chain hydrocarbons that contain a carbon–carbon triple bond and have the general formula C_nH_{2n-2}.

Positional isomers differ in the location of a multiple bond.

An *unsaturated* hydrocarbon is a hydrocarbon to which additional hydrogen atoms can be added.

Dehydration (as used in chemistry) is the elimination of a molecule of water from some other molecule. *Hydration* is the addition of water to another molecule. *Hydrogenation* is the addition of hydrogen to another molecule.

A *reactive intermediate* is a real chemical substance that can be observed and sometimes isolated, but that is generally unstable. A *transition state* is the chemical species at the point of highest energy for any elementary process. The transition state does not represent an isolable chemical substance but is a fleeting state in which some bonds are partly formed and others partly broken.

Activation energy is the energy required to go from reactants to the transition state.

Arrow formalism is the use of curved arrows to show the movement of electrons in a reaction.

A *polymer* (from the Greek for *many* and *parts*) is a giant molecule made up of small units called *monomers*. An *addition polymer* is formed by the combination of monomers, with all atoms of the monomer being incorporated into the polymer.

An *oxonium ion* is an ion that has a positively charged oxygen atom.

The *resonance theory* says that some molecules, such as benzene, cannot be adequately represented by a single structure; the real molecule is a *resonance hybrid* of two or more structures. *Kekulé structures* are resonance structures that represent the benzene molecule.

When an electron pair is spread over several atoms through a network of adjacent and parallel *p* orbitals, the electrons are said to be *delocalized*.

Aromatic hydrocarbons contain a benzene ring. *Aliphatic hydrocarbons* are hydrocarbons that do not contain a benzene ring.

A *nucleophile* is an atom, an ion, or a molecule that seeks out positive charge. An *electrophile* is an atom, an ion, or a molecule that seeks out negative charge.

An *electrophilic aromatic substitution* is a reaction that involves the addition of an electrophile to a benzene ring, with the subsequent elimination of a proton.

CHAPTER SUMMARY

The *alkenes* are hydrocarbons that have a carbon–carbon double bond consisting of one σ and one π bond. The double-bonded carbon atoms are in the sp^2 hybridization state, and the arrangement of atoms about those carbons is planar triangular.

In the IUPAC system, the names of alkenes are obtained by changing the *-ane* suffix of the corresponding alkane to *-ene*. A number designates the position of the double bond.

The characteristic reaction of alkenes is *addition* to the double bond—for example, the addition of hydrogen gas, the hydrogen halides, water, or bromine or chlorine, as well as addition polymerization. According to *Markovnikov's* rule, when HX is added to an alkene or an alkyne, the H adds to the carbon that already bears the greater number of H atoms. A mechanism that is consistent with this rule (and with other experimental evidence such as rate studies) involves the formation of a carbonium ion as an intermediate in the rate-determining step of the reaction. Tertiary carbonium ions are more stable than secondary carbonium ions, which in turn are more stable than primary carbonium ions. Use of the *arrow formalism,* which shows the movement of electrons, is helpful in following reaction mechanisms.

Alkynes are open-chain hydrocarbons with a carbon–carbon triple bond that consists of one σ and two π bonds. The triple-bonded carbons are in the *sp* hybridization state, and the arrangement of atoms about those carbons is linear. The chemistry of the alkynes is essentially the same as that of the alkenes, with addition to the triple bond the characteristic reaction.

In the IUPAC system the names of alkynes are obtained by changing the *-ane* suffix of the corresponding alkane to *-yne*. A number designates the position of the triple bond.

Aromatic hydrocarbons are hydrocarbons that contain the *benzene ring*. The benzene ring is surprisingly stable. This special stability is explained by the resonance theory, which says that no single structure adequately represents benzene. The real molecule is a *resonance hybrid* of two structures called *Kekulé structures,* and has carbon–carbon bonds that are intermediate between single and double bonds. The π bond electrons of benzene are delocalized around the ring through a network of adjacent and parallel *p* orbitals.

The characteristic reaction of benzene is *electrophilic aromatic substitution.* Benzene does not readily undergo addition reactions.

PROBLEMS

15-1. Name each of the following according to the IUPAC system:

(a) $CH_3-\overset{CH_3}{\overset{|}{C}}H-CH=CH_2$

(b) $CH_2=CH-CH=CH-CH_3$

(c) $CH_3-C\equiv C-CH_2-CH_3$

15-2. Draw structural formulas for

(a) 2,3-dimethyl-2-butene
(b) 3-ethyl-2-pentene
(c) 2-chloro-1-butene
(d) cyclopentene
(e) 3-hexyne.

15-3. Draw structural formulas for all isomers of C_3H_5Cl.

15-4. Which of the following alkenes can exist in *cis* and *trans* form?

(a) $CH_3-\overset{CH_3}{\overset{|}{C}}=CH-CH_3$

(b) $CH_3-CH=CH_2$

(c) $CH_2=CHCl$

(d) $ClCH=CHCl$

(e) $Cl-C\equiv C-H$

15-5. How do you account for the fact that cyclohexene (A) and cyclohexa-1,3-diene (B) are relatively stable compounds, whereas cyclohexa-1,2-diene (C) has never been prepared? (*Hint:* consider the bond angles for the different hybridization states of carbon.)

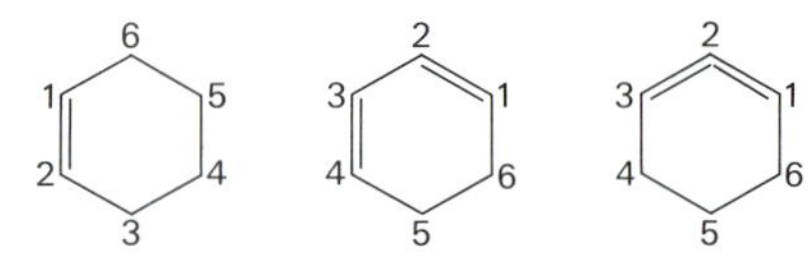

A B C

15-6. Write the structural formula of the alkene that

would be produced by the dehydration of 1-pentanol ($CH_3CH_2CH_2CH_2CH_2OH$).

15-7. Write an equation for the hydrogenation of methyl propene.

15-8. State Markovnikov's rule, and write a reaction that illustrates its application.

15-9. Write equations for the addition of hydrogen bromide to
(a) 2-methyl-2-pentene
(b) 2-pentene.

15-10. Write equations for the addition of hydrogen chloride (2 moles) to 1 mole of 1-pentyne. Use curved arrows to show the mechanism of the reaction, and state whether any carbonium ions formed as intermediates are primary, secondary, or tertiary.

15-11. Write a structural formula for a primary, a secondary, and a tertiary carbonium ion.

15-12. Describe a simple method by which you could distinguish between (a) hexane and hexene; (b) hexane and hexyne.

15-13. The π bond of an alkene can be regarded as a Lewis base, because alkenes react with H^+ to form carbonium ions. Arrange the following alkenes in order of increasing base strength, clearly indicating the strongest and weakest base:
(a) $CH_2{=}CH_2$
(b) $CH_3{-}CH{=}CH_2$
(c) $CH_3{-}C(CH_3){=}C(CH_3){-}CH_3$
(d) $CH_3{-}CH{=}CH{-}CH_3$

15-14. Write structural formulas of the products of the following reactions:

(a) 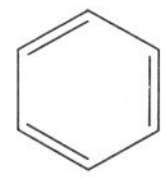 + Cl_2

(b) 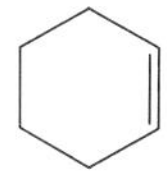 + Cl_2

(c) $CH_3{-}CH{=}CH_2 + Cl_2$

(d) 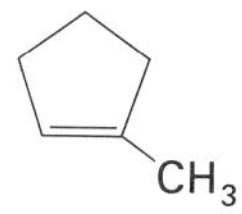 + HCl

(e) $CH_3{-}CH_2{-}CH(OH){-}CH_2{-}CH_3 \xrightarrow{H_2SO_4} C_5H_{10} + H_2O$

15-15. How would you synthesize

(a) $CH_3{-}CH_2{-}CH(Cl){-}CH_3$

from $CH_3{-}CH_2{-}CH{=}CH_2$

(b) $CH_3{-}CH_2{-}CH_2{-}CH_3$

from $CH_3{-}CH_2{-}C{\equiv}CH$

(c) $CH_3{-}CH(Br){-}CH(Br){-}CH_3$

from $CH_3{-}CH{=}CH{-}CH_3$

Use any simple chemicals in addition to the ones given, and indicate any catalyst necessary.

15-16. (a) Alkenes and alkynes undergo addition reactions with chlorine. Write a structural formula for the product of the following reaction:

$$CH_2{=}CH{-}CH_2{-}CH_3 + Cl_2 \longrightarrow \text{Product}$$

(b) This reaction depends on the breakup of chlorine molecules into ions as the first step:

$$:\ddot{\underset{..}{Cl}}{-}\ddot{\underset{..}{Cl}}: \longrightarrow :\ddot{\underset{..}{Cl}}^{+} + :\ddot{\underset{..}{Cl}}:^{-}$$

In the second step, which leads to product, a reactive intermediate has been observed:

$$:\ddot{\underset{..}{Cl}}^+ + CH_2{=}CH{-}CH_2{-}CH_3 \longrightarrow \text{Intermediate}$$

Write a structural formula for the intermediate.

(c) When the reaction is carried out in the presence of bromide ions, Br^-, a second product is also formed. Write the structural formula of this other product.

15-17. The reaction of chlorine with cyclohexene is an addition reaction that depends on the breakup of chlorine molecules into ions:

$$:\ddot{\underset{..}{Cl}}{-}\ddot{\underset{..}{Cl}}: \longrightarrow :\ddot{\underset{..}{Cl}}^+ + :\ddot{\underset{..}{Cl}}:^-$$

An intermediate has been observed in the addition reaction:

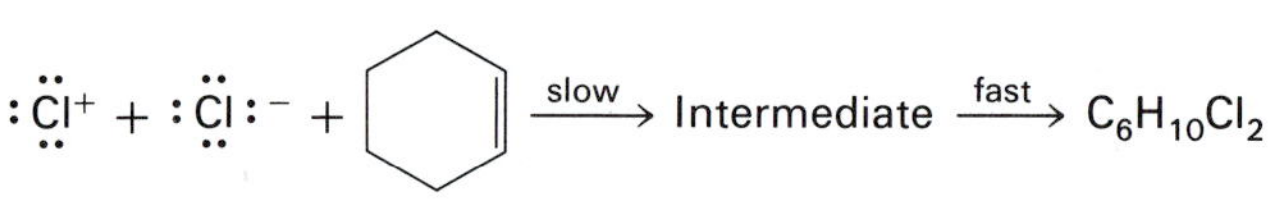

Write the structural formulas of the reactive intermediate and the product.

15-18. (a) Draw the structure of the product of the reaction between the ethyl carbonium ion ($CH_3\overset{+}{C}H_2$) and benzene.

(b) Write equations that show a reasonable mechanism for the reaction of part a, and indicate any nucleophile(s) and electrophile(s).

15-19. A compound of general formula C_nH_{2n+2} can have no ring and no π bond. A compound of general formula C_nH_{2n} can have one ring or one π bond. A compound of general formula C_nH_{2n-2} can have a triple bond, two rings, two π bonds, or one ring *and* one π bond. Write structures that illustrate each of these possibilities.

15-20. The compound iodine chloride (ICl) is sometimes used to determine the iodine number of fats and oils. Write the structure of the main product from the addition of ICl to propene.

15-21. Consider the experiment in which bromine is added to propene. When chloride ion (Cl^-)—say, from LiCl—is also present in the reaction mixture (in addition to Br_2 and $CH_2{=}CH{-}CH_3$), *two* products are obtained. Write their structures. Explain how this experiment supports the mechanism proposed in Section 15-9 for the addition of halogens to an alkene.

15-22. The structure of an oil is shown below. Calculate the iodine number of the oil—that is, the maximum number of grams of iodine that could be added to 100 g of the oil. The molecular weight of the oil is 878 g/mole.

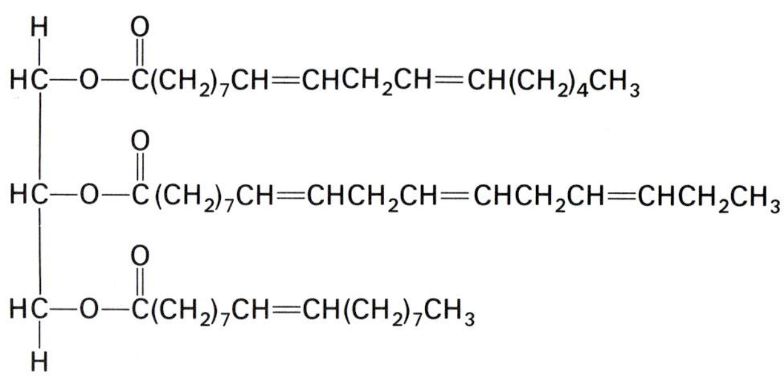

15-23. In 1938, DuPont chemists were performing experiments with fluorine-containing gases. One day, chemist Roy J. Plunkett filled several small cylinders with tetrafluoroethylene ($CF_2{=}CF_2$). When the cylinders were filled, he sealed their valves and stored them overnight. The next morning, he noticed that the pressure gauge had dropped to zero on one cylinder. But when he examined the tank, he found the seal still tight, and the tank weighed as much as when it was filled.

(a) What type of chemical reaction had occurred?

(b) When chemists cracked open the cylinder, what substance do you suppose they found? (Draw its structure.)

15-24. Draw the structure of a monomer that could be polymerized by a free-radical addition process.

15-25. Toy "superballs" bounce so high because they contain polybutadiene. In this synthetic rubber, butadiene ($CH_2{=}CH{-}CH{=}CH_2$) is polymerized end-to-end, just as in natural rubber. Ignoring geometrical isomers, draw a reasonable structure for polybutadiene, showing at least two repeating units of the polymer chain.

15-26. A sample of polystyrene has an average molecular weight of 12,000 g/mole. What, approximately, is the number of monomers in an "average" molecule of polystyrene? Styrene is

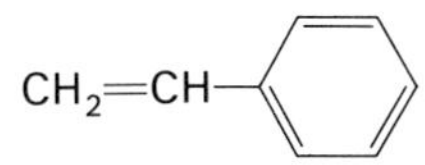

15-27. Draw the structures of monomers from which each of the following polymers could be constructed:

(a) 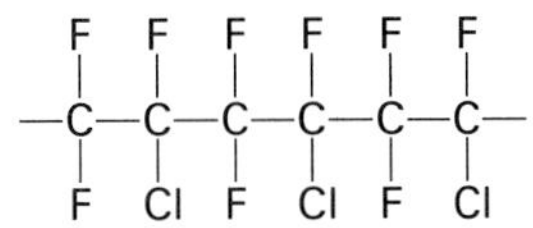

(b) 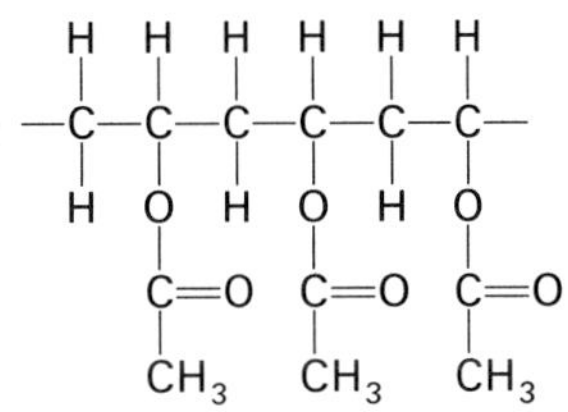

On a laboratory vacuum line like this one, organic compounds can be synthesized in an inert atmosphere, such as nitrogen or argon, at atmospheric or reduced pressure.

16

Reactions of Alkyl Halides

Organic compounds that contain halogens are rare in nature. In the laboratory, however, halogen-containing organic compounds are important starting materials in the synthesis of a wide variety of substances. In this chapter, alkyl halides illustrate the mechanisms of two important classes of organic reactions: substitution at a saturated carbon atom, and elimination.

16-1 SUBSTITUTION REACTIONS AT SATURATED CARBON ATOMS

Unsubstituted alkanes are very unreactive; combustion is the only reaction they undergo with ease. The substitution of a halogen atom for a hydrogen atom in an alkane, however, provides a functional group that makes the molecule more reactive. For example, *tert*-butane, $(CH_3)_3CH$, does not react with hydroxide ion, but *tert*-butylbromide reacts with hydroxide to yield an alcohol:

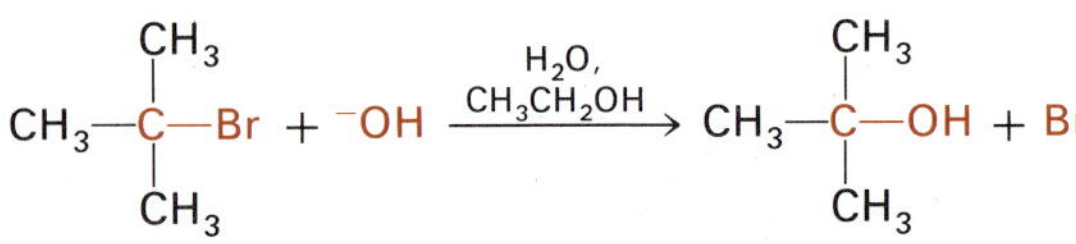

(16-1)

***tert*-butyl bromide** **tert-butyl alcohol**

(The H_2O and CH_3CH_2OH written above the arrow here designate the solvent in which the reaction is carried out. A mixture of water and alcohol is used because both reactants are soluble in it.)

In the following discussion, it is useful to define alkyl halides as primary (1°), secondary (2°), or tertiary (3°). In a primary alkyl halide, the carbon bonded to the halogen is bonded to only *one* other carbon atom. Ethyl chloride is an example:

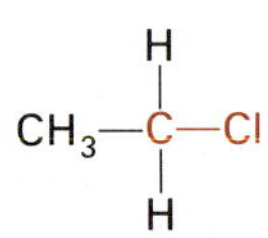

ethyl chloride (1° alkyl halide)

In a secondary alkyl halide, the carbon bonded to the halogen is bonded to *two* other carbon atoms. An example is 2-iodopropane:

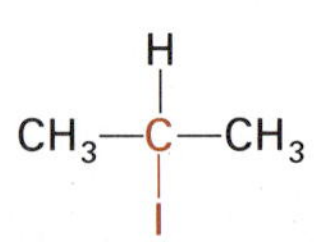

2-iodopropane (2° alkyl halide)

In a tertiary alkyl halide, the carbon bonded to the halogen is bonded to three other carbon atoms. An example is *tert*-butyl bromide, a reactant in Equation 16-1. We shall soon see that the mechanisms of reactions of alkyl halides depend on whether the alkyl halide is primary, secondary, or tertiary.

The substitution of a hydroxide ion for a halide ion in an alkyl halide is not always the best way to prepare an alcohol in the laboratory, but this reaction provides an excellent illustration of an important class of organic reaction that we shall encounter again and again—the **substitution reaction.** Here we consider substitutions at saturated carbon atoms, in contrast to substitution in aromatic compounds (discussed in Chapter 15). The *structure* of the alkyl halide determines the mechanism of the substitution reaction. Two different mechanisms have been determined. Both depend upon the polar nature of the carbon–halogen bond:

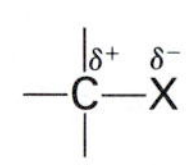

When the alkyl halide is *tertiary* (as in Equation 16-1), kinetic studies show that the *rate* of the reaction depends *only* on the concentration of the *tert*-butyl bromide, and *not* on the concentration of the hydroxide ion:

$$\text{Rate} = k[(CH_3)_3CBr]$$

where k is some constant particular to this reaction at some specific temperature (see Section 10-10). The following proposed mechanism is consistent with the rate law because the slow step involves only *tert*-butyl bromide as a reactant:

Step 1: Formation of carbonium ion

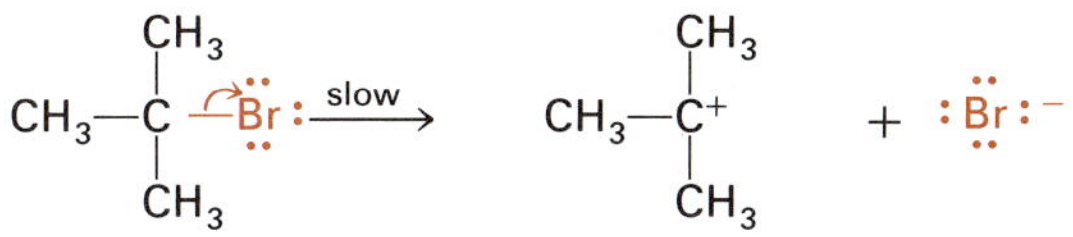

3° carbonium ion

Step 2: Addition of OH^-

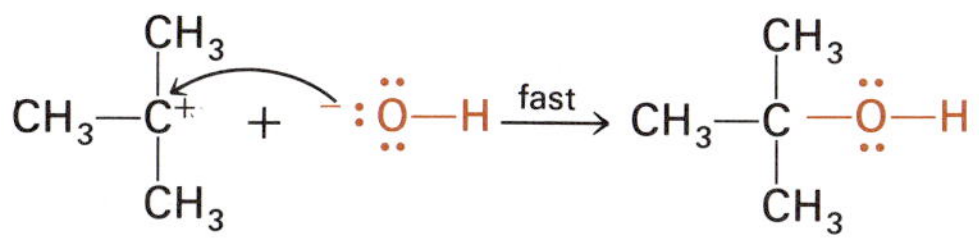

In the first and rate-determining step, the carbon–bromine bond is broken to yield a relatively stable tertiary carbonium ion. In the rapid second step, the nucleophilic hydroxide ion attacks the carbonium ion to produce the alcohol. The *net* reaction (the *sum* of the two steps) is just the substitution of OH^- for Br^- in the tertiary bromide (Equation 16-1).

Because the formation of the activated complex in the rate-determining step involves only *one* reactant, the reaction mechanism is said to be **unimolecular.** The name given to this mechanism is S_N1 for *substitution, nucleophilic, unimolecular.* The S_N1 mechanism is typical of substances that can form tertiary or other relatively stable carbonium ions.

EXERCISE 16-1 When 3-chloro-3-methylpentane reacts with I^-, the main product is 3-iodo-3-methylpentane.

(a) Write equations that show a likely mechanism for the reaction. (Use curved arrows to show electron movement.)

(b) Write an equation for the *net* reaction.

In the case of *primary* alkyl halides, kinetic studies show that a different mechanism applies. For example, consider the substitution of hydroxide ion for bromide ion in ethyl bromide:

$$CH_3CH_2{-}Br + OH^- \longrightarrow CH_3CH_2{-}OH + Br^-$$

The rate is found to depend on the concentrations of *both* the ethyl bromide and the hydroxide ion:

$$\text{Rate} = k[CH_3CH_2Br][OH^-]$$

The rate law shows that *both* ethyl bromide *and* the hydroxide ion are involved in the rate-determining step of the reaction. The mechanism consists of a single step:

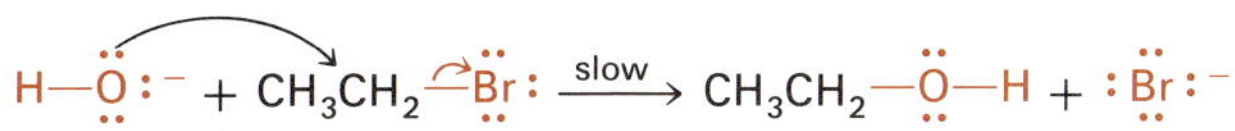

In this one-step reaction, the hydroxide ion is added and the bromide ion leaves *simultaneously*. The new bond (C—OH) is formed at the same time the old bond (C—Br) is broken. Unlike the mechanism for the reaction of hydroxide ion with the tertiary bromide, *the reaction of this primary bromide does not involve a carbonium ion.* The carbonium ion that might form from ethyl bromide would be a primary carbonium ion, which is not stable. So, instead of a carbonium-ion intermediate, there is a fleeting transition state with the geometry shown here:

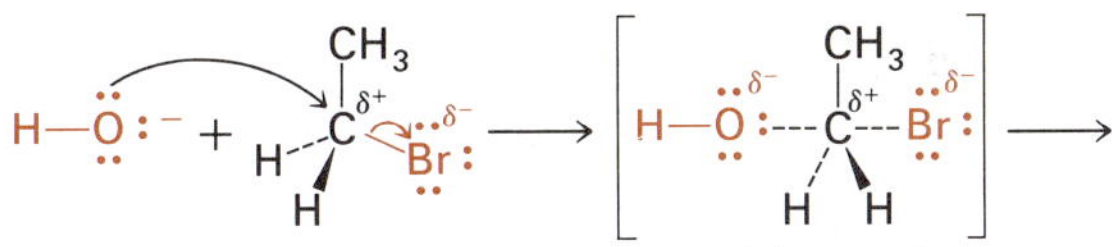

transition state

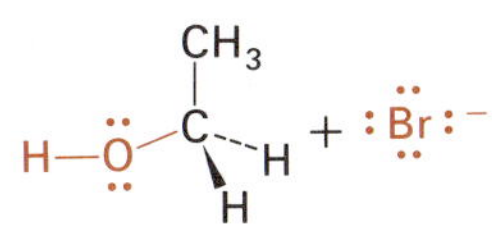

The nucleophilic OH^- attacks the partially positively charged carbon of the C—Br bond. The tetrahedron about the central carbon atom flattens out as the hydroxide ion approaches, and simultaneously the bromide ion leaves from the other side. The hydroxide ion is called an **entering group:** an atom (or group of atoms) that joins a molecule during a reaction as a result of the formation of a new chemical bond. The bromide ion is called a **leaving group:** an atom (or group of atoms) that leaves a molecule during a reaction as a result of the breaking of a chemical bond. At some point, the carbon is equally bonded to both the entering (OH^-) and the departing (Br^-) groups. Then the central carbon and the three nonreacting groups attached to it are planar. As the bromine departs further, the tetrahedron is reformed about the central carbon atom. During the reaction, the negative charges are thus spread out over the regions between the carbon and the OH and Br, and the carbon never bears a full positive charge (that is, a carbonium ion is *not* formed).

Because the formation of the activated complex in the rate-determining step (the *only* step) involves two reactants, the reaction is said to be **bimolecular.** The name given to the mechanism is S_N2 for *substitution, nucleophilic, bimolecular.* The S_N2 reaction is typical of *primary* alkyl halides.

EXERCISE 16-2 When 1-chloropropane reacts with Br^-, the main product is 1-bromopropane.

(a) Write a likely mechanism for the reaction, showing the configuration of the transition state.

(b) Write an equation for the *net* reaction.

We have seen that tertiary halides react by S_N1 mechanisms with the formation of stable tertiary carbonium ions; and primary halides react by S_N2 mechanisms via a transition state in which one bond is partially broken and another partially formed. *Secondary* halides react by either mechanism. If the reaction is run in a polar solvent, which helps stabilize the carbonium ion, the S_N1 mechanism is favored. A low concentration of the substituting nucleophile also favors the unimolecular mechanism. The bimolecular mechanism (S_N2) is favored by the use of a nonpolar solvent or by a high concentration of the nucleophile, which results in a large number of collisions between the substrate and the nucleophile.

16-2 STEREOCHEMISTRY OF SUBSTITUTION REACTIONS

In the unimolecular reaction of Equation 16-1, a tertiary carbonium ion is the reactive intermediate. The positively charged carbon and the three groups attached to it are in one plane because the positively charged carbon (being bonded to only three groups) is in the sp^2 hybridization state:

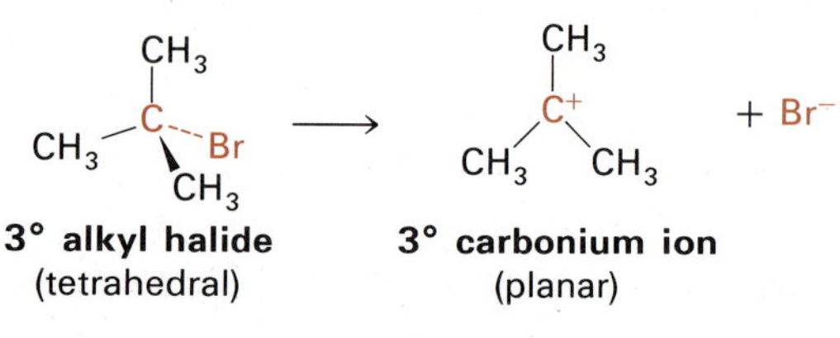

(Carbon can readily change hybridization state during chemical reactions. Here it has changed from sp^3 in the alkyl bromide to sp^2 in the carbonium ion.) The attacking nucleophile (OH^-) can approach the C^+ from either side of the plane. In the case of the carbonium ion shown, where all three groups bonded to the C^+ are the same, the product is the same in either case; $(CH_3)_3COH$ cannot exist as a pair of enantiomers. When an optically active compound such as 2-bromobutane undergoes a unimolecular reaction (S_N1), *both* optical enantiomers are formed—that is, the product is a racemic mixture of enantiomers:

Step 1: Formation of carbonium ion

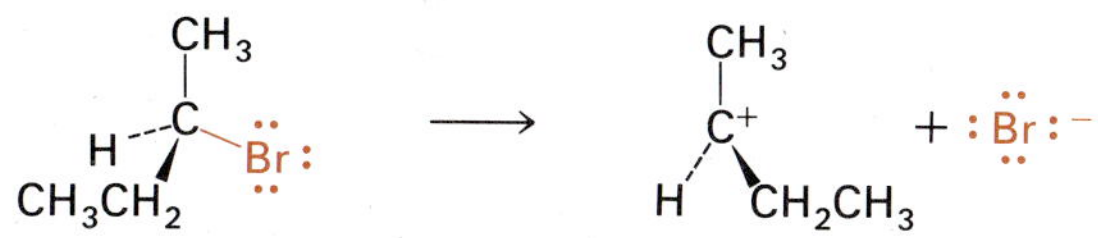

optically active compound **planar carbonium ion**

Step 2: Addition of OH^-

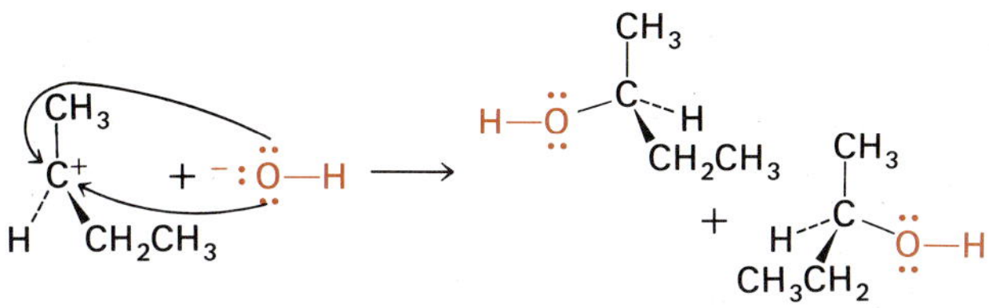

planar carbonium ion **racemic mixture of enantiomers**

In step 2, the nucleophile (OH^-) can attack from either side. Equal numbers of molecules of the two enantiomers are formed—a racemic mixture of enantiomers.

In the bimolecular mechanism (S_N2), the attacking nucleophile always approaches from the side opposite the leaving group (called a back-side attack). If the original compound is optically active, the product will also be optically active, but with an *inverted configuration,* like that of an umbrella turned inside out in a strong wind. This inversion of configuration occurs in the reaction of the optically active 2-bromobutane with hydroxide ion by an S_N2 mechanism:

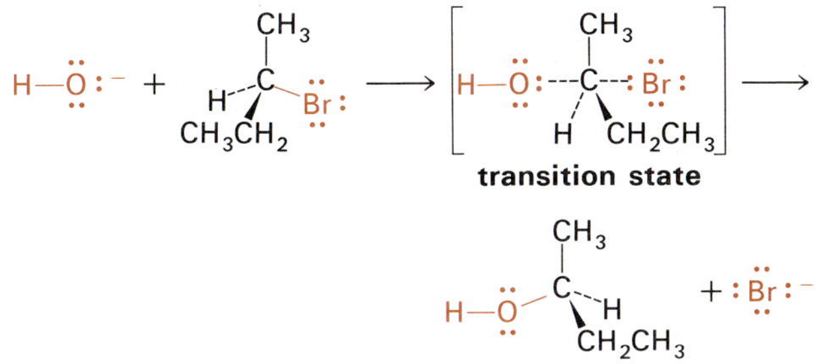

The product is the mirror image of the substrate with OH substituted for Br. Organic chemists use such changes in the optical activity of optically active compounds to determine whether a reaction is S_N1 or S_N2. If an optically active reagent yields a racemic mixture of enantiomers, the mechanism is S_N1. If there is an inversion of configuration leading to an optically active product, then the mechanism is S_N2.

16-3 RATES OF SUBSTITUTION REACTIONS

Steric effect. From the relative rates of substitution reactions of alkyl halides given in Tables 16-1 and 16-2, we see that the *structure* of the alkyl group has a significant effect on the rate.

Methyl chloride is relatively exposed to attack by the OH^- at the rear of the C—Cl bond. But, as the hydrogens of methyl chloride are replaced by methyl groups, the rate of the S_N2 reaction decreases. These bulky methyl groups interfere with the approach of the attacking OH^- to the rear of the C—Cl bond, and thus they slow the rate of the reaction. This change in reactivity, due to the *size* of groups attached at (or near) the carbon where reaction occurs, is called a **steric effect.** In order of decreasing reactivity via an S_N2 mechanism,

$$\underset{\text{methyl}}{CH_3\text{—}X} > \underset{1°}{CH_3CH_2\text{—}X} > \underset{2°}{(CH_3)_2CH\text{—}X} >> \underset{3°}{(CH_3)_3C\text{—}X}$$

TABLE 16-1
Relative rates of S_N2 reactions

Alkyl halide	Reaction	Relative rate
Methyl halide	$OH^- + CH_3Cl \xrightarrow{S_N2} CH_3OH + Cl^-$	1000

Primary halide	$OH^- + CH_3CH_2{-}Cl \xrightarrow{S_N2} CH_3CH_2OH + Cl^-$	100

Secondary halide	$OH^- + (CH_3)_2CH{-}Cl \xrightarrow{S_N2} (CH_3)_2CHOH + Cl^-$	1
	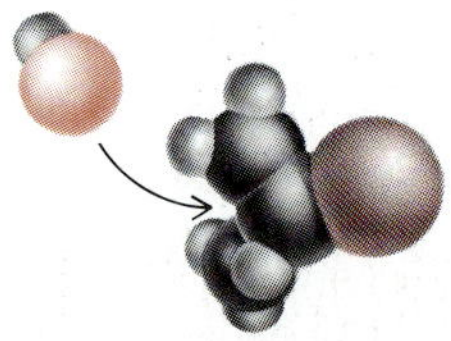	
Tertiary halide	$OH^- + (CH_3)_3C{-}Cl \xrightarrow{S_N2}$ No reaction via S_N2	0
	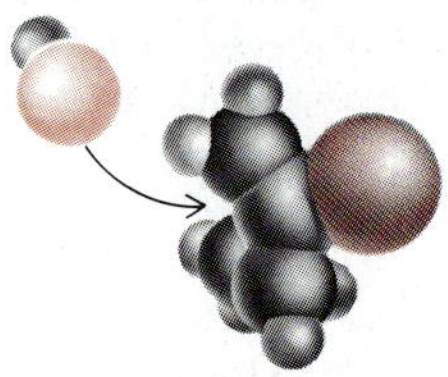	

TABLE 16-2
Relative rates of S_N1 reactions

Alkyl halide	Reaction	Relative rate
Primary halide	$CH_3CH_2{-}Cl + OH^- \xrightarrow{S_N1}$ No reaction via S_N1	0
Secondary halide	$(CH_3)_2CH{-}Cl + OH^- \xrightarrow{S_N1} (CH_3)_2CH{-}OH + Cl^-$	1
Tertiary halide	$(CH_3)_3C{-}Cl + OH^- \xrightarrow{S_N1} (CH_3)_3C{-}OH + Cl^-$	1000

The tertiary halide, in which this steric effect is greatest, does not undergo substitution by the S_N2 mechanism. Instead, it reacts rapidly by an S_N1 mechanism in which the rate-determining step is the formation of a relatively stable *planar* tertiary carbonium ion. This results in less crowding in the activated complex of the rate-determining step, and thus it accelerates the reaction.

Effect of nucleophile on rate. The rate of an S_N2 reaction depends on the strength of the nucleophile as well as on the structure of the alkyl halide. If a less reactive nucleophile (say, H_2O) is used instead of OH^-, the reaction rate decreases, because the rate-determining step involves the nucleophile.

The rate of an S_N1 reaction, on the contrary, does not depend on the reactivity of the nucleophile, because the rate-determining step (the formation of a carbonium ion) does not involve the nucleophile. This difference gives another way of distinguishing the unimolecular and bimolecular mechanisms experimentally.

Effect of leaving group on rate. In either S_N1 or S_N2 nucleophilic substitution reactions, the alkyl iodides react most rapidly; then the alkyl bromides; and the alkyl chlorides react the least rapidly (all other conditions being the same). Any group that is the conjugate base of a strong acid and that can sustain a negative charge makes a good leaving group in substitution reactions. Other examples (in addition to Cl^-, Br^-, and I^-) are NO_3^- and HSO_4^-.

16-4 ELIMINATION REACTIONS

We saw in Chapter 15 that alkenes can be prepared by the elimination of hydrogen bromide from an alkyl bromide, or by elimination of water from an alcohol. *These elimination reactions can occur at the same time, and in competition with, the substitution reactions we have just seen.* For example, a primary alkyl halide can be converted to an alcohol by an S_N2 *reaction, or to an alkene in an elimination reaction:*

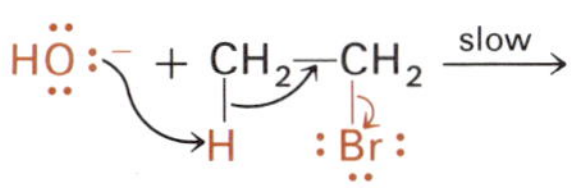

$$CH_2{=}CH_2 + H\ddot{O}{-}H + :\ddot{Br}:^- \qquad (16\text{-}2)$$

The mechanism of this reaction consists of a single step: a hydroxide ion removes a proton and, *simultaneously,* a bromide ion leaves and a new carbon–carbon bond (a π bond) is formed. Note that the two atoms eliminated (H and Br) come from adjacent carbon atoms. The only step in the reaction involves two reactants, so this is a **bimolecular elimination (E-2) mechanism.**

Thus *primary* alkyl halides react with a nucleophile via an S_N2 mechanism to yield a substitution product, or via an E-2 mechanism to yield an alkene. Neither of these reactions involves the formation of a carbonium ion intermediate: each involves a fleeting transition state in which a new bond is partially formed and an old one partially broken. In the following reaction, S_N2 and E-2 mechanisms proceed side by side; thus some molecules of starting material yield alcohol and some alkene:

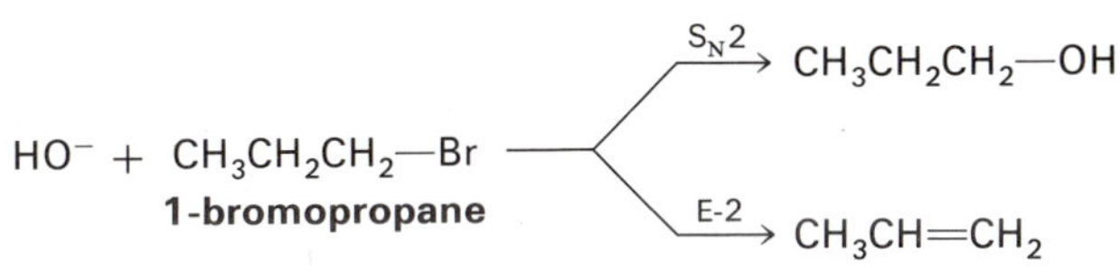

(Following the custom of organic chemists, we do not balance the equation. We show only the reactants and the products of interest.)

Thus, when 1-bromopropane reacts with hydroxide ion, there is not only one product, but a mixture of products. This is the case with most organic reactions. However, by controlling the conditions of the reaction, it often is possible to favor the desired product over others. The elimination reaction is favored by high temperatures, by the use of a strongly basic nucleophile, or by the presence of large, bulky side chains near the reaction site, which hinder the approach of the nucleophile in the substitution reaction.

Elimination can also accompany a substitution reaction of tertiary alkyl halides. In the conversion of *tert*-butyl bromide to an alcohol by an S_N1 mechanism, some methyl propene is also formed:

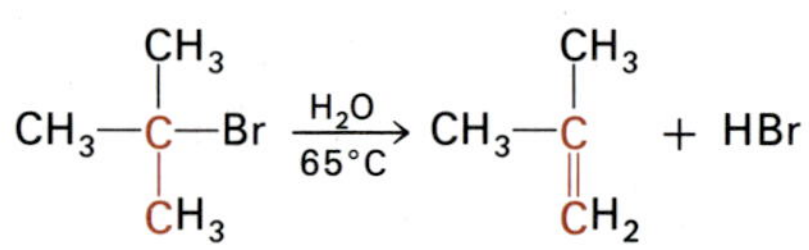

(16-3)

The mechanism for this reaction consists of two steps:

Step 1: Formation of carbonium ion

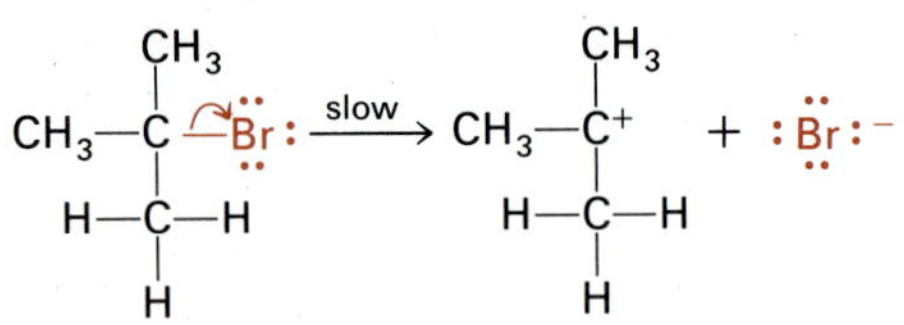

Step 2: Elimination of H^+

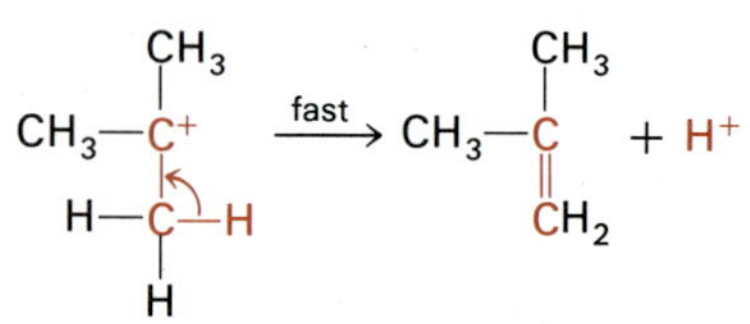

The *net* reaction, the sum of the two steps, is the reaction shown in Equation 16-3—the elimination of HBr from *tert*-butyl bromide. The first and rate-determining step is the same as that for the substitution reaction of a tertiary alkyl halide: the formation of a tertiary carbonium ion. But instead of adding a hydroxide ion in the second step, a proton is eliminated and a new carbon–carbon bond (a π bond) is formed, to produce an alkene.

In this example, although one hydrogen atom has been singled out, all nine hydrogens are identical. It makes no difference which is eliminated—the product is always the same. Note that, as with the E-2 mechanism, the two atoms that are eliminated come from adjacent carbons. This reaction is unimolecular, because only one reactant is involved in the rate-determining step. The mechanism is named **E-1,** for *elimination, unimolecular.*

Thus we have seen that *tertiary* alkyl halides can react with a nucleophile via an S_N1 mechanism to yield a substitution product such as an alcohol, or via an E-1 mechanism to yield an alkene. Each reaction involves the formation of a tertiary carbonium ion as an intermediate:

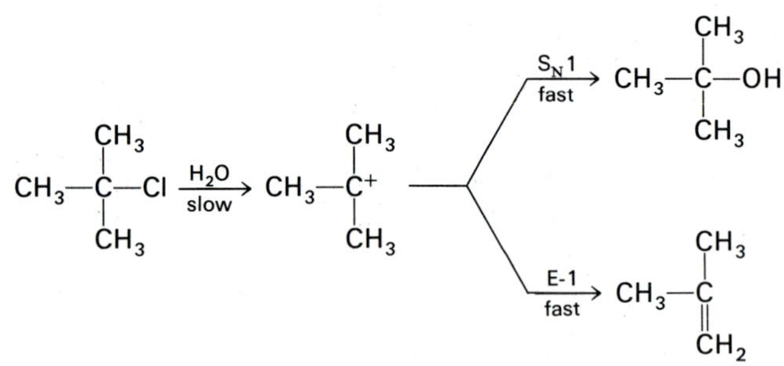

Again, a mixture of products is obtained. The relative amounts of each depend strongly on the specific reactants and on the reaction conditions. In the presence of a strong base such as OH^-, elimination via the E-2

mechanism occurs along with the E-1 and S_N1 reactions.

Secondary alkyl halides can react by all four mechanisms: S_N1, S_N2, E-1, and E-2. Which reaction is favored depends on the specific reactants and the reaction conditions.

EXERCISE 16-3 (a) When 1-chlorobutane is heated in the presence of OH^-, one of the products is an alkene. Write the structural formula of the alkene, and name the type of mechanism the reaction follows (S_N1, S_N2, E-1, or E-2).

(b) When 3-iodo-3-ethylpentane is heated with Br^-, one of the products is an alkene. Write the structural formula of the alkene, and name the type of mechanism the reaction follows.

Each elimination reaction considered so far produces only *one* alkene. In a more general case such as the one shown at the top of the next column, more than one elimination product is possible. Of the three possible products, which do we get? The answer is that all three are formed, but in different amounts. The relative amounts of the products depend on the specific reactant and on the conditions under which the reaction is run. There are rules to help predict the relative amounts, but we shall not discuss them here.

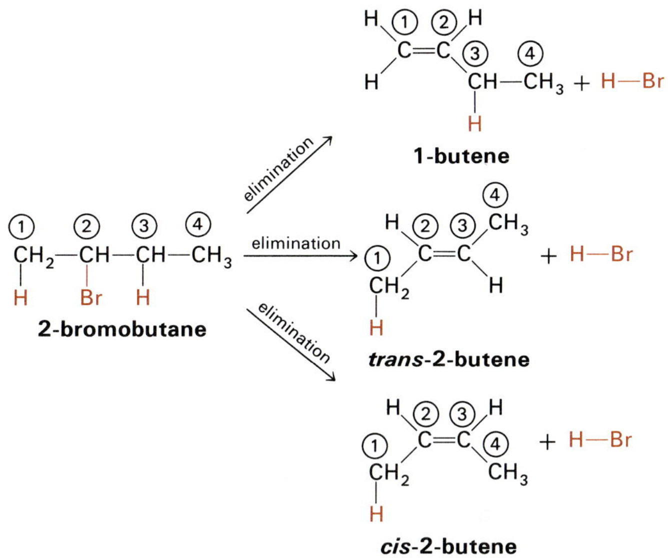

EXERCISE 16-4 Draw structural formulas of the product or products of the elimination of HCl from (a) 1-chloropentane, (b) 2-chloropentane, and (c) 3-chloropentane.

KEY TERMS

In a *primary* (1°) alkyl halide, the carbon atom bonded to the halogen is bonded to only one other carbon atom. In a *secondary* (2°) alkyl halide, the carbon bonded to the halogen is bonded to two other caron atoms. In a *tertiary* (3°) alkyl halide, the carbon bonded to the halogen is bonded to three other carbon atoms.

When the formation of the activated complex in the rate-determining step of a reaction involves only *one* reactant, the mechanism of the reaction is said to be *unimolecular*. When the formation of the activated complex in the rate-determining step of a reaction involves *two* reactants, the mechanism of the reaction is said to be *bimolecular*.

An S_N1 mechanism (*substitution, nucleophilic, unimolecular*) is typical of substances that can form tertiary carbonium ions, which are relatively stable. An S_N2 mechanism (*substitution, nucleophilic, bimolecular*) is typical of primary alkyl halides, which cannot form stable carbonium ions.

An E-1 mechanism (*elimination, unimolecular*) is typical of substances such as tertiary alkyl halides that can form tertiary carbonium ions. An E-2 mechanism (*elimination, bimolecular*) is typical of primary alkyl halides, which cannot form stable carbonium ions.

An *entering group* is an atom (or group of atoms) that joins a molecule during a reaction as a result of the formation of a new chemical bond. A *leaving group* is an atom (or group of atoms) that leaves a molecule during a reaction as a result of the breaking of a chemical bond.

A change in reactivity due to the *size* of groups attached at or near the carbon where reaction occurs is called the *steric effect*.

CHAPTER SUMMARY

Substitution reactions of *tertiary* alkyl halides usually occur by an S_N1 mechanism; the rate-determining step is the formation of a relatively stable tertiary carbonium ion. Substitution reactions of *primary* alkyl halides, on the contrary, usually occur by an S_N2 mechanism. The rate-determining step is the simultaneous formation of a new bond to the entering group and the breaking of the old bond to the leaving group. Secondary alkyl halides can react by either an S_N1 or an S_N2 mechanism. The relative rates of substitution reactions depend on steric effects and on the specific entering and leaving groups.

Elimination reactions of alkyl halides to produce alkenes can occur at the same time (and in competition with) substitution reactions. Tertiary alkyl halides usually react by an E-1 mechanism, with the formation of a tertiary carbonium ion. Primary alkyl halides usually react by an E-2 mechanism, with a fleeting transition state in which a new bond is partially formed and an old bond partially broken.

PROBLEMS

16-1. Write the structure of the main product of each of the following substitution reactions. For reactions a, c, d, e, and f, state whether the mechanism would more likely be S_N1 or S_N2. What can you say about the mechanism of reaction b?

(a) $CH_3I + Br^-$

(b) 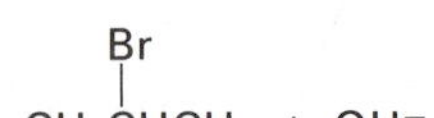

(c) $CH_3(CH_2)_4CH_2Cl + HS^-$
(the product is a thiol)

(d) 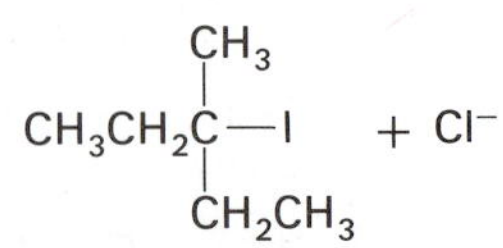

(e) 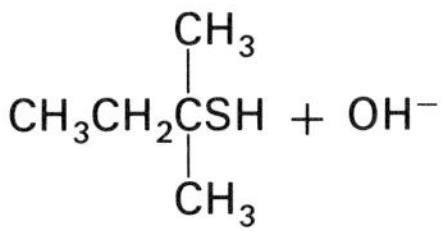

(f) 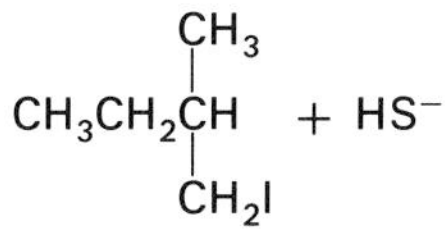

16-2. Write the stucture of an alkyl halide you could use to prepare each of the following alcohols:

(a) 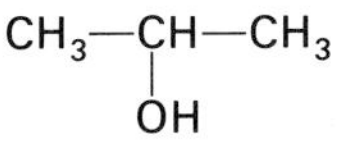

(b) 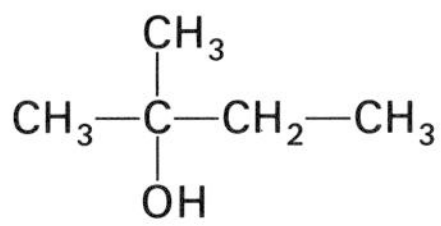

(c) 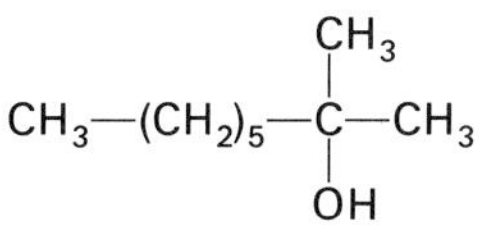

16-3. Which of the following two reactions proceeding by an S_N2 mechanism would be more rapid? Explain.

(a) 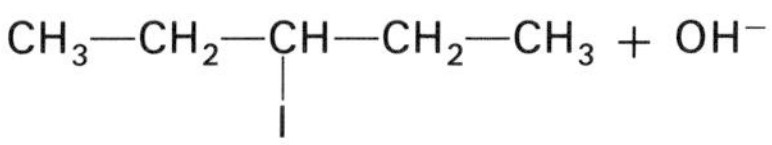

(b) CH_3—$(CH_2)_3$—$CH_2I + OH^-$

16-4. Which of the following two reactions proceeding by an S_N1 mechanism would be more rapid? Explain.

(a) 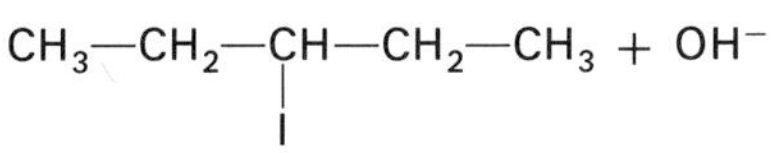

(b) 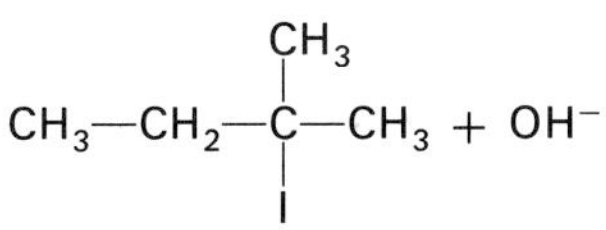

16-5. In one of the two following reactions, a single substitution product is formed. In the other, a racemic mixture of enantiomers is produced. Tell which reaction gives only one product and which gives the racemic mixture of enantiomers, and in each case explain why.

(a) $CH_3(CH_2)_4CH_2SH + Cl^-$

(b) 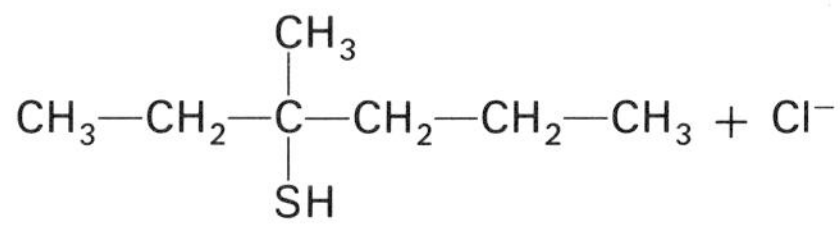

16-6. Write the structure of an alkyl bromide from which you could prepare each of the following alkenes:

(a) $CH_3CH_2CH{=}CHCH_3$

(b) 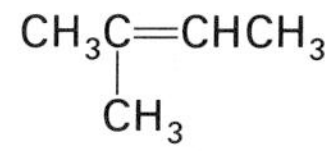

(c) 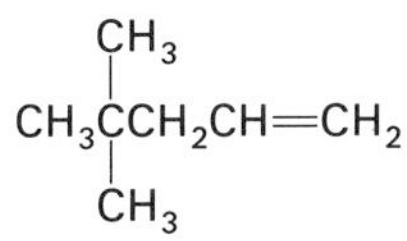

(d) 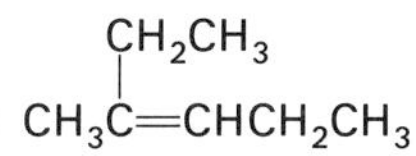

16-7. When 2-iodobutane is heated in the presence of OH^-, there are three elimination products and two substitution products. Write the structures of all the products.

16-8. When propyl iodide ($CH_3CH_2CH_2I$) is heated in the presence of OH^-, there are two products. Write mechanisms for both reactions using curved arrows.

16-9. (a) Write structures for the elimination and substitution products formed when 3-bromo-3-methylpentane is heated with hydroxide ion. Does the substitution reaction yield a pair of enantiomers?
(b) Write a mechanism, using curved arrows, for the substitution reaction of part a.
(c) Write a mechanism, using curved arrows, for one of the elimination reactions of part a.

16-10. Genetic changes in houseflies have produced some species that are resistant to the insecticide DDT. These species produce an enzyme that catalyzes the elimination of HCl from DDT. The product is a compound called DDE, which is estimated to be between 1/100 and

1/200 as toxic as DDT. Here is DDT:

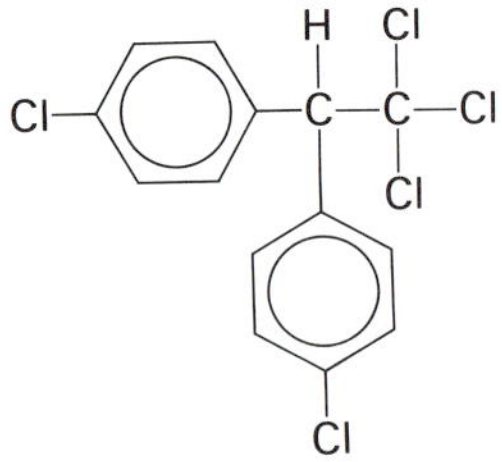

Write the strueture of DDE.

16-11. When 1-iodobutane is heated with aqueous NaOH solution, the major product formed is 1-butene.

$$CH_3CH_2CH_2CH_2{-}\ddot{\underset{..}{I}}{:} + {}^{-}{:}\ddot{\underset{..}{O}}{-}H \longrightarrow CH_3CH_2CH{=}CH_2 + {:}\ddot{\underset{..}{I}}{:}^{-} + H{-}\ddot{\underset{..}{O}}{-}H$$

Experiments show that the rate at which 1-butene is formed depends on the concentration of OH^- as well as on that of butyl iodide.
(a) Write a mechanism for the formation of 1-butene from 1-iodobutane that is consistent with these facts. Show any intermediates and use curved arrows to indicate electron shifts.
(b) Using the shorthand notation for reaction mechanisms (for instance, S_N1), how would you designate the reaction in part a?
(c) A small amount of 1-butanol is also formed in the reaction of 1-iodobutane with OH^-:

$$CH_3CH_2CH_2CH_2{-}\ddot{\underset{..}{I}}{:} + {}^{-}{:}\ddot{\underset{..}{O}}{-}H \longrightarrow CH_3CH_2CH_2CH_2{-}\ddot{\underset{..}{O}}{-}H + {:}\ddot{\underset{..}{I}}{:}^{-}$$

Name the *specific* mechanism by which this product most logically arises.

16-12. The following reaction of compound A takes place at room temperature:

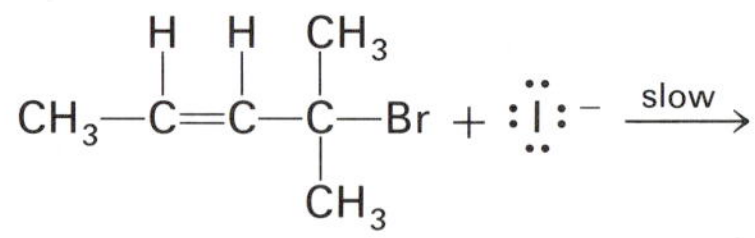

A

$$\text{Intermediate } \mathbf{B} \xrightarrow{\text{fast}} \mathbf{C} + \mathbf{D} + {:}\ddot{\underset{..}{Br}}{:}^{-}$$

The two isomeric products C and D are each formed from intermediate B.
(a) What *general* type of organic chemical reaction does this process belong to?
(b) What is the structure of the intermediate B? Is the rate-determining step unimolecular or bimolecular?
(c) What are the structures of products C and D? (At least one of these should be obvious. The other one has the iodine on a different carbon—see if you can figure out which one.)
(d) How do you account for the formation of two products from a single reactive intermediate? Be explicit and use formulas to support your answer.
(e) Very small amounts of a third product F are also formed in the above reaction from B:

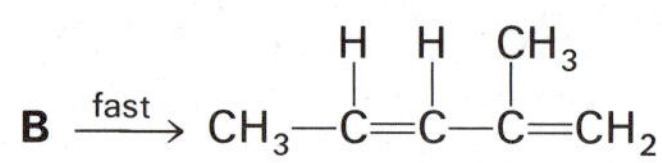

F

By what *specific* reaction mechanism is F formed from A? Use the shorthand notation (S_N1, for instance) to designate your answer.

Redwood wine tanks. These tanks were built in 1933; each one is 9 feet in diameter and $14\frac{1}{2}$ feet high and holds 24,500 gallons of wine.

17

Alcohols, Phenols, and Ethers

Alcohols, phenols, and ethers are characterized by oxygen-containing functional groups: C—OH in alcohols, —OH in phenols, and C—O—C in ethers. These functional groups are common, both in synthetic and in natural products. In fact, one of the earliest chemical processes developed by human beings was the preparation of ethyl alcohol from grains and other carbohydrates by fermentation. In this chapter, we discuss the structures, properties, and chemical reactions of alcohols, phenols, and ethers.

17-1
STRUCTURE AND PROPERTIES OF ALCOHOLS

Alcohols are compounds that contain the hydroxyl functional group (—OH), and it is this group that determines their chemistry. We can think of alcohols as derivatives of water in which an alkyl group replaces one of the hydrogen atoms to give a compound with the general formula R—OH. This resemblance is not purely formal; we shall see that alcohols have some of the properties of water.

The structure of methyl alcohol (CH_3OH, the simplest alcohol) is shown in Figure 17-1. In methyl alcohol, both the carbon and the oxygen atoms are in the sp^3 hybridization state, and all the bond angles are close to the 109.5° expected for sp^3 hybridization. The O—H bond is polar. Hydrogen is bonded to a small, highly electronegative atom (oxygen); so, like water, alcohols can participate in hydrogen bonds (see Section 7-11). Hydrogen is bonded to one oxygen atom by a strong covalent bond, and to another oxygen atom by the weaker hydrogen bond:

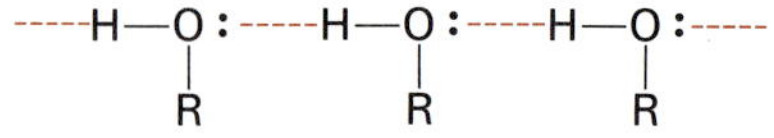

The C—H bonds are essentially nonpolar and do not participate in hydrogen bonds. The hydrogen bonds that form between neighboring molecules in the liquid make it more difficult to separate the molecules from one another—that is, to vaporize the liquid. Additional energy must be provided to break the hydrogen bonds. As a result, alcohols have higher boiling points than the corresponding alkanes (Table 17-1). In fact, although the first *four* alkanes are gases at room temperature and atmospheric pressure, even methyl alcohol is a liquid.

Actually, the difference in boiling point between an alcohol and the corresponding alkane—say, between methyl alcohol and methane—is the result of two factors: hydrogen bonding and van der Waals forces. As we saw in Chapter 7, van der Waals forces increase as the number of electrons in the molecule and the surface area of the molecule increase. Two molecules of

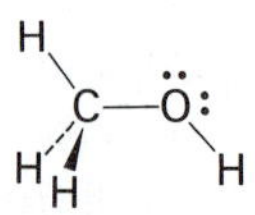

Bond length (nm)		Bond angle	
C—H	0.11	H—C—H	109°
O—H	0.096	H—C—O	110°
C—O	0.14	C—O—H	109°

FIGURE 17-1
Structure of methanol.

TABLE 17-1
Boiling points of some alkanes and alcohols

Name	Structure	Molecular weight	Boiling point (°C)
Methane	CH_4	16	−161
Methyl alcohol	CH_3—OH	32	65
Ethane	CH_3CH_3	30	−88
Ethyl alcohol	CH_3CH_2—OH	46	78
Propane	$CH_3CH_2CH_3$	44	−42
n-Propyl alcohol	$CH_3CH_2CH_2$—OH	60	97
n-Butane	$CH_3CH_2CH_2CH_3$	58	−0.5

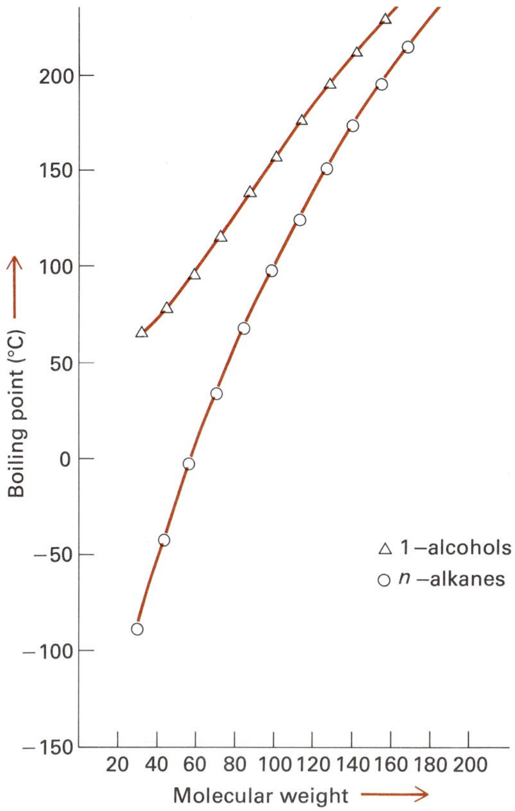

FIGURE 17-2

Boiling points of unbranched alcohols and alkanes.

the same molecular weight are roughly the same in size and in number of electrons, and so they have van der Waals forces of about the same strength. Thus, if we compare the boiling point of methyl alcohol (molecular weight 32) with that of ethane (molecular weight 30), we can say that the much higher boiling point of the alcohol is due almost exclusively to hydrogen bonding and not to a difference in van der Waals forces. Similarly, the boiling point of ethyl alcohol is much higher than that of propane, although they are about the same in molecular weight. Figure 17-2 graphs boiling points against molecular weights for alcohols and alkanes. Note that in the case of the heavier molecules (where the hydroxyl group constitutes a smaller *fraction* of the molecule), the boiling points of the alcohols are only slightly higher than those of alkanes of the same molecular weight.

We can also see the tremendous effect of hydrogen bonding on boiling point if we consider the *number* of hydrogen atoms in a molecule that can participate in hydrogen bonds. For example, Table 17-2 compares the boiling points of water (which has *two* hydrogens that can form hydrogen bonds), alcohols (with only *one* hydrogen that can form a hydrogen bond), and an ether (with *no* hydrogens that can form hydrogen bonds). We see that even though it is the smallest of these molecules, water has the highest boiling point. And dimethyl ether (CH_3—O—CH_3) is a gas at room temperature and boils 101 degrees below ethyl alcohol, with which it is isomeric.

TABLE 17-2
Boiling points of water, alcohols, and ether

Name	Structure	Number of H atoms that can hydrogen-bond	Molecular weight	Boiling point (°C)
Dimethyl ether	CH_3—O—CH_3	0	46	−23
Methyl alcohol	CH_3—O—H	1	32	65
Ethyl alcohol	CH_3CH_2—O—H	1	46	78
Water	H—O—H	2	18	100

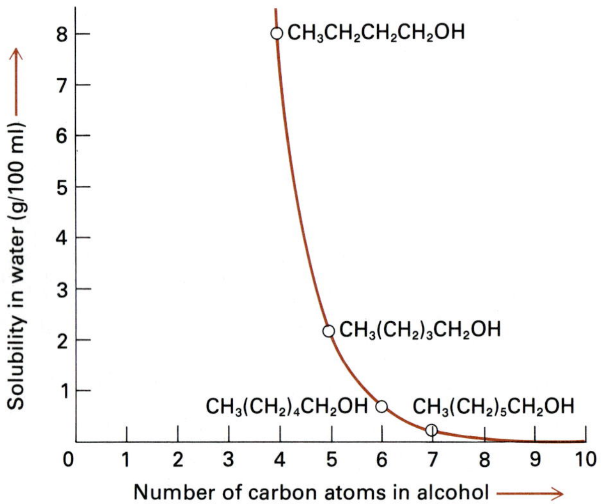

FIGURE 17-3

Water solubilities of alcohols.

Because they can form hydrogen bonds with water molecules, the small alcohols are very soluble in water—in contrast to the alkanes, which are essentially insoluble. As the size of the nonpolar alkyl group increases, the solubility of the alcohol in water decreases (Figure 17-3).

EXERCISE 17-1 Consider the molecules

(1) $CH_3CH_2CH_2CH_3$ (2) $CH_3CH(OH)CH_3$

(3) $CH_3CH_2OCH_3$

(a) Which is the least soluble in water?
(b) Which has the highest boiling point?
(c) Which has the lowest boiling point?

Like water, alcohols are very weak acids, dissociating to a small extent in water to yield alkoxide ions and hydronium ions:

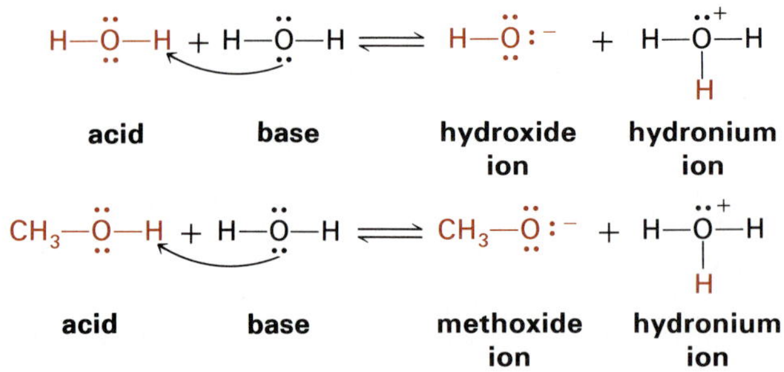

Methyl and ethyl alcohols are comparable to water in acid strength, and the higher alcohols are even less acidic.

Alcohols are also weak Brønsted–Lowry and Lewis bases—just as water is:

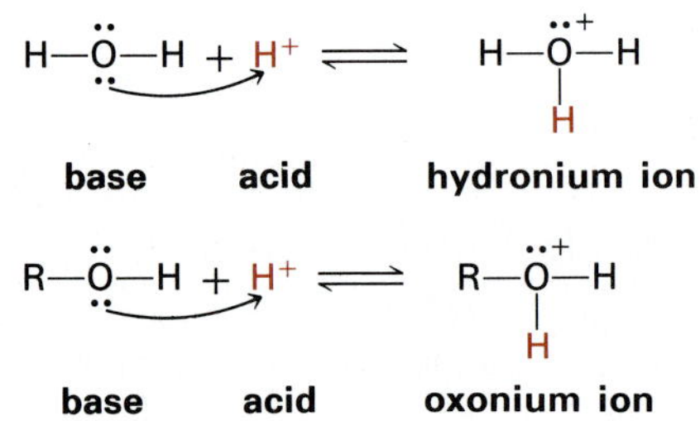

We shall see many examples of the formation of **oxonium ions.** In both hydronium and oxonium ions, the oxygen bears a positive charge because it has, in effect, lost a half-share of the electron pair that it donated to the proton.

So far in writing organic structures, we have usually included lone pairs of electrons when writing formulas of compounds containing atoms such as oxygen or the halogens. In general, we hereafter omit these lone pairs of electrons in writing formulas, except in cases where we want to show how they enter into the mechanism of a reaction.

Like alkenes, alcohols are prevalent in nature. The following structures of some naturally occurring alcohols illustrate the wide variety of molecules in which the —OH group is found:

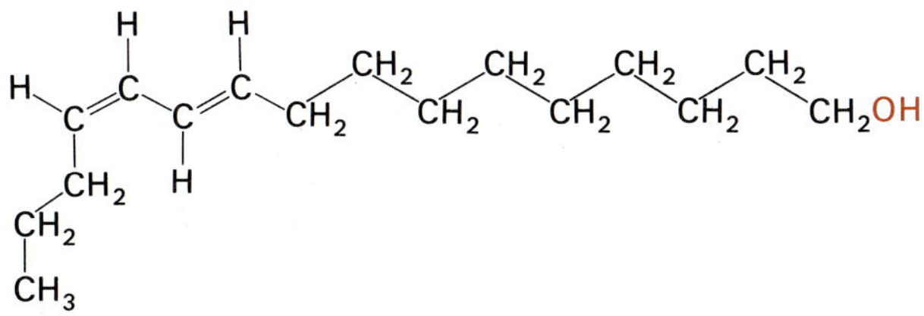

bombykol
(10-*trans*,12-*cis*-hexadecadien-1-ol; sex attractant of silkworm moth, physiologically active at concentrations of 10 μg per ml of air)

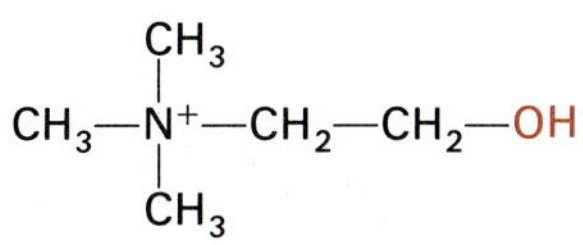

choline
(building block for the neurotransmitter acetylcholine)

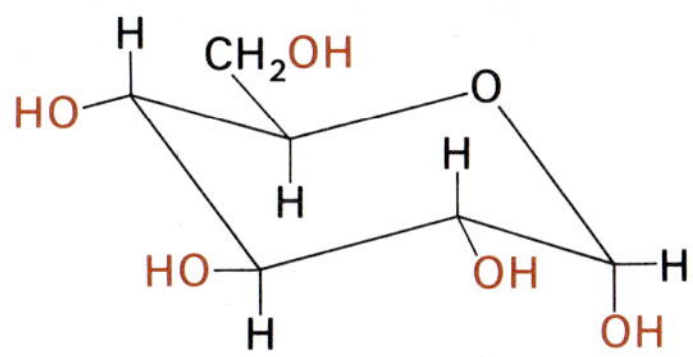

α-D-glucose
(a simple sugar)

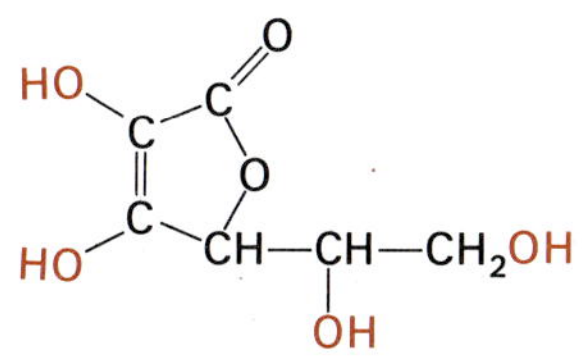

ascorbic acid
(vitamin C)

Phenols are aromatic alcohols—compounds with the hydroxyl group attached to a benzene ring. The following are structures of some simple phenols:

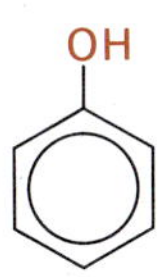

phenol

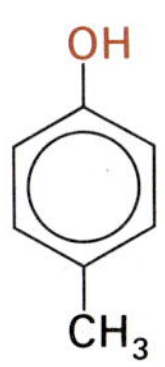

***p*-cresol**

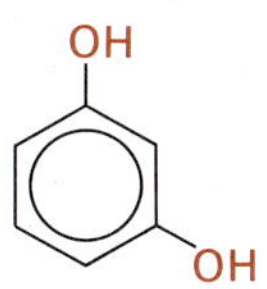

resorcinol

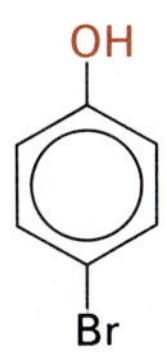

***p*-bromophenol**

The parent compound (phenol) itself is a low-melting solid, only slightly soluble in water at room temperature but very soluble in alcohol, ether, and other organic solvents. Phenols are much stronger acids than alcohols. The pK_a for phenol is 10; that of ethyl alcohol, in contrast, is 16:

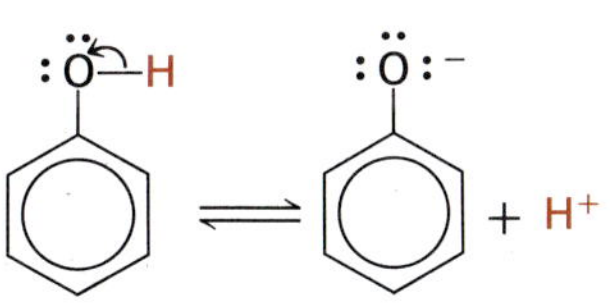

$K_a = 10^{-10}$ p$K_a = 10$

$$CH_3CH_2{-}\ddot{O}{-}H \rightleftharpoons CH_3CH_2{-}\ddot{O}{:}^- + H^+$$

$K_a = 10^{-16}$ p$K_a = 16$

Phenol, sometimes called carbolic acid, was the first widely used antiseptic (a substance that kills microorganisms on living tissue) and disinfectant (a substance that kills microorganisms on inanimate objects such as floors and walls). Joseph Lister used it as an operating room disinfectant and for antiseptic surgery in 1867. However, phenol can burn the skin and is poisonous, so it is no longer used as an antiseptic. Phenol and the cresols (*ortho-*, *meta-*, and *para-*) are used in dilute solution in commercial disinfectants. Phenols are extensively used in industrial processes such as the synthesis of dyes and plastics, and they occur widely in nature:

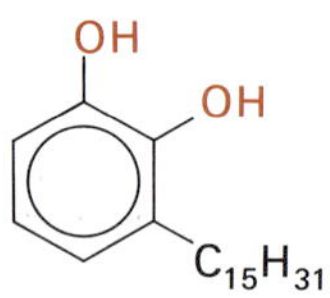

urushiol
(irritant in poison ivy)

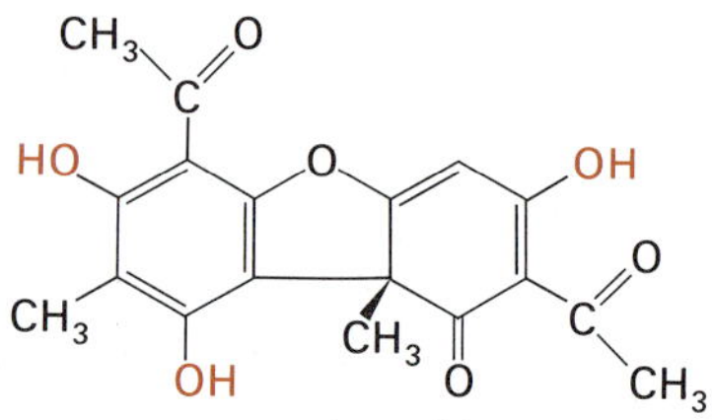

usnic acid
(common yellow lichen pigment)

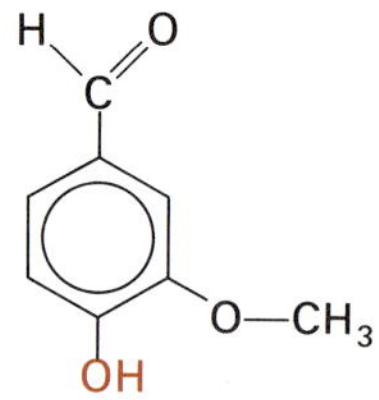

vanillin
(fragrant component of vanilla bean, responsible for the flavor of "vanilla flavoring" and of extract of vanilla)

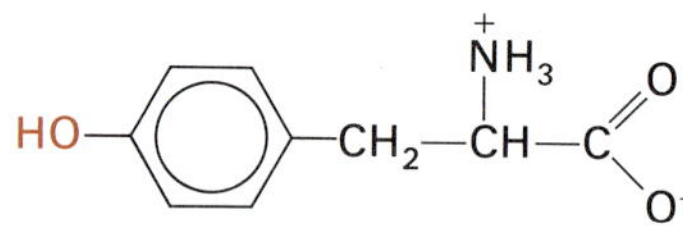

tyrosine
(an α-amino acid found in proteins; the structure shown is that at physiological pH)

17-2 NOMENCLATURE OF ALCOHOLS

Alcohols are classified as primary (1°), secondary (2°), or tertiary (3°), depending on the number of *other* carbons bonded to the carbon bearing the hydroxyl group:

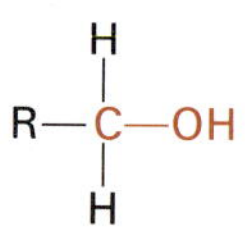

1° alcohol

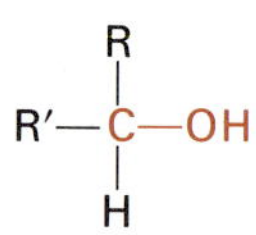

2° alcohol

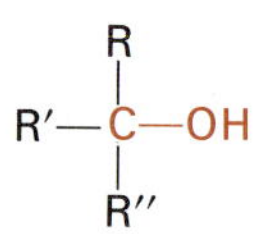

3° alcohol

The following rules govern the naming of alcohols according to the IUPAC system.

1. Alcohols are named by removing the *-e* ending from the corresponding alkane, and substituting the ending *-ol*:

CH_3—OH
methanol
(methyl alcohol)

CH_3—CH_2—OH
ethanol
(ethyl alcohol)

Common names are given in parentheses under the IUPAC names.

2. The position of the hydroxyl group is given by a number prefixed to the name. The numbering begins at the end of the carbon chain closer to the —OH group:

CH_3—CH_2—CH_2—OH
1-propanol
(propyl alcohol)

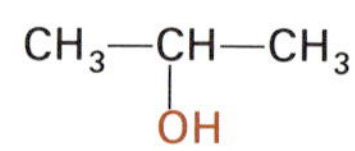

2-propanol
(isopropyl alcohol)

Condensed formulas: $CH_3CH_2CH_2OH$ $CH_3CHOHCH_3$

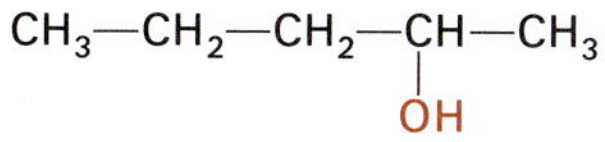

2-pentanol

Condensed formula: $CH_3CH_2CH_2CHOHCH_3$

3. The name of a branched alcohol is based on the longest unbranched chain of carbons that contains the hydroxyl group, and the numbering begins at the end of the chain closer to the hydroxyl group:

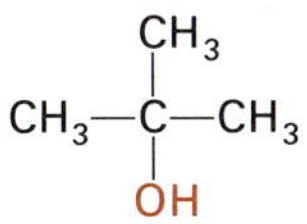

2-methyl-2-propanol
(*tert*-butyl alcohol)

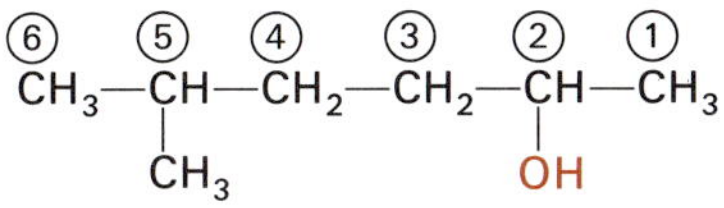

5-methyl-2-hexanol

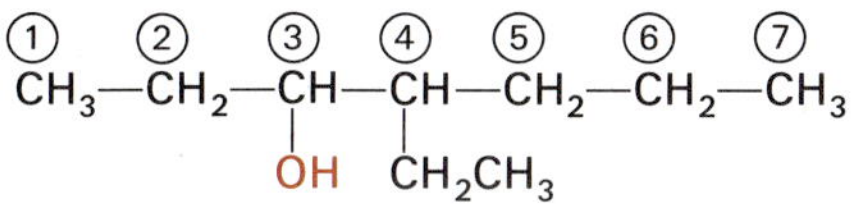

4-ethyl-3-heptanol

4. Compounds with two hydroxyl groups are called **diols;** compounds with three are called **triols:**

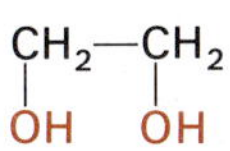

1,2-ethanediol
(ethylene glycol)

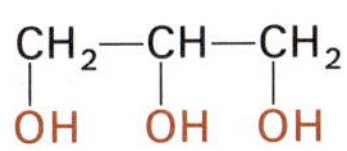

1,2,3-propanetriol
(glycerol)

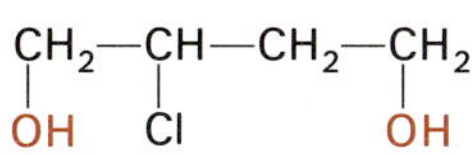

2-chloro-1,4-butanediol

(Compounds with two hydroxyl groups on the *same* carbon are almost never isolated, although we shall see that they sometimes occur as reactive intermediates.) Ethylene glycol is used as the basis for permanent antifreeze. It is infinitely soluble in water, lowers the freezing point of the water (as does any solute, Chapter 9), and does not boil off. Because of the *two* hydroxyl groups that can form hydrogen bonds, ethylene glycol boils at the high temperature of 198°C. Glycerol is used in hand lotions and as an additive to food and tobacco—the *three* hydroxyl groups form hydrogen bonds with water molecules, thereby helping maintain moistness. Glycerol plays a significant biological role as the backbone of two major classes of lipid: phospholipids (found in all cell membranes) and fats and oils (used by cells for energy storage; see Chapter 24). Glycerol occurs in the intestine as a product of fat digestion.

EXERCISE 17-2 (a) Write structural formulas for (1) 2-pentanol, (2) 3-methyl-2-butanol, and (3) 1,2-propanediol.

(b) Name the following compounds:

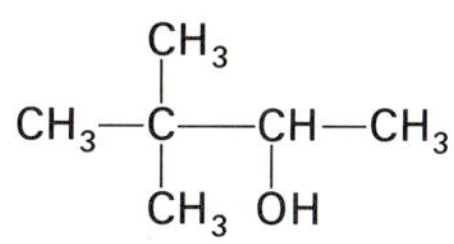

(4)

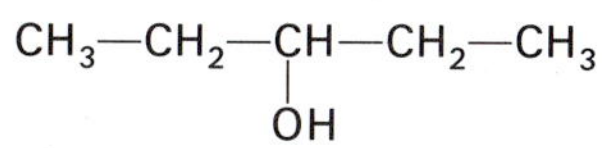

(5)

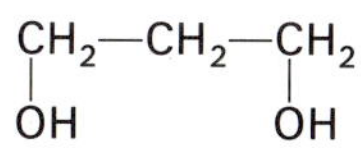

(6)

17-3
PREPARATION OF ALCOHOLS

Methyl alcohol was once called wood alcohol because it was prepared by the distillation of wood in the absence of air. It is now prepared by the catalytic hydrogenation of carbon monoxide at high temperature and pressure:

$$CO(g) + 2H_2(g) \xrightarrow{ZnO—CrO_3} CH_3OH$$

Methyl alcohol is an important laboratory reagent and solvent; it is also used to denature ethyl alcohol (make it unfit to drink). Methyl alcohol is extremely toxic; small amounts can cause nausea, blindness, and even death.

Ethyl alcohol (the intoxicating component of alcoholic beverages) is only mildly toxic. It is, therefore, often used as a solvent for flavorings and drugs. In addition, it is used as rubbing alcohol, as an antiseptic, and as a laboratory solvent. Ethyl alcohol can be prepared by the fermentation of sugars by enzymes produced by living yeast cells. The enzymes act as catalysts in the reaction

$$\underset{\text{glucose}}{C_6H_{12}O_6} \xrightarrow{\text{enzymes}} \underset{\text{ethyl alcohol}}{2CH_3CH_2OH} + 2CO_2(g)$$

If the sugar comes from fruit, the product is wine. The yeast needed to produce the catalytic enzymes is killed when the alcohol content of the mixture reaches 10% to 15% alcohol by volume, limiting the extent of the reaction. Stronger "spirits" are obtained by distillation or by the addition of more ethanol. Brandy is obtained by distilling wine, and whiskey by distilling fermented grain. Any kind of carbohydrate can be used to produce ethyl alcohol—grains and sugar cane, for example. The carbohydrates are first broken down and converted to glucose by enzymes.

The energy crisis has promoted interest in using **gasohol,** a mixture of 10% ethyl alcohol and 90% unleaded gasoline in place of "pure" gasoline as automobile fuel. The production of gasohol is now subsidized by the federal government and by some state governments, but it is not yet clear whether gasohol produces more energy than is consumed in its production (that there is a *net* gain in energy in the production and use of gasohol). Because the energy used to produce the gasohol can be in the form of nonpetroleum sources such as coal, however, the use of gasohol definitely is a way to conserve petroleum.

Alcohols also are prepared by other methods such as the hydration of alkenes (Section 15-8). Isopropyl alcohol, for example, is formed by the addition of water to propene:

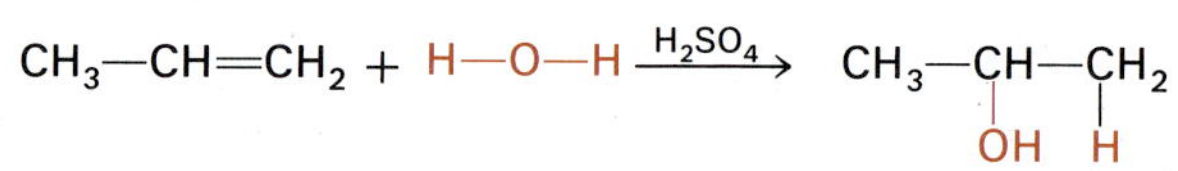

propene **isopropyl alcohol**

Like ethanol, isopropyl alcohol in a 70% solution is used as rubbing alcohol and as an antiseptic to cleanse the skin before a blood sample is taken or an injection is given.

Hydration of an alkene takes place in the citric acid (or Krebs, or tricarboxylic acid) cycle, the central pathway for the aerobic oxidation of fuel molecules in the body (see also Section 26-3). In one step of the cycle, fumarate is hydrated to yield malate:

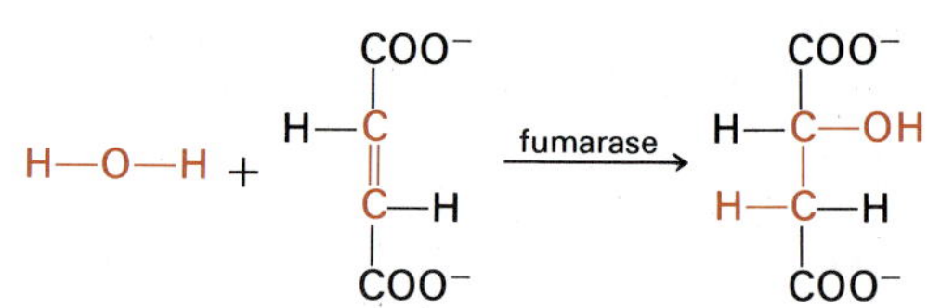

fumarate **malate**

The enzyme fumarase catalyzes the reaction. Although the molecules are somewhat more complex than those we have discussed so far, the reaction and its mechanism are the same that we saw in Section 15-8.

17-4 REACTIONS OF ALCOHOLS: REACTIONS WITH ACTIVE METALS

Alcohols, like water, react with alkali metals to liberate hydrogen gas and leave an alkoxide. In this reaction, the O—H bond of the alcohol is broken:

$$2Na + 2CH_3CH_2\text{—}O\text{—}H \longrightarrow 2CH_3CH_2O^- + Na^+ + H_2(g)$$

sodium ethoxide

An alkali metal reacts less vigorously with an alcohol than with water, and scraps of sodium are disposed of in the laboratory by dropping them in isopropyl alcohol. The alkoxide ions are strong bases, as strong as or stronger than the hydroxide ion. They are commonly used in reactions requiring a strong base in a nonpolar solvent, because alkoxide ions are soluble in nonpolar solvents and hydroxide ions are not.

Alkoxide ions react with water to yield the alcohol and the hydroxide ion:

$$CH_3CH_2CH_2O^- + H_2O \rightleftharpoons CH_3CH_2CH_2OH + OH^-$$

This is a Brønsted–Lowry acid–base reaction. In the forward reaction, water is the acid (the proton donor) and the alkoxide ion is the base (the proton acceptor). (What is the acid, and what the base, in the reverse reaction?)

EXERCISE 17-3 Write an equation for the reaction between potassium metal and isopropyl alcohol.

17-5 REACTIONS OF ALCOHOLS: SUBSTITUTION

Like the alkyl halides (Chapter 16), alcohols undergo substitution reactions by two different mechanisms, depending on the alcohol's structure. In these reactions the C—OH bond is broken, and another atom or group is substituted for the hydroxyl group. With tertiary alcohols, as with tertiary halides, the mechanism is S_N1, as shown for the reaction between *tert*-butyl alcohol and hydrogen bromide. (The first step, a protonation of the OH group, will be discussed shortly.)

Step 1: Protonation of OH

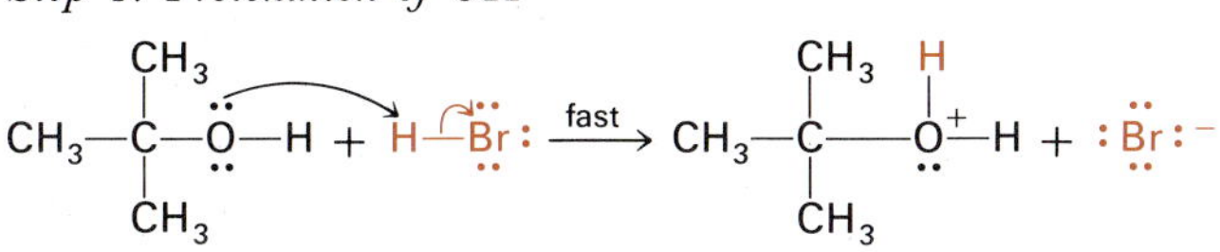

oxonium ion

Step 2: Formation of carbonium ion

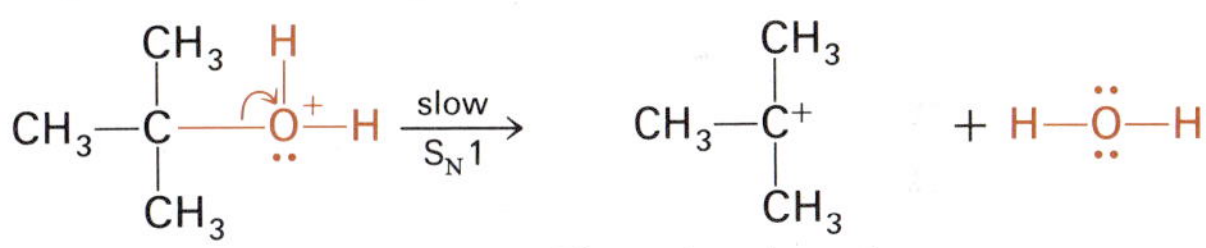

3° carbonium ion

Step 3: Addition of Br⁻

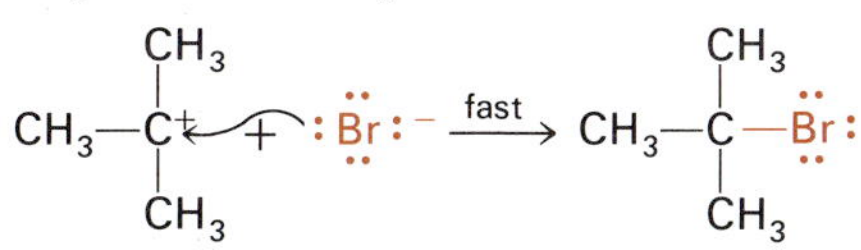

The *net* reaction is just the substitution of Br^- for OH^-:

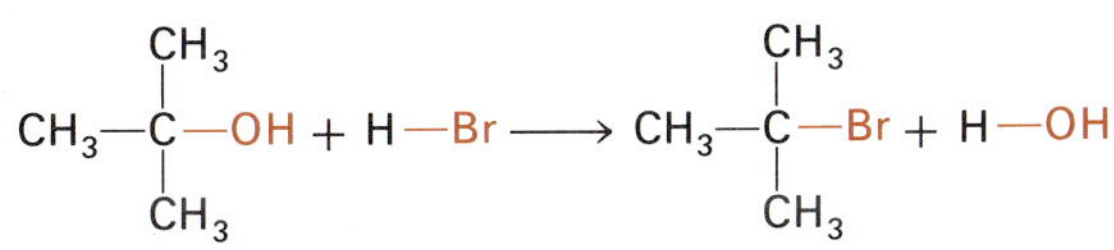

The rate-determining step (step 2) involves only the oxonium ion as a reactant, so the mechanism is unimolecular—S_N1. The mechanism is essentially the same as that we saw earlier for the substitution of a tertiary alkyl halide (Section 16-1). The only difference is the additional (rapid) first step—the protonation of the OH group to yield an oxonium ion.

Why is this additional step (the formation of the oxonium ion) needed? Why isn't the carbonium ion formed immediately, as is the case with *tert*-butyl bromide? The answer is that the hydroxide ion is a strong base—in contrast to the bromide ion, which is a weak base. As a result, in the reaction

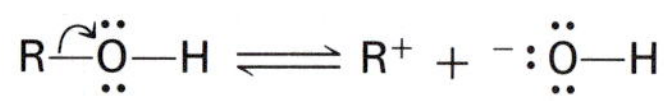

the position of equilibrium lies far to the left. (The reverse reaction is favored.) Very little carbonium ion is formed.

In the reaction,

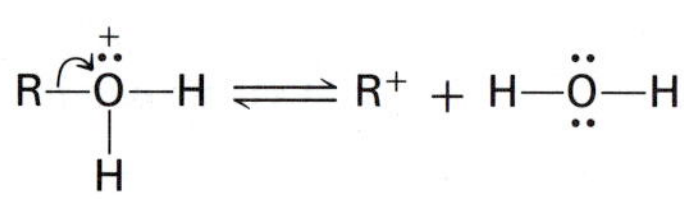

the position of equilibrium is more favorable for the formation of the carbonium ion, because water is a much weaker base than the hydroxide ion.

In these reactions, water and the hydroxide ion are called leaving groups (Section 16-1), and *water is a much better leaving group than the hydroxide ion.* By protonation, the hydroxide ion (which is a poor leaving group) is converted into the better leaving group, water. (Thus, substitution reactions of alcohols are always carried out in acid solution. In the example given, the proton was provided by the HBr, which is a strong acid.) In general, *weak bases are good leaving groups.* Examples of good leaving groups are Cl^-, Br^-, I^-, and H_2O.

Primary alcohols, like primary halides, undergo substitution by an S_N2 mechanism. The primary carbonium ion is unstable and is not formed. An example is the reaction between ethyl alcohol and hydrogen bromide:

Step 1: Protonation of OH

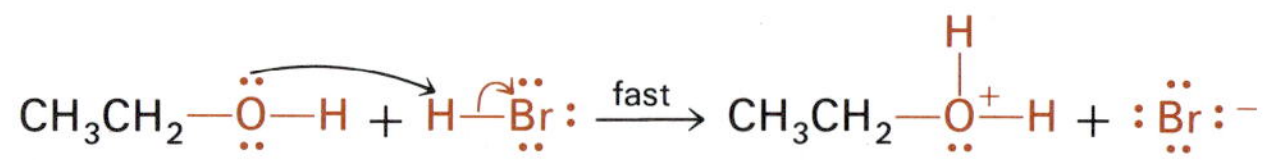

Step 2: Substitution of Br^- for H_2O

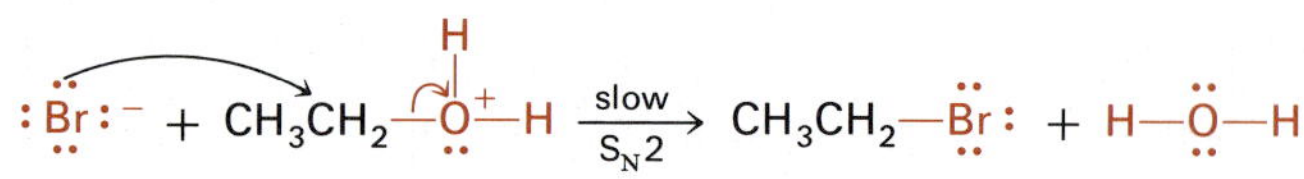

As with the tertiary alcohol, the first step is the protonation of the OH group to convert it to the better leaving group, H_2O. The second step is rate-determining, and is bimolecular because it involves two reactants—the oxonium ion and bromide ion. The transition state for the second step has the geometry shown, with the C—Br bond partially formed and the C—O bond partially broken:

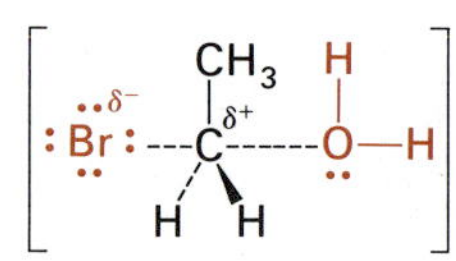

Like secondary alkyl halides, secondary alcohols can react by *both* S_N1 and S_N2 mechanisms.

The rate of the substitution is influenced by the structure of the alcohol. Tertiary alcohols are rapidly converted to halides at room temperature, but primary alcohols require heating for several hours. A common laboratory test that distinguishes between primary, secondary, and tertiary water-soluble alcohols is based on these facts. In the Lucas test, an aqueous solution of HCl and $ZnCl_2$ (a catalyst) is added to the alcohol. Tertiary alcohols react instantly to produce the alkyl chloride, secondary alcohols react in five to ten minutes, and primary alcohols take hours. Alkyl chlorides are insoluble in water, so their presence is easily detected by a cloudiness in the solution or by the separation of a second layer.

EXERCISE 17-4 (a) What reagent would you use to prepare 2-iodo-2-methylpentane from 2-methyl-2-pentanol? Write a mechanism for the reaction, using arrows to show electron movements. (Be sure to indicate which step is rate determining.)

(b) What reagent would you use to prepare 1-chloropentane from 1-pentanol? Write a mechanism for the reaction, using arrows.

17-6 REACTIONS OF ALCOHOLS: DEHYDRATION

We saw in Chapter 15 that alcohols can be converted to alkenes by **dehydration** (the elimination of a molecule of water), which occurs in the presence of an acid catalyst. This dehydration (elimination) can accompany substitution. For example, in the reaction between *tert*-butyl alcohol and hydrogen bromide, one product (as we have seen) is *tert*-butyl bromide. But *another* product, an alkene, is formed by elimination:

Step 1: Protonation of OH

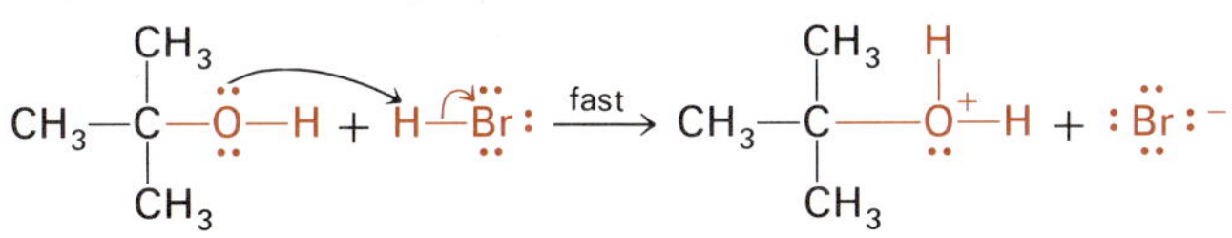

oxonium ion

Step 2: Formation of carbonium ion

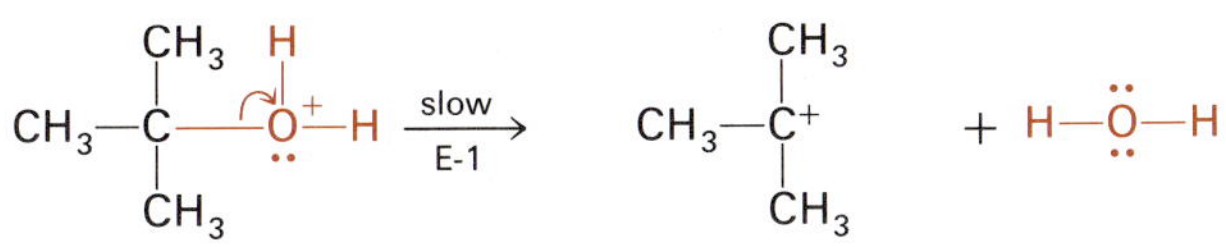

3° carbonium ion

Step 3: Elimination of H^+

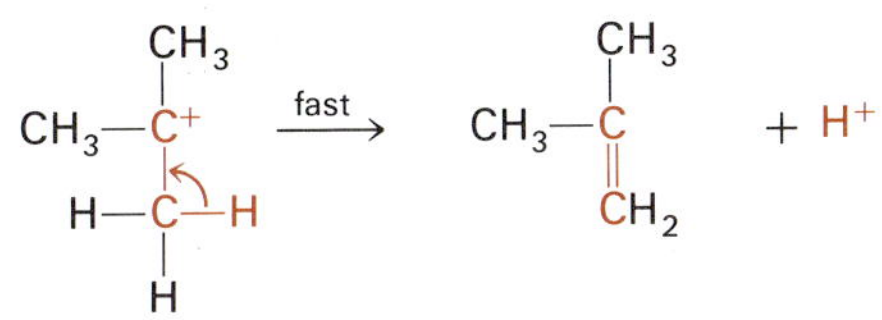

methyl propene

Again, the rapid first step is the protonation of the hydroxyl group to yield the better leaving group, water. The second (and rate-determining) step involves only the oxonium ion as reactant, so the mechanism is unimolecular, E-1. In a final, rapid step, the carbonium ion loses a proton to yield the alkene. Note that the first two steps of the E-1 mechanism are identical to those of the S_N1 mechanism (Section 17-5). The same carbonium ion is formed in both cases. In the S_N1 mechanism, the product is an alkyl halide or other substitution product; in the E-1 mechanism, the product is an alkene.

The net reaction, which is the reverse of the hydration of an alkene, is

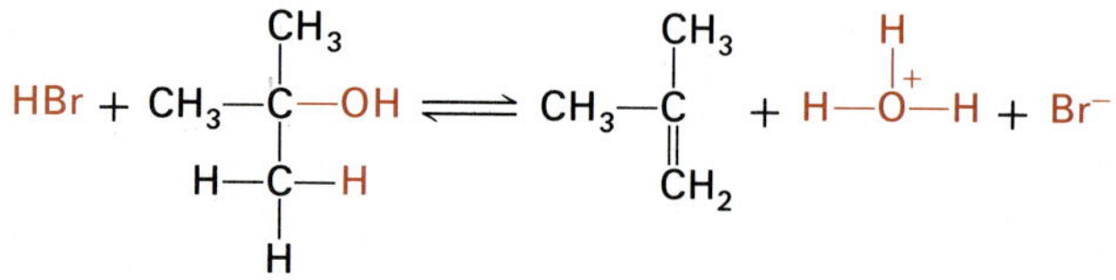

The forward reaction, the dehydration of an alcohol, is favored by the removal of either water or the alkene from the reaction mixture. The reverse reaction, the hydration of an alkene, is favored by the presence of a large quantity of water. The mechanism of the dehydration is essentially the same as that for the elimination of HX from an alkyl halide—the only difference is the addition of the proton to convert the poor leaving group (OH^-) to the good leaving group (H_2O).

Because tertiary carbonium ions are the most stable and primary are the least, we would expect dehydration to occur most readily with tertiary alcohols and least readily with primary alcohols—and this is the case. The following are other examples of the dehydration of alcohols (a fairly strong acid such as H_3PO_4 or a strong acid such as H_2SO_4 is used):

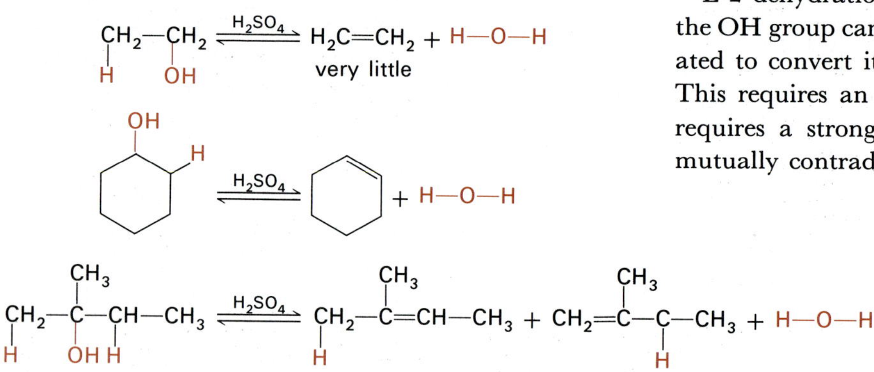

Dehydration reactions also occur in biological systems. An example is the dehydration of citrate to yield *cis*-aconitate, another step in the citric acid cycle:

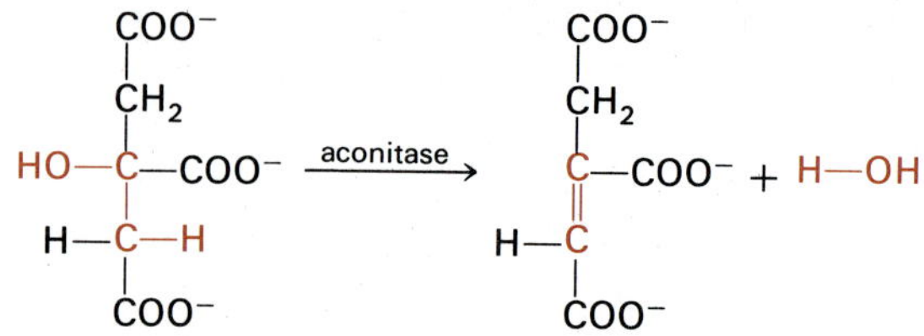

citrate **cis-aconitate**

The reaction is catalyzed by the enzyme aconitase.

EXERCISE 17-5 What reagent would you use to dehydrate cyclopentanol? Draw the structural formula of the product of the reaction.

E-2 dehydrations of alcohols do not occur. Before the OH group can be removed, it must first be protonated to convert it to the better leaving group, H_2O. This requires an *acid* solution; but the E-2 reaction requires a strong *base*—and these requirements are mutually contradictory.

17-7 REACTIONS OF ALCOHOLS: OXIDATION

Primary and secondary alcohols are *partially* oxidized by common oxidizing agents such as potassium permanganate ($KMnO_4$) or potassium dichromate ($K_2Cr_2O_7$). Primary alcohols are oxidized to aldehydes, which are rapidly oxidized further to organic acids unless the aldehyde can be removed from the reaction mixture—say, by distillation.

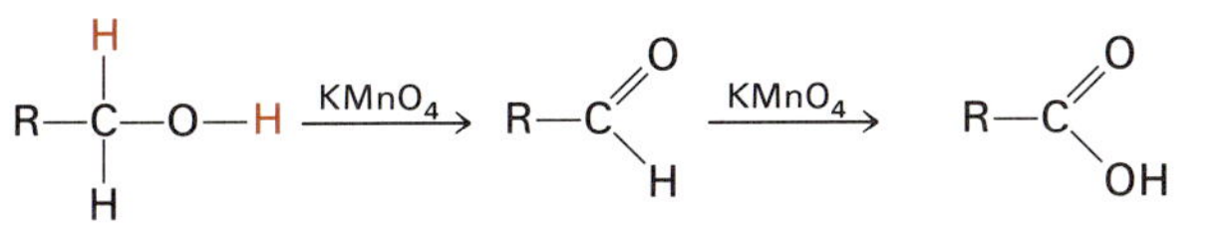

1° alcohol **aldehyde** **carboxylic acid**

Note that in the first step, the oxidation of the alcohol, one hydrogen is removed from the OH group, and the other from the carbon bearing the OH group. The mechanism of the reaction is not completely understood.

Secondary alcohols are oxidized to ketones, which are not easily oxidized further:

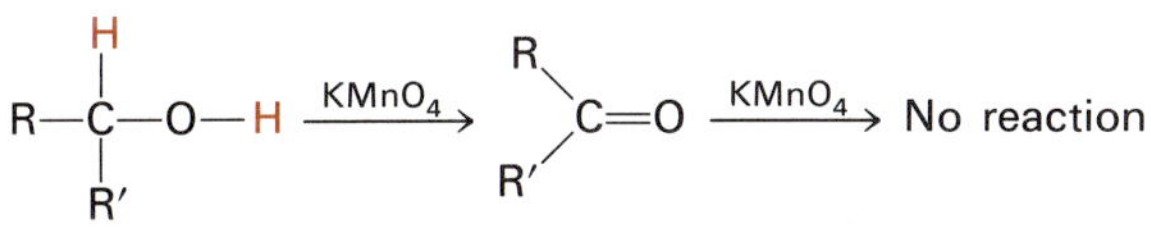

2° alcohol **ketone**

Again, one hydrogen is removed from the OH group, and the other from the carbon bearing the OH group.

Tertiary alcohols are resistant to oxidation by these oxidizing agents:

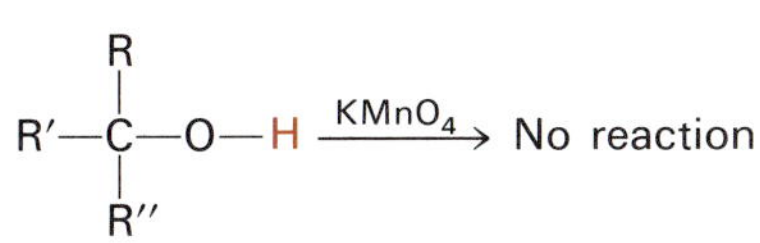

3° alcohol

In the tertiary alcohol, the carbon bearing the OH group has no additional hydrogen atoms to be removed on oxidation; thus tertiary alcohols are not easily oxidized. All alcohols, however, can be oxidized to CO_2 and H_2O by strong oxidizing agents—for example, by burning in air.

The oxidation of ethanol to acetic acid by dichromate ion in acid solution is the reaction that provides the breath analyzer test used for drunk-driving suspects:

$$16H^+ + \underset{\text{(yellow-orange)}}{2Cr_2O_7^{2-}} + 3CH_3CH_2OH \longrightarrow 3CH_3C(=O)OH + 11H_2O + \underset{\text{(green)}}{4Cr^{3+}}$$

The solution changes from the yellow-orange of the $Cr_2O_7^{2-}$ to the green of the Cr^{3+} as the reaction progresses, and the color change is used to give a quantitative measure of the alcohol present in the breath.

Oxidation of ethanol to acetic acid (the principal ingredient of vinegar) by oxygen in the air is responsible for the souring of wine (wine turning to vinegar). The reaction is catalyzed by enzymes produced by certain bacteria:

$$O_2(g) + \underset{\textbf{ethanol}}{CH_3CH_2OH} \xrightarrow{\text{enzymes}} \underset{\textbf{acetic acid}}{CH_3C(=O)OH} + H_2O$$

In the body, ethanol is oxidized to acetaldehyde by enzymes in the liver:

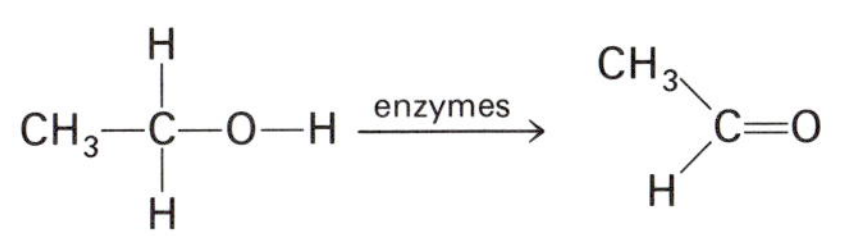

ethanol **acetaldehyde**

Acetaldehyde is readily oxidized further to acetic acid, which is a normal constituent of cells and is not toxic. When methanol is ingested, it is also oxidized by enzymes in the liver; in this case to formaldehyde:

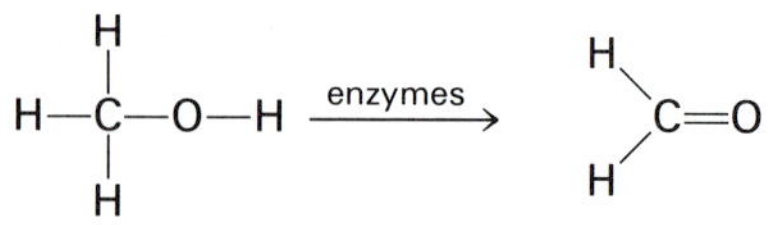

methanol **formaldehyde**

Formaldehyde reacts rapidly; it causes the lenses of the eyes to become opaque, producing blindness, and it denatures enzymes. The antidote to methanol poisoning is ethanol administered intravenously. The ethanol saturates the liver enzymes and prevents them from oxidizing the methanol. The body then slowly excretes the methanol.

Many biological reactions are oxidation–reduction reactions. Yet another step in the citric acid cycle is the oxidation of a 2° alcohol (malate) to a ketone (oxaloacetate):

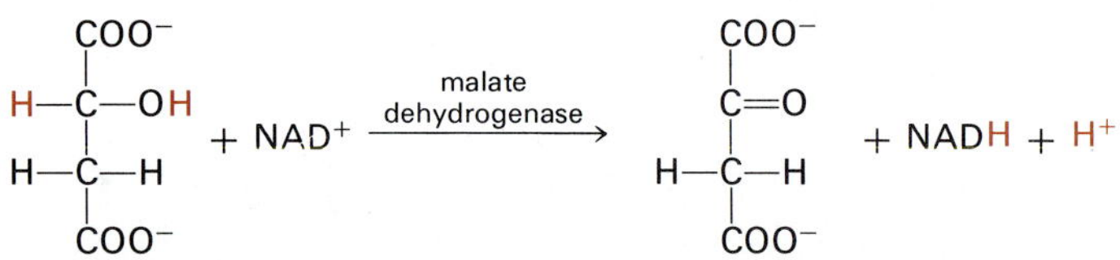

malate **oxaloacetate**

The oxidizing agent is NAD^+, the oxidized form of nicotinamide adenine dinucleotide (Section 12-4). The structure of NAD^+ is given in Chapter 26 together with a discussion of the citric acid cycle, of which we have seen several steps in this chapter. The reaction is catalyzed by the enzyme malate dehydrogenase.

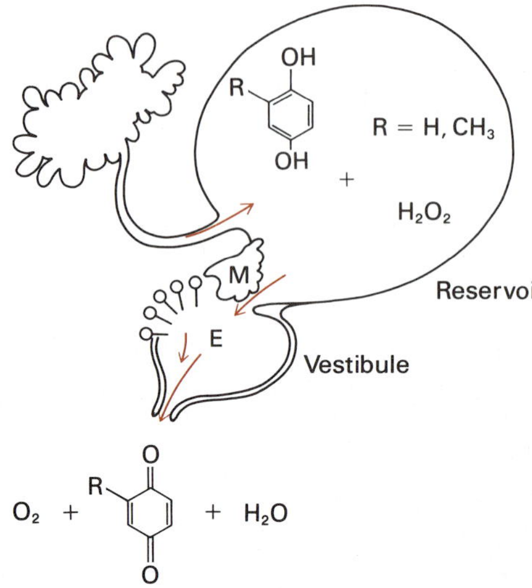

FIGURE 17-4

Diagram of gland of bombardier beetle. The reservoir contains hydroquinones and hydrogen peroxide obtained from a duct that drains secretory tissue. The enzymes (E) are secreted in the vestibule. The muscle (M) controls the valve between the two compartments. In quinone R = H, and in methyl quinone R = CH_3.

An example from nature of the oxidation of a phenol to a ketone is seen in the extraordinary defense mechanism of certain beetles of the genus *Brachinus,* commonly called bombardier beetles. These beetles eject from the tip of the abdomen a defensive spray composed of a hot (100°C) aqueous solution of two diketones, quinone and methyl quinone (Figure 17-4). The quinones are strongly repellent to many predators.

The abdominal reservoir of the beetle contains a solution of hydroquinones and the oxidizing agent, hydrogen peroxide, which is present in excess. When attacked, the beetle squeezes some of this solution into the vestibule, where enzymes catalyze both the oxidation of the hydroquinones to quinones by H_2O_2 and the decomposition of the excess peroxide:

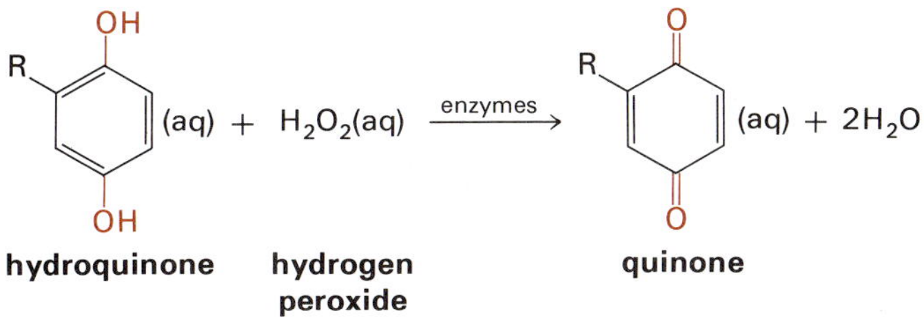

$$\Delta H = -203 \text{ kJ/mole}$$

The heat evolved is sufficient to heat the spray to the boiling point and to vaporize about one-fifth of it. The pressure of the water vapor and the oxygen gas causes an explosive ejection of the hot spray. Figure 17-5 shows a beetle defending itself against a tree frog.

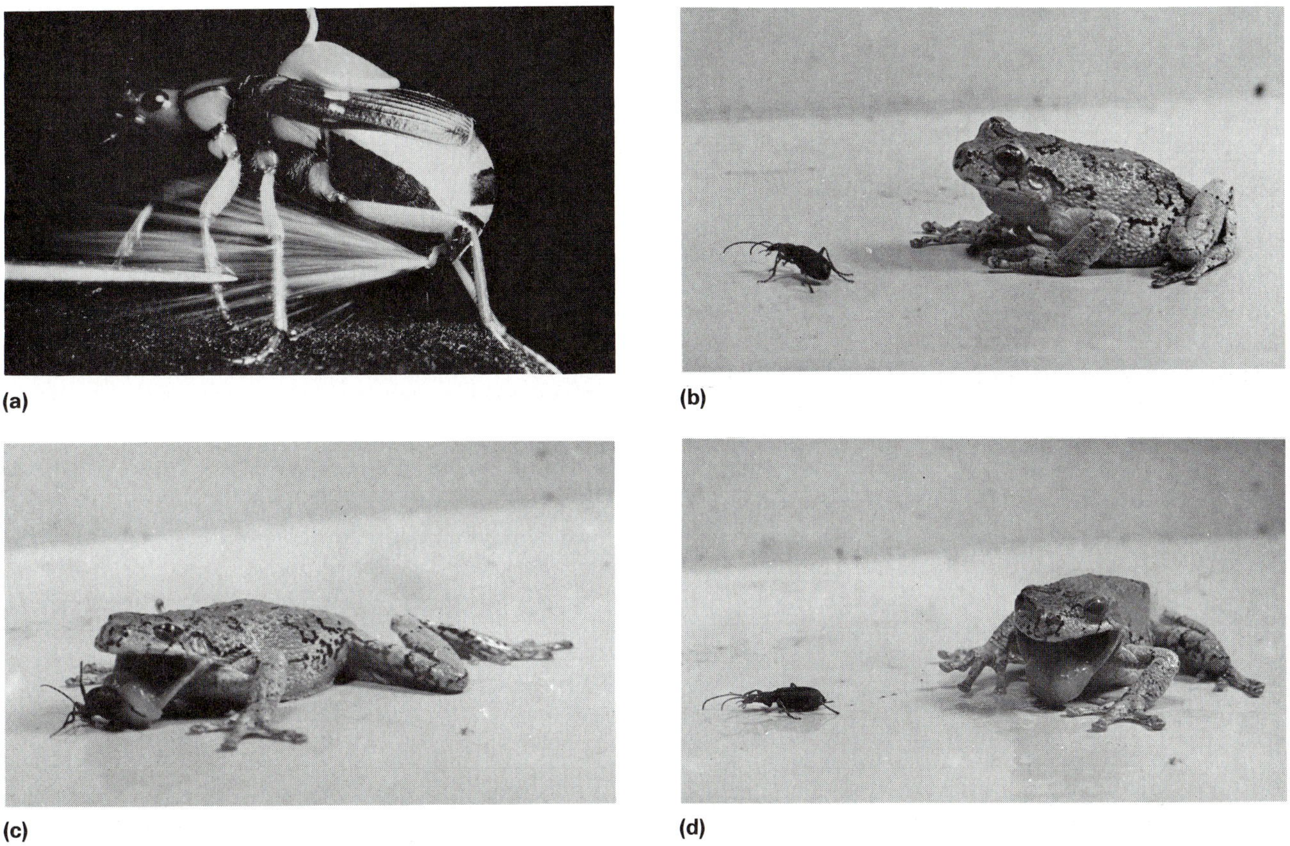

FIGURE 17-5

A bombardier beetle in action. **(a)** Bombardier beetle (*Stenaptinus ignitus*) firing in response to stimulation of its left front leg with forceps. The forceps are seen as a horizontal bar extending into the picture from the lower left. **(b)** A tree frog (*Hyla versicolor*) watches a bombardier beetle. **(c)** The frog strikes with its tongue. **(d)** The frog rejects the beetle. [(a) courtesy of D. Aneshansley and T. Eisner; (b), (c), and (d) courtesy of T. Eisner and C. Walcott.]

EXERCISE 17-6 Draw structural formulas for the products of the following reactions:

(a) 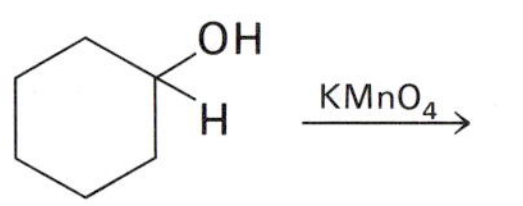

cyclohexanol

(b) 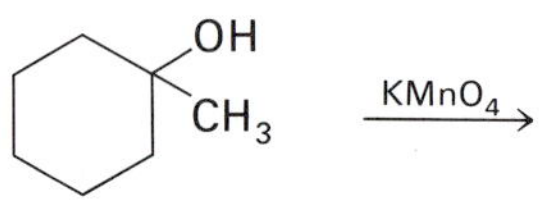

1-methylcyclohexanol

(c) 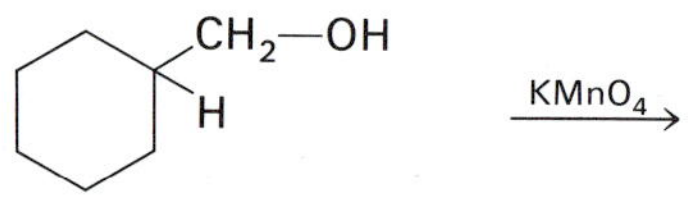

(hydroxymethyl)cyclohexane

Reduction is the reverse of oxidation. A reduction reaction forms, for example, an alcohol from an aldehyde:

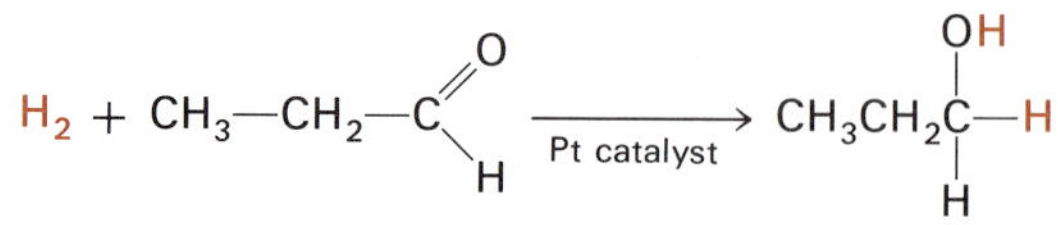

propanal (aldehyde) **1-propanol (alcohol)**

This reaction is just like the hydrogenation of an alkene. The hydrogenation of ethylene, for example, is the addition of H_2 to the π bond of $CH_2{=}CH_2$ to yield ethane ($CH_3{-}CH_3$). Here we added H_2 to the π bond of an aldehyde to yield an alcohol.

17-8 STRUCTURE AND PROPERTIES OF ETHERS

Ethers are much less reactive than alcohols, almost as unreactive as alkanes, and so are widely used as solvents. Ethers have the general formula R—O—R′, and are isomeric to alcohols. For example, dimethyl ether (CH_3OCH_3) and ethyl alcohol (CH_3CH_2OH) are structural isomers, both having the molecular formula C_2H_6O.

Ethers are named by naming the two alkyl groups attached to the oxygen, and adding the word *ether*. If the two alkyl groups are the same, the prefix di- often is used (but it sometimes is omitted, so dimethyl ether and methyl ether refer to the same compound).

$CH_3CH_2{-}O{-}CH_3$
methyl ethyl ether

$CH_3CH_2{-}O{-}CH_2CH_3$
diethyl ether (or ethyl ether)

$CH_3CH_2CH_2{-}O{-}CH_3$
methyl propyl ether

All the hydrogen atoms in ethers are bonded to carbon, and so cannot participate in hydrogen bonds. Ethers therefore boil at much lower temperatures than the alcohols of which they are isomers (see Table 17-3).

Although ethers have no hydrogen atoms that can

TABLE 17-3
Boiling points of some alcohols and isomeric ethers

Formula	Name	Structure	Boiling point (°C)
C_2H_6O	Dimethyl ether	CH_3OCH_3	−25
	Ethyl alcohol	CH_3CH_2OH	78
C_3H_8O	Methyl ethyl ether	$CH_3OCH_2CH_3$	11
	n-Propyl alcohol	$CH_3CH_2CH_2OH$	97
$C_4H_{10}O$	Diethyl ether	$CH_3CH_2OCH_2CH_3$	35
	n-Butyl alcohol	$CH_3CH_2CH_2CH_2OH$	118

participate in hydrogen bonds, the oxygen atom can participate in a hydrogen bond with a molecule that can provide the hydrogen atom—water, for example:

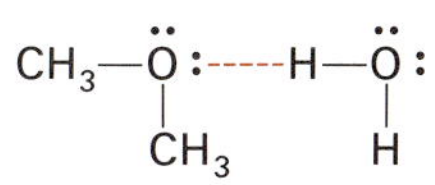

Ethers, therefore, have approximately the same solubilities in water as alcohols of comparable sizes.

Ether molecules are slightly polar because of the C—O bonds. The individual bond dipoles partially cancel each other to give a small net molecular dipole:

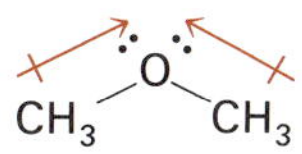
bond dipoles

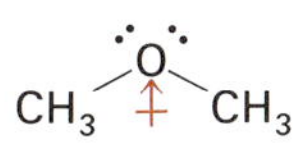
net molecular dipole

This small dipole moment gives ethers their special usefulness as solvents. Because of it, ethers can dissolve *both* hydrocarbons (which are nonpolar) *and* organic compounds with polar functional groups. Some cyclic ethers that are common laboratory solvents are

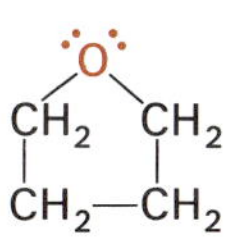
tetrahydrofuran

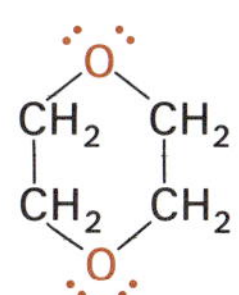
dioxane

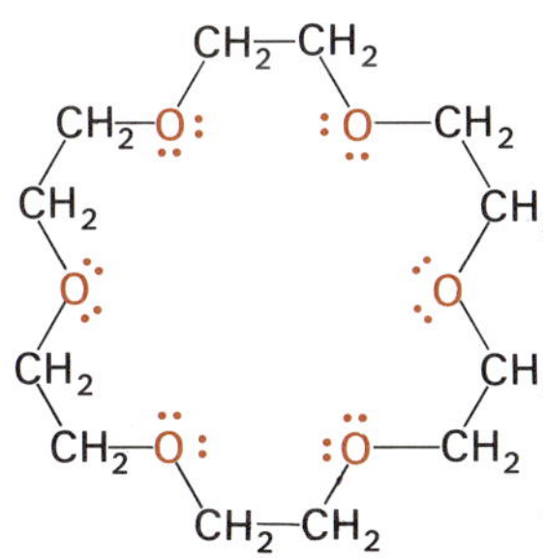
18-crown-6

If even a small amount of the "crown" ether 18-crown-6 is added to the nonpolar solvent benzene, the mixture is able to dissolve an inorganic salt such as $KMnO_4$, though the salt is almost totally insoluble in pure benzene. The K^+ interacts with the oxygens and occupies the center of the ring. Because most chemical reactions occur in solution, ethers play an important role in bringing together polar and nonpolar molecules for reaction.

Diethyl ether is most familiar as the general anesthetic usually called simply "ether"; it has been widely used since the mid-1840s. Diethyl ether is a colorless liquid with a high vapor pressure at room temperature (see Section 4-4). When its vapors are inhaled, ether acts as a depressant on the central nervous system. Because the effective dose is much smaller than the lethal one, ether is a relatively safe anesthetic, but it may produce nausea as a side effect. It must be used with care, both in hospitals and in laboratories, because of its high volatility and flammability.

The following two chlorine-containing ethers are strongly suspected of being human carcinogens; their use is strictly controlled by the Occupational Safety and Health Administration (OSHA).

$ClCH_2—O—CH_3$
chloromethyl methyl ether

$ClCH_2—O—CH_2Cl$
bis(chloromethyl)ether

In the name bis(chloromethyl)ether, *bis* means that there are *two* chloromethyl groups.

Ethers occur widely in nature. Here are a few examples:

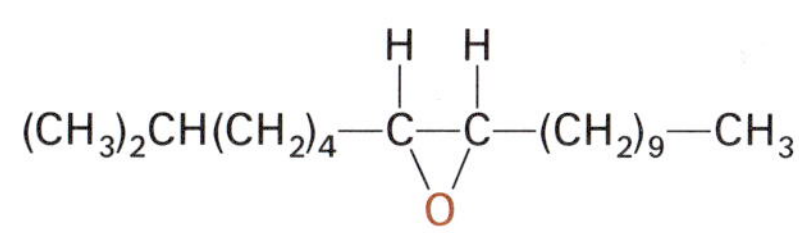
disparlure
(gypsy moth sex attractant)

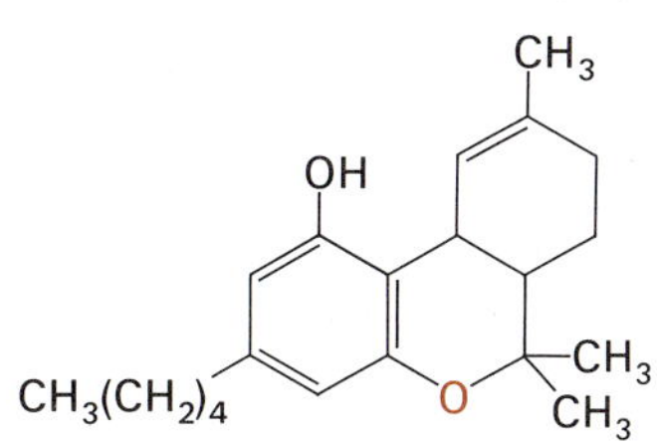
tetrahydrocannabinol
(marihuana)

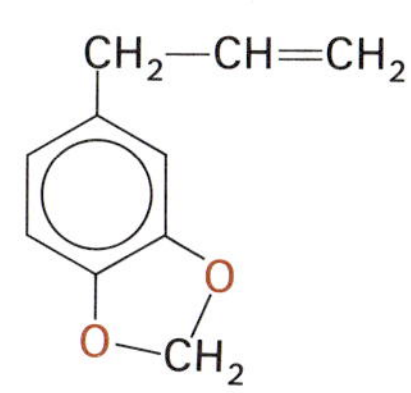
saffrole
(oil of sassafras)

17-9
PREPARATION OF ETHERS

Ethers can be prepared by elimination of a molecule of water from *two* molecules of alcohol, using an acid catalyst (in contrast to the elimination of a molecule of water from *one* molecule of alcohol, which yields an alkene):

$$CH_3CH_2OH + HOCH_2CH_3 \xrightarrow[\text{heat}]{H^+} CH_3CH_2{-}O{-}CH_2CH_3 + H_2O$$

Thus, we now see that heating ethyl alcohol with HBr can yield *three* products: CH_3CH_2Br by an S_N2 mechanism; $CH_2{=}CH_2$ by an *intra*molecular dehydration; and $CH_3CH_2{-}O{-}CH_2CH_3$ by an *inter*molecular dehydration. The relative amounts of the three products depend on the experimental conditions and on the specific alcohol and acid used. With primary alcohols it is possible to obtain a high yield of the ether, whereas tertiary alcohols yield primarily the alkene.

The mechanism for the formation of ethers consists of three steps:

Step 1: Protonation of OH

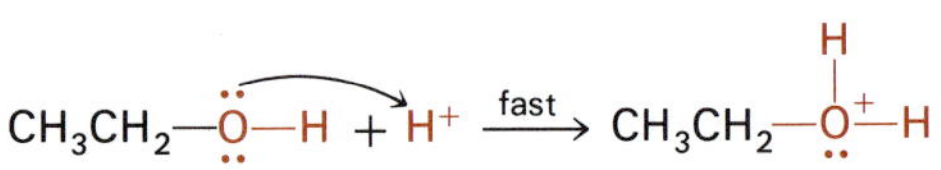

oxonium ion

Step 2: Substitution of ROH for H_2O

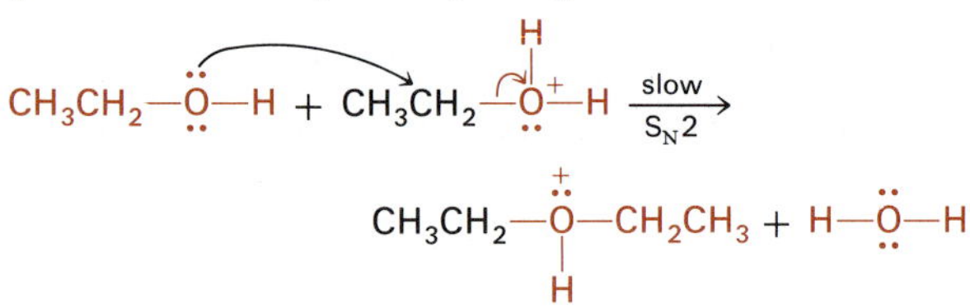

oxonium ion

Step 3: Elimination of H^+

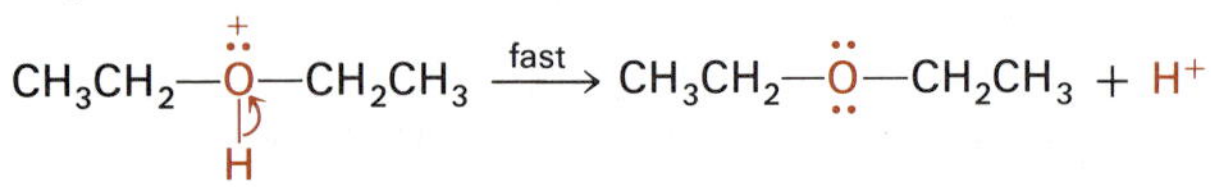

The first step is the addition of a proton to the OH group of the alcohol to convert it to H_2O^+. The second step is an S_N2 reaction in which a second, unprotonated, CH_3CH_2OH molecule is the nucleophile and H_2O is the leaving group. A proton departs in the third step to yield the ether and regenerate the catalyst. The reaction is most satisfactory when a symmetrical ether is desired—otherwise a mixture of products is obtained, as seen in the following exercise.

EXERCISE 17-7 Three different ethers are obtained when a mixture of methanol and ethanol is heated with sulfuric acid. Draw the structural formulas of the ethers.

Ethers can also be synthesized by the Williamson synthesis, a substitution reaction in which a metal alkoxide such as sodium ethoxide reacts with an alkyl halide:

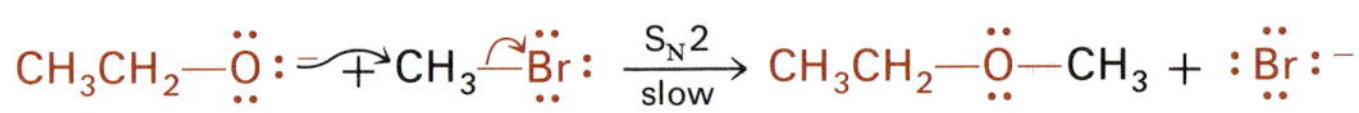

This method is preferable to the acid-catalyzed intermolecular dehydration when an unsymmetrical ether is needed. Because the mechanism is classic S_N2, with the substitution of an alkoxide ion for a halide ion, the alkyl halide should be primary. The alkoxide ion, on the contrary, may be derived from a primary, secondary, or tertiary alcohol:

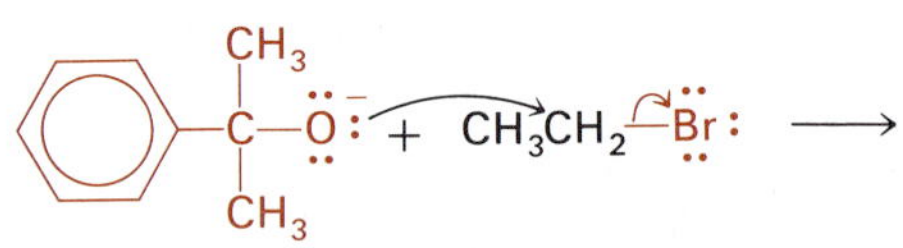

3° alkoxide **1° alkyl halide**

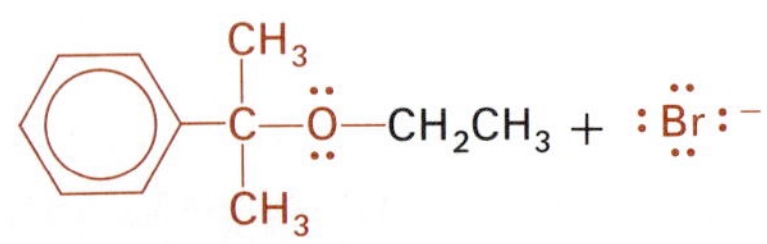

EXERCISE 17-8 What reagent(s) would you need to prepare ethyl *tert*-butyl ether, starting with potassium *tert*-butoxide?

KEY TERMS

An *alcohol* has the functional group OH. Alcohols are classified as *primary* (1°), *secondary* (2°), or *tertiary* (3°) depending on the number of other carbons bonded to the carbon bearing the hydroxyl group.

A *phenol* is an aromatic alcohol; the OH group of a phenol is bonded directly to the benzene ring.

A *diol* has two OH groups; a *triol* has three OH groups.

An *ether* has the functional group C—O—C.

Dehydration is the elimination of a molecule of water—for example, the elimination of H—OH from an alcohol to yield an alkene.

The *citric acid* (or *Krebs* or *tricarboxylic acid*) *cycle* is the series of reactions that comprise the final common pathway for the aerobic oxidation of fuel molecules in the body—amino acids, fatty acids, and carbohydrates.

CHAPTER SUMMARY

Alcohols are compounds that contain the OH functional group. Because of the formation of hydrogen bonds between alcohol molecules, alcohols boil at higher temperatures than alkanes of the same molecular weight. Hydrogen bonding also accounts for the fact that the small alcohols are soluble in water.

Alcohols are very weak acids and also are weak Brønsted–Lowry and Lewis bases. They react with active metals to produce alkoxide ions (RO^-) and hydrogen gas. Alkoxide ions are strong bases. Phenols are much stronger acids than are alcohols.

A laboratory method of synthesizing alcohols is by *substitution* of OH^- for X^- in an alkyl halide, R—X. For tertiary alkyl halides, the mechanism is S_N1, and the rate-determining step is the formation of a tertiary carbonium ion. For primary halides, the mechanism is S_N2, and the single step is the *simultaneous* formation of the HO—R bond and the breaking of the R—X bond, without the formation of an unstable primary carbonium ion.

Substitution reactions in acid solution are characteristic of the alcohols—tertiary alcohols react by the S_N1 mechanism, and primary alcohols by the S_N2 mechanism. The OH group is always protonated first to convert it to the better leaving group H_2O.

Alcohols can be dehydrated to yield alkenes by an

E-1 mechanism. Again, the OH group is first protonated to convert it to H_2O. Dehydration occurs most readily with tertiary alcohols because they can form stable tertiary carbonium ions.

Primary alcohols can be oxidized to aldehydes and acids by common oxidizing agents such as $KMnO_4$ or $K_2Cr_2O_7$; and secondary alcohols can be oxidized to ketones. Tertiary alcohols are not readily oxidized by mild oxidizing agents.

Ethers have the general formula R—O—R′. Because they have no hydrogen atoms that can form hydrogen bonds, they boil at lower temperatures than the alcohols of which they are isomers. They are good solvents, because they can dissolve both hydrocarbons and organic compounds with polar functional groups. Ethers can be prepared by the *intermolecular* dehydration of alcohols, and also by the substitution of an alkoxide ion (RO^-) for X^- in an alkyl halide.

PROBLEMS

17-1. Copy the following structures and label all alcohol, phenol, and ether functional groups:

(a) Epinephrine (adrenaline), a hormone secreted by the adrenal glands

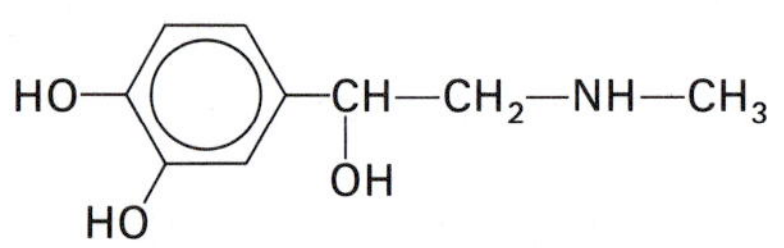

(b) Thyroxine, a hormone secreted by the thyroid gland

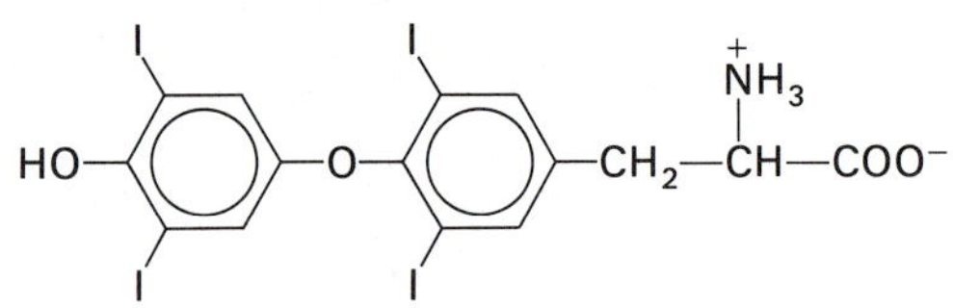

(c) D-Ribose, a simple five-carbon sugar

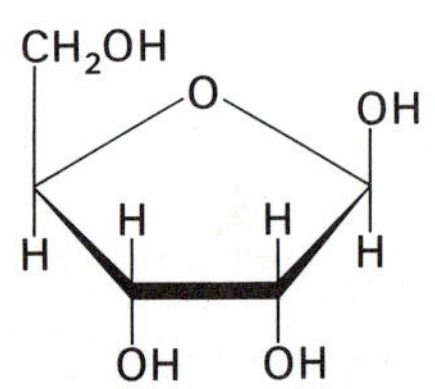

17-2. In hot, dry climates, octadecanol ($C_{18}H_{37}OH$) is sometimes thrown on reservoirs, where it forms a film one molecule thick on the surface of the water and prevents the water from evaporating. Explain this phenomenon. Could *n*-propyl alcohol be used instead of octadecanol? Explain why, or why not.

17-3. Why is ethylene glycol a more satisfactory antifreeze than the cheaper methyl alcohol?

17-4. (a) Write the structural formulas of all alcohols having the molecular formula $C_4H_{10}O$, and give the IUPAC name of each.
(b) State whether each of the alcohols is primary, secondary, or tertiary.

17-5. Write the structural formulas of (and name) all ethers that are isomeric with the alcohols in Problem 17-4.

17-6. Which has the highest boiling point?
(a) $CH_3(CH_2)_4CH_3$
(b) $CH_3—O—(CH_2)_3CH_3$
(c) $CH_3CH_2—O—(CH_2)_2CH_3$
(d) $CH_3(CH_2)_3CH_2OH$
(e) $CH_2═CH—CH═CH—CH_2—CH_3$

17-7. Write structural formulas for
(a) 3-ethyl-2-pentanol
(b) *tert*-butyl alcohol
(c) methyl isopropyl ether
(d) 1,3-butanediol
(e) potassium *n*-propoxide
(f) sodium methoxide
(g) cyclopentanol
(h) 1,3-cyclopentanediol.

17-8. In the reaction of HBr with 2-methylpropene using ethyl alcohol as solvent, the major product is the tertiary bromide, as predicted by Markovnikov's rule (Section 15-5):

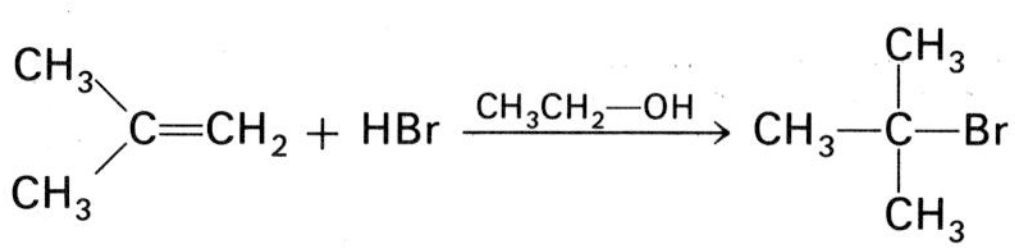

Careful analysis reveals, however, that due to side reactions with the solvent, three other products are formed in low yield:

(a) CH_3CH_2Br (b) $CH_2{=}CH_2$ (c) 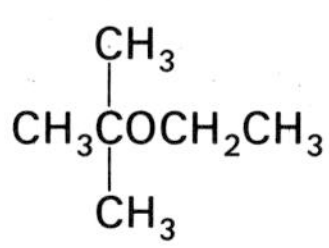

For each of these three byproducts, draw a reasonable mechanism to account for its formation, and classify the mechanism according to type (S_N1, S_N2, etc.).

17-9. (a) Draw the structures of the products of the following nucleophilic substitution reactions:

(1) $CH_3CH_2—\ddot{\underset{..}{O}}:^- + CH_3—\ddot{\underset{..}{I}}: \longrightarrow$

(2) $\ddot{N}H_3 + CH_3CH_2—\ddot{\underset{..}{Br}}: \longrightarrow$

(3) 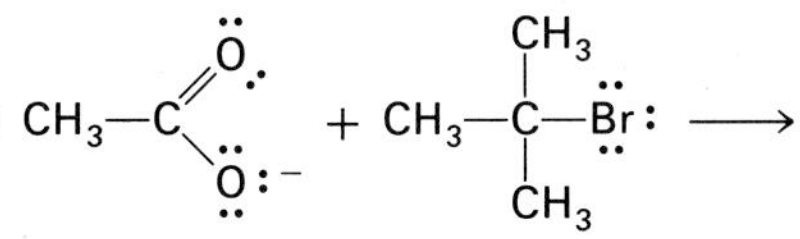

(b) Using arrows to show electron movement, depict the mechanism of each of the above reactions.

17-10. A compound C_3H_8O reacts only very slowly (many hours at room temperature) with HCl and $ZnCl_2$. However it does undergo the S_N2 reaction with HBr to form C_3H_7Br. Write the correct structures of the compounds C_3H_8O and C_3H_7Br?

17-11. What laboratory test could you use to distinguish the two alcohols below?

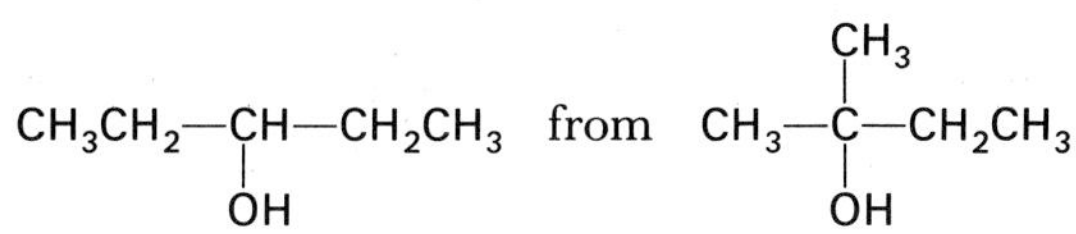

(A correct test causes one compound to react, but not the other.)

17-12. If you wanted to make some *cis*- and *trans*-2-pentene by dehydration of an alcohol, what would be the best alcohol to use? Why?

17-13. How could you prepare

(a) $CH_3(CH_2)_2CH{=}CH_2$ from $CH_3(CH_2)_3CH_2OH$

(b) 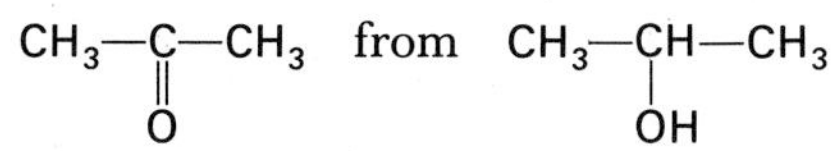

(c) $CH_3(CH_2)_2COOH$ from $CH_3(CH_2)_2OH$

(d) 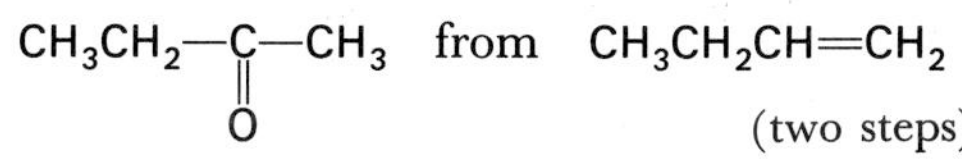

(two steps)

In each case, use any other simple chemicals you need. Write equations for the reactions, and label each as an oxidation, hydration, or dehydration.

17-14. Draw structures of the products of the following reactions:

(a) $CH_3CH_2\text{—}\underset{\large H}{\overset{\large OH}{C}}\text{—}CH_2CH_3 \xrightarrow{H_2SO_4} C_5H_{10}$

(b) $CH_3CH_2\text{—}\ddot{\underset{..}{O}}{:}^- + CH_3\text{—}\ddot{\underset{..}{I}}{:} \longrightarrow C_3H_8O + {:}\ddot{\underset{..}{I}}{:}^-$

(c) 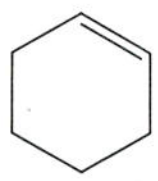 $+ H_2 \xrightarrow{Ni} C_6H_{12}$

17-15. When the alcohol $CH_3CH_2CH_2OH$ is heated with sulfuric acid, two products are obtained: C_3H_6 and $C_6H_{14}O$. Draw the structures of these two products.

17-16. List two properties of ethers that make them especially useful solvents in the chemical laboratory.

17-17. (a) The diol ethylene glycol (CH_2OHCH_2OH) is toxic. If ingested, it is oxidized by enzymes in the liver to oxalic acid. The oxalate crystallizes in the kidneys as the insoluble salt calcium oxalate and causes damage to the kidneys. Write the structure of oxalic acid.
(b) Another diol, propylene glycol ($CH_3CHOHCH_2OH$), is nontoxic. Propylene glycol is used as a solvent for drugs and as a moisturizing agent for foods. Propylene glycol is oxidized by enzymes in the liver to pyruvic acid, an intermediate in the citric acid cycle. Write the structure of pyruvic acid.

17-18. Write a net unbalanced equation for each of the following classes of reaction: (a) oxidation of an alcohol to an aldehyde; (b) oxidation of an alcohol to a carboxylic acid; (c) dehydration of an alcohol; (d) hydration of an alkene.

A silver mirror is formed when an aldehyde reacts with silver ions (page 476).

18

Aldehydes and Ketones

The carbonyl functional group (C=O) is of unique importance in organic chemistry because of the variety of reactions that it undergoes. The principal reactions of aldehydes and ketones (both of which contain the carbonyl group) are addition to the C=O bond and condensation reactions in which two molecules are combined. The carbonyl group occurs in many biological molecules—most notably in sugars, which contain both hydroxyl and carbonyl groups.

Starting with simple aldehydes and ketones, the synthetic chemist can construct an endless assortment of compounds with unbranched or branched chains of carbon atoms. Reactions of the carbonyl group are also important in the biosynthesis of large molecules such as cholesterol, the steroid hormones, and glycogen from simple, small molecules.

18-1 STRUCTURE AND PROPERTIES

Aldehydes and ketones are compounds that contain the **carbonyl functional group** >C=O, and that have the general formulas:

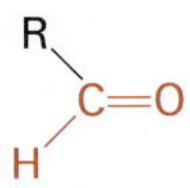

aldehydes (RCHO)

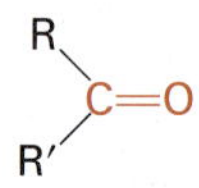

ketones (RCOR′)

In an **aldehyde,** at least *one* of the groups bonded to the carbonyl carbon is an H atom. The other group (R) can be H (in formaldehyde) or any alkyl or aromatic group. A carbon atom of the R group is bonded to the carbonyl carbon. The condensed formula for an aldehyde is RCHO:

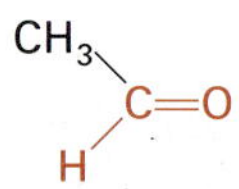

acetaldehyde
CH_3CHO

Naming of aldehydes and ketones is discussed in Section 18-2.)

In a **ketone,** the carbonyl carbon is bonded to *two* other carbon atoms—one in each R group. The condensed formula for a ketone is RCOR′:

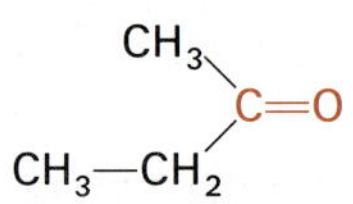

ethyl methyl ketone
$CH_3COCH_2CH_3$

The carbonyl carbon is bonded to three other atoms and is thus in the sp^2 hybridization state, with three σ bonds and one π bond, as shown in Figure 18-1 for the simplest aldehyde, formaldehyde. The molecule is planar, with bond angles close to the 120° expected for sp^2 hybridization.

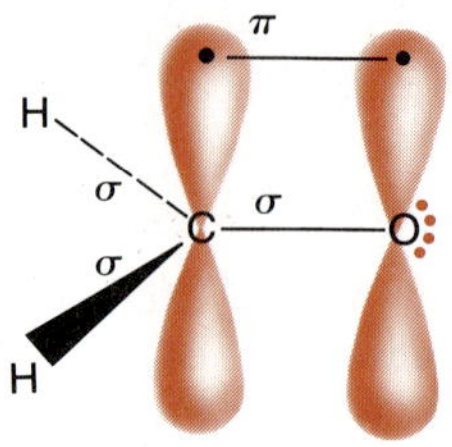

Bond length (nm)	
C=O	0.12
C–H	0.11
Bond angle	
H–C–O	122°
H–C–H	117°

FIGURE 18-1
Structure of formaldehyde.

The carbon–oxygen double bond is polar, because oxygen is more electronegative than carbon and attracts the shared electrons more strongly. This effect can be represented by the use of the ↦ or $^{\delta+}{}^{\delta-}$ notation—or by drawing resonance structures, one of which places a negative charge on the oxygen:

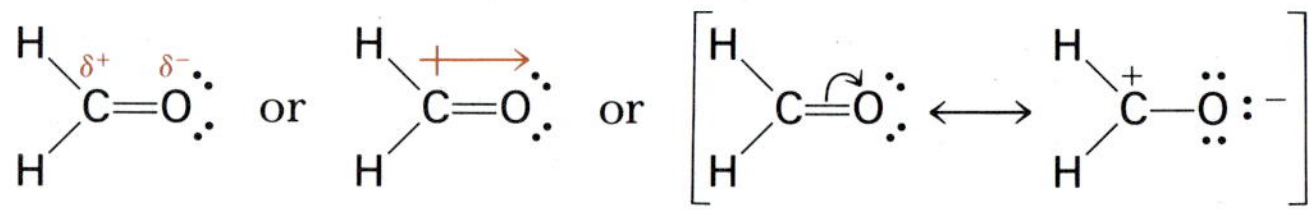

The resonance notation means that the real structure of the molecule corresponds to *neither* of the structures shown, but is a composite of the two.

As we shall see, much of the chemistry of carbonyl compounds is determined by this polarity of the C=O bond. The partial positive charge on the carbon atom makes it susceptible to attack by nucleophiles (represented by the symbol Nu:). The partially negative oxygen is vulnerable to electrophilic attack (usually by H^+).

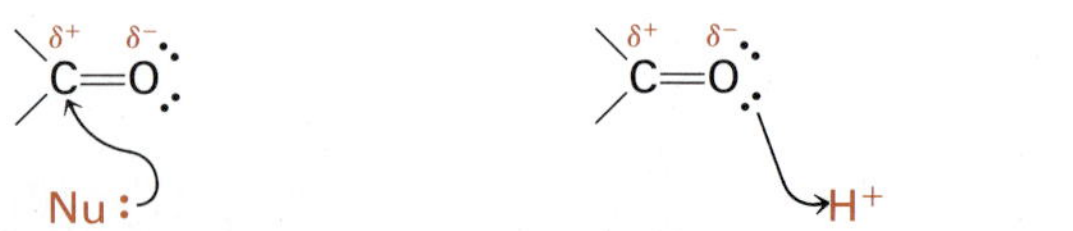

nucleophile attacks C **electrophile (H^+) attacks O**

TABLE 18-1
Boiling and melting points of aldehydes and ketones

Common name (IUPAC name)	Condensed formula	Molecular weight	Boiling point (°C)	Melting point (°C)
Formaldehyde (methanal)	HCHO	30	−21	−92
Acetaldehyde (ethanal)	CH_3CHO	44	21	−121
Propionaldehyde (propanal)	CH_3CH_2CHO	58	49	−81
Butyraldehyde (butanal)	$CH_3CH_2CH_2CHO$	72	76	−99
Acetone (propanone)	CH_3COCH_3	58	56	−95
Ethyl methyl ketone (butanone)	$CH_3CH_2COCH_3$	72	80	−86
Methyl *n*-propyl ketone (2-pentanone)	$CH_3(CH_2)_2COCH_3$	86	102.4	−78
Diethyl ketone (3-pentanone)	$CH_3CH_2COCH_2CH_3$	86	102.0	−40

Boiling and melting points of some aldehydes and ketones are given in Table 18-1. Figure 18-2 compares the boiling points of aldehydes with those of alcohols and alkanes of the same molecular weights. (Ketones are not shown on the graph, but their boiling points lie close to the line of the aldehydes.) The aldehydes boil at higher temperatures than alkanes (but at lower temperatures than alcohols) of the same molecular weights. The boiling points of aldehydes and ketones are higher than those of alkanes because attractive forces between dipoles of neighboring molecules make it more difficult to separate them from one another in the conversion from liquid to gas.

The cohesive forces between neighboring dipoles are not nearly as strong as hydrogen bonds. Aldehydes and ketones have no hydrogen atoms that can participate in hydrogen bonds, so they boil at temperatures lower than the boiling points for alcohols of comparable molecular weight. The effect of these dipole–dipole

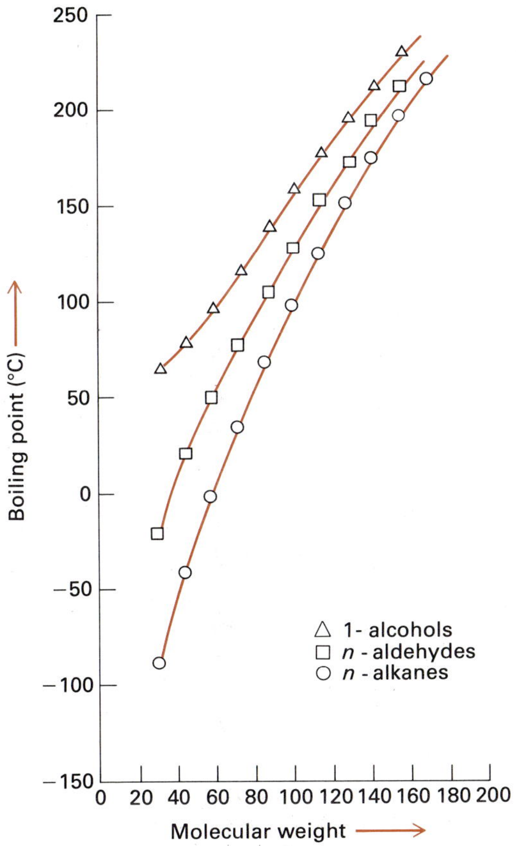

FIGURE 18-2
Boiling points of unbranched alcohols, aldehydes, and alkanes.

attractive forces becomes less important as the nonpolar hydrocarbon part of the molecule becomes larger. As a result, the larger aldehydes and ketones boil at temperatures only slightly higher than those for alkanes of comparable molecular weight.

EXERCISE 18-1 List the following compounds in order of decreasing boiling point:

(a) $CH_3(CH_2)_4CH_3$

(b) $CH_3(CH_2)_2-\overset{O}{\overset{\|}{C}}-CH_3$

(c) $CH_3(CH_2)_2\overset{OH}{\overset{|}{C}}-CH_3$

(d) $CH_3-\overset{O}{\overset{\|}{C}}-CH_2CH_2OH$

(e) $CH_3CH_2-O-(CH_2)_2CH_3$

Many natural products are aldehydes and ketones—the following are a few examples:

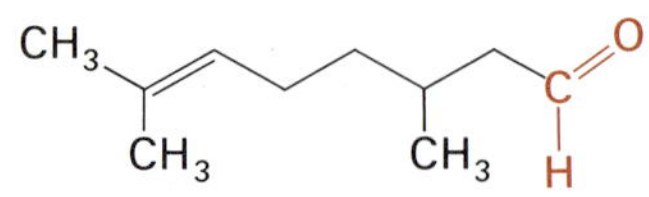

citronellal
(oil of lemon)

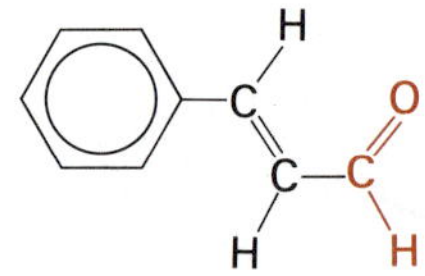

cinnamaldehyde
(oil of cinnamon)

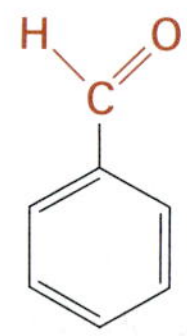

benzaldehyde
(oil of bitter almond)

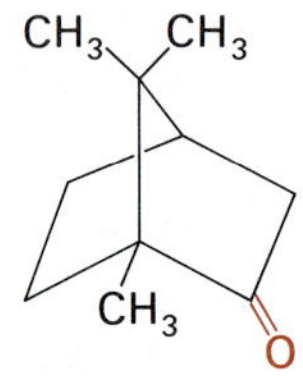

camphor
(oil of camphor tree)

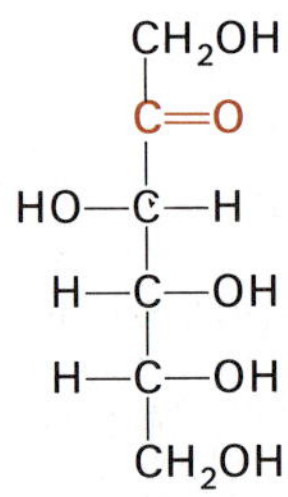

fructose
(a simple sugar found in honey and fruit juices)

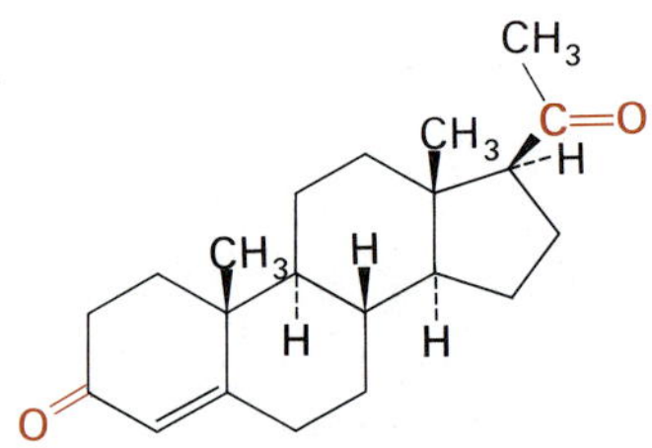

progesterone
(female sex hormone)

18-2 NOMENCLATURE

In the IUPAC system, aldehyde names are obtained by dropping the final *-e* from the name of the corresponding alkane and adding *-al:*

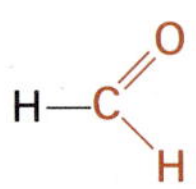

methanal
(formaldehyde)

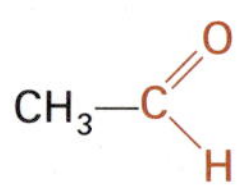

ethanal
(acetaldehyde)

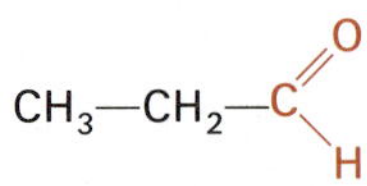

propanal
(propionaldehyde)

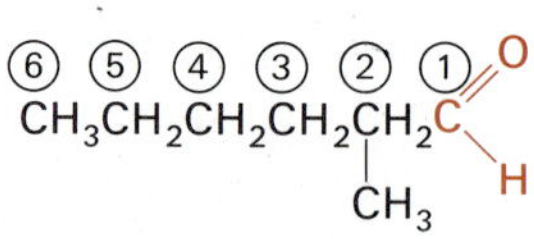

2-methylhexanal

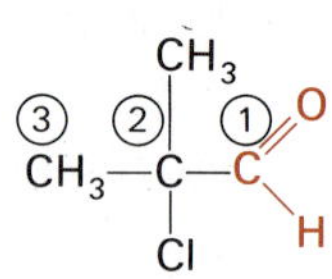

2-chloro-2-methylpropanal

(Common names are given in parentheses under the IUPAC names.)

Because the carbonyl carbon is always at one end of the chain, the positions of side groups are designated by numbering from this end, with the carbonyl carbon assigned the number 1. Common names are given for the simple aldehydes. Formaldehyde, at room temperature and atmospheric pressure, is a gas with a pungent odor. It is used industrially in the production of Bakelite, the first synthetic plastic. We saw (Section 17-7) that formaldehyde produced by the oxidation of methyl alcohol in the liver is responsible for the toxicity of methyl alcohol.

In the *common* (not in the IUPAC) naming system, positions of attached groups are designated by assigning the Greek letters α, β, γ, and so on. The lettering begins with the carbon atom *next to the carbonyl group.* (The rationale for this convention is that there can be no substituent on the carbonyl carbon of an aldehyde.)

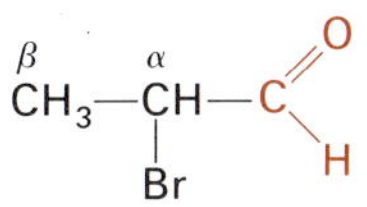

α-bromopropionaldehyde
(IUPAC name: 2-bromopropanal)

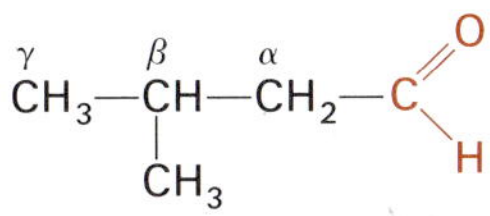

β-methylbutyraldehyde
(IUPAC name: 3-methylbutanal)

In the IUPAC system, the names of ketones are obtained by replacing the final *-e* in the name of the corresponding alkane with *-one.* The position of the carbonyl group is specified by a number prefix. The longest carbon chain containing the carbonyl group is numbered from the end closer to the carbonyl carbon:

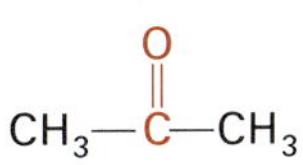

propanone
(acetone, or dimethyl ketone)

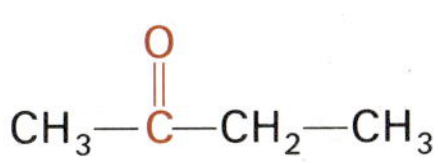

butanone
(ethyl methyl ketone)

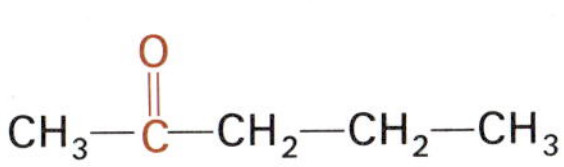

2-pentanone
(methyl *n*-propyl ketone)

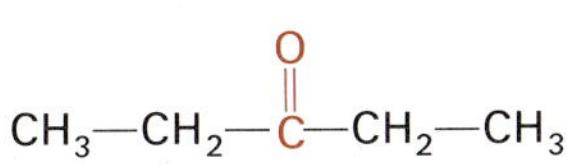

3-pentanone
(diethyl ketone)

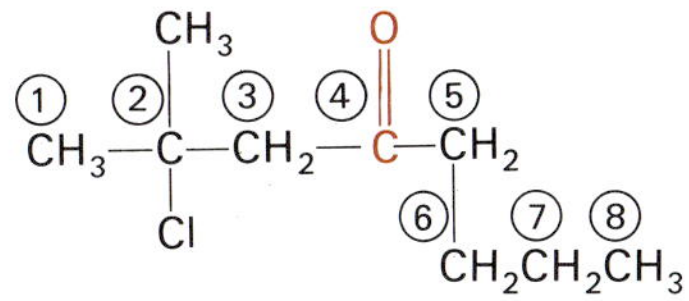

2-chloro-2-methyl-4-octanone

In the common system, the name *ketone* is prefixed by the names of the two alkyl groups attached to the carbonyl carbon, as just shown in parentheses for some of the simple ketones. As with aldehydes, the positions of attached groups are designated by Greek letters:

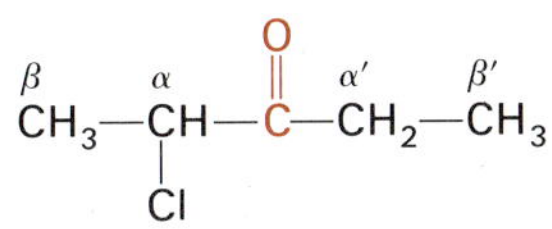

α-chlorodiethyl ketone
(2-chloro-3-pentanone)

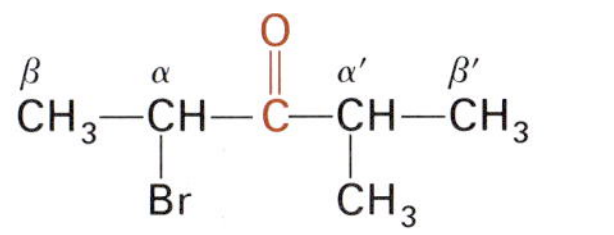

α-bromo-α'-methyl-diethyl ketone
(2-bromo-4-methyl-3-pentanone)

(IUPAC names are given in parentheses.)

Acetone is an important organic solvent. It dissolves a wide variety of organic substances, but it is also infinitely soluble in water. A small amount of acetone is normally present in the blood. About 20 mg per day are formed in the liver (as a product of the digestion of fats) and excreted by the kidneys. During starvation, diets very low in carbohydrates, and some disorders of carbohydrate metabolism such as diabetes mellitus, acetone accumulates in the blood and urine; sometimes it can even be detected in the breath (see also Section 27-3).

EXERCISE 18-2 Name the following compounds according to the IUPAC system:

(a) 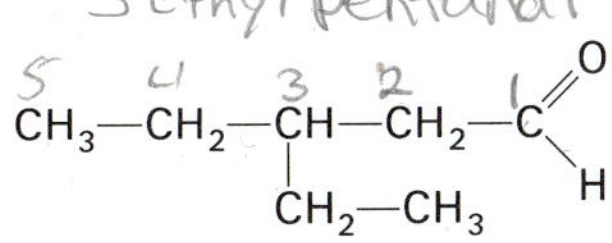

(b) 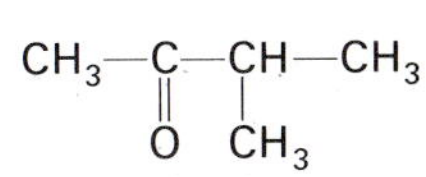

(c) 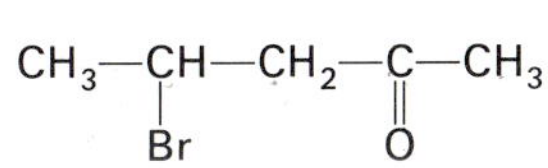

18-3
PREPARATION

As we saw in Chapter 17, aldehydes and ketones can be prepared by the oxidation of primary and secondary alcohols, respectively, using oxidizing agents such as dichromate ($Cr_2O_7^{2-}$) or permanganate (MnO_4^-):

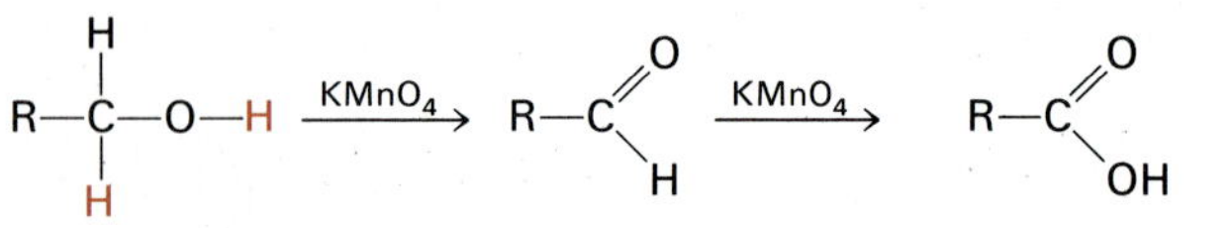

1° alcohol **aldehyde** **carboxylic acid**

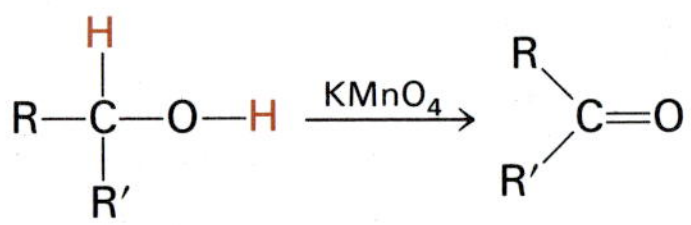

2° alcohol **ketone**

The ketone is not easily oxidized further, and this reaction provides the ketone in good yield. The aldehyde, on the contrary, is rapidly oxidized further to the carboxylic acid, so this method is useful only for aldehydes with low boiling points, which can be removed from the reaction mixture by distillation. (Aldehydes, as we have seen, boil at lower temperatures than the corresponding alcohols. Thus if the temperature of the reaction is kept above the boiling point of the aldehyde but below that of the alcohol, aldehydes can be distilled off as they are formed.) Aldehydes that boil at temperatures above normal distilling temperatures can be prepared by a dehydrogenation process in which the alcohol vapor is passed over hot copper metal:

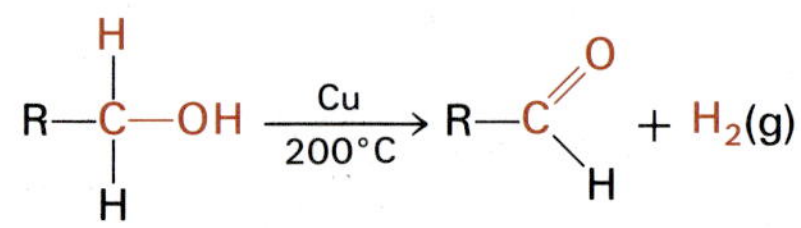

EXERCISE 18-3 Draw the structural formula of the alcohol from which you could prepare each of the following compounds, and write an equation for each reaction:

(a) cyclohexanone, 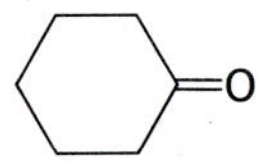

(b) acetone, 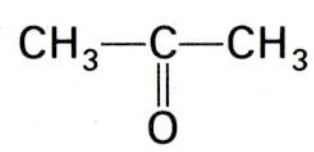

(c) butanal, 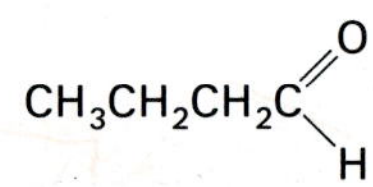

18-4
REACTIONS: ADDITION OF H—X

A characteristic reaction of aldehydes and ketones involves **addition** of H—X across the carbon–oxygen double bond. A generalized mechanism is given here for an addition reaction initiated by attack on the partially positively charged carbonyl carbon by a nucleophile (Nu:⁻). We are assuming a nucleophile with a negative charge (OH^-, for example), although it could also be a neutral one such as NH_3.

Step 1: Addition of nucleophile

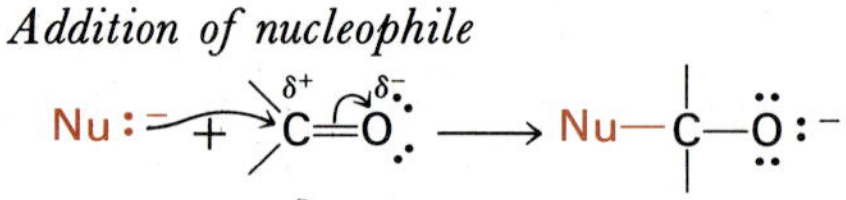

Step 2: Addition of H^+

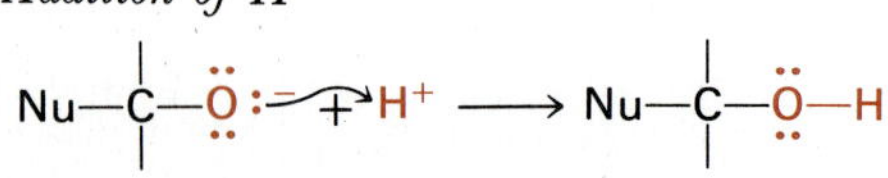

In the first step the nucleophile attacks the positive end of the carbonyl group, forming a new bond Nu—C. A proton is then added to the negatively charged oxygen. *The net reaction is just the addition of Nu—H* (as Nu:$^-$, H$^+$):

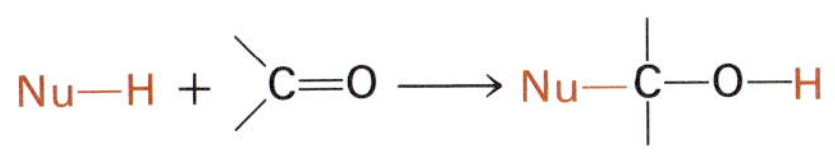

A typical example is the addition of hydrogen cyanide (HCN), which is initiated by the addition of a cyanide ion (as the nucleophile) to the carbonyl carbon:

Step 1: Addition of nucleophile

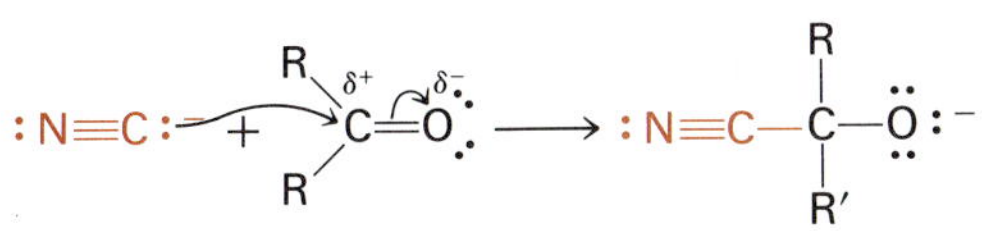

Step 2: Addition of H$^+$

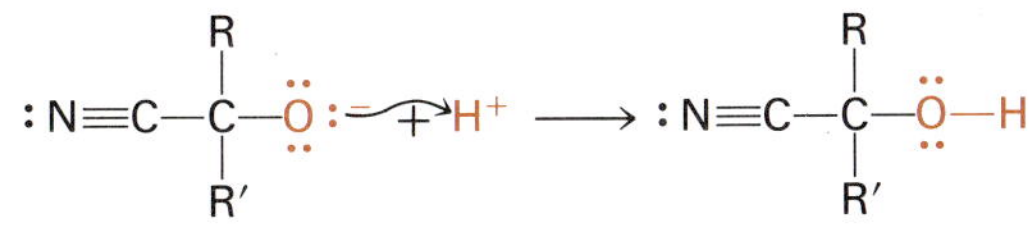

a cyanohydrin

Note that the H$^+$ has been added to the oxygen, and the nucleophile (NC$^-$) to the carbon. Cyanohydrins are used in sugar chemistry to lengthen carbon chains and thus make longer sugars from shorter ones (see Problem 23-16).

EXERCISE 18-4 Write the structure of the addition product of acetaldehyde with HCN.

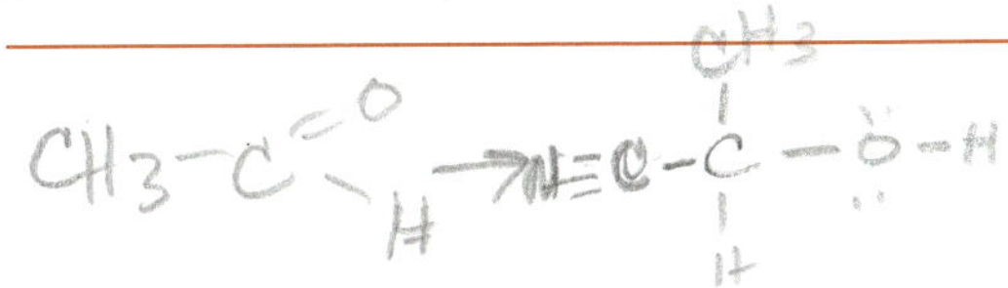

18-5 REACTIONS: ADDITION OF WATER AND OF ALCOHOLS

The **addition of water** to aldehydes and ketones is not an important synthetic reaction, but it does illustrate the principles of carbonyl addition reactions. In pure water, ketones are normally hydrated only to a small extent—but formaldehyde is over 99.9% hydrated, and other aldehydes are about 50% hydrated. The addition of water to formaldehyde occurs by the following mechanism:

Step 1: Addition of nucleophile

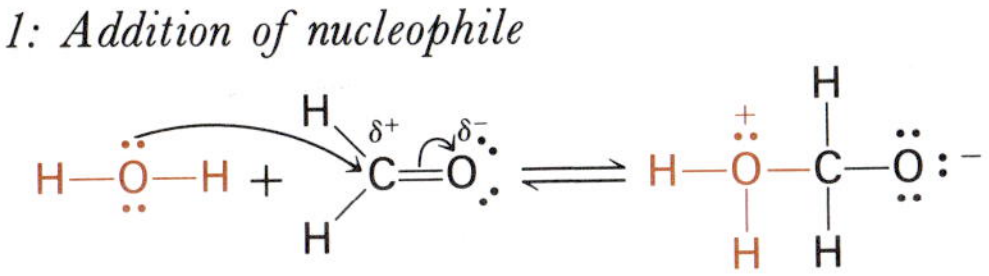

Step 2: Proton transfer

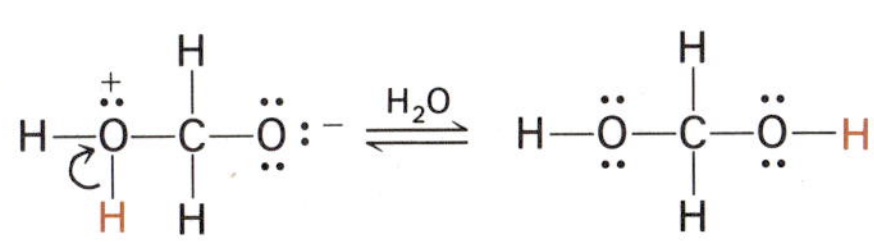

a 1,1-diol, or geminal (gem) diol

Water, acting as a nucleophile, attacks the carbonyl carbon. A proton is then rapidly transferred to produce the *diol*. The net reaction is the *addition* of the nucleophile (OH$^-$) to the carbon, and of H$^+$ to the oxygen. A saturated solution of formaldehyde gas in water (called "formalin") is about 37% by weight formaldehyde. The formaldehyde is essentially all in the 1,1-diol form shown. Formalin solutions are used as disinfectants and as biological preservatives.

This reaction is slow in pure water, but faster in the presence of small amounts of acid or base, either of which can act as a catalyst. In the acid-catalyzed reaction, the first step is the protonation of the oxygen of the carbonyl group:

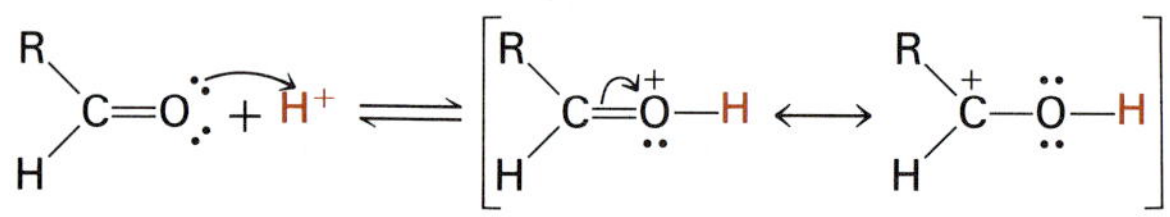

resonance structures

This protonation increases the positive charge on the carbonyl carbon, making it more susceptible to attack by the nucleophilic H_2O molecule:

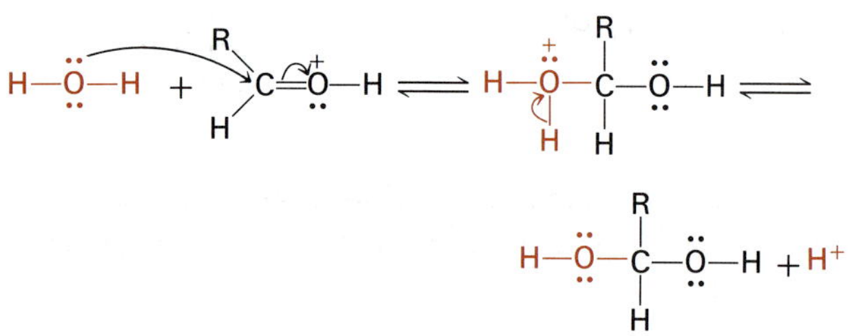

The reaction is also catalyzed by OH^-, which is a better nucleophile than H_2O:

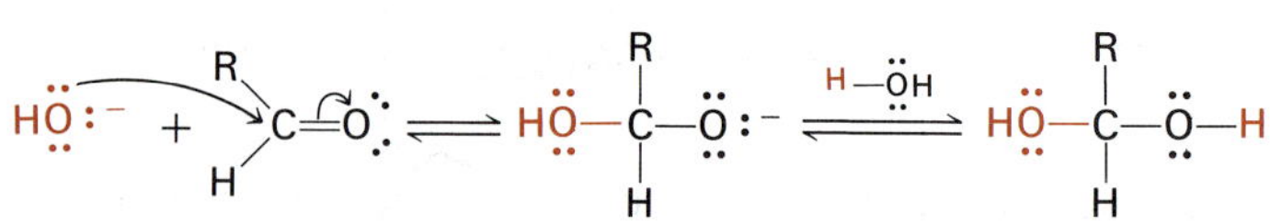

The alkoxide ion produced in the first step rapidly abstracts a proton from a water molecule to yield the 1,1-diol.

The hydrated forms of aldehydes and ketones—1,1-diols, or geminal (gem) diols—generally cannot be isolated because they readily lose water to regenerate the original aldehyde or ketone. Thus, these compounds exist only in water solution. The analogous **addition of alcohols** to aldehydes and ketones, on the contrary, is of significance in carbohydrate chemistry (see Chapter 23). The addition of one mole of alcohol (R—OH) to one mole of an aldehyde or a ketone yields a **hemiacetal** or **hemiketal,** respectively. This reaction, like that with water, is catalyzed by acid or base, and the *net* reaction is the *addition* of RO^- to the carbonyl carbon and H^+ to the oxygen:

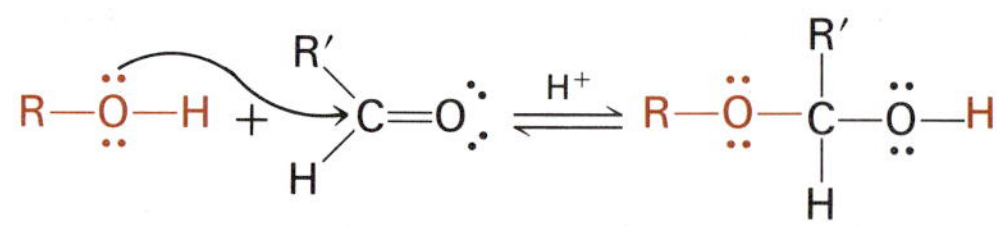

alcohol **aldehyde** **a hemiacetal**

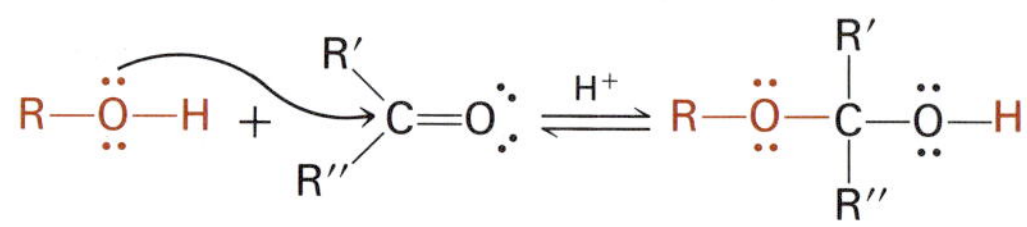

alcohol **ketone** **a hemiketal**

The *mechanism* for the addition of an alcohol to an aldehyde or a ketone is essentially identical to that just shown for the acid-catalyzed addition of water. Note that the addition of water (or of alcohol) to the C═O bond is the same in principle as the addition described in Section 18-4. These are simply further examples of the same principles.

EXERCISE 18-5 Write a likely mechanism for the addition of a molecule of methanol to a molecule of acetaldehyde.

Simple sugars are molecules in which both the alcohol and carbonyl functional groups are present. When these groups come close together, they can react. Thus, the simple sugar glucose ($C_6H_{12}O_6$) occurs in nature as an open-chain molecule in equilibrium with a cyclic hemiacetal formed by the *intramolecular* addition of an alcohol to an aldehyde:

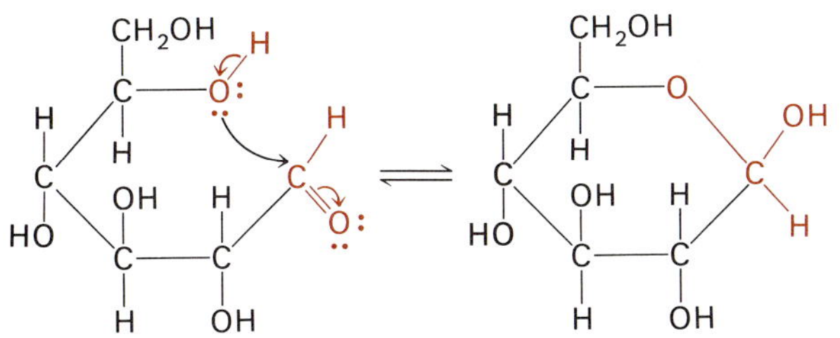

D-glucose (open-chain form) **β-D-glucose** (a cyclic hemiacetal)

Again, the R—O⁻ (from the alcohol) is added to the carbonyl carbon, and H^+ to the carbonyl oxygen atom. The cyclic hemiacetal is a six-membered ring whose most stable configuration is the chair form (Section 14-6):

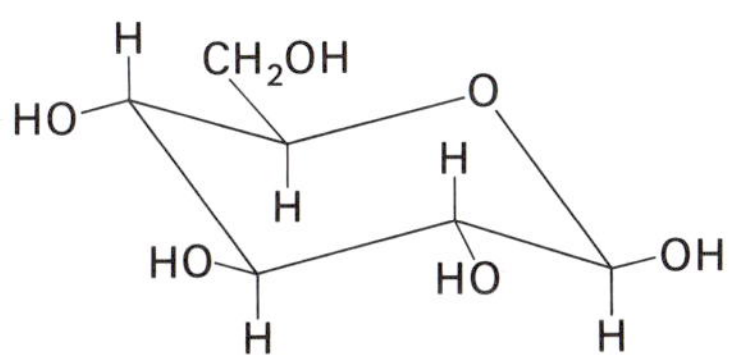

β-D-glucose

(In Section 23-3, we shall see that there are actually two cyclic forms of D-glucose.)

Full acetals and ketals are formed by *substituting* —OR for —OH in the hemiacetal or hemiketal. An acid catalyst is used to protonate the OH group and convert it to the better leaving group, water. The *net* reactions are

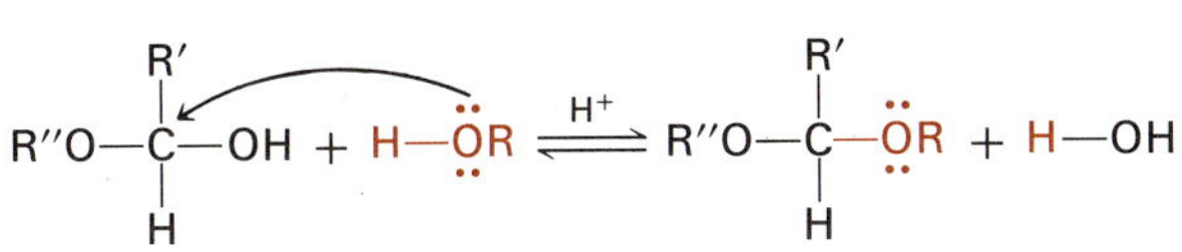

a hemiacetal **an acetal**

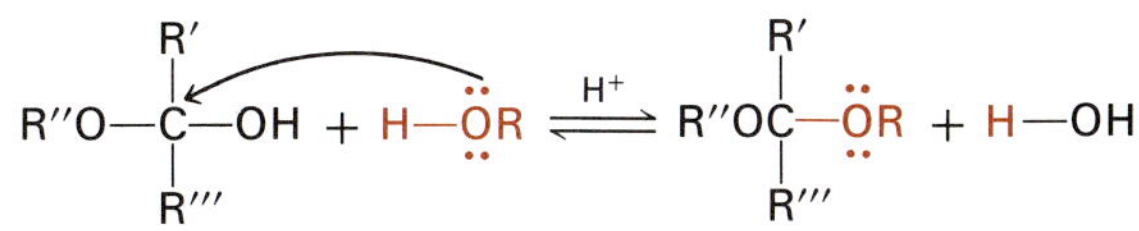

a hemiketal **a ketal**

Thus, reaction between *one* molecule of an alcohol and one molecule of an aldehyde or ketone yields a hemiacetal or hemiketal. Reaction between *two* molecules of an alcohol and one molecule of an aldehyde or ketone yields an acetal or a ketal.

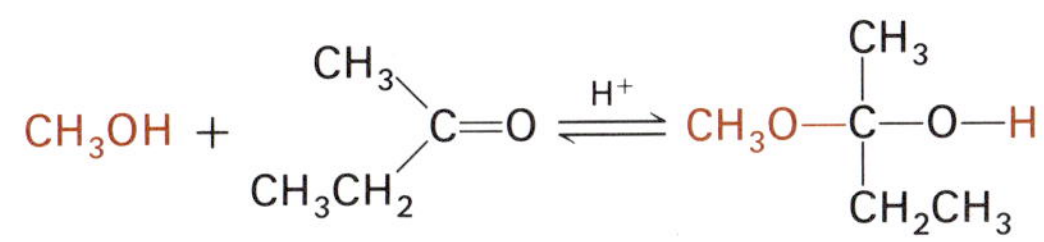

a hemiketal

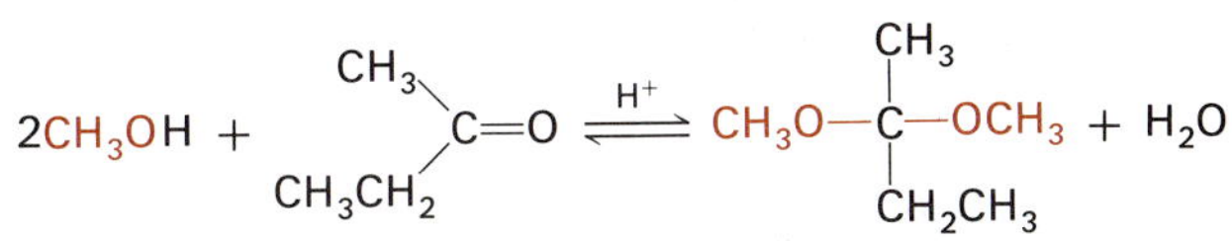

a ketal (18-1)

Note that both the hemiketal and the ketal have carbon bonded to two oxygen atoms. The same is true of the hemiacetal and acetal:

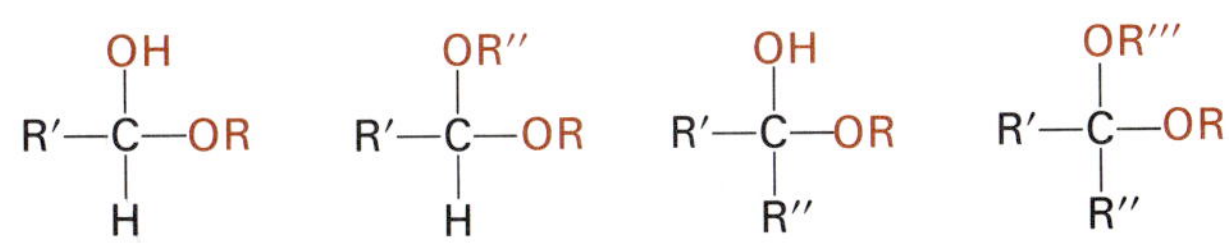

hemiacetal **acetal** **hemiketal** **ketal**

In both the hemiacetal and hemiketal, there is one alcohol and one ether group; in both the acetal and ketal, there are two ether groups. In acid solution, acetals and ketals can be readily **hydrolyzed**—broken up by water with the addition of the components of water (H and OH). The water adds across (and splits) the C—O bonds to regenerate the aldehyde and alcohol. The net reaction for the hydrolysis of a ketal is the reverse reaction of Equation 18-1. We shall learn more about these compounds and their reactions when we study sugars and carbohydrates in Chapter 23.

EXERCISE 18-6 (a) Draw the structural formula of the product of the addition of one molecule of ethanol to one molecule of acetaldehyde.

(b) Draw the structural formula of the product of the reaction of *two* molecules of ethanol with one molecule of acetaldehyde.

(c) Identify each of the following compounds as a hemiacetal, hemiketal, acetal, or ketal:

(1)

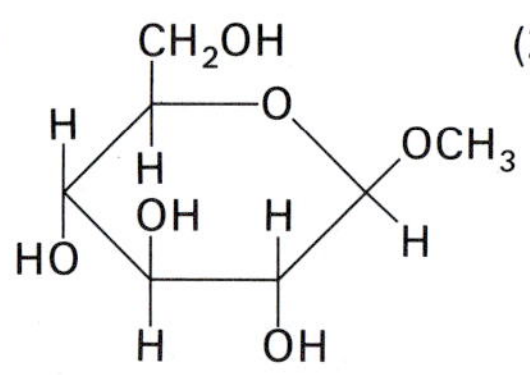

methyl β-D-glucoside

(2)

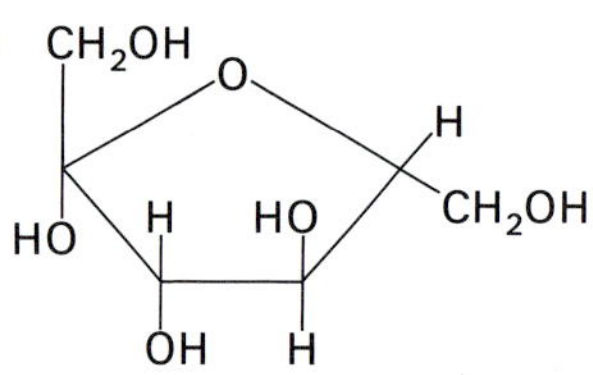

α-D-fructose

(3)

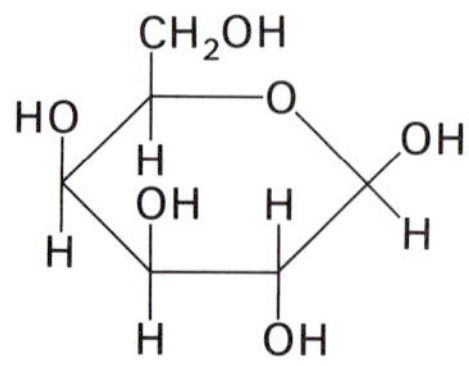

β-D-galactose

(4)

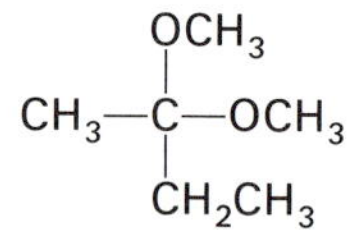

2,2-dimethoxybutane

(d) Write the structures of the alcohol and the aldehyde or ketone from which each of the following compounds were synthesized:

(1)

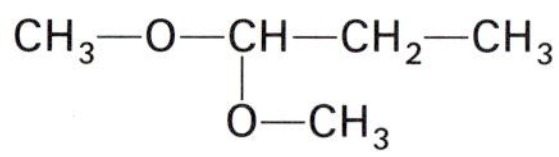

(2)

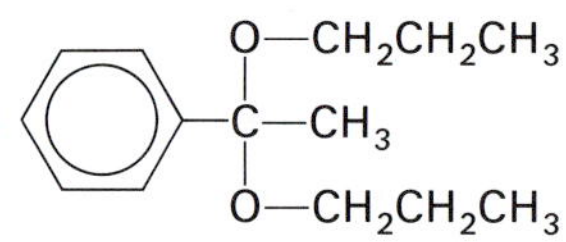

18-6 REACTIONS: ADDITION OF AMMONIA AND OF AMMONIA DERIVATIVES

The nucleophile ammonia (a Lewis base) also can add to an aldehyde or ketone. The addition product formed then eliminates a molecule of water to yield an **imine:**

Step 1: Addition of nucleophile

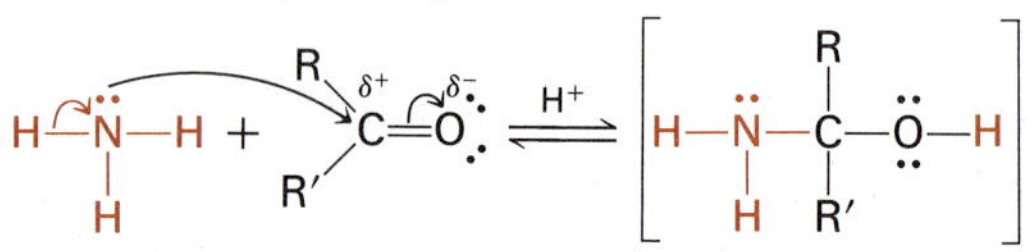

addition product

Step 2: Elimination of H_2O

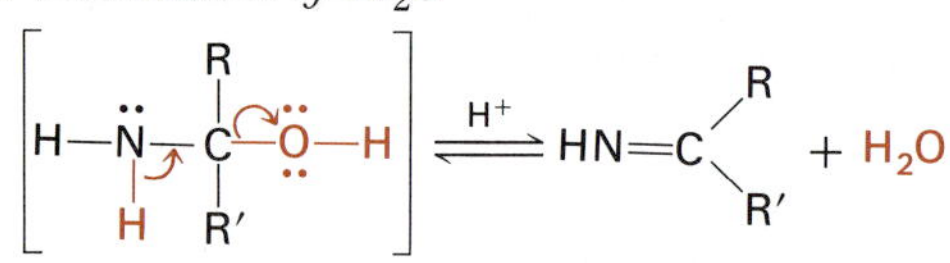

an imine

The first step is the addition of NH_3 as NH_2^- to the carbonyl carbon and H^+ to the oxygen. The second step is the elimination of H_2O to yield the imine. The final product is unstable; it hydrolyses by the addition of H_2O to the C═N bond of the imine in the reverse of the above reactions to yield the aldehyde or ketone.

When instead of ammonia, a primary amine ($R—NH_2$) is the nucleophile, then the product (a substituted imine) is called a **Schiff base:**

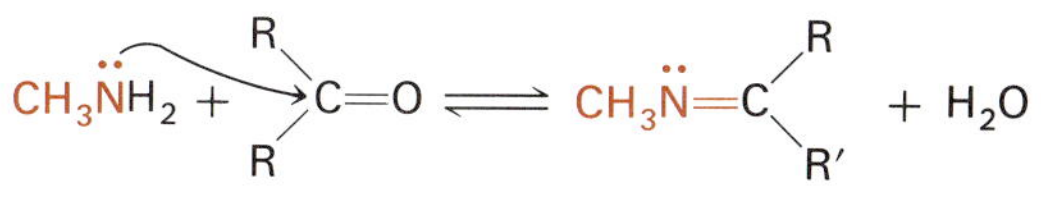

a Schiff base

Schiff bases are important in biochemical processes in which amino groups are transferred. (We shall meet a Schiff base again in our discussion of coenzymes in Chapter 22.)

Another interesting and important reaction, one that can be used to help identify aldehydes and ketones, involves 2,4-dinitrophenylhydrazine, an ammonia derivative:

Step 1: Addition of nucleophile

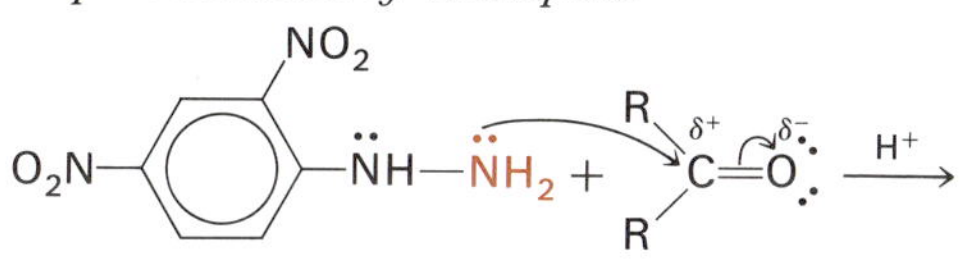

2,4-dinitrophenylhydrazine

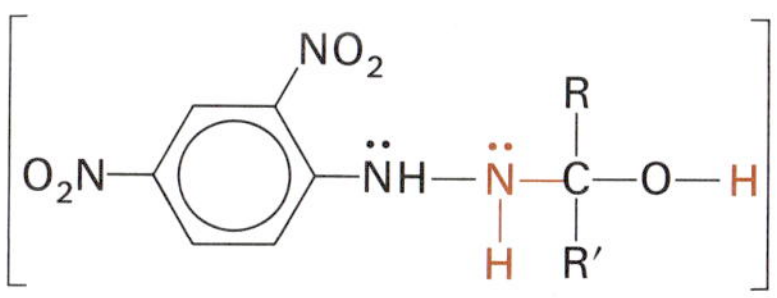

addition product

Step 2: Elimination of H_2O

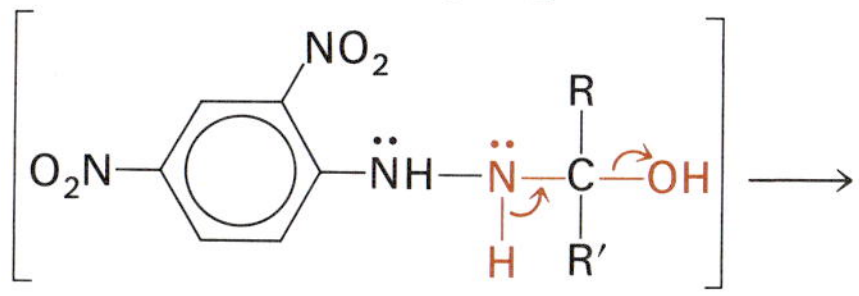

addition product

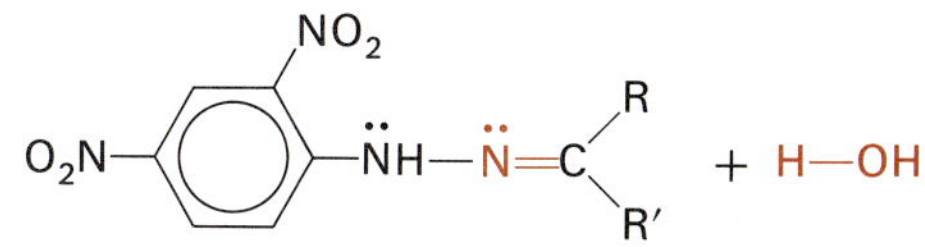

a 2,4-dinitrophenylhydrazone
(2,4-DNP; yellow or orange solid)

The intermediate addition product formed in step 1 eliminates a molecule of water (step 2) to yield a stable, brightly colored, crystalline solid product—a dinitrophenylhydrazone (or a 2,4-DNP) derivative. This product can readily be purified by recrystallization, and its melting point can be used to help identify the original aldehyde or ketone if that is unknown.

EXERCISE 18-7 Write the structure of the final product (after elimination of water) of the addition of phenylhydrazine ($H_2N—NH—$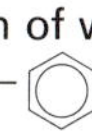) to acetone.

18-7 REACTIONS: ADDITION OF GRIGNARD COMPOUNDS (ORGANOMAGNESIUM COMPOUNDS)

Almost all of the organic reactions we have studied thus far interconvert functional groups but do not affect the carbon skeleton of the molecule. The addition of CN^- to the carbonyl group (Section 18-4) is the only reaction we have seen that lengthens the carbon chain. Not many reactions are available to the organic chemist for synthesizing new carbon–carbon bonds; that is why the work of Victor Grignard is so important. Grignard, a French chemist, was awarded the Nobel Prize in 1912 for his discovery and investigations of the reactions of organomagnesium compounds of the general formula R—MgX, where R is an alkyl or an **aryl** (aromatic) group, and X is a halogen atom. *The addition of a Grignard reagent to a carbonyl compound forms new carbon–carbon bonds.*

The reagents are prepared by reaction between magnesium metal and an organic halide. For example, ethyl bromide in ether (an inert solvent) reacts with metallic magnesium to yield ethyl magnesium bromide:

$$CH_3CH_2Br + Mg(s) \xrightarrow{\text{ether}} CH_3CH_2MgBr$$

ethylmagnesium bromide
(a Grignard reagent)

The reaction is carried out in dry ether because Grignard reagents react rapidly with water to yield hydrocarbons:

$$CH_3CH_2MgBr + HO—H \longrightarrow CH_3CH_3 + HO—MgBr$$

EXERCISE 18-8 Write the structure of the Grignard reagent produced by reaction between magnesium metal and *ortho*-bromotoluene,

Magnesium is less electronegative than carbon (C has an electronegativity of 2.5, and Mg of 1.2). Thus we can think of the Grignard reagent as having a highly polar R—Mg bond, which we can represent by the use of the $\delta^+\delta^-$ notation:

$$\overset{\delta^-}{R}—\overset{\delta^+}{Mg}Br$$

Grignard reagents provide carbanions as nucleophiles in addition reactions with carbonyl compounds. The addition product has the general structure shown here, in which the carbanion of the Grignard reagent (R^-) adds to the carbonyl carbon, and the MgBr adds to the oxygen:

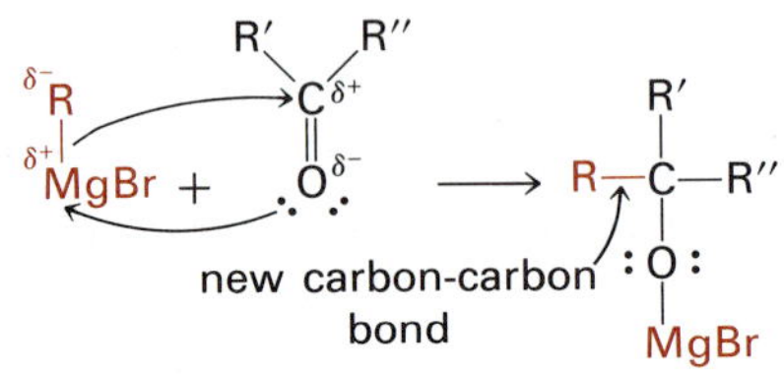

The addition product can be converted to an alcohol by acid **hydrolysis,** as shown for the addition of propylmagnesium bromide to acetaldehyde to yield 2-pentanol:

Step 1: Addition of nucleophile

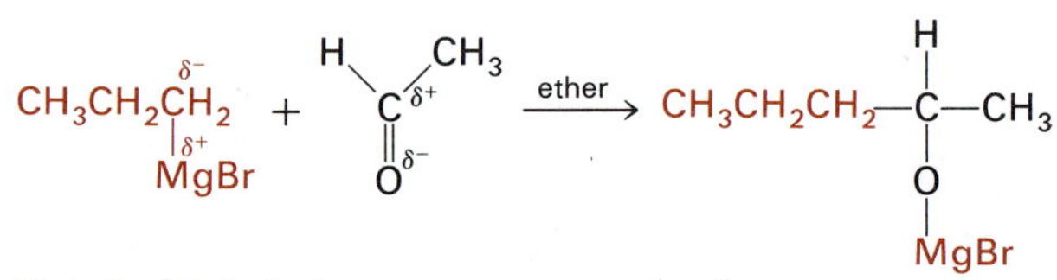

Step 2: Hydrolysis

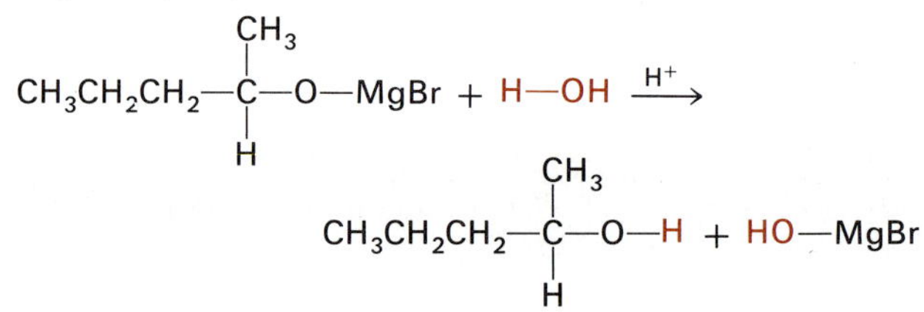

2-pentanol

The net result of the two steps is to add R—H (an alkane; in this case, $CH_3CH_2CH_2$—H) to the C═O bond. A carbanion (here $CH_3CH_2CH_2^-$), which is provided by the Grignard reagent, adds to the carbonyl carbon atom. Then, in the acid hydrolysis, a proton adds to the oxygen. Because alkanes are so unreactive, they cannot add directly to the C═O bond as do HX, H_2O, ROH, and NH_3 (Sections 18-4 through 18-6). Instead, an indirect method such as the Grignard reaction, which converts the alkane to a carbanion, is required.

If you study the reactions of Equations 18-2 and 18-3, you can see that reaction between a Grignard reagent and an aldehyde yields a 2° alcohol. Convince yourself that reaction between a Grignard reagent and formaldehyde yields a 1° alcohol; and reaction between a Grignard reagent and a ketone yields a 3° alcohol.

A secondary or tertiary alcohol typically can be produced by more than one combination of Grignard reagent and carbonyl compound. For example, the 2° alcohol 2-butanol has two alkyl groups attached to the hydroxyl-bearing carbon: an ethyl group and a methyl group. 2-Butanol can be prepared from either of two combinations of aldehyde and Grignard reagent, as shown in Equations 18-2 and 18-3:

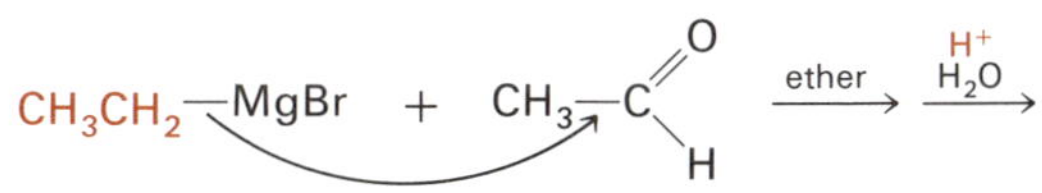

ethylmagnesium bromide **acetaldehyde**

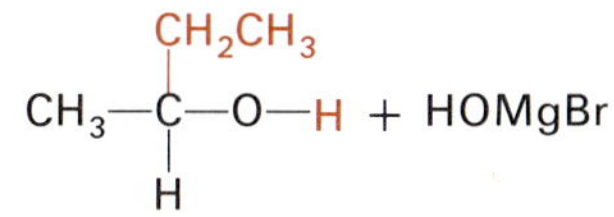 (18-2)

2-butanol

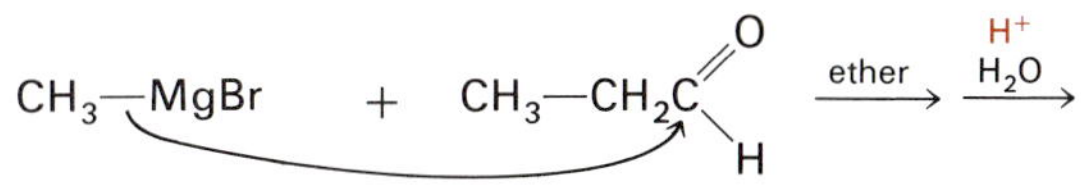

methylmagnesium bromide **propionaldehyde**

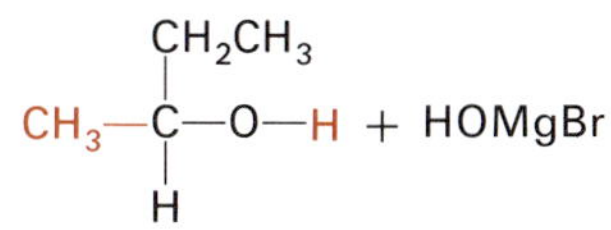 (18-3)

2-butanol

EXERCISE 18-9 Give *two* combinations of Grignard reagent and carbonyl compound that can be used to produce the tertiary alcohol

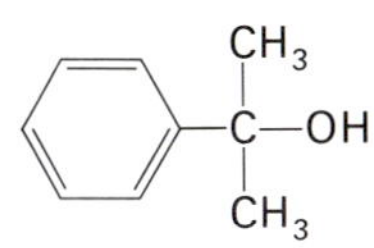

18-8 REACTIONS: OXIDATION AND REDUCTION

We have already seen that aldehydes are readily oxidized further to carboxylic acids but that ketones are resistant to further oxidation. This fact is the basis of two simple laboratory tests that use mild oxidizing agents to distinguish aldehydes from ketones—in both tests, *aldehydes are oxidized, but ketones fail to react.* In Fehling's test, the cupric ion (Cu^{2+}) in basic solution (in the form of a complex ion with tartrate; see Section 13-5) is the oxidizing agent. The formation of a red precipitate of Cu_2O indicates an aldehyde:

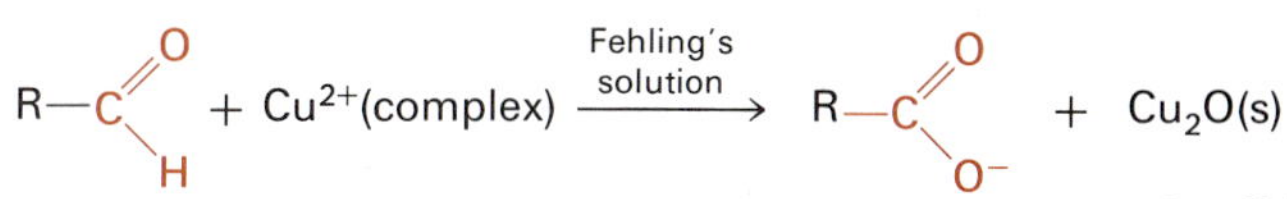

blue solution **salt of acid** **red solid**

The aldehyde is oxidized to the acid, and the copper (II) is reduced to copper(I).

Because the solution is basic, a salt of the acid (the sodium salt, say, if NaOH was the base) is formed rather than the acid itself. Benedict's test is identical to Fehling's, except that it uses Cu^{2+} complexed with citrate instead of with tartrate.

In Tollens' test, also called the silver-mirror test, silver ion in the form of an ammonia complex is reduced to metallic silver, which deposits as a mirror on the walls of a test tube, or as a black precipitate. The overall reaction is

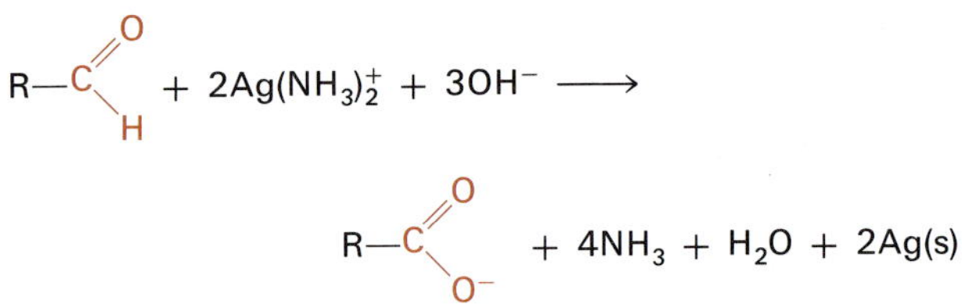

salt of acid

The same reaction—the reduction of Ag^+ by formaldehyde—is used commercially in the manufacture of silver mirrors. In Chapter 23 we shall see that Benedict's, Fehling's and Tollens' tests are commonly used to distinguish between reducing and nonreducing sugars.

Aldehydes and ketones can be reduced by catalytic hydrogenation to the corresponding alcohols. Commonly used catalysts are platinum, palladium, and nickel. Aldehydes produce primary alcohols, and ketones produce secondary ones. This is the exact reverse of the oxidation of alcohols to aldehydes and ketones.

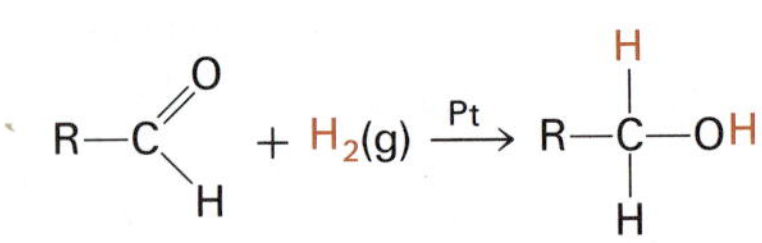

aldehyde **1° alcohol**

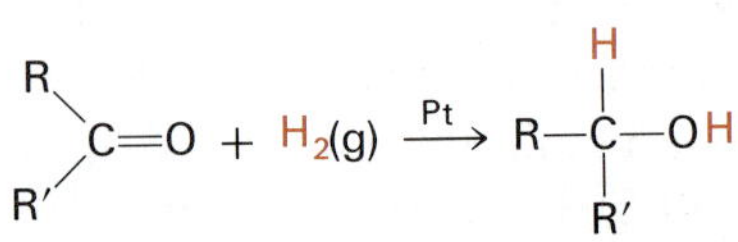

ketone **2° alcohol**

If the aldehyde or ketone also contains carbon–carbon double or triple bonds, the π bonds are hydrogenated simultaneously:

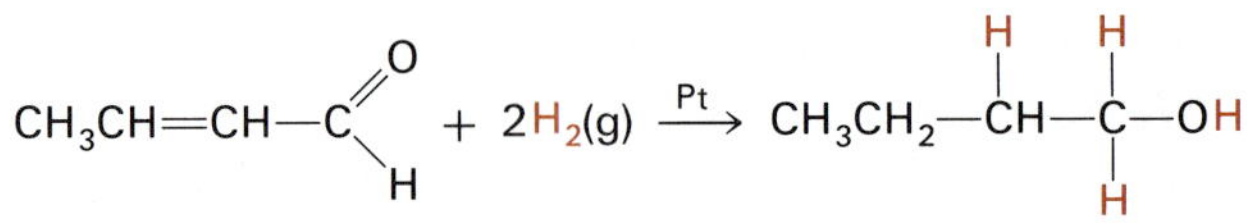

Remember that the benzene ring is especially stable, so it is possible to hydrogenate double bonds in groups attached to a benzene ring without hydrogenating the π bonds of the benzene ring:

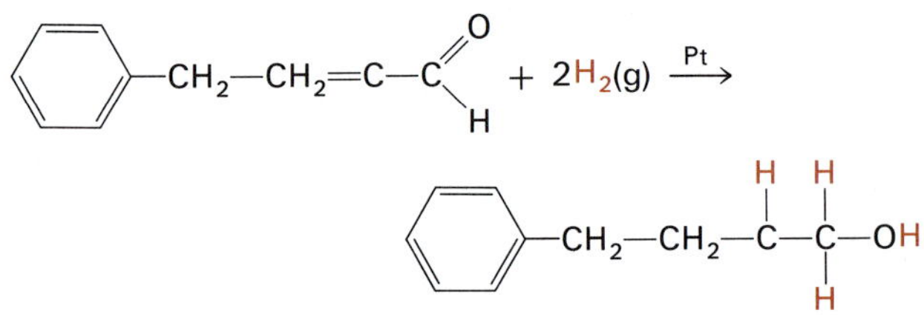

EXERCISE 18-10 Write equations for the catalytic hydrogenation of the following compounds. Show the structures of the products.

(a) Benzaldehyde, 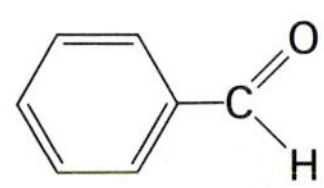

(b) Diethyl ketone, $CH_3CH_2COCH_2CH_3$

(c) Propenal, 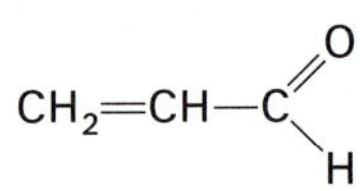

Special reagents are available to reduce the carbonyl group in a compound without affecting a carbon–carbon double bond. Metal hydrides—such as lithium aluminum hydride ($LiAlH_4$) or sodium borohydride ($NaBH_4$)—reduce carbonyl groups without reducing carbon–carbon double or triple bonds. The reduction by $LiAlH_4$ usually is carried out in ether solution. Acid hydrolysis then produces an alcohol from the aldehyde or ketone:

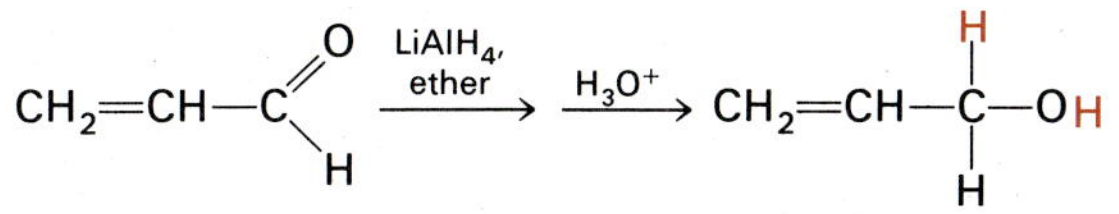

Lithium aluminum hydride and sodium borohydride reduce the carbonyl group by delivering a hydride ion ($H:^-$) to carbon; a proton is added to the oxygen:

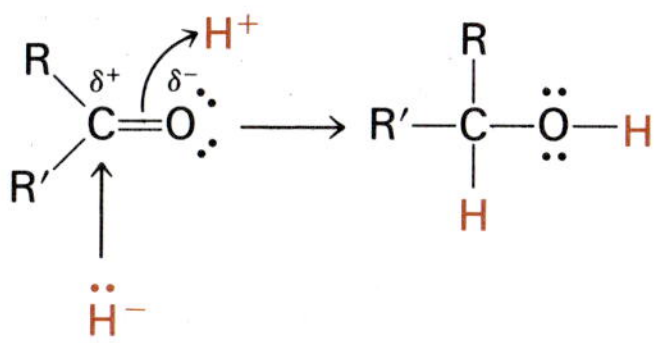

Like $LiAlH_4$ or $NaBH_4$, the biological reducing agent NADH reacts by delivering a hydride ion (in the form of H^+ and $2e^-$). The oxidized form of NADH (NAD^+) is a major electron acceptor in the oxidation of fuel molecules in the citric acid cycle (see Section 12-4). In the final step in the conversion of glucose to ethanol by fermentation, NADH reduces acetaldehyde to ethanol:

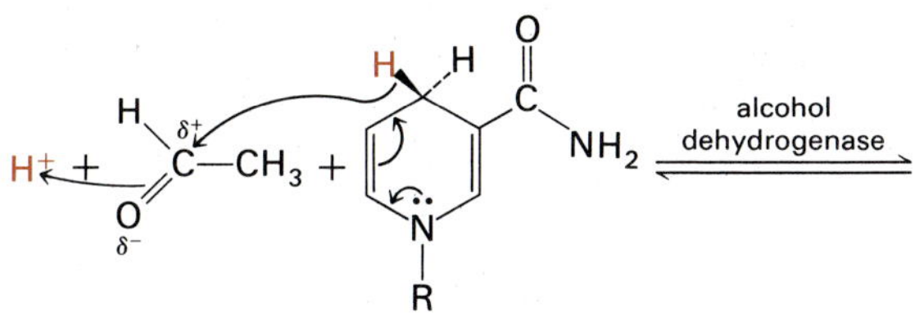

NADH
(reducing agent)

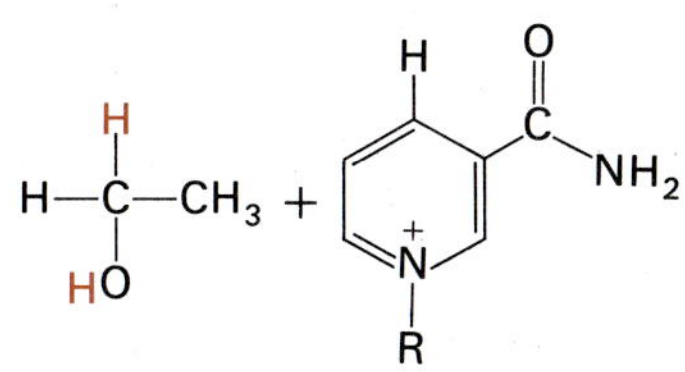

NAD+
(oxidizing agent)

The enzyme *alcohol dehydrogenase* catalyzes the reaction. (The complete structures of NADH and NAD^+ are given in Chapter 26, where bioenergetics and metabolism are discussed.)

In human muscle tissue, NADH reduces pyruvate to lactate, with the enzyme lactate dehydrogenase as catalyst. The same reduction can be accomplished in the laboratory with $NaBH_4$:

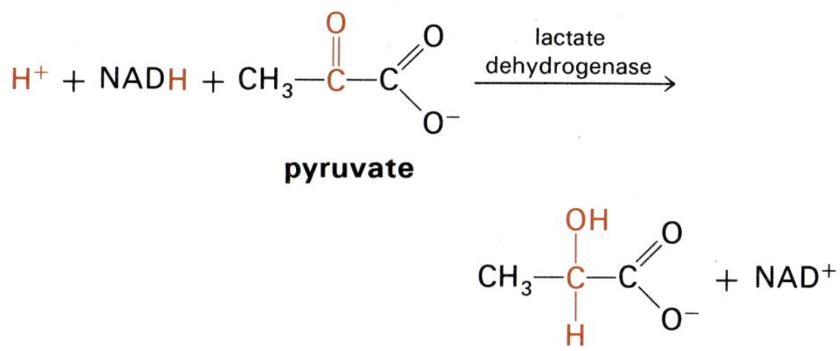

pyruvate **lactate**

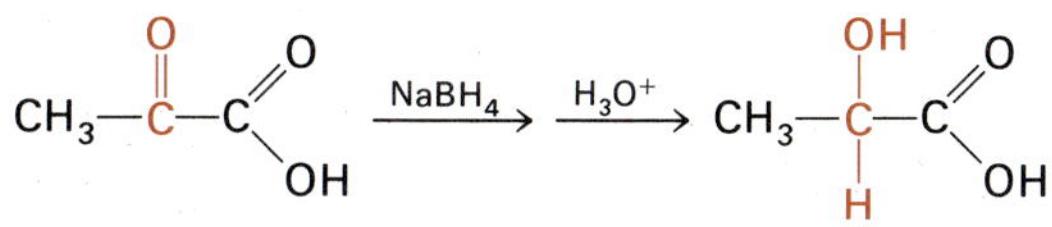

pyruvic acid **lactic acid**

In the biological reaction, which occurs at physiological pH, pyruvic and lactic acids are in the form of anions. Note that in both reactions, only the ketone C=O group is reduced; the carboxylic acid C=O is not. The reduction of pyruvate to lactate in muscle is important because it regenerates NAD^+ and thus enables the process of glycolysis (the anaerobic catabolism of carbohydrates to pyruvic acid with production of energy; see Chapter 26) to continue.

Unlike the metal hydrides, NADH reacts very slowly with oxygen in the absence of specific enzyme catalysts. This stability is essential for the biological function of NADH because it permits enzymes to control the flow of energy and reducing power.

18-9
ACIDITY: INDUCTIVE AND RESONANCE EFFECTS

The reactions of aldehydes and ketones that we have considered thus far all involve reactions of the carbonyl functional group itself. But the presence of this polar functional group also causes changes in the α-carbon atom (the carbon atom adjacent to the carbonyl carbon). The carbonyl carbon pulls electrons toward itself, causing the α-hydrogen atoms to be more acidic than normal C—H hydrogens:

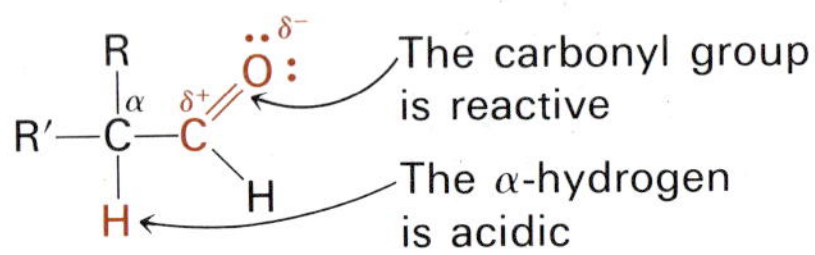

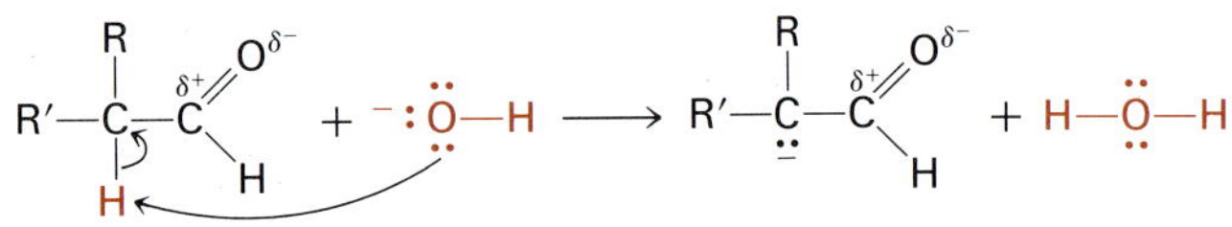

One consequence of the reactivity of the α-hydrogen is that aldehydes and ketones exist in solution as an equilibrium mixture of two isomeric forms called the keto form and the enol form:

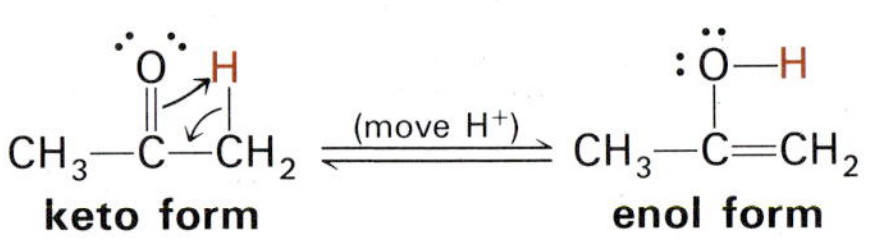

This transfer of hydrogen can involve any of the six hydrogens in acetone because they are all equivalent.

When compounds exist in structural arrangements of this type (which differ only in the distribution of electrons and the position of a hydrogen atom), the isomeric structures are called **tautomers.** The phenomenon itself is called **tautomerism.** In later chapters, we see that tautomerism plays an important role in the structures of some biologically important molecules.

The tautomerism is catalyzed by hydroxide ion, which reacts (in an acid–base reaction) with an α-hydrogen to produce water and an anion called an **enolate ion.** Here we see the tautomerism of acetone:

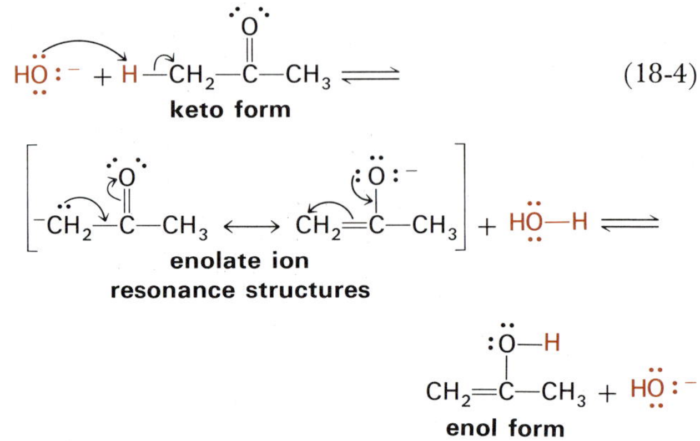

(18-4)

If a proton adds to the negative *carbon* of the enolate ion, the keto form is regenerated. If it adds instead to the negative *oxygen* of the enolate ion, the enol form is produced. Note the difference between tautomers and resonance structures. Tautomers are distinct chemical species that differ both in the position of a hydrogen atom and in the distribution of electrons. Resonance structures, which differ only in electron distribution, correspond to only one chemical species—the real structure is intermediate between that of the resonance structures.

Although acetone is not nearly so strong an acid as hydrochloric acid or acetic acid, and it is a weaker acid even than ethanol, it is still a much stronger acid than ethane:

ethane: $CH_3CH_3 \rightleftharpoons CH_3CH_2^- + H^+$

$K_a \cong 10^{-50}$ $\quad$ $pK_a \cong 50$

acetone: $CH_3-\overset{\overset{\Large O}{\|}}{C}-CH_3 \rightleftharpoons CH_3-\overset{\overset{\Large O}{\|}}{C}-CH_2^- + H^+$

$K_a \cong 10^{-20}$ $\quad$ $pK_a \cong 20$

ethanol: $$CH_3CH_2OH \rightleftharpoons CH_3CH_2O^- + H^+$$

$$K_a \cong 10^{-16} \qquad pK_a \cong 16$$

acetic acid: $$CH_3C(=O)OH \rightleftharpoons CH_3C(=O)O^- + H^+$$

$$K_a \cong 10^{-5} \qquad pK_a \cong 5$$

How can we account for the fact that acetone is so much stronger an acid than ethane? Two separate factors are responsible for this enhanced acidicity. First (but less important) is the **inductive effect** of the polar, electron-withdrawing carbonyl group, which pulls electrons away from the α-carbon toward the highly electronegative oxygen (Figure 18-3). The carbonyl carbon has a partial positive charge that pulls on the electrons of the C—C and the C—H bonds. This pulling of electrons through the σ bonds (indicated by the arrows in the diagram) results in a lower density of electrons in the C—H bond, weakening the bond and making it easier to remove a proton. The inductive effect of the carbonyl group thus makes aldehydes and ketones more acidic than alkanes.

The principal reason for the enhanced acidity of these compounds, however, is resonance stabilization of the carbanion that is formed. The anion that is produced by the ionization of ethane is a carbanion, with the negative charge *localized* on the carbon. The enolate ion produced by the ionization of acetone is stabilized by *resonance,* which delocalizes the electrons, as we saw in Equation 18-4.

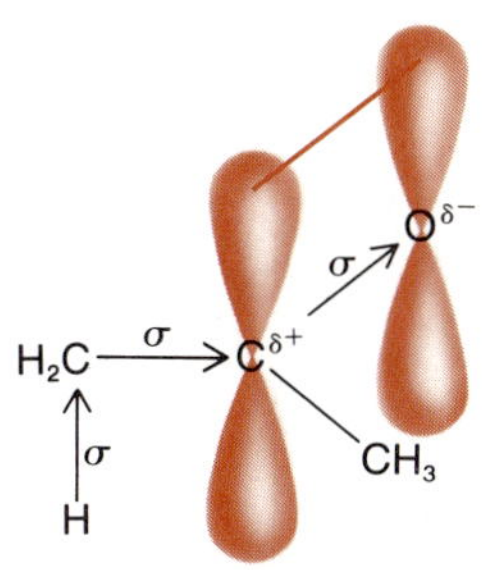

FIGURE 18-3

Inductive effect of carbonyl group. Electrons are pulled through σ bonds, toward the carbonyl group.

The two carbon atoms shown are in the sp^2 hybridization state, so each has an unhybridized p orbital available for a π bond. (The α-carbon rehybridizes from the sp^3 to the sp^2 state upon losing a proton.) The oxygen atom also has an unhybridized p orbital available for a π bond. Thus, one of the conditions required for resonance—that there be *adjacent* parallel p orbitals available—is satisfied. There are four π electrons (electrons in π bonds or lone pairs of electrons) that can drift through this chain of three parallel p orbitals. Just as with benzene, this delocalization of negative charge stabilizes the anion. Note that *in the resonance structures, there is no rearrangement of atoms but only rearrangement of electrons.* The structure of the real anion is a composite of the two resonance structures, but it is closer to the one that has the negative charge on the oxygen, because oxygen is more electronegative than carbon. Reactions of this anion can take place as though *both* the α-carbon and the oxygen have a negative charge—*both can act as nucleophiles.*

The inductive and resonance effects operate together to account for the enhanced acidity of aldehydes and ketones. The inductive effect causes a weakening of the α-carbon–hydrogen bond, and the resonance effect (which is by far the more important) stabilizes the anion. We can show that the resonance effect dominates by considering acetaldehyde. In acetaldehyde, the inductive effect should operate on both the hydrogen attached to the carbonyl carbon and the α-hydrogen. Yet only the α-hydrogen is acidic:

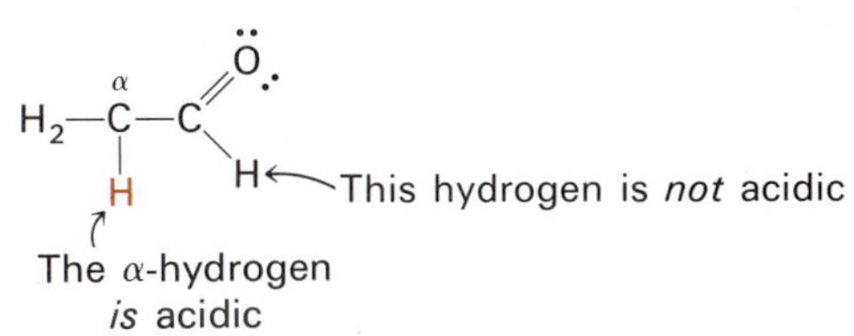

The anion formed when an α-hydrogen is removed from acetaldehyde is stabilized by resonance; the anion formed when the hydrogen bonded to the carbonyl carbon is removed is not stabilized by resonance:

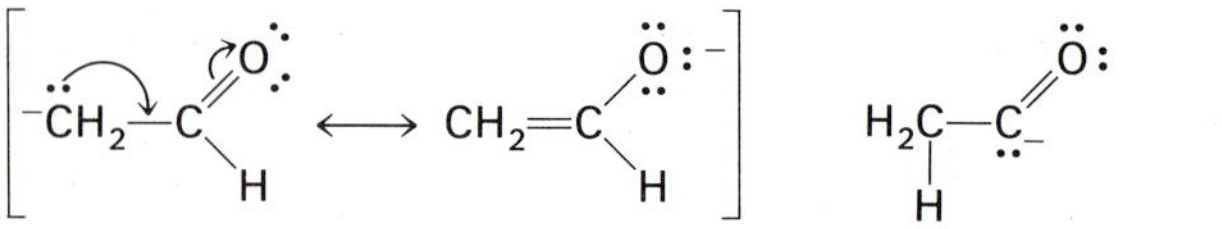

(resonance structures after loss of α-hydrogen)

(no resonance structures after loss of C═O hydrogen)

With an unsymmetric ketone such as ethyl methyl ketone, two different anions are formed in basic solution, depending on which α-hydrogen is removed:

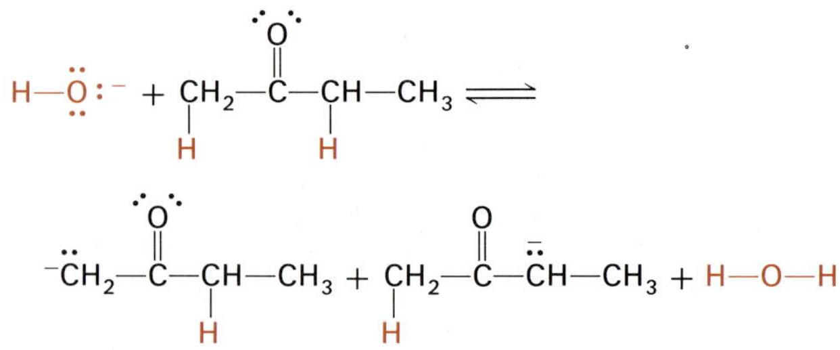

structure A

structure B

These anions are isomers of each other, and each has an additional resonance structure. (As an exercise, write the resonance structures for A and B.)

EXERCISE 18-11 (a) Which would you expect to be most acidic: CH_4, CH_3Cl, CH_2Cl_2, or $CHCl_3$? Explain.

(b) Write two resonance structures for the anion formed when sodium hydroxide solution is added to propionaldehyde (CH_3CH_2CHO).

18-10 THE ALDOL CONDENSATION

A consequence of the acidity of α-hydrogens of aldehydes and ketones is a **self-addition** that occurs in basic solution. For example, when acetaldehyde is treated with dilute sodium hydroxide, a 50% yield of 3-hydroxybutanal (an **aldol** from *ald*ehyde and alco*hol*) is formed:

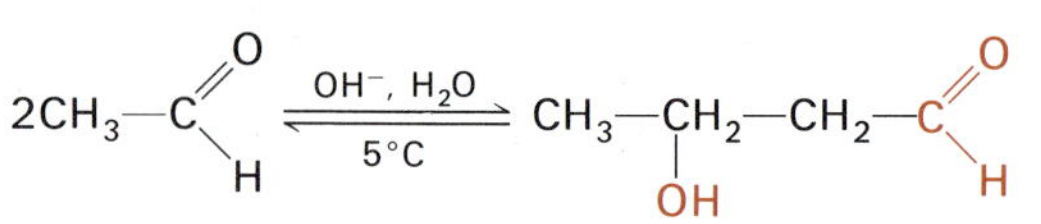

an aldol

Reactions of this type are called **aldol condensations.**

Such reactions result in the formation of new carbon–carbon bonds and longer carbon chains. A possible mechanism for the reaction is

Step 1: Formation of enolate ion

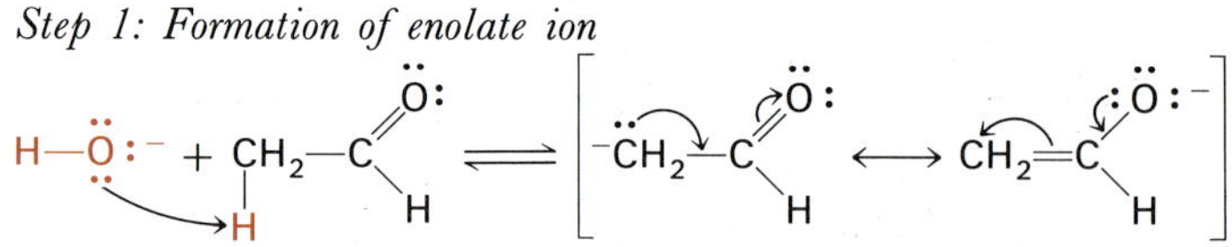

enolate ion resonance structures

Step 2: Addition of nucleophile

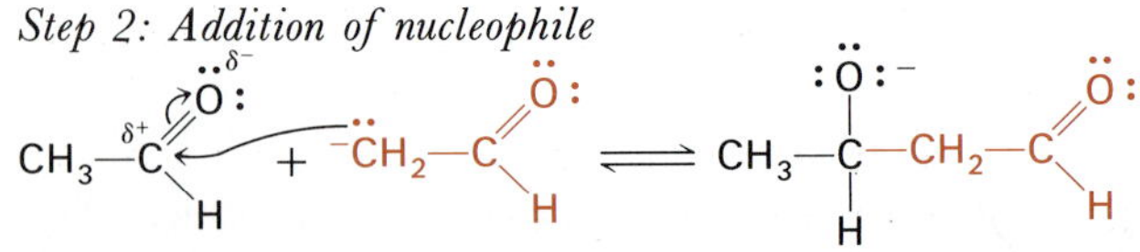

Step 3: Addition of H^+

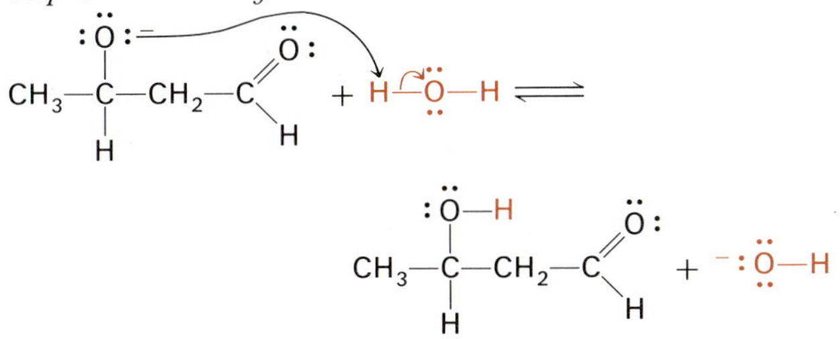

In the first step, an enolate anion is formed. The second step is simply the addition of a nucleophile (the carbanion form of the enolate ion) to the carbonyl carbon of acetaldehyde. Finally, a proton is transferred from a water molecule to produce an alcohol and regenerate the catalyst (OH^-). The final step occurs because this alcohol is a weaker acid than water; thus the anion of the alcohol is a stronger base than the hydroxide ion.

Note that in step 2, the negative carbon rather than the negative oxygen attacks, although oxygen (being more electronegative than carbon) has the greater negative charge. One reason for this is that in organic solvents, the oxygen is likely to have associated with it the metal ion of the base—say, Na^+ from NaOH. This *protects* the oxygen, and prevents it from reacting. Reactions are, therefore, more likely to occur at the negatively charged carbon. Furthermore, the compound formed by the addition of the carbanion is more stable than that formed by the addition of the resonance structure with the negatively charged oxygen.

Cross-condensation between two different aldehydes or ketones that both have α-hydrogens leads to a complex mixture of products and so is not useful for synthesis. (There are four products, for example, of the reaction between acetaldehyde and propanal.) But an aldehyde or ketone that does not form an enolate ion because it has *no* α-hydrogen can be cross-condensed with one that has an α-hydrogen—for example, benzaldehyde with acetone:

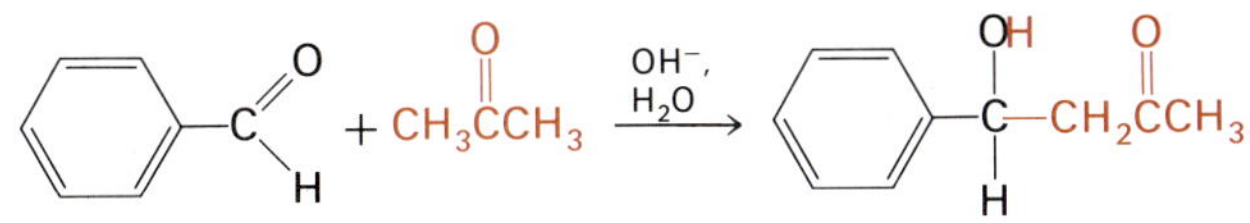

Because benzaldehyde has no α-hydrogen, the only anion formed is from acetone. The acetone carbanion adds to the carbonyl carbon of benzaldehyde. In general, ketones are less reactive than aldehydes, so the main reaction is the cross-condensation of the benzaldehyde and acetone, and not the self-condensation of the acetone. However, some small amount of the self-condensation product also is formed.

EXERCISE 18-12 Draw the structure of the product of the self-condensation of acetone.

We shall see that aldol and aldol-type condensations occur widely in the biosynthesis of large molecules such as fats and the steroid hormones (Chapter 24). The aldol condensation, for example, is the reaction that cross-links the fibers in collagen—the major structural protein of skin, bone, teeth, cartilage, and tendon (Chapter 21). A step in the biosynthesis of the six-carbon sugar fructose is the aldol condensation of the two three-carbon units dihydroxyacetone phosphate and glyceraldehyde 3-phosphate. The reaction is catalyzed by the enzyme aldolase:

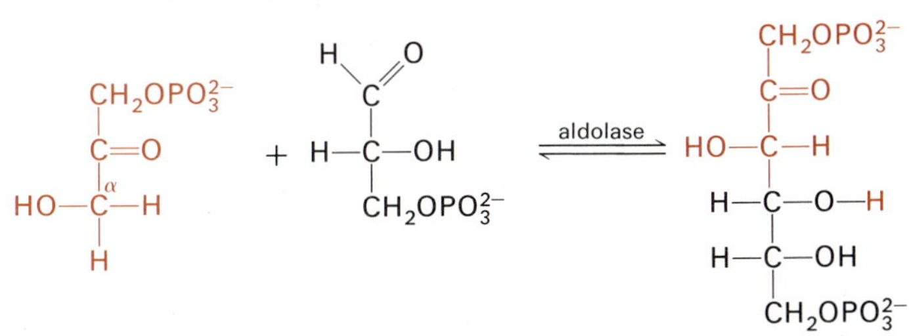

dihydroxyacetone phosphate **glyceraldehyde 3-phosphate** **fructose 1,6-diphosphate**

The reverse of this reaction, also catalyzed by aldolase, is a step in glycolysis. Figure 18-4 is a photomicrograph of crystals of aldolase.

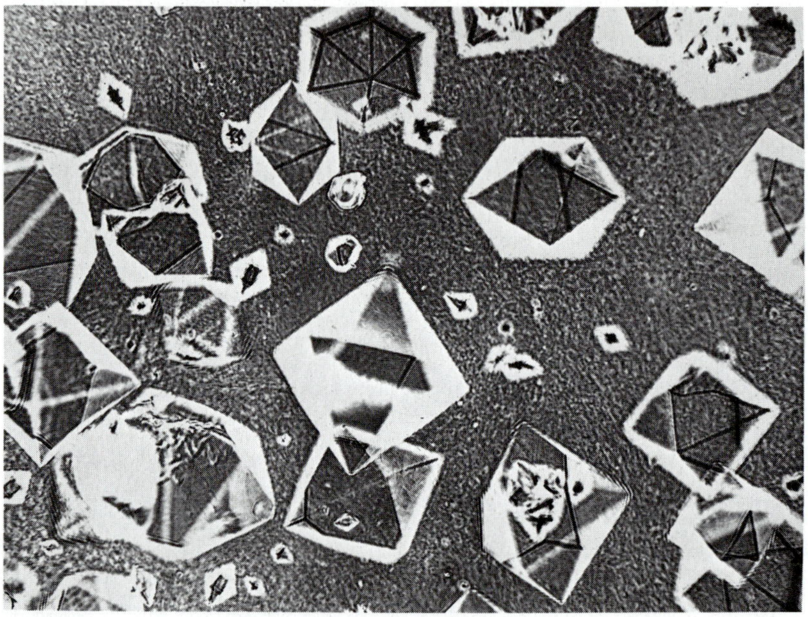

FIGURE 18-4

Photomicrograph of crystals of aldolase. [Courtesy of David Eisenberg.]

KEY TERMS

A *Grignard reagent* is an organomagnesium compound of general formula R—MgX, where X is a halogen atom and R is an alkyl or an aryl (aromatic) group.

Tautomers are isomers that differ only in the distribution of electrons and in the position of a hydrogen atom.

The polar, electron-withdrawing carbonyl group of an aldehyde or ketone pulls electrons from the α-carbon toward the highly electronegative oxygen. This pulling of electrons through σ bonds is called the *inductive effect*.

The *aldol condensation* is a self-addition of aldehydes or ketones that occurs in basic solution.

Hydrolysis is the breaking up of a molecule to form two or more smaller ones by the addition of the components of water (H and OH).

CHAPTER SUMMARY

Aldehydes and ketones contain the *carbonyl functional group* ($>C=O$). Aldehydes and ketones can undergo a variety of chemical reactions. Both can be reduced to alcohols; aldehydes (but not ketones) can be oxidized to carboxylic acids. Both aldehydes and ketones undergo *addition to the carbon–oxygen double bond:* a nucleophile adds to the partially positively charged carbonyl carbon, and an electrophile (usually H^+) adds to the partially negatively charged oxygen. Compounds that form addition products with carbonyl compounds include HCN, ROH, 2,4-DNP, and Grignard reagents.

Because of the acidity of the α-hydrogen atoms, *enolate anions* of aldehydes and ketones are formed in basic solution. The acidity of α-hydrogens is due primarily to the resonance stabilization of the anion, and to a lesser degree to the *inductive effect* of the carbonyl group. The enolate anions participate in *aldol condensation* reactions, and also in substitution reactions.

New carbon–carbon bonds are formed in the *Grignard addition reactions,* in the aldol condensations, and in substitution reactions of the enolate anions. These carbon–carbon bond-forming reactions change the basic carbon framework of the molecule, lengthening or branching the carbon chain.

PROBLEMS

18-1. Which of the following compounds will have the lowest (and which the highest) solubility in water?

(a) $CH_3CH_2OCH_2CH_3$ (b) $CH_3-C(=O)-CH_2CH_3$

(c) $CH_3CH_2CH_2CH_2OH$ (d) $CH_3-C(=O)-CH_2OH$

18-2. Rank the following compounds in order of increasing boiling point:

(a) $CH_3OCH_2CH_3$ (b) $CH_3CH_2CH_2OH$

(c) $CH_3CH_2CH_2CH_3$ (d) $HOCH_2CH_2OH$

18-3. Name the following compounds according to the IUPAC system:

(a) $CH_3CH(Cl)C(=O)CH_3$ (b) $CH_3CH(CH_3)CH(Br)CHO$

(c) $CH_3CH_2C(=O)CCl_3$

18-4. From what alcohol could you prepare each of the following compounds?

(a) 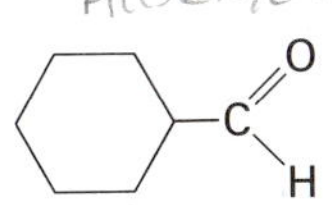

(b) $CH_3-CH(CH_3)-C(=O)-CCl_3$

(c) $CH_3-C(=O)-CH_2-C(=O)-CH_3$

18-5. Write the structures of the products of the following reactions:

(a) between 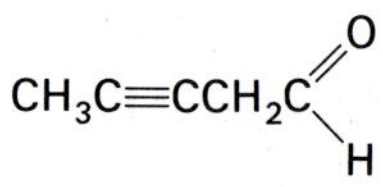 and an excess of $H_2(g)$, with a platinum catalyst;

(b) between 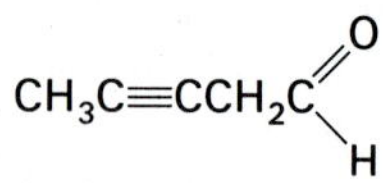 and $LiAlH_4$.

18-6. Write the structures of the products of the following reactions:

(a) 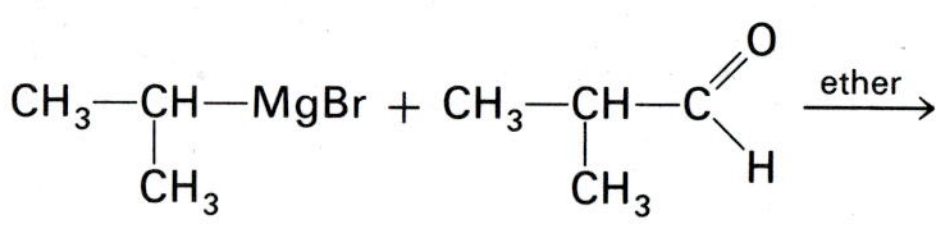

Addition product

(b) Water added to the product of part a.

(c) 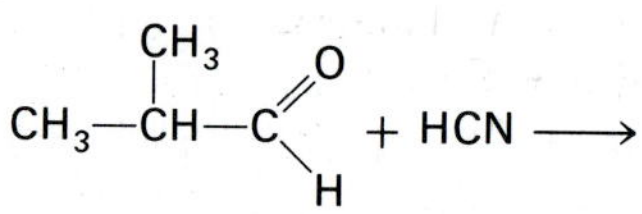

(d) 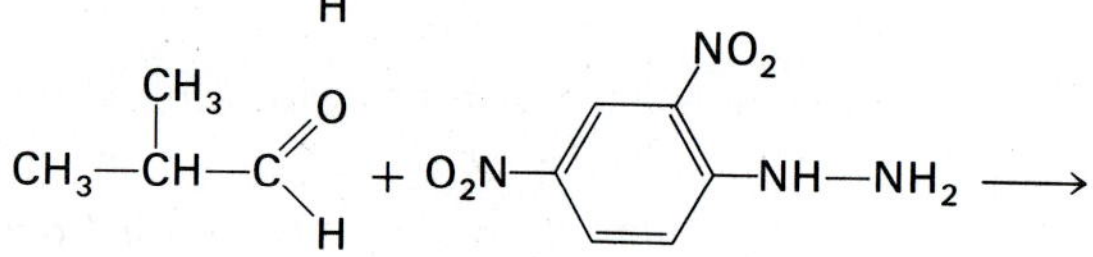

(e) 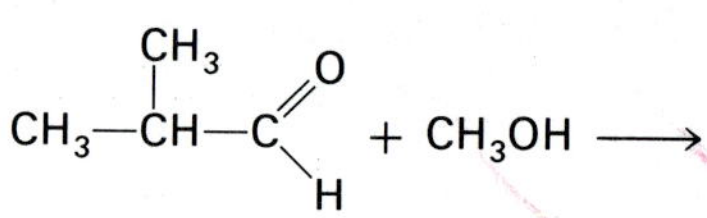

(f) 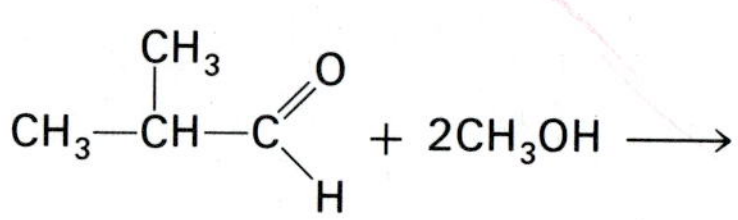

18-7. What alcohol and what carbonyl compound would you use to prepare the following compounds?

(a) the hemiketal 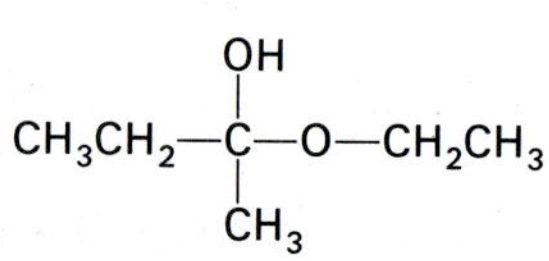

(b) the hemiacetal 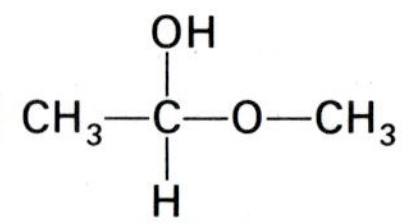

18-8. A certain millipede uses a compound called mandelonitrile as its defensive secretion. This substance readily breaks down in the presence of an enzyme to form only two products: a poisonous gas and a substance C_7H_6O that is a tissue irritant:

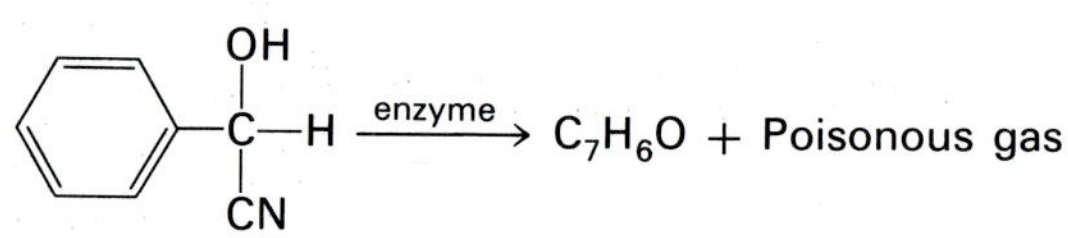

mandelonitrile

(a) Write a simple mechanism (using curved arrows for electron-pair movement) to show what reaction is occurring.
(b) Draw the structures of C_7H_6O and the poisonous gas.

18-9. The melon fly poses a serious problem in Hawaii. With increased travel to that part of the world, this insect threatens the general U.S. agricultural community. Recently the following compound was discovered to be an effective male melon fly attractant, and its use could eradicate this pest:

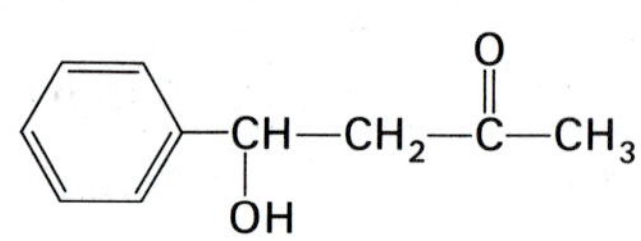

Show how you could prepare this compound using benzaldehyde

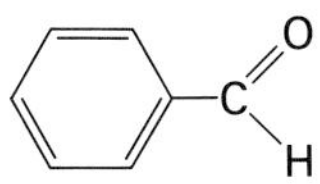

and any other simple compounds you need. Be sure to show your steps clearly.

18-10. Perfume companies all around the world are interested in the following alcohol, which has the pleasant aroma of freshly cut red roses:

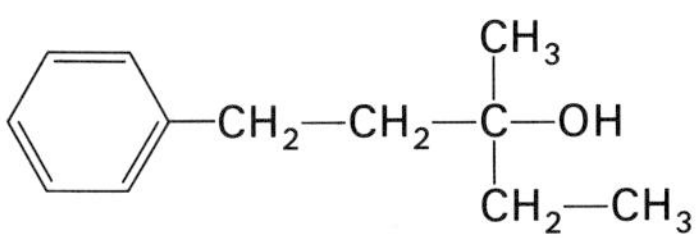

There are at least three simple ways to make this compound, all based on the same technique. Show *one* way in detail, using magnesium and any simple organic compounds you need. Use curved arrows when appropriate.

18-11. One of the most common visual pigments in the vertebrate eye, rhodopsin, is formed by joining a molecule of oxidized vitamin A to one NH_2 group of a protein in the eye (called opsin):

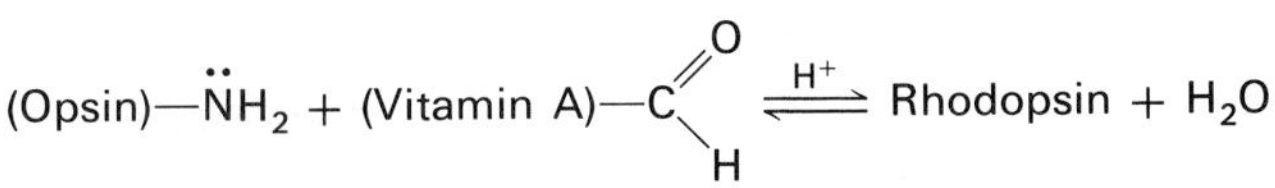

Draw a reasonable mechanism for this reaction (which actually takes place in the eye) and clearly show the new bond that is formed between opsin and vitamin A aldehyde.

18-12. When mammalian tissues consume oxygen to burn energy, carbon dioxide (CO_2) is released. Hemoglobin, an important blood protein, uses its amino groups to transport CO_2 from the tissues to the lungs, where it is released. This reversible reaction of hemoglobin with CO_2 to form X accounts for 60% of the total CO_2 exchanged by red blood cells during respiration:

$$\text{Hemoglobin—}\ddot{N}H_2 + CO_2 \rightleftharpoons X$$

(a) Draw the structure of X, clearly showing the new bond formed in the reaction between hemoglobin and CO_2. Draw a reasonable mechanism for this reaction, using curved arrows.

(b) Which of the three general types of organic chemical reactions we have studied (addition, substitution, elimination) is this?

18-13. Resonance stabilizes carbanions, carbon radicals, and carbonium ions. Resonance can also stabilize other atoms such as nitrogen and oxygen, which in compounds can have lone pairs of electrons in *p* orbitals. Notice the remarkable similarity of electronic structure between a carboxylate ion (A) and a carboxylic amide (B). Then draw one other resonance structure of B that follows logically from this comparison:

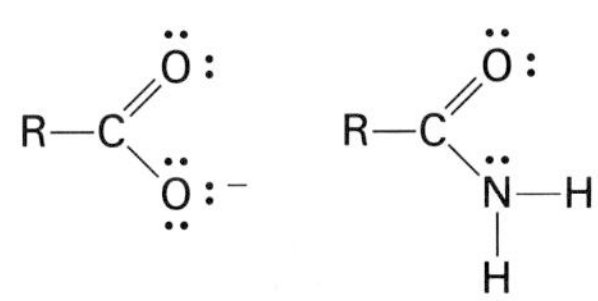

A B

18-14. (a) Draw one other resonance structure for

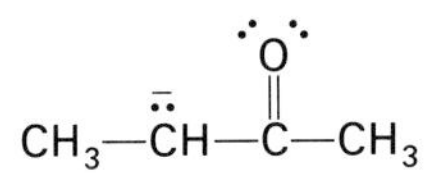

(b) Draw two additional resonance structures for

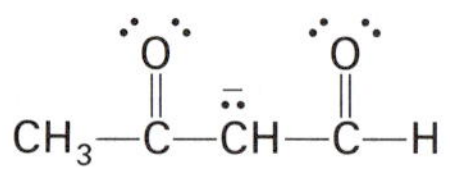

(c) Draw two additional resonance structures for

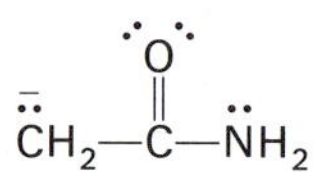

(d) Draw three additional resonance structures for

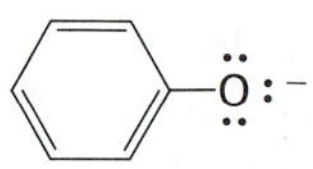

(e) Draw four additional resonance structures for

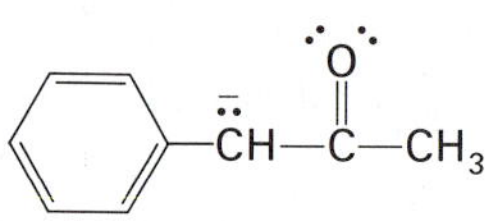

(f) Draw one other resonance structure for

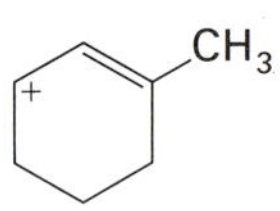

18-15. Which of the following compounds would you expect to be the stronger acid? Explain.

(a) $CH_3\overset{\overset{O}{\|}}{C}CH_2\overset{\overset{O}{\|}}{C}CH_3$ (b) $CH_3CH_2\overset{\overset{O}{\|}}{C}CH_2CH_3$

18-16. (a) Draw the structures of the products of the following reactions:

(1) 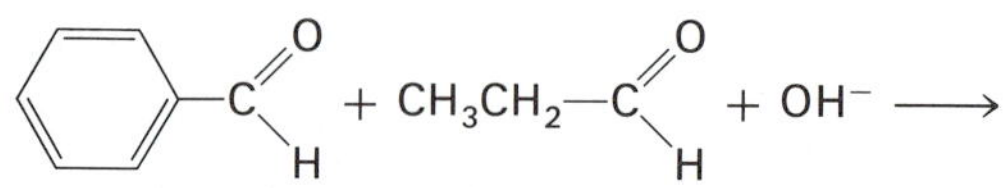

(2) 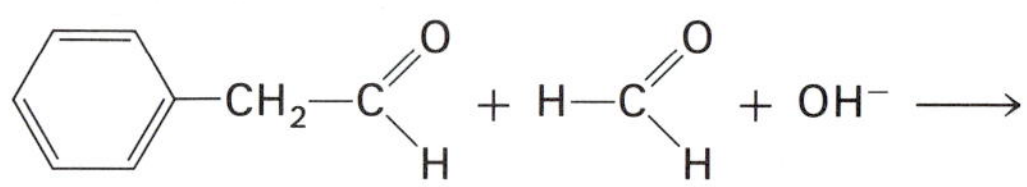

(b) The products of the reactions of part a readily lose water to yield a compound containing a carbon–carbon double bond. Draw the structures of the products of the elimination of water from the compounds produced in part a. (*Hint:* Remember that the α-hydrogens are the most acidic.)

18-17. Suggest a synthesis for each of the following compounds. Use equations, but you need not describe mechanisms. Use any additional simple chemicals you need.

(a) 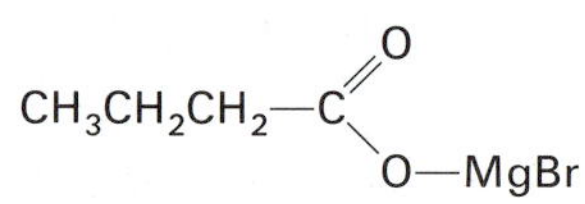 from $CH_3CH_2CH_2Br$ and Mg

(b) 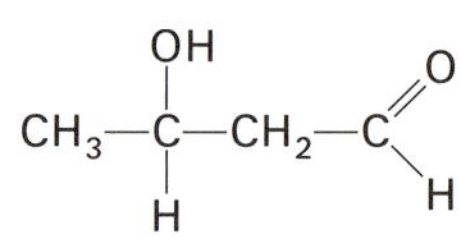 from 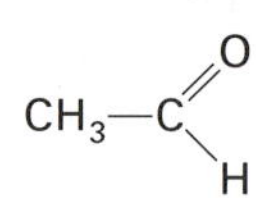

(c) 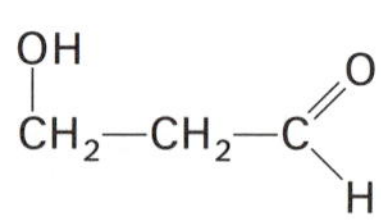 from 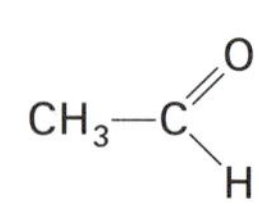

18-18. Draw the structures of the products of the following reactions. Using curved arrows, show the mechanism of each reaction.

(a) 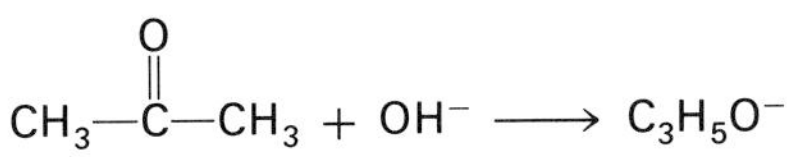

(b) Product of part a + CH_3I ⟶

I^- + Two isomeric organic products

(c) Product of part a + 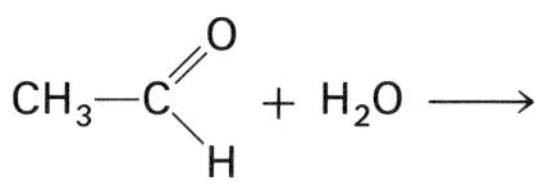

(*Hint:* Consider the resonance structures of the product of part a.)

18-19. The compound 5-aminohexanaldehyde

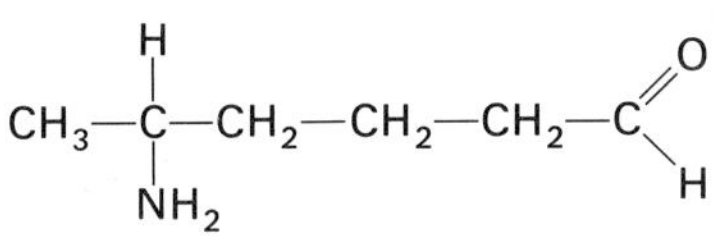

undergoes an intramolecular addition reaction to form a six-membered ring. Draw the structure of the cyclic product or products.

18-20. Describe a laboratory test (or reagent) that could be used to distinguish the following compounds:

(a) $CH_3CH_2CH_2OH$ and 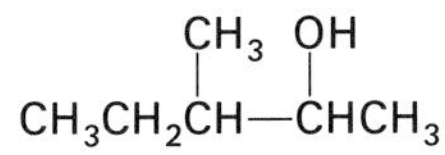

(b) 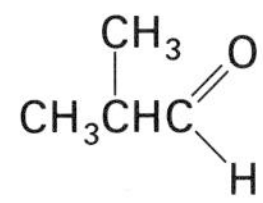 and 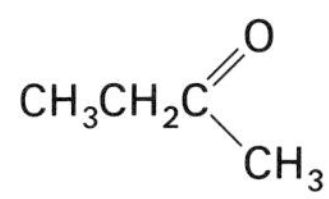

18-21. Consider the molecular formula C_3H_6O.
(a) Draw a structure with only σ bonds.
(b) Draw a structure with a C—C π bond.
(c) Draw a structure with a C—O π bond.
(d) In your structure for part c, what is the hybridization of the carbon or carbons attached to the oxygen?

18-22. We saw that in most cases it is impossible to isolate hydrates of aldehydes and ketones from solution. An exception is chloral hydrate (the hydrate of trichloroacetaldehyde, Cl_3CCHO), which is a stable, crystalline solid. Chloral hydrate is used as a sedative and as a sleep-inducing drug. Write its structure.

18-23. In the reaction of Section 18-10 between dihydroxyacetone phosphate and glyceraldehyde 3-phosphate to yield fructose 1,6-diphosphate, an NH_2 group of the aldolase reacts with the carbonyl group of the dihydroxyacetone phosphate to yield a Schiff base (Section 18-6). The Schiff base is protonated (has an extra proton) and thus is positively charged. Using E—NH_2 to represent the enzyme, write the structure of the Schiff base.

Aspirin consumption in the United States is 12,000 tons per year—150 tablets for every man, woman, and child.

19

Carboxylic Acids and Esters

Carboxylic acids are weak compared to the strong inorganic acids such as H_2SO_4 and HCl. But carboxylic acids are typical acids—they taste sour, turn blue litmus red, react with magnesium to produce hydrogen gas, and react with bases to form salts and water. The acetic acid of vinegar, the citric acid of lemon juice, and the lactic acid of sour milk are familiar carboxylic acids.

Esters are derivatives of carboxylic acids, formed by reaction between an acid and an alcohol. Esters are plentiful in nature. They are responsible for many of the pleasant odors of fruits and flowers. Fats and oils also are esters. Dacron is a polyester. Another important derivative of carboxylic acids, the amide, is discussed in Chapter 20.

19-1 STRUCTURE AND PROPERTIES OF CARBOXYLIC ACIDS

Carboxylic acids have the **carboxyl** functional group:

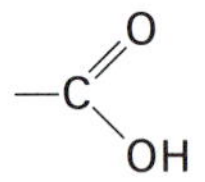

carboxyl group

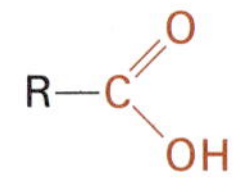

general formula of a carboxylic acid

The R group can be H (in formic acid, the simplest carboxylic acid) or any alkyl or aromatic group. A carbon atom of the R group is bonded to the carboxyl carbon atom. The condensed formula for carboxylic acids is RCOOH or RCO_2H:

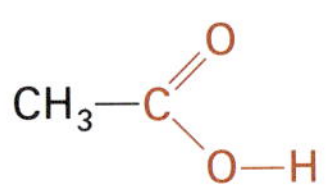

acetic acid

Condensed formula: CH_3COOH or CH_3CO_2H

The carboxyl carbon atom is in the sp^2 hybridization state and forms three σ bonds (one to each oxygen and one to the R group) and one π bond. Figure 19-1 shows the bonding in formic acid (HCOOH). The three sp^2 orbitals lie in a plane; the bond angles about the carboxyl carbon are close to the 120° angles of sp^2 hybridization. Note that, like carbon–carbon bonds, the carbon–oxygen double bond is shorter than the carbon–oxygen single bond.

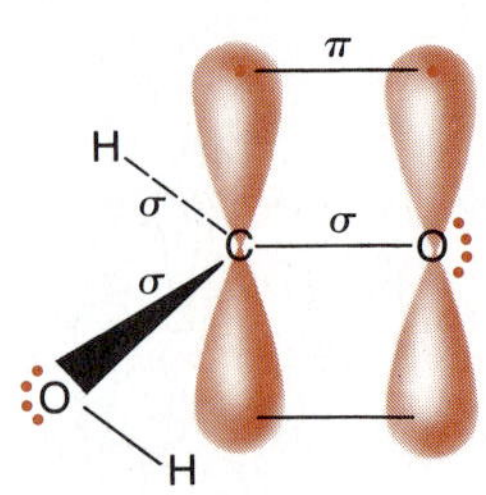

Bond length (nm)	
C=O	0.12
C–O	0.13
C–H	0.11
O–H	0.093
Bond angle	
H–C=O	124°
O–C=O	125°
H–C–O	111°
H–O–C	106°

FIGURE 19-1

Structure of formic acid

Table 19-1 lists the names, melting points, boiling points, and water solubilities of some unbranched saturated carboxylic acids. The boiling points of the carboxylic acids are unexpectedly high for their molecular weights—even higher than those of the corresponding alcohols (Figure 19-2). The high boiling points of the acids are attributed primarily to the fact that in the liquid two molecules are held together by hydrogen bonds:

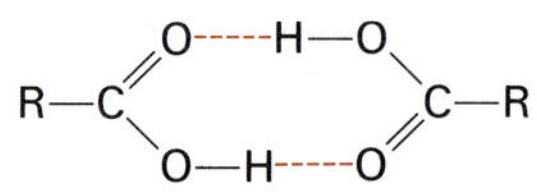

A unit like this is called a **dimer** because it is composed of two molecules joined together; a **trimer** consists of three and a **polymer** of many small molecules joined together. Even in the vapor state, carboxylic acids exist to some extent as dimers. These larger units cause much higher boiling points than would be expected for the monomers.

The acids of low molecular weight (through octanoic acid) are liquids at room temperature, and the first four are infinitely soluble in water because of hydrogen bonds that form between the carboxyl group and water molecules. As the length of the nonpolar hydrocarbon chain increases, the water solubility decreases.

EXERCISE 19-1 List the following compounds in order of decreasing boiling point:

(a) $CH_3CH_2OCH_2CH_3$

(b) $CH_3CH_2\overset{O}{\overset{\|}{C}}CH_3$

(c) $CH_3CH_2\overset{O}{\overset{\|}{C}}OH$

(d) $CH_3CH_2\overset{OH}{\overset{|}{C}}HCH_3$

TABLE 19-1
Physical properties of carboxylic acids

Formula	Name: Common	Name: IUPAC	Melting point (°C)	Boiling point (°C)	Solubility at 20°C (g/100 g H_2O)
HCOOH	Formic acid	Methanoic acid	8	101	∞
CH_3COOH	Acetic acid	Ethanoic acid	17	118	∞
CH_3CH_2COOH	Propionic acid	Propanoic acid	−21	141	∞
$CH_3(CH_2)_2COOH$	Butyric acid	Butanoic acid	−5	164	∞
$CH_3(CH_2)_3COOH$	Valeric acid	Pentanoic acid	−34	186	4.97
$CH_3(CH_2)_4COOH$	Caproic acid	Hexanoic acid	−3	205	0.968
$CH_3(CH_2)_6COOH$	Caprylic acid	Octanoic acid	17	239	0.068
$CH_3(CH_2)_8COOH$	Capric acid	Decanoic acid	32	270	0.015
$CH_3(CH_2)_{10}COOH$	Lauric acid	Dodecanoic acid	44	299	0.0055
$CH_3(CH_2)_{12}COOH$	Myristic acid	Tetradecanoic acid	54	Decomposes	0.0020
$CH_3(CH_2)_{14}COOH$	Palmitic acid	Hexadecanoic acid	63	Decomposes	0.00072
$CH_3(CH_2)_{16}COOH$	Stearic acid	Octadecanoic acid	70	Decomposes	0.00029

Red ants carry *formic acid* (HCOOH) in their jaw glands, and the acid is partly responsible for the irritation caused by ant bite. The "sting" of the stinging nettle is also partly due to formic acid. *Acetic acid* (CH_3COOH) is the acid in vinegar and is an oxidation product of the ethanol obtained by fermentation of fruits and grains. As we shall see in Chapter 24, acetyl coenzyme A (which may be regarded as a derivative of acetic acid) is the fundamental unit in the biosynthesis of a variety of large molecules such as the fatty acids and the steroid hormones. *Butyric acid,*

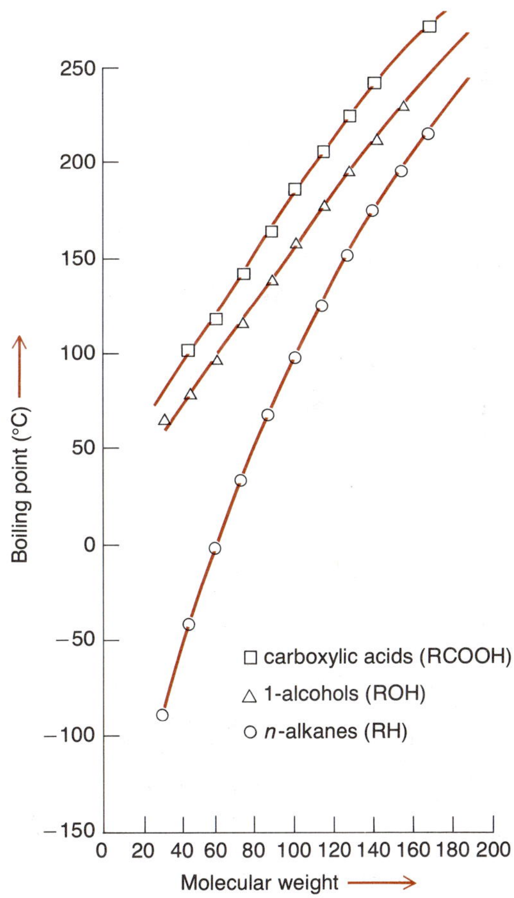

FIGURE 19-2
Boiling points of unbranched carboxylic acids, alcohols, and alkanes.

$CH_3(CH_2)_2COOH$, produces the odor of rancid butter. Penetrating and unpleasant odors also characterize the larger carboxylic acids—*caproic, caprylic,* and *capric acids* are said to be reminiscent of goats. Their names, indeed, are derived from the Latin *caper,* which means goat. Of course, the vapor pressures decrease with increasing molecular weight, and so the odors decrease in intensity.

Carboxylic acids occur in combined form in vegetable oils and animal fats: laurel and coconut oils are rich sources of *lauric acid,* coconut oil of *palmitic acid,* and beef fat of *stearic acid.* The straight-chain carboxylic acids that can be obtained from fats and oils are often called **fatty acids.** Fatty acids are used in the manufacture of soaps and detergents (see Chapter 24). Because fats and oils are synthesized in nature from the two-carbon units of acetic acid, essentially all the carboxylic acids found in nature that are larger than caproic acid (the six-carbon acid) have even numbers of carbon atoms.

The following are some other naturally occurring carboxylic acids:

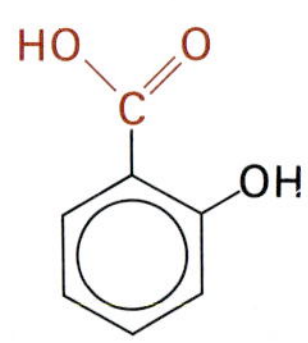

salicylic acid
(from the bark of willow trees; reduces fever and swelling)

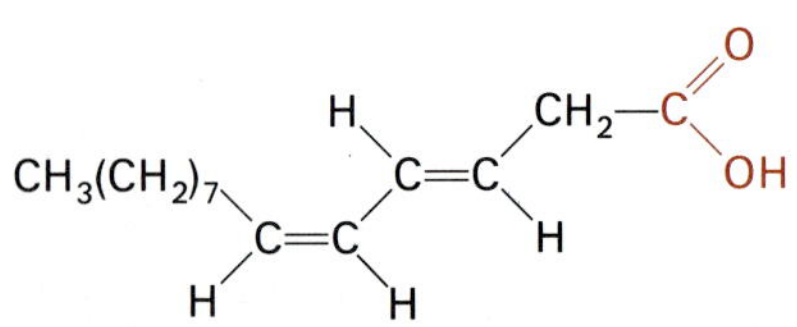

3-*trans*-5-*cis*-tetradecadienoic acid
(sex attractant of black carpet beetle)

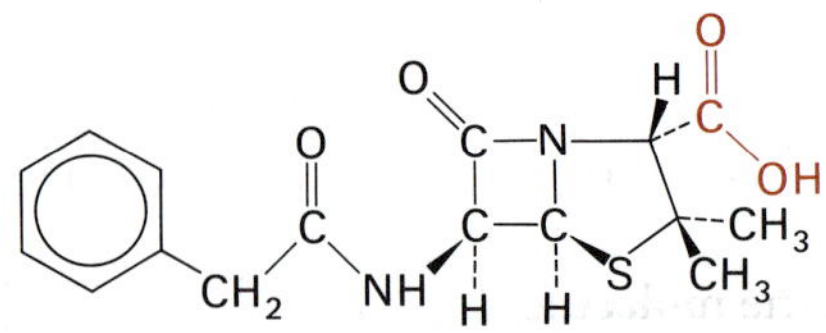

penicillin G
(secretion of a blue-green mold; antibiotic)

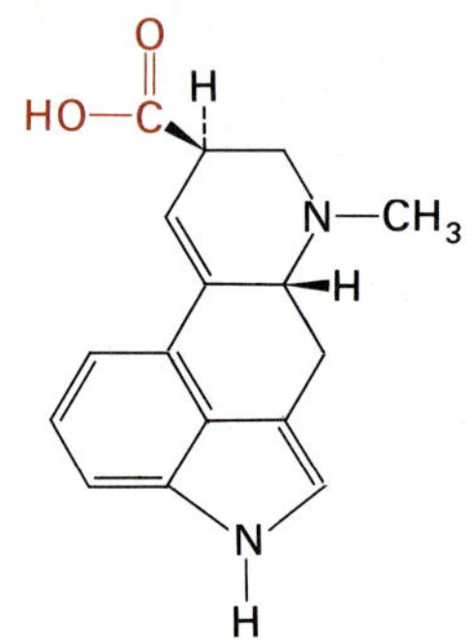

lysergic acid
(from ergot, a fungus that grows on grain and hops)

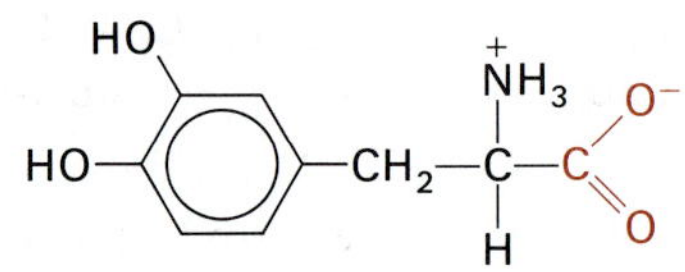

DOPA (3,4-dihydroxyphenylalanine)
(an amino acid—the L form is produced in the adrenal gland and nerve terminals; used in treatment of Parkinson's disease)

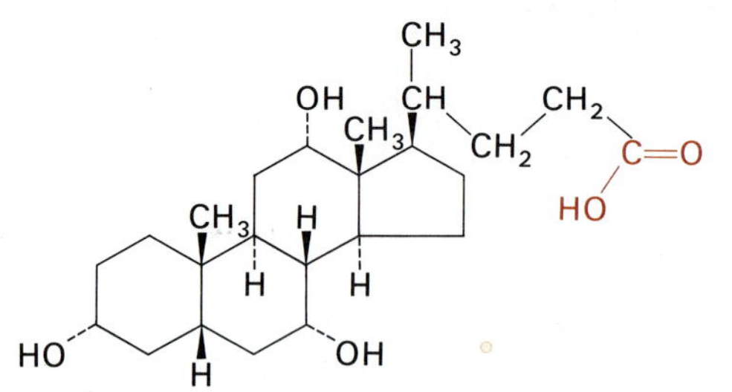

cholic acid
(a bile acid; aids in absorption of fats)

Many of these substances, like many we have seen in earlier chapters, have more than one functional group. Each functional group displays its own properties, modified somewhat by the presence of other groups in the molecule. In biochemical reactions, each functional group of a substrate can interact with a specific group on the enzyme that catalyzes the reaction. (We shall see an example of this in the discussion of the enzyme chymotrypsin in Chapter 22.)

19-2 NOMENCLATURE OF CARBOXYLIC ACIDS

The small carboxylic acids are generally known by their common names (Table 19-1). When using these names, as with the common names for aldehydes and ketones, the positions of substituents are designated by the Greek letters α, β, γ, and so on, *beginning with the carbon next to the carboxyl carbon:*

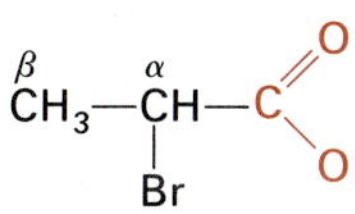

α-bromopropionic acid

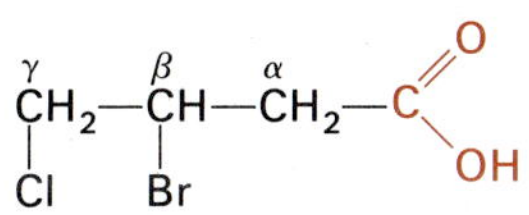

β-bromo-γ-chlorobutyric acid

In the IUPAC system, carboxylic acids are named by dropping the final *-e* from the name of the corresponding alkane and adding the ending *-oic acid.* The positions of substituents are designated by number. The carboxyl carbon is assigned the number 1:

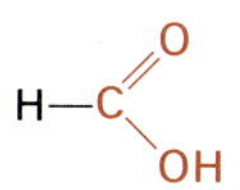

methanoic acid

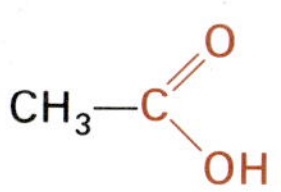

ethanoic acid

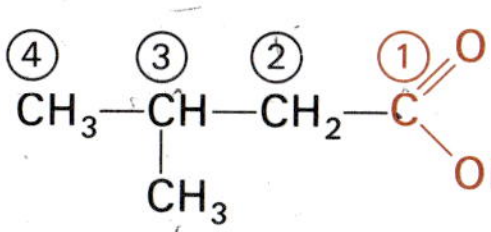

3-methylbutanoic acid

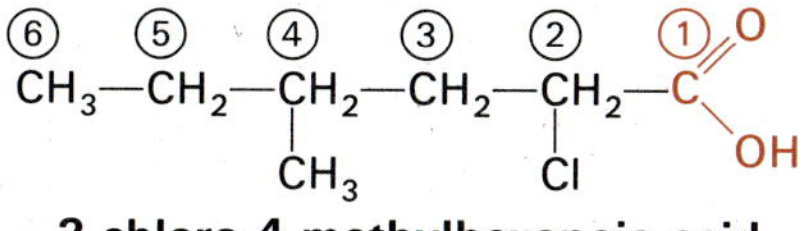

2-chloro-4-methylhexanoic acid

The simplest aromatic acid is benzoic acid:

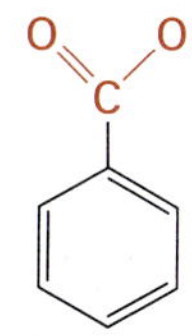

benzoic acid

Some compounds have two or more carboxyl groups:

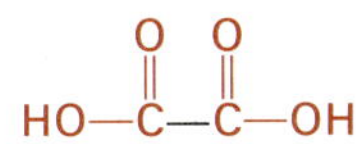

oxalic acid

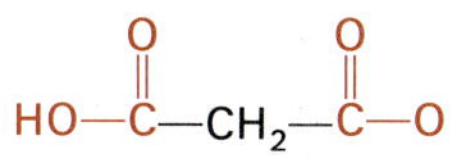

malonic acid

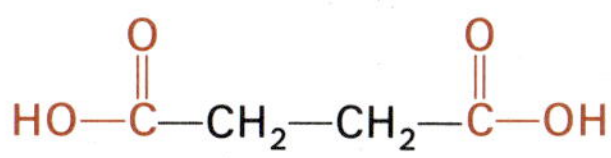

succinic acid

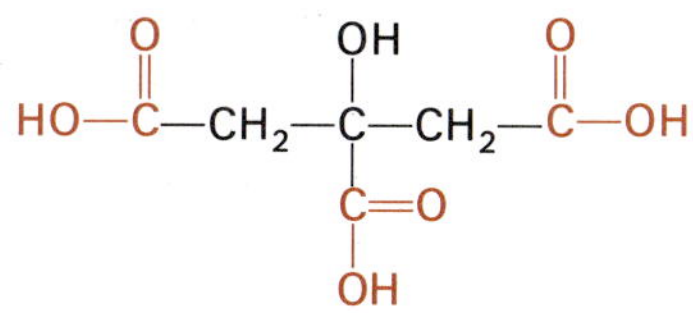

citric acid

Oxalic acid is a constituent of rhubarb and spinach; citric acid is found in citrus fruits. In ionized form, succinic acid (as succinate) and citric acid (as citrate) play essential rules as metabolites in the citric acid cycle, a central energy-producing pathway found in all aerobic cells (Chapter 26).

EXERCISE 19-2 Name the following compounds according to the IUPAC system:

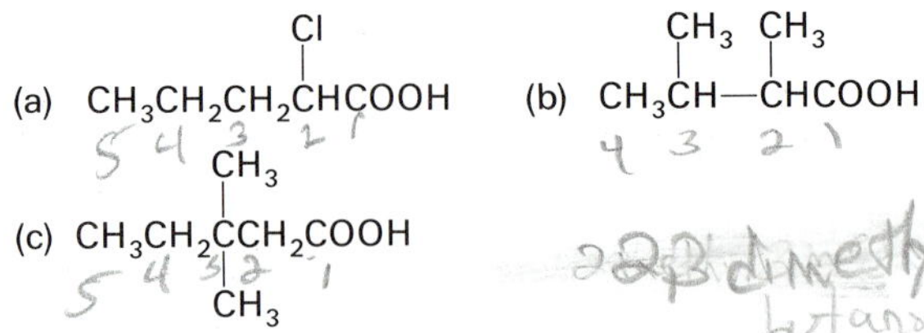

19-3 ACIDITY OF CARBOXYLIC ACIDS

Carboxylic acids are weak acids that dissociate in water. For example, acetic acid dissociates as follows:

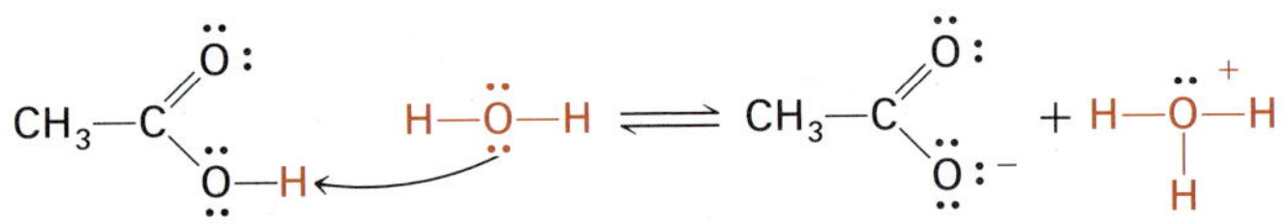

Note that it is the hydroxyl hydrogen atom that dissociates.

Although carboxylic acids are much weaker than inorganic acids such as HCl and HNO_3, they are among the strongest organic acids—roughly 10^{11} times as acidic as the common alcohols and 10^{15} times as acidic as aldehydes and ketones:

acetic acid: $CH_3COOH \rightleftharpoons CH_3COO^- + H^+$

$K_a \cong 10^{-5} \quad pK_a \cong 5$

ethanol: $CH_3CH_2OH \rightleftharpoons CH_3CH_2O^- + H^+$

$K_a \cong 10^{-16} \quad pK_a \cong 16$

propionaldehyde: 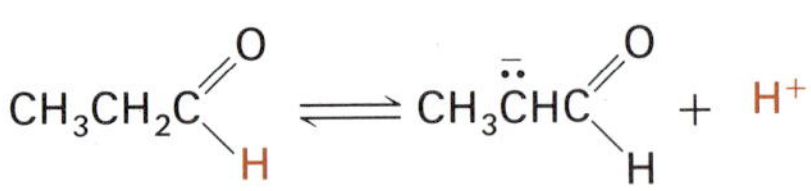

$K_a \cong 10^{-20} \quad pK_a \cong 20$

Why is acetic acid so much stronger an acid than ethanol? The explanation lies primarily in the resonance stabilization of the acetate ion, which delocalizes the negative charge over two oxygen atoms. In the ethoxide ion, on the contrary, the negative charge is localized on the single oxygen atom—there is no resonance stabilization:

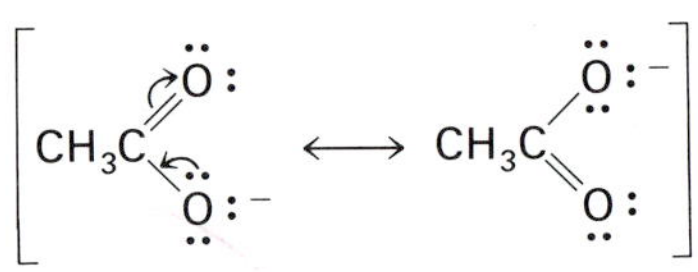

acetate ion resonance structures

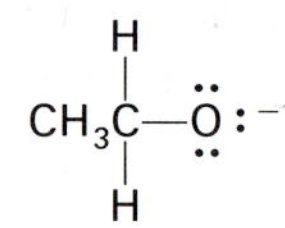

ethoxide ion (no resonance structures)

Ethanol is therefore a much weaker acid than acetic acid; correspondingly, the ethoxide ion is a much stronger base than the acetate ion.

The actual structure of the acetate ion is intermediate between the two resonance structures shown; the negative charge is distributed equally between the two oxygen atoms, and the bond lengths of the two carbon–oxygen bonds are identical (intermediate between the normal length of carbon–oxygen double and single bonds). The structure of a carboxylate anion is often represented as

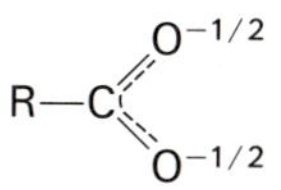

Spreading the negative charge over several atoms rather than having it concentrated on one atom results in a more stable structure.

An electronegative substituent such as —Cl in the structure of a carboxylic acid pulls electrons away from the —COOH group; this weakens the O—H bond and stabilizes the $RCOO^-$ ion by spreading the negative charge. The proton, then, dissociates more easily in water. That is, the acid is stronger, with a larger dissociation constant (K_a). As can be seen in Table 19-2, substitution of the electronegative atom chlorine, bromine, or iodine increases the strength of the acid. Adding a second carboxyl group, with its highly polar carbonyl group that has a strong attraction for electrons, also increases the strength of the acid. Thus, malonic acid ($HOOC—CH_2—COOH$) is a much stronger acid than acetic acid. The substitution of *three* chlorine atoms, instead of only one, increases the strength of the acid tremendously—trichloroacetic acid has a K_a of 0.23 M, much higher than any others given in the table. The electronegative substituents act indirectly, by the inductive effect (operating through a chain of σ bonds), to stabilize the carboxylate ion.

EXERCISE 19-3 (a) List the following substances in order of decreasing strength as acids:

CH_3CH_2COOH $\quad CH_3CHCOOH$ (with Br on the CH) $\quad CH_3CH_2CHO$ and $CH_3CH_2CH_2OH$

(b) List the following substances in order of decreasing strength as bases:

$CH_3CH_2COO^-$ $\quad CH_3CHCOO^-$ (with Br on the CH) $\quad CH_3\bar{C}HCHO$ and $CH_3CH_2CH_2O^-$

TABLE 19-2
Acidity of some substituted acetic acids

Name	Formula	K_a(M)	pK_a
Acetic acid	CH_3COOH	1.8×10^{-5}	4.76
Iodoacetic acid	ICH_2COOH	67×10^{-5}	3.18
Bromoacetic acid	$BrCH_2COOH$	130×10^{-5}	2.90
Chloroacetic acid	$ClCH_2COOH$	140×10^{-5}	2.86
Malonic acid	$HOOCCH_2COOH$	140×10^{-5}	2.86
Dichloroacetic acid	$Cl_2CHCOOH$	5000×10^{-5}	1.30
Trichloroacetic acid	Cl_3CCOOH	$23{,}000 \times 10^{-5}$	0.64

19-4
SALTS OF CARBOXYLIC ACIDS

Neutralization of a carboxylic acid by a strong base (such as sodium hydroxide) results in the formation of a carboxylate salt and water:

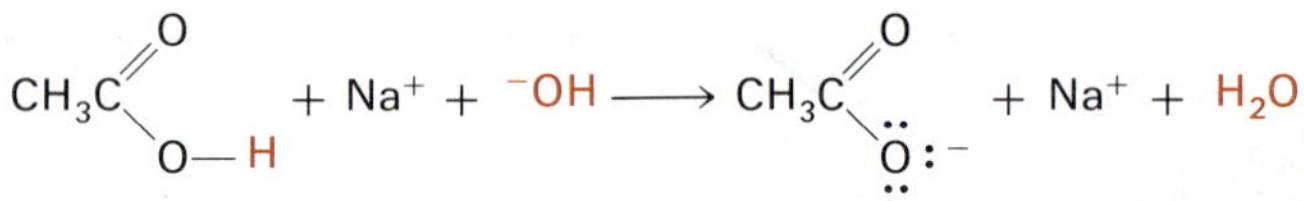

acetic acid **sodium acetate solution**

The Na^+ is a spectator ion and can be omitted from the equation. Carboxylic acids that have relatively large and nonpolar alkyl groups (and are only slightly soluble in water) can often be converted to soluble salts by neutralization with sodium or potassium hydroxide. These ionic salts are much more soluble in water than the corresponding neutral acid molecules. (Remember that carboxylic acids are weak acids, existing mostly as the undissociated molecule in water.)

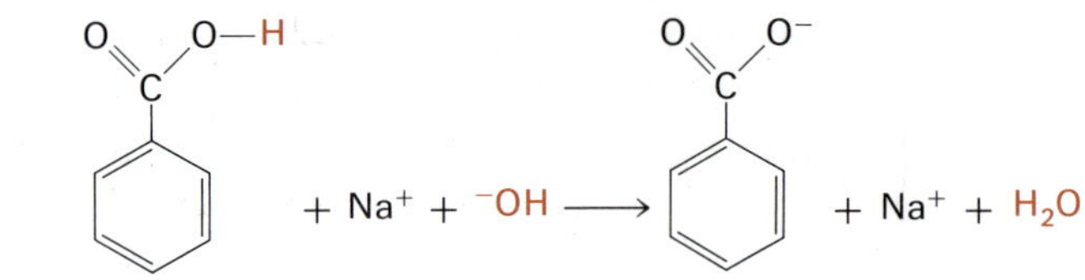

benzoic acid (slightly soluble; 0.27 g/100 g H_2O)

sodium benzoate (very soluble; 66 g/100 g H_2O)

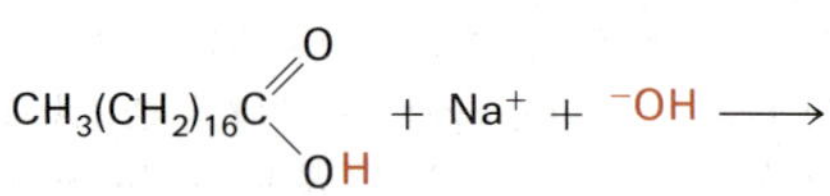

stearic acid (slightly soluble; 0.00029 g/100 g H_2O)

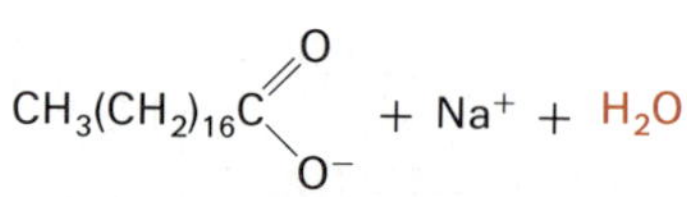

sodium stearate (very soluble; a soap)

The salts of carboxylic acids, like inorganic salts, are crystalline substances composed of positively charged metal ions and negatively charged carboxylate ions. The sodium salts are the most common and are soluble in water. Sodium benzoate, sodium and calcium propionates, and sodium and calcium citrates are added to foods to inhibit the growth of molds and bacteria. The sodium salts of the long-chain carboxylic acids (such as stearic acid) are soaps.

The anions of the carboxylic acids are named by removing the *-ic* ending from the name of the acid and substituting *-ate*. Common (and IUPAC) names for some carboxylate ions are:

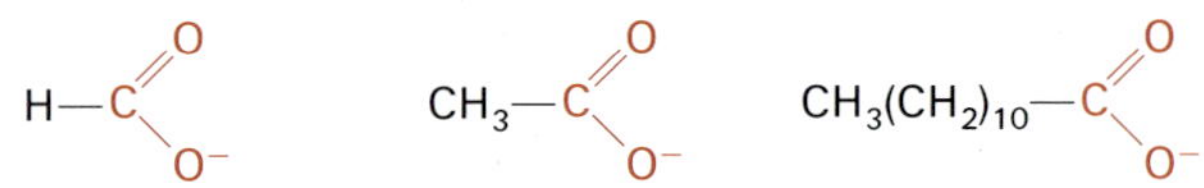

formate ion (methanoate ion) **acetate ion** (ethanoate ion) **laurate ion** (dodecanoate ion)

The salts are named by placing the name of the metal ion first and following it with the name of the carboxylate ion. For example, $Na^+CH_3COO^-$ is sodium acetate.

In the laboratory, slightly soluble carboxylic acids often can be separated from other organic compounds by treatment with a solution of sodium hydroxide. For example, a mixture of the liquids octanoic acid and *n*-octane (each of which is only slightly soluble in water) can be separated by the scheme outlined in Figure 19-3.

The mixture is first shaken with dilute NaOH in a separatory funnel (Figure 19-4). Two layers form—the upper organic layer is composed of the octane, and the lower aqueous layer contains the salt sodium octanoate. The two layers are separated, and the aqueous layer is then made acidic with HCl. This regenerates the organic acid, which is only slightly soluble in water and so again forms a separate layer. (In general, organic substances that are not highly halogenated are less dense than water, so the organic layer is usu-

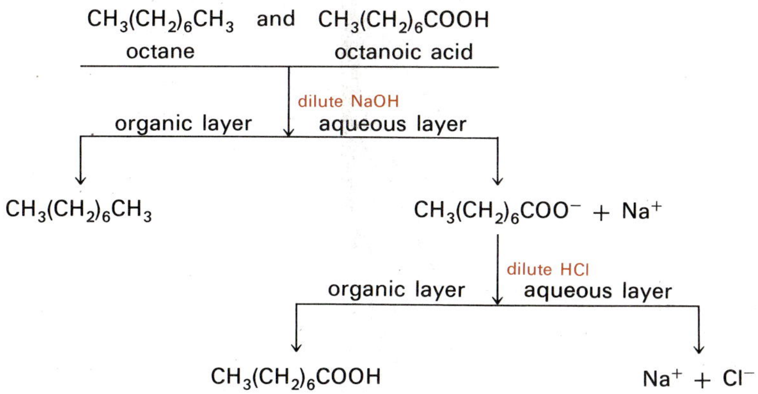

FIGURE 19-3

Separation of a carboxylic acid from a neutral organic substance.

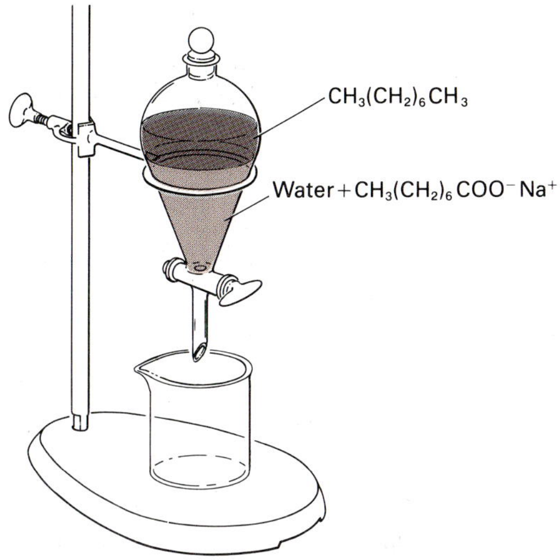

FIGURE 19-4

A separatory funnel. The upper (organic) layer is octane. The lower (aqueous) layer contains the salt, sodium octanoate.

ally the upper layer.) The net result of the whole process, therefore, is to separate the octanoic acid and the *n*-octane—we now have them in two separate containers. This technique is used regularly by organic chemists to help separate the desired product from a reaction mixture.

EXERCISE 19-4 Would you be able to separate octanoic acid from *n*-octanol by the method just described? Explain.

19-5 PREPARATION OF CARBOXYLIC ACIDS

We saw in Chapter 17 that primary alcohols can be oxidized first to aldehydes and then to acids, using potassium permanganate or potassium dichromate:

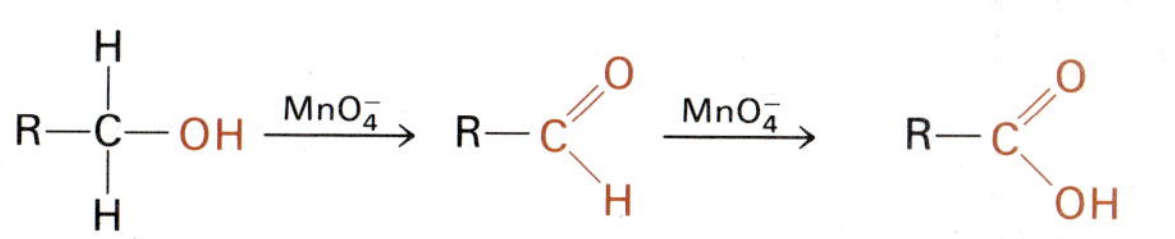

1° alcohol **aldehyde** **carboxylic acid**

Aldehydes can be oxidized with milder oxidizing agents such as Fehling's solution or Tollens' solution (see Section 18-8):

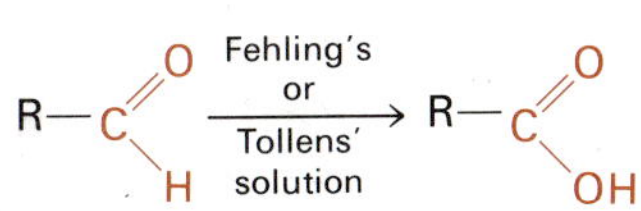

Another method of preparing carboxylic acids is by the *addition* of Grignard reagents to carbon dioxide, with subsequent *hydrolysis* of the addition product. (If you solved Problem 18-17a, you have already written a mechanism for this reaction. The mechanism is the same as that for addition of Grignard reagents to aldehydes and ketones.) The following example shows the formation of acetic acid from carbon dioxide and methylmagnesium bromide:

Addition:

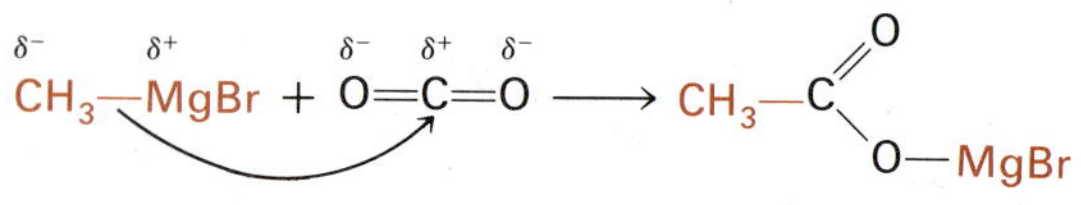

addition product

Hydrolysis:

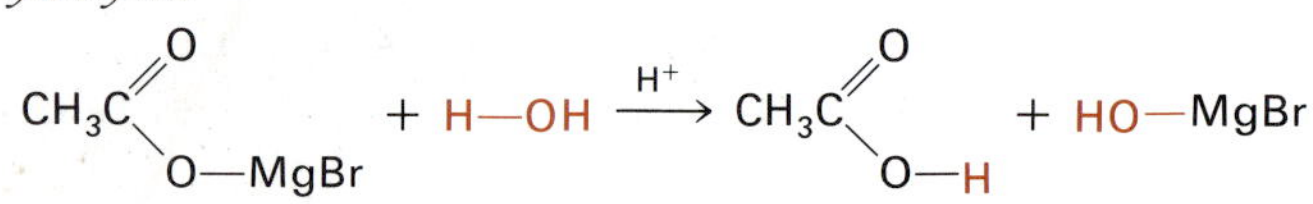

Note that this reaction leads to the formation of a carboxylic acid that has one carbon more than the R group of the Grignard reagent. This is a typical carbonyl addition reaction, like those we saw earlier for aldehydes and ketones.

EXERCISE 19-5 Write equations that show how you could prepare benzoic acid, starting with a Grignard reagent. Use any other simple chemicals you need.

19-6 ESTERS AND ESTERIFICATION

Carboxylic acids react readily with alcohols, in the presence of strong acids as catalysts, to yield derivatives called **esters,** in which the —OH group of the acid is replaced by an —OR group. The general formula for esters is

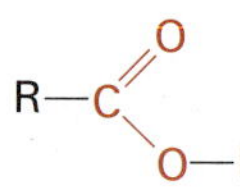

condensed formula: RCOOR′ or RCO_2R'

Because they have no hydrogen atoms that can participate in hydrogen bonds, esters boil at lower temperatures than the carboxylic acids of which they are isomers. For example, the ester methyl formate ($HCOOCH_3$) boils at 32°C; its isomer acetic acid (CH_3COOH) boils at 118°C.

Esters are widely distributed in nature, providing many of the pleasant odors of fruits and flowers. Fats and oils are esters of the long-chain fatty acids and glycerol ($CH_2OHCHOHCH_2OH$). The following are some naturally occurring esters:

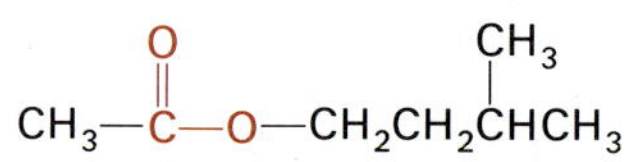

isoamyl acetate
(banana oil)

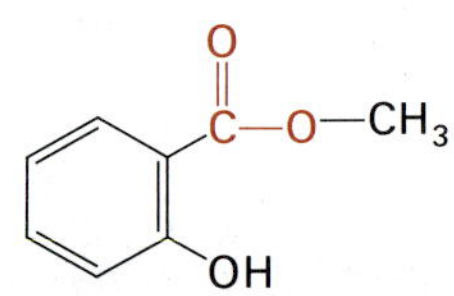

methyl salicylate
(oil of wintergreen)

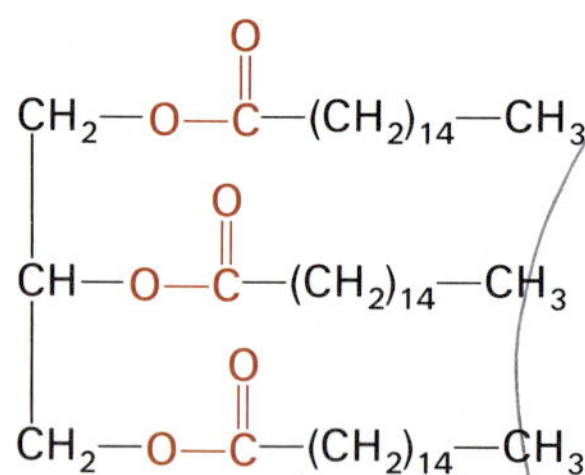

tripalmitin
(a fat)

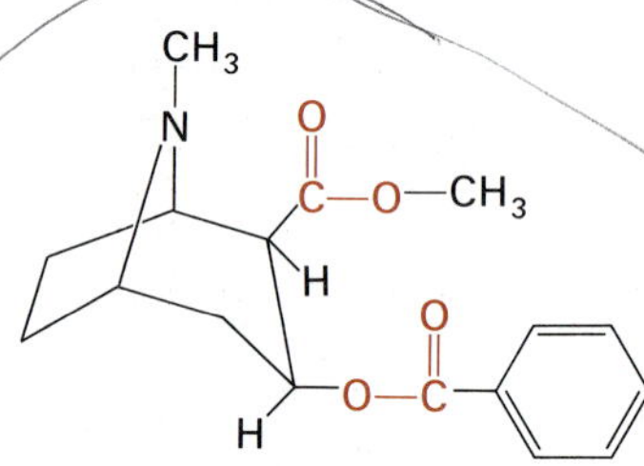

cocaine
(from leaves of coca plant; anesthetic and stimulant)

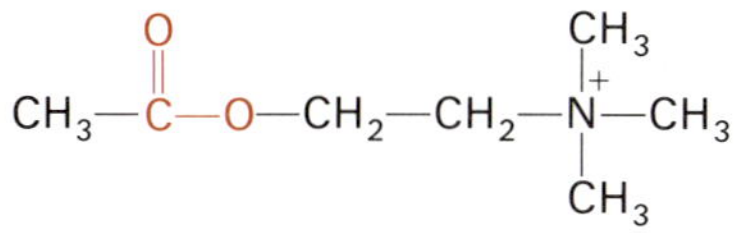

acetylcholine
(neurotransmitter; synthesized in nerve cells)

The *net* equation for a typical **esterification**—the reaction between an acid and an alcohol to yield an ester—is the following, for the reaction between acetic acid and methyl alcohol to yield the ester methyl acetate:

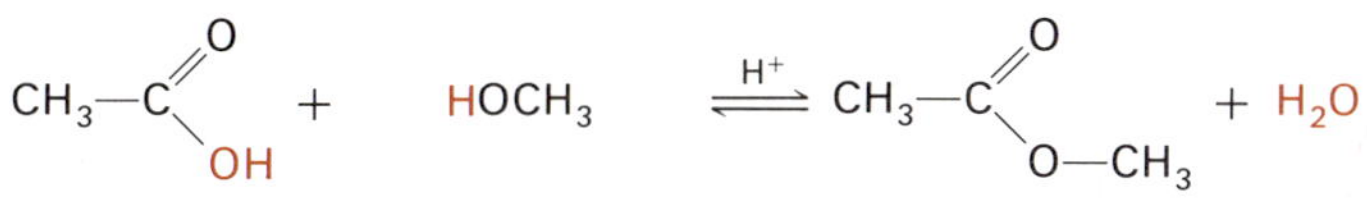

acetic acid **methyl alcohol** **methyl acetate**

The proposed mechanism for the reaction consists of many steps, several of which are proton transfers. The following is an abbreviated mechanism, with most of the proton transfers omitted:

Step 1: Addition of alcohol to carboxyl group

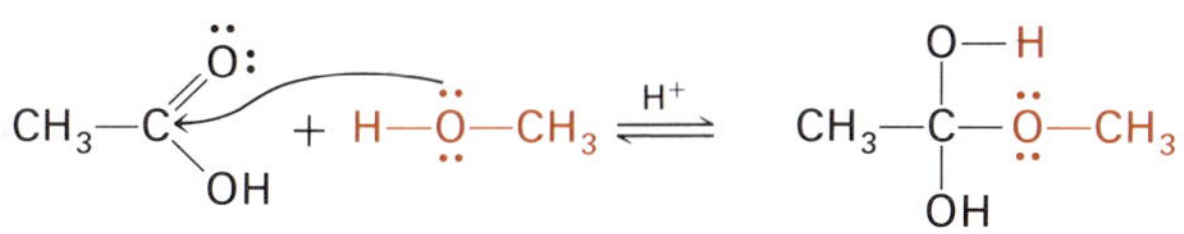

addition intermediate

Step 2: Addition of H^+ to convert OH to better leaving group, H_2O

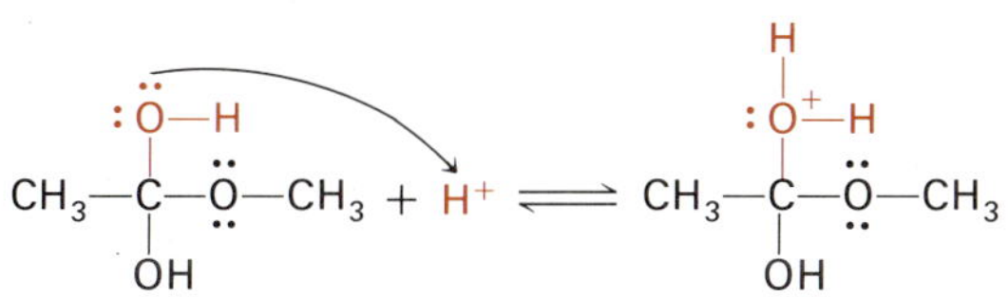

Step 3: Elimination of H_2O and H^+

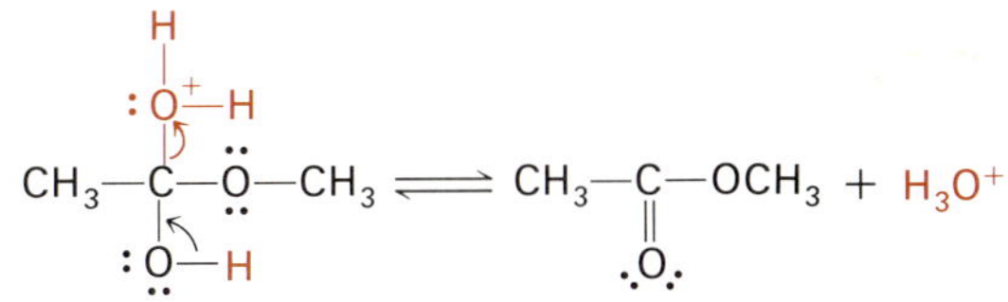

The addition intermediate is like that formed in the addition of an alcohol to an aldehyde or a ketone to form a hemiacetal or a hemiketal (see Section 18-5). The overall reaction is just the *substitution* of $—OCH_3$ for —OH in the acid; but it results from *addition* of the alcohol to the carbonyl group, followed by the *elimination* of water.

The mechanism of esterification was established in part by studies using the radioactive isotope of oxygen, ^{18}O. Ordinary benzoic acid was heated with methanol "labeled" with the oxygen isotope: $CH_3^{18}OH$. When the products of the reaction were collected and separated, the ester was found to be radioactive, and the water was not. The water produced in the reaction contained only ^{16}O; all the ^{18}O was in the ester. This experiment showed that the C—OH bond of the acid and the O—H bond of the alcohol are broken in the reaction:

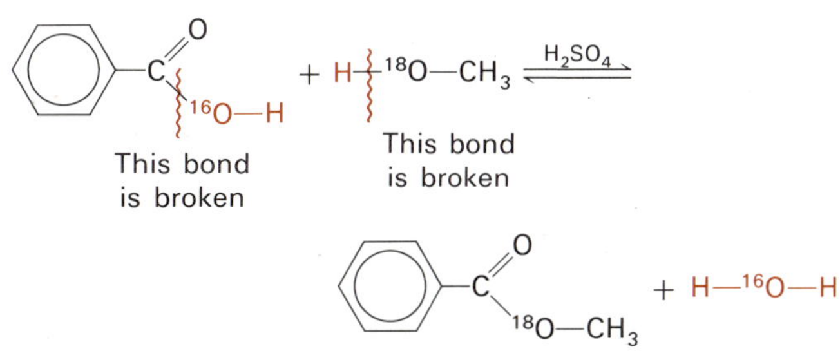

The mechanism given earlier for the process of esterification is consistent with this fact. Note that the part of the ester molecule indicated in color comes from the alcohol, and the rest comes from the carboxylic acid.

Esters are easy to name if you first determine from what alcohol and what acid the ester could be synthesized:

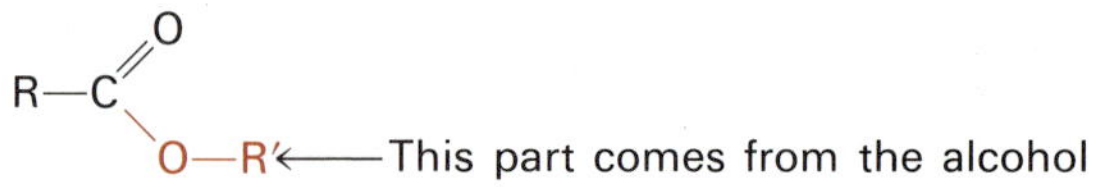

Name the alkyl group of the alcohol; then name the acid as if it were the anion. Common and (IUPAC) names for some esters are:

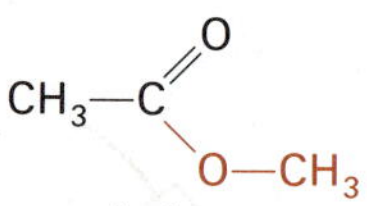

methyl acetate (methyl ethanoate)

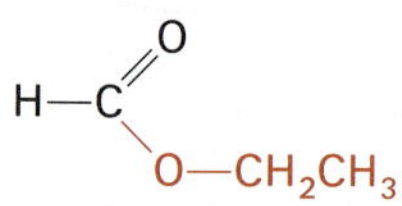

ethyl formate (ethyl methanoate)

Equilibrium constants for esterifications are usually of the order of magnitude of 1; that is, neither reactants nor products are strongly favored (see Chapter 10). For example, the equilibrium constant is 3.38 for the reaction between acetic acid and ethyl alcohol to yield ethyl acetate and water:

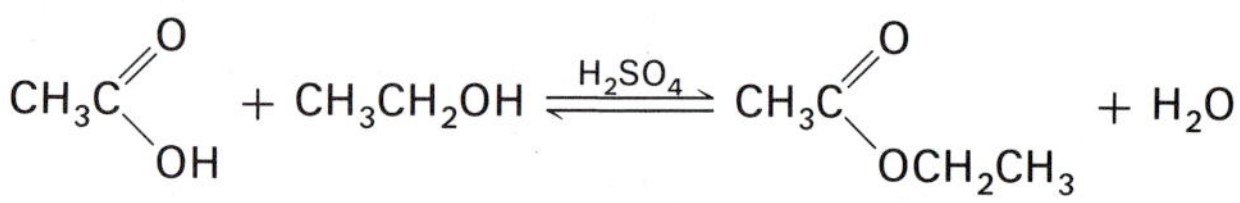

$$K = \frac{[CH_3COOC_2H_5][H_2O]}{[CH_3COOH][CH_3CH_2OH]} = 3.38$$

Note that water is not the solvent but is only a product. Therefore, its concentration is not constant and must be included in the equilibrium constant expression. For equilibria in dilute aqueous solution (Chapter 11), the concentration of water is essentially constant and is included in the equilibrium constant.

If equal amounts of the acid and the alcohol are mixed at the beginning of the reaction, the conversion to products is only 65% at equilibrium (see Table 19-3). In order to obtain a high yield of the ester, a large excess of one of the reactants (usually the less expensive one) may be used. Table 19-3 shows that there is 97% conversion of acid to the ester when the initial ratio of alcohol to acid is 10 moles to 1, and there is 99% conversion when the ratio is 100:1. The forward reaction is also favored by removing water from the reaction mixture—say, by boiling it off as it is formed. This is practical if the boiling points of the reactants and the ester are all above 100°C.

EXERCISE 19-6 (a) Write structures for an alcohol and an acid from which you could synthesize acetylcholine (see structure given near the beginning of this section).

(b) Write a net equation for the reaction between salicylic acid

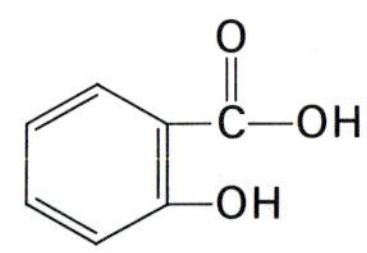

and methanol to yield an ester. Show the structure of the product.

(c) Write a net equation for the reaction between salicylic acid and acetic acid to yield an ester. Show the structure of the product. (The product is aspirin, acetyl salicylic acid.)

TABLE 19-3
Percentage conversion of acetic acid to ester (ethyl acetate)

Experiment	Concentrations measured at	Acetic acid	+	Ethyl alcohol	⇌	Ethyl acetate	+	Water	Conversion (%)
		Concentrations (moles)							
1	Start	1.00		1.00		0		0	
	Equilibrium	0.35		0.35		0.65		0.65	65
2	Start	1.00		10.00		0		0	
	Equilibrium	0.03		9.03		0.97		0.97	97
3	Start	1.00		100		0		0	
	Equilibrium	0.01		99.01		0.99		0.99	99

Note: The net reaction involved here is $CH_3C(=O)OH + CH_3CH_2OH \xrightleftharpoons{H_2SO_4} CH_3C(=O){-}O{-}CH_2CH_3 + H_2O$

Source: Reprinted with permission of Macmillan Publishing Company from *Introduction to Organic Chemistry* by A. Streitweiser, Jr., and C. H. Heathcock. Copyright © 1976, Macmillan Publishing Co., Inc.

19-7 HYDROLYSIS OF ESTERS

The *reverse* of the esterification process is an example of hydrolysis—the acid-catalyzed reaction between an ester and water to yield the corresponding acid and alcohol. The *net* reaction is the addition of the components of water across the C—OR bond to yield an alcohol and a carboxylic acid:

$$CH_3{-}C(=O){-}OCH_3 + H{-}OH \xrightleftharpoons{H^+} CH_3{-}C(=O){-}OH + H{-}OCH_3$$

The probable mechanism for the reaction is just the reverse of the process whose abbreviated mechanism appears in Section 19-6. (To see the mechanism of the hydrolysis, start with step 3 of the esterification process, and read each reaction in the reverse direction.) We have seen that the esterification reaction is favored by the removal of water from the reaction mixture; hydrolysis is favored by the use of a large excess of water.

The hydrolysis of an ester also is catalyzed by base; a typical reaction is

$$\underset{\text{methyl acetate}}{CH_3C(=O)OCH_3} + Na^+ + OH^- \longrightarrow \underset{\text{sodium acetate}}{CH_3{-}C(=O)O^-} + Na^+ + \underset{\text{methyl alcohol}}{CH_3OH}$$

Leaving out proton transfers, the mechanism can be summarized in two steps—addition of the nucleophilic OH^-, and elimination of an alcohol:

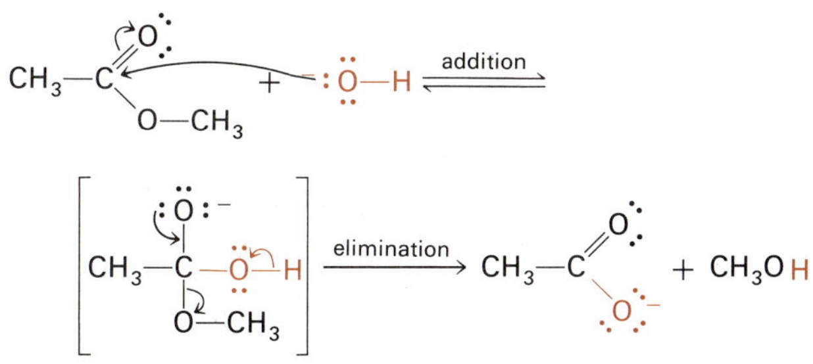

Instead of acetic acid, this reaction yields the sodium salt, sodium acetate. Because the reaction mixture contains NaOH and is basic, any acetic acid formed reacts immediately with base to give the salt and water. Thus the NaOH is used not just in small, catalytic amounts; instead, equal numbers of moles of the ester and the NaOH react. The OH^- is a reagent as well as a catalyst, because it is consumed in the reaction. This reaction, unlike the acid-catalyzed hydrolysis, is essentially irreversible because the acetate ion is quite stable and has no tendency to react with the alcohol. The reaction, therefore, goes to completion.

Saponification, used through the ages to produce soaps, is the hydrolysis of fats (triesters of glycerol and fatty acids) in basic solution:

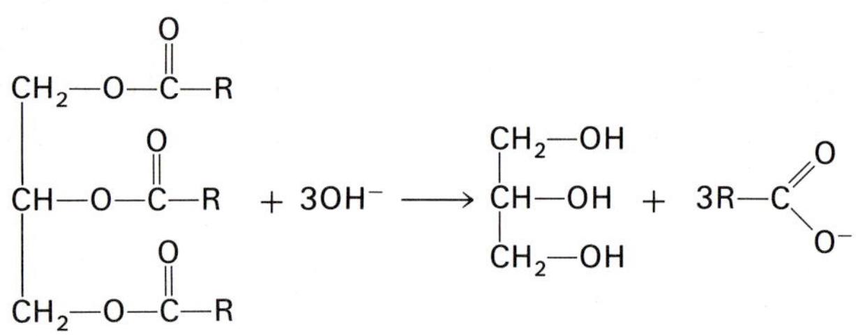

fat (triglyceride) **glycerol** **salt of acid**

The R groups in the fat are not necessarily all the same. They consist primarily of saturated or unsaturated hydrocarbons with chains 12 to 20 carbons long. (Fats and soaps, which are the sodium or potassium salts of long-chain fatty acids, are discussed in more detail in Chapter 24.)

EXERCISE 19-7 Draw the structures of the products of the acid hydrolysis of isoamyl acetate,

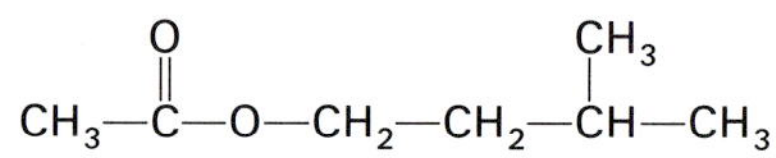

19-8 THE CLAISEN CONDENSATION

Esters undergo a carbonyl *addition* reaction called the **Claisen condensation** that is like the aldol condensations we saw in Section 18-10. As with aldehydes and ketones, the first step involves the removal of an α-hydrogen atom by a strong base or an enzyme. The α-hydrogens of esters are less acidic than those of aldehydes and ketones, so a very strong base such as sodium ethoxide is used in the laboratory. The *net* reaction for the self-condensation of the ester ethyl acetate is

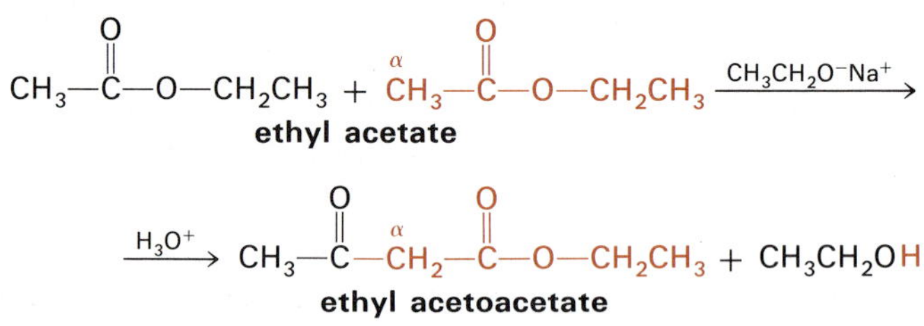

The mechanism consists of familiar steps: (1) removal of an α-hydrogen by a strong base; (2) addition of a

nucleophile to the carbonyl carbon; (3) elimination; (4) and (5) acid–base reactions.

Step 1: Removal of α-hydrogen by a strong base

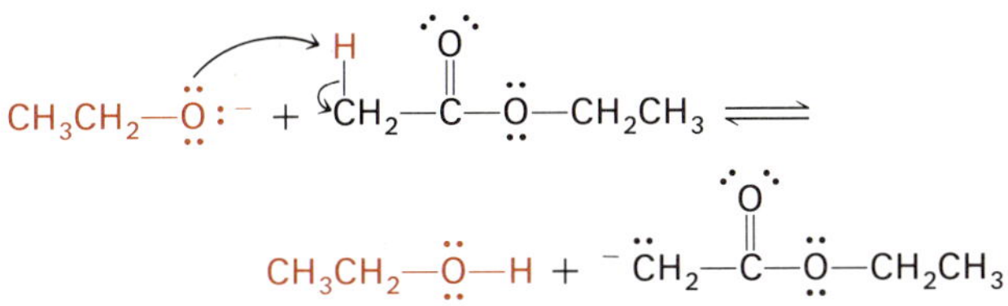

Step 2: Addition of nucleophile to carbonyl carbon

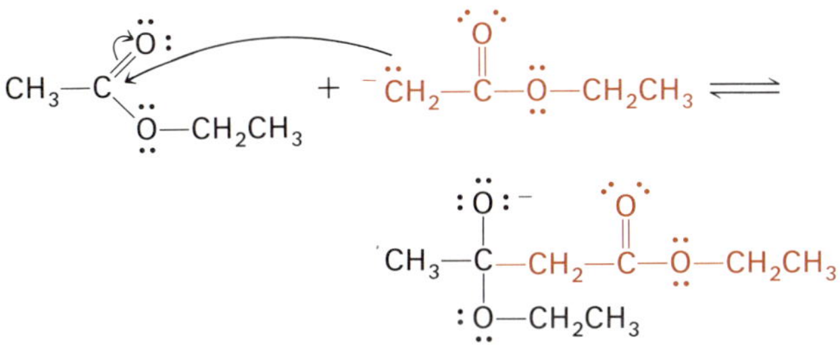

addition intermediate

Step 3: Elimination of ethoxide ion

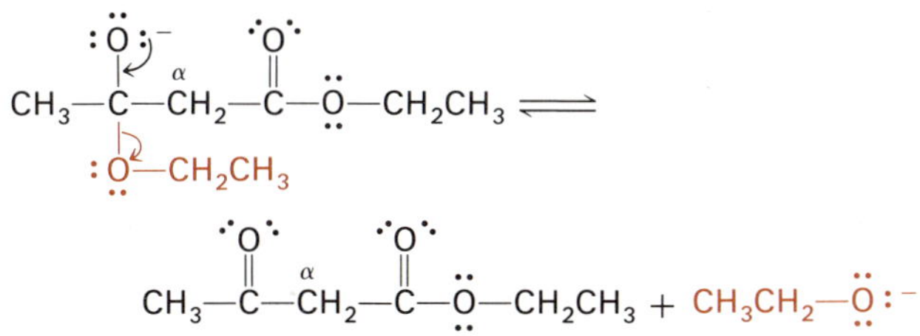

ethyl acetoacetate

The ethoxide ion is a very strong base. It instantly attacks the acidic hydrogen atoms that are α to *both* carbonyl groups; step 4 is an essentially irreversible step that provides the driving force for the reaction.

Step 4: Acid–base reaction

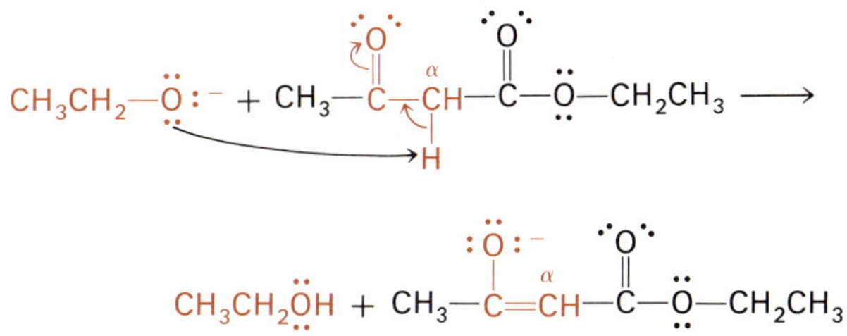

The product is the *sodium* salt of ethyl acetoacetate (if sodium ethoxide was used as the base), which is only slightly soluble and precipitates out of solution. The salt can be separated by filtration, and the β-ketoester (ethyl acetoacetate) can be obtained in step 5 by dissolving the salt in dilute acid solution.

Step 5: Acid–base reaction to yield ethyl acetoacetate

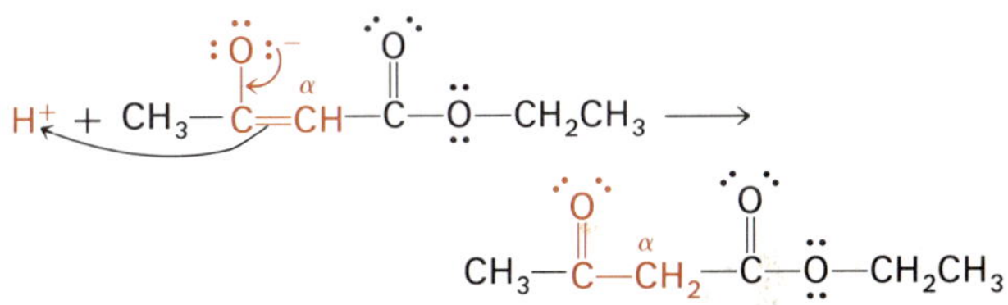

ethyl acetoacetate

The pure, liquid product can be obtained by extraction with ether, followed by drying and distillation. (Because ethyl acetoacetate is more soluble in ether than it is in water, it can be separated from the reaction mixture by extracting it with ether.)

In the laboratory, this reaction can be carried out only with a very strong base. In nature, very similar ester condensations are carried out, with enzymes as catalysts, at physiological pH. We shall see an example in the biosynthesis of fatty acids in Chapter 24.

19-9
ACYL HALIDES AND ACID ANHYDRIDES

We have thus far seen one kind of carboxylic acid derivative, the ester. We now consider two other carboxylic acid derivatives, the acyl halide and the acid anhydride. (We discuss still another, the amide, in Chapter 20.) The structures of the acyl halide and anhydride are

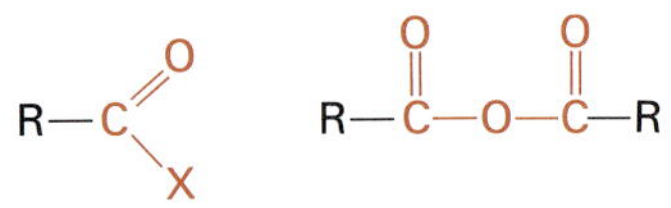

acyl halide **acid anhydride**

In the acyl halides, a halogen (X) replaces the OH of the carboxyl group. In the acid anhydrides, an ester group (RCOO—) replaces the —OH of the carboxyl group.

Acyl halides are produced by reaction between a carboxylic acid and phosphorus trichloride (PCl_3), phosphorus tribomide (PBr_3), or thionyl chloride ($SOCl_2$):

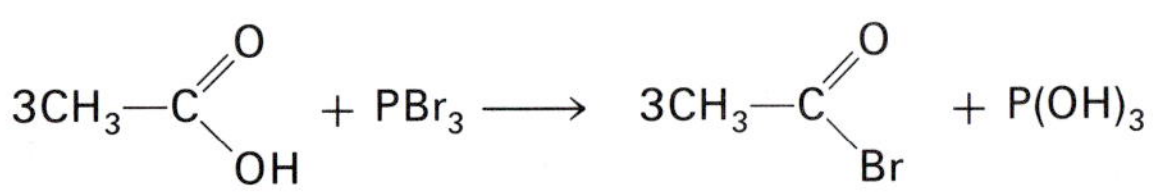

acetyl bromide

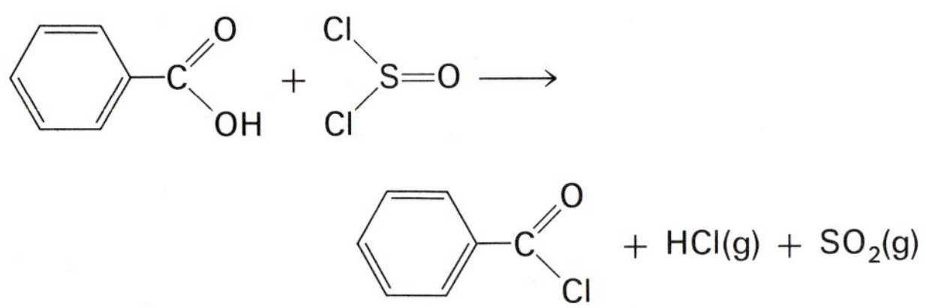

benzoyl chloride

Because acyl halides are highly reactive (more reactive than the acids themselves), they are useful intermediates in the preparation of esters:

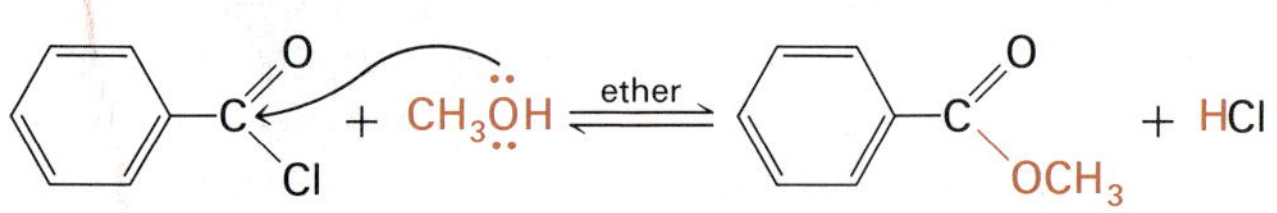

Acyl chlorides also react with water, some vigorously, to give hydrogen chloride, and to regenerate the acid:

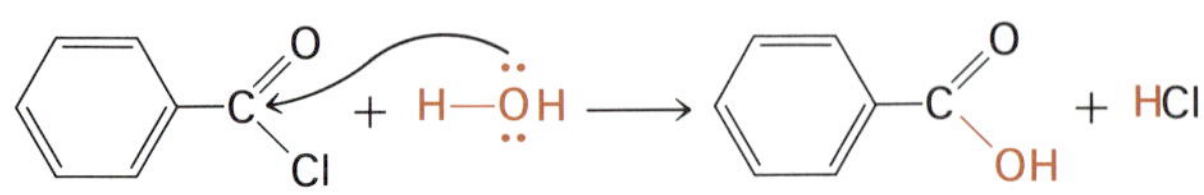

Acid anhydrides are prepared by reaction of a carboxylic acid with a dehydrating agent such as phosphorus pentoxide (P_2O_5). *One* molecule of water is removed from each *two* molecules of the acid:

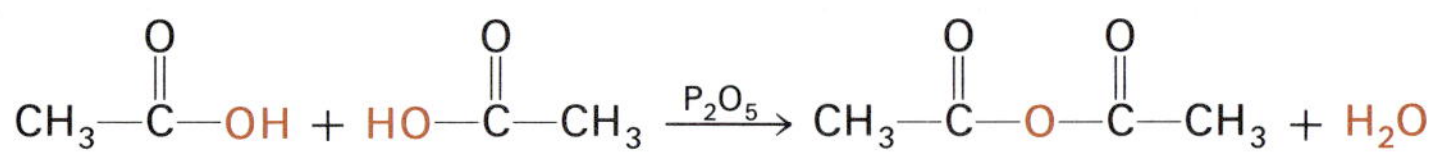

acetic anhydride

The water combines with the P_2O_5 to form phosphoric acid (H_3PO_4). Like the acyl halides, acid anhydrides are more reactive than the acids from which they derive. The anhydrides react readily with alcohols to produce one molecule of an ester and one molecule of the acid that corresponds to the anhydride used:

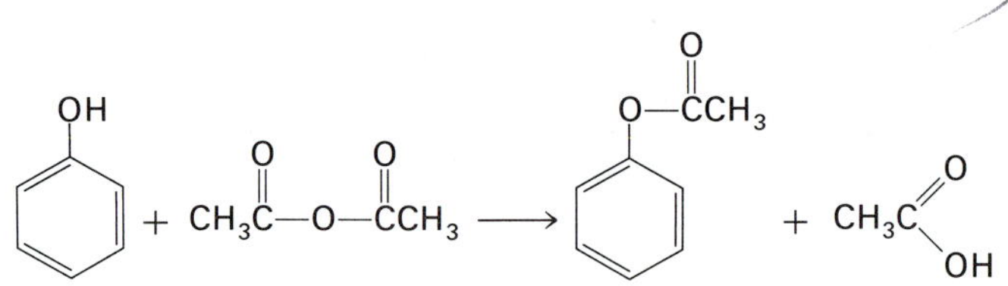

phenol **acetic anhydride** **phenyl acetate** **acetic acid**

Because the acetic acid boils at a lower temperature than do the reactants and the phenylacetate, it can be removed from the reaction mixture by distillation, driving the reaction to completion. Anhydrides also react vigorously with water to regenerate two molecules of acid—the reaction is just the reverse of the formation of the anhydride:

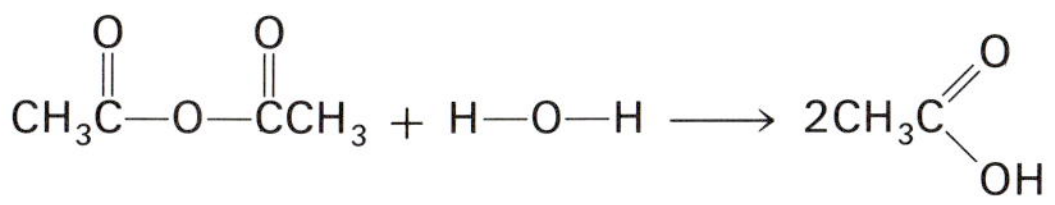

EXERCISE 19-8 Draw the stuctures of the products of the following reactions:

(a) 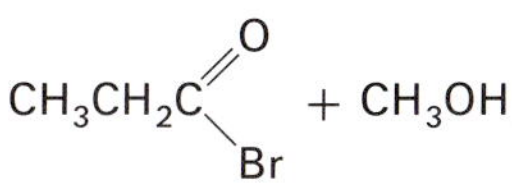

(b) 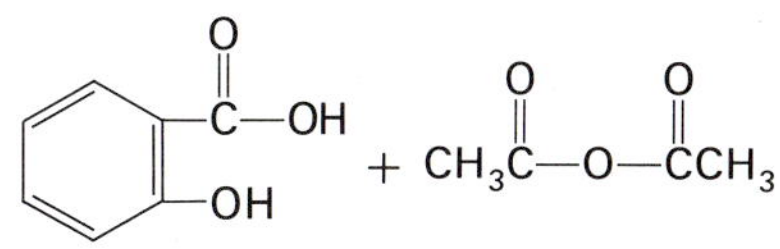

(c) 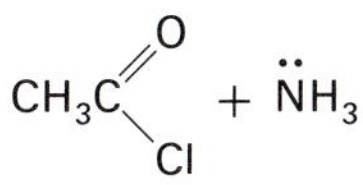

19-10 POLYESTERS

The polyesters of which so many consumer products are made are polymers in which monomers are joined by ester linkages. The process by which polyesters are formed is **condensation polymerization,** in which *monomers combine through the elimination of a small molecule such as water, ethanol, or hydrogen chloride.* This is in contrast to addition polymerization (Section 15-10), in which all atoms of the monomer are incorporated into the polymer. In condensation polymerization, each monomer must have *two* functional groups so that the chain can grow. A typical example is the formation of the fiber Dacron. Terephthalic acid and ethylene glycol condense with the elimination of water:

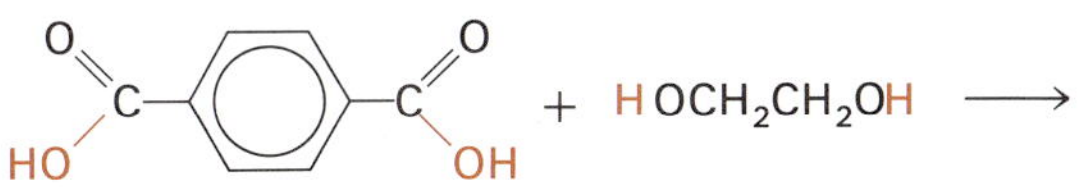

terephthalic acid **ethylene glycol**

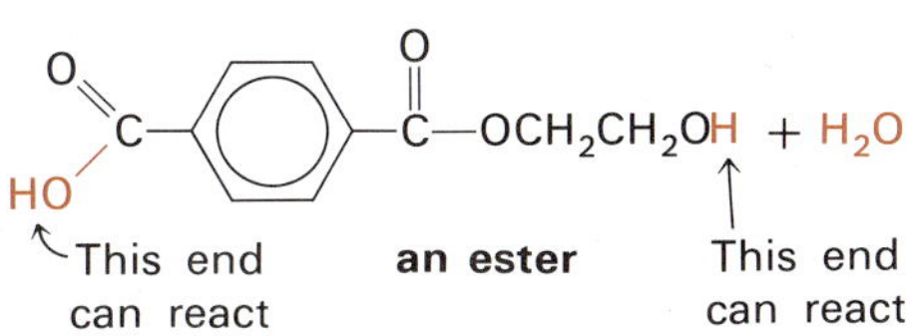

The reaction is a simple esterification. The product of the reaction is an ester that has an alcohol and an acid group and these groups can react further. The COOH can react with a molecule of ethylene glycol, and the OH can react with a molecule of terephthalic acid, to yield the following *tri*ester:

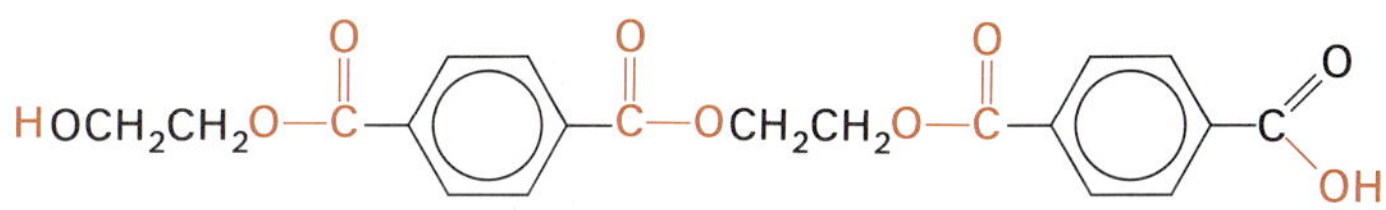

The product again has functional groups at both ends, and so can react further, until finally a polyester is formed with the general structure

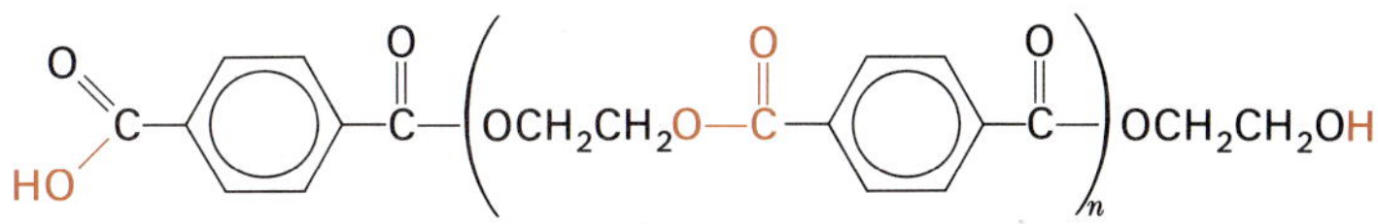

Dacron

Dacron is an example of a **copolymer,** a polymer made up of two or more different monomers. Dacron, often combined with natural fibers such as cotton or wool, is widely used in both woven and knitted fabrics. In medicine, Dacron fabrics are used for sutures and to replace damaged or diseased sections of blood vessels and of the esophagus.

EXERCISE 19-9 Acetate rayon, one of the first synthetic polymers, is a polyester. It is made by reaction between cellulose (Chapter 23), which has many hydroxyl groups, and acetic anhydride,

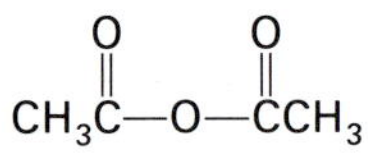

Write a structure for acetate rayon, and tell what small molecule is eliminated in its formation. Represent cellulose by the structure.

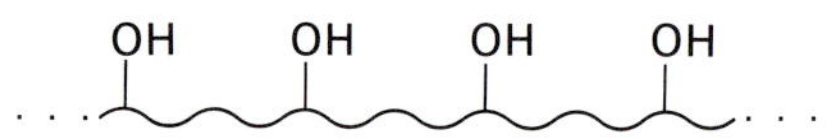

If one of the monomers used in the preparation of a polymer is trifunctional (that is, has three functional groups), it may form branched chains or cross-links between polymer chains. This can be seen in Glyptal® resin, a polymer formed of glycerol (a *tri*hydroxyl alcohol) and orthophthalic acid (a dicarboxylic acid). The first stage in the polymerization leads to chains with some unreacted hydroxyl functional groups:

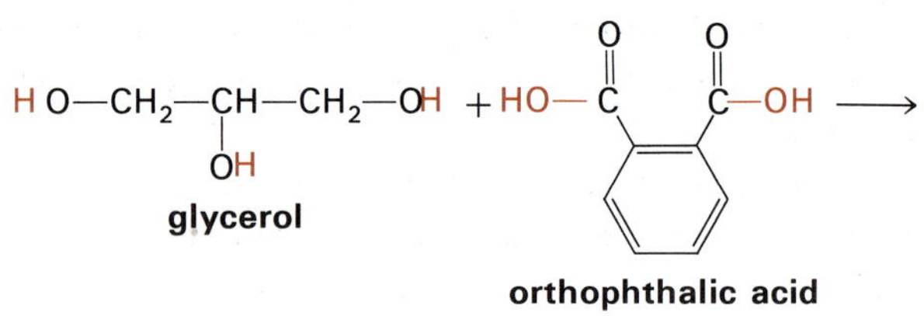

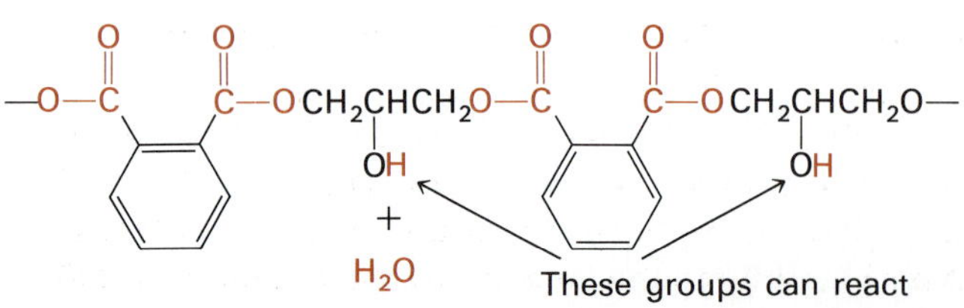

A branch forms when one of these unreacted hydroxyl groups reacts with a molecule of the diacid:

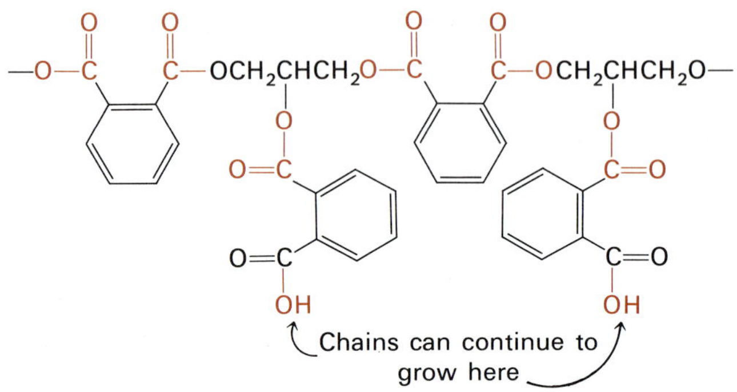

If, instead, polymer chains with free hydroxyl groups are aligned as shown below, the chains can be cross-linked by a molecule of the diacid:

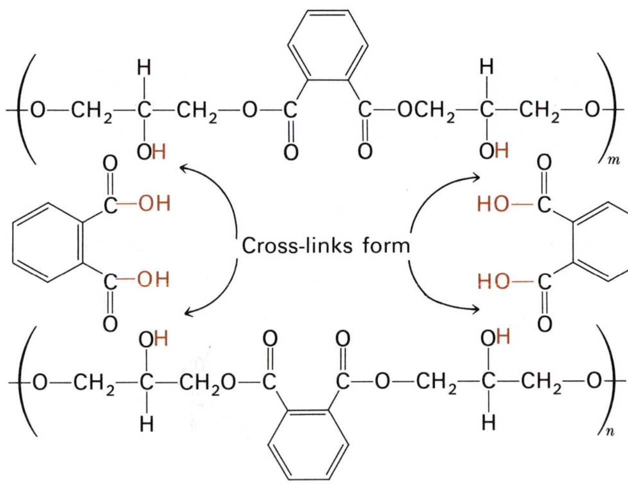

In this way, as the mixture is heated further, polymer chains are linked together in a giant network (Figure 19-5). The network is strong and rigid, and the strands of polymer can no longer slide along each other. The greater the degree of cross-linking, the stronger and more rigid the polymer.

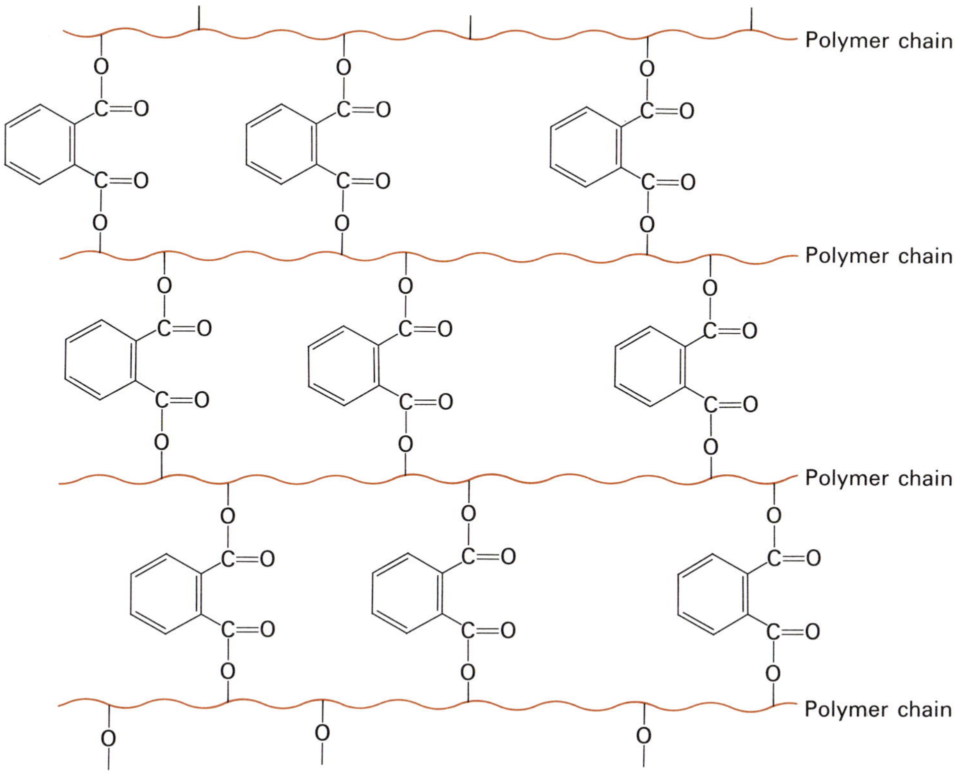

FIGURE 19-5

Glyptal resin, a highly cross-linked polymer. An entire sample of glyptal is one giant molecule.

19-11 SYNTHESIS OF AN ESTER: ASPIRIN

Aspirin (acetylsalicylic acid, an ester of salicylic acid and acetic acid) was first synthesized for medicinal use in 1893 by Felix Hoffmann, a German chemist working for F. Bayer and Company.

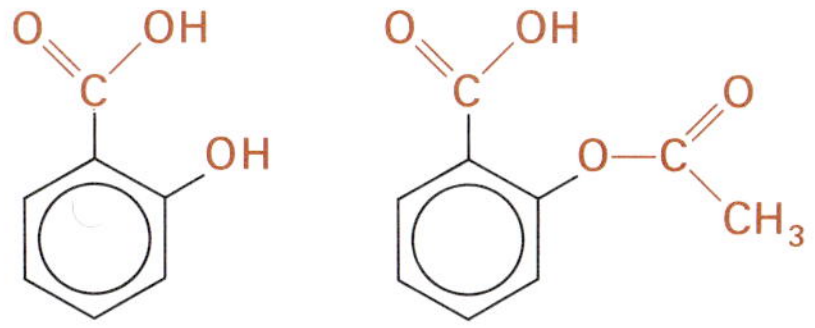

salicylic acid **acetylsalicylic acid (aspirin)**

Salicylates, salts of salicylic acid extracted from willow and other plants, had been used from the 1700s. (Hippocrates around 400 B.C. recommended chewing willow bark to relieve the pain of childbirth.) Salicylates and salicylic acid, however, have dreadful tastes and are extremely irritating to the lining of the stomach. Hoffmann's discovery of aspirin was the result of attempts to find a form of salicylate without these disadvantages.

The long-established uses of aspirin include the reduction of fever, and alleviation of pain and swelling caused either by injury or by illness such as arthritis. A new use of aspirin—now the subject of intensive study—is to prevent blood clots that may cause heart attacks and strokes.

Despite the discovery of a variety of prescription pain killers, aspirin is still the leading pain killer and the standard treatment to reduce fever and swelling. Aspirin consumption in the United States is 12,000 tons per year—the equivalent of 150 tablets for every man, woman, and child. The worldwide consumption of aspirin is 100,000 tons per year.

Although aspirin can be prepared from salicylates extracted from plants, that is not how it is prepared commercially. We shall now see how benzene (a component of coal tar, which is obtained by distilling coal in the absence of air) is converted to aspirin. In each step of the synthesis, only one atom or functional group of the molecule reacts, while the rest of the molecule remains unchanged.

In the first step of the reaction, benzene reacts with chlorine in the presence of a catalyst such as aluminum chloride, to produce chlorobenzene:

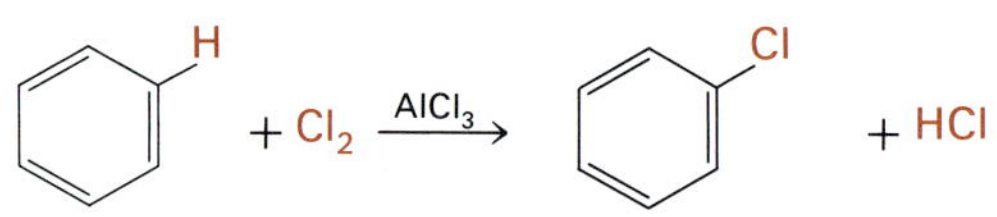

benzene **chlorobenzene**

Chlorobenzene is much more reactive than benzene and can be converted to phenol by heating with sodium hydroxide:

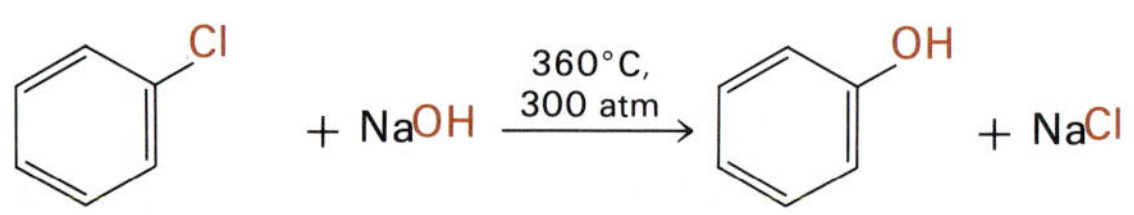

chlorobenzene **phenol**

In a third step, the hydrogen of the hydroxyl group of the phenol, which is acidic, is neutralized by the base sodium hydroxide to form the salt sodium phenolate:

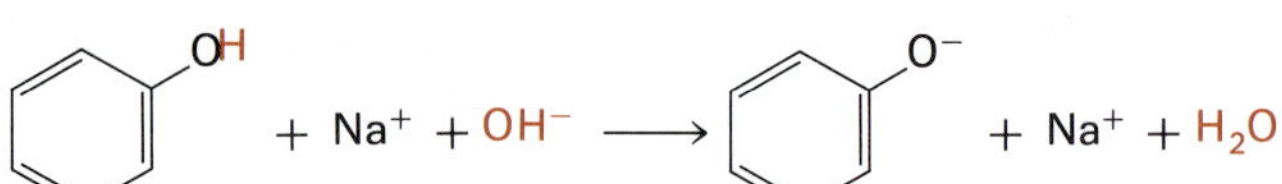

phenol **sodium phenolate**

The sodium phenolate is treated with carbon dioxide gas under pressure to yield the sodium salt of salicylic acid:

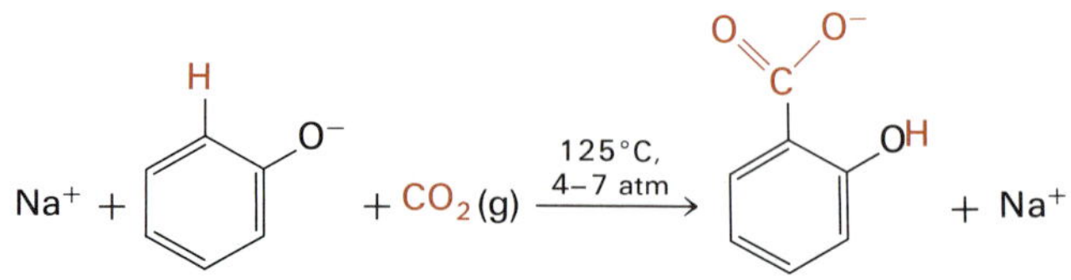

phenolate **sodium salicylate**

Treatment with hydrochloric acid converts the sodium salt to salicylic acid:

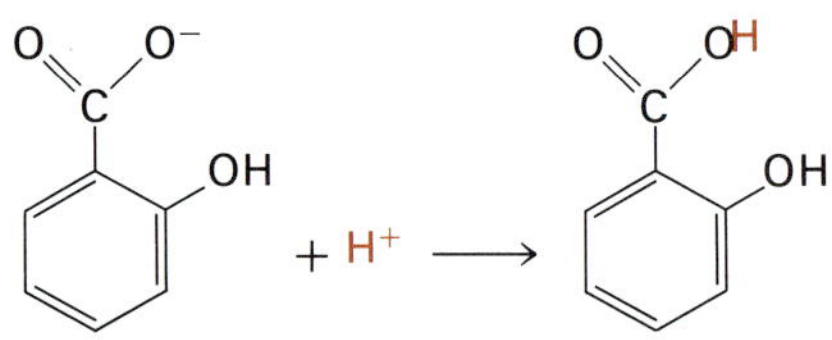

salicylate **salicylic acid**

The hydroxyl group of salicylic acid is esterified to form acetylsalicylic acid:

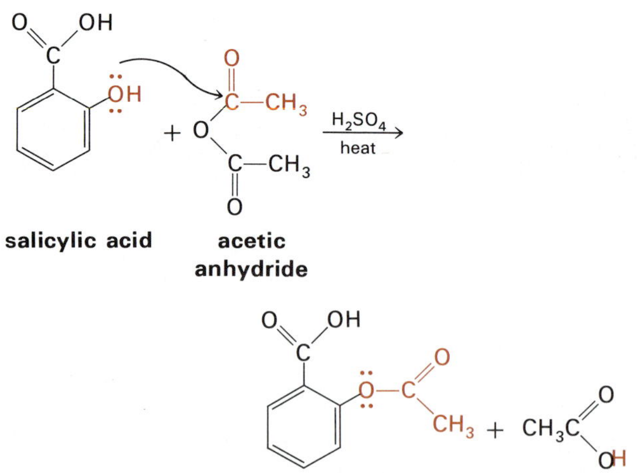

salicylic acid **acetic anhydride**

acetylsalicylic acid (aspirin) **acetic acid**

In the last step of the synthesis of aspirin, salicylic acid is treated with acetic anhydride, which reacts with the phenolic OH group of the salicylic acid to form an ester. If, instead, salicylic acid is treated with an alcohol, an ester is formed with the acid group of the salicylic acid. For example, if salicylic acid is treated with methanol, the product of the esterification reaction is methyl salicylate or oil of wintergreen:

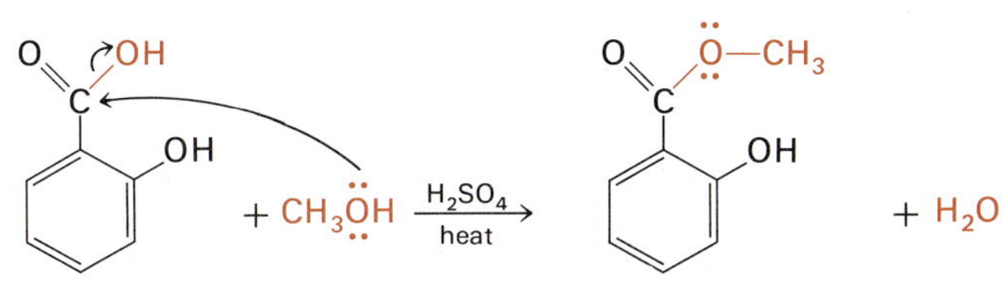

salicylic acid **methyl salicylate (oil of wintergreen)**

In the synthesis of an organic compound, the yield of product is always less than 100%, and it may be very low—especially for a multistep synthesis. At any step in the synthesis, the position of equilibrium may be such that the extent of conversion of reactants to products is low. There may be side reactions that give other products in addition to the desired one. If an intermediate must be isolated and purified before going on to the next step in the synthesis, some product is lost in the isolation process, and more in the purification of the final product. These losses can occur at each step. Thus, the more steps, the lower the yield is likely to be. For example, in the two-step reaction A $\longrightarrow$ B and B $\longrightarrow$ C, if the yield for the first step is 40% of the theoretical, and that for the second step is 60% of the theoretical, then the yield of product C is only 24% of the theoretical amount for a given amount of starting material A. For an industrial process, it is of crucial importance to find the synthesis that involves the smallest possible number of steps and the best experimental conditions. The ideal conditions give the greatest possible yield at each step (that is, the conditions shift the position of equilibrium to the right), while at the same time providing a reasonable rate of reaction.

19-12 ANHYDRIDES AND ESTERS OF PHOSPHORIC ACID

Because carboxylic acid anhydrides react vigorously with water (Section 19-9), they cannot exist in the aqueous medium of living organisms. Anhydrides of phosphoric acid, (an inorganic acid), on the contrary, are stable in water and are the principal carriers of energy in biochemical reactions. Phosphoric acid (H_3PO_4; often called *ortho*phosphoric acid), has a strong tendency to condense, eliminating a molecule of water to form the anhydride *pyro*phosphoric acid ($H_4P_2O_7$):

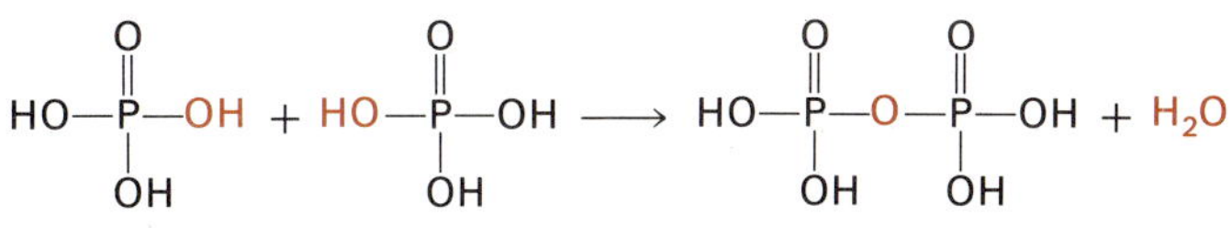

orthophosphoric acid (H_3PO_4) **pyrophosphoric acid ($H_4P_2O_7$)**

Pyrophosphoric acid is an anhydride of orthophosphoric acid; its structure is similar to that of a carboxylic acid anhydride:

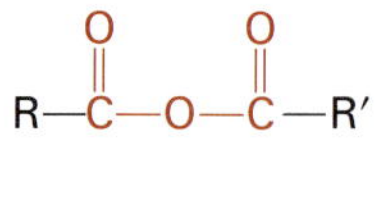

carboxylic acid anhydride

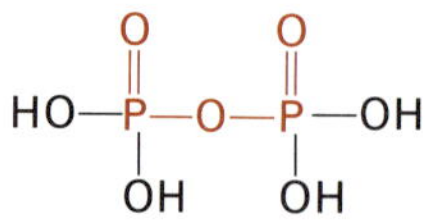

phosphoric acid anhydride (pyrophosphoric acid)

Both orthophosphoric acid and pyrophosphoric acid react with strong bases to form salts. Orthophosphoric acid can lose one, two, or three protons, forming the salt NaH_2PO_4, Na_2HPO_4, or Na_3PO_4. Similarly, pyrophosphoric acid can lose up to four protons to form salts.

Phosphoric acid and anhydrides of phosphoric acid can also form esters. For example, the first step in glycolysis—the anaerobic catabolism (or breaking down) of glucose—is the formation of the ester glucose 6-phosphate:

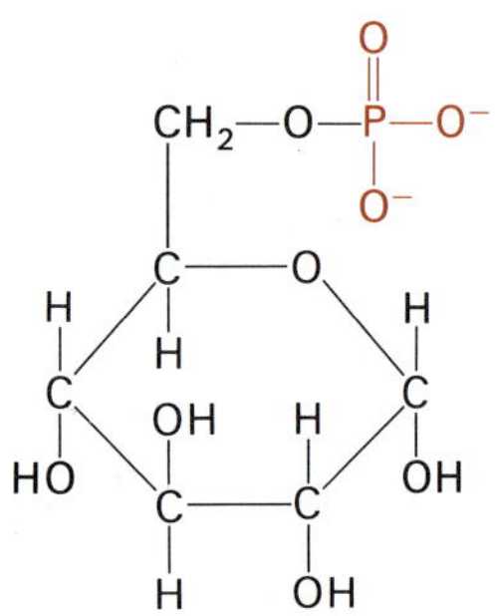

glucose 6-phosphate

At physiological pH, the phosphate group has charge of 2−.

Isopentenyl pyrophosphate (an ester of pyrophosphate) is a building block in the synthesis of cholesterol and other large biological molecules (Section 24-6):

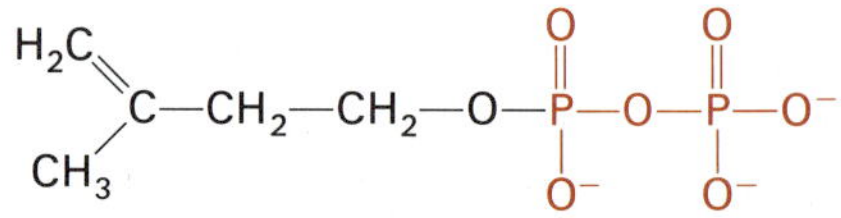

isopentyl pyrophosphate

The pyrophosphate is shown with a 3− charge, its charge at physiological pH.

Adenosine triphosphate (ATP), the special carrier of energy within living cells, is a triphosphate ester of adenosine, whose structure we shall study in Chapter 25:

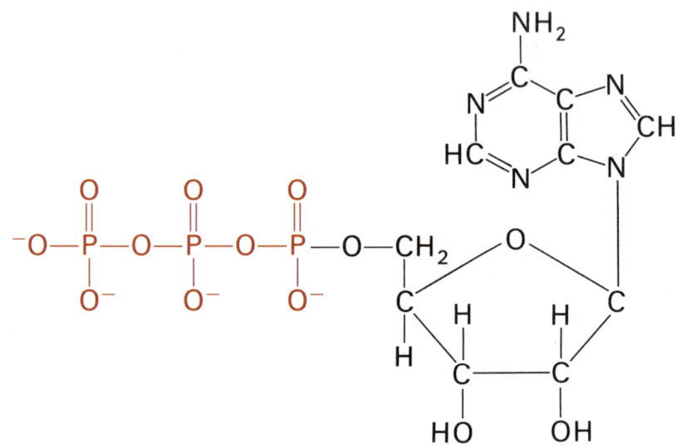

ATP

The triphosphate group has a 4− charge at physiological pH. Note that ATP contains two phosphoric acid anhydride linkages as well as a phosphate ester linkage. ATP is a product of the catabolism of fats and carbohydrates (Chapter 26). When ATP is hydrolyzed under typical cellular conditions to yield adenosine diphosphate (ADP) and phosphate, approximately 50 kJ of energy is released per mole of ATP hydrolyzed:

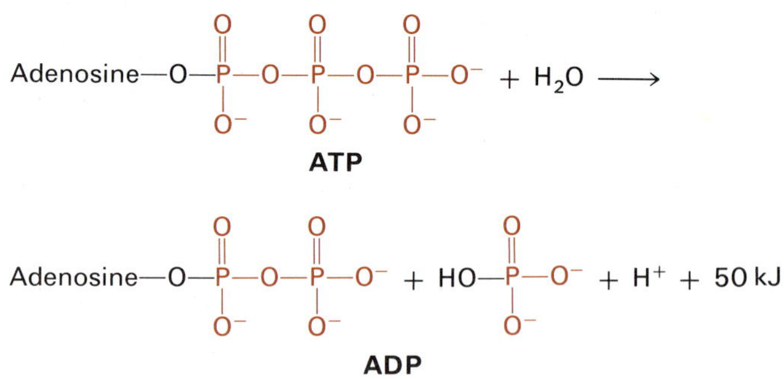

The energy produced in this and similar reactions is used to do mechanical work such as muscle contraction, to transport ions and molecules across cell membranes, and to synthesize large molecules from small ones. We shall learn more about the formation and reactions of ATP and ADP when we discuss bioenergetics and metabolism in Chapter 26. In Chapter 25, we shall see that nucleic acids are polymers whose backbones consist of phosphate esters.

KEY TERMS

A *dimer* is composed of two molecules joined together.

Esterification is the process in which a carboxylic acid and an alcohol react, eliminating a molecule of water, to yield an ester. The *hydrolysis of an ester* is the reverse of esterification: an ester reacts with water to produce an alcohol and an organic acid.

Fats are triesters of long-chain acids (called fatty acids) and glycerol.

Saponification is the base-catalyzed hydrolysis of fats to yield soaps and glycerol. Soaps are the sodium salts of long-chain fatty acids.

An *acyl halide* is an acid derivative in which a halogen atom (X) replaces the OH of the carboxyl group.

An *acid anhydride* is an acid derivative in which RCOO replaces the OH of the carboxyl group.

A *condensation polymer* is formed by the combination of monomers, with the elimination of a small molecule.

A *copolymer* is a polymer composed of two or more different monomers.

CHAPTER SUMMARY

Organic acids have the *carboxyl functional group* (—COOH). The carboxylic acids boil at even higher temperatures than alcohols of the same molecular weight because of the formation of *dimers* held together by hydrogen bonds.

Carboxylic acids are much weaker than the strong inorganic acids such as HCl and HNO_3, but they are the strongest organic acids—much stronger acids than alcohols, aldehydes, or ketones. The strength of the carboxylic acids is attributed primarily to the resonance stabilization of the carboxylate anion. Carboxylic acids that are only slightly soluble in water can often be separated from neutral organic compounds by converting them into soluble salts through neutralization with a strong base such as sodium or potassium hydroxide.

Acids can be prepared by oxidation of primary alcohols or aldehydes, or by the addition of Grignard reagents to carbon dioxide.

Carboxylic acids, *acid anhydrides,* and *acyl halides* react with alcohols to yield *esters*. Esters have no hydrogen atoms that can form hydrogen bonds, so their boiling points are lower than those of the carboxylic acids of which they are isomers.

The acid *hydrolysis* of an ester yields the carboxylic acid and the alcohol of which the ester is composed. Basic hydrolysis yields the alcohol and a salt of the carboxylic acid. Esters also undergo a reaction called the *Claisen condensation* in which two ester molecules combine. This reaction is analogous to the aldol condensations of aldehydes and ketones. Ester condensations are common in nature in the build-up of large biological molecules.

PROBLEMS

19-1. Name each of the following compounds according to the IUPAC system.

(a) 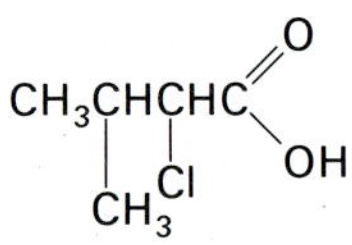

(b) 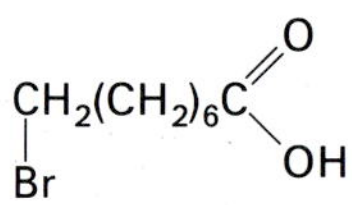

(c) 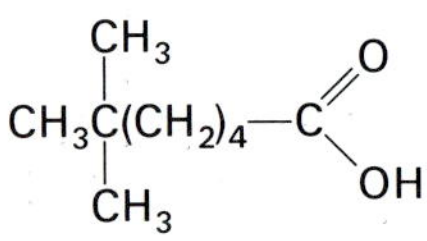

(d) 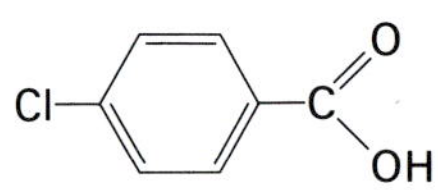

(e) 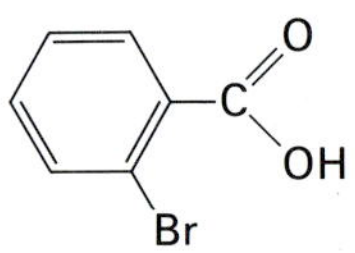

19-2. List the following compounds in order of decreasing boiling points:

(a) $CH_3—CH{=}CH_2$ (b) $CH_3CH_2CH_2OH$
(c) CH_3COOH (d) $HOCH_2CH_2OH$
(e) $HOCH_2—CH(COOH)—COOH$

19-3. List the following compounds in order of increasing boiling points:

(a) $CH_3CH_2OCH_2CH_3$ (b) CH_3CH_2COOH
(c) $CH_3(CH_2)_3OH$ (d) $CH_3(CH_2)_3CH_3$
(e) $CH_3—C(=O)—O—CH_3$

19-4. Explain why propionic acid is about 10^{10} times as strong an acid as propyl alcohol.

19-5. Rank the following compounds in order of *decreasing* strength as acids.

(a) $CH_3CH_2CH_2C(=O)CH_3$ (b) $CH_3CH_2CH_2CH_3$
(c) CH_3CH_2COOH (d) $CH_3CH_2CH_2OH$
(e) H_2O

19-6. Rank the following compounds in order of *increasing* strength as bases.

(a) OH^- (b) CH_3COO^-
(c) $CH_3CH_2CH_2O^-$ (d) $CH_3CH_2\ddot{C}H_2$
(e) $CH_3CH_2\ddot{C}HC(=O)H$

19-7. Which of the following compounds is the weakest acid, and which the strongest?

(a) $Cl—CCl_2—COOH$ (b) $ClCH_2—COOH$
(c) $CH_3—COOH$ (d) $I—CI_2—COOH$

19-8. Which of the following compounds is the strongest acid?

(a) $HOOCCH_2COOH$ (b) CH_3CH_2COOH

(c) 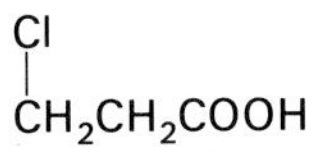 (d) CH_3CH_2OH

19-9. Outline a method for separating a mixture (a) of acetic acid and decanoic acid (*Hint:* consult Table 19-1); (b) of hexanoic acid and cyclohexane.

19-10. Describe a reagent or test you could use to distinguish between

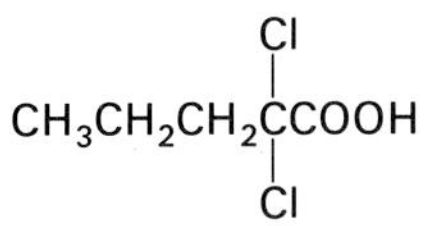 and 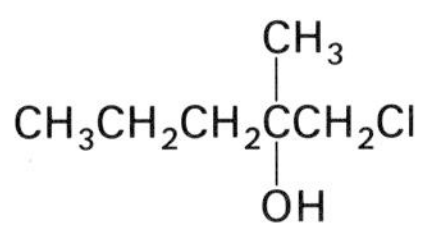

19-11. Which would have the lower freezing point: a 6 M solution of acetic acid (CH_3COOH) or a 6 M solution of HCl? Explain.

19-12. Fragrances or flavors of some esters are given together with their names and structures. Write the structure of the acid and the alcohol from which each ester could be synthesized.

(a) Isobutyl formate 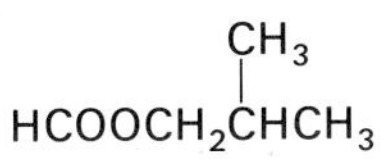 raspberry

(b) *n*-Octyl acetate $CH_3COO(CH_2)_7CH_3$ orange

(c) Ethyl *n*-butyrate $CH_3(CH_2)_2COOCH_2CH_3$ pineapple

(d) Benzyl acetate 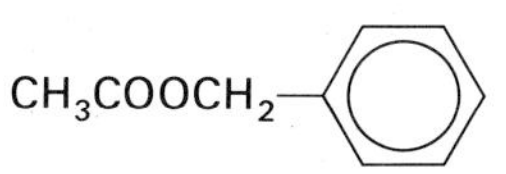 jasmine

(e) Coumarin 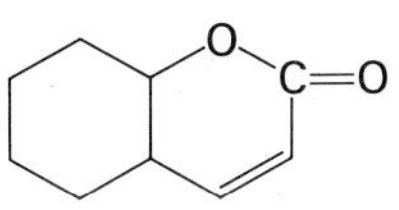 sweet clover

19-13. Write an equation for the reaction between each of the following compounds and a dilute solution of NaOH:
(a) CH_3CH_2COOH
(b) $HOOC(CH_2)_2COOH$
(c) $CH_3(CH_2)_{14}COOH$

19-14. Glyceraldehyde 3-phosphate, an intermediate in the catabolism of glucose in glycolysis, is an ester of phosphoric acid and glyceraldehyde. The structure of glyceraldehyde 3-phosphate is

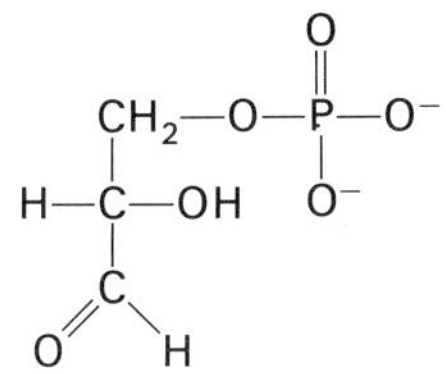

Write the structure of glyceraldehyde.

19-15. Describe a reagent or test you could use to distinguish between

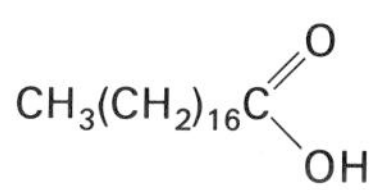 and 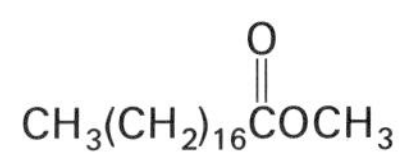

stearic acid **methyl stearate**

19-16. Of the two isomers shown, which will be more soluble in water? Explain.

(a) 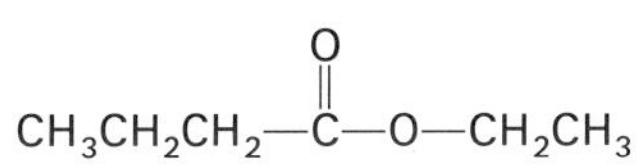
ethyl propionate

(b) 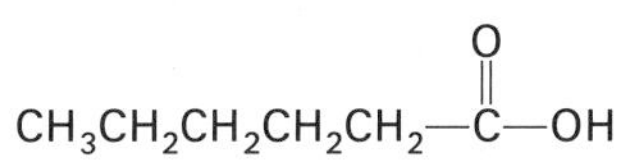
hexanoic acid

19-17. Write equations that describe how you could prepare each of the following compounds. Use any additional simple compounds you need.

(a) $\mathrm{CH_3CH_2\overset{\overset{O}{\|}}{C}{-}OH}$ (starting with Mg and an alkyl halide)

(b) $\mathrm{CH_3CH_2\overset{\overset{O}{\|}}{C}{-}Br}$ (starting with an alcohol)

(c) $\mathrm{CH_3CH_2{-}\overset{\overset{O}{\|}}{C}{-}O{-}\overset{\overset{O}{\|}}{C}{-}CH_2CH_3}$ (starting with an acid)

(d) $\mathrm{HO{-}\overset{\overset{O}{\|}}{C}{-}CH_2{-}\overset{\overset{O}{\|}}{C}{-}OH}$ (starting with an alcohol)

19-18. (a) Write the structures of an acyl halide and an alcohol you could use to prepare the ester,

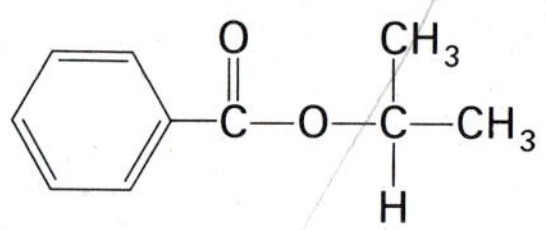

(b) Write the structures of an acid anhydride and an alcohol you could use to prepare the ester,

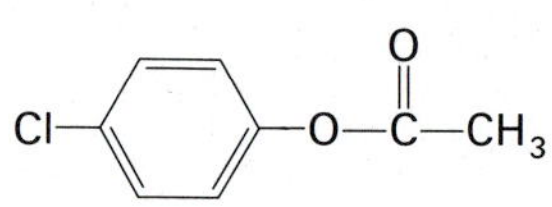

19-19. Draw the structures of the products of the following reactions

(a) 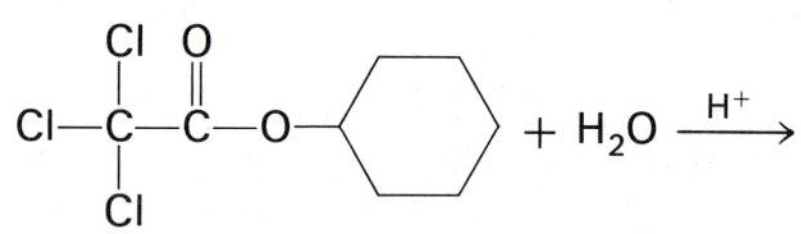

(b) 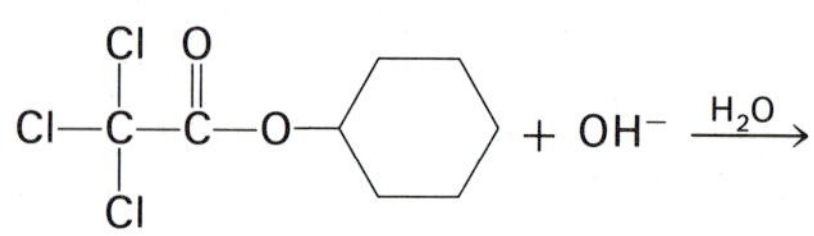

(c) 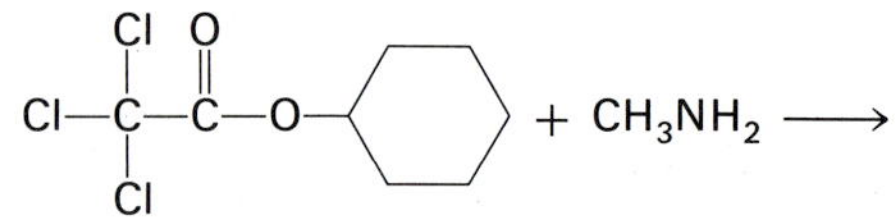

(d) $\mathrm{HOCH_2CH_2\overset{\overset{O}{\|}}{C}OH + HOCH_2CH_2\overset{\overset{O}{\|}}{C}OH \xrightarrow{H^+}}$

(e) $\mathrm{Cl{-}\overset{\overset{O}{\|}}{C}{-}(CH_2)_4{-}\overset{\overset{O}{\|}}{C}{-}Cl + 2CH_3OH \longrightarrow}$

19-20. Write the structure of the fat produced when three moles of lauric acid—a fatty acid obtained from the oil of the laurel (bay) tree, $CH_3(CH_2)_{10}COOH$—react with a mole of glycerol.

19-21. Write structures of monomers from which the following polymer could be synthesized:

$$\mathrm{{-}O(CH_2)_3O\overset{\overset{O}{\|}}{C}(CH_2)_4\overset{\overset{O}{\|}}{C}O(CH_2)_3O\overset{\overset{O}{\|}}{C}(CH_2)_4\overset{\overset{O}{\|}}{C}O{-}}$$

19-22. Write the structure of a molecule that can copolymerize with adipic acid, the six-carbon diacid, $HOOC(CH_2)_4COOH$, to yield a highly cross-linked polyester.

19-23. Write the structure of a compound that can polymerize by itself to form a polyester.

Water droplets on a nylon fishline. Because nylon is largely hydrocarbon (Section 20-7), it is not wet by water. In the color photograph on the cover, you can see that the water droplets act like tiny prisms and divide the sunlight into a continuous spectrum—a rainbow.

20

Amines and Amides

Amines are the most important bases of organic chemistry and biochemistry. They are, in particular, the bases present in nucleic acids, the giant molecules that store genetic information. The amino functional group $-NH_2$, together with the carboxyl group, is present in α-amino acids—the building blocks of proteins—and in a wide variety of other natural substances. In this chapter, we study the reactions that amines undergo and also the reactions of amides, derivatives of amines and carboxylic acids.

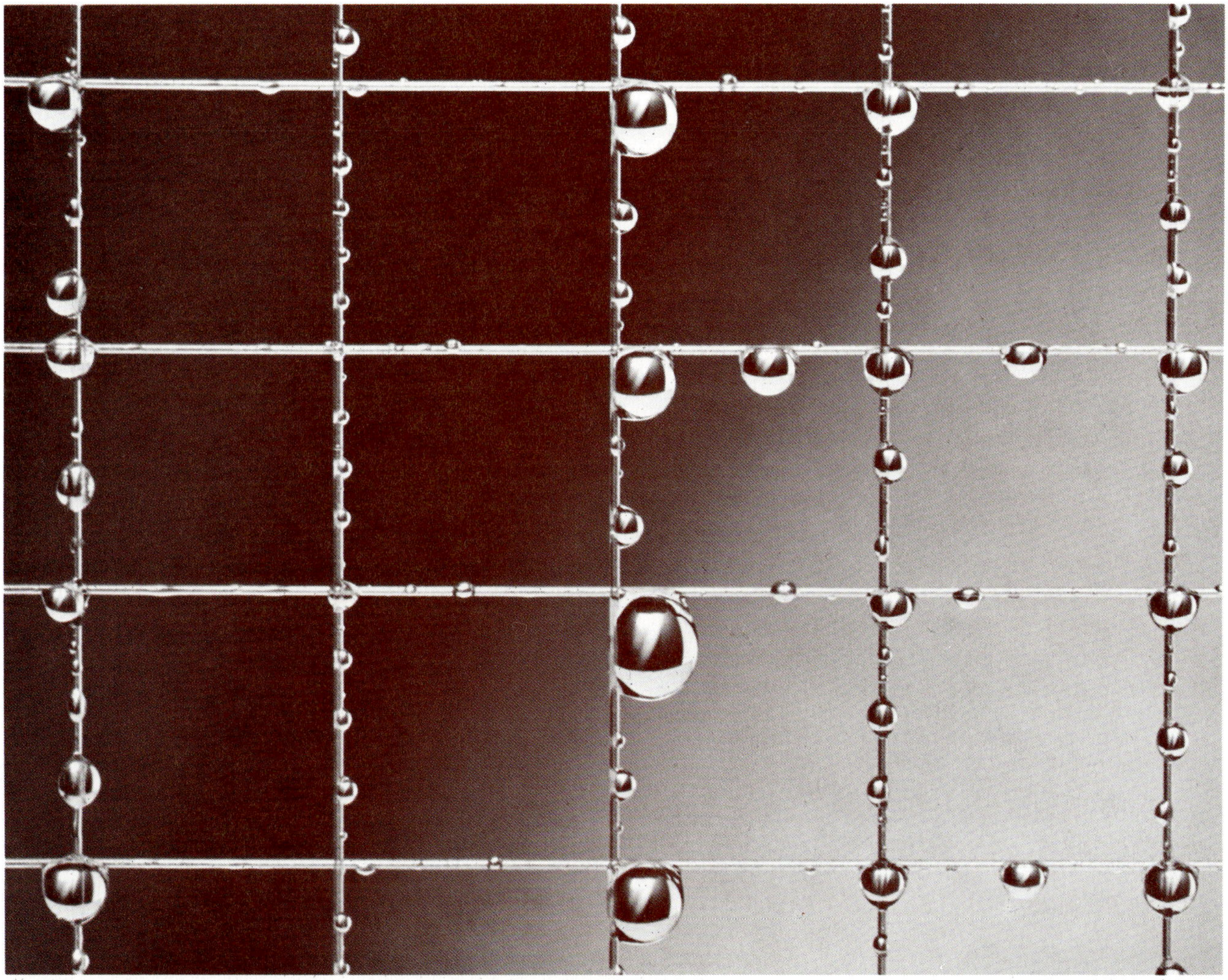

20-1 STRUCTURE AND PROPERTIES OF AMINES

Alcohols and ethers can be regarded as water molecules in which one or both of the hydrogen atoms are replaced by alkyl groups. Similarly, amines can be regarded as ammonia molecules in which one, two, or all three hydrogen atoms are replaced by alkyl groups. Amines are classified as primary, secondary, or tertiary—depending on the number of alkyl groups attached to the *nitrogen*. Primary amines have the general formula RNH_2, secondary amines the general formula RR′NH, and tertiary amines RR′R″N:

1° amine: RNH_2 Example: $CH_3CH_2CH_2\text{—}\ddot{N}H_2$
propylamine

2° amine: RR′NH Example: $CH_3CH_2\text{—}\ddot{N}(H)\text{—}CH_2CH_3$
diethylamine

3° amine: RR′R″N Example: $CH_3\text{—}\ddot{N}(CH_2CH_3)\text{—}CH_2CH_2CH_3$
methylethylpropylamine

Note that the R groups of a 2° or 3° amine can be the same or different.

Similarly, there are quaternary ammonium salts that are analogous to the NH_4^+ ion, with alkyl groups in place of hydrogens. Although quaternary ammonium salts are ionic solids, they have melting and boiling points much lower than those of typical inorganic salts.

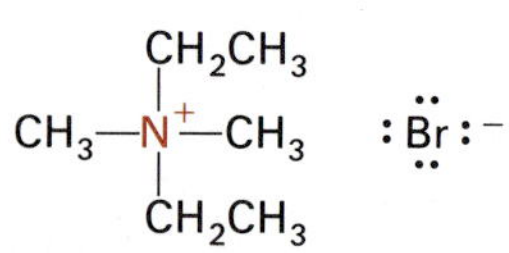

dimethyldiethylammonium bromide
(a quaternary ammonium salt)

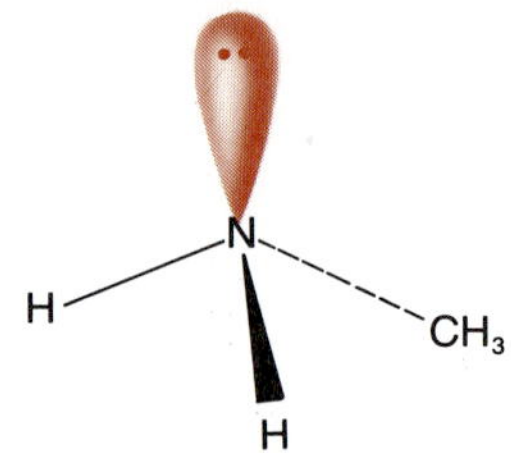

Bond length (nm)
N—H 0.10
C—N 0.15

Bond angle
H—N—H 106°
H—N—C 113°

FIGURE 20-1
Structure of methylamine.

Remember that the ammonia molecule is pyramidal, with nitrogen in the sp^3 hybridization state (Section 7-8). We can also think of the molecule as having nitrogen at the center of a tetrahedron, with hydrogen atoms at three corners of the tetrahedron, and a lone pair of electrons occupying the fourth corner. In amines, one or more of the hydrogen atoms is replaced by an alkyl group, but the structure of the molecule is still similar to that of ammonia. Figure 20-1 shows the structure of the simplest amine, methylamine (CH_3NH_2). The bond angles about the nitrogen atom are close to the expected tetrahedral angle of 109.5°.

There are also cyclic amines, in which the nitrogen is incorporated into the ring structure. These compounds, which have one or more ring atoms that are not carbon, are called heterocyclic compounds (see also Section 15-15). Examples of heterocyclic amines are piperidine (which has the properties of a typical secondary amine) and pyridine (an analogue of benzene, with properties of an aromatic compound):

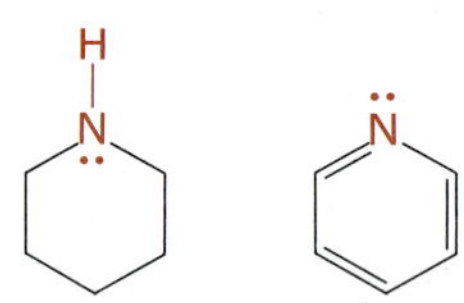

piperidine **pyridine**

The melting and boiling points of some amines are

TABLE 20-1
Properties of amines

Name	Structure	Melting point (°C)	Boiling point (°C)
Ammonia	NH_3	−78	−33
Methylamine	CH_3NH_2	−94	−6.3
Ethylamine	$CH_3CH_2NH_2$	−81	16.6
Dimethylamine	$(CH_3)_2NH$	−93	7.4
Diethylamine	$(CH_3CH_2)_2NH$	−48	56.3
Trimethylamine	$(CH_3)_3N$	−117	2.9
Triethylamine	$(CH_3CH_2)_3N$	−114	89.3

given in Table 20-1. Like alcohols, primary and secondary amines have hydrogen atoms that can participate in hydrogen bonds, and thus their boiling points are higher than those of alkanes of corresponding molecular weight (Figure 20-2). Nitrogen is less electronegative than oxygen, so the N—H---N hydrogen bond is weaker than the O—H---O hydrogen bond. As a result, amines boil at lower temperatures than alcohols of corresponding molecular weights. Hydrogen bonding is most important for primary amines (which have two hydrogen atoms that can participate in hydrogen bonds) and is not possible for tertiary amines (which have no hydrogens available for hydrogen bonds). Because amines can also form hydrogen bonds to water, the smaller amines (those with up to five carbons) are soluble in water.

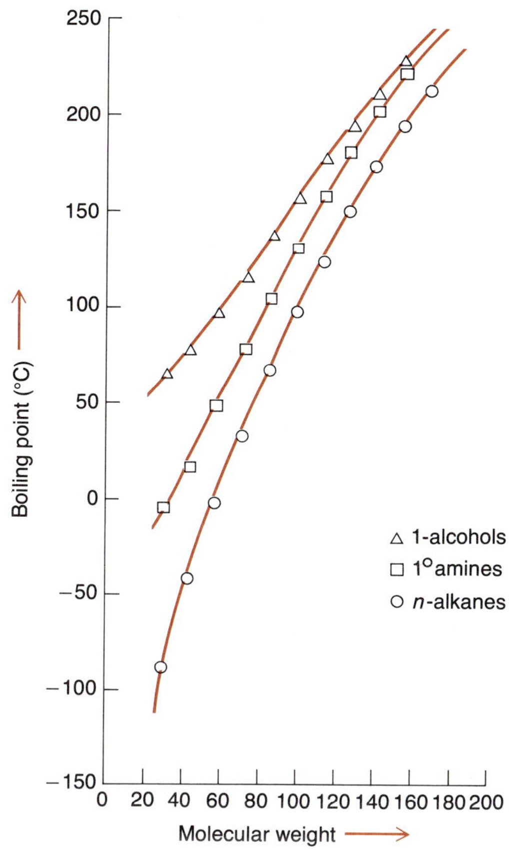

FIGURE 20-2
Boiling points of unbranched alcohols, primary amines, and normal alkanes

The following are some naturally occurring amines, including heterocyclic amines:

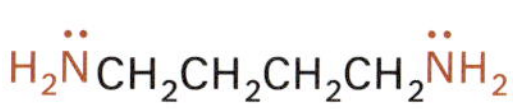

1,4-diaminobutane, or **putrescine** (responsible for odor of decaying fish)

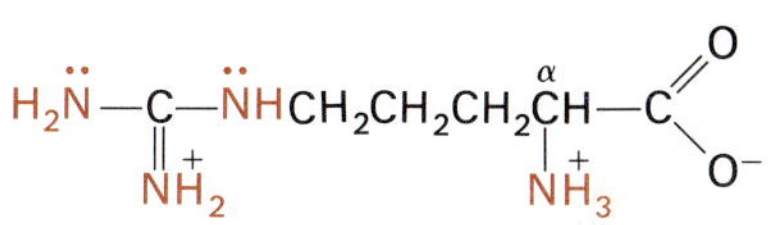

arginine (an α-amino acid; the structure shown is that at physiological pH)

EXERCISE 20-1 Rank the following compounds in order of increasing boiling point:
(a) $CH_3CH_2NHCH_3$ (b) $CH_3CH_2CH_2NH_2$
(c) $(CH_3)_3N$ (d) $(CH_3)_3NH^+Cl^-$

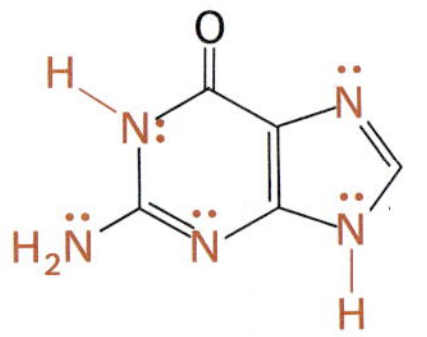

guanine
(one of the bases in DNA and RNA)

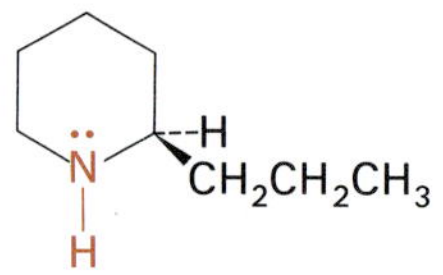

coniine
(poison that Socrates drank; from hemlock plant)

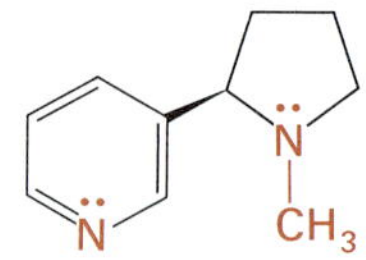

nicotine
(poison; from tobacco plant)

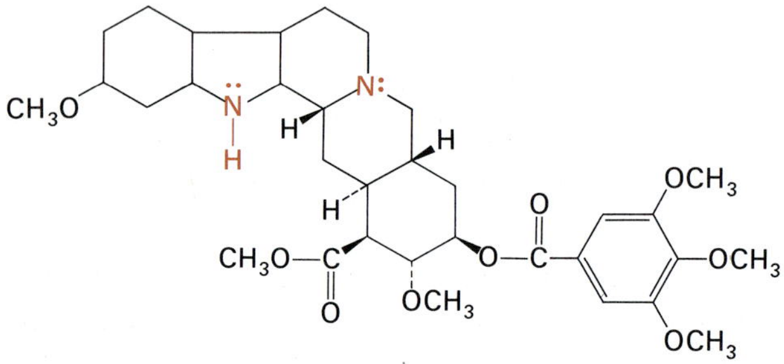

reserpine
(from Indian snakeroot; used as tranquilizer)

Ammonia and the small amines have sharp, penetrating odors. Somewhat larger amines smell like decaying fish—putrescine is an appropriate common name for 1,4-diaminobutane; and 1,5-diaminopentane is aptly called cadaverine.

EXERCISE 20-2 Label all the amino groups in nicotine and reserpine as primary, secondary, or tertiary or aromatic.

20-2 NOMENCLATURE OF AMINES

In the IUPAC system, amines are named as derivatives of the parent hydrocarbon:

CH_3NH_2

aminomethane

$CH_3CH_2CH_2CH_2NH_2$

1-aminobutane

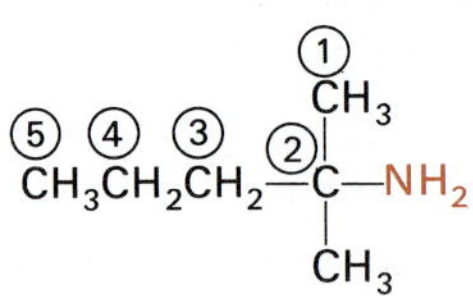

2-amino-2-methylpentane

The simple amines usually are known by their common names; all the alkyl groups are named, and "amine" is added as a suffix. The entire name is one word:

CH_3NH_2

methylamine

$CH_3CH_2NH_2$

ethylamine

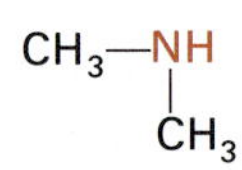

dimethylamine

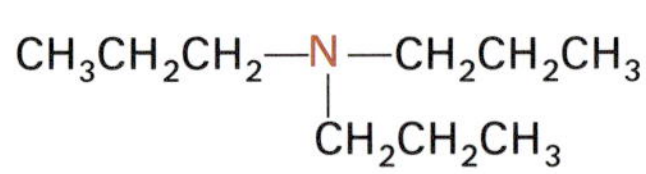

tri-*n*-propylamine

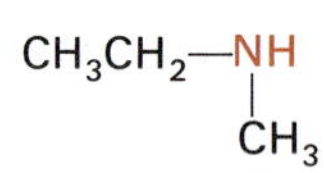

methylethylamine

The simplest aromatic amine is aniline, and its derivatives are named after it:

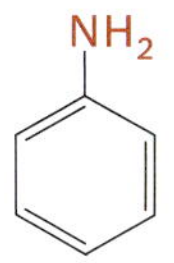
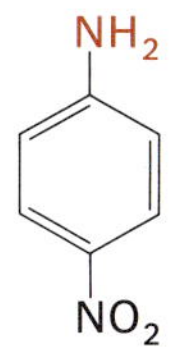
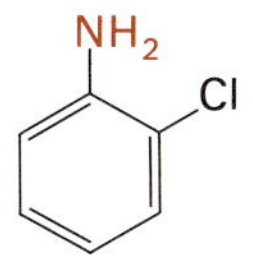

aniline **para-nitroaniline** **ortho-chloroaniline**

EXERCISE 20-3 Name the following compounds, using either common or IUPAC names:

(a) $CH_3CH_2CH_2—NH_2$

(b) $H_2N—CH_2CH_2CH_2CH_2CH_2—NH_2$

(c)

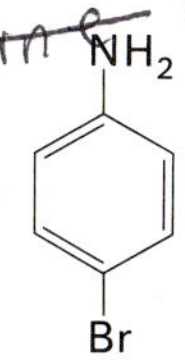

20-3 PREPARATION OF AMINES

There are many ways to prepare amines, and we consider a few here. Section 20-8 describes one more method of preparation. The most direct, though not always the most satisfactory, method of preparation of amines is the nucleophilic substitution of ammonia or an amine for a halogen atom in an alkyl halide (see Section 16-1). An example is the preparation of ethylamine from ethyl bromide:

Step 1: Substitution of NH_3 for Br^-

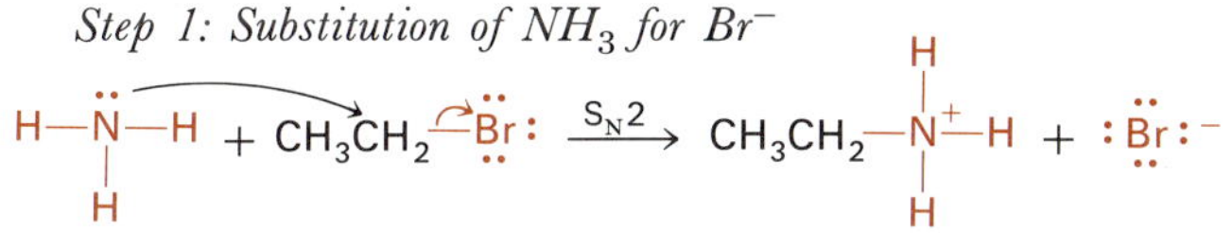

ethylammonium bromide

In this first step, an ammonium salt is produced. If the solution is then made basic with sodium hydroxide, the product is ethylamine:

Step 2: Elimination of H^+

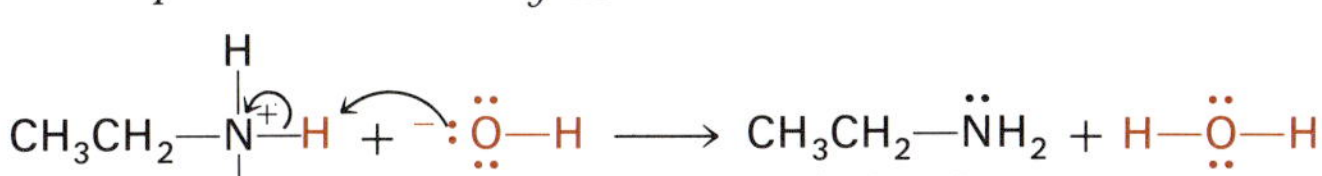

ethylamine

In practice, however, the process is complicated by the fact that the ethylamine is also a good nucleophile and can displace the bromine from ethyl bromide to yield diethylamine:

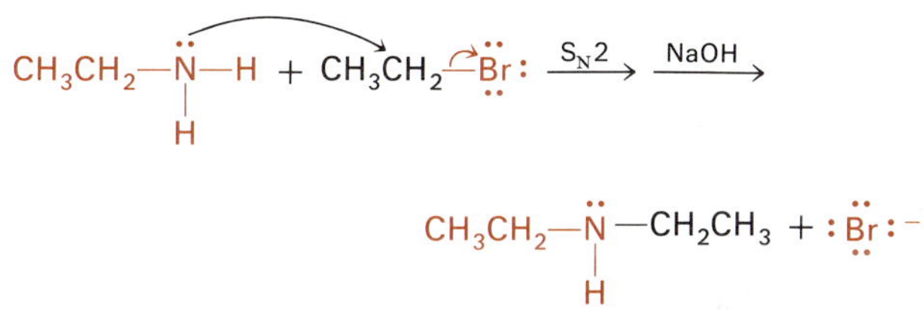

diethylamine

The reaction can continue in this way, to yield triethylamine and even tetraethylammonium bromide, $(CH_3CH_2)_4N^+Br^-$. The result often is a complex mixture of products. By using a large excess of ammonia (or amine) and carefully controlling the reaction conditions, it is possible to obtain a moderate yield of the desired product.

A second method of preparation of amines (one that occurs in the biosynthesis of amino acids) is **reductive amination,** a combination of two reactions we have seen before:

1. The addition of ammonia or of a primary amine to a carbonyl carbon to yield an imine, C=NH or C=NR′ (Section 18-6);

2. The addition of H_2 to a double bond (reduction)—in this case, the H_2 is added to the C=N double bond of the imine. We saw the addition of H_2 to C=C and C=O bonds in Sections 15-4 and 18-8.

As an example, we consider the reductive amination of benzaldehyde to yield benzylamine.

Step 1: Amination of benzaldehyde to yield an imine

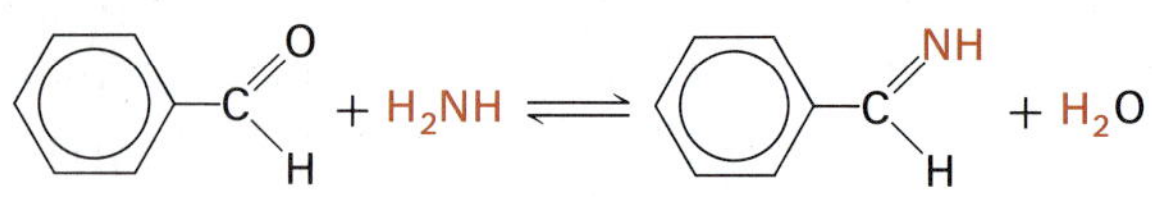

an imine

A likely mechanism for the reaction is given in Section 18-6. The imine is unstable, and it cannot be isolated from the reaction mixture. The reduction is carried out by the addition of H_2 using a catalyst (Ni, for example).

Step 2: Reduction of the imine to yield an amine

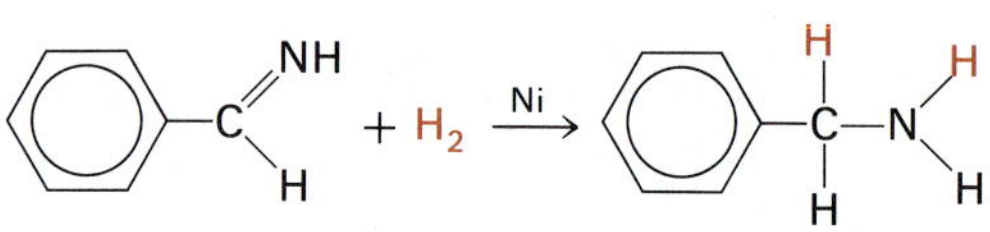

imine **benzylamine**

The net result of the two steps is the replacement of the carbonyl oxygen with an amino group and the addition of a hydrogen atom to the carbonyl carbon.

A key step in the biosynthesis of amino acids is the reductive amination of α-ketoglutarate to yield glutamate, an α-amino acid. The reducing agent is not H_2, but is one of the biological reducing agents NADH or its phosphate ester NADPH (see also Chapter 26). The reaction is catalyzed by the enzyme glutamate dehydrogenase:

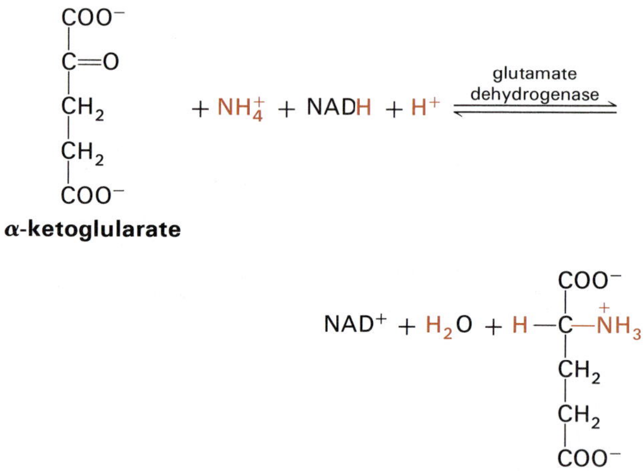

α-ketoglularate **glutamate**

Aromatic amines are prepared from the corresponding aromatic nitro compounds by catalytic hydrogenation, or by use of reducing agents such as metals and acid (for example, iron or zinc and hydrochloric acid):

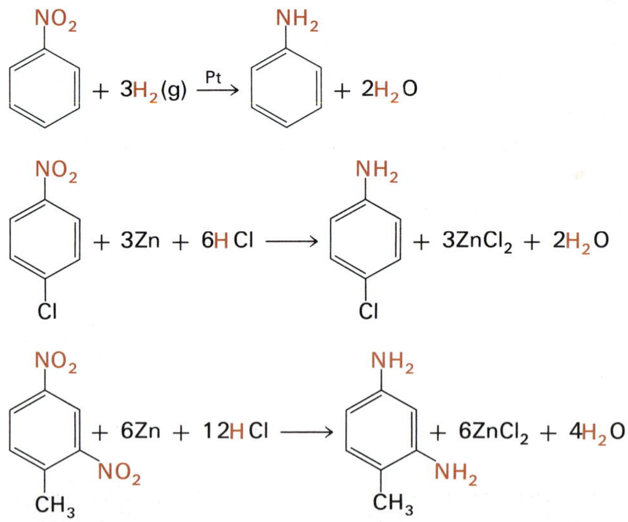

20-4
BASICITY OF AMINES

Amines, with a lone pair of electrons on the nitrogen, are Brønsted–Lowry and Lewis bases. Nitrogen is less electronegative than oxygen, so amines have a greater tendency to *donate* electrons (and are stronger bases) than alcohols or water.

Aqueous solutions of amines are basic because of the equilibrium

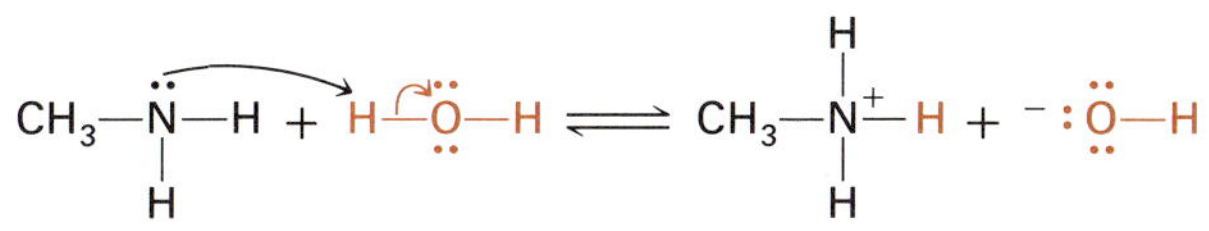

$$K = \frac{[OH^-][CH_3\overset{+}{N}H_3]}{[CH_3NH_2]} = 4.2 \times 10^{-4}\ \text{M}$$

This is analogous to the equilibrium in an aqueous solution of ammonia:

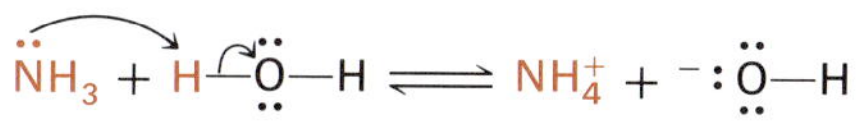

$$K = \frac{[OH^-][NH_4^+]}{[NH_3]} = 1.8 \times 10^{-5}\ \text{M}$$

The K for methylamine is typical for an alkyl amine and is somewhat greater than that of ammonia—that is, methylamine is a somewhat stronger base than ammonia.

Aromatic amines, on the contrary, are weaker bases than ammonia. The lone pair of electrons of nitrogen is delocalized over the benzene ring:

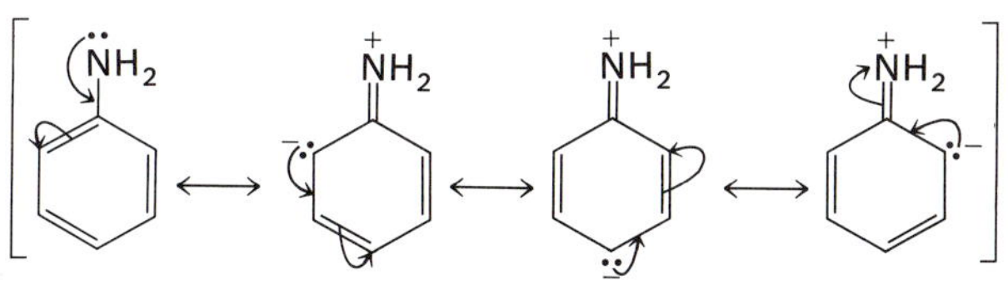

This delocalization means that the nitrogen's lone pair of electrons is less accessible for bonding than is the lone pair on the nitrogen of an aliphatic amine, for which there are no resonance structures. Hence the lower basicity of aromatic amines.

Because they are bases, amines react with acids to form salts that are analogues of inorganic ammonium salts:

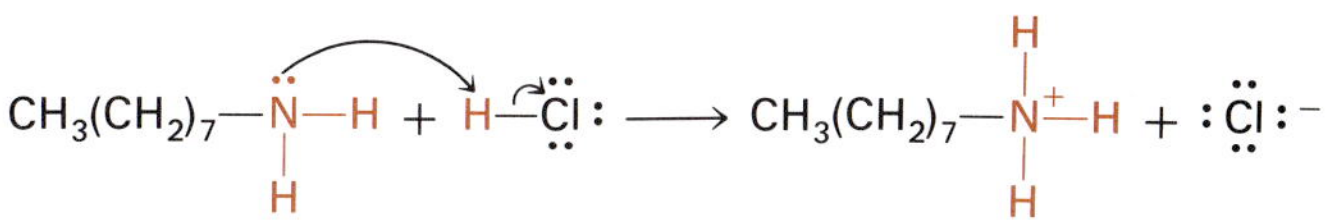

The *salts* are ionic, and so are often soluble in water even if the corresponding amine is essentially insoluble. Therefore, amines can sometimes be separated from neutral or acidic organic compounds by treating the mixture with a strong acid such as hydrochloric acid. For example, a mixture of the liquids *n*-octylamine and *n*-octane (each only slightly soluble in water) can easily be separated by the scheme shown in Figure 20-3.

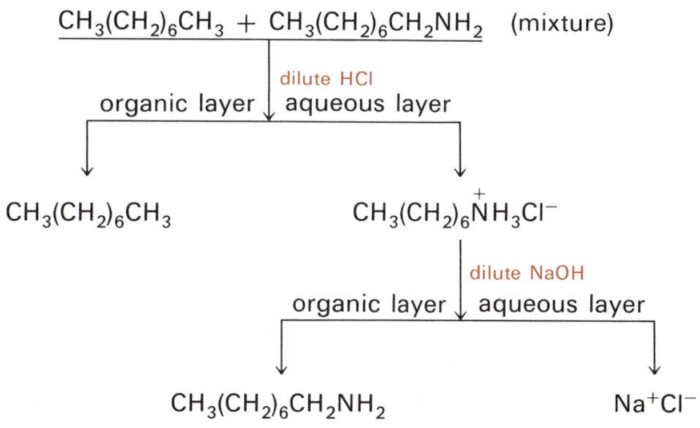

FIGURE 20-3

Separation of an amine from a neutral organic substance.

The mixture is shaken with dilute HCl in a separatory funnel, and the two layers are separated. The aqueous layer, which contains the salt *n*-octylammonium chloride, is treated with dilute NaOH until the solution is basic. This procedure regenerates the amine, which is only slightly soluble in water and forms a separate layer.

EXERCISE 20-4 Which of the following reagents could be used to separate aniline, from cyclohexanol? Explain your answer.

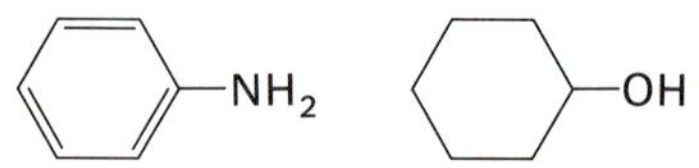

aniline **cyclohexanol**

(a) ethyl alcohol (b) diethyl ether
(c) dilute H_2SO_4 (d) H_2O (e) dilute KOH

20-5 REACTIONS OF AMINES WITH NITROUS ACID

A variety of reactions occur between nitrous acid (HNO_2) and amines. The particular reaction depends on the structure of the amine. With *primary alkyl* amines, an unstable diazonium salt is produced:

$$\underset{\textbf{1° amine}}{CH_3CH_2{-}NH_2} + HNO_2 + HCl \longrightarrow \underset{\textbf{diazonium salt}}{CH_3CH_2{-}\overset{+}{N}{\equiv}N} + Cl^- + 2H_2O$$

The amine must be primary for this reaction to occur because *two* H atoms must be removed from the N atom to form the N≡N bond. The diazonium salt rapidly decomposes to give nitrogen gas, and a variety of substitution and elimination products from the organic portion with which we shall not be concerned:

$$CH_3CH_2{-}\overset{+}{N}{\equiv}N \longrightarrow N_2(g) + \text{Organic products}$$

The reaction is not useful in synthesis because of the mixture of products, but it is a good test for primary amines. Neither secondary nor tertiary amines produce nitrogen gas with nitrous acid because they do not have two H atoms on the amino nitrogen. Thus, the evolution of nitrogen gas is a qualitative test for primary alkyl amines. If the nitrogen is collected and its volume measured, a quantitative determination of primary amines can be made. For each mole of primary amine, one mole of nitrogen gas is produced (with one N contributed by the amine and the other by the nitrous acid, as seen in the equation).

EXERCISE 20-5 A mixture of amines is treated with HNO_2, and the N_2 gas that evolves is collected. If 2.24 ℓ of $N_2(g)$ are collected at STP, how many moles of primary amines were contained in the mixture?

With secondary amines, nitrous acid forms oily compounds, usually yellow, called *N*-nitrosoamines (or *N*-nitrosamines):

$$\underset{\substack{\textbf{diethylamine}\\ \text{(2° amine)}}}{CH_3CH_2{-}\underset{\displaystyle CH_2CH_3}{\underset{|}{N}}{-}H} + HNO_2 \longrightarrow \underset{\substack{\textbf{\textit{N}-nitroso-diethylamine}\\ \text{(a yellow oil)}}}{CH_3CH_2{-}\underset{\displaystyle CH_2CH_3}{\underset{|}{N}}{-}N{=}O} + H_2O$$

The appearance of a yellow oil upon the addition of an unknown amine to a solution of HNO_2 is a test for secondary amines. Because *N*-nitroso compounds are now known to be carcinogens, this test is not recommended. It has been suggested that nitrates (NO_3^-) and nitrites (NO_2^-) added to retard the spoilage of hot dogs, bacon, and sandwich meats lead to the formation of *N*-nitroso compounds in the stomach and thus to an increased incidence of stomach cancer. Hydrochloric acid in the stomach can react with NO_2^- to give HNO_2, and the HNO_2 can react with secondary amines produced by the breakdown of foods. (NO_3^- can be reduced to NO_2^- by bacteria in the stomach, so addition of NO_3^- to foods is equivalent to the addition of NO_2^-.) Whether or not these compounds should be used as food additives is now a topic of debate in the United States, but evidence is insufficient to prohibit their use at this time. It is quite possible that their use will be prohibited in the future. Because tertiary amines have no hydrogen atoms on the nitrogen, they do not form nitroso compounds.

When primary *aromatic* amines react with nitrous acid, the diazonium ions produced are relatively stable if maintained at about 0°C, and they are important intermediates in a variety of organic synthesis. The benzene diazonium ion is produced by the reaction of aniline with nitrous acid:

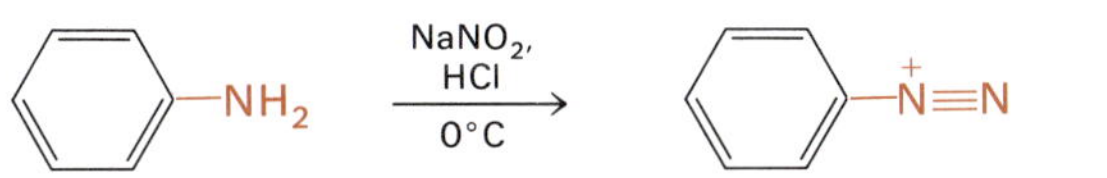

aniline (1° aromatic amine) **benzene diazonium ion**

(Because HNO_2 is unstable, it is usually produced in solution from HCl and $NaNO_2$.)

Aromatic diazonium ions readily undergo substitution reactions in which nucleophiles such as Cl^-, Br^-, I^-, CN^-, ROH, and H_2O are substituted for nitrogen. Copper (I) salts are used as catalysts.

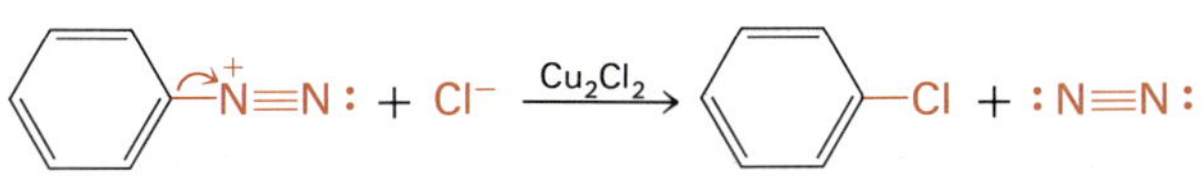

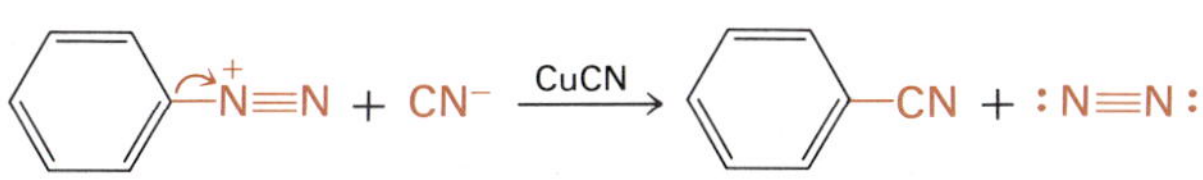

When the nucleophile is water, the product is a phenol:

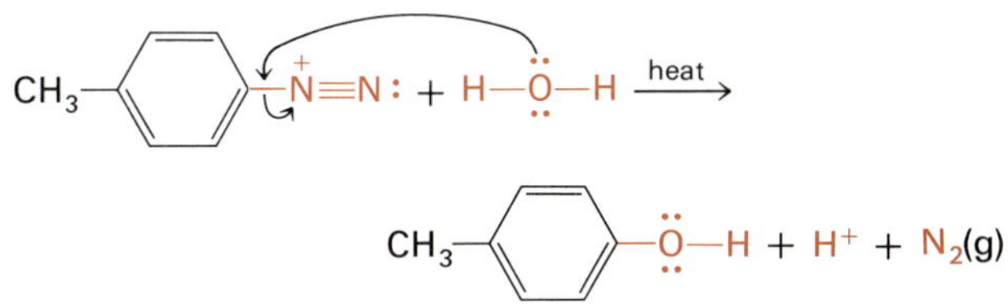

Because aromatic amines are relatively easy to prepare, this is a good way to synthesize a variety of substituted aromatic compounds.

EXERCISE 20-6 Ortho-chloroaniline,

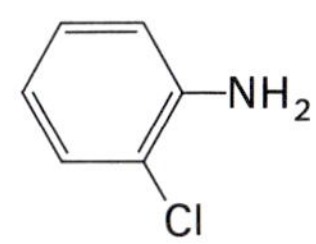

is treated with sodium nitrite ($NaNO_2$) and hydrogen iodide (HI) at 5°C. Potassium iodide is then added, and the solution is heated to 25°C. Nitrogen gas is given off. Draw the structure of the other product.

Besides participating in substitution reactions, aromatic diazonium ions also can couple with another electron-rich aromatic compound such as a phenol or an aromatic amine to yield a dye that contains the **azo group** (—N=N—):

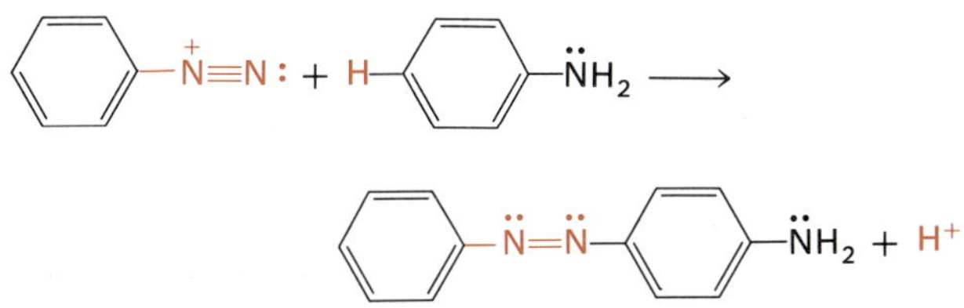

***para*-aminoazobenzene**
(orange)

The coupling occurs at the *para* position of the aromatic compound. Most of the substituted azo compounds are colored; they form a large class of dyes called azo dyes. Some also are used as indicators in the laboratory. Examples of diazo dyes are shown at the bottom of the page. (Butter yellow was formerly used as yellow food coloring, but it too is now suspected of being carcinogenic and so is no longer used.)

The structural feature of azo compounds responsible for their color is the extensive system of alternating double and single bonds. These alternating bonds are called **conjugated double bonds.** Such molecules undergo electronic transitions of the π electrons that correspond in energy to wavelengths of light in the visible region of the spectrum (400 to 700 nm; Chapter 6). The electrons *absorb* light in going to higher energy levels, and thus they remove certain colors from ordinary white light, which is a continuous spectrum of all colors. Only the light that is *not* absorbed reaches the eye as reflected or transmitted light (Table 20-2). Molecules without long conjugated systems also absorb light in undergoing electronic transitions, but these transitions require more energy and thus are not in the visible region of the spectrum.

TABLE 20-2
Color and wavelength of light

Wavelength absorbed (nm)	Color absorbed	Observed color
400	Blue	Yellow
500	Green	Red
600	Yellow	Blue
700	Red	Green

The green color of chlorophyll (see structure, Figure 13-1) and the red of carotene (bottom of page) are also due to the absorption of light in the visible region by π electrons in the long conjugated system of double and single bonds.

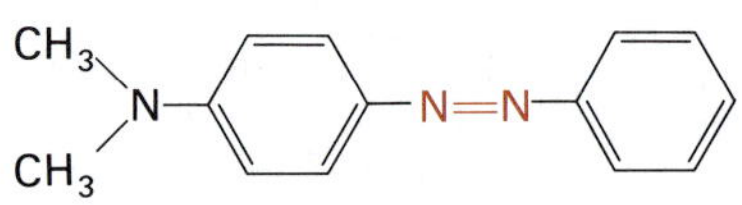

butter yellow

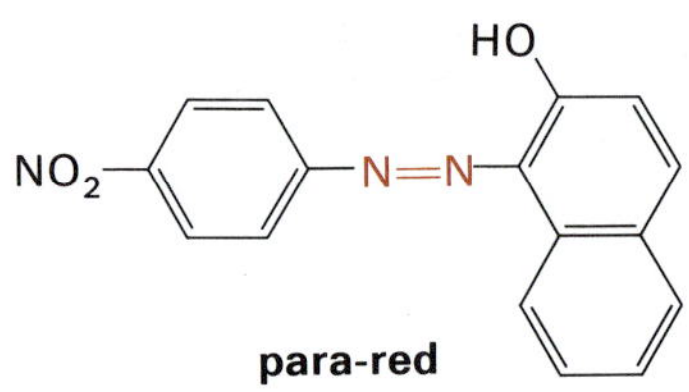

para-red

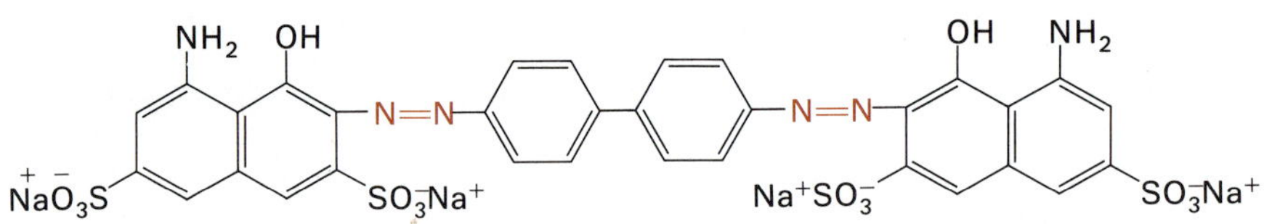

direct blue 2B

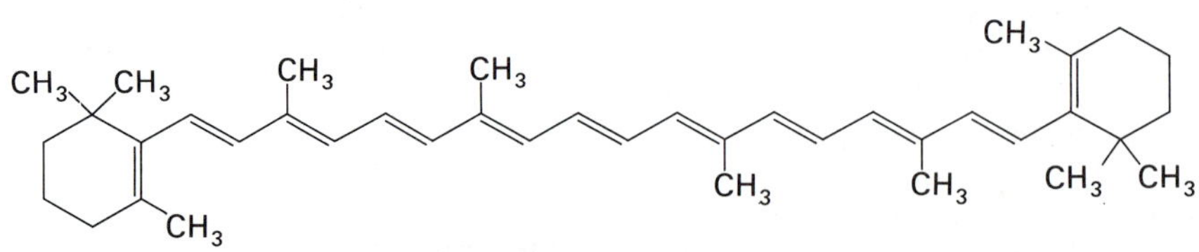

β-carotene
(precursor of vitamin A)

20-6
STRUCTURE AND PROPERTIES OF AMIDES

Amides are compounds in which an amino group replaces the OH group of a carboxylic acid. In a simple amide, the two remaining bonds to nitrogen are to hydrogen atoms; in a substituted amide, one or both of the remaining bonds to nitrogen are to alkyl or aryl groups:

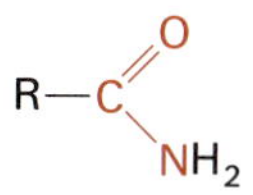

$RCONH_2$
simple amide

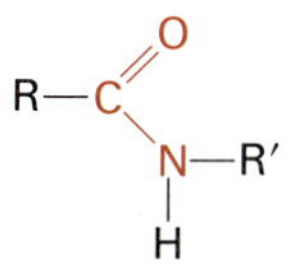

RCONHR′

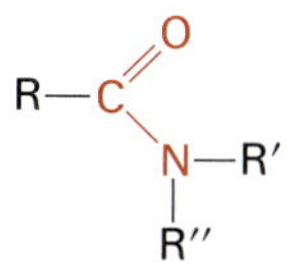

RCONR′R″
substituted amides

The following are examples of amide structures:

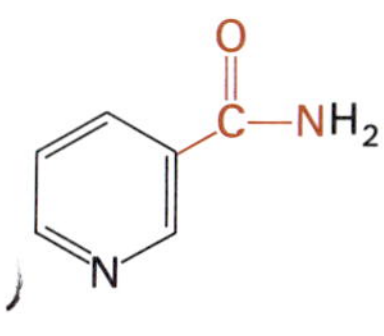

nicotinamide
(a B vitamin)

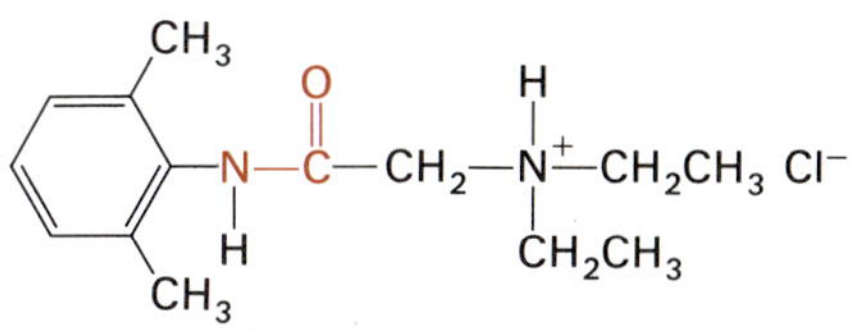

lidocaine hydrochloride
(Xylocaine; a synthetic local anesthetic)

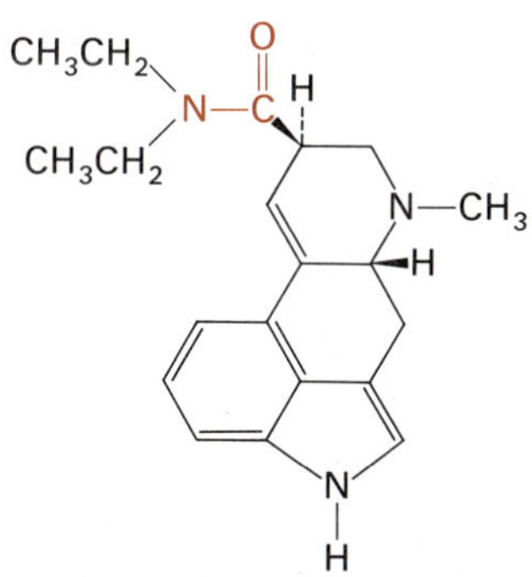

lysergic acid diethylamide
(LSD; hallucinogen; synthetic variation on the naturally occurring acid)

Urea is of historical interest as the first organic compound to be synthesized. Urea is a diamide, a product of amino acid catabolism in humans:

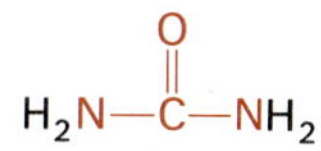

urea

Formamide (Figure 20-4) is the simplest amide. The carbon is in the sp^2 hybridization state, and the bond angles about the carbon atom are approximately the 120° angles expected for sp^2 hybridization. The entire molecule is planar and relatively rigid, because rotation about the carbon–nitrogen bond is restricted. The carbon–nitrogen bond in formamide is also shorter (0.14 nm) than the carbon–nitrogen bond in methylamine (0.15 nm; Figure 20-1). The rigidity and the shorter length of the carbon–nitrogen bond in formamide are attributed to a partial double-bond character of the carbon–nitrogen bond, as seen in the resonance structures

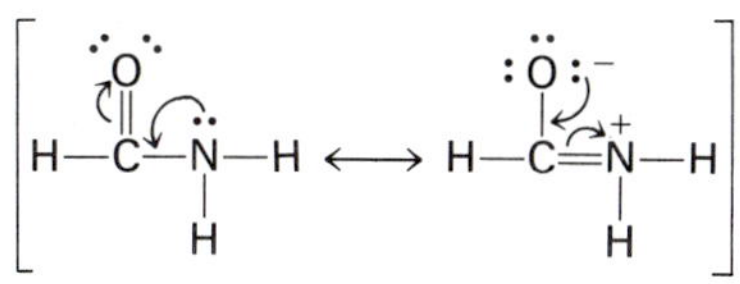

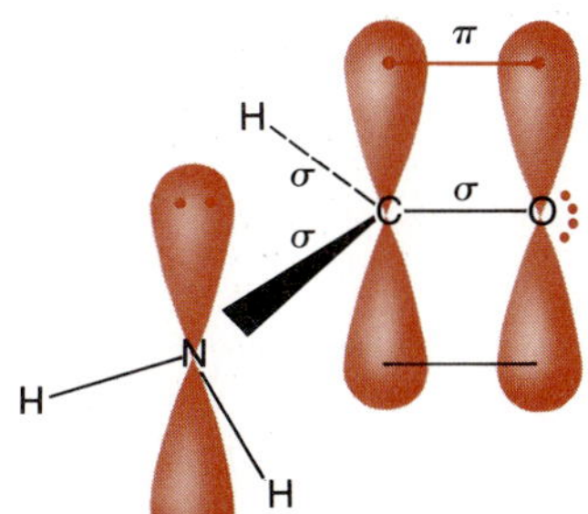

Bond length (nm)	
C=O	0.12
C–N	0.14
C–H	0.11
Bond angle	
H–C=O	123°
H–C–N	113°
N–C=O	124°

FIGURE 20-4
Structure of formamide.

In Chapter 21, we shall see that the basic functional group of proteins is the amide group (called the **peptide group** in protein chemistry). Many special properties of proteins are due to the rigidity of the C—N bond in the amide group.

Except for formamide, which melts at 3°C, the amides are solid at room temperature. Hydrogen bonds between molecules in the solid and liquid account for the relatively high melting and boiling points of amides:

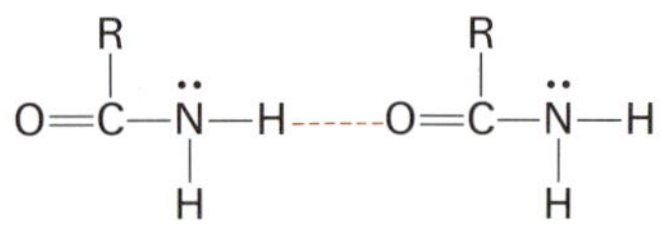

Large numbers of hydrogen bonds must be broken to melt the solid or boil the liquid. Because amides can also form hydrogen bonds with water, the small amides (those with up to five carbons) are soluble in water. Amides are named by dropping the *-ic* or *-oic* ending from the name of the acid and adding the suffix *-amide*. If the amide is substituted, alkyl groups attached to the nitrogen are named; the fact that they are bonded to nitrogen is indicated by using the prefix *N*- (for nitrogen):

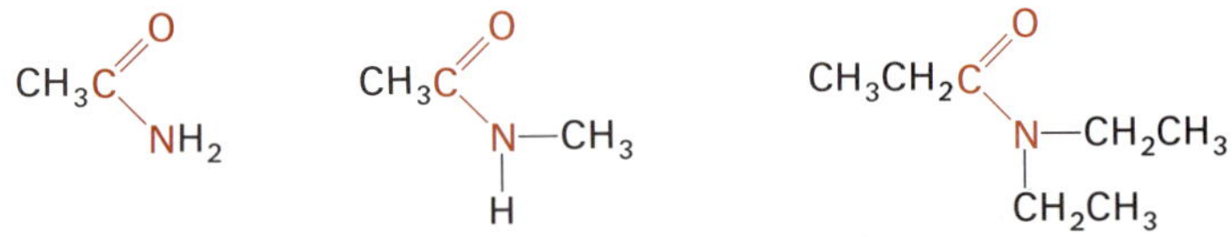

acetamide **N-methylacetamide** **N,N-diethylpropionamide**

20-7 PREPARATION OF AMIDES

Amides can be prepared by reaction between (1) ammonia or a primary or secondary amine and (2) a carboxylic acid, acyl halide, or acid anhydride:

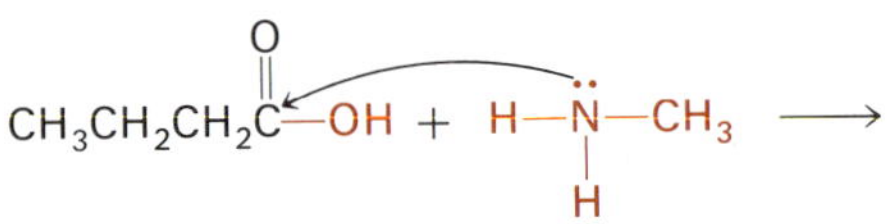

butyric acid **methylamine**

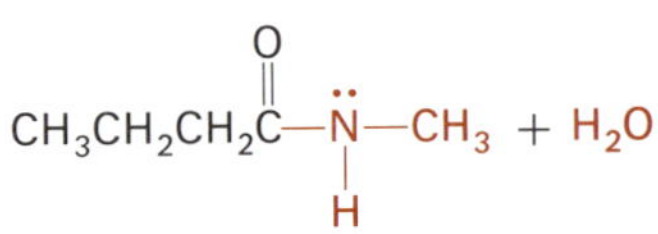

***N*-methylbutyramide**

When an acyl or acid anhydride is used to produce an amide, an acid is also produced in the reaction. The HCl formed in the following reaction is produced as the salt, ammonium chloride ($HCl + NH_3 \rightarrow NH_4^+Cl^-$):

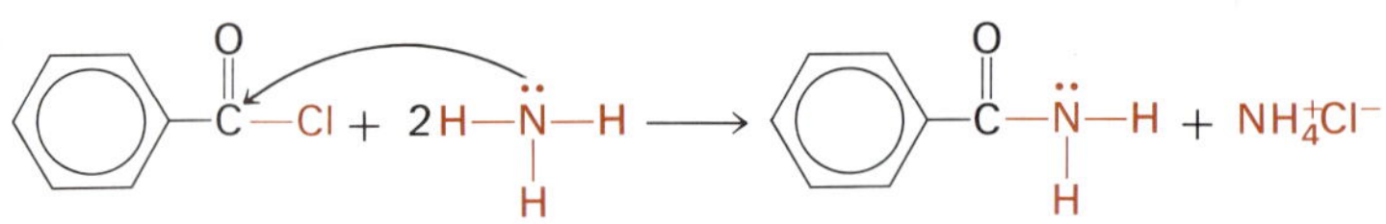

benzoyl chloride **benzamide**

Similarly, amides can be prepared from esters. The products are an amide plus the alcohol of the ester:

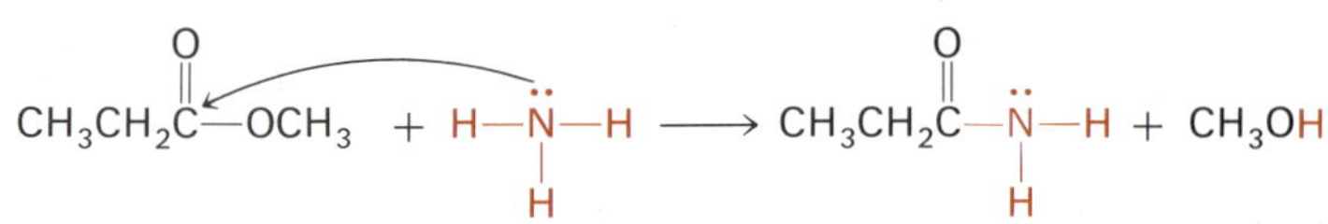

methyl propionate **ammonia** **propionamide** **methyl alcohol**

Note that in the formation of the amide bond, a hydrogen atom bonded to nitrogen is eliminated from ammonia or the amines. Tertiary amines, which have three alkyl groups and *no* hydrogen atom bonded to nitrogen, thus cannot react to form amides.

The mechanisms of all these reactions to prepare amides are like those described for the formation of esters in Section 19-6. The only difference is that the nucleophile added in the first step of the formation of amides is ammonia or an amine (in the formation of esters, the nucleophile is an alcohol). The final elimination reactions are the same for amides as for esters.

EXERCISE 20-7 Write out the mechanism for the reaction in which propionamide is formed from methyl propionate and ammonia.

Proteins, including silk and wool, are **polyamides** (or **polypeptides**)—polymers in which the monomers are joined by amide groups. The discovery of nylon, a synthetic polyamide, resulted from attempts to produce an artificial silk. Synthetic polyamides, like polyesters (Section 19-10), are produced by condensation polymerization. An example is Nylon 66, the most common form of nylon. Adipic acid (a dicarboxylic acid) and hexamethylene diamine condense with the elimination of water:

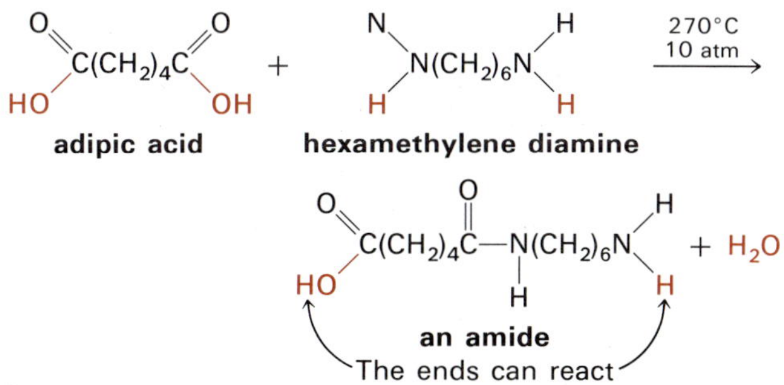

The reaction is just the formation of an amide from a carboxylic acid and an amine. The product has a functional group at each end, so it can react further. The COOH group can react with a molecule of the diamine, and the NH_2 group can react with a molecule of adipic acid, to yield the following product:

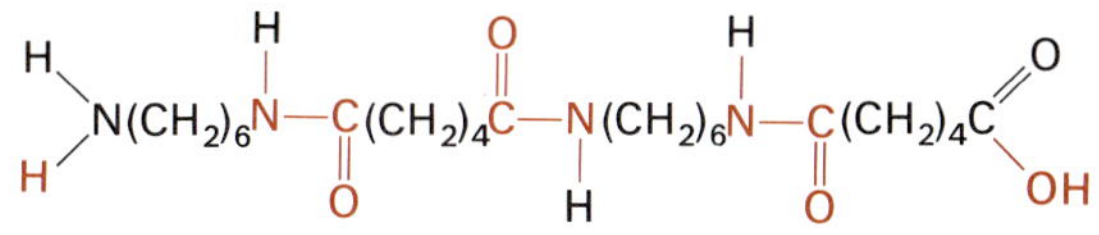

Again, the product has functional groups at both ends, and so can react further, until finally a polyamide is formed with the structure:

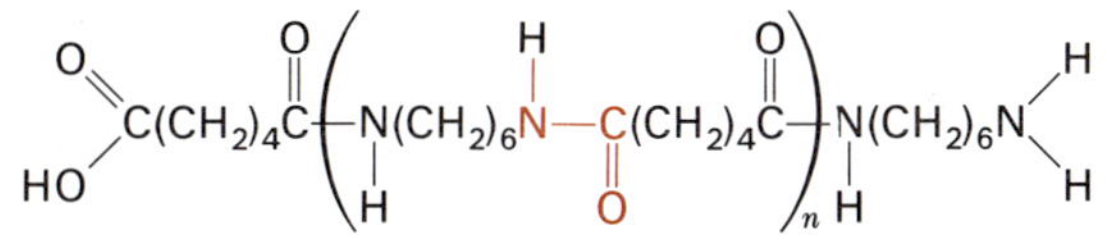

Nylon 66

(The 66 in the name Nylon 66 means that it is composed of a six-carbon acid and a six-carbon amine.)

Instead of adipic acid, adipyl chloride (an acid chloride) will also react with hexamethylene diamine.

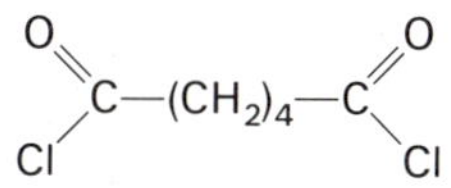

adipyl chloride

The product is still Nylon 66, but the small molecule eliminated in the reaction is hydrogen chloride instead of water. Figure 20-5 shows a laboratory synthesis of nylon from adipyl chloride and hexamethylene diamine.

When nylon is spun and drawn out into fibers, the polymer assumes a regular (almost crystalline) structure, with the long chains parallel to each other. Hydrogen bonds form between the N—H and C=O bonds of neighboring molecules, as shown in Figure 20-6. These hydrogen bonds, each of which is only about one-tenth as strong as a covalent bond, account for the strength of nylon. The hydrogen bonds are strong enough to hold the chains together in fibers, but weak enough that the fibers can be drawn out in the first place. When we discuss protein structure in Chapter 21, we shall see that the β-pleated sheet structure of proteins is very similar to that of nylon.

FIGURE 20-5

The synthesis of nylon from hexamethylenediamine and adipyl chloride. The polymer is formed at the junction of the two layers. [Photograph by Richard Byrnes.]

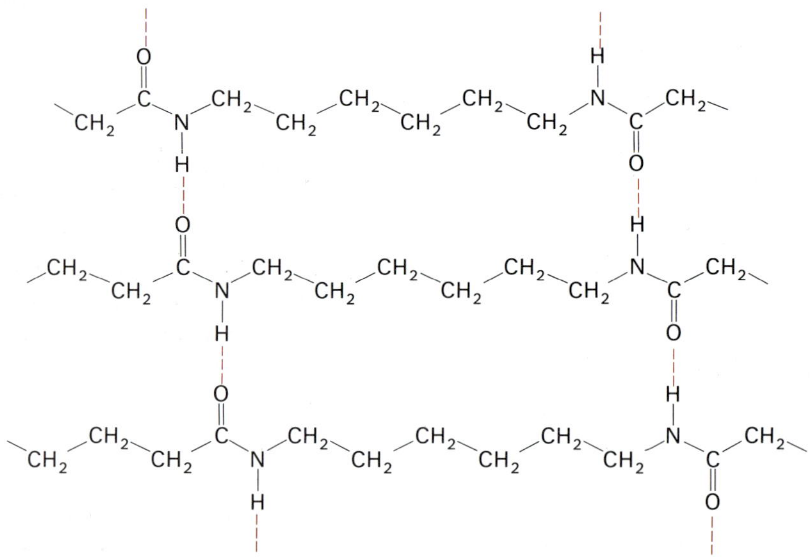

FIGURE 20-6

Hydrogen bonds between neighboring polymer chains account for the high tensile strength of nylon.

20-8
REACTIONS OF AMIDES

The reduction of amides by lithium aluminum hydride ($LiAlH_4$) is a good method of preparing aliphatic amines:

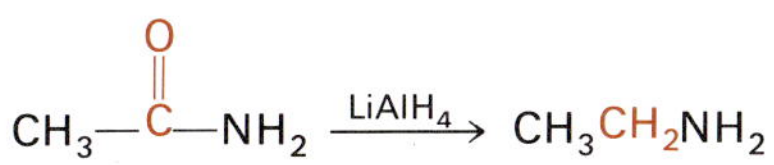

acetamide **ethylamine**

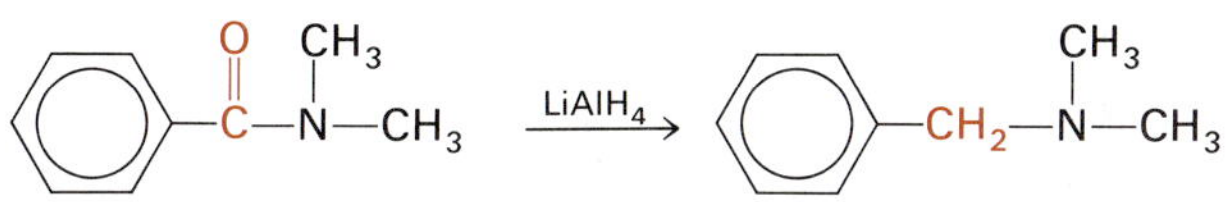

***N,N*-dimethylbenzamide** ***N,N*-dimethylbenzamine**

Amides, like esters, can be hydrolyzed in acidic or basic solution, but the reaction is much slower with amides than with esters. To break amide bonds, it is necessary to heat for several hours with a moderately concentrated solution of a *strong* acid or base. The acid hydrolysis of benzamide yields benzoic acid and the ammonium salt of the strong acid. (The NH_3 formed in the following reaction immediately reacts with excess HCl to yield NH_4^+.)

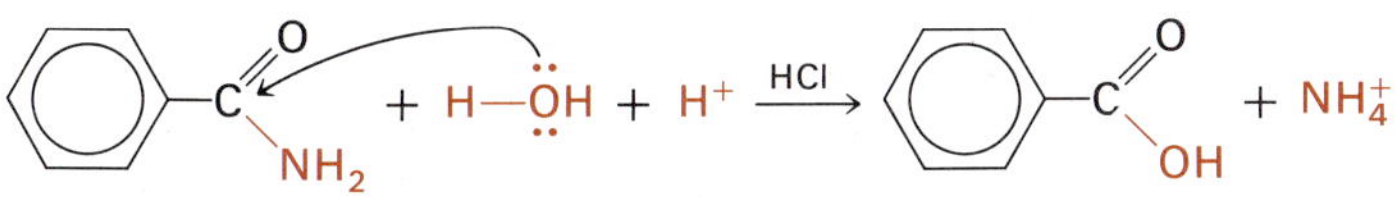

benzamide **benzoic acid**

The NH_3 reacts primarily with the strong acid instead of with the much weaker benzoic acid.

The basic hydrolysis of benzamide yields a salt of the benzoate ion (sodium benzoate if the base is NaOH) and ammonia. Benzoic acid formed in the hydrolysis immediately reacts with the strong base to yield the benzoate ion:

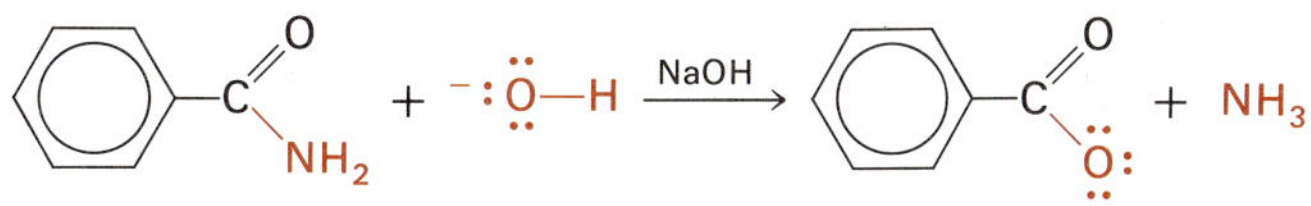

benzamide **benzoate ion**

These reactions are essentially irreversible; the NH_4^+ ion formed in the acid-catalyzed hydrolysis and the benzoate ion formed in the base-catalyzed hydrolysis are both very stable and have little tendency to react. In living cells, amide groups of proteins are hydrolyzed in the presence of enzyme catalysts—without strong acids or bases, and at normal body temperatures.

We have seen that the carboxylic acid derivatives (acyl halides, acid anhydrides, esters, and amides) react with nucleophiles such as water (in hydrolysis), alcohols, or amines. The acyl chlorides and acid anhydrides react vigorously with water; esters can be made to react by heating with acid or base; and amides can be made to react by heating even longer and harder with moderately concentrated solutions of a strong acid or base. The mechanism is the same for all these hydrolysis reactions; the first step is attack by the nucleophile (say, water) on the C=O carbon of the acid derivative. This carbon bears a partial positive charge:

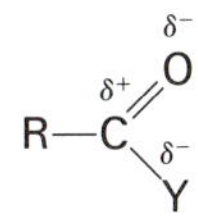

Here, —Y represents the —Cl in an acyl chloride; —OOCR′ in an acid anhydride; —OR′ in an ester; or —NH_2, —NHR′ or —NR′R″ in an amide. All of these groups are electron-withdrawing groups that pull electrons away from the carbonyl carbon and make its partial positive charge greater.

The greater the partial positive charge on the C=O carbon, the more readily it is attacked by a nucleophile. Because chlorine is more electronegative than oxygen or nitrogen, chlorine is the best of the above electron-withdrawing groups, and the acyl chlorides are the most reactive of the carboxylic acid derivatives. The acid anhydride, which has two oxygen atoms bonded to each C=O carbon atom, is the next strongest electron-withdrawing group, and the acid anhydrides are the next-most-reactive derivatives. Because oxygen is more electronegative than nitrogen, esters are stronger electron-withdrawing groups than amides; and therefore esters are hydrolyzed more readily than amides.

In order of decreasing reactivity with water or other nucleophiles,

acyl chlorides > acid anhydrides > esters > amides

Acyl chlorides and carboxylic acid anhydrides are so reactive that they cannot exist in the watery medium of living organisms. Esters, which are less reactive, do occur in nature—in fats, for example. Amides, the least reactive of the carboxylic acids, form the backbones of proteins—the subject of the next chapter.

KEY TERMS

Amines are classified as primary, secondary, or tertiary—depending on the number of alkyl groups attached to the nitrogen. A 1° amine has the general formula RNH_2; a 2° amine, RR′NH; and a 3° amine, RR′R″N.

Any amine that has a benzene ring bonded to nitrogen is an *aromatic amine.*

A *quaternary ammonium salt* has four alkyl groups bonded to nitrogen; the nitrogen is positively charged. The salts are ionic.

Reductive amination is the addition of ammonia or an amine to a carbonyl carbon to yield an imine (—C=N—). This is followed by reduction of the C=N bond. The net reaction is the substitution of an amino group for the oxygen of a carbonyl group.

Conjugated double bonds are alternating double and single bonds.

Amides are derivatives of carboxylic acids in which the carboxyl —OH group is replaced by $—NH_2$, —NHR, or —NRR′.

Polyamides are polymers in which the monomers are joined by amide groups. Polyamides are condensation polymers. Proteins are polyamides, or *polypeptides.*

CHAPTER SUMMARY

Primary amines (which have two hydrogen atoms that can form hydrogen bonds) and *secondary amines* (which have one such hydrogen) boil at higher temperatures than do alkanes of the same molecular weight, but at lower temperatures than do alcohols.

Amines can be prepared by the substitution of ammonia or an amine for a halogen atom in an alkyl halide, by reductive amination, or by the hydrolysis of amides. Aromatic amines are prepared by hydrogenation of nitro compounds.

The aliphatic amines are generally slightly stronger bases than ammonia, but the aromatic amines are weaker, due to the delocalization of the nitrogen lone pair of electrons over the benzene ring. Amines react with acids to form ammonium salts. These salts usually are more soluble in water than the amines, and treatment with a strong inorganic acid is a useful method of separating an amine from a neutral or an acidic organic compound.

When primary aromatic amines react with nitrous acid, *aromatic diazonium ions* (which are resonance stabilized) are produced. These aromatic diazonium ions readily undergo a variety of substitution reactions. Such an ion also can be coupled with another aromatic compound to yield a colored compound called an azo dye. Primary aliphatic amines react with HNO_2 to produce N_2 gas; and 2° amines to produce nitroso amines.

Amides are compounds in which an amino group replaces the hydroxyl group of a carboxylic acid. The amide group is planar and rigid due to the partial double-bond nature of the amide carbon-nitrogen bond.

Amides are prepared by reactions of carboxylic acids, acyl halides, acid anhydrides, or esters with ammonia or amines. Amides can be hydrolyzed, but the reaction is much slower than the hydrolysis of esters, and it requires heating for several hours with a moderately strong acid or base.

PROBLEMS

20-1. Write structural formulas for the eight amines having molecular formula $C_4H_{11}N$, and label each as primary, secondary, or tertiary.

20-2. Name each of the following compounds using either common or IUPAC names:

(a) 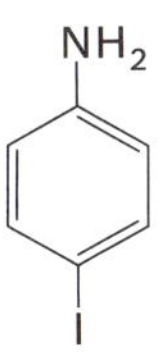

(b) 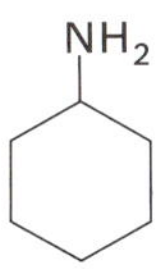

(c) 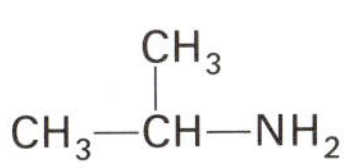

(d) 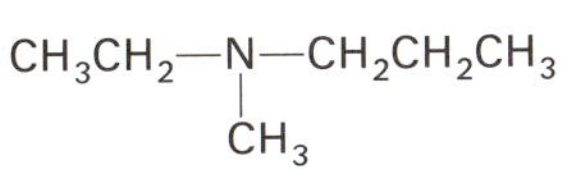

20-3. Rank the following compounds in order of increasing boiling point.

(a) $CH_3(CH_2)_3COOH$ (b) $CH_3(CH_2)_4NH_2$

(c) 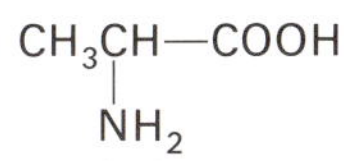

(d) $(CH_3CH_2)_2NCH_3$

20-4. Explain how lemon juice can eliminate "fishy" odors due to amines.

20-5. Calculate the concentration of OH^- and of H_3O^+ in a 1 M solution of methylamine in water:

$$CH_3NH_2 + H_2O \rightleftharpoons CH_3NH_3 + {}^-OH$$
$$K = 4 \times 10^{-4}$$

20-6. Draw the structures of the products of the following reactions:

(a) 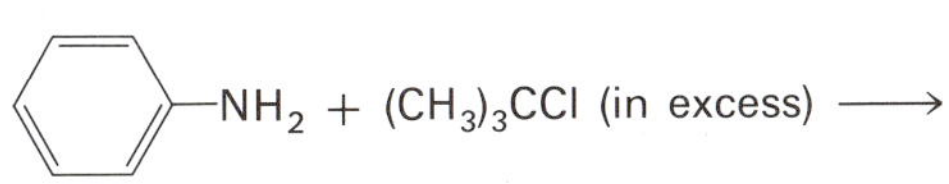

(b) 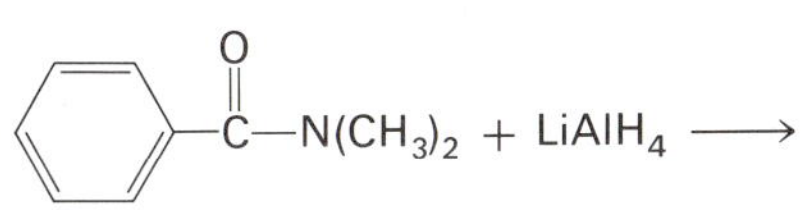

(c) 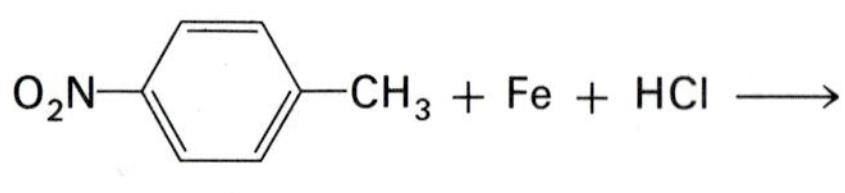

20-7. Write equations that show how you could prepare the following compounds.

(a) $CH_3-CH(CH_3)-\ddot{N}H_2$ from $CH_3-CH(Cl)-CH_3$

(b) 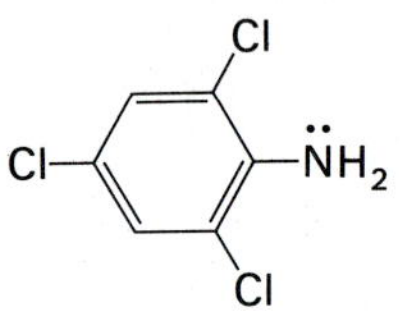 from 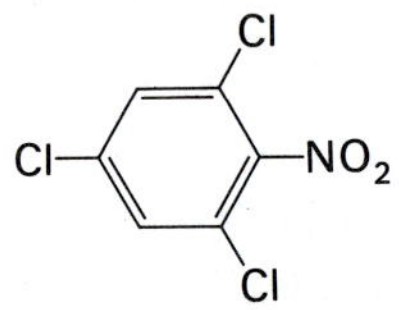

(c) $CH_3CH_2CH_2\ddot{N}(CH_3)_2$ from $CH_3CH_2\overset{O}{\overset{\|}{C}}-\ddot{N}(CH_3)_2$

20-8. Describe a method by which you could separate di-*n*-propylamine $(CH_3CH_2CH_2)_2NH$ from di-*n*-propyl ketone $(CH_3CH_2CH_2)_2CO$. Write equations for any reactions that occur.

20-9. Draw one additional resonance structure for the compound

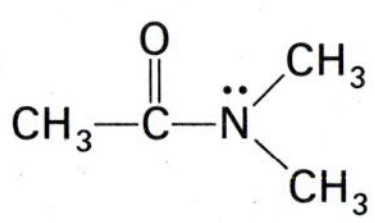

20-10. Which would you expect to be the stronger base, cyclohexylamine or aniline? Explain your answer.

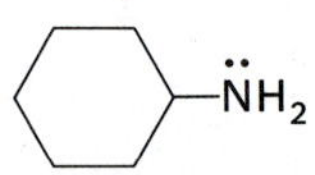

cyclohexylamine

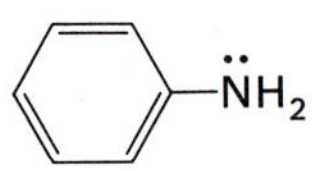

aniline

20-11. Draw structures of the products of the following reactions:

(a) 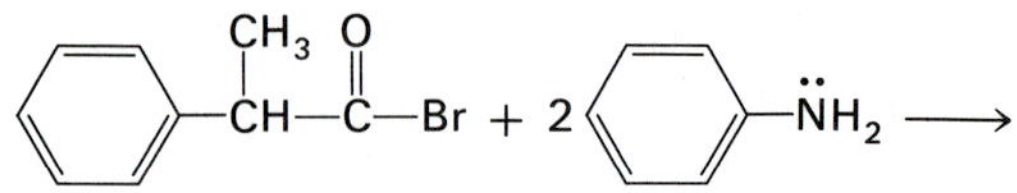

(b) $CH_3-\overset{O}{\overset{\|}{C}}-O-\overset{O}{\overset{\|}{C}}-CH_3 + 2(CH_3)_2\ddot{N}H \longrightarrow$

(c) $CH_3CH_2\overset{O}{\overset{\|}{C}}O(CH_2)_2CH_3 + CH_3CH_2CH_2\ddot{N}H_2 \longrightarrow$

(d) $CH_3CH_2C(=O)O^-NH_4^+ + NaOH \longrightarrow$

20-12. Draw the structures of an ester and an amine you could use to prepare

(a) $CH_3CH_2CH_2-\overset{O}{\overset{\|}{C}}-\ddot{N}(CH_3)_2$

(b) 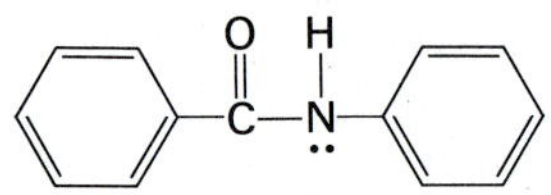

20-13. Draw the structures of the products of the following reactions.

(a) The hydrolysis (with HCl) of formamide

$H-\overset{O}{\overset{\|}{C}}-\ddot{N}H_2$

(b) The hydrolysis (with KOH) of nicotinamide

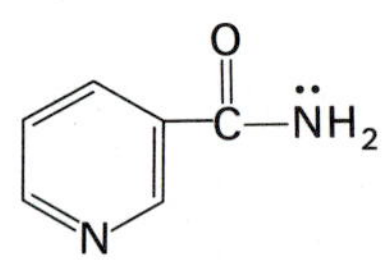

(c) The hydrolysis (with NaOH) of LSD (see structure, Section 20-6).

(d) The hydrolysis (with H_2SO_4) of lidocaine (see structure, Section 20-6).

20-14. Draw the structures of the two compounds that can be coupled (by reaction with HNO_2 and subsequent heating) to give
(a) butter yellow (see structure, Section 20-5)
(b) para-red (see structure, Section 20-5)
(c) direct blue 2B (see structure, Section 20-5).

20-15. Draw the structures of the products of the following reaction

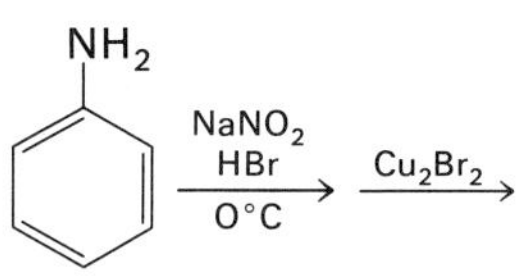

20-16. Draw the structure of an intermediate likely to be formed in the reaction between methyl acetate,

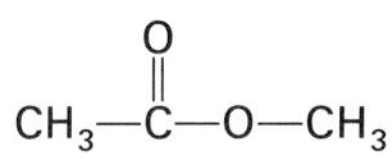

and ammonia to yield an amide and an alcohol.

20-17. Draw the structures of the products of the acid-catalyzed hydrolysis of
(a) caffeine:

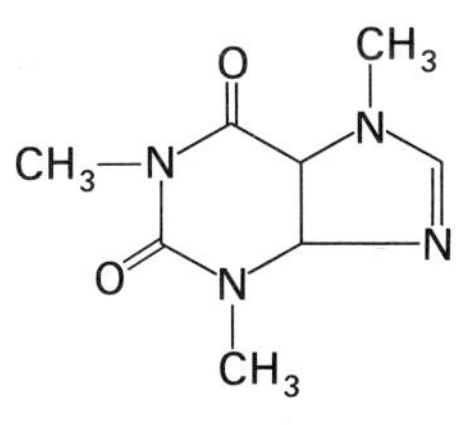

(b) acetylaminofluorene (AAF; a herbicide; thought to cause liver cancer):

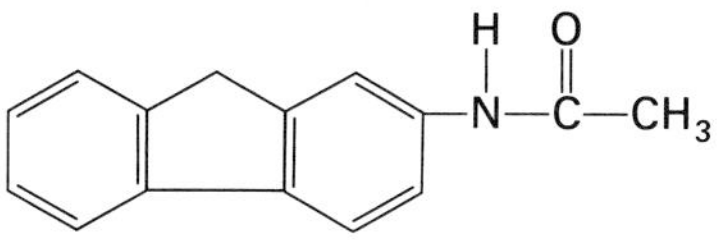

20-18. If a drop of hydrochloric acid is spilled on a nylon stocking, a hole appears. Explain.

20-19. Draw the structure of a compound that can polymerize by itself to form a polyamide.

20-20. Draw the structure of a molecule that can copolymerize with hexamethylenediamine $H_2N—(CH_2)_6—NH_2$ to yield a highly cross-linked polymer.

20-21. Write the structure of a monomer from which the following polymer could be constructed:

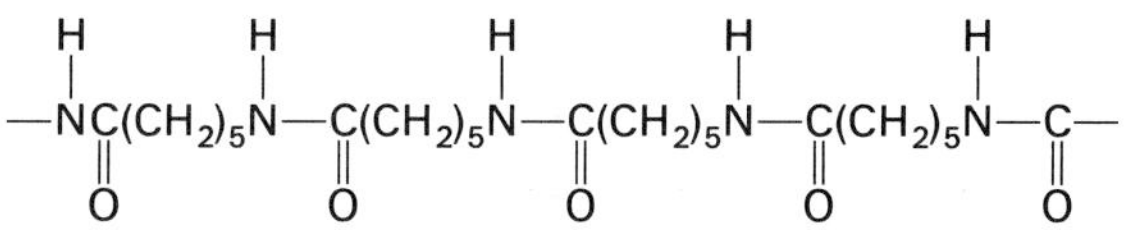

20-22. Quinine, an antimalarial drug obtained from the bark of the cinchona tree, is not very soluble in water. Suggest a way to make quinine more soluble in water.

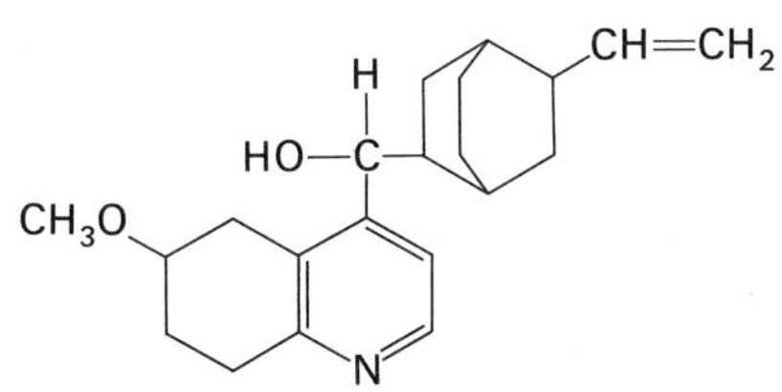

quinine

20-23. Another way to prepare amines is by reduction of a nitrile. Write an equation for a reaction by which $CH_3CH_2CH_2—C{\equiv}N$ could be converted to an amine.

IV

BIOLOGICAL CHEMISTRY

Collagen fibrils from human skin. Transmission electron micrograph, ×40,000.

21

Introduction to Peptide and Protein Structure

Amino acids are molecules of which peptides and proteins are formed. They occur in a number of other structures as well. This chapter examines properties of amino acids, together with some examples of the roles of amino acids in biological processes. We shall see how amino acids are joined to form peptides and proteins with precise three-dimensional structures. The biological activity of a protein is critically dependent on this structure.

21-1
COMMON PROPERTIES OF AMINO ACIDS

The amino acids are an extremely versatile set of biochemical compounds. Hundreds of distinct proteins, polymers of amino acids, occur in nature with very diverse functions. Each of these proteins has a unique structure that is defined by the order of amino acid units linked together. Amino acids also occur in molecules other than proteins—in hormones and antibiotics and as structural components of cell-surface polymers. For certain amino acids, a direct biological role can be demonstrated. Glutamic acid and phenylalanine, for example, are known to be involved in brain function.

The 20 principal amino acids from which proteins are synthesized have a number of common features. Each possesses an amino group, $—NH_2$ (which is basic), and a carboxylic acid group, —COOH. In all natural amino acids, these are both bonded to the same carbon atom:

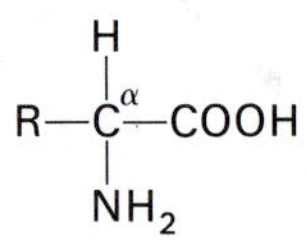

Because it is the first carbon after the carboxyl group, this central carbon atom is called the **alpha carbon** (α-carbon). Compounds with this arrangement are called α-amino acids. The other bonds to the α-carbon are occupied by a hydrogen atom and an R group that is specific for each of the amino acids. The exact structure for an amino acid depends on the pH, because three distinct forms of the carboxyl and amino groups can exist:

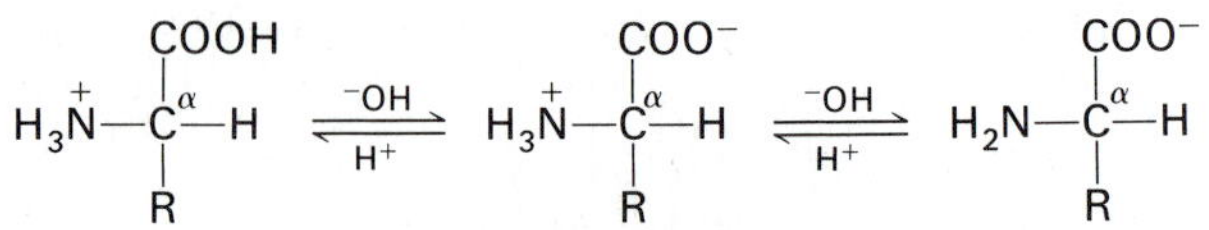

positively charged; predominates at low pH — zwitterion; predominates at pH 7 — negatively charged; predominates at high pH

(The R group may also have different forms under different conditions of pH.)

Although the three forms are in equilibrium and some of each is always present, in neutral solution (pH near 7) amino acids are predominantly in the form of **zwitterions** (from the German *zwitter,* which means "hermaphrodite"). In this form the carboxyl group is unprotonated ($—COO^-$), and the amino group is protonated ($—NH_3^+$). In acid solution (say pH = 1) the carboxyl group is un-ionized (—COOH), whereas the amino group remains ionized ($—NH_3^+$). In basic solution (say pH = 11) the amino group is un-ionized ($—NH_2$), whereas the carboxyl group is ionized ($—COO^-$).

The pure amino acids also are in the zwitterion form. Thus, like salts, they are solids at room temperature. They are insoluble in nonpolar solvents, and they are relatively soluble in water.

EXERCISE 21-1 Draw the structure of alanine that predominates at pH 7, at pH 1, and at pH 11. The R group of alanine is $—CH_3$.

The simplest amino acid, glycine, has two hydrogens attached to the α-carbon: $H_2N—CH_2—COOH$. Each of the other amino acids has four different substituents on the α-carbon. They are therefore optically active and exist in enantiomorphic forms (mirror images) known as D and L (see also Chapter 13). For the amino acid alanine, the configurations corresponding to D and L are

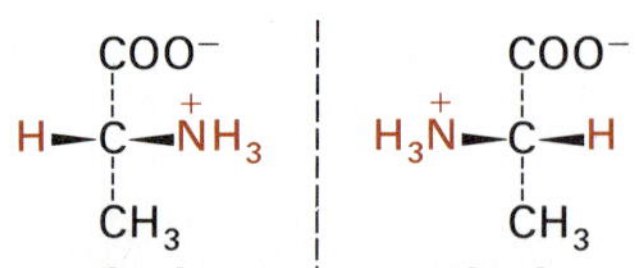

D-alanine | **L-alanine**

Hypothetical mirror plane

All of the optically active amino acids of proteins occur only in the L form. Unless specific reference is made to D-amino acids, we are referring to the L forms in all the following discussions of amino acids.

21-2
THE AMINO ACIDS OF PROTEINS

The 20 amino acids from which proteins are synthesized can be classified on the basis of the chemical properties of their R groups. An R group is often called a side chain. In the following structures, the R groups are shown in color.

1. *Alkyl R groups.* The simplest amino acid is **glycine,** whose R group is a hydrogen atom. **Alanine, valine, leucine, isoleucine,** and **proline** have alkyl groups as side chains:

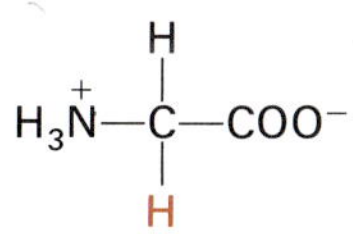

glycine
(Gly)

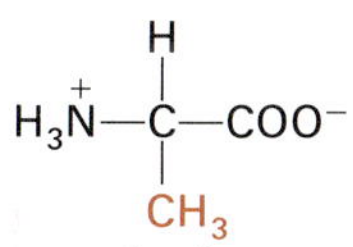

alanine
(Ala)

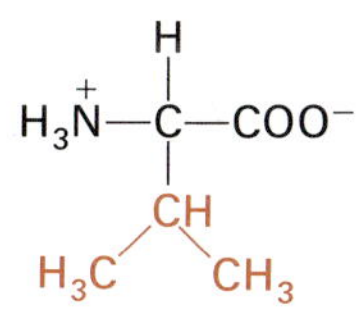

valine
(Val)

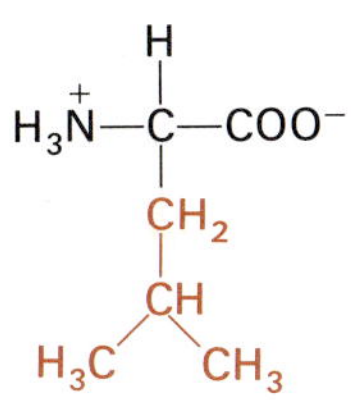

leucine
(Leu)

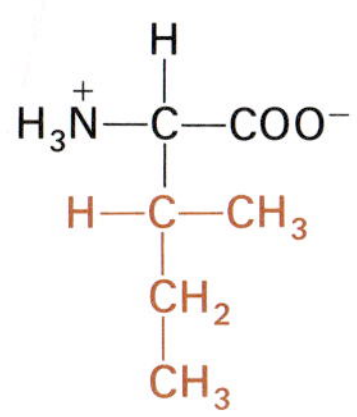

isoleucine
(Ile)

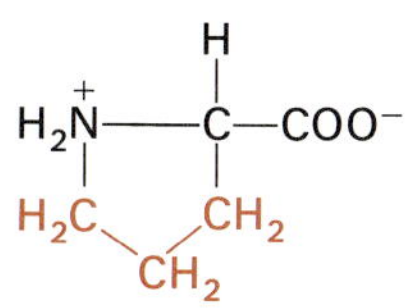

proline
(Pro)

In proline, a three-carbon chain bonds both to the α-carbon and to the amino group, forming a cyclic 2° amine. These six amino acids have nonpolar side chains. Because these side chains are insoluble in water, these amino acids are also known as **hydrophobic** (from the Greek *hydro,* water, and *phobos,* aversion).

2. *Aromatic alkyl groups.* **Phenylalanine, tyrosine,** and **tryptophan** have aromatic side chains:

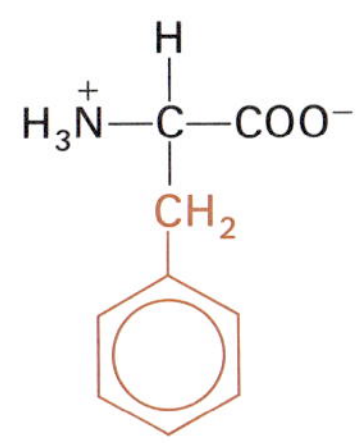

phenylalanine
(Phe)

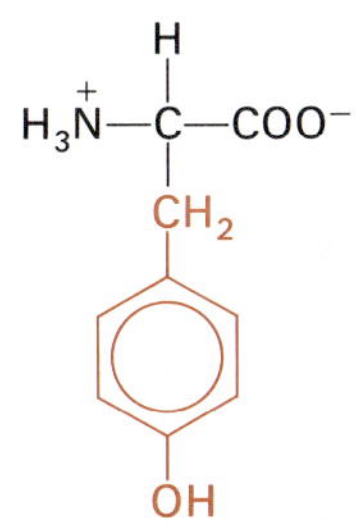

tyrosine
(Tyr)

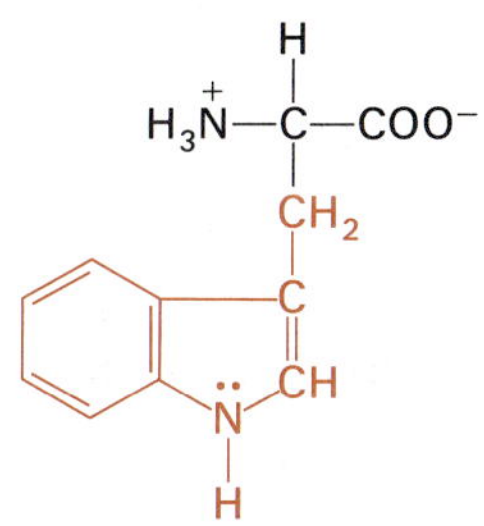

tryptophan
(Trp)

The side chain of phenylalanine is nonpolar and thus hydrophobic. The side chain of tryptophan also is essentially nonpolar because the lone pair of electrons of the ring nitrogen is delocalized over the aromatic system of the rings. (Several resonance structures can be written for tryptophan. A similar delocalization of nitrogen electrons over an aromatic ring is described in Section 20-4). Tyrosine, on the contrary, has a polar side chain that can participate in hydrogen bonds. The phenol group of tyrosine is **hydrophilic** (from the Greek for *water* and *loving*).

3. *Hydroxyl-containing R groups.* **Serine** and **threonine** are in this class together with tyrosine, which we have already listed with the aromatic amino acids.

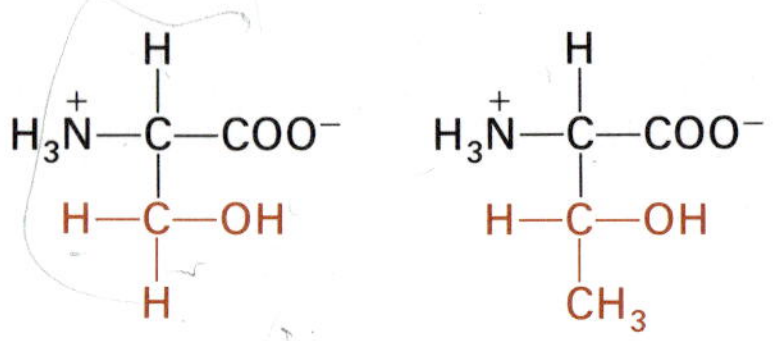

serine (Ser) **threonine (Thr)**

The side chains of these amino acids are polar and hydrophilic. The fact that they can participate in hydrogen bonds has important implications for protein structure, as we shall see.

4. *Basic R groups.* **Lysine, arginine,** and **histidine** have basic R groups:

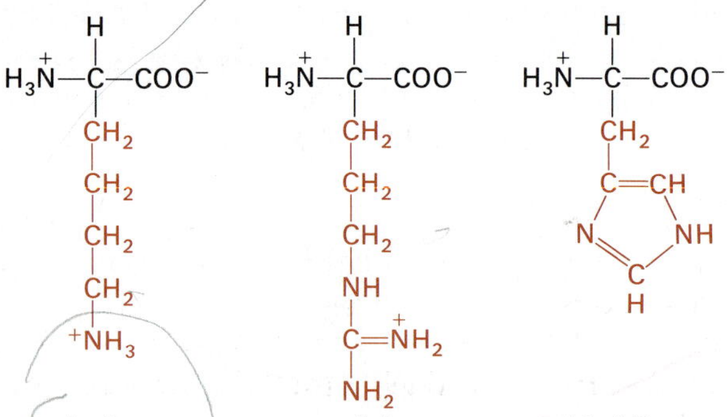

lysine (Lys) **arginine (Arg)** **histidine (His)**

The R groups of the first three types (alkyl, aromatic, and hydroxyl-containing) are uncharged at neutral pH. Lysine and arginine, on the contrary, have strongly basic side chains that are positively charged in neutral solution. The ring of histidine is much less basic than the side chains of lysine and arginine, and near pH 7 both the charged and the uncharged forms are present in significant concentration:

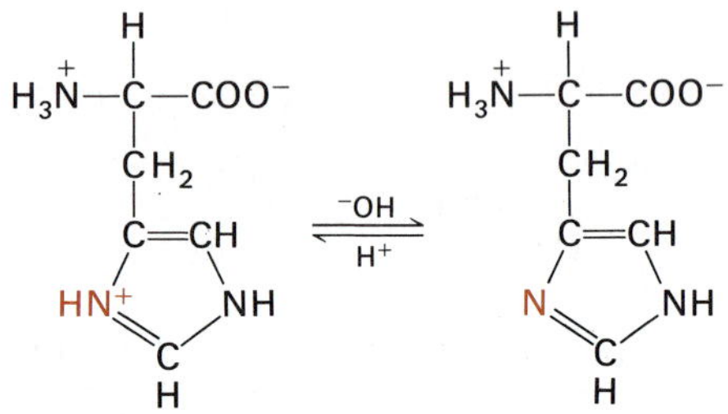

Both forms of histidine occur in proteins, and interconversions between them play a role in certain enzyme mechanisms.

5. *Acid R groups.* **Aspartic acid** and **glutamic acid** are the two amino acids that have negatively charged side chains in neutral solution. Because they exist as negative ions at physiological pH, they are usually called aspartate and glutamate:

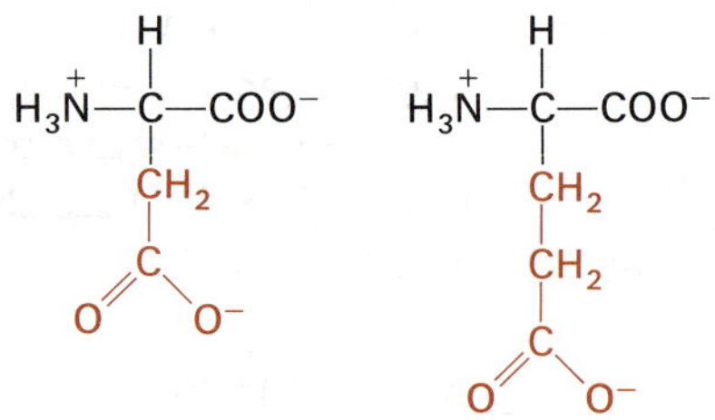

aspartate (Asp) **glutamate (Glu)**

6. *Amide-containing R groups.* **Asparagine** and **glutamine** are derivatives of aspartic acid and glutamic acid, respectively. The side chains of asparagine and glutamine contain polar amide groups instead of the

terminal carboxylate groups of aspartate and glutamate:

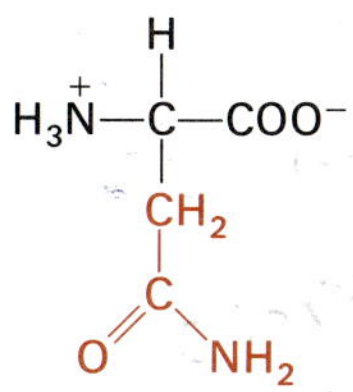

asparagine (Asn)

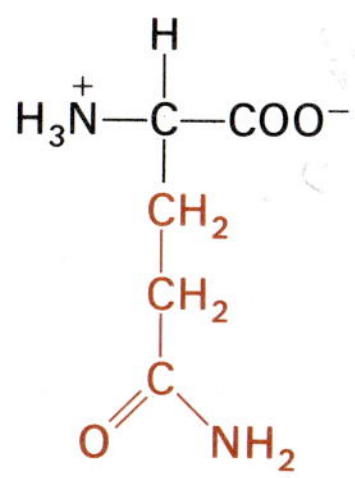

glutamine (Gln)

7. *Sulfur-containing R groups.* **Cysteine** and **methionine** are the two amino acids that contain sulfur:

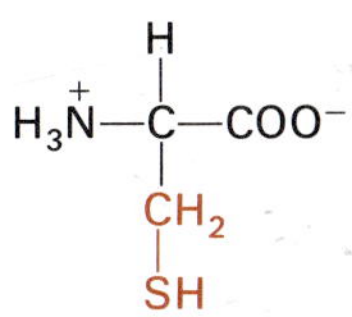

cysteine (Cys)

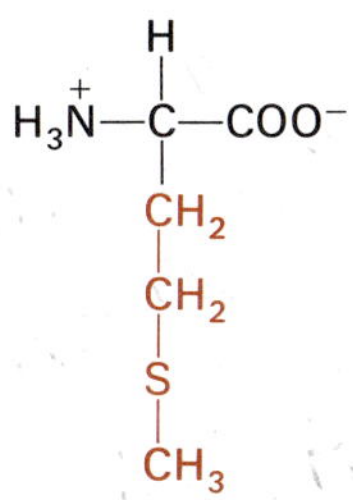

methionine (Met)

The SH group of cysteine is a **thiol** (or **sulfhydryl**) group, the sulfur analog of an alcohol group. The —SH group is a weak acid that ionizes to the $—S^-$ form at pH greater than 8. We shall see shortly that cysteine is unique among the amino acids in its ability to form covalent cross-links. Methionine contains a thioether functional group, the sulfur analog of an ether.

Although proteins are synthesized from only the 20 amino acids we have just discussed, some amino acids are modified after having been incorporated into a protein. In Section 21-6, we shall see that collagen—the fibrous protein of skin, bone, cartilage, tendon, and teeth—contains derivatives of proline and lysine. And an iodinated derivative of tyrosine, **thyroxine,** is found in the protein thyroglobulin, which is synthesized by the thyroid gland:

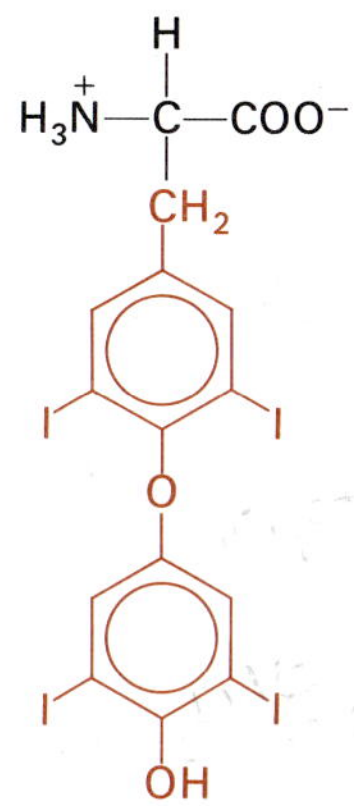

thyroxine

The amino acids can also be characterized from a nutritional point of view. Eight of the amino acids must be supplied to adult human beings in the diet, and they are called **essential** amino acids. The other amino acids can be synthesized by tissues of the body from available metabolic precursors. Table 21-1 lists the 20 amino acids from which proteins are synthesized, and indicates those that are essential amino acids.

Although the main functions of amino acids involve their incorporation into peptides and proteins, certain amino acids have other biological roles as well. One of these is glutamate, which as "monosodium glutamate" (MSG) is a widely used flavor additive, especially common in seasoning Chinese cooking. Too much MSG in food can cause severe headache and facial numbness along with other characteristic symptoms that together have come to be called "Chinese restaurant syndrome." Such observations indicate that glutamic acid may be active in the nervous sys-

TABLE 21-1
The 20 principal amino acids

Name	Abbreviation	Need in human diet
Alanine	Ala	——
Arginine	Arg	Essential to children only
Asparagine	Asn	——
Aspartic acid	Asp	——
Cysteine	Cys	——
Glutamic acid	Glu	——
Glutamine	Gln	——
Glycine	Gly	——
Histidine	His	——
Isoleucine	Ile	Essential
Leucine	Leu	Essential
Lysine	Lys	Essential
Methionine	Met	Essential
Phenylalanine	Phe	Essential
Proline	Pro	——
Serine	Ser	——
Threonine	Thr	Essential
Tryptophan	Trp	Essential
Tyrosine	Tyr	——
Valine	Val	Essential

tem, a hypothesis supported by the fact that normal levels of glutamic acid and of glutamine in the brain are about ten times larger than those of any other amino acids. This and other information on stimulation of the nervous system by glutamic acid suggest that glutamic acid is a natural chemical transmitter within the brain. Ingestion of large amounts may disturb the normal levels of glutamate and related compounds in the brain and thereby cause the symptoms of distress. Fortunately, the symptoms generally disappear shortly after eating and can be avoided altogether by reducing intake of MSG.

A condition associated with the amino acid phenylalanine cannot be so readily reversed unless detected very early. A genetic inability to metabolize phenylalanine properly causes a disease called phenylketonuria (abbreviated PKU). PKU results in the accumulation of high levels of phenylalanine and related compounds in the body. In infants, the excess phenylalanine leads to irreversible brain damage and mental retardation. However, if phenylketonuria is diagnosed at birth, a special diet low in phenylalanine can be designed that will prevent brain damage during the early years of life. After the age of eight, the special diet can be relaxed because the brain is no longer sensitive to excess phenylalanine. Most hospitals now routinely test newborn infants for abnormal levels of phenylalanine to identify cases of PKU early in life when treatment will be successful. The incidence of PKU in newborns is about 1 in 10,000. Many other genetic diseases involving abnormal amino acid metabolism have also been identified, but most occur even less frequently than does PKU.

21-3 ACID–BASE PROPERTIES OF AMINO ACIDS

We have said that amino acids exist predominantly as zwitterions at physiological pH. The zwitterion can act either as an acid or a base in the Brønsted–Lowry system (Section 11-6):

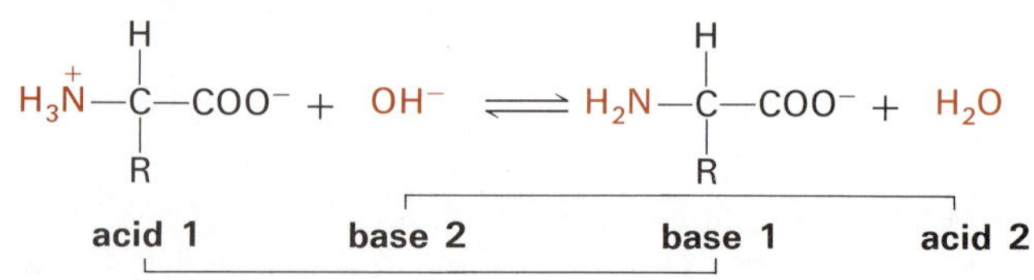

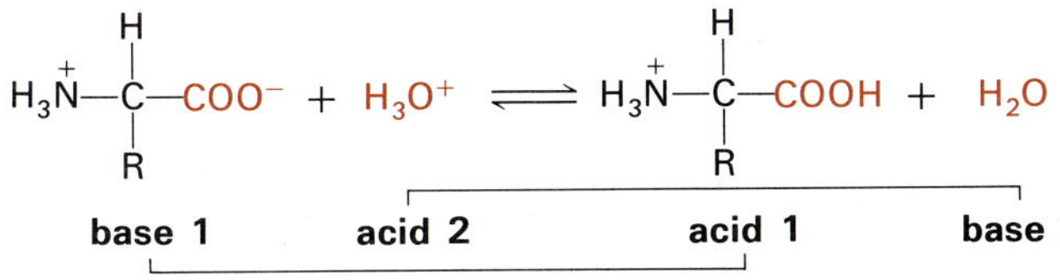

(21-1)

In strongly acid solution, the position of equilibrium of Equation 21-1 lies to the right, and the positively charged form of the amino acid predominates. This positively charged species is a weak acid like acetic acid and has properties typical of a weak acid. For example, it has a characteristic K_a (and pK_a) that reflects its acid strength. Alanine has $K_a = 5 \times 10^{-3}$ M:

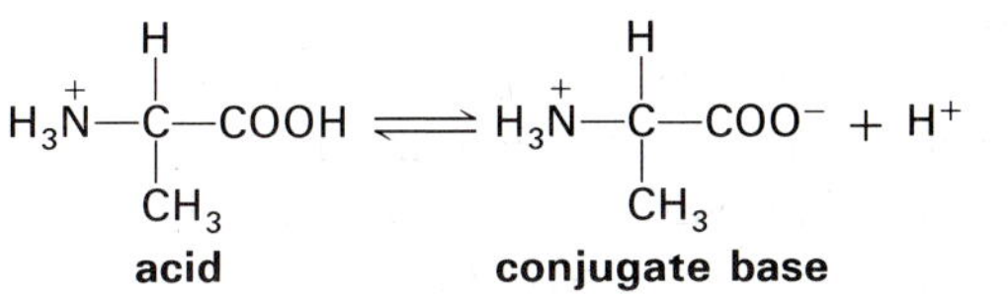

$$K_a = \frac{[H^+][Base]}{[Acid]} = 5 \times 10^{-3}\ \text{M} \qquad pK_a = 2.3$$

The K_a of acetic acid, in contrast, is about 10^{-5}, so the positively charged form of alanine is a much stronger acid than acetic acid. In Section 11-11 we saw that when a weak acid and its conjugate base are both present in significant concentration in a solution, the solution is buffered. Then the pH of the solution remains almost constant in spite of the addition of limited amounts of strong acid or base. When equal concentrations of the positively charged form of alanine and its conjugate base (the zwitterion) are present in solution,

$$K_a = \frac{[H^+][Base]}{[Acid]} = [H^+] = 5 \times 10^{-3}\ \text{M}$$

and $pH = pK_a = 2.3$

In the pH region near pH = 2.3, alanine is a buffer.

Similarly, the zwitterion form of alanine is a weak acid with a characteristic pK_a, different from that of the positively charged form:

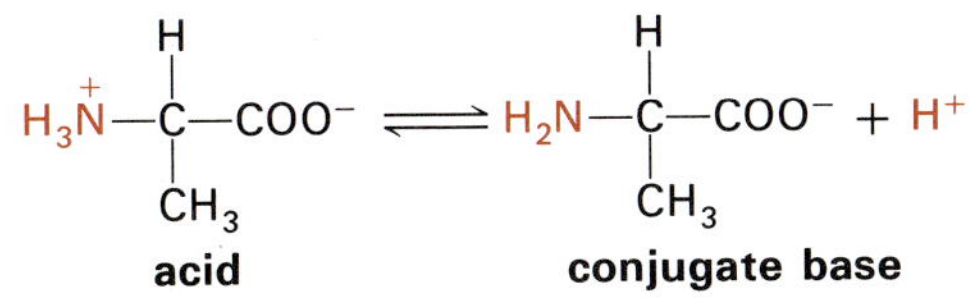

$$K_a = \frac{[H^+][Base]}{[Acid]} = 2 \times 10^{-10}\ \text{M} \qquad pK_a = 9.7$$

When equal concentrations of the acid (the zwitterion) and its conjugate base are present in solution,

$$[H^+] = K_a = 2 \times 10^{-10}\ \text{M}$$

and $$pH = pK_a = 9.7$$

Alanine is also a buffer in the pH region near pH 9.7. In the pH region below pH 2.4, the positively charged species predominates; in the pH region above pH 9.7, the negatively charged species predominates; and in the region between pH 2.4 and 9.7, the zwitterion form predominates.

Table 21-2 gives pK_a values for several amino acids. The amino acids with ionizable side chains have a third pK_a that represents the dissociation of H^+ from the acid form of the side chain. Thus, for lysine, which has a strongly basic side chain, the pK_a for the side chain represents the equilibrium

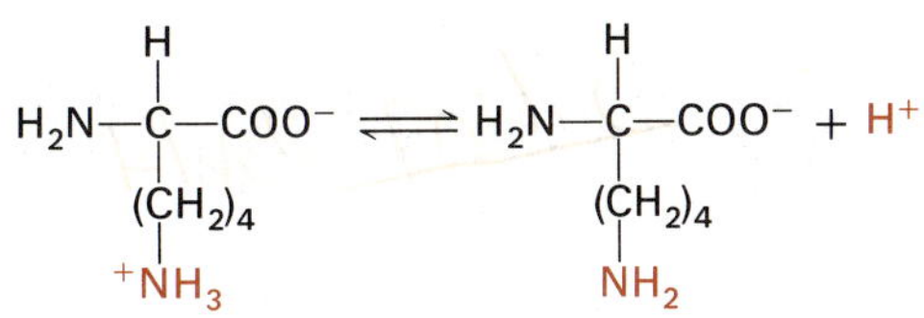

$pK_a = 10.5$

And for aspartic acid, the pK_a for the side chain represents the equilibrium

TABLE 21-2
The pK_a values for some amino acids (25°C)

Amino acid	$pK_a(\alpha\text{—COOH})$	$pK_a(\alpha\text{—NH}_3)$	pK_a(side chain)
Glycine	2.3	9.6	
Alanine	2.3	9.7	
Serine	2.2	9.2	
Aspartic acid	2.1	9.8	3.9
Glutamic acid	2.2	9.7	4.3
Histidine	1.8	9.2	6.0
Lysine	2.2	9.0	10.5
Arginine	2.2	9.0	12.5
Cysteine	1.7	10.8	8.3

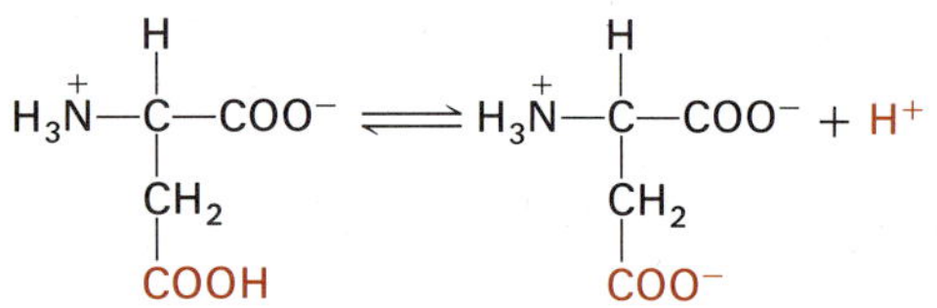

$pK_a = 3.9$

The pK_a for the dissociation of the proton from histidine's side chain is 6.0:

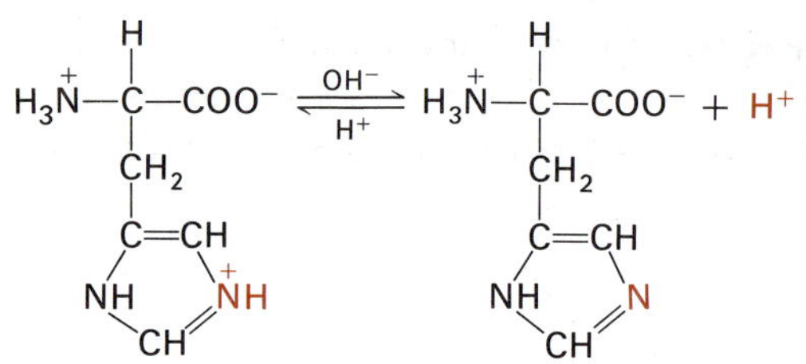

$pK_a = 6.0$

Histidine is, therefore, a buffer in the pH region near 6. Histidine is the only amino acid that has a pK_a close to the physiological pH (see Table 21-2), and thus is the only amino acid that buffers in this pH region. An important consequence of this fact is that proteins with a high percentage of histidine are good buffers in living organisms. In hemoglobin, for example, histidine makes a significant contribution to the buffering capacity of blood for this very reason.

EXERCISE 21-2 (a) In what three pH regions is aspartic acid a buffer?

(b) Which is the stronger acid, aspartic acid or glutamic acid?

(c) Which is the stronger base, lysine or arginine?

21-4 POLYPEPTIDE CHAINS

Amino acids can be linked together by **amide bonds** between the carboxyl group of one amino acid and the amine group of another. Between two amino acids, this amide linkage is also called a **peptide bond.** Although peptide-bond formation in living cells involves activation of the carboxyl group to enhance reactivity, peptide-bond formation can formally be considered as a condensation reaction between the amino group of one amino acid and the carboxyl group of another amino acid, with the elimination of water (see Section 20-7):

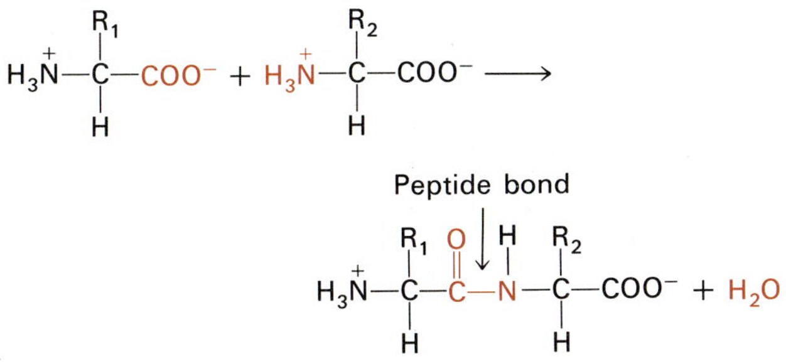

Peptide bonding can give rise to two-unit **dipeptides,** three-unit **tripeptides,** or longer chains. Chains of about 50 amino acid units or less are called peptides; longer chains are called **polypeptides.** A protein is composed of one or more polypeptide chains. Large peptides and polypeptides are synthesized on ribosomes in living cells in sequences specified by the DNA of the genes (Chapter 25). Small peptides are synthesized by specific enzyme systems or can be formed as cleavage products from larger peptides synthesized on ribosomes.

When amino acids are linked in peptides or proteins via peptide bonds, they are properly termed **amino acid residues,** to reflect the fact that their α-amino and α-carboxyl groups are no longer free. For example, the hormone angiotensin II (which causes constriction of blood vessels) is an octapeptide, composed of eight amino acid residues (Figure 21-1). The **backbone** (indicated by color) is made up of regularly repeating units. The **side chains** branch from the backbone.

Instead of writing a complete structure for a peptide or protein, chemists almost always use a shorthand notation that describes the structure in terms of the abbreviations given in Table 12-1. In this notation, the structure of angiotensin is

NH_3^+-Asp–Arg–Val–Tyr–Ile–His–Pro–Phe-COO^-

or simply

Asp–Arg–Val–Tyr–Ile–His–Pro–Phe

By convention, the residue with the free amino group is taken to be the start of the peptide chain; the residue with the free carboxyl group is, then, the end of the chain. The sequence of amino acids is always written from left to right, starting with the amino-terminal residue and finishing

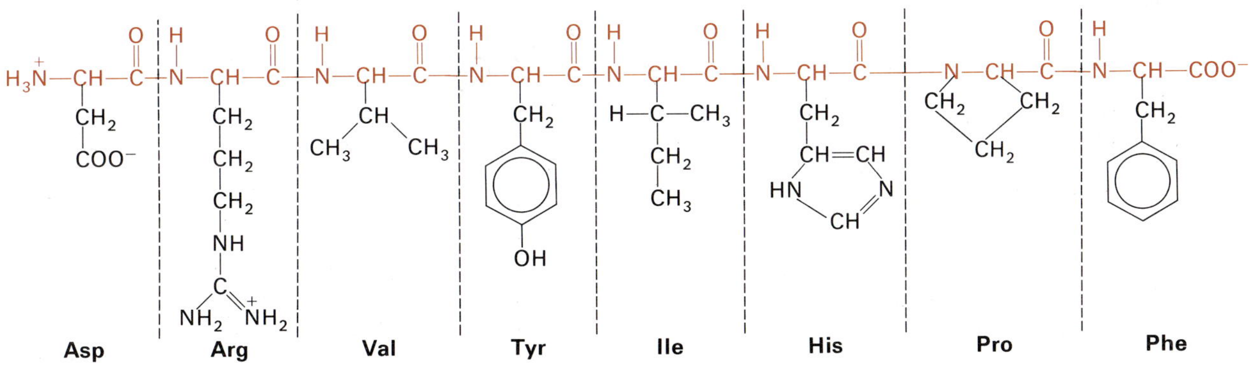

FIGURE 21-1

The hormone angiotensin II, an octapeptide (composed of eight amino acid residues).

TABLE 21-3
Selected hormones

Name	Number of amino acid units	Origin	Action
Angiotensin II	8	Blood plasma	Constriction of blood vessels
Oxytocin	9	Pituitary	Stimulates contraction of the uterus and other smooth muscles
Vasopressin	9	Pituitary	Antidiuretic (decreases the volume of urine excreted)
Somatostatin	14	Hypothalamus	Inhibits production of human growth hormone
Gastrin	17	Stomach	Stimulates acid secretion in stomach
Glucagon	29	Pancreas	Stimulates glycogen metabolism in liver
Adrenocorticotropic hormone (ACTH)	39	Pituitary	Stimulates production of adrenal hormones
Insulin	51 (in 2 chains: 21 + 30)	Pancreas	Controls carbohydrate metabolism
Human growth hormone (HGH)	188	Pituitary	General effects; bone growth
Prolactin	198	Pituitary	Stimulates lactation
Follicle stimulating hormone (FSH)	202 (89 + 113)	Pituitary	Stimulates sperm production and follicle maturation

with the carboxyl-terminal residue, as we have just done. In precise terminology, the names of the amino acid residues should be modified to include a *-yl* suffix, to reflect the incorporation into peptide linkages. Thus, the third position of the angiotensin II peptide is occupied by a "valyl residue." However, in common usage the distinction between amino acid residues and amino acids is often ignored, and we speak of "valine at the third position." Table 21-3 describes hormones that range in size from 8 to 202 amino acid residues.

The physiological properties of angiotensin and all other peptides and proteins depend not only on what amino acids are present, but also on the particular **sequence,** the order in which the amino acid residues are linked together. Because they have two different functional groups, any two amino acids can combine to form two different dipeptides. Histidine and phenylalanine, for example, can form the dipeptides Phe–His and His–Phe:

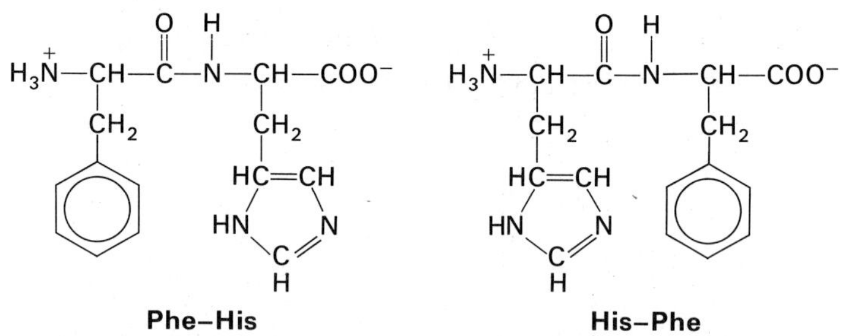

Phe–His (melting point: 173°C)

His–Phe (melting point: 214°C)

These dipeptides are different compounds, with different chemical and physical properties.

As the number of amino acid residues in a peptide increases, the number of possible variations in sequence increases tremendously. Three different amino acids, for example, can form six different tripeptides. Although sometimes a change in the amino acid sequence of a protein has little or no effect on its ability to function properly, at other times an error in a single amino acid residue can have serious consequences. Sickle-cell anemia is caused by a single mistake in the amino acid sequence of hemoglobin (Section 28-3).

EXERCISE 21-3 Using shorthand notation to describe the amino acid sequence, write the structures of the six different tripeptides that can be formed from valine, leucine, and tyrosine.

A particularly interesting class of peptides with biological activity are the enkephalins (from the Greek, "in the head"). This recently discovered group of peptides binds at the opiate receptors of the brain, the sites in the brain that bind morphine and other sleep inducers and pain relievers. Thus, the enkephalins may be nature's own painkillers. Enkephalins, composed of five amino acid residues, have been isolated in two forms, with either methionine or leucine at the carboxyterminal position:

Methionine-enkephalin:

$$NH_3^+\text{–Tyr–Gly–Gly–Phe–Met–COO}^-$$

Luecine-enkephalin:

$$NH_3^+\text{–Tyr–Gly–Gly–Phe–Leu–COO}^-$$

Many other opiate-type peptides have now been discovered or synthesized in the laboratory; peptides with opiate activity have been given the general name **endorphins.**

EXERCISE 21-4 (a) Name the following structure using the shorthand notation for naming peptides:

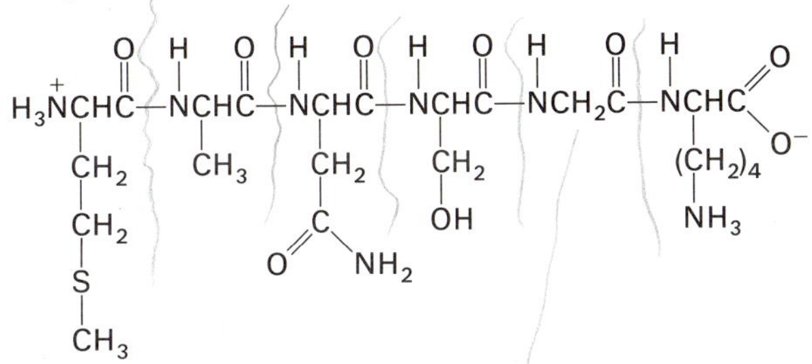

(b) Write the structure of leucine-enkephalin showing all peptide bonds and side chains.

21-5 DISULFIDE BONDS

Although peptides and proteins are formed by covalent linkages of amino acids principally via peptide bonds, a second type of covalent bonding is found between cysteine residues. **Disulfide bonds,** also known as S—S bonds or S—S bridges, are formed by the oxidation of two cysteine residues:

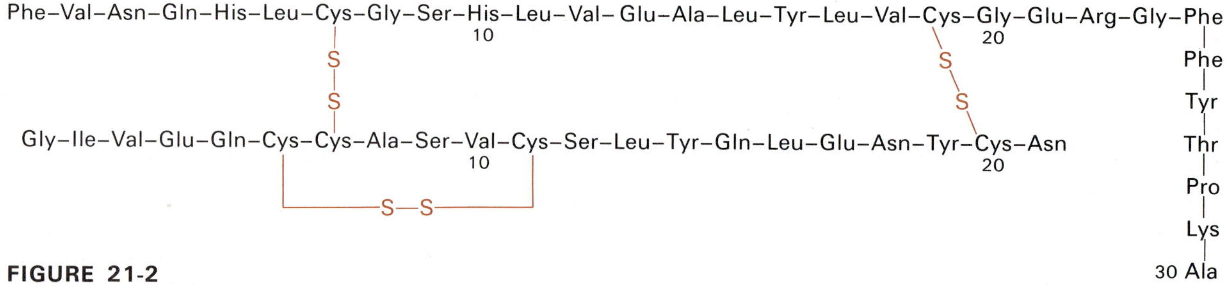

FIGURE 21-2

The amino acid sequence of insulin.

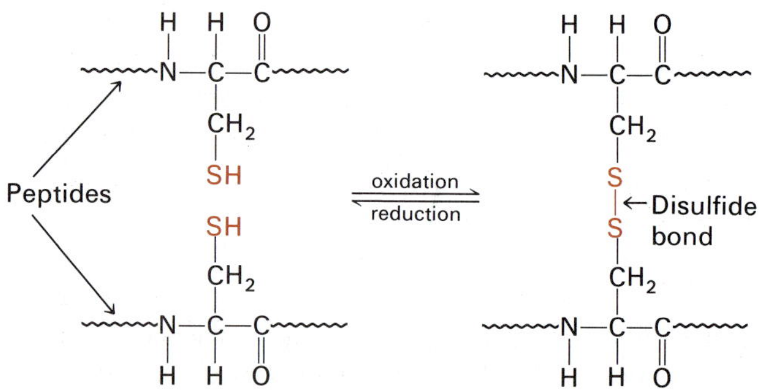

When a disulfide bond is formed between two molecules of the free amino acid cysteine, the product is **cystine.**

Disulfide bonds are found in several hormones, in many digestive enzymes, and in antibody molecules; most proteins, however, do not contain disulfide bonds. When present, disulfide bonds contribute to the three-dimensional structure or folding pattern of the protein chain. They generally involve cysteine residues some distance apart in the linear sequence of amino acids or on different polypeptide chains. Figure 21-2 shows that amino acid residues in the two polypeptide chains of insulin are linked by two disulfide bridges. A third disulfide bridge links two cysteine residues within a single chain.

EXERCISE 21-5 The pituitary hormone oxytocin, which causes contractions of the uterus in childbirth, contains a disulfide linkage. The amino acid sequence of oxytocin is Cys–Tyr–Ile–Glu–Arg–Cys–Pro–Leu–Gly · NH_2. Write the complete structure of oxytocin, clearly showing the disulfide bridge. The abbreviation Gly · NH_2 means that glycine is present as the amide rather than as the acid.

21-6 PROTEIN STRUCTURE

Polypeptide chains with more than about 50 amino acids are considered proteins. The proteins encountered in biochemistry vary enormously in size, shape, and function. They range in size from insulin with 51 amino acids to giant proteins such as the enzyme β-galactosidase, which is composed of four identical polypeptide chains, each with 1021 amino acids, to give a total molecular weight of about 465,000.

We shall see that there is a close relationship between the three-dimensional structure of a protein and its biological activity. Enzymes (proteins that catalyze biochemical reactions) typically are compact and roughly spherical, a shape that is termed **globular.** Hemoglobin also is globular. Certain other proteins—such as keratin in hair and collagen in tendons and bone, which function as structural elements and provide mechanical support—are long, rodlike molecules with a shape that is termed **fibrous.** In some cases, the same protein molecule may have regions that are globular and others that are fibrous.

Before we consider the factors that determine the shapes of proteins, we briefly review the special properties of the peptide bond.

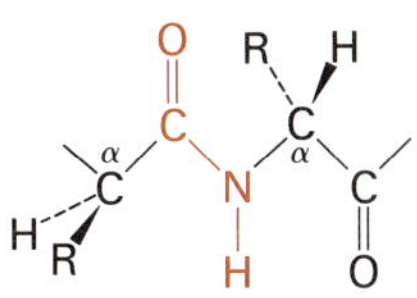

planar peptide group

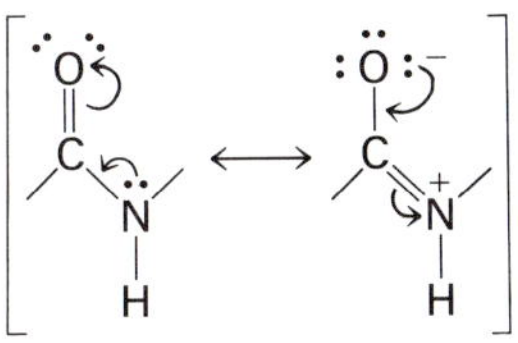

resonance structures

Because the peptide group (CONH) can exist in the two resonance structures shown, both the bond between the carbonyl carbon and oxygen and that between the carbonyl carbon and nitrogen are intermediate between double and single bonds. And because both bonds have some double-bond character, *the peptide group is planar, and rotation about the carbon–nitrogen bond is restricted.* The bond between the α-carbon and the carbonyl carbon and that between the α-carbon and nitrogen, on the contrary, are ordinary single bonds. There is relatively free rotation of groups about these bonds.

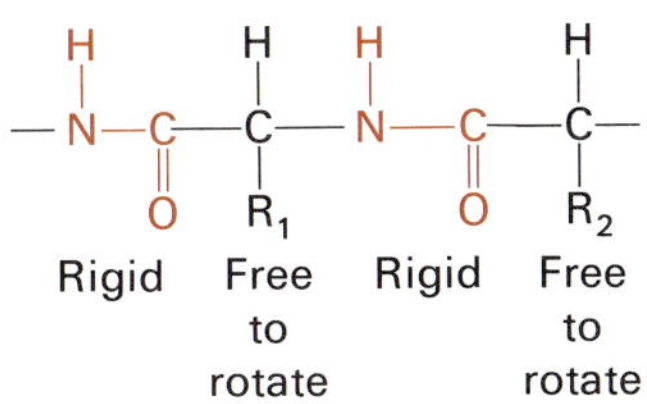

These characteristics of the peptide group are important in determining the three-dimensional structures of proteins.

Because of the great diversity in their sizes and shapes, it is difficult to generalize about protein structures. Nevertheless, certain observed patterns suggest some generalizations. X-ray crystallography (Section 8-5) has been particularly valuable in discovering the structural features common among proteins. This technique has been used to determine the detailed three-dimensional structure (at the atomic level) of dozens of proteins. Apparently each protein has a distinct three-dimensional structure, arising from a particular folding of its polypeptide chain. Some flexibility in the conformation of the protein may be possible (and may even be essential to its function), but the basic pattern of chain folding is rather rigid and is distinct for each protein.

Once we recognize that a protein has a unique conformation, we can begin to specify levels of structure. The most elementary level is the sequence of the amino acid residues in the polypeptide chain. For any particular protein, the sequence of amino acids and the location of any disulfide bridges constitute the **primary structure.** The primary structure thus describes all the covalent bonds that link the amino acid residues.

But the long peptide chain of a protein does not exist in a completely random conformation; instead, much of the chain is folded or wound in a particular, regular way into coils or sheets. **Secondary structure** refers to these regular patterns of organization. The principal examples of secondary structure encountered in proteins are the **alpha** (α) **helix** and the **beta pleated** (β-pleated) **sheet.** (The details of these structures are described in Section 21-8.)

A polypeptide chain, which may be composed of regions of α helix, β-pleated sheet, and various irregular regions, is itself folded, bent, or wrapped into a three-dimensional **tertiary structure.** Tertiary structure describes the spatial relationships of amino acid residues that are widely separated in the extended polypeptide

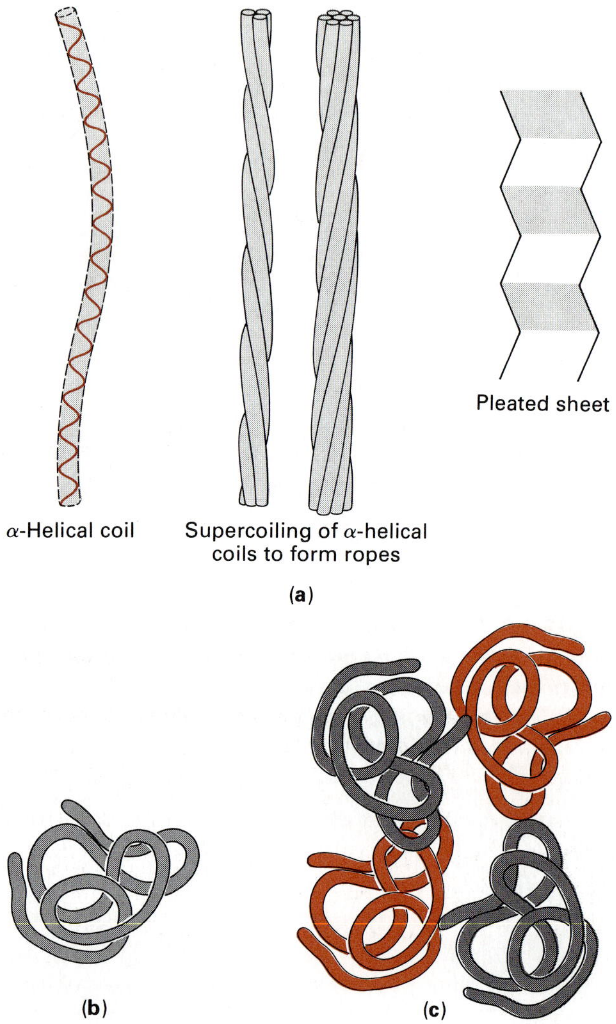

FIGURE 21-3

A schematic illustration of secondary, tertiary, and quaternary structure of a protein. **(a)** Secondary structure refers to a regular coil or sheet arrangement of polypeptide chains. **(b)** Tertiary structure refers to the folding of a coil into a compact globular shape. **(c)** Quaternary structure refers to the spatial arrangement of the separate polypeptide chains of an oligomeric protein.

chain, but that are close together when the chain is folded to form a globular protein. The dividing line between secondary and tertiary structure is arbitrary, and these terms are sometimes ambiguous—the general idea, however, is illustrated schematically in Figure 21-3.

Many proteins contain more than one polypeptide chain; these are known as **oligomeric** proteins. In many cases, the several polypeptide chains that make up the protein are identical, as they are in β-galactosidase, but there are also many important proteins that contain different polypeptide chains. The three-dimensional arrangement of the separate polypeptide chains of a multichain protein is the protein's **quaternary structure.** An oligomeric protein described later in this chapter is hemoglobin, which contains two polypeptide chains **(subunits)** of a type designated alpha, plus two subunits of a type designated beta.

We have said that the covalent bonds linking amino acid residues in a protein describe the primary structure, but we have not yet discussed the forces that account for the secondary, tertiary, or quaternary structure of a protein. These forces, often called **noncovalent forces,** are weaker than covalent bonds. They include hydrogen bonds, ionic forces called **salt linkages** that are the same as forces that exist between Na^+ and Cl^- ions, and hydrophobic interactions. **Hydrophobic interactions** result from the cohesiveness of water—the tendency of water molecules to form hydrogen bonds with each other and to exclude nonpolar molecules with which water cannot form hydrogen bonds. Nonpolar side chains of polypeptides minimize their contact with water by clustering together. The polypeptide chain folds in such a way that the nonpolar groups are located in the internal region of a globular protein. Thus hydrophobic amino acids like valine, leucine, and phenylalanine tend to be located in the relatively nonpolar interior of a protein. Water is largely excluded from this region.

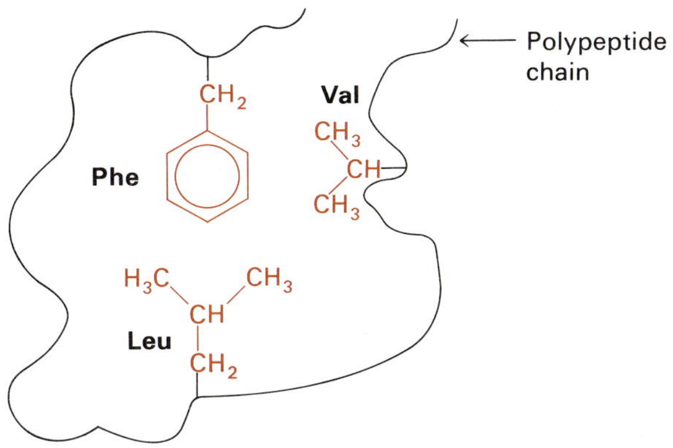

hydrophobic interactions

In Section 21-8, we shall see that hydrophobic forces are important in determining the shape of myoglobin, the oxygen-binding protein of muscle.

Although a single hydrogen bond is not very strong, the tremendous numbers of hydrogen bonds that occur in proteins account for the stabilities of the α helix and the β-pleated sheet. The secondary structure of a protein depends almost exclusively on hydrogen bonds between the C=O of one peptide group and the N—H of another. Hydrogen bonds can also form between any two side chains that have a hydrogen donor group such as OH or NH_2 and a hydrogen receptor (an oxygen or nitrogen):

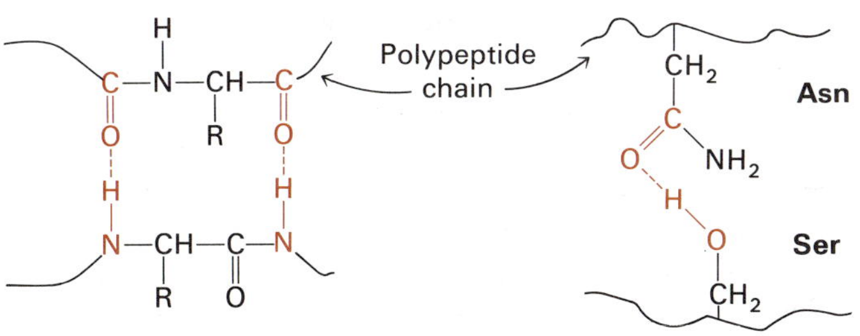

Hydrogen bonds between peptide groups

A hydrogen bond between two side chains

In contrast to amino acids with hydrophobic side chains, which are buried inside a protein, those with polar hydrophilic side chains that can form hydrogen bonds with water often are found on the surface of proteins:

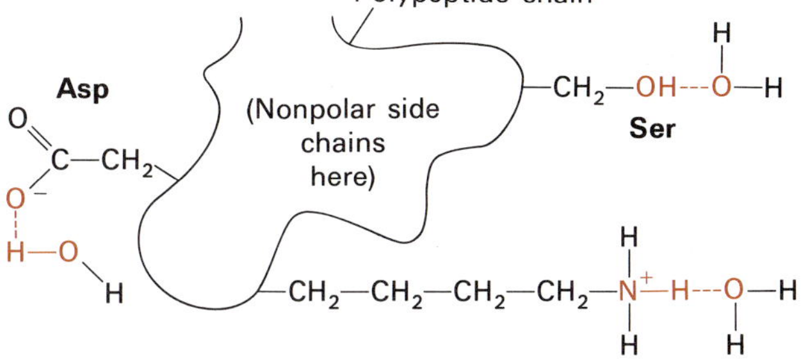

EXERCISE 21-6 (a) List three amino acids (different from those given in the text) that are likely to be found on the inside of a globular protein.

(b) List three amino acids that are likely to be found on the outside of a globular protein.

Salt linkages, which are ionic bonds, can form between a side chain with a positive charge and one with a negative charge—between aspartate and lysine, for example:

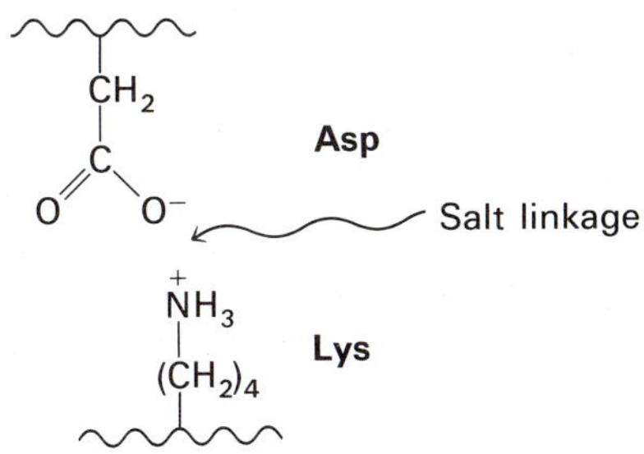

The oppositely-charged groups attract each other just as Na^+ and Cl^- do. In contrast, side chains with like charges repel each other.

EXERCISE 21-7 List two pairs of amino acids (other than the pair just given) that can participate in salt linkages.

21-7 HYDROLYSIS AND DENATURATION OF PROTEINS

The primary structure of a protein is destroyed in the body by digestion (hydrolysis) with enzymes, and in the laboratory by hydrolysis with acids and bases. As we saw in Section 20-8, an amide can be hydrolyzed by heating with a moderately concentrated solution of a strong acid or base. If the pentapeptide leucine-enkephalin is hydrolyzed with 6 M HCl, the products are the acid forms of the amino acid residues:

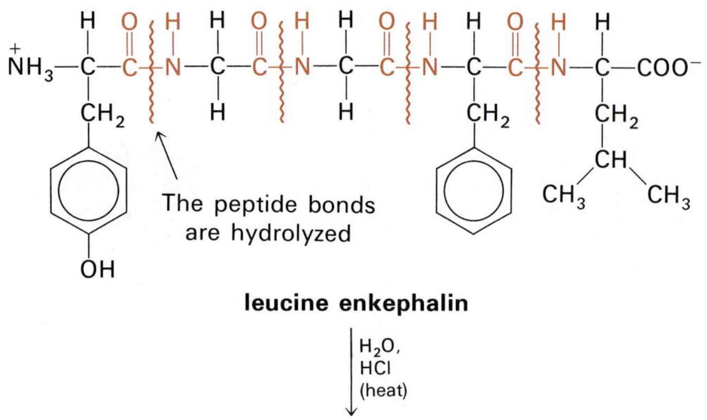

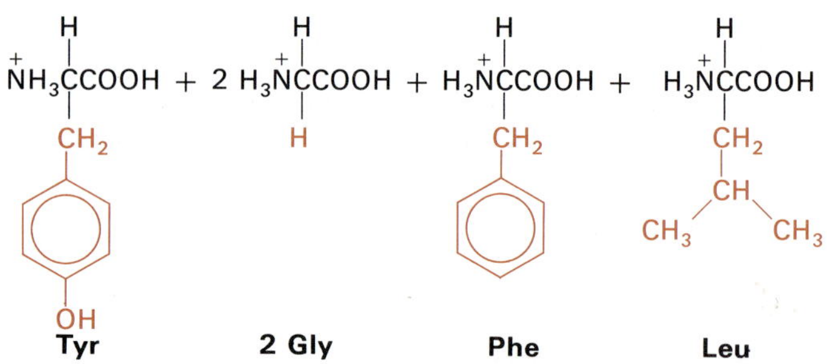

If the forces that stabilize protein secondary, tertiary, and quaternary structure are counteracted, a general unfolding of the protein structure occurs. This process called **denaturation** converts the ordered three-dimensional structure of a protein into a disordered structure and results in loss of the protein's biological activity. In a fully denatured protein, the polypeptide chain has no ordered three-dimensional structure, but is a **random coil.** The primary structure of a protein, however, is retained on denaturation. Treatments that tend to denature proteins are high temperature, pH extremes, and certain solvents such as solutions of alcohol, acetone, urea, or detergents (which interfere with hydrogen bonds or salt linkages). However, there are many exceptions to this generalization. Some proteins, for example, are stable at pH extremes or at high temperature. Certain bacteria, for example, live in hot sulfur springs at a temperature of 95°C. Still other proteins are denatured by low temperature. In the disordered or denatured state, proteins are generally less soluble than they are in the native state, and they may precipitate out of solution. For very concentrated solutions of proteins, denaturation can lead to formation of a solid gel—as in the hard boiling of eggs, where high temperature denatures the proteins of the egg.

EXERCISE 21-8 Many of the solvents that denature proteins are thought to operate by disrupting hydrogen bonds. When 70% ethyl alcohol is used as a disinfectant, for example, it kills bacteria by denaturing protein. Explain how ethyl alcohol might disrupt the hydrogen bonds of a protein.

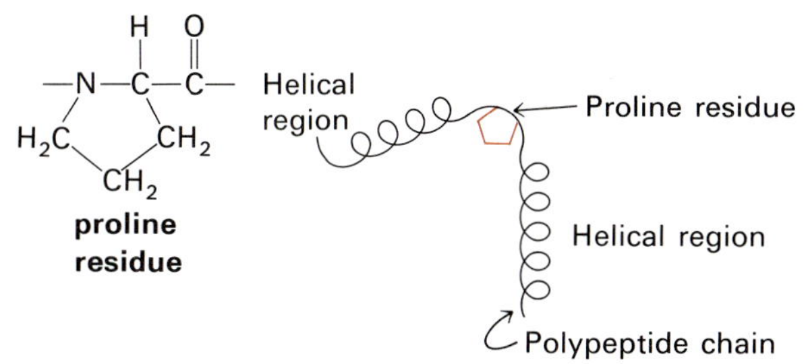

One of the key ideas of contemporary protein chemistry is that the primary structure of a protein dictates its secondary, tertiary, and quaternary structure. In other words, once a protein is synthesized in a cell, the chain of amino acid residues spontaneously folds into a particular secondary and tertiary structure determined by noncovalent interactions between the various portions of the polypeptide chain. As the individual polypeptide chains assume their natural conformation, disulfide bonds may form—this sets up further associations and may bind the individual chains in a quaternary structure. One of the major challenges to modern protein chemists is to try to understand the interactions that determine conformation well enough to be able to predict the secondary and tertiary structure from knowledge of primary structure. Some progress has been made in this direction, primarily based on the recognition that each amino acid has a characteristic tendency either to participate in or to destabilize an α helix or a β sheet. It is known that glutamate is very likely to occur in an α helix but is not likely to be found in a β-pleated sheet. Tyrosine, on the contrary, is likely to occur in a sheet, but not in a helix. And proline breaks the regular structure of both the sheet and the helix. Its cyclic structure limits the possible conformations of the polypeptide chain around the prolyl residue.

The idea that the primary structure of proteins determines their secondary, tertiary, and quaternary structures is supported by the fact that some proteins, after being denatured or unfolded into random coils with no biological activity, spontaneously return to their original, biologically active forms (Figure 21-4).

We next consider in more detail the regular structures of the α helix and the β-pleated sheet, and of a third regular structure—the triple helix of collagen.

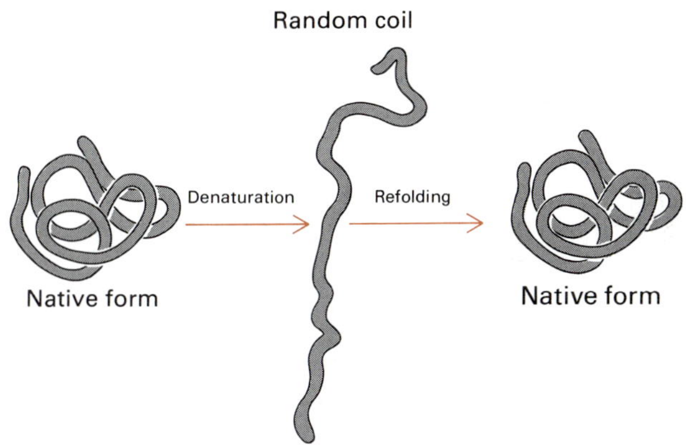

FIGURE 21-4

Denaturation and renaturation of a protein. Some proteins spontaneously resume their native shapes after having been denatured.

21-8
SECONDARY STRUCTURE OF PROTEINS: THE α HELIX, THE β-PLEATED SHEET, AND THE TRIPLE HELIX

Even before the first three-dimensional structure of a protein was determined by X-ray analysis, chemists proposed two possible regular structures for polypeptides: the α helix and the β-pleated sheet. The **α helix** (Figure 21-5) is a right-handed corkscrewlike structure in which the polypeptide chain winds around an axis in a spiral. One of the major features of the α helix is the presence of *hydrogen bonds* between the peptide nitrogen atoms on one turn of the helix and the carbonyl oxygen on the turn above. The C=O group of each amino acid residue is hydrogen-bonded to the NH group of the amino acid residue four residues ahead, as shown at the top of the next page. Only these groups are in proper position for hydrogen bonding. Although a single hydrogen bond is relatively weak, the cumulative effect of the large number

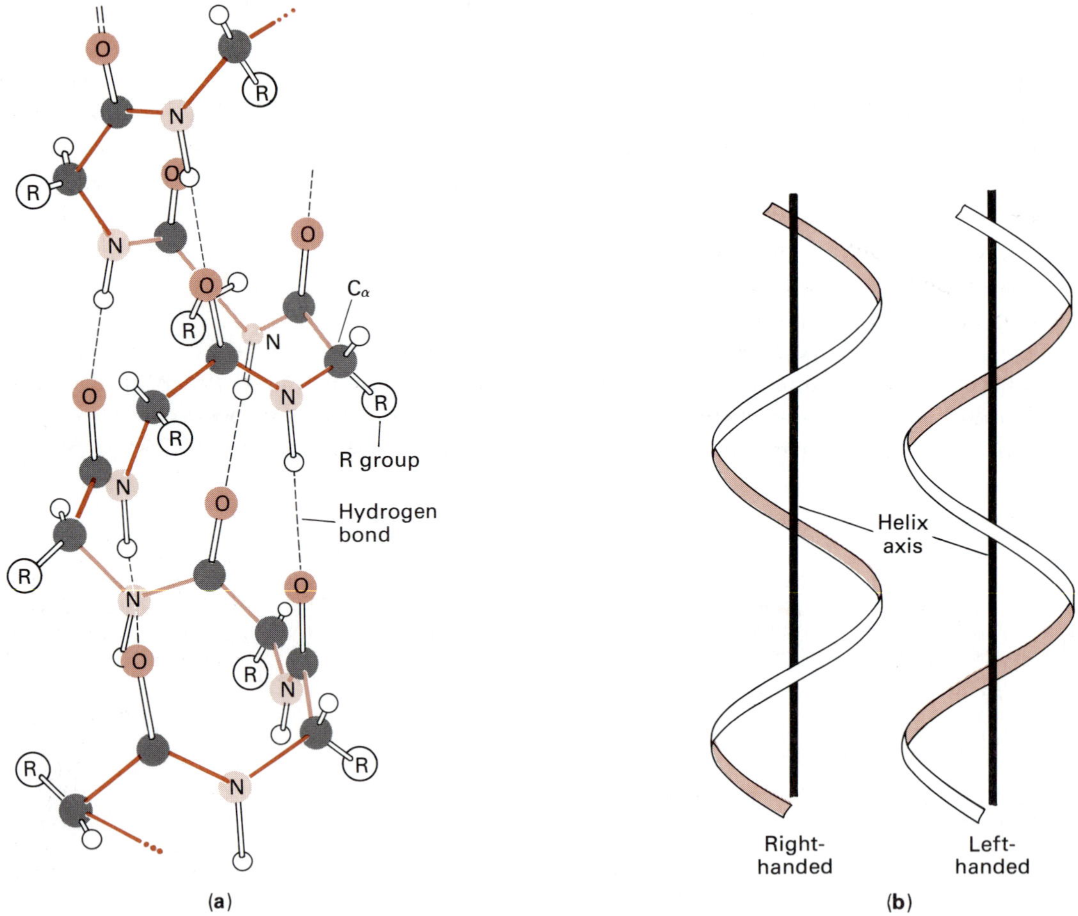

FIGURE 21-5

Helical structures. **(a)** The α helix (right-handed) **(b)** Helix handedness.

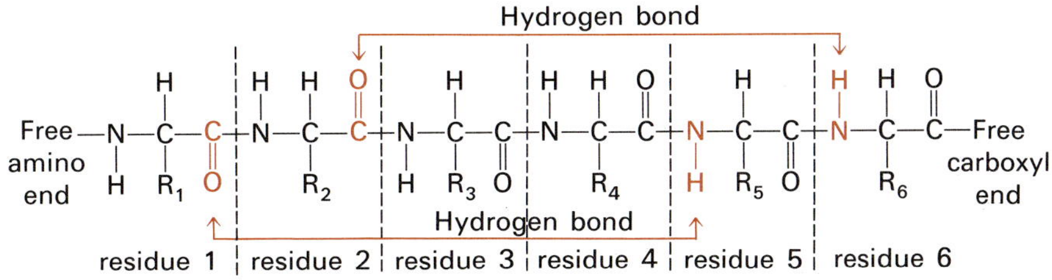

of hydrogen bonds in the α helix accounts for the strength and stability of this helix. Although α helices of both the left-handed and right-handed variety can exist in principle, only the right-handed helix places all of the side chains of L-amino acids on the outside of the helix. The left-handed form, in contrast, places the side chains of L-amino acids on the inside of the helix, where there is not enough room for them.

The first structure of a globular protein to be solved by X-ray crystallography was achieved in 1959 for **myoglobin,** the oxygen-binding protein of muscle. Myoglobin consists of 153 amino acid residues, about 75% of them arranged in eight stretches of α helix (Figure 21-6). The regions between the helical ones are random coil. (There are four proline residues in myoglobin; they account for some, but not all, of the breaks in the helix.) The heme group, a nonprotein group that binds the oxygen, fits into a crevice in the myoglobin. The iron atom in the heme is bonded to a histidine in the myoglobin. Heme is a **prosthetic group**— a nonprotein molecule that is tightly bound to a protein and is required for the protein's biological activity. We shall see other examples of prosthetic groups when we discuss vitamins and coenzymes in Chapter 22.

The myoglobin molecule is folded in such a way that essentially all side chains in the interior of the molecule, including those close to the heme group, are nonpolar ones such as leucine, valine, and tryptophan. The only exceptions are two histidines: one of them is

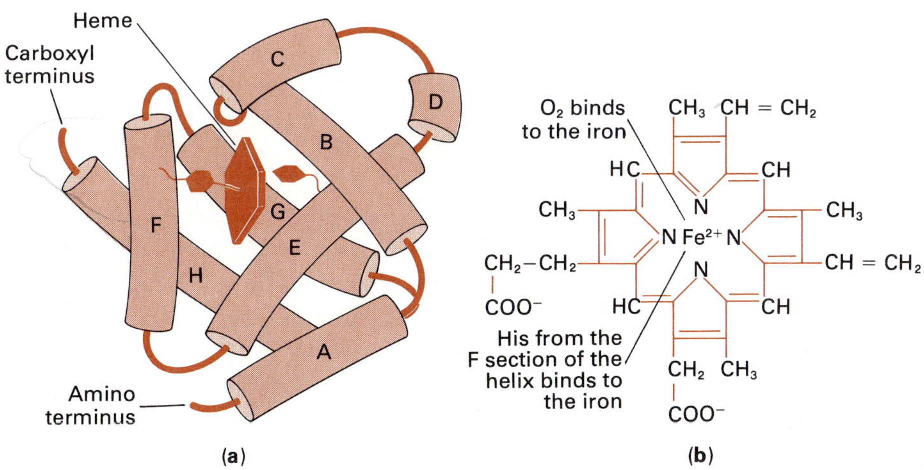

FIGURE 21-6

The structure of myoglobin. **(a)** The eight separate α helices in myoglobin are labeled A through H. The heme group and two histidine residues from the polypeptide chain are shown in color. **(b)** The structure of heme. The heme group, with its system of conjugated double and single bonds, is responsible for the red color of blood.

bonded to the iron atom of the heme, and the other is close to the heme (Figure 21-6a). Charged hydrophilic groups—including glutamate, aspartate, lysine, and arginine—are on the outside of the molecule where they can hydrogen-bond to water molecules. Thus, the way the myoglobin molecule is folded (its tertiary structure) is determined by these noncovalent interactions.

Hemoglobin, the oxygen-carrying protein of the blood, consists of four subunits held together in a particular quaternary structure by noncovalent interactions (Figure 21-7). Hydrophobic interactions are most important in holding the subunits together, but hydrogen bonds also contribute, and there are several salt linkages. There are two each of two types of polypeptide chain, called alpha and beta. Both the alpha and the beta subunits are very similar to myoglobin.

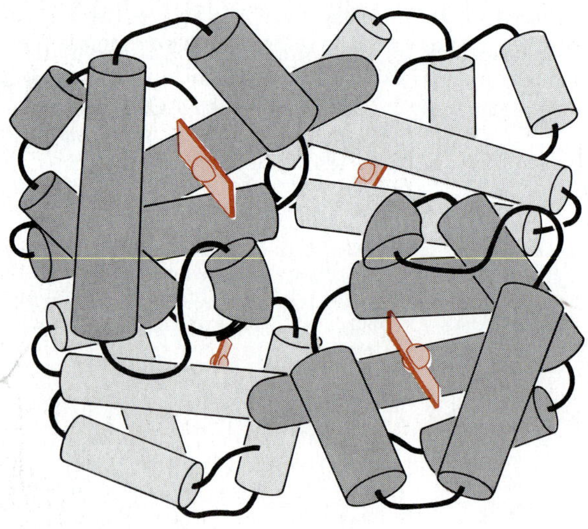

FIGURE 21-7

Arrangement of the four subunits of hemoglobin. The molecule is compact and nearly spherical. Each subunit is similar to the myoglobin molecule.

Each of the four subunits contains a heme group. Chapter 22 discusses the binding and release of oxygen by hemoglobin and myoglobin.

The first studies of protein structure suggested that the α helix is the principal type of secondary structure, but later studies revealed the α helix to be less common than first thought. For example, the first enzyme structure to be determined (that of lysozyme, an antibacterial substance found in mucus and tears) showed only a small fraction of residues in α helices. Large stretches of β-pleated sheet were more prominent.

The α helix places hydrogen bonds between atoms of residues that are near neighbors along the polypeptide chain. In contrast, the **β-pleated sheet** structure involves hydrogen bonds between groups that are widely separated in the polypeptide chain (Figure 21-8). The β-pleated sheet structure can also involve entirely separate polypeptide chains. Usually the regions of the polypeptide chain that participate in the β-sheet structure are aligned in opposite directions (antiparallel) like those shown in Figure 21-8.

Large regions of the β-pleated sheet structure occur in fibroin, the protein of silk. The strength of silk fibers is due to the tremendous number of hydrogen bonds that hold the chains together. Silk is resistant to stretching because the polypeptide chains are in an extended or stretched conformation rather than coiled in a helix. The most common form of silk fibroin has glycyl side chains (H atoms) on one side of the sheet, and primarily alanyl side chains (CH_3 groups) on the other. The β-sheet structure of silk is very similar to the structure of nylon (Figure 20-5).

Collagen, an important fibrous protein, is found in connective tissue, skin, bone, and teeth. It is the most abundant protein in mammals, comprising over one-quarter of the total protein. The protein collagen is made up of structural units of a protein called tropocollagen. A tropocollagen molecule consists of three polypeptide chains, each of about a thousand amino acid residues, twisted together in a long **triple helix.**

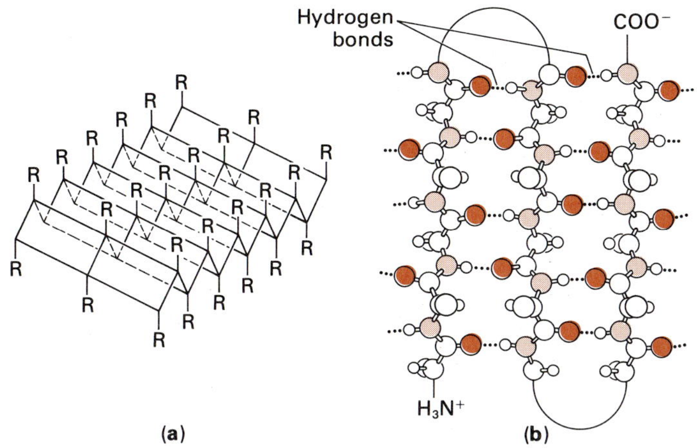

FIGURE 21-8

The β-pleated sheet. **(a)** A schematic representation of three parallel polypeptide chains, showing the pleated-sheet arrangement. **(b)** A ball-and-stick model showing hydrogen bonds between polypeptide chains, or between segments of the same polypeptide chain. Oxygen atoms are shown in color, and nitrogen atoms are shaded.

Figure 21-9 is an electron micrograph of collagen fibrils from skin, and Figure 21-10a illustrates the triple helix.

Glycine residues, which constitute nearly one-third of collagen, are critical for the structure of tropocollagen because only the H atom (glycine's side chain) is small enough to fit inside the triple helix. The three strands of the triple helix are held together by hydrogen bonds (and in old collagen, also by covalent cross-links). If a solution of collagen is heated, the hydrogen bonds break, and the triple helix disintegrates into randomly coiled polypeptide chains. Heating denatures collagen and converts it to gelatin.

The triple helices of tropocollagen molecules associate in the pattern shown in Figure 21-10b to form tough, rigid collagen fibers that are strengthened by covalent cross-links. The greater the number of cross-links, the more rigid the fiber. Collagen fibers are so strong that a fiber only 1 mm in diameter can support a load of more than 10 kg. Adjacent parallel rows of tropocollagen molecules are displaced about one-quarter of the length of a single molecule to form the staggered array shown in Figure 21-10b. The individual tropocollagen molecules in a row are not bonded together end-to-end. Instead there is a gap between the ends of two adjacent molecules. This gap is thought to be involved in bone formation. Bone consists primarily of collagen and hydroxylapatite,

FIGURE 21-9

Electron micrograph of collagen fibrils obtained from skin. [Courtesy of Jerome Gross.]

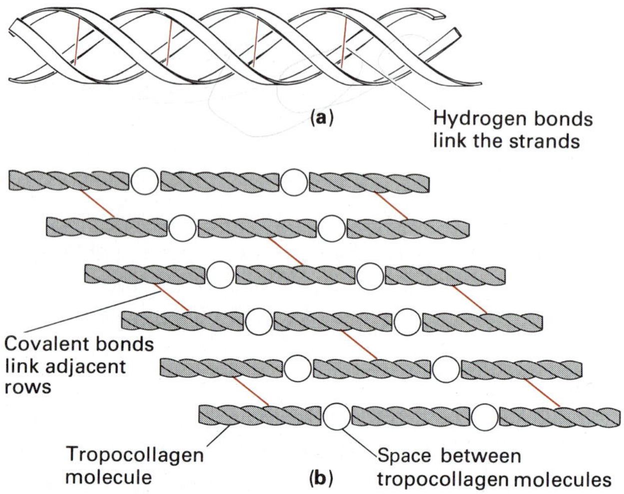

FIGURE 21-10

The structure of collagen. **(a)** A tropocollagen molecule with triple-stranded structure. Hydrogen bonds link the stands. **(b)** A schematic illustration of a collagen fiber. Each ropelike section represents a tropocollagen triple helix. Covalent cross-links join adjacent rows. The circles represent gaps between the tropocollagen molecules. The staggered arrangement of these gaps gives rise to the spiral grooves evident in the fibrils of Figure 21-9.

$Ca_{10}(PO_4)_6(OH)_2$ (see Section 10-8). It may be that the gaps between the tropocollagen molecules are sites where the hydroxylapatite begins to crystallize.

Collagen contains two unusual amino acids that are not incorporated when the protein is synthesized. Instead, hydroxyproline and hydroxylysine are formed by the hydroxylation of proline and lysine after the polypeptide chain is formed.

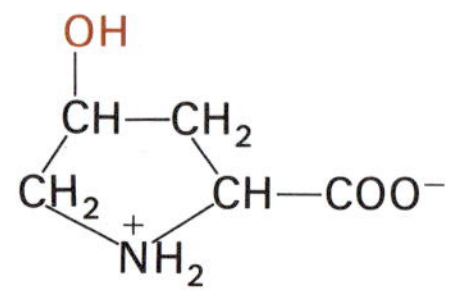

hydroxyproline

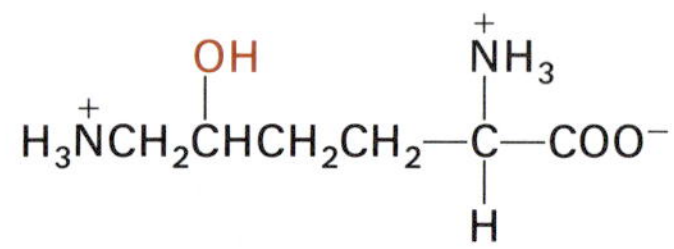

hydroxylysine

Studies with radioactive C-labeled proline and hydroxyproline established the fact that the proline is hydroxylated after incorporation in the polypeptide chain. If hydroxyproline containing radioactive ^{14}C is fed to rats, none is found in the collagen. If, however, proline containing ^{14}C is fed to rats, the collagen is radioactive.

A reducing agent such as ascorbic acid (vitamin C) is required for the hydroxylation of proline and lysine. Scurvy—a disease characterized by bleeding gums, spotting of the skin, and prostration—results from vitamin C deficiency, which prevents proper formation of collagen.

21-9 DETERMINATION OF AMINO ACID COMPOSITION OF PROTEINS

We have discussed the structures of several peptides and proteins, but we have not yet seen how their amino acid compositions and sequences are determined. An important step in characterizing proteins and peptides isolated from biological sources is that of finding which amino acids are present in the sample and in what quantities. For such a determination, the protein must first be hydrolyzed to break the peptide bonds and give the free amino acids. There are several methods for hydrolyzing peptide bonds. Among them are treatment with strong acid or strong base and digestion with proteins that cleave peptide bonds (**proteolytic enzymes**). Each method has advantages and disadvantages related to yields and to the destruction of certain amino acids. The most widely used method is hydrolysis in 6 N HCl at 100°C for 24 hours, in the absence of oxygen. Oxygen is excluded in order to minimize oxidative destruction of the amino acids. Once the free amino acids are produced, their identity and quantity can be determined.

Column chromatography (Section 8-1) is used to separate the amino acids. The column is filled with an ion-exchange resin—a relatively inert polymer such as polystyrene that is bonded to charged groups. The charged groups can be either positive or negative, and for amino acid analysis a resin with a negatively charged group such as sulfonate (SO_3^-) is used. A cation (usually Na^+) is associated with the SO_3^-:

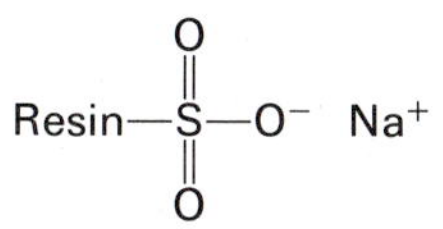

The acidic amino acid mixture from the protein hydrolysis is applied to the column. The amino acids are in their acid form and so are positively charged. The amino acids displace some of the Na^+ ions and stick to the column:

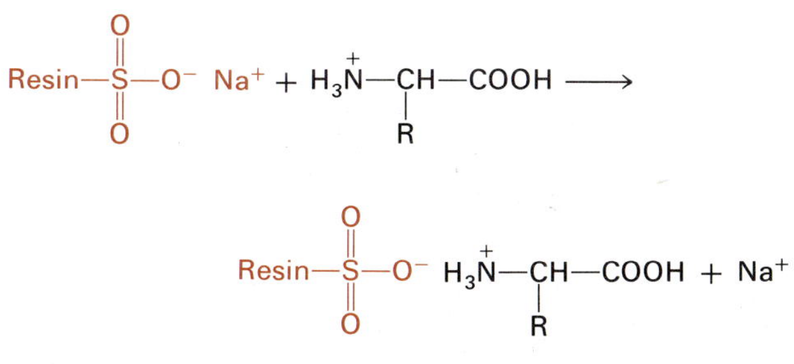

The amino acids with basic side chains (lysine and arginine) are most tightly bound to the resin; and the amino acids with acidic side chains (glutamic and aspartic acids) are least firmly bound.

Several buffers are passed in turn through the column—each of slightly higher pH than the last. This procedure causes the bound amino acids to dissociate protons from their carboxylic acid groups to yield carboxylate ions ($—COO^-$). Each amino acid dissociates protons at a slightly different pH. When the negatively charged carboxylate ion is formed, the positive charge of the amino group binding that amino acid to the resin is counteracted, and the amino acid is released and flushed from the column.

Thus the different amino acids emerge from the column in a particular order—each at a particular pH. The strongest acid (aspartic acid) comes off first, and the strongest base (arginine) comes off last. The amount of each amino acid in the column washings is determined by spectrophotometric analysis of the purple product of ninhydrin reacting with the amino acid:

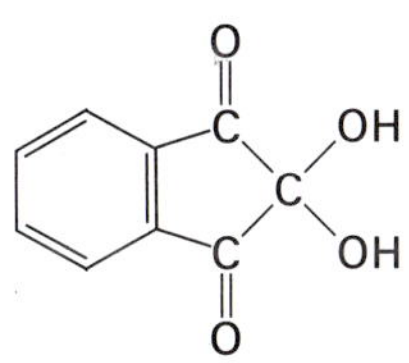

ninhydrin

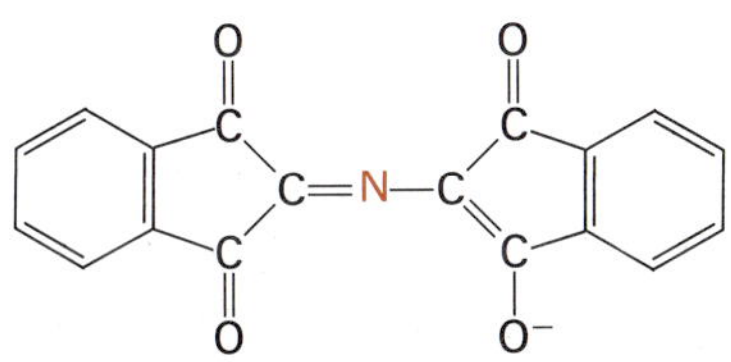

purple product (the N atom comes from the amino acid)

(The system of conjugated double and single bonds is responsible for the purple color of the product; see Section 20-5.) The absorbance of light in the blue region of the spectrum, measured by spectrophotometer, is proportional to the concentration of amino acid in the solution. The entire procedure has been automated, and a complete amino acid analysis can be completed in only a few hours in a device called an amino acid analyzer. Figure 21-11 illustrates the output of an amino acid analyzer for the peptide bradykinin, a nonapeptide (found in the blood) that helps regulate blood pressure.

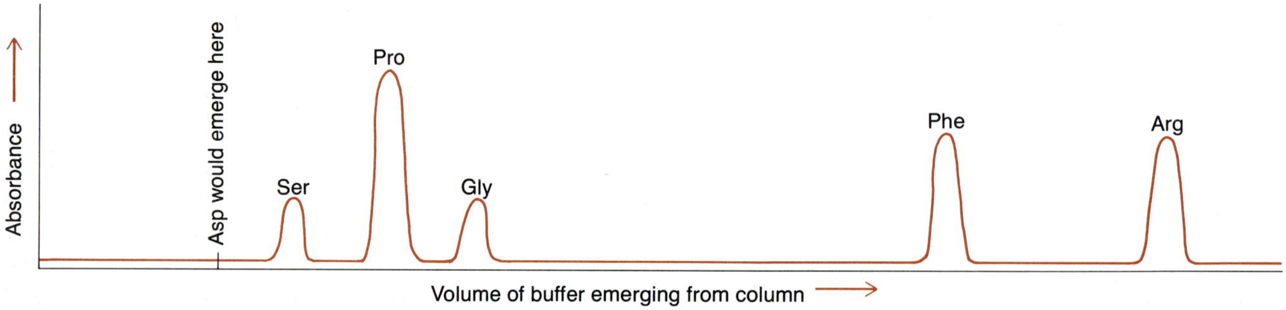

FIGURE 21-11

Amino acid analysis for the nonapeptide bradykinin:

Arg–Pro–Pro–Gly–Phe–Ser–Pro–Phe–Arg

The area under the curve for each amino acid is proportional to the amount of that amino acid in the mixture.

21-10 DETERMINATION OF AMINO ACID SEQUENCE

Hydrolysis of a protein for analysis of its amino acid content yields only the *composition* of the protein. In order to solve many important biological problems, the *sequence* of the amino acids in the protein (the primary structure) is also required. Sequence information is more difficult to obtain, especially for large proteins, because it requires breaking the polypeptide chain into smaller peptide units, determining the amino acid sequence of each of these peptides, and then deducing the order of the peptides in the original protein. The first complete protein sequence determination was accomplished by Frederick Sanger in 1956 for the hormone insulin, which has 51 amino acid residues. Since Sanger's work, hundreds of other protein sequences have been determined, and a number of valuable techniques have been developed that make sequencing a relatively straightforward process.

The first step in sequencing a large protein is to cleave it into smaller peptides. Specific chemicals or enzymes are used, and we soon shall see how they are used. The smaller peptides are separated by chromatographic methods, and the sequence of amino acids in each peptide is determined by the **Edman method** of sequencing proteins. This method uses a special reagent that removes amino acids in sequence, one residue at a time, from the amino-terminal end of a peptide. The reagent, phenylisothiocyanate, reacts with the amino-terminal end of the peptide to form a phenylthiocarbamoyl (PTC) derivative:

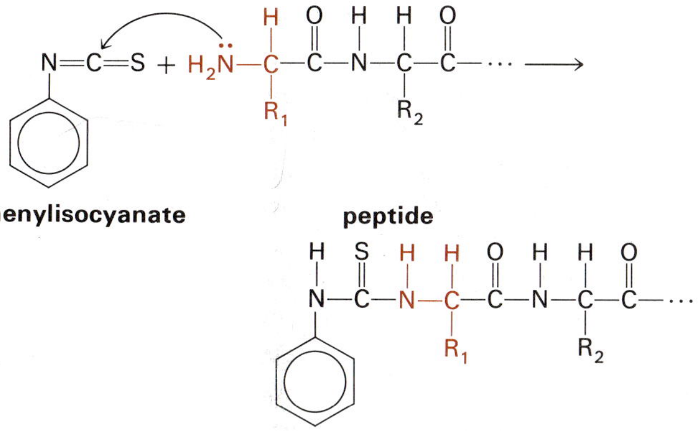

phenylisocyanate

peptide

PTC derivative

The PTC derivative can be hydrolyzed in dilute acid, under conditions that do not affect the peptide bonds. The products are a cyclic phenylthiohydantoin (PTH) derivative of the amino acid residue at the amino-terminal end of the peptide chain, plus the remainder of the peptide intact:

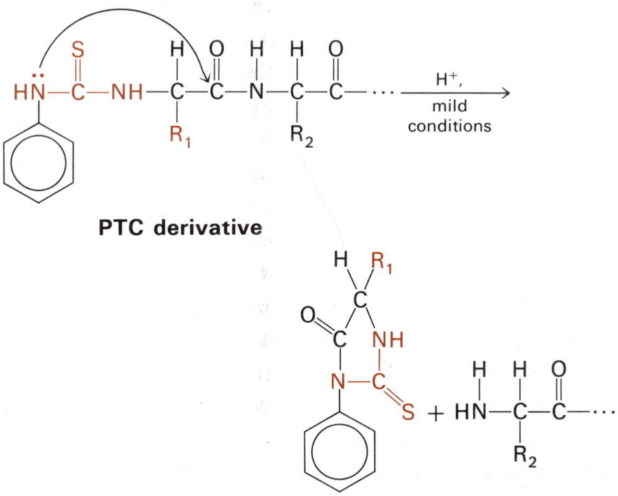

PTC derivative

PTH-amino acid

remainder of peptide

The PTH can be identified by chromatographic methods, and the remaining peptide can be subjected to another cycle of the Edman reagent to identify the second amino acid. The cycles of reaction and cleavage to remove the N-terminal amino acid residue can be repeated many times. In this way the sequence of peptides up to 40 or more amino acids in length can be determined. Such analyses are now carried out routinely with an automated device known as a sequenator.

How is a large protein cleaved into smaller peptides that can be sequenced by the Edman method? One way is to use particular proteolytic enzymes (or **proteases**) that hydrolyze polypeptide chains at specific residues. The two most commonly used proteases are the pancreatic enzymes *trypsin* and *chymotrypsin.* Both have high degrees of specificity for particular amino acids. Trypsin, for example, catalyzes the hydrolysis of polypeptides only on the carboxyl side of arginine and lysine residues:

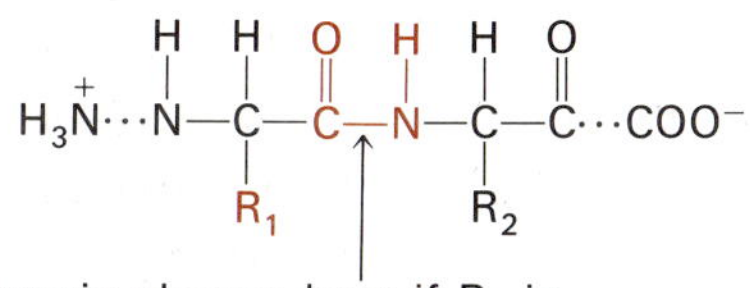

Trypsin cleaves here if R_1 is the side chain of Lys or Arg; R_2 can be any amino acid side chain.

Similarly, chymotrypsin cleaves mainly peptide bonds on the carboxyl side of one of the aromatic amino acids phenylalanine, tyrosine, or tryptophan:

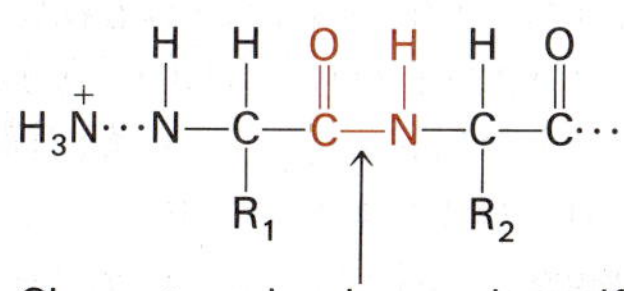

Chymotrypsin cleaves here if R_1 is the side chain of Phe, Tyr, or Trp; R_2 can be any amino acid side chain.

A number of other proteases that are less specific than trypsin and chymotrypsin also are available. Carboxypeptidases cleave peptide bonds at the carboxyl-terminal end of peptides; aminopeptidases cleave peptide bonds at the amino-terminal end of the peptides. A chemical reagent, cyanogen bromide (CNBr), specifically hydrolyzes peptide bonds on the carboxyl side of methionine residues.

Now we are ready to look at the entire process by which a large protein can be sequenced. First, the protein is cleaved into smaller peptides—say, by trypsin—and the amino acid sequence of each peptide is determined. But the order of the separate peptides in the whole protein is not yet known. If a second set of peptides is obtained from the original protein by a different protease—say, chymotrypsin—then *overlaps* in the composition of the two sets of peptides can be used to establish their order. The following example illustrates how this is done.

EXAMPLE 21-1

A nonapeptide is hydrolyzed and found to be composed of one each of the following amino acid residues (listed in alphabetical order): Ala, Arg, Ile, Leu, Lys, Met, Phe, Ser, Tyr. Hydrolysis with trypsin gives two smaller peptides and a free amino acid:

Peptide 1: Leu–Phe–Met–Arg

Peptide 2: Ile–Ser–Ala–Lys

Free amino acid: Tyr

Hydrolysis with chymotrypsin gives two smaller peptides:

Peptide 3: Met–Arg–Tyr

Peptide 4: Ile–Ser–Ala–Lys–Leu–Phe

What is the amino acid sequence of the original nonapeptide?

From the tryptic hydrolysis, we conclude that Tyr must be at the carboxyl-terminal end of the original nonapeptide. Tyr was left as a free amino acid following hydrolysis at the carboxyl side of Arg or Lys. But we cannot yet tell whether peptide 1 or peptide 2 is at the amino-terminal end of the original nonapeptide.

Chymotryptic peptide 4 overlaps the tryptic peptides and thus provides the additional information needed to establish the amino acid sequence of the original nonapeptide:

tryptic peptide 2 ⟵⟶ tryptic peptide 1 ⟵⟶

$H_3\overset{+}{N}$–Ile–Ser–Ala–Lys–Leu–Phe–Met–Arg–Tyr–COO^-

⟵⟶ chymotryptic peptide 4 ⟵⟶ chymotryptic peptide 3

(Chymotryptic peptide 3, together with tryptic peptide 1, confirms that Tyr is at the carboxyl-terminal end.)

In practice, the amino acid sequence of a peptide (of up to 40 or 50 amino acid residues) would be determined by the Edman method. The principles illustrated in the example, however, are those used to determine the sequence of peptides in a large protein. One of the first proteins to be sequenced by the methods outlined above was the enzyme **ribonuclease,** an enzyme that hydrolyzes RNA. Its sequence with 124 amino acids and four disulfide bonds (Figure 21-12) was completed in 1962. Since then, the sequences of hundreds of other proteins have been determined, including some extremely long ones. The primary structure of the enzyme β-galactosidase (which contains 1021 amino acids in each of four identical polypeptide chains) was reported in 1977.

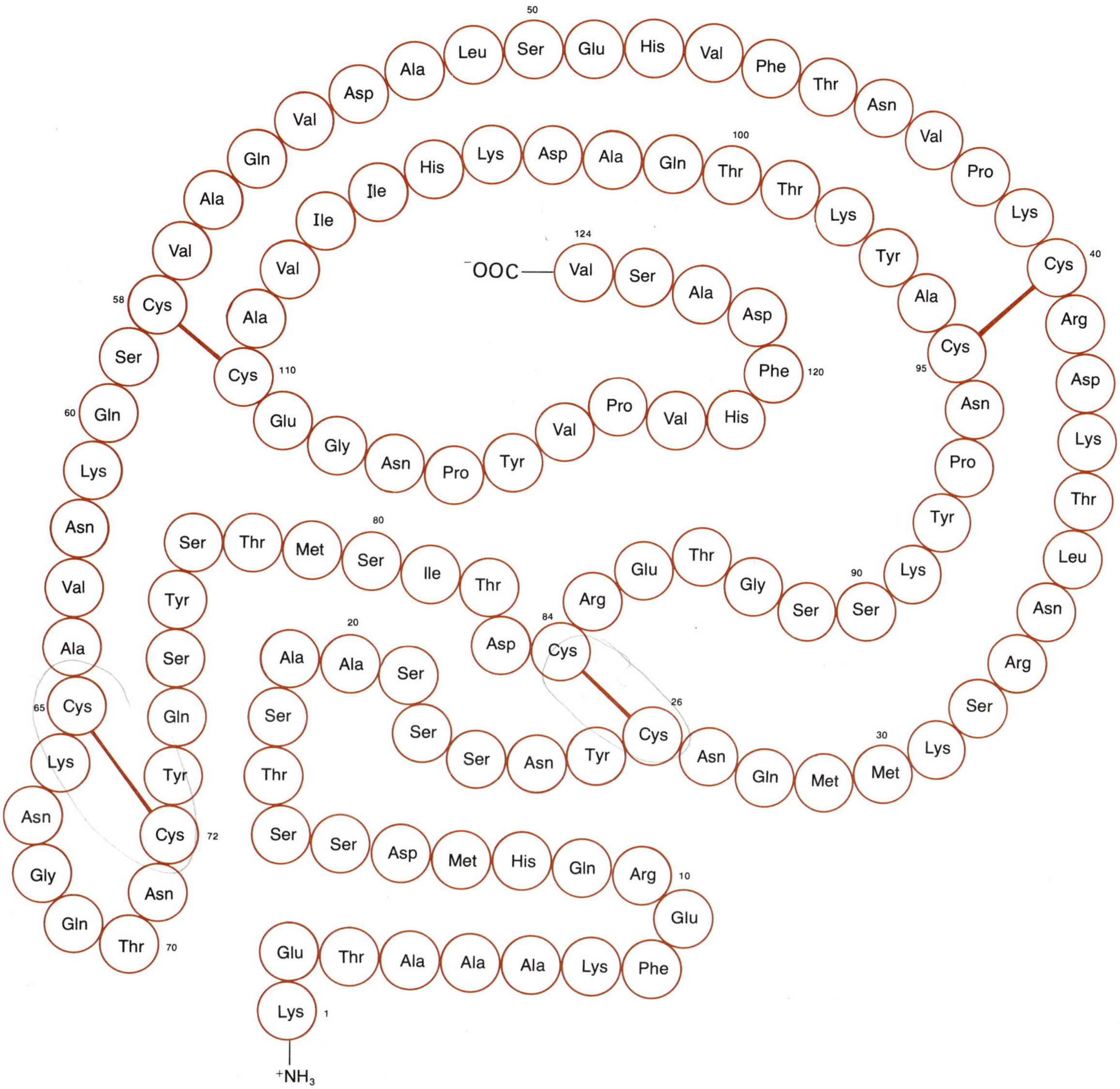

FIGURE 21-12

Amino acid sequence of ribonuclease, an enzyme that hydrolyzes RNA.

Once the amino acid composition and the primary structure of a protein are known, then the secondary, tertiary, and quaternary structures are studied. As we have said, X-ray diffraction (described in Section 8-5) is used to determine the three-dimensional positions of most of the atoms. A crystalline sample of the protein is required for the X-ray studies, and often it is very difficult to grow protein crystals. Crystals of myoglobin like those shown in Figure 21-13 are obtained by adding a high concentration of a salt (often ammonium sulfate) to a solution of the protein. This procedure reduces the solubility of the protein and causes it to crystallize. Using the methods we have described and similar ones, the three-dimensional structures of about a hundred proteins have been established.

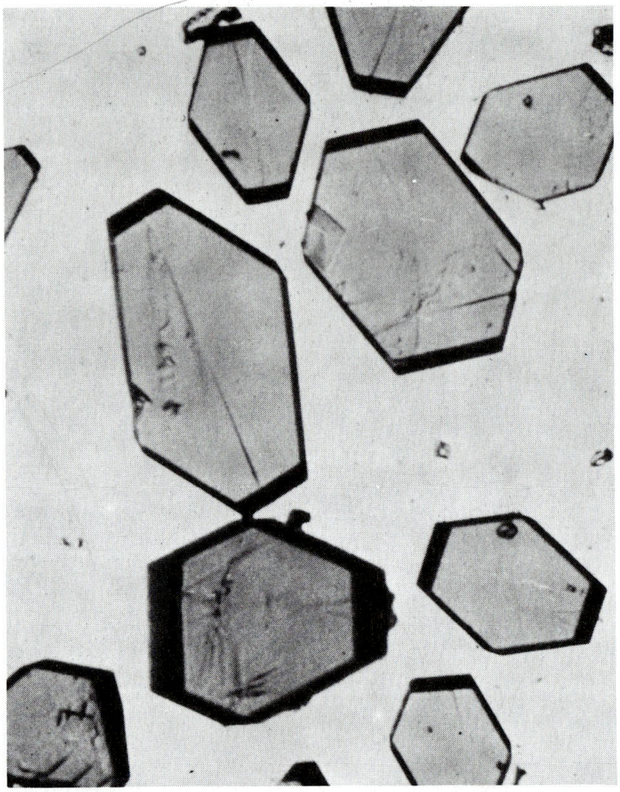

FIGURE 21-13
Crystals of myoglobin. [Courtesy of John Kendrew.]

EXERCISE 21-9 (a) What smaller peptides would be produced by the digestion of the following peptides with trypsin and with chymotrypsin?

Lys–Cys–Tyr–Val–Arg–His–Glu–Trp–Ala–Ser

(b) A peptide yields the following smaller peptides upon digestion with trypsin and chymotrypsin:

Trypsin:
Val–Arg and Leu–Gly–Asp–Phe–Ala–Tyr–Glu

Chymotrypsin:
Ala–Tyr–Glu and Val–Arg–Leu–Gly–Asp–Phe

What is the sequence of amino acids in the peptide?

21-11 CHEMICAL SYNTHESIS OF PEPTIDES AND PROTEINS

The problem of synthesizing in the laboratory even a "small" protein such as insulin (51 amino acid residues) is awesome. There are, however, many reasons why laboratory synthesis of proteins is important. Additional information about the structure and function of peptides and proteins can be obtained by synthesizing similar molecules in the laboratory. By producing peptides and proteins that differ only slightly in structure from the natural peptides, we can gain insight into the roles of particular amino acid residues. Chemical synthesis also serves to confirm the sequence determination. If the synthesized molecules are indistinguishable in properties from the naturally occurring forms, then the proposed structure must be correct. Laboratory synthesis is also emerging as a viable means of producing hormones for medical use. Several peptide hormones have been synthesized in the laboratory. The protein hormone insulin has been synthesized in laboratories in this country and in China. The enzyme ribonuclease and portions of several other proteins have also been synthesized.

The main principle of peptide synthesis is an adage that might be spoken by the line coach of a football team: "block and react." In other words, the strategy is to block all the functional groups of amino acids to be joined, except for those particular groups that will participate in the peptide bond. For example, if we were to begin the synthesis of insulin, as a start the first two residues of the smaller chain could be joined to give Gly–Ile. If glycine and isoleucine were simply mixed under conditions that favor peptide-bond formation, an assortment of peptides of different lengths (and containing glycine and isoleucine in varying order) would be obtained:

$$\text{Gly} + \text{Ile} \longrightarrow \begin{array}{l}\text{Gly–Gly}\\ \text{Ile–Gly}\\ \text{Gly–Ile}\\ \text{Ile–Ile}\\ \text{Gly–Ile–Gly, and so on}\end{array}$$

If only the dipeptide Gly–Ile is to be produced, the reaction mixture must contain glycine with its amino group blocked so that it cannot react, and isoleucine with its carboxyl group blocked. Then the two derivatives can react only to give the Gly–Ile peptide (bottom of page). (As in Part III, we show only the main reactants and products; the equation is not balanced.)

Removal of the blocking groups then yields the free Gly–Ile peptide. The dipeptide can be blocked at one of its ends and combined with another suitably blocked amino acid derivative or peptide derivative to produce a large peptide. In this way, fragments of the protein can be built up individually and joined by appropriate "block-and-react" strategies until the final peptide or protein is completed. This is the approach that was successfully used in the synthesis of hormones, including insulin.

Blocked synthesis:

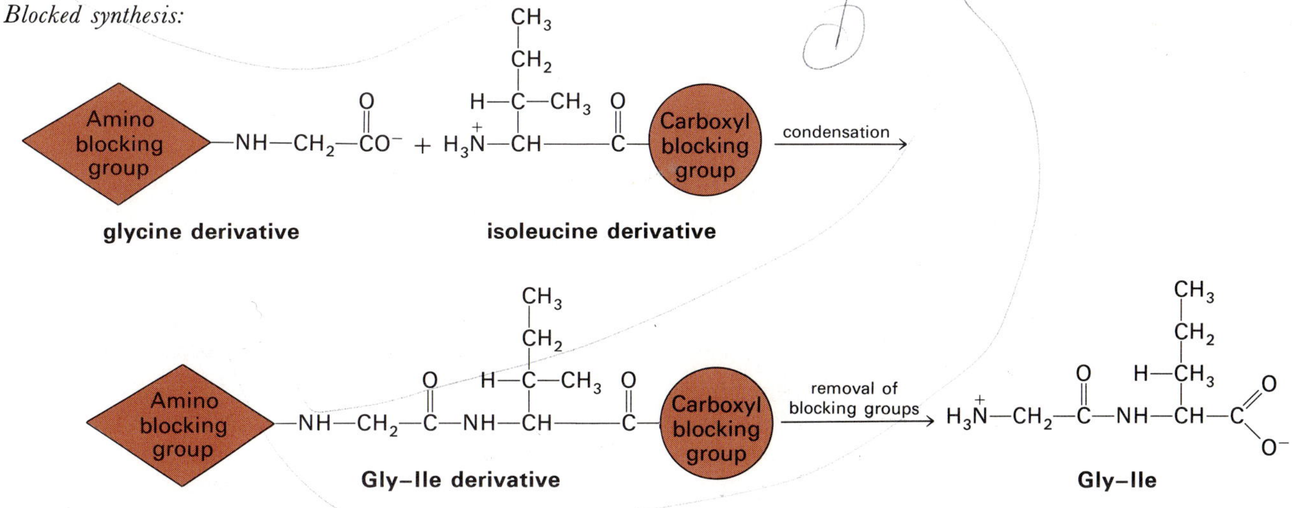

KEY TERMS

An *amino acid* is a compound that contains an amino group and a carboxyl group. In an *α-amino acid,* both functional groups are bonded to the α-carbon—the first carbon after the carboxyl group. *Essential amino acids* cannot be synthesized in the body and so must be provided in the diet.

A *Zwitterion* is a compound with both a positive and a negative charge, such as an amino acid at neutral pH.

A *hydrophobic* substance is nonpolar and thus is insoluble in water. A *hydrophilic* substance is polar and is soluble in water.

The *peptide bond* is the covalent bond between the carboxyl carbon of one amino acid and the amino nitrogen of another amino acid. A *peptide* is a chain of about 50 or fewer amino acids linked by peptide bonds. A *polypeptide* is a peptide that contains more than about 50 amino acids.

Amino acid residues are amino acids that are in peptide chains and thus do not have free α-amino or α-carboxyl groups.

The *backbone* of a peptide chain consists of regularly repeating units. The *side chains* of the amino acid residues form branches from the backbone.

A *disulfide bond* can be formed between the sulfur atoms of two cysteine units to produce cystine.

Globular proteins are folded into a compact, almost spherical shape, Enzymes are examples of globular proteins. *Fibrous proteins* are long, rodlike molecules that function as structural elements and provide mechanical support. Collagen in tendons and bone is a fibrous protein.

An *oligomeric protein* contains more than one polypeptide chain. Each chain of an oligomeric protein is called a *subunit.*

Noncovalent forces are any forces other than covalent bonds that exist between molecules. They include hydrogen bonds, van der Waals forces, and *salt linkages* (or ionic forces).

Denaturation (caused by heating or other harsh treatment) converts the ordered three-dimensional structure of a protein into a disordered one and thus results in loss of the protein's biological activity.

The *α helix* is a common conformation of portions of polypeptide chains in proteins. Adjacent turns of the spirally wound chain are stabilized by hydrogen bonds between amide hydrogen atoms and carbonyl oxygen atoms four residues along the polypeptide chain.

The *β sheet* is a common conformation of portions of polypeptide chains in proteins, in which two extended stretches of the chains interact with each other through hydrogen bonds between amide hydrogen atoms and carbonyl oxygen atoms.

A *prosthetic group* is a nonprotein molecule that is tightly bonded to a protein molecule.

In *ion-exchange chromatography,* a mixture of compounds is bound to a charged resin. The compounds are eluted from the column in an order based on their charges.

Proteolytic enzymes (or *proteases*) cleave peptide bonds; many of them cleave only peptide bands adjacent to a specific amino acid residue.

CHAPTER SUMMARY

Amino acids are the building blocks of proteins and peptides. Twenty principal amino acids are found in proteins, each with a central carbon atom that is bonded to an amino group, a carboxylic acid group, a hydrogen atom, and a characteristic R group. The amino acids usually are found in the L form of stereoisomers (except for one, glycine, which is not optically active). The R groups vary in size and shape, in chemical reactivity, in electric charge, and in ability to form hydrogen bonds. The tremendous variety of functions performed by proteins results from the characteristic properties of these R groups. At physiological pH, amino acids exist in the form of *zwitterions.*

In peptides or proteins, amino acids are joined in linear sequences via peptide bonds formed by a condensation between the carboxylic acid group of one amino acid and the amino group of the adjacent amino acid. Covalent bonds can form also between the sulfhydryl groups of cysteine amino acids to give disulfide bonds.

A protein is a large molecule composed of one or more polypeptide chains. The sequence of amino acids and the disulfide bridges in a protein describe its *primary structure.* Every protein has a specific three-dimensional structure known as its *conformation.* The conformation may include portions of the polypeptide chain in regular structural arrangements such as α helix, β-pleated sheet, or the collagen triple helix. *Secondary structure* describes these regular patterns of organization. The polypeptide chain (which may be composed of regions of α helix, β sheet, and irregular regions) can be folded into a compact shape. *Tertiary structure* describes this folding. The globular protein myoglobin provides an example. If a protein contains more than one polypeptide chain (as hemoglobin does), the three-dimensional arrangement of these subunits is the protein's *quaternary structure.* Noncovalent forces (including hydrogen bonds, hydrophobic forces, and salt linkages) account for the secondary, tertiary, and quaternary structure of a protein.

To determine the complete three-dimensional structure of a protein, several steps are necessary. (1) The amino acid composition is determined by hydrolyzing the protein and separating and identifying the amino acids by ion-exchange chromatography. The amount of each amino acid is determined spectroscopically. (2) The protein is cleaved into peptides (often by proteolytic enzymes), and the amino acid sequence of each peptide is determined by the Edman method. (3) The order of peptides in the protein is established by using two (or more) proteases that yield overlapping peptides. (4) The three-dimensional arrangement of the atoms in space is determined by X-ray crystallography.

Proteins can be synthesized from amino acids, using special methods to ensure that the amino acids can bond only in the order desired.

PROBLEMS

21-1 Consider the following seven amino acids: aspartic acid, cysteine, isoleucine, lysine, serine, phenylalanine, and proline. Select one of these amino acids that exhibits each of the following characteristics:

(a) an amino acid with an aromatic R group
(b) an amino acid with a sulfhydryl group
(c) an amino acid with a hydroxyl group
(d) an amino acid that is a secondary amine
(e) an amino acid with a positively charged R group
(f) an amino acid with a negatively charged R group
(g) an amino acid with more than one chiral carbon atom.

21-2. (a) Draw the structures of the following amino acids in the forms that predominate at pH 1, at pH 7, and at pH 12: cysteine, phenylalanine, and glutamic acid. (See Table 21-2.)
(b) Draw the structures of aspartic acid at pH 1, at pH 3, and at pH 7, and those of lysine at pH 1, at pH 7, and at pH 10.

21-3. Which of the following are likely to be on the outside of a globular protein? (a) leucine; (b) aspartic acid; (c) threonine; (d) glycine; (e) glutamic acid.

21-4. Which of the common α-amino acids does not have an enantiomer?

21-5. Large concentrations of urea interfere with the hydrogen bonds, ionic forces, and van der Waals interactions in a solution of a protein. What types of protein structure (1°, 2°, 3°, or 4°) would you expect to be affected by urea? (Urea is H_2NCONH_2.)

21-6. Why are globular proteins generally water soluble, but fibrous proteins generally water insoluble?

21-7. Draw the complete structure of the tripeptide Ala–Gly–Ser.

21-8. For the following pairs of amino acids, pick the one most likely to have the property indicated:

Pair of amino acid residues

a. Glutamic acid or phenylalanine
b. Lysine or leucine
c. Alanine or proline
d. Serine or glycine
e. Asparagine or aspartic acid

Property

A. Found on the surface of a protein
B. Buried in the interior of a protein
C. Present in an α-helical portion of a protein
D. Participating in a hydrogen bond involving a side-chain hydrogen
E. Participating in an ion-pair interaction

21-9. Designate the stucture of the peptide methionine-enkephalin in an abbreviated form:

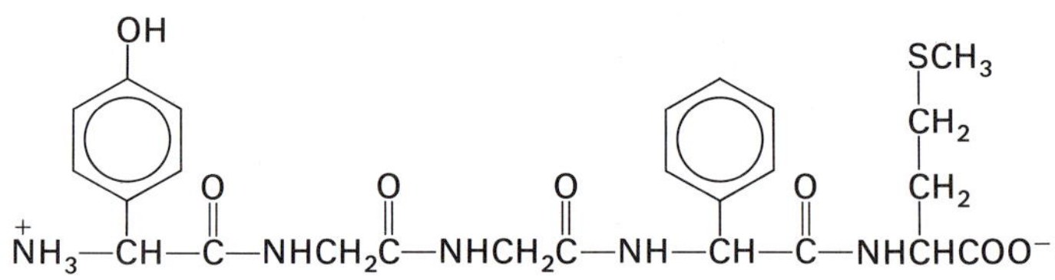

21-10. Could the following peptide form a cyclic structure without making an additional peptide bond? If so, draw a probable structure.

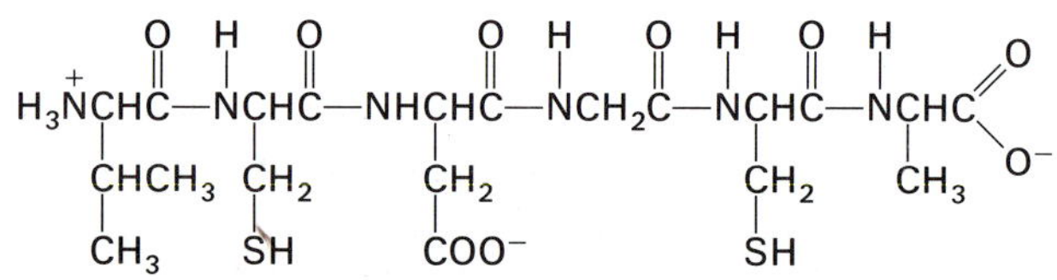

21-11. The tripeptide glutathione, which is found in all cells and plays an important role in plant respiration, has the structure:

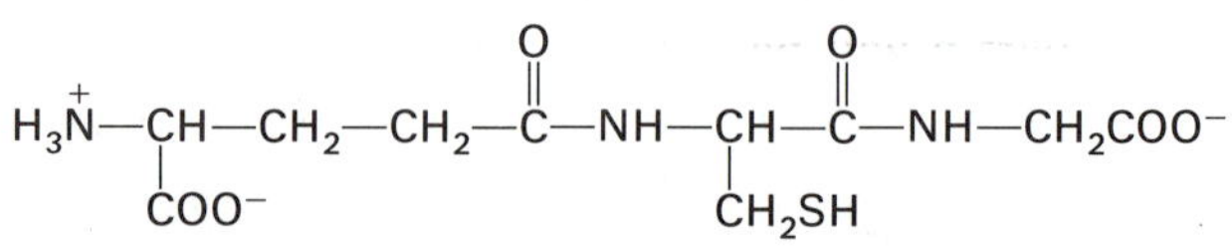

What is unusual about the structure of glutathione?

21-12. (a) Histones are basic proteins that serve as the polypeptide "wrappers" for DNA. On the basis of this fact, name two amino acids that you would expect to find highly concentrated in these proteins.
(b) Histones in their native conformation contain a very high proportion of α-helical structure. Knowing this, which amino acid would you expect to be *absent* from these proteins?

21-13. Laboratory-produced samples of each pure α-amino acid usually are mixtures that contain identical amounts of the L and D enantiomers. These mixtures are called **racemic mixtures.**
(a) What will be the effect on plane-polarized light of passage through a racemic mixture of synthetic phenylalanine?
(b) When the tripeptide Leu–Ala–Val is artificially synthesized using racemic mixtures of each α-amino acid, how many isomers (including enantiomers) of the tripeptide will form?
(c) Will this artificially made sample of tripeptide (in part b) be optically active?

21-14. Leucine-enkephalin, a painkiller produced in minute amounts in the body, is a pentapeptide with the amino acid sequence Tyr–Gly–Gly–Phe–Leu. (a) How many asymmetric carbons are there in this specific pentapeptide?
(b) How many different structures are theoretically possible for this enkephalin (including all enantiomers)? (c) Recalling that only one enantiomer of each optically active α-amino acid is produced in nature, how many of those different pentapeptide isomers calculated in part b will be present in the naturally occurring enkephalin? Is the naturally occurring substance optically active?

21-15. (a) If trypsin were used to hydrolyze the following peptide, what smaller peptides would be produced?

Gly–Ala–Lys–Gly–Arg–Gly

(b) What smaller peptides would be produced by digestion with chymotrypsin?

21-16. If chymotrypsin were used to hydrolyze the following peptide, what smaller peptides would be produced?

Gly–Tyr–Ala–Val–Phe–Trp

21-17. A newly discovered protein hormone from elephants is shown to have four identical subunits in its overall structure. One hundred grams of the intact protein in 1 ℓ of water is agitated under conditions which break apart the quaternary structure. The resulting solution exerts an osmotic pressure of 7.6 mm Hg at 300 K. Calculate the molecular weight of the complete, intact protein hormone (refer to Chapter 9, if necessary).

21-18. A funny thing happened to a protein chemist who tried to determine the molecular weight of a new enzyme X by measuring its freezing-point depression. When 10 g of X were dissolved in 1000 g of H_2O, the solution froze at $-0.000186°C$.
(a) Calculate the molecular weight of X from these data. (The molal freezing-point depression of water is 1.86°C; see Chapter 9.)
(b) Now the chemist decided to double-check the result by measuring the freezing point of an acetone solution of the enzyme. Surprisingly, the value of the molecular weight of X came out to be 25,000 g/mole! What, if anything, do these experiments tell you about the enzyme's structure?

21-19. A mixture of the following four peptides in acid solution is applied to an ion-exchange column (the ion-exchange resin is bonded to SO_3^- groups):
(a) Arg–Lys–Gly–Ala
(b) Ala–Gly–Gly–Glu
(c) Glu–Gly–Ala–Asp
(d) Arg–Ala–Ala–Gly
In what order would you expect the peptides to be eluted (washed out) from the column when the column is flushed with buffers of increasing pH?

21-20. An octapeptide gives the following fragments when hydrolyzed by trypsin and chymotrypsin:

Trypsin:
Ala–Ala–Trp–Gly–Lys and Thr–Val–Lys

Chymotrypsin:
Gly–Lys and Thr–Val–Lys–Ala–Ala–Trp

What is the amino acid sequence of the octapeptide?

21-21. A nonapeptide gives the following fragments when hydrolyzed by trypsin and chymotrypsin:

Trypsin:
Trp–Lys and Ser–Tyr–Ile–Val–His and Gly–Arg

Chymotrypsin:
Lys–Gly–Arg–Ser–Tyr and Ile–Val–His and free Trp

What is the amino acid sequence of the nonapeptide?

21-22. (a) Describe a reasonable tertiary structure for a peptide that contains ten aspartate residues, followed by three glycine residues, and then ten lysine residues.
(b) Describe a reasonable tertiary structure for a peptide that contains ten lysine residues alternating with ten threonine residues and then followed by six leucine residues.

21-23. What would be the charge at physiological pH of each of the following peptides? (Consult Table 21-2, and assume that the pK values of any amino acids not listed are the same as the pK values of glycine. Remember that in a peptide the α-amino and α-carboxyl groups of all except the terminal amino acids are tied in peptide bonds.)
(a) Leucine-enkephalin:
Tyr–Gly–Gly–Phe–Leu
(b) Oxytocin: Cys–Tyr–Ile–Glu–Arg–Cys
(c) Angiotensin:
Asp–Arg–Val–Tyr–Ile–His–Pro–Phe
(d) Bradykinin:
Arg–Pro–Pro–Gly–Phe–Ser–Pro–Phe–Arg
(e) Which of these peptides would bind most strongly to an anion-exchange resin?

A network of capillaries associated with muscle fibers in the rat. These capillaries deliver nutrients to the muscle fibers—the protein hemoglobin transports the oxygen. Scanning electron micrograph, $\times 650$.

22

Protein Function

In their role as enzymes, proteins catalyze virtually all the biochemical reactions of living organisms. In addition, proteins play a number of other essential roles: as muscle proteins, they generate movement; as transporters, they deliver molecules to the cells for metabolism; and as fibrous proteins, they are a principal structural material of living organisms.

In this chapter we discuss the general properties of enzymes and look at the enzyme chymotrypsin in detail. We briefly consider some other protein functions and see how myoglobin and hemoglobin transport oxygen in the body.

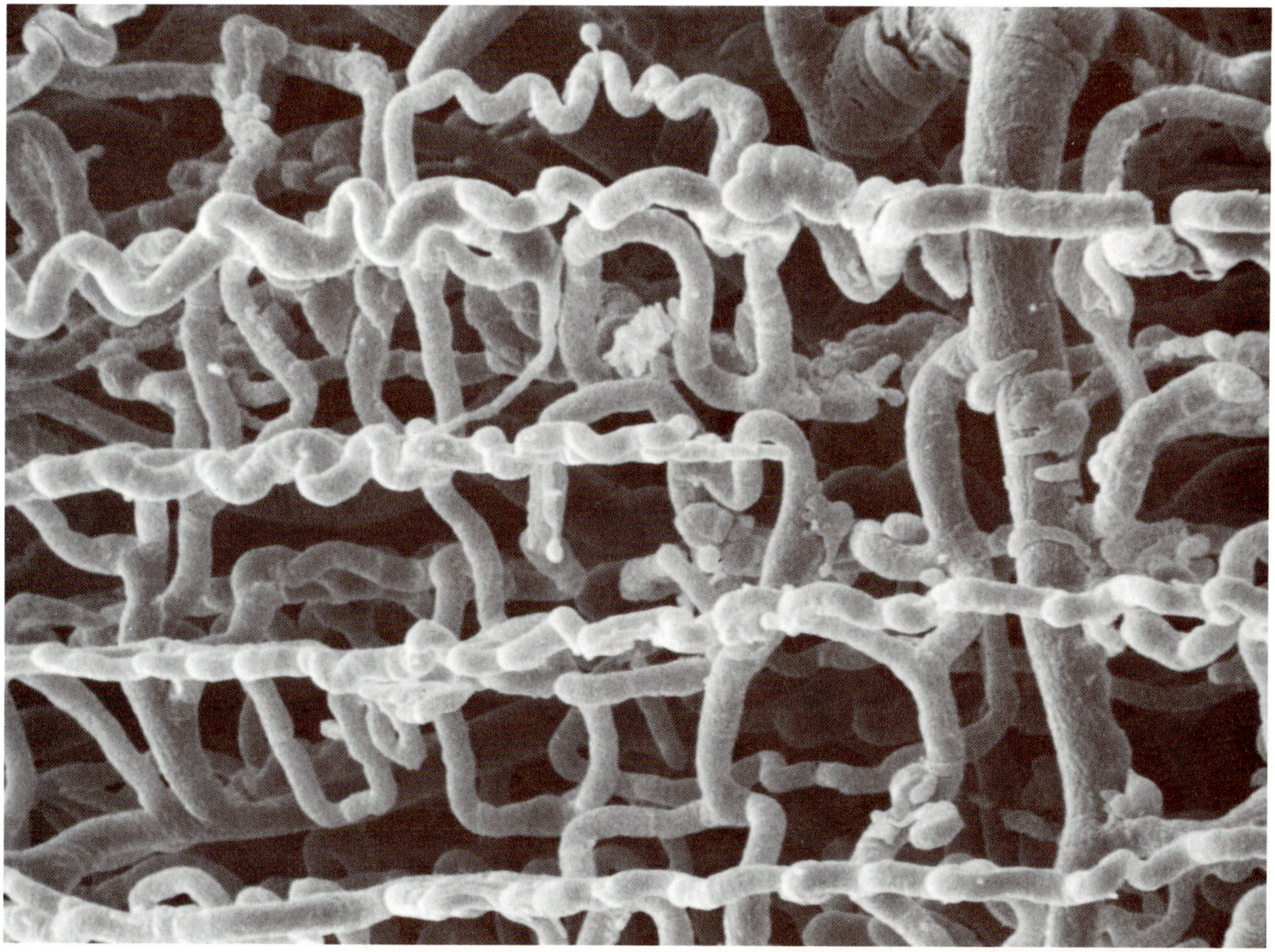

22-1
GENERAL PROPERTIES OF ENZYMES

Proteins are found in every cell and are of crucial importance to virtually every biological process. This chapter describes many protein functions; still others are discussed in later chapters. The largest class of proteins is the enzymes, a group of extraordinarily efficient biological catalysts. An enzyme might speed a reaction by a factor of billions—a reaction that would otherwise require years might occur in a fraction of a second with the enzyme present. About two thousand different enzymes are known, and many of them have been purified. We now consider some properties enzymes have in common before looking more carefully at one enzyme—chymotrypsin, an enzyme that catalyzes the hydrolysis of particular peptide bonds.

Although each enzyme has a lengthy official name that describes its function in precise detail, simpler (older) names and shorthand versions of the lengthy names are widely used by biochemists, and we shall use them exclusively. A common way to name an enzyme is simply to name the **substrate,** the compound on which the enzyme acts, and then add the suffix *-ase*. Urease, for example, catalyzes the hydrolysis of urea to ammonia and carbon dioxide:

$$H_2O + \underset{\textbf{urea}}{H_2N{-}\overset{\overset{\displaystyle O}{\|}}{C}{-}NH_2} \xrightleftharpoons{\text{urease}} 2NH_3 + CO_2 \qquad (22\text{-}1)$$

Urea is a substrate of the enzyme urease.

Fumarase catalyzes the addition of water to the double bond of fumarate to yield malate:

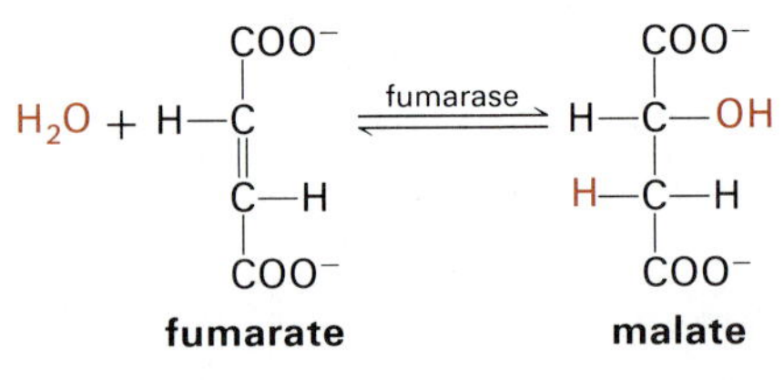

(22-2)

Fumarate is a substrate of the enzyme fumarase. The reaction is a step in the citric acid cycle (Section 26-3).

Other enzyme names refer to the type of reaction catalyzed. Thus, dehydrogenases catalyze the removal of hydrogen atoms, and oxygenases catalyze the addition of oxygen atoms. Malate dehydrogenase catalyzes the removal of two hydrogen atoms from malate to yield oxaloacetate:

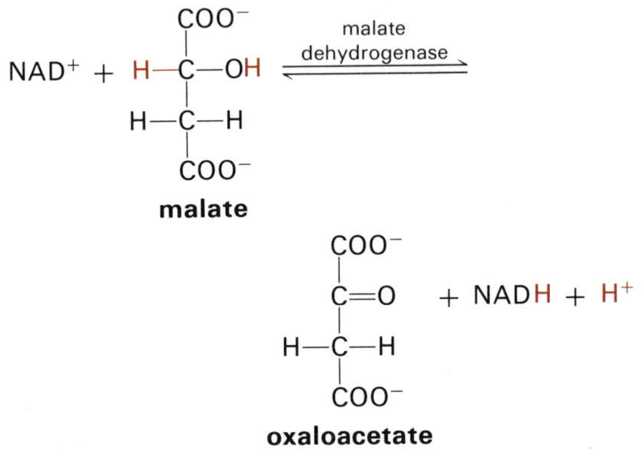

(22-3)

NAD^+ is the oxidizing agent. The reaction of Equation 22-3 follows that of Equation 22-2 in the citric acid cycle.

Like all catalysts, enzymes participate in reactions and increase the reaction rates without themselves being consumed in the process. Enzymes do not alter the relative amounts of products and reactants present at equilibrium. We saw in Chapter 10 that the rate of a reaction is partly determined by the height of the energy barrier between reactants and products: the higher the activation energy, the slower the reaction. The activation energy in turn depends on the structure of the transition state, the chemical form through which reactants must pass to be converted to products (Section 15-6). A catalyst speeds both forward and reverse reactions by providing a transition state of lower energy (Figure 22-1).

Enzymes have a high degree of **specificity:** each enzyme recognizes only one or a small number of substrates and catalyzes only one or a limited number of reactions. We shall shortly see why this is so. Urease is

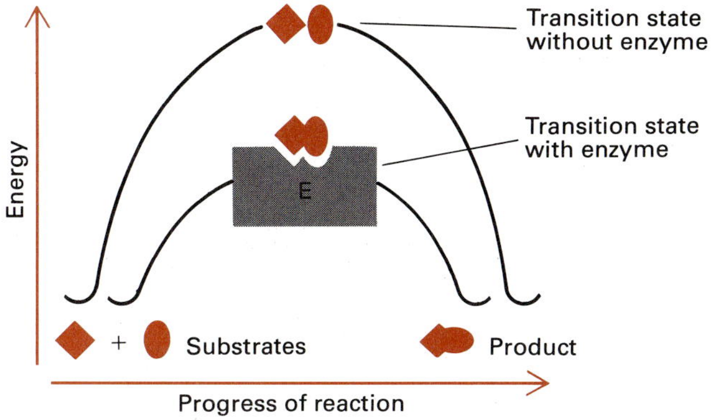

FIGURE 22-1

Diagram of enzyme catalysis. The enzyme provides a transition state of lower energy.

an extreme example. It catalyzes a single reaction (and its reverse): the hydrolysis of the peptide bonds of urea (Equation 22-1). Urease has no effect on the rate of hydrolysis of other amides.

Many enzymes are stereospecific, acting only on one of a pair of enantiomers. For example, in the last step of the urea cycle (in which urea is synthesized from the breakdown of amino acids), the enzyme arginase catalyzes the hydrolysis of L-arginine, but not that of D-arginine:

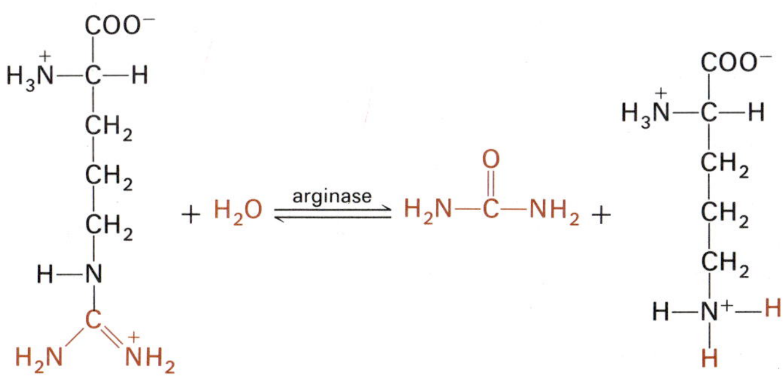

L-arginine **urea** **L-ornithine**

The digestive enzyme chymotrypsin is less specific than either urease or arginase. It catalyzes primarily the hydrolysis of peptide bonds on the carboxyl side of the aromatic amino acid residues of phenylalanine, tyrosine, and tryptophan; but it also acts on ester linkages. Furthermore, any bulky hydrophobic group can replace the aromatic rings of these substrate amino acids without changing the effectiveness of chymotrypsin. Even the α-amino group is not required. For example, chymotrypsin catalyzes the hydrolysis of the ester

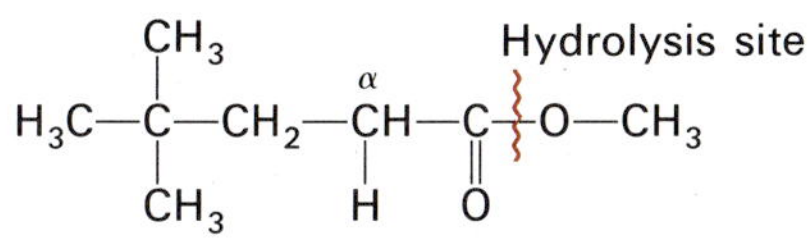

methyl 4,4-dimethylpentanoate

which has a bulky, hydrophobic tertiary butyl group instead of an aromatic ring, and has a hydrogen atom instead of an α-amino group.

Enzymes are responsible for catalyzing virtually all biochemical reactions, so an important element in the regulation of cellular chemistry is the activation and deactivation of enzymes as the biochemical environment changes. A common mechanism by which enzyme activity is regulated is the synthesis of the enzyme in an inactive precursor form called a **zymogen,** or **proenzyme.** Cleavage of one or a few peptide bonds of the zymogen yields the active enzyme. The digestive enzymes pepsin, trypsin, and chymotrypsin and many of the enzymes involved in blood clotting are produced and activated in this way. Trypsinogen, the inactive form of trypsin, is produced in the pancreas and activated in the small intestine by hydrolysis of a single peptide bond. The products are the active enzyme (trypsin) plus a hexapeptide:

$$\text{Trypsinogen} + H_2O \xrightarrow{\text{enterokinase}} \text{Trypsin} + \text{Hexapeptide}$$

The reaction is catalyzed by the enzyme enterokinase. In the serious disease called pancreatitis, digestive enzymes are released in active form in the pancreas. There they digest the pancreas itself and its blood vessels.

A second method of enzyme regulation is **feedback inhibition:** the enzyme that catalyzes the first step of a series of reactions is inhibited (made inactive) by the final product of the reactions. This inhibition prevents the accumulation of extra product in the cell. Feedback inhibition occurs in the biosynthesis of isoleucine from threonine:

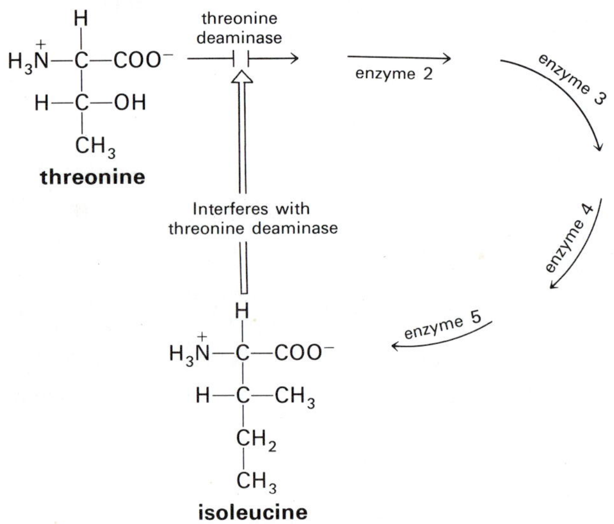

Threonine deaminase, the enzyme for the *first* of the five reactions by which threonine is converted to isoleucine, is inhibited by isoleucine, the final product of the reaction series. Thus, the reaction slows when the concentration of isoleucine becomes high enough, and no more isoleucine is produced. When the concentration of isoleucine falls to a low enough level, the enzyme becomes more active, and more isoleucine is synthesized. This feedback prevents the cell from wasting energy by making products that are not needed. We shall see in Section 22-3 *how* the product can cause the inhibition.

EXERCISE 22-1 (a) In the biosynthesis of isoleucine from threonine, it is the first enzyme in the series (rather than some later one) that is subject to feedback inhibition. Explain how this is advantageous for the cell.

(b) The following branched biosynthetic pathway is subject to feedback inhibition. Enzyme 1 is inhibited when concentrations of *both* products, D and H, are high. Which enzyme is inhibited by a high concentration of D only, and which by a high concentration of H only? (The numbers over the arrows are the enzyme numbers.)

$$\mathrm{A} \xrightarrow{1} \mathrm{B} \xrightarrow{2} \mathrm{C} \xrightarrow{4} \mathrm{E} \xrightarrow{5} \mathrm{F} \xrightarrow{6} \mathrm{G} \xrightarrow{7} \mathrm{H}$$
$$\mathrm{C} \xrightarrow{3} \mathrm{D} \ \text{(downward branch)}$$

22-2 REACTION RATE AND THE ENZYME–SUBSTRATE COMPLEX

We have discussed some general properties of enzymes, but we have not yet seen how they select their substrates nor how they accomplish increases in reaction rates. The general principles of reaction kinetics (Section 10-10) apply also to enzymes. Consider the simple, uncatalyzed, one-step reaction $A \rightarrow B$. The rate law, determined by experiment and graphed in Figure 22-2a, is

$$\text{Reaction rate} = k_T[\text{A}]$$

where k_T is the rate constant for this particular reaction at temperature T. According to the rate law, the rate of the reaction is directly proportional to the concentration of A.

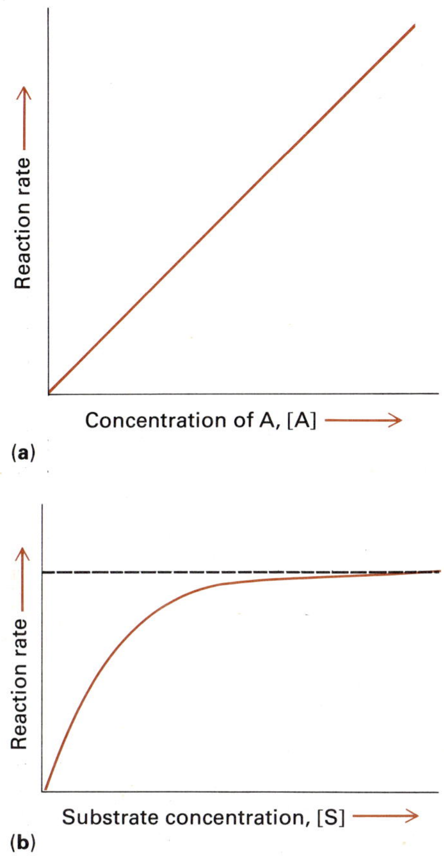

FIGURE 22-2

Reaction rates. **(a)** For the one-step reaction A $\longrightarrow$ B, the reaction rate is directly proportional to the concentration of A: Reaction rate $= k_T[\text{A}]$. **(b)** For an enzyme-catalyzed reaction (at a fixed enzyme concentration under defined conditions), a maximal reaction rate is observed. When the concentration of substrate [S] is low, the reaction rate increases with [S]. At high values of [S], the reaction rate is independent of [S].

We might expect to obtain the same graph for the simple enzyme-catalyzed reaction

$$\text{Substrate} \xrightarrow{\text{enzyme}} \text{Product}$$

Instead, when we graph reaction rate against substrate concentration [S], we obtain the results shown in Figure 22-2b. At low concentrations of substrate the reaction rate *is* nearly proportional to [S], as expected. Increasing [S] increases the reaction rate, but only up to some maximal rate. After the maximal rate is reached, further increases in [S] have no effect on the rate. The enzyme is said to be **saturated** with substrate. All enzymes show this saturation effect.

EXERCISE 22-2 (a) We have seen that when an enzyme is saturated with substrate, the reaction rate cannot be increased by further addition of substrate. Suggest two possible ways of further increasing the reaction rate. (Refer to Chapter 10, if necessary.)

(b) Write a rate law for the reaction E + S $\rightleftharpoons$ ES. Assume that it occurs in one step.

(c) What would be the effect on the reaction rate of doubling [E] when [S] is in large excess and can be assumed to be constant?

How can we explain the saturation effect? A simple model has been proposed (and widely accepted) to account for these results. According to this model, for a reaction with only one substrate, the enzyme E combines with substrate S to form an enzyme–substrate complex E—S. The enzyme–substrate complex can dissociate to product P and free enzyme:

$$\textit{Step 1:} \quad \text{E} + \text{S} \rightleftharpoons \text{E—S}$$

$$\textit{Step 2:} \quad \text{E—S} \rightleftharpoons \text{P} + \text{E}$$

Both steps are assumed to be reversible.

The substrate binds to a region of the enzyme called the **active site.** When all the active sites are filled, the enzyme is saturated with substrate, and the reaction rate is at a maximum—it cannot be further increased by increasing the substrate concentration. All enzymes for which the active site has been studied so far are globular or partially globular proteins. Their active sites are clefts or crevices formed by the folding of the polypeptide chain.

The amino acid residues that line the cleft are almost entirely nonpolar, and they come from widely separated parts of the polypeptide chain. We saw an example of this in the formation of the heme-binding pocket of myoglobin (Section 21-8). The active site typically contains a few polar or charged side chains that are needed for bonding to the substrate and for catalyzing the breaking and making of bonds. The active site thus provides a place where all the reactants are brought together in positions favorable for the reaction. (We shall see an example of this when we study chymotrypsin in Section 22-6.) The forces that hold the substrate in the active site are the same ones responsible for the secondary, tertiary, and quaternary structures of proteins: hydrophobic interactions, hydrogen bonds, and ionic interactions.

The specificity of an enzyme for a particular substrate depends on the precise three-dimensional arrangement of atoms in the active site. Only a substrate with a complementary shape can fit into the active site, as a hand fits a glove or a key fits a lock. The **lock-and-key model** of the interaction between an enzyme and its substrate (Figure 22-3), first suggested by Emil Fischer in 1890, is still a useful and essentially correct way to picture the specificity of an enzyme for a substrate. The active sites of some enzymes, however, appear not to have a rigid structure. Instead, the active site complements the substrate only *after* the substrate is bound (Figure 22-4). According to this model, called the **induced-fit model,** the enzyme changes its shape to fit the substrate.

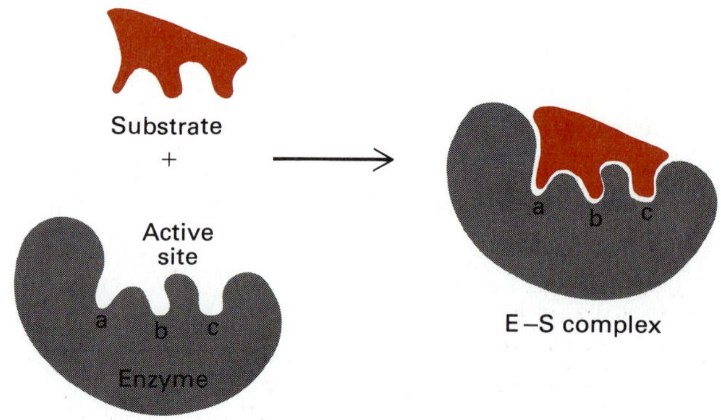

FIGURE 22-3

Lock-and-key model of an enzyme and its substrate. The shapes of the substrate and the active site are complementary.

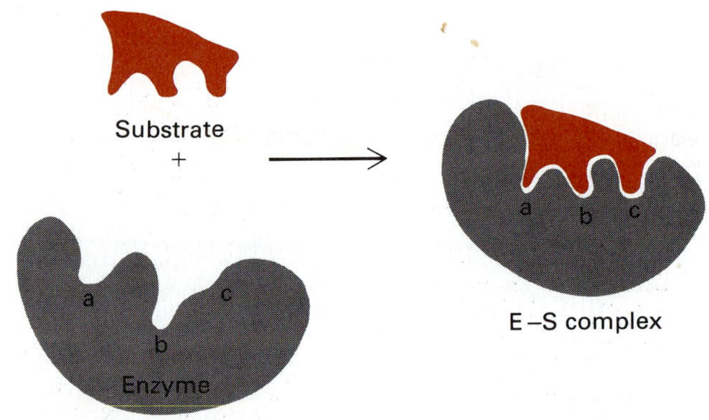

FIGURE 22-4

Induced-fit model of an enzyme and its substrate. The shape of the enzyme changes to fit the substrate when the substrate binds to the enzyme.

We have seen that many principles of reaction kinetics apply to enzyme-catalyzed reactions, but that enzyme-catalyzed reactions show a saturation effect. This observation led to the model of the enzyme–substrate complex and the active site. There are some additional respects in which enzyme-catalyzed reac-

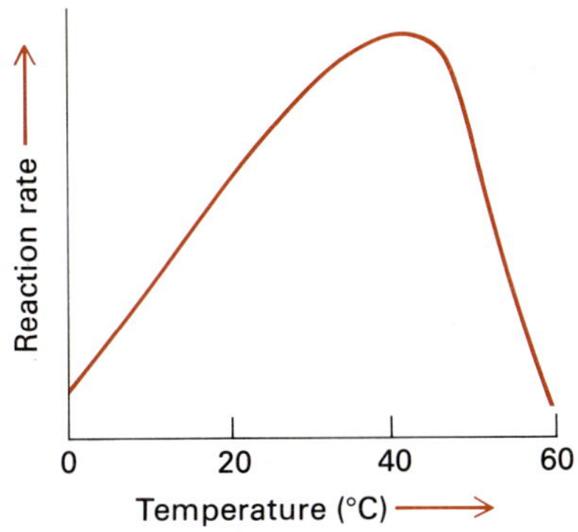

FIGURE 22-5

The effect of temperature on enzyme activity. As the temperature is raised, the activity increases—up to a certain temperature, at which the enzyme is denatured.

tions differ from those studied earlier. We know that the rate of a reaction depends on the temperature (Chapter 10); increasing the temperature increases the reaction rate because a greater number of molecules collide with enough energy to reach the top of the activation energy barrier. This holds true for enzymes as well, but only up to a certain temperature; above that temperature, the reaction rate falls sharply (Figure 22-5). This is not surprising if we recall that enzymes are proteins and thus may be denatured at high temperatures. When an enzyme is denatured, the very precise conformation of the active site is destroyed and the enzyme inactivated. Most enzymes are denatured at temperatures above 50°C, and many enzymes that catalyze reactions in the human body are denatured at temperatures above about 41°C (106°F).

The rates of many enzyme-catalyzed reactions are also dependent on the pH. Lysozyme, an enzyme found in egg white and in tears, catalyzes the hydrolysis of particular bonds in polysaccharides (polymers of sugar molecules; Chapter 24). When these polysaccharides are in the cell walls of bacteria, the effect of hydrolysis is to cause the cell to burst because of its high osmotic pressure. Figure 22-6 graphs the rate of hydrolysis of a polysaccharide against pH. The reaction rate is greatest at pH 5; the rate drops rapidly both at higher pH and at lower pH. The lysozyme is active only when the aspartate side chain is ionized (COO^-) and the glutamic acid side chain is un-ionized (COOH). The decrease on the acid side is caused by protonation of the side chain of an aspartate residue at the active site. The decrease on the basic side is due to the dissociation of a proton from the side chain of a glutamic acid residue at the active site. The optimal pH for most enzymes in the human body is close to the physiological pH (7.4). Trypsin, for example, has an optimal pH of about 8. However, pepsin (a digestive enzyme that hydrolyzes peptide bonds) is most effective at a pH of about 2, which is the pH of the stomach, where pepsin operates.

A property of an enzyme called the turnover number reflects the speed to which the enzyme accelerates a reaction. The **turnover number** is defined as the number of molecules of substrate converted to product per minute by each molecule of enzyme. It can be determined when the enzyme concentration and the

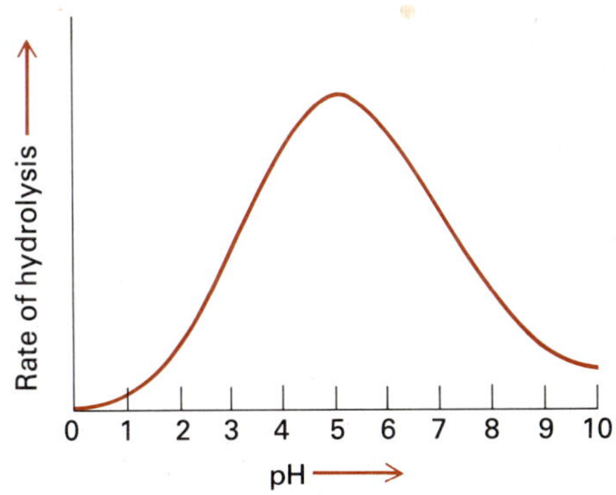

FIGURE 22-6

The rate of hydrolysis of a polysaccharide by lysozyme as a function of pH.

TABLE 22-1
Turnover numbers of some enzymes

Enzyme	Reaction catalyzed	Turnover number (per minute)
Carbonic anhydrase	$CO_2 + H_2O \rightleftharpoons H_2CO_3$	36,000,000
Catalase	$2H_2O_2 \rightleftharpoons 2H_2O + O_2$	5,600,000
Lactate dehydrogenase	Pyrruvate + NADH + $H^+ \rightleftharpoons$ Lactate + NAD^+	60,000
Chymotrypsin	Hydrolysis of specific peptide bonds	6,000
DNA polymerase I	Addition of nucleotide monomers to DNA chains (Chapter 25)	900
Lysozyme	Hydrolysis of specific polysaccharide bonds	30

maximal rate of reaction are known. The highest turnover number known is $3.6 \times 10^7\ min^{-1}$ for carbonic anhydrase, an enzyme in red blood cells that catalyzes the combination of CO_2 and H_2O to form H_2CO_3:

$$CO_2 + H_2O \xrightleftharpoons{\text{carbonic anhydrase}} H_2CO_3$$

The turnover number of 3.6×10^7 means that each molecule of carbonic anhydrase catalyzes the reaction of 36 million CO_2/H_2O pairs per minute. In other terms, each mole of enzyme catalyzes the reaction of 3.6×10^7 moles of substrate. This also means that a 10^{-7} M solution of enzyme catalyzes the formation of 3.6 M H_2CO_3 every minute. More common turnover numbers for the enzymes of cellular metabolism are closer to $10^3\ min^{-1}$, or 1000 reactions per minute. Nevertheless, even such lower numbers give a substantial rate of product formation; such values dramatize the speed with which a small number of enzyme molecules can transform a large number of substrate molecules. Table 22-1 gives the turnover numbers of several enzymes.

EXERCISE 22-3 The turnover number of lactate dehydrogenase is 60,000 *min*$^{-1}$. What is the rate of formation of lactate in moles/ℓ per minute when the concentration of enzyme is 10^{-5} M and the substrate (pyruvate) is present in excess? (This reaction occurs in glycolysis; Chapter 26.)

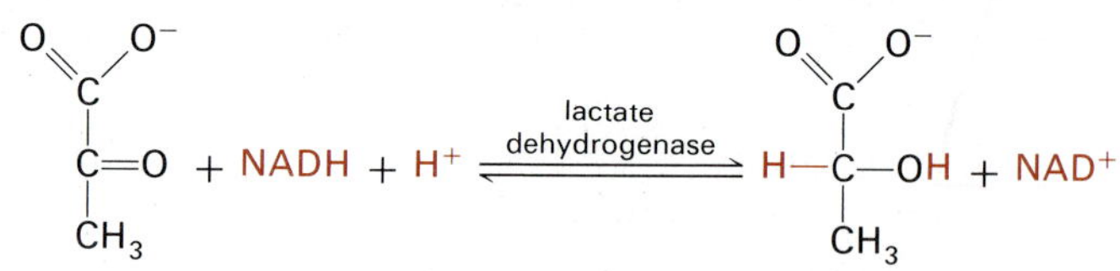

pyruvate **lactate**

22-3
ENZYME INHIBITION

We have seen that certain enzymes can be regulated by feedback inhibition, in which the final product of a reaction pathway inhibits the first enzyme of the pathway. But, more generally, how does any **inhibitor** (a molecule that inteferes with the catalysis of a reaction) interact with an enzyme to stop the reaction? In order to characterize inhibitors, we must first describe some properties of an enzyme. Figure 22-7a is the same graph of reaction rate against substrate concentration for a fixed concentration of enzyme that was shown in Figure 22-2b. The maximal rate, usually called the maximal velocity (V_{max}) by biochemists, is indicated on the graph; the half-maximal velocity, $\frac{1}{2} V_{max}$, is also indicated. The substrate concentration required to give the half-maximal velocity is a quantity called K_M, the **Michaelis constant.** The value of V_{max} is often given in moles/ℓ per minute, and that of K_M in moles/ℓ.

The following example shows that V_{max} and K_M can be determined by a series of experiments in which the initial reaction rate is measured for different concentrations of substrate. (It is the initial rate of reaction that is measured because substrate is consumed as the reaction proceeds, and the reaction rate decreases with time. It is the initial reaction rate that corresponds to the given substrate concentration; see Section 10-10.)

EXAMPLE 22-1

For a certain enzyme, the rate of product formation as a function of substrate concentration is observed to vary as shown below. Find V_{max} and K_M for the enzyme.

Substrate concentration (μM)	Initial reaction rate (μmoles of product formed per minute)
0.3	2.7
0.5	4.2
1.0	6.7
1.5	7.7
2.0	8.2
2.5	8.4
3.0	8.4

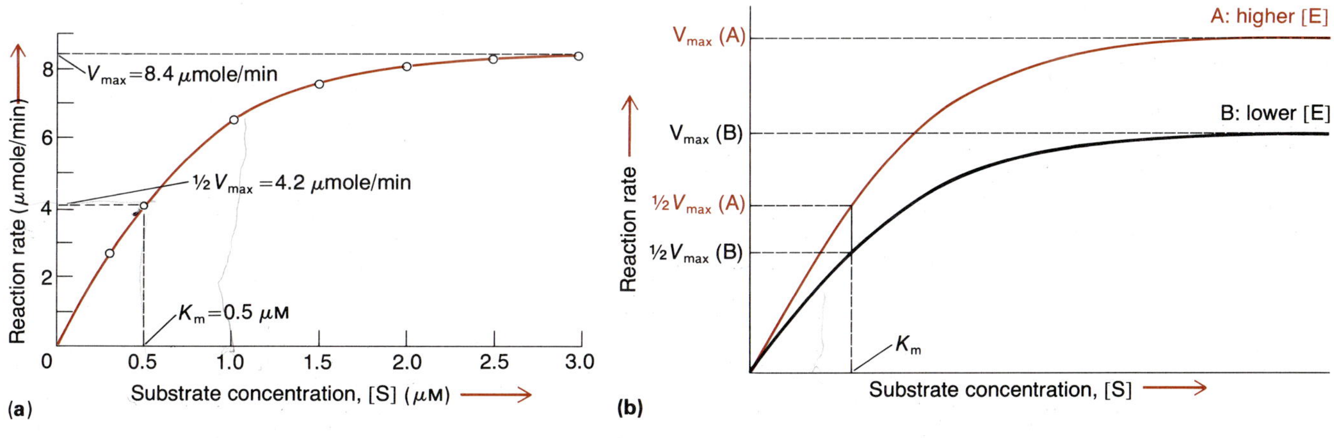

FIGURE 22-7

Reaction rate as a function of substrate concentration. **(a)** The value V_{max} is the maximal rate of reaction; K_M is the concentration of substrate that gives a reaction rate equal to $\frac{1}{2} V_{max}$. **(b)** Reaction rate as a function of substrate concentration for two different enzyme concentrations. The value of V_{max} is greater at the higher enzyme concentration, but K_M is the same for both.

(a) The first step is to plot reaction rate as a function of substrate concentration. The graph is shown in Figure 22-7.

(b) The value of V_{max} (in μmoles of product per minute) can be determined readily from the graph. We see that V_{max} is 8.4 μmole/min.

(c) The value of K_M is the substrate concentration at which the reaction rate is $\frac{1}{2} V_{max}$, or 4.2 μmole/min. From the graph, we see that K_M is 0.5 μM (μmole/ℓ).

Although V_{max} depends on the concentration of enzyme, K_M does not; *K_M is characteristic of the particular enzyme and substrate, but is independent of the enzyme concentration.* Thus, for a particular substrate, if the enzyme concentration is increased, V_{max} increases, but K_M remains the same. These ideas are illustrated schematically in Figure 22-7b, which shows V_{max} and K_M for two different enzyme concentrations. The value of V_{max} is greater at the higher enzyme concentration, but K_M is the same for both enzyme concentrations.

For most enzymes, K_M is approximately equal to the equilibrium constant for dissociation of the enzyme–substrate complex:

$$ES \rightleftharpoons E + S \qquad K_M = \frac{[E][S]}{[ES]}$$

Like other equilibrium constants, K_M is constant for a particular reaction at fixed temperature (as we saw in Figure 22-7b). We see from the equilibrium constant expression that when $[S] = K_M$, then $[E] = [ES]$. That is, one-half of the enzyme is present as free enzyme E, and one-half is in the form of the ES complex. Thus under these conditions, only one-half of the enzyme is "working" at any moment—only half the active sites are filled. (This is consistent with the fact that K_M is the substrate concentration when the reaction rate is $\frac{1}{2} V_{max}$.)

The larger the value of K_M, the more readily ES dissociates to E and S, and the weaker the bonding of the ES complex. The smaller the value of K_M, the less readily ES dissociates to E and S, and the stronger the bonding of the ES complex. Enzymes vary widely in the values of K_M for their particular substrates, but values of 10^{-5} to 10^{-4} M are typical. The smaller K_M, the more effective is the enzyme in catalyzing the reaction at low substrate concentrations. The enzyme of Example 22-1 has a K_M of 0.5 μM (5×10^{-7} M). This enzyme is effective at catalyzing its reaction (that is, one-half or more of the enzyme sites are filled) when the substrate concentration is 0.5 μM or greater. When the substrate concentration is much lower than 0.5 μM, only a small fraction of the enzyme sites are occupied.

22-4 COMPETITIVE AND NONCOMPETITIVE INHIBITION

The quantities V_{max} and K_M help to characterize the different types of enzyme inhibitors. The two principal types of reversible enzyme inhibition are competitive inhibition and noncompetitive inhibition. With competitive inhibition, the inhibitor molecule typically resembles the substrate and can bind reversibly at the active site of the enzyme, thereby preventing the true substrate from binding. For example, the reaction catalyzed by the enzyme succinate dehydrogenase converts succinate to fumarate (along with the reduction of the hydrogen acceptor, NAD^+):

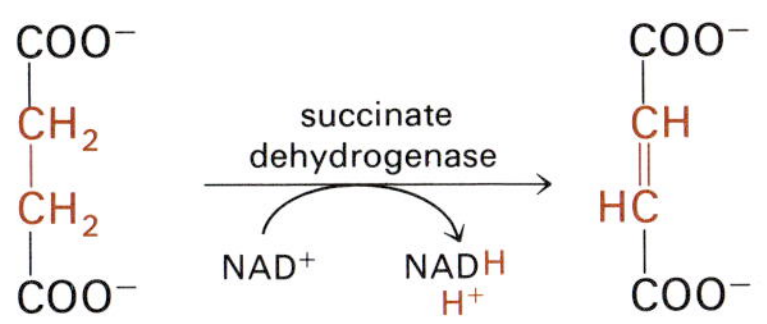

succinate **fumarate**

(Remember that the arrow

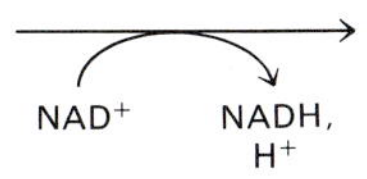

means that NAD^+ is a reactant, and NADH and H^+ are products).

A compound that inhibits the enzymatic reaction competitively is malonate, a molecule very similar to succinate, lacking only a —CH_2—unit:

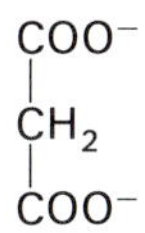

malonate

As the name implies, **competitive inhibition** means that the substrate and the inhibitor vie for binding at the active site. Because some inhibitor will succeed in binding the enzyme, those sites become unavailable for binding the substrate, and the reaction rate is lower than it would be in the absence of inhibitor. However, for a given inhibitor concentration, as the concentration of substrate is increased, more and more of the sites on the enzyme become occupied by substrate. Finally, at very high substrate concentrations, the number of substrate molecules is so much in excess of inhibitor molecules that virtually all enzyme sites are occupied by substrate. Then V_{max} has the same value as it has in the absence of inhibitor. Therefore, one of the criteria used to identify cases of competitive inhibition is the following: for a fixed concentration of inhibitor, high concentrations of substrate can overcome the inhibition. Figure 22-8 shows a plot of velocity against substrate concentration. The presence of a competitive inhibitor shifts the curve so that the apparent K_M is higher, but the V_{max} achieved at high substrate concentrations remains the same. Hence, a competitive inhibitor can be characterized as a compound that causes a change in K_M but causes no change in V_{max}.

The concept of competitive inhibition is put to practical use in the treatment of poisoning. For example, about 50 deaths occur each year in the United States due to the accidental drinking of ethylene gly-

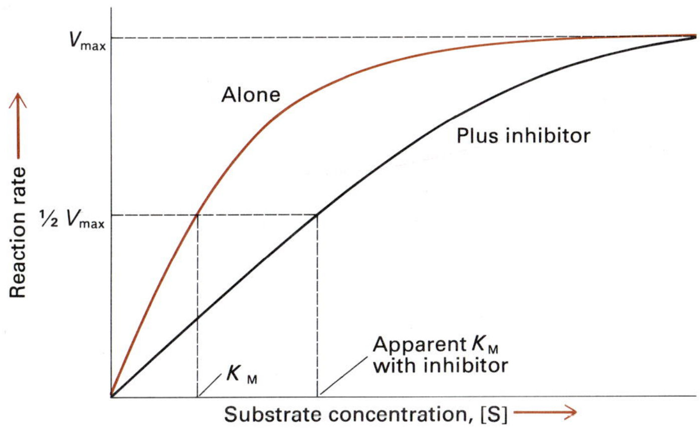

FIGURE 22-8

Reaction rate as a function of substrate concentration in the absence and in the presence of a competitive inhibitor. The value of V_{max} is the same in both cases; K_M is greater in the presence of the inhibitor.

col, the main ingredient in automobile antifreeze. Death occurs not from the ethylene glycol directly but rather from the oxalic acid produced in the oxidation of ethylene glycol by enzymes of the body:

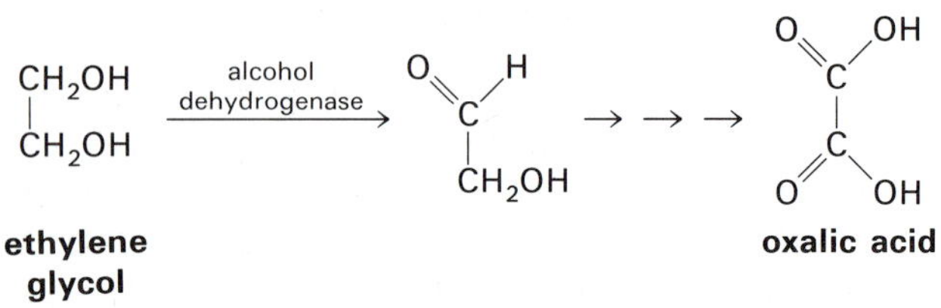

The poisoning by ethylene glycol can be counteracted by administering nearly intoxicating quantities of ethanol. The ethanol acts as a competitive inhibitor for the alcohol dehydrogenase. In this way, the oxidation to oxalic acid is prevented, and the ethylene glycol is excreted in the urine. In Chapter 17 we saw that this same treatment, administration of large quantities of ethanol, is used to prevent the oxidation of methanol to formic acid in cases of methanol poisoning.

The action of a **noncompetitive inhibitor** cannot be overcome by high concentrations of substrate, because noncompetitive inhibitors bind at a site other than the active site on the enzyme molecule. Noncompetitive inhibitors typically do not resemble the substrate, but they react with a portion of the enzyme to destroy its catalytic power by changing the properties of the active site. Mercury is a noncompetitive inhibitor for some enzymes. It destroys the catalytic power of enzymes by reacting with sulfhydryl groups at or near the active site. One mercury atom can cross-link two protein chains or segments of a protein chain:

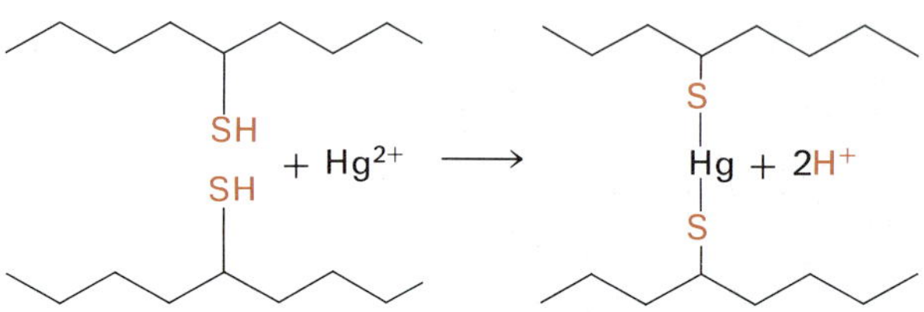

It is because of mercury's action as an enzyme inhibitor that there is concern about the polluting effects of mercury from industrial wastes. The mercury is concentrated in fish, which then may be eaten by human beings. Similarly, the destruction of enzymes by reaction between Pb^{2+} and sulfhydryl groups gives rise to the effects of lead poisoning.

In the presence of a noncompetitive inhibitor, a certain number of enzyme molecules are inactivated; the effect is as if some fraction of the enzyme had been removed from solution. A graph of reaction rate against substrate concentration (Figure 22-9) shows

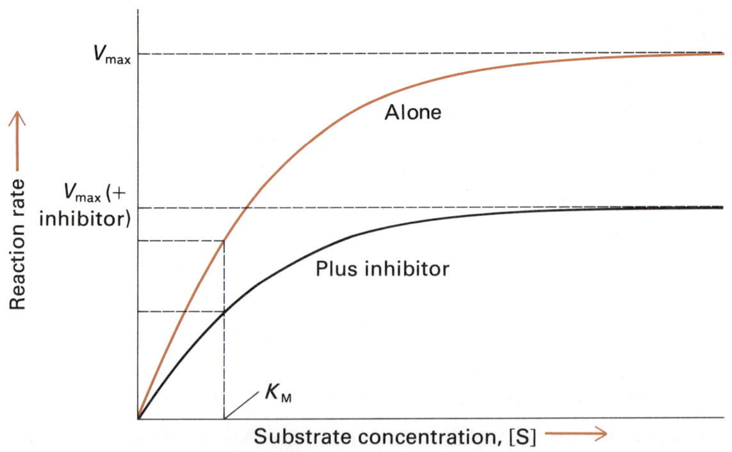

FIGURE 22-9

Reaction rate as a function of substrate concentration in the absence and in the presence of a noncompetitive inhibitor. The value of K_M is the same in both cases; V_{max} is lower in the presence of the inhibitor.

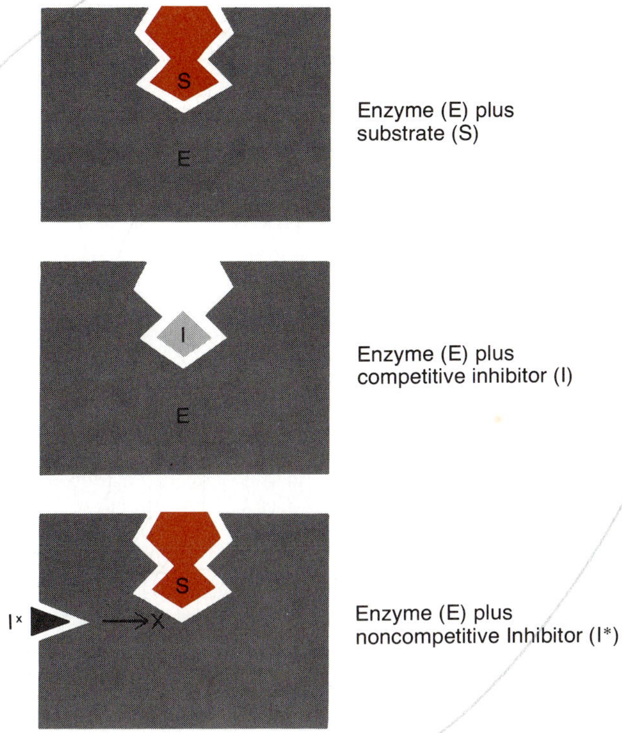

FIGURE 22-10

Schematic illustration of competitive and noncompetitive inhibition. The competitive inhibitor prevents substrate from binding. The noncompetitive inhibitor does not prevent the substrate from binding, but it does produce a defect (X) that impairs the ability of the enzyme to catalyze the reaction.

that the result is to lower the apparent V_{max} while leaving the K_M unchanged, because the remaining enzyme molecules are normal. (Compare Figure 22-9 with Figure 22-7b.)

The difference between competitive and noncompetitive inhibition can be illustrated schematically as in Figure 22-10. The competitive inhibitor resembles the substrate and binds at the same site. In contrast, the noncompetitive inhibitor bears no resemblance to the substrate and binds at a different site.

Both competitive and noncompetitive inhibition are reversible. The inhibitor (I) is bound rapidly and reversibly to the enzyme by relatively weak noncovalent forces to form an enzyme–inhibitor complex:

$$E + I \rightleftharpoons E\text{---}I$$

Some enzymes, however, are *irreversibly* inactivated by agents that bind covalently to the enzyme, permanently destroying its activity. The classic example of an irreversible inhibitor is the nerve gas diisopropylfluorophosphate, abbreviated DIFP. It is one of a class of highly toxic organophosphates developed during World War II in Germany by I. G. Farbenindustrie, first as insecticides and later as potential agents of chemical warfare. DIFP inactivates an enzyme involved in nerve transmission, acetylcholinesterase, by reacting with a serine residue at the active site and thereby preventing transmission of nerve impulses. The chemistry of the DIFP reaction with the key serine residue can be summarized as

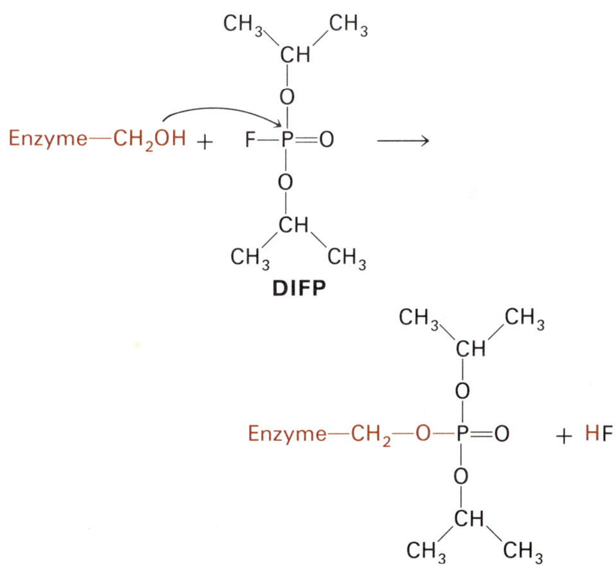

An especially interesting form of enzyme inhibition, with potential medical applications, concerns the so-called **suicide inhibitors.** These molecules are not inhibitors until modified by the catalytic activity of the target enzyme. Once activated, suicide inhibitors generally bond covalently with the enzyme to cause irreversible inhibition. Thus, these molecules are substrates that become inhibitors only after they are converted to products. Certain strains of *Neisseria gonorrhoeae* resistant to the antibiotic penicillin pose a serious threat. Their antibiotic resistance can be overcome by a method based on this type of inhibition. Penicillin-resistant strains of these microorganisms produce penicillinase, an enzyme that inactivates penicillin by cleaving the amide bond in the four-membered ring:

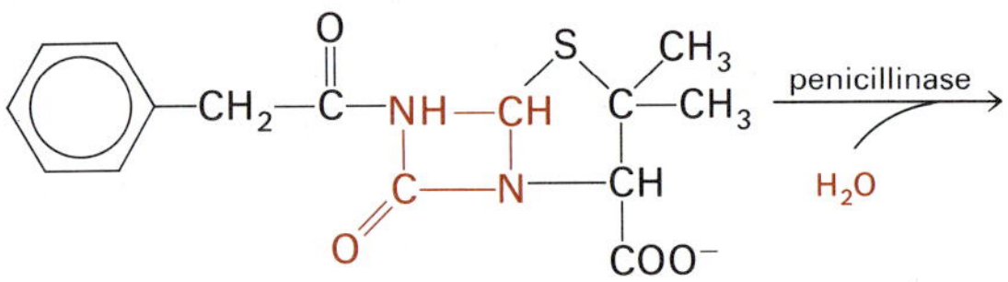

penicillin

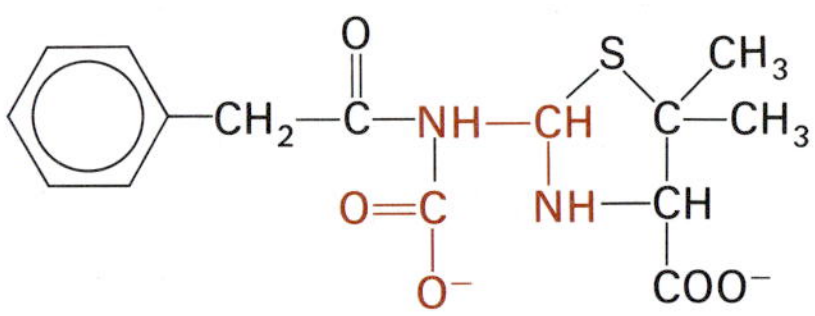

penicilloic acid (anion)

The product, penicilloic acid, is inactive as an antibiotic.

A suicide inhibitor of the penicillinase has been discovered; it is clavulinic acid:

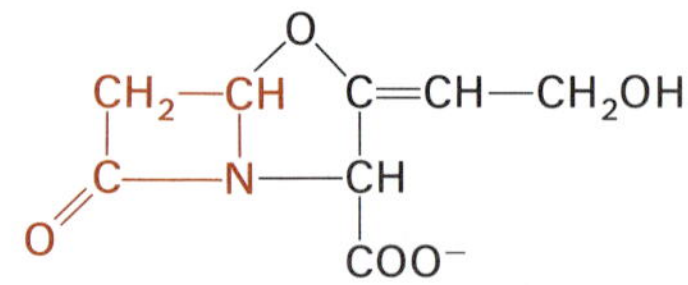

clavulinic acid (anion)

The enzyme penicillinase acts on clavulinic acid to cleave its four-membered ring. This cleavage product forms a covalent bond with penicillinase and inhibits irreversibly the catalytic activity of penicillinase. Inhibitors of this type can enhance the effectiveness of penicillin against resistant strains of bacteria.

Enzyme-inhibition studies provide valuable information on the mechanism of catalysis, on the shape of the active site, and on the identities of the functional groups in the active site responsible for binding substrate and for catalytic action.

Inhibition of enzymes by metabolic products is important in the regulation of metabolic pathways. A reaction product sometimes closely resembles the normal substrate and thus acts as a competitive inhibitor. However, in the natural regulation of enzymes in living cells, control usually is achieved by a more complicated process known as allosteric regulation, which is discussed in the following section.

22-5 ALLOSTERIC REGULATION OF ENZYME ACTIVITY

Very sensitive regulation of enzyme activity is needed to keep metabolic pathways functioning at optimal rates. The enzyme threonine deaminase, which catalyzes the first step in the biosynthesis of isoleucine from threonine (Section 22-1), is subject to feedback inhibition by the final product, isoleucine. Because the structure of isoleucine is quite different from that of threonine, isoleucine is not a competitive inhibitor:

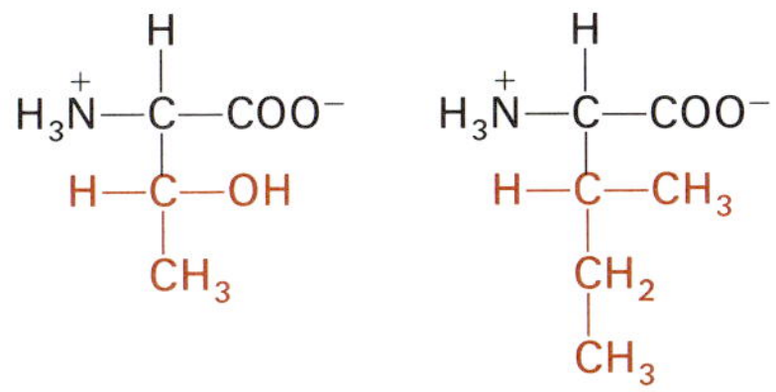

threonine **isoleucine**

Inhibition by isoleucine occurs at a site on the enzyme distinct from the one that binds threonine. This second site, called the allosteric site, specifically recognizes isoleucine. Its presence there triggers a change in the conformation of the enzyme to a form that binds threonine poorly. The binding of isoleucine therefore has an inhibiting effect on the enzyme activity. Enzyme-activity regulation by this type of noncompetitive inhibition is known as **allosteric regulation.**

A simple model for allosteric regulation proposes two conformations of the enzyme (Figure 22-11). One conformation, called T (for *tense*), binds the inhibitor strongly but the substrate only weakly. The other conformation, called R (for *relaxed*), binds the substrate strongly but the inhibitor only weakly. In the absence of inhibitor, the substrate keeps the enzyme in the R state, which has high catalytic activity. However, when inhibitor binds to the enzyme, the conformation of the enzyme shifts to the T state and its catalytic activity drops.

Every allosteric enzyme is composed of two or more polypeptide chains, or subunits. The enzyme illustrated in Figure 22-11 is a dimer; each subunit contains a site for substrate and a site for inhibitor. In addition to the interaction between substrate and inhibitor sites already described, the active sites of two subunits interact in an important way: binding of substrate to one subunit makes it easier for another substrate molecule to bind to the other subunit. The two sites **cooperate,** and this cooperative behavior can also be explained by the T and R states. If the enzyme is initially in the T state, binding of the first substrate molecule can switch *both* subunits to the R form (Figure 22-12). A second molecule of substrate can then easily bind to the remaining free site.

The cooperative behavior exhibited by allosteric enzymes accounts for the shape of the curve of their reaction rate against substrate concentration (Figure 22-13). The curve has an S-shape, quite different from

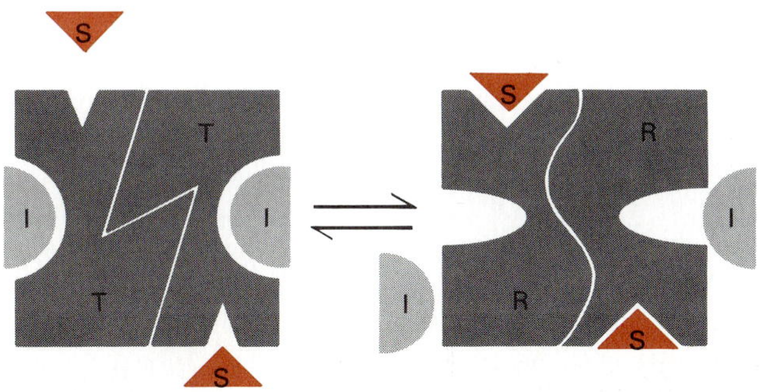

FIGURE 22-11

Allosteric regulation via R and T states.

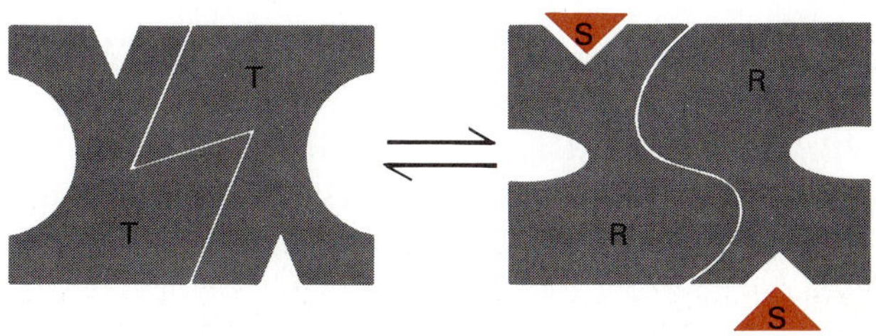

FIGURE 22-12

Cooperativity in an allosteric enzyme. Both subunits switch from the T state to the R state when one molecule of substrate is bound.

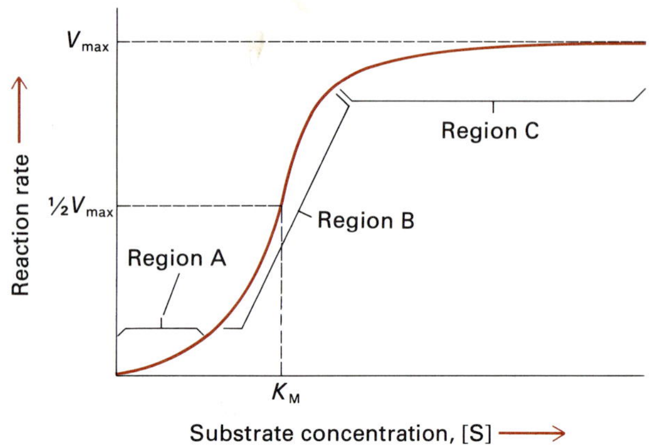

FIGURE 22-13

Reaction rate as a function of substrate concentration for an allosteric enzyme. The reaction rate increases only gradually when the value of [S] is low, but it increases very dramatically at intermediate values of [S].

that obtained for non-allosteric enzymes (Figure 22-7). At low substrate concentrations (region A in Figure 22-13) a small change in [S] has little effect on the reaction rate—the rate increases only gradually as [S] is increased. This is because when a substrate molecule encounters an enzyme molecule in the T state, the substrate can bind only with difficulty. But when it does bind to one subunit, it causes both subunits to shift to the R state, which binds substrate strongly. Thus, at intermediate values of [S] (region B in Figure 22-13), most enzyme molecules are in the R state; a small increase in [S] causes a large increase in the reaction rate because nearly all the added substrate molecules bind rapidly to enzyme. At high values of [S], the enzyme is nearly saturated with substrate, and the rate is close to its maximum.

Both the cooperative interactions between substrate sites and the ability to be inhibited by allosteric inhibitors are important factors in the regulation of metabolic pathways. In addition, for some enzymes, allosteric activators also exist; they bind preferentially to the R state and counteract the effects of the inhibitors. In Section 22-8, we shall see allosteric regulation of the oxygen-carrying capacities of myoglobin and hemoglobin.

22-6 A DETAILED LOOK AT ONE ENZYME: CHYMOTRYPSIN

One enzyme that has been studied extensively is the protease chymotrypsin, which catalyzes the hydrolysis of peptide bonds on the carboxyl side of phenylalanine, tryptophan, and tyrosine (Chapter 21). The enzyme is synthesized in the pancreas in the inactive zymogen form, chymotrypsinogen, thus preventing digestion of the cells in which it is synthesized. Like trypsin (Section 22-1), chymotrypsinogen becomes active chymotrypsin in the intestine after certain peptide bonds are hydrolyzed.

The single, inactive polypeptide chain of chymotrypsinogen contains 245 amino acid residues cross-linked by five disulfide bridges. In the intestine, this chain is split into three distinct smaller polypeptide units (designated A, B, and C) that together are catalytically active. The activation of chymotrypsinogen is summarized in Figure 22-14a. The first step is cleavage of the bond between arginine-15 (Arg^{15}) and isoleucine-16 (Ile^{16}). This step is catalyzed by trypsin, another digestive enzyme found in the intestine.

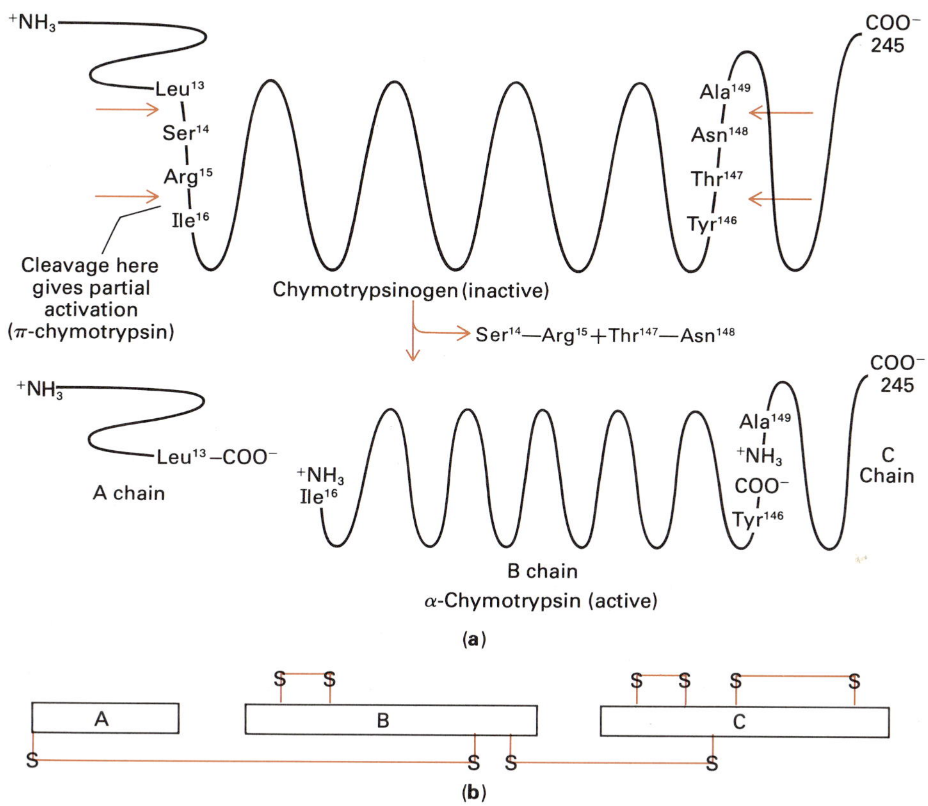

FIGURE 22-14

Activation of chymotrypsinogen. **(a)** The peptide bonds that are hydrolyzed. **(b)** The disulfide linkages that stabilize the structure.

Cleavage of the Arg—Ile bond produces an active form of the enzyme known as π-chymotrypsin. This molecule cleaves other π-chymotrypsin molecules to produce the final (stable) form of the enzyme, α-chymotrypsin.

As seen in Figure 22-14b, the three polypeptide chains of α-chymotrypsin are connected by two disulfide bridges. Chymotrypsin contains regions of β-pleated sheet structure, especially in the B chain, but very little α helix. The molecule is folded so that all the charged groups are on the outside, except for three that have a key role in catalysis. The substrate fits into a nonpolar pocket in the enzyme.

Three amino acid residues at the active site collaborate to provide the catalytic action of chymotrypsin. The three residues involved are aspartate-102, histidine-57, and serine-195 (Figure 22-15). These three residues form a *charge-relay network* in which the nega-

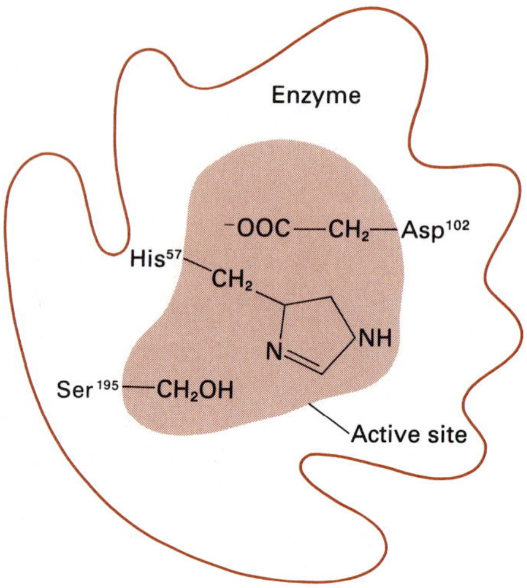

FIGURE 22-15
Charge-relay system at the active site of chymotrypsin.

tive charge on Asp^{102} can rapidly be transferred to Ser^{195} by shifting two protons a short distance:

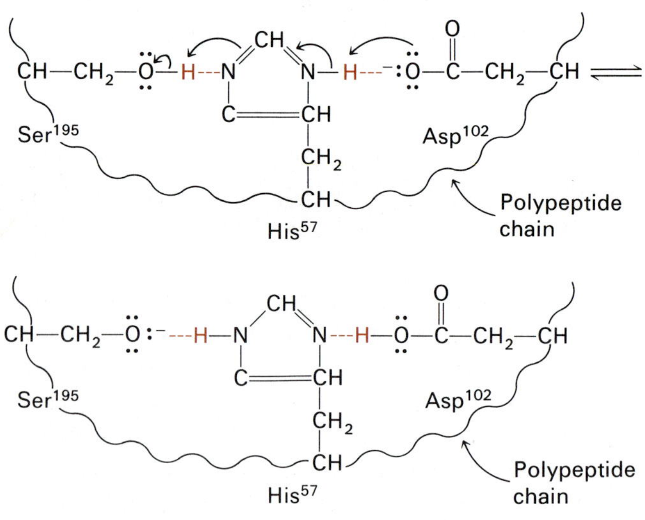

The negatively charged serine residue is a very strong nucleophile.

A plausible catalytic mechanism for chymotrypsin consists of two stages. The first stage, in which the peptide bond of the substrate is cleaved, starts with a nucleophilic attack by the serine oxygen on the carbonyl carbon of the peptide bond. The product of this step, an addition intermediate, is formed by a mechanism like those we saw for the hydrolysis of esters and amides (Sections 19-7 and 20-8):

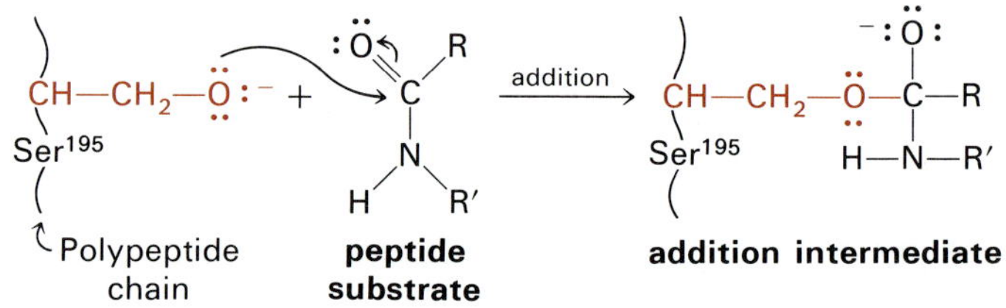

(The Ser—O^- must be produced by the charge-relay network after the substrate is bound and in concert with the attack on the peptide bond. Because Ser—O^- is such a strong nucleophile, it cannot exist free at physiological pH.)

In a second step, an elimination reaction, the peptide bond is cleaved, freeing one product (the amino component of the substrate, H_2NR'). The carbonyl carbon of the substrate is now in an ester linkage with Ser^{195} of the enzyme.

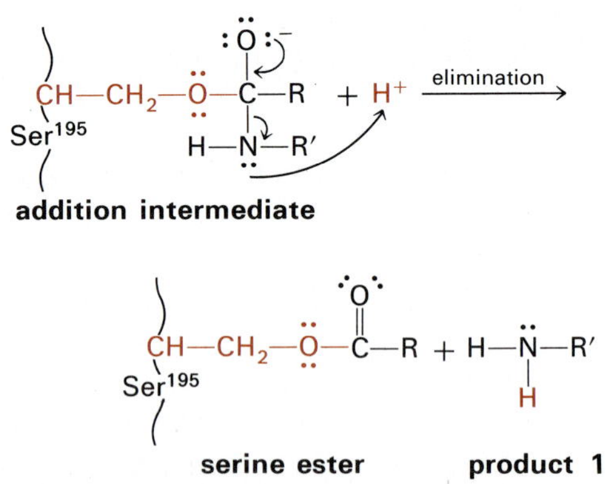

(22-4)

(The H^+ is furnished by His^{57}.)

In the second stage of the reaction, this serine ester is hydrolyzed. The mechanism consists of two steps that are the same as those just discussed, with H_2O substituting for the amine H_2NH'. In the first step, the nucleophilic oxygen of a water molecule attacks the carbon of the ester to form an addition intermediate. Then the C—O bond of the ester is cleaved in an elimination reaction, freeing the second product, the acid component of the substrate, RCOOH.

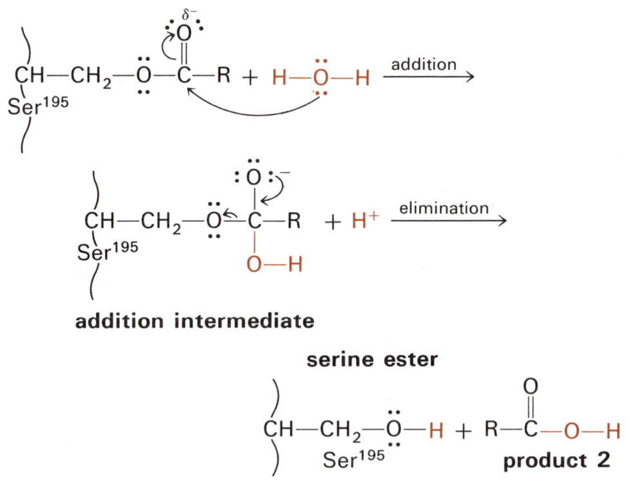

(Protons produced or consumed in the reactions are transferred to and from His[57].) The enzyme is now in its original state, and it can catalyze the hydrolysis of another substrate molecule.

The function of the enzyme is to bring together all the reactants in positions favorable for reaction, and to provide a very strong nucleophile (Ser—O^-) for the attack on the carbonyl carbon of the peptide bond. When an amide or peptide bond is hydrolyzed in the absence of a catalyst, the nucleophile that attacks the carbonyl carbon is water, a much weaker nucleophile than Ser—O^-. As we saw in Chapter 20, to achieve a reasonable rate of hydrolysis of an amide in the laboratory, it is necessary to boil the reaction mixture for hours using a strong acid or base as catalyst.

In contrast, chymotrypsin is effective at body temperature at physiological pH. Each molecule of chymotrypsin catalyzes the hydrolysis of 6000 peptide bonds every minute. The charge-relay system of aspartate, histidine, and serine is thought to increase the rate of catalysis by a factor of about 1000. The importance of this set of three amino acid residues is emphasized by its occurrence in many enzymes, even some that evidently evolved from very different origins.

22-7 PARTICIPATION OF COENZYMES IN ENZYMATIC REACTIONS

The structure of chymotrypsin and its mechanism of action illustrate how several amino acid residues can interact to provide properties that no single amino acid residue could produce. However, for many of the complex reactions that must be catalyzed by enzymes in various biochemical pathways and cycles, even the versatility provided by the various side chains of the 20 principal amino acids is inadequate. In particular, reactions that involve oxidation–reduction and the making or breaking of carbon–carbon bonds require other chemical species called **cofactors.** Enzymes that require cofactors are inactive without them. The cofactor may be a nonprotein organic molecule or a metal ion such as iron (Fe^{2+}), copper (Cu^{2+}), zinc (Zn^{2+}), magnesium (Mg^{2+}), or manganese (Mn^{2+}). Alcohol dehydrogenase, for example, which reduces pyruvate to lactate, requires Zn^{2+} as a cofactor; and NADH dehydrogenase, an enzyme that oxidizes NADH to NAD^+, requires Fe^{2+}.

When the cofactor is a nonprotein organic sub-

TABLE 22-2
Vitamins and coenzymes

Vitamin	Coenzyme derived from vitamin	Function
B_1: Thiamin	Thiamine pyrophosphate	Aldehyde reactions and decarboxylation (elimination of CO_2)
B_2: Riboflavin	Flavine adenine dinucleotide; flavin mononucleotide	Oxidation–reduction
B_6: Pyridoxine	Pyridoxal phosphate	Amino group reactions
B_{12}: Cobalamin	Cobamide coenzyme	C—C bond rearrangements
Niacin (Nicotinate)	Nicotinamide adenine dinucleotide (NAD^+)	Oxidation–reduction
Pantothenate	Coenzyme A	Acyl (RCO—) reactions
Biotin	Biotin	CO_2 reactions; carboxylation and decarboxylation
Folic acid	Tetrahydrofolate	Reactions of one-carbon units

stance, it is called a **coenzyme.** The coenzymes have a wide variety of structures, specially adapted to the reactions in which they participate. Coenzymes are closely associated with the enzymes with which they act, either by covalent bonds or by noncovalent interactions. Many coenzymes are derived from vitamins; Table 22-2 lists the B-complex vitamins with the names of their derived coenzymes and descriptions of the reactions in which they participate. Because the structures of most of the coenzymes are highly complex, they are not given in the table; many include both carbohydrate and nucleotide portions (Chapters 23 and 25). But the coenzyme pyridoxal phosphate, a derivative of vitamin B_6, has a relatively simple structure:

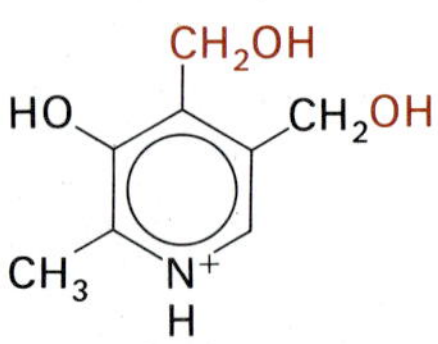

pyridoxine
(vitamin B_6)

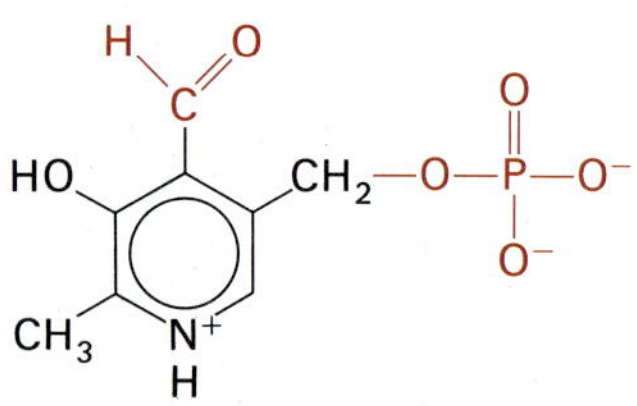

pyridoxal phosphate
(coenzyme form)

Pyridoxal phosphate is a versatile coenzyme that binds to many different kinds of enzymes. It is a prosthetic group (Section 21-6) of all transaminases, enzymes that catalyze the transfer of an amine group from an α-amino acid to an α-keto acid. For example, it is a prosthetic group of the enzyme glutamate transaminase that catalyzes the conversion shown in Equation 22-5; this process occurs in the degradation of surplus amino acids in the body:

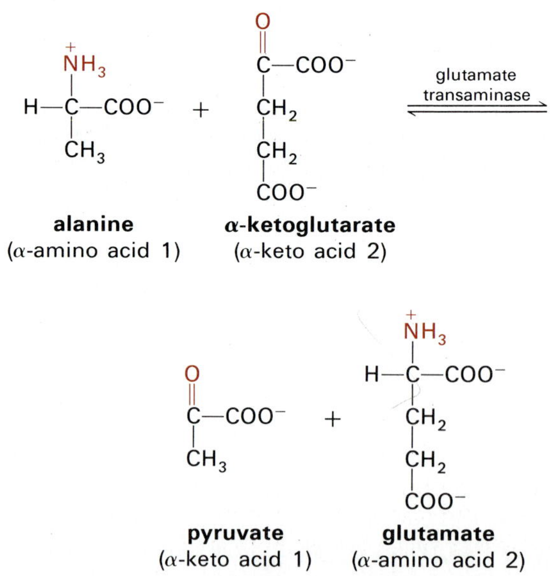

(22-5)

The amino group of glutamate is eventually converted to urea.

In the absence of substrate, the aldehyde group of pyridoxal phosphate forms a Schiff base (Section 18-6) with the side-chain amino group of a lysine residue at the active site of the enzyme:

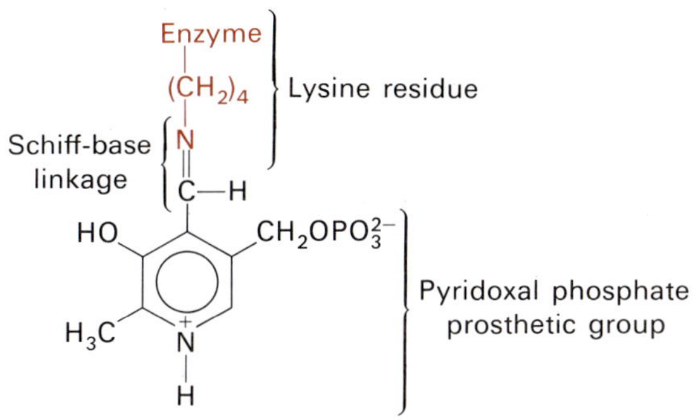

During catalysis, the Schiff-base bond to the enzyme is broken, and a new one is formed with the α-amino acid:

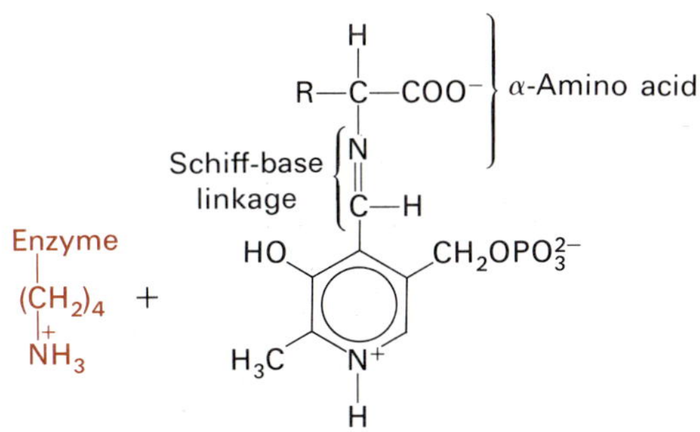

Noncovalent forces keep the new Schiff base tightly bound to the enzyme during the catalytic reaction. At the end of the reaction, the pyridoxal phosphate is once again in Schiff-base linkage with the enzyme.

Transamination is only one of many kinds of reactions in which pyridoxal phosphate acts as a coenzyme. Other reactions catalyzed by pyridoxal phosphate enzymes include decarboxylation reactions (removal of CO_2), deaminations (removal of amino groups), and racemizations (the formation of a D-amino acid from an L-amino acid).

Sulfa drugs function by preventing the formation of a coenzyme by harmful bacteria. The bacteria synthesize folic acid from *para*-aminobenzoic acid (PABA):

***p*-aminobenzoic acid (PABA)**

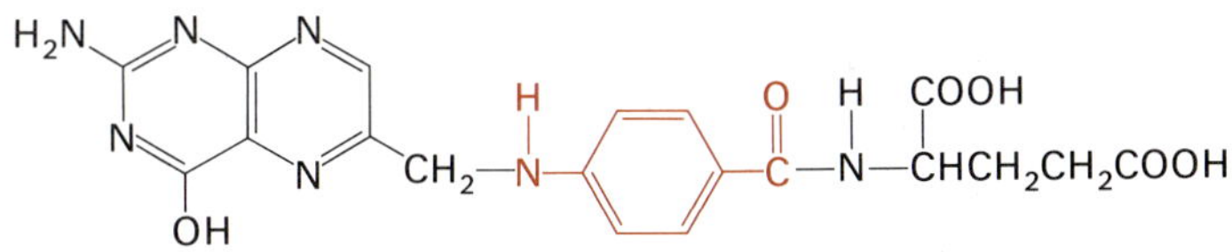

folic acid

The folic acid is then converted to the coenzyme. Sulfa drugs are similar in structure to *para*-aminobenzoic acid:

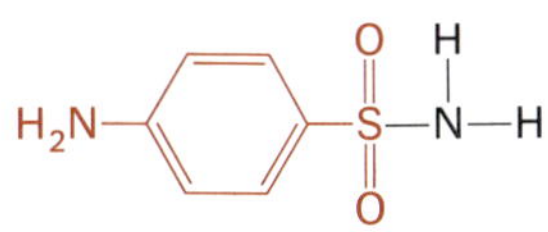

sulfanilamide

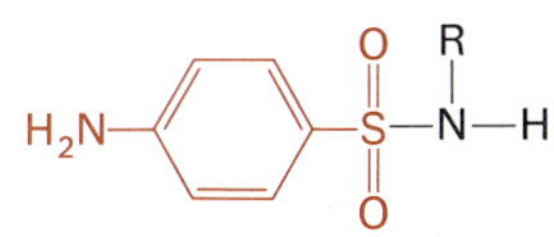

general formula of a sulfa drug

The bacteria mistake the sulfa drug for PABA and incorporate the drug into "folic" acid in place of PABA. This results in a compound that cannot be used to synthesize the needed coenzyme, and so the bacteria die. Sulfa drugs do not interfere with the biosynthesis of the coenzyme in humans because humans cannot synthesize their own folic acid; folic acid is obtained in the diet.

22-8
NONENZYMATIC PROTEINS: TRANSPORT PROTEINS

Although the enzymes constitute the largest class of proteins, many proteins are not enzymes at all, and a few have other functions in addition to being enzymes. In this section and the next, we consider (briefly) some special functions performed by nonenzymatic proteins and see how the proteins myoglobin and hemoglobin transport oxygen.

Many of the small molecules and ions required by cells for metabolism are carried by **transport proteins.** These proteins share with enzymes their specificity for other molecules, but they only bind these molecules and do not catalyze any reaction. Hemoglobin (the major constituent of red blood cells) transports oxygen from the lungs to the tissues, and myoglobin transports oxygen in muscle cells. About three-quarters of the solutes in blood plasma, the noncellular component of the blood, are proteins. Many of them are transport proteins; they carry nutrients from the intestine to the liver and other organs, and carry waste products to the kidneys for excretion. Examples of plasma proteins include serum albumin (which transports fatty acids), lipoproteins (which transport fats; see Section 24-7), and transferrin (which carries iron).

Some transport proteins, instead of carrying molecules from one part of the body to another, are stationed in the cell membrane. There the protein molecule binds a small molecule such as an amino acid. When the amino acid is bound, the conformation of the transport protein is altered in such a way that the small molecule can cross the membrane and enter the cell. A protein that binds glucose molecules and transports them across the cell membrane has recently been isolated from the surface of red blood cells.

We shall now describe how the transport proteins myoglobin and hemoglobin function. As we saw in Chapter 21, myoglobin is a globular protein folded in a particular way to accommodate a heme molecule. It is the iron atom of the heme that binds the oxygen. Hemoglobin consists of four subunits, two of type α and two of type β. Both the α and β subunits are similar to myoglobin, and each subunit contains a heme group (Figure 22-16). The fact that hemoglobin is an allosteric protein gives it important transport properties different from those of myoglobin.

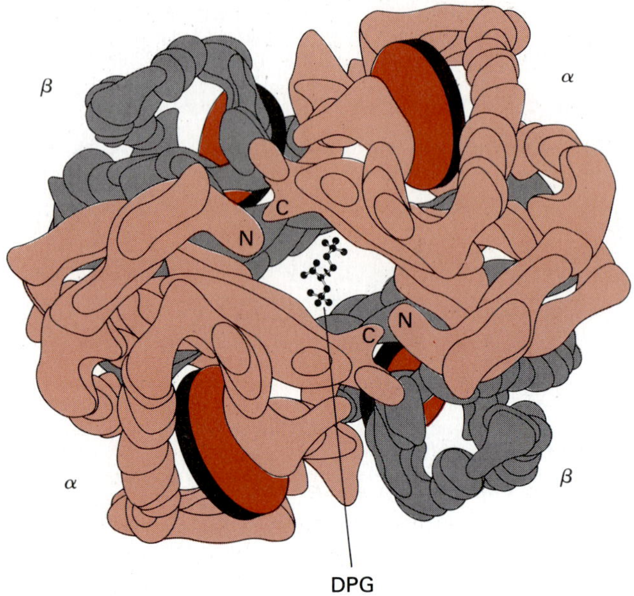

FIGURE 22-16

The arrangement of the four subunits of hemoglobin. The central cavity is occupied in deoxyhemoglobin by a molecule of diphosphoglycerate (DPG). [After M. F. Perutz, "The Hemoglobin Molecule." Copyright © 1964 by Scientific American, Inc. All rights reserved.]

Figure 22-17 plots the percentage of oxygen sites filled (the **oxygen affinity**) against the partial pressure of oxygen, P_{O_2} (essentially the concentration of O_2; see Chapter 3), for myoglobin and for hemoglobin. These curves are like those we discussed for enzymes, except that we are plotting percentage of O_2 sites filled rather than reaction rate. The curve for hemoglobin is S-shaped, like that of an allosteric enzyme (Figure 22-

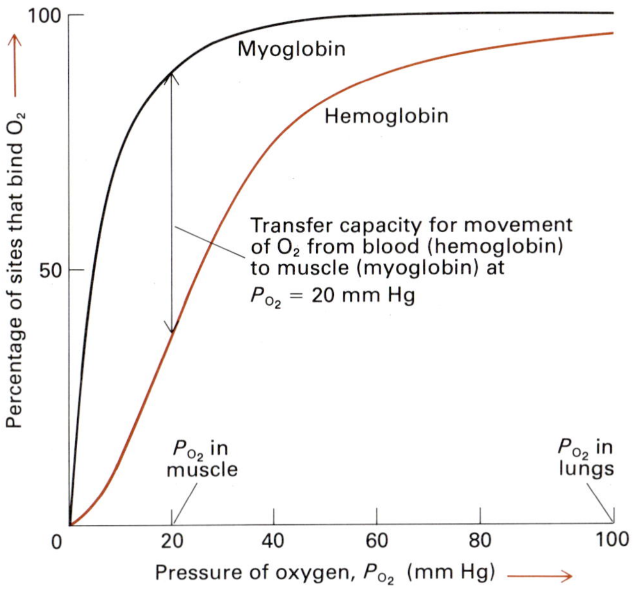

FIGURE 22-17

The percentage of oxygen sites filled as a function of the partial pressure of oxygen P_{O_2} for myoglobin and hemoglobin. Typical values of P_{O_2} in the lungs and in muscle tissue are indicated.

13). The curve for myoglobin is like that of a nonallosteric enzyme (Figure 22-7). The S-shaped curve means that there is communication between the four subunits of heme, just as there is between the subunits of an allosteric enzyme. The binding of O_2 by heme is cooperative. That is, the binding of an O_2 molecule by the heme group of *one* subunit changes the conformations of the other subunits and makes it easier for O_2 to bind to the other three heme groups. The cooperativity exhibited by the subunits of hemoglobin, like that exhibited by the subunits of allosteric enzymes, can be explained in terms of the conversion of a T state that binds oxygen poorly to an R state that binds oxygen well. The shift from the T state to the R state requires breaking salt bridges (Section 21-6) that help hold the four subunits together.

Figure 22-17 shows that at all partial pressures of O_2 below about 100 mm Hg, the percentage of sites occupied by O_2 is higher in myoglobin than in hemoglobin. Myoglobin has a greater affinity for oxygen than hemoglobin has. Typically, the partial pressure of oxygen in the lungs is about 100 mm Hg. Thus, when blood saturated with O_2 in the lungs reaches the muscles where the oxygen partial pressure is about 20 mm Hg, a large amount of oxygen is transferred from hemoglobin to the myoglobin in the muscles, as indicated in Figure 22-17. The figure also shows that myoglobin releases O_2 only when the partial pressure of O_2 is very low. In actively exercising muscle, P_{O_2} is close to zero, and oxygen release proceeds readily. Because P_{O_2} in the tissues (which require oxygen for metabolism) is also much less than 100 mm Hg (typically about 40 mm Hg), oxygen is released by hemoglobin as the blood circulates through the body. The oxygen diffuses from the capillaries into the cells.

The binding of oxygen by hemoglobin is partly regulated by the concentrations of H^+ and CO_2. Increasing the acidity of the blood causes hemoglobin to release oxygen (Figure 22-18). Increasing the concentration of CO_2 at constant pH also favors the release of oxygen by hemoglobin. Both CO_2 and H^+ are produced by actively metabolizing cells, so both these factors enhance the release of oxygen by hemoglobin to the cells where oxygen is needed. Although the equilibrium is really much more complicated, we can summarize it with the simple equation

$$HbO_2 + H^+ + CO_2 \rightleftharpoons CO_2\text{—}Hb\text{—}H^+ + O_2$$

where Hb represents one subunit of hemoglobin. The binding of H^+ and CO_2 by hemoglobin causes it to release oxygen. In the lungs the reverse reaction takes place, and H^+ and CO_2 are released as oxygen binds to the hemoglobin.

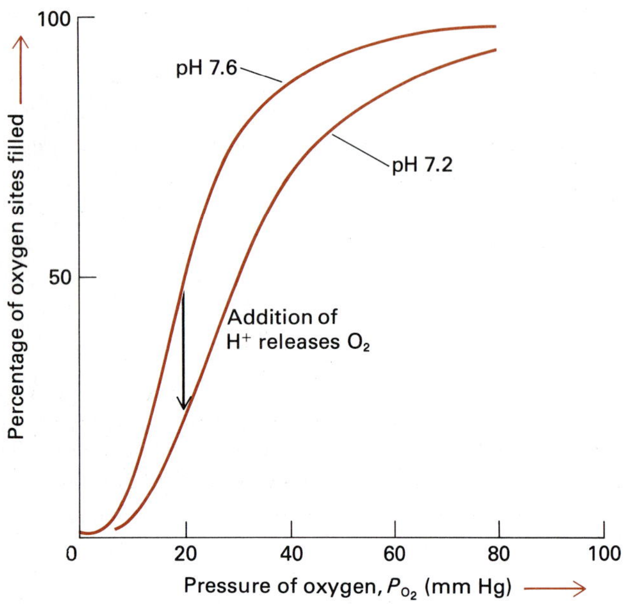

FIGURE 22-18

The effect of pH on the oxygen affinity of hemoglobin. Lowering the pH causes oxygen to be released. (Physiological pH is 7.4.)

EXERCISE 22-4 CO_2 binds to hemoglobin by reacting with uncharged amino groups. In Section 18-6 we discussed reactions between amino groups and aldehydes or ketones. Consider CO_2 as a form of ketone, and write a mechanism for the reaction by which it binds to the amino group. Use the formula R—NH_2 to represent a subunit of hemoglobin.

The binding of O_2 by hemoglobin is regulated by a small organic molecule diphosphoglycerate (DPG) as well as by H^+ and CO_2.

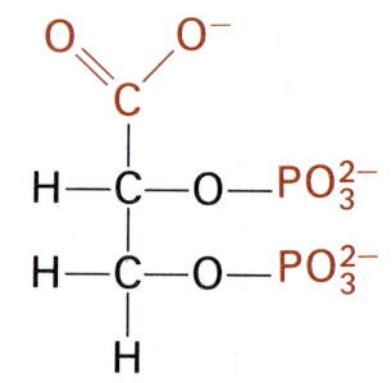

DPG

As shown in Figure 22-16, the DPG fits into a cavity between the β subunits of hemoglobin. The negative charges of the DPG form salt links with positively charged amino acid residues in the β chains, but only in **deoxyhemoglobin** (hemoglobin without any bound oxygen). When hemoglobin is oxygenated, these salt linkages must be broken—an energy-requiring process. Thus, DPG lowers the oxygen affinity of hemoglobin. **Oxyhemoglobin** (the oxygenated form of hemoglobin) does not bind DPG. The change in the conformations of the subunits that occurs when hemoglobin is oxygenated brings the subunits closer together, leaving no space for the DPG between the β chains. The equilibrium between hemoglobin, O_2, and DPG can be represented by the equation

$$Hb_4\text{—}DPG + 4O_2 \rightleftharpoons Hb_4(O_2)_4 + DPG$$

where Hb_4 represents the four subunits of the hemoglobin molecule, and $Hb_4(O_2)_4$ represents a completely oxygenated hemoglobin molecule.

DPG plays a role in the adaptation of the body to disorders in the delivery of oxygen to the tissues. In emphysema (a respiratory disease that causes a reduced flow of air in the bronchioles), hemoglobin in the lungs is not fully saturated with oxygen. In partial compensation for this, the DPG concentration in an emphysema patient is higher than normal. This reduces the oxygen affinity of the hemoglobin and causes it to release more O_2 to the tissues than it would at a normal DPG concentration. The concentration of

DPG in the blood also increases as a person accustomed to living at sea level acclimatizes to a high altitude. At the higher altitude, where the atmospheric pressure is lower, less O_2 is bound by the hemoglobin in the lungs. The DPG concentration begins to increase as soon as a person gains in altitude, causing an increase in the amount of oxygen delivered to the tissues. (A significant increase in DPG concentration occurs in less than one day.)

We have seen that allosteric regulation of oxygen binding in hemoglobin by H^+, CO_2, and DPG permits hemoglobin to transport O_2 from the lungs to the tissues where it is needed for metabolism. In the reverse direction, hemoglobin transports H^+ and CO_2 from the tissues to the lungs, where the CO_2 is exhaled and the H^+ is taken up by HCO_3^-, the major blood buffer. The allosteric regulation results from interactions between separate binding sites on the polypeptide chains. These interactions cause changes in the conformations of the chains and in the quaternary structure of the molecule. There is also cooperativity between the O_2-binding sites of the separate subunits, giving the S-shaped curve of oxygen affinity against partial pressure of oxygen.

EXERCISE 22-5 What is the effect of each of the following changes on the equilibrium concentration of oxyhemoglobin [HbO_2] in a sample of blood? (a) decrease [H^+]; (b) decrease [CO_2]; (c) decrease [DPG].

22-9 OTHER NONENZYMATIC PROTEIN FUNCTIONS

In Chapter 21 we saw that there are several types of **fibrous proteins** that play a purely structural role. The α-keratin of hair, hooves, nail, horn, feathers, and wool has properties that depend to a large extent on its α-helical structure. Three or seven helices can coil together to form ropes stabilized by disulfide bridges. The more disulfide cross-links, the more rigid the material. Silk, which has the β-pleated sheet structure, is also a fibrous protein. The most important fibrous protein is collagen—the structural protein of skin, bone, and teeth. As noted in Chapter 21, tropocollagen molecules are triple helices; in collagen, these molecules are cross-linked in a way that gives a strong, rigid material.

Some proteins have a **protective function.** These include the blood-clotting substances (thrombin and fibrinogen) and **antibodies** that give protection against foreign organisms such as bacteria or viruses. Antibodies are also involved in the rejection of tissue from another person (in organ transplants). Most antibodies are molecules known as γ-globulins, which have two chains of low molecular weight and two of high molecular weight (Figure 22-19). These chains are arranged in such a way that they provide two binding sites for a specific antigen. **Antigens** are substances that stimulate the synthesis of antibodies. Any large molecule, such as a protein, that possesses on its surface an arrangement of atoms different from those on the surfaces of molecules of the host is an antigen. The immune system of an organism can distinguish the organism's own molecules from foreign ones by these differences on the molecule's surfaces. When an

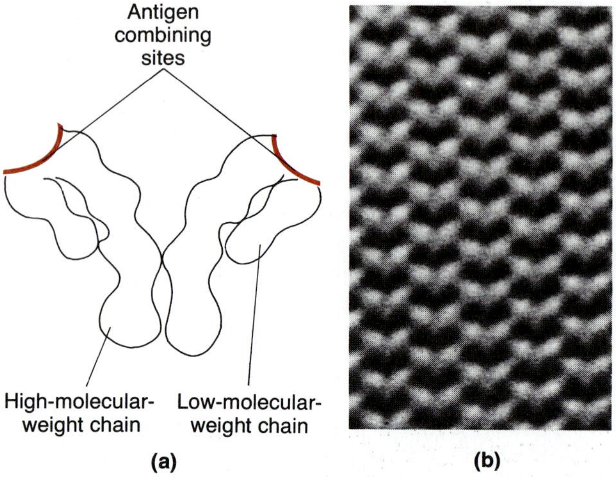

FIGURE 22-19

Antibody molecules. **(a)** Schematic representation of the structure of an antibody molecule. **(b)** Electron micrograph of a crystal of antibody molecules. (Note the Y shapes of the molecules.) [Courtesy of David R. Davies.]

antibody binds an antigen (say a virus), the antigen is inactivated so that it cannot cause any damage. In Section 23-9 we discuss various antibody–antigen reactions that can occur when blood transfusions are given.

Receptor proteins share with enzymes and transport proteins their specificity for other molecules. They do not catalyze any reaction, but they can trigger an event after the particular molecule is bound. For example, when a receptor at a nerve junction binds the small molecule acetylcholine

$$CH_3\overset{\overset{\displaystyle O}{\|}}{C}—O—CH_2—CH_2—\overset{+}{N}(CH_3)_3$$

acetylcholine

the conformation of the receptor changes. This change alters the permeability of the nerve cell membrane to allow sodium ions to enter the cell and trigger the transmission of a nerve impulse.

Proteins also are responsible for **muscle contraction.** The two main proteins that serve this function are actin (which has a long filamentous structure) and myosin (which has a globular head and a rodlike tail). Their structures are shown in Figure 22-20. Muscle proteins perform the important function of transforming chemical energy into mechanical work. Myosin has an unusual structure uniquely suited to causing this transformation. Myosin combines features of a fibrous protein and an enzyme; the globular heads of myosin act as enzymes for the hydrolysis of ATP, which provides the energy for muscle contraction; and the rodlike tail does mechanical work. In muscle fibers, actin and the tails of myosin are arranged in parallel rows that slide along each other during muscle contractions (Figure 22-21).

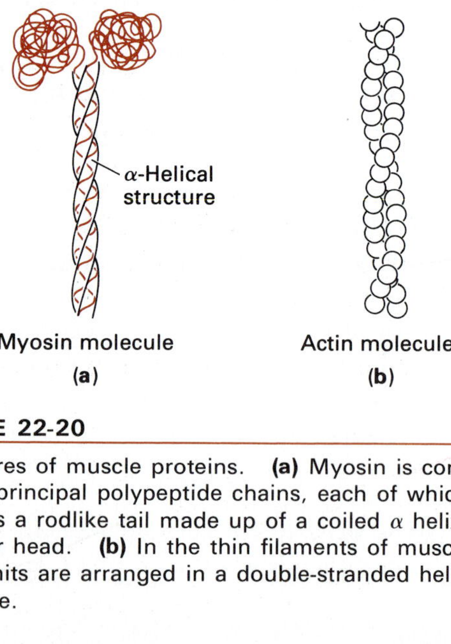

FIGURE 22-20

Structures of muscle proteins. **(a)** Myosin is composed of two principal polypeptide chains, each of which contains a rodlike tail made up of a coiled α helix and a globular head. **(b)** In the thin filaments of muscle, actin units are arranged in a double-stranded helical structure.

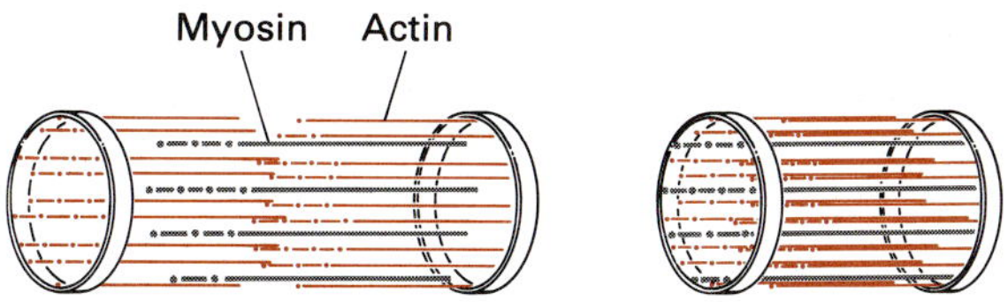

FIGURE 22-21

Interaction of myosin and actin in muscle contraction. The parallel rows of myosin and actin molecules slide along each other during muscle contraction.

KEY TERMS

Enzymes are globular proteins that catalyze biochemical reactions. A *substrate* is a specific compound that is transformed into a product by a particular enzyme. A *zymogen* or *proenzyme* is an inactive precursor form of an enzyme.

The *active site* of an enzyme contains amino acid residues that bind the substrate and catalyze the reaction. An enzyme is said to be *saturated* with substrate when all the active sites are occupied.

The *turnover number* is the number of molecules of product formed in one minute by one molecule of enzyme. The *maximal velocity,* V_{max}, is the maximal rate of conversion of reactants to products for a given concentration of enzyme under defined conditions. The *Michaelis constant,* K_M, is the substrate concentration required to produce half-maximal reaction rate for an enzyme under defined conditions.

Competitive inhibition is inhibition of the catalytic activity of an enzyme through the action of a compound that resembles the substrate in structure and displaces it from the active site of the enzyme; the inhibitor effect can be overcome by increased concentrations of the substrate. *Noncompetitive inhibition* is inhibition of the catalytic activity of an enzyme by a compound that does not resemble the substrate structurally; the inhibitor effect cannot be overcome by increasing the concentration of substrate. A *suicide inhibitor* is a substrate that the enzyme converts to an inhibitor for the enzyme itself.

In an *allosteric enzyme,* one active site can interact with another active site in the molecule (typically in another subunit). The enzyme activity can also be regulated by inhibitors that bind to sites other than the active site. In *feedback inhibition,* the final product of a reaction sequence inhibits the allosteric enzyme that catalyzes the first step of the sequence.

Coenzymes are relatively small compounds, derived from or related to vitamins, that associate with enzymes and participate in catalytic processes.

Antibodies are protein molecules of the immune system that bind foreign compounds called antigens. *Antigens* are substances that elicit an immune response to produce specific antibodies that bind to the antigens and inactivate them.

CHAPTER SUMMARY

Many proteins function in living organisms as enzymes. Virtually every biochemical reaction has its particular enzyme. *Enzymes* are highly specific catalysts; each acts on only a small number of biochemical compounds. The compounds transformed by an enzyme are known as its *substrates*. Enzymes function by lowering the activation energy needed to reach the transition state for the conversion of substrate to product. Enzymes are commonly named after the substrates of the reactions they catalyze, with the suffix *-ase*. They vary in the rates at which they catalyze the transformations of their substrates and in the concentration of substrate needed for maximal efficiency. The highest rate that a given amount of enzyme can achieve is called its *maximal velocity*. The concentration of substrate required to give half-maximal velocity is called the *Michaelis constant* (K_M); it is characteristic of the enzyme under defined conditions and is independent of the enzyme concentration. The *turnover number* is defined as the number of molecules of substrate converted to product per minute by each molecule of enzyme; the values for some enzymes are as high as 10^7, although 10^3 is more typical. Enzymes that are capable of degrading other molecules in cells are often synthesized in an inactive form known as a *zymogen*. Chymotrypsin is one such protein, and its inactive zymogen is called chymotrypsinogen.

Enzymes can be inhibited by various compounds, including molecules that resemble their substrates and are called *competitive inhibitors*. The ability of enzymes to catalyze reactions is related to the properties of amino acid residues at the active site of the enzyme. In the case of the enzyme chymotrypsin, the close proximity of three amino acid side chains (Asp, His, and Ser) leads to an unusually active Ser–OH that plays an essential role in the reaction catalyzed by the enzyme. For certain enzymes, compounds known as *coenzymes* (related to vitamins) must bind to the enzymes in order to activate the enzyme for its catalytic process.

In the class of enzymes known as *allosteric enzymes*, the catalytic activity of the enzyme is regulated by key biochemical compounds that need not resemble the substrates for the enzymes. In this case, the regulating compounds bind at sites distinct from the substrate binding site and influence enzyme activity by shifting the equilibrium between relatively active conformations (designated R) and relatively inactive conformations (designated T). When reaction rate is plotted against substrate concentration for an allosteric enzyme, an S-shaped curve is obtained.

Other proteins are concerned only with binding molecules, but not with any catalytic transformation of the molecules bound. These include antibodies and hemoglobin. Hemoglobin is an allosteric protein; its oxygen-binding capacity is regulated by the concentrations of H^+, CO_2, and diphosphoglycerate. There also is cooperativity between the oxygen-binding sites on the different subunits of the hemoglobin molecule, causing the curve of oxygen affinity against partial pressure of oxygen to be S-shaped.

PROBLEMS

22-1. For each pair of terms on the left, pick the one that most closely fulfills the definition on the right.

(a) Allosteric inhibitor or competitive inhibitor	(a) An enzyme inhibitor most likely to resemble the substrate for the enzyme

(b) α helix or β sheet	(b) A regular structure involving hydrogen bonds between amino acid residues about four residues apart on the polypeptide chain
(c) Coenzyme or antigen	(c) A small molecule that participates with an enzyme in catalyzing a biochemical reaction
(d) Antibody or zymogen	(d) The inactive form of a protein, activated by cleaving key peptide bonds
(e) K_M, or V_{max}	(e) The property of an enzyme that indicates the concentration of substrate needed to achieve one-half the maximal rate of catalysis
(f) Secondary structure or quaternary structure	(f) The aspect of protein structure related to the arrangement of polypeptide chains in an oligomeric protein

22-2. The charge-relay system of chymotrypsin and related enzymes involves the residues Asp, His, and Ser. Considering the R groups of these residues, which amino acids (if any) could substitute for Asp, for His, or for Ser?

22-3. Two enzymes, A and B, are observed to have the plots of reaction rate against substrate concentration shown below. On the basis of this

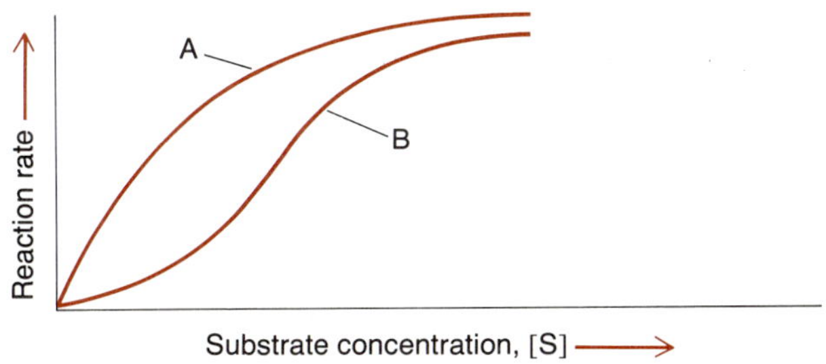

information, what inferences (if any) can be made about whether enzyme A or B is more likely to be composed of subunits?

22-4. Data are given in the table below for a series of experiments on the rate of product formation against substrate concentration for (1) a solution containing a certain enzyme E alone; (2) a solution containing enzyme E plus compound X; and (3) a solution containing enzyme E plus compound Y. The concentration of E is the same in all the experiments.

Substrate concentration (μM)	Product formed (μmoles/min)		
	Exp. 1 (E only)	Exp. 2 (E + X)	Exp. 3 (E + Y)
0.5	4.2	2.3	2.2
1.0	6.7	4.5	3.8
1.5	7.7	5.5	5.2
2.0	8.2	6.1	6.5
2.5	8.4	6.5	7.5
3.0	8.4	6.5	8.4

(a) Find V_{max} and K_M for each of the three sets of experiments.
(b) What conclusions can be drawn about compounds X and Y?
(c) Which compound, X or Y, is more likely to bear a close structural relationship to the substrate?

22-5. Draw the graph for experiment 1 of Problem 22-4. Draw also the graphs you would expect to obtain if the experiment were repeated (a) at a higher temperature (assume that the enzyme is not denatured at this new temperature); and (b) with a lower concentration of enzyme. Show clearly on your graphs how these changes affect V_{max} and K_M.

22-6. Some types of leukemia (a cancer of the white blood cells) require asparagine as an essential nutrient. These cancers can be suppressed by the intravenous administration of asparaginase, an enzyme that catalyzes the reaction

$$\text{Asparagine} + H_2O \rightleftharpoons \text{Aspartate} + NH_4^+$$

Aspartate cannot be used as a nutrient by the cancer cells. Asparaginases from different sources were found to vary widely in their K_M for asparagine. The concentration of asparagine in the blood is very low. Which asparaginase would you expect to be more effective against cancer, one with a K_M of 10^{-3} M or one with a K_M of 10^{-6} M? Explain.

22-7. The enzyme acetylcholinesterase (which catalyzes the hydrolysis of acetylcholine, a transmitter of nerve impulses) contains serine, aspartate, and histidine at its active site, just as chymotrypsin does. Explain why acetylcholinesterase is inactive in acidic solution but is active in basic solution up to a pH of over 10.

22-8. Trypsin is produced in the pancreas in the form of a zymogen. The pancreas also contains a trypsin inhibitor. Explain why it is important to have a trypsin inhibitor in the pancreas.

22-9. Diphosphoglycerate mutase catalyzes the isomerization of 1,3-diphosphoglycerate to 2,3-diphosphoglycerate, a regulator of oxygen transport in red blood cells:

1,3-DPG **2,3-DPG**

The enzyme is inhibited by the product, 2,3-diphosphoglycerate. Describe an experiment that would determine whether the inhibition is competitive or noncompetitive. Which type of inhibition would you expect this product to exhibit?

22-10. (a) Some bacteria are resistant to penicillin because they produce the enzyme penicillinase, which hydrolyzes an amide bond and inactivates the antibiotic. The molecular weight of penicillinase is 30,000. A solution containing 0.3 μg of penicillinase catalyzes the hydrolysis of 0.6 μmole of penicillin per minute. What is the turnover number of penicillinase? (Assume one active site per molecule.)

(b) The table below gives the results of a series of experiments in which the rate of hydrolysis of penicillin by penicillinase was measured as a function of the concentration of penicillin. What is V_{max} for this experiment, and what is K_M?

Penicillin concentration (moles/ℓ)	Rate of hydrolysis (moles/min)
2×10^{-6}	2.0×10^{-10}
4×10^{-6}	4.0×10^{-10}
6×10^{-6}	5.0×10^{-10}
10×10^{-6}	7.0×10^{-10}
20×10^{-6}	9.5×10^{-10}
40×10^{-6}	11.0×10^{-10}
100×10^{-6}	12.0×10^{-10}
200×10^{-6}	12.0×10^{-10}

(c) From the data in the table, would you deduce that penicillinase consists of only one subunit or of more than one subunit?

22-11. The enzyme acetylcholinesterase has a turnover number of 1,500,000 min^{-1}. How many moles of acetylcholine molecules are hydrolyzed per minute by a solution containing 10^{-6} mole of acetylcholinesterase? How many seconds does it take for one molecule of enzyme to cleave one acetylcholine molecule?

22-12. The hydrolysis of pyrophosphate ($P_2O_7^{4-}$) to phosphate (PO_4^{3-}), is an important source of energy for many biosynthetic reactions. In the bacterium *Escherichia coli,* this reaction is catalyzed by the enzyme pyrophosphatase, which has a turnover number of 37,500 min^{-1}. The molecular weight of the enzyme is 120,000, and it consists of six subunits, each of which contains one active site. How many moles of pyrophosphate are hydrolyzed per minute by 1 mg of pyrophosphatase?

22-13. The hormone progesterone contains as its only functional groups two ketone groups and one double bond. Not much is known about the properties of the receptor protein that binds progesterone. What types of noncovalent forces might help bind progesterone to its receptor? Name two amino acid residues that could help bind progesterone to the receptor.

22-14. All enzymes known to be subject to feedback inhibition are allosteric enzymes. Under natural conditions they are not subject to competitive inhibition, although in the laboratory competitive inhibitors can be found. Explain why the final product of a sequence of reactions might be expected to be an allosteric rather than a competitive inhibitor of the first enzyme of the sequence.

22-15. Crystals of deoxyhemoglobin shatter when exposed to air. Suggest an explanation for this phenomenon.

22-16. What is the effect of each of the following changes on the equilibrium concentration of oxyhemoglobin [HbO_2] in a blood sample? (a) increase pH; (b) increase [CO_2]; (c) decrease [DPG]; (d) add a denaturing agent that dissociates the hemoglobin into subunits but does not interfere with the structure of the subunits.

22-17. What change in concentration of DPG would you expect to occur when a person moves from a mountaintop to sea level? Explain.

22-18. Suggest three amino acid residues that could participate in salt links with the DPG in hemoglobin.

22-19. A positively charged histidine residue that is involved in the binding of DPG in adult hemoglobin is replaced by a neutral serine residue in fetal hemoglobin. Explain why this change is of advantage to the fetus.

Sheep red blood cells bound to a human lymphocyte. Scanning electron micrograph, ×7000.

23

Carbohydrates

Carbohydrates are the principal source of energy for living organisms. They include simple sugars (monosaccharides) such as glucose, an important constituent of blood; disaccharides (two simple sugars bonded together) such as sucrose; and polysaccharides, polymers of simple sugars. Polysaccharides include the important energy-storage substances starch and glycogen, as well as cellulose, the major structural material of plants. New research is showing that carbohydrates enable one cell to "recognize" another—a process of special importance in embryo development.

23-1 THE IMPORTANCE OF CARBOHYDRATES

Carbohydrates are one of the four major classes of molecules in living cells: carbohydrates, proteins, lipids, and nucleic acids. They were studied extensively in the early stages of biochemistry. In fact, the origins of biochemistry can be traced to studies by Louis Pasteur and others on the fermentation of the simple carbohydrates (sugars) to form alcohol:

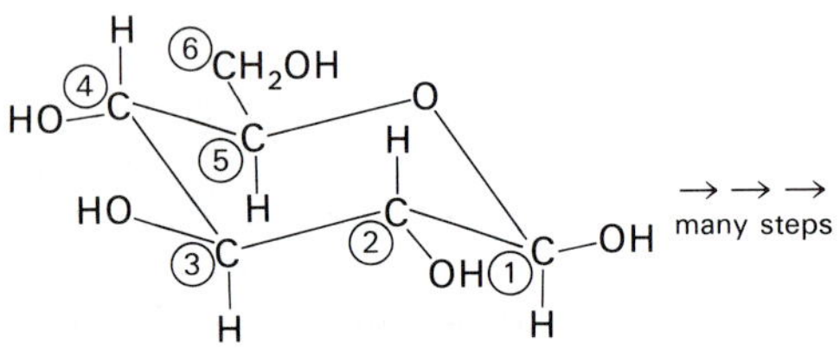

D-glucose
($C_6H_{12}O_6$)

$$2\ CH_3CH_2OH + 2CO_2 \quad (23\text{-}1)$$
ethanol

Fermentation has been known since antiquity. Reaction 23-1 was described as early as 1810 by Joseph Gay-Lussac. By 1837 several investigators had demonstrated that fermentation requires living yeast cells and had assumed that the process of fermentation is intimately linked to the life of these small organisms. A major breakthrough occurred in 1897 when Eduard Büchner was able to obtain a nonliving, cell-free extract of yeast that would convert glucose to ethanol. In some respects, this event can be viewed as the trigger that led to the flourishing of biochemistry.

The more recent discoveries concerning DNA and proteins have somewhat overshadowed carbohydrate chemistry, but the carbohydrates have an essential role: they provide energy to cells and are important building blocks of other cellular components, including DNA. Today, carbohydrates are again taking a key role in research with the discovery that carbohydrates on the cell surfaces are critically involved in cell-to-cell communication and may be partly responsible for the abnormal behavior of cells in the diseases collectively known as cancer.

The name carbohydrate means a "hydrate of carbon," or a combination of carbon and water. This name arises from the fact that the relative proportions of C, H, and O in the simple sugars or monosaccharides is $C_n(H_2O)_n$—that is, one atom of carbon for each set of the atoms that form water. The formula for glucose, for example, is $C_6H_{12}O_6$ or $C_6(H_2O)_6$. These formulas are not very illuminating because they do not show that glucose contains five hydroxyl groups and an aldehyde group. We can see this if we write glucose as an open-chain structure:

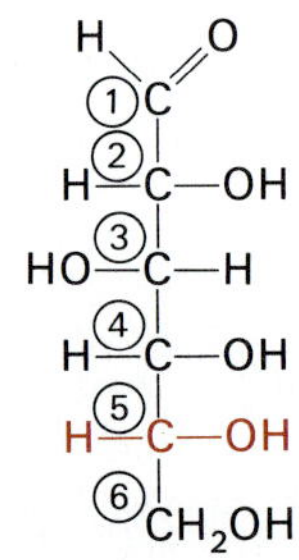

D-glucose

A small amount of glucose in the open-chain structure exists in equilibrium with the cyclic structure shown earlier; we shall learn more about this equilibrium shortly (see also Section 18-5). Carbohydrates are either polyhydroxy aldehydes or ketones, or substances that can be hydrolyzed to yield these.

The simple carbohydrates (those that cannot be hydrolyzed to smaller units) are called **monosaccharides.** Those that can be hydrolyzed to yield two monosaccharide units are called **disaccharides.** The disaccharide maltose, for example, can be hydrolyzed to give two molecules of glucose (see Section 23-5). Those carbohydrates that can be hydrolyzed to yield many monosaccharide units are called **polysaccharides.** We first consider the structures and reactions of monosaccharides, and then we see how they combine to form larger molecules.

23-2 MONOSACCHARIDES

A typical monosaccharide contains from three to seven carbon atoms. Monosaccharides are classified according to the number of carbon atoms. The ending *-ose* designates a carbohydrate, and prefixes are added to indicate the number of carbons. Thus, a *triose* contains three carbon atoms, a *tetrose* contains four carbon atoms, a *pentose* contains five carbon atoms, a *hexose* contains six carbon atoms, and a *heptose* contains seven carbon atoms. For example, D-glucose is a hexose because it is a six-carbon sugar.

Because carbohydrates may be aldehydes or ketones, a further classification defines the carbohydrates as **aldoses** (aldehydes) or **ketoses** (ketones). The prefixes aldo- and keto- are often used together with the names that designate the number of carbon atoms:

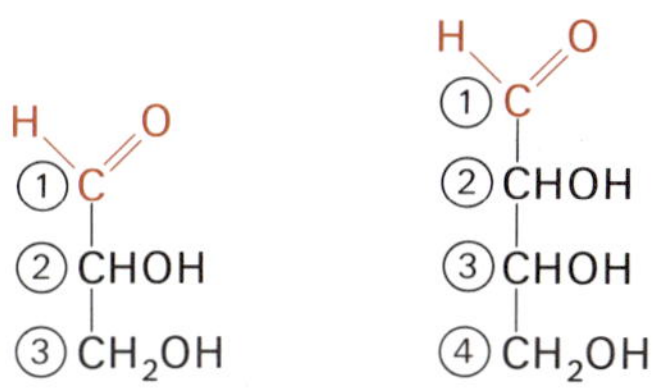

glyceraldehyde (an aldotriose) **erythrose** (an aldotetrose)

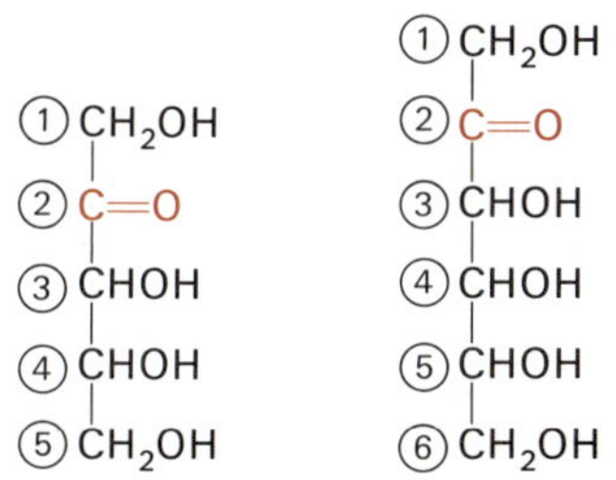

ribulose (a ketopentose) **fructose** (a ketohexose)

The carbons are numbered starting at the carbonyl carbon end of the chain for aldoses or at the end closest to the carbonyl carbon for ketoses.

The simplest monosaccharides are the trioses, glyceraldehyde and dihydroxyacetone, both of which are derived from glycerol (by oxidation):

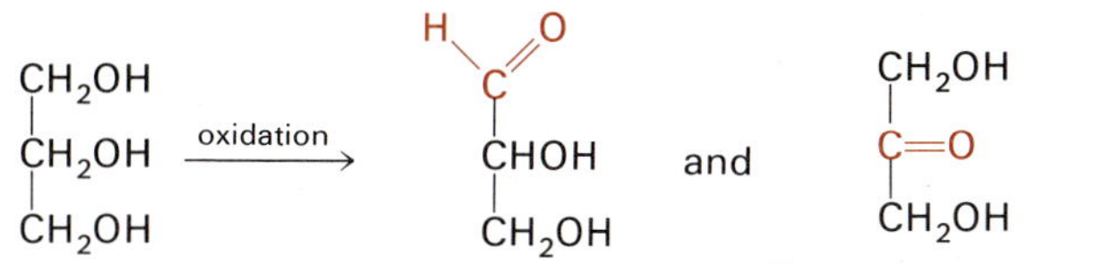

glycerol **glyceraldehyde** **dihydroxyacetone**

These two distinct compounds, both of which are carbohydrates of molecular formula $C_3H_6O_3$, are structural isomers. One is an aldose; the other is a ketose. Both are produced as phosphate esters (Section 19-12) in the course of glycolysis, the first metabolic pathway in the production of energy from glucose. Glycolysis is discussed in Section 26-2.

Glyceraldehyde has a chiral carbon atom and therefore exists as a pair of enantiomers:

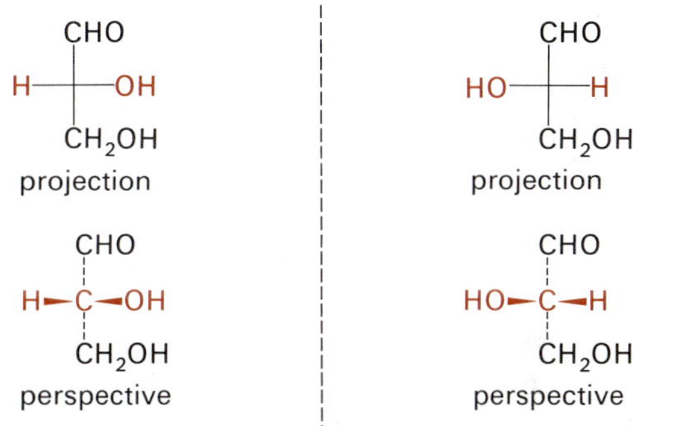

D-(+)-glyceraldehyde $[\alpha]_D = +8.7°$ Mirror **L-(−)-glyceraldehyde** $[\alpha]_D = -8.7°$

Fischer projection formulas, given here along with the perspective formulas, commonly are used to show stereochemistry of sugars. In Fischer projections, each intersection of lines represents a chiral C atom. Sometimes the chiral carbon atoms are shown, but usually they are not. Horizontal lines represent bonds projecting above the plane of the paper; vertical lines represent bonds projecting below the plane of the paper. The Fischer projections are especially useful when there are several chiral carbons.

The quantity $[\alpha]_D$ is the **specific rotation,** the rotation of the plane of polarized light measured when sodium light passes through a solution containing 1 g of solute per 100 g of water for a distance of 10 cm (see Section 13-5). D-Glyceraldehyde is the (+) or dextrorotatory isomer. The D and L convention designates the stereochemistry of carbohydrates: all D sugars have the same configuration as D-glyceraldehyde at the next-to-last carbon atom. Sugars with the opposite configuration belong to the L family.

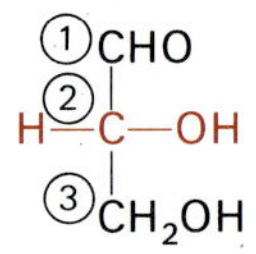

D-(+)-glyceraldehyde

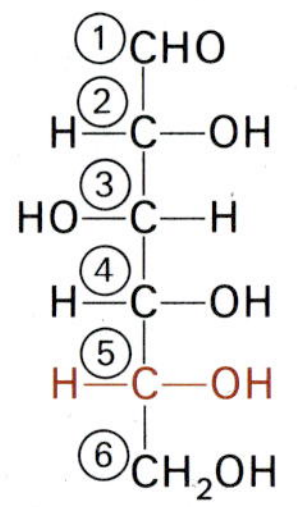

D-(+)-glucose

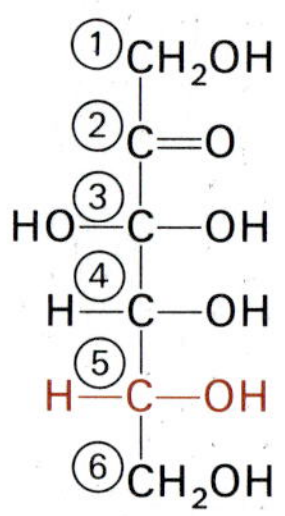

D-(−)-fructose

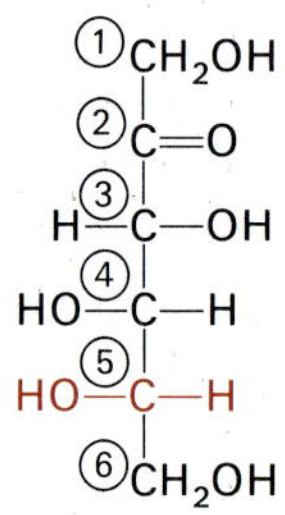

L-(+)-fructose

(Although chiral carbons are usually shown in projection formulas as intersections, they are shown explicitly here.) The D sugars predominate in nature. Note that these D and L notations give the stereochemistry but do not imply any specific direction for the rotation of polarized light; D-glucose is (+), and D-fructose is (−).

The important aldoses with more than three carbon atoms can be related to D-glyceraldehyde by imagining that the larger aldoses are synthesized starting with glyceraldehyde without disturbing carbon atoms number two and three. (Such a synthesis can be accomplished in the laboratory; a method is outlined in Problem 23-16). If we extend the chain by adding one carbon to the carbonyl end of D-glyceraldehyde, the products will be two four-carbon sugars known as D-erythrose and D-threose:

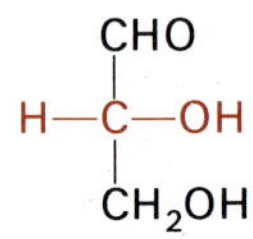

D-glyceraldehyde

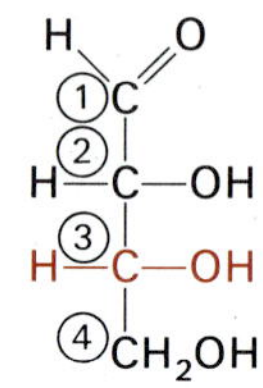

D-erythrose
$[\alpha]_D = -14.8°$

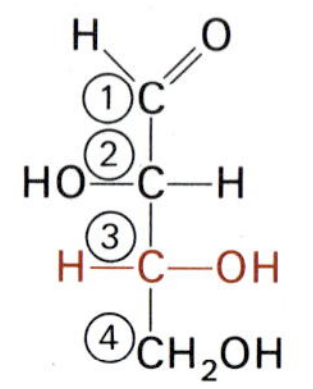

D-threose
$[\alpha]_D = +19.6°$

The two compounds retain the D designation because the next-to-last carbon atom (carbon-3) has the same configuration originally present in D-glyceraldehyde. The two compounds differ, however, in the configuration about the second carbon atom, which is also chiral, and hence they are not enantiomers.

D-Erythrose and D-threose are stereoisomers, isomers with the same skeletal structures but different three-dimensional arrangements of the atoms in space. As we saw in Sections 13-4 and 13-5, stereoisomers include geometric isomers and optical isomers. D-Erythrose and D-threose have different specific rotations, and thus are optical isomers; but they are not mirror images, and so are not enantiomers. (If this is not clear, molecular models will help.) Such substrates, which are optical isomers but not enantiomers, are called **diastereomers.** Optical isomers thus include enantiomers and diastereomers. One of a pair of diastereomers can be converted into the other only by changing the configuration at one or more chiral centers. Although enantiomers are identical in properties except for their optical rotations, diastereomers differ in their chemical and physical properties.

EXERCISE 23-1 Both D-erythrose and D-threose have enantiomers. Write the structures of the enantiomers, and give their specific rotations.

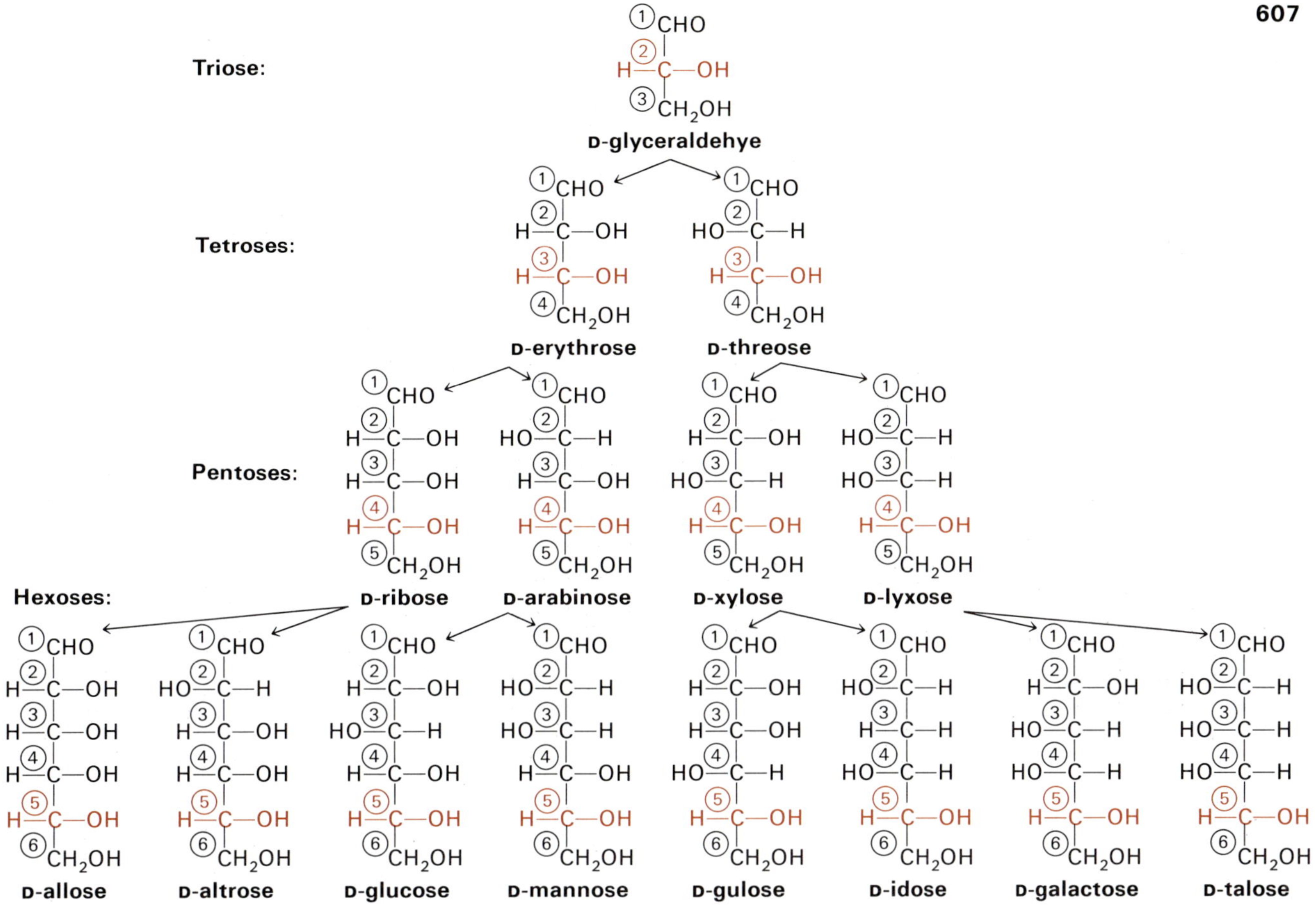

FIGURE 23-1

Aldoses related to D-glyceraldehyde.

EXERCISE 23-2 Which pairs from among the following four compounds are (a) enantiomers or (b) diastereomers:

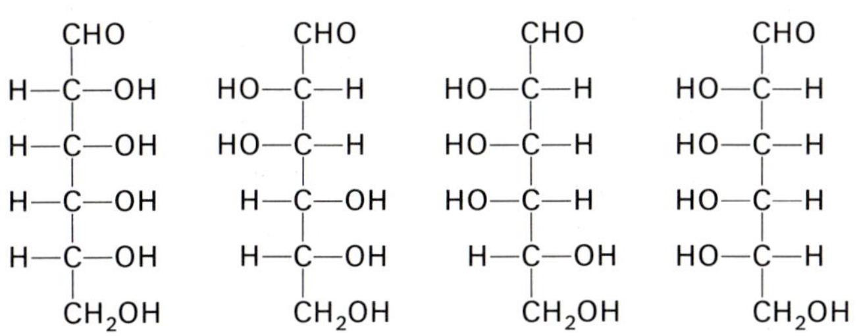

A B C D

If the imaginary synthesis is repeated, with D-erythrose and then D-threose as the starting materials, two five-carbon sugars are produced from each to give a total of four diastereomers that are D-pentoses. Their structures are summarized in Figure 23-1, along with those of the eight hexoses (six-carbon sugars) that could be synthesized from the four pentoses. The eight hexoses are all diastereomers. All of the molecules shown in Figure 23-1 have the same configuration about the next-to-last carbon atom as D-glyceraldehyde, and thus they are members of the D-family.

Each of the D-aldoses has an enantiomer in the L-family. Thus, there are a total of 4 tetroses (two in the D family and two in the L family), 8 pentoses, and 16 hexoses. The maximal number of stereoisomers possible for a compound having n chiral atoms is 2^n. Glyceraldehyde, with one chiral carbon atom, has $2^1 = 2$ stereoisomers; an aldotetrose with two chiral carbon atoms has $2^2 = 4$ stereoisomers, and so on.

EXERCISE 23-3 There is a family of ketoses that is analogous to the family of aldoses in Figure 23-1. In the ketoses, the first optically active compound is D-erythrose, a four-carbon ketose with the same configuration at the next-to-last carbon atom as D-glyceraldehyde.

(a) Write the structures of D-erythrose and of the two D-ketopentoses. (Carbon-2 is the carbonyl carbon.)

(b) How many chiral carbons are there in a ketopentose? How many stereoisomers are there for a ketopentose?

23-3 RING STRUCTURES OF MONOSACCHARIDES

The structures of aldoses and ketoses can be conveniently represented as straight chains. However, in solution, aldoses with four or more carbon atoms and ketoses with five or more carbons cyclize to form five- and six-membered rings. As we saw in Section 18-5, aldehydes and ketones can react with alcohols to form hemiacetals and hemiketals:

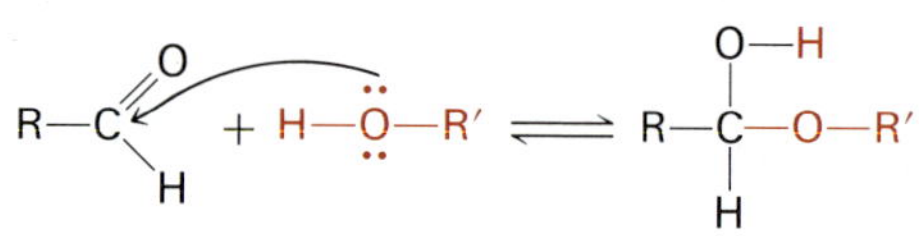

aldehyde **alcohol** **hemiacetal**

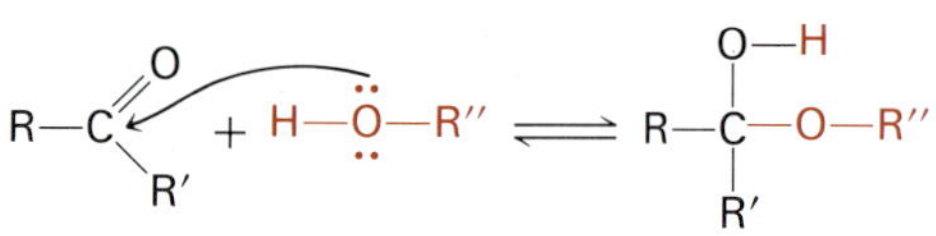

ketone **alcohol** **hemiketal**

In an intramolecular reaction, the carbonyl carbon (carbon-1, or C-1) of glucose reacts with the C-5 hydroxyl group to form a cyclic hemiacetal:

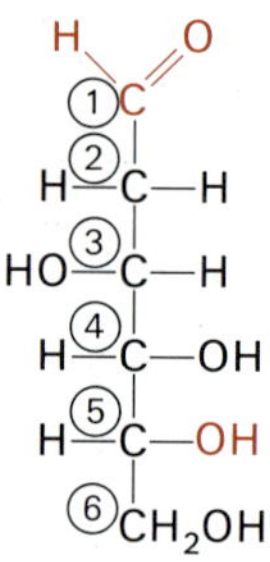

D-glucose (open-chain form)

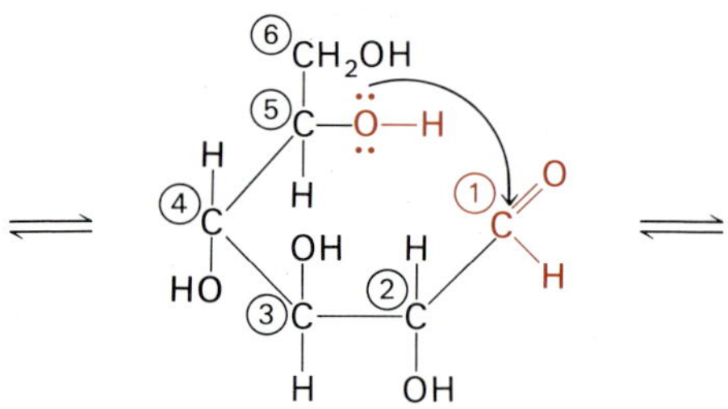

(open-chain form in a conformation favorable for reaction)

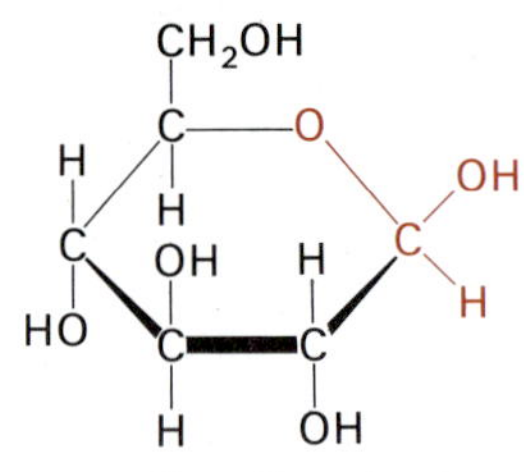

β-D-glucose (a cyclic form of glucose, a hemiacetal)

Similarly, the hexoketose D-fructose cyclizes to form

a five-membered ring that is a hemiketal. The carbonyl carbon (C-2) reacts with the hydroxyl group of C-5:

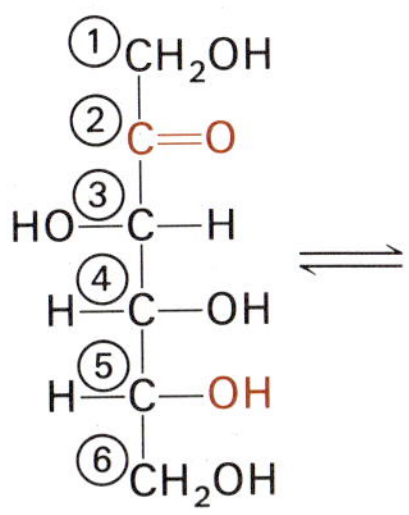

D-fructose (open-chain form)

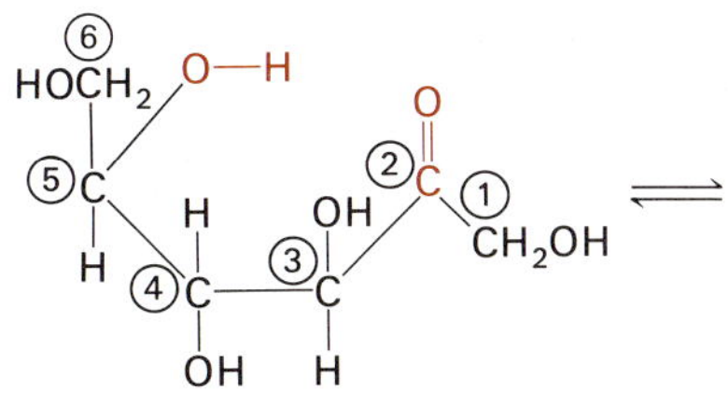

(open-chain form in a conformation favorable for reaction)

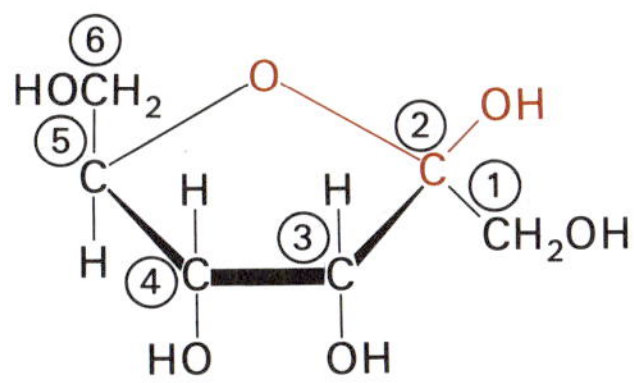

β-D-fructose (a cyclic form of fructose, a hemiketal)

In these ring structures, the ring is considered to be approximately planar and perpendicular to the plane of the paper, with the heavy line toward the reader. The substituents project up and down from the plane of the ring.

EXERCISE 23-4 The best way to visualize both the Fischer projection and the cyclic form of a carbohydrate is with molecular models.

(a) Construct a model of the Fischer projection of D-glucose. First make the carbon chain with the —CHO group at one end and the —CH_2OH group at the other end. Suspend the chain vertically with the CHO group at the top. Now fold the CH_2OH group *away* from you, and bring it up to touch the CHO group. Hold both groups in the same hand, forming a rather rigid ring. Look at each carbon atom in turn starting with C-2, and attach the OH and H groups as they are seen in the Fischer projection. Note that when you focus on one carbon atom, the OH and H groups project toward you, and the carbon–carbon bonds project away from you.

(b) To convert the open-chain model of D-glucose into the cyclic form, join the oxygen atom of the C-5 OH group to the aldehyde carbon (C-1). The ring is approximately planar. Note that when the ring is in a horizontal plane with the ring oxygen directed away from you and the CH_2OH group pointing up, groups that are on the left in the Fischer projection are directed upward, and groups that are on the right in the Fischer projection are directed downward.

When glucose cyclizes, an additional asymmetric center is created: C-1 (which was originally the carbonyl carbon) is now a chiral carbon atom. Thus, two distinct cyclic structures are formed with different configurations of the H and OH groups about C-1. Which configuration is formed depends on the orientation of the carbonyl group when the bond is formed:

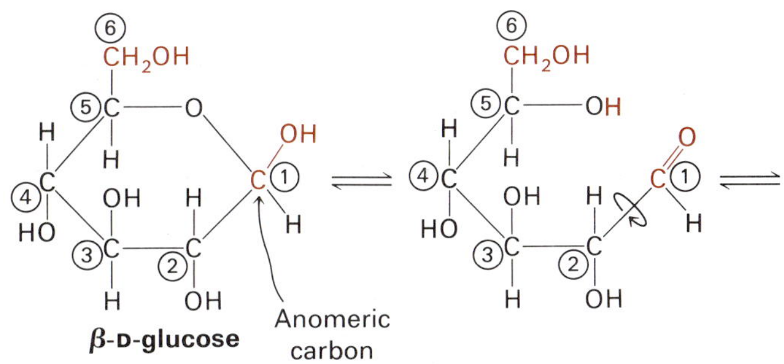

β-D-glucose

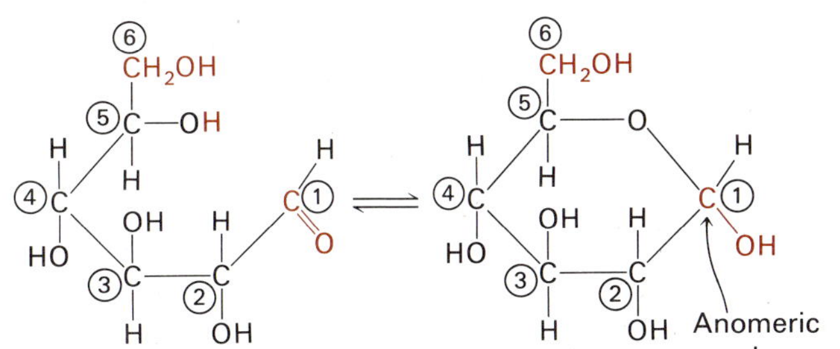

α-D-glucose

The OH group of C-1 can be above or below the ring. When the OH group of C-1 is pointing up (in the same direction as C-6), the compound is β-D-glucose; when the OH group is pointing down, the compound is α-D-glucose. The α and β forms of a cyclic carbohydrate are called **anomers,** and the new chiral carbon atom (the one formed when the open chain cyclizes) is the **anomeric carbon atom.** According to these conventions, the two anomers of D-fructose are β-D-fructose and α-D-fructose:

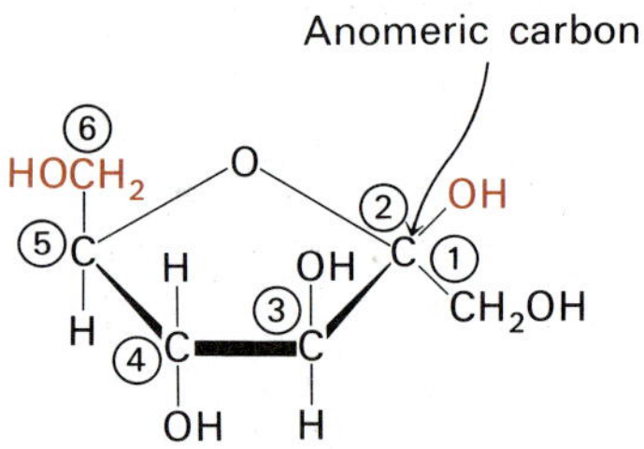

β-D-fructose

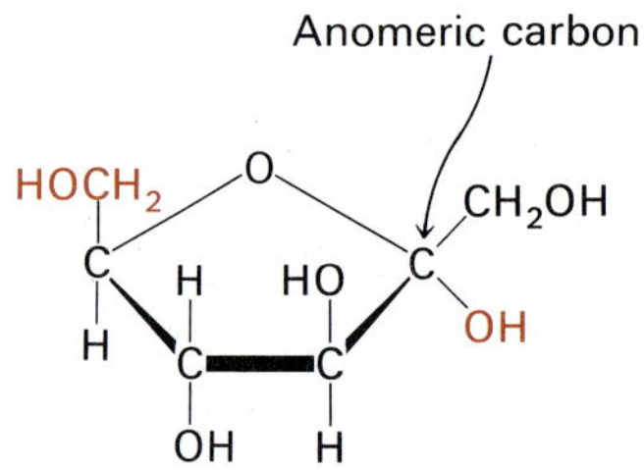

α-D-fructose

The cyclic structures are often written in the simplified form we have used for many organic structures, in which corners represent carbon atoms and all bonds not shown are to hydrogen atoms:

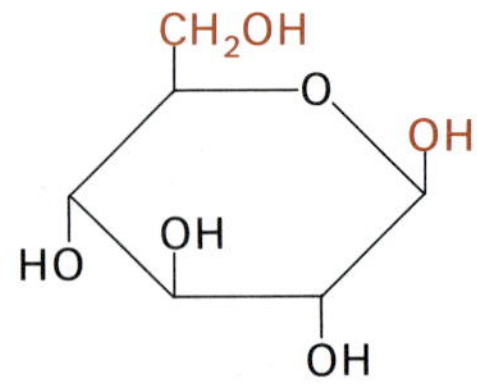

β-D-glucose

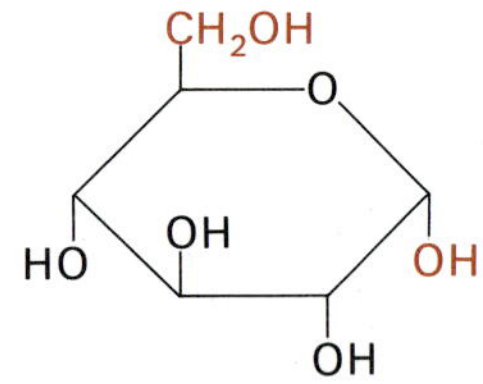

α-D-glucose

The cyclic structures we have been using are convenient and give an approximate picture of the orientations of the hydrogen atoms and hydroxyl groups in the ring forms of glucose. However, as we noted in Section 14-6, six-membered rings tend to assume nonplanar conformations that can be described in terms of their resemblance to a "chair" or to a "boat." These conformations are also assumed by the six-membered rings of carbohydrates (Figure 23-2). In the chair conformation, the substituents on each carbon atom of the ring are described as either **equatorial** or **axial,** according to their orientation with respect to the approximate plane of the chair. Note that in the chair conformation for β-D-glucose, all of the bulky substituents (—OH and —CH_2OH) are equatorial, an arrangement that minimizes crowding between these groups. Among all the aldohexoses, only glucose can

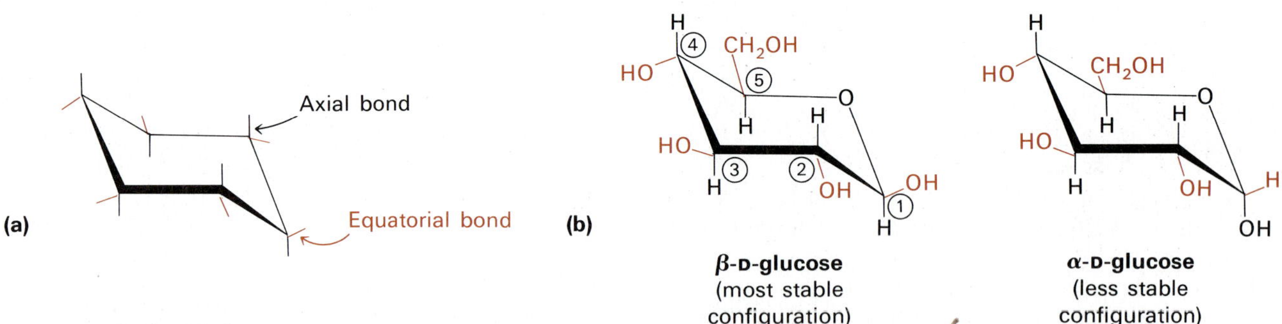

FIGURE 23-2

Chair conformation of D-glucose. **(a)** General representation of a cyclohexane ring (equatorial bonds are indicated by color). **(b)** Chair conformation for D-glucose (equatorial bonds are indicated by color).

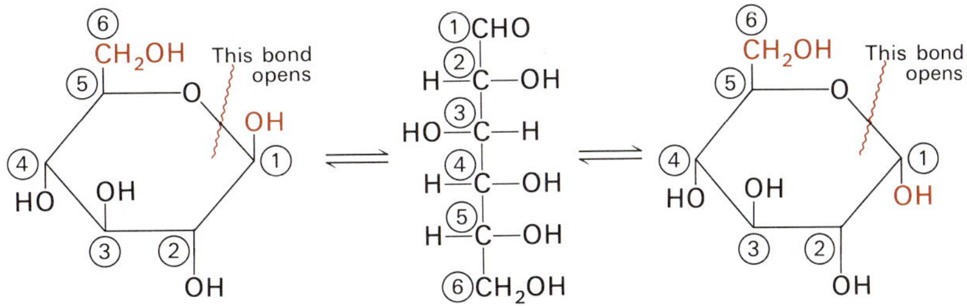

β-D-glucose (63.6%) (0.024%) **α-D-glucose** (36.4%)

have *all* bulky substituents in equatorial positions; others have at least one bulky substituent in an axial position. This distinction may help account for the fact that β-D-glucose is by far the most abundant aldohexose in nature. Chair structures can easily be written for all the aldohexoses of Figure 23-1 by noting how they differ from β-D-glucose, which has all its substituents in equatorial positions. For example, D-galactose differs from D-glucose only in the configuration of groups about C-4.

EXERCISE 23-5 A Fischer projection for D-galactose, a component of lactose (milk sugar), is given in Figure 23-1. Write the structure of β-D-galactose in the chair conformation.

EXERCISE 23-6 The Fischer projection for the aldopentose D-ribose is given in Figure 23-1. This sugar, which occurs in ribonucleic acids (RNA), cyclizes to form a five-membered ring. Write structures for the α and β form of cyclic D-ribose.

Pure crystalline D-glucose can be obtained in either the α form (which has a melting point of 146°C and a specific rotation $[\alpha]_D$ of +18.7°) or the β form (which has a melting point of 150°C and a specific rotation of +112°). If *either* form is dissolved separately in water, the optical rotation of the solution changes until an equilibrium value of +52.7° is reached. This phenomenon is called **mutarotation.** It results from the interconversion of the two forms in solution. The unstable cyclic hemiacetal opens to give the open-chain form, which then closes again to give either the α-form or the β-form cyclic structures (top of page). At equilibrium, the solution contains 63.6% of the more stable β form, 36.4% of the α form, and only about 0.02% of the open-chain form.

23-4 REACTIONS OF CARBOHYDRATES: OXIDATION–REDUCTION REACTIONS

In studying energy transformations in cells (Chapter 26), we shall encounter many reactions that involve the stepwise oxidation of carbohydrates to CO_2. The energy generated by the oxidation is used for the synthesis of other important biochemical compounds. The fact that carbohydrates can be oxidized in steps is an essential property of carbohydrate metabolism in cells, and the oxidation and reduction of carbohydrates is a useful analytic tool for carbohydrate chemists.

As we saw in Section 18-8, aldehydes are readily oxidized to carboxylic acids with mild oxidizing agents: in the Tollens' (or silver-mirror) test, silver ions (Ag^+) are reduced to metallic silver; in both the Benedict's and Fehling's tests, copper(II) ions are reduced to copper(I). These same tests can be used to detect aldehyde groups in sugars. Benedict's test, for example, is often used to detect glucose in urine. Urine normally does not contain glucose, but in certain diseases (such as diabetes) glucose is present. Benedict's reagent contains a citrate complex of Cu^{2+} in basic solution:

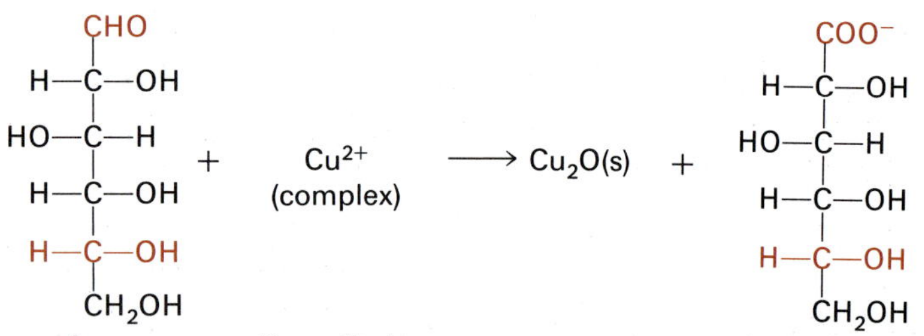

D-glucose — Benedict's reagent (blue solution) — red precipitate — D-gluconic acid (anion)

If no glucose is present in the urine, the solution remains bright blue. A positive test for glucose varies in color from green to yellow-orange to bright red, depending on the concentration of glucose. In a positive test, the aldose is oxidized to the anion of a carboxylic acid, and the Cu^{2+} is reduced to Cu^{1+}. Thus, carbohydrates that give a positive Benedict's test (or Fehling's or Tollens' test) are called **reducing sugars.**

We would expect aldoses to give positive tests because they are aldehydes in their open-chain forms, but ketoses do too. How can we explain this? Ketones in general are not readily oxidized by such mild reducing agents. The reason is that ketoses (like aldoses) have α-hydrogen atoms that are acidic and are subject to keto–enol tautomerism. Tautomers differ only in the distribution of electrons and in the position of a hydrogen atom (Section 18-9). If a ketose is dissolved in basic solution (as in the reducing sugar tests), an equilibrium is set up in which some ketose, some enol, and some aldose is present:

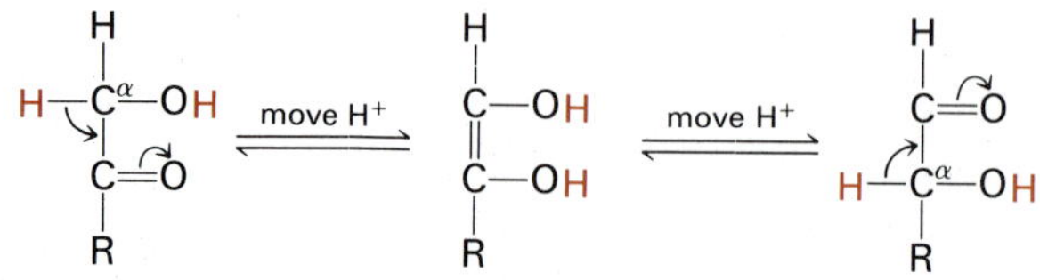

ketose — enol form — aldose

Because there is always some aldose present in a basic solution of a ketose, ketoses give positive tests for reducing sugars. (As the aldose is consumed in the test, the equilibrium shifts to replenish it, and so a strong test is obtained.)

Mild oxidation of the aldehyde group of D-glucose yields D-gluconic acid. This can be done in the laboratory—say, with Cu^{2+} or Ag^+, as in the tests for reducing sugars. In the body, the oxidation is catalyzed by a particular enzyme. D-Gluconic acid spontaneously forms an intramolecular cyclic ester by reaction between the COOH group and the OH group of C-5:

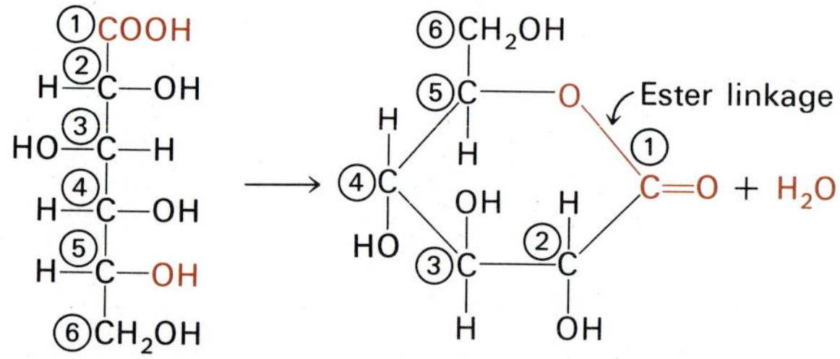

D-gluconic acid — D-gluconolactone

Internal esters are called **lactones,** and D-gluconic acid in this form is called D-gluconolactone.

Another lactone that has been widely publicized in recent years for its role in human nutrition is L-ascorbic acid, or vitamin C. It is a lactone with an enediol structure:

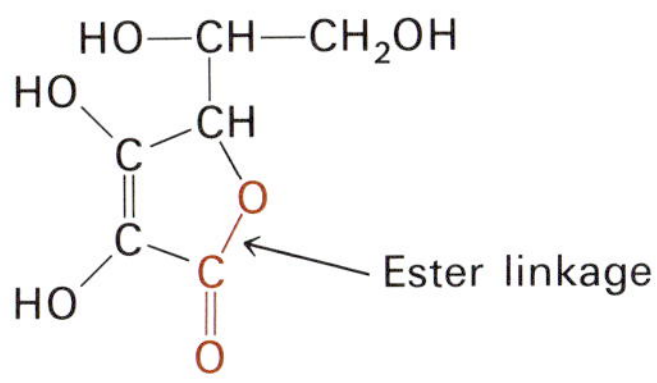

L-ascorbic acid

Other oxidation reactions of carbohydrates also occur, both in living cells and in the laboratory. There are enzymes, for example, that catalyze oxidation of the primary hydroxyl groups of carbohydrates instead of the aldehyde groups. When the primary alcohol group of D-glucose is oxidized, the product is D-glucuronic acid:

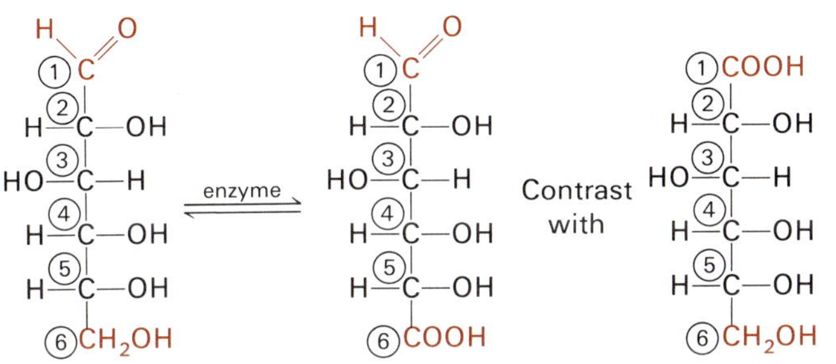

D-glucose **D-glucuronic acid** **D-gluconic acid**

Both D-gluconic and D-glucuronic acids are produced in the body, but each reaction is catalyzed by its own specific enzyme. A phosphate ester of D-gluconic acid is an important intermediate in metabolism. D-Glucuronic acid is a component of the coats of particular cells and also is a component of the extracellular fluid of humans. More extensive oxidation reactions involve the cleavage of carbon–carbon bonds in living cells; these are discussed in the chapter on metabolic transformations.

Carbohydrate-reduction reactions are also encountered in various chemical or enzymatic processes. The simplest one is the reduction of an aldehyde to an alcohol, as in the conversion of an aldose to the corresponding **sugar alcohol.** D-sorbitol is the sugar alcohol produced by reduction of D-glucose with hydrogen gas using a metal catalyst:

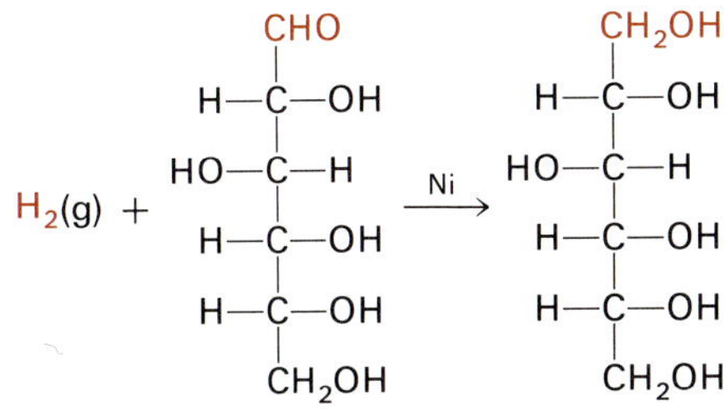

D-glucose **D-sorbitol**

Because of its large number of hydroxyl groups that can form hydrogen bonds to water, D-sorbitol is often used to help keep moisture in candies (such as caramels) that tend to dry out. Glycerol, which is a component of fats and oils (Chapter 24), is also a sugar alcohol:

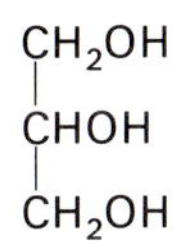

glycerol

A reduction of the five-carbon sugar D-ribose forms one of the principal components of the genetic material, DNA (see Chapter 25). This important carbohydrate is 2-deoxyribose:

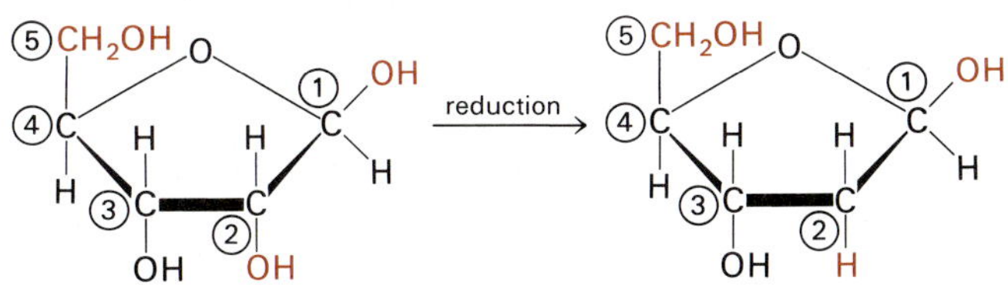

β-D-ribose **β-2-deoxyribose**

2-Deoxyribose is produced from ribose in a series of reactions catalyzed by enzymes that have evolved specially for this purpose.

23-5
REACTIONS OF CARBOHYDRATES: REACTIONS WITH ALCOHOLS

In Section 18-5 we saw that a hemiacetal can react with an alcohol in acid solution to yield an acetal:

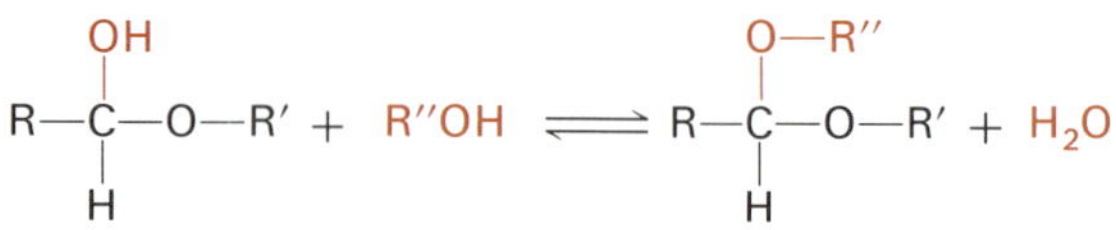

hemiacetal **alcohol** **acetal**

When an aldohexose in the cyclic (hemiacetal) form reacts with an alcohol in acid solution, the products are **glycosides** (as is the custom with biochemists, we do not balance the equation):

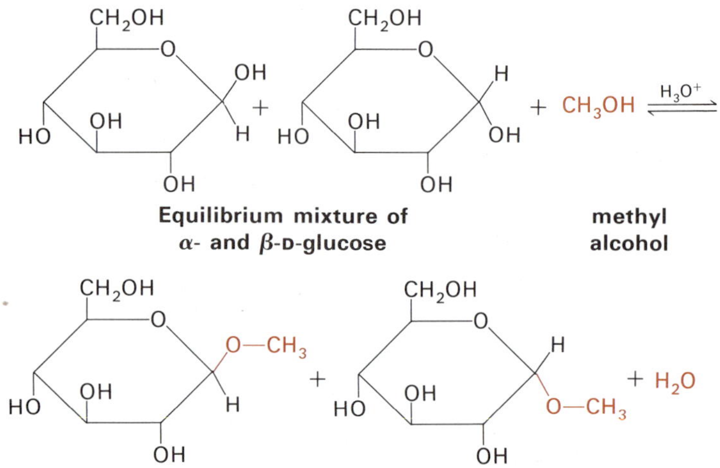

Equilibrium mixture of α- and β-D-glucose **methyl alcohol**

methyl-β-D-glucoside **methyl-α-D-glucoside**

Glycosides, then, are acetals formed by reaction of the hydroxyl group of C-1 of the ring form of an aldohexose with an alcohol. The new carbon–oxygen bond is a **glycosidic bond.** In Section 23-7, we shall see that two monosaccharides can react to form a disaccharide when the hemiacetal OH group of one surgar molecule reacts with an OH group of a second sugar molecule. The disaccharide product is a glycoside. In polysaccharides, the individual monosaccharides are also linked principally through glycosidic bonds.

Although carbohydrate hemiacetals are unstable and are in equilibrium with open-chain forms of the monosaccharides, the glycosides (acetals) are stable. Glycosides of monosaccharides, therefore, are not reducing sugars—they do not react with Tollens', Benedict's, or Fehling's solutions.

EXERCISE 23-7 (a) Write structures for the products of the reaction of D-galactose with ethanol in acid solution.

(b) Which of the following compounds would react with Tollens′ solution?

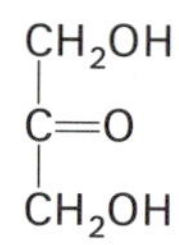

(1) **dihydroxyacetone**

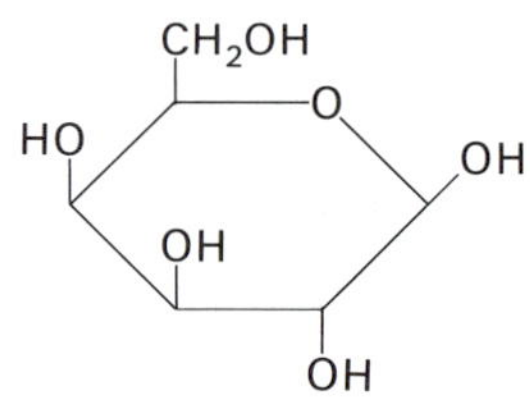

(2) **β-D-galactose**

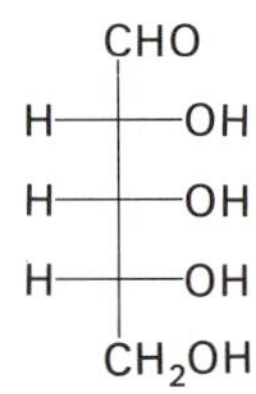

(3) **D-ribose**

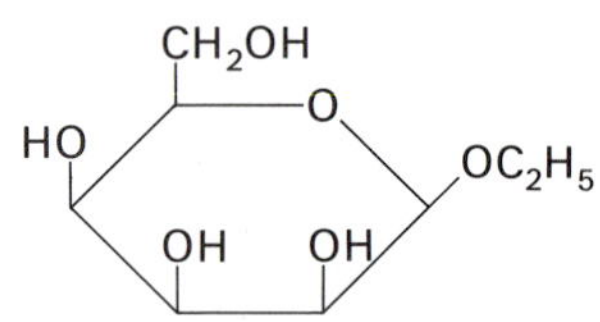

(4) **ethyl-β-D-mannoside**

23-6 REACTIONS OF CARBOHYDRATES: DERIVATIVES INVOLVING OTHER ELEMENTS

A variety of carbohydrate derivatives encountered in biochemical reactions involve atoms other than carbon, hydrogen, and oxygen, Most of the transformations of carbohydrates in metabolic reactions involve a carbohydrate present as a sugar phosphate, an ester of phosphoric acid and a carbohydrate. Glucose, for example, which circulates freely in the blood, is phosphorylated in the cells before it participates in metabolic reactions. The first step in glycolysis (Chapter 26) is the phosphorylation of glucose by ATP (adenosine triphosphate, see Section 19-12) to yield the phosphate ester, glucose 6-phosphate:

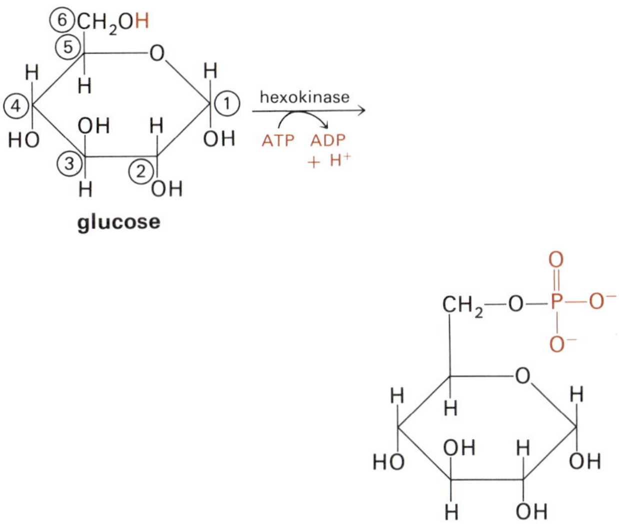

glucose

glucose 6-phosphate

The reaction is catalyzed by hexokinase, an enzyme that transfers a phosphate group from ATP to a variety of six-carbon sugars. The reaction is essentially irreversible under physiological conditions.

In some cases, two phosphate groups are present on the same molecule. An example is fructose 1,6-diphosphate, which is produced in another step in glycolysis:

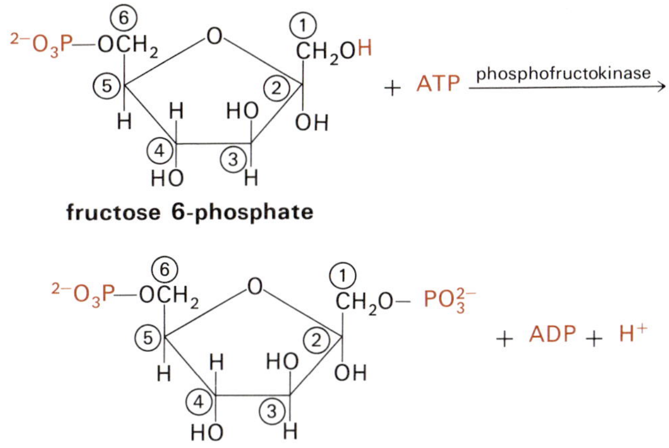

fructose 6-phosphate

fructose 1,6-phosphate

The backbone of nucleic acid polymers contains phosphodiester linkages in which a single phosphoric acid molecule forms esters with two sugar molecules. The sugar molecule in RNA (ribonucleic acid) is ribose; and that in DNA (deoxyribonucleic acid) is deoxyribose. In RNA a phosphate group links C-3 of one ribose molecule to C-5 of the adjacent ribose molecule:

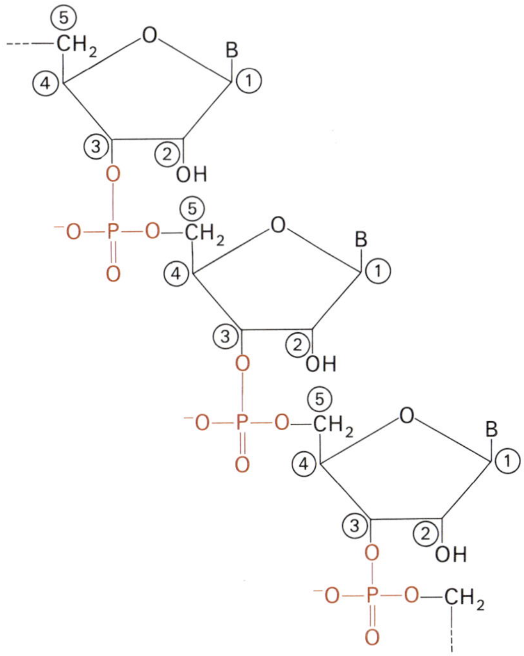

In the nucleic acids, each ribose molecule is also bonded to a specific base represented in the above structure as B and discussed fully in Chapter 25.

Other carbohydrate derivatives involving nitrogen-containing functional groups are the **amino sugars,** found in large quantities in the polymers that form bacterial cell walls. In an amino sugar, one of the hydroxyl groups of a sugar is replaced by an amino group (present in the charged form at neutral pH), as in D-glucosamine:

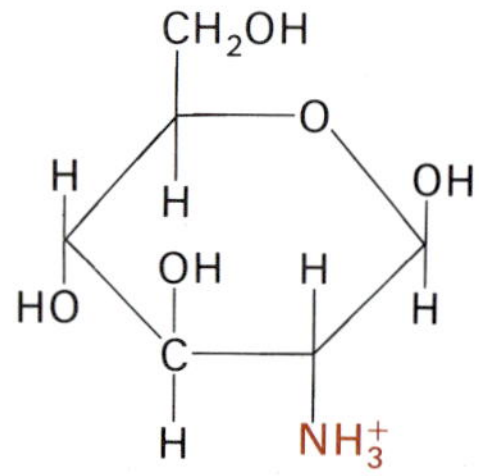

D-glucosamine

Amino sugars are often in the form of an *N*-acetyl derivative, an amide of the amino sugar and acetic acid. Such a derivative of D-glucosamine is *N*-acetyl-D-glucosamine (abbreviated NAG):

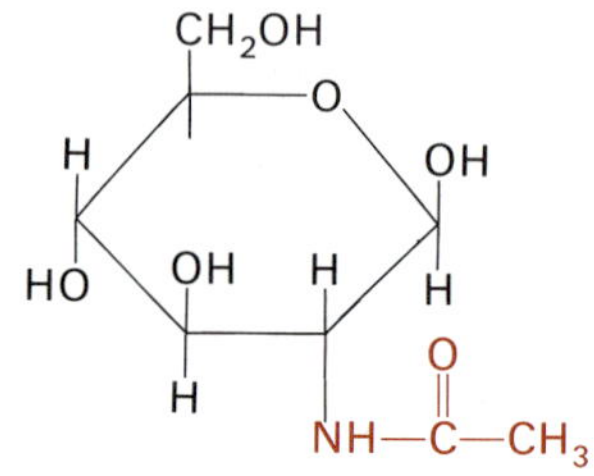

NAG

23-7 DISACCHARIDES

When an alcohol reacts with the hemiacetal group of a sugar in its cyclic form, the bond formed is called a glycosidic linkage, and the new molecule is a glycoside:

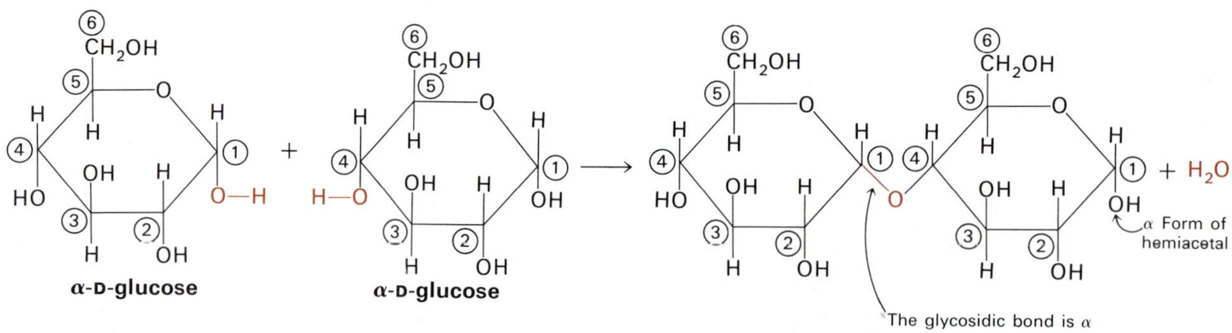

$+ H_2O$ (23-2)

α-maltose (a glycoside)

When, as here, two sugar units are linked, the glycoside formed is called a disaccharide. Maltose is the disaccharide composed of two units of D-glucose joined in a glycosidic linkage from the anomeric carbon (C-1) of one ring to C-4 of the other ring. The righthand ring of maltose is a hemiacetal and can, like glucose, undergo mutarotation; it can open to the free aldehyde and then reclose in the β form:

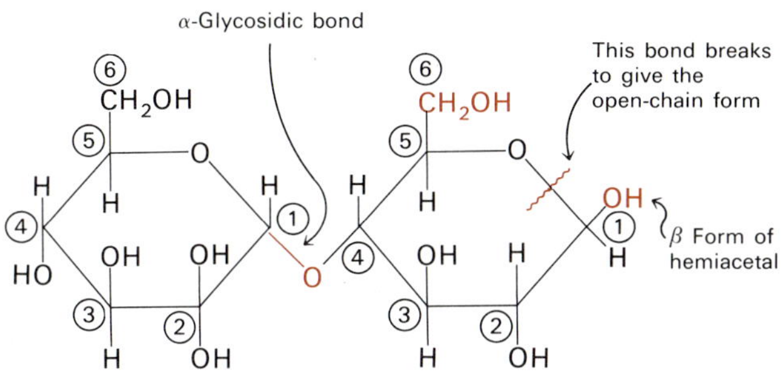

β-maltose

Note that the α and β prefixes for maltose refer to the hydroxyl group of the free anomeric carbon (C-1 of the righthand molecule), not to the glycosidic bond (which is alpha in both cases). Because maltose has a hemiacetal group that can undergo mutarotation, it is a reducing sugar. The glycosidic bond is α because it was formed from the α form of D-glucose. Because glycosides are stable, the glycosidic bond is fixed in the α position. If the two D-glucose units were instead joined in a β-glycosidic linkage, the disaccharide would not be maltose but a different compound, cellobiose (see Problem 23-8). We shall see (Section 23-6) that cellulose is composed of D-glucose units joined by β-glycosidic linkages.

The bond between the two glucose units in maltose is designated $\alpha(1\rightarrow4)$ because the glycosidic linkage is alpha, and the bond is between C-1 of the sugar that forms the glycosidic bond and C-4 of the other sugar. Maltose is formed as a partial breakdown product of the carbohydrate polymer called starch (Section 23-6). Maltose can be further hydrolyzed by the enzyme maltase (or by dilute acids) to yield two molecules of D-glucose. The reaction is the reverse of that in Equation 23-2.

Lactose, a disaccharide of galactose and glucose, is the principle carbohydrate of milk. The anomeric carbon (C-1) of galactose is joined by a β-glycosidic bond to C-4 of glucose:

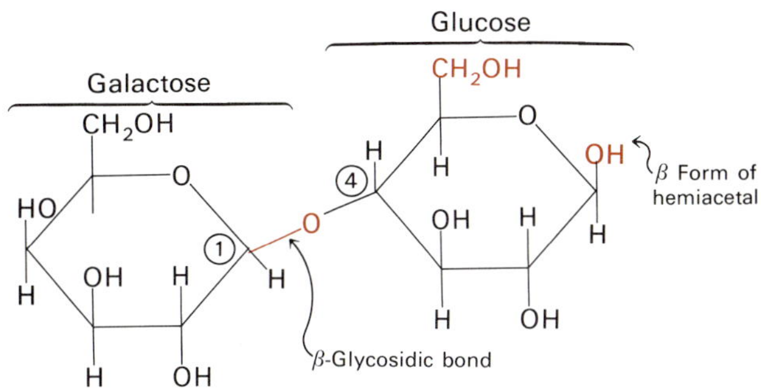

β-lactose

Human beings generally cannot digest glucose units joined by β-glycosidic bonds. However, a specific enzyme—lactase, a β-galactosidase (an enzyme that splits β-galactosides)—present in the small intestine splits lactose into glucose and galactose units for further metabolic processing. This enzyme is absent in certain groups of people around the world, causing an inability to digest milk, known as **lactose intolerance.**

A serious inherited disease called **galactosemia** results from the absence of an enzyme that catalyzes one step in the metabolism of galactose. Infants born with this condition cannot tolerate milk; the consumption of milk causes vomiting, diarrhea, jaundice, and sometimes mental retardation. The treatment for galactosemia is the exclusion of galactose from the diet.

EXERCISE 23-8 Write the structure of α-lactose. Is lactose a reducing sugar? Explain.

Sucrose, common table sugar, is a disaccharide composed of glucose and fructose. The α anomer of carbon (C-1) of glucose is joined in a glycosidic linkage with the β anomer of carbon (C-2) of fructose:

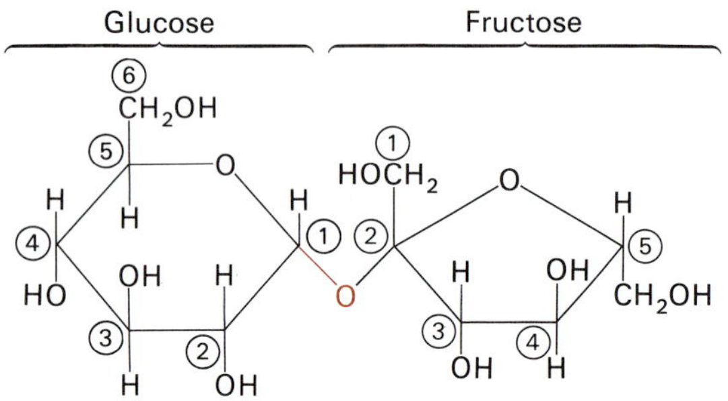

sucrose

(Note that C-2 of fructose is on the left, instead of on the right as in the structure given in Section 23-3. The molecule has been flipped over in such a way that groups that extend above the plane of the ring in the structure given in Section 23-3 extend below the plane of the ring here, and vice versa.)

There is no hemiacetal group in sucrose, so sucrose is *not* a reducing sugar. The reason for this is that the hemiacetal group of glucose has been converted to a stable glycoside (acetal), and the hemiketal group of fructose has been converted to a stable ketal.

Sucrose is usually isolated from sugar cane or beet sugar, although it occurs throughout the plant kingdom.

EXERCISE 23-9 Write the structures of the products of the hydrolysis of sucrose.

23-8 CARBOHYDRATE POLYMERS

Carbohydrates in living organisms are present principally as polysaccharides, polymers of monosaccharides. In the process of photosynthesis, plants make glucose from CO_2 and H_2O. Some of the glucose is converted to starch for storage. The starch provides an energy reserve for the plant (and also for any consumer of the plant).

Starch occurs in two forms: **amylose,** the unbranched form, consists of long chains of D-glucose units in an $\alpha(1\rightarrow4)$ linkage (the linkage of maltose); and **amylopectin,** the branched form, consists of $\alpha(1\rightarrow4)$ chains linked together via $\alpha(1\rightarrow6)$ glycosidic bonds at the points of branching (Figure 23-3). Amylopectins generally contain one $\alpha(1\rightarrow6)$ branch point every 20 to 30 glucose units along the chain. A number of digestive enzymes in saliva and in the digestive tract catalyze hydrolysis of starch into free maltose units and chains of intermediate length called **dextrins,** and ultimately into glucose. Figure 23-4 illustrates the hydrolysis of amylopectin by β-amylase, an enzyme that cleaves away maltose units, beginning at the nonreducing ends of the chains. Although each starch molecule has one reducing end, it is such a small fraction of the molecule that starches give negative tests with Fehling's, Benedict's, and Tollens' solutions.

Starch is the storage form of glucose in plants; but glycogen is the storage form of glucose in animals. **Glycogen,** found in the liver and in muscle, resembles amylopectin but is more highly branched, with one branch point for approximately every ten glucose units. Like starch molecules, glycogen molecules vary in size; a typical molecule consists of about 25,000 glucose units. Glycogen is stored in the cells in the form of

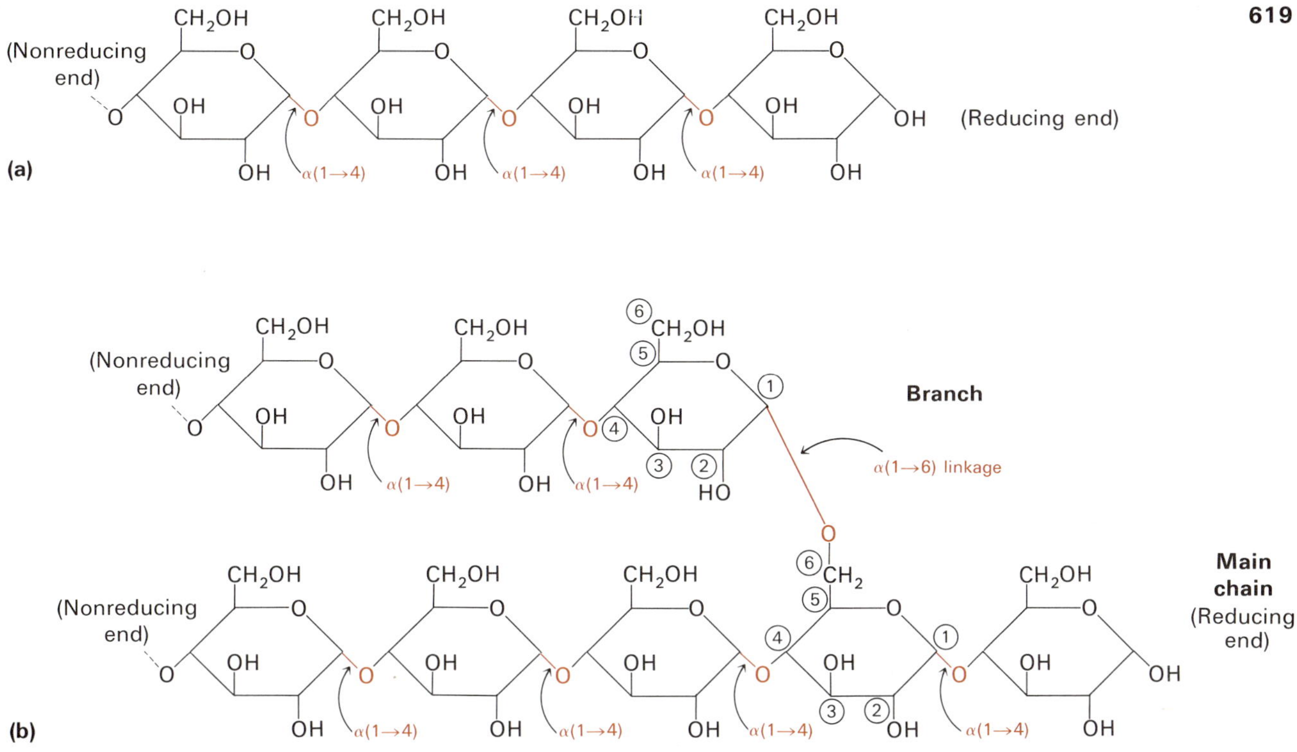

FIGURE 23-3

Segments of amylose and amylopectin molecules. **(a)** α-Amylose: all of the glycosidic linkages are $\alpha(1 \rightarrow 4)$. There may be from 25 to 3000 D-glucose units per chain. **(b)** Amylopectin: $\alpha(1 \rightarrow 4)$ chains linked by $\alpha(1 \rightarrow 6)$ glycosidic bonds. There may be as many as 600,000 D-glucose units per molecule.

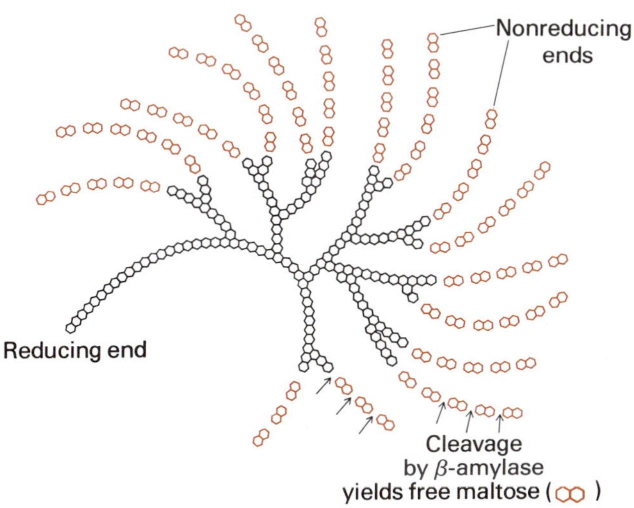

FIGURE 23-4

The enzyme β-amylase cleaves maltose units beginning at the nonreducing ends of the chains. The enzyme is called β-amylase to distinguish it from α-amylase, which cleaves amylopectins at random. Thus, in the case of amylase, the designation α or β does not refer to the configuration of the anomeric carbon at the glycosidic bond that is cleaved.

granules that can be seen in the electron micrograph of Figure 23-5.

The liver helps to maintain a relatively constant concentration of glucose in the blood, normally approximately 100 mg of glucose per 100 ml of blood. Glycogen in the liver is hydrolyzed by specific enzymes to release glucose between meals and during

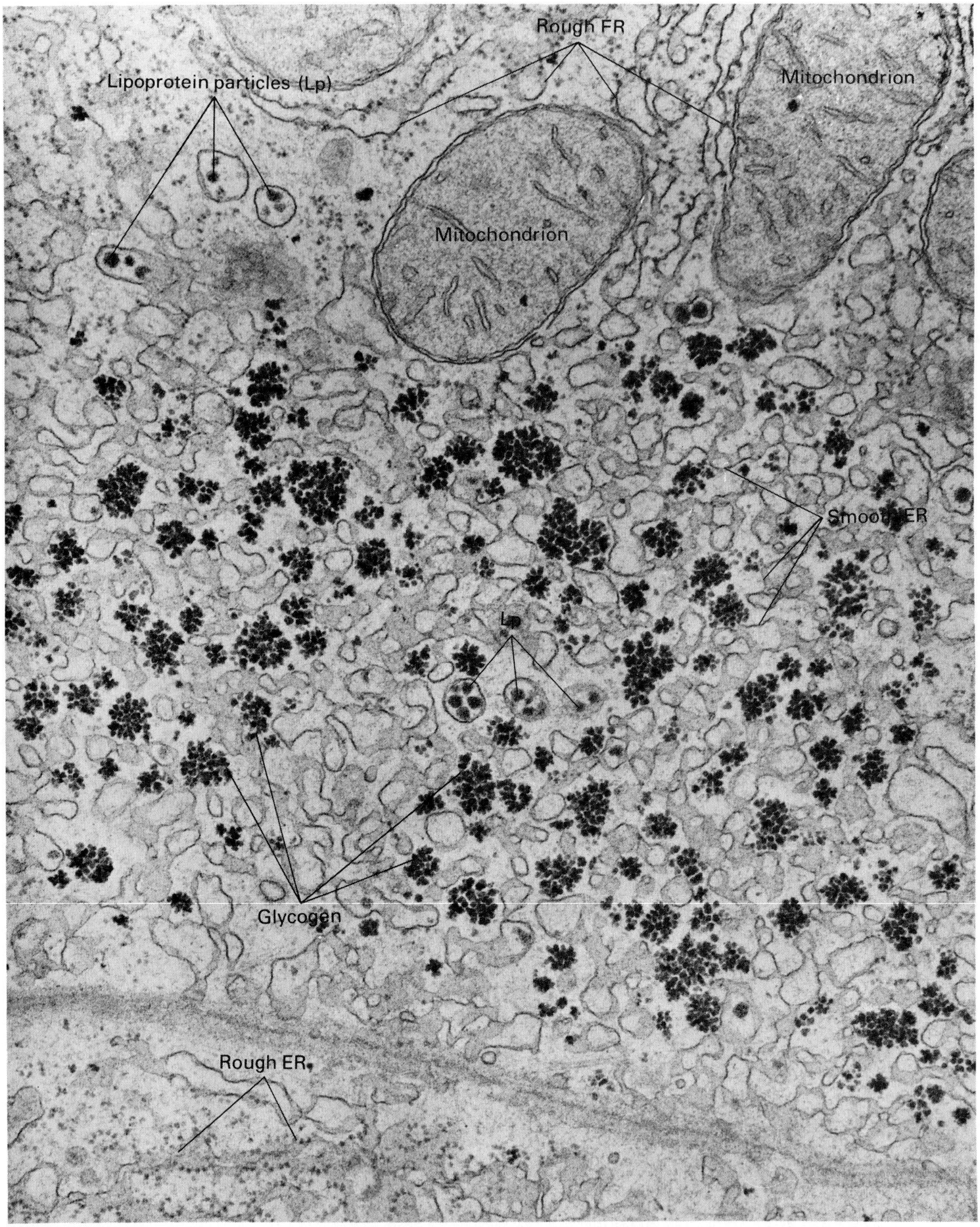

FIGURE 23-5

Electron micrograph of a liver cell. The small dark particles in clusters are glycogen granules. ER = endoplasmic reticulum. The parts of a cell are discussed in Section 27-1. [Courtesy of George Palade.]

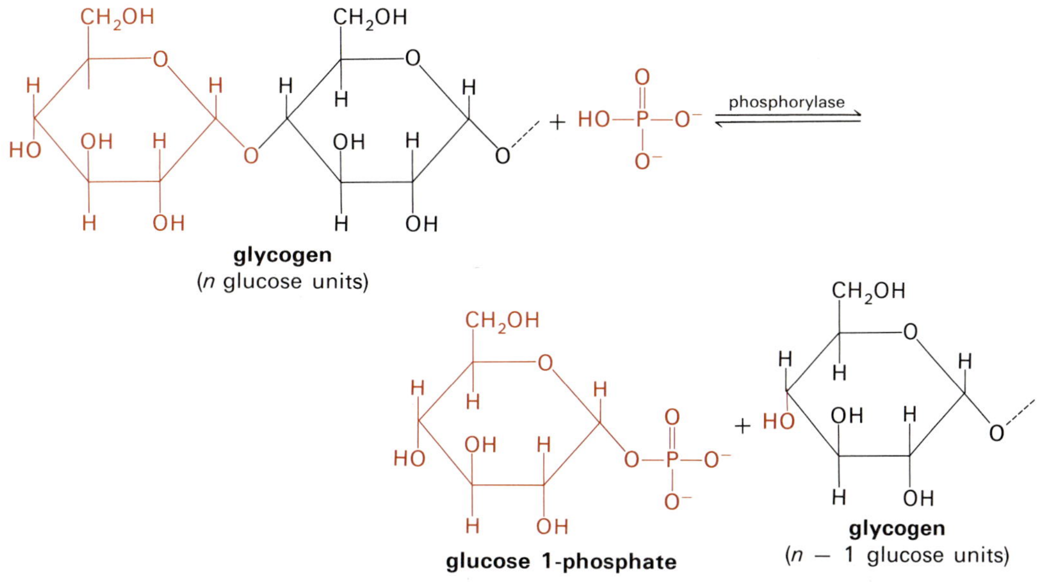

strenuous muscular activity. Also, glycogen is synthesized from glucose when the blood sugar concentration is high. (In Chapter 24 we shall see that glucose can also be converted to fats for storage.) In diabetes, when very high concentrations of glucose may occur in the blood, the kidneys excrete some of the excess glucose in the urine.

Several different genetically determined glycogen-storage diseases are known. Type VI, or Hers' disease, is characterized by an enlargement of the liver and low blood sugar levels between meals. Patients with this disease lack the enzyme phosphorylase, which catalyzes the first step in the breakdown of glycogen to glucose (top of page).

Carbohydrate polymers also are a key structural material in plants. The major component of wood is **cellulose.** The total weight of cellulose accounts for more than one-half of the organic matter present in living things on earth. Cellulose is a polymer of D-glucose with individual monosaccharide units joined by $\beta(1 \rightarrow 4)$ glycosidic bonds (Figure 23-6). The great strength of the woody substances of plants arises from cellulose molecules that are organized into bundles or fibrils. The fibrils are in parallel arrangement

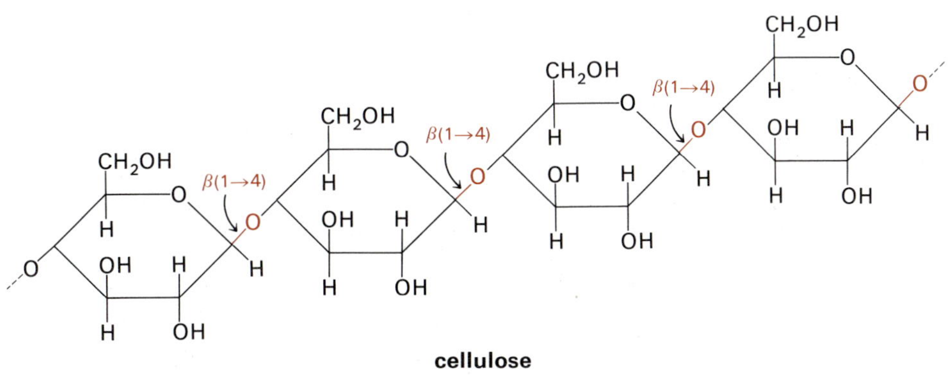

FIGURE 23-6

Cellulose is a polymer of D-glycose with $\beta(1 \longrightarrow 4)$ glycosidic bonds.

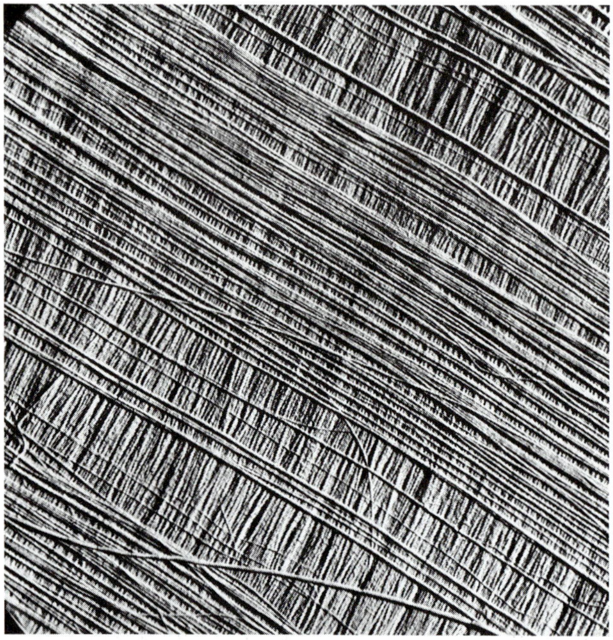

FIGURE 23-7

Electron micrograph of the cell wall of an alga, *Chaetomorpha melagonium.* The fibrils are in parallel arrangement within a layer; the layers cross each other. [Courtesy of Eva Frei and R. D. Preston.]

within a layer, and the layers cross each other, as illustrated in Figure 23-7 for the cell wall of an alga. The human body does not form cellulases, enzymes that can digest cellulose. Cellulase-producing microbes that are harbored in the digestive tracts of cattle and termites do allow these animals to hydrolyze cellulose.

EXERCISE 23-10 The exoskeletons of insects and of crustacea are composed largely of the polysaccharide chitin. Chitin is very similar to cellulose; it consists of NAG in $\beta(1 \rightarrow 4)$ linkage. The structure of NAG is given in Section 23-6. Write a structure for chitin. Illustrate the $\beta(1 \rightarrow 4)$ linkage clearly.

The polysaccharides we have considered thus far consist of glucose molecules only. There are other polymers that contain both sugars and amino acids. The cell walls of some bacteria, for example, consist of networks of polysaccharide chains cross-linked by peptides. Polysaccharides are often called **glycans,** and these cell-wall polymers are called **peptidoglycans.** A bacterial cell wall must be very strong to withstand the osmotic pressures of up to 20 atm that result from the high concentrations of metabolites inside the cell. The peptidoglycan network provides this mechanical support. Figure 23-8 shows an isolated bacterial cell wall. This huge single molecule maintains the distinctive sacklike shape of the bacterium it once surrounded.

The peptidoglycan consists of three basic units.

1. The backbone is a linear polysaccharide chain in which two derivatives of glucose alternate in a

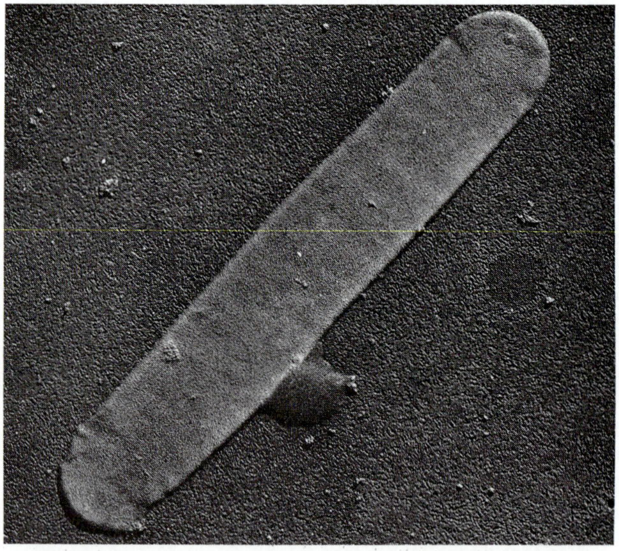

FIGURE 23-8

Electron micrograph of a peptidoglycan bacterial cell wall. [Courtesy of Nathan Sharon.]

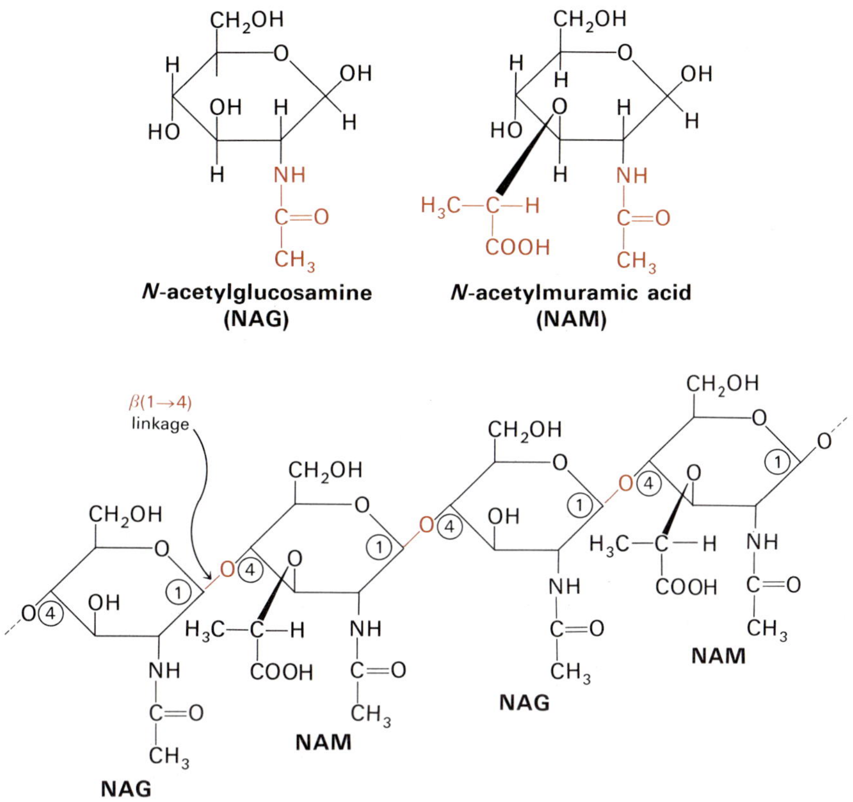

FIGURE 23-9

The peptidoglycan cell walls consist of alternating NAG and NAM units in $\beta(1 \longrightarrow 4)$ glycosidic linkage.

$\beta(1\rightarrow4)$ glycosidic linkage (Figure 23-9). The two glucose derivatives are *N*-acetylglucosamine (NAG) and *N*-acetylmuramic acid (NAM), whose structures are shown in the figure.

2. A tetrapeptide of L-alanine, D-glutamine, L-lysine, and D-alanine is bonded via an amide bond to the carboxyl groups of the NAM residues.

3. A peptide of five glycine residues cross-links the NAM tetrapeptides on neighboring polypeptide chains (Figure 23-10).

The final bond formed in the cross-linking process is the peptide bond between a glycine residue of the glycine pentapeptide and a D-alanine residue of the NAM tetrapeptide (Figure 23-10). Penicillin acts as an antibiotic through its inhibition of this particular step in the formation of the bacterial cell wall. Several methods are available to biochemists who need to break these strong cell walls in order to study cell components. The enzyme lysozyme, for example, hydrolyzes the glycosidic bond of the polysaccharide chains between NAM and NAG. Lysozyme can be used to

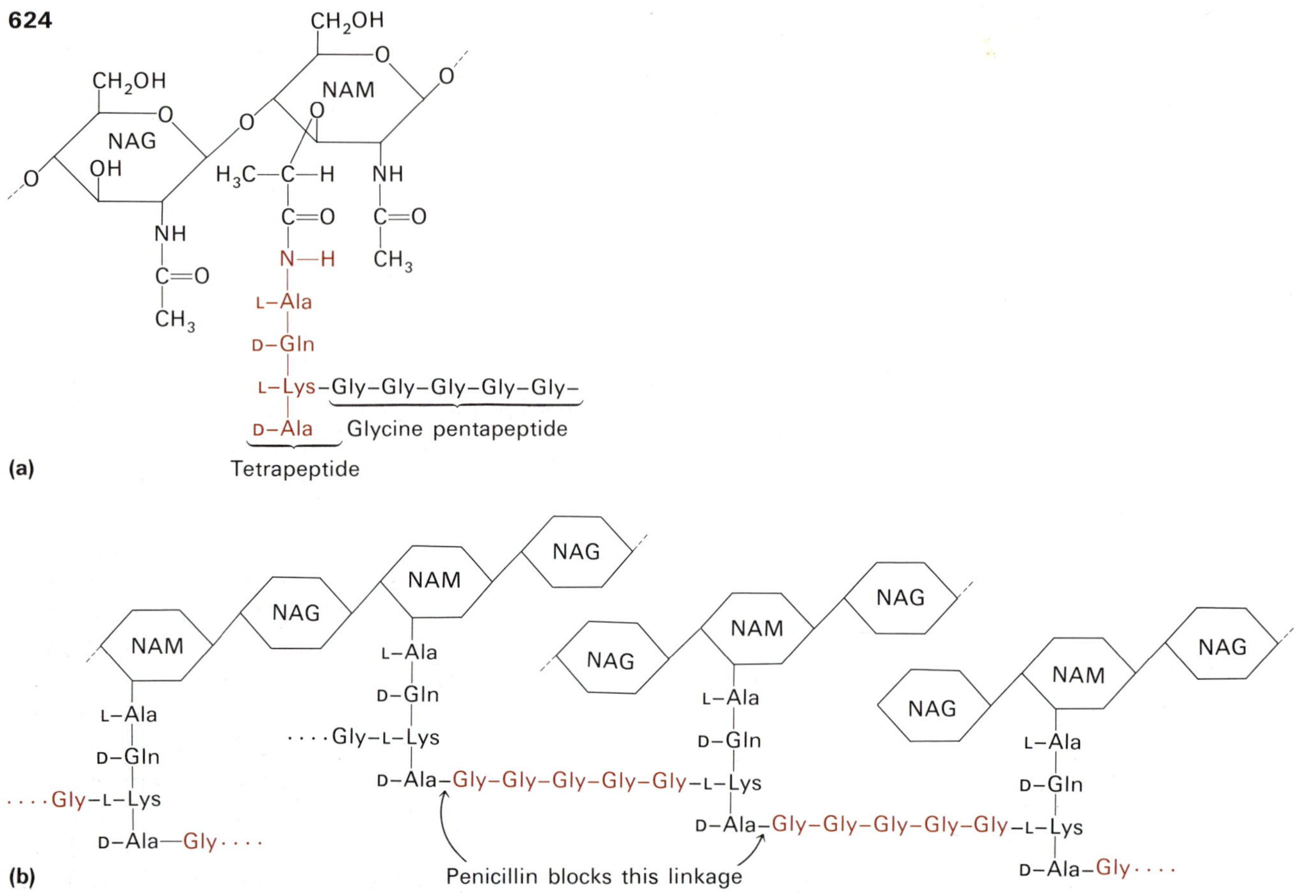

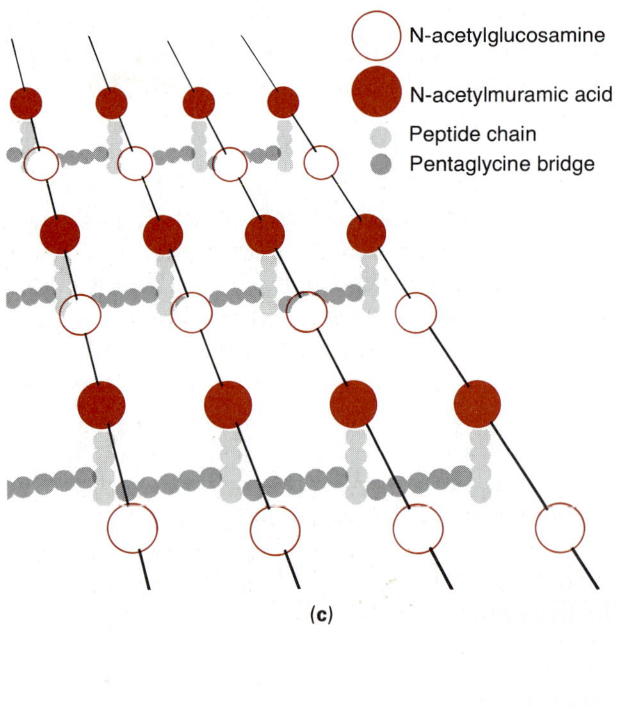

FIGURE 23-10

A peptidoglycan cell wall. **(a)** The basic repeating unit. **(b)** The glycine pentapeptides that cross-link two polysaccharide chains are indicated by color. Penicillin blocks the linkage between D-alanine and the glycine pentapeptide. **(c)** Schematic diagram showing many crosslinks. [From Nathan Sharon, "The Bacterial Cell Wall."]

make nicks in the cell wall; then, when the cells are poured into a dilute salt solution, the osmotic pressure of the cell interior bursts the walls. Another common method of splitting the cell walls is to use a "French press." This device forces the cells at high pressure through a very small opening; when the cells emerge into the lower pressure of the atmosphere, they explode.

23-9
GLYCOPROTEINS AT THE SURFACES OF ANIMAL CELLS

We have seen that peptidoglycans are polymers that consist of polysaccharide chains cross-linked by peptides. **Glycoproteins,** in contrast, are macromolecules that consist primarily of proteins, but that have oligosaccharide (two to about ten sugar units) or polysaccharide side chains. Both the sequence of sugars and the sequence of amino acids in a glycoprotein are specific to the particular glycoprotein.

The high degree of specificity of the oligosaccharide units is illustrated by the blood-group glycoproteins. Karl Landsteiner first recognized in 1900 that human beings can be classified on the basis of blood groups, with the earliest classification based on groups called O, A, B, and AB. Landsteiner observed that something in the serum of group B causes the red cells of group A to clump together. (The serum is the noncellular component of blood.) Similarly, red cells from group B are aggregated (or agglutinated) by something in the serum of group A. Thus, type-A blood contains an anti-B factor but not an anti-A factor, and type-B blood contains an anti-A factor but not an anti-B factor. These observations are explained by a model in which type-A antigens (Section 22-9) on the red blood cells are captured by anti-A antibodies in the serum; and, similarly, type-B antigens are captured by anti-B antibodies. These antigens on the cell wall are glycoproteins. The aggregation of the blood cells results from the interaction between antigens and antibodies. The antigen and antibody characteristics of the O, A, B, and AB blood types are summarized in Table 23-1.

TABLE 23-1
ABO blood groups

Blood type	Antigens on cells	Antibodies in serum	Reaction of red cells	
			Anti-A	Anti-B
O	None	Anti-A and anti-B	−	−
A	A	Anti-B	+	−
B	B	Anti-A	−	+
AB	A + B	None	+	+

The main danger in transfusion arises when the recipient possesses antibodies against antigens on the surfaces of the transfused blood cells. Normally only a small number of these antibodies are present in the recipient's blood. However, a transfusion of cells containing antigens stimulates the immune system of the recipient to produce tremendous numbers of antibodies against these antigens. This causes the transfused cells to agglutinate. It is important then that the recipient not possess antibodies for the antigens on the transfused blood cells. By consulting Table 23-1, we can see which combinations should be safe and which dangerous. Suppose a patient has type-A blood—the blood cells have type-A antigens and the serum contains anti-B antibodies. Blood cells with B antigens must, therefore, not be transfused; so neither B nor AB blood may be used, but either type A or type O would be safe. If the recipient has type-B blood, she can be transfused with either type B or type O. In contrast, a type-O recipient has both anti-A and anti-B antibodies, and so can safely be transfused only with type-O blood in which the red blood cells have neither A nor B antigens. Type-O blood, sometimes called the "universal donor," can thus be used for patients of all blood types—type-O cells react with neither anti-A nor anti-B antibodies in the recipient's serum.

In contrast to the care needed with whole blood, a patient can be given serum of any type. Clumping of the recipient's cells by antibodies in the transfused serum is negligible because the concentration of antibodies in the serum is small and no additional antibodies can be produced. Furthermore, the serum is immediately diluted by a factor of 10 to 12 for a one-pint transfusion (the blood volume of an adult is 10 to 12 pints). (The serum is always checked for antibody concentration, however, because occasionally the serum of a donor contains an unusually high concentration of antibodies.)

EXERCISE 23-11 Use information in Table 23-1 to explain why a type-AB recipient can be given a transfusion of whole blood of group A, B, or AB. Such people are sometimes called "universal recipients."

We now know enough about the molecular structure of blood cells to relate differences in red cells of types A, B, and O to the chemistry of their surface glycoproteins, which act as the antigens. Remarkably, the differences revolve around a single monosaccharide unit. The structure of the oligosaccharide portion of the A-type glycoprotein is shown in Figure 23-11. The O-type antigen differs solely in the absence of the two NAG units indicated. Blood types A and B differ only in the fact that type B has galactose in place of the two NAG units. Thus, the major complication in transfusion caused by type-A serum of the recipient mixing with type-B red cells results simply from the presence or absence of an acetyl amino group.

The specificity of the blood-group substances illustrates a more general phenomenon of cell identification at the biochemical level. Many of the complex processes of differentiation of a fertilized egg into an adult organism require some mechanism by which one cell type can recognize another. The oligosaccharides of glycoproteins may serve to "label" the surface of cells. In addition to the glycoprotein labels, cells are postulated to possess receptors, units that recognize the labels of other cells. In this way, cells could both "recognize" and "be recognized." Such recognition would take the observable form of adhesion between cells. Evidence for this theory is given by the facts that under certain conditions, isolated cells grown in culture will stick together, and that it is possible to interfere with adhesion by altering the surface glycoproteins. Figure 23-12 shows sheep red blood cells

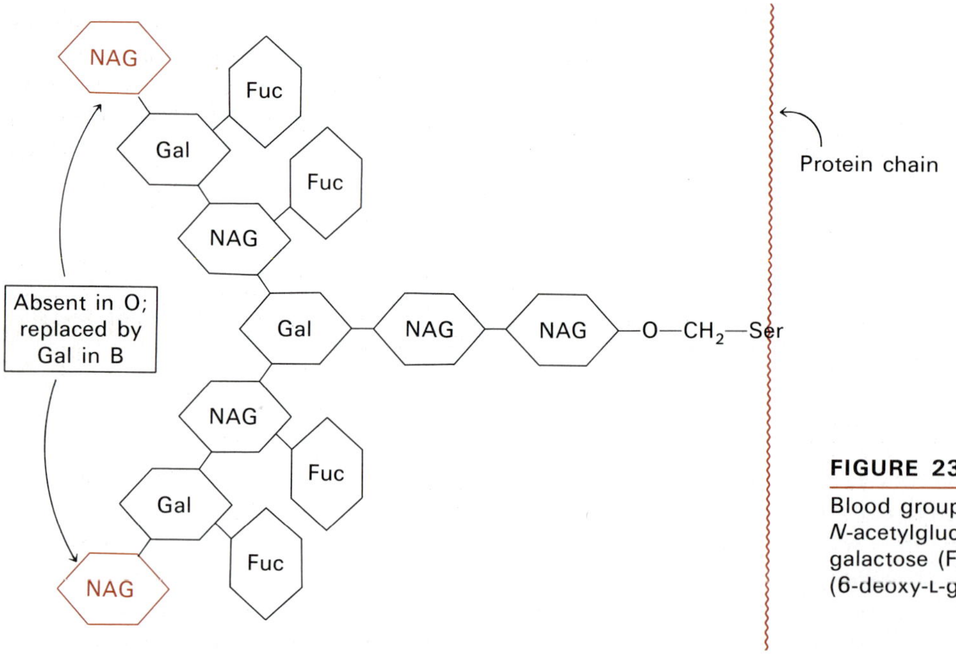

FIGURE 23-11

Blood group "A" oligosaccharide. NAG is *N*-acetylglucosamine (Figure 23-9). Gal is galactose (Figure 23-1). Fuc is fucose (6-deoxy-L-galactose).

FIGURE 23-12

Scanning electron micrograph of sheep red blood cells bound to a human lymphocyte (*center*). The surface of the lymphocyte contains glycoprotein receptors specific for antigens on the surface of the sheep red cells. [Courtesy of Jerry Thornthwaite and Robert Leif.]

bound to a human lymphocyte. Receptors on the lymphocyte's surface recognize antigens on the sheep blood cells.

Special interest in the role of cell-surface glycoproteins has been generated in recent years because of a possible relationship to cancer. Cells isolated from a cancerous tumor and grown in culture tend to grow to a more crowded condition than normal cells. They also have a more rounded shape. Normal cells can be transformed to a cancerlike condition by treatment with certain chemical agents (carcinogens) or with tumor-producing viruses. One result of such transformation is the loss of a major cell-surface glycoprotein. When this protein (fibronectin) is purified and reintroduced into transformed cells, the cells are restored to a more normal, flat shape.

KEY TERMS

A *monosaccharide* is a single carbohydrate molecule; it contains from three to seven carbon atoms and has the general formula, $C_n(H_2O)_n$. A *disaccharide* is two monosaccharide units linked together; an *oligosaccharide* is from two to about ten monosaccharide units linked together; and a *polysaccharide* is many monosaccharide units linked together.

An *aldose* is a carbohydrate with an aldehyde functional group. A *ketose* is a carbohydrate with a ketone functional group.

An *anomer* is one of the two possible structures (designated α or β) that can occur when a carbohydrate in the open-chain form undergoes cyclization to form a hemiacetal. *Mutarotation* is the process of interconversion between the α and β ring forms of a monosaccharide via the open-chain form.

A *glycoside* is a carbohydrate in the ring form that has reacted with an alcohol. The linkage between carbohydrate units in disaccharides and polysaccharides is called a *glycosidic bond.*

A *reducing sugar* is a sugar that gives a positive reaction with Benedict's, Fehling's, or Tollens' reagent.

A *sugar alcohol* is a sugar in which the aldehyde group has been reduced to an alcohol. An *amino sugar* is a sugar derivative in which one of the hydroxyl groups has been replaced with an amino group. *Phosphate derivatives* of sugars are esters of phosphoric acid and one of the hydroxyl groups of the sugar.

A *peptidoglycan* is a polysaccharide with peptide side chains. A *glycoprotein* is a protein with oligosaccharide or polysaccharide side chains.

CHAPTER SUMMARY

Simple carbohydrates fall into two categories, aldoses and ketoses. The *aldoses* contain an aldehyde functional group, and the *ketoses* contain a ketone functional group. Carbohydrates can exist as single sugar units (*monosaccharides*) or joined together in molecules ranging from a few units (*oligosaccharides*) to thousands of units (*polysaccharides*).

For aldoses with three or more carbons, stereoisomers can exist, including compounds that are not enantiomers and that are designated *diastereomers.* A family of four-, five-, and six-carbon aldoses can be synthesized, all of which share the configuration of D-glyceraldehyde at the next-to-last carbon atom. The most common sugar, D-glucose, is one of the six carbon members of this family. A similar family of ketoses can be prepared.

Although carbohydrate structures are often drawn as linear chains for convenience, in solution the larger aldoses and ketoses exist principally as cyclized structures of five- and six-membered rings that include an oxygen atom in the ring. Ring closure introduces a new center of asymmetry, at the anomeric carbon. The two distinct structures that can form as a result of ring closure are called α or β *anomers.* Interconversions can take place between the two forms, and the process of interconversion is called *mutarotation.* The cyclic structures of carbohydrates tend to assume nonplanar arrangements, such as the "chair" conformation of six-membered rings.

Interconversion of aldoses and ketoses can be accomplished in dilute alkali in a reaction called *enolization* because intermediates with an enol arrangement are formed. *Glycosides* are formed when a sugar in cyclic form reacts with an alcohol. Oxidation reactions can convert the aldehyde group or the primary alcoholic groups to carboxylic acids. Reduction of the aldehyde group leads to *sugar alcohols.* Elements other than carbon, hydrogen, and oxygen are found in certain carbohydrate derivatives—for example, phosphorus in sugar-phosphate esters, and nitrogen in amino sugars.

Cetain disaccharides are present in living organisms in appreciable quantities, and the most common compound in this category is sucrose, the common table sugar. Polysaccharides exist in many varieties in different forms of life; they include starch and glycogen (the storage polymers of plants and animals, respectively) and cellulose (the major component of wood).

All three of these polymers are composed of D-glucose. Other polysaccharides are composed of carbohydrate derivatives, such as amino sugars, and fulfill a number of structural roles, especially at the surfaces of cells. Carbohydrates may also be found in association with proteins to give a class of molecules called *glycoproteins*. Glycoproteins at the surfaces of cells are involved in the chemical identification of cells, as in the ABO blood types and in certain changes associated with cells becoming cancerous.

PROBLEMS

23-1. For each substance on the left, pick a description from the list on the right that best describes its structure:

(a) D-glucose	(A) a sugar alcohol
(b) D-glucoside	(B) a disaccharide
(c) glycerol	(C) a hemiacetal
(d) glycogen	(D) an acetal
(e) maltose	(E) a polysaccharide

23-2. The simplest ketose in the D-family is the ketotetrose, D-erythulose:

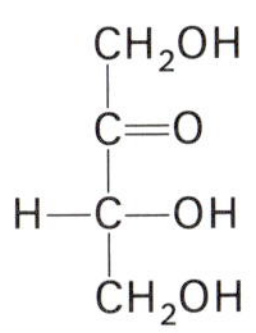

(a) Write Fischer projection formulas for the two ketopentoses in the D family. (The carbonyl carbon is C-2.)

(b) Write Fischer projection formulas for the four ketohexoses in the D family. (The carbonyl carbon is C-2.)

(c) Write the cyclic structure of *one* of the ketohexoses.

(d) Write a Fischer projection formula for an enantiomer of one of the ketohexoses.

23-3. Write the structures of α- and β-D-mannose in the cyclic form. (A Fischer projection formula for D-mannose is given in Figure 23-1.)

23-4. Draw D-mannose in the chair conformation.

23-5. Write structures of the products of oxidation of D-galactose (a) at C-1 and (b) at C-6.

23-6. Write the structures of galactose 6-phosphate and galactose 1-phosphate in the cyclic form.

23-7. Write the structure of the disaccharide composed of two units of mannose joined by an $\alpha(1\rightarrow4)$ glycosidic bond.

23-8. Cellobiose is a disaccharide produced in the hydrolysis of cellulose by cellulases. Write the structure of cellobiose, and describe its glycoside linkage.

23-9. Which of the following would give a positive test with Benedict's reagent? (a) sucrose; (b) α-maltose; (c) methyl-β-L-glucoside; (d) β-lactose; (e) glycogen.

23-10. Write Fischer projection formulas for all possible stereoisomers of an aldotetrose. Tell which are enantiomers and which are diastereomers.

23-11. The following family of diaminopentoses has been discovered on a distant planet:

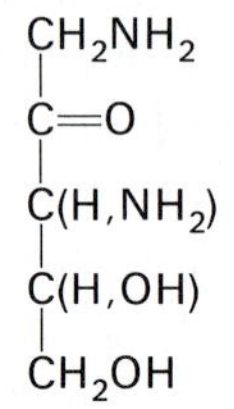

(a) Copy the formula above, and label each asymmetric carbon with an asterisk.
(b) How many different structures are possible for this family (include all enantiomers)? Write the structures.
(c) Indicate which structures are enantiomers.

23-12. How many *different* chemical compounds are shown here?

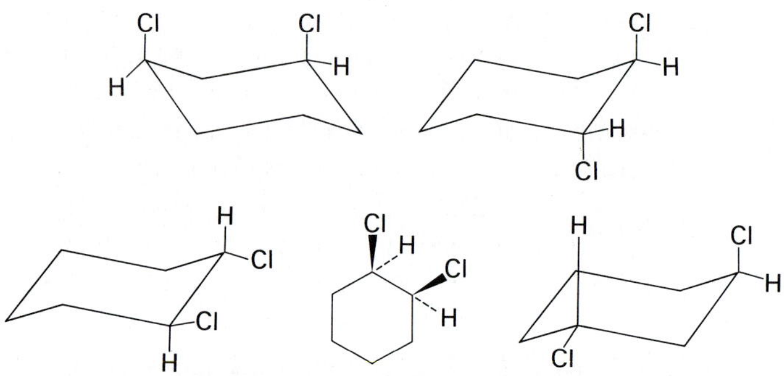

23-13. The polyhydroxy compound sorbitol, shown here in Fischer projection, is the sweetener used in "sugarless" gum.

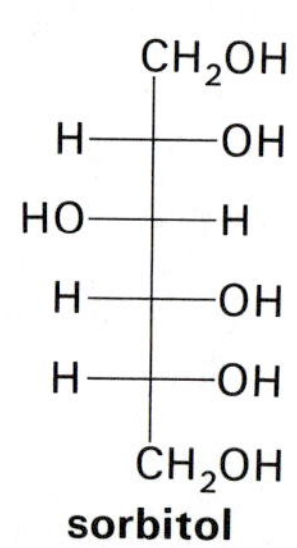

sorbitol

Which sugar (or sugars from the following list will form sorbitol upon reduction (by H_2 with a nickel catalyst)? This question requires careful analysis.

(a)
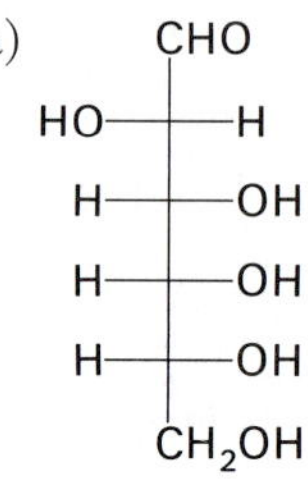

(b)
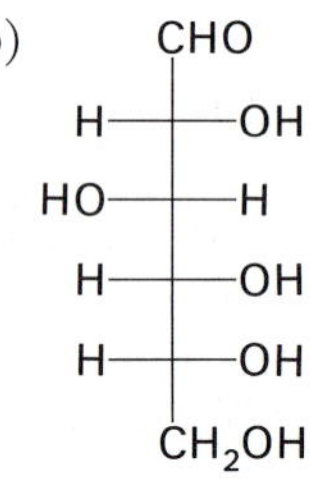

(c)
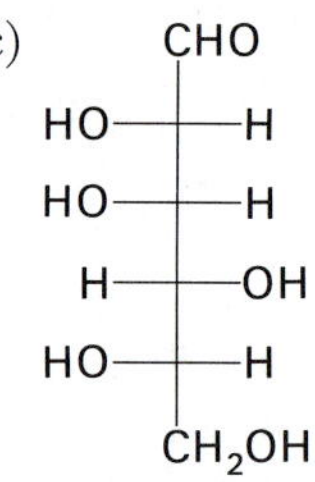

(d)
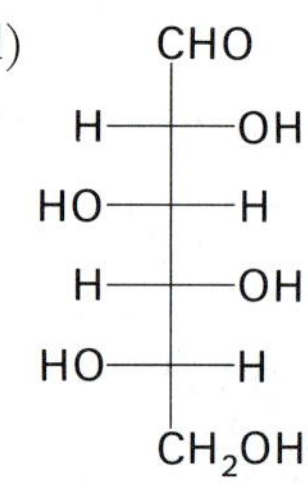

(e)
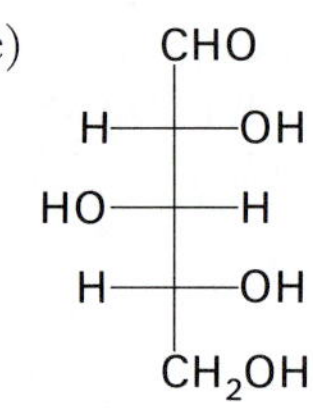

23-14. The compound 5-aminohexanaldehyde can be separated into its enantiomers. The D-enantiomer, shown here in a Fischer projection, rotates plane-polarized light $+80°$ when it has been freshly prepared. However, upon standing for one or two days at room temperature, its rotation slowly changes until it becomes constant at $+20°$. Paper chromatography on this "aged" sample shows three spots, one of which corresponds to the original pure D-5-aminohexanaldehyde.

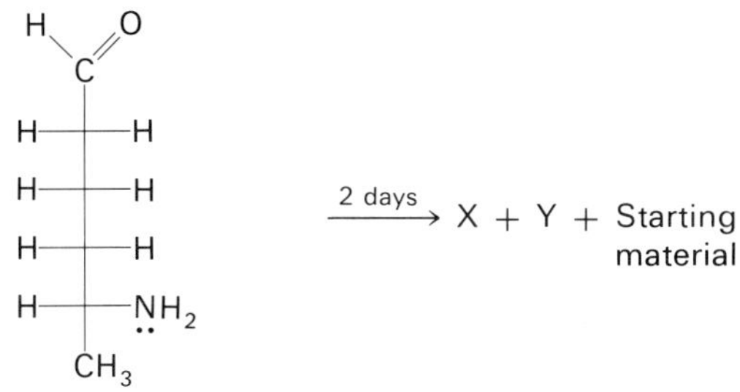

D-5-aminohexanaldehyde

(a) Propose structures for the two new compounds X and Y that have appeared. Write their structures.
(b) Are these new compounds optically active?

23-15. A controversy is raging over the alleged anticancer properties of the drug Laetrile™, although tests in the United States do not substantiate such therapeutic activity. This rather simple substance contains a β-glycoside bond reminiscent of polysaccharides.

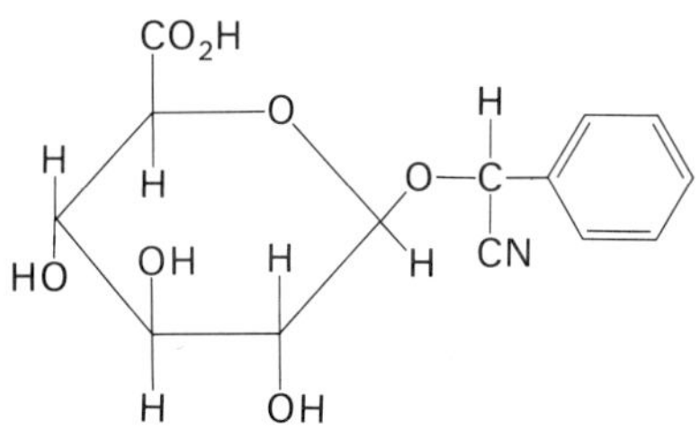

Laetrile™

Scientists in other countries say that Laetrile™ does not act directly but requires activation first. They claim that enzymes that act on polysaccharides catalyze the breakdown of Laetrile™ at the cancer site to form a very reactive poisonous gas that stops the cancer growth.
(a) Propose a series of reactions that explains this breakdown, and name the poisonous gas. (Consult Section 18-4, if necessary.)
(b) Two other products are formed in the breakdown, $C_6H_{10}O_7$ and C_7H_6O. Write their structures.

23-16. We said in Section 23-2 that the D family of aldoses can be synthesized from D-glyceraldehyde without disturbing the stereochemistry of C-2. The Kiliani–Fischer synthesis is the method by which this was first accomplished. We shall see how this method can be used to synthesize D-erythrose and D-threose from D-glyceraldehyde.
(a) The first step is the addition of HCN to the carbonyl carbon of D-glyceraldehyde. Write the structures of the two diastereomers formed in this step.
(b) Acid hydrolysis of the products of step a yields two carboxylic acids. (Carbon-1, which was the CN carbon, is the carboxyl carbon.) Write the structures of the two products of the hydrolysis.
(c) Write a reaction that shows how D-erythrose and D-threose can be obtained from the carboxylic acids formed in step b.

Human adipocytes, or fat-storing cells. Each cell contains a single large vacuole of liquid lipid. Scanning electron micrograph, ×800.

24

Lipids

Lipids, biological compounds insoluble in water but soluble in nonpolar solvents, can be divided into two major subdivisions: the complex lipids, which are esters or amides of fatty acids, and the simple lipids (such as steroids), which do not contain fatty acids. The main functions of complex lipids are fat storage and formation of cell membranes. Steroids also play a role in membranes, and in addition they are involved in endocrinology (the male and female sex hormones are steroids) and nutrition (vitamin D is a steroid).

24-1
CLASSIFICATION OF LIPIDS

Lipids are substances from biological sources that are insoluble in water but can be dissolved in nonpolar solvents such as ether and chloroform. Crude preparations of lipids generally contain a large variety of different compounds, but they can be divided into two groups based on whether or not they yield soaps upon hydrolysis in basic solution. Production of "soap," the sodium or potassium salts of fatty acids (long-chain carboxylic acids), is one of the earliest examples of practical chemistry. Recipes for making soap from animal fats (which are largely lipids) and wood ash (which contains the base KOH) can be found in the writings of the ancient Greeks. The name for this process of soap production by alkaline hydrolysis is **saponification.** Lipids that hydrolyze to yield fatty acids are called **complex lipids.** We shall see shortly that complex lipids are esters or amides. The complex lipid tristearin is composed of three molecules of stearic acid, an 18-carbon fatty acid, esterified to glycerol. A saponification reaction of tristearin may be summarized as

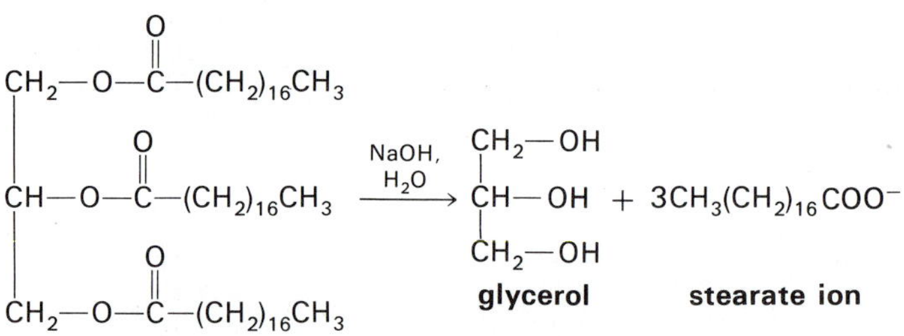

tristearin **glycerol** **stearate ion**

The mechanism of this reaction, the hydrolysis of an ester, is discussed in Section 19-7.

The complex lipids are further classified on the basis of the compound to which the fatty acids are bonded: glycerol, glycerol 3-phosphate, sphingosine, or a fatty alcohol (a monohydroxyl alcohol with a long hydrocarbon chain).

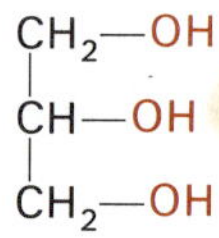

glycerol

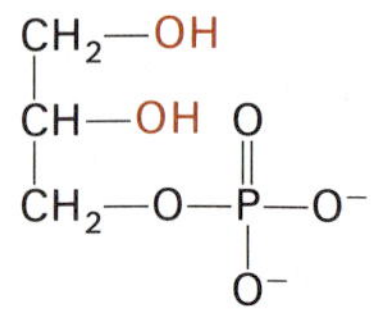

glycerol 3-phosphate

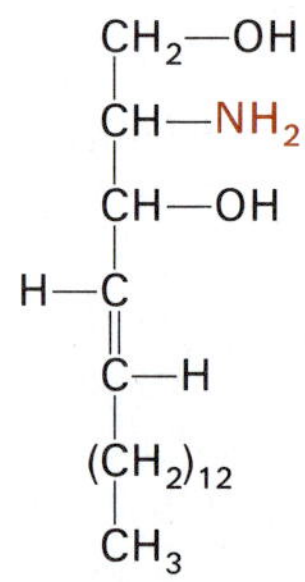

sphingosine

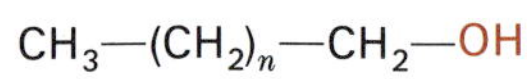

fatty alcohol

The groups to which the fatty acids are esterified are shown in color. The fatty acid forms an amide with the NH_2 group of sphingosine instead of an ester with an hydroxyl group (Section 24-4). We discuss all these classes of lipids in the following sections.

Lipids that are nonsaponifiable are called **simple lipids.** The simple lipids include terpenes, steroids, and prostaglandins, whose structures we describe in Section 24-6.

24-2 FATTY ACIDS

We have already studied some of the properties and reactions of fatty acids and fats: the hydrogenation of unsaturated fatty acids and the determination of the degree of unsaturation, or iodine number (Chapter 15); and the hydrolysis of fats and oils to yield soaps (Chapter 19).

The major fatty acids range in size from 14 to 22 carbon atoms; the most abundant saturated ones are palmitic acid (with 16 carbon atoms) and stearic acid (with 18 carbon atoms). Oleic acid, with 18 carbons, is the most abundant unsaturated fatty acid. The structures and melting points of a representative selection of fatty acids are presented in Table 24-1. The naturally occurring unsaturated fatty acids all have the *cis* configuration at the double bonds.

The presence or absence of double bonds in fatty acids has a major impact on their physical properties. The hydrocarbon tails of saturated fatty acids can assume a straight, extended conformation that permits close packing (Figure 24-1). In contrast, the *cis* double bond leads to kinks in the hydrocarbon chain that hinder close packing. The space-filling models shown in the figure are of the anions of the acids—the predominant species at physiological pH. Because they cannot pack tightly, the van der Waals attractive forces between neighboring hydrocarbon chains are weaker for the unsaturated than for the saturated fatty acids. As a result, the unsaturated fatty acids form less stable solids and have lower melting points (see Table 24-1). We saw an example of this in the copepods, the tiny sea creatures whose fats contain a ratio of unsaturated to saturated fatty acids that is higher for specimens taken from colder water (Section 15-4). A similar observation has been made with the bacterium *E. coli*: the ratio of its unsaturated to saturated fatty acids increases when the temperature at which it is grown is decreased. Cell membranes are largely composed of lipids, and we shall see that cell

TABLE 24-1
Fatty acids

Name	Structure	Melting point (°C)
Saturated fatty acids:		
Lauric	$CH_3—(CH_2)_{10}—COOH$	44
Myristic	$CH_3—(CH_2)_{12}—COOH$	54
Palmitic	$CH_3—(CH_2)_{14}—COOH$	63
Stearic	$CH_3—(CH_2)_{16}—COOH$	70
Unsaturated fatty acids:		
Oleic	$CH_3—(CH_2)_7—CH{=}CH—(CH_2)_7—COOH$	13
Linoleic	$CH_3—(CH_2)_4—CH{=}CH—CH_2—CH{=}CH—(CH_2)_7—COOH$	−5
α-Linolenic	$CH_3—CH_2—CH{=}CH—CH_2—CH{=}CH—CH_2—CH{=}CH—(CH_2)_7—COOH$	−11
γ-Linolenic	$CH_3—(CH_2)_4—CH{=}CH—CH_2—CH{=}CH—CH_2—CH{=}CH—(CH_2)_4—COOH$	−14
Arachidonic	$CH_3—(CH_2)_4—(CH{=}CH—CH_2)_4—(CH_2)_2—COOH$	−50

Note: Most of the saturated and unsaturated fatty acids can be synthesized in the bodies of mammals from simple dietary components. Linoleic and γ-linolenic acids are exceptions that must be provided in the diet, and they are known as **essential fatty acids.** The essential fatty acids are required for the synthesis of certain other fatty acids and of prostaglandins, compounds that we discuss in Section 24-6.

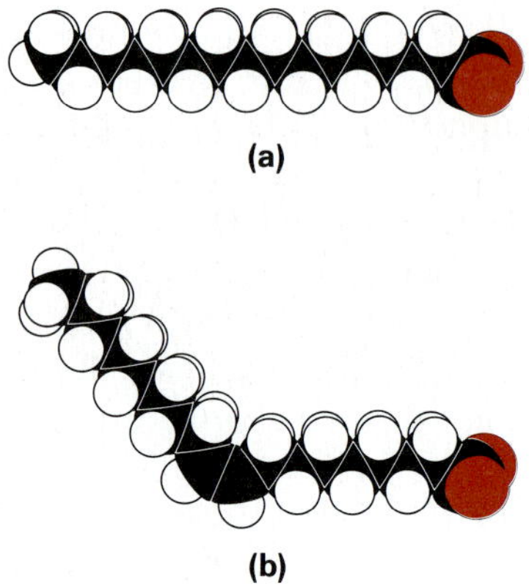

FIGURE 24-1

Space-filling models of anions of **(a)** palmitate, the 16-carbon saturated fatty acid, and **(b)** oleate, an 18-carbon unsaturated fatty acid. The anions are shown because they predominate at physiological pH. [From Lubert Stryer, *Biochemistry,* 2nd ed., W. H. Freeman and Company. Copyright © 1981.]

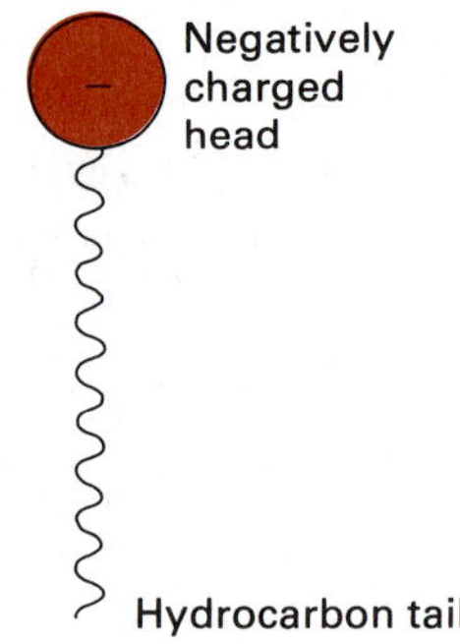

FIGURE 24-2

Schematic representation of a fatty acid anion.

membranes must be fluid to perform their functions (Section 24-5). Thus, an increase in the proportion of unsaturated fatty acids keeps cell membranes from becoming too stiff at the lower temperatures.

The data given in Table 24-1 also show that the melting points of fatty acids increase as the length of the carbon chain is increased. We would expect this, because the van der Waals forces also increase with increasing chain length.

We have noted that the sodium and potassium salts of fatty acids are soaps, but we have not yet seen how soaps work. In water at neutral pH, fatty acids are present predominantly as anions. The anion consists of a negatively charged head region and a long hydrocarbon tail. This general structure can be represented conveniently by a circle for the COO^- group connected to a wavy line (Figure 24-2). Because of the long nonpolar hydrocarbon chain, these anions do not really dissolve in water. Instead, the fatty acid anions spontaneously associate to form spheres (called **micelles**) with the negatively charged heads on the outside where they can form hydrogen bonds with the hydrogen atoms of water molecules. The nonpolar hydrocarbon tails are inside the micelle (Figure 24-3). These micelles, as we shall see, are similar in many ways to cell membranes.

A micelle is held together by the hydrophobic interactions of the nonpolar hydrocarbon tails and by the hydrogen bonds between the ionic heads and the

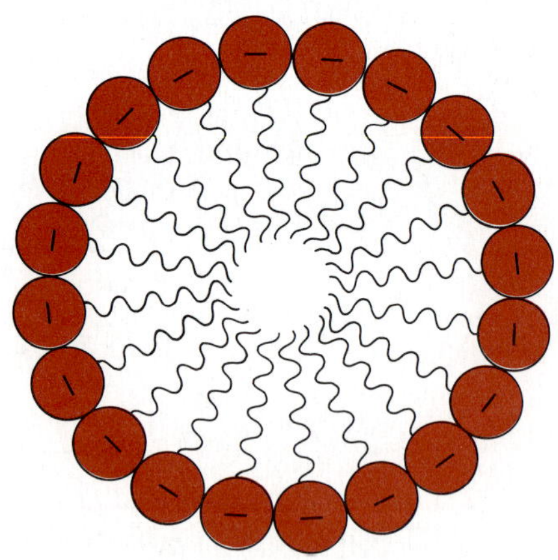

FIGURE 24-3

Cross section of a soap micelle.

water molecules. Van der Waals forces between neighboring hydrocarbon tails also contribute to the stability of a micelle. These are the same interactions that we saw are responsible for the folding of a globular protein.

Greasy dirt on skin and clothing will not wash off with water because grease and oils are nonpolar and thus are insoluble in water. When soap is added to the water, the hydrocarbon tails penetrate the grease, and the ionic heads dissolve in water. This breaks the greasy dirt into small particles and disperses it through the water in micelles (Figure 24-4). The surface of a single micelle, which may contain from hundreds to thousands of fatty acid anions, has a high concentration of negative charge. Repulsive forces between the charges on individual micelles keep them apart and suspended in the water.

Hard water contains metal ions such as Mg^{2+} and Ca^{2+}. The "soap scums" that form in hard water are insoluble salts of metal ions and fatty acid anions:

$$Ca^{2+} + \underset{\substack{\textbf{palmitate soap}\\ \text{(soluble)}}}{2CH_3(CH_2)_{14}COO^-} + 2Na^+ \longrightarrow \underset{\substack{\textbf{calcium palmitate}\\ \text{insoluble ``soap scum''}}}{Ca[CH_3(CH_2)_{14}COO^-]_2(s)} + 2Na^+$$

Synthetic detergents have structures similar to those of soaps. Most synthetic detergents are salts of linear alkanesulfonates:

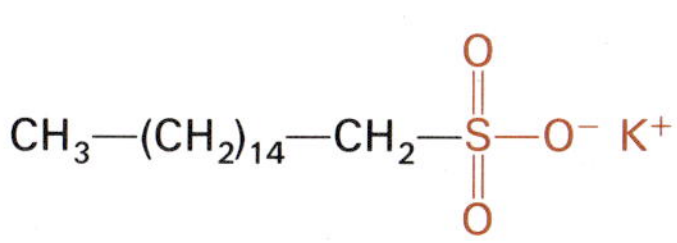

a synthetic detergent

Because detergents form soluble complexes with the metal ions in hard water, they do not form scums.

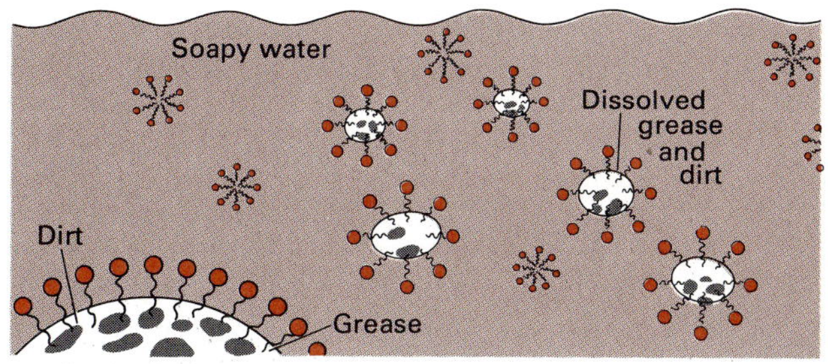

FIGURE 24-4

Schematic diagram of the action of soap in dissolving grease and dirt.

EXERCISE 24-1 (a) A disadvantage of soaps is that in acidic solution (tap water is usually acidic) the soap precipitates out of solution as a scum, even in soft water. Suggest a reason for this.

(b) Name one or two compounds that could be combined with the soap to prevent the formation of scum.

In nature fatty acids occur almost exclusively with an even number of carbon atoms. We shall now see that this is a result of the way they are synthesized. The chain-lengthening step in the biosynthesis of fatty acids is the condensation of two thioesters (esters in which a sulfur atom replaces the singly bonded oxygen atom of an ester group):

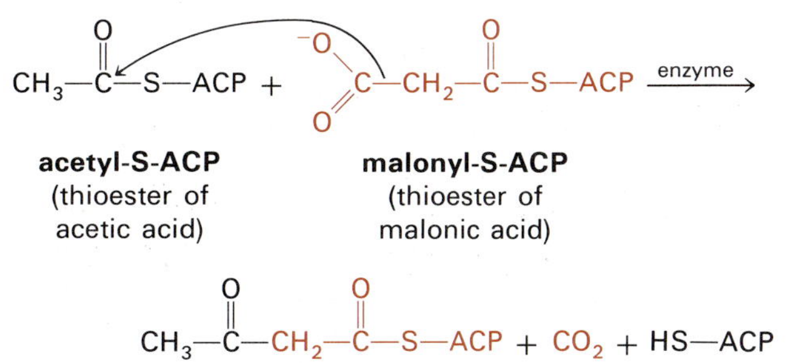

acetyl-S-ACP (thioester of acetic acid)

malonyl-S-ACP (thioester of malonic acid)

acetoacetyl-S-ACP (thioester of acetoacetate)

(24-1)

We shall not be concerned here with the structure of the ACP (acyl carrier protein), which is a polypeptide chain that contains a thiol (—SH) group, but only in the fact that it forms reactive thioesters with carboxylic acids. (The condensation of the two thioesters is very similar to the Claisen condensation of two esters discussed in Section 19-8.) In the reaction of Equation 24-1, a four-carbon unit is formed from a two-carbon unit and a three-carbon unit, with the elimination of CO_2 (a decarboxylation).

The condensation reaction is followed by (1) reduction of the ketone to an alcohol; (2) dehydration of the alcohol to give a carbon–carbon double bond; and (3) further reduction to yield a four-carbon (or butyrate) thioester. All of these reactions are catalyzed by enzymes, and the reducing agent is NADPH (nicotinamide adenine dinucleotide phosphate; structure, Figure 26-3).

1. Reduction of the ketone to yield an alcohol:

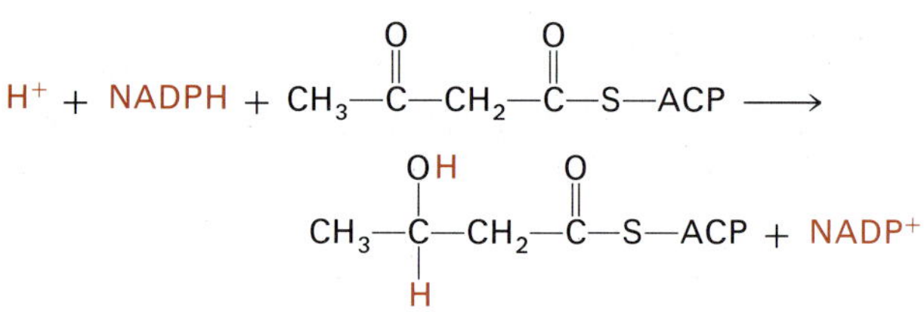

2. Dehydration of the alcohol to yield a carbon–carbon double bond:

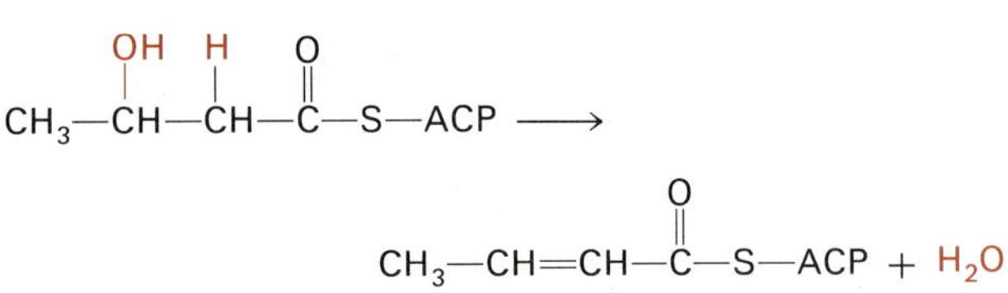

3. Reduction of the carbon–carbon double bond to yield butyryl-S-ACP:

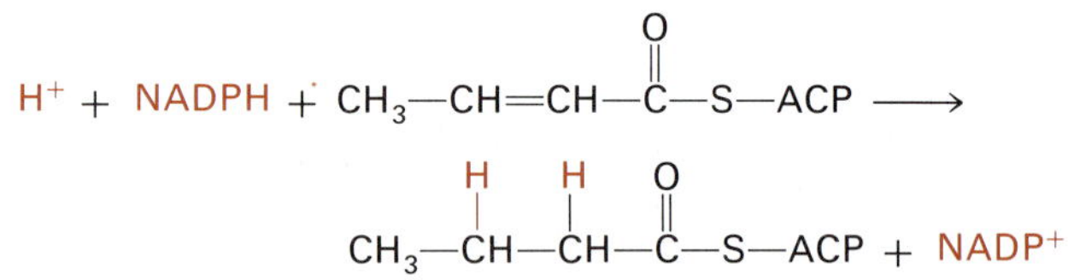

butyryl-S-ACP

This completes the first cycle of the fatty acid synthesis; a two-carbon thioester has now been converted to a four-carbon thioester. The butyryl-S-ACP can undergo condensation with another molecule of malonyl-S-ACP, and so on, increasing the carbon chain by two carbons with each round of reactions. Chains of up to 16 carbon atoms are built up in this way. (There are other mechanisms for making longer chains.) The thioesters are hydrolyzed to yield fatty acids:

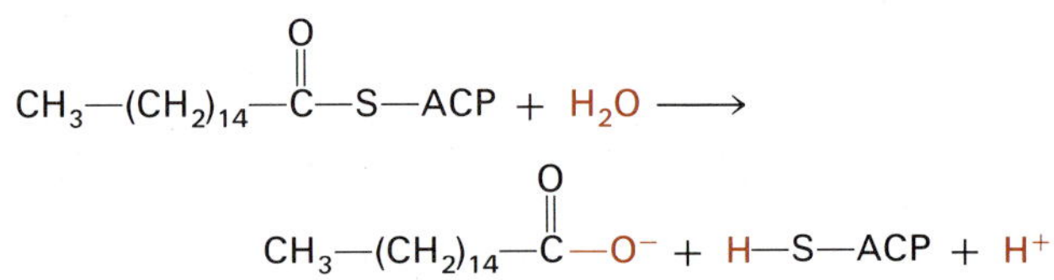

palmitate

Thus we see that the fatty acid is synthesized by starting with an acetate unit (two carbons) and adding two carbons at a time. This path by which most naturally occurring fatty acids are synthesized explains why most of them contain even numbers of carbon atoms.

EXERCISE 24-2 Some marine organisms synthesize fatty acids with odd numbers of carbon atoms. The synthesis starts with propionyl-S-ACP, which condenses with malonyl-S-ACP to yield a five-carbon chain:

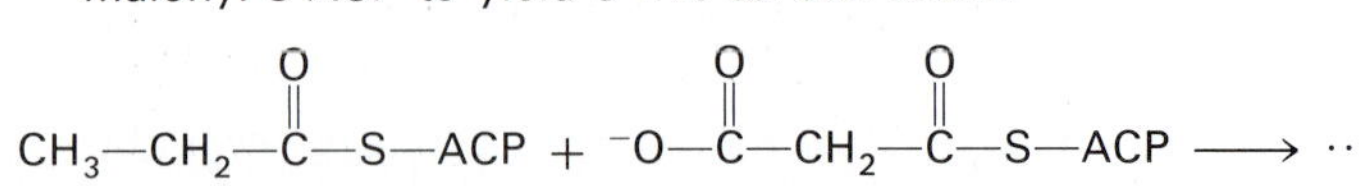

propionyl-S-ACP **malonyl-S-ACP**

Write the structures of the product of this reaction and of the products of the first reduction, dehydration, and second reduction reactions that follow this reaction to complete the first round in the biosynthesis.

24-3 TRIACYLGLYCEROLS (TRIGLYCERIDES)

Only small amounts of fatty acids are found free in nature; most are combined in complex lipids. The **triacylglycerols** or **triglycerides,** triesters of the fatty acids and glycerol, constitute the most abundant class of lipids. Fats and oils (Section 19-6), the fuel molecules that provide the principal energy reserves in plants and animals, are triglycerides. In **simple triglycerides,** the fatty acids from which the triester was formed are identical, as in tripalmitin:

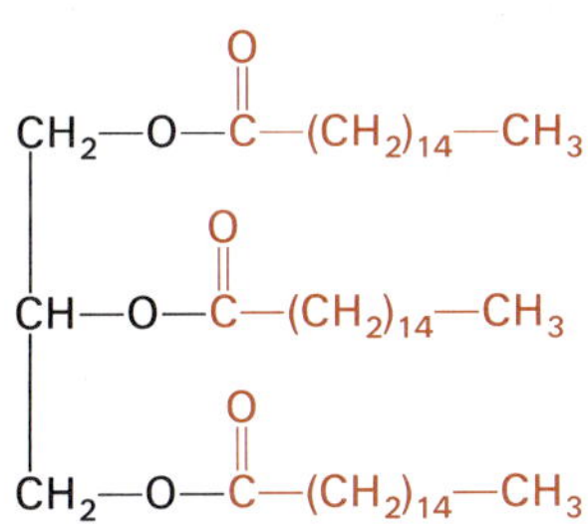

tripalmitin

In the more common **mixed triglycerides,** there are two or three *distinct* fatty acid residues in the same molecule, often including both saturated and unsaturated fatty acids. Figure 24-5 shows a space-filling model of a typical triglyceride; it contains myristic acid (14 carbons) esterified to the hydroxyl group of the glycerol C-1, and palmitic acid (16 carbons) esterified to the hydroxyl groups of the other two carbons. The melting points of triglycerides generally increase

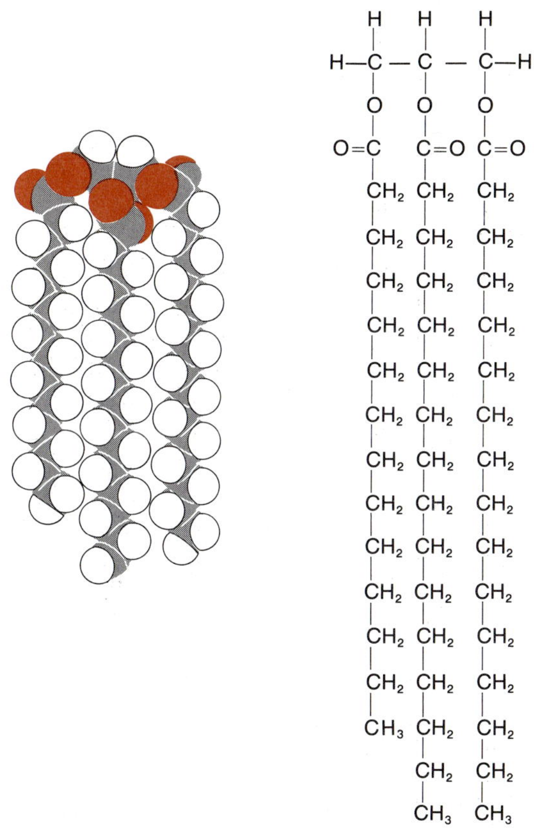

FIGURE 24-5

A space-filling model and the structural formula of the mixed triglyceride 1-myristoyldipalmitoyl glycerol.

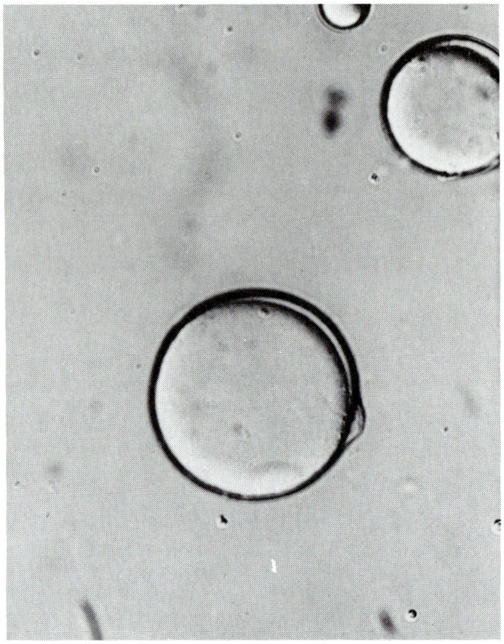

FIGURE 24-6

Photomicrograph of an adipose (fat storage) cell. A large globule of fat is surrounded by a thin ring of cytoplasm (the living part of a cell, excluding the nucleus) and a bulging nucleus. [Courtesy of Pedro Cuatrecasas.]

with increasing lengths of the fatty acid substituents, and they decrease with increasing degree of saturation, as suggested by the data of Table 24-1.

We saw in Chapter 23 that some energy is stored in the liver and in muscle as glycogen, a polysaccharide. Glycogen is used as a ready source of energy between meals and during strenuous exercise. Most of the body's energy reserves, however, are stored in the form of fats. Figure 24-6 is a photomicrograph of a fat cell—a large globule of fat occupies most of the cell volume. In these cells, the fats are synthesized from small molecules, are stored, and (when needed as an energy source) are degraded again into two-carbon units. A person weighing 60 kg (130 lb) typically might have energy reserves of about 85,000 kcal in triglycerides, 500 kcal in glycogen, and 35 kcal in glucose. (Because the joule is not yet as widely used in biochemistry as it is in chemistry, we use the calorie as the unit of energy in this section of the book. As noted in Section 4-10, 1 cal = 4.184 J.)

Energy is much more highly concentrated in fats than in carbohydrates or proteins. When 1 g of fat is oxidized to CO_2 and H_2O, 9 kcal of energy are released. In contrast, the energy yield from 1 g of anhydrous carbohydrate or protein is only 4 kcal. The difference is even greater than it first appears, because fats (which are nonpolar) are stored in anhydrous form; but glycogen (which is polar) is stored in hydrated form, with 2 g of water for every 1 g of glycogen. Thus, 1 g of stored, hydrated glycogen contains only one-third of 4 kcal, or 1.3 kcal. This is to be contrasted with the 9 kcal of energy obtained from 1 g of fat. A gram of fat, therefore, stores more than six times as much energy as a gram of glycogen.

EXERCISE 24-3 Write the structural formula of a triglyceride formed from lauric, oleic, and linoleic acids.

EXERCISE 24-4 (a) A person weighing 60 kg typically has a fuel reserve of 85,000 kcal stored as triglycerides. What is the total weight of the triglycerides?

(b) If the same amount of energy were stored instead as hydrated glycogen, how much more would the person weigh?

24-4
PHOSPHOLIPIDS, SPHINGOLIPIDS, AND WAXES

The **phospholipids,** or **phosphoglycerides,** are derivatives of glycerol in which fatty acids are esterified to the hydroxyl groups of C-1 and C-2 of glycerol, and phosphoric acid is esterified to the third hydroxyl group. Phosphoglycerides are important constituents of cell membranes (Section 24-5). The simplest phosphoglyceride is phosphatidic acid (or phosphatidate, the ionized form that predominates at physiological pH):

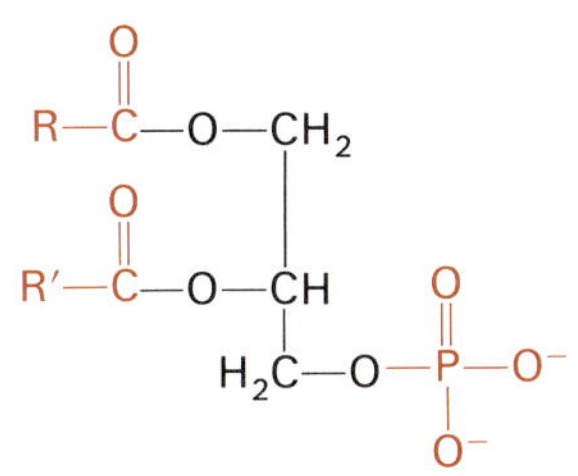

phosphatidate

Here R and R′ represent the hydrocarbon chains of the fatty acids.

Phosphatidate is only a minor component of the naturally occurring phospholipids; the phosphoric acid is usually esterified to a second alcohol. A typical phosphoglyceride can be represented schematically as

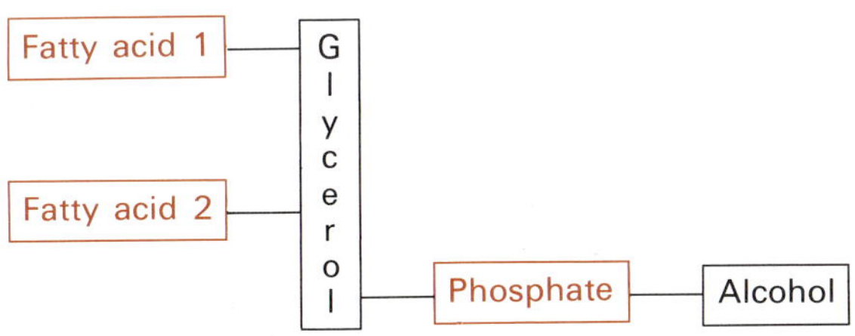

The common alcohols that are components of phosphoglycerides are ethanolamine, choline, serine, glycerol, and inositol. At physiological pH they exist as:

$HO-CH_2-CH_2-\overset{+}{N}H_3$

ethanolammonium ion

$HO-CH_2-CH_2-\overset{+}{N}(CH_3)_3$

choline

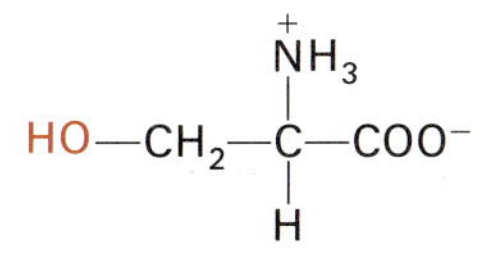

serine

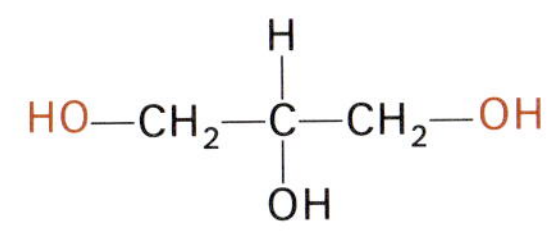

glycerol

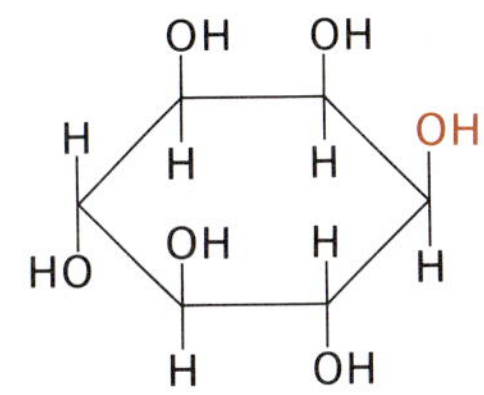

inositol

Phosphatidyl ethanolamine and phosphatidyl choline are the two most abundant phosphoglycerides in plant and animal cells. The structure of a typical phosphatidyl choline is

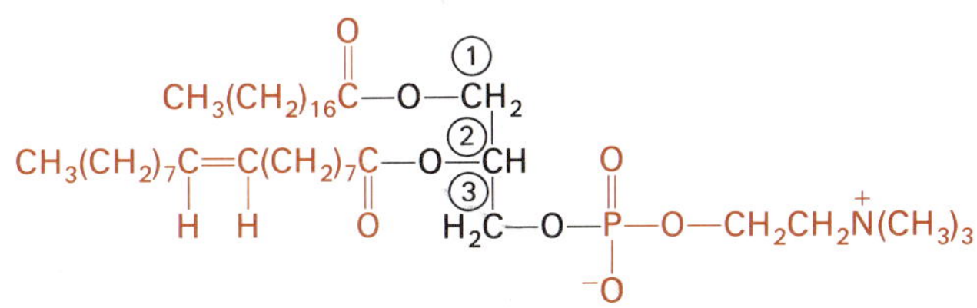

1-stearoyl-2-oleoyl-phosphatidyl choline

A space-filling model of this molecule is illustrated in Figure 24-7; structural formulas of other phosphoglycerides are given in Figure 24-8.

Sphingolipids are complex lipids derived from sphingosine, an amino alcohol with a long hydrocarbon tail. At physiological pH, it exists as an ammonium ion:

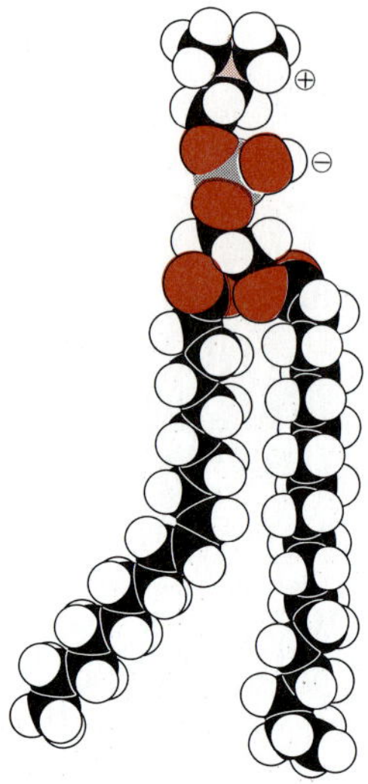

FIGURE 24-7

A space-filling model of 1-stearoyl-2-oleoyl-phosphatidyl choline. [From Lubert Stryer, *Biochemistry,* 2nd ed., W. H. Freeman and Company. Copyright © 1981.]

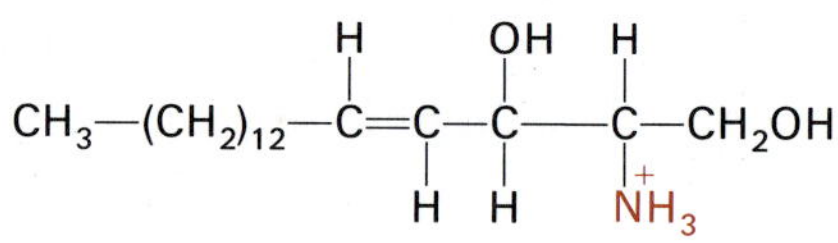

sphingosine

In all sphingolipids, the amino group of sphingosine is linked via an amide bond to a fatty acid. The fatty acid–sphingosine unit is called a **ceramide:**

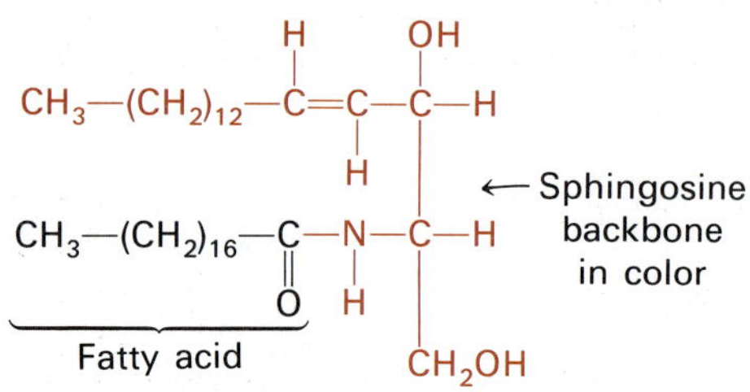

a ceramide

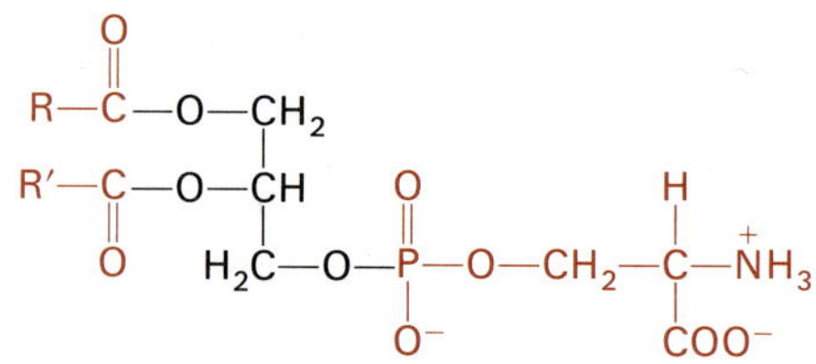

phosphatidyl serine

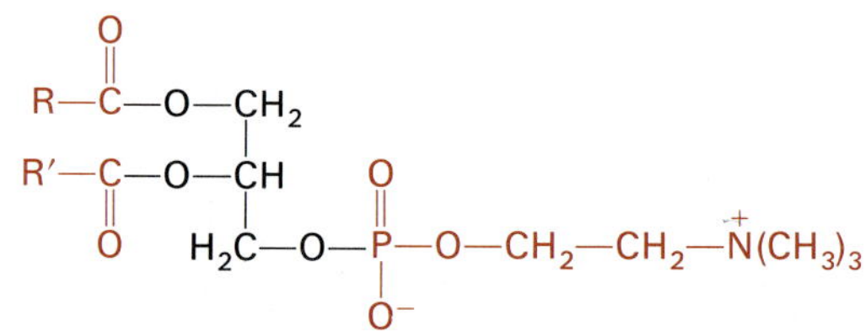

phosphatidyl choline

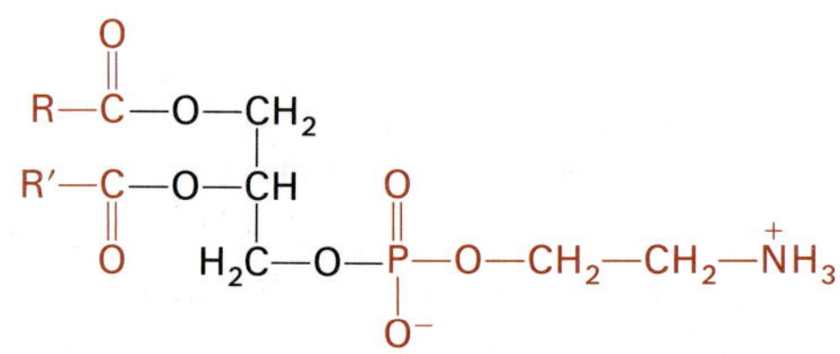

phosphatidyl ethanolamine

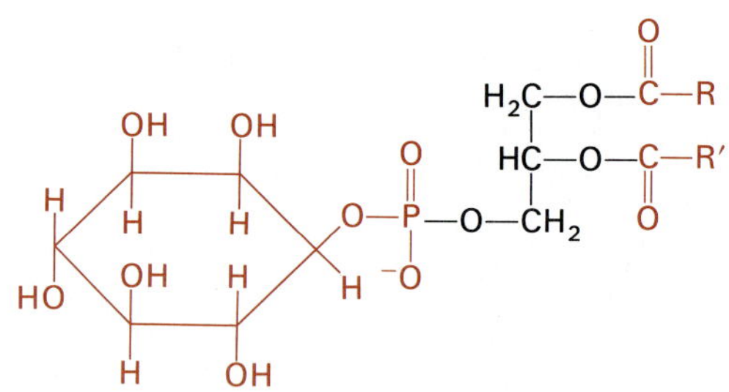

phosphatidyl inositol

FIGURE 24-8

Structural formulas of some phosphoglycerides.

The primary hydroxyl group of sphingosine is also substituted in sphingolipids. **Sphingomyelin,** a phospholipid found in cell membranes, has a choline phosphate group attached to the primary hydroxyl group of the ceramide:

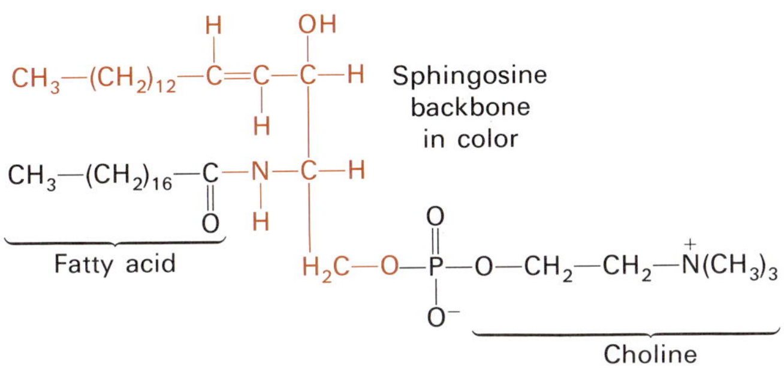

sphingomyelin

The structure is written in a form that shows the similarities between the conformation of sphingomyelin and that of phosphatidylcholine (see also Figure 24-9).

Glycosphingolipids, or **glycolipids,** are sphingolipids that contain one or more sugar units. A fatty acid is linked to the amino group of sphingosine as in ceramide and sphingomyelin; but instead of the choline phosphate group of sphingomyelin, a sugar is attached to the primary hydroxyl group of ceramide. The simplest glycolipid is cerebroside, in which a single glucose or galactose unit is bonded to the primary hydroxyl of sphingosine:

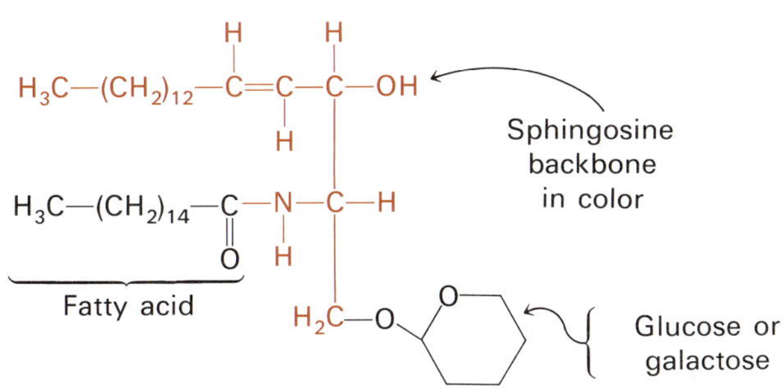

a cerebroside

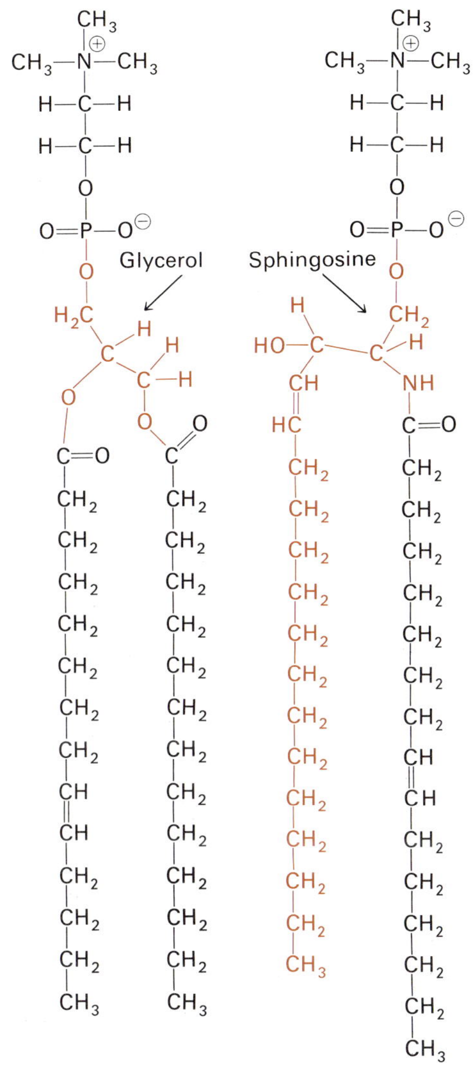

phosphatidylcholine **sphingomyelin**

FIGURE 24-9

Structural formulas of phosphatidyl choline and sphingomyelin. Note that both molecules have hydrophobic hydrocarbon tails and charged, hydrophilic heads. The fatty acid components of the nonpolar tails can vary. (A space-filling model of a phosphatidyl choline is shown in Figure 24-7.)

In **gangliosides,** the most complex glycolipids, an oligosaccharide chain of up to seven monosaccharides is attached to the primary hydroxyl group of ceramide. At least one of the sugars is an acidic one such as *N*-acetylneuramic acid (see Section 23-8). Gangliosides are concentrated in the nervous system, particularly in the gray matter of the brain. Tay–Sachs disease, a genetic disease that is usually fatal by the age of three, arises from the absence of an enzyme required for the degradation of a particular ganglioside. As a result, the ganglioside content of the brain of an infant with Tay–Sachs disease is abnormally high, and some parts of the brain are greatly swollen.

Glycolipids are prevalent on the surfaces of cell membranes where they contribute to cell-to-cell recognition and thus play a crucial role in the development of tissues and in the detection of foreign cells by the immune system.

The triglycerides, phospholipids, and sphingolipids are all complex lipids: they can all be saponified. The **waxes,** water-insoluble esters of fatty acids and fatty alcohols, constitute the remaining family of complex lipids. Waxes provide protective coatings for fur, skin, leaves, and fruits, and they are a major component of beeswax.

Copepods store both fat and wax as an energy reserve. The fats are usually distributed through the body and are used as a quick source of energy. The waxes are usually stored for long-term use in a special sac (Figure 24-10). The waxes synthesized by copepods that live in arctic or antarctic waters are highly unsaturated, and they remain liquid even at the low ocean temperatures there. Wax may constitute as much as 70% of the dry weight of a copepod that lives in cold water, and the animal can live for as long as ten months on the energy stored in the wax.

EXERCISE 24-5 A wax found in a copepod that lives in arctic waters consists of a 16-carbon straight-chain fatty alcohol and a 20-carbon fatty acid that contains five double bonds. Write a possible structure for the wax. (The double bonds are always separated by at least one $—CH_2—$ group.)

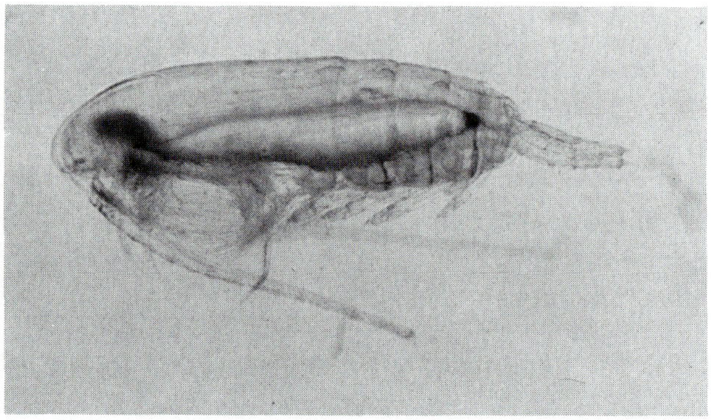

FIGURE 24-10

A large sac filled with liquid wax is visible in this photograph of a copepod of the species *Calanus plumchrus,* which lives in the waters of British Columbia. Half of the dry weight of the copepod is wax. [Courtesy of A. A. Benson.]

24-5
BIOLOGICAL MEMBRANES

The phospholipids, sphingomyelin, and the glycolipids are of primary importance in biological membranes. The central role of these lipids in membranes results from their characteristic structures: they have nonpolar hydrocarbon tails, and charged or polar heads, as shown in Figure 24-9 for phosphatidylcholine and sphingomyelin. We can represent the membrane lipids schematically as we did soaps—with a circle for the charged or polar head and a wavy line for each hydrocarbon tail (Figure 24-11).

In water these lipids can form micelles as soaps do. But instead, a different arrangement is favored and is formed spontaneously—a double sheet called a **lipid bilayer** (Figure 24-12). The bilayer is a sandwichlike sheet with the hydrophobic tails of the lipids at the center (from which water is excluded) and the hydrophilic head groups on both surfaces in contact with the aqueous environment.

The lipid bilayer, like soap micelles and the folding of proteins, is stabilized primarily by hydrophobic interactions. Hydrogen bonds between water molecules and the polar or charged heads of the lipid molecules, and van der Waals attractions between the hydrocarbon tails, also help to hold the bilayer together. The cell membrane is a lipid bilayer that completely envelops the cell and separates two distinct watery environments: the cell interior and the cell exterior (Figure 24-13).

The phospholipid bilayer structure of cell membranes provides a barrier against diffusion of molecules in and out of cells. The bilayers are impermeable to most ions and polar molecules except in membranes that possess specific receptor and transport proteins permitting the passage of such molecules. Water is an exception; although it is a polar molecule, it traverses the membrane relatively freely.

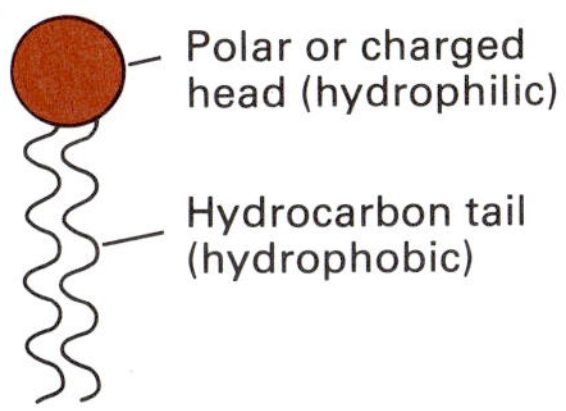

FIGURE 24-11
Schematic representation of a membrane lipid.

The cell membranes contain many globular proteins as well as lipids. The membrane proteins can be visualized as floating in a sea of lipids. This view, summarized as the **fluid mosaic model** of membranes, pictures the cell membrane as a fluid consisting of proteins and lipids in constant lateral motion within the bilayer (Figure 24-14). The membrane is fluid in the sense that a small lipid molecule might diffuse laterally (within the bilayer) the entire length of a bacterium in one second—that is a distance of about 2×10^{-4} cm. Proteins, which are much larger than lipids, diffuse much more slowly. There is almost no tendency for a protein or a lipid to flip from the inside to the outside of the bilayer or vice versa. It has been estimated that a small lipid flips from one side of the membrane to the other no more often than once a month; and large protein molecules flip over much less often than that.

Two classes of protein are associated with cell membranes. **Integral proteins** are partly embedded in the hydrophobic region of the bilayer. Each integral protein studied so far spans the entire width of the bilayer and is exposed to the aqueous regions on both sides of the bilayer. As we would expect, the exposed parts of the protein contain hydrophilic amino acid residues on their surfaces. The parts of the protein that are buried inside the bilayer have hydrophobic amino acid residues on their surfaces. Integral proteins can be removed only by disrupting the bilayer—by adding detergent, for example. **Peripheral proteins** are located on the surface of the membrane and are rather easily removed—say, by increasing the salt concentration. This interferes with the salt linkages and hydrogen bonds that bind the protein. Each of the periph-

(a)

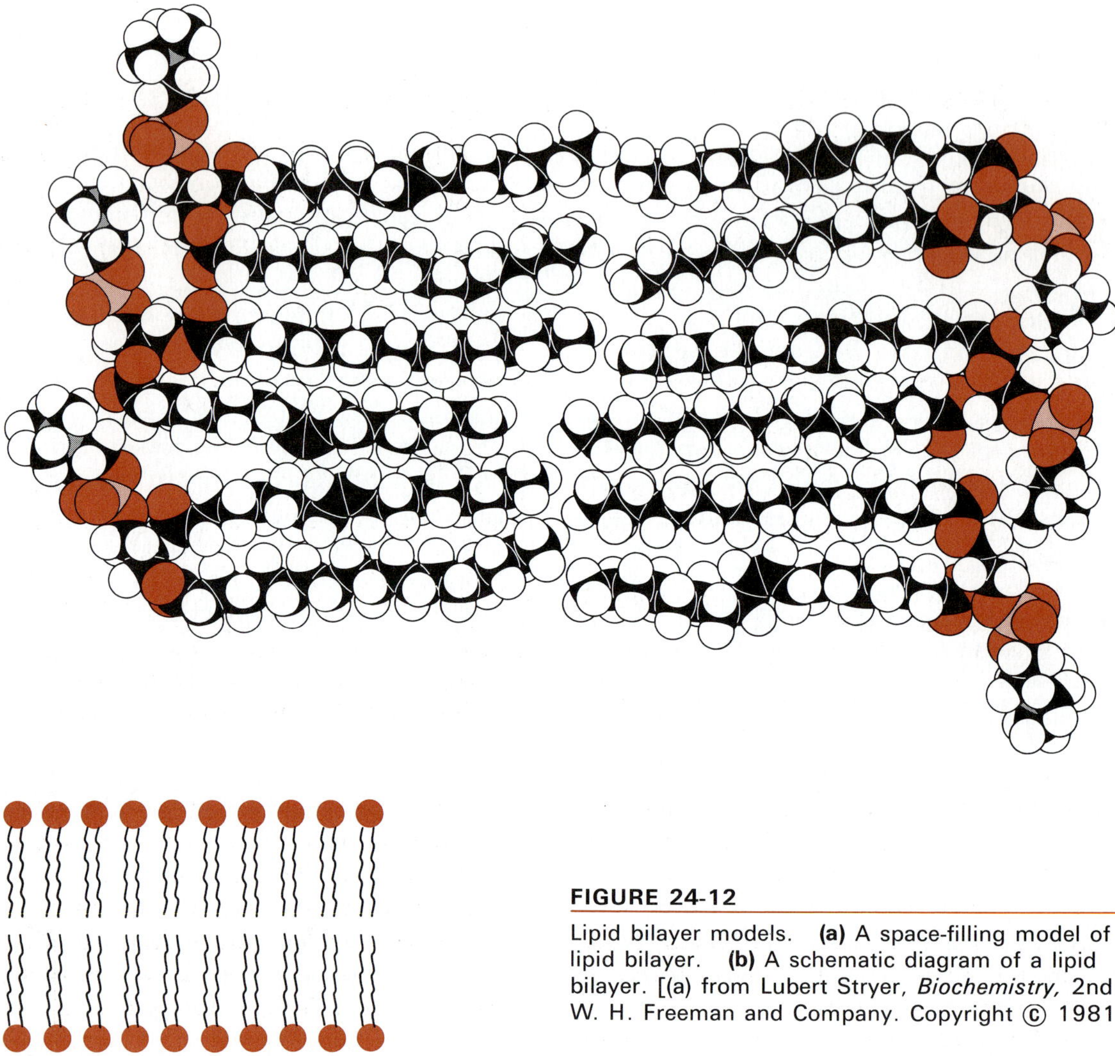

(b)

FIGURE 24-12

Lipid bilayer models. **(a)** A space-filling model of a lipid bilayer. **(b)** A schematic diagram of a lipid bilayer. [(a) from Lubert Stryer, *Biochemistry,* 2nd ed., W. H. Freeman and Company. Copyright © 1981.]

eral proteins studied so far is bound to an integral protein, and most of them are located on the inside surface of the cell membrane.

All integral proteins are inserted into the membrane asymmetrically, so the two sides of the membrane are different. And every molecule of a particular protein has the same orientation in the bilayer. This orientation of a protein with respect to the inside and outside of the membrane is related to its specific transport or receptor function. For example, a protein responsible

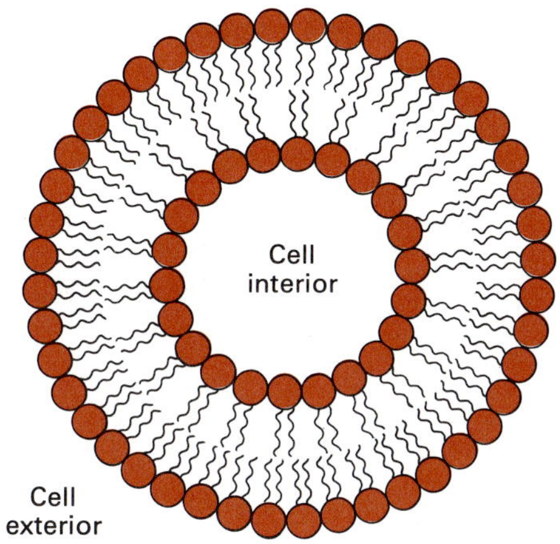

FIGURE 24-13

Schematic diagram of a bilayer membrane. Both the interior and the exterior of the cell are aqueous solutions.

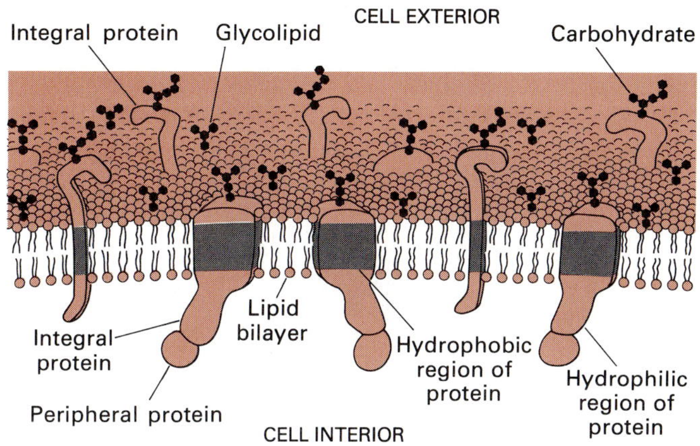

FIGURE 24-14

The fluid mosaic model of a cell membrane. The membrane includes proteins as well as lipids. Sugar units are attached to some of the lipids (glycolipids) and to some of the proteins (glycoproteins). The cell membrane is asymmetric. The sugar units are always found on the exterior surface of the membranes, and the peripheral proteins are almost always found on the interior. [From H. F. Lodish and James E. Rothman, ''The Assembly of Cell Membranes.'' Copyright © 1978 by Scientific American, Inc. All rights reserved.]

for transporting K^+ *into* the cell must be oriented in a certain way—if it were flipped over, it would instead transport K^+ *out* of the cell. The sugar units of glycolipids and of glycoproteins (proteins with attached carbohydrate groups; Section 23-9) are always on the outer surfaces of cell membranes. It may be that one function of the hydrophilic sugar residues in a glycoprotein is to keep the protein properly oriented in the cell membrane.

Some antibiotics work by making a bacterial cell membrane permeable to certain ions. This interferes with metabolic reactions and kills the bacterium. Valinomycin, for example, makes the membrane permeable to K^+. Valinomycin is a cyclic molecule that consists of three different amino acid residues, each repeated three times. Its structure is illustrated schematically in Figure 24-15a. The K^+ ion is loosely bound to six polar oxygen atoms located on the inside of the molecule. The outside of the valinomycin molecule is nonpolar and thus is soluble in the hydrophobic interior of the membrane. Figure 24-15b illustrates how the valinomycin molecule can carry K^+ across a cell membrane. The antibiotic gramicidin A also renders the membrane permeable. Gramicidin A is an open-chain polypeptide of 15 amino acid residues. The gramicidin molecule forms a tunnel that extends across the cell membrane. The exterior of the molecule is hydrophobic and is soluble in the interior of the cell membrane. The interior of the tunnel formed by the molecule contains polar groups that bind ions and permit them to pass through the membrane (Figure 24-15c).

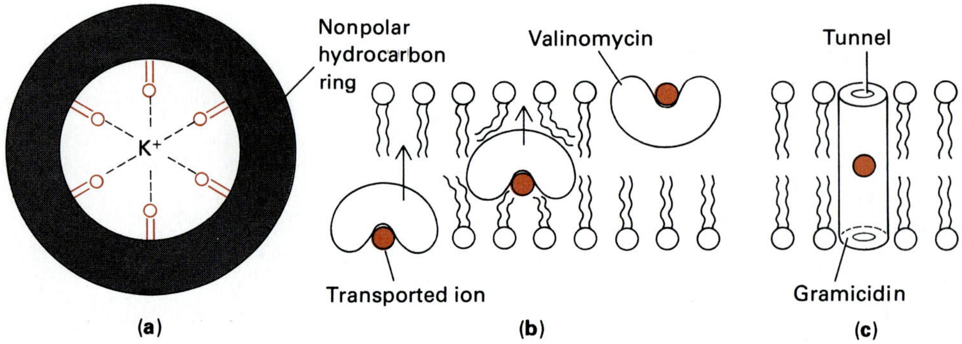

FIGURE 24-15

Antibiotics that act on membranes. **(a)** A schematic diagram illustrating how the valinomycin molecule binds K^+. **(b), (c)** Schematic diagrams illustrating how valinomycin and gramicidin transport ions across cell membranes.

24-6 SIMPLE LIPIDS: TERPENES

The complex lipids—the triglycerides, the waxes, the phospholipids, and the sphingolipids—all can be hydrolyzed to yield fatty acids. In contrast, the simple lipids—the terpenes, the steroids, and the prostaglandins—cannot be saponified.

The **terpenes** are natural products abundant in the oils obtained from plants and flowers. Many terpenes are responsible for the characteristic odors or flavors of the plants from which they are obtained. The terpenes can be pictured as composed of two or more five-carbon isoprene units:

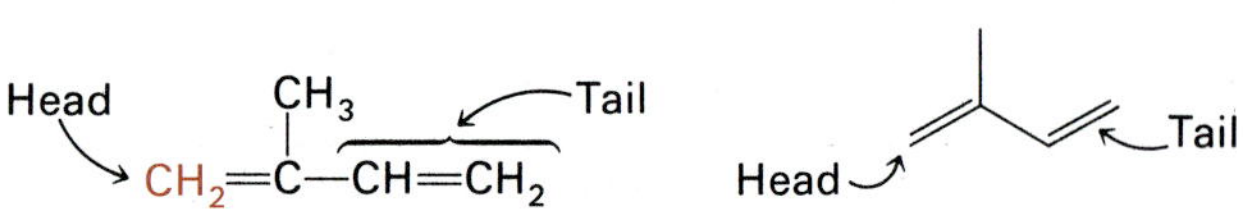

isoprene **isoprene**

We saw in Section 15-10 that natural rubber and gutta percha are addition polymers of isoprene. However, in nature terpenes are not synthesized from isoprene itself, but rather from a pyrophosphate ester of isoprene. The isoprene units of terpenes are usually connected head-to-tail, but occasionally they are arranged tail-to-tail:

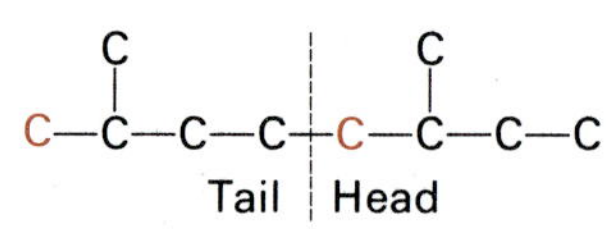

Head-to-tail arrangement

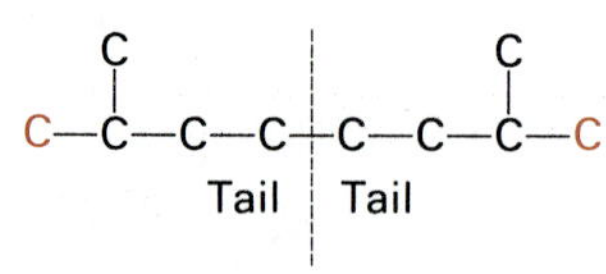

Tail-to-tail arrangement

The simplest terpenes contain two isoprene units (ten carbon atoms) and are called **monoterpenes.** An example is geraniol, found in the oil of geranium:

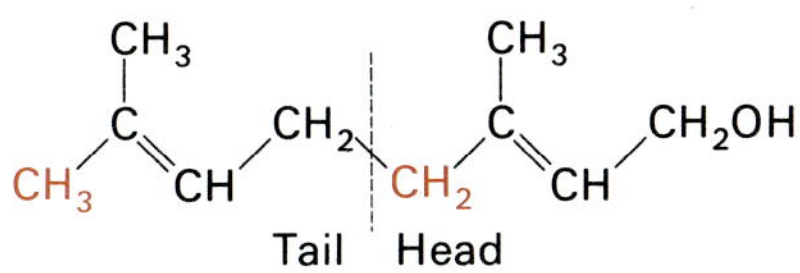

geraniol (a monoterpene)

Farnesol, which consists of three isoprene units (15 carbon atoms), is a **sesquiterpene:**

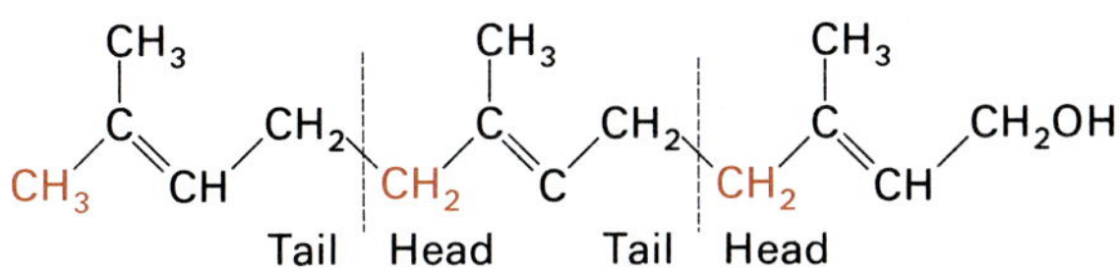

farnesol (a sesquiterpene)

Phytol, a component of chlorophyll, a **diterpene** composed of four isoprene units (20 carbon atoms), is shown below. Shorthand structures are often used to represent terpenes; in this form, phytol is

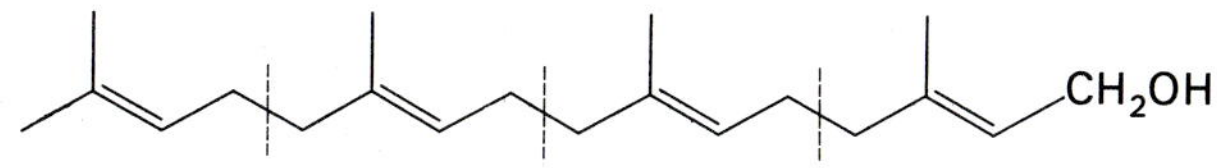

phytol

Squalene, a precursor of cholesterol and of many other important biological molecules, is a **triterpene** composed of six isoprene units (30 carbon atoms). Squalene is synthesized from isoprene in the following reaction sequence:

$$C_5 \longrightarrow C_{10} \longrightarrow C_{15} \text{ (a sesquiterpene)} \qquad (24\text{-}2a)$$

$$C_{15} + C_{15} \longrightarrow C_{30} \text{ (squalene)} \qquad (24\text{-}2b)$$

The structure of squalene is shown at the bottom of the page. Within each sesquiterpene section of squalene, the isoprene units are arranged head-to-tail, as usual. The two sesquiterpene units, however, are in the unusual tail-to-tail arrangement. We shall look at the structure of squalene again when we discuss the steroids because the basic steroid structure is derived from squalene.

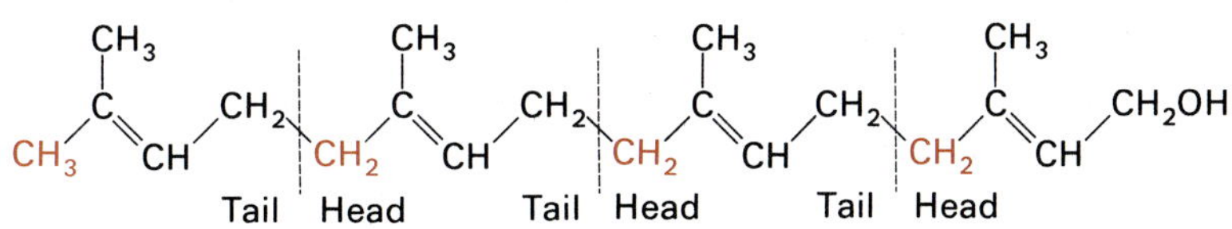

phytol (a diterpene)

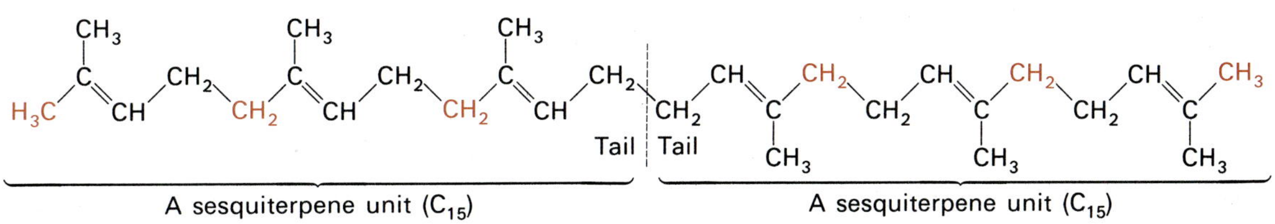

squalene (a triterpene)

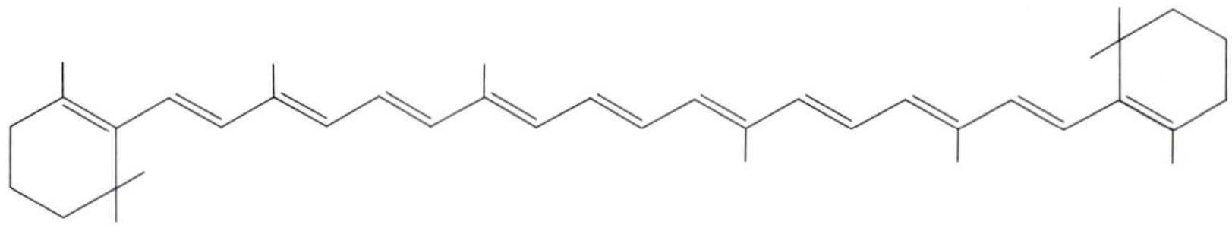

β-carotene
(a precursor of vitamin A)

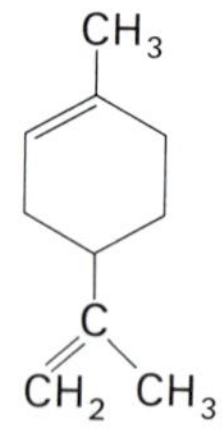

limonene
(lemon oil)

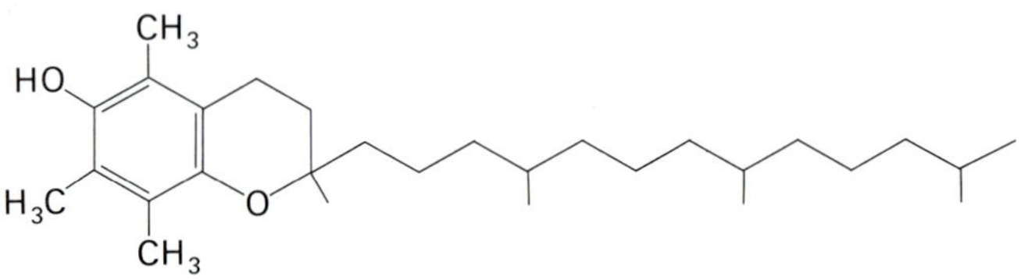

vitamin E
(α-tocopherol; prevents oxidation of unsaturated fatty acids of cell membranes)

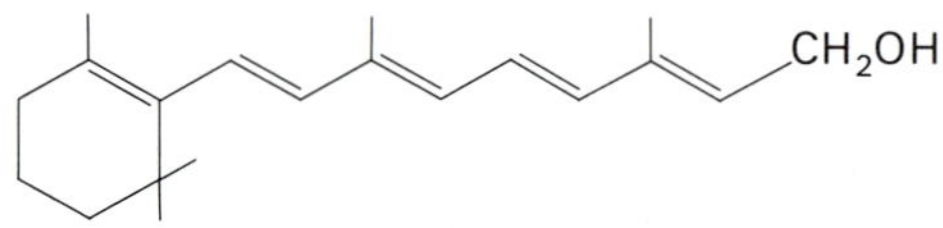

vitamin A$_1$
(retinol; plays an important role in the visual cycle in vertebrates)

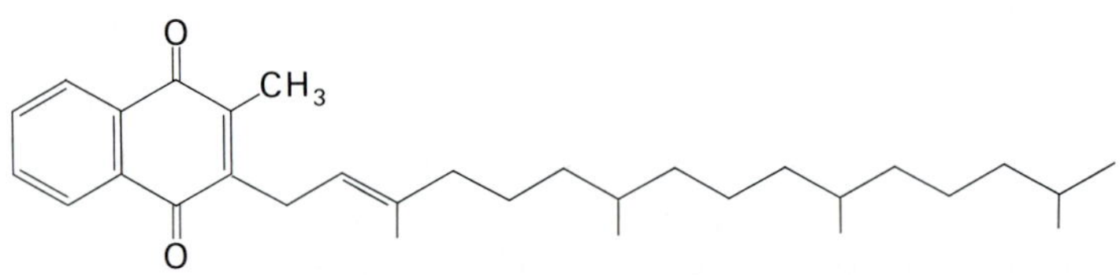

vitamin K$_1$
(required for blood clotting)

FIGURE 24-16
Structures of some terpenes.

All of the terpenes we have considered up to now are open-chain molecules, but many terpenes are cyclic, and many have both cyclic and acyclic parts:

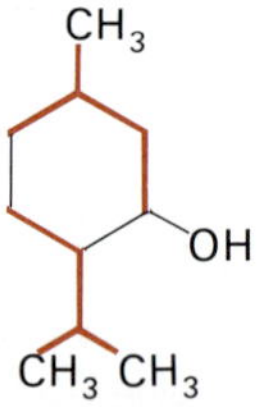

menthol
(peppermint oil)

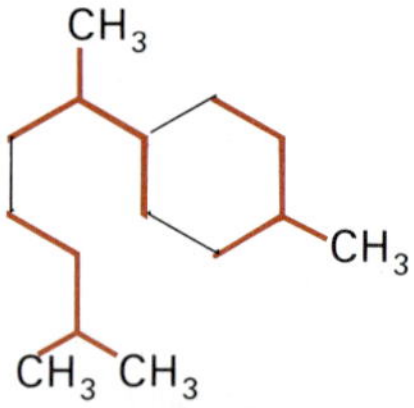

zingiberene
(ginger oil)

The colored bands highlight the individual Y-shaped isoprene units. Figure 24-16 gives structures of some additional terpenes, including those of three of the four fat-soluble vitamins (A, E, and K). Many terpenes are more complex than those we have considered, consisting of two or more rings fused together.

EXERCISE 24-6 Copy the structures of the terpenes given in Figure 24-16, and mark each isoprene unit. Indicate any that are in the unusual tail-to-tail arrangement.

24-7
SIMPLE LIPIDS: STEROIDS AND PROSTAGLANDINS

The **steroids** are tetracyclic natural products synthesized from the terpene squalene. The basic steroid skeleton has the structure

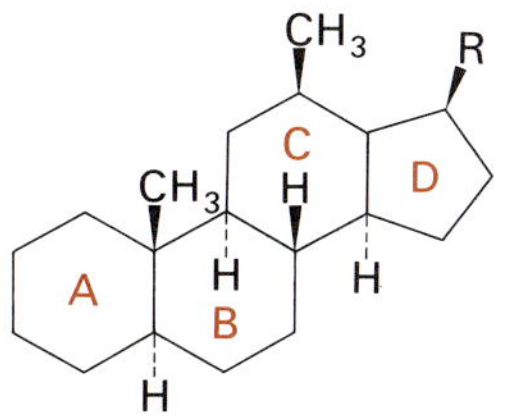

The rings are labeled A, B, C, and D as shown. The stereochemistry of the asymmetric carbon atoms is almost always that shown, and we shall not usually indicate it in structures.

Cholesterol is the most abundant steroid in animals. Cholesterol is present in the brain and in many cell membranes; it is the major component of gallstones, and it is present in the fat deposits that line hardened arteries. It has not yet been established, however, whether its presence there is one of cause or effect. An adult body typically contains one-third to one-half pound of cholesterol.

Cholesterol is synthesized in the body from the 30-carbon terpene, squalene. The synthesis starts with the condensation of two molecules of acetyl coenzyme A (acetyl-S-CoA), a thioester with an active acetyl group. The reaction is a biological Claisen-type condensation similar to the one involved in fatty acid synthesis (Equation 24-1):

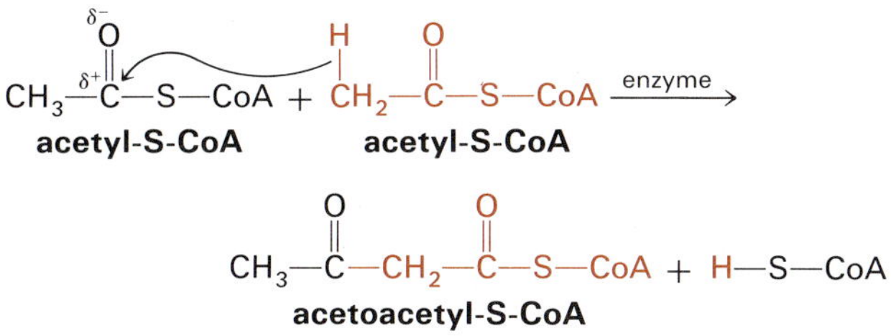

The second step is a biological aldol condensation (Section 18-10):

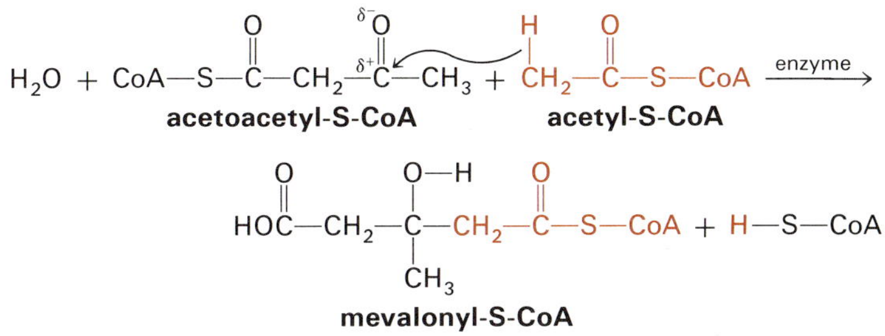

The condensation is accompanied by hydrolysis of a thioester bond of the acetoacetyl-S-CoA. The six-carbon compound mevalonyl-S-CoA is converted in a series of reactions that includes a decarboxylation to a five-carbon compound, isopentenyl pyrophosphate:

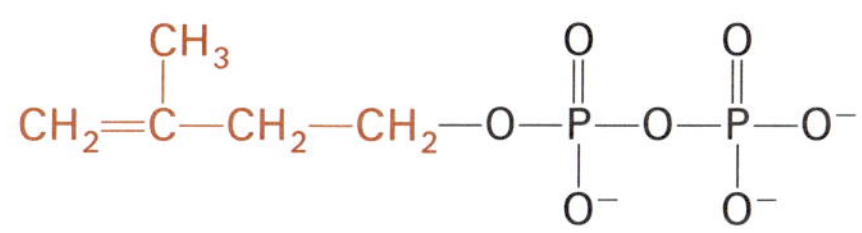

isopentenyl pyrophosphate

The carbon skeleton of isopentenyl pyrophosphate is essentially that of isoprene. This is the compound from which the terpenes are actually synthesized in nature, and squalene is formed as shown in Equation 24-2 (Section 24-6).

The final stage of the cholesterol biosynthesis includes the addition of an oxygen atom, the closing of rings, electron movement, migration of a methyl group, removal of three methyl groups, and reduction. We simply summarize these reactions here, starting with squalene in a conformation that illustrates the ring formations (colored dashed lines indicate the original isoprene units):

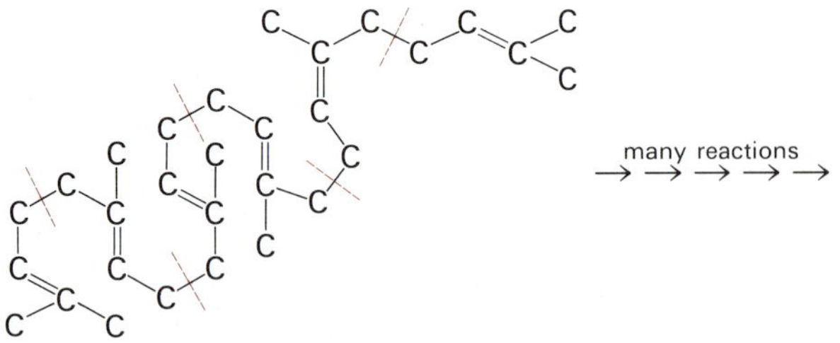

squalene (30 carbons)

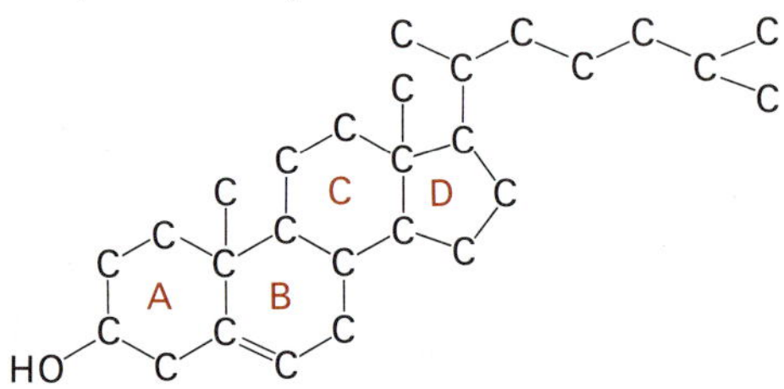

cholesterol (27 carbons)

Cholesterol is a steroid alcohol, or **sterol,** by virtue of the hydroxyl group on the A ring. It is the biological precursor for other steroids compounds—among them the **bile acids,** which play a role in the absorption of fats in the intestine; the **steroid sex hormones; cortisone,** which is involved in the metabolism of carbohydrates; and one **steroid vitamin** (D). The structures of cholesterol and of some other steroids are given in Figure 24-17.

The sex hormones are a particularly interesting class of compounds because of the influence they exert in the differentiation of male and female secondary sexual characteristics. The striking difference between males and females is triggered by the male sex hormone **testosterone** and the female sex hormone **estradiol,** but the structures of these hormones are remarkably similar (Figure 24-17). A second female sex hormone, **progesterone,** which is secreted by the ovaries, prepares the lining of the uterus for an ovum and helps to maintain pregnancy.

Synthetic steroid derivatives closely related to progesterone and estradiol make up "the pill," **oral contraceptive drugs** that represent one of the major impacts of chemistry on modern society. Scientists working in the 1950s reasoned that progesterone-like steroids would serve as contraceptive agents by suppressing ovulation, just as progesterone itself suppresses ovulation during pregnancy. Trials began in Puerto Rico in 1955 with Enovid. A tablet composed principally of 10 mg of norethynodrel (see Figure 24-17) was taken from day 5 to day 25 of the menstrual cycle. The results were highly successful: no pregnancies occurred when the pills were taken on schedule. Studies of the progesterone-like steroids used revealed that some were contaminated with other steroids, particularly a methyl ether of ethynyl estradiol called **mestranol** (Figure 24-17). This contamination was fortunate; it was later found that estrogen-like compounds increase the effectiveness of the progesterone-like compounds in suppressing ovulation. By combining the two types of compounds, it is possible to construct pills with much lower dosages (thereby reducing side effects). One product, Enovid-E, contains daily doses of 2.5 mg norethynodrel (one-tenth the original dose) and 0.1 mg mestranol. Other pills on the market contain related compounds in varying proportions.

Vitamin D (also known as the "sunshine" vitamin) was discovered as an agent that can alleviate the symptoms of rickets. Exposure to ultraviolet light enhances the potency of certain forms of vitamin D. Moreover, exposure of patients with rickets to sunlight improves their condition. The current understanding of these processes is that a derivative of cholesterol is the precursor of vitamin D_3 (the naturally occurring form of the vitamin in human beings) and that the conversion is stimulated by ultraviolet light.

The final simple lipids to be described, the **prostaglandins,** are derived from arachidonic acid, a fatty acid, via an intermediate, prostanoic acid (Figure 24-

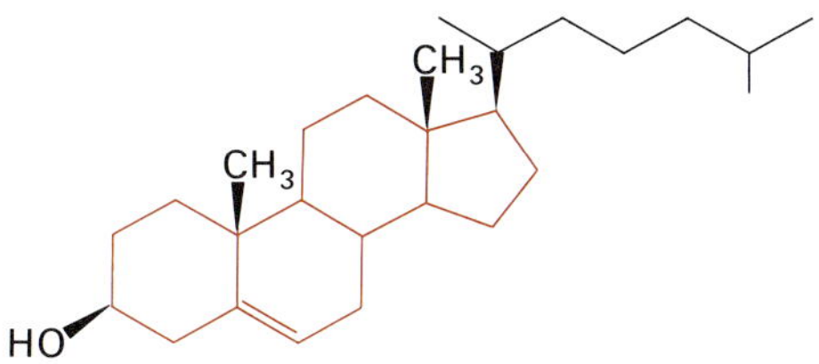

cholesterol

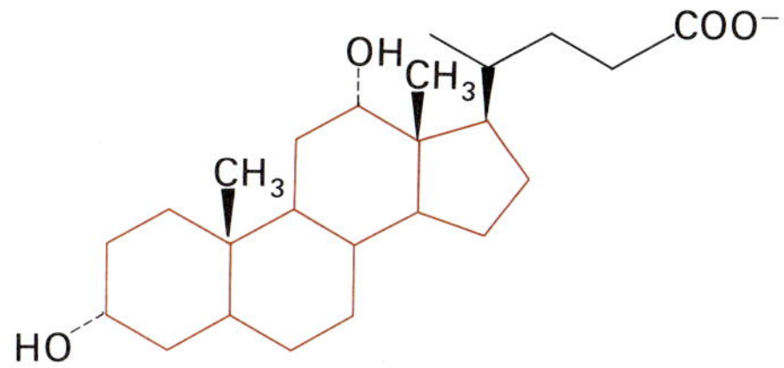

deoxycholic acid
(a bile acid; anionic form present at physiological pH)

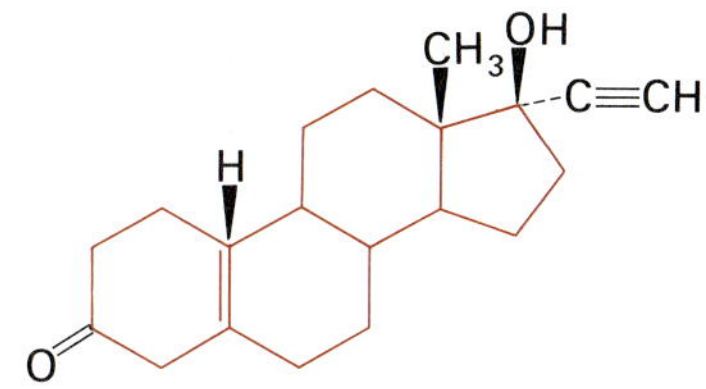

norethynodrel
(synthetic hormone)

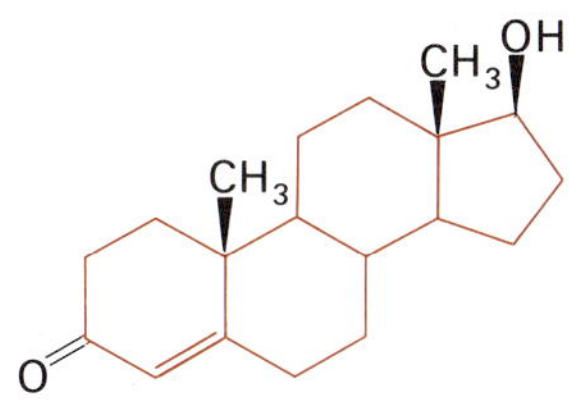

testosterone
(male sex hormone)

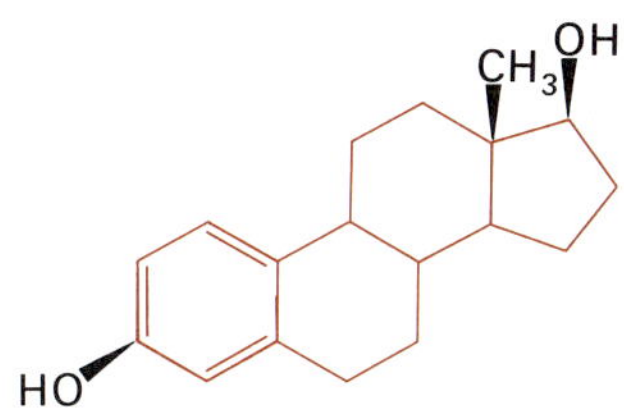

β-estradiol
(female sex hormone)

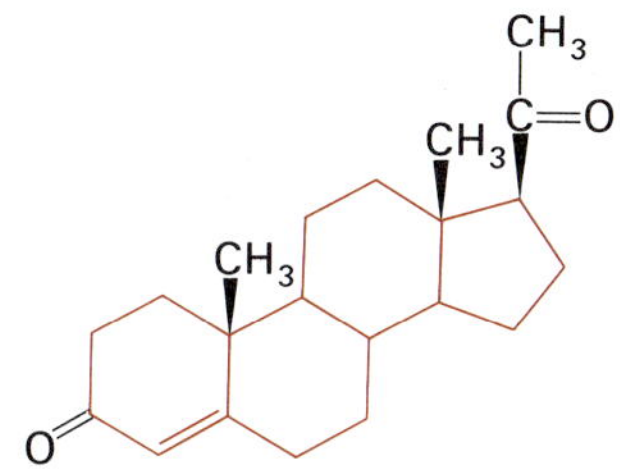

progesterone
(female sex hormone)

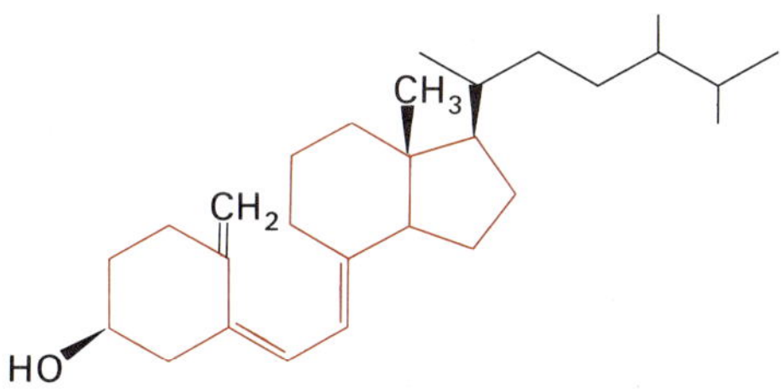

cholecalciferol
(vitamin D_3)

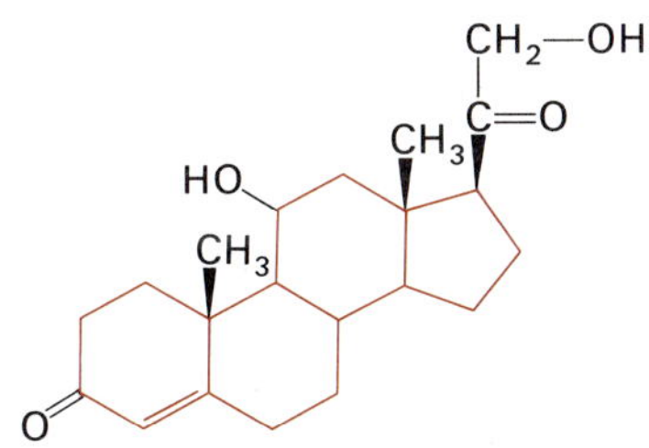

corticosterone
(involved in control of carbohydrate metabolism)

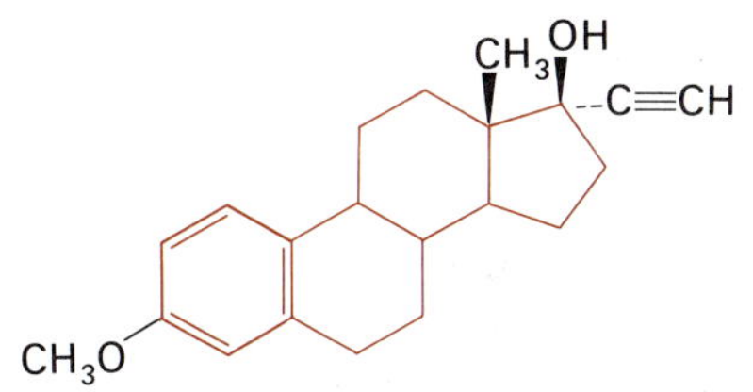

mestranol
(synthetic hormone)

FIGURE 24-17

Structures of some steroids. Vitamin D_3 can be regarded as a steroid in which one ring has been ruptured.

18). Prostaglandins are oxygenated 20-carbon fatty acids that contain a five-membered ring. The prostaglandins were first isolated from the prostate, a gland of the male reproductive system—hence their name—but they have since been found in cells throughout the body, and have also been synthesized in the labora-

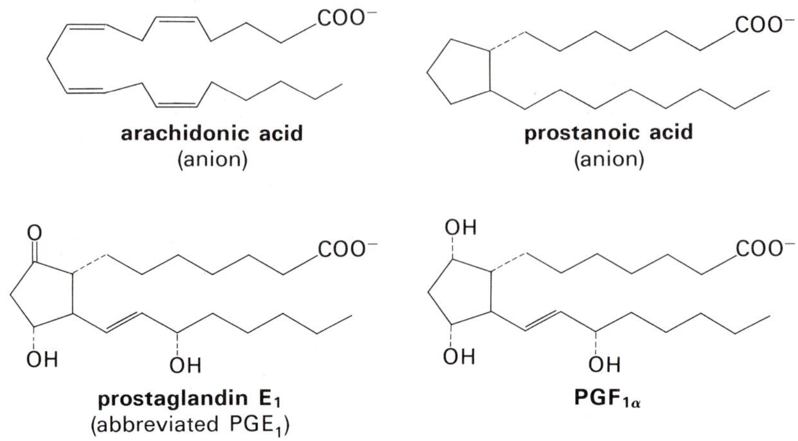

FIGURE 24-18

Structures of arachidonic acid, prostanoic acid, and prostaglandins.

tory. Prostaglandins in low concentration (as low as 10^{-7} M) have profound physiological effects; they appear to modulate the action of hormones, although their mode of action is not understood.

Prostaglandins have a wide range of physiological effects, including the induction of swelling, redness, fever, and increased sensitivity to pain. They cause headaches when injected into the blood, and they cause fever when injected into the brain. In addition, they stimulate smooth muscle contractions, decrease blood pressure, and modulate the transmission of nerve impulses. Clinical studies with pregnant women show that prostaglandins can induce labor when administered late in pregnancy or induce abortion when administered early in pregnancy. A new understanding of the action of aspirin in reducing fever, pain, and swelling suggests that aspirin operates by inhibiting the biosynthesis of prostaglandins. The aspirin apparently interferes with the biosynthetic reactions in which oxygen is inserted into the prostanoic acid.

24-8 LIPOPROTEINS

Among the proteins associated with membranes, many are tightly bound lipids. Such protein–lipid complexes, **lipoproteins,** are also present in blood plasma. The lipoproteins of the blood are presumably involved in lipid transport throughout the body. The lipoproteins are classified according to the ratio of lipid to protein: the very low density lipoproteins (abbreviated VLDL) are about 90% lipid and 10% protein; the low-density lipoproteins (abbreviated LDL) are about 75% lipid and 25% protein; and the high density lipoproteins (abbreviated HDL) are about 40% lipid and 60% protein. The average values of HDL are 45 mg/100 ml for men, and 55 mg/100 ml for women.

In certain diseases, the amounts of the various plasma lipoproteins are altered. For example, studies in the 1950s and early 1960s revealed low levels of HDL in men with coronary heart disease. The idea of a relationship between HDL and heart disease lay dormant until about 1975, when several new studies also uncovered evidence for low levels of HDL (and high levels of LDL) in heart disease. One large study in Framingham, Massachusetts, indicated that every decrease of 5 mg/100 ml in blood HDL below the average corresponds to an increase of roughly 25% in the risk of heart attack. The higher average HDL value for women may be related to the fact that women have a lower rate of heart disease than men. It now appears that HDL is a more accurate indicator of heart-disease risk than is the level of blood cholesterol.

The concentration of HDL appears to rise substantially as a consequence of regular vigorous exercise. In a Stanford University study for men over 35 years old who ran at least 15 miles per week, HDL values were found to average 65 mg per 100 ml of blood. Even higher levels, 80 to 100 mg HDL/100 ml, were reported in the blood of alcoholics by a Wisconsin study; this may be related to other data linking moderate drinking (one to two drinks per day) with reduced risk of heart disease. One current hypothesis is that HDL protects against coronary disease, possibly by eliminating excess cholesterol from the blood stream and so reducing the opportunity for the deposits on the walls of the arteries that are characteristic of atherosclerosis.

KEY TERMS

Lipids are substances from biological sources that are insoluble in water but are soluble in nonpolar solvents. *Complex lipids* are saponifiable. They yield *soaps* (salts of fatty acids) upon hydrolysis in basic solution. *Simple lipids* are nonsaponifiable.

Soaps are not truly soluble in water but are dispersed in the form of charged spheres called *micelles*. The negatively charged heads of the fatty acid anions are on the outside of the micelle, and the nonpolar hydrocarbon tails are on the inside. A *lipid bilayer* is the arrangement of phospholipids found in membranes, a "sandwich" structure with the nonpolar hydrocarbon tails of the fatty acids at the center and the polar head groups of the phospholipids on the two outside surfaces.

Triacylglycerols, or *triglycerides,* are triesters of fatty acids and glycerol. *Phospholipids,* or *phosphoglycerides,* are derivatives of glycerol in which fatty acids are esterified to C-1 and C-2 and phosphoric acid is esterified to C-3 of glycerol. *Sphingolipids* are complex lipids derived from sphingosine, an amino alcohol with a long hydrocarbon tail. *Glycolipids* are sphingolipids that contain one or more sugar residues. *Waxes* are esters of fatty acids and fatty alcohols.

Terpenes are composed of two or more five-carbon isoprene units. *Steroids* are natural products synthesized from a 30-carbon terpene, squalene; steroids have a basic skeletal structure that consists of four fused rings. *Prostaglandins* are simple lipids derived from a fatty acid, arachidonic acid; they have a variety of physiological functions.

CHAPTER SUMMARY

Lipids are biological substances that dissolve in nonpolar solvents but not in water. They can be divided into two categories on the basis of whether or not they yield fatty acids upon alkaline hydrolysis. Lipids that yield fatty acids are called *complex lipids;* those that do not are called *simple lipids.*

The *fatty acids* are carboxylic acids with long hydrocarbon chains. They range in size from about 14 to 22 or more carbon atoms and generally contain an even number of atoms because their biosynthesis proceeds by addition of two-carbon units. The melting points of fatty acids increase as the chain length is increased and as the number of double bonds is decreased.

In most of the complex lipids, the fatty acids are esterified to alcohols, commonly glycerol or glycerol-3-phosphate. In the case of glycerol, three fatty acids can be esterified to each glycerol to yield *triacylglycerols,* or *triglycerides.* These compounds are the most abundant class of lipids, comprising the bulk of the energy reserves in plants and animals. When isolated, they are either solids or liquids known as *fats* or *oils,* respectively.

Complex lipids formed with glycerol-3-phosphate can have a maximum of two fatty acids esterified with the glycerol. These compounds are called *phospholipids* and have a number of interesting properties related to the presence of a polar head group and a nonpolar tail group on the same molecule. One special role of phospholipids is the formation of *bilayer structures* of cell membranes. Phospholipids are usually found with the phosphate group esterified to another small compound, such as ethanolamine or choline.

In another class of complex lipids, fatty acids are bonded via an amide linkage to sphingosine (to give a *sphingolipid*). When a sugar is also attached to the compound, it is called a *glycolipid.* When fatty acids are esterified to fatty alcohols, the resulting compounds are called *waxes.*

The simple lipids are derived from fatty acids, as in the compounds known as *prostaglandins,* or from multiples of five-carbon isoprene units. Compounds with two or more isoprene units are known as *terpenes.* Squalene (a terpene) is a precursor for the biosynthesis of steroids such as cholesterol. Other terpenes include vitamins A, E, and K. The *steroids,* derivatives of fused-ring terpenes, include vitamin D and the male and female sex hormones.

When lipids occur in association with proteins, such molecules are called *lipoproteins.* Low levels of the blood lipoproteins called HDL (high density lipoproteins) *appear* to correlate with a high incidence of heart disease.

PROBLEMS

24-1. For each compound in the first column, list the substance from the second column of which it is a component.

(a) glycerol	(A) squalene
(b) choline	(B) wax
(c) isoprene	(C) sphingomyelin
(d) fatty alcohol	(D) fat

24-2. Saponification of compound X yields glycerol, lauric acid, oleic acid, and stearic acid. Write a structure of compound X. How many structures are possible for X (including enantiomers)?

24-3. How many fatty acid units are esterified to

each unit of (a) glycerol, (b) glycerol-3-phosphate, (c) sphingosine, (d) fatty alcohol?

24-4. What do stearic and oleic acids have in common? How do they differ, and what effects on their melting points can be expected from the differences?

24-5. Write structures for a simple and a mixed triglyceride.

24-6. A sample of lard and a sample of corn oil are hydrolyzed, and their fatty acid compositions are determined. The fatty acid composition of one is 30% palmitic, 25% stearic, 40% oleic, and small amounts of several others. The fatty acid composition of the other is 10% palmitic, 30% oleic, 55% linoleic, and small amounts of several others. Tell which fatty acid composition is that of the corn oil, and explain how you chose your answer.

24-7. Bile salts act as detergents, helping to disperse fats in the upper intestine. This process aids in the digestion of fats. Explain how bile salts act to provide a detergent action. (The structure of one bile salt, deoxycholic acid, is given in Figure 24-17.)

24-8. How many ml of 1 M KOH are required to saponify 100 g of the triglyceride tripalmitin?

24-9. A cell membrane contains about 60% by weight of protein and about 40% by weight of phosphoglycerides. Calculate the number of protein molecules and the number of phospholipid molecules in 100 g of cell membrane if the proteins have an average molecular weight of 50,000, and the lipids have an average molecular weight of 750.

24-10. Cardiolipin, a lipid first isolated from heart muscle, has the structure

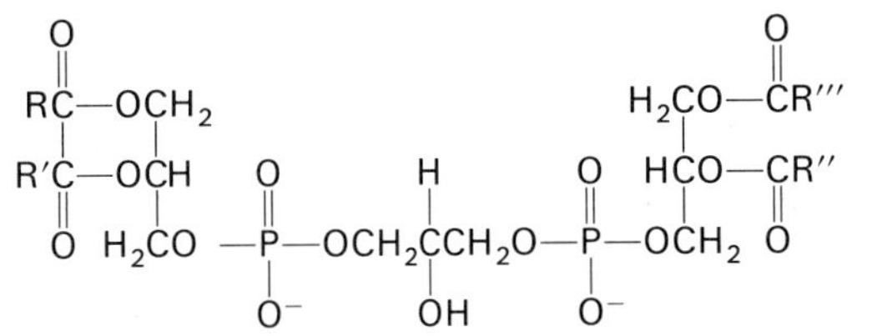

Write the structures of all the products formed when cardiolipin is hydrolyzed in basic solution.

24-11. Elaidic acid, the *trans* isomer of oleic acid, does not occur naturally. Elaidic acid is formed from oleic acid during the hydrogenation of vegetable oils to produce margarine. Which of the isomers (the *cis* or the *trans*) would you expect to have the higher melting point? Explain.

24-12. Write the structure of the cerebroside that contains stearic acid as its fatty acid and has glucose as its sugar.

24-13. Which pair of carbon atoms is added last in the biosynthesis of palmitate: C-1 and C-2, or C-15 and C-16? (The numbering in a carboxylic acid starts with the carboxyl carbon.)

24-14. One of the waxes found in beeswax is myricyl palmitate:

$$CH_3\text{—}(CH_2)_{14}\text{—}COO\text{—}CH_2\text{—}(CH_2)_{28}\text{—}CH_3$$

Write the structures of the fatty acid and the fatty alcohol of which myricyl palmitate is composed.

24-15. Spermaceti wax, which is obtained by cooling oil from the head of a sperm whale, was once widely used in cosmetics, soap, and candles. Spermaceti is an ester of palmitic acid and cetyl alcohol, $CH_3(CH_2)_{14}CH_2OH$. Write the structure of spermaceti.

24-16. Proteins that occur in cell membranes can be removed and studied. Peripheral proteins are soluble in water, but integral proteins usually are insoluble in water. What conclusions about the proteins' stuctures can be draw from these observations?

24-17. Leukotriene C is a biologically potent molecule that causes allergic responses in asthma patients. Its synthesis and release in the lungs are triggered by antibodies produced by the body's immune system against allergens (substances that cause allergic reactions).

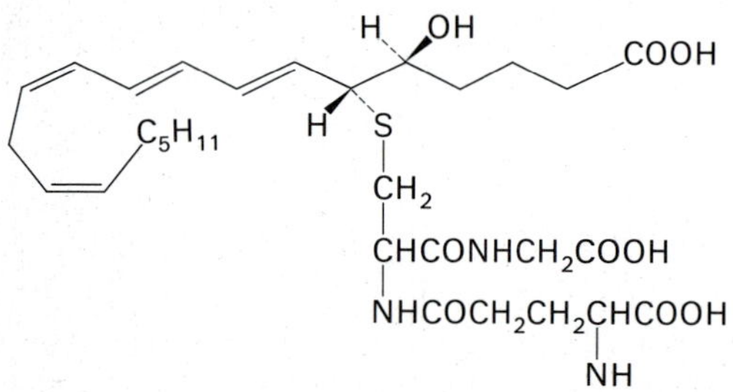

leukotriene C

(a) Of what fatty acid is leukotriene C a derivative?

(b) Write the structures of the products of the hydrolysis of leukotriene C in basic solution, and name as many of these products as you can.

(c) Would leukotriene C be classified as a complex lipid or as a simple lipid?

A single DNA molecule is released from the ruptured head of an *E. coli* T2 bacteriophage. This single DNA molecule has a molecular weight of 1.2×10^8 and a length of 0.052 mm. The intact bacteriophage has a length of 18×10^{-6} mm, only $\frac{1}{3000}$ as great as the length of the DNA it contained ($\times$1,000,000).

25

Nucleic Acids and Protein Synthesis

In two different forms, nucleic acids play two essential biological roles. One form is DNA—the genetic material that stores the information carried in the genes. The other form is RNA, which uses the information stored in the DNA to direct the synthesis of the thousands of different proteins produced by each cell.

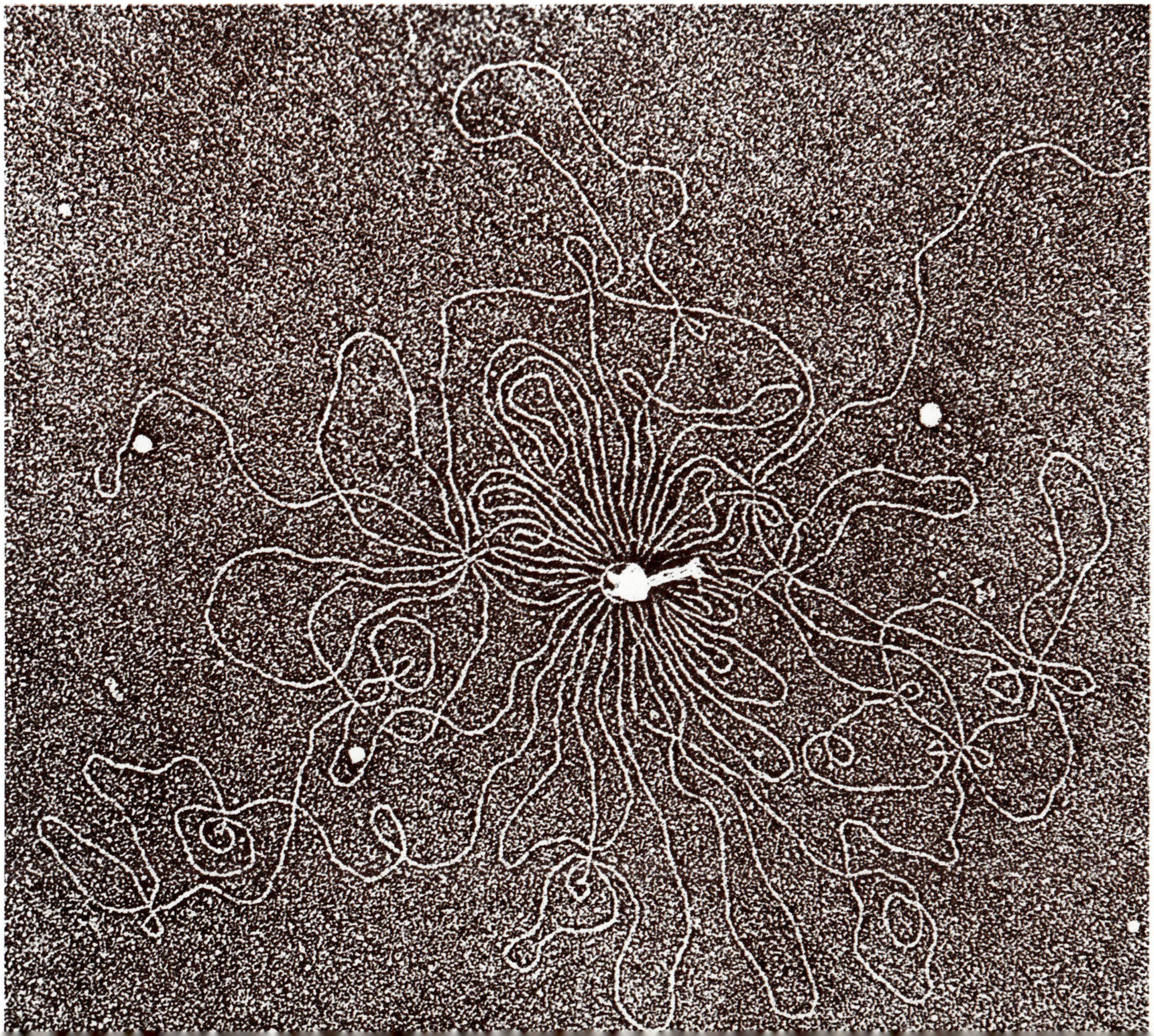

25-1
THE CHEMICAL UNITS OF NUCLEIC ACIDS

We have seen in Chapters 21 and 22 that living organisms produce thousands of different proteins and that these protein molecules are both huge and highly sensitive to "errors" in structure. Replacement of just one amino acid out of thousands can render an enzyme useless or harmful. How does a cell store the information needed to spell out proteins hundreds of thousands of units long? And how does it manufacture the needed proteins with such dependable accuracy? These two questions are at the heart of modern biochemistry. The answer to both lies in the structure and function of the complex polymers known as nucleic acids.

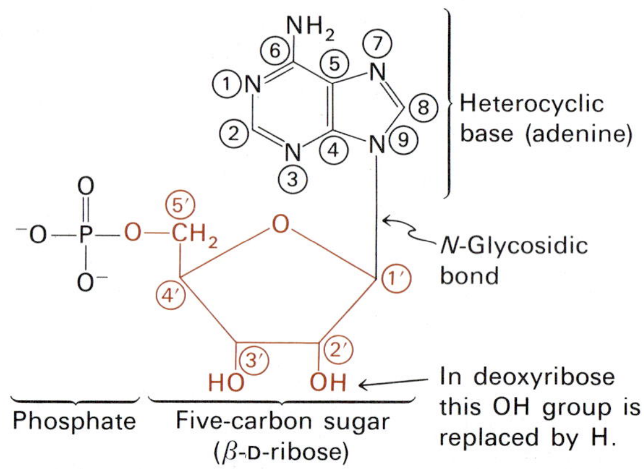

(25-1)

Nucleic acids, so called because they were first isolated from cell nuclei about 100 years ago, are of two general types: deoxyribonucleic acid (DNA) and ribonucleic acid (RNA). In the mid-1940s, DNA was shown to be the carrier of the genetic information that is passed from parent organism to its offspring. This information is, in large part, prescriptions for synthesizing the proteins essential to life processes. This finding narrowed the problem of information storage and transmission to one of understanding what it is about the DNA structure that lets it store information, pass the information along accurately, and eventually express that information as usable proteins.

In one of the most dramatic developments in the history of biology, James D. Watson and Francis Crick in 1953 deduced a structure for DNA that immediately suggested an answer to one of these questions and pointed the way to answering the others. To understand their contribution, we must first know the chemical composition of nucleic acids.

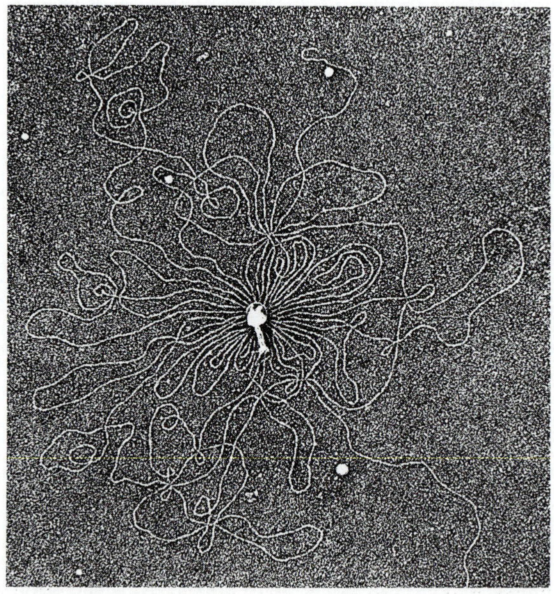

FIGURE 25-1

Electron micrograph of a bacteriophage, surrounded by its single molecule of DNA. The head of the bacteriophage was split open by osmotic shock, releasing the DNA. The DNA normally is circular, but it has been broken at one point; the two free ends can be seen at the top and bottom of the picture. [From A. K. Kleinschmidt, D. Lang, D. Jacherts, and R. K. Zahn, *Biochim. Biophys. Acta* 61:857 (1962).]

Like proteins and polysaccharides, **nucleic acids** are huge molecules (Figure 25-1). Each nucleic acid consists of one or two chains of **nucleotides.** Each nucleotide, in turn, is composed of three covalently bonded units: a heterocyclic base, a five-carbon sugar, and a phosphate. An example in which the heterocyclic base is adenine and the sugar is β-D-ribose follows:

A nitrogen atom of the heterocyclic base (at position 9 for adenine) is bonded to the anomeric carbon (C-1) of the sugar by an *N*-glycosidic bond (see Section 23-4). The phosphate unit is bound to carbon-5 of the sugar by an ester linkage. To distinguish the positions of the atoms of the sugar from the atoms of the heterocyclic base, a prime (′) is added to the numbers of the sugar atoms. Thus the nucleotide in Expression 25-1 is called a 5′-phosphate ester.

The phosphate ester bond can be visualized as forming between phosphoric acid and the sugar hydroxyl, with the elimination of water:

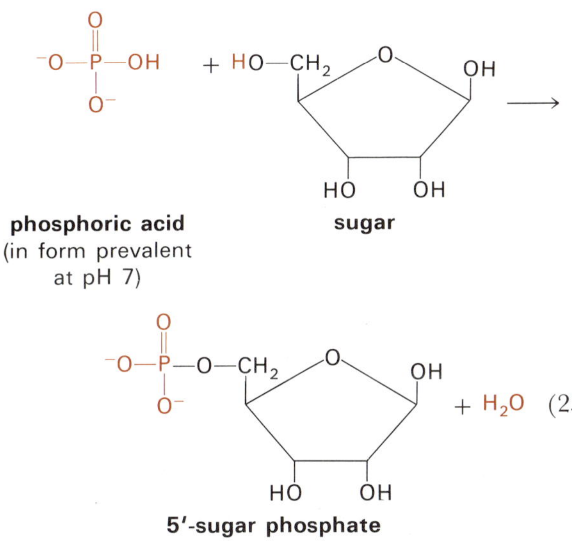

(25-2)

(At neutral pH, only one of the oxygen atoms of phosphoric acid is protonated.) In cells, the phosphate ester bond to sugar is not actually formed by a simple condensation reaction of the type shown, but rather is formed by specialized mechanisms described in Chapter 26. The simple esterification reaction (Equation 25-2) is given here only to illustratc why thc phosphate bond to the sugar is designated an ester.

In nucleic acids, individual nucleotide units are joined by a second ester bond between the phosphate group and the 3′-position of an adjacent nucleotide unit. This type of phosphate bridge is called a 3′-5′-phosphodiester linkage.

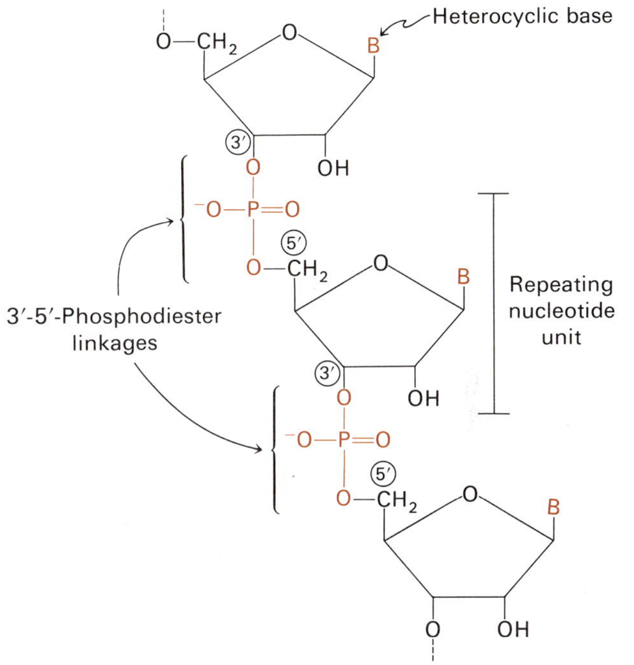

Chains of nucleotides joined by 3′-5′-phosphodiester linkages in this way are called **polynucleotides,** or simply **nucleic acids.**

In the first nucleotide structure presented (Expression 25-1), adenine is the heterocylic base and β-D-ribose is the sugar. The sugar unit may instead be deoxyribose. In naturally occurring nucleic acids, the sugar units are either all ribose or all deoxyribose. When the sugar is deoxyribose the nucleic acid is **deoxyribonucleic acid,** or **DNA.** Of the five principal heterocyclic bases that exist in cells, only four are found in DNA: **adenine, guanine, cytosine,** and **thymine** (see Figure 25-2).

The bases often are abbreviated in the shorthand form by their first letters, so the bases of DNA are

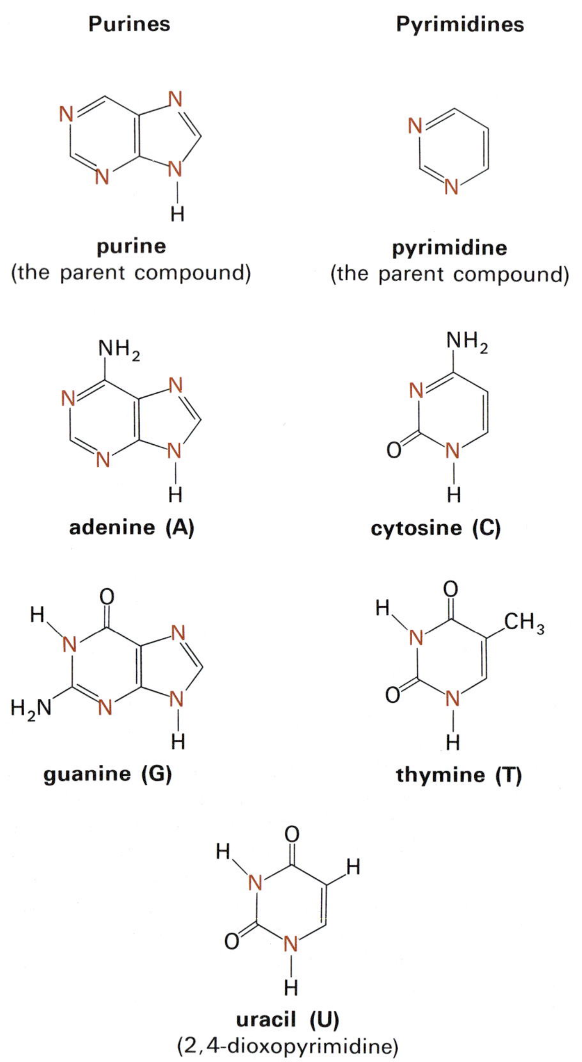

FIGURE 25-2

The principal bases of nucleic acids.

called A, G, C, and T. When the sugar is ribose, the nucleic acid is **ribonucleic acid,** or **RNA.** Of the four principal bases found in RNA, three also are found in DNA: adenine, guanine, and cytosine. The fourth RNA base, **uracil,** occurs only in RNA, just as thymine occurs only in DNA.

Note how closely thymine resembles uracil, differing only in the presence of a methyl group. The bases of RNA are called A, G, C, and U. The bases A and G are derivatives of purine and, therefore, called **purines;** the bases C, T, and U are called **pyrimidines** because of their resemblance to pyrimidine (Figure 25-2). Figure 25-3 shows hypothetical "alphabetical" stretches of DNA and RNA.

Although the five purine and pyrimidine compounds just described are present in organisms mainly as parts of nucleic acids, significant amounts also are found in the free or unpolymerized forms. Molecules composed of just a heterocyclic base and a sugar (but no phosphate) are called **nucleosides.** These nucleosides are named according to their bases (Figure 25-4).

Thymidine is found exclusively with deoxyribose as the sugar. The other nucleosides can occur with either ribose or deoxyribose as the sugar. If the sugar is deoxyribose, the full name includes a *deoxy-* prefix (or the letter *d* preceding the letter for the base in the abbreviated form), as in deoxyadenosine (abbreviated dA):

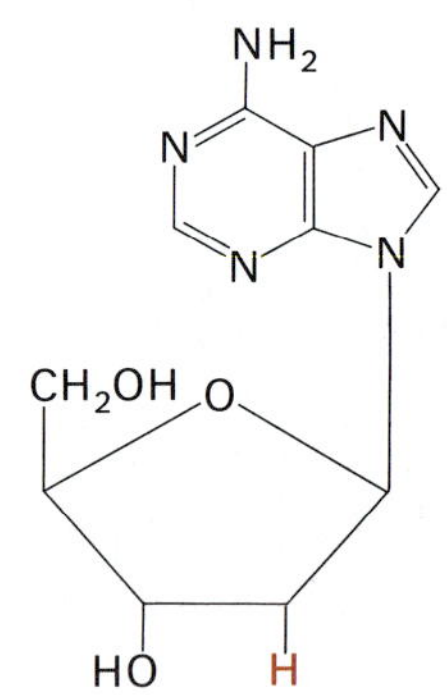

deoxyadenosine (dA)

In nucleic acids, uridine occurs only in the ribose form, but in cells it is also present as deoxyuridine monophosphate, a precursor of deoxythymidine monophosphate.

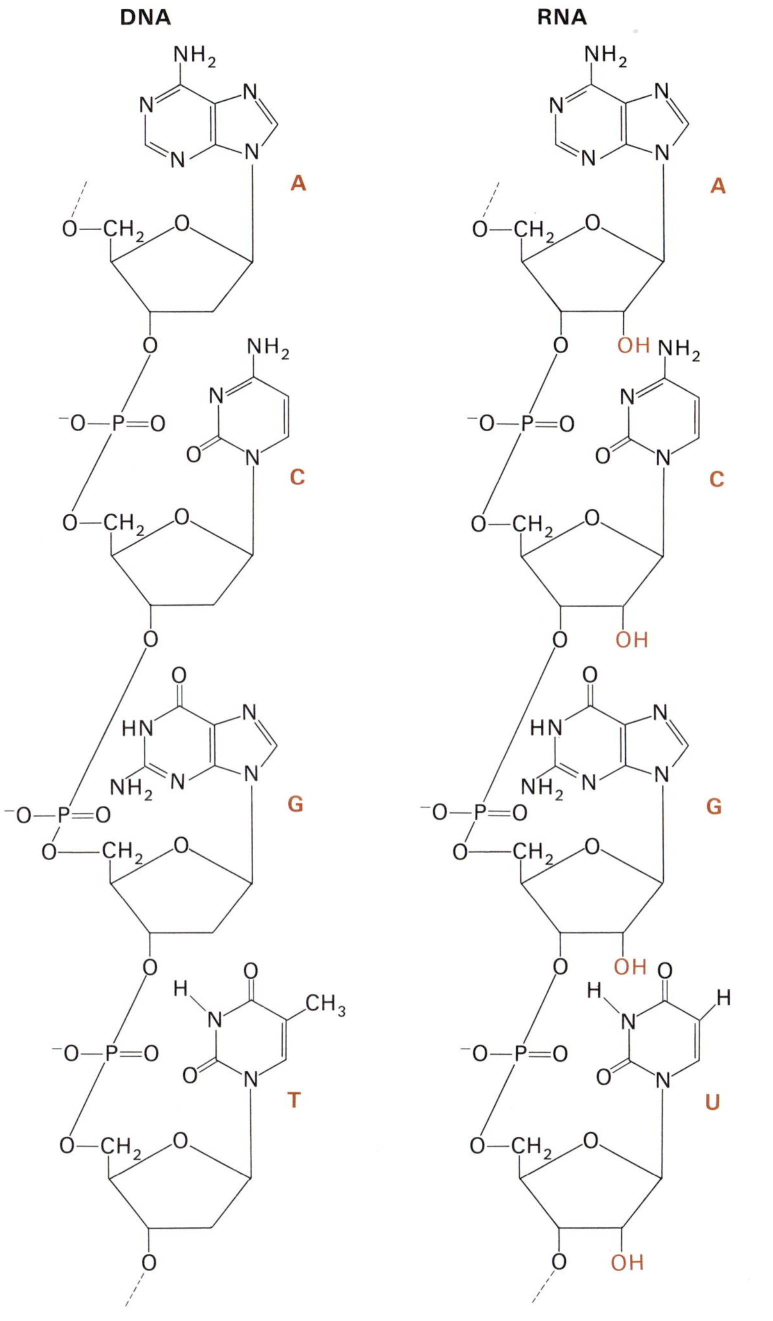

FIGURE 25-3

Hypothetical "alphabetical" stretches of DNA and RNA, illustrating the principal bases. In true DNA or RNA molecules, the four types of bases characteristic of DNA or RNA would occur in various sequences and combinations.

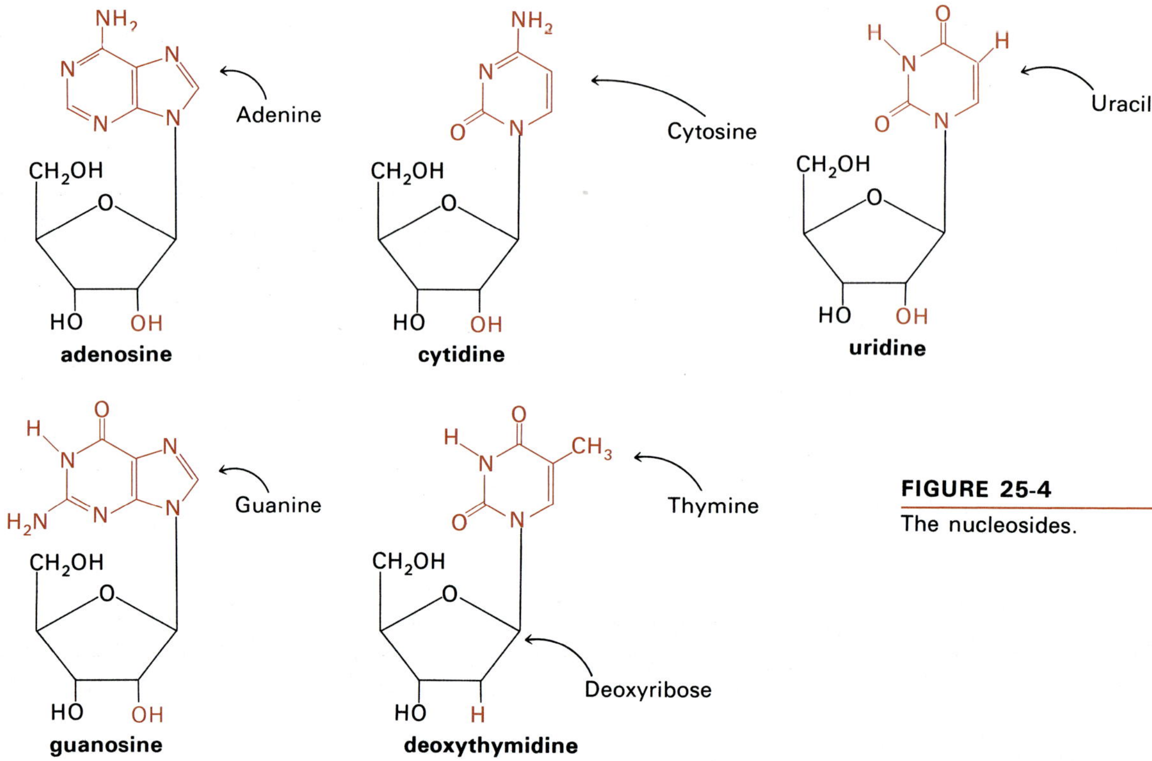

FIGURE 25-4
The nucleosides.

The nucleotides composed of adenosine with a single phosphate are named adenosine monophosphate (AMP) or deoxyadenosine monophosphate (dAMP) if the sugar is deoxyribose:

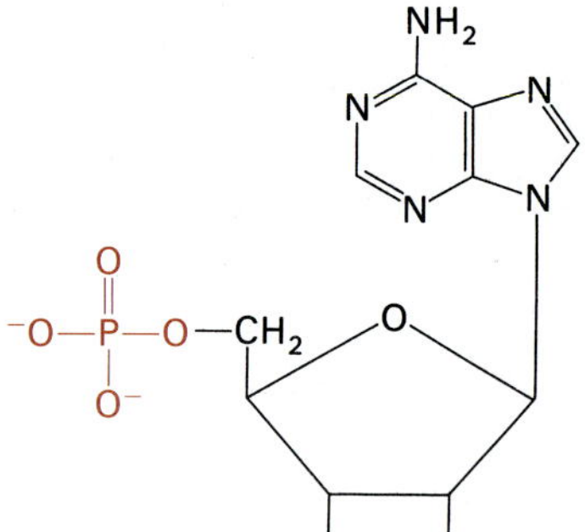

adenosine monophosphate (AMP)

Compounds with two or three phosphate groups also occur, and they are named with the appropriate nucleoside designation, plus diphosphate or triphosphate:

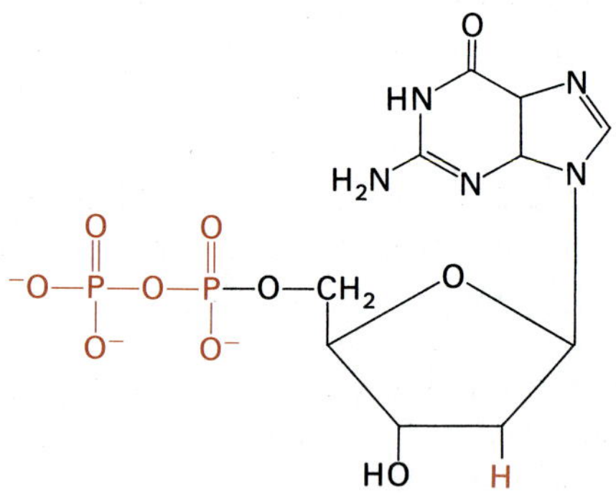

deoxyguanosine diphosphate (dGDP)

Adenosine triphosphate (ATP) is a very common nucleoside derivative, central to cellular energetics (Chapter 26). Although adenosine triphosphate (ATP) is the most abundant nucleoside triphosphate

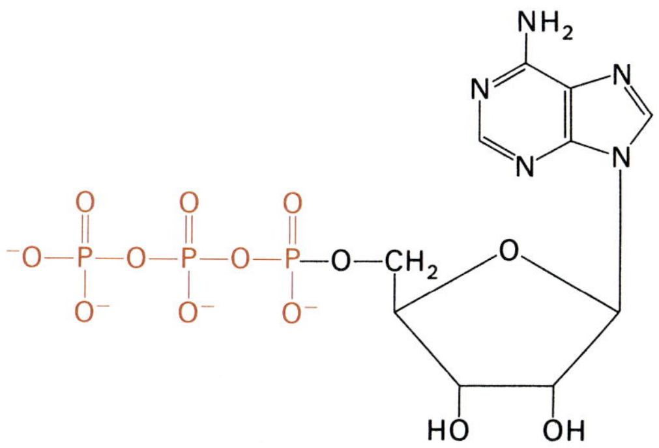

adenosine triphosphate (ATP)

in cells, appreciable amounts of GTP, CTP, and UTP also are present, because these molecules are used in the synthesis of RNA. Deoxynucleoside triphosphates (dATP, dGTP, dCTP, and dTTP) are used in the synthesis of DNA, but there is generally much less DNA in cells than RNA, so the deoxyribose forms are found in much smaller amounts than the ribose forms.

EXERCISE 25-1 Draw the structures indicated by the following abbreviated names: UMP, dCDP, TTP, GTP, dADP.

25-2 THE DNA DOUBLE HELIX

Watson and Crick's great contribution was to explain how the DNA chain of nucleotides, with its backbone of alternating phosphate and sugar groups and its variable array of bases, is arranged in space. Several critical clues led Watson and Crick to propose a double helix for DNA.

X-ray data revealed three characteristic features of DNA: fibers of DNA have a helical structure; one full turn of the helix occurs at a spacing of 3.4 nm along the fiber axis; and a basic structural unit repeats every 0.34 nm along the fiber axis. A second line of evidence concerned the relative amounts of the four kinds of bases in samples of DNA from many species. Measurement of the amounts of A, T, G, and C revealed a striking pattern: in every sample of DNA, the amount of A is the same as the amount of T, and the amount of G is the same as the amount of C. This A = T and G = C relationship prevails although the absolute percentages of the bases vary widely. For example, in the DNA from the bacterium *Sarcina lutea,* A is 13% but in human cells A is 30% (see Table 25-1). Yet in both species, the percentage of T is the same as that of A. Similarly, G and C vary from 37% in *Sarcina lutea* to 20% in humans, but are always equal. Table 25-1 illustrates the equalities A = T and G = C for several other species.

TABLE 25-1
Composition of DNA of various species (in percentages of total bases)

	Mole percentage			
Species	A	T	G	C
Sarcina lutea (a bacterium)	13	13	37	37
Human	30	30	20	20
Escherichia coli (a bacterium)	25	25	25	25
Wheat	27	27	23	23
Bovine (cattle)	28	28	22	22
Saccharomyces cerevisiae (brewer's yeast)	32	32	18	18

Watson and Crick put all of this information together in a brilliant scientific insight that has been

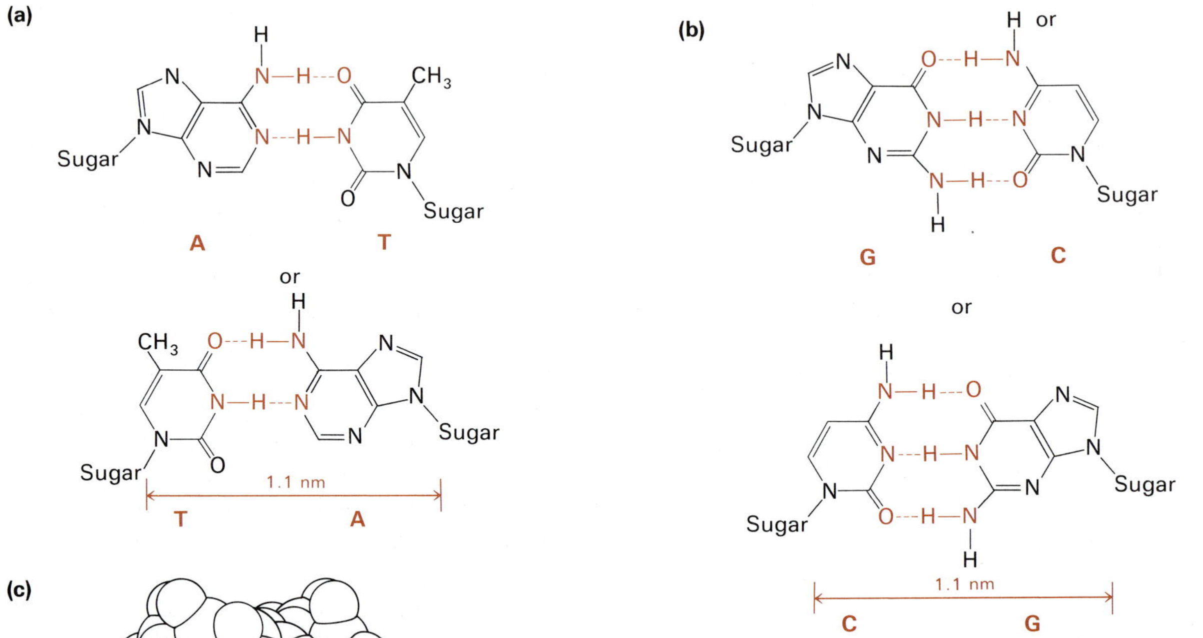

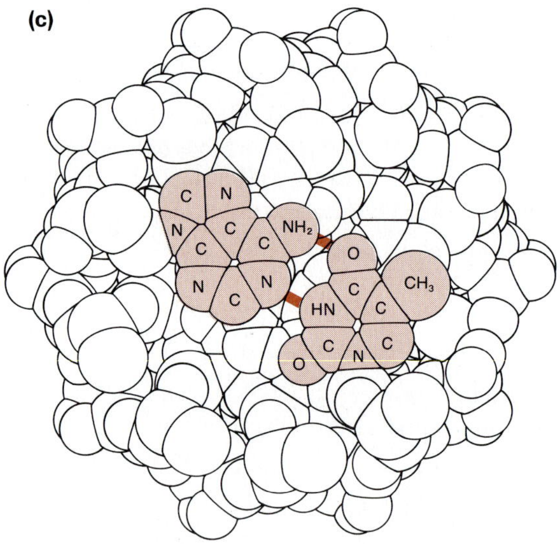

FIGURE 25-5

Base pairs. **(a)** The A–T pair. **(b)** The G–C pair. **(c)** Space-filling model of a base pair. This is an end-on view of a DNA double helix, with the first base pair (an A–T pair) colored and labeled. Hydrogen bonds are indicated by color. [The help of Gene Dykes in preparing the DNA models from which (c) was drawn is gratefully acknowledged.]

described in intimate detail by Watson in the book *The Double Helix*. The main breakthrough was the realization that the bases must somehow be paired in the structure of DNA: A always with T, and G always with C. Moreover if, as they believed, this pairing involves hydrogen bonds between the bases, the most likely arrangement would be a double-stranded structure with one base from each strand contributing to each pair. The key to the double helix lies in the special pattern of complementary hydrogen bonding (Figure 25-5) that permits two hydrogen bonds between A and T and three hydrogen bonds between G and C.

The structure of DNA rests on the fact that suitable

base pairs can form only between A and T and between G and C. Two pyrimidines would be too small to form a base pair together, because they could not bridge the gap across the strands of DNA. Therefore, possible pairs such as C–C, T–T, or C–T are eliminated. Similarly, pairs involving two purines would be too large to fit between the strands, eliminating possible pairs such as A–A, G–G, and A–G.. The remaining pairs, each involving one purine and one pyrimidine, that would be the right size are A–C and G–T. These base pairs cannot form because the various atoms are not in suitable positions to form complementary hydrogen bonds:

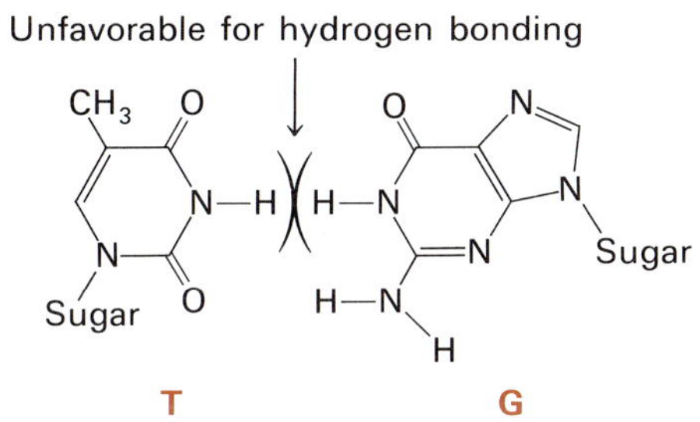

Thus only A–T and G–C base pairs can form, and this fact accounts for the fundamental structure of DNA. The two strands are held by a structural complementary—a kind of lock-in-key fit—between units of the strands.

The DNA helix arises from the two strands twisting about one another. The phosphate unit links the 5′-position of one nucleotide to the 3′-position of the next along the whole chain, and this defines a direction for each DNA chain. The two strands of a DNA molecule always run in opposite directions; the chains are said to be **antiparallel.** This arrangement is illustrated in Figure 25-6 for an unwound stretch of the structure. When the helical twist is included, the full structure can be represented as in Figure 25-7. The strands twist around each other as in a spiral staircase. The base pairs stack at a spacing of 0.34 nm, and each base pair is rotated by 36° from its neighbor. Therefore in each ten base pairs, the structure rotates $10 \times 36°$ or 360°, which corresponds to a full circle or one complete turn (Figure 25-7c,d). Thus as indicated by the X-ray data, the helical structure has a full repeat at a spacing corresponding to ten bases pairs, or 3.4 nm. Also apparent in the structure are a wide groove and a narrow groove between the strands (Figure 25-7a,b).

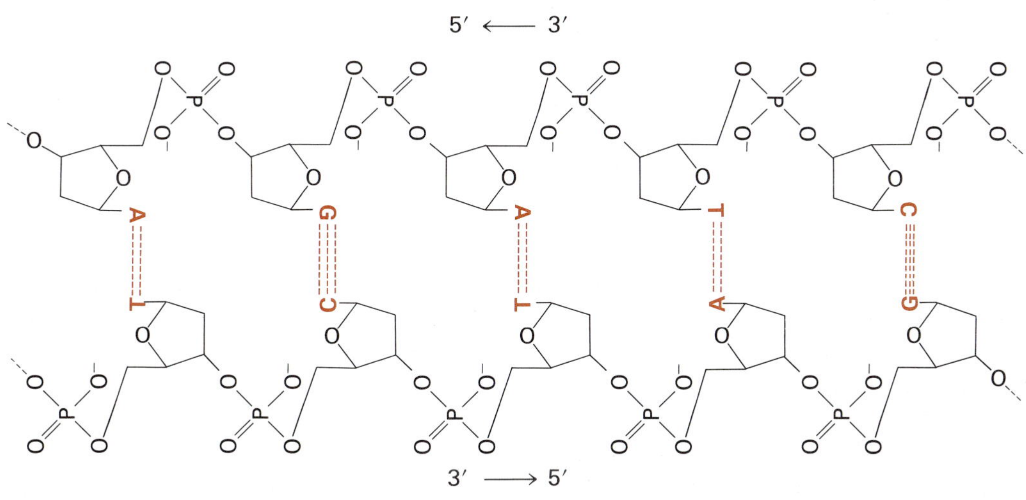

FIGURE 25-6

Antiparallel arrangement of the strands of DNA.

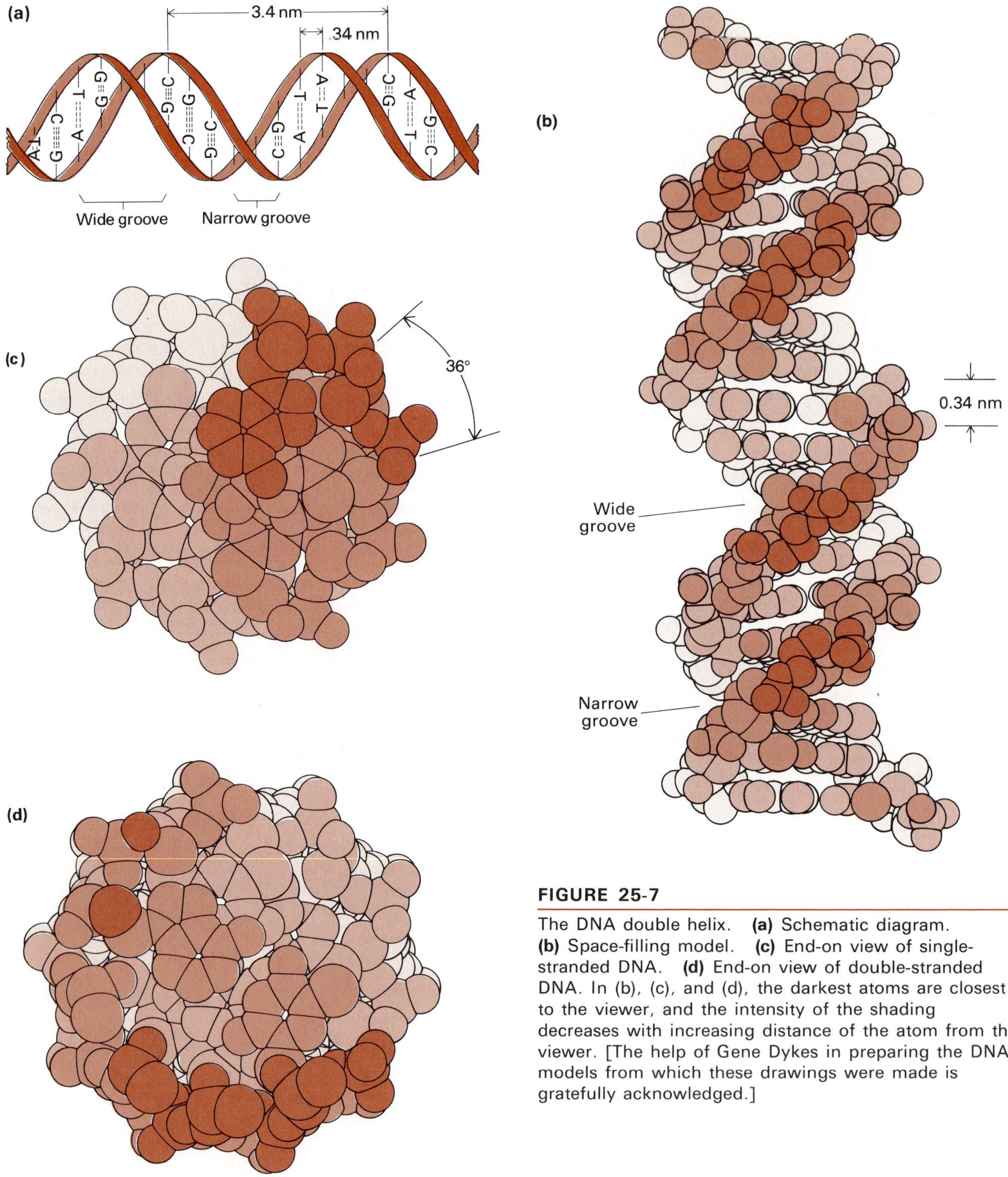

FIGURE 25-7

The DNA double helix. **(a)** Schematic diagram. **(b)** Space-filling model. **(c)** End-on view of single-stranded DNA. **(d)** End-on view of double-stranded DNA. In (b), (c), and (d), the darkest atoms are closest to the viewer, and the intensity of the shading decreases with increasing distance of the atom from the viewer. [The help of Gene Dykes in preparing the DNA models from which these drawings were made is gratefully acknowledged.]

25-3
THE REPLICATION OF DNA

The double-helical structure for DNA proposed by Watson and Crick was very satisfying in terms of its ability to accommodate all of the existing data, but its most exciting feature was that it immediately suggested how DNA could be **replicated,** or reproduced, each time a cell divided. It had long been established that the genetic information is passed on from cell to cell during the course of development of a complex organism from a single fertilized egg. One of the central mysteries of biology had been how this genetic material replicates each time a cell divides to insure that each daughter cell has a complete copy.

From the double-helical structure, replication can be readily visualized. Suppose the individual strands of the helix separate. The pairing of bases always follows the pattern of A with T and G with C, so each strand alone carries all of the information needed to specify its complementary strand. The "missing" strand therefore can be replaced in biochemical processes of the cell by matching a T for the new strand to an A of the existing strand, matching a C to a G, and so on. For each replication of DNA, the strands are assumed to separate, with each strand providing the "template" (or pattern) for its new complementary strand, thereby forming two double helices where there had been one (Figure 25-8). Each new DNA double helix thus contains one strand from a parent molecule and one strand synthesized by that parent strand acting as a template, using the A–T and G–C pairing pattern. Such replication is termed **semi-conservative,** because one-half of the parent DNA continues to exist in each new molecule.

The actual replication process involves many separate biochemical steps and many enzymes, principally DNA polymerases. The main lines of the mechanism of replication have been deduced during the last 20

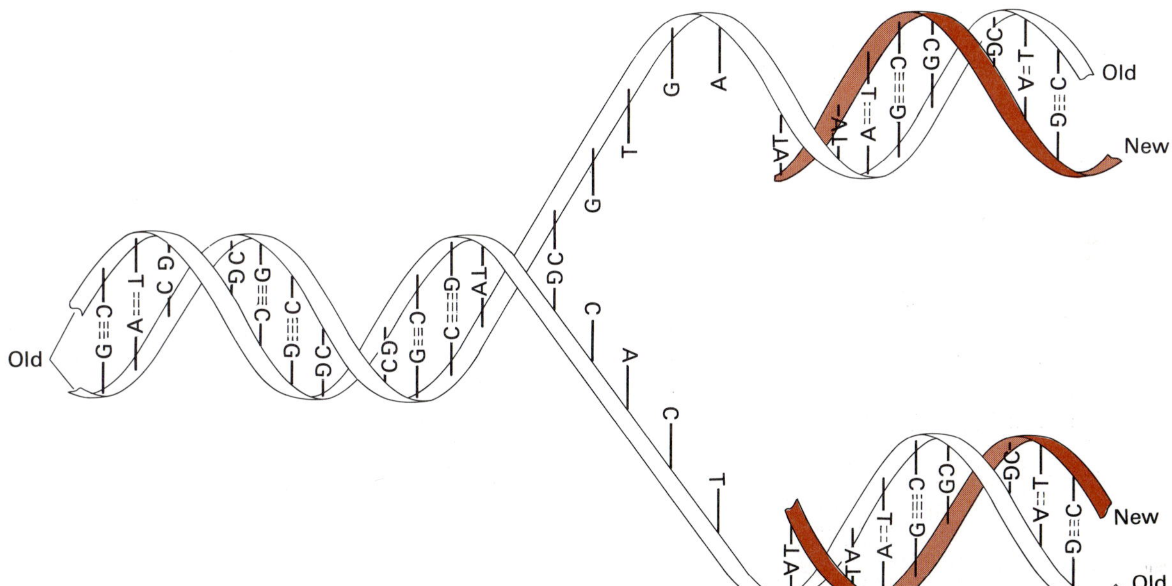

FIGURE 25-8

Semi-conservative replication.

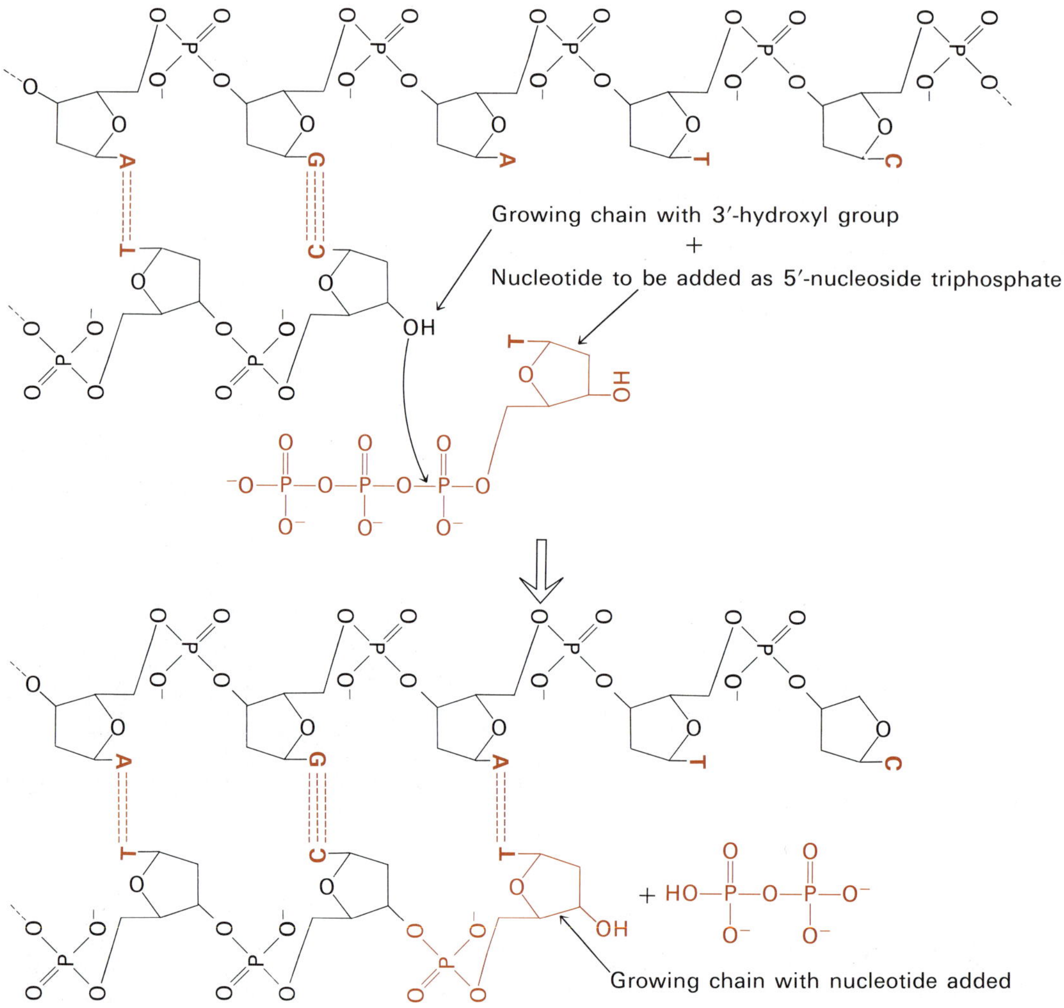

FIGURE 25-9

DNA synthesis.

years, following the critical first experiments by Arthur Kornberg. He demonstrated that DNA replication takes place by the addition of nucleoside triphosphate molecules. As the triphosphate reacts, the two terminal phosphate groups split off, leaving the nucleotide unit attached (Figure 25-9). This mechanism requires that replication always occurs with a 5′-nucleoside phosphate attaching to a free 3′-hydroxyl group. This pattern of synthesis, known as **5′→3′ replication,** has been found in all species examined.

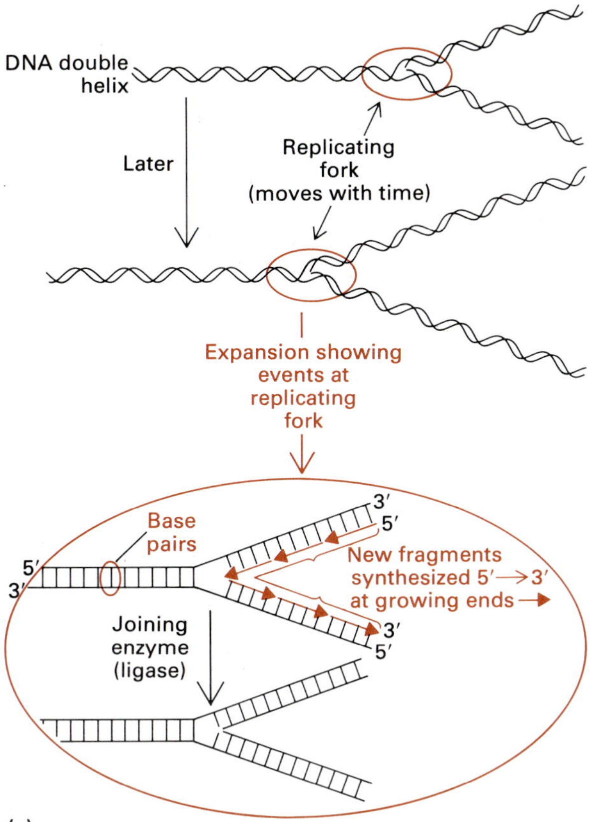

(a)

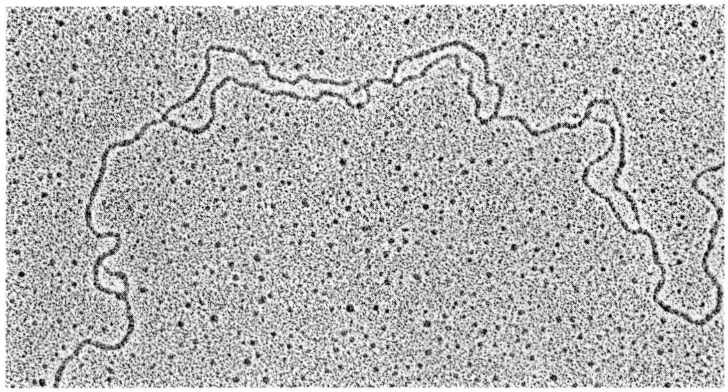

(b)

FIGURE 25-10

The replicating fork. **(a)** Schematic diagram. **(b)** Electron micrograph of replicating chromosomal DNA. [Courtesy of David Hogness.]

The strands of DNA separate, as indicated in Figure 25-8, creating a place of active DNA synthesis called the **replicating fork** (Figure 25-10). As illustrated in the detail of the figure, the new strands are formed in short pieces that are synthesized in a direction toward the replicating fork for one new strand and away from the replicating fork for the other new strand. The short segments are then joined by special enzymes called ligases to give the complete strands.

EXERCISE 25-2 The base sequences of the complementary strands of a hypothetical region of DNA are as follows:

. . .-A-G-G-G-T-T-C-T-G-A-A-T-G-A-T-. . .

. . .-T-C-C-C-A-A-G-A-C-T-T-A-C-T-A-. . .

To convince yourself that you know how DNA replicates, cover up the lower strand, and see if you can reproduce its sequence by examining just the upper strand.

25-4
THE ROLE OF RNA AND THE CENTRAL DOGMA

The double helix elegantly explains the structure of DNA and its replication, but it offers little insight into how a sequence of nucleotides in the DNA dictates (encodes) the sequence of amino acids of a protein. Yet such coding is as important as the mechanism of DNA replication because the function of large segments of the DNA molecules (the *genes*) is to specify the primary structure of a corresponding protein. The problem has two phases: (1) finding the mechanism by which the information to specify the sequence of protein is stored in the DNA, and (2) finding the sequence by which the stored information is turned into actual protein—that is, the mechanism by which the protein is assembled. It was impossible to devise any direct complementary relationships that would permit a bonding of specific amino acids with specific sequences of nucleotides. Nucleotides would be required to fit amino acids of diverse sizes and shapes, with differences (as between leucine and valine) that were far too subtle to be distinguished by differences among the nucleotides themselves.

In 1958, Crick suggested a solution to this dilemma. He proposed the "adaptor hypothesis," according to which DNA sequence is translated into a special set of adaptor nucleic acid molecules that then are linked to the needed amino acids. There were already reasons to believe that such molecules were RNA, with at least one type of RNA for each amino acid. We now know that protein synthesis does indeed take place via adaptor molecules of ribonucleic acid, but it involves three different RNA intermediates. We shall summarize the events of protein synthesis briefly here in order to place the substances and processes involved in context. The next few sections describe assembly of protein molecules in more detail.

The information specifying the primary structure of a protein is encoded in the sequence of nucleotides of DNA. The first class of RNA, **messenger RNA (mRNA)**, includes many copies of that sequence made by a process like that of DNA replication but using ribonucleotides rather than deoxyribonucleotides. This copying of genetic information from DNA onto RNA is called **transcription.** Molecules of the second class of RNA, **transfer RNA (tRNA)**, function as the adaptor molecules proposed by Crick. They are attached to the individual amino acids and assemble them in the proper sequence along molecules of messenger RNA. The third class of RNA, **ribosomal RNA,** participates in the large cell structures called ribosomes where the actual bonding of amino acids into the protein takes place. The process of building protein molecules from the information in messenger RNA is called **translation.**

This idea—that DNA acts as the template for production of RNA (transcription) and that RNA directs synthesis of protein (translation)—has so dominated the progress of biochemistry in the last two decades that it has come to be called the "central dogma." It is often summarized as

$$\text{DNA} \xrightarrow{\text{transcription}} \text{RNA} \xrightarrow{\text{translation}} \text{Protein}$$

With few exceptions (among some tumor-causing viruses), this simple expression describes the flow of genetic information in all living cells.

25-5 THE GENETIC CODE

Knowing something about the roles of transfer RNA and messenger RNA, we can set about learning how the sequence of nucleotides in DNA actually codes for amino acids. The only difference between DNA molecules lies in their heterocyclic bases. There are 4 bases and 20 amino acids, so the correspondence obviously cannot be between a single base and a single amino acid. Scientists therefore believed that these 4 bases—guanine, cytosine, adenine, and thymine—act as a kind of four-letter alphabet to spell out the amino acids and proteins. Similarly, sequences of two bases are insufficient because the four bases, taken in pairs, could write only 16 different two-letter words (AA, AC, AG, AT, CA, CC, CG, CT, etc.)—still not enough to code for all 20 amino acids. The four bases taken as 3 letter words, or **triplets,** give more than enough combinations (64) to code uniquely for every amino acid, and early experiments confirmed the triplet nature of the genetic code.

For the specific assignment of triplet sequences to each of the amino acids, several experimental approaches were used. For the most part they involved using various forms of mRNA, synthesized in the laboratory with known composition, to direct protein synthesis. Eventually the full "dictionary" of the genetic code was obtained. (Table 25-2).

All 64 distinct codons are presented in a convenient grid with the first base along the left edge, the second base across the top, and the third base along the right edge. Certain amino acids are represented by several different codons. Leucine, for example, is represented by the following six triplets: UUA, UUG, CUU, CUC,

TABLE 25-2
The genetic code

First base	Second base: U		C		A		G		Third base
U	UUU	Phe	UCU	Ser	UAU	Tyr	UGU	Cys	U
	UUC	Phe	UCC	Ser	UAC	Tyr	UGC	Cys	C
	UUA	Leu	UCA	Ser	UAA	***	UGA	***	A
	UUG	Leu	UCG	Ser	UAC	***	UGG	Trp	G
C	CUU	Leu	CCU	Pro	CAU	His	CGU	Arg	U
	CUC	Leu	CCC	Pro	CAC	His	CGC	Arg	C
	CUA	Leu	CCA	Pro	CAA	Gln	CGA	Arg	A
	CUG	Leu	CCG	Pro	CAG	Gln	CGG	Arg	G
A	AUU	Ile	ACU	Thr	AAU	Asn	AGU	Ser	U
	AUC	Ile	ACC	Thr	AAC	Asn	AGC	Ser	C
	AUA	Ile	ACA	Thr	AAA	Lys	AGA	Arg	A
	AUG	Met	ACG	Thr	AAG	Lys	AGG	Arg	G
G	GUU	Val	GCU	Ala	GAU	Asp	GGU	Gly	U
	GUC	Val	GCC	Ala	GAC	Asp	GGC	Gly	C
	GUA	Val	GCA	Ala	GAA	Glu	GGA	Gly	A
	GUG	Val	GCG	Ala	GAG	Glu	GGG	Gly	G

Note: The codons are read in the $5' \rightarrow 3'$ direction. The symbol *** indicates a termination codon.
Source: F. H. C. Crick, "The Genetic Code: III." Copyright © 1966 by Scientific American, Inc. All rights reserved.

CUA, and CUG. Most amino acids are specified by at least two codons, but two amino acids (methionine and tryptophan) are specified by unique codons. In addition, three codons are reserved for punctuation. Like the period at the end of a sentence, they are a termination signal, indicating that the protein sequence is complete. How particular codons came to specify particular amino acids is a mystery related to the earliest events in the origin of life—one that we may never unravel.

25-6 TRANSCRIPTION

In the first step of protein synthesis, many copies of the coding sequences in DNA are made in the form of messenger RNA. In this way, the information in the genes is amplified, making it possible for protein to be synthesized more rapidly.

Transcription also provides the genetic information in a single-stranded form more suitable for directing amino acid assembly. For each gene of the DNA that codes for the amino acid sequence of a protein, one DNA strand contains the information, and the other strand is the complement of the informational strand. Synthesis of mRNA uses the complementary strand of the DNA as a template to form a polynucleotide that resembles the information strand of the DNA (with the difference that T in DNA is replaced by U in RNA). Transcription is catalyzed by enzymes called RNA polymerases and, like DNA synthesis, proceeds in the 5′→3′ sense (Figure 25-11).

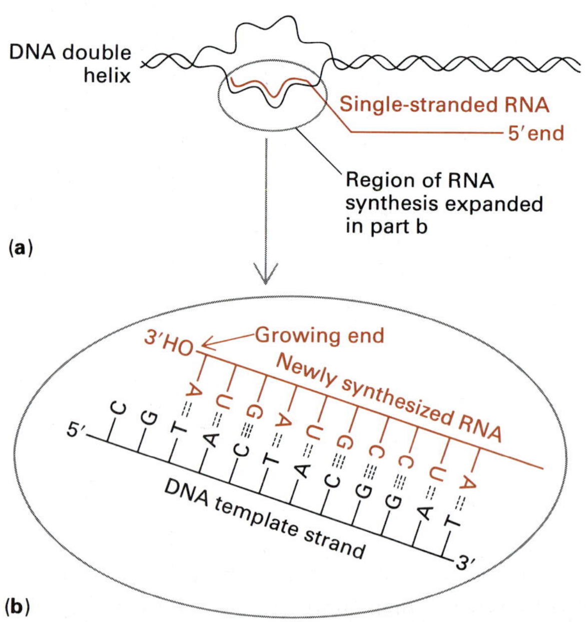

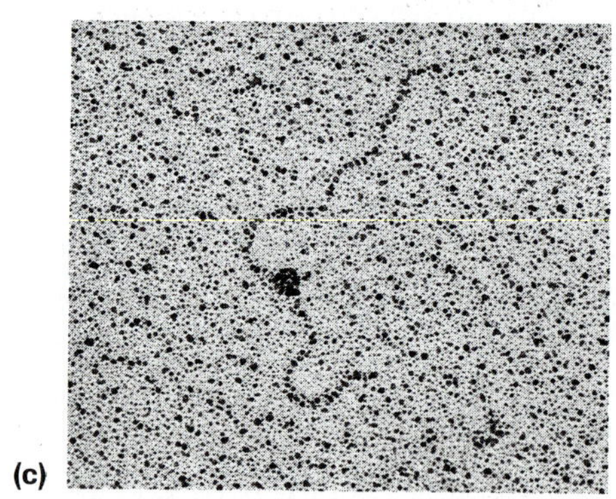

FIGURE 25-11

Transcription. **(a)** DNA double helix with region transcribed. **(b)** Expanded diagram of the region of RNA synthesis. **(c)** Electron micrograph of RNA polymerase attached to a promoter site on a fragment of DNA from the bacterium *E. coli*. [Courtesy of Laurien Polder.]

Synthesis of mRNA must begin on the correct strand of DNA and exactly at the start of the gene that spells out the needed protein. To assure that mRNA synthesis begins at the correct place, there are certain sequences of nucleotides in the DNA called **promoters** that are recognized by the enzyme RNA polymerase. These recognition sites can also serve in the control of protein synthesis. When the sites are blocked, initiation of mRNA formation is prevented. In this way, individual genes or clusters of genes can be "turned on" or "turned off" at this very first step of protein synthesis.

One well-studied example of this type of regulation concerns the enzyme β-galactosidase in the bacterium *Escherichia coli.* This enzyme splits lactose into glucose and galactose for further metabolic processing. If there is no lactose available to the cells, the enzyme is not needed, and a specific protein called a **repressor** binds to the DNA at the promoter site of the gene for β-galactosidase, blocking transcription. If lactose is present, it binds to the repressor and converts it to an inactive form that no longer binds to the DNA. With the repressor removed, transcription of the β-galactosidase gene then occurs. In higher organisms, the processes that activate certain genes at certain times in the development of the organism are not yet well understood.

EXERCISE 25-3 In order to visualize the process of transcription, repeat Exercise 25-2, but this time cover the lower strand, and use the upper strand to specify the mRNA sequence that would be formed using the upper strand as a template.

25-7 TRANSLATION INTO PROTEIN

Once a gene has been activated and its corresponding mRNA synthesized, the tRNA present in the cell comes into play. Each cell contains at least one set of tRNA molecules for each amino acid. In addition, it contains a set of 20 enzymes, called *aminoacyl-tRNA synthetases,* one capable of recognizing each amino acid and its corresponding tRNA. In this way, the recognition of slightly different amino acids is accomplished by enzymes, which we know to be highly specific, rather than by the nucleotides themselves. The tRNA molecules are relatively small, each with fewer than 100 nucleotides.

Each tRNA forms a characteristic structure in which regions of the RNA fold back on themselves to form three stretches of double helix, with a loop at the end. These loops, plus a "stem" consisting of a double helix formed by the two free ends, give the molecules a characteristic cloverleaf structure. The tRNA from yeast that is specific for alanine was the first tRNA molecule fully analyzed in terms of its nucleotide sequence (Figure 25-12). The amino acid is attached to the free 3′—OH of the stem. At the opposite end of the molecule, there are three bases that by complementary pairing recognize the codon nucleotide triplet sequence corresponding to alanine. These three bases in the tRNA that recognize the codon are called the **anticodon,** and the loop in which they are found is called the anticodon loop.

The process of linking each amino acid with its tRNA occurs in two steps (Figure 25-13). First each aminoacyl-tRNA synthetase catalyzes a reaction between its amino acid and ATP to yield an "activated" amino acid derivative, called aminoacyl-AMP:

$$\text{Amino acid} + \text{ATP} \longrightarrow \text{Aminoacyl—AMP} + \text{PP}_i$$

The effect of this reaction is to form a mixed anhydride between a phosphate of ATP and the carboxyl

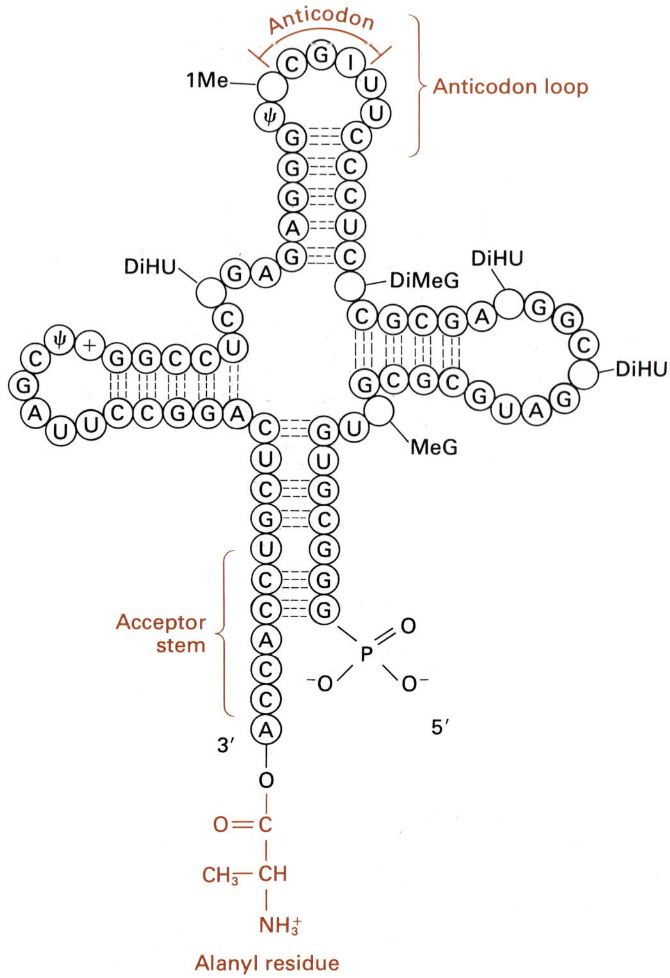

FIGURE 25-12

Yeast alanine tRNA with alanyl residue attached at the 3′-end. In addition to the usual bases found in RNA—A, C, G, and U—a number of modified bases are also present, as indicated by the abbreviations DiHU (for dihydrouridine) ψ (for pseudouridine), DiMeG (for N-dimethyl guanosine), 1ME (for methylguanosine) and I (for inosine). These bases are formed by enzymatic modification after the tRNA is synthesized.

group of the amino acid. The two remaining phosphates are released together in an inorganic form called pyrophosphate (PP_i). The aminoacyl-AMP derivative is said to be activated because the highly reactive mixed anhydride bond permits the second step of the aminoacyl-tRNA synthetase reaction to proceed with a high yield:

$$\text{Aminoacyl—AMP} + \text{tRNA} \longrightarrow \text{Aminoacyl—tRNA} + \text{AMP}$$

In this second step, an ester bond is formed between the free 3′—OH of the corresponding tRNA and the carbonyl carbon of the amino acid to give an aminoacyl-tRNA, with the release of AMP (Figure 25-13). The aminoacyl-tRNA is thus the key adaptor molecule: it can recognize a codon sequence with its anticodon loop, and it can deliver the amino acid that corresponds to that codon.

Now that we have identified the agents directly involved in the flow of information from DNA to protein—namely, tRNA and mRNA—we can understand more about how this flow of information is actually coordinated in cells. In fact, the process of aligning aminoacyl-tRNA molecules along the appropriate sequence of mRNA is a very intricate process that involves one of the most complicated pieces of molecular machinery of the cell, the ribosome.

Ribosomes are huge particles by molecular standards, with diameters of about 20 nm and molecular

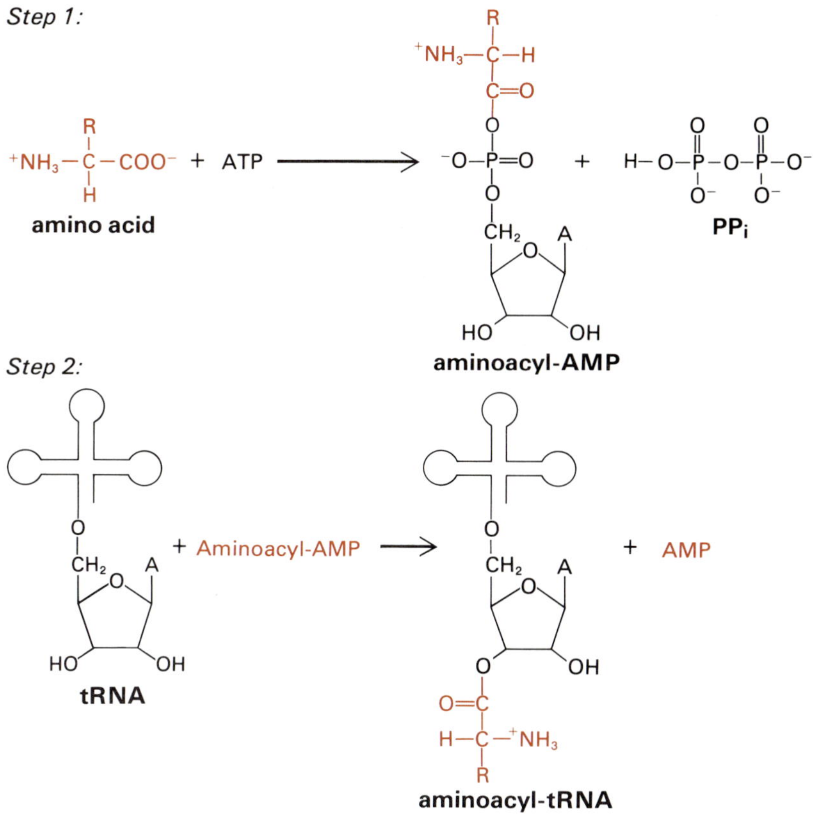

FIGURE 25-13

The two steps in the reactions catalyzed by aminoacyl-tRNA synthetase. Step 1: activation of the amino acid. Step 2: attachment of the amino acid to the tRNA specific for that amino acid. Both steps are catalyzed by the same enzyme, which is specific for the amino acid.

weights of about 3 million. Each ribosome is composed of over 50 different proteins, as well as a specialized form of RNA called **ribosomal RNA** (**rRNA**). As far as we know, rRNA is not informational; it has none of the coding functions of a messenger RNA. As soon as a portion of the mRNA has been synthesized in transcription, the decoding process called translation begins at the ribosome (Figure 25-14).

In translation, the message is aligned at the starting position on the ribosome, and the mRNA is read beginning near the free 5′-end. Protein synthesis always starts with a methionine residue, coded by AUG. In many proteins, however, the N-terminal Met is clipped off after protein synthesis has begun. When the mRNA is properly aligned on the ribosome, a methionyl-tRNA with the CAU anticodon binds to the AUG codon. As the codon and the anticodon form base pairs with the mRNA and tRNA aligned antiparallel, they mimic the base pairs that exist in the corresponding region of the DNA double helix (Figure 25-15).

A ribosome has two distinct sites for protein synthesis, the D (for "donor") site and the A (for "acceptor") site (Figure 25-14). Molecules of tRNA bearing their amino acids enter at the A site to match with the proper codon of the mRNA. The mRNA–tRNA complex then shifts to the D site, and the next aminoacyl-tRNA indicated by the codon of the mRNA enters at the A site. The amino acid attached to the tRNA in the D site is then added to the amino acid attached to

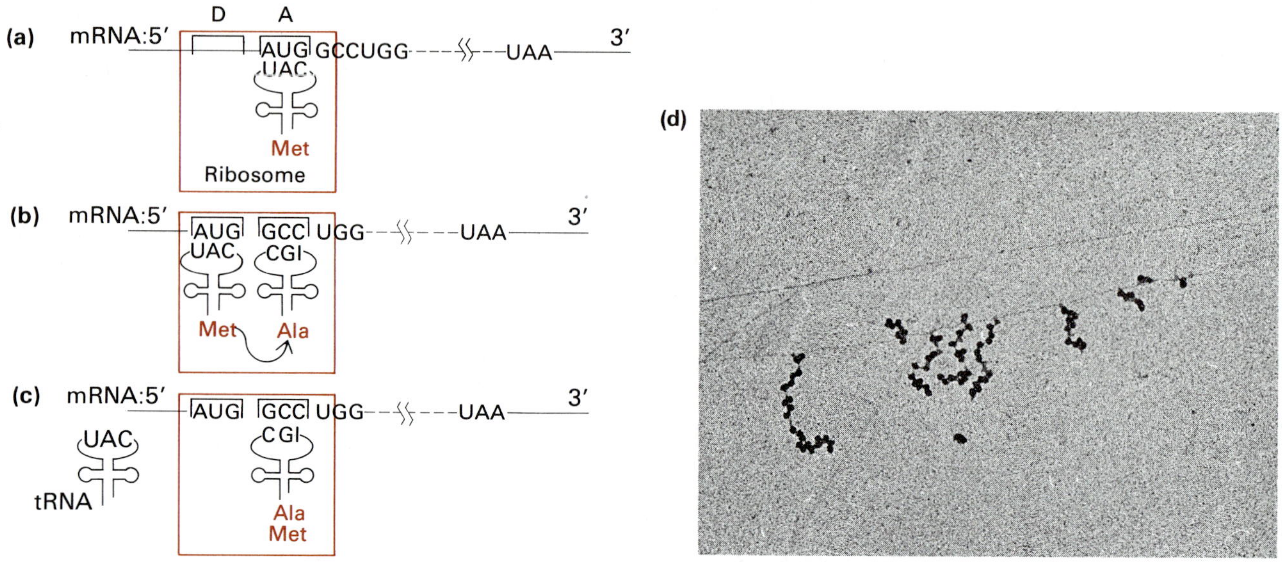

FIGURE 25-14

Translation on ribosomes. **(a)** Initiation: the mRNA is aligned on the ribosome with the AUG codon in the A (acceptor) site, and a methionyl-tRNA is bound to the codon. **(b)** The mRNA-tRNA complex shifts to the D (donor) site, and an alanyl-tRNA binds to the mRNA codon in the A site. **(c)** Donation: a peptide bond is formed between the Met and the Ala. The Met is released from its tRNA, and the free tRNA is released from the D site. **(d)** Electron micrograph showing transcription and translation of DNA from the bacterium *E. coli*. As each mRNA strand is synthesized and peels off the DNA, ribosomes (dark particles) attach to the growing strand. Each cluster of ribosomes is reading the same mRNA strand; and each ribosome translates the mRNA into a complete polypeptide chain. The micrograph shows several mRNA strands being formed simultaneously. The longest one, at the lefthand end, is the first to be transcribed from the DNA. [**(d)** from O. L. Miller, Jr., Barbara A. Hamkalo, and C. A. Thomas, Jr., *Science* 169:392 (1970). Copyright © 1970 by the American Association for the Advancement of Science.]

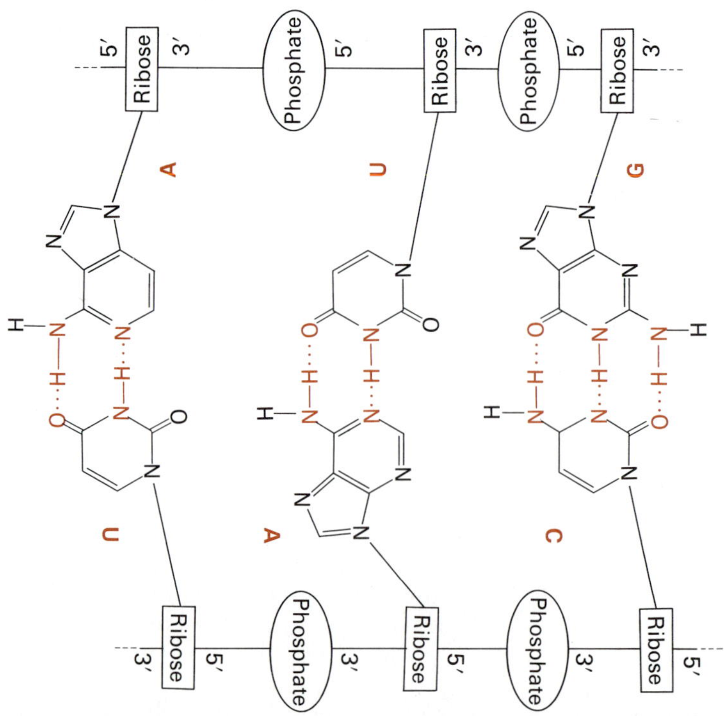

FIGURE 25-15

Base pairs between the AUG codon for methionine and its anticodon. The RNA strands corresponding to the mRNA and the tRNA are presented in the upper and lower parts of the figure, respectively. For the mRNA the 5′ ⟶ 3′ direction is to the right, while for the tRNA the 5′ ⟶ 3′ direction is to the left. Thus the two strands are antiparallel, permitting typical base pairs to form in the pattern A with U, U with A, and G with C.

the tRNA in the A site by peptide-bond formation (Figure 25-16). The entire unit (the tRNA with the dipeptide and mRNA attached) is then shifted to the D site, and the next aminoacyl-tRNA enters at the A site. The process repeats as the dipeptide in the D site is bonded to the amino acid in the A site. The cycle continues to repeat until the complete polypeptide is formed. The process stops when one of the three termination codons (UAA, UAG, or UGA; see Table 25-2) is encountered on the mRNA. Then the polypeptide chain is released from the last tRNA.

In the base pairing between codons of the mRNA and anticodons of the tRNA, the precise base-pairing relationship (G with C, and A with U, the replacement for T in RNA) is relaxed slightly for the pairing involving the third base of the codon. Evidently there is some flexibility in the codon–anticodon fit at this position. This looseness of fit, referred to as "wobble," allows pairs other than the usual A–U and G–C to occur. The most extreme form of wobble occurs with **inosine** (abbreviated I), a modified form of guanosine found in the anticodon of yeast alanine tRNA (Figure 25-12). When in the position of the anticodon that pairs with the third codon base, I can pair with U, C, or A (see Figure 25-17). As a result, some tRNA molecules can bind to more than one codon sequence. In yeast, for example, alanine tRNA can bind to GCA, GCU, or GCC codons.

EXERCISE 25-4 For the following sequence in a mRNA molecule, write the sequence of amino acids you would expect to be produced by translation:

· · · –A–U–G–A–C–G–A–G–A–G–C–U–U–A–C–

C–U–U–A–U–A–G–C–G–G–G–G–U–A–G– · · ·

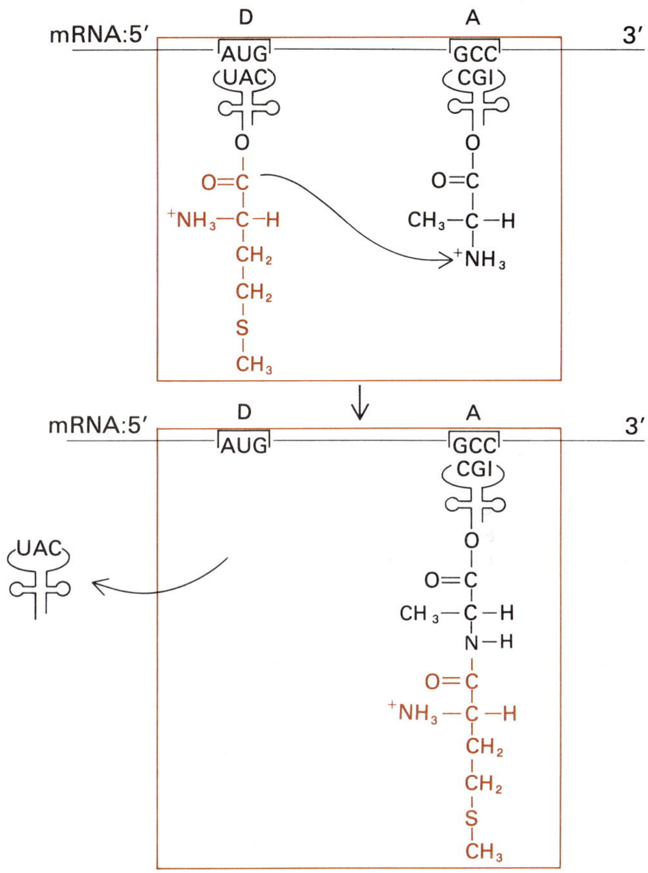

FIGURE 25-16

Donation step: formation of peptide bond between units in A and D sites. The initiator tRNA with methionine attached is in the D site and a tRNA bearing the next amino acid enters at the A site (upper portion of figure). A peptide bond is formed between the amino group of the amino acid in the A site and the carboxyl group of the methionine in the D site. As a result the methionine is released from the tRNA in the D site. The free tRNA of the D site is then released.

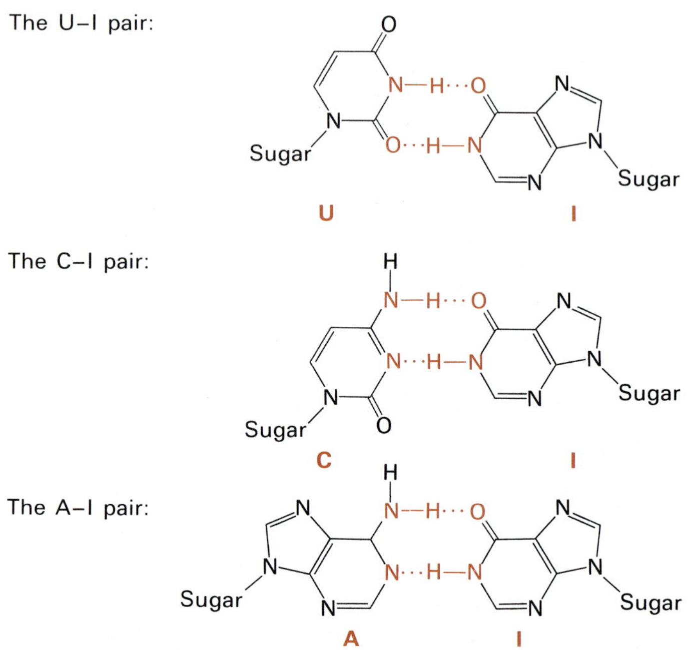

FIGURE 25-17

Base pairs of I with U, C, and A. According to the wobble hypothesis, I can form irregular base pairs with bases of RNA other than G, when they occur at the third position of the codon. In each case two hydrogen bonds are involved.

25-8 THE NONENZYMATIC BASE-CLEAVAGE METHOD FOR DNA SEQUENCING

In the last several years, a number of major developments have greatly increased our understanding of the structure and action of nucleic acids. They hint at the enormous potential of artificial genes and other modifications of DNA in solving health-related problems. One of the most important developments has been a dramatic breakthrough in determining the base sequence of DNA. Using nonenzymatic methods, Allan M. Maxam and Walter Gilbert worked out a rapid and convenient procedure for sequencing DNA (first published in 1977). In previous studies of nucleic acids—such as the work that established the sequence of the tRNA (Figure 25-10)—the approach was like the one described for protein sequencing: nucleic acids were cleaved at specific bases through the action of specialized enzymes called **nucleases.** Many small fragments, called oligonucleotides, were produced; these were isolated and sequenced individually. Eventually, by studying overlapping oligonucleotides, a full sequence could be determined. However, the procedures required were laborious, and years of work were required for longer sequences. Maxam and Gilbert's method is revolutionary by comparison; it enables scientists to determine sequences at a rate of approximately 100 nucleotides per day. In fact, DNA sequencing has become so easy that it is now preferable to find the sequence of a large protein by sequencing the DNA of the corresponding gene and decoding the base sequence, rather than by sequencing the protein itself.

Several factors were involved in developing the new method of DNA sequencing. One factor was the use of a technique called **gel electrophoresis,** which had reached the stage where polynucleotides differing in size by only a single nucleotide could be distinguished. This technique involves a gelatinous substance (the gel) made from a synthetic polymer such as polyacrylamide. When a mixture of singled-stranded polydeoxynucleotides of various sizes is applied in a band at one end of the gel (the position called the origin) and an electric field is applied between the ends of the gel, the various DNA fragments—all negatively charged because of the abundance of phosphate groups—migrate to the positive electrode. (Hence the name electrophoresis from electro- plus the Greek word *phoresis* for "something carried"). Each fragment moves at a rate that depends upon its size: the larger the fragment, the more it is held by the gel network. Thus by carefully choosing conditions (such as the length of time of the electrophoresis and the density of the gel), polynucleotides can be separated into a series of bands, differing in size by as little as one nucleotide. The different molecules are separated in order of length—the smallest having traveled the greatest distance from the origin.

In itself, the resolution of DNA fragments at the level of one nucleotide was a notable achievement. However, by itself, resolution of DNA fragments by gel electrophoresis was still a long way from sequencing them. Three obstacles remained to be overcome by Maxam and Gilbert

The first obstacle was that of obtaining specific fragments of DNA. Although gel electrophoresis can resolve DNA fragments differing in size by a single nucleotide, this resolution can be achieved only for fragments containing up to a few hundred nucleotides. However, even the smallest DNA molecules—those found in certain viruses—contain tens of thousands of nucleotides. Even the simplest cells, such as bacteria, have DNA molecules with millions of nucleotides. In higher organisms, the DNA is wrapped in the basic histone proteins and is packaged in compact structures called **chromosomes.** A human cell contains 46 chromosomes, each made up of a large DNA molecule that may consist of as many as 100 million nucleotides. Thus some way was needed to produce small specific pieces of DNA for eventual sequencing.

Maxam and Gilbert used enzymes called **restriction endonucleases** for this purpose. These endonucleases cleave DNA at specific sequences of base pairs. The sequence required to cause cleavage is generally four or more bases long, a sequence unlikely to appear very often in the overall sequence of DNA (Figure 25-18).

Thus digestion of large DNA molecules with restriction enzymes (of which many are now available that cleave DNA at different nucleoside sequences) produces well-defined fragments of a manageable size. By using different restriction endonucleases, it is possible to produce overlapping fragments of DNA like those produced by cleavage of a protein with different proteases (see Section 21-10). Finally, the strands of the double-stranded fragment can be separated for sequence determination on single strands. For strand separation, gel electrophoresis may also be used. At high pH the double helix is destabilized, causing the strands to unwind and come apart.

The second obstacle faced by Maxam and Gilbert was that of obtaining cleavage specific for each type of base. Obtaining fragments of DNA containing several hundred bases or fewer was a necessary first step, but some way of cleaving these fragments specifically at each type of base was needed in order to determine the full sequence. Many chemists had tried to develop specific reactions for splitting selectively at A, G, C, or T, but none had been fully successful. Maxam and Gilbert were not fully successful either, but they did advance the chemistry of nonenzymatic cleavage to the point where it could be used in sequencing.

(a)

↓
G–C–C–G–C–A–G–C–T–G–G–C–C–G–C
C–G–G–C–G–T–C–G–A–C–C–G–G–C–G
↑

(b)

↓ ↓
. . . G–C–A–G–C–T–G–C–C–A–T–T–G–C–C–T–T–G–A–C–A–T–A–G–T–A–G–C–T–G–A–T–C . . . . .
. . . C–G–T–C–G–A–C–G–G–T–A–A–C–G–G–A–A–C–T–G–T–A–T–C–A–T–C–G–A–C–T–A–G . . . . .
↑ ↑

↓ Hind III

A–G–C–T–G–C–C–A–T–T–G–C–C–T–T–G–A–C–A–T–A–G–T
C–G–G–T–A–A–C–G–G–A–A–C–T–G–T–A–T–C–A–T–C–G–A

↓ Strand separation

A–G–C–T–G–C–C–A–T–T–G–C–C–T–T–G–A–C–A–T–A–G–T
+
C–G–G–T–A–A–C–G–G–A–A–C–T–G–T–A–T–C–A–T–C–G–A

FIGURE 25-18

Restriction endonuclease cleavage of DNA. The restriction endonuclease, Hind III, was isolated from the bacterium *Hemophilus influenzae.* **(a)** Hind III cleaves the DNA strand adjacent to the two A bases in the sequence A–G–C–T in each strand. **(b)** Treatment of the DNA molecule with Hind III splits out a *restriction fragment,* which can then be separated into single strands at high pH.

The principle involved is to establish different conditions that cleave each of the four types of bases from its deoxyribose unit. Removing the base from the sugar destabilizes the phosphate ester bonds at that sugar, and the polynucleotide chains can be readily split at those weakened points by heating in basic solution.

Two types of chemical reactions were employed to selectively weaken the base-to-sugar bonds. Methylation by dimethyl sulfate, $(CH_3)_2SO_4$, facilitates cleavage of purines from their sugars; treatment with hydrazine, $H_2N—NH_2$, facilitates cleavage of pyrimidines from their sugars. Normally the methylation reaction proceeds more rapidly at the purine guanidine than at adenine, so the reaction is partially specific (G > A). However, the specificity can be reversed by treatment with dilute acid, which cleaves methylated A more rapidly than methylated G. The pyrimidine cleavage is normally equally effective at C and T, but conditions were found (2 M NaCl) that prevent reaction of T, so that cleavage can be limited to C. Thus if a single-stranded fragment of DNA is isolated, it can be divided into four samples and subjected to the four cleavage procedures. The result is to split the chain in four ways:

Reaction 1: (A > G): Splitting mainly at A (with some splitting at G) by methylation followed by dilute-acid treatment before heating with base.

Reaction 2: (G > A): Splitting mainly at G (with some splitting at A) by methylation followed by heating with base.

Reaction 3: (C) Splitting only at C by treatment with hydrazine in 2 M NaCl followed by heating with base.

Reaction 4: (C + T): Splitting at both pyrimidines by treatment with hydrazine followed by heating with base.

Possible points of cleavage

↓ ↓ ↓ ↓ ↓ ↓

Strand: A–G–C–T–G–C–C–A–T–T–G–C–C–T–T–G–A–C–A–T–A–G–T

Products:
A–G
A–G–C–T–G
A–G–C–T–G–C
A–G–C–T–G–C–C–A–T–T–G
A–G–C–T–G–C–C–A–T–T–G–C
A–G–C–T–G–C–C–A–T–T–G–C–C–T–T–G–A
T–G
T–G–C
T–G–C–C–A–T–T–G
T–G–C–C–A–T–T–G–C
T–G–C–C–A–T–T–G–C–C–T–T–G–A
T–G–C–C–A–T–T–G–C–C–T–T–G–A–C–A–T–A–G–T
C–A–T–T–G
C–A–T–T–G–C
C–A–T–T–G–C–C–T–T–G–A
C–A–T–T–G–C–C–T–T–G–A–C–A–T–A–G–T
A–T–T–G
A–T–T–G–C
A–T–T–G–C–C–T–T–G–A
A–T–T–G–C–C–T–T–G–A–C–A–T–A–G–T
C–T–T–G–A
C–T–T–G–A–C–A–T–A–G–T
T–T–G–A
T–T–G–A–C–A–T–A–G–T

FIGURE 25-19

Products formed by partial splitting of a single-stranded DNA at C. The cleavage points are indicated by the arrows. All possible products that can arise from incomplete splitting are shown.

None of the reactions is complete, so for each set of conditions a mixture of products is formed, depending on how many potential cleavage points are actually attacked. For example, splitting one of the strands shown in Figure 25-18 at C (reaction 3) gives the family of products illustrated in Figure 25-19. When all four reaction conditions are used, four families of products are formed, each ending at the base adjacent to the chemical attack. However, before this procedure could be used to determine the sequence, a third factor had to be introduced by Maxam and Gilbert.

The third obstacle was that of limiting the number of cleavage products detected. Even though a length of DNA can be split at specific bases and the products separated by size on gel electrophoresis, the results would be too confusing to interpret without one additional step. The problem is that for each starting strand of DNA of a certain size—say, the 23 bases in Figure 25-19—any of the specific cleavage reactions could lead to many products of the same size.

For example, cleavage of the sequence of Figure 25-19 at C could lead to three different five-nucleotide units: A–G–C–T–G (from positions 1 to 5), A–T–T–G–C (from positions 8 to 12), and C–T–T–G–A (from positions 13 to 17). This happens because the reactions are incomplete, and not all links at C are destroyed by chemical attack. Because these different products all contribute to the band corresponding to five nucleotides, they cannot be distinguished by electrophoresis.

The key used by Maxam and Gilbert to overcome this problem was radioactive labeling of one end of the starting fragment. By following the radioactive label during electrophoresis, the portions of the DNA that retain the terminal nucleotide after chemical attack can be distinguished, and a unique set of fragments is produced.

For example, for splitting at C in the example of Figure 25-19, suppose we label the first nucleotide (A). Then only the first six products are tagged. Maxam and Gilbert labeled the terminal nucleotide with the radioactive phosphorus isotope ^{32}P by a method we shall not describe here. When the fragments of the labeled DNA are separated, the fragments that contain the terminal A nucleotide can be identified by their radioactivity. After electrophoresis, the bands on the gel are made visible as an **autoradiograph.** The radioactive emissions of the ^{32}P produce dark patches on photographic film at each band containing the labeled A. Only the end-labeled fragments contribute to the autoradiograph. The other (unlabeled) fragments are effectively invisible.

If end-labeled DNA from Figure 25-19 is subjected to the four reaction conditions, a series of oligonucleotides is produced. The relatively small products so formed are shown in Figure 25-20. The products of each of the four reactions are then placed in adjoining bands at the origin of a gel and subjected to electrophoresis. An autoradiograph of the gel will have the features represented by Figure 25-21.

A band appears on the autoradiograph under the origin of each of the four types of reactions if a susceptible base occurs at the corresponding position. For example, the band farthest from the origin (hence, the smallest fragment) appears in the second column. This means that it was produced by cleavage at G. (A faint band appears at the same level in the first column, because the G > A specificity is only partial.) Because there is no smaller fragment, this band must correspond to a single nucleotide, so and the unknown first base, indicated by X, must be followed by G. We now know this much of the structure: X–G.

The next closest bands to the origin (that is, the next

Labeled (*) end of the fragment:	*A–G–C–T–G–C–C–A···	
Products of reaction 1 (A > G):	*A–G–C–T–G–C–C	(and larger fragments)
Products of reaction 2 (G>A):	*A	
	*A–G–C–T	(and larger fragments)
Products of reaction 3 (C):	*A–G	
	*A–G–C–T–G	
	*A–G–C–T–G–C	(and larger fragments)
Products of reaction 4 (C + T):	*A–G	
	*A–G–C	
	*A–G–C–T–G	
	*A–G–C–T–G–C	(and larger fragments)

FIGURE 25-20

The four chemical cleavage reactions applied to a DNA fragment. Only the fragments that retain an end-label are considered, and only the shorter fragments are shown.

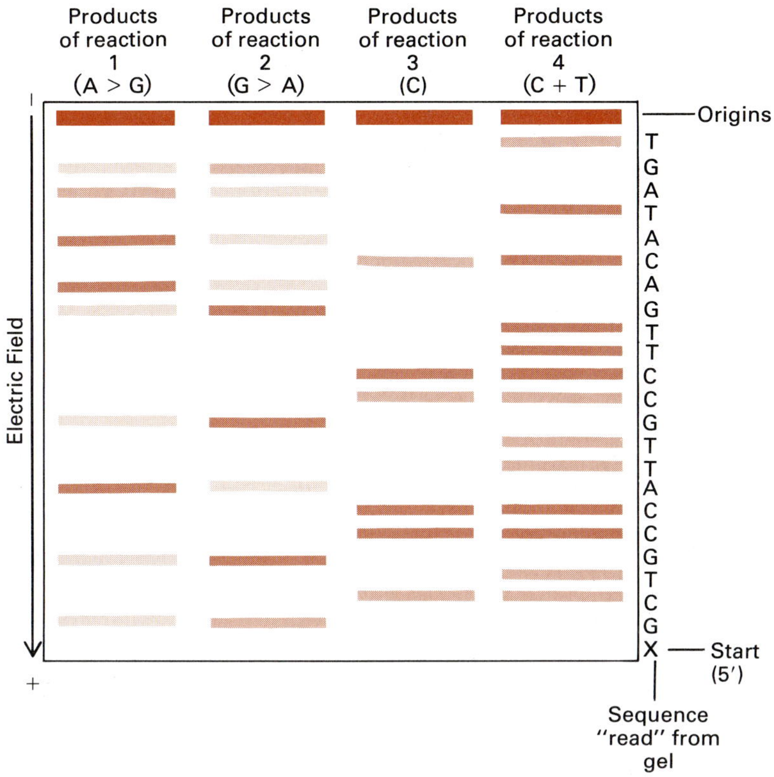

FIGURE 25-21

Representation of a gel used in DNA sequencing. The products of each of the four cleavage reactions are placed at the top of a channel (origin). Application of an electric field results in migration of the cleavage products in each channel according to size (the smallest migrates the farthest). The sequence of the original DNA fragment can be "read" from the gel.

smallest fragments, two nucleotides long) appear in the third and fourth columns. Because it appears in both columns, the fragment must arise from cleavage at C, indicating that the G is followed by a C: we now have X–G–C. At the next higher position is a band in the C + T column but not in the C column. The sequence must therefore contain a T at the next position: X–G–C–T. Continuing in this way, the autoradiograph represented can be "read" to give the sequence X–G–C–T–G–C–C–A · · ·.

In an actual autoradiograph a sequence with a hundred or more bases can be "read." The other strand of the original fragment (Figure 25-18) can also be sequenced in the same way. The assignment of the sequence can then be checked by finding if the sequence of the partner strand agrees with the predictions from A–T and G–C base pairing. Sequencing of fragments produced by other restriction endonucleases further confirms the accuracy of the structure and gives the identity of the first base that we have called X.

EXERCISE 25-5 For the following fragment of DNA, list all of the subfragments that would be produced by chemical cleavage at G. Which of these would be labeled by ^{32}P attached at the left (5′) end of the fragment? What labeled fragments would be produced by cleavage at T?

C–C–T–T–G–A–A–G–T–C–C–T–G–C–T

25-9
RECENT DEVELOPMENTS IN NUCLEIC ACID TECHNOLOGY

The ease and speed of the DNA-sequencing method has caused an "explosion" in DNA-sequence determinations. This new information has produced several major surprises in our understanding of DNA organization in genes and of transcription and translation to form proteins. The DNA for a particular gene can often be isolated if a purified mRNA for that gene is available.

For example, certain blood cells are rich in the mRNA for globins, the protein portion of hemoglobin. If restriction-endonuclease fragments for the total DNA of an organism such as the mouse are run in gel electrophoresis, regions of the gel containing the DNA for the globin genes can be identified by using radioactively labeled mRNA as a **probe** in a method called **hybridization.** The method is based on heating the DNA so that it "melts" and the base-paired strands come apart. When the radioactive mRNA probe is added and the mixture cooled, a DNA–RNA hybrid double helix forms between the mRNA and its complementary DNA. After the nonhybridizing mRNA is washed out of the gel, an autoradiograph will reveal the location of the labeled RNA and hence of the complementary DNA corresponding to the gene. With some additional steps to "clean up" and amplify the DNA, the necessary fragments of DNA were obtained to provide the full nucleotide sequence of the globin genes.

The results of sequencing the globin gene (and information emerging on other genes) provided one of the biggest surprises since the description of the double helix in 1953. When the nucleotide sequence was compared with the known amino acid sequence, the correspondence between nucleotide triplets and amino acids was just as expected for portions of the gene. However, these portions are interrupted by **intervening sequences** that have no relation to the protein. Two intervening sequences appear in the globin genes (Figure 25-22).

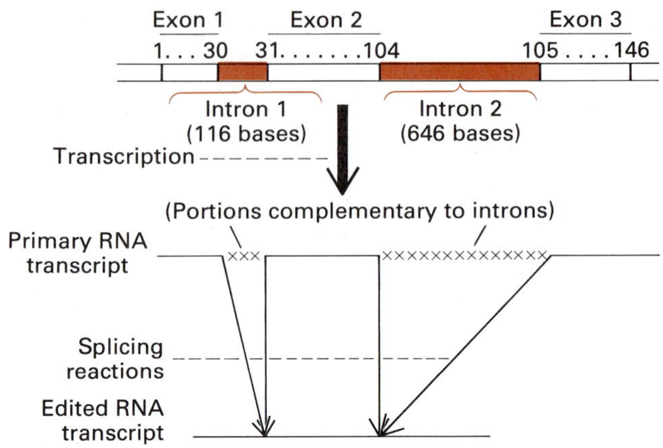

FIGURE 25-22

The β-globin gene. The entire base sequence of the gene contained in the DNA, with both exons and introns, is transcribed onto the primary RNA. From this primary RNA the intron sections are recognized and cut away so that the mRNA is "spliced" to form the "edited" transcript.

Intervening sequences have now been discovered in about a dozen other genes, including those for insulin and several other hormones, and those for antibodies. The terms **exons** and **introns** have been coined to denote the regions of the DNA that are **ex**pressed in the amino acid sequence (**ex**ons) and the **int**ervening regions that are not expressed (**int**rons).

The concept of "split" genes (genes split by introns) has a number of major implications. First, transcription is more complicated than originally formulated, because the mRNA must be processed before it is translated. Initially, a full RNA transcript of the gene containing both exons and introns is made. Then special "splicing" enzymes recognize the RNA sequences corresponding to introns and eliminate them. How the splicing enzymes recognize the RNA sequences corresponding to introns is not fully understood, but one clue is that all introns yet studied start with the sequence GT and end with the sequence AG. These sequences may be part of the signals recognized by the splicing enzymes, but there may also be other signals in the sequence that are not yet identified. The splicing process of the globin genes is summarized in Figure 25-22.

The significance of the intervening sequences is currently under active debate. They may have something to do with regulating which genes are expressed at various times in the life of a cell. Alternatively, introns may provide some advantage in evolution by permitting new proteins to be formed rapidly as exons are combined in new ways, rather than relying on single base changes and replacement of individual amino acids to alter gene sequences.

A surprise perhaps even greater than intervening sequences emerged when the complete nucleotide sequence of a bacterial virus was determined by Frederick Sanger and colleagues. This virus, designated ϕX174, contains a single-stranded DNA. Using an enzyme cleavage method, Sanger's group determined the sequence of all 5375 nucleotides of the DNA. The sequence reveals that several of the proteins specified by the DNA are coded by **overlapping genes.** In other words, some portions of the DNA code for more than one protein.

For example, in Figure 25-23, the gene designated B is contained entirely within the A gene. The start of the A gene is at a certain position on the DNA marked by an ATG initiation codon. Another ATG initiation

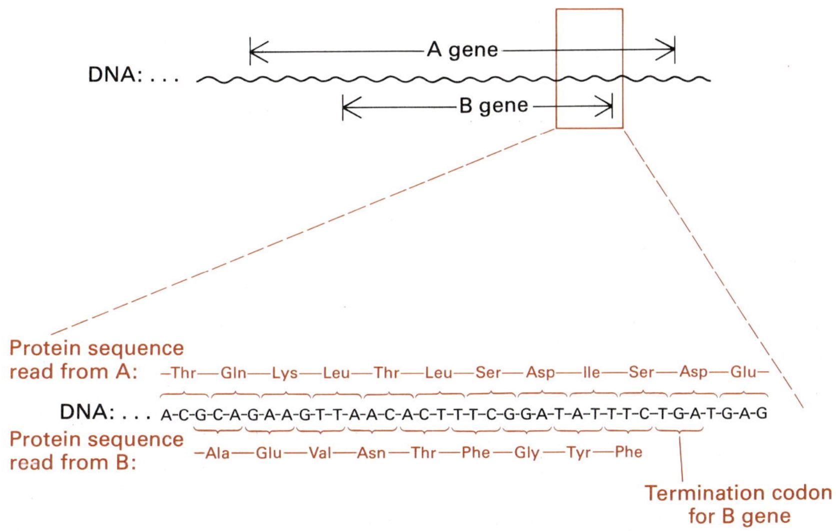

FIGURE 25-23

Overlapping genes. In the portion of the DNA represented, the A and B genes overlap as indicated in the upper portion of the figure. In the enlarged area of the genes represented in the lower portion of the figure, the base sequence of the DNA near the end of the B gene is given. The corresponding amino acid sequences for the A and B proteins, read in different reading frames, are presented above and below the DNA sequence. The base sequence includes a termination codon for the B gene, but because of the difference in reading frame, this region of the DNA specifies a continuing series of amino acid residues for the A chain.

codon that occurs 1091 bases later signals the start of the B gene. Because 1091 bases correspond to 363 codons plus two nucleotides ($1091/3 = 363\frac{2}{3}$), the B gene will be read differently than the very same sequence of nucleotides as part of the A gene. Thus overlapping genes can code for distinct proteins by using the DNA in different reading frames.

In Figure 25-23, the codon triplets coding for the A protein are indicated above the DNA, and codon triplets coding for the B protein are shown below the DNA. The A–C–G triplet at the beginning of the portion of the DNA codes for Thr in the A protein, but its third nucleotide starts the G–C–A codon that specifies Ala in the B protein. The two proteins are specified in this way until the T–G–A termination codon is reached for the B protein after a Phe residue. Because these three nucleotides are read as the end of the Ser and beginning of the Asp codons for the A protein, synthesis of the A protein continues until a termination codon in its reading frame is encountered some 31 codons later.

Thus the overlapping genes reveal a wholly unexpected "economy" in which the same DNA is used for two proteins by a shift in reading frames. This means that the two proteins must somehow have evolved together, because a mutation in one protein would also affect the other in the overlapping regions. In fact, certain mutant forms of the virus that have been isolated prove this point. For example, in one mutant form analyzed, the sixth nucleotide (G) in the portion summarized in Figure 25-23 is replaced by a C. As a result the A protein changes from Gln to His at that position, while the B protein changes from Glu to Gln. This degree of coordination suggests that compactness of the DNA may have had certain advantages for this virus that do not necessarily extend to other organisms. Nevertheless, this discovery of overlapping genes reveals a new dimension in the versatility of the DNA coding scheme that has evolved in life on earth.

EXERCISE 25-6 Another portion of the sequence coding for the A and B proteins is summarized below. Deduce the corresponding amino acid sequence. Note that the first three bases comprise the first codon of this portion for the A gene, but the first full codon of the B gene begins with the third base (G).

· · · –A–T–G–C–G–C–G–C–T–T–C–G–A–T–A–
A–A–A–A–T–G–A–T–T–G–G–C–G–T–A– · · ·

25-10 SYNTHETIC GENES

Current research directed toward synthesis of artificial DNA genes is generating considerable excitement. With this approach, it now seems likely that bacteria programmed with artificial genes will be used in the near future to produce insulin and other scarce and expensive hormones on a large scale. The bacteria are programmed in several different ways.

1. Fragments of the DNA of a higher organism (produced by restriction endonucleases, or even by simply breaking up the DNA mechanically) can be inserted directly into bacterial chromosomes.

2. Molecules of mRNA can be isolated, and the complementary DNA (called cDNA) can be formed

by the action of the enzyme reverse transcriptase from certain RNA tumor viruses. This enzyme forms DNA from the corresponding RNA.

3. The DNA sequence needed can be synthesized from individual nucleotides, using methods analogous to the ones used to synthesize proteins, in which certain functional groups are blocked, and catalysts are added to join units in the desired sequence.

In one recent approach used to synthesize the gene for insulin, 20 nucleotide triplets corresponding to each of the 20 natural amino acids were first synthesized. Then triplets were joined according to the sequence of amino acids in insulin. Overall, a great deal of activity can be expected in this area, and drug companies have already begun commercial ventures. Growing proteins like insulin in bacteria is a very attractive alternative to the current mode of production (isolating it in small amounts from the pancreas of slaughtered pigs or cattle).

Very recently, scientists have reported production of **interferon** in bacteria. This protein is believed to have anticancer properties, but it is present in such small quantities in humans that it had been impossible to study its properties adequately. Now there is some hope that its properties can be checked in experiments with cancer patients, and some dramatic medical outcome may result.

KEY TERMS

Deoxyribonucleic acid (DNA) is the genetic material; molecules of DNA store the information that is carried in the *chromosomes* of cells. The DNA molecules are divided into functional units called *genes.*

Ribonucleic acid (RNA) is involved in the translation and transcription of the information stored in DNA. There are three major classes of RNA: transfer RNA (tRNA), messenger RNA (mRNA), and ribosomal RNA (rRNA).

Nucleotides are the building blocks of the nucleic acids DNA and RNA. A nucleotide is composed of a *heterocyclic base* (related to *purine* or *pyrimidine*), a five-carbon sugar (ribose or deoxyribose), and a phosphate group. Compounds containing only a heterocyclic base and a sugar are called *nucleosides.*

A *phosphodiester linkage* is the bridge provided by phosphate groups between adjacent nucleotide units in nucleic acids. The phosphate group is bonded to C-3′ of one nucleotide and C-5′ of the neighboring nucleotide.

The structure of DNA is a *double helix,* with two antiparallel strands joined by base pairing through hydrogen bonds; the strands complete one turn every ten base pairs along the helix.

Base pairs are the specific interactions between bases that are found in the DNA double helix (A with T, and G with C) or the related interactions in RNA (A with U, and G with C).

Replication is the process by which DNA is duplicated during cell division. The two strands separate to give a *replicating fork,* and each strand provides a template for the synthesis of the missing strand (*semiconservative* replication). *DNA polymerases* are the enzymes responsible for the replication of DNA.

The *genetic code* describes the particular amino acid specified by each of the 64 possible base triplets (*codons*).

RNA polymerases are the enzymes responsible for the synthesis of RNA during the transcription process; the synthesis begins at specific sites on the DNA called *promoters*, and it can be blocked by proteins called *repressors*.

Transcription is the process of RNA synthesis in which a sequence of bases in a region of DNA specifies the sequence of bases in the RNA. *Messenger RNA* (*mRNA*) is the class of RNA that contains a transcript of the base sequence of a gene in DNA; mRNA programs ribosomes to synthesize proteins.

Translation is the process in which base sequences of RNA are used to specify the amino acid sequences of proteins. *Transfer RNA* (*tRNA*) molecules are charged with specific amino acids; they bind to ribosomes in response to the presence of specific codons of mRNA, which are recognized by the *anticodons* of the tRNA.

The *central dogma* describes the flow of information within cells from DNA to RNA to proteins.

Ribosomes are the particles on which protein synthesis takes place; they are composed of many proteins and several molecules of *ribosomal RNA* (*rRNA*).

Nucleases are enzymes that cleave nucleic acids; an important class of such enzymes are the *restriction endonucleases*, which cleave double-stranded DNA at specific base sequences.

Electrophoresis is the technique for the separation of molecules on the basis of their migration in an electric field. *Autoradiographs* are the images that radioactively labeled molecules produce on photographic film by the emissions from the radioactive isotopes. *Hybridization* is a technique of forming mixed RNA–DNA molecules from DNA that has been melted and its complementary RNA.

Intervening sequences are stretches of DNA found between coding regions of genes in higher organisms. These intervening sequences are also called *introns*, and the regions that code for proteins are called *exons*. *Overlapping genes* are regions of DNA that code for two distinct proteins by decoding the DNA in different triplet sequences.

CHAPTER SUMMARY

Nucleic acids play fundamental roles in the storage and replication of the genetic information and in its expression as sequences of amino acids in proteins. *Nucleic acids* are polymers made up of units called *nucleotides*. Three distinct chemical entities occur in a nucleotide: a heterocyclic base related to purine or pyrimidine, a five-carbon sugar (ribose or deoxyribose), and phosphate. The base is attached to the sugar by an *N*-glycosidic bond, and the sugar units are esterified to the phosphate. Each phosphate can participate in a second ester linkage, and successive nucleotides are linked in the polymers through phosphodiester bridges. The polymers are named according to the sugar units present: *ribonucleic acid* (RNA) if the sugar unit is ribose, or *deoxyribonucleic acid* (DNA) if the sugar unit is deoxyribose. Four types of heterocyclic bases are found in DNA: two *purines,* adenine (A) and guanine (G); and two *pyrimidines,* cytosine (C) and thymine (T). In RNA the bases A, G, and C also occur, but T is replaced by uracil (U). Units involving only a base and a sugar are called *nucleosides,* and they may occur as monophosphates, diphosphates, or triphosphates. The bonds between the phosphates in these compounds are anhydrides with certain high-energy properties that are important in bioenergetics.

DNA is the genetic material. It exists in cells in a two-stranded form known as the *double helix*. The two strands are stabilized by hydrogen bonds between

A–T base pairs and G–C base pairs. The composition of bases in DNA always reflects this pairing pattern: A = T, and G = C. Replication of DNA during cell division also relies on this pairing scheme.

The sequence of nucleotides in DNA is used by cells to specify the sequence of amino acids in proteins. The flow of information from DNA to proteins is coordinated by three types of molecules of RNA. Copies of the nucleotide sequence in DNA are made in an RNA form called *messenger RNA* (mRNA) in the process known as *transcription*. The next stage of protein synthesis involves molecules called *transfer RNA* (tRNA). A different class of tRNA molecules exists for each of the 20 amino acids, and specific enzymes link the amino acids with their corresponding tRNA molecules. Molecules of tRNA with their attached amino acids are aligned along the sequence of bases of the mRNA according to a three-base (triplet) *genetic code*. For each amino acid, there are one or more specific base triplets on the mRNA (*codons*) and a complementary set of bases on the tRNA (*anticodons*) that form specific complexes. These complexes arise from base pairs related to the A–T and G–C base pairs of DNA, with A–U base pairs in the case of mRNA–tRNA interactions. The actual alignment of mRNA with tRNA units bearing amino acids in order to synthesize proteins is called *translation* and takes place on ribosome particles that contain a third class of RNA, *ribosomal RNA* (rRNA).

Much of the understanding of nucleic acids and their roles in the genetic code and protein synthesis is based on information obtained about the sequences of bases in DNA and RNA. A chemical cleavage method has recently been developed for determining the sequence of bases in DNA rapidly and conveniently. Studies of DNA sequences have produced startling new discoveries. One recent discovery is that the regions of DNA coding for a single protein can be interrupted by *intervening sequences*. These sequences appear in the primary RNA transcript, but they are excised in the final mRNA used for translation. The role of these intervening sequences is still uncertain. Another recent discovery is that two distinct genes can *overlap* on the same region of DNA by using different reading frames for decoding the base sequence during protein synthesis. Progress in the chemical synthesis of genes is occurring at a rapid pace and may soon lead to the commercial production of insulin and other hormones in bacteria programmed with artificial genes for these substances.

PROBLEMS

25-1. For each pair on the left, choose the term that corresponds to the definition on the right.

(a) Nucleoside or nucleotide	(a) Composed of a heterocyclic base, a sugar, and a phosphate.
(b) Thymine or uracil	(b) Found in DNA but not in RNA
(c) Adenine or cytosine	(c) A purine derivative
(d) Guanine or guanosine	(d) Contains an *N*-glycosidic bond
(e) A–C or A–T	(e) One of the base pairs found in DNA
(f) Repressor or promoter	(f) A protein that regulates transcription
(g) AUG or UAG	(g) One of the termination codons

(h) A site or D site	(h) Where aminoacyl-tRNA molecules first bind to the ribosome
(i) I–G or I–C	(i) One of the extra base pairs that occur in codon–anticodon interactions because of "wobble"
(j) Cleavage at A or cleavage at T	(j) Where cleavage is enhanced in the Maxam–Gilbert DNA-sequencing method by dilute-acid treatment
(k) Introns or exons	(k) Always begin with G–T and end with A–G
(l) Reverse transcriptase or RNA polymerase	(l) Can catalyze synthesis of RNA from DNA templates

25-2. The first step in linking alanine with its tRNA is formation of the aminoacyl-AMP derivative of alanine. Draw the structure of this compound.

25-3. Name the following compound, and give its three-letter abbreviation:

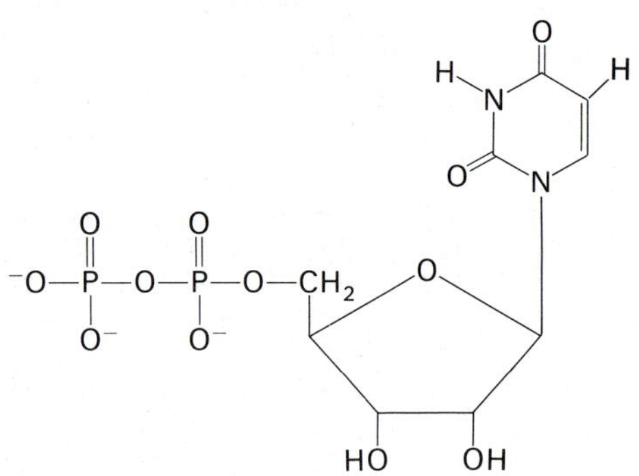

25-4. A portion of a nucleic acid chain contains the base sequence –A–C–T–. Draw the structure for this region of the nucleic acid.

25-5. The following is one-half of a base pair. Draw the missing half and name the two bases involved.

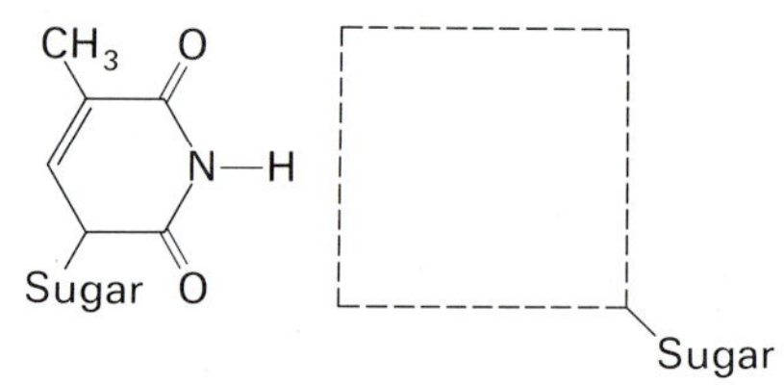

25-6. For the sequence

–C–T–A–G–G–T–C–A–T–T–G–A–G–C–T–

provide (a) the complementary strand expected in DNA; and (b) the complementary strand that would occur in mRNA.

25-7. For the following short DNA sequence, write the corresponding amino acid sequence:

–A–T–G–G–T–A–A–C–T–C–T–A–T–G–C–T–
A–A–C–C–T–T–C–G–C–C–A–A–C–T–G–A–

Could this sequence be harboring two short overlapping genes?

25-8. In the course of a Maxam–Gilbert DNA-sequence determination, the following end-labeled oligonucleotide is subjected to treatment with hydrazine in 2 M NaCl. What is the smallest end-labeled fragment that will be produced?

*A–G–T–A–A–G–C–T–A–T–G–T–A–A–G

Beef heart mitochondria in cross section. Energy is stored in mitochondria in the form of ATP during cellular respiration. Transmission electron micrograph, ×80,000.

26

Bioenergetics and Metabolism

In a series of enzymatic reactions, cells can gradually extract the energy stored in carbohydrates, fats, and proteins. This energy is used to form ATP, which in turn is used to do work or to synthesize needed compounds. In plant cells, light energy can be used directly to synthesize ATP.

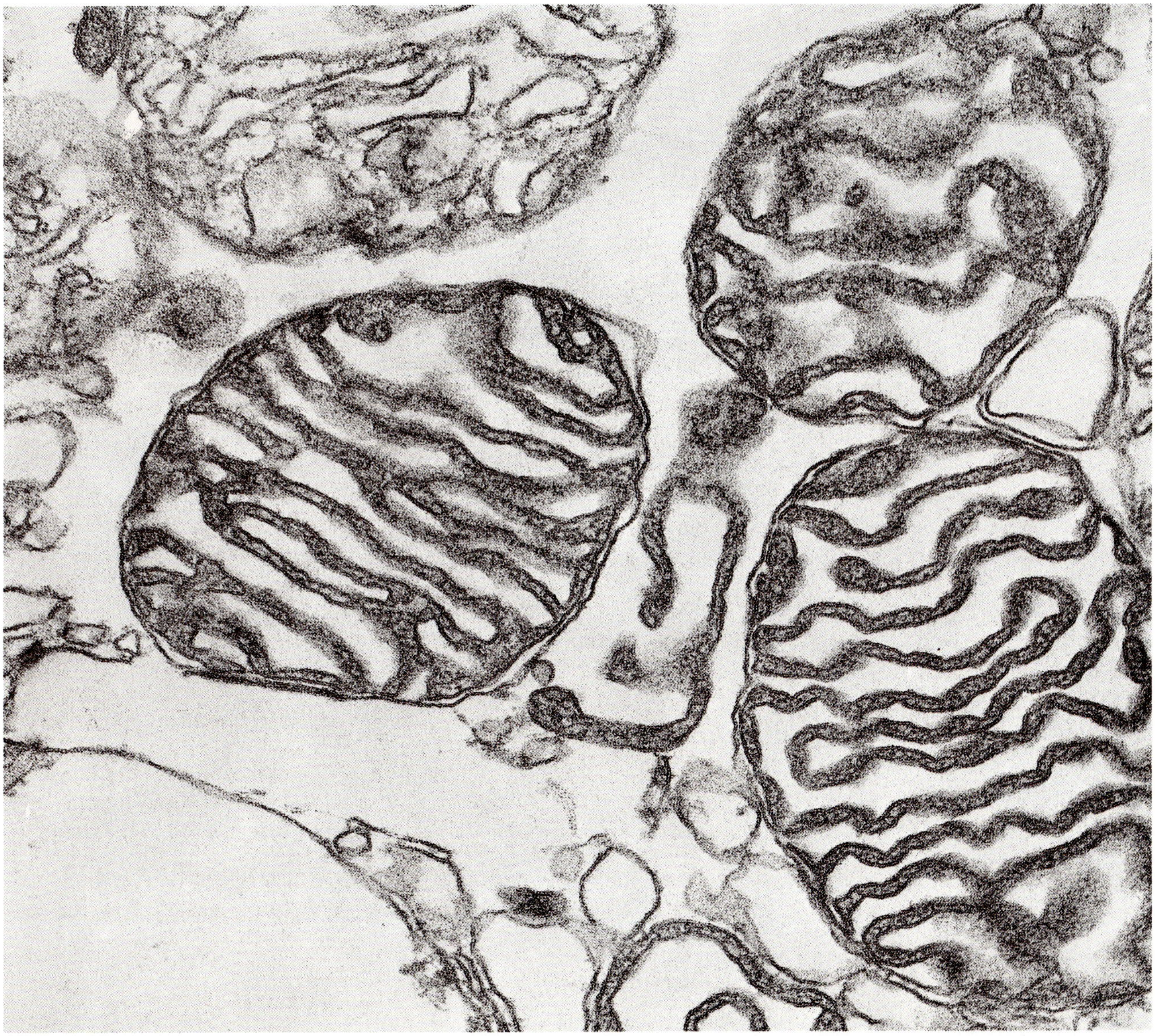

26-1 THE OVERALL PATTERN OF BIOENERGETICS AND METABOLISM

Thus far we have discussed the various biochemical components of living organisms as individual chemical entities. In this chapter, we see how the biochemical processes of cells are used to metabolize these compounds, extracting energy when it is needed or using energy to synthesize new molecules required for growth. On the most general level, the processes of bioenergetics can be viewed as the accumulation and exploitation of **reducing power.**

The business of accumulating reducing power is largely the work of plants. In the process of photosynthesis, green plants convert CO_2 and H_2O from the environment into glucose and oxygen:

In plants:

$$\underset{\text{sun}}{\overset{\text{Energy}}{\text{from}}} + 6CO_2 + 6H_2O \longrightarrow \underset{\text{glucose}}{C_6H_{12}O_6} + 6O_2$$

The carbon is reduced from the 4+ oxidation state in CO_2 to the zero oxidation state in $C_6H_{12}O_6$. (See Chapter 12 for a review of oxidation and reduction, and of oxidation numbers.) In effect, plants store energy from sunlight in the form of chemical energy in the relatively "reduced molecule" $C_6H_{12}O_6$. Such a reduced, energy-rich molecule is a valuable biological commodity. It can react with oxygen to release energy for the various energy-requiring tasks of living organisms: the synthesis of biomolecules such as DNA and proteins, the contraction of muscle, and the transport of ions and molecules across the cell membranes.

Glucose is the main carbohydrate of metabolic processes. When animals eat glucose, they reverse the process of plants, oxidizing glucose back to CO_2 and H_2O, and releasing energy:

In animals:

$$C_6H_{12}O_6 + 6O_2 \longrightarrow 6CO_2 + 6H_2O + \text{Energy}$$

The same total amount of energy is given off in the biological oxidation of the carbon-containing fuel molecules in cells as in the combustion of those fuel molecules outside the cell. We shall see, however, that in cells the energy is released in small packets. This permits a more efficient use of the energy by the cell than would be possible if the glucose were oxidized to CO_2 in one step.

The various steps of glucose formation by plants and oxidation by animals involve ATP as a form of "metabolic cash" that is a ready store of energy. Thus cellular metabolism revolves around the formation and use of ATP. The special role for this nucleoside triphosphate (one of the four principal nucleoside triphosphates discussed in Chapter 25) results from the unique properties of phosphate anhydride bonds. The high density of negative charges on the oxygen atoms of the three phosphate groups introduces repulsive forces that make hydrolysis of the phosphate anhydride bonds very favorable. The charge distribution on ATP can be represented most accurately in terms of partial charges:

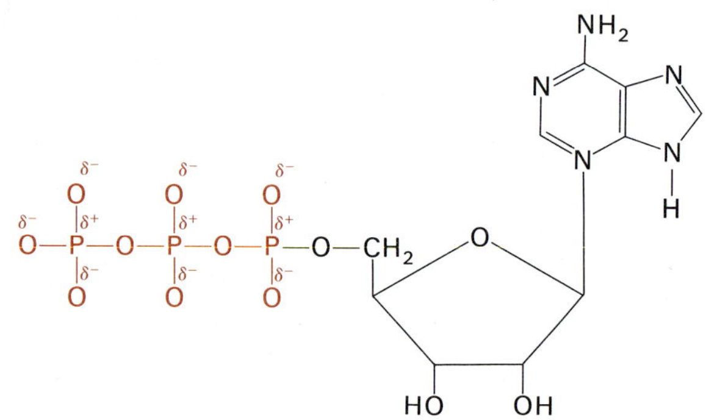

adenosine triphosphate (ATP)

Hydrolysis is favored also because the free phosphate ion (but not the phosphate tied up in anhydride bonds) is stabilized by being a resonance hybrid:

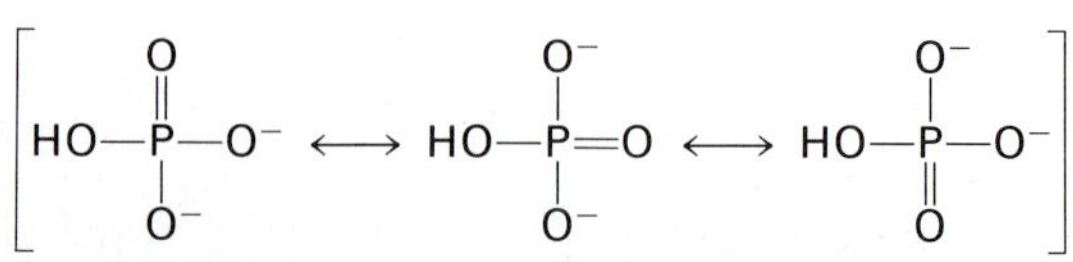

When a phosphate anhydride bond is hydrolyzed, a large amount of energy is released. Thus the phosphate anhydride bonds can be considered "high-energy" bonds, sometimes symbolized with a squiggle (∼). The structure of ATP can be represented as:

$$\text{AMP} \sim \text{P} \sim \text{P}$$

The squiggle emphasizes the fact that a large amount of energy is stored in each of the two phosphate anhydride bonds, and P represents a phosphate group, not just a phosphorus atom. Phosphate groups (HPO_4^{2-}) not part of an organic compound are called **inorganic phosphate,** abbreviated P_i; the pyrophosphate group ($HP_2O_7^{3-}$) is abbreviated PP_i. (The term "high-energy bond" as it is used here means a chemical bond that is relatively weak and easily broken. This should not be confused with the term "high bond energy." As we saw in Chapter 7, a high bond energy means that the chemical bond is strong and that a large amount of energy is required to break it.)

EXERCISE 26-1 Write an equation for the hydrolysis of ATP to yield ADP and P_i. Write the complete structures of all the reactants and products. (If necessary, refer to Section 19-12.)

One of the main roles of ATP in metabolic reactions involves the **coupling** of ATP hydrolysis with less favorable reactions, thus providing "driving" energy for the less favorable reactions. For example, in the metabolism of glucose, one of the first steps is the production of a phosphorylated derivative of glucose, glucose 6-phosphate. Formation of this compound from glucose and phosphate directly requires energy and thus is an unfavorable reaction:

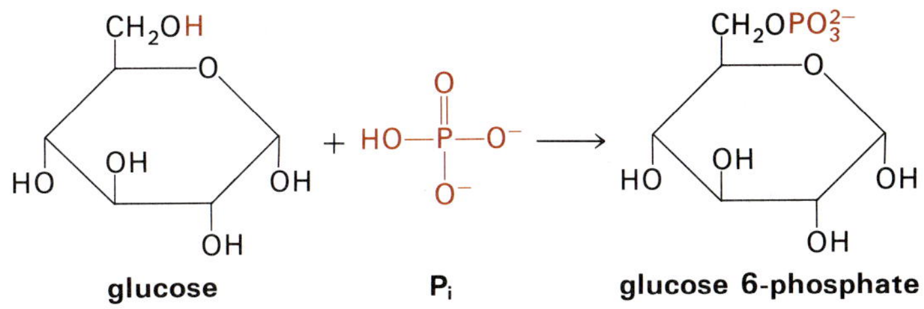

$$\Delta G^{0'} = +3.3 \text{ kcal/mole}$$

(See Section 10-9 for a review of ΔG and of the conditions for spontaneous reaction.) The free energy value is given here in terms of $\Delta G^{0'}$, the standard free energy change at pH 7.0, rather than ΔG^0, the standard free energy change at $[H^+] = 1.0$ M. Because $\Delta G^{0'}$ for the reaction is positive, virtually no glucose 6-phosphate forms spontaneously from glucose and phosphate. On the other hand, hydrolysis of ATP to form ADP and P_i is favored by a $\Delta G^{0'}$ value of -7.3 kcal/mole:

$$\text{ATP} \longrightarrow \text{ADP} + \text{P}_i \qquad \Delta G^{0'} = -7.3 \text{ kcal/mole}$$

So, if the phosphate group is provided by ATP, then the favorable ATP-hydrolysis reaction is effectively coupled to the unfavorable glucose-phosphorylation reaction, and production of glucose 6-phosphate occurs to an appreciable extent:

	$\Delta G^{0'}$ (kcal/mole)
Unfavorable reaction:	
Glucose + P_i → Glucose 6-P	+3.3
Very favorable reaction:	
ATP → ADP + P_i	−7.3
Net reaction (favorable):	
Glucose + ATP → Glucose 6-P + ADP	−4.0

As shown, the $\Delta G^{0'}$ values sum to give the favorable $\Delta G^{0'}$ value of -4.0 kcal/mole for the overall reaction.

In addition to nucleoside triphosphates, several other "high-energy" compounds are important in biochemical processes. Mixed anhydrides can form be-

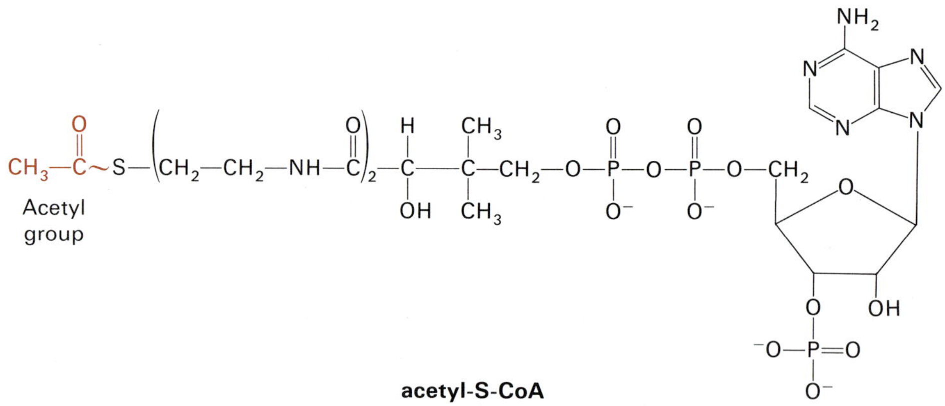

acetyl-S-CoA

tween phosphoric acid and carboxylic acid groups. Such compounds are called **acyl phosphates.** One class of acyl phosphates includes the aminoacyl-AMP derivatives formed in the first step of the reactions catalyzed by the aminoacyl-tRNA synthetases (see Chapter 25). Another important acyl phosphate is 1,3-diphosphoglycerate, the first high-energy intermediate produced in glucose metabolism:

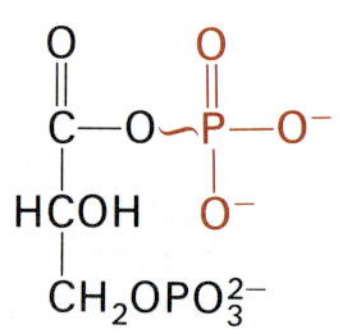

1,3-diphosphoglycerate

Another high-energy intermediate of glucose metabolism is phosphoenolpyruvate, which has high-energy properties because of a phosphate group that holds the structure in the enolic form. Hydrolysis of the phosphate permits the product to assume the more stable keto form:

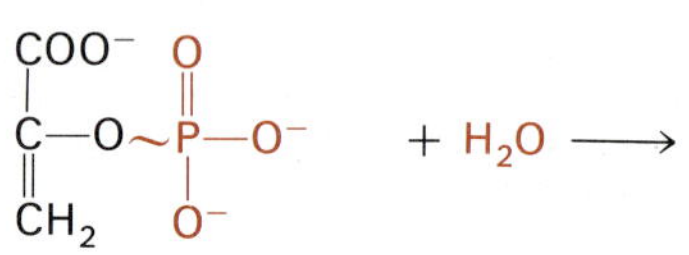

phosphoenolpyruvate

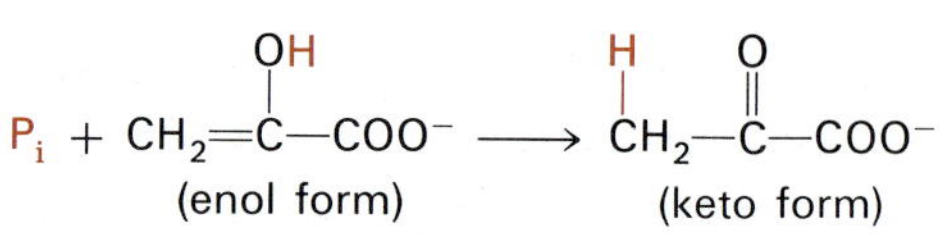

pyruvate

Creatine phosphate, a high-energy storage compound of muscle, has a high-energy bond of the guanidinium phosphate type (the guanidinium group is the functional group of the arginine side chain):

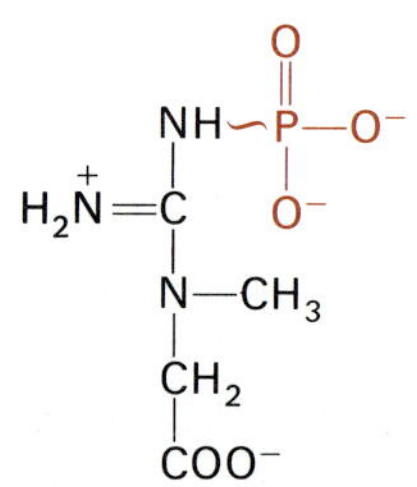

creatine phosphate

Thioesters, such as the acetylated form of coenzyme A (Acetyl-S-CoA), also have a highly favorable free energy of hydrolysis (structure at top of page). We shall see in Section 26-5 that acetyl-S-CoA is a reactive molecule that provides acetyl groups in the synthesis of the long hydrocarbon chains of fatty acids.

Table 26-1 gives the reactions and the free energies of hydrolysis of some phosphorylated compounds. Note that some of these compounds, such as phosphoenolpyruvate and creatine phosphate, have a greater tendency to hydrolyze (a more negative $\Delta G^{0'}$) than does ATP. These molecules spontaneously transfer phosphate groups to ADP to yield ATP. We shall see that this is one way ATP is produced in the catabolism of sugars. In contrast, ATP spontaneously transfers phosphate groups to molecules such as glucose and

TABLE 26-1
Standard free energies of hydrolysis of some phosphorylated compounds

Compound	Reaction	$\Delta G^{0'}$ kcal/mole	$\Delta G^{0'}$ kJ/mole
Phosphoenolpyruvate (PEP)	PEP $\xrightleftharpoons{H_2O}$ Pyruvate + P_i	−14.8	−61.9
Creatine phosphate	Creatine ~P $\xrightleftharpoons{H_2O}$ creatine + P_i	−10.3	−43.1
Acyl phosphate	1,3-DPG $\xrightleftharpoons{H_2O}$ 3-phosphoglycerate + P_i	−10.3	−43.1
Pyrophosphate	PP_i $\xrightleftharpoons{H_2O}$ 2 P_i	− 8.0	−33.4
ATP	ATP $\xrightleftharpoons{H_2O}$ ADP + P_i	− 7.3	−30.5
Glucose 1-phosphate	Glucose 1-P $\xrightleftharpoons{H_2O}$ glucose + P_i	− 5.0	−20.9
Fructose 6-phosphate	Fructose 6-P $\xrightleftharpoons{H_2O}$ fructose + P_i	− 3.8	−15.9
Glucose 6-phosphate	Glucose 6-P $\xrightleftharpoons{H_2O}$ glucose + P_i	− 3.3	−13.8
Glycerol 1-phosphate	Glycerol 1-phosphate $\xrightleftharpoons{H_2O}$ glycerol + P_i	− 2.2	−20.2

fructose. The phosphates of these molecules have less tendency to hydrolyze (less-negative $\Delta G^{0'}$) than does ATP. The intermediate position of ATP in the table enables ATP to function very effectively as a carrier of phosphate groups in biochemical reactions. Figure 26-1 summarizes the ATP cycle that functions in cells. The ATP provides chemical energy for the energy-requiring tasks of cells; it is then regenerated from ADP and P_i using energy released by the degradation of glucose.

Oxidation–reduction reactions and reactions involving ATP are the basis of bioenergetics. In plants, light is used to form ATP, which is then used to drive the synthesis of glucose and other fuel molecules. In the process, hydrogen atoms are removed from water, freeing O_2. Animal cells use the oxygen released by plants to oxidize glucose to generate energy and to provide chemical building blocks for the synthesis of other molecules. Thus oxygen is an essential component of the global metabolic economy.

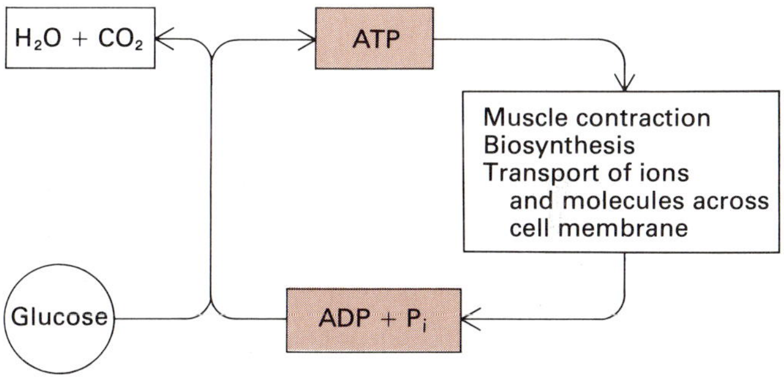

FIGURE 26-1
The ATP cycle. Energy released in the hydrolysis of ATP is used to do work. Energy released by the oxidation of glucose is used to regenerate ATP.

Before discussing in detail how energy is extracted from fuel molecules and stored in ATP, we shall take a brief overview of the main steps in the process: glycolysis, the citric acid cycle, and oxidative phosphorylation. Most cells have a simple glucose-metabolizing system that does not require oxygen. This system, which splits six-carbon glucose molecules into three-carbon units, is named **glycolysis** for this glucose-splitting function. Glycolysis may be a biochemical "fossil," left over from an early period of life when there was little or no oxygen in the atmosphere. However, it is still an important use of glucose, especially in muscle. There it is often necessary to generate large amounts of energy faster than oxygen can be supplied to the muscle. In addition, the anaerobic reactions of glycolysis are the first step in the metabolism of glucose in aerobic cells. Products of glycolysis are further processed in the aerobic phase of metabolism to CO_2 and H_2O. It is in the aerobic phase that most of the energy is extracted.

Glycolysis accounts for only a small fraction of the total ATP produced in the aerobic catabolism of glucose. In aerobic cells, the products of glycolysis are further degraded, first to acetyl-S-CoA, and finally (via a series of energy-yielding reactions) to CO_2 and H_2O. In the **citric acid cycle,** acetyl groups from acetyl-S-CoA are oxidized to CO_2 (each acetyl group produces $2CO_2$). Four pairs of electrons are transferred to the coenzymes NAD^+ (nicotinamide adenine dinucleotide) and FAD (flavin adenine dinucleotide) for each acetyl group that is oxidized.

The reduced coenzymes $NADH + H^+$ and $FADH_2$ produced in the citric acid cycle are oxidized in a series of oxidation–reduction reactions that are coupled with the formation of ATP. In this series of reactions, which is called **oxidative phosphorylation,** or the **respiratory chain,** or the **electron-transport chain,** electrons are ultimately transferred to molecular oxygen (which is reduced to H_2O). Most of the ATP produced by the oxidation of glucose is formed in this stage of metabolism.

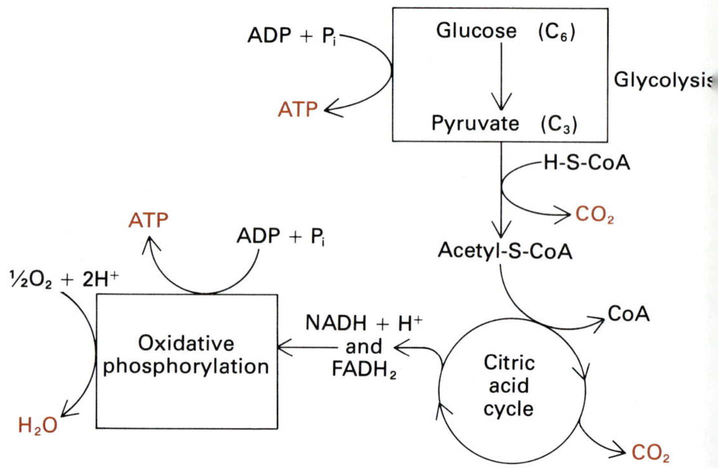

FIGURE 26-2

Main stages in the oxidation of glucose: glycolysis, the citric acid cycle, and oxidative phosphorylation.

The main stages in the oxidation of glucose to CO_2 and H_2O are summarized in Figure 26-2. We shall now consider each stage in greater detail, starting with glycolysis.

26-2 GLYCOLYSIS

The overall result of glycolysis in anaerobic cells is to convert each molecule of glucose processed into two molecules of lactate. Everyone has personal experience with lactate; it is the compound that causes muscle fatigue and cramps during vigorous exercise. Along with production of lactate, glycolysis achieves the net synthesis of two molecules of ATP from ADP and P_i. Thus glycolysis may be summarized as

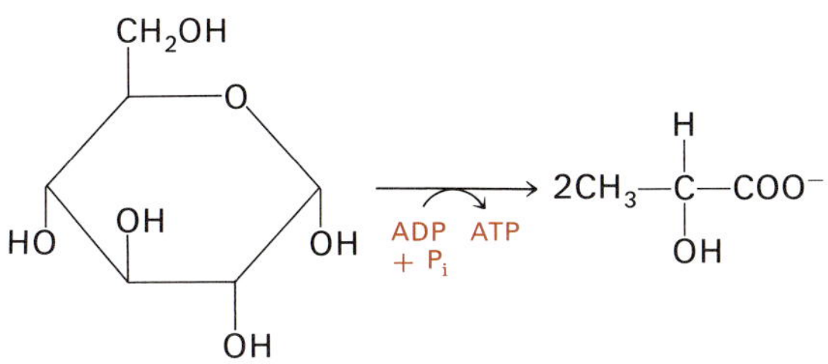

glucose **lactate**

The glucose itself may be obtained directly from plant substances, from the breakdown of storage polymers such as starch or glycogen, or by the conversion of other sugars such as galactose. Glycolysis requires 11 distinct reactions; they can be divided into three phases:

1. an "investment" phase, in which two molecules of ATP are consumed in phosphorylation reactions that prepare the hexose units for splitting into triose units;

2. a "recovery" phase, in which energy is extracted from the first high-energy compound (1,3-diphosphoglycerate) to pay back the ATP consumed in the first phase;

3. a "profit" phase, in which energy is extracted from the second high-energy compound (phosphoenolpyruvate) to provide the net gain from glycolysis of 2ATP.

In the first (investment) phase of glycolysis, glucose is phosphorylated by ATP to form glucose 6-phosphate and ADP. We have discussed this reaction in our explanation of coupling of ATP hydrolysis to unfavorable phosphorylation reactions. The phosphorylation reaction is followed by an isomerization reaction in which glucose 6-phosphate is converted to fructose 6-phosphate. A second phosphorylation reaction yields the compound that will be split into triose units, fructose 1,6-diphosphate. (As usual, we shall not balance the equation).

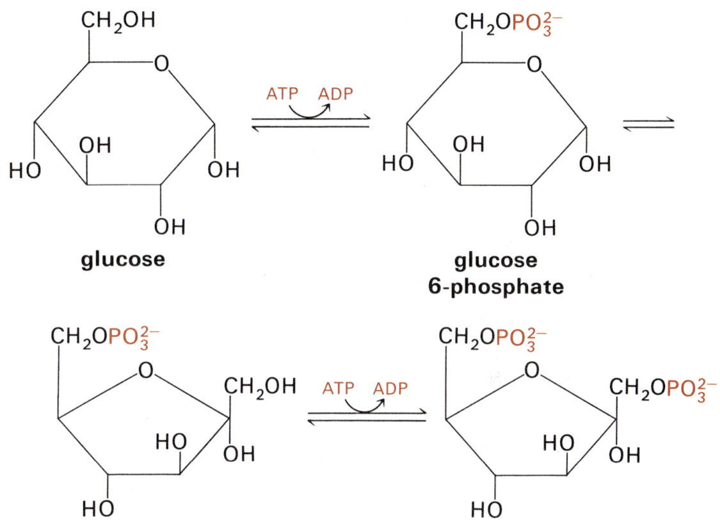

glucose **glucose 6-phosphate** **fructose 6-phosphate** **fructose 1,6-diphosphate**

In the last reaction of this first preparatory phase of glycolysis, fructose 1,6-diphosphate is split into two distinct triose units, glyceraldehyde 3-phosphate and dihydroxyacetone phosphate. The origin of the atoms in these two compounds is most readily visualized in terms of the open-chain representation:

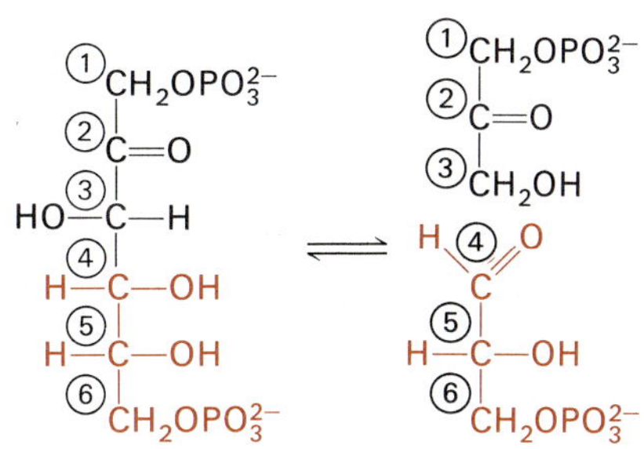

fructose 1,6-diphosphate

dihydroxyacetone phosphate

glyceraldehyde 3-phosphate

The second (recovery) phase of glycolysis begins with the conversion of glyceraldehyde 3-phosphate to the first high-energy compound of glycolysis, 1,3-diphosphoglycerate. For each molecule of fructose 1,6-diphosphate split, *two* molecules of 1,3-diphosphoglycerate are formed from glyceraldehyde 3-phosphate, because dihydroxyacetone phosphate also is converted to glyceraldehyde 3-phosphate in an isomerization reaction:

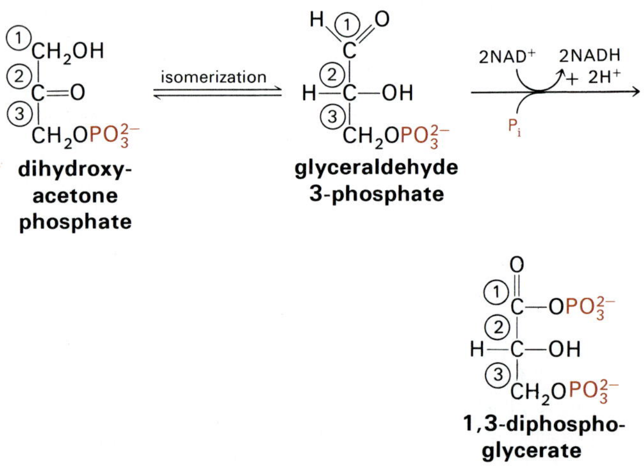

The reaction to form 1,3-diphosphoglycerate is complex, involving oxidation of the aldehyde group of glyceraldehyde 3-phosphate to a carboxylic acid and formation of an acyl phosphate (a "high-energy" bond). There is a simultaneous reduction of the principal oxidation–reduction coenzyme, nicotinamide adenine dinucleotide (NAD^+). The structures of NAD^+ and NADH and a general formula for the reaction are given in Figure 26-3. (We have seen examples of oxidation–reduction reactions involving NAD^+ in Sections 17-7 and 18-8.)

This second phase of glycolysis is completed when the phosphate group introduced in the formation of 1,3-diphosphoglycerate is donated in the next reaction to ADP, forming ATP and 3-phosphoglycerate:

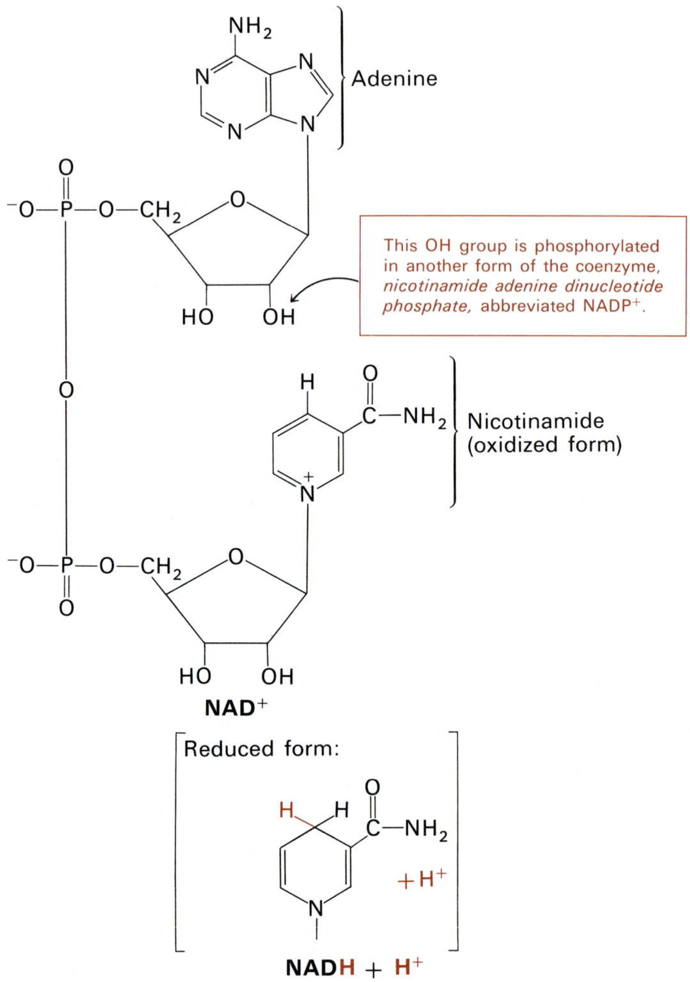

FIGURE 26-3

Nicotinamide adenine dinucleotide (NAD^+). A typical NAD^+ reaction is

$$\text{Substrate–}H_2 + NAD^+ \longrightarrow \text{Substrate} + NADH + H^+$$

NAD^+ reactions always involve removal of two H atoms from the reduced substrate, although only one of the H atoms (and two electrons) appears in the structure of NADH, the reduced form of NAD^+. Therefore, a free proton (H^+) is always produced together with the NADH.

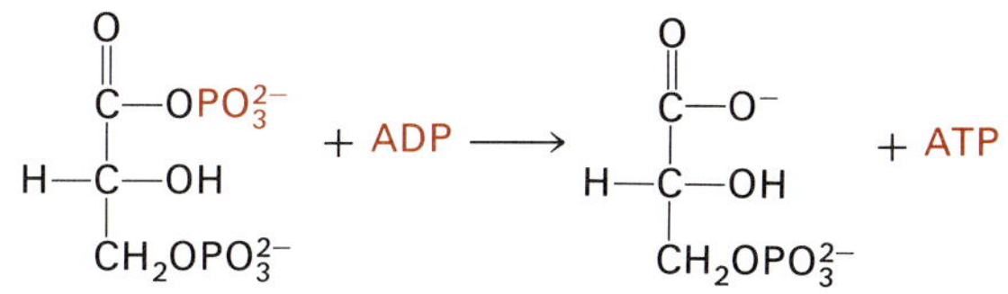

1,3-diphospho-glycerate **3-phospho-glycerate**

Because there are *two* molecules of 1,3-diphosphoglycerate for each molecule of glucose that enters glycolysis, *two* molecules of ATP are formed in this reaction per original glucose molecule. Thus the ATP formed here exactly "repays" the two molecules of ATP consumed in the initial phosphorylation reactions leading to fructose 1,6-diphosphate. Therefore, with respect to ATP consumption or production, glycolysis is even at this point.

The third and final (profit) phase of glycolysis begins with formation of the high-energy compound phosphoenolpyruvate from 3-phosphoglycerate in a transformation that takes place via the intermediate compound 2-phosphoglycerate:

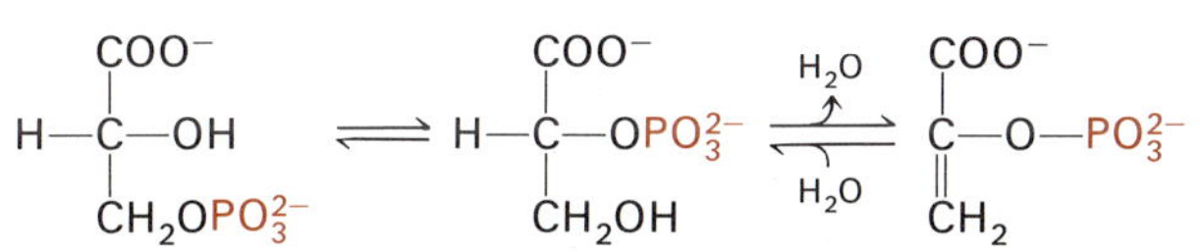

3-phospho-glycerate **2-phospho-glycerate** **phospho-enolpyruvate**

In the next step, phosphoenolpyruvate reacts with ADP to form pyruvate and ATP:

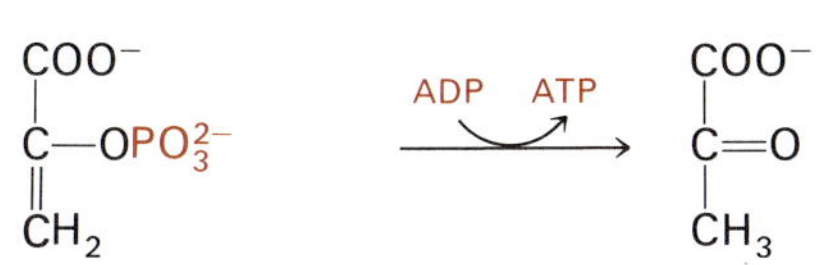

phosphoenolpyruvate **pyruvate**

The series of reactions in which glucose is transformed to pyruvate is summarized:

$$\underset{\text{glucose}}{C_6H_{12}O_6} + 2ADP + 2P_i + 2NAD^+ \longrightarrow \underset{\text{pyruvate}}{2C_3H_3O_3^-} + 2ATP + 2NADH + 2H^+$$

Two molecules of ATP are produced for each molecule of glucose. This represents the net yield of ATP in glycolysis. The individual reactions of glycolysis are summarized in Figure 26-4.

In order for glycolysis to continue, a constant supply of NAD^+ is required for the oxidation of glyceraldehyde-3-phosphate. Under anaerobic conditions, the NAD^+ is regenerated in a final reaction of glycolysis with the reduction of pyruvate to lactate:

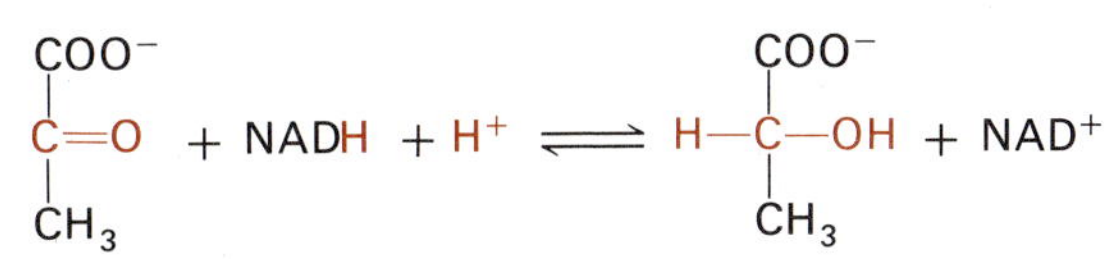

pyruvate **lactate**

The reaction for the conversion of glucose to lactate is summarized:

$$\underset{\text{glucose}}{C_6H_{12}O_6} + 2ADP + 2P_i \longrightarrow \underset{\text{lactate}}{2C_3H_5O_3^-} + 2ATP$$

When lactate is formed, typically during strenuous exercise under conditions of "oxygen debt," it accumulates in the muscles and causes an acidification that we associate with cramps or soreness. When activity is reduced, enough oxygen becomes available. Then the aerobic reactions take over, removing pyruvate and stopping production of lactate. The excess lactate that is left from the period of oxygen debt remains in the bloodstream until it is absorbed by the liver. In the liver, lactate is converted back to glucose in a process discussed more fully in Chapter 27.

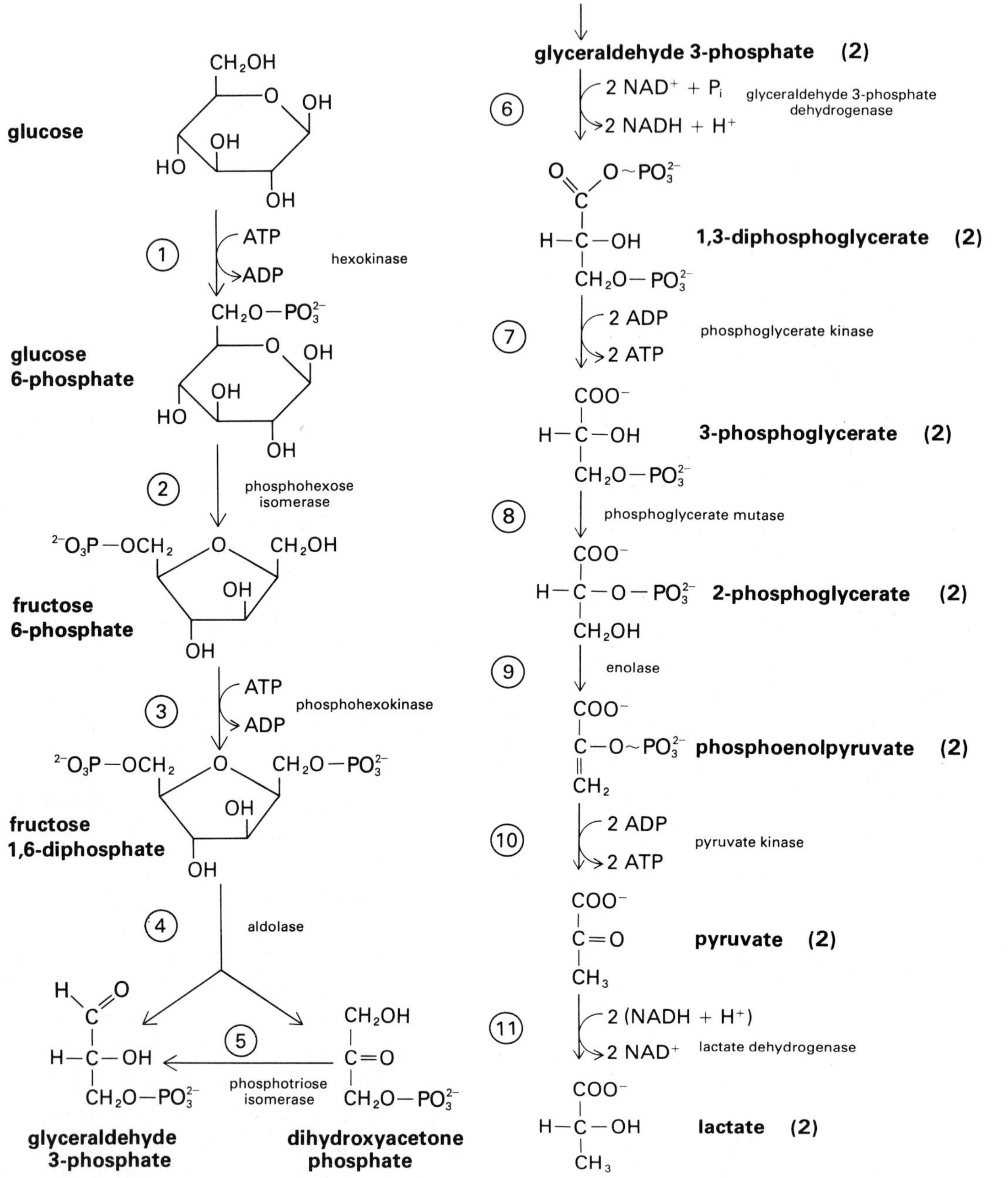

FIGURE 26-4

Glycolysis. ATP is consumed in reactions 1 and 3 (−2 ATP). ATP is produced in reactions 7 and 10 (+4 ATP). There is a net gain of 2 ATP. For each molecule of glucose that reacts, there are *two* molecules of each of the substances in reactions 6 through 11.

In some microorganisms such as yeast, pyruvate is converted anaerobically to ethanol rather than to lactate. The pyruvate is decarboxylated to yield acetaldehyde and CO_2; then the acetaldehyde is reduced to ethanol, simultaneously converting the NADH + H^+ back to NAD^+:

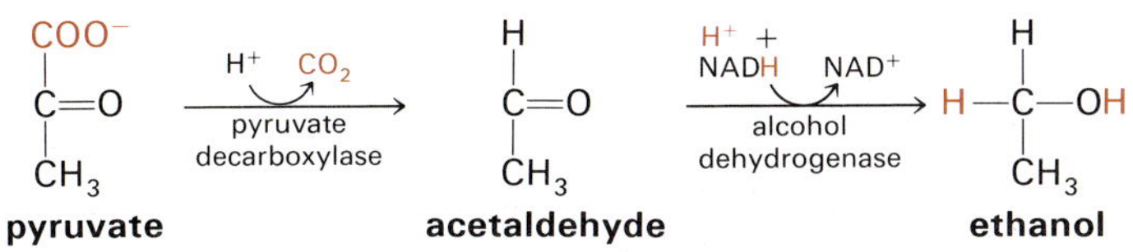

The conversion of pyruvate to ethanol is known as **fermentation.**

In aerobic cells, pyruvate is "siphoned off" for processing in the citric acid cycle, and the NAD^+ is regenerated when NADH + H^+ ultimately donates an electron pair to O_2 in the respiratory chain. We shall discuss these reactions in Section 26-3.

We can now judge the overall efficiency of glycolysis. First, we can estimate what fraction of the energy potentially available in glucose is released by glycolysis. The standard free energy for the complete oxidation of glucose to CO_2 and H_2O is about −700 kcal/mole. That for oxidation of lactic acid is about −325 kcal/mole.* The free energy available from the conversion of glucose to lactic acid is the difference between the glucose value and twice the lactic acid value:

$$\Delta G^0(\text{glucose}) - 2\Delta G^0(\text{lactic acid}) = (-700) - (2 \times -325) = -50 \text{ kcal/mole}$$

Thus of the total free energy potentially available from glucose, only a relatively small proportion (about 7%) is tapped in glycolysis. The remainder of the energy is extracted in the reactions of the citric acid cycle (Section 26-3).

Concerning the efficiency of glycolysis, the 50 kcal/mole lost from glucose results in the net synthesis of 2 moles of ATP, corresponding to a gain of about (2 × 7.3) kcal/mole, or about 15 kcal per mole of glucose consumed. The capture of 15 kcal/mole out of a potential 50 kcal/mole represents an efficiency of about 30%, which is typical of biochemical processes. The energy not captured in ATP formation is dissipated as heat.

EXERCISE 26-2 Chemical methods can be used to "label" any of the carbon atoms of glucose with the radioactive isotope of carbon, ^{14}C. The reactions of glycolysis can be carried out in cells or in extracts, and the lactic acid produced can be analyzed to determine which position(s) contain the ^{14}C label. Which positions of glucose must contain ^{14}C in order to produce lactate with ^{14}C in the methyl group?

A disease called fructose intolerance that is due to a disturbance of glycolysis has been recognized in some individuals. This disease often appears for the first time when infants are weaned from mother's milk (which contains lactose as the sugar) to formulas supplemented with cane sugar containing sucrose, a disaccharide of fructose and glucose. At that time, these infants receive large amounts of fructose, and individuals with fructose intolerance begin to display symptoms such as vomiting, weight loss, liver enlargement, and (in extreme cases) convulsions. (Similar symptoms occur in lactose intolerance, described in Chapter 23.)

This syndrome arises because these babies lack an enzyme (fructose 1-phosphate aldolase) that splits

*In SI units, ΔG^0 for glucose is about −2930 kJ/mole, and ΔG^0 for lactic acid about −1360 kJ/mole. (1 cal = 4.18 J.)

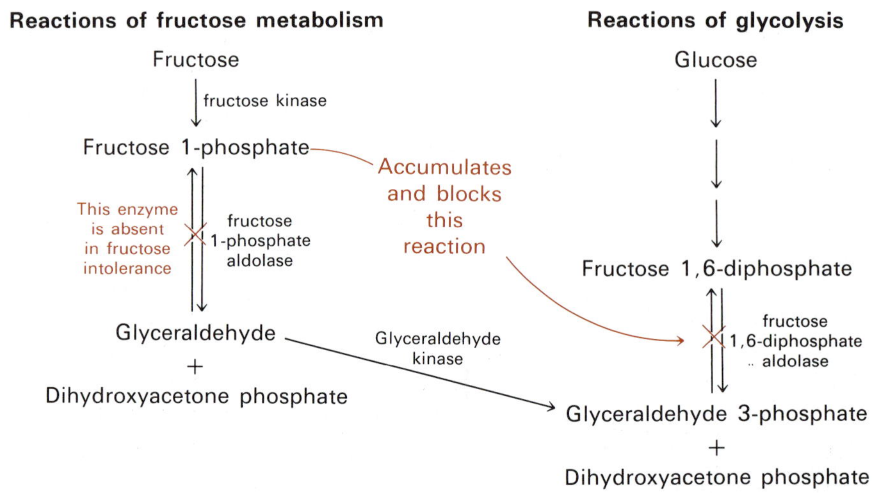

FIGURE 26-5
The metabolic basis of fructose intolerance.

fructose 1-phosphate into dihydroxyacetone phosphate and glyceraldehyde. If this enzyme is absent, fructose 1-phosphate accumulates and inhibits the fructose 1-6-diphosphate–splitting step of glycolysis (Figure 26-5). When individuals with fructose intolerance are placed on a diet low in fructose, recovery is rapid.

EXERCISE 26-3 Which reaction or reactions of glycolysis can be classified as each of the following reaction types?

(a) Transfer of phosphate group; (b) isomerization; (c) dehydration; (d) aldol cleavage (the reverse of an aldol condensation); (e) oxidation–reduction. (In your answer, refer to the reaction numbers given in Figure 26-4.)

26-3 THE CITRIC ACID CYCLE AND OXIDATIVE PHOSPHORYLATION

The energy left in the products of glycolysis can be released in an additional series of reactions. These reactions occur in two phases. First, in the citric acid cycle, the fuel molecules are oxidized to CO_2. Energy contained in the reduced fuel molecules is converted to the temporary storage form of reduced coenzymes—either $NADH + H^+$ or a related oxidation–reduction coenzyme, flavin adenine dinucleotide (abbreviated FAD, or in the reduced form $FADH_2$). FAD resembles NAD^+, but it contains a triple ring system in place of the nicotinamide. The ring system is derived from vitamin B_2 (riboflavin) and is shown here in the oxidized and reduced forms:

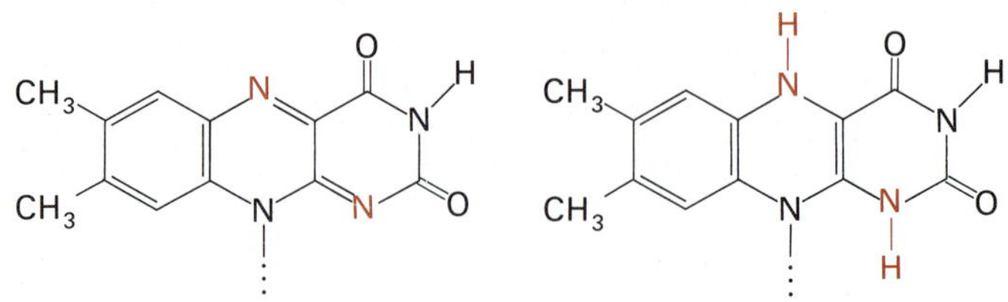

oxidized form reduced form

flavine adenine dinucleotide

In the second phase, the "reducing power" contained in the reduced coenzymes is extracted in a series of oxidation–reduction reactions. In this cascade of reactions, the reduced coenzymes gradually lose their reducing power, with the loss coupled to the synthesis of ATP from ADP and P_i. Because electron pairs from the reduced coenzymes can be visualized as being transported from one ATP-synthesizing site to another, this cascade of reactions is called the **electron-transport chain.** There are three sites of ATP synthesis in the electron-transport chain, and *at the end of the chain the electrons are used to reduce oxygen to water, thus completing the catabolism of glucose to* CO_2 *and* H_2O. Because phosphorylation of ADP to give ATP in this system is tightly linked to the consumption of oxygen, it is called **oxidative phosphorylation.** Phosphorylation reactions that occur in the absence of oxygen (for example, those of glycolysis) are called **substrate-level phosphorylations.**

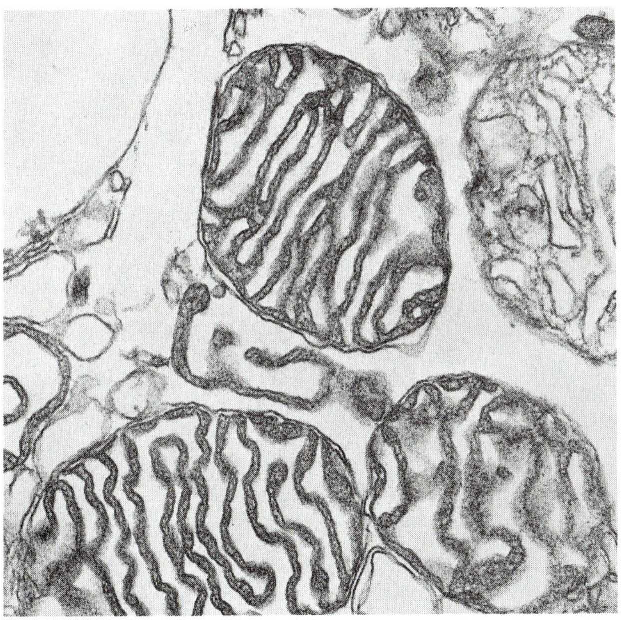

FIGURE 26-6

Cross section of mitochondria. [Courtesy of John Telford, Cornell University.]

The reactions of the citric acid cycle and oxidative phosphorylation take place in plant and animal cells in specialized subcellular structures, or organelles, called **mitochondria** (Figure 26-6). Many of the enzymes involved are attached to the surface of the inner mitochondrial membranes. In contrast, the enzymes involved in glycolysis are freely dispersed throughout the inner contents of the cell, known generally as the **cytoplasm.** (Cell structures are described more fully in Section 27-1 and Figure 27-1.)

The specific reduced compound processed by the citric acid cycle is acetyl-S-CoA (Section 26-1). We have noted that the compounds of glycolysis are diverted at the stage of pyruvate. This siphoning-off occurs by a complex reaction that involves decarboxylation of pyruvate, reduction of NAD^+, and formation of acetyl-S-CoA from CoA-SH:

$$\underset{\textbf{pyruvate}}{CH_3{-}\overset{\displaystyle O}{\overset{\|}{C}}{-}COO^-} + NAD^+ + CoA{-}SH \xrightarrow[\text{dehydrogenase}]{\text{pyruvate}}$$

$$\underset{\textbf{acetyl-S-CoA}}{CH_3{-}\overset{\displaystyle O}{\overset{\|}{C}}{\sim}S{-}CoA} + NADH + H^+ + CO_2$$

The citric acid cycle processes one acetyl-S-CoA unit in each "turn" of the cycle, with a release of the C atoms as CO_2 and a capture of the reducing power as reduced coenzymes. At the time of its description in 1937 by Hans Krebs, the elucidation of the citric acid cycle was a remarkable insight that ranks as one of the major achievements of biochemistry. In honor of its discoverer, the reactions are sometimes called the **Krebs cycle.** (It is also frequently called the **tricarboxylic acid cycle**).

The overall pattern of the cycle is to form six-carbon citrate molecules from four-carbon oxaloacetate and the two-carbon acetyl units of acetyl-S-CoA. The six-carbon units then are processed to five-carbon and four-carbon units, with release of two molecules of CO_2, reduction of coenzymes, and formation of one high-energy phosphate anhydride bond. Figure 26-7 gives an overview of the citric acid cycle. In the $C_4 + C_2 \rightarrow C_6$ step, oxaloacetate condenses with acetyl-S-CoA to form citrate; the citrate is then isomerized to isocitrate:

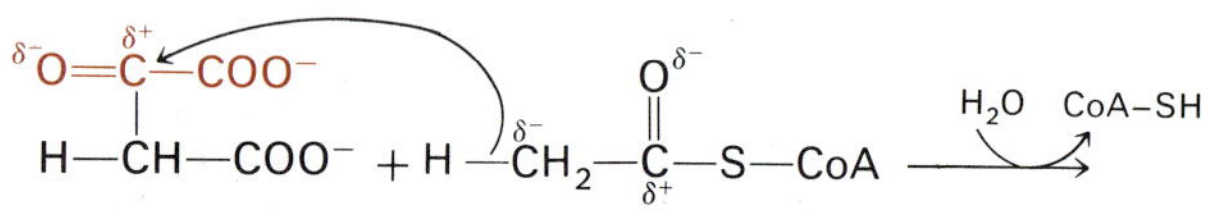

oxaloacetate (four carbons) **acetyl-S-CoA** (two carbons)

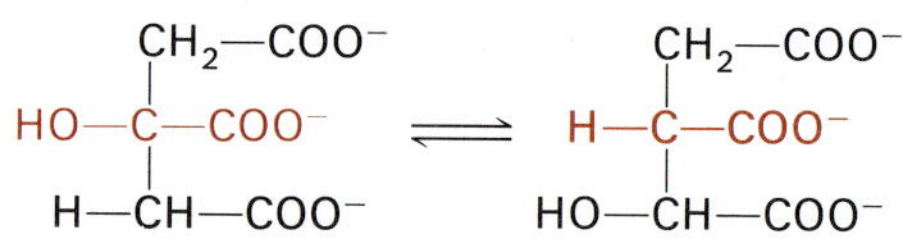

citrate (six carbons) **isocitrate**

In the first of the decarboxylation reactions, isocitrate becomes α-ketoglutarate, with the concomitant reduction of NAD^+ to $NADH + H^+$:

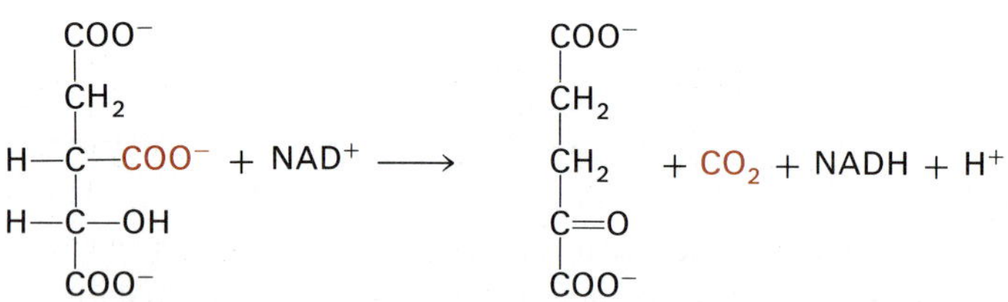

isocitrate **α-ketoglutarate**

This reaction is followed by the second decarboxylation, which forms succinate from α-ketoglutarate, along with reduction of another molecule of NAD^+. This reaction also involves the only substrate-level phosphorylation of the citric acid cycle, with

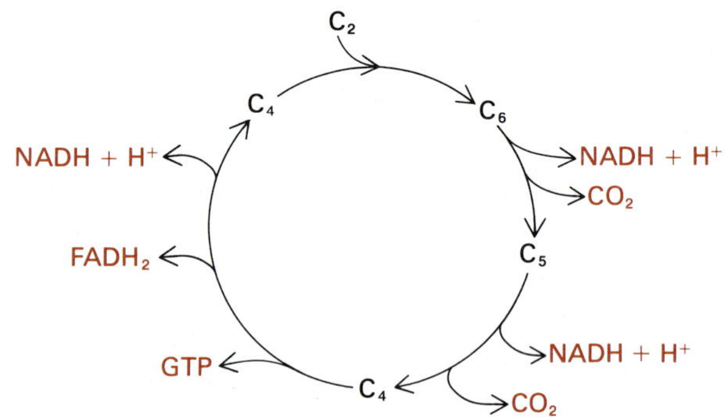

FIGURE 26-7

An overview of the citric acid cycle. For each acetyl group that enters the cycle, two CO_2, one GTP, three ($NADH + H^+$), and one $FADH_2$ are produced.

$GDP + P_i$ forming GTP (energetically, the high-energy phosphate anhydride bonds of GTP are equivalent to those of ATP):

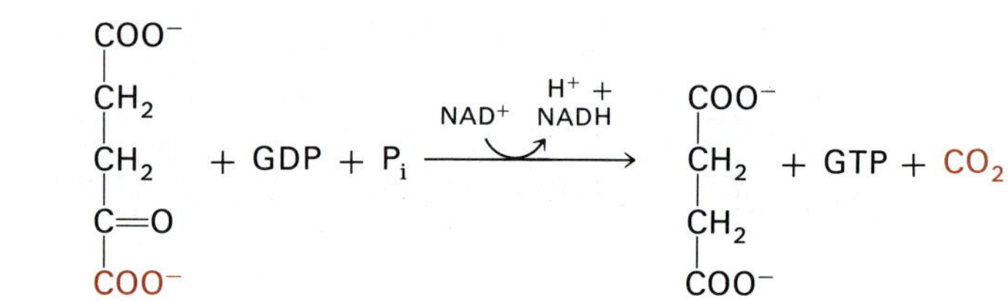

α-ketoglutarate **succinate**

With the formation of succinate, all of the decarboxylation reactions of the cycle have been completed. The remainder of the reactions of four-carbon compounds serve to extract additional reducing power and to form oxaloacetate in preparation for another turn of the cycle. These reactions begin with oxidation of succinate to form fumarate, with the reduction of FAD to $FADH_2$. A "rehydration" step then occurs in which a water molecule is added to fumarate, forming malate. The malate, in turn, is oxidized to form oxalo-

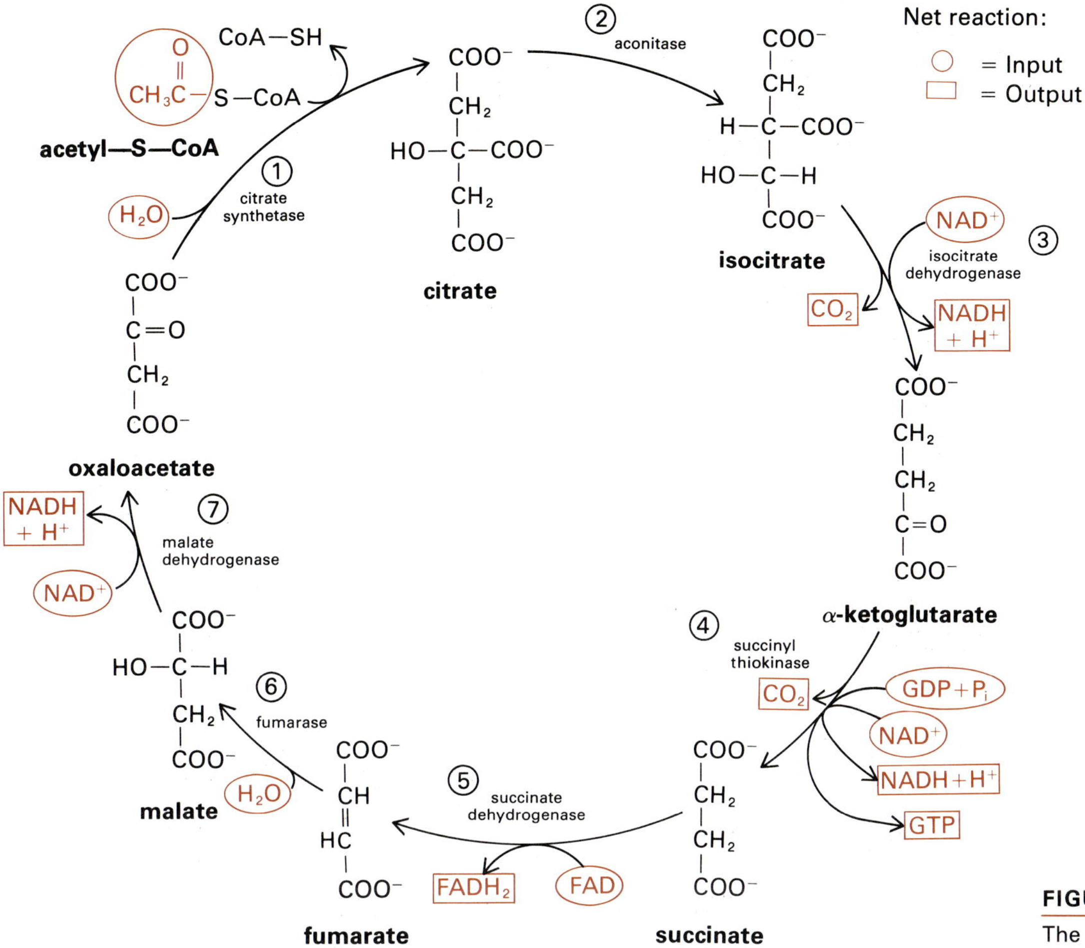

FIGURE 26-8

The citric acid cycle.

acetate, with the reduction of the final molecule of NAD^+:

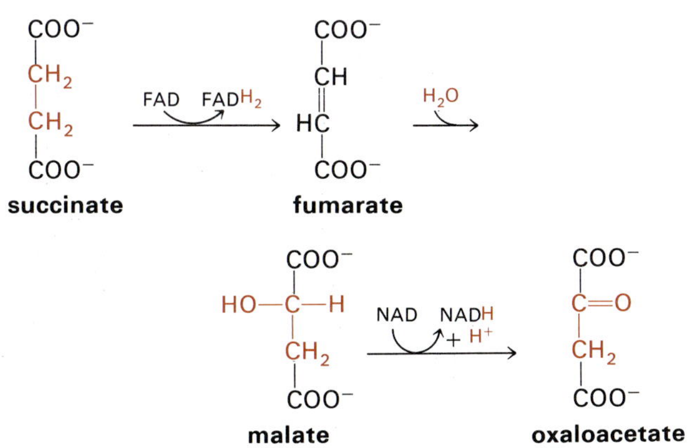

The reactions of the citric acid cycle are summarized in Figure 26-8:

$$\text{Acetate} + 3NAD^+ + FAD + GDP + P_i + 2H_2O \longrightarrow 2CO_2 + 3NADH + 3H^+ + FADH_2 + GTP$$

The oxidations that occur in the cycle are actually dehydrogenations in which H atoms are donated to NAD^+ or FAD; O_2 does not participate directly. Oxygen is required for the continued operation of the cycle, however, because NAD^+ and FAD can be regenerated only by the transfer of electrons from the reduced coenzymes to molecular oxygen through the electron-transport chain; that is, by the reduction of O_2 to form H_2O. Only one substrate-level phosphorylation reaction occurs in the citric acid cycle (to form GTP), but the reduced coenzymes, NADH + H^+ and $FADH_2$, can be "redeemed" for many more ATP molecules in the reactions of oxidative phosphorylation, which we now consider.

EXERCISE 26-4 Each reaction of the citric acid cycle can be described by one (or by a combination of two or more) of the following reaction types: condensation, isomerization, hydration, decarboxylation, oxidation–reduction, or substrate-level phosphorylation. Consult Figure 26-8, and classify each of the reactions as to reaction type.

The net result of oxidative phosphorylation is to donate protons and electron pairs from the reduced coenzymes NADH + H^+ and $FADH_2$ to O_2 to form H_2O. A large amount of energy is released in the process, and this energy is used to generate ATP. This is the main source of ATP in aerobic cells. In Section 12-4, the standard reduction potential $\mathcal{E}^{0'}$ for the net reaction in which O_2 is reduced to H_2O by NADH + H^+ is calculated to be 1.14 V. This corresponds to a $\Delta G^{0'}$ of −52.6 kcal per mole of H_2O formed:

$$O_2 + 2NADH + 2H^+ \longrightarrow 2NAD^+ + 2H_2O$$
$$\mathcal{E}^{0'} = 1.14\text{ V} \qquad \Delta G^{0'} = -52.6\text{ kcal/mole } H_2O$$

The reaction does not occur in one step, but in a series of steps that (like those of glycolysis and the citric acid cycle) release the energy in small packets. This permits the energy to be more efficiently coupled to the formation of ATP than would be possible if the energy were released all at once. In this series of oxidation–reduction reactions, electrons are passed from the reducing agents NADH + H^+ and $FADH_2$ to progressively stronger oxidizing agents, and ultimately to the very strong oxidizing agent O_2.

The electron carriers include flavins, nonheme iron compounds, quinones, and hemes. Some steps in the electron-transport chain are shown in Figure 26-9. The first step is catalyzed by NADH dehydrogenase, a flavoprotein that contains nonheme iron. A common form of coenzyme Q has a tail consisting of ten isoprene units:

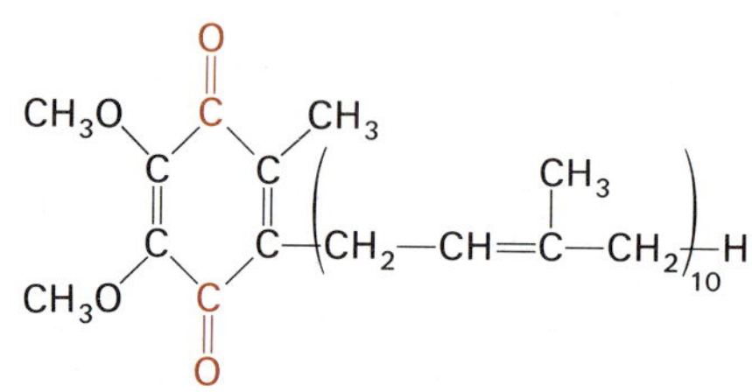

coenzyme Q_{10} (CoQ)$_{10}$
(oxidized form)

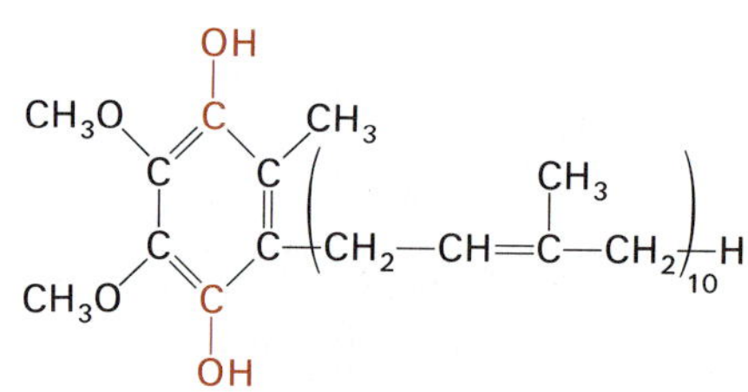

coenzyme Q_{10}
(reduced form)

(Coenzyme Q is also called ubiquinone because it is ubiquitous in living organisms.) The cytochromes are proteins containing heme groups like those in hemoglobin. The iron atoms in the cytochromes alternate between Fe^{2+} (reduced state) and Fe^{3+} (oxidized state) as electrons pass through the chain. Cytochrome a_3 also contains a copper atom that alternates between Cu^+ and Cu^{2+} as it transfers electrons to O_2. Table 26-2 gives reduction potentials for some half-reactions of the electron transport chain. The strongest oxidizing agent in the table is O_2, which has the greatest tendency to gain electrons. The strongest reducing agent is NADH + H^+, which has the greatest tendency to lose electrons.

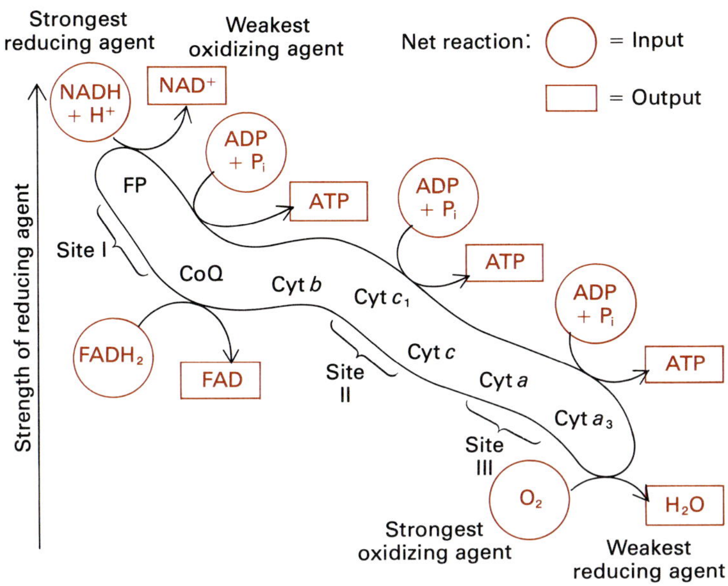

FIGURE 26-9

The electron-transport chain. In addition to the cytochromes, components of the chain include flavoprotein (FP) and coenzyme Q (CoQ).

TABLE 26-2
Standard reduction potentials of some reactions in the electron-transport chain

Half-reaction	$\mathcal{E}^{0\prime}$ (V)
$O_2 + 4H^+ + 4e^- \longrightarrow 2H_2O$	+0.82
Cytochrome *a* (Fe^{3+}) + e^- $\longrightarrow$ cytochrome *a* (Fe^{2+})	+0.39
Cytochrome *c* (Fe^{3+}) + e^- $\longrightarrow$ cytochrome *c* (Fe^{2+})	+0.22
CoQ (oxidized) + $2H^+ + 2e^-$ $\longrightarrow$ CoQ (reduced)	+0.10
Cytochrome *b* (Fe^{3+}) + e^- $\longrightarrow$ cytochrome *b* (Fe^{2+})	+0.07
$NAD^+ + 2H^+ + 2e^- \longrightarrow NADH + H^+$	−0.32

EXERCISE 26-5 Calculate $\mathcal{E}^{0\prime}$ for the reaction

$$\text{cytochrome } b(Fe^{2+}) + \text{cytochrome } c(Fe^{3+}) \longrightarrow \text{cytochrome } b(Fe^{3+}) + \text{cytochrome } c(Fe^{2+})$$

Should this reaction occur spontaneously? (Refer to Section 12-4, if necessary.)

The series of oxidation–reduction reactions that constitutes the electron-transport chain is coupled to a series of three phosphorylations at three different sites (Figure 26-9). Three molecules of ATP are produced for each pair of electrons donated by NADH + H^+. The way this coupling is achieved is fairly complex; only the general outlines are presented here. In some manner not yet well understood, when electrons are passed along the electron-transport chain of reactions, protons (H^+ ions) are simultaneously expelled from the mitochondria.

As a result of the expulsions of protons, the H^+ concentration becomes higher on the outside of the membrane than on the inside. Because molecules always tend to distribute homogeneously in a given volume, this difference in H^+ concentration creates a force that drives H^+ into the mitochondria. This force can be harnessed for chemical work. In this case, the work is the phosphorylation of ADP. Proton movement is controlled by a protein–membrane complex called the **coupling factor.** The coupling factor contains a channel through which the protons move to equalize the H^+ concentrations on the inside and outside. In addition, the coupling factor contains sites that bind ADP and P_i. By some unknown mechanism, the inward movement of protons is coupled to the condensation of ADP and P_i to form ATP.

Each pair of electrons originating in $NADH + H^+$ is passed along the electron-transport chain in a series of reactions leading to the expulsion of a quantity of protons that corresponds to the synthesis of three molecules of ATP. The reduced coenzyme $FADH_2$ is a weaker reducing agent than $NADH + H^+$ and donates its electrons partway along the electron-transport chain. As a result, the quantity of protons expelled when the electrons from $FADH_2$ are passed along the electron-transport chain corresponds to the synthesis of only two molecules of ATP. Thus each molecule of $NADH + H^+$ generates three molecules of ATP, but each molecule of $FADH_2$ generates only two molecules of ATP.

Some poisons act as inhibitors of electron transport by blocking the chain at specific points. Rotenone, an extremely toxic substance obtained from the root of a plant and used as an insecticide, interferes with the enzyme NADH dehydrogenase, and thus blocks ATP synthesis at site 1. The poisons CN^- and CO block the electron flow between cytochrome $(a + a_3)$ and O_2, inhibiting phosphorylation at site 3. Cyanide binds to Fe^{3+} and carbon monoxide binds to Fe^{2+}; when the iron is complexed it cannot be as readily oxidized and reduced.

Looking again at the net reaction for the citric acid cycle, we can now evaluate its effectiveness:

$$\text{Acetate} + 3\text{NAD}^+ + \text{FAD} + \text{GDP} + \text{P}_i + 2\text{H}_2\text{O} \longrightarrow 2\text{CO}_2 + 3\text{NADH} + 3\text{H}^+ + \text{FADH}_2 + \text{GTP}$$

For each molecule of glucose processed by glycolysis, two acetate units enter the citric acid cycle. Each leads to the production of three molecules of $(NADH + H^+)$ or a total of six $(NADH + H^+)$. Each of these in turn produces three ATP in oxidative phosphorylation, for a total of 18 ATP from this source. To this, we can add two ATP from each of the $FADH_2$ molecules produced per acetate unit, for an additional four ATP for each glucose molecule. Finally, the molecules of GTP formed by substrate-level phosphorylation are equivalent to another two ATP per glucose $(GTP + ADP \rightleftharpoons ATP + GDP)$ for an overall total of $18 + 4 + 2 = 24$. Thus a great deal more ATP is produced by the citric acid cycle and oxidative phosphorylation than by glycolysis.

If a glucose molecule is converted by glycolysis to the point of pyruvate and then diverted to enter the citric acid cycle, the two molecules of $NADH + H^+$ produced in the conversion of two molecules of glyceraldehyde 3-phosphate to 1,3-diphosphoglycerate cannot be reoxidized by forming lactic acid. Therefore these $NADH + H^+$ molecules are also available to donate electrons to the electron-transport chain and generate additional ATP. However, these molecules of $NADH + H^+$ must first be transported from the cytoplasm where glycolysis occurs to the inside of the mitochondria. One molecule of ATP is consumed in the transport across the mitochondrial membrane. Thus these "imported" molecules of $NADH + H^+$

TABLE 26-3
ATP yield from oxidation of glucose to CO_2 and H_2O

Source	ATP formed per glucose molecule
Glycolysis (in the cytosol):	
ATP yield	+2
2 NADH are also formed	
Conversion of pyruvate to acetyl-S-CoA (inside mitochondria):	
2 NADH	
Citric acid cycle (inside mitochondria):	
2 molecules of GTP	+2
6 NADH	
2 $FADH_2$	
Oxidative phosphorylation (inside mitochondria):	
2 NADH from glycolysis (each yields 2 ATP, not 3, because of cost of shuttle)	+4
2 NADH from conversion of pyruvate to acetyl-S-CoA (each yields 3 ATP)	+6
2 $FADH_2$ formed in citric acid cycle (each yields 2 ATP)	+4
6 NADH formed in citric acid cycle (each yields 3 ATP)	+18
Total yield:	+36 ATP per glucose

can be credited only with a net production of two ATP, or an additional four ATP per glucose.

The pyruvate decarboxylation, which transforms pyruvate to acetyl units for entry into the citric acid cycle, generates two additional "internal" NADH + H^+ per glucose in the mitochondria, corresponding to three ATP each, or an additional six ATP.

If we tabulate the ATP molecules arising from the various sources (Table 26-3), we find that 36 molecules of ATP are formed for each molecule of glucose consumed. Therefore, the overall reaction of glucose metabolism may be summarized as follows:

$$\text{Glucose} + 6O_2 + 36\text{ADP} + 36P_i + 36H^+ \longrightarrow 6CO_2 + 36\text{ATP} + 42H_2O$$

In this case the full 700 kcal/mole of glucose is consumed to produce 36 ATP, equivalent to about 260 kcal/mole. Thus the overall efficiency is about 40%, near the 30% value found for glycolysis.

26-4 PHOTOSYNTHESIS

Now that we have seen how energy from the oxidation of glucose to CO_2 and H_2O is extracted and stored in ATP, it will be interesting to consider the similarities and contrasts in the production of glucose from CO_2 and H_2O in **photosynthesis.** As the name implies, an important ingredient is light, which is captured in the green **chlorophyll** pigments of plants (Figure 26-10). These pigments are located in the plant-cell organelles

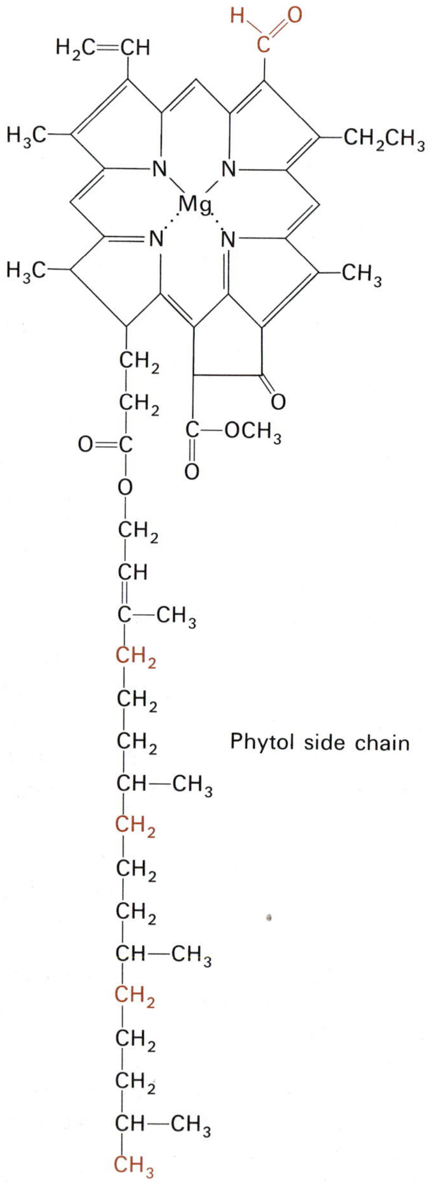

FIGURE 26-10

The structure of chlorophyll *b*. Chlorophyll *a* has a methyl group in place of the aldehyde group shown here in color. The phytol side chain is composed of four isoprene units—their heads are shown in color.

called **chloroplasts.** Although this light trapping apparatus has many unique features not found in animal cells, it is helpful to consider its similarities to the processes already described for animal cells.

Fundamentally, photosynthesis operates on principles like those of oxidative phosphorylation. In both cases, the main activity is a flow of electrons along an electron-transport chain, with the simultaneous formation of a difference in H^+ concentrations across a membrane. In the case of oxidative phosphorylation, the flow of electrons begins by the donation of electrons from NADH + H^+ and ends with the acceptance of the electrons by oxygen.

In the case of photosynthesis, the electrons are obtained by absorbing light and harnessing light energy to split water into oxygen plus protons and electrons (in what is effectively a reversal of the last stage of oxidative phosphorylation). These electrons have a high level of reducing power and are used in two ways. First, they initiate a flow of electrons between cytochromes that causes proton movement. The proton movement across membranes is used by coupling factors (very like those found in mitochondria) to form ATP from ADP and P_i. Second, they reduce $NADP^+$ (the analogue of NAD^+ with the extra phosphate group at the position noted in Figure 26-3) to NADPH + H^+ to provide the hydrogen atoms needed in the reduction of CO_2 to form glucose. To emphasize the role of light in the production of ATP in plants, it is called **photophosphorylation.**

The ATP and NADPH + H^+ generated by photophosphorylation are used to synthesize glucose in a series of reactions first described by Melvin Calvin. The reactions generally are called the **Calvin cycle** in his honor. The first step is the combination of CO_2 with a five-carbon sugar derivative, ribulose 1,5-diphosphate; the reaction involves a splitting of the hypothetical six-carbon product (shown in brackets) into two three-carbon 3-phosphoglycerate units:

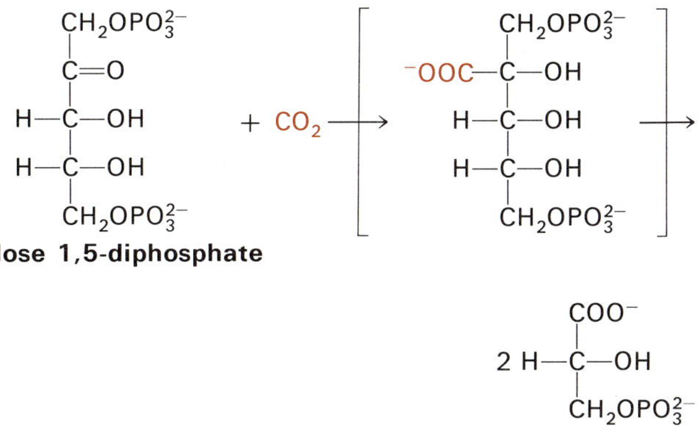

ribulose 1,5-diphosphate

3-phosphoglycerate

The 3-phosphoglycerate is converted to glyceraldehyde 3-phosphate in the important reduction step of the cycle. This step uses the NADPH + H^+ formed in photophosphorylation and also requires ATP hydrolysis:

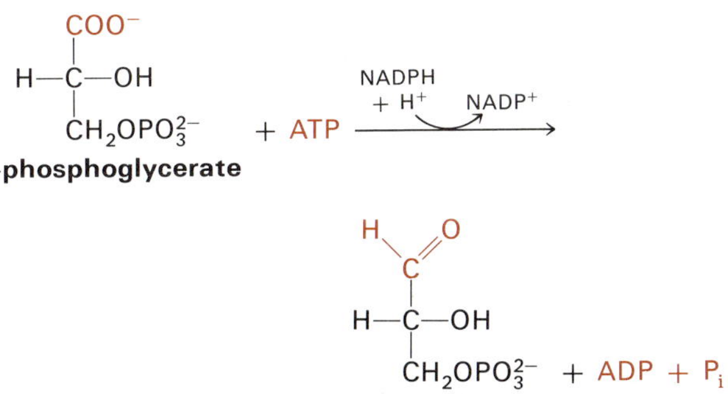

3-phosphoglycerate

glyceraldehyde 3-phosphate

The glyceraldehyde 3-phosphate feeds into a complex series of reactions known as the **pentose shunt.** For every 12 molecules of glyceraldehyde 3-phosphate that enter the shunt, rearrangement of the atoms produces one glucose molecule, six ribulose 5-phosphate molecules, and six P_i. A phosphorylation of ribulose 5-phosphate by ATP to give ribulose 1,5-diphosphate (+ADP) completes the cycle.

The reactions of the Calvin cycle are summarized in Figure 26-11. In this figure the molecules are shown in the numbers that correspond to the production of one molecule of glucose. Thus there are six CO_2, twelve 3-phosphoglycerate, etc. The overall net reaction is

$$6CO_2 + 18ATP + 12NADPH + H^+ \longrightarrow \text{Glucose} + 18ADP + 18P_i + 12NADP^+$$

If we consider that each molecule of NADPH + H^+ (which has essentially the same reducing power as NADH + H^+) is equivalent to 3 ATP by conversion in mitochondria, then the 12 NADPH + H^+ consumed by the Calvin cycle correspond to 36 ATP. This amount plus the 18 ATP consumed directly by the Calvin cycle add up to a total of 54 ATP required for the synthesis of each molecule of glucose. Recall that in oxidative phosphorylation, 36 molecules of ATP are produced by oxidation of one molecule of glucose. Thus in a sense, plants invest 54 ATP in each glucose, but animals can extract only 36 ATP, or two-thirds of the initial value. The moral is that in metabolic trades, the energy invested in the synthesis of a molecule can never be fully regained in its oxidation.

EXERCISE 26-6 Although ^{14}C is commonly used in metabolic studies (and was invaluable in working out the details of the Calvin cycle), the nonradioactive isotope ^{13}C is also very useful. It can be used in studies on humans for which the use of ^{14}C is precluded by its radiation hazards. In nature, ^{13}C constitutes about 1% of the carbon atoms in any material containing carbon. However, enriched compounds can be obtained by certain isolation and separation procedures. The amount of ^{13}C in various positions of organic molecules can be determined by mass spectroscopy or by nuclear magnetic resonance.

An extract of plant cells is exposed to light and to CO_2 enriched with ^{13}C to a level of 10%. The 3-phosphoglycerate is then isolated and analyzed for the amount of ^{13}C it contains and for the location of the ^{13}C atoms. In which position of the 3-phosphoglycerate would you expect the ^{13}C to appear, and at what percentage abundance?

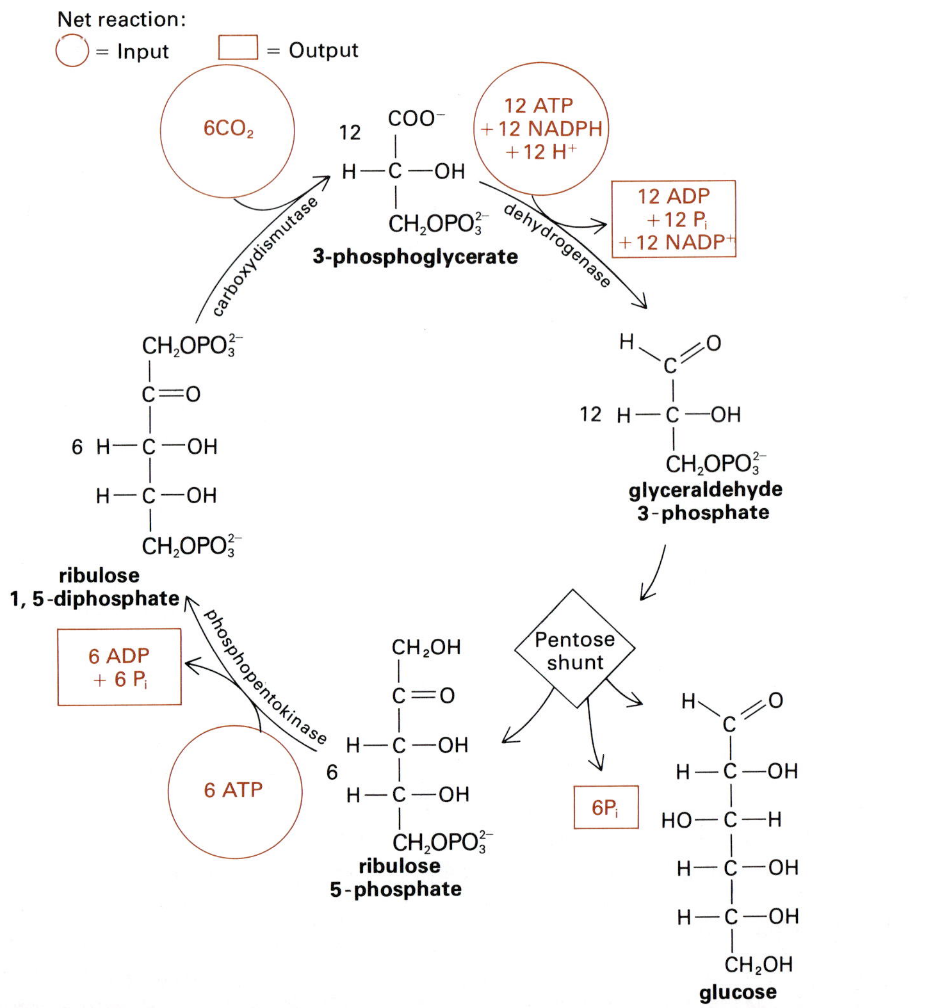

FIGURE 26-11

The Calvin cycle.

26-5
FATS AND PROTEINS AS ENERGY SOURCES

Energy can be derived from the fats and proteins as well as from carbohydrates. Fats can provide more than twice as much energy per gram as glucose because they are more highly reduced and thus yield more energy upon oxidation. Fats also have the advantage that they can be stored in a more compact form than carbohydrate polymers (which are stored with large amounts of water). However, fats have the disadvantage that they cannot be mobilized for oxidation as quickly as carbohydrates when energy is needed.

Fats are oxidized by a series of enzymes that first cleave the ester bonds of the triglycerides to free fatty acids and then cleave the fatty acids into two-carbon units of acetyl-S-CoA. This cleavage begins with a CoA-S– derivative of the fatty acid:

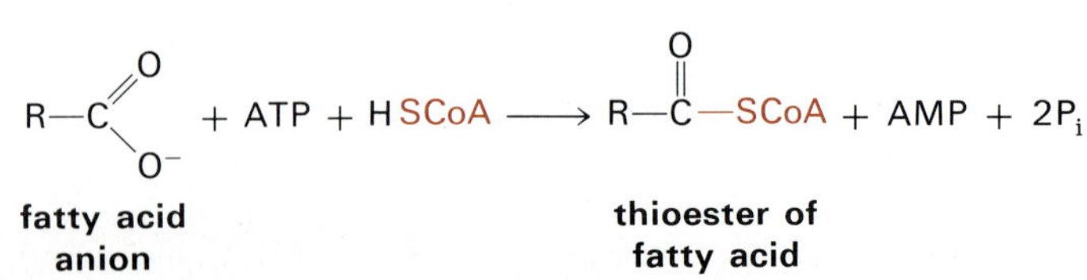

(Note that *two* high-energy phosphate bonds are hydrolyzed in this reaction.) For palmitic acid, a 16-carbon fatty acid, the first reaction is formation of a double bond between C-3 and C-4, with reduction of FAD:

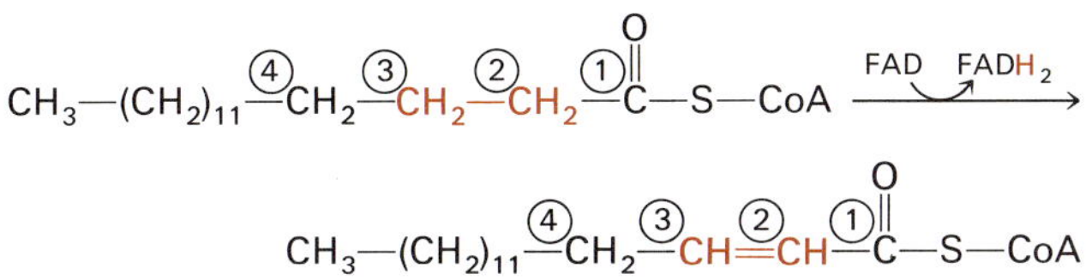

Next a hydration reaction occurs:

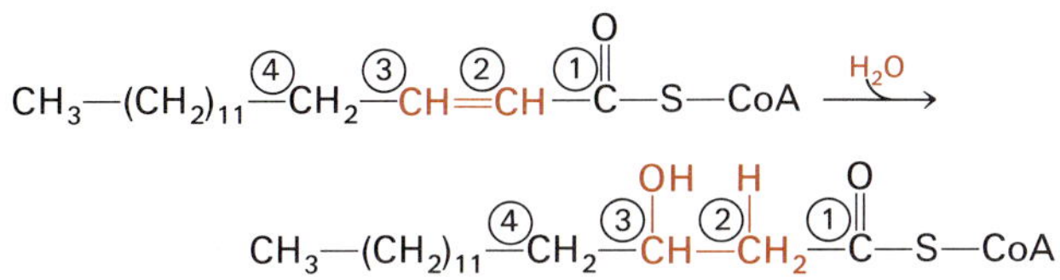

The hydration is followed by another oxidation–reduction reaction, with the oxidation of the hydroxyl group to a carbonyl group and reduction of the coenzyme NAD^+:

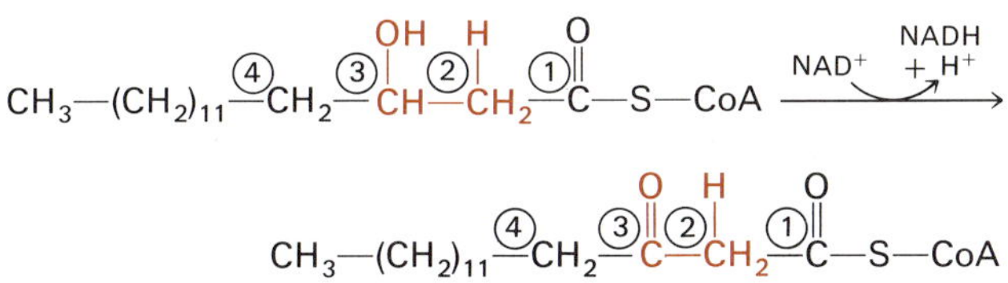

The keto configuration at C-3 can now be attacked by another molecule of CoA-SH, forming a 14-carbon fatty acid derivative (two carbons shorter than the starting compound) plus a unit of acetyl-S-CoA:

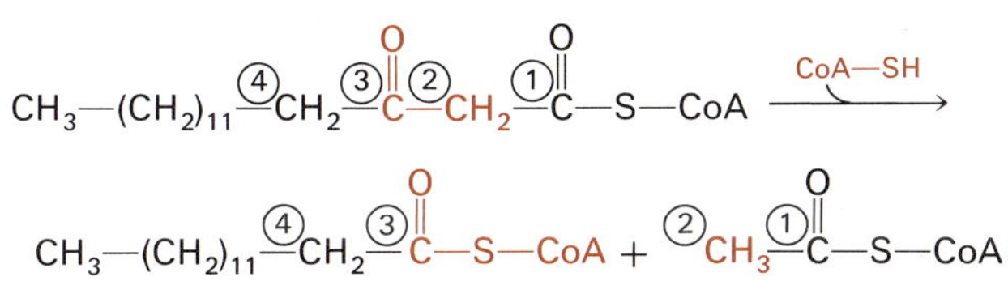

Thus each passage of a fatty acyl-S-CoA derivative through these reactions removes a two-carbon acetyl-CoA unit to leave a fatty acyl-S-CoA derivative with two fewer carbon atoms. For complete oxidation of palmitic acid, seven repeats through the cycle would be necessary, forming eight molecules of acetyl-S-CoA (two acetyl-CoA units would be formed on the last passage through the reactions). The overall reaction (omitting H_2O) for palmitic acid is

$$\text{Palmitoyl—S—CoA} + 7\text{FAD} + 7\text{NAD}^+ + 7\text{CoA—SH} \longrightarrow 8\text{Acetyl—S—CoA} + 7\text{FADH}_2 + 7\text{NADH} + 7\text{H}^+$$

To find the amount of ATP derived from fatty acid oxidation, we can tabulate the amount of ATP generated by each of the reduced coenzymes and acetyl-CoA units. The reactions of fatty acid oxidation occur within the mitochondria, so the direct conversion rates used for the reduced coenzymes in oxidative phosphorylation may be applied: 3 ATP for each $NADH + H^+$ and 2 ATP for each $FADH_2$, plus 12 ATP for each acetyl-S-CoA unit processed in the citric acid cycle. Thus the seven reduced coenzymes yield 21 ATP (from $NADH + H^+$) plus 14 ATP (from $FADH_2$), or 35 ATP, and the eight acetyl-S-CoA units yield 96 ATP, for a grand total of 131 ATP per palmitoyl-S-CoA.

Two high-energy phosphate anhydride bonds are consumed in the initial reaction of CoA-SH with palmitic acid, so the true yield of ATP per fatty acid is 129 (see Table 26-4). One way to evaluate this number is to compare it to the production of ATP from glucose in terms of the ATP produced per carbon atom. The 36 ATP per glucose correspond to a yield of 6 ATP per carbon atom, whereas about 8 ATP (129 ATP $\div$ 16 carbons) are produced by each carbon atom from the more highly reduced fatty acids.

It is more difficult to generalize about oxidation of proteins as energy sources because proteins are first

TABLE 26-4
ATP formation from fatty acid oxidation

Source	ATP formed per palmitic acid
From (7) $FADH_2$ (2 ATP per $FADH_2$)	14
From (7) NADH + H^+ (3 ATP per NADH + H^+)	21
	35
	− 2*
	33
From (8) Acetyl—S—CoA:	
3 NADH + H^+ per acetyl—S—CoA (24 NADH + H^+ total)	72
1 $FADH_2$ per acetyl—S—CoA (8 $FADH_2$ total)	16
1 GTP per acetyl—S—CoA (8 GTP total)	8
	Total: 129

*Minus 2 for thioester formation between palmitic acid and CoA—S—H, in which ATP is split to AMP and $2P_i$ (so that two high-energy bonds are lost).

digested to individual amino acids, and each amino acid has its distinct oxidation pathway and characteristic yield in terms of ATP production. However, on the average, the energy obtained from proteins is similar to that obtained from an equal weight of carbohydrates. In general the amino groups of the amino acid are removed first and combined with carbonate in a series of reactions that ultimately leads to urea, H_2NCONH_2, which is less toxic than free ammonia and is the main nitrogen-excretion compound in humans. Following deamination, the carbon skeletons of the amino acids are metabolized further by conversion to intermediates of glycolysis or the citric acid cycle.

KEY TERMS

High-energy compounds are those that contain a bond (such as the anhydride bonds of ATP) that yields a large release of free energy upon hydrolysis.

ATP (adenosine triphosphate) is the principal high-energy compound of cells and has a variety of roles in energy-requiring reactions. *Acyl phosphates* are a class of high-energy compounds with mixed anhydride bonds of phosphoric acid and a carboxylic acid—for example, 1,3-diphosphoglycerate, an intermediate in glycolysis. *Thioesters* are a class of high-energy compounds in which an ester is formed with a sulfur atom in place of the oxygen atom that would normally be derived from a hydroxyl group. One example of a high-energy thioester is acetyl-S-CoA.

Glycolysis is the name given to the reactions that extract energy from glucose anaerobically by degrading it to lactic acid. *Fermentation* is a process that resembles glycolysis but diverts intermediates at the stage of pyruvate to form ethanol and CO_2.

The *citric acid cycle* is the series of reactions for incorporating acetyl units into citrate, with the subsequent release of the acetyl carbons as CO_2 and the accompanying reduction of molecules of NAD^+ and FAD for use in ATP synthesis by oxidative phosphorylation.

Oxidative phosphorylation is the process of ATP synthesis from ADP and P_i in reactions that require oxygen and lead to the reduction of oxygen to form water. *Substrate-level phosphorylation* is the process of ATP synthesis from ADP and P_i (or an organic phosphate donor) that can occur in the absence of oxygen, as in ATP formation in glycolysis.

Mitochondria are the subcellular organelles responsible for the citric acid cycle, fatty acid oxidation, and oxidative phosphorylation.

Cytoplasm is the term for the overall intracellular fluid of the cell. The enzymes of glycolysis are dispersed freely throughout the cytoplasm.

Coupling factors are the key proteins of oxidative phosphorylation that harness the energy arising from proton movement across the mitochondrial membrane to synthesize ATP from ADP and P_i.

Photosynthesis refers to the process of glucose formation from CO_2 and H_2O in plants, using energy harnessed from light. *Chloroplasts* are the subcellular organelles unique to plant cells that house the molecular machinery of photosynthesis. *Chlorophyll* is the green light-trapping pigment of photosynthesis.

The *Calvin cycle* is the series of reactions that involve the synthesis of glucose from CO_2 and H_2O, using energy from the light-dependent reactions of photosynthesis.

CHAPTER SUMMARY

The main principle of bioenergetics is the conversion of atoms in glucose and other nutrient molecules to more oxidized forms. The energy that is liberated by this process is trapped by formation of ATP from ADP and P_i. The energy stored in the ATP molecules can then be used to drive various biochemical reactions. The role of ATP as an energy-storage molecule is related to the anhydride bonds between its phosphate groups; these bonds have a high energy of hydrolysis that can be coupled to drive less favorable reactions. Other "high-energy" compounds also participate in biochemical reactions, including acyl phosphates, guanidinium phosphates, phosphoenols, and thioesters.

The reactions that extract the energy available in glucose occur in two phases: a phase that does not require oxygen, and a phase that does require oxygen. The phase that can take place in the absence of oxygen is called *glycolysis;* it converts each molecule of glucose processed into two molecules of lactic acid through the action of 11 distinct enzyme-catalyzed reactions. In the course of these reactions, there is a net synthesis of two molecules of ATP from ADP and P_i for each molecule of glucose consumed. Fermentation, a variation of glycolysis, occurs in certain microorganisms, such as yeast, in which glucose is transformed into ethanol and CO_2.

For processing in the oxygen-requiring phase of glucose metabolism, pyruvate (the precursor of lactic acid) is decarboxylated to give an acetyl unit linked to coenzyme A via a thioester. The acetyl unit enters the *citric acid cycle,* where its carbon atoms are released as

CO_2 and the energy stored in the unit is used principally to reduce oxidation–reduction coenzymes (NAD^+ and FAD). The reduced coenzymes then donate their electrons to an *electron-transport chain.* The passage of the electrons along the chain is coupled to the synthesis of ATP from ADP and P_i to provide the bulk of the net synthesis of ATP derived from glucose metabolism.The reactions leading to ATP formation from the electron-transport chain are termed *oxidative phosphorylation.* Along with the reactions of the citric acid cycle, they occur in the subcellular organelles called *mitochondria.* Overall, 34 molecules of ATP are formed (for each molecule of glucose consumed) from the oxidative reactions involving the tricarboxylic acid cycle, in addition to the two molecules of ATP generated by glycolysis.

In plant cells, the energy to drive metabolism is obtained directly from light and is trapped by subcellular organelles called *chloroplasts.* The light is absorbed by the pigment *chlorophyll,* and the energy derived is used to split water. The electrons derived from water are captured in the reduction of $NADP^+$, and the oxygen atoms are liberated as O_2 gas. Energy from the electrons is coupled to the synthesis of ATP from ADP and P_i. In contrast to the synthesis of ATP in mitochondria (oxidative phosphorylation), the synthesis of ATP in chloroplasts is termed *photophosphorylation,* a portion of the process called *photosynthesis.* The other part of photosynthesis concerns the formation of glucose from CO_2. In a series of reactions called the *Calvin cycle* (requiring ATP and reduced $NADP^+$), a molecule of glucose is formed for every six molecules of CO_2 consumed.

Nutrients other than glucose also can be metabolized in energy-yielding reactions, and fatty acids in particular can be oxidized with a high yield in terms of ATP formed. The fatty acids are digested by cleaving pairs of atoms as acetyl units that can be processed in the tricarboxylic acid cycle. Oxidation of proteins occurs with the production of carbon skeletons that can be fed into glycolysis or the tricarboxylic acid cycle, while the amino groups are excreted as urea.

PROBLEMS

26-1. Which of the following compounds may be considered to have high-energy bonds?
(a) adenosine monophosphate
(b) adenosine diphosphate
(c) adenosine triphosphate
(d) glyceraldehyde 3-phosphate
(e) 1,3-diphosphoglycerate
(f) 3-phosphoglycerate
(g) phosphoenolpyruvate
(h) glucose 6-phosphate
(i) acetyl-S-CoA

26-2. If glucose is synthesized with isotopically labeled carbon in one or more of the six positions of carbon, which positions would lead to the appearance of the label in CO_2 if the glucose were metabolized by fermenting yeast?

26-3. Calculate $\mathcal{E}^{0'}$ for the reaction

$$\text{NADH} + \text{H}^+ + 2\ \text{cytochrome}\ c\ (\text{Fe}^{3+}) \longrightarrow \text{NAD}^+ + 2\text{H}^+ + 2\ \text{cytochrome}\ c\ (\text{Fe}^{2+})$$

26-4. The structure of lauric acid, a 12-carbon fatty

acid, is CH_3—$(CH_2)_{10}$—COOH. Write the overall reaction for the oxidation of lauric acid in the mitochondria. For each molecule of lauric acid oxidized, how many molecules of ATP are produced? Although we have not described the synthesis of fatty acids in detail, would you expect the synthesis of lauric acid to require more or less ATP than is produced by its oxidation?

26-5. (a) Write the equation for the hydrolysis of ATP to AMP and pyrophosphate (PP_i). Write the structures of all reactants and products.
(b) Pyrophosphate (PP_i) hydrolyzes to yield $2P_i$. Would you expect this reaction to occur spontaneously under typical cellular conditions? Explain.

26-6. Use data given in Table 26-1 to show (a) that phosphoenolpyruvate should spontaneously transfer a phosphate group to ADP; and (b) that ATP should spontaneously transfer a phosphate group to glycerol.

26-7. In Chapter 22 we saw that 2,3-diphosphoglycerate (2,3-DPG) regulates the binding of O_2 to hemoglobin by stabilizing deoxyhemoglobin. The 2,3-DPG is synthesized and degraded in a detour from the pathway of glycolysis:

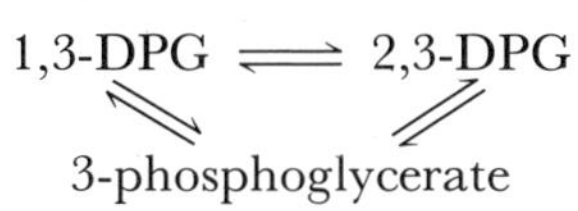

Predict the effect on the O_2 affinity of hemoglobin of (a) a deficiency of the enzyme hexokinase; and (b) a deficiency of the enzyme pyruvate kinase. (Consult Figure 26-4 to see which steps in glycolysis are catalyzed by these enzymes.)

26-8. Is there a net oxidation in glycolysis in the transformation of glucose (a) to pyruvate, and (b) to lactate?

26-9. (a) What is the yield of ATP when one molecule of pyruvate is oxidized to CO_2 and H_2O via the citric acid cycle and oxidative phosphorylation?
(b) What is the yield of ATP when one molecule of fructose 1,6-diphosphate is oxidized to CO_2 and H_2O as in part a?

26-10. The administration of nitrite (an oxidizing agent) is an effective treatment for cyanide poisoning. Suggest a possible basis for the effectiveness of nitrite as an antidote.

26-11. List the following substances in order of increasing tendency to donate electrons: (a) cytochrome *c* (Fe^{2+}); (b) NADH + H^+; (c) coenzyme Q (reduced); (d) cytochrome *b* (Fe^{2+}).

26-12. List the following substances in order of increasing tendency to accept electrons: (a) cytochrome *c* (Fe^{3+}); (b) NAD^+; (c) coenzyme Q (oxidized); (d) cytochrome *b* (Fe^{3+}).

Air sacs in the lungs are surrounded by networks of capillaries through which oxygen and carbon dioxide diffuse. Scanning electron micrograph of capillary network surrounding an alveolar sac, $\times 950$.

27

Integration of Biochemical Processes

Cells are the fundamental units of life. Two kinds of cells occur in nature. Prokaryotic cells are small and relatively simple, with all their molecular contents in a single cytoplasmic compartment. The other kind, eukaryotic cells, are large and relatively complex, with their biochemical processes compartmentalized among a variety of subcellular organelles. Plants and animals are composed of eukaryotic cells. Integration of biochemical processes occurs between cells of higher organisms, as illustrated by the perfected functions of a trained athlete and the malfunctions of a diabetic.

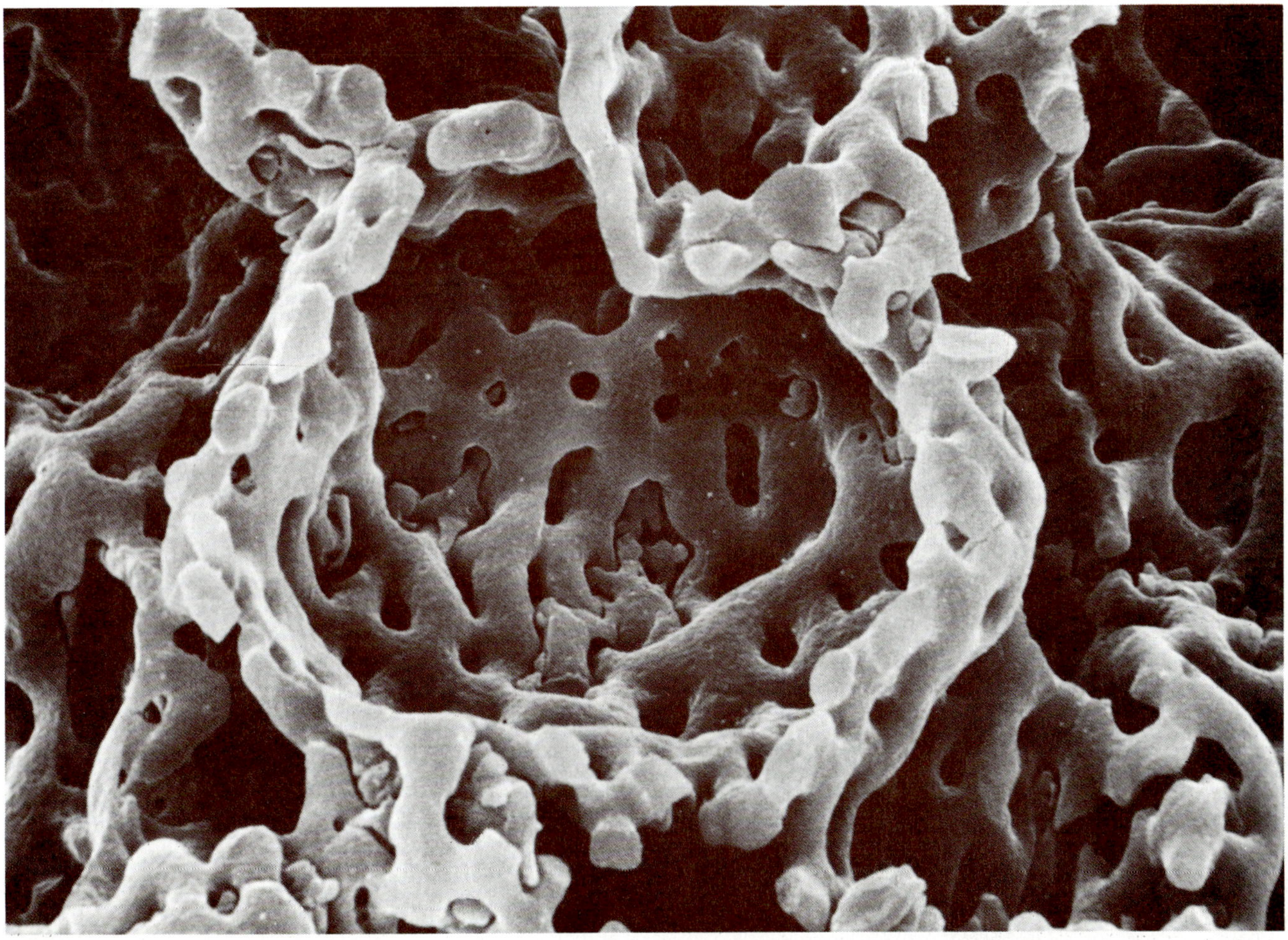

27-1
CHARACTERIZATION OF CELLS

All cells can be divided into two broad categories on the basis of the organization of their chromosomes. In cells of higher organisms (such as plants and animals) and in certain complex microorganisms (such as yeast, amoeba, and paramecium), the chromosomes are contained in a **nucleus.** The nucleus is a subcellular organelle with its own envelope, called the **nuclear membrane.**

Cells possessing a nucleus are termed **eukaryotic** (from the Greek *eu* for "true" and *karyon* for "nucleus"). Eukaryotic cell nuclei contain many distinct chromosomes, often in pairs. (Human cells, for example, contain 46 chromosomes in 22 pairs plus a pair of X chromosomes in females or a set of X and Y chromosomes in males.)

In contrast, the cells of bacteria and blue-green algae lack a nucleus and are termed **prokaryotic.** In prokaryotic cells, the chromosomal material (principally one long circular molecule of DNA) is distributed throughout the cellular contents, or **cytoplasm.** The absolute amount of DNA is much smaller in a prokaryotic cell than in a eukaryotic cell. In general, eukaryotic cells are about a thousand times larger in volume and contain about a thousand times more DNA than prokaryotic cells. For example, a typical bacterium, *Escherichia coli* (which inhabits the human gut), contains about a million base pairs of DNA in each cell, whereas a human cell contains several billion base pairs of DNA.

Prokaryotic cells and eukaryotic cells also differ in several other important respects. In addition to the nucleus, eukaryotic cells contain a number of other organelles, such as the mitochondria where the reactions of oxidative phosphorylation occur. Similar reactions are carried out in prokaryotic cells without such compartmentalization. In fact, mitochondria resemble bacteria in certain respects, and it has been suggested that mitochondria may have arisen from bacterial cells that were incorporated into larger cells and evolved a symbiotic (mutually beneficial) relationship with them.

Mitochondria are thus, in a sense, tiny cells within cells; they possess their own DNA, ribosomes, and other machinery for protein synthesis. Several subunits for the coupling factor of oxidative phosphorylation, as well as several subunits for the enzyme cytochrome-*c* oxidase (that catalyzes the reduction of oxygen to form water), are encoded by the DNA of mitochondria.

In plant cells, photophosphorylation takes place in chloroplast organelles that may have arisen from blue-green algae. Like mitochondria, chloroplasts also have their own DNA and protein-synthesis machinery.

Many eukaryotic cells also contain a complicated network of membranes called the **endoplasmic reticulum,** or ER. Proteins that will ultimately be secreted are first injected into the interior of the ER. From there they are incorporated into small membrane-bound sacs, called **vesicles,** and transported to the **Golgi apparatus** (another membranous network) for secretion. Proteins destined to follow this secretion pathway contain a **signal sequence** of amino acids at the N-terminus that recognize receptors on the ER. In this way, the ribosomes synthesizing such proteins are led to the surface of the ER, and the newly synthesized protein is fed directly into the ER. Once inside the ER, these signal sequences (usually 15 to 30 amino acid residues, including several hydrophobic residues) are cleaved from the rest of the protein and hence do not appear in the secreted product.

The various components of eukaryotic cells described so far are illustrated in Figure 27-1. This figure also includes a number of other important structures, such as **lysosomes**—large vesicles containing degradative enzymes that must be kept in isolation to protect the rest of the cell contents and only used when some cell digestion occurs (as during fasting). Animal cells also contain structures called **centrioles,** or **centrosomes,** which participate in the separation of chromosomes during cell division.

When cells are examined at high magnification, a

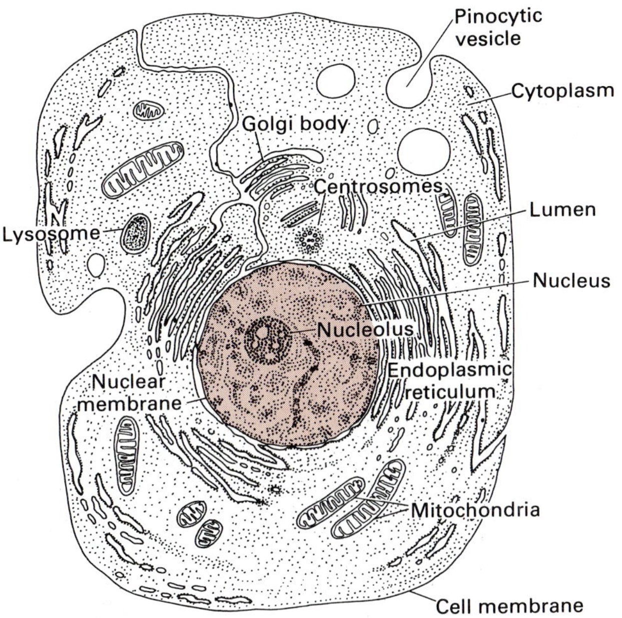

FIGURE 27-1

A "typical" eukaryotic cell. [After J. Brachet, "The Living Cell." Copyright © 1961 by Scientific American, Inc. All rights reserved.]

variety of smaller filamentous structures are revealed that fill the cell with a lacy network. This network includes **microfilaments** (about 50 nm in diameter), somewhat larger **intermediate filaments,** and hollow, tubular structures called **microtubules** (about 25 nm in diameter). These various filamentous structures are collectively called the **cytoskeleton;** they endow cells with motility and influence their shape.

Within a eukaryotic cell, events must be coordinated between organelle and cytoplasmic enzymes, as in carbohydrate metabolism (Chapter 26). We noted that NADH + H^+ can be used to reduce pyruvate to form lactic acid, or the NADH + H^+ can be used by mitochondria to form ATP in oxidative phosphorylation if sufficient oxygen is present. In the body, coordination also occurs between events in diverse tissues such as the liver and muscle. The processes involved in integrating events within and between cells function at maximal efficiency during certain athletic exercise that places great stress on the systems involved. Therefore, for a fuller exposition of the integration of biochemical processes, we turn to a close examination of an especially demanding form of exercise.

27-2 MAXIMIZING BIOCHEMICAL PROCESSES: A DAY IN THE LIFE OF A MARATHONER

Marathon running is probably the form of athletic activity that best illustrates the integration of the various biochemical processes of the human body. In the well-trained athlete, the biochemical functions are close to optimal levels, and the various factors that contribute to marathon performance touch upon most of these functions. For a premier marathon runner such as Bill Rodgers, dedication, drive, and determination are obviously crucial ingredients in success that cannot be evaluated in biochemical terms. However, if we follow the bioenergetic processes that occur during a marathon, we can learn a great deal about how reactions in various parts of the body are integrated.

As the runners stand at the starting line, tension

FIGURE 27-2

Start of the Finger Lakes Marathon, October 8, 1978. [Courtesy of Raymond Pompilio.]

builds until the gun fires and the runners are swept along on a wave of excitement. Although the competitors have at least two hours of running ahead of them, the elite marathoners sprint away from the starting line at a very fast pace (Figure 27-2). For the first 10 to 20 seconds, they can rely on short-term stores of energy already present in the muscles, particularly in the form of creatine phosphate. The creatine phosphate is used to phosphorylate ADP to form ATP in a reaction catalyzed by the enzyme creatine kinase:

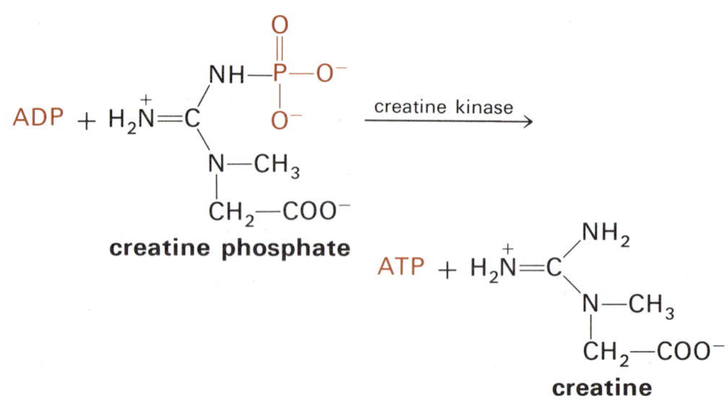

The hydrolysis of ATP back to ADP and P_i by the muscle protein myosin provides the driving force for muscle contraction and propulsion of the runner's body.

As the available creatine phosphate is consumed in these first seconds of the race, the next source of energy is tapped—muscle glycogen. In general, glucose is stored in the body in the form of the branched polymer glycogen (see Chapter 23) both in muscle and in the liver. The glucose available in the muscles can be tapped for additional energy by processing in glycolysis with the formation of lactic acid. However, this source of energy is likely to provide useful work for only an additional two or three minutes, until the full process of oxidative phosphorylation takes over.

The leading marathoners usually run the first mile in times much faster than the average time of about five minutes per mile for the full 26.2 miles. These quickly expended forms of energy, creatine phosphate and muscle glycogen, are responsible for this burst, but they leave a quantity of lactic acid to be dealt with. The lactic acid escapes into the blood and is then salvaged by the liver and used to regenerate glucose.

The formation of glucose from lactic acid by the liver is called **gluconeogenesis** (meaning "creation of glucose anew") and illustrates integration of the cells of the liver and muscles. The two systems cooperate, so that while the muscle is busy producing lactic acid, the burden of lactic acid accumulation can be relieved by the liver. The reactions of gluconeogenesis are essentially a reversal of glycolysis, with the addition of several new reactions to circumvent those reactions that are not readily reversible (principally the ones involving phosphorylations and having a large free energy increase).

Thus various biochemical processes contribute to the bioenergetics of the first few minutes of a marathon, but as these short-term energy sources are expended, the main sustaining reactions take over. Basically these reactions are the breakdown of glycogen to glucose and the processing of glucose by glycolysis to the point of pyruvate, followed by the reactions of the citric acid cycle and oxidative phosphorylation. This process is initiated by **epinephrine** (adrenaline), one of the important small molecules produced by the nervous system.

Epinephrine, which is generated in greater amounts in times of excitement, is released by the nerve cells and reaches the cells containing glycogen. There it triggers a reaction common to the action of many chemicals of the nervous system and hormones. This reaction is activation of the enzyme embedded in the membranes of the cells and called **adenyl cyclase.** This enzyme catalyzes the formation of a derivative of ATP, 3′,5′-cyclic AMP (abbreviated cAMP):

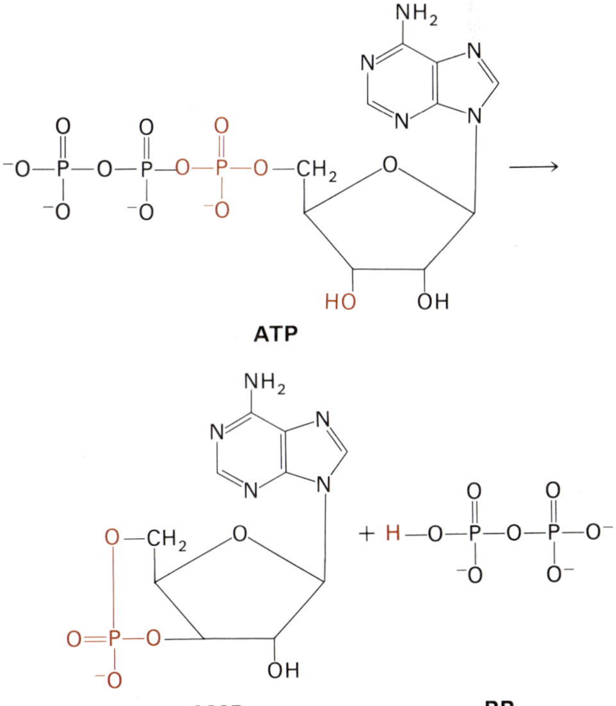

ATP

cAMP

PP_i

A great many cellular processes are triggered in a similar manner—the hormone controlling the process triggers formation of cAMP, and the internal machinery of the cell then responds to the cAMP as the "second messenger" to alter biochemical reactions. In the case of glycogen-rich cells, the cAMP formed in this way following epinephrine stimulation causes a change in the enzyme responsible for breaking down glycogen in glucose units. This enzyme is called **glycogen phosphorylase,** because the glucose units split from glycogen are phosphorylated in the process. Thus the reaction (which is described in Section 23-8) may be summarized as

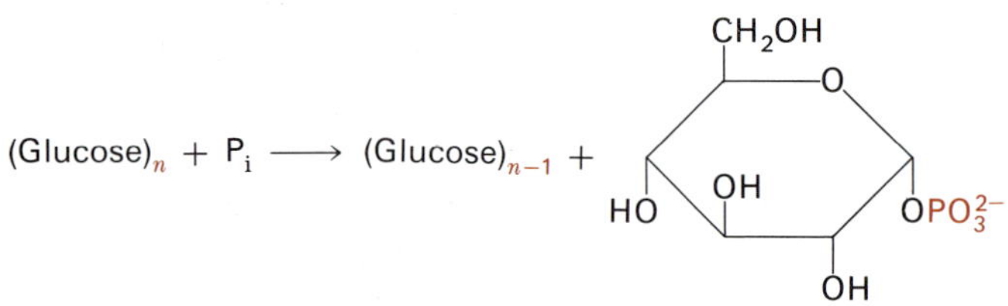

glycogen **glycogen** **glucose 1-phosphate**

The glucose 1-phosphate is converted to glucose 6-phosphate for processing in glycolysis by an enzyme called phosphoglucomutase. Glycogen phosphorylase can exist in two forms, a highly active "A" form and a relatively inactive "B" form. The A form is derived from the B form by the phosphorylation of a key serine residue at the expense of ATP. The increased levels of cAMP produced by epinephrine stimulate enzymes that carry out the phosphorylation of the key serine residue. In this way, the excitement of the marathon in the runner's mind leads to formation of epinephrine, which via cAMP produces increased formation of glucose.

During the long haul of the marathon, glucose derived from glycogen stores will provide the bulk of the energy for the runners. Some contributions are also made from fatty acid oxidation, especially in the later phases of the marathon, but glycogen levels can be the main factor in determining success or failure.

This important role for carbohydrates is widely recognized, and some coaches view the dietary preparations for a marathon as one of the critical factors in a runner's performance. By and large, the more carbohydrate reserves, the better will be the running performance in a race of the duration of marathons. Special efforts are often made to deplete the body of carbohydrates and then "load up" just before the race. The belief is that the depletion phase will cause an "overshoot" in the loading phase, to give exceptionally high levels of glycogen at the time of the race. There is some scientific evidence in support of this belief and a good deal of folklore among runners about it. As a result, consumption of spaghetti and other dishes high in starches is part of the premarathon preparation.

This emphasis on loading up with carbohydrates is related to the fact that no net production of glucose can be achieved from fats. Glucose can be formed from amino acids under certain conditions, but this route does not contribute significantly in healthy individuals. Therefore, the only realistic way to increase body carbohydrate is to eat carbohydrate. (The body has no trouble making fats from carbohydrates, because fats are formed from acetyl-S-CoA units; see Chapter 24.) What contributes to making marathons such dramatic events is that even with carbohydrate loading, the best estimates are that only enough glycogen is present to carry a runner about 20 miles. Thus, it is during the last six miles that will power (and fatty acids) take over.

Some fatty acid oxidation occurs throughout the marathon, but in the final miles (as glycogen stores are depleted) fatty acids become the principal energy source. It has been suggested that women are especially suited to marathon events (there is a long standing tradition of successful women marathon swimmers) because of higher fat levels in women than in men.

As noted in Chapter 26, fatty acids are rich sources of ATP and can be readily metabolized in mitochondria. One reason for the success of certain runners of marathons and ultramarathons (races of 50 to 100 miles) may be that they have especially efficient fat-utilization mechanisms. One side effect of reliance on fats for energy is the formation of acetone and related compounds, called **ketone bodies.** Formation of ketone bodies can become prominent during fasting, in diabetes (Section 27-3), and during carbohydrate-free diets for weight reduction.

Ketone bodies accumulate when metabolism relies heavily on fatty acids. With intensive fatty acid metabolism, the citric acid cycle is overloaded, acetyl-S-CoA units accumulate, and pairs of acetyl-S-CoA molecules condense to form acetoacetyl-S-CoA:

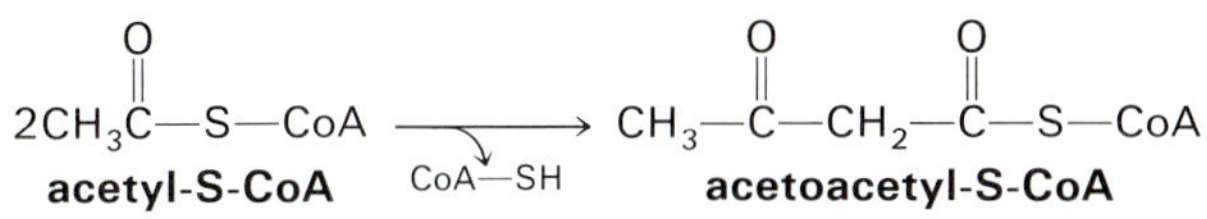

The acetoacetyl-S-CoA is then hydrolysed to yield acetoacetate, which can be further metabolized by reduction or decarboxylation to yield β-hydroxybutyrate or acetone:

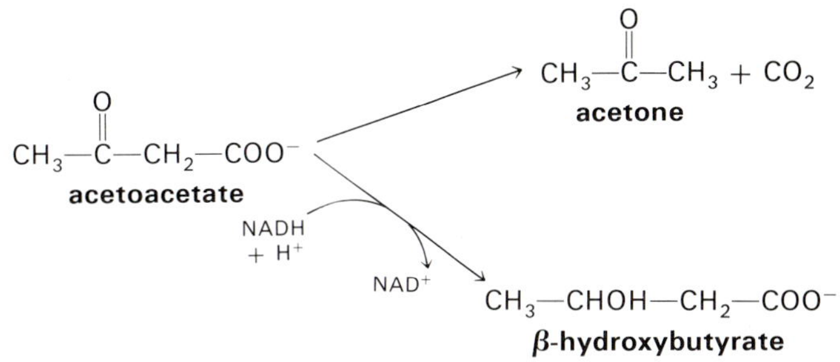

These three substances (acetoacetate, acetone, and β-hydroxybutyrate) are known collectively as ketone bodies. The large amounts of these substances that accumulate in the later stages of a marathon irritate the joints and cause some of the distress associated with long-distance running. Their effects on the brain may alter consciousness, so they may be partly responsible for the experience known as "runner's high."

With the production of ketone bodies described, we have completed the description of the various phases of metabolism experienced by the long-distance runner. Other aspects of endurance events, such as the fight against dehydration and loss of salts, play a role in success or failure but are not considered further here. Instead we turn to a major metabolic problem that also involves ketone bodies.

27-3 METABOLIC DISTURBANCES ASSOCIATED WITH DIABETES

It has long been recognized that humans can suffer a debilitating disease associated with excretion of sugar in the urine. Water is lost with the glucose, so in the extreme phases individuals suffer both hunger and thirst. This disease, called **diabetes,** was identified as arising from a deficiency of the hormone **insulin.** The link between diabetes and insulin was established in the 1920s by F. G. Banting and C. H. Best, two of Canada's most famous scientists, working in Toronto. Following an earlier observation that dogs without pancreases show symptoms similar to those of diabetes in humans, Banting and Best prepared extracts of pancreas that reversed the symptoms in dogs.

In 1922 the first treatment of a human was carried out on a 14-year-old boy, emaciated from the disease that was causing him to excrete daily about 4 ℓ of urine highly concentrated in glucose. The outlook for his survival was grim. However, the adverse symptoms were dramatically reversed by administration of the active extract prepared by Banting and Best. Eventually the active ingredient, insulin, was purified and characterized (see Table 21-3), and convenient commercial preparations from slaughterhouse materials were developed. Currently, there is considerable interest in the synthesis of human insulin in bacteria using synthetic genes (Chapter 25).

Administration of insulin greatly aids the management of diabetes, but it rarely is possible to control the insulin levels with the same degree of "fine tuning" as is present in normal individuals. As a result, glucose levels fluctuate outside their usual limits, and high levels of free glucose may cause damage to many of the organs and tissues known to be involved in the complications of diabetes.

An illustration involving hemoglobin of the way high levels of free glucose can alter the natural constituents of the body has recently been discovered. Normally about 5% of hemoglobin contains a glucose molecule covalently attached to the amino-terminal position of the β chains (a Val residue). Apparently, the glucose units become attached by first binding to form a Schiff base (Section 18-6), and then undergoing rearrangement to form a ketoamine:

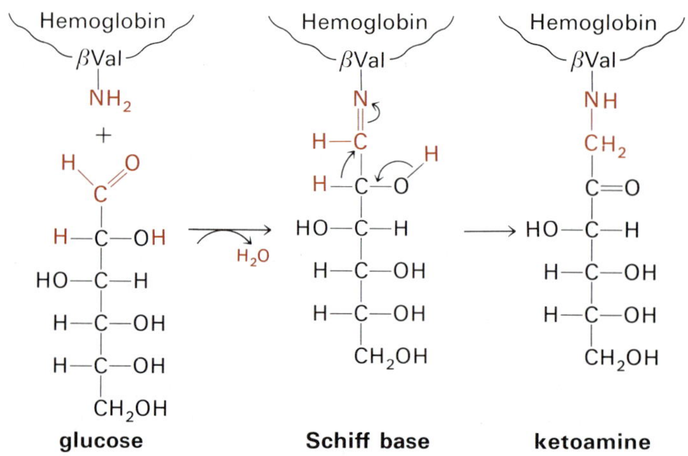

In diabetics, the fraction of glycosylated hemoglobin (hemoglobin that is bound to glucose) is 2 or 3 times that in normal individuals. Thus glycosylation may be used as a convenient indicator of average blood glucose levels. In addition, the glycosylation reaction may itself provide insights into the complications associated with diabetes, because some of these complications may be due to glycosylation of other critical proteins.

How does insulin regulate glucose levels? One of the main functions of insulin is to facilitate the uptake of glucose by cells. In the absence of insulin, glucose cannot be transported into cells. Instead it accumulates in the blood and is excreted in the urine. Glycogen levels become depleted because glucose cannot be taken up by the liver cells to restore glycogen. The body overreacts to this situation by making more glucose at the expense of proteins. In the absence of glycogen, the body must rely on fats for energy, causing an accumulation of acetyl-S-CoA and the formation of ketone bodies. "Acetone breath" is one of the noticeable signs of diabetes.

KEY TERMS

The *nucleus* is the organelle of the cells of higher organisms that contains the genetic material, DNA, packaged into structures called *chromosomes.*

A *eukaryotic* cell is one with a true nucleus, in contrast to a *prokaryotic* cell, which is smaller and has its genetic material dispersed throughout its contents, or *cytoplasm.*

The *endoplasmic reticulum* (ER) is the complicated network of membranes involved in secretion (along with the other membraneous network, the *Golgi apparatus*).

A *signal sequence* is a string of hydrophobic amino acids found at the end of proteins that are destined to pass through a membrane.

Lysosomes are vesicles in which enzymes that could be damaging to the cells are packaged until needed.

The *cytoskeleton* is a fibrous network of proteins that includes *microfilaments, intermediate filaments,* and *microtubules.*

Gluconeogenesis is the process of glucose synthesis from lactic acid.

Cyclic AMP (3′,5′-cyclic adenosine monophosphate, or cAMP) is commonly the second messenger formed within cells in response to a hormone that triggers its formation from ATP by stimulating the enzyme *adenyl cyclase.*

Glycogen phosphorylase converts stored glycogen into glucose 1-phosphate units for further metabolic processing. It is controlled by the interconversion of a highly active A form and a relatively inactive B form.

Ketone bodies are the compounds (such as acetone) produced by conditions (such as endurance athletics or diabetes) that place great reliance on oxidation of fatty acids for energy.

Diabetes is the disease produced by insufficient insulin.

CHAPTER SUMMARY

Cells can be grouped into two categories: prokaryotic cells (such as bacteria) and eukaryotic cells (such as those that make up plants and animals). *Prokaryotic cells* are small and simple, with the genetic material distributed throughout the cell; *eukaryotic cells* are large and complex, with many compartmented organelles. The largest organelle of eukaryotic cells typically is the *nucleus,* which contains the genetic material packaged into *chromosomes.* Other organelles of eukaryotic cells include *mitochondria* (and *chloroplasts* in plant cells). Eukaryotic cells may also contain an *endoplasmic reticulum,* a *Golgi apparatus,* and a *cytoskeleton.*

Metabolic events are coordinated between the organelles and cytoplasm of individual cells, as well as between the cells of different tissues. Aspects of this coordination can be revealed by considering the processes that take place during the running of a marathon. Initial energy is provided by high-energy stores of the muscle, such as creatine phosphate. Next, glucose is tapped to produce energy. Finally, fatty acids take over as the principal source of energy. Breakdown of glycogen into glucose is stimulated by epinephrine, through the enhancement of cyclic AMP levels in liver cells. The high levels of cyclic AMP lead to a conversion of the enzyme responsible for glucose formation from glycogen (glycogen phosphorylase) to a relatively active form.

Reliance on fatty acids for energy can lead to the production of *ketone bodies* (compounds such as acetone). Ketone-body formation occurs in the metabolic disease *diabetes,* which results from a deficiency in insulin production. In the absence of insulin, glucose cannot be effectively utilized, and it accumulates in the blood and is excreted in the urine. Diabetes can cause death, but administration of commercially-produced insulin can greatly aid in control of the disease. Some of the complications associated with diabetes may arise from nonenzymatic glycosylation of proteins.

A red blood cell from a patient with sickle-cell disease. Scanning electron micrograph, $\times$14,000.

28

Biochemistry and Society

The active intervention of human beings in the biochemical processes of life began on a large scale with drugs such as penicillin and other antibiotics that fight infectious disease. More recent studies have focused on certain genetic diseases and major current causes of mortality—heart disease and cancer. Biochemistry is also impinging on the realm of human behavior through the development of psychoactive drugs. On a global scale, the insights of biochemistry may also contribute to reducing the world food and energy crises.

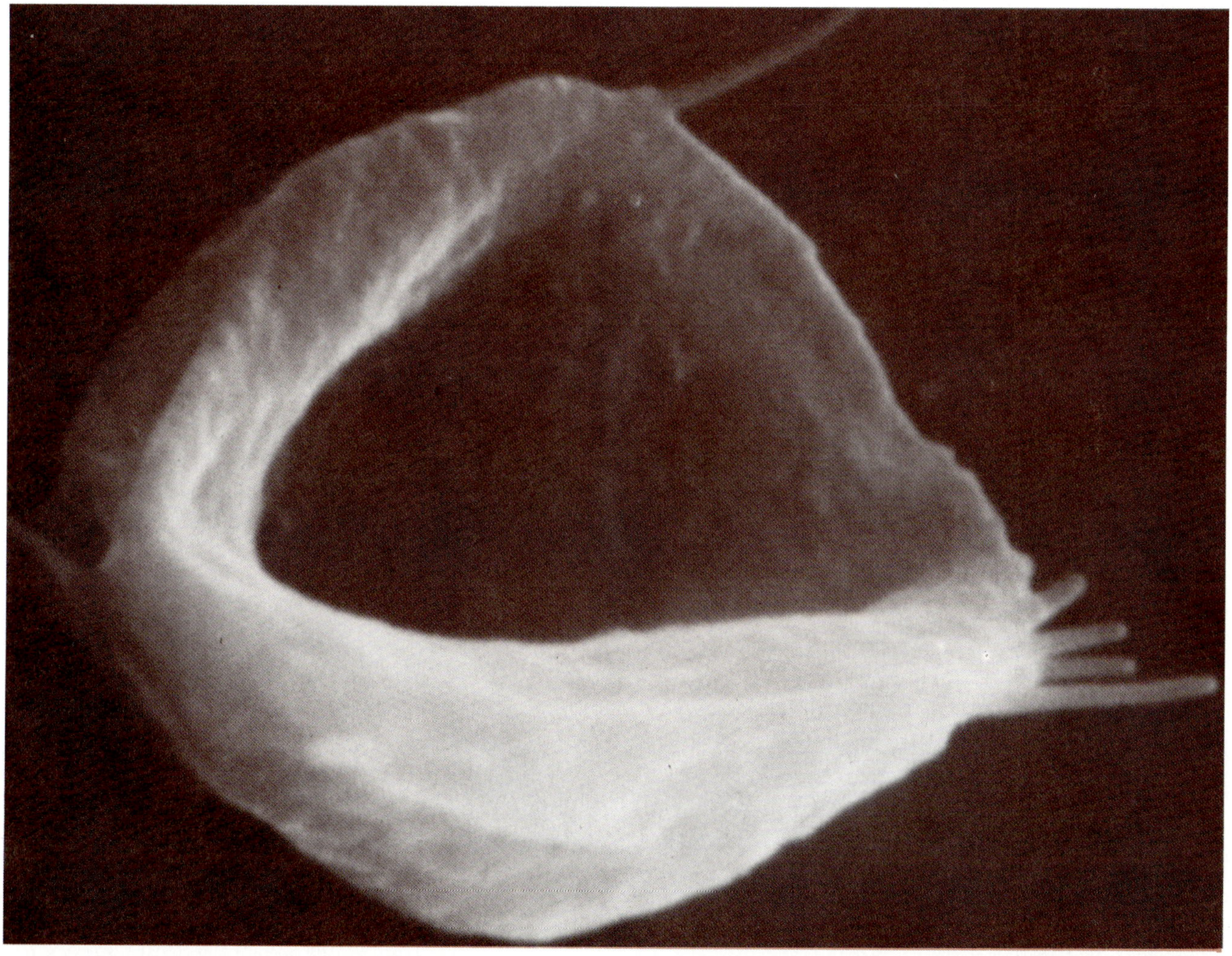

28-1
BIOCHEMISTRY AND MEDICINE

The discovery of molecules that counteract the diseases of human beings has origins in the prescientific traditions of folk medicine. In many parts of the world, herbs and other plant products have been used medicinally for centuries. In some cases, the active ingredient has been isolated and its chemical structure established. One of the most complete of these stories concerns **quinine,** a potent antimalarial agent from the bark of the cinchona tree native to Peru.

The powers of the bark were reported by a Jesuit in South America in 1633, and some years later a "Roman connection" was established to ship the bark to the Vatican, where it was given to pilgrims from all over Europe to speed its distribution. By 1820, the active ingredient, quinine, was isolated by two French scientists, and the availability of the drug in relatively pure form increased its usefulness in the treatment of malaria. The discovery of quinine was a great benefit to civilization; malaria, with its recurring chills and fevers, was an unpleasant and often fatal disease prevalent in many parts of the world. The disease is carried by mosquitos that inject the protozoan parasite, *Plasmodium,* that causes the disease. The parasites multiply in the red blood cells.

It is ironic to note that sickle-cell disease, the most prevalent of genetic diseases yet discovered, owes its high incidence to the fact that it provides some resistance to malaria. Thus in regions where malaria was epidemic, the carrier form of the disease (arising when the sickle-cell gene is inherited from only one parent) lessened the chances of death from malaria. More and more carriers survived, until in some parts of Africa the incidence of the sickle-cell gene is near 50%. A more complete description of sickle-cell disease is presented in Section 28-3.

The structure of quinine (next column) was proposed in 1907, and its validity was established through a complete synthesis of quinine by Robert Woodward and William Doering in 1945. The synthetic material was found to have full antimalarial activity.

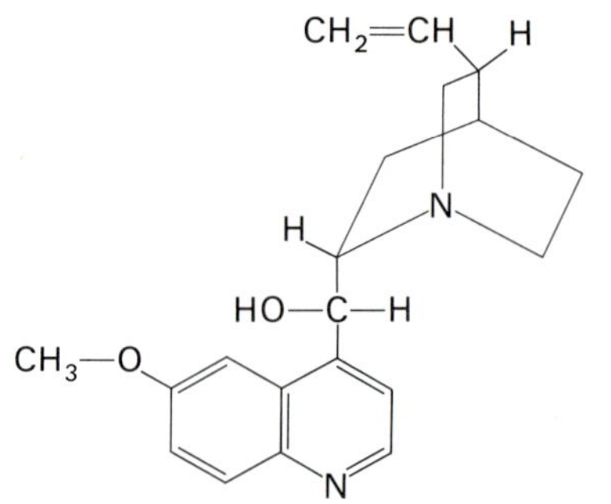

quinine

The age of modern chemical warfare against disease began when **sulfanilamide,** a chemical used in the dye industry, was found to have toxic effects on the bacteria responsible for a number of diseases, including pneumonia, tuberculosis, scarlet fever, and plague. Sulfanilamide acts through its resemblance to *p*-aminobenzoic acid, a component of folic acid (see Section 22-6). Although still useful in treating certain bacterial infections, sulfanilamide and related compounds have been replaced to a large extent by penicillin and other antibiotics.

An **antibiotic** is defined as a chemical, produced by one type of microorganism, that kills or inhibits growth of another type of microorganism. The discovery of the first important antibiotic, **penicillin,** can be traced to Alexander Fleming in 1929. He noticed that a culture of bacteria had been contaminated with a mold, and that the bacteria had been destroyed in regions near the mold. The mold, called *Penicillium notatum,* was evidently producing a substance toxic to the bacteria. The possibility that this material might be useful in treating infection did not occur to Fleming, but in 1940 two British scientists followed up the report of his work and set about isolating the active compound, penicillin whose structure is discussed in Section 22-3.

Other antibiotics discovered more recently have also come into prominence, beginning with **streptomycin** (discovered in 1943) and progressing to the currently popular **tetracycline** family of antibiotics.

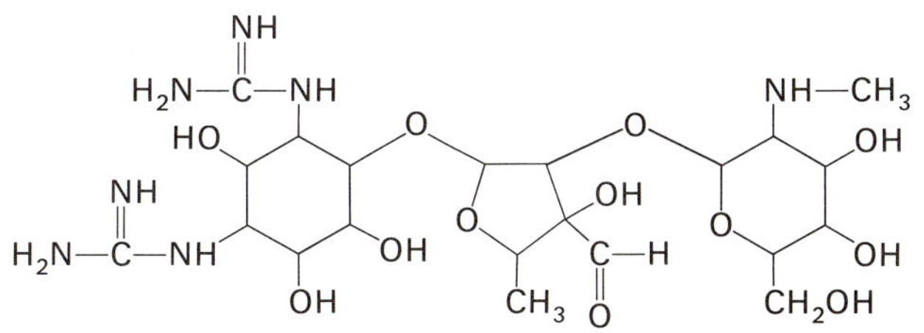

streptomycin

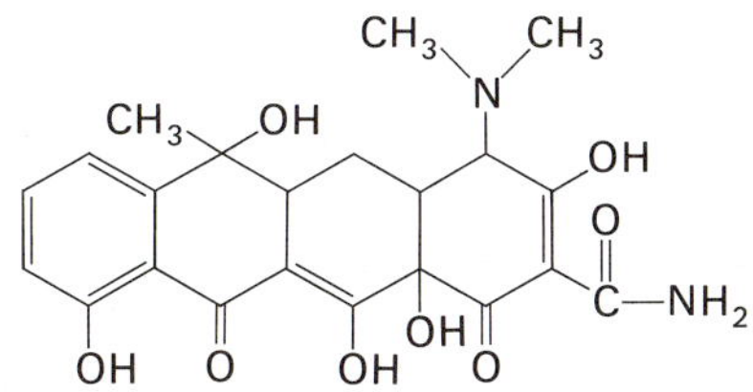

tetracycline

28-2 THE FIGHT AGAINST VIRAL DISEASES

Throughout history, the most dangerous enemies of human beings have become smaller and smaller. In the early stages of human existence, the greatest threats came from large predatory animals and other human beings. Although the threat from other human beings is still real, the dangers from animals are now far less than the danger human beings pose to animals.

However, as people became more and more crowded in cities and towns, the infectious diseases transmitted by bacteria became a major cause of death and illness. The development of antibiotics reduced much of the risk from bacteria, leaving the danger from even smaller enemies, the **viruses.**

Viruses are parasites of cells; a virus is made up of small chromosomes contained within a shell of protein. The viruses are incapable of growing by themselves, but if they invade a cell, the machinery of the cell can be commandeered to replicate the viral chromosome and transcribe and translate the viral genes into viral proteins. Depending on the specific characteristics of the virus and the host cell, the virus may multiply and destroy the cell, or it can coexist with the cell in such a way that the virus chromosome is incorporated into the chromosome of the cell. In this later case, the cell is often **transformed** by the virus and takes on entirely new properties.

Among the early diseases that were identified as being caused by viruses were smallpox and polio. For these diseases, which can be transmitted from one individual to another, it has been possible to develop vaccines. A **vaccine** is an inactive form of the virus with which people can be inoculated to stimulate their immune systems to produce antibodies against the virus. If inoculated individuals later encounter active virus, their immune systems will already have the antibodies to fight it. This process, called **immunization,** has been highly successful in eliminating such diseases as polio and smallpox.

As infectious diseases have been brought under control and the life expectancy of human beings has been extended, the threats from the class of diseases known as **cancer** have become more apparent. A viral cause for cancer has been suggested, but a role of viruses in most human cancers has not been established. Cancer cells are characterized by the loss of normal growth-control ability, leading to formation of a mass of rapidly growing cells (a tumor) or other abnormalities. It has been possible to mimic the transformation from normal to cancer-type behavior in cells isolated from human beings and grown in the laboratory by infecting the cells with certain viruses known to cause cancer in animals. These results suggest a role of virus in human cancer, but conclusive proof is lacking.

A complicating factor is the evidence that cancer can be caused also by exposure to radiation or to certain toxic chemicals (**carcinogens**), as in the case of lung cancer associated with cigarette smoking. In studies with a simple bacterial system introduced by Bruce Ames, it has been shown that virtually all carcinogens are **mutagens**—that is, they cause alterations in the base sequence of DNA. Many compounds have now been identified as mutagens and thus presumably are carcinogens. The compounds include items that come into widespread contact with people, such as hair dyes and a flame retardant known as TRIS used in childrens' sleepwear. In the case of the latter, mounting concerns led to the prohibition of the material.

Efforts to identify viruses that may be involved in cancer are continuing actively, but there is increasing support for the view that cancer is triggered by chemicals in the environment. This view is strengthened by studies showing that cancer rates of all kinds are highest in regions where industrial activities that produce chemical pollution of the environment are concentrated. Many questions remain to be answered in the area of cancer research, and some pessimism may be warranted about the outlook for reducing cancer rates, particularly in the case of lung cancer. The relationship between cigarette smoking and lung cancer has been established beyond any reasonable doubt, yet cigarette smoking continues at a surprisingly high level.

28-3 THE ULTIMATE FIGHT: GENETIC DISEASES

As one level of biological threat after another is brought closer to elimination—predatory animals, then microbes, viruses, and toxic chemicals of the environment—we are left with diseases arising from the basic properties of biochemical functioning. *Homo sapiens* arose through the process of evolution, which is based on variations in the DNA of the chromosomes. During replication of DNA, the new strands are formed according to the A–T and G–C basepairing scheme described in Chapter 25. Although this process is extraordinarily precise, a very low level of error is present: one mistake occurs for roughly every billion base pairs formed. Gross rearrangements of genetic material may also occur, but at an even lower rate.

Single base changes may have no effect on the amino acid sequence if the mutation involves a change from one codon to another codon specifying the same amino acid. For example, an error substituting A for G in the codon UCG would not change the amino acid for which the triplet codes, because both UCG and UCA code for serine. Mutations can also be neutral, changing an amino acid that is not critical for the function of the protein—such as a change from a lysine to an arginine on the surface of the protein. In some cases the mutation will have a deleterious effect, producing a protein that functions poorly and impairs the reproductive ability of the organism. In a small fraction of the cases, a single mutation can be fatal, so that the organism produced with this mutation never reaches full development. Occasionally a mutation will improve the ability of the organism to function in its environment. Then the organism will thrive and reproduce abundantly, and a new line of evolution may be launched.

Thus the biochemical basis of life leads to constant change, continuous flux. Certain fundamental aspects appear to be "locked in"—the genetic code itself, for example, and the basic patterns of metabolism seem to be universal and pretty much constant. In the amino acid sequences of individual proteins, however, changes are always taking place. For example, over 200 mutant forms of human hemoglobin have been discovered, most with changes in single amino acids that can be related to a single base change in the DNA.

Mutant forms of enzymes are also known. Some of these mutants cause functional deficiencies that impair the health of the individual and lead to the broad range of afflictions called **genetic diseases.** Different genetic diseases have been identified that affect virtually every pathway of metabolism. To understand the general problems of genetic diseases, we now consider the most thoroughly studied of the genetic diseases, **sickle-cell disease.**

The incidence of most genetic diseases is very low, for two reasons. First, two copies of each gene are inherited, one from the mother and one from the father, so most genetic diseases are expressed only if a mutant form of a gene in inherited from *both* parents, and such coincidences are rare. The exception to this situation is for genetic diseases on the X or Y chromosomes. Each of these chromosomes is present singly in the male. Hemophilia (or bleeder's disease), for example, is the result of a defect on the X chromosome inherited from the mother.

For most genetic diseases there is no benefit from having the mutant form of the gene. The special feature of sickle-cell disease is that, as already noted, the sickle-cell gene does afford some resistance to malaria. Thus individuals with the sickle-cell gene on one chromosome benefit from the malaria resistance but are also virtually free of any harmful effects of sickle-cell disease. The selective advantage of having one sickle-cell gene led to a rise in the ability of these individuals to survive and reproduce, and after many generations in certain parts of Africa almost every adult had one sickle-cell gene.

Malaria is no longer a major problem, because the use of pesticides has controlled the mosquitos that transmitted the disease. However, population groups descended from Africans still have a high incidence of the sickle-cell gene. In the case of black Americans, about 10% have a single sickle-cell gene, and about 50,000 individuals have sickle-cell disease. Other ethnic groups have particular genetic diseases (such as Tay-Sachs disease among European Jews, and Cooley's anemia among Italians and Greeks), but usually the incidence is much lower because none of these diseases carries any survival advantage. Because of the relatively high incidence of sickle-cell disease, it has received considerable attention, and many of the molecular steps in the disease can be described.

Sickle-cell disease has been recognized for many centuries in Africa. It even has names in several West African tribal languages, with onomatopoetic qualities that suggest the recurring painful crises associated with the disease: *chewechweechwe* in Ga, *ahotutuo* in Twi, and *nuidudui* in Ewe.

In the West, sickle-cell disease was first reported by J. B. Herrick in 1910. He observed the characteristic sickle shape (Figure 28-1) in the red blood cells of an anemic patient in Chicago. The molecular nature of the disease was established in 1949 by Linus Pauling and three colleagues, who observed that the hemoglobin isolated from sickle-cell patients has a different charge from that of normal hemoglobin (Figure 28-2). The molecular basis of this charge difference was established by V. M. Ingram in 1956. He observed that in hemoglobin from patients with sickle-cell disease, the two α chains are normal, but the two β chains are altered. The difference in the β chains is substitution of a neutral valine residue at the sixth position in place of the negatively charged glutamate residue normally found there.

The change from glutamate (Glu) to valine (Val) corresponds to a change in a single base in the DNA: a

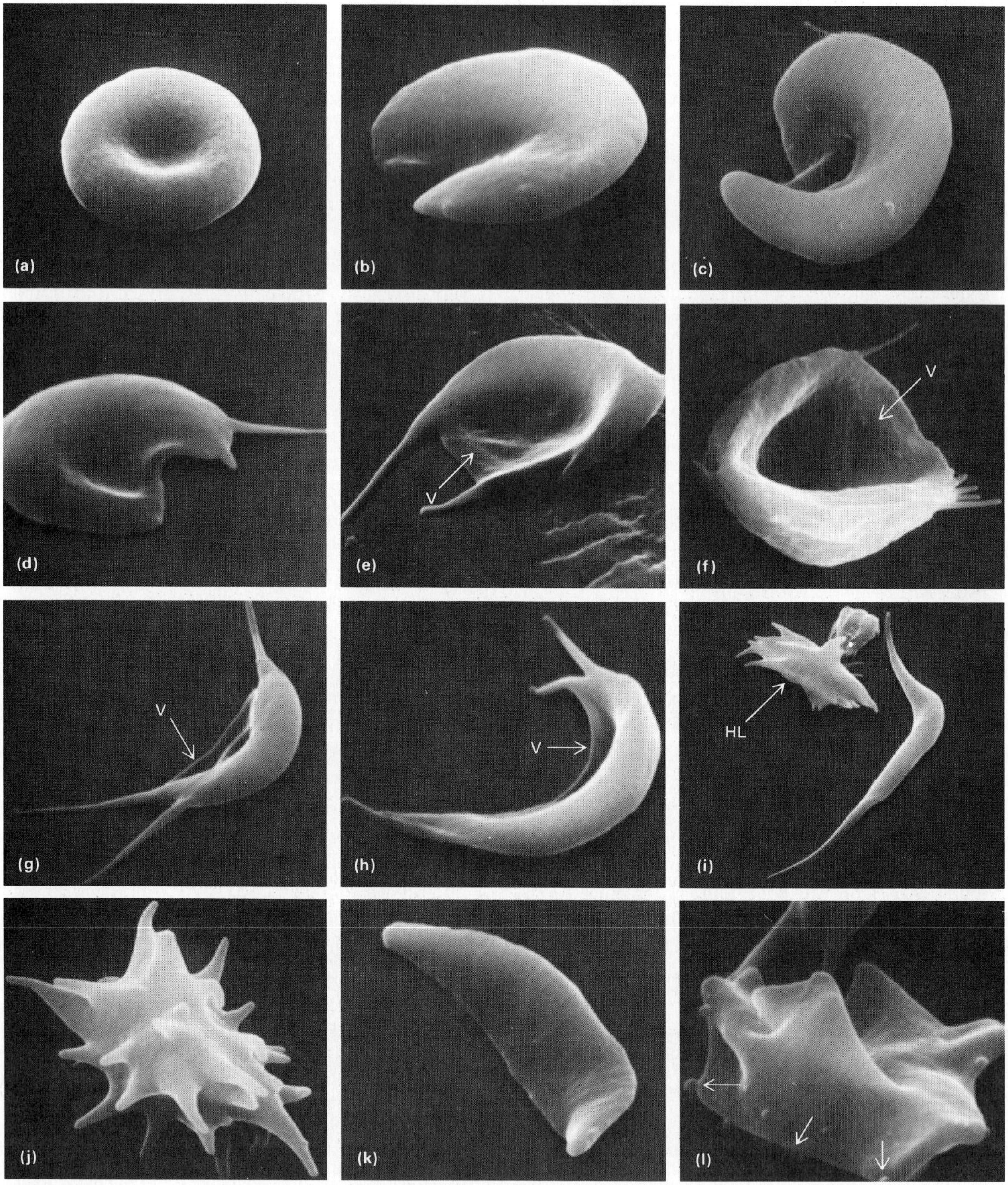

FIGURE 28-1

Scanning electron micrographs of normal (a) and sickled (b–l) red cells. The various sickled forms illustrated are representative of the shapes seen in patients with sickle-cell disease. These include cells with a very thin membrane veil (V in e–h) and the holly-leaf form (HL in i). [Courtesy of James G. White.]

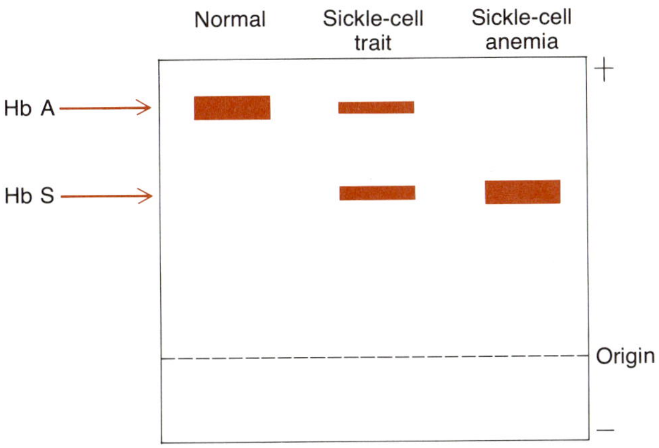

FIGURE 28-2

Electrophoresis of normal and sickle-cell hemoglobins. [From Lubert Stryer, *Biochemistry,* 2nd ed., W. H. Freeman and Company. Copyright © 1981.]

change from an A to a T converts the GAA codon for Glu to the GTA codon (in the DNA, or GUA in the RNA) for Val. The change from a charged residue (Glu) to a hydrophobic one (Val) lowers the charge on the molecule, reducing its solubility. At the very high concentrations of hemoglobin in the red blood cells, the less soluble sickle-cell hemoglobin molecules (called **hemoglobin S**) stick together to form long fibers (Figure 28-3) that force the cell into the abnormal sickle-shaped form. These misshapen cells clog the capillaries of the body, preventing oxygen delivery and leading to the many symptoms of the disease (swollen joints, leg ulcers, and brain and heart damage that may lead to death).

So much is known about sickle-cell disease that it presents a challenge to biochemistry and medicine to develop a treatment. For a brief period, cyanate was administered as a treatment. It carbamylates the α-amino groups of the protein (top of next column) and appeared to lessen sickling. However, carefully controlled studies revealed that the antisickling effects are marginal, whereas the treatment causes some serious

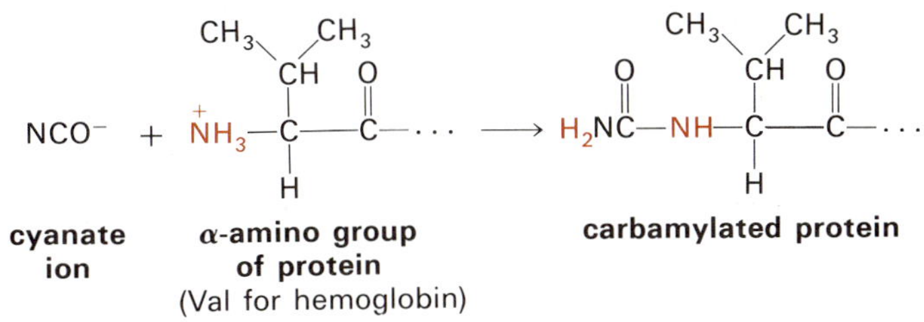

neurological complications as side effects. Thus cyanate treatment was dropped, and at the present time no other clinically useful antisickling agents have been developed.

In the case of another genetic disease, **phenylketonuria,** detection at birth can lead to successful treat-

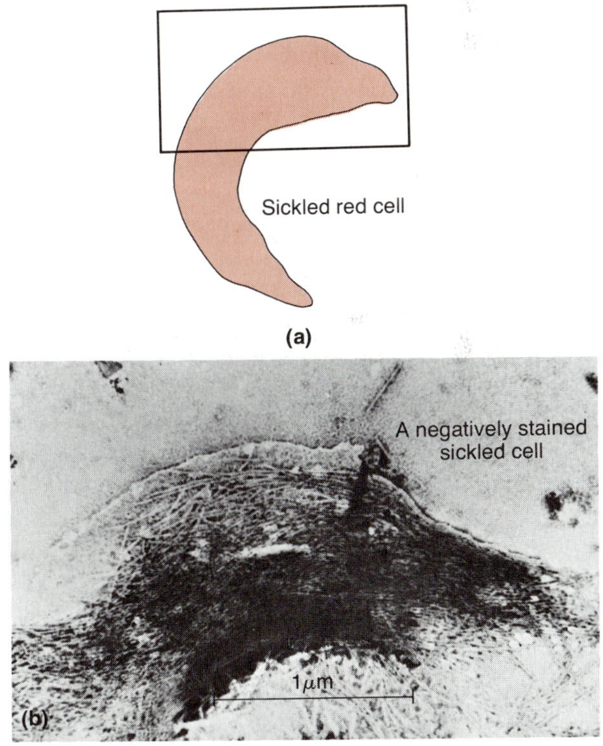

FIGURE 28-3

Fibers within sickled cell.

ment. This disease arises from the absence of an enzyme that oxidizes the amino acid phenylalanine (see Chapter 21). As a result, phenylalanine accumulates, and the consequences can include permanent mental retardation. However, if the disease is detected at birth, the infant can be placed on a diet low in phenylalanine, and this will permit normal development. In many states, children are now routinely screened at birth for this disease.

Certain other diseases can even be detected before birth, at about the fourteenth week of pregnancy, by a process called **amniocentesis.** Cells are withdrawn from the amnionic fluid around the fetus and cultured. Examination of the chromosomes of these cultured cells can reveal the presence of certain diseases (such as Down's syndrome) that involve chromosomal abnormalities. In addition, testing of the cultures for certain enzymes can also reveal genetic diseases involving enzyme deficiencies. In cases where a severe genetic disease is detected, the fetus is still in an early enough period of the pregnancy that a therapeutic abortion can be considered.

28-4 GENETIC ENGINEERING AND RECOMBINANT DNA

Increased knowledge of biochemical processes and their abnormalities has led to specific chemical treatments that lessen the impact of disease, but there has been speculation that intervention at the level of DNA may someday be possible. The idea is that some form of genetic manipulation (along the lines that achieved the synthesis of the hormone insulin in bacteria, described in Chapter 25) could be used to alter the genetic makeup of human beings. The thought is awesome, and the necessary technology is still so far away that there is yet no convincing evidence that such "genetic engineering" will ever be achieved. Nevertheless, biochemists have started down a road of laboratory experimentation that could ultimately lead to genetic manipulation at the human level.

For the present, the goals are more limited—to understand more fully the nature of eukaryotic gene structure by incorporating portions of chromosomes in the bacterium *Escherichia coli* or other suitable hosts. The strains of bacteria with specific portions of foreign DNA incorporated are known as "clones," and the overall line of experimentation with recombining DNA from different organisms is sometimes called **gene cloning,** or **gene splicing.** (This is to be distinguished from organism cloning—the production of identical individuals from single cells—that has been popularized in science fiction for humans and has actually been achieved for plants and lower animals such as frogs.)

Although there is no question about the potential benefits from recombinant DNA research in terms of advances in basic biology and in practical applications, there are also (according to some geneticists) possible risks. For example, if studies were carried out to clone a portion of the DNA of a cancer virus in *E. coli,* and if the particular clone of *E. coli* escaped from the laboratory, and if it infected human beings, a "cancer epidemic" could be imagined as the result. Because of their concerns, in 1974 a group of scientists working on recombinant DNA launched an appeal for a moratorium on cloning experiments involving eukaryotic DNA until a set of guidelines could be established to minimize risks. This step was unprecedented in the history of science, and it stimulated a debate on the issue that is still raging.

The moratorium had several consequences. First, it was successful; research on recombinant DNA from eukaryotic sources stopped during the period when guidelines were being formulated. Second, committees were constituted to set guidelines, and in 1976 such guidelines were published by the National Institutes of Health. One of the requirements set down in the guidelines is the development of a special "crippled" strain of *E. coli* for use in such work, a strain incapable of growing in the human gut. Such a strain was created and is now in wide use in recombinant DNA work. The guidelines also defined levels of risk from P1 for the least risk to P4 for highest risk.

Public discussion may have reached a peak in 1976 when several scientists (mainly individuals not working with recombinant DNA themselves) claimed that recombinant DNA experiments were still unsafe, even within the conditions of the guidelines, and could have disastrous effects on life as we know it. A number of municipalities (most notably Cambridge, Massachusetts) began debating whether to permit such research within their boundaries. On the other side, many scientists (including some who first joined in the call for the moratorium) have recently expressed the view that the guidelines are unnecessarily strict and that the moratorium movement may itself have raised concerns that exaggerate the potential dangers. Recently some relaxation of the guidelines has taken place as experimental evidence has indicated that the risks are less serious than originally feared.

28-5 BIOCHEMISTRY AND BEHAVIOR

The biochemical basis of the genetic and bioenergetic processes of life has been revealed, but the relationship between biochemistry and behavior is still only vaguely understood. The brain is composed of neurons and other types of cells that can be described in biochemical terms, but the ways in which these cells interact to account for memory and various forms of behavior are unclear.

One aspect that has been partially unraveled is the role of **neurotransmitters.** These small compounds are released from nerve cells at the junctions with other nerve cells. These junctions are called **synapses.** The neurotransmitters diffuse across the synapse, where they bind to the receptors on the neighboring neuron and trigger electrical changes. The structures of several of the known neurotransmitters follow.

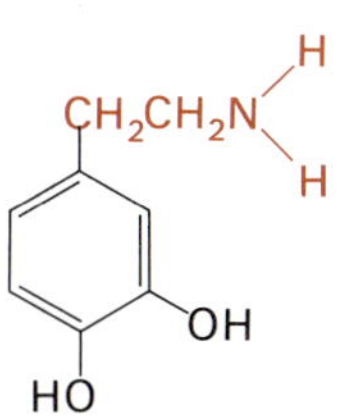

dopamine

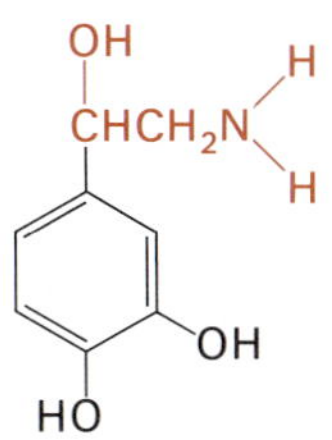

norepinephrine

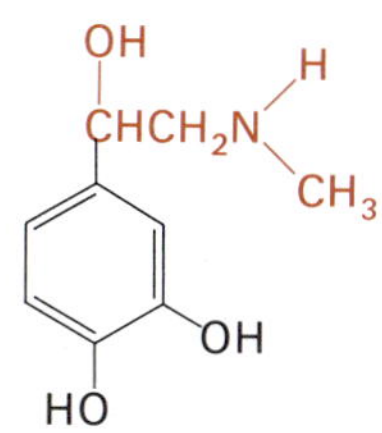

epinephrine

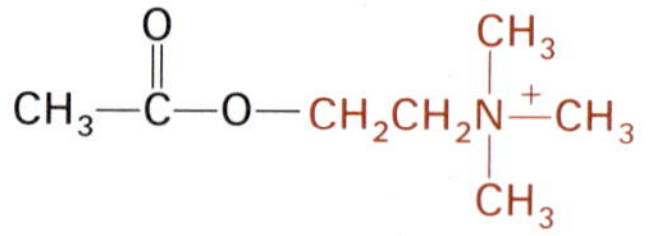

acetylcholine

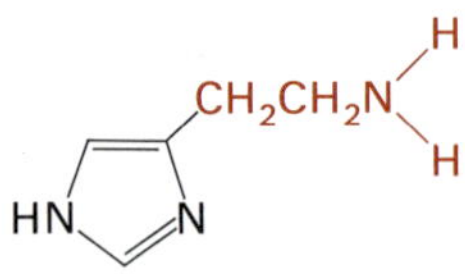

histamine

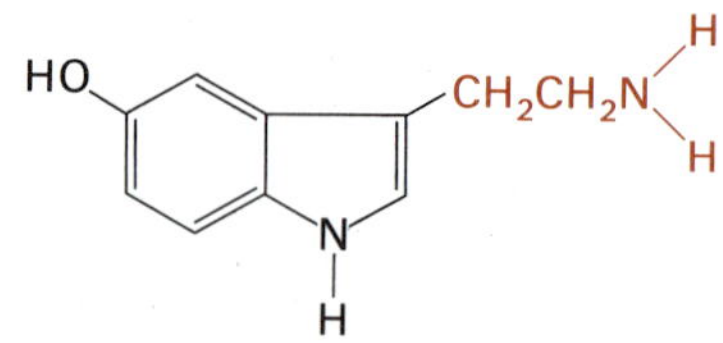

seratonin

The involvement of biochemistry in behavior occurs largely in relation to the neurotransmitters. Chemicals have been discovered that apparently perturb neurotransmitters to alter behavior. The **psychoactive drugs** such as marihuana or LSD are examples of such chemicals. The effects of alcohol may also fall into this class, but they are not as clearly related to neurotransmitters. **Pharmacological agents** (such as barbiturates, amphetamines, and morphine) and compounds structurally related to neurotransmitters that can be administered directly (such as L-dopa, used in the treatment of Parkinson's disease) are other examples.

One of the clearest demonstrations of the relationships between biochemistry and behavior concerns the psychoactive drugs. The structures of these compounds:

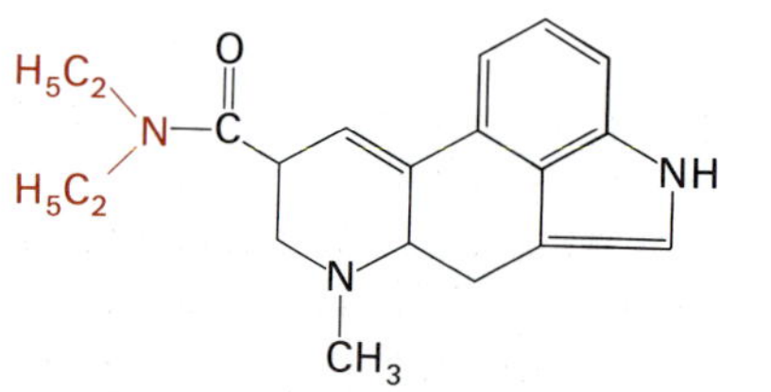

D-lysergic acid diethylamide
(LSD)

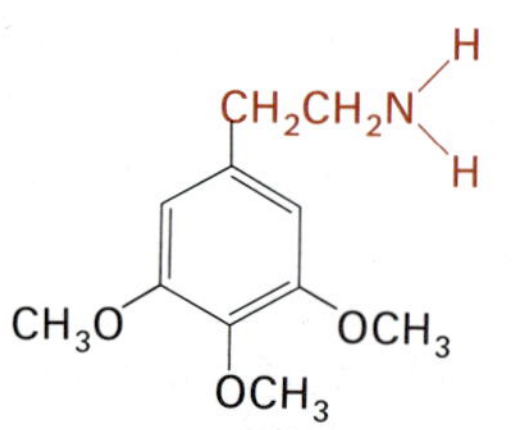

mescaline

resemble neurotransmitters, and their mode of action apparently involves some mimicking of neurotransmitter effects. For example, **LSD** (one of the most powerful psychoactive drugs) has features that suggest a possible similarity with epinephrine, seratonin, and histamine. **Mescaline** (the active ingredient of peyote, the substance from cactus used in the religious ceremonies of certain native North American peoples) has a close similarity to the neurotransmitter dopamine.

The biochemical basis of behavior is becoming more clearly understood as a result of the use of a variety of pharmacological drugs, such as tranquilizers and drugs used to control mental illnesses such as schizophrenia. One of the most effective drugs used in mental illness is the simple salt lithium chloride. Other compounds used for their effect on the brain are **barbiturates** (such as phenobarbital) that induce sleep, stimulants (such as **amphetamine**), and pain-relieving compounds (such as **morphine**):

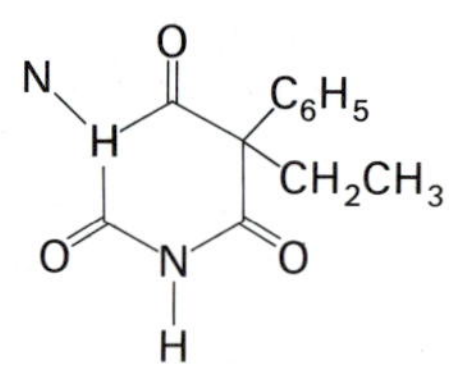

phenobarbital

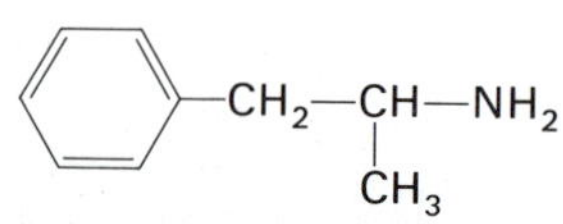

amphetamine

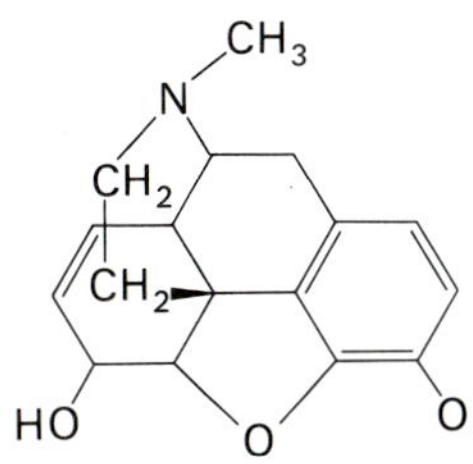

morphine

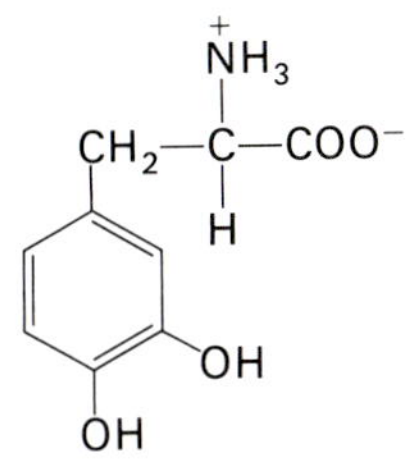

L-dopa

In Parkinson's disease, administration of the compound L-dopa has been very helpful. L-Dopa is the precursor of dopamine and leads to elevation of dopamine levels. Dopamine itself is not absorbed into the brain as readily as L-dopa is.

The development of new drugs and the understanding of the mode of action of drugs that alter behavior present one of the greatest challenges for future biochemists.

28-6 BIOCHEMICAL PROBLEMS ON A GLOBAL SCALE

Certain problems with a biochemical basis occur on a scale that involves the entire planetary environment. Two such problems are the food problem and the energy problem. The human population of this planet has grown to such proportions that providing the basic food needs of the people strains the planetary resources. The issue of energy, although less a matter of life and death than the food problem, is also a major concern. The energy problem relates to the food problem via the highly technological processes used for fertilizing and harvesting food. Thus the "reducing power" (see Chapter 26) that forms the basis of the fuel for both people and machines must be considered in the context of the ability to maintain life on this planet.

Most of the concern related to the crisis in food supply is directed at proteins. Proteins make up about 20% of the human body, and they occur in active tissues that are constantly being broken down and built up. Therefore a continuous need for amino acids exists. Of the twenty principal amino acids, eight cannot be synthesized by human beings and must be supplied in the diet. These essential amino acids (see Chapter 21) must be provided in relatively large amounts (compared to other essential substances such as vitamins). The essential amino acids and their approximate daily requirements are listed in Table 28-1.

A human being obtains the essential amino acids by eating protein, which is broken down to amino acids in the digestive tract. The amino acids are then absorbed through the walls of the intestine into the bloodstream and move throughout the body. The critical feature of amino acid nutrition is that the amino acids can be utilized only to the extent that the least abundant essential amino acid is supplied.

In certain protein sources (such as egg), the amino acid composition is close to the proportions of the essential amino acids required in the diet, and utilization is near 100%. In contrast, protein from soybeans has only a 70% utilization because it is low in methio-

TABLE 28-1
Essential amino acid requirements

Amino acid	Requirement (g/day)
Isoleucine	0.7
Leucine	1.1
Lysine	0.8
Methionine	1.1
Phenylalanine	1.1
Threonine	0.5
Tryptophan	0.3
Valine	0.8

nine, and protein from wheat flour has only a 50% utilization due to a deficiency in lysine. As a result of their varied amino acid composition, different proteins are able to complement each other. Although soy protein is deficient in methionine, it has an "excess" of lysine in comparison with the proportions of essential amino acids needed. Similarly, although wheat protein is deficient in lysine, it has an "excess" of methionine. Thus if soy protein and wheat protein are eaten together, the deficiency of each is complemented, and the utilization of the combination is near 100%.

Various national groups have, over the centuries, developed diets in which complementary proteins are eaten together. Mexican foods, for example, often combine rice (low in lysine but rich in methionine) with beans (low in methionine but rich in lysine). In India, the rice is often complemented by lentils.

Obtaining an adequate supply of amino acids is generally not a problem for individuals on a normal varied diet that includes meat. (Vegetarians must take special precautions to ensure an adequate diet. A major problem can arise if animal products such as eggs and milk also are avoided, because vitamin B_{12} is not obtained, and pernicious anemia can result.) However, many of the underdeveloped countries of the world are severely pressed to provide their populations with sufficient protein sources to supply the essential amino acids above a starvation level.

Millions of individuals starve to death each year, and many of the deaths (especially among children) are caused by protein deficiency. The pot-bellied child suffering from **kwashiorkor** (a West African word that means "the sickness the child develops when another baby is born") has a protein deficiency that is often fatal. The name reflects the plight of the child forced to be weaned because of the birth of a sibling, and thus deprived of the reliable protein source—mother's milk.

Two factors must be considered in seeking remedies to the problems of the world supply of protein: (1) making use of protein supplies for direct human consumption versus their use as livestock food; and (2) providing proteins in appropriate combinations or with appropriate amino acid supplements to ensure complete utilization. The issue of using protein for humans or for livestock concerns the heavy reliance in developed countries on meat as a protein source. It has been estimated that about 20 pounds of feed protein are required for each pound of beef produced for human consumption. Somewhat less feed is required for pork, chicken, or dairy products. The feed is principally grain products.

Well over one-half of the corn crop, oat crop, barley crop, and unexported soybean crop of the United States is used to provide animal protein for human consumption. Thus from the more than 20 million tons of protein feed used annually in the United States, only about 2 million tons of protein for human consumption are obtained. The effective "loss" of 18 million tons of protein has been estimated to be close to the yearly world protein deficiency. Assuming an average daily protein requirement of 40 g, the 18 million tons of protein could feed 1 billion people for a year. The estimated requirement of 40 g of protein per day is that for an adult male weighing 150 pounds.

The protein would go even farther if adjustments are made for the lesser amounts needed by children.

Whether used for humans or for livestock, the existing supplies of proteins can be used more efficiently if steps are taken to ensure full utilization. One approach to this goal is to mix feeds, just as complementary proteins are mixed in human diets. For example soy and corn are often mixed in animal feeds to compensate for the methionine deficiency of soy and the lysine deficiency of corn. A second approach has been to breed corn and barley strains that produce lysine-rich grains. These grains are used primarily as animal feed; high-lysine corn is particularly effective in raising hogs. Similarly, addition of 0.2% lysine to wheat flour nearly doubles the utilizability of its protein, and addition of 0.5% methionine to soy protein renders its protein value close to the value of eggs. Work on identifying the gene that controls methionine production in soybeans is underway. Eventually the principles of genetic engineering may be applied to developing a strain of high-methionine soybeans that are a source of complete protein.

Thus in the current world crisis of population and resources, protein supply is an important factor. Humanitarian or economic factors may eventually limit the production of meat, so that the grains now consumed as animal feed can be used more efficiently to meet human food needs. Careful evaluation of the utilization of essential amino acids will ensure that protein is used most effectively.

KEY TERMS

Antibiotics are chemicals that selectively kill the bacterial agents responsible for infectious diseases.

Immunization is the process of disease prevention in which the immune system is stimulated by an inactivated form of a potential disease-causing agent (a *vaccine*), so that if an active agent is later brought into the body, sufficient antibodies are already present to prevent infection.

Cancer is the constellation of cellular disorders related to uncontrolled cell growth and tumor formation. A chemical that causes cancer is called a *carcinogen*. A *mutagen* is a chemical agent that increases the frequency of mutation in DNA. A *transformed cell* is a cell rendered cancerous, by either chemical or viral treatment.

A *genetic disease* is a disease related to a malfunctioning gene; it can be transmitted only by inheritance.

Amniocentesis is the process of withdrawing cells from the amnionic fluid surrounding a fetus and analyzing the cells to detect possible birth defects.

Genetic engineering includes the broad category of phenomena related to the manipulation of the genetic material, DNA.

In genetic engineering, *cloning* refers to the growth of large amounts of DNA from any species, including human, by incorporating the DNA into a microorganism such as a bacterium.

Recombinant DNA is DNA artificially inserted into an organism for experimental purposes.

Neurotransmitters are compounds found at the synapses of nerve cells; they are released by one cell, diffuse across the synapse, and then trigger an electrical change in the adjacent cell.

Psychoactive drugs are those that alter perception or behavior.

The *essential amino acids* are those amino acids that cannot be synthesized by the body and must be supplied in the diet, usually in the form of proteins.

CHAPTER SUMMARY

The impact of biochemistry in medicine has gone through many stages, beginning with the discovery of natural products such as quinine. Later, chemical agents were synthesized that were found to have antibacterial activity—for example, the dye sulfanilamide. More dramatic antibacterial activites were discovered for the compounds produced by microorganisms and known as *antibiotics*. *Penicillin* is a prominent representative of the antibiotics.

Steps in the treatment of diseases caused by viruses were taken with the development of *vaccines,* inactivated forms of the viruses that can be injected into individuals to provide protection against later infections of an active virus. The inactive virus stimulates production of antibodies that later can attack an active virus; this approach to dealing with viral disease is called *immunization*. One of the pressing medical problems at the present time is *cancer,* which may be caused by viruses, although environmental chemicals (*carcinogens*) clearly also play a role.

As more and more has been learned about biochemisry, a new class of diseases has been discovered, the *genetic diseases*. In these diseases, a malfunctioning gene that produces an abnormal protein is inherited. One well-characterized example is *sickle-cell disease,* which arises from an alteration in just one of the amino acids of the β chains of hemoglobin. At the present time, there is no satisfactory treatment for sickle-cell disease. In the case of another genetic disease, *phenylketonuria,* detection at birth can lead to circumvention of the symptoms of the disease if the infant is placed on a special diet. For certain diseases, detection can take place before birth through a process called *amniocentesis*.

Future advances in biochemical approaches to medicine may come from research on recombinant DNA, in which portions of the DNA from human beings are linked to the DNA of the bacterium *E. coli* to facilitate study of the human DNA. Biochemistry is also coming to play an expanding role in the realm of human behavior through the development of psychoactive drugs and through increasing knowledge of the chemistry of the brain. On a global scale, biochemistry provides insights into the food crisis, especially in regard to the requirements for amino acids in relation to total protein consumption.

APPENDIX 1

Tables

TABLE A-1
Units in the SI system

Quantity measured	Name of unit	Symbol
Length	meter	m
Mass	kilogram	kg
Time	second	sec
Electric current	ampere	A
Temperature	kelvin	K
Quantity of substance	mole	mole
Energy	joule	J
Electric charge	coulomb	C
Electric potential difference	volt	V
Frequency	hertz	Hz

TABLE A-2
Prefixes for decimal fractions or multiples of SI units

Fraction	Prefix	Symbol	Multiple	Prefix	Symbol
10^{-1}	deci-	d	10^{1}	deka-	da
10^{-2}	centi-	c	10^{2}	hecto-	h
10^{-3}	milli-	m	10^{3}	kilo-	k
10^{-6}	micro-	μ	10^{6}	mega-	M
10^{-9}	nano-	n	10^{9}	giga-	G
10^{-12}	pico-	p	10^{12}	tera-	T

TABLE A-3
Conversion factors

1 ton (2000 lb)	= 907.2 kg	1 kg	= 2.205 pounds
1 pound	= 453.6 g	1 km	= 0.621 mile
1 inch	= 2.54 cm	1 m	= 39.37 inches
1 mile	= 1.609 km	1 cm	= 0.3937 inch
1 quart	= 0.946 ℓ	1 ℓ	= 1.057 quarts

1 electron volt (ev) $= 1.602 \times 10^{-19}$ J
1 calorie = 4.184 J
1 electron volt/atom = 23.06 kcal/mole = 96.48 kJ/mole
1 atmosphere (atm) = 760 mm Hg

TABLE A-4
Fundamental constants

Avogadro's number:	$N = 6.0225 \times 10^{23}$ molecules/mole
Gas constant:	$R = 8.206 \frac{\ell \text{ atm}}{\text{mole deg}} = 8.31 \frac{\text{J}}{\text{mole deg}} = 1.99 \frac{\text{cal}}{\text{mole deg}}$
Velocity of light:	$c = 3.00 \times 10^{8}$ m/sec $= 3.00 \times 10^{10}$ cm/sec
Planck's constant:	$h = 6.6256 \times 10^{-34}$ J sec
Electron mass:	$m_{\text{e}} = 9.1091 \times 10^{-28}$ g
Electron charge:	$q_{\text{e}} = 1.6021 \times 10^{-19}$ C
Proton mass:	$m_{\text{p}} = 1.6725 \times 10^{-24}$ g
Neutron mass:	$m_{\text{n}} = 1.6748 \times 10^{-24}$ g

APPENDIX 2

Exponents and Logarithms

EXPONENTS

In chemistry, many of the numbers we must use are very small (say, the mass of a water molecule, which is 0.00000000000000000000003 g) or large (say, Avogadro's number, N, 602,250,000,000,000,000,000,000 molecules/mole). It is impractical to work with such numbers, but they may be expressed compactly in exponential form, in which 10^x represents a 1 with x zeros after it, and $10^{-x} = 1/10^x$:

$$
\begin{aligned}
10^6 &= 1{,}000{,}000\\
10^5 &= 100{,}000\\
10^4 &= 10{,}000\\
10^3 &= 1{,}000\\
10^2 &= 100\\
10^1 &= 10\\
10^0 &= 1\\
10^{-1} &= 1/10^1 = 0.1\\
10^{-2} &= 1/10^2 = 0.01\\
10^{-3} &= 1/10^3 = 0.001\\
10^{-4} &= 1/10^4 = 0.0001\\
10^{-5} &= 1/10^5 = 0.00001
\end{aligned}
$$

When a number is expressed in scientific notation, the decimal point is placed so that the number lies between 1 and 10, and the new number is multiplied by the proper power of 10 to maintain its original value. For example,

$$
\begin{aligned}
204 &= 2.04 \times 100 = 2.04 \times 10^2\\
1974 &= 1.974 \times 1000 = 1.974 \times 10^3\\
320{,}678 &= 3.20678 \times 100{,}000 = 3.20678 \times 10^5\\
0.91 &= 9.1 \times 0.1 = 9.1 \times 10^{-1}\\
0.03 &= 3 \times 0.01 = 3 \times 10^{-2}\\
0.000271 &= 2.71 \times 0.0001 = 2.71 \times 10^{-4}
\end{aligned}
$$

Multiplication by 10^x means the decimal point should be moved x places to the right to express the number in nonscientific notation. Multiplication by 10^{-x} means the decimal point should be moved x places to the left to express the number in nonscientific notation. Thus,

$5.31 \times 10^2 = 531$ (decimal point moved 2 spaces to right)

$6.1 \times 10^{-3} = 0.0061$ (decimal point moved 3 spaces to the left)

In multiplying exponential numbers, the exponents are added:

$$2 \times 10^2 \times 3 \times 10^6 = 6 \times 10^{(2+6)} = 6 \times 10^8$$

$$4 \times 10^2 \times 5 \times 10^{-6} = 20 \times 10^{(2+(-6))} = 20 \times 10^{-4} = 2.0 \times 10^{-3}$$

$$6 \times 10^{-6} \times 7 \times 10^4 = 42 \times 10^{(-6+4)} = 42 \times 10^{-2} = 4.2 \times 10^{-1}$$

In dividing exponential numbers, the exponents are subtracted:

$$\frac{6 \times 10^8}{3 \times 10^4} = 2 \times 10^{(8-4)} = 2 \times 10^4$$

$$\frac{8 \times 10^4}{2 \times 10^8} = 4 \times 10^{(4-8)} = 4 \times 10^{-4}$$

$$\frac{9 \times 10^4}{3 \times 10^{-4}} = 3 \times 10^{(4-(-4))} = 3 \times 10^8$$

If an exponential number is squared, the exponent is doubled:

$$(2 \times 10^3)^2 = (2 \times 10^3)(2 \times 10^3) = 2^2 \times 10^{(3+3)} = 4 \times 10^6$$

In taking a square root, the exponent is halved:

$$\sqrt{4 \times 10^{-10}} = (4 \times 10^{-10})^{1/2} = 4^{1/2} \times 10^{-10/2} = 2 \times 10^{-5}$$

When taking a square root, if the exponent is not divisible by 2, rewrite the number so that the exponent *is* divisible by 2:

$$\sqrt{2.5 \times 10^{13}} = \sqrt{2.5 \times 10 \times \frac{10^{13}}{10}} = \sqrt{25 \times 10^{12}} = 5 \times 10^6$$

In the second step, the number was multiplied by 10 and divided by 10, giving an *even* exponent while maintaining the original value of the number.

When adding or subtracting numbers with powers of 10, the powers of 10 must be the same; if they are not the same, the numbers must be rewritten so the powers of 10 *are* the same:

$$3 \times 10^2 + 2.0 \times 10^3 = 3 \times 10^2 + 20 \times 10^2 = 23 \times 10^2 = 2.3 \times 10^3$$

LOGARITHMS, pH, AND pK_a

The logarithm of a number is the power (exponent) to which 10 must be raised to give that number. For example, the logarithm of 1000 is 3, because $1000 = 10^3$; and the logarithm of 0.01 is -2, because $0.01 = 10^{-2}$. To find the logarithm of a number such as 300, it is first necessary to express the number in scientific notation:

$$300 = 3 \times 10^2$$

The logarithm of a product is the sum of the logarithms:

$$\log(a \times b) = \log a + \log b$$

$$\begin{aligned}\log 300 &= \log(3 \times 10^2)\\ &= \log 3 + \log 10^2\\ &= 0.48 + 2 = 2.48\end{aligned}$$

We find the value of log 3 in a table of logarithms such as Table A-5.

TABLE A-5
Two-place logarithms

Number	Logarithm
1.0	0.00
2.0	0.30
3.0	0.48
4.0	0.60
5.0	0.70
6.0	0.78
7.0	0.85
8.0	0.90
9.0	0.95
10.0	1.00

We can use logarithms to find the pH of 6×10^{-4} M HCl. Because HCl is a strong acid, $[H^+] = 6 \times 10^{-4}$ M.

$$\begin{aligned}pH &= -\log[H^+] \qquad \text{where } [H^+] \text{ is in moles}/\ell\\ &= -\log(6 \times 10^{-4})\\ &= -(\log 6 + \log 10^{-4})\\ &= -[0.78 + (-4)] = 3.22\end{aligned}$$

We can also use logarithms to find the pK_a of acetic acid, which has $K_a \cong 2 \times 10^{-5}$.

$$\begin{aligned}pK_a &= -\log K_a\\ &= -\log(2 \times 10^{-5})\\ &= -(\log 2 + \log 10^{-5})\\ &= -[0.30 + (-5)] = 4.70\end{aligned}$$

We can also find $[H^+]$, given the pH of a solution. For example, given a solution of pH 8.30:

$$pH = -\log[H^+] = 8.30$$

$$\log[H^+] = -8.30 = +0.70 - 9$$

Here we have rewritten the logarithm, −8.30, as +0.70 − 9, so that we can look up the **antilog** of 0.70 (the number whose logarithm is 0.70) in Table A-5. (We can look up the antilog of a number if its logarithm is expressed as a *positive* number between 0 and 1.)

$$\begin{aligned}[H^+] &= \text{antilog } 0.70 \times \text{antilog } (-9)\\ &= 5.0 \times 10^{-9} \text{ M}\end{aligned}$$

Similarly, given that the pK_a of HCN is 9.40, we can calculate K_a:

$$\begin{aligned}pK_a &= -\log K_a = 9.40\\ \log K_a &= -9.40 = +0.60 - 10\\ K_a &= \text{antilog } 0.60 \times \text{antilog } (-10)\\ &= 4 \times 10^{-10}\end{aligned}$$

APPENDIX 3

Significant Figures

Suppose someone weighs the same penny several times on a centigram balance, obtaining the weights 3.14 g, 3.13 g, 3.12 g, and 3.13 g. What is the "true" weight of the penny? * In all of the weighings, the first two digits are identical, but there is an uncertainty in the third digit. The penny weighs between 3.12 g and 3.14 g; the average weight is 3.13 g with an uncertainty of ±0.01 g (that is, all the measurements lie between 3.13 − 0.01 g and 3.13 + 0.01 g). The measurement "3.1**3** g" is said to have three significant figures: two digits that are certain, and a third that is still significant although subject to some uncertainty. The uncertain digit is shown in dark type.

The number of significant figures in a measurement depends upon the sensitivity of the measuring instrument. When the penny is weighed on bathroom scales that weigh to the nearest pound, the penny appears to weigh nothing (0 pounds); when weighed on an analytical balance, it weighs 3.134**7** g, with an uncertainty of ±0.0002 g. In this last series of measurements, the average weight was 3.134**7** g, and all of the measurements lay between 3.1345 g and 3.1349 g. The uncertainty in a measurement can never be entirely eliminated, no matter how elaborate the measuring instrument.

In performing calculations with measured quantities, it is important to maintain the proper number of significant figures—it is easy, especially when using a calculator, to obtain a result with more digits than are significant, but the additional figures are meaningless. For example, suppose you have weighed a rock to the nearest gram, and found it to weigh 454 g ± 1 g, that is, between 453 g and 455 g. You have also weighed a pebble on an analytical balance, and found it to weigh 3.1529 g ± 0.0001 g. What is the combined weight of the rock plus the pebble? The total weight is expressed as 457 g ± 1 g, and not as 457.1529 g. Because we know the weight of the rock only within ±1 g, we cannot know the combined weight more precisely than ±1 g. If we wrote the total weight as

*We are assuming that the balance is operating properly—that is, that it is giving the correct weight within the limits of uncertainty of the balance.

457.1529 g, that would imply *erroneously* that the weight of the rock, as well as that of the pebble, is known to four decimal places.

Here are some rules for the use of significant figures.

1. Zeros that appear to the left of all other digits in the number are not counted as significant. For example, 0.0125 has three significant figures, and 0.007 has only one significant figure. The zeros serve only to place the decimal point. In scientific notation, these numbers are 1.25×10^{-2} and 7×10^{-3}.

2. Zeros that appear to the right of all other digits in the number may or may not be significant. The deepest gorge in the world is Hell's Canyon (Snake River, Idaho), which is 7,900 ft deep. Here it is not clear how precisely the depth has been measured, and we do not know whether the uncertainty is ±100 ft or ±1 ft; that is, we do not know whether the gorge is 7900 ± 100 ft or 7900 ± 1 ft deep. We may assume that here, as is usually the case when numbers are not expressed in scientific notation, the zeros are not intended to be significant but only to show the position of the decimal point, and hence that the gorge is 7.9×10^3 ft ± 100 ft deep. If the uncertainty is ±1 ft, the measurement should be written 7.900×10^3 ft, where the zeros at the right (after the decimal point) are then taken to be significant.

3. Zeros in the middle of a number are significant: the measurement 200.35 has five significant figures.

4. When adding or subtracting measured quantities, the result should have one uncertain digit. For example, when adding 48.**2** g and 3.2**8** g, the sum is

$$\begin{array}{r} 48.\mathbf{2} \\ 3.2\mathbf{8} \\ \hline 48.\mathbf{48} \end{array}$$

The sum should be rounded off to give only *one* of the uncertain digits. If the first of the digits to be discarded is 5 or greater, the last significant figure is increased by 1; if the first of the digits to be discarded is less than 5, the last significant figure is not changed. Thus, the sum 48.**48** should be rounded off to 48.5; 2.0**332** is rounded off to 2.03; and 7.8**563** is rounded off to 7.8**6**.

5. In the multiplication and division of measured quantities, the result should have the same number of significant figures as the *least* precise measurement. Thus, the product $6.02 \times 10^{23} \times 3.267 = 19.\mathbf{667} \times 10^{23}$ should be rounded off to three significant figures, giving the result $19.\mathbf{7} \times 10^{23}$, or 1.97×10^{24}. (This is a rule of thumb, adequate for most purposes. A more precise statement is that the percentage uncertainty in the result is at least as great as that in the least precise of the measurements.)

APPENDIX 4

Answers to Exercises and Even-Numbered Problems

CHAPTER 1

EXERCISES

1-1. (a) 9×10^5 cm (b) 6×10^3 ml
1-2. 10.5 g/cm^3
1-3. (a) −38°F (b) 40.6°C
1-4. 13% protein; 36% fat; 51% carbohydrate. Not well balanced: too little carbohydrate and too much fat.
1-5. $\frac{\text{mass O}}{\text{mass C}} = 1.33$, and $\frac{\text{mass O}}{\text{mass C}} = 2.66$;
ratio 1:2

PROBLEMS

1-2. (a) He (b) N (c) Br (d) Au (e) Ag (f) As (g) Cu (h) K
1-6. 6.7×10^8 miles/hour
1-8. (a) 350 g (b) 3.50×10^3 g (c) 0.708 g/ml
1-10. 6.2×10^4 g (137 lb)
1-12. 327°C
1-14. 38°C
1-16. 2.5×10^3 ml; 2.6 qt
1-18. (a) 29°F (b) 113°F
1-20. Measure: mass of tin, and mass of tin oxide
Calculate: mass of oxygen; and % oxygen =

$$\frac{\text{mass oxygen}}{\text{mass tin oxide}} \times 100$$

1-22. (a) 39.3% (b) 100 g (c) 865 g
1-24. $\frac{\text{mass O}}{\text{mass N}} = 1.14$, and $\frac{\text{mass O}}{\text{mass N}} = 2.28$;
ratio of $\frac{\text{mass O}}{\text{mass N}}$ in two compounds = ½

CHAPTER 2

EXERCISES

2-1. C = 8 atoms; H = 18 atoms
2-2. (a) 323 (b) 180 (c) 180 (d) 194

2-3. (a) 6.02×10^{23} (b) 2, 1 (c) 2, 1; 2, 16 (d) 3 (e) 3, 18.1×10^{23}
2-4. (a) 0.0100 mole (b) 6.02×10^{21} atoms (c) 3.27×10^{-22} g
2-5. (a) $2ZnS + 3O_2 \rightarrow 2ZnO + 2SO_2$
(b) $N_2 + 3H_2 \rightarrow 2NH_3$
(c) $6CO_2 + 6H_2O \rightarrow C_6H_{12}O_6 + 6O_2$
(d) $COCl_2 + H_2O \rightarrow CO_2 + 2\,HCl$
(e) $2CCl_4 + O_2 \rightarrow 2COCl_2 + 2Cl_2$
2-6. (a) O_2 (b) 2 moles C (c) 6 moles CO
2-7. $C_4H_{10}O$
2-8. (a) $Fe_2O_3 + 3CO \rightarrow 2Fe + 3CO_2$
(b) 5.6×10^7 g (5.6×10^4 kg)
(c) 4.2×10^7 g (46 tons)
2-9. (a) $Fe_3O_4 + 4H_2 \rightarrow 3Fe + 4H_2O$
(b) 12 molecules (c) 0.30 mole (d) H_2 is limiting reagent; Fe_3O_4 in excess (e) 0.90 mole (f) 50 g

PROBLEMS

2-2. (a) 0.050 mole (b) 3.0×10^{22} atoms (c) 1.5×10^{22} atoms
2-4. 2 g He
2-6. CH_4
2-8. (a) 0.35% (b) 3.5 g Fe (c) 0.062 mole (d) 3.7×10^{22} atoms
2-10. (a) 5×10^{12} molecules/ℓ (b) 1×10^{18}
2-12. 72 kg
2-14. (a) $CH_4 + 2O_2 \rightarrow CO_2 + 2H_2O$ (b) 4 g (c) 16 g (d) 11 g
2-16. (a) $2C_8H_{18} + 25O_2 \rightarrow 16CO_2 + 18H_2O$
(b) 10 kg (c) 4 g
2-18. (a) $2NH_3 \rightarrow N_2 + 3H_2$ (b) 1.5 moles (c) 2.7 moles
2-20. (a) $C_3H_5(OH)_3 + 3HNO_3 \rightarrow C_3H_5(NO_3)_3 + 3H_2O$
(b) HNO_3 (c) 3.0×10^2 g (d) 67%

CHAPTER 3

EXERCISES

3-1. (a) $P_{N_2} = 0.742$ atm; $P_{CO_2} = 4 \times 10^{-4}$ atm; $P_{H_2O\ vap} = 0.062$ atm
(b) $P_{total} = P_{atm} = 1.000$ atm
3-4. 33 atm
3-5. 327°C
3-6. 1832 psi
3-7. The final volume is ¼ the initial volume.
3-8. NH_3

PROBLEMS

3-3. (a) 0.500 atm (b) 0.50 atm (c) 0.92 atm (d) 0.50 atm (e) 1.05 atm (f) 1.500 atm
3-4. The final volume is ⅔ the initial volume.
3-6. (a) 8 ℓ (b) ½ atm (c) 10 ℓ
(d) They contain equal numbers of molecules.
(e) 3×10^{-4} mole
3-8. (a) The pressure is multiplied by 1.26.
(b) The pressure is multiplied by 5.
(c) The pressure is multiplied by 6.30.
3-10. 67 ℓ
3-12. 13 moles, 5.6×10^2 g
3-14. 45 minutes
3-16. (a) 1.34 g/ℓ (b) 0.090 g/ℓ (c) 1.29 g/ℓ
3-18. 0.082 ℓ atm/mole deg
3-20. $C_6H_{12}O$
3-22. 4.8×10^7 atoms/ℓ
3-24. C_3H_8

CHAPTER 4

EXERCISES

4-1. Height of Hg column = 600 mm
4-2. The liquid is stable at 0°C and the vapor at 40°C.

4-3. (a) $\cong 40°C$ (b) $\cong 50$ mm (c) liquid
(d) vapor
4-4. The vapor pressure would increase; the ice would sublime.
4-5. The melting point of benzene increases when the pressure is raised; that of water decreases.
4-6. about 60°C
4-7. (a) 418 kJ (4.18×10^5 J) (b) 200 kcal
4-8. 0.46 J/g deg
4-9. 245 kJ
4-10. 2.1×10^3 kJ
4-11. (a) 14 kJ

PROBLEMS

4-2. It sublimed.
4-4. When the flask was cooled, some of the vapor inside the flask condensed, so the pressure inside was less than 1 atm. At reduced pressure, water boils at a lower temperature.
4-6. (a) 2.8×10^3 kJ (b) 3.7×10^3 kJ
4-8. (a) $\cong 78°C$ (b) $\cong 40$ mm
4-12. The beaker containing ether was empty.
4-14. (a) 100 g steam (b) 100 g steam
4-16. (a) 0°C (c) 99 kJ (d) 0.50 J/g deg
4-18. Average K.E. is the same for all three.
Average P.E. is greatest for gas; least for solid.
Entropy is greatest for gas; least for solid.
4-22. 17 kJ per g carbohydrate or protein; 39 kJ per g fat
4-24. (a) 100 kcal (b) 263 kcal (c) 25.1 kcal
(d) 140 kcal (e) 50.0 kcal (f) 126 kcal

CHAPTER 5

EXERCISES

5-1. $^{205}_{81}Tl$; 30% ^{203}Tl and 70% ^{205}Tl
5-2. (a) 92 protons, 143 neutrons, 92 electrons
(b) 92 protons, 146 neutrons, 92 electrons
5-3. $^{226}_{88}Ra \rightarrow ^{222}_{86}Rn + ^{4}_{2}\alpha$
5-4. $^{16}_{7}N \rightarrow ^{16}_{8}O + ^{0}_{-1}e$ (β^- particle)
5-5. 3.13 mg
5-6. $^{238}_{92}U + ^{12}_{6}C \rightarrow ^{246}_{98}Cf + 4^{1}_{0}n$
5-7. 2.74×10^9 kJ/mole

PROBLEMS

5-2. (a) -9.63×10^4 coulombs
(b) $+9.63 \times 10^4$ coulombs
5-4. Pb stops α, β, and γ rays.
5-6. See Section 5-6.
5-8. (a) $1 \times 10^5:1$ (b) $1 \times 10^{15}:1$
5-10. (a) 11 p^+; 10 e^-; 12 n (b) 13 p^+; 10 e^-; 14 n
(c) 17 p^+; 18 e^-; 18 n (d) 16 p^+; 18 e^-; 16 n
5-12. 208
5-14. No, they may have different numbers of protons and different numbers of neutrons. It is only necessary that the *sum* $(n + p)$ be 40.
5-16. 40. The atomic mass is close to 40, so only small amounts of the lighter isotopes can be present.
5-18. (a) $^{3}_{1}H \rightarrow ^{3}_{2}He + ^{0}_{-1}e$ (β^- particle)
5-20. $^{212}_{86}Rn \rightarrow ^{208}_{84}Po + ^{4}_{2}He^{2+}$ (α particle) + 2 e^-
5-22. (a) 82 (b) 207
5-24. 8
5-26. 11,140 years
5-28. none of A; 0.25 mole B; none of C; 0.75 mole D
5-30. (a) positron (b) about 3×10^7 sec (1 year)

CHAPTER 6

EXERCISES

6-1. (a) $K(g) \rightarrow K^+(g) + e^-$ (b) Li
6-3. $K(s) + H_2O \rightarrow K^+ + OH^- + \frac{1}{2} H_2(g)$
6-4. $Ba(s) + Cl_2(g) \rightarrow BaCl_2(s)$
6-5. (a) Na (b) Ba
6-6. (a) $Ba(s) + F_2(g) \rightarrow BaF_2(s)$
(b) $2Cs(s) + Cl_2(g) \rightarrow 2CsCl(s)$
(c) $H_2(g) + I_2(s) \rightarrow 2HI(g)$

6-7. (a) Cl (b) Xe (c) Na
6-8. 3×10^4 cm
6-9. 986 kJ/mole
6-10. 32; $4s^2\,4p^6\,4d^{10}\,4f^{14}$
6-11. 20
6-12. K^+, Ca^{2+}, Cl^-
6-13. (a) Na (b) Br (c) Cl^-

PROBLEMS

6-2. alkali metals: Cs and Li; alkaline earths: Be and Ca; halogens: Br and F; noble gas: Ne
6-6. (a) Cs (b) He (c) I_2
6-8. (a) Pb (b) Pb
6-10. (a) S (b) K (c) Si (d) Cl (e) Ca (f) As (g) Te
6-16. (a) The one with the higher I.P. has a noble gas electron configuration.
(b) The one with the higher I.P. is chemically similar to Ar.
6-18. 12 and 56.
6-20. (a) I^-, Cs^+, Ba^{2+} (b) S^{2-}, Cl^-, K^+, and Ca^{2+}

CHAPTER 7

EXERCISES

7-1. (a) Ca: + 2:$\ddot{\text{Br}}$· → Ca^{2+} + 2:$\ddot{\text{Br}}$:$^-$

(b) 2K· + :$\dot{\text{O}}$· → $2K^+$ + :$\ddot{\text{O}}$:$^{2-}$

7-2. :$\ddot{\text{I}}$:$\ddot{\text{I}}$:

7-3. (a) $\overset{+\!\!\longrightarrow}{\text{H—Cl}}$ (b) $\overset{\longleftarrow\!\!+}{\text{Br—I}}$

7-4. 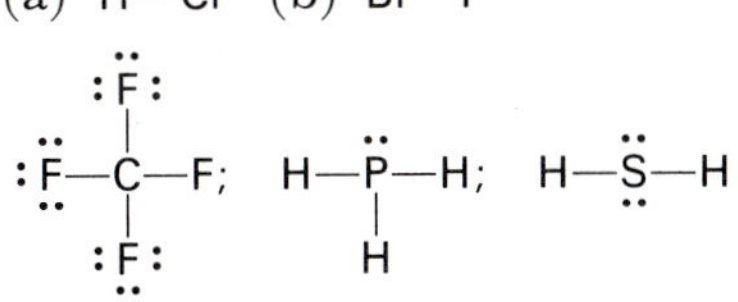

7-5. 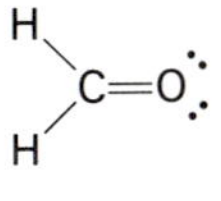

7-6. 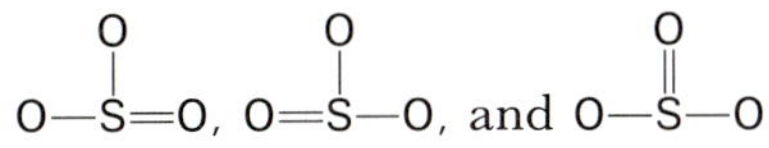

7-7. (a) C═C bond is the shortest and strongest
(b) C≡C bond is the shortest and strongest
7-8. (a) linear (b) bent
7-9. (a) The —C≡C— bond consists of 1 σ + 2 π bonds. These two C atoms are using sp hybrid orbitals. The other C atom is using sp^3 hybrid orbitals.
(b) The C═C bond consists of 1 σ + 1 π bond. These two C atoms are using sp^2 hybrid orbitals. The other two C atoms are using sp^3 hybrid orbitals.

PROBLEMS

7-2. (a) Al_2O_3 (b) K_3N (c) Ba_3N_2 (d) Rb_2S (e) $CaAt_2$ (f) $Ca(NO_3)_2$ (g) $Ba_3(PO_4)_2$ (h) Na_2SO_4 (i) $(NH_4)_2CO_3$ (j) $Cr(OH)_3$ (k) $KHCO_3$ (l) $Zn(CH_3CH_2COO)_2$
7-4. H^+, H:$^-$, :C≡N:$^-$
7-6. $n = 2$

7-8. (a) 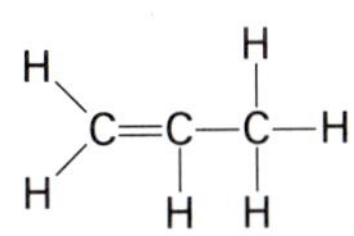 (b) 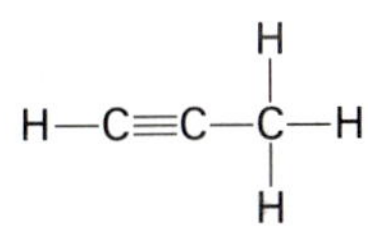

7-10. Si $1s^2\,2s^2\,2p^6\,3s^2\,3p^2$; 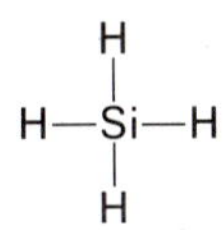

tetrahedral; sp^3 hybrid orbitals (atomic number 14)

P $1s^2\,2s^2\,2p^6\,3s^2\,3p^3$; 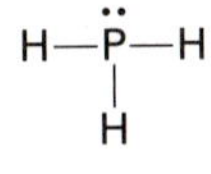

tetrahedral; sp^3 hybrid orbitals (atomic number 15)

7-12. (a) C has 5 bonds (b) N has 4 bonds (c), (d) H has 2 bonds
7-14. CHF_3, CH_2F_2, CH_3F
7-16. sp^2 hybrid orbitals used by carbon atoms (Double bond consists of one σ and one π bond)
7-18. 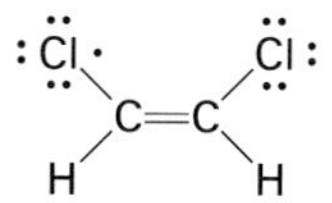

7-20. C_2H_5OH highest; C_2H_{16} lowest
7-22. Hydrogen bonding between HF molecules makes its boiling point unexpectedly high.

CHAPTER 8

EXERCISES

8-1. CH_2O
8-2. $C_2H_4O_2$
8-3. an alcohol ($\lambda = 3.2 \times 10^3$ nm)
8-4. $(CH_3)_2S$ has one peak; CH_3CH_2SH has three (ratio of areas = $3CH_3:2CH_2:1SH$)
8-5. metallic, Zn; ionic, KBr; molecular, Cl_2; network, Si

PROBLEMS

8-2. 4.4 g CO_2; 1.26 g H_2O
8-4. (a) $C_{10}H_{12}NO$ (b) $C_{20}H_{24}N_2O_2$
8-6. (a) 2 peaks, 1:1 (b) 2 peaks, 3:1 (c) 1 peak (d) 2 peaks, 3:1
8-10. (a) Hg (b) Ge (c) PH_3 (d) BaS
8-12. covalent bonding; diamond structure
8-14. For ionic bonding, one needs a combination of an atom that has a great tendency to gain electrons with an atom that loses electrons relatively easily.

CHAPTER 9

EXERCISES

9-1. 0.25 M
9-2. (a) 0.500 mole (b) 90 g
9-3. (a) 0.400 ml (b) 0.50 ℓ
9-4. (a) 18 g (b) 2.5%
9-5. (a) 0.025 M KCl (b) 0.200 M KCl
9-6. 77.55°C
9-7. 1.00 molal (2.00 molal in ions)
9-8. 6.8 atm

PROBLEMS

9-2. (a) Weigh 0.30 moles NaCl (17.6 g). Transfer to 2 ℓ volumetric flask and fill to the line with water.
(b) Weigh 3.0 moles (176 g) NaCl and make up as in part a.
(c) 1 ℓ. Measure out 100 ml of 1.5 M NaCl solution (by pipet or graduated cylinder). Transfer to 1 ℓ volumetric flask, and fill to the line with water.
(d) 50 ml
9-4. (a) 2.5×10^2 g (b) 5.0×10^2 g
9-6. (a) 360 g (b) 171 g (c) 0.88% NaCl (d) 0.25 M (e) 9.9 g
9-8. (a) 0.05 molal (b) 137 g
9-10. Nonelectrolytes: sucrose, methanol, glycerol
9-14. (a) Yes, it's polar. (b) *n*-Hexane, because only a small fraction of the stearic acid is polar; most of the molecule is nonpolar. (We'll learn more about this in Chapter 19.)
9-22. 100 g/mole
9-24. 5.7×10^3 g/mole
9-26. (a) 11 atm (b) 13 atm (c) Solution b has the higher boiling point and the lower freezing point. (d) 22 atm

CHAPTER 10

EXERCISES

10-1. (a) −2816 kJ/mole (b) −16.5 kJ/g sucrose, −15.6 kJ/g glucose

10-4. (a) decrease $[HbO_2]$ (b) $[HHb^+]$ increased, $[HbO_2]$ decreased (c) $[HbO_2]$ greater in lungs; $[HHb^+]$ greater in tissues

10-5. $[I_2] = 0.20$ M, and $[HI] = 1.4$ M

10-6. The reverse reaction ($2\,HI \rightarrow H_2 + I_2$) will occur spontaneously.

10-7. (a) $K = \dfrac{[\text{fructose}][\text{glucose}]}{[\text{sucrose}]}$ (b) $K = \dfrac{1}{[O_2]^3}$

10-8. $K_{sp} = [Al^{3+}][OH^-]^3$

10-9. more soluble in pure water

10-10. Would expect a precipitate, because $[Ca^{2+}]^3[PO_4^{3-}]^2 > K_{sp}$. (The K_{sp} for $Ca_3(PO_4)_2$ in blood at 37°C may be larger than that in water at 25°C. The concentrations, $[Ca^{2+}]$ and $[PO_4^{3-}]$, may also be inaccurate.)

10-11. The rate would be multiplied by $(3)^2$, or 9.

10-12. (a) [A] = 1 M; [B] = 2 M; [C] = 1 M
(b) [A] = 2 M; [B] = 4 M; [C] = 2 M
(c) more rapid in the 500 ml flask; 8 times as fast

10-13. (a) $2\,HI\,(g) \rightleftharpoons H_2\,(g) + I_2\,(g)$
(b) H_2I_2 and I (c) Rate = $k[HI]^2$

10-14. (a) decrease (1/16 of original rate)
(b) increase (16 × original rate)
(c) increase (4 × original rate)
(d) decrease (e) increase

PROBLEMS

10-4. 5.64×10^3 kJ

10-6. (a) 178 kJ/mole $CaCO_3$
(b) −283 kJ/mole CO (c) −395 kJ/mole C

10-8. −640 kJ/mole

10-10. (a) −9966 kJ/mole (b) −38.9 kJ/g

10-12. (a) 4×10^{-7} M (b) 1.1×10^{-5} M
(c) Yes (though carbonates and phosphates, which are cheaper, are usually used).

10-14. (a) no effect (b) less CO (c) no effect
(d) less CO

10-16. (a) increase (b) decrease (c) decrease
(d) no effect (e) increase

10-18. no effect

10-20. $[Fe^{3+}][OH^-]^3$; $[Mg^{2+}]^3[PO_4^{3-}]^2$; $[Pb^{2+}][I^-]^2$

10-22. no; $[Ba][OH^-]^2 < K_{sp}$

10-24. (a) $\Delta G = 992$ kJ/mole (b) no
(c) rate is too slow

10-26. iron filings

10-30. (a) Rate = $k[H_2][NO]^2$ (b) 144 times

10-32. (a) 21,402 atm^{-1} (b) Reaction occurs spontaneously in forward direction. (c) high pressure and low temperature

CHAPTER 11

EXERCISES

11-1. (a) $H^+ + OH^- \rightarrow H_2O$
(b) $CH_3COOH + OH^- \rightarrow CH_3COO^- + H_2O$

11-2. (a) acidic (b) basic (c) 1×10^{-5} M

11-3. (a) $[H^+] = 1 \times 10^{-3}$ M; $[OH^-] = 1 \times 10^{-11}$ M; pH = 3
(b) $[OH^-] = 1.0$ M; $[H^+] = 1 \times 10^{-14}$ M; pH = 14
(c) $[H^+] = 10$ M; $[OH^-] = 1 \times 10^{-15}$ M; pH = −1

11-4. (a) between 4 and 5, but close to 5
(b) between 1×10^{-10} and 1×10^{-11}, but closer to 1×10^{-11}

11-5. pH = 13

11-6. (a) 4.5×10^{-13} M (b) 1.0×10^{-7} M
(c) 0.018 M

11-7. (a) products favored (b) reactants favored
(c) $PO_4^{3-} + H_2O \rightleftharpoons HPO_4^{2-} + OH^-$

11-8. $NH_3, CH_3OCH_3, CH_3CH_2OH, CH_3CH_2NH_2$, PH_3, and H_2S
11-10. HF strongest, and HCN weakest
11-11. 4.2×10^{-3} M
11-12. (a) 1.0×10^{-3} M (b) 4.3 (between 4 and 5)
11-13. Thymol blue: yellow to red
Erythrosine: red to yellow
11-14. (a) $NaHSO_4$ and Na_2SO_4
(b) $[H^+] = 1.3 \times 10^{-2}$; pH = 1.89

PROBLEMS

11-2. (a) (1) 4 N (2) 2 N (3) 36 N (4) 0.5 N (5) 0.20 N
(b) (1) 3 M (2) 0.5 M (3) 0.025 M (4) 0.05 M (5) 0.1 M
(c) (1) 0.15 mole, 0.15 eq (2) 0.025 moles, 0.050 eq (3) 2.5×10^{-3} mole, 5.0×10^{-3} eq (4) 2×10^{-4} mole, 2×10^{-4} eq (5) 0.25 mole, 0.50 eq
11-6. (a) $H^+ + NH_3 \rightarrow NH_4^+$
(b) $CH_3COOH + NH_3 \rightarrow CH_3COO^- + NH_4^+$
(c) $H^+ + OH^- \rightarrow H_2O$
(d) $CH_3COOH + OH^- \rightarrow CH_3COO^- + H_2O$
11-8. H_2SO_4, $NaHSO_4$, NaCN, Na_2S
11-10. (a) 75 ml (b) H_2O, Na^+, Cl^- (c) pH = 7
11-12. (a) 10^{-4} M (b) 10^{-8} M (c) 10^{-4} M (d) 10 M
11-14. (a) 5 (4.83) (b) 4 (3.75) (c) 12 (12.43) (d) 1 (0.64) (e) 7 (7.40)
11-16. pH = −0.3
11-18. 6×10^{-8} M
11-20. (a) pH = 2 (b) pH = 11 (c) pH = 7 (d) pH = 5
11-22. 2×10^{-6} M
11-24. (a) products favored (b) reactants favored (c) products favored (d) reactants favored
11-26. 6.1×10^{-10} M
11-28. $HCO_3^- + OH \rightarrow H_2O + CO_3^{2-}$
$HCO_3^- + H^+ \rightarrow H_2O + CO_2$
11-30. (a) 64.9 kJ/mole (b) 1.55°C
11-34. (a) 0.015 M (b) 7.18
(c) pH of HCl solution = 2

CHAPTER 12

EXERCISES

12-1. Oxidizing agents (a) O_2 (b) Fe^{2+} (c) Ag^+ (d) H^+ Reducing agents (a) C_2H_5OH (b) Al (c) Fe (d) Mg
12-2. (a) +8/3, +4, +2, +3, +7, +6
(b) (1) $\overset{-2}{CH_3OH} + \overset{+5}{NO_3^-} \rightarrow \overset{+4}{CO_2} + \overset{0}{N_2}$;
CH_3OH oxidized, NO_3^- reduced
(2) $\overset{-2}{H_2S} + \overset{0}{O_2} \rightarrow \overset{+6-2}{[SO_4]^{2-}}$;
H_2S oxidized, O_2 reduced
(3) $\overset{+6}{SO_4^{-2}} + \overset{-2}{H_2S} \rightarrow \overset{0}{S}$;
H_2S oxidized, SO_4^{2-} reduced
12-3. (a) silver (b) anode, Zn; cathode, Ag (c) Zn (anode)
12-4. Zn, Cr, Fe
12-5. (a) yes (b) no
12-6. (b) 2.09 V
(c) $Mg(s) + Co^{2+} \rightarrow Mg^{2+} + Co(s)$
(d) no (e) yes
12-7. more than 1.50 V
12-8. $Cr_2O_7^{2-} + 6Br^- + 14H^+ \rightarrow 2Cr^{3+} + 3Br_2(l) + 7H_2O$
12-9. $5CH_3OH + 6NO_3^- + 6H^+ \rightarrow 3N_2 + 5CO_2 + 13H_2O$

PROBLEMS

12-2. Cl_2 is the oxidizing agent; Br^- is the reducing agent.

12-4. (a) $\overset{0}{C}_6H_{12}O_6 + \overset{+6}{S}O_4^{2-} \rightarrow H_2\overset{-2}{S} + \overset{+4}{C}O_2$

(b) $\overset{0}{Cl_2} \rightarrow HO\overset{+1}{Cl} + \overset{-1}{Cl^-}$;

(Cl$_2$ gets oxidized and reduced)

(c) $\overset{0}{Cd} + \overset{+3}{Ni}_2O_3 \rightarrow \overset{+2}{Cd}(OH)_2 + \overset{+2}{Ni}(OH)_2$

12-6. $\overset{-4}{C}H_4$, $\overset{-2}{C}H_3OH$, $\overset{0}{C}H_2O$, $H\overset{+2}{C}O_2H$, $\overset{+4}{C}O_2$

12-8. (a) Al is oxidized; Fe^{3+} is reduced (b) 13.5 g
12-10. (b) 432 g (c) cathode
12-12. Using Table 12-1, calculate $\mathcal{E}^0$ for the various metals dissolving in acid ($2H^+ + 2e^- \rightarrow H_2(g)$). Soil water is slightly acidic.
12-14. (a) no (b) yes
12-16. $3Ag_2S + 2Al(s) \pm 6Ag + 2Al^{3+} + 3S^{2-}$
12-20. Li, K, Ca
12-22. 112 g
12-24. isocitrate + $NAD^+ \rightarrow$ α-ketoglutarate + CO_2 + NADH + H^+ $\mathcal{E}^{0'} = 0.06$ V
12-26. $CH_3CHO + NADH + H^+ \rightarrow CH_3CH_2OH + NAD^+$
$\mathcal{E}^{0'} = 0.12$ volts
12-28. (a) $12H^+ + 4MnO_4^- + 5CH_3OH \rightarrow 4Mn^{2+} + 5HCO_2H + 11H_2O$
(b) $2CH_3CH_2OH + 2O_2 \rightarrow 2CH_3CO_2H + 2H_2O$
(c) $CH_3CO_2H + 2H_2 \xrightarrow{Pt} CH_3CH_2OH + H_2O$

CHAPTER 13

EXERCISES

13-1. (a) The staggered conformation is most stable; the eclipsed conformation is least stable.

(b) 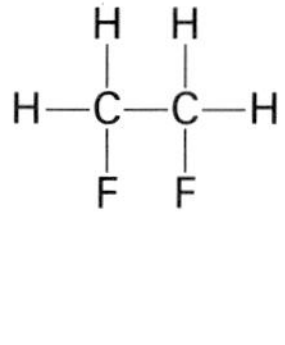 CH_2FCH_2F or FCH_2CH_2F

13-2. $CH_3CH_2CH_2CH_2CH_3$; 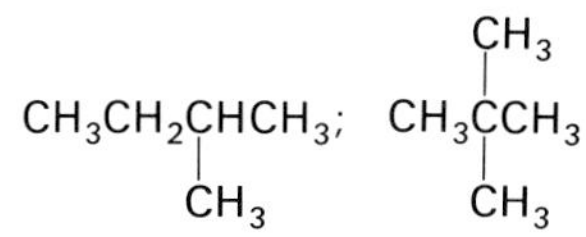

13-3. In both compounds, one of the double-bonded carbon atoms is bonded to two of the same atoms (two H atoms in both cases).

13-4. (a) 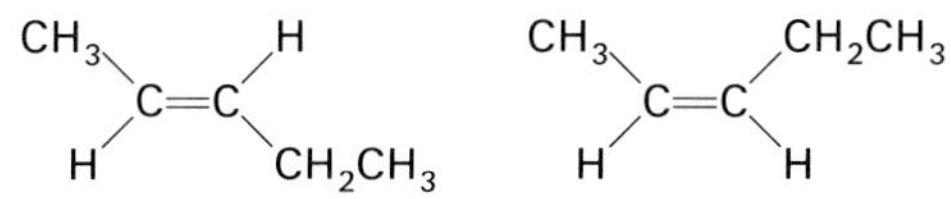

***trans*-2-pentene** ***cis*-2-pentene**

13-5. alanine, glyceraldehyde, glucose

PROBLEMS

13-2. *sp*^{2}

13-6. $CH_3—CH_2—CH_2—CH_2Cl$; 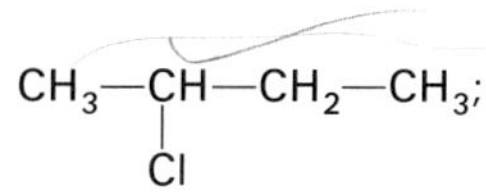

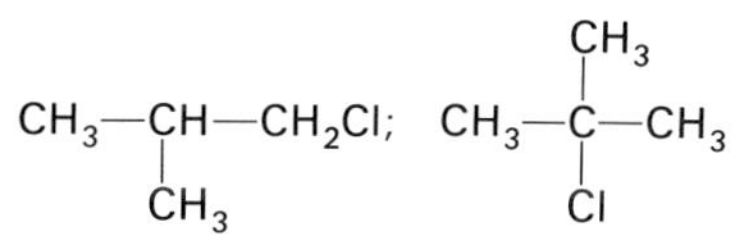

13-10. $H_2C{=}CH—CH_2Cl$; 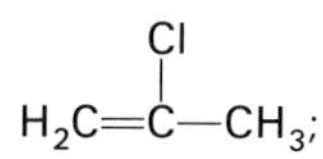

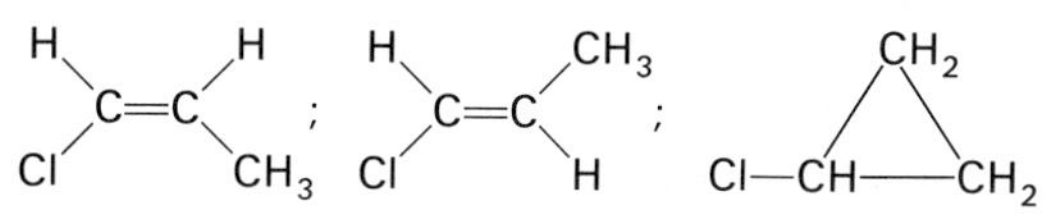

(*cis*-isomer) (*trans*-isomer)

13-16. (a) 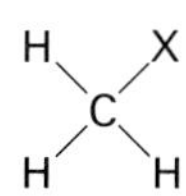

(b) 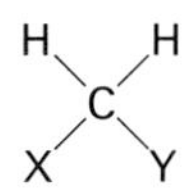 and 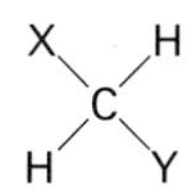

(c) 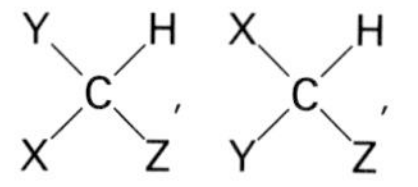 and

13-18. $H_2C{=}CHCH_2CH_2Cl$; $H_2C{=}CH{-}\overset{Cl}{\overset{|}{\overset{*}{C}}}HCH_3$ (and enantiomer); $H_2C{=}\overset{Cl}{\overset{|}{C}}CH_2CH_3$; $ClCH{=}CHCH_2CH_3$ (*cis* and *trans*); $CH_3CH{=}CHCH_2Cl$ (*cis* and *trans*) $CH_3CH{=}CClCH_3$ (*cis* and *trans*); $CH_2{=}\underset{CH_3}{\underset{|}{C}}{-}CH_2Cl$;

$ClCH{=}C(CH_3)_2$; 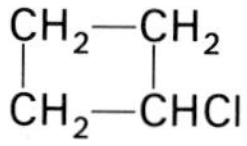

13-20. (a) sp^2 (b) 109.5° (c) 109.5° (d) none (e) none (f) free

13-22. (a) 1, 4, and 6 (b) 1

CHAPTER 14

EXERCISES

14-1. (a) 4-ethyl-2,4-dimethylhexane and 5-ethyl-2,3,4-trimethylheptane

(b) 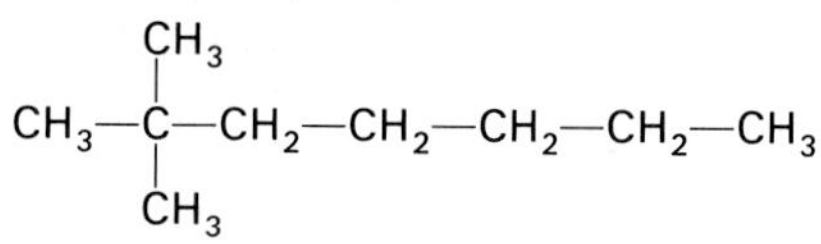

14-3. (b) 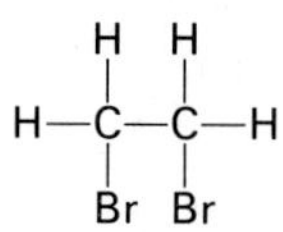 and 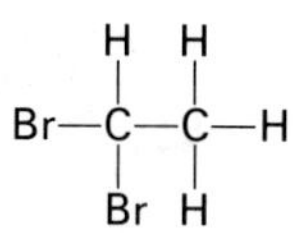

14-4. (a) 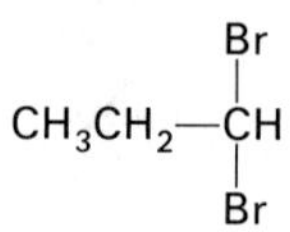

1,1-dibromopropane

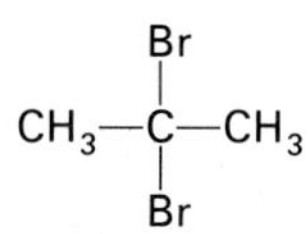

2,2-dibromopropane

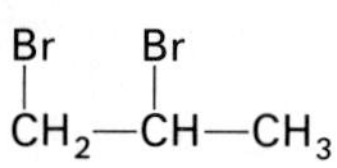

1,2-dibromopropane

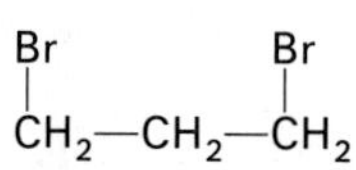

1,3-dibromopropane

14-5. 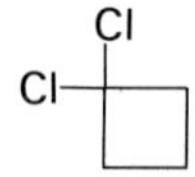

1,1-dichloro-cyclobutane

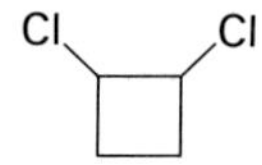

1,2-dichloro-cyclobutane (*cis* and *trans*)

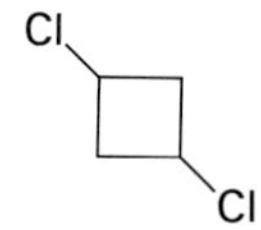

1,3-dichloro-cyclobutane (*cis* and *trans*)

PROBLEMS

14-2. (a) 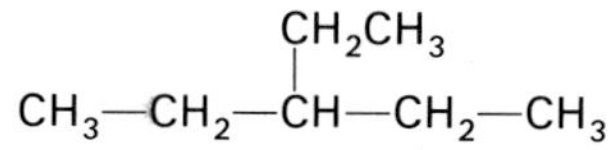

(b) 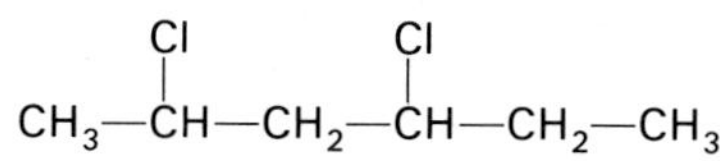

(c) 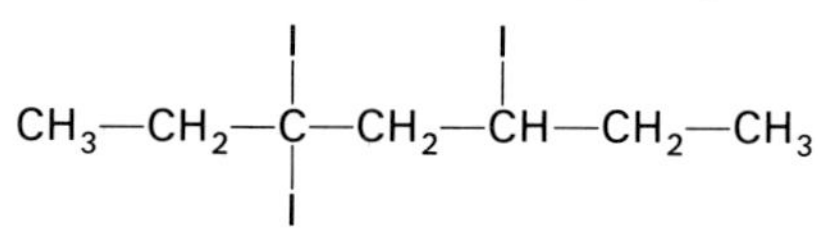

(d) 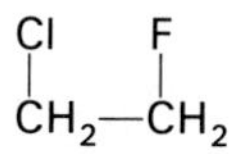

(e) 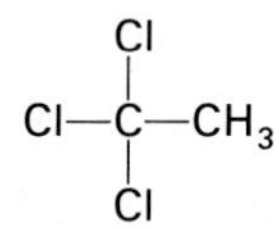

(f) 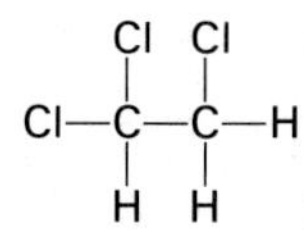

(g) 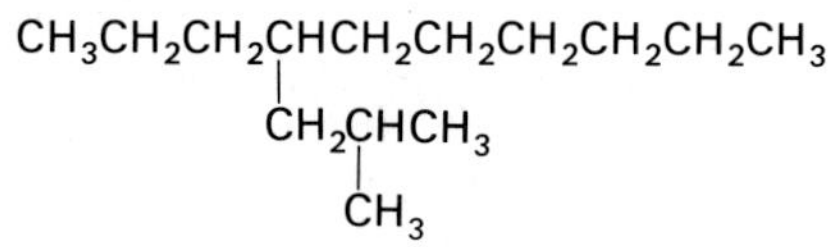

(h) 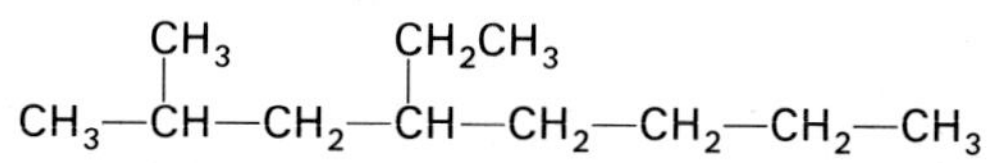

(i) 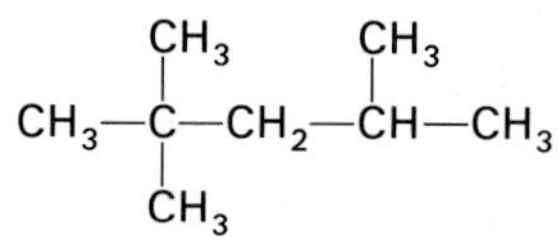

14-4. $CH_3{-}(CH_2)_4{-}CH_3$

***n*-hexane**

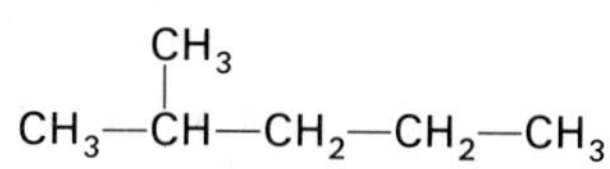

2-methylpentane

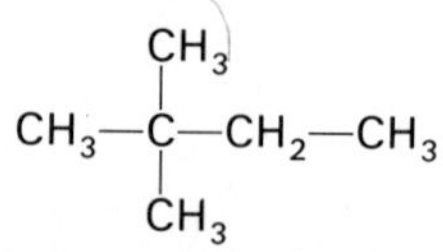

2,2-dimethylbutane

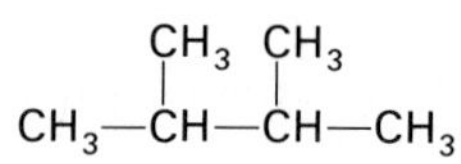

2,3-dimethylbutane

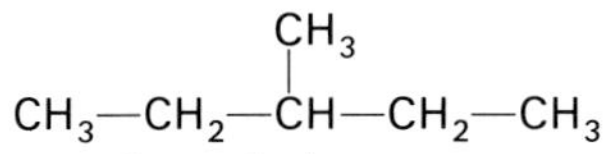
3-methylpentane

14-6. cyclopentane; methylcyclobutane; 1,1-dimethylcyclopropane; *cis*- and *trans*-1,2-dimethylcyclopropane.

14-8. chair form with bromine in equatorial position

14-10. (a) A and B, and A and C are isomers. (b) B and C are different conformations of the same molecule.

14-12. (a) —C≡C— (b) 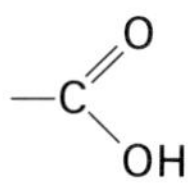 (c) —OH

(d) 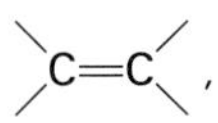, —OH (e) —Cl, —OH, 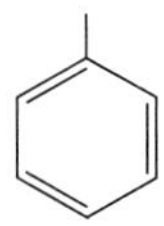

(f) 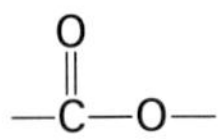 (g) —O— (h) $-NH_2$, 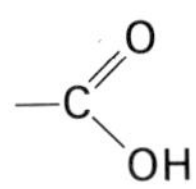

(i) —SH

CHAPTER 15

EXERCISES

15-1. $CH_2{=}CHCH_2CH_2CH_3$ **1-pentene**; $CH_3CH{=}CHCH_2CH_3$ **2-pentene** (*cis* and *trans* isomers)

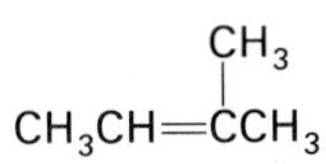
2-methyl-2-butene

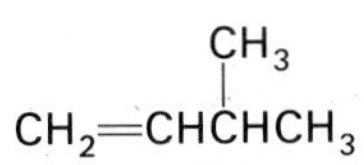
3-methyl-1-butene

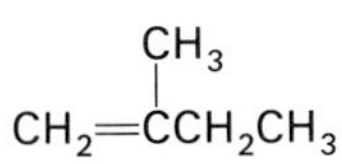
2-methyl-1-butene

15-2. (a) 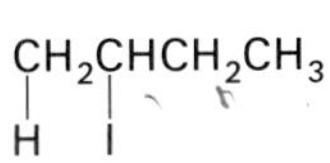 (b) 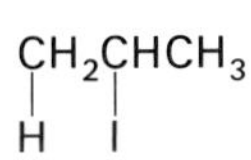

15-3. 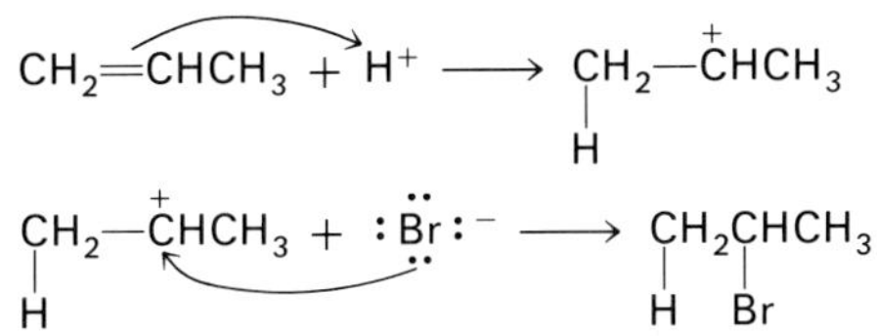

15-4. 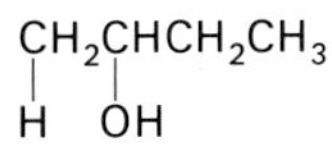

15-5. 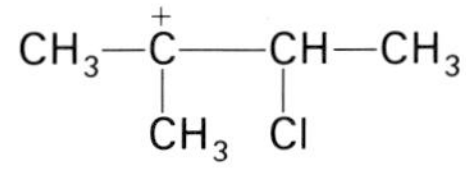

15-6. 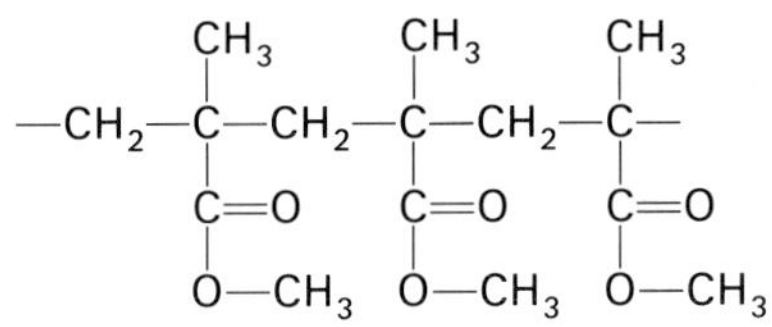

15-7. A high concentration of initiator leads to more free radicals. Thus chain-terminating steps occur more frequently, before the chains grow very long.

15-8. 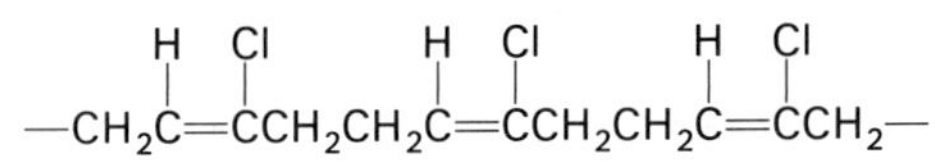

15-9. 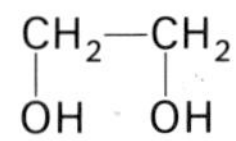

15-10. (a) 3-bromo-6-methyl-4-nonyne
(b) 5-bromo-6-chloro-2-methyl-3-heptyne

15-11. (a) 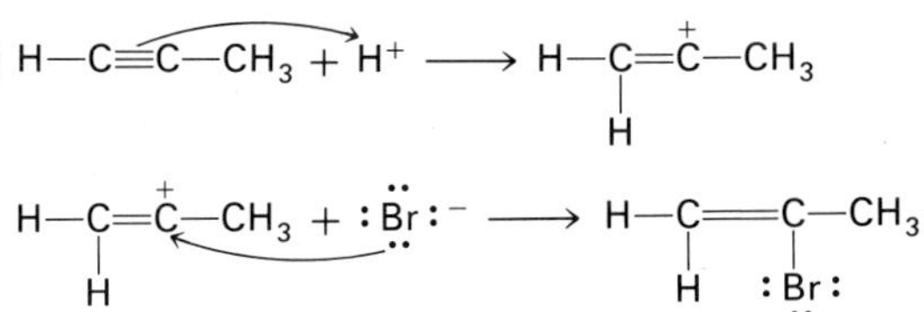

(b) 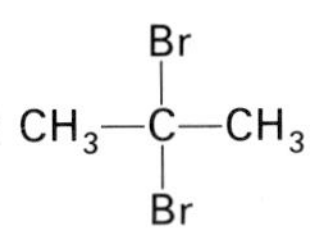

15-12. 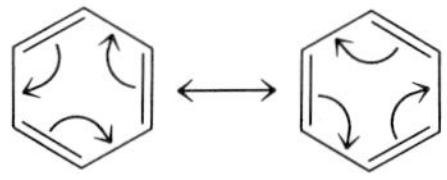

15-13. 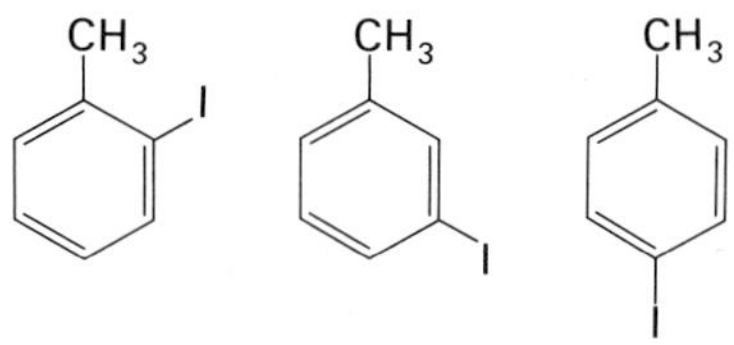

ortho *meta* *para*

PROBLEMS

15-2. (a) 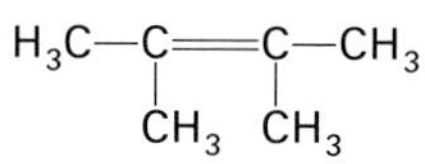

(b) 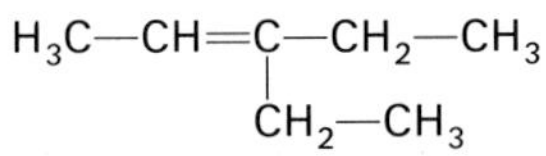

(c) 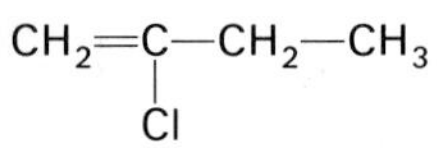 (d) 

(e) $H_3C—CH_2—C≡C—CH_2—CH_3$

15-4. d only

15-6. $CH_3CH_2CH_2CH=CH_2$

15-10. 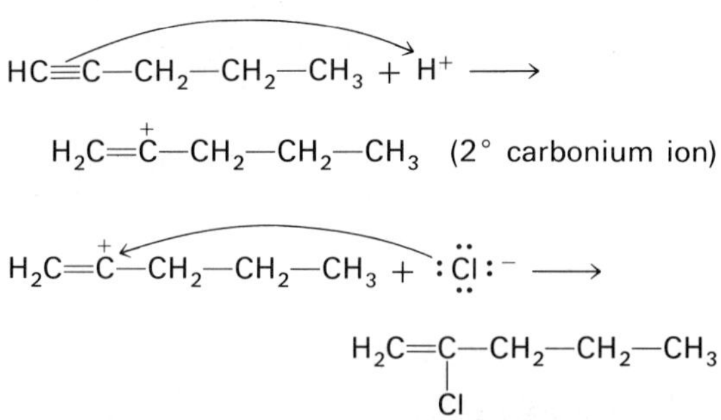

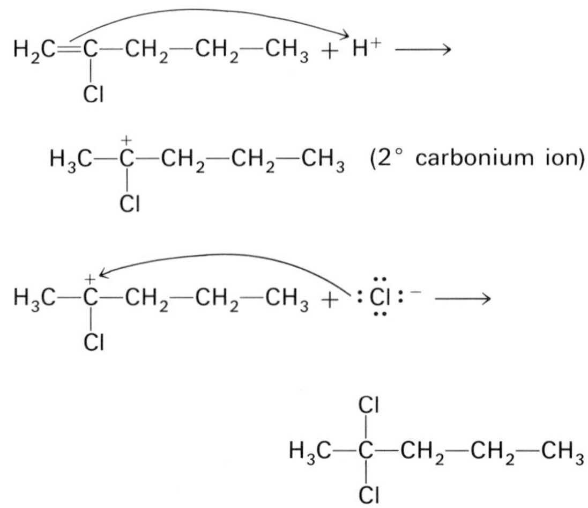

15-12. (a) Hexene will react with $KMnO_4$ and with Br_2, and hexane will not. (b) Hexyne will react with $KMnO_4$ and with Br_2, and hexane will not.

15-14. (a) 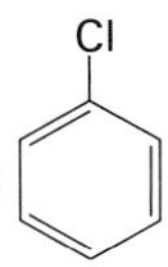 (b) 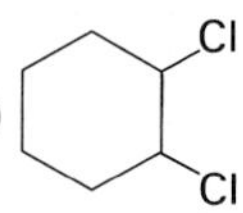 (c) 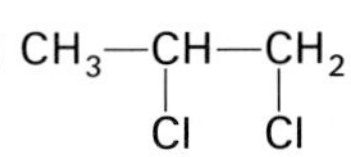

(d) 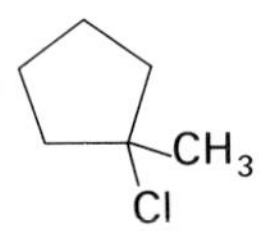 (e) $CH_3—CH=CH—CH_2—CH_3$

15-16. (a) 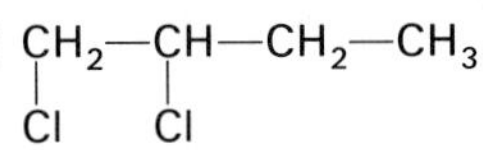

(b) 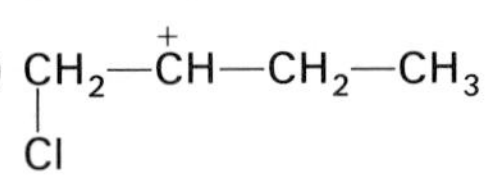 (a 2° carbonium ion)

(c) 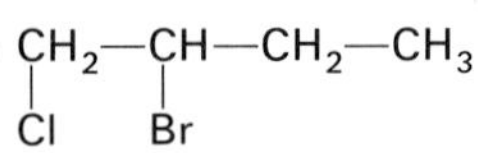

15-18. (a) 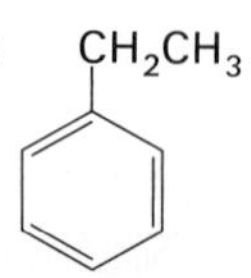

(b) 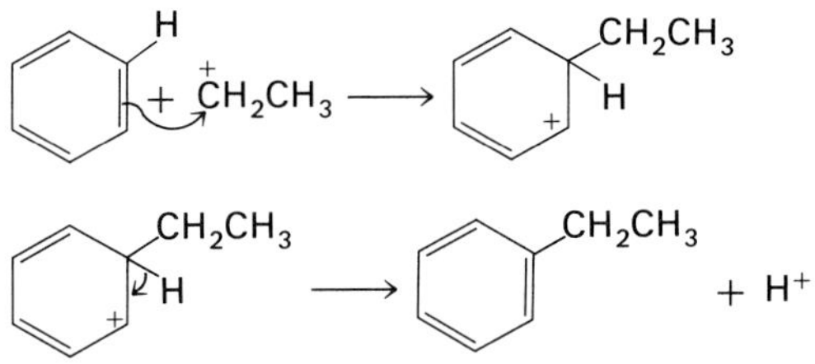

15-20. 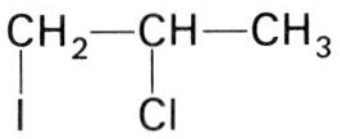

15-22. 87 g of iodine could be added to 100 g of the oil.

15-24. Any compound with one or more carbon–carbon double bonds

15-26. 115 monomers

CHAPTER 16

EXERCISES

16-1. (a) 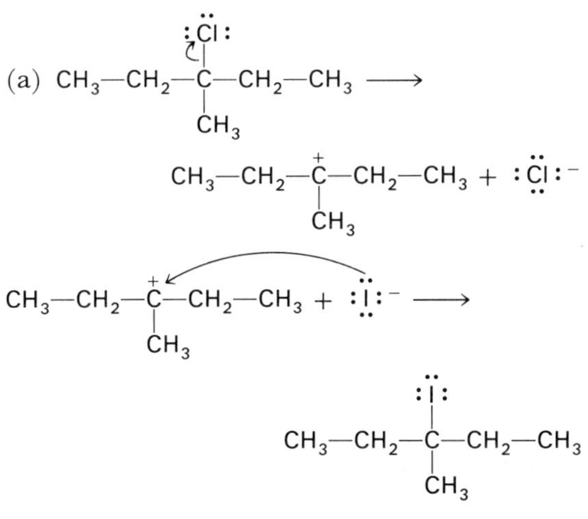

(b) 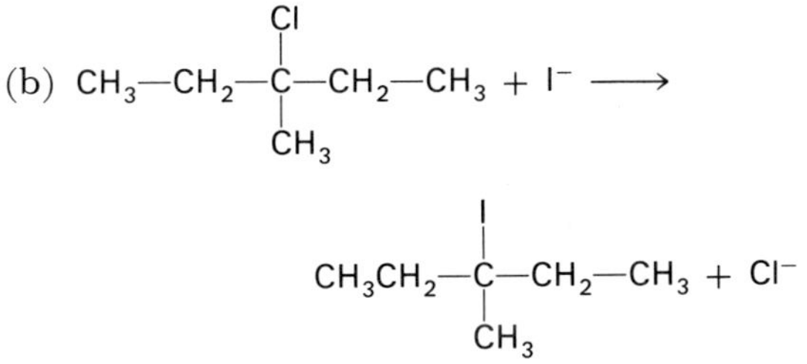

16-2. (a) 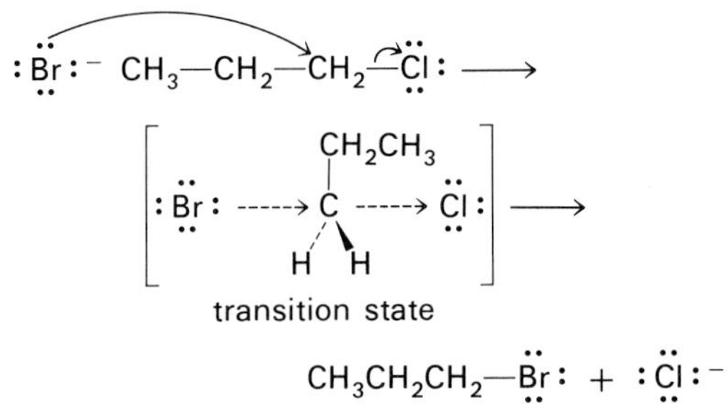

(b) $CH_3CH_2CH_2{-}Cl + Br^- \longrightarrow CH_3CH_2CH_2{-}Br + Cl^-$

16-3. (a) $CH_2{=}CH{-}CH_2{-}CH_3$, E-2

(b) 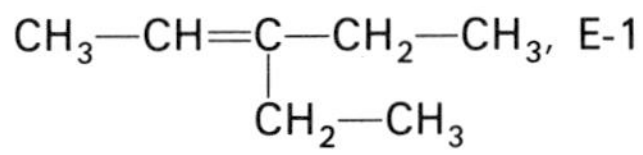 E-1

16-4. (a) $CH_2{=}CH{-}CH_2CH_2CH_3$

(b) $CH_3{-}CH{=}CH{-}CH_2CH_3$ (*cis* and *trans*); $CH_2{=}CH{-}CH_2CH_2CH_3$

(c) $CH_3{-}CH{=}CH{-}CH_2CH_3$ (*cis* and *trans*)

PROBLEMS

16-2. (a) 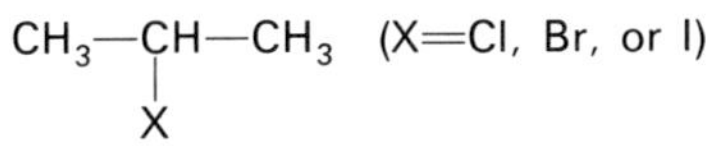 (X=Cl, Br, or I)

(b) 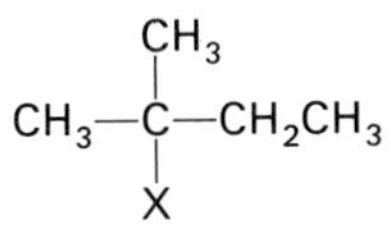 (c) 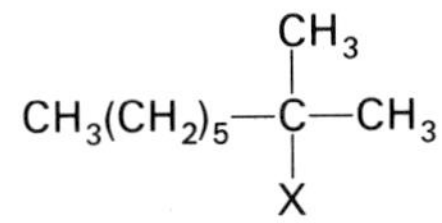

16-4. (b) would be faster, because the 3° carbonium ion is more stable than the 2° carbonium ion that would be formed in (a).

16-6. (a) 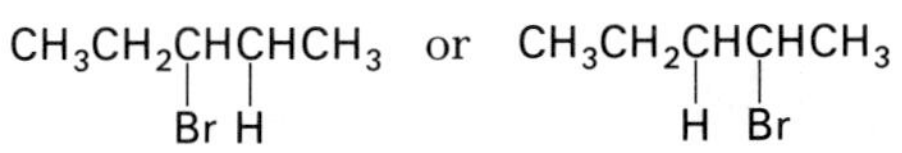

(b) 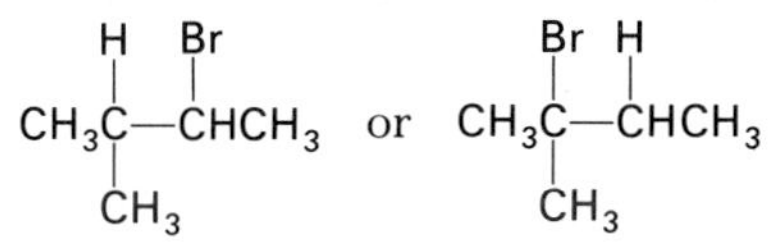

(c) 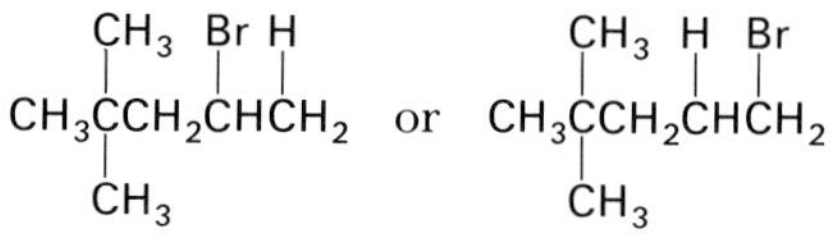

(d) 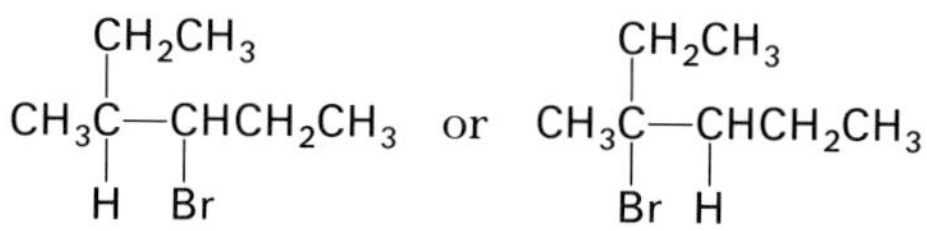

16-8. 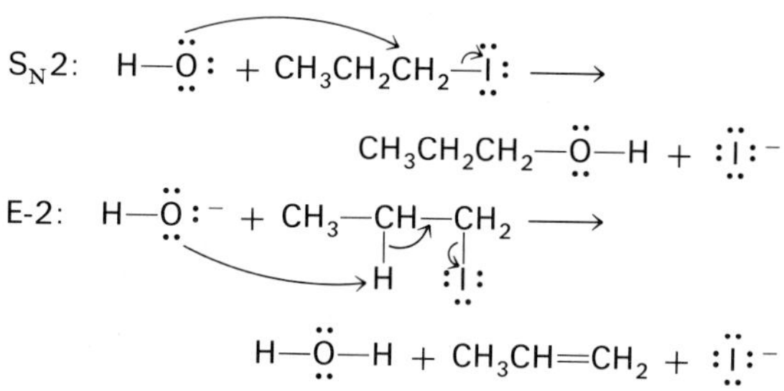

16-10. 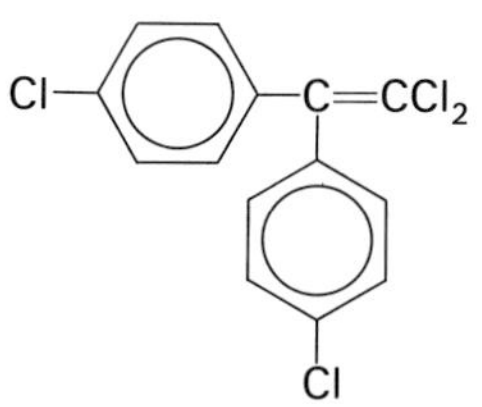

16-12. (a) Substitution

(b) 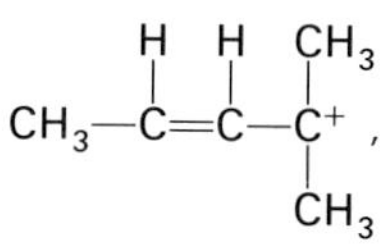 , unimolecular

(c) 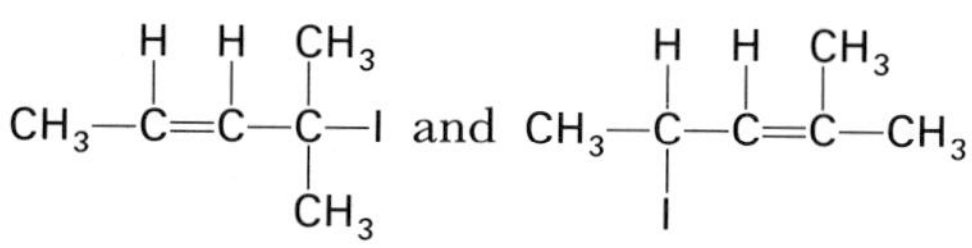

(d) Resonance

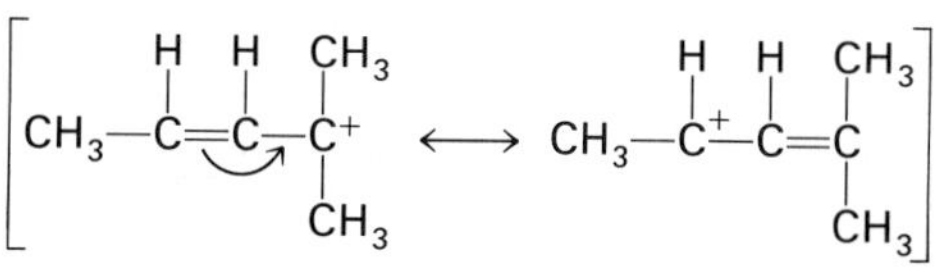

(e) E-1

CHAPTER 17

EXERCISES

17-1. (a) $CH_3CH_2CH_2CH_3$ (b) 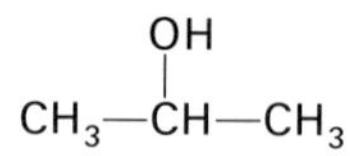

(c) $CH_3CH_2CH_2CH_3$

17-2. (a) 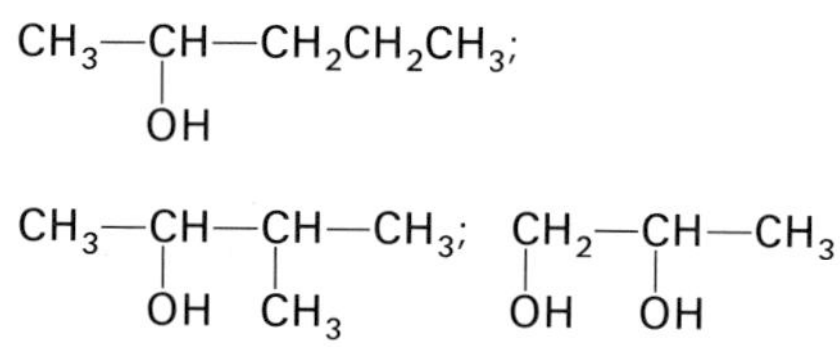

(b) 3,3-dimethyl-2-butanol; 3-pentanol; 1,3-propanediol

17-3. 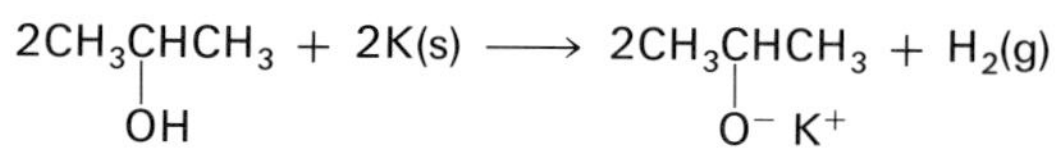

17-4. (a) HI by an S_N1 mechanism in 3 steps:

Step 1:

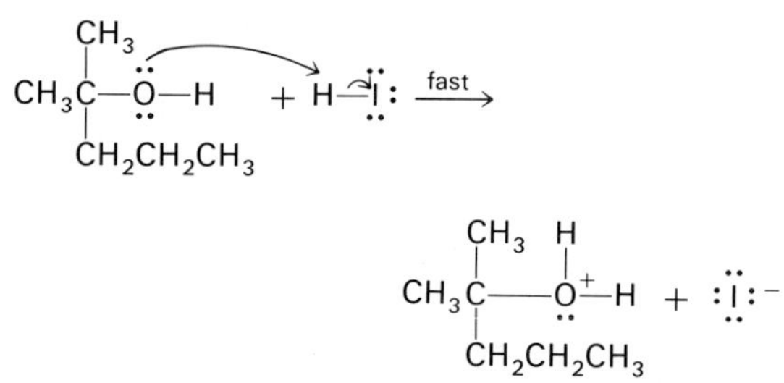

Step 2:

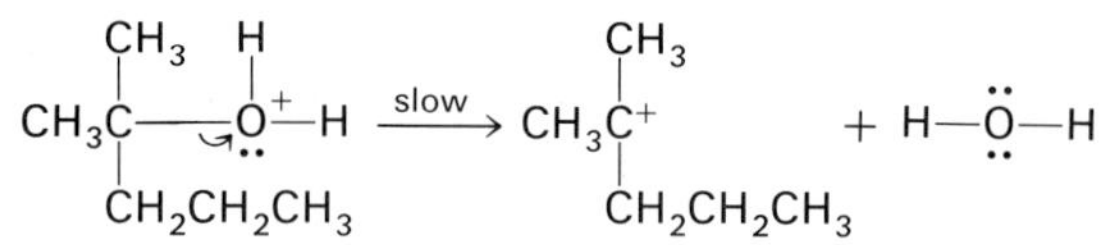

(rate-determining)

Step 3:

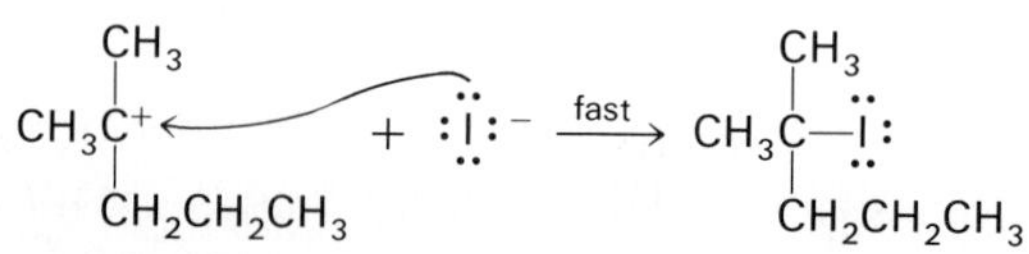

(b) HCl by an S_N2 mechanism in 2 steps:

Step 1:

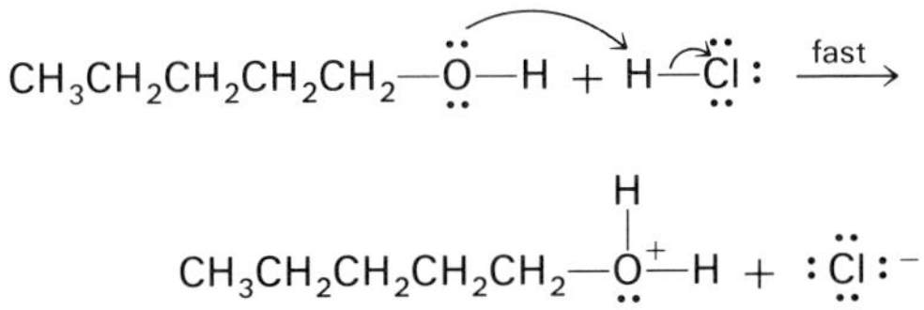

Step 2:

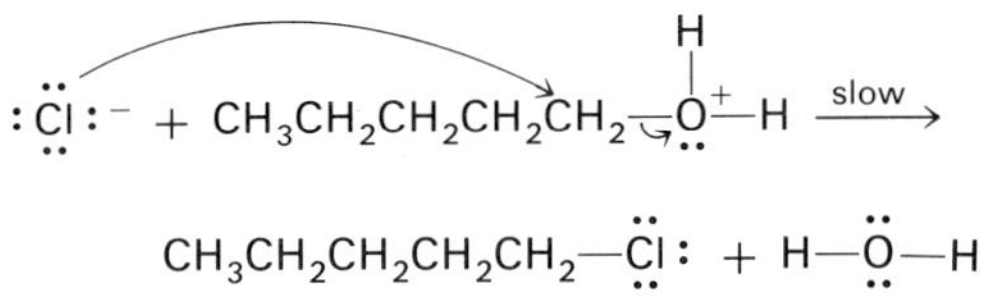

17-5. A strong acid, 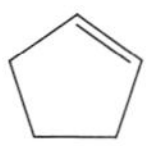

17-6. (a) 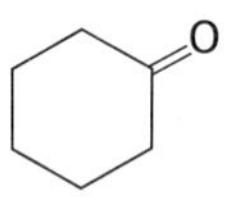

cyclohexanone

(b) no reaction

(c) 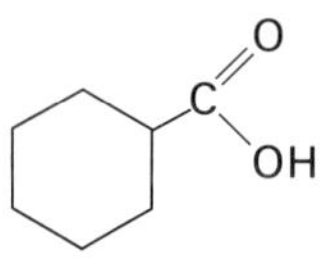

cyclohexanecarboxylic acid

17-7. $CH_3-O-CH_2CH_3$, $CH_3CH_2-O-CH_2CH_3$, and CH_3-O-CH_3

17-8. an ethyl halide

PROBLEMS

17-2. *n*-Propyl alcohol is soluble in water, and so could not be used.

17-4. $CH_3-CH_2-CH_2-CH_2-OH$ 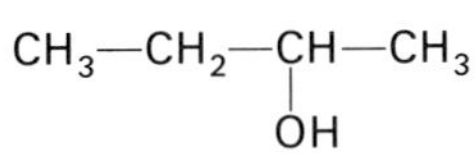

1-butanol (1°) **2-butanol** (2°)

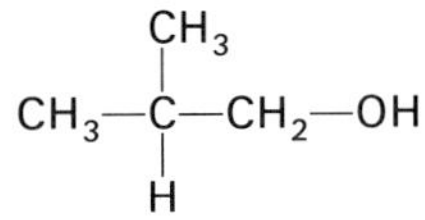 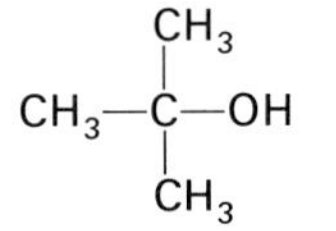

2-methyl-1-propanol (1°) **2-methyl-2-propanol** (3°)

17-6. d

17-8. (a) 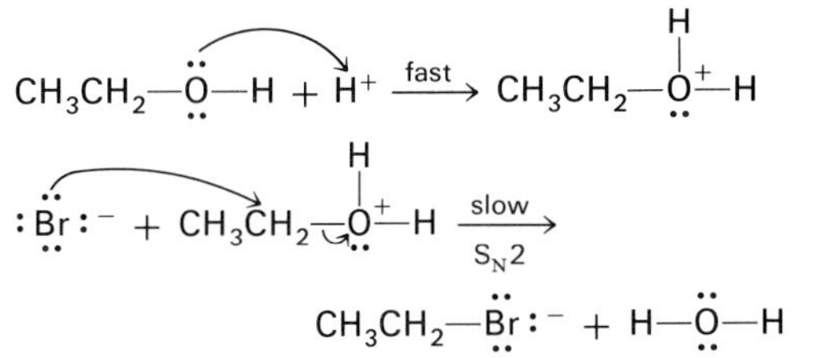

(b) 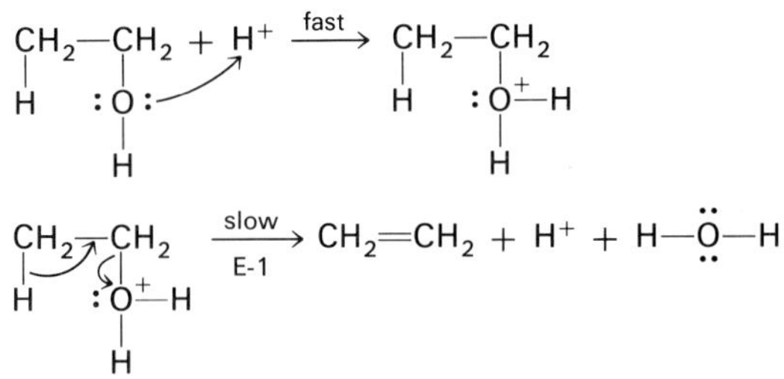

(c) 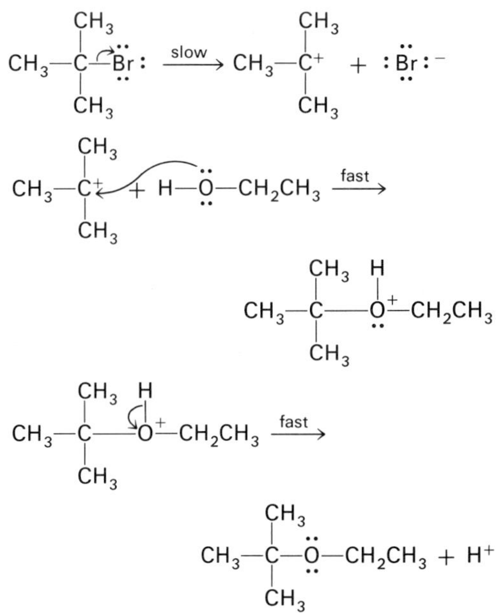

S_N1 (followed by elimination of proton)

17-10. $CH_3CH_2CH_2OH$ and $CH_3CH_2CH_2Br$
(C_3H_8O is a 1° alcohol)

17-12. $CH_3CH_2CH(OH)CH_2CH_3$ gives *cis*- and *trans*-2-pentene as only products (no 1-pentene)

17-14. (a) $CH_3CH{=}CH{-}CH_2CH_3$ (*cis* and *trans*)
(b) $CH_3{-}\ddot{O}{-}CH_2CH_3$
(c) 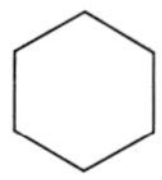

17-16. lack of reactivity, ability to dissolve both polar and nonpolar substances

CHAPTER 18

EXERCISES

18-1. d (highest), c, b, e, a (or d > c > b > e > a)

18-2. (a) 3-ethylpentanal
(b) 3-methyl-2-butanone
(c) 4-bromo-2-pentanone

18-3. (a) 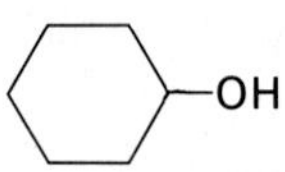 (b) $CH_3{-}CH(OH){-}CH_3$
(c) $CH_3CH_2CH_2CH_2OH$

18-4. $:N{\equiv}C{-}C(H)(CH_3){-}OH$

18-5. 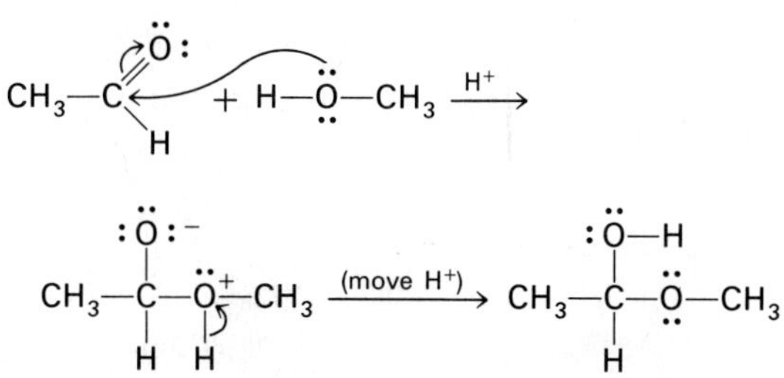

18-6. (a) $CH_3{-}C(H)(OH)(OCH_2CH_3)$ (b) $CH_3{-}C(H)(O{-}CH_2CH_3)_2$
(c) (1) acetal (2) hemiketal
(3) hemiacetal (4) ketal
(d) (1) CH_3OH and CH_3CH_2CHO
(2) $CH_3CH_2CH_2OH$ and 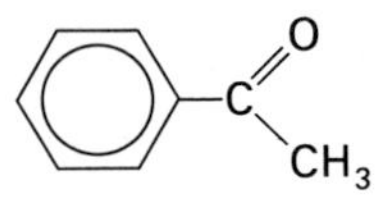

18-7. 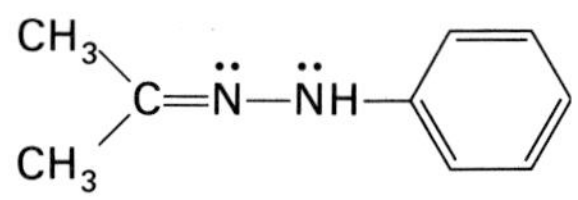

18-8. 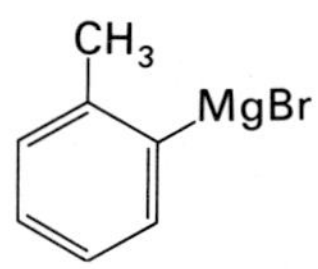

18-9. (a) 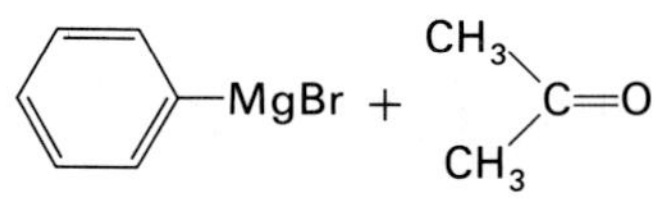
(b) CH_3MgBr + 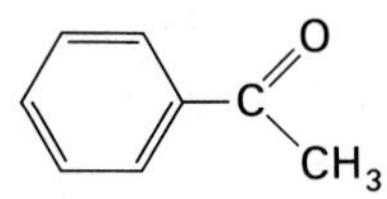

18-10. (a) 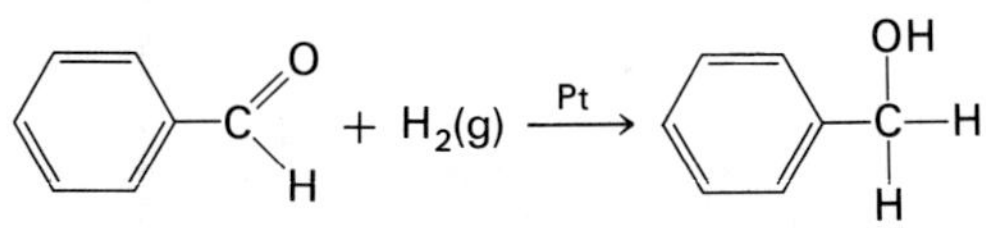
(b) $(CH_3CH_2)_2C{=}O + H_2(g) \xrightarrow{Pt} CH_3CH_2C(OH)(H)CH_2CH_3$
(c) 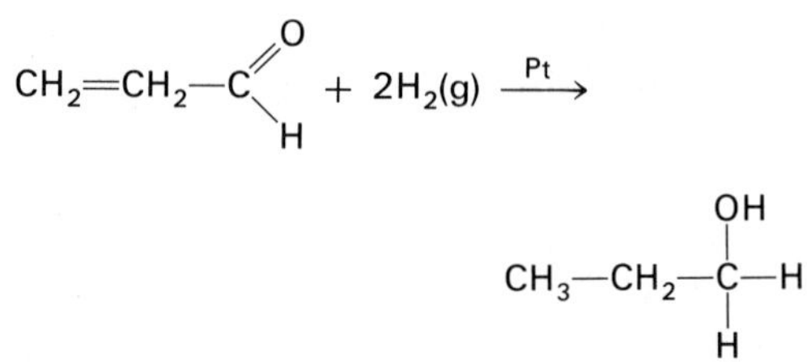

18-11. (a) $CHCl_3$ (inductive effect greatest)

(b) 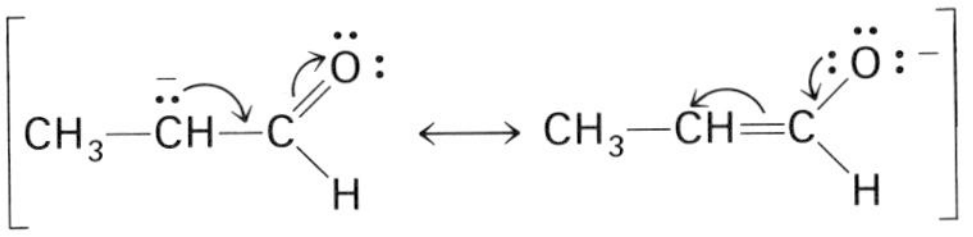

18-12. 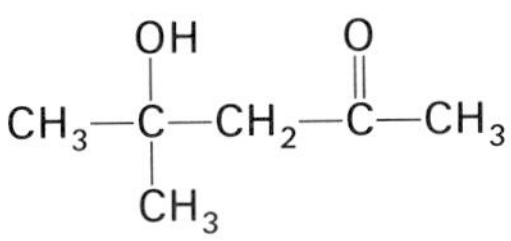

PROBLEMS

18-2. c < a < b < d

18-4. (a) 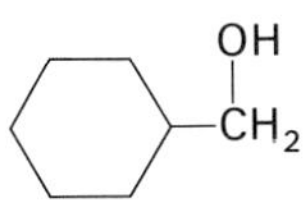 (b) 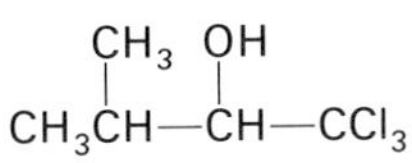

(c) 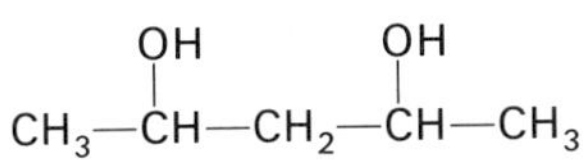

18-6. (a) 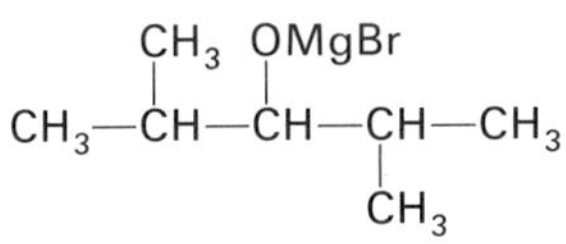

(b) 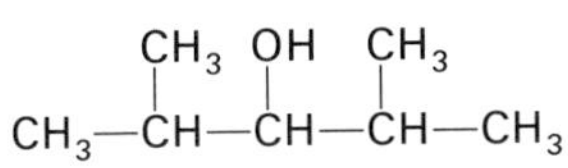

(c) 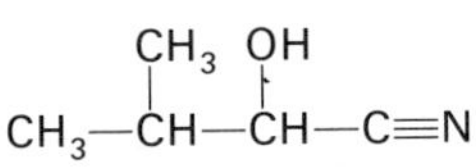

(d) 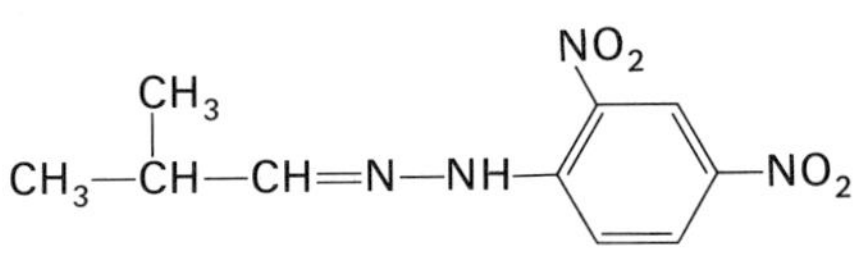

(e) 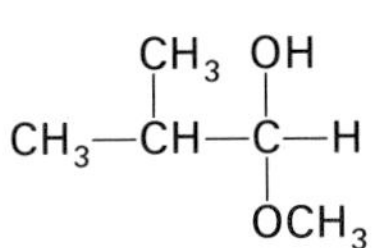 (f) 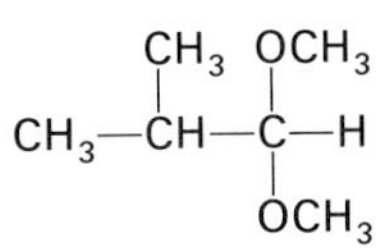

18-8. (a) 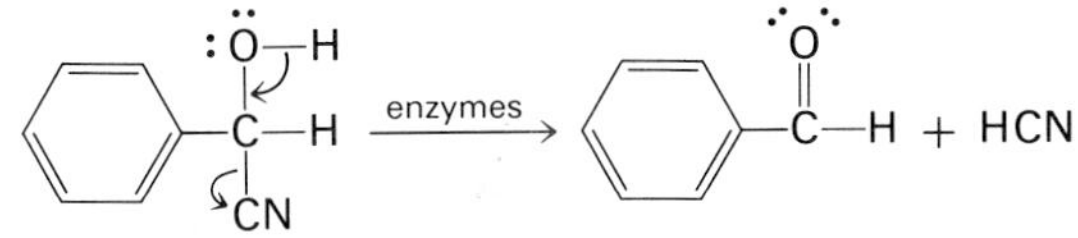

(reverse of
HCN addition)

(b) 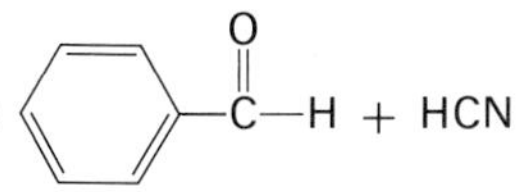

18-10. 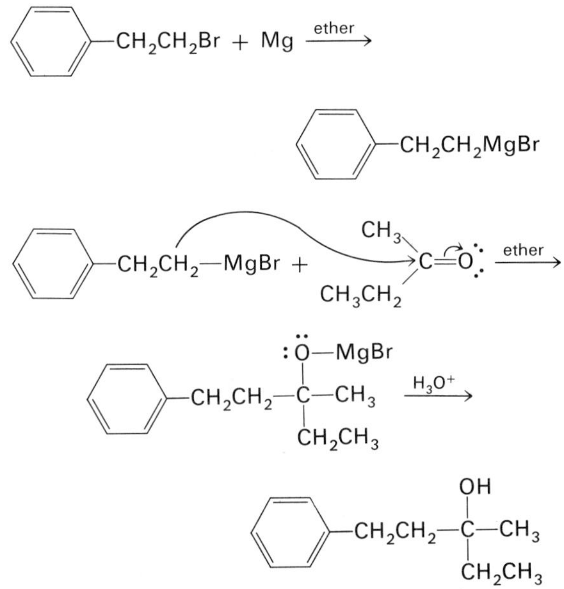

Other possible combinations are:

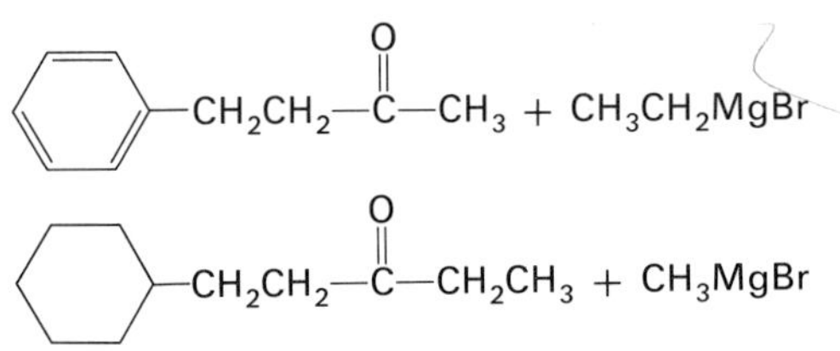

18-12. 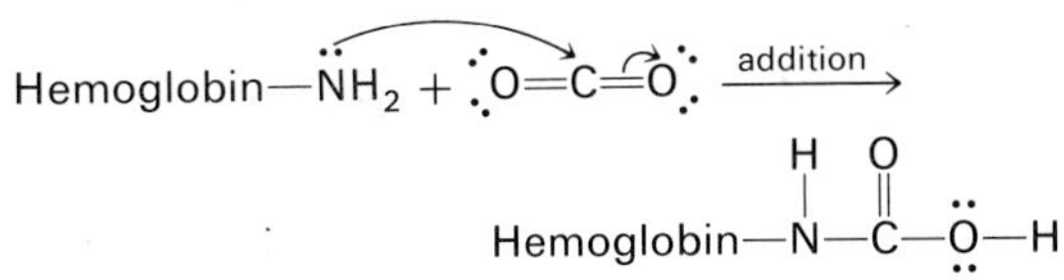

18-14. (a) 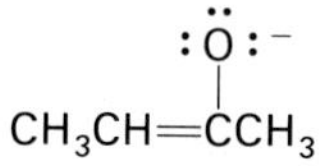

(b) 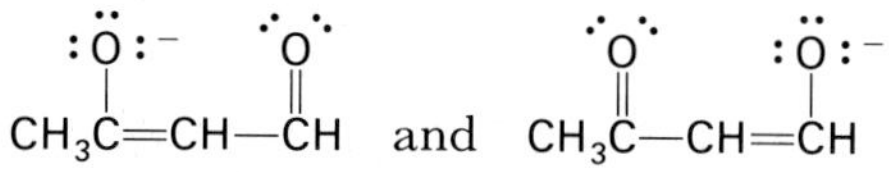

(c) 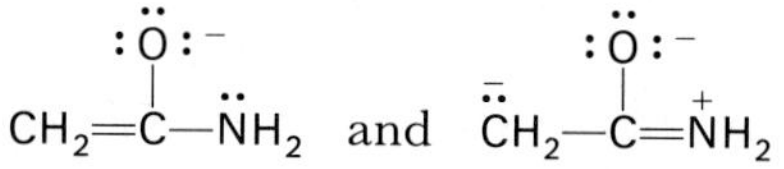

(d) 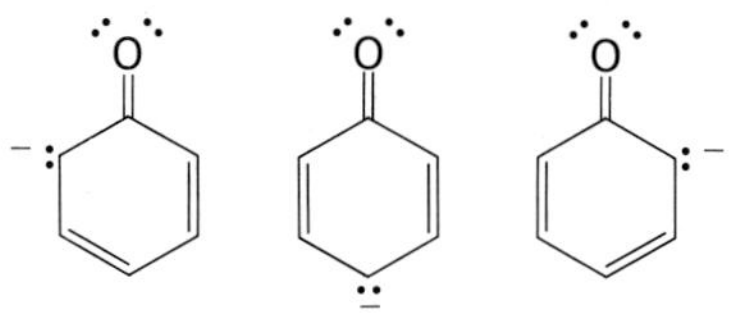

(e) 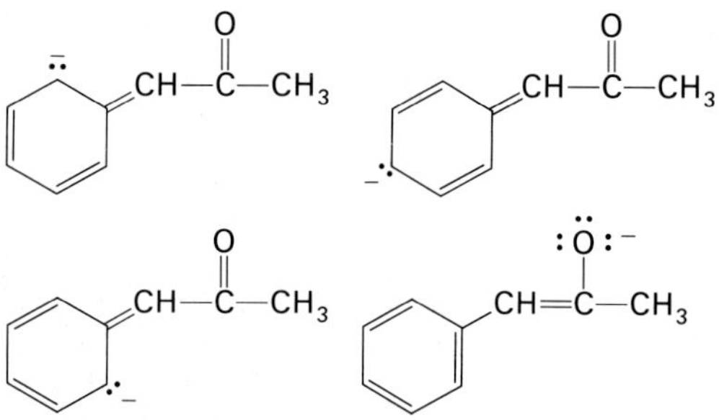

(f) 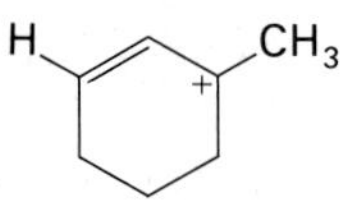

18-16. (a) (1) 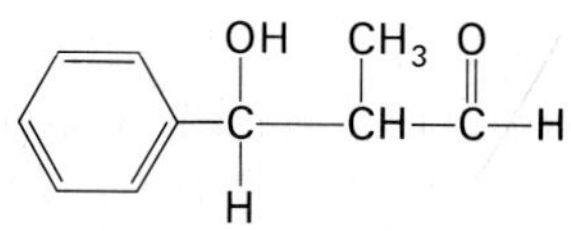

(2) 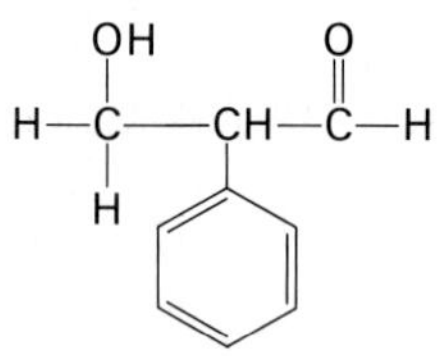

(b) (1) 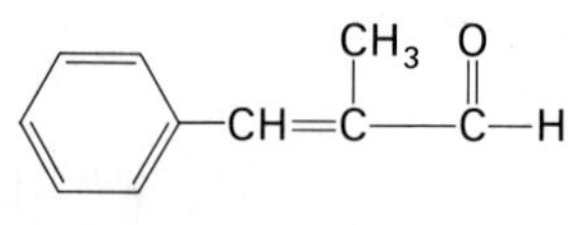

(2) 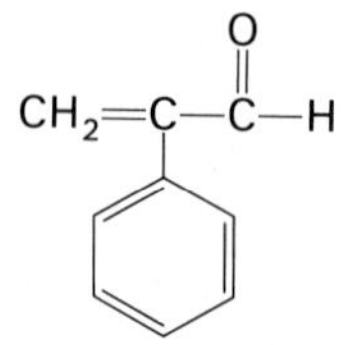

18-18. (a) 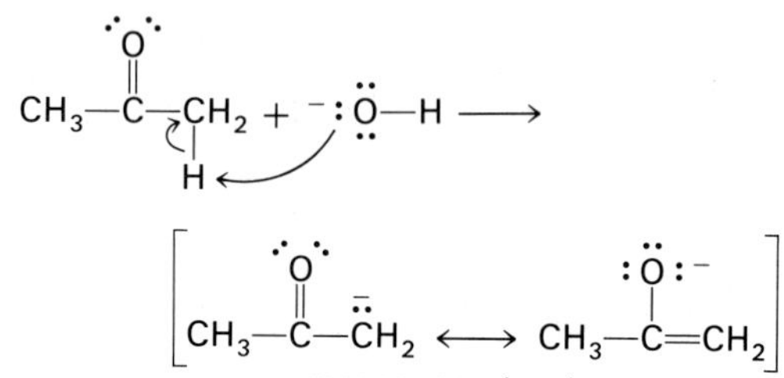

(b) 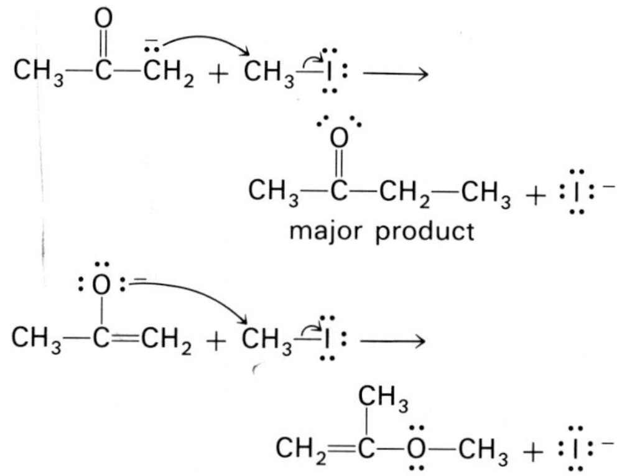

(c) 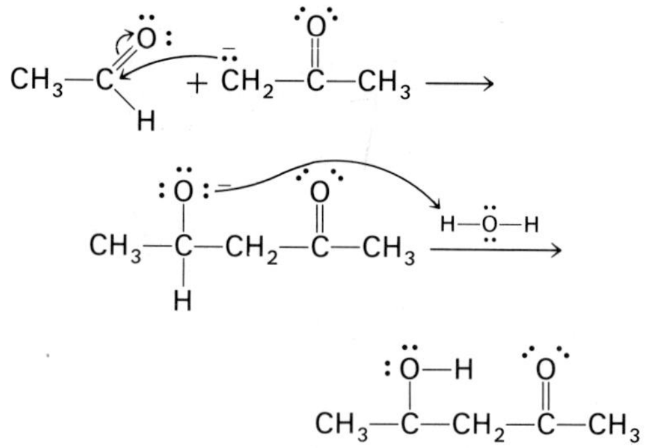

18-20. (a) solubility in water; Lucas test (Section 17-5)

(b) Tollens' test; Fehling's test; $K_2Cr_2O_7$ (or $KMnO_4$) and H_2SO_4.

18-22. 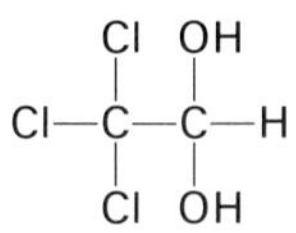

CHAPTER 19

EXERCISES

19-1. c > d > b > a

19-2. (a) 2-chloropentanoic acid
(b) 2,3-dimethylbutanoic acid
(c) 3,3-dimethylpentanoic acid

19-3. (a) $CH_3-\underset{Br}{\underset{|}{CH}}-COOH > CH_3CH_2COOH > CH_3CH_2CH_2OH > CH_3CH_2CHO$

(b) $CH_3\overset{\ominus}{C}HCHO > CH_3CH_2CH_2O^{\ominus} > CH_3CH_2COO^{\ominus} > CH_3-\overset{Br}{\overset{|}{CH}}-COO^{\ominus}$

19-4. Yes—octanoic acid is a *much* stronger acid than octanol. Octanoic acid would dissolve in NaOH solution, and octanol would not dissolve.

19-5. 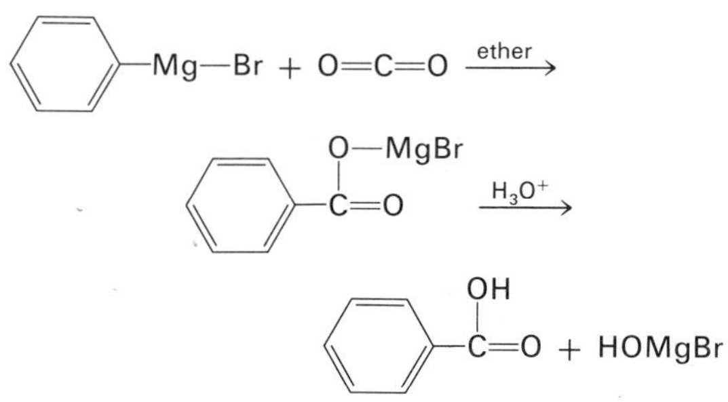

19-6. (a) $CH_3-\overset{O}{\overset{\|}{C}}-OH + HOCH_2CH_2-\overset{CH_3}{\overset{|}{\underset{CH_3}{\underset{|}{N^+}}}}-CH_3$

(b) 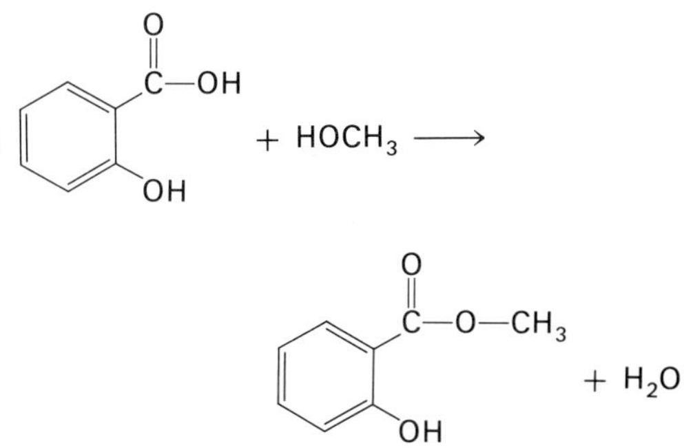

methyl salicylate
(oil of wintergreen)

(c) 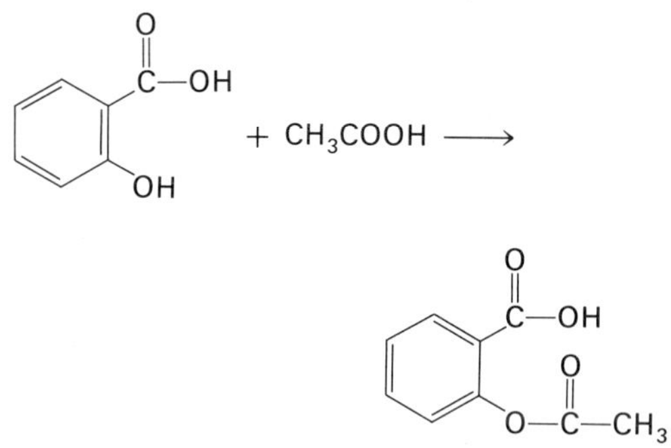

acetylsalicylic acid
(aspirin)

19-7. $CH_3-\overset{O}{\overset{\|}{C}}-OH$ and $HO-CH_2CH_2-\overset{CH_3}{\overset{|}{CH}}-CH_3$

19-8. (a) $CH_3CH_2\overset{O}{\overset{\|}{C}}-OCH_3 + HBr$

(b) 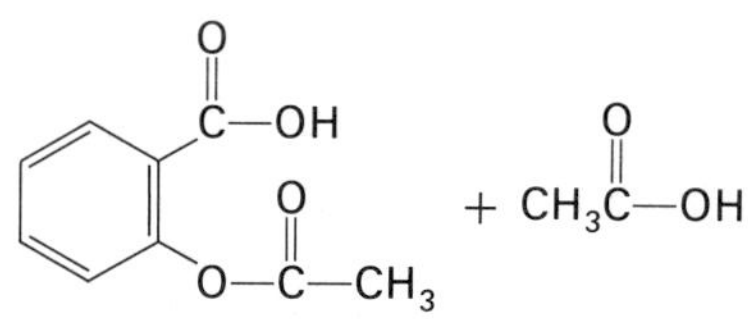

acetylsalicylic acid
(aspirin)

(c) $CH_3\overset{O}{\overset{\|}{C}}-NH_2 + HCl$

19-9. 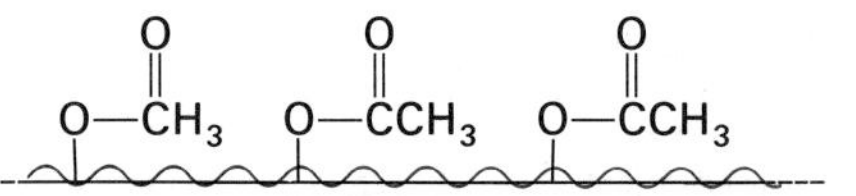

The molecule eliminated is CH_3COOH (acetic acid).

19-20. 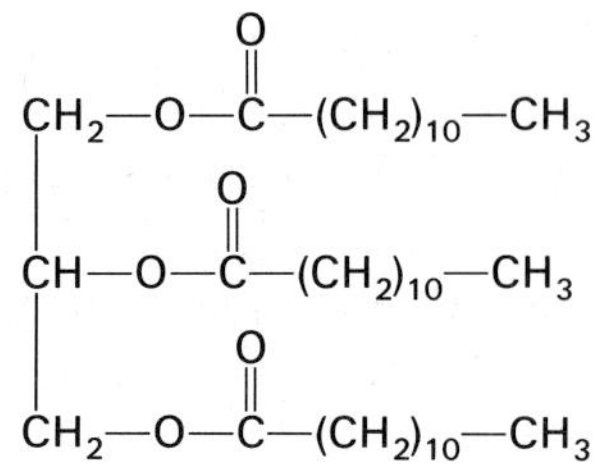

19-22. $HOCH_2-CH(OH)-CH_2OH$, or any trihydroxy alcohol

PROBLEMS

19-2. e > b > c > d > a
19-6. b < a < c < e < d
19-8. a
19-10. Blue litmus paper would turn red with the acid.

19-12. (a) HCOOH and $CH_3-CH(CH_3)-CH_2OH$
(b) CH_3COOH and $CH_3(CH_2)_7OH$
(c) $CH_3CH_2CH_2COOH$ and CH_3CH_2OH
(d) CH_3COOH and 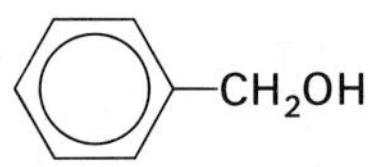
(e) 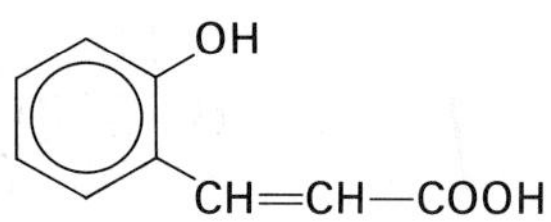

19-14. 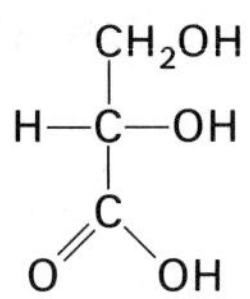

19-16. The acid is more soluble in water.

19-18. (a) 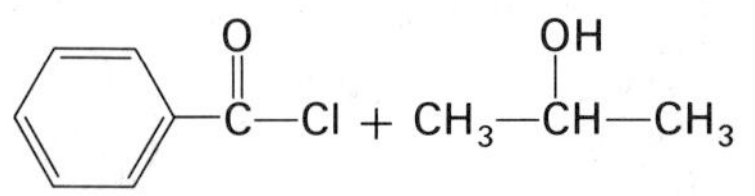
(b) 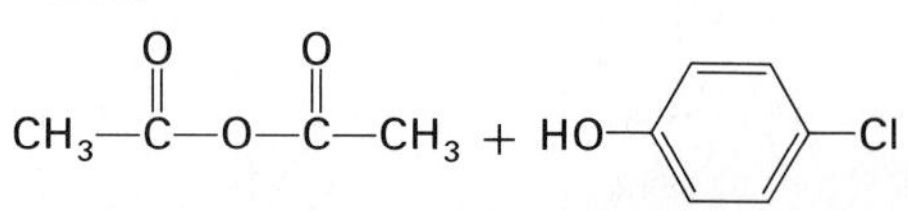

CHAPTER 20

EXERCISES

20-1. c < a < b < d
20-2. Nicotine: one aromatic; one 3°. Reserpine: one 2°, one 3°.
20-3. (a) 1-aminopropane (b) 1,5-diaminopentane (c) para-bromoaniline
20-4. c
20-5. 0.1 mole
20-6. 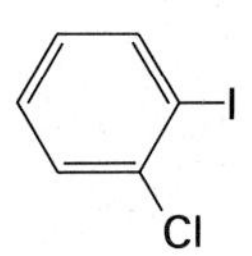

PROBLEMS

20-2. (a) para-iodoaniline (b) cyclohexylamine (c) isopropylamine (d) ethylmethylpropylamine
20-4. Lemon juice contains citric acid. This can react with the basic amines to yield citrates—ionic salts that are odorless.

20-6. (a) 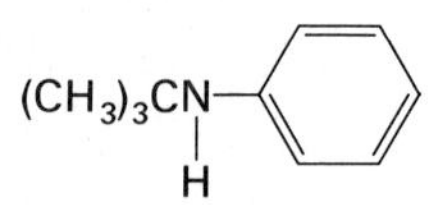 (b) 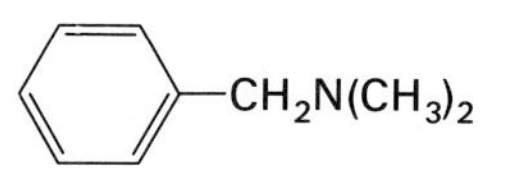
(c) 

20-8. Add dilute HCl. The amine will go into solution as an ammonium salt. The ketone will remain as a separate organic layer. Separate the layers, and add dilute NaOH solution to the aqueous layer until it is basic. This regenerates the amine, which will separate out as an organic layer because it is only slightly soluble in water.

20-10. Cyclohexylamine is the stronger base. The lone pair of electrons on the N in aniline is delocalized through the benzene ring.

20-12. (a) $CH_3CH_2CH_2\overset{O}{\overset{\|}{C}}-O-CH_2CH_3$ and $NH(CH_3)_2$

(b) 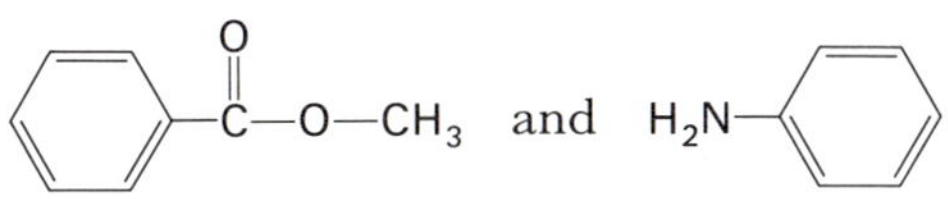

20-14. (a) 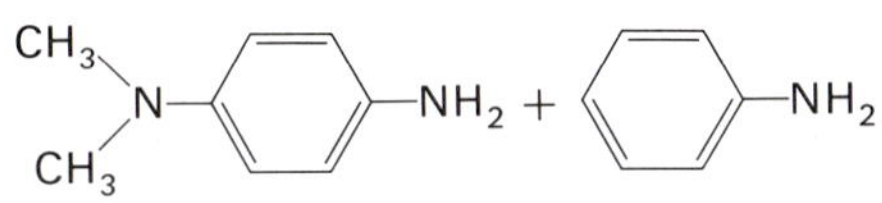

(b) 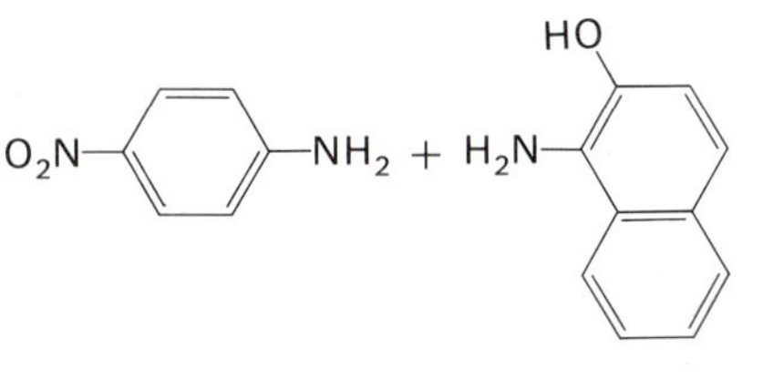

(c) 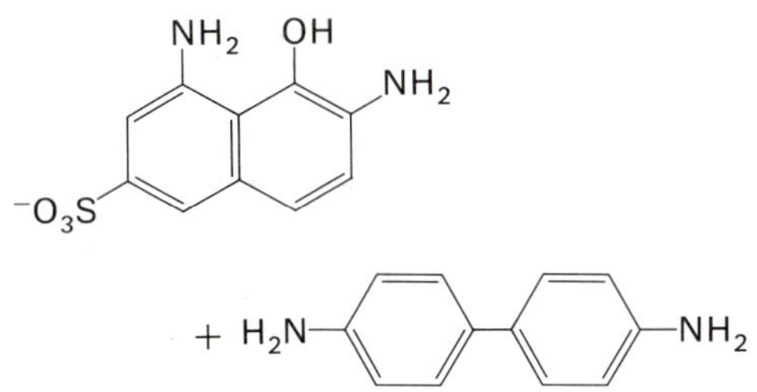

20-16. 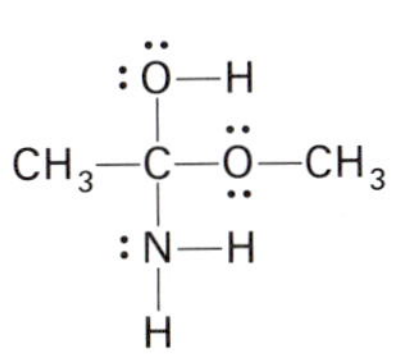

20-18. The amide linkages are hydrolyzed.

20-20. $HOOC-CH_2-\underset{COOH}{\underset{|}{CH}}-CH_2-COOH$

a molecule that has three COOH groups (or COCl or acid anhydrides)

20-22. Convert the amine to a salt by adding an acid such as HCl or H_2SO_4. The salt is ionic and more soluble in water.

CHAPTER 21

EXERCISES

21-1. 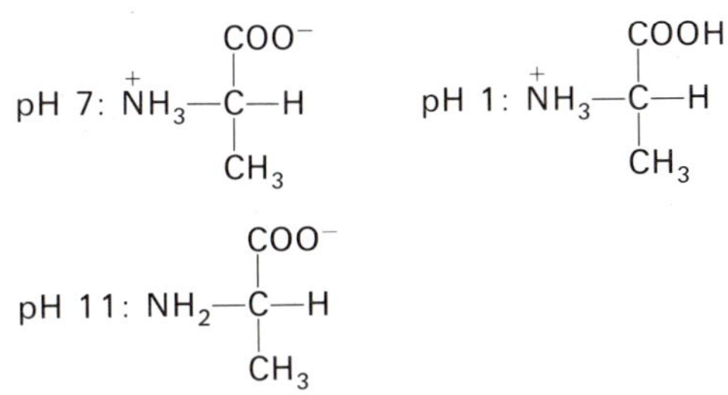

21-2. (a) Aspartic acid is a buffer at pH values of approximately 2, 4, and 10. (b) Aspartic acid is a stronger acid than glutamic acid. (c) Arginine is a stronger base than lysine.

21-3. Val–Leu–Tyr; Val–Tyr–Leu; Leu–Val–Tyr; Leu–Tyr–Val; Tyr–Val–Leu; Tyr–Leu–Val

21-4. (a) Met–Ala–Asn–Ser–Gly–Lys

(b) 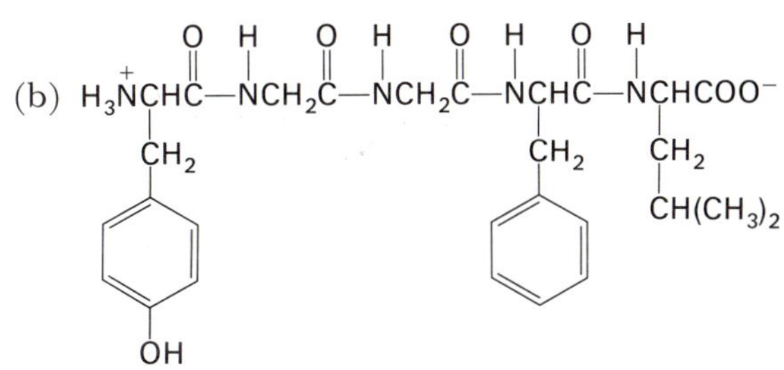

21-5. See the structure at the bottom of the page.
21-6. (a) Any of the nonpolar amino acids.
(b) Any of the polar or charged amino acids.
21-7. Arg–Asp; Arg–Glu; Lys–Glu; charged form of His with Asp or Glu
21-8. Both the O and an H of C_2H_5OH can form hydrogen bonds with side chains of amino acid residues in the protein.
21-9. (a) Trypsin: Lys; Cys–Tyr–Val–Arg; His–Glu–Trp–Ala–Ser
Chymotrypsin: Lys–Cys–Tyr; Val–Arg–His–Glu–Trp; Ala–Ser
(b) Val–Arg–Leu–Gly–Asp–Phe–Ala–Tyr–Glu

PROBLEMS

21-2. (a) pH 1: 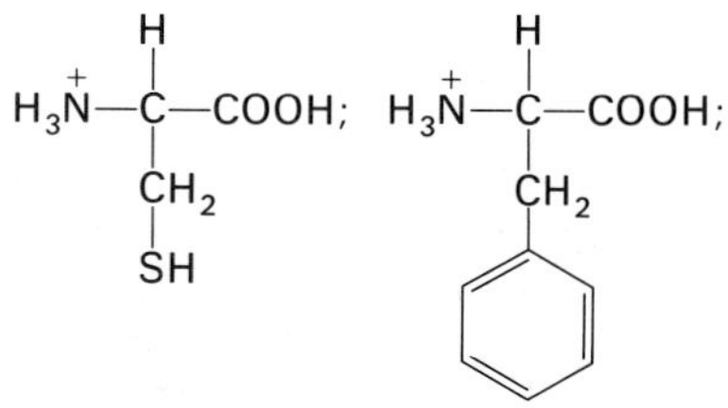

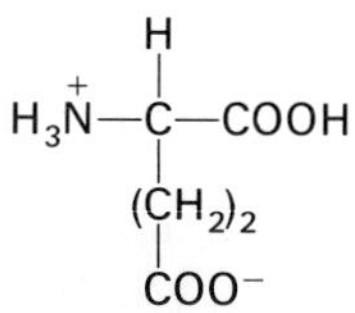

pH 7: 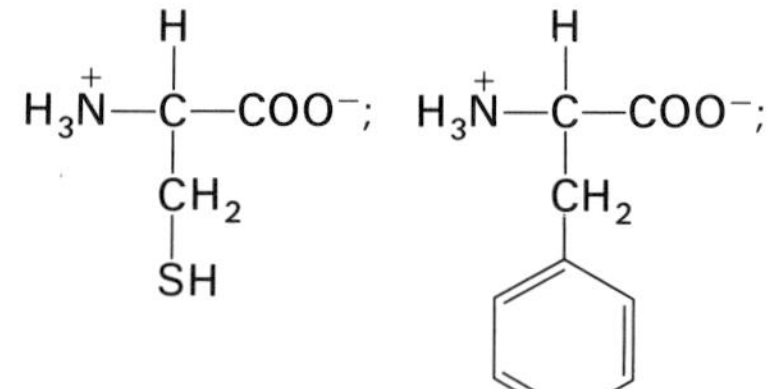

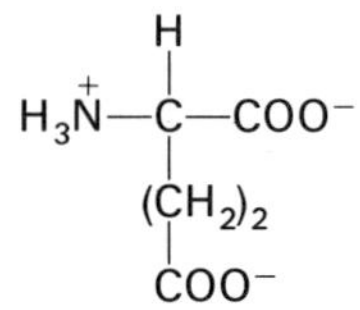

pH 12: 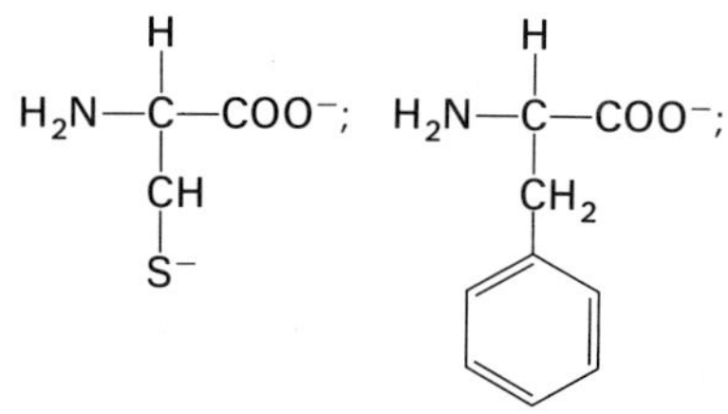

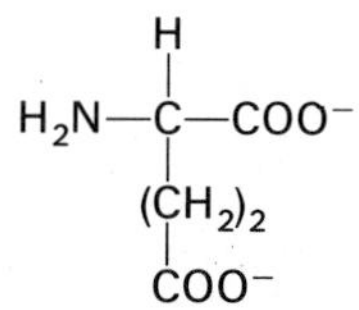

(b) Asp: 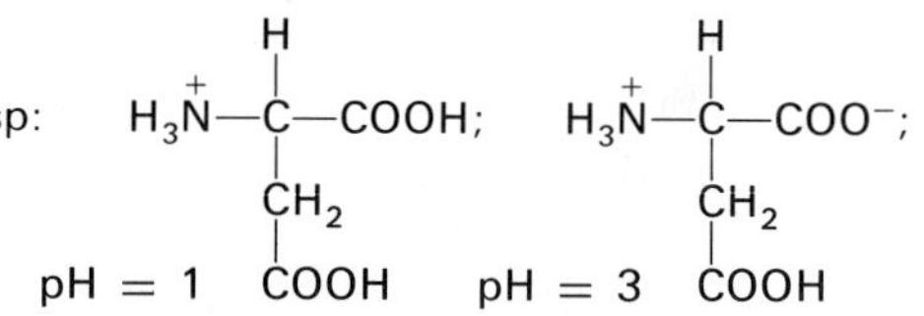

21-5. 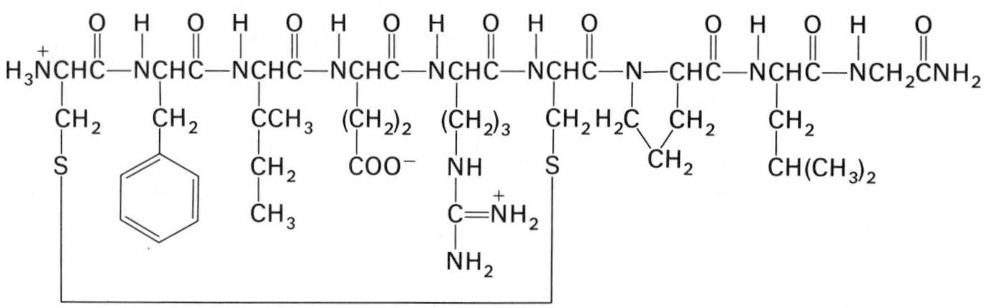

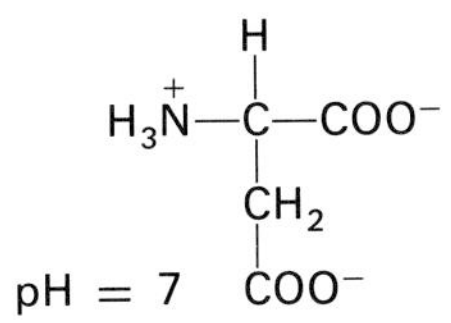

Lys: 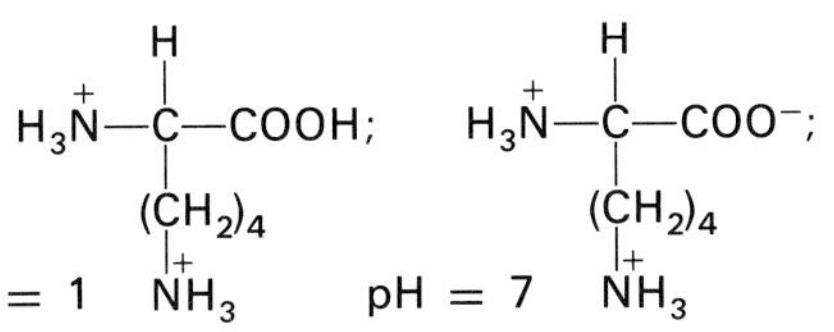

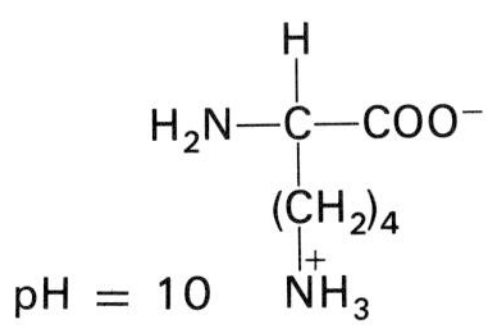

21-4. Gly
21-6. In globular proteins, hydrophilic groups are on the outside and hydrophobic groups are on the inside. In fibrous proteins, both hydrophilic and hydrophobic groups are on the outside.
21-8. (a) Glu (b) Leu (c) Ala (d) Ser (e) Asp
21-10. A disulfide bond forms a cyclic structure.
21-12. (a) Lys, Arg, and His (b) Pro
21-14. (a) three (b) eight (as in Problem 21-13) (c) one; optically active
21-16. Gly–Tyr; Ala–Val–Phe; Trp
21-18. (a) 100,000 g/mole (b) The enzyme consists of four subunits. Acetone disrupts the quaternary structure.
21-20. Thr–Val–Lys–Ala–Ala–Trp–Gly–Lys
21-22. (a) Folded in a U-shape so that salt linkages can form between Asp and Lys; with the glycines at the bend. (b) The Leu residues would probably be on the interior; the Thr and Lys residues would be folded so as to hydrogen bond with each other and with water.

CHAPTER 22

EXERCISES

22-1. (a) The cell does not waste energy making substances that are not needed. (b) Enzyme 3 is inhibited by D, and enzyme 4 by H.
22-2. (a) increasing the enzyme concentration; increasing the temperature (b) Rate $= k[E][S]$ (c) The rate would be doubled.
22-3. 0.6 mole/ℓ min
22-4. 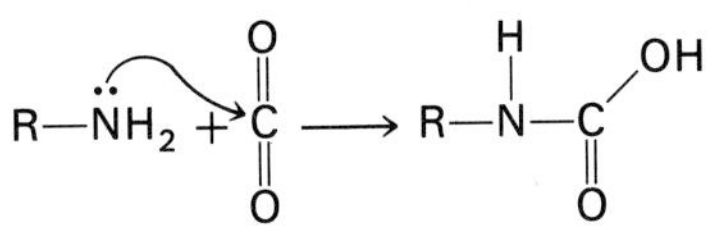
22-5. (a) increase $[HbO_2]$ (b) increase $[HbO_2]$ (c) increase $[HbO_2]$

PROBLEMS

22-2. possibly Glu for Asp; none for His; possibly Thr for Ser
22-4. (a) $V_{max} = 8.4, 6.5, 8.4$ (b) X is a noncompetitive inhibitor; Y is a competitive inhibitor. (c) Y
22-6. Asparaginase with $K_M = 10^{-6}$ M. This asparaginase is effective at concentrations as low as 10^{-6} M.
22-8. If trypsin were produced in the pancreas, the trypsin might digest proteins of the pancreatic cells.
22-10. (a) 60,000 per minute (b) $V_{max} = 12.0 \times 10^{-10}$ mole/min; $K_M \cong 8 \times 10^{-6}$ M; probably one subunit
22-12. 2×10^{-3} moles/min
22-14. The product of a series of reactions (in contrast to the product of a single reaction) is usually quite different in structure from the reactant of the first step in the series.

22-16. (a) $[HbO_2]$ increases (b) $[HbO_2]$ decreases (c) $[HbO_2]$ increases (d) Increase $[HbO_2]$. (This situation would be similar to that of myoglobin, which binds O_2 better than hemoglobin.)

22-18. His, Lys, Arg

CHAPTER 23

EXERCISES

23-1.

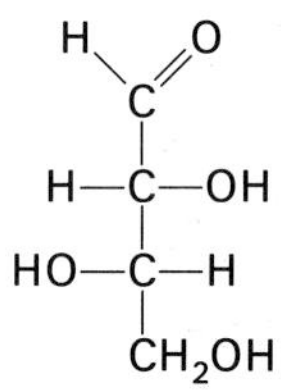

L-erythrose
$[\alpha]_D = +14.8°$

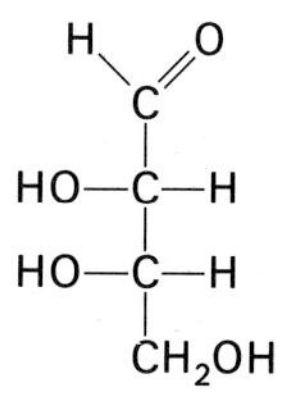

L-threose
$[\alpha]_D = -19.6°$

23-2. (a) A and D are enantiomers.
(b) diastereomers: B and C; A and B; A and C; B and D; and, C and D

23-3. (a)

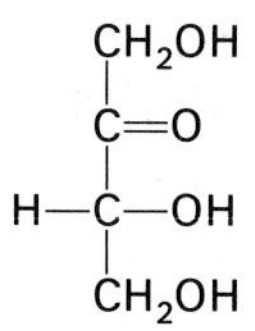

D-erythrose

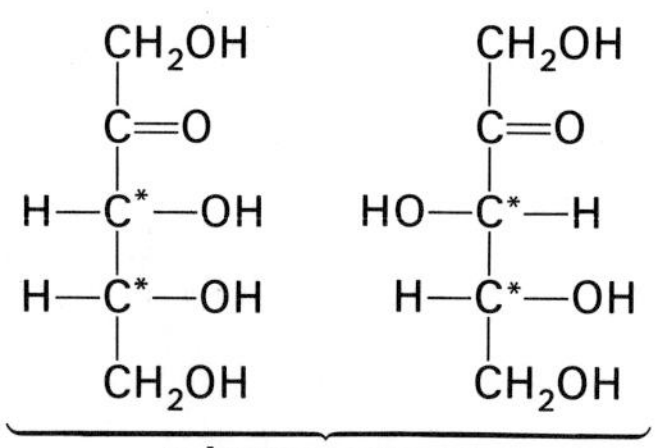

D-ketopentoses

(b) There are 2 chiral carbons in a ketopentose; and 4 stereoisomers.

23-5. The chair conformation for D-galactose is identical to that for β-D-glucose except that the H and OH groups on C-4 are interchanged.

23-6.

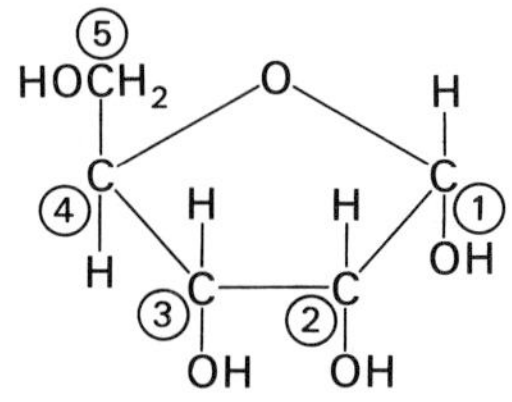

α-D-ribose

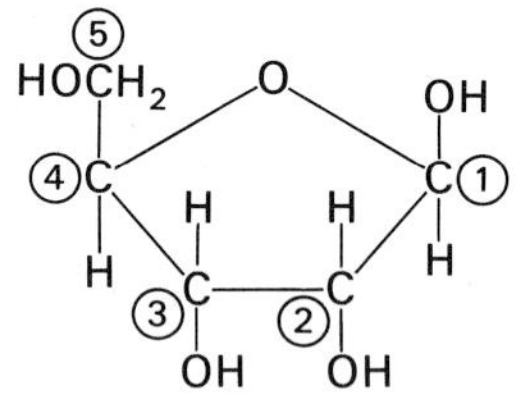

β-D-ribose

23-7. (a)

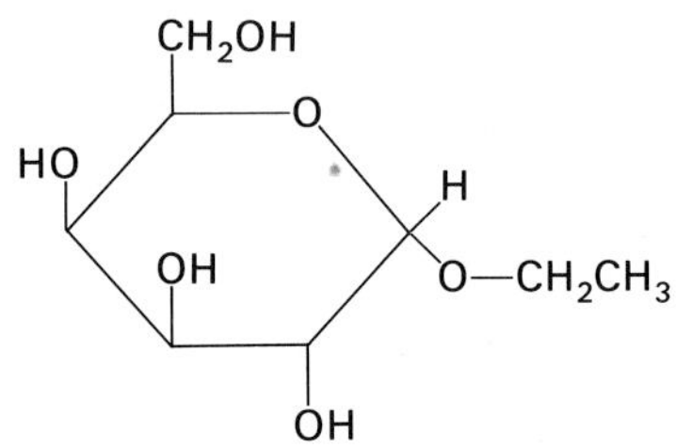

and

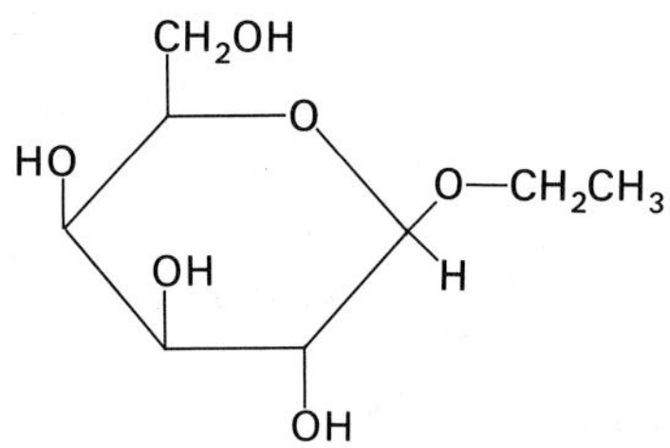

(b) 1, 2, 3

23-8.

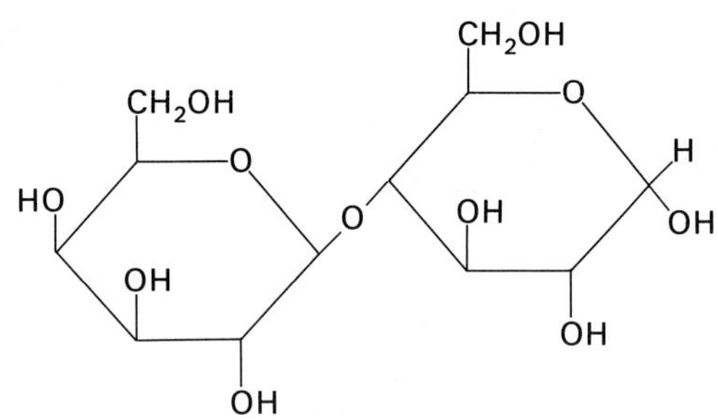

α-lactose

Lactose is a reducing sugar because the right-hand ring can open to form an aldehyde.

23-9.

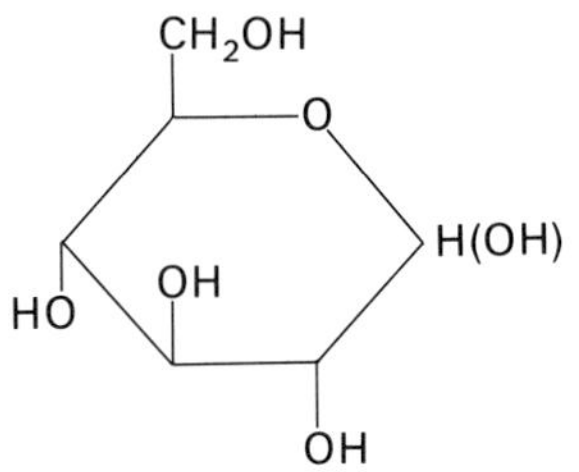

α- and β-D-glucose

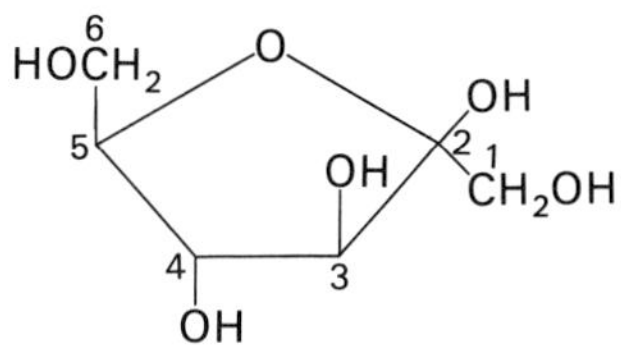

β-D-fructose

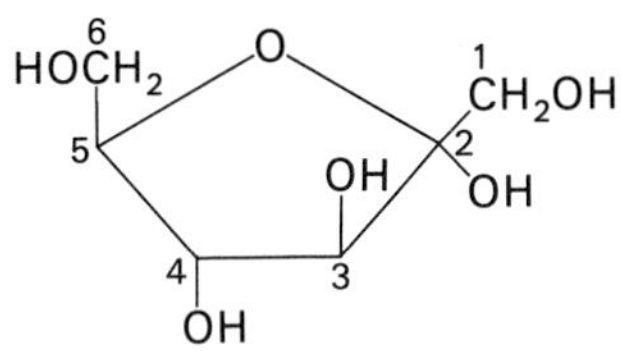

α-D-fructose

23-10.

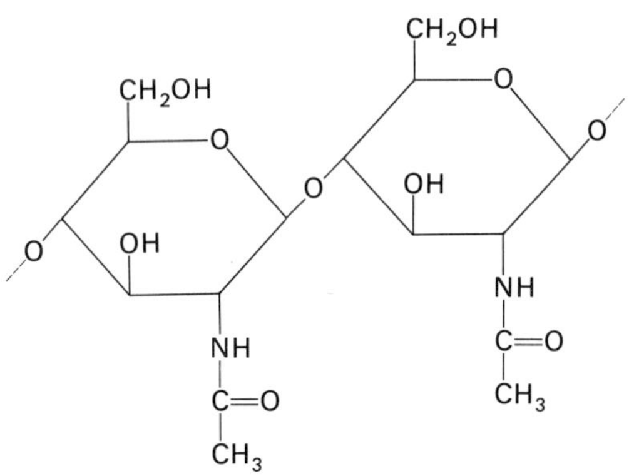

chitin

23-11. An AB recipient has neither A-type nor B-type antibodies, and can therefore receive blood cells containing both A and B antigens.

PROBLEMS

23-2. (a)

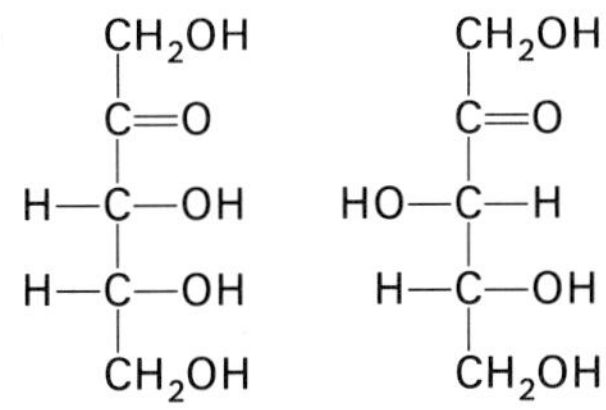

(b)

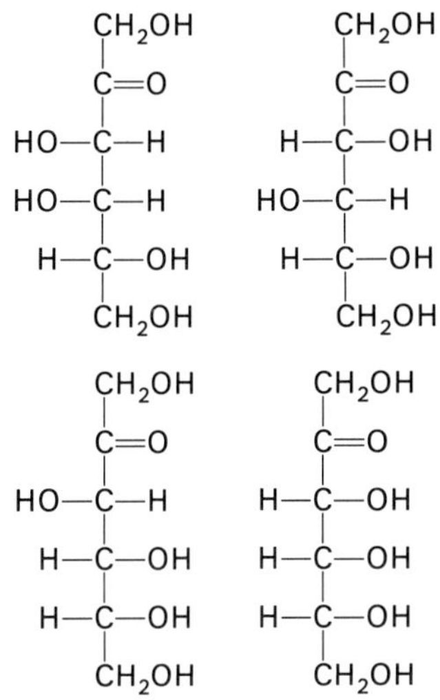

(c)

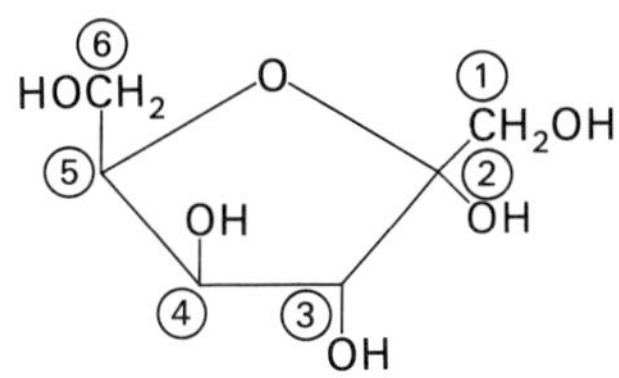

(d) The structure should have the L configuration about C-5: HO—C—H

23-4. The structure is the same as that for D-glucose except that the H and OH groups of C-2 are interchanged.

23-6. 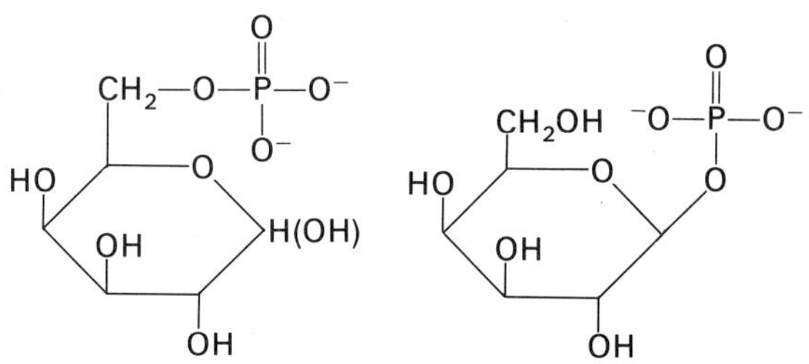

23-8. 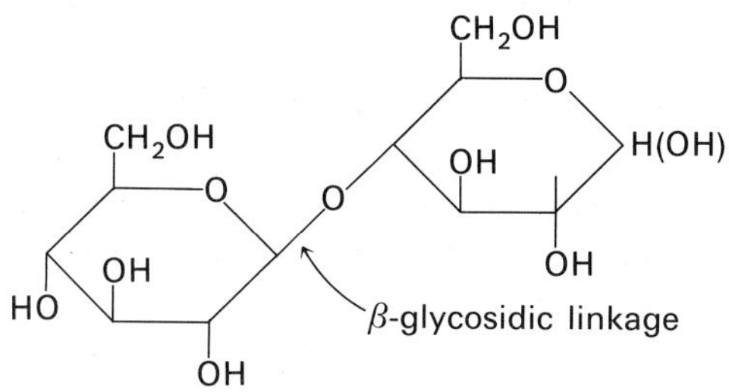

cellobiose

23-10. 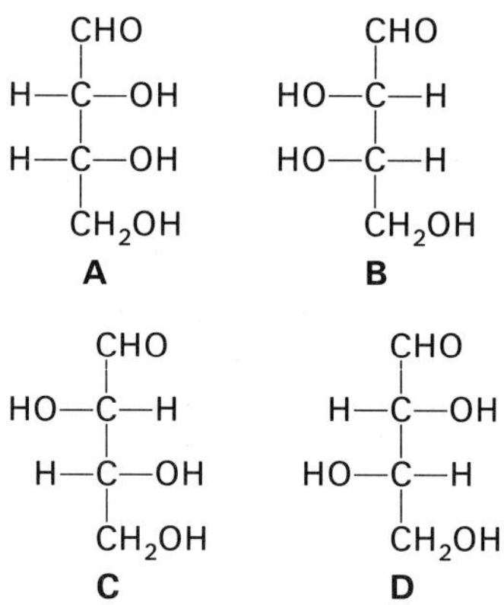

Enantiomers: A and B; C and D
Diastereomers: A and C; A and D; B and C; B and D

23-12. Four different compounds. There are two representations of *trans*-1,2-dichlorocyclohexane.

23-14. 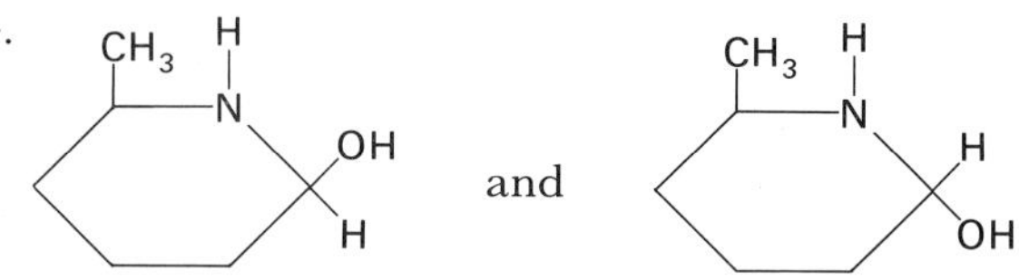

They are optically active.

23-16. (a) 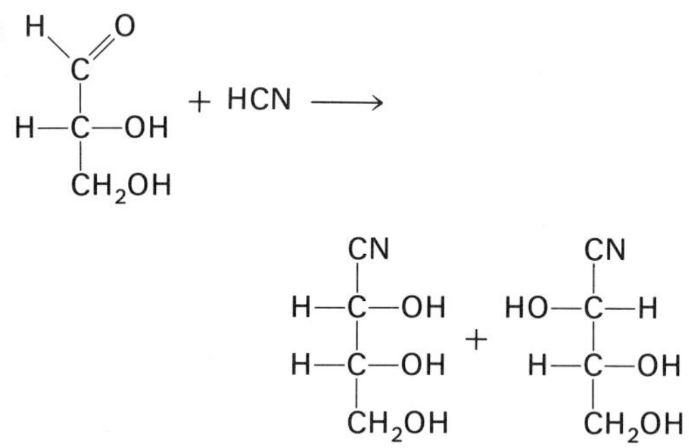

(b) 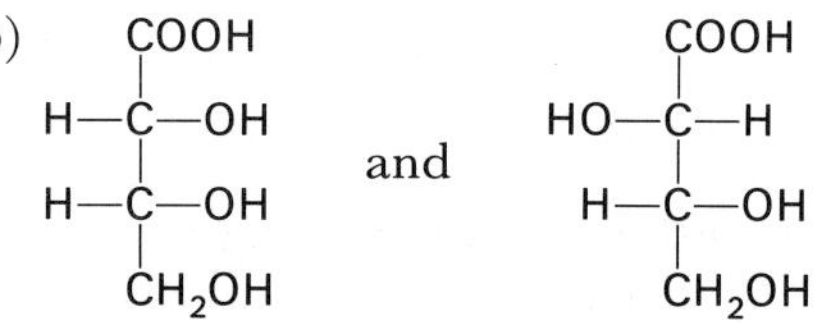

(c) 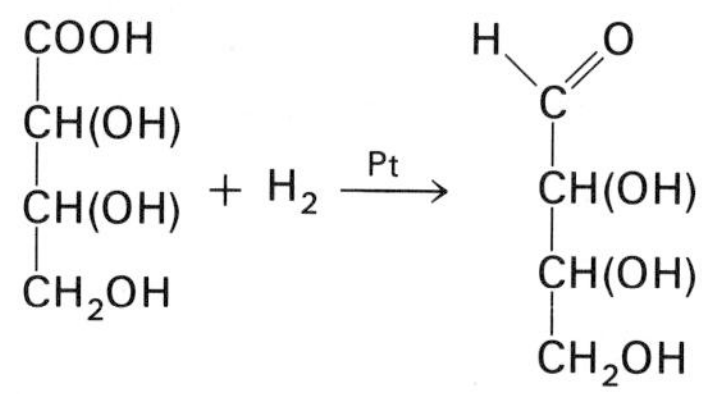

CHAPTER 24

EXERCISES

24-1. (a) The fatty acids, R—COOH, are not soluble in water; they are soluble only in the form of their sodium or potassium salts.
(b) Basic substances such as Na_2CO_3 or Na_3PO_4.

24-2. 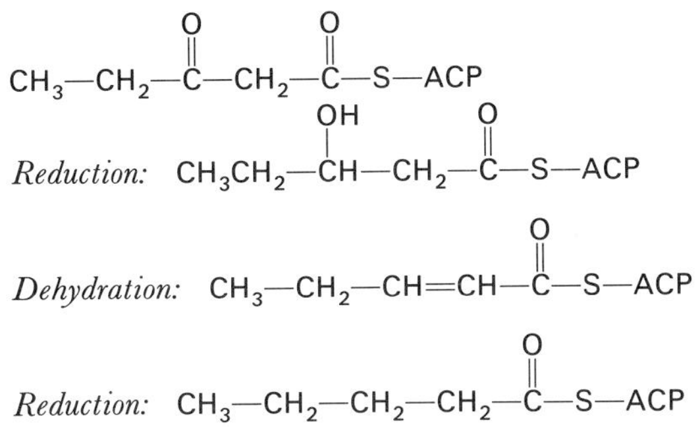

24-3. One structure is

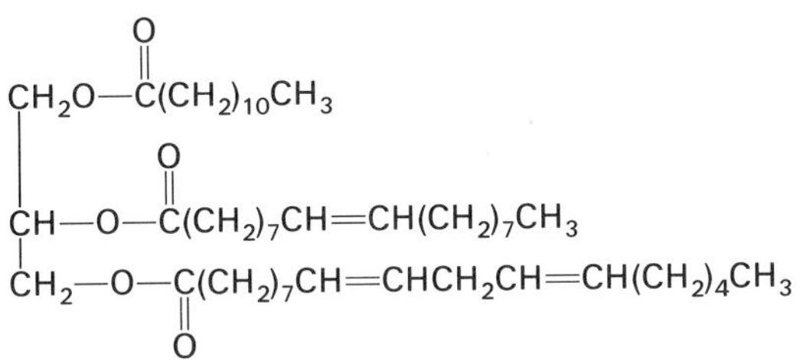

(Two other structures are possible with the positions of the fatty acids interchanged.

24-4. (a) 9.5 kg of fats. (b) 60 kg of glycogen. The difference is about 50 kg.

24-5. 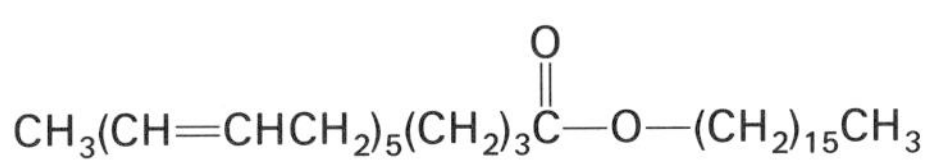

There are several others having different positions of the double bonds.

24-6. 2 in limonene; 8 in β-carotene, tail-to-tail arrangement in center of molecule; 4 in vitamin A; 4 (plus an additional one in lefthand ring) in vitamin E; 4 (plus one in rings) in vitamin K.

PROBLEMS

24-2. 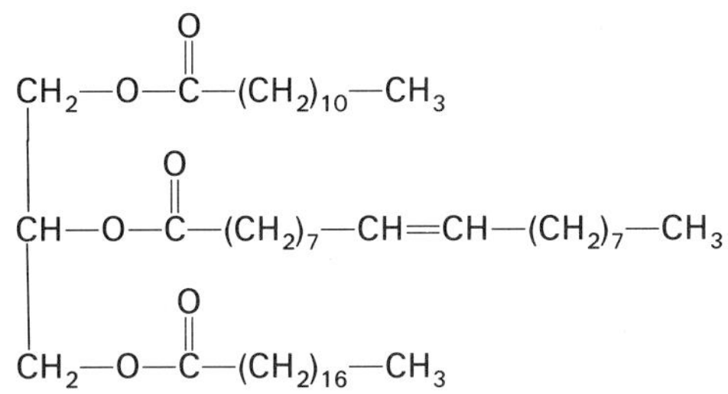

A total of 6 including enantiomers.

24-4. Same number of carbons. Oleic has a lower melting point because it has a double bond.

24-6. The corn oil (which has a lower melting point than lard) contains a higher percentage of unsaturated fatty acids.

24-8. 900 ml (0.9 ℓ, to one significant figure)

24-10. CH_2OH–$CHOH$–CH_2OH; RCOOH; R′COOH; R″COOH; R‴COOH; PO_4^{3-}

24-12. 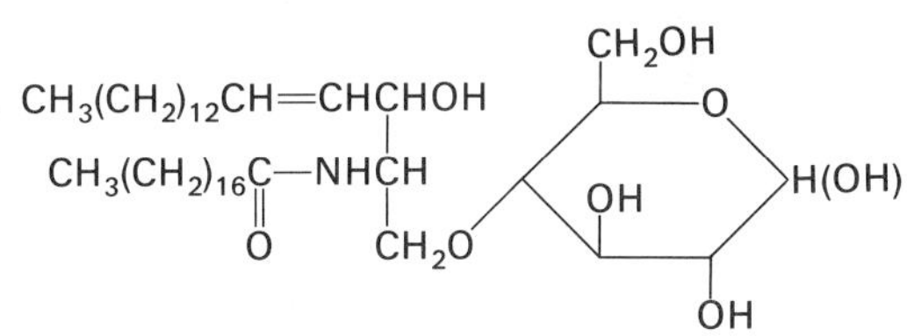

24-14. $CH_3(CH_2)_{14}COOH$ and $CH_3(CH_2)_{28}CH_2OH$

24-16. Peripheral proteins have (primarily) polar or charged amino acid residues on their outer surfaces. Integral proteins have (primarily) hydrophobic amino acid residues on their outer surfaces.

CHAPTER 25

EXERCISES

25-1.

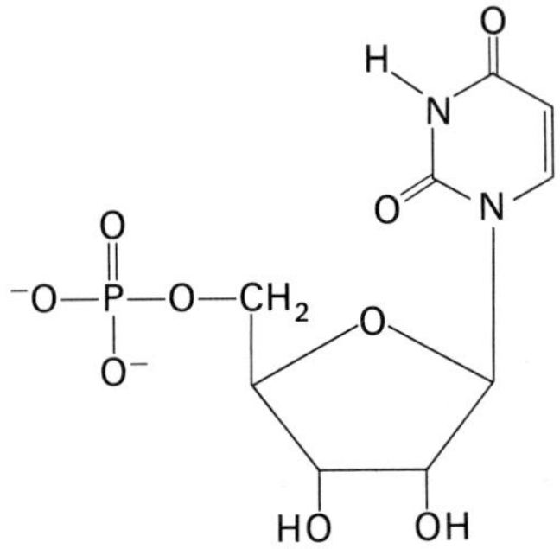

UMP

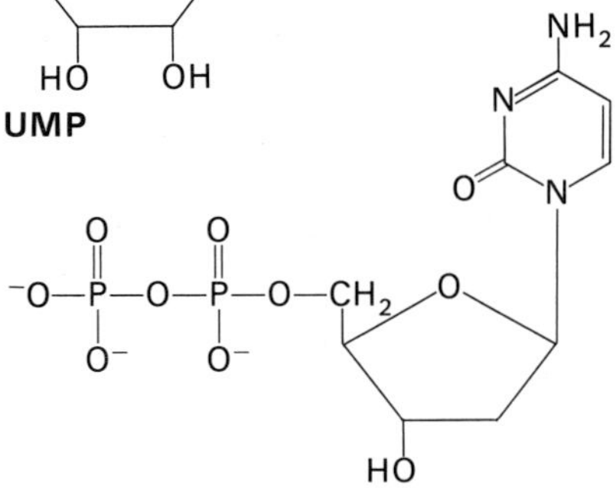

dCDP

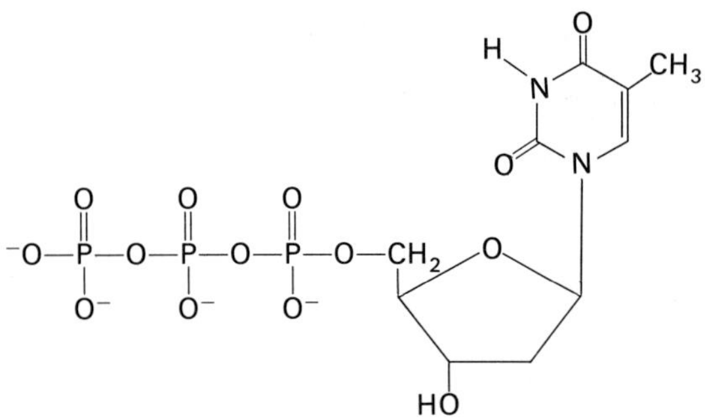

TTP

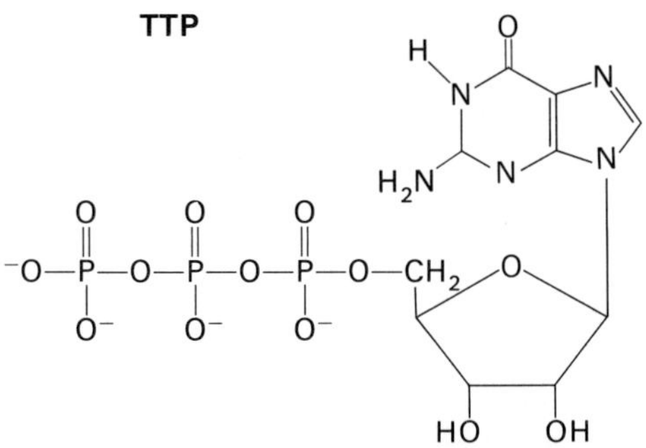

GTP

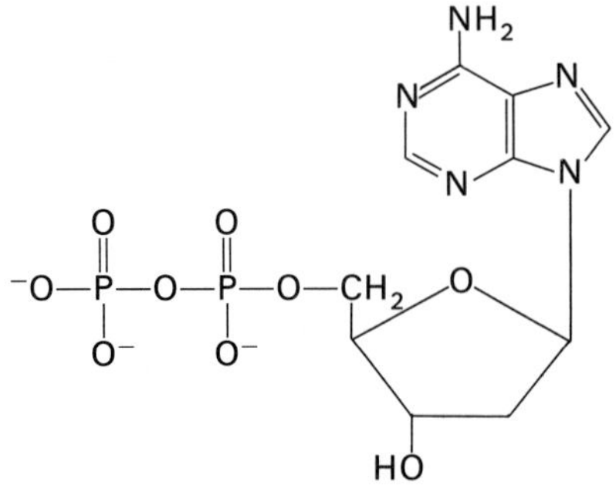

dADP

25-2. Answer contained in the exercise.

25-3. . . . –A–G–G–G–T–T–C–T–G–A–
A–A–T–G–C–C–A–T–G–A–T– . . .

. . . –U–C–C–C–A–A–G–A–C–U–
U–U–A–C–G–G–U–A–C–U–A– . . .

25-4. Met–Thr–Arg–Ala–Ile–Leu–Ile–Ala–Gly

25-5. C–C–T–T–G–A–A–G–T–C–C–T
C–C–T–T–G–A–A
C–C–T–T
A–A–G–T–C–C–T–G–C–T
A–A–G–T–C–C–T
A–A
T–C–C–T–G–C–T
C–T

Only the first three fragments in the list would be labeled by ^{32}P attached at the 5′-end. Labeled fragments produced by cleavage at T would be

C–C
C–C–T
C–C–T–T–G–A–A–G
C–C–T–T–G–A–A–G–T–C–C
C–C–T–T–G–A–A–G–T–C–C–T–G–C

25-6.

A protein: – Met – Arg – Ala – Ser – Ile – Lys – Met – Ile – Gly – Val –

. . .–A–T–G–C–G–C–G–C–T–T–C–G–A–T–A–A–A–A–A–T–G–A–T–T–G–G–C–G–T–A–. . .

B protein: – Ala – Arg – Phe – Asp – Lys – Asn – Asp – Trp – Arg –

PROBLEMS

25-2.

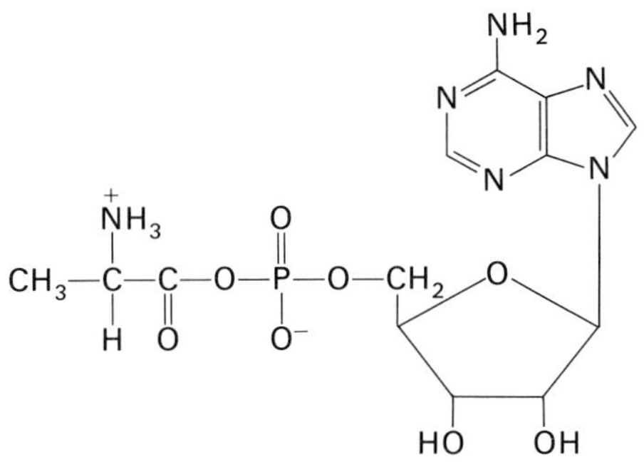

alanyl-AMP

25-4.

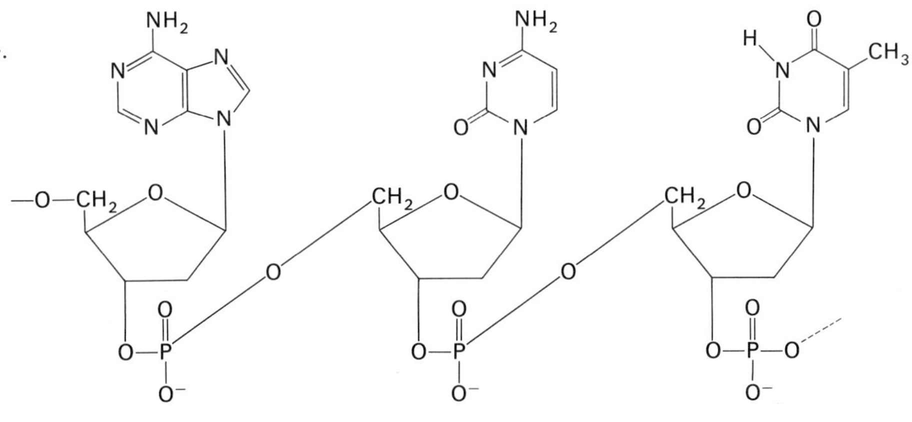

25-6. –C–T–A–G–G–T–C–A–T–T–G–A–G–C–T–

DNA: –G–A–T–C–C–A–G–T–A–A–C–T–C–G–A–

RNA: –G–A–U–C–C–A–G–U–A–A–C–U–C–G–A–

25-8. A–G–T–A–A–G

CHAPTER 26

EXERCISES

26-1.

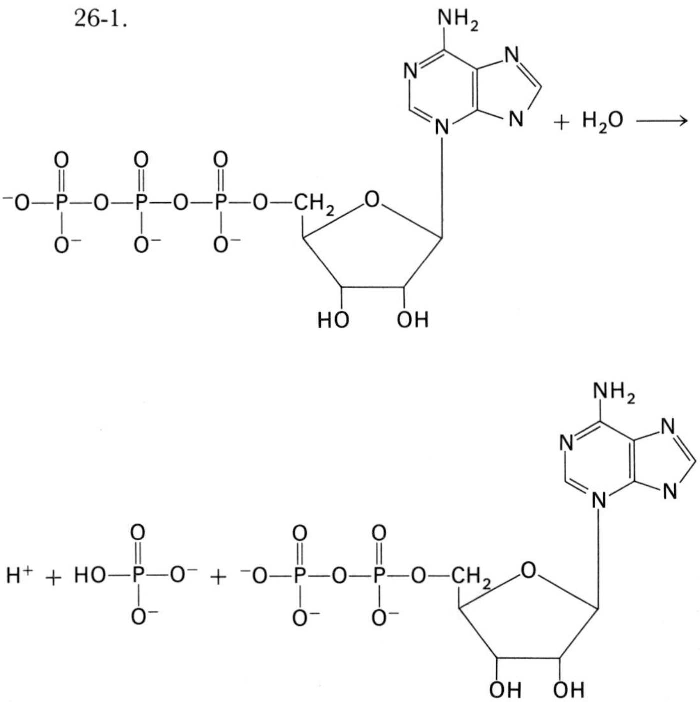

26-2. positions 1 or 6

26-3. (a) Reactions 1, 3, 6, 7, and 10. In reaction 8, a phosphate group shifts positions within a molecule.
(b) Reactions 2 and 5 (c) Reaction 9
(d) Reaction 4 (e) Reactions 6 and 11

26-4. condensation, reaction 1; isomerization, reaction 2; hydration, reaction 6; decarboxylation, reactions 3 and 4; oxidation, reactions 3, 5, and 7; phosphorylation, reaction 4.

26-5. $\mathcal{E}^{0'} = +0.22 - 0.07 = +0.15$ V; spontaneous

26-6. position 1 (carboxylic acid group); abundance 5% because only one-half of the molecules contain the C atom that was originally in CO_2

PROBLEMS

26-2. positions 3 and 4

26-4. Lauric acid + ATP + CoA–SH →
Lauryl-S-CoA + ADP + P_i

Lauryl–S–CoA + 5FAD + $5NAD^+$
+ 5CoA–SH + $5H_2O$ → 6Acetyl–S–CoA
+ $5FADH_2$ + 5NADH + H^+

ATP produced:
15 from 5NADH + H^+
10 from $5FADH_2$
72 from 12 acetyl–S–CoA
− 1 consumed in forming lauryl–S–CoA

Total 96

26-6. (a) PEP + ADP → Pyruvate + ATP
$\Delta G^{0'} = -7.5$ kcal/mole
(b) ATP + Glycerol →
ADP + Glycerol 1-phosphate
$\Delta G^{0'} = -5.1$ kcal/mole

26-8. (a) There is a net oxidation in glucose → pyruvate.
(b) There is no net oxidation in glucose → lactate.

26-10. The nitrite oxidizes Fe^{2+} to Fe^{3+}, which does not complex with CN^-.

26-12. $b < d < c < a$ (greatest tendency to accept electrons)

INDEX

ATOMIC WEIGHTS OF THE ELEMENTS

Element	Symbol	Atomic number	Atomic weight	Element	Symbol	Atomic number	Atomic weight
Actinium	Ac	89	(227)	Europium	Eu	63	151.96
Aluminum	Al	13	26.9815	Fermium	Fm	100	(257)
Americium	Am	95	(243)	Fluorine	F	9	18.9984
Antimony	Sb	51	121.75	Francium	Fr	87	(223)
Argon	Ar	18	39.948	Gadolinium	Gd	64	157.25
Arsenic	As	33	74.9216	Gallium	Ga	31	69.72
Astatine	At	85	(210)	Germanium	Ge	32	72.59
Barium	Ba	56	137.33	Gold	Au	79	196.967
Berkelium	Bk	97	(247)	Hafnium	Hf	72	178.49
Beryllium	Be	4	9.0122	*Hahnium	Ha	105	(262)
Bismuth	Bi	83	208.980	Helium	He	2	4.0026
Boron	B	5	10.81	Holmium	Ho	67	164.930
Bromine	Br	35	79.904	Hydrogen	H	1	1.0079
Cadmium	Cd	48	112.40	Indium	In	49	114.82
Calcium	Ca	20	40.08	Iodine	I	53	126.904
Californium	Cf	98	(251)	Iridium	Ir	77	192.22
Carbon	C	6	12.011	Iron	Fe	26	55.847
Cerium	Ce	58	140.12	Krypton	Kr	36	83.80
Cesium	Cs	55	132.905	Lanthanum	La	57	138.91
Chlorine	Cl	17	35.453	Lawrencium	Lr	103	(260)
Chromium	Cr	24	51.996	Lead	Pb	82	207.2
Cobalt	Co	27	58.9332	Lithium	Li	3	6.941
Copper	Cu	29	63.546	Lutetium	Lu	71	174.97
Curium	Cm	96	(247)	Magnesium	Mg	12	24.305
Dysprosium	Dy	66	162.50	Manganese	Mn	25	54.9380
Einsteinium	Es	99	(252)	Mendelevium	Md	101	(258)
Erbium	Er	68	167.26	Mercury	Hg	80	200.59

The atomic weights are based on the carbon-12 standard: $^{12}C = 12.000...$ exactly. Parentheses denote a radioactive element; the number given is the mass number of the most stable isotope.

*Elements 104 and 105 have not yet been officially named.